PreK–2 Standards
NUMBER AND OPERATIONS
In prekindergarten through grade 2 all students should—

Understand numbers, ways of representing numbers, relationships among numbers, and number systems and should—
- Count with understanding and recognize "how many" in sets of objects;
- Use multiple models to develop initial understandings of place value and the base-ten number system;
- Develop understanding of the relative position and magnitude of whole numbers and of ordinal and cardinal numbers and their connections;
- Develop a sense of whole numbers and represent and use them in flexible ways, including relating, composing, and decomposing numbers;
- Connect number words and numerals to the quantities they represent, using various physical models and representations;
- Understand and represent commonly used fractions, such as $\frac{1}{4}$, $\frac{1}{3}$, and $\frac{1}{2}$

Understand meanings of operations and how they relate to one another and should—
- Understand various meanings of addition and subtraction of whole numbers and the relationship between the two operations;
- Understand the effects of adding and subtracting whole numbers;
- Understand situations that entail multiplication and division, such as equal groupings of objects and sharing equally

Compute fluently and make reasonable estimates and should—
- Develop and use strategies for whole-number computations, with a focus on addition and subtraction;
- Develop fluency with basic number combinations for addition and subtraction;
- Use a variety of methods and tools to compute, including objects, mental computation, estimation, paper and pencil, and calculators

ALGEBRA
In prekindergarten through grade 2 all students should—

Understand patterns, relations, and functions and should—
- Sort, classify, and order objects by size, number, and other properties;
- Recognize, describe, and extend patterns such as sequences of sounds and shapes or simple numeric patterns and translate from one representation to another;
- Analyze how both repeating and growing patterns are generated

Represent and analyze mathematical situations and structures using algebraic symbols and should—
- Illustrate general principles and properties of operations, such as commutativity, using specific numbers;
- Use concrete, pictorial, and verbal representations to develop an understanding of invented and conventional symbolic notations

Use mathematical models to represent and understand quantitative relationships and should—
- Model situations that involve the addition and subtraction of whole numbers, using objects, pictures, and symbols

Analyze change in various contexts and should—
- Describe qualitative change, such as a student's growing taller;
- Describe quantitative change, such as a student's growing two inches in one year

PreK–2 Standards
GEOMETRY
In prekindergarten through grade 2 all students should—

Analyze characteristics and properties of two- and three-dimensional geometric shapes and develop mathematical arguments about geometric relationships and should—
- Recognize, name, build, draw, compare, and sort two- and three-dimensional shapes;
- Describe attributes and parts of two- and three-dimensional shapes;
- Investigate and predict the results of putting together and taking apart two- and three-dimensional shapes

Specify locations and describe spatial relationships using coordinate geometry and other representational systems and should—
- Describe, name, and interpret relative positions in space and apply ideas about relative position;
- Describe, name, and interpret direction and distance in navigating space and apply ideas about direction and distance;
- Find and name locations with simple relationships such as "near to" and in coordinate systems such as maps

Apply transformations and use symmetry to analyze mathematical situations and should—
- Recognize and apply slides, flips, and turns;
- Recognize and create shapes that have symmetry

Use visualization, spatial reasoning, and geometric modeling to solve problems and should—
- Create mental images of geometric shapes using spatial memory and spatial visualization;
- Recognize and represent shapes from different perspectives;
- Relate ideas in geometry to ideas in number and measurement;
- Recognize geometric shapes and structures in the environment and specify their location

MEASUREMENT
In prekindergarten through grade 2 all students should—

Understand measurable attributes of objects and the units, systems, and processes of measurement and should—
- Recognize the attributes of length, volume, weight, area, and time;
- Compare and order objects according to these attributes;
- Understand how to measure using the nonstandard and standard units;
- Select an appropriate unit and tool for the attribute being measured

Apply appropriate techniques, tools, and formulas to determine measurements and should—
- Measure with multiple copies of units of the same size, such as paper clips laid end to end;
- Use repetition of a single unit to measure something larger than the unit, for instance, measuring the length of a room with a single meterstick;
- Use tools to measure;
- Develop common referents for measures to make comparisons and estimates

PreK–2 Standards

DATA ANALYSIS AND PROBABILITY

In prekindergarten through grade 2 all students should—

Formulate questions that can be addressed with data and collect, organize, and display relevant data to answer them and should—

- Pose questions and gather data about themselves and their surroundings;
- Sort and classify objects according to their attributes and organize data about their objects;
- Represent data using concrete objects, pictures, and graphs

Select and use appropriate statistical methods to analyze data and should—

- Describe parts of the data and the set of data as a whole to determine what the data show

Develop and evaluate inferences and predictions that are based on data and should—

- Discuss events related to student's experiences as likely or unlikely

Grades 3–5

NUMBER AND OPERATIONS

In grades 3–5 all students should—

Understand numbers, ways of representing numbers, relationships among numbers, and number systems and should—

- Understand the place-value structure of the base-ten number system and be able to represent and compare whole numbers and decimals;
- Recognize equivalent representations for the same number and generate them by decomposing and composing numbers;
- Develop understanding of fractions as parts of unit wholes, as parts of a collection, as locations on number lines, and as divisions of whole numbers;
- Use models, benchmarks, and equivalent forms to judge the size of fractions;
- Recognize and generate equivalent forms of commonly used fractions, decimals, and percents;
- Explore numbers less than 0 by extending the number line and through familiar applications;
- Describe classes of numbers according to characteristics such as the nature of their factors

Understand meanings of operations and how they relate to one another and should—

- Understand various meanings of multiplication and division;
- Understand the effects of multiplying and dividing whole numbers;
- Identify and use relationships between operations, such as division and the inverse of multiplication, to solve problems;
- Understand and use properties of operations, such as the distributivity of multiplication over addition

Compute fluently and make reasonable estimates and should—

- Develop fluency with basic number combinations for multiplication and division and use these combinations to mentally compute related problems, such as 30×50;
- Develop fluency in adding subtracting, multiplying, and dividing whole numbers;
- Develop and use strategies to estimate the results of whole-number computations and to judge the reasonableness of such results

Grades 3–5

NUMBER AND OPERATIONS (continued)

In grades 3–5 all students should—

- Develop and use strategies to estimate computations involving fractions and decimals in situations relevant to students;
- Use visual models, benchmarks, and equivalent forms to add and subtract commonly used fractions and decimals;
- Select appropriate methods and tools for computing with whole numbers from among mental computation, estimation, calculators, and paper and pencil according to the context and nature of the computation and use the selected method or tool

ALGEBRA

In grades 3–5 all students should—

Understand patterns, relations, and functions and should—

- Describe, extend, and make generalizations about geometric and numeric patterns;
- Represent and analyze patterns and functions, using words, tables, and graphs

Represent and analyze mathematical situations and structures using algebraic symbols and should—

- Identify such properties as commutativity, associativity, and distributivity and use them to compute with whole numbers;
- Represent the idea of a variable as an unknown quantity using a letter or a symbol;
- Express mathematical relationships using equations

Use mathematical models to represent and understand quantitative relationships and should—

- Model problem situations with objects and use representations such as graphs, tables, and equations to draw conclusions

Analyze change in various contexts and should—

- Investigate how a change in one variable relates to a change in a second variable;
- Identify and describe situations with constant or varying rates of change and compare them

GEOMETRY

In grades 3–5 all students should—

Analyze characteristics and properties of two- and three-dimensional geometric shapes and develop mathematical arguments about geometric relationships and should—

- Identify, compare, and analyze attributes of two- and three-dimensional shapes and develop vocabulary to describe the attributes;
- Classify two- and three-dimensional shapes according to their properties and develop definitions of classes of shapes such as triangles and pyramids;
- Investigate, describe, and reason about the results of subdividing, combining, and transforming shapes;
- Explore congruence and similarity;
- Make and test conjectures about geometric properties and relationships and develop logical arguments to justify conclusions

Specify locations and describe spatial relationships using coordinate geometry and other representational systems and should—

- Describe location and movement using common language and geometric vocabulary;
- Make and use coordinate systems to specify locations and to describe paths;
- Find the distance between points along horizontal and vertical lines of a coordinate system

Instructor's Edition

Mathematics for Elementary School Teachers

Second Edition

Phares O'Daffer
Illinois State University

Randall Charles
San Jose State University

Thomas Cooney
University of Georgia

John Dossey
Illinois State University

Jane Schielack
Texas A&M University

Boston San Francisco New York
London Toronto Sydney Tokyo Singapore Madrid
Mexico City Munich Paris Cape Town Hong Kong Montreal

Senior Acquisitions Editor: Anne Kelly
Marketing Manager: Becky Anderson
Senior Production Supervisor: Peggy McMahon
Production Services: Elm Street Publishing Services, Inc.
Manufacturing Buyer: Evelyn Beaton
Senior Prepress Supervisor: Caroline Fell
Cover Designer: Barbara T. Atkinson
Cover Photo: © Photonica/Zesa
Text Design: Elm Street Publishing Services, Inc.
Photo Acknowledgments: page xxii © John Henley/Corbis Stock Market **page 58** © AP/Wide World Photos **page 122** © AP/Wide World Photos **page 188** © H. David Seawall/CORBIS **page 230** © AP/Wide World Photos **page 276** © Erik Freeland/CORBIS SABA **page 344** © Tony Freeman/PhotoEdit **page 390** © AP/Wide World Photos **page 452** © Gary Holscher/Stone **page 502** © Corbis **page 584** PhotoDisc © 2001 **page 646** PhotoDisc © 2001 **page 702** PhotoDisc © 2001

The Library of Congress has already cataloged the student edition as follows:

Library of Congress Cataloging-in-Publication Data

Mathematics for elementary school teachers / Phares O'Daffer ... [et al.]—2nd ed.
p. cm.
Includes bibliographical references and index.
ISBN 0-201-69951-6 (alk. paper)
1. Mathematics—Study and teaching (Elementary) I. O'Daffer, Phares G.

QA135.6 .M38 2001
510'.24'372—dc21

2001035787

Instructor's Edition *Mathematics for Elementary School Teachers*
ISBN 0-201-73535-0

1 2 3 4 5 6 7 8 9 10—QWT—04030201

Contents

CHAPTER 6 Rational Number Operations and Properties 276

CHAPTER 7 Proportional Reasoning 344

CHAPTER 8 Analyzing Data 390

Preface

INTRODUCTION

In *Mathematics for Elementary School Teachers*, Second Edition, we offer students a unique opportunity to develop a clear understanding of mathematical concepts, procedures, and processes. We also provide students with the tools to communicate these ideas to others and to apply these ideas to the real world.

We have designed this book to help students develop mathematical power, so that they will come to better understand what mathematics is and what it means to "do mathematics." We think this can be done by focusing on the five fundamental processes of **Making Connections, Communicating, Problem Solving, Mathematical Reasoning,** and **Representation.**

We believe that these fundamental processes are often best developed as students become actively involved in their own learning. We believe this active involvement should take place in an environment rich in real-world models, technology that promotes understanding, and opportunities to communicate with others in small groups.

In revising the first edition of *Mathematics for Elementary School Teachers*, we have utilized extensive feedback from thousands of teachers and students who have used this text. Our goal is to present a thorough development of mathematical content in a way that is understandable to students. We have also revised the material so that it powerfully embodies the new *Principles and Standards for School Mathematics* of the National Council of Teachers of Mathematics.

We urge instructors using this text to build their courses around active student involvement using the five fundamental processes as they deal with specific mathematics content and procedures. Here is a look at how each of these processes is integrated into *Mathematics for Elementary School Teachers.*

MAKING CONNECTIONS

Content Connections

Each chapter begins with three distinct features: **Chapter Perspective, Connection to the NCTM Principles and Standards,** and **Connection to the PreK–8 Classroom.** All are designed to help students make connections that enable them to see how the content of the chapter relates to real-world applications, to mathematics content in other chapters, and to how children learn mathematics. We believe this opening material is very important in giving students a "road map" of where they have been and where they are going, as well as providing motivation for learning the chapter concepts and explaining why it might be interesting to do so (see pp. 276–277; 321).

Modeling

We further help students make connections by using real-world models to develop and illustrate mathematical concepts and by using models and technology to promote discovery of key mathematical patterns and relationships. We also highlight important real-world applications of mathematics to demonstrate how the real world and the various ideas and relationships in mathematics fit together. For example, on p. 218, see how the model of "Orbiting Spacecrafts" is used to help students understand the concept of least common multiples. And on p. 107, see how a base-ten block model is used in understanding numeration. In Section 5.1, see how models are used to help students understand how to add, subtract, and multiply integers.

COMMUNICATING

We believe students who are able to communicate effectively what they have learned have truly mastered the mathematical concepts studied. This is particularly important for future teachers who must feel comfortable with mathematical ideas and be able to communicate them competently to their students. We facilitate communication by providing numerous situations in which students are encouraged to talk, write, and work in groups; create pictures; and make tables, graphs, and charts to help explain mathematical ideas.

Section 1.1, Mathematics as Communication, examines the various roles of communication in perceiving, understanding, and using mathematics (see pp. 1–14).

Communication is also enhanced by the following activities:

- Students develop communication skills in the *Communicating and Connecting Ideas* exercises (see p. 319).
- Student communication is fostered through the *Your Turn* parts of all text examples (see p. 280).
- Mini-Investigations are intended to promote student communication and to provide an opportunity for active involvement. These optional activities encourage students to explore ideas individually or in small groups. Students are asked to Write About; Talk About; Make a Graph, Chart, or Table; or Draw a Diagram or Picture to communicate their findings or observations (see p. 289).
- Many of the *Deepening Understanding* and *Reasoning and Problem Solving* exercises also require communication of mathematical ideas (see pp. 304–305).
- Exercises designed for group activity are designated as such by the icon (see p. 294).

PROBLEM SOLVING

In Chapter 1, we present problem solving as a central part of mathematics. We first engage students in the development of a reliable problem-solving process, then suggest a number of useful strategies such as making a list or table, drawing a picture, creating a graph, etc., to help facilitate problem solving. We then continue to highlight the problem-solving process and strategies as students move on to solve a large number of thought-provoking problems throughout the text (see Section 1.3, Mathematics as Problem Solving, pp. 37–47).

Problem-Solving Examples and Exercises

Problem-solving examples are included throughout the text to help students better understand the problem-solving process. These examples, as seen on pp. 322–324, identify the following components of the problem-solving process and provide an ongoing emphasis on the role each plays in problem solving:

- Understand the problem
- Develop a plan
- Implement the plan
- Look back

Other emphases on problem solving in the text include the following, which appear throughout the text:

- Students solve problems in the *Reasoning and Problem Solving* portion of the exercise sets at the end of each section (see p. 337).
- Student problem-solving opportunities are integrated into many of the Mini-Investigations (see p. 284).
- Students are often involved in problem solving in the *Reflect* part of the in-text examples (see p. 280).
- Students read and try out examples in whole sections devoted to problem solving (see Section 10.2, Solving Problems in Geometry, pp. 525–541).
- Students learn to ask what-if... questions to create and solve new problems (see pp. 574–576).

MATHEMATICAL REASONING

In Chapter 1, Section 1.2, some of the basic ideas of inductive and deductive reasoning are introduced and experienced. Then these processes are utilized in a variety of ways throughout the text, as indicated below.

- Students use mathematical reasoning in special Mini-Investigations labeled "Using Mathematical Reasoning" (see p. 296).
- Students use mathematical reasoning in exercises in the *Reasoning and Problem Solving* sections of the exercise sets (see p. 318).
- Students develop mathematical reasoning through the "alternative ways of thinking" examples that show students different ways of reasoning through problem situations (see p. 333). These examples show students how they can construct their own problem-solving strategies and that frequently there are several ways to solve a problem. The *Your Turn* parts of each example encourage students to put their thinking into practice and then to reflect on the techniques and concepts they have just learned (see p. 298).
- Students use reasoning processes in numerous real-world estimation applications (see pp. 283 and 302). Problems in the exercise sets involving estimation are identified by the icon ≈ (see p. 293).

Using Inductive Reasoning

Inductive reasoning is developed in Chapter 1 (see Section 1.2) and utilized extensively throughout the book in selected Mini-Investigations and as students study patterns and relationships in other situations to discover generalizations. (See p. 198, Table 4.3, for the use of inductive reasoning in number theory and p. 545 for the use of inductive reasoning in geometry.)

Using Deductive Reasoning

Deductive reasoning is also developed in Chapter 1 (see Section 1.2) and is utilized throughout the book. In the second edition of *Mathematics for Elementary School Teachers*, more attention has been given to using deductive reasoning to verify important theorems, both in the text development and in the reasoning exercises. (See the following pages for examples: pp. 151–152, pp. 194–195, p. 264, p. 313, pp. 358–359, pp. 571–572.)

REPRESENTATION

This text provides ongoing reinforcement of the use of different representations to organize, record, and communicate mathematical ideas. Various representations are provided throughout the text in verbal, numerical, pictorial, graphic, and symbolic forms.

Using Technology

Explanations of the way technology provides opportunities for learning mathematics are integrated throughout this text. The use of technology is incorporated as an enhancement to the text discussion and exercises. Although strongly encouraged, use of technology is an optional feature of this text. We point out the technology option in some of the Mini-Investigations, in exercises and examples, and in a special *Spotlight on Technology* feature (see Chapter 1, p. 6.) We have also included useful *Graphing Calculator* and *Geometry Exploration Software* appendices to provide instruction on the basic functions and features of these valuable tools. Exercises involving technology have been included and are designated by the icon (see p. 305).

Finding Patterns

In Chapter 1, Section 1.2, the techniques for searching for patterns are introduced. Then these procedures are utilized in a variety of ways throughout the text, as indicated below.

- "Finding a Pattern" Mini-Investigations encourage students to use verbal, numerical, pictorial, graphic, and symbolic representations to organize data in order to find patterns (see p. 284).
- Students discover patterns in order to form generalizations in various content areas. (For example, in Chapter 4, on Number Theory, see the pattern search in Table 4.2 on p. 195.)

REVIEW AND ASSESSMENT

Exercises play an important role in student conceptualization of mathematical ideas. Our exercises have been carefully chosen to help students learn basic concepts and procedures and to extend their understanding to broader concepts. In addition, the exercises further encourage students to develop their ability to solve problems, use mathematical reasoning, communicate ideas, and relate mathematics to the real world.

Section Exercises

The section exercises have been divided developmentally into four categories (see pp. 292–294):

- *Reinforcing Concepts and Practicing Skills* exercises focus on basic concepts and skills.
- *Deepening Understanding* exercises stimulate further thinking about key topics.
- *Reasoning and Problem Solving* exercises involve the use of higher-level thinking processes.
- *Communicating and Connecting Ideas* exercises enhance student ability to communicate mathematically and make important connections to history and to other mathematical ideas.

End-of-Chapter Material

The *Chapter Summary* allows students to revisit the mathematical concepts, basic techniques, and higher-level processes they have studied (see pp. 339–342):

- *Key Ideas: Questions and Answers* provide succinct summaries of each topic discussed in the chapter. These questions and answers serve as a self-check for student understanding and can be used as a valuable reference tool.
- *Key Terms and Generalizations* provide a handy reference (including page numbers) with which students can look up specific mathematical terms and ideas.

The *Chapter Review Exercises* can be used to assess student understanding of the chapter's important ideas or to serve as a source of additional review exercises. The items in the Chapter Review are divided developmentally into three categories (see pp. 342–343):

- *Concepts and Skills* exercises assess student understanding of basic concepts and procedures developed in the chapter.
- *Reasoning and Problem Solving* exercises assess student ability to use higher-level thinking processes.
- *Alternative Assessment* items provide opportunities for a more open-ended, observational approach to assessment.

SUPPLEMENTS

For the Instructor

Instructor's Edition (ISBN 0-201-73535-0)
The IE includes a complete answer section consisting of answers to nearly all text exercises.

Instructor's Resource Manual (ISBN 0-201-73536-9)
This manual highlights important points to cover in each chapter and key exercises to use. It also provides guidance on using the Mini-Investigations and includes sample student responses. Technology options are further expanded in this manual as well. Answers to the *Your Turn* sections of examples are also included. Finally, the Instructor's Resource Manual contains blackline masters for the instructor's use.

Instructor's Solutions Manual (ISBN 0-201-73537-7)
This manual provides solutions to all problems contained in the text.

TestGen-EQ with QuizMaster-EQ Windows/Macintosh (Dual-Platform CD-ROM, ISBN 0-201-73691-8)
TestGen-EQ's user-friendly graphical interface enables instructors to easily view, edit, and add questions, transfer questions to test, and print tests in a variety of fonts and forms. Search and sort features let the instructor quickly locate questions and arrange them in a preferred order. Six question formats—short-answer, true/false, multiple-choice, essay, matching, and bimodal—are available. A built-in question editor gives the user power to create graphs, import graphics, insert mathematical symbols and templates, and insert variable numbers or text. Computerized testbanks include algorithmically defined problems organized according to each textbook. An "Export to HTML" feature lets instructors create practice tests for the web.

QuizMaster-EQ enables instructors to create and save tests using TestGen-EQ so students can take them for practice or a grade on a computer network. Instructors can set preferences for how and when the tests are administered. QuizMaster-EQ automatically grades the exams, stores results on disk, and allows the instructor to view or print a variety of reports for individual students, classes, or courses. Consult your Addison-Wesley sales representative for details.

Instructor's Testing Manual (ISBN 0-201-73538-5)
This manual contains prepared tests for each chapter. In addition, it provides guidelines and items for open-ended alternative assessments of the major ideas in each chapter by briefly summarizing the section objectives and then posing a few questions that students should be able to answer when they have gained an understanding of the basic concepts covered in that section.

Web site (www.aw.com/odaffer)
The web site for this text provides additional resources for both instructors and students. For example, *Interactivate* provides technology-enhanced explorations indexed to the text which promote learner-centered, collaborative, open-ended, and inquiry-based experiences.

For the Student

Cooperative Learning Activities Manual with Manipulatives and Technology (ISBN 0-201-73689-6)
This manual contains activities for group investigations to involve students in an interactive learning and discovery process, using manipulatives and/or technology when appropriate and reinforcing or extending ideas covered in the text.

Student's Solutions Manual (ISBN 0-201-73690-X)
This manual provides detailed, worked-out solutions to nearly every odd-numbered problem in the text.

InterAct Math® Tutorial Software (CD-ROM, ISBN 0-201-74889-4)
InterAct Math Tutorial Software has been developed and designed by professional software engineers working closely with a team of experienced mathematics educators. The software includes exercises that are linked with every objective in the textbook and require the same computational and problem-solving skills as their companion exercises in the text. Each exercise has an example and interactive guided solutions that are designed to involve students in the solution process and to help them identify precisely where they are having trouble. It recognizes common student errors and provides students with appropriate customized feedback. With its

sophisticated answer-recognition capabilities, InterAct Math Tutorial Software recognizes appropriate forms of the same answer for any kind of input. It also tracks student activity and scores for each section that can then be printed out. The software is free to qualifying adopters or can be bundled with books for sale to students.

The Geometer's Sketchpad™
The Geometer's Sketchpad™, from Key Curriculum Press, is a dynamic geometry construction and exploration tool for students and teachers alike. With Sketchpad™ you can construct precise figures and manipulate them interactively, while preserving geometric relationships. Dynamic interaction gives you the power to explore, analyze, and understand mathematics as never before. This amazing software is available for a nominal charge with the purchase of *Mathematics for Elementary School Teachers*, Second Edition.

Addison-Wesley Math Tutor Center
The AWL Math Tutor Center is staffed by qualified mathematics instructors who provide students with tutoring on examples and exercises from the textbook. Tutoring is available via toll-free telephone, fax, or e-mail five days a week, seven hours a day.

ACKNOWLEDGMENTS

We wish to express our appreciation to the following individuals who reviewed the text and provided valuable suggestions for improvement.

Ellen Ansell, University of Pittsburgh
Susan Bradley, Angelina College
Andrea DeCosmo, Waubonsee Community College
Maria Diamantis, Southern Connecticut State University
Deborah Farro-Lynd, SUNY at Oneonta
Richard Francis, Southeast Missouri State
Barb Gentry, Parkland College
Peter L. Glidden, West Chester University
Joseph Harkin, SUNY College at Brockport
Kimberly Kelton, Austin Community College
Russ Killingsworth, Seattle Pacific University
Christopher Kribs Zaletta, University of Texas at Arlington
Richard E. Pfiefer, San Jose State University
Janice Richardson, Elon College
Eric Rowley, Utah State University
Jerry Stonewater, Miami University
Joseph Walker, Elizabethtown College

We also would like to express our gratitude to the preK, elementary, and junior high school pre-service teachers who have been in our classes and have given us valuable insight that has been of significant help to us in writing this book. We also gratefully acknowledge our colleagues who have taught this course and who have shared very helpful ideas with us. We especially appreciate the input of Roger Day at Illinois State University, who supplied valuable ideas and material for improving the technology dimension of the first edition of this text.

A special thanks to the dedicated editors at Addison Wesley Longman: Bill Poole (in the early stages), Anne Kelly, Peggy McMahon, Becky Anderson, and others who gave dedicated effort to making this revision of *Mathematics for Elementary School Teachers* a quality textbook.

We also acknowledge the quality work done by Sandi Goldstein, Editorial Consulting Services, and Susan Gallier at Elm Street Publishing Services, Inc., at various stages of this project.

Finally, we want to thank our spouses and families for their continued affirmation and helpfulness during the writing of this text.

Phares O'Daffer, Randy Charles, Tom Cooney, John Dossey, Janie Schielack

Making Connections

Each **Chapter Perspective** shows the connection between the content of the chapter and how it applies to real-world situations. It also helps students to see how the concepts they are studying in each chapter connect to other mathematical topics. Additionally, it provides an overview of the chapter and functions as an advance organizer.

6

Rational Number Operations and Properties

Chapter Perspective

Both fractions and decimals describe a comparison between two quantities that is the basis for rational numbers. Measuring the worth of stocks requires describing quantities between whole number values. Stock prices, which have been reported in fraction form, are now commonly represented by decimals. For example, the price $58\frac{3}{8}$ is now reported as 58.375. An understanding of rational numbers and their related symbols provides an important foundation for extending the use of addition, subtraction, multiplication, and division to solve problems involving measurement, probability, and statistics.

In this chapter, you can develop a greater sense of the importance of the rational number system by exploring the symbolization used to represent rational numbers, the use of operations with rational numbers to solve problems, and the ordering and density properties of rational numbers.

276

The **Communicating and Connecting Ideas** exercises, including special *Historical Pathways* exercises, help students to communicate mathematically and make important connections to history and other mathematical concepts.

D. Communicating and Connecting Ideas

35. a. Find the product of $\frac{1}{2}$ and $\frac{1}{4}$ in fraction form. Then write the fractions $\frac{1}{2}$ and $\frac{1}{4}$ in decimal form and find the product in decimal form. Which product is more accurate? Explain your answer.
   b. Find the product of $\frac{1}{3}$ and $\frac{2}{3}$ in fraction form. Then write the fractions $\frac{1}{3}$ and $\frac{2}{3}$ in decimal form and find the product in decimal form. Which product is more accurate? Explain your answer.

36. **Historical Pathways.** Decimal notation was firmly established in the sixteenth century, largely owing to demands from new developments in astronomy, navigation, trade, engineering, and war that computations be performed more quickly and accurately. With a partner, discuss the advantages and disadvantages of multiplication in both fraction and decimal form. Write a brief paragraph summarizing your conclusions.

The **Connection to the NCTM Standards,** through specific curriculum recommendations and quotes, helps the students see that the concepts they are studying are important topics recommended by curriculum experts.

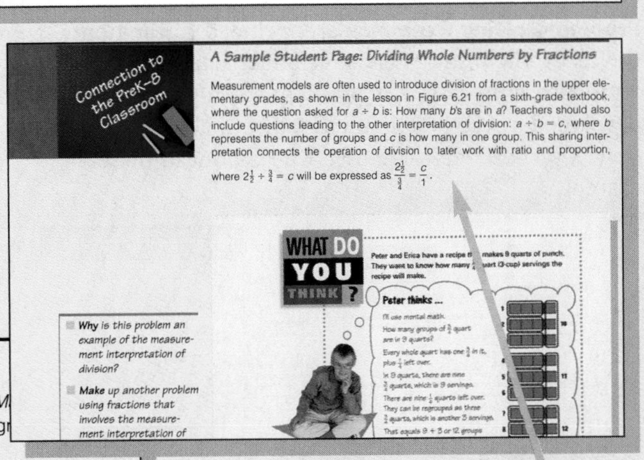

Connection to the PreK–8 Classroom

A Sample Student Page: Dividing Whole Numbers by Fractions

Measurement models are often used to introduce division of fractions in the upper elementary grades, as shown in the lesson in Figure 6.21 from a sixth-grade textbook, where the question asked for $a \div b$ is: How many b's are in a? Teachers should also include questions leading to the other interpretation of division: $a \div b = c$, where b represents the number of groups and c is how many in one group. This sharing interpretation connects the operation of division to later work with ratio and proportion, where $2\frac{1}{4} \div \frac{3}{4} = c$ will be expressed as $\frac{2\frac{1}{4}}{\frac{3}{4}} = \frac{c}{1}$.

Connection to the NCTM Principles and Standards

The NCTM *Principles and Standards for M*... mentary school mathematics curriculum g... rational numbers so that students can

- understand and represent commonly used fractions, suc... *(p. 78);*
- develop understanding of fractions as parts of unit whole... tion, as locations on number lines, and as divisions of wh... and
- work flexibly with fractions, decimals, and percents to so...

The **Connection to the PreK–8 Classroom** relates the concepts to their development and application for children in Grades PreK–8. Student Pages are included to show how the mathematics is actually introduced to the PreK–8 student.

COMMUNICATING

The **Your Turn—Practice and Reflect** part of each text example is another way we promote student communication of the mathematical ideas being studied.

Your Turn

Practice: If the length of the line segment

____

represents 1 unit of length, use ideas of halves, thirds, and so on to describe th lowing lengths.

a. ______

b. ______________

c. ____________________________

Reflect: In each of your answers, identify the number you used to repr the part and the number you used to represent the whole. Is the number tha resents the part necessarily less than the number that represents the w Explain. ■

Exercises designed for group activity are identified by the icon shown.

32. Making Connections. Form a group with three other classmates. See who can find the magazine or newspaper article that contains the greatest use of rational number ideas. With the rest of the class, discuss the kinds of articles you looked for and maybe some surprises that you found.

Students are asked to **Write About; Talk About; Make a Graph, Chart or Table;** or **Draw a Diagram or Picture** to communicate their findings or observations from Mini-Investigations.

Talk with other classmates about what they remember about learning to use these symbols and compare your experiences.

Mini-Investigation 6.1 **Communicating**

What symbols are used to describe quantities between 0 and 1?

PROBLEM SOLVING

Problem-Solving Examples are included throughout the text to help students better understand the problem-solving process. Key problem-solving strategies developed in Chapter 1—make a list, make a table, draw a picture, etc.—are used throughout in these examples.

The continual focus on the **Problem-Solving Process** (Understand the Problem; Develop a Plan; Implement the Plan; Look Back) helps students develop techniques that become an integral part of their approach to solving problems.

Example 6.15 **Problem Solving: Producing Salsa**

Suzanne's special salsa recipe makes $4\frac{3}{4}$ quarts of salsa. The local market wants to sell it in decorative jars that hold $\frac{2}{3}$ quart. How many recipes would Suzanne have to make to get full jars with no salsa left over?

Working toward a Solution

Understand the problem	*What does the situation involve?*	Filling jars with salsa and trying to make the recipe amounts and the amount needed to fill the jars come out even.
	What has to be determined?	How many recipes must be made to produce a whole number of full jars?
	What are the key data and conditions?	The recipe makes $4\frac{3}{4}$ quarts. Each jar holds $\frac{2}{3}$ quart.
	What are some assumptions?	The recipes always come out with exactly the same amount; each jar holds exactly $\frac{2}{3}$ quart.

C. Reasoning and Problem Solving

26. The Inventory Problem. A spool of ribb $25\frac{1}{3}$ yards of brocade ribbon. The ribbon evenly among three stores. One store sold bon. Another store sold $\frac{1}{2}$ of its ribbon. Th was closed for inventory and sold none c How much of the original ribbon is left?

a. Use rational numbers and operations equation or set of equations to describe t

b. Can you think of a different equation o tions that could be used? If so, write it. explain why not.

c. Use the equation or set of equations t question.

The **Reasoning and Problem Solving** exercises in each section reinforce students' problem-solving skills.

Student problem-solving opportunities are integrated into many of the Mini-Investigations.

Mini-Investigation 6.8 **Solving a Problem**

How can you use square tiles of three colors (blue, yellow, and green) to make a rectangle that is $\frac{1}{2}$ green, $\frac{1}{3}$ blue, and $\frac{1}{6}$ yellow?

MATHEMATICAL REASONING

Most examples present **alternative ways of thinking** to show students that they can construct their own procedures, reason their way to a solution, and come to understand that frequently there is more than one way to accomplish a mathematical task.

Sharon's thinking: I drew a rectangle to represent the entire job. Then I separated the rectangle into sections to represent the parts done by John and the part done by Gretchen.

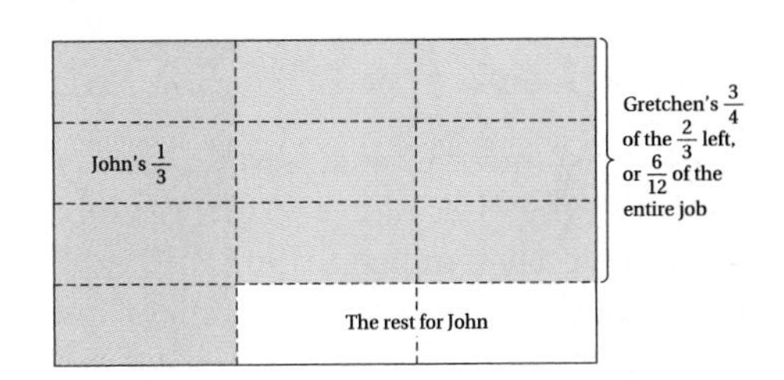

John left Gretchen $1 - \frac{1}{3}$, or $\frac{2}{3}$ of the job to do. She did $\frac{3}{4}$ of the $\frac{2}{3}$. The diagram shows the parts of the whole in terms of twelfths. Thus, Gretchen did $\frac{6}{12}$, or $\frac{1}{2}$, of the job and she should receive $\frac{1}{2} \times \$150 = \$75$.

Marco's thinking: I used the equation $\frac{3}{4}(1 - \frac{1}{3}) \times \$150 = \$75$.

Both **Inductive Reasoning** and **Deductive Reasoning** are developed in Chapter 1 and utilized throughout the book. Additionally, **Mini-Investigations** offer students opportunities to study patterns and relationships and to make generalizations.

Mini-Investigation 6.4 **Using Mathematical Reasoning**

When you divide to find a decimal for a fraction, how do you know that every nonterminating decimal must repeat a certain set of digits over and over?

Estimation helps students develop a better sense of size and magnitude, and fosters higher-level thinking.

Estimation Strategies

Products of rational numbers in either fraction or decimal form can be estimated by using the same techniques, based on place value, that are used to estimate products of whole numbers. Some of the techniques discussed in Chapter 3 were rounding, front-end estimation, and substituting compatible numbers. Because decimal notation is based on place value, these estimation techniques work just as they do with whole numbers. In fact, using estimation techniques with decimal notation can verify where the decimal point should be placed in the product, as shown in Example 6.14.

Estimation Application

How You Spend Your Time

Estimate the fraction of your life so far that you've spent

a. sleeping,

b. eating,

c. watching TV, and

d. going to school.

Describe how you made your estimates.

Special exercises involving estimation are identified by the ≈ icon as shown.

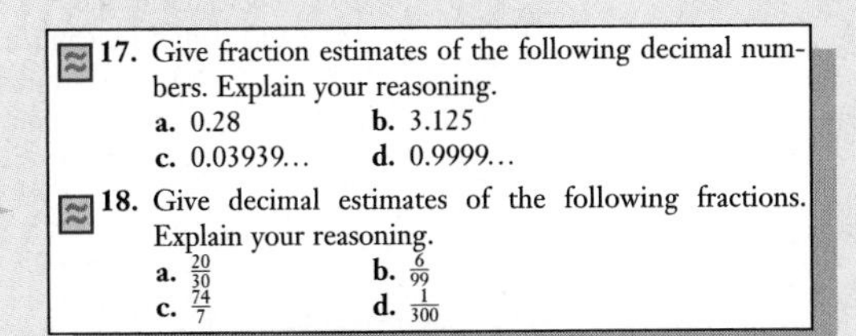

17. Give fraction estimates of the following decimal numbers. Explain your reasoning.

a. 0.28 **b.** 3.125

c. 0.03939... **d.** 0.9999...

18. Give decimal estimates of the following fractions. Explain your reasoning.

a. $\frac{20}{30}$ **b.** $\frac{6}{99}$

c. $\frac{74}{7}$ **d.** $\frac{1}{300}$

REPRESENTATION

"Finding a Pattern" Mini-Investigations encourage students to use verbal, numerical, pictorial, graphical, and symbolic representations to organize data in order to find patterns.

> **Mini-Investigation 6.3** **Finding a Pattern**
>
> What pattern can you discover in Figure 6.6 that would help you produce a set of equivalent fractions associated with the point on the number line that is one-fourth of the way from 1 to 2?

The **Spotlight on Technology** feature suggests opportunities for optional incorporation of technology.

> **Spotlight on Technology** **Computer Spreadsheets**
>
> A computer spreadsheet is a grid of rows and columns used to store, organize, and manipulate numbers and other data. The first spreadsheet for a personal computer was VisiCalc. Released in 1979, this software helped popularize desktop computing. Many publishers now offer spreadsheet software for a variety of computers.
>
> The rows and columns of a computer spreadsheet are labeled with numbers and letters, as shown in Figure 1.2. The labels provide an easy way to identify spreadsheet cells. For example, the number 3.8 is in cell B3, the cell where column B and row 3 intersect. Note that spreadsheet cells can contain words or numbers.

Technology is also incorporated into the text as an enhancement to the discussion and exercises, but remains a completely optional feature. Technology exercises are identified with the 💻 icon as shown.

> **22.** A student used her fraction calculator to examine the following pattern.
>
> | Key Sequence | Display |
> |---|---|
> | [+] 2 [/] 5 [=] | 2/5 |
> | [=] | 4/5 |
> | [=] | 6/5 |
> | [=] | 8/5 |
> | [=] | N/D→n/d 10/5 |
>
> **a.** What will appear on the screen when she presses the equals sign again?
> **b.** Including the five times indicated here, how many times must she press the equals sign until 42/5 appears on the display?

Graphing Calculator and Geometry Exploration Software appendices further provide instruction on the basic functions and features of these valuable tools.

SECTION EXERCISES

These exercises focus on basic concepts and skills.

These exercises stimulate further thinking about key topics.

These exercises extend opportunities to use higher-level thinking processes.

These exercises enhance student ability to communicate mathematically and make important connections to history and to other mathematical ideas.

> **A. Reinforcing Concepts and Practicing Skills**
>
> **1.** Use an **area model** to illustrate the quantity $\frac{3}{5}$. (Indicate clearly the corresponding representation for 1.)
> **2.** Use a **linear model** to illustrate the quantity $\frac{3}{5}$. (Indicate clearly the c...
> **3.** Use a **discr**... quantity $\frac{3}{5}$. (... tation for 1...
> **4.** Use an **area**... clearly the c...

> **B. Deepening Understanding**
>
> **14.** For each of the following rational number models, show a corresponding model for 1.
>
> a. c. e.

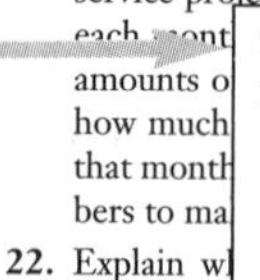

> **C. Reasoning and Problem Solving**
>
> **21. The Service Project Problem.** A group of local junior high school students meets once a month to plan a service project. Different numbers of students show up each mont... amounts o... how much... that month... bers to ma...
> **22.** Explain w... denominat...

> **D. Communicating and Connecting Ideas**
>
> **30.** A certain telephone company commercial asserts that large discounts advertised by other companies may not be a better deal for the consumer. In the commercial, the statement is made that "They say you save $\frac{1}{5}$ of the cost, when really you save only $\frac{1}{20}$ of the cost." Describe a situation in which both fractions, in a sense, could give correct information.

CHAPTER SUMMARY

CHAPTER SUMMARY

Key Ideas: Questions and Answers

Section 6.1

- **What language and ideas are needed to explain rational numbers?** *(pp. 277–281)* After identifying a *whole* and separating the whole into equal parts, we use two integers to describe a portion of the whole. For example, three of four equal parts is $\frac{3}{4}$ of the whole.
- **What are rational numbers?** *(pp. 281–282)* A rational number may be thought of as the relationship represented by an infinite set of ordered pairs, each of which describes the same quantity.
- **What are fractions and how can they be used to represent rational numbers?** *(pp. 282–284)* A fraction is a symbol, $\frac{a}{b}$, with a called the numerator and b called the denominator. Two fractions are equivalent ($\frac{a}{b} = \frac{c}{d}$) if

Key Ideas: Questions and Answers provide succinct summaries of each topic discussed in the chapter.

Key Terms and Generalizations provide a handy reference (including page numbers) with which students can look up specific mathematical terms and ideas.

Key Terms, Concepts, and Generalizations

Section 6.1
Whole (p. 278)
Equal parts (p. 278)
Ordered pair (p. 278)
Rational number (p. 281)
Fraction (p. 282)
Numerator (p. 282)
Denominator (p. 282)
Adding and subtracting in fraction form (p. 299)
Closure property of addition (p. 300)
Identity property of addition (p. 300)
Commutative property of addition (p. 300)
Associative property of addition (p. 300)
Zero property of multiplication (p. 312)
Commutative property of multiplication (p. 312)
Associative property of multiplication (p. 312)
Distributive property of multiplication over addition (p. 312)

CHAPTER REVIEW EXERCISES

These exercises assess student understanding of basic concepts and procedures developed in the chapter.

These exercises assess student ability to use higher-level thinking processes.

CHAPTER REVIEW

Concepts and Skills

1. Draw a diagram to show the meaning of each rational number.
   a. $\frac{5}{8}$ **b.** $2\frac{3}{5}$ **c.** 0.125
2. State the fundamental law of fractions and give an example of it.
3. Complete the following chart of symbols for the rational numbers shown.

Fraction form	*Decimal form*
$\frac{5}{12}$	?
?	0.125125 …
?	0.56

Reasoning and Problem Solving

11. Is 2.357357… a rational number? Use the definition of rational number to justify your answer.
12. Describe the connections between rational numbers and division.

Chapter Review Exercises end with **Alternative Assessment.** These items provide opportunities for a more open-ended, observational approach to assessment.

Alternative Assessment

19. Write a paragraph, or design a flow chart, to describe an efficient process for deciding whether any two rational numbers are equivalent. (Remember, the rational numbers might be given in either decimal or fraction form or both.)
20. When young students first encounter multiplication and division of rational numbers, they are often confused because their generalizations that multiplication

1 Viewing Mathematics

Chapter Perspective

When asked, *What is mathematics?* some people will focus on what mathematics is used for, that is, how mathematics can be used in science or in everyday applications. Other people will think of numbers or geometric shapes. Still others will think of mathematics as reasoning or problem solving. For example, in the picture above an architect is showing blueprints to a family. The creation of the blueprint involves many mathematical ideas, including scaling, ratio and proportion, geometry, and many other topics addressed in the following chapters.

In this chapter we invite you to consider what mathematics means to you. The chapter also highlights the processes of communicating with mathematics and technology, reasoning mathematically, searching for patterns, solving problems, and making connections. These mathematical processes will be introduced in this chapter and used throughout the book.

Connection to the NCTM Principles and Standards

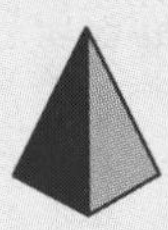

The National Council of Teachers of Mathematics (NCTM) produced the document *Principles and Standards for School Mathematics* (2000) that provides curricular, teaching, and learning guidelines for school mathematics grades PreK–12. This document outlines standards for different content areas and for the following mathematical processes: problem solving, reasoning and proof, communication, connections, and representation. These processes are fundamental to doing mathematics. A sample taken from the *Principles and Standards* that illustrates the meaning of these processes is given below.

Instructional programs should enable students to

- solve problems that arise in mathematics and in other contexts *(p. 52);*
- recognize reasoning and proof as fundamental aspects of mathematics *(p. 56);*
- communicate their mathematical thinking coherently and clearly to peers, teachers, and others *(p. 60);*
- understand how mathematical ideas interconnect and build on one another to produce a coherent whole *(p. 64);* and
- create and use representations to organize, record, and communicate mathematical ideas *(p. 67).*

The processes emphasized in this chapter and throughout the text are based on these NCTM standards.

Connection to the PreK–8 Classroom

In grades PreK–2, children experience mathematics as an exploratory activity in which they can investigate numbers and shapes. Their mathematical experiences are based on inductive reasoning grounded in the use of concrete materials.

In grades 3–5, children develop various strategies for solving problems and use different types of reasoning processes.

In grades 6–8, students use technology to solve problems and use mathematics to communicate ideas. They are familiar with a variety of strategies to solve problems.

Section 1.1 MATHEMATICS AS COMMUNICATION

- Views and Attitudes about Mathematics
- Communicating Mathematically
- Using Technology to Communicate Mathematically

In this section, we examine the various roles of communication in perceiving, understanding, and using mathematics. We ask you to communicate about the kinds of experiences you have had in doing mathematics, to be open-minded about your attitude toward mathematics, and to begin to formulate a revised view—if appropriate—of what you consider mathematics to be. We also consider how technology can be used to display data in ways that can help you explore mathematical ideas, solve mathematical problems, and communicate the results.

Views and Attitudes about Mathematics

What does *mathematics* mean to you? How do you feel about mathematics? We consider both questions in this section.

Views about Mathematics. Some people view mathematics as solving problems—much like solving a mystery—whereby they are challenged to find a solution. Other people view mathematics as carrying through systematic procedures, such as calculations, with a goal of doing the procedure correctly to get a correct answer. Still others view mathematics as a process of reasoning, in which things fit together logically and theorems are proved—leaving no room for contradictions or inconsistencies. And then there are those who view mathematics as a search for patterns, often using technology, with an inexhaustible source of new relationships to discover and communicate to others.

For some, mathematics is a game with rules for play given; for others, mathematics is a work of art—a thing of beauty. Whether mathematics is seen as a finite, static body of material to be learned or as an ever-growing body of ideas where new discoveries are made every day, you can be sure that there is a broad spectrum of popular ideas about mathematics.

Certainly, adults' views of mathematics are influenced by their school experiences with mathematics. If their school experiences were primarily computational in nature, it is not surprising that people might view mathematics as a subject that consists primarily of doing computations. But, if their mathematical experiences focused on solving problems, they are likely to view mathematics primarily as problem solving.

Attitudes toward Mathematics. You may have known people who get a great deal of satisfaction from meticulously completing a long-division problem to arrive at the correct answer, as well as others who might be totally bored with doing so. These examples illustrate that people have varying attitudes about mathematics. Nevertheless, research shows that mathematics is a favorite subject for many elementary school children.

Our attitudes about mathematics are influenced not only by our school experiences, but also by the media. In *Radio Days*, Woody Allen portrays mathematics negatively by asking a schoolmate to recite mathematical formulas to illustrate what a disagreeable character the classmate really is. On the other hand, TV shows in the past decade, such as "Square One," have made an explicit effort to help children develop a positive attitude toward mathematics.

The attitudes of family members and friends about mathematics are also influential. You may have heard someone say, in an almost bragging way, "I was never any good at mathematics." But hardly anyone ever brags about not being able to read. Mathematics is such a crucial part of life that we cannot afford to encourage children to develop the attitude that not being as good as they can be at doing mathematics is "okay."

Mini-Investigation 1.1 will help you analyze your attitude toward mathematics.

Draw a picture showing your temperature. State what you like and dislike about mathematics, express your feelings about your ability to do mathematics, and relate what has most influenced your attitude toward mathematics.

Mini-Investigation 1.1 **Communicating**

How high would your temperature be on this "mathematics thermometer"?

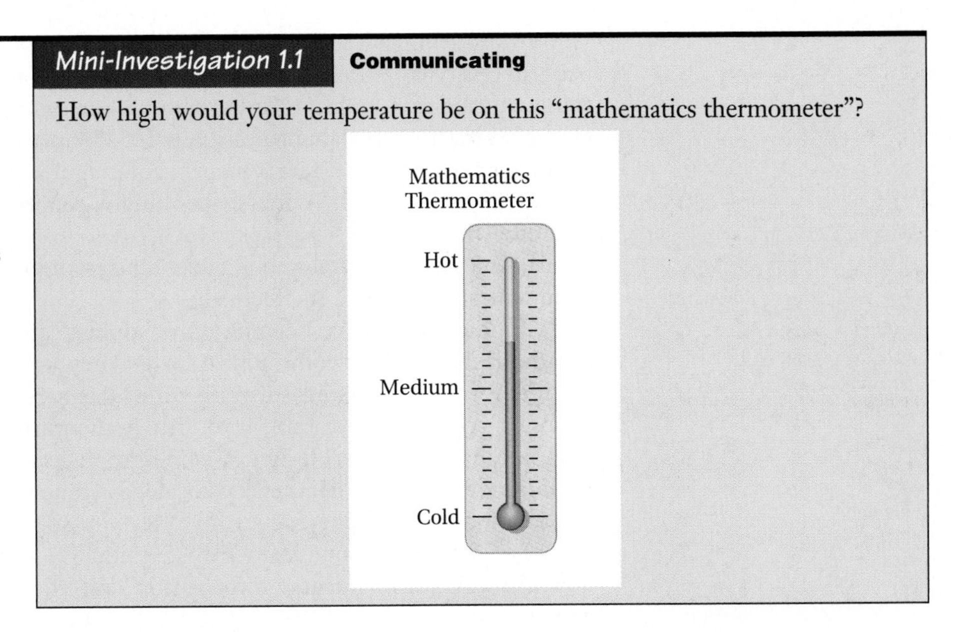

Communicating Mathematically

When people think of communication, they typically think about activities such as writing, drawing, speaking, or using body language. The message of this section is that words, symbols, graphs, tables, numerical data, and pictures can be used to represent and communicate ideas. Often these representations are produced and processed using technology. We first look at the role of mathematical words and symbols in communication.

Communicating with Mathematical Words and Symbols. Mathematics is a language of words and symbols that is used to communicate ideas of number, space, and real-world phenomena. For example, the language of mathematics may be used to describe how to get from one store to another by traveling *parallel* or *perpendicular* to certain roads. Or the words *similar* and *congruent* may be used to explain the relationship between pairs of geometric figures.

In both everyday and mathematical situations you have probably used words such as *square*, *diagonal*, and *circle*, which have commonly understood meanings. You may also have stated some relationships, such as *intersect*, *inside*, *on*, or *tangent to*. As you begin to use the language of mathematics meaningfully, you reinforce your understanding of mathematics as you learn to communicate your ideas to others.

Symbols are often used to represent mathematical ideas and can play an important role in communicating those ideas clearly and efficiently. For example, the symbols $E = mc^2$ may stimulate a scientist to think about significant ideas involved in Einstein's theory of relativity. Seeing the symbol π may cause you to think about a relationship involving a circle. Ideas can be represented by symbols, and the symbols can stimulate thinking about those ideas. As you think about familiar mathematical words and symbols and what they mean, you will see that symbols and carefully chosen words can economically represent ideas in mathematics. This use of words and symbols not only allows you to communicate ideas more efficiently, but makes discovering new ideas and important relationships easier. Exercises 6 and 27 in Section 1.1 will help you understand the importance of communication in mathematics.

Communicating with Data and Graphs. Data and interpretations of data are used in everyday situations and in newspapers, magazines, and television to represent and communicate information. For example, you might say that the average salary for a certain profession is $45,000 per year or that the median height of students in a class is 5 feet 3 inches. You might also say that the scores on a test ranged from 74 to 98. A newspaper might report a high correlation between cigarette smoking and lung cancer or between exercise and mental health. In the media, data are often displayed in tables, charts, or graphs in order to condense and communicate information effectively.

Graphic artists have become quite sophisticated in their ability to produce just the right graph to communicate what they want the reader to perceive. For example, word processing software often has a feature that allows the user to choose a graph and customize it to display the information they want to emphasize. Because the clever design of some graphs can disguise or visually alter the meaning or impact of data, everyone needs to be able to interpret correctly the information such graphs convey. We develop several techniques for examining graphs in Chapter 8, but Mini-Investigation 1.2 provides an initial example of how a graph communicates information and allows you to interpret properly what you see.

Write a one-paragraph interpretation of the graph and compare your interpretation with those of others in your class.

Mini-Investigation 1.2 **Communicating**

What message is communicated by this graph?

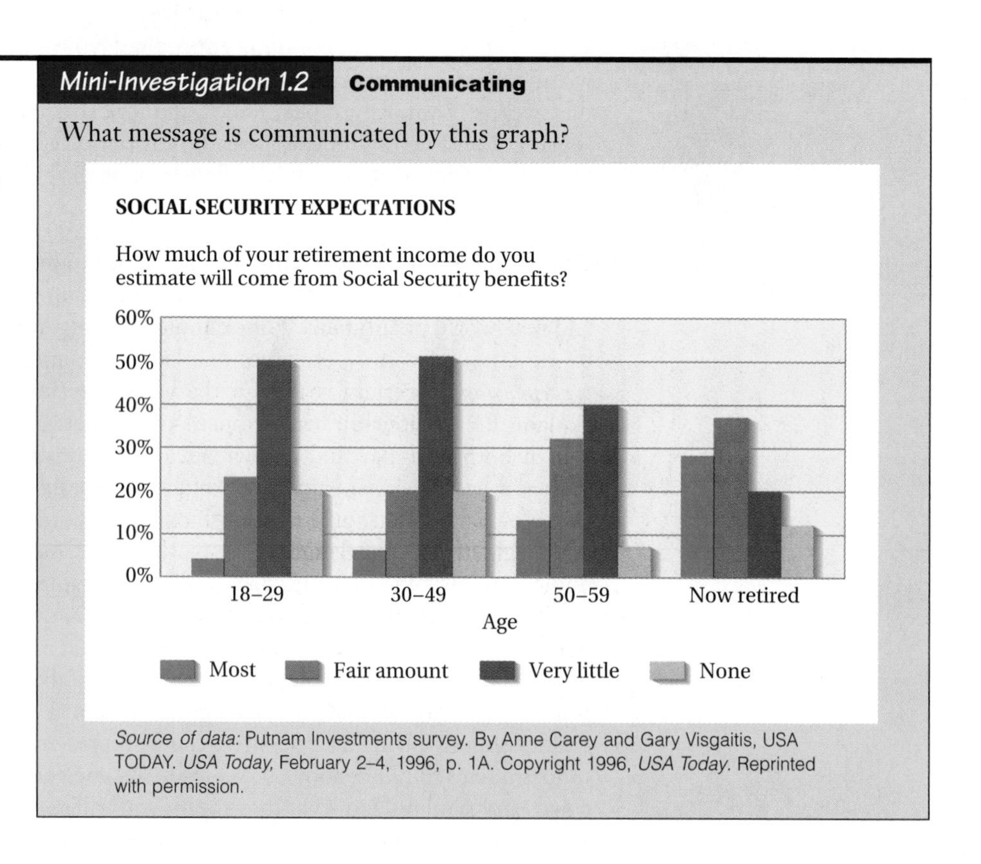

Source of data: Putnam Investments survey. By Anne Carey and Gary Visgaitis, USA TODAY. *USA Today,* February 2–4, 1996, p. 1A. Copyright 1996, *USA Today.* Reprinted with permission.

Another example of how graphs communicate information is shown in Figure 1.1. This circle graph visually conveys information about how teenagers spend their time during a typical weekday.

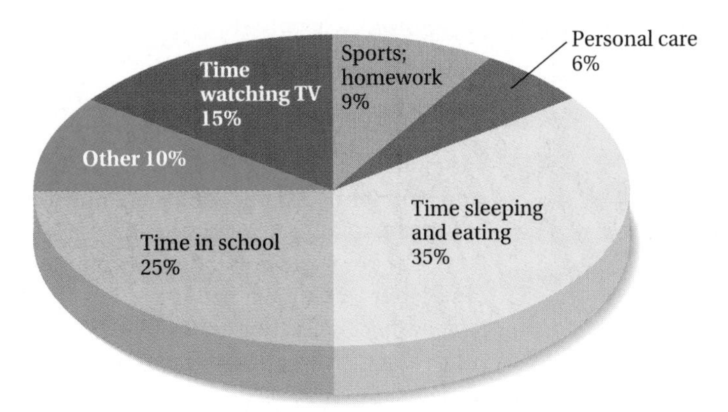

FIGURE 1.1 | Circle graph showing the weekday distribution of time for a typical teenager.

Graphs are an important way of displaying the results of classifying data efficiently and visually. In Chapter 8, we discuss in detail various methods of data analysis and types of graphs used to display data.

One of the important aspects of doing mathematics in the elementary grades is to encourage students to communicate their mathematical ideas and thinking. The NCTM *Principles and Standards* call for the use of "mathematical communities" in the classroom where the emphasis is on students talking and writing about mathematics. In part, the goal is to encourage the acquisition of the language of mathematics, but, more fundamentally, it is to emphasize the importance of learning that doing mathematics involves communicating and representing mathematical ideas.

Using Technology to Communicate Mathematically

The use of technology plays a central role in representing and communicating mathematical ideas. Technology can be used to show a graph of an equation, to produce and vary a chart containing related data, or to solve and present a solution to a problem. These various forms of representation can be used to communicate ideas.

Because many useful and relatively inexpensive types of technology have been developed, technology plays an increasingly important communication role in the development and application of mathematics. People using mathematics in many different professions now regularly utilize the computer in their daily activities. The computer also helps them communicate with each other to analyze data, share conjectures, and find and present solutions to problems.

Technology is also being used more and more to help people learn mathematics and to communicate with other learners about mathematics. Initially, educators debated the wisdom of allowing students to use calculators while learning mathematics. The evidence now seems clear that the use of calculators not only does not impede learning of mathematics but can even enhance students' abilities to solve problems and think conceptually. In the following subsections, we explore three forms of technology that are commonly used in mathematics and that will be referred to throughout this book: computer spreadsheets, calculators, and geometry exploration software.

Spotlight on Technology **Computer Spreadsheets**

A computer spreadsheet is a grid of rows and columns used to store, organize, and manipulate numbers and other data. The first spreadsheet for a personal computer was VisiCalc. Released in 1979, this software helped popularize desktop computing. Many publishers now offer spreadsheet software for a variety of computers.

The rows and columns of a computer spreadsheet are labeled with numbers and letters, as shown in Figure 1.2. The labels provide an easy way to identify spreadsheet cells. For example, the number 3.8 is in cell B3, the cell where column B and row 3 intersect. Note that spreadsheet cells can contain words or numbers.

The user can enter a formula into a spreadsheet cell to instruct the computer to calculate values from other cells. For example, we could select cell B6 and enter the formula = sum (B3 + B4 + B5). The sum of the three cells in the column above B6 will then be calculated and recorded in cell B6. The value of a computer spreadsheet is obvious when we type in new values in cells B3, B4, and B5. The formula automatically recalculates the sum and replaces the old sum with it in cell B6.

	A	B	C	D	E
1	Projected Sales by City (millions of dollars)				
2	City	1996	1997	1998	
3	Denver	3.8	4.1	5.1	
4	Miami	7.6	8.9	8.9	
5	Phoenix	4.7	5.1	5.1	
6	Total Sales				
7					
8					

FIGURE 1.2 | Portion of a computer spreadsheet showing row and column entries.

Because a computer spreadsheet allows cell values to be changed and formulas used to recalculate solutions, a spreadsheet is an excellent tool for mathematical problem solving. Computer spreadsheets are also being widely used in personal recordkeeping, in small businesses, and in a variety of complex technical applications.

Using a Computer Spreadsheet. Using a computer spreadsheet for problem solving allows the user to manipulate data and try alternative solutions to the problem quickly and easily. To gain an initial understanding of how a spreadsheet works, let's consider the following problem-solving situation.

> A club is planning a car wash day to make money. The members think that a charge of \$6 for washing a car is reasonable, but they have found that the cost will be \$25 to rent space for the day and \$2 per car for water and equipment. They now want to know how many cars they would have to wash before they begin to make a profit and how many cars they would have to wash to make a profit of \$100.

To find the answers to these questions, we could use pencil and paper to set up a table. In the first column of the table, we would list the number (n) of cars to be washed. In the next two columns, we would calculate the expense and income for each car washed. We know that the expenses are $2 for each car washed and $25 no matter how many are washed. To calculate the expense, we would use the formula Expense $= 2n + 25$ in the second column of the table. To calculate the income at $6 per car washed, we would use the use the formula Income $= 6n$ in the third column of the table.

Number of Cars (n)	**Expense $= (2n + 25)$**	**Income $= 6n$**
1	$(2 \times 1) + 25 = 27$	$6 \times 1 = 6$
2	$(2 \times 2) + 25 = 29$	$6 \times 2 = 12$
⋮	⋮	⋮
10	$(2 \times 10) + 25 = 45$	$6 \times 10 = 60$
⋮	⋮	⋮

Instead, we could use a computer spreadsheet that finds and displays the answers quickly and easily. We would set up the spreadsheet to display the number of cars in the cells of the first column. Then we would instruct the computer to apply the formulas we used in the table to the cells in the second and third columns. The results are shown in Figure 1.3.

	A	B	C
1	Number of Cars	Expense ($)	Income ($)
2	1	27	6
3	2	29	12
4	3	31	18
5	4	33	24
6	5	35	30
7	6	37	36
8	7	39	42
9	8	41	48
10	9	43	54
11	10	45	60

FIGURE 1.3 | Spreadsheet printout showing part of the expense and income for a proposed car-wash project.

By looking at such a spreadsheet, the club members could easily see that they would begin to make a profit when they had washed the seventh car. If they looked at more rows of the spreadsheet (not shown in Figure 1.3), they would find that they have to wash 32 cars to make a profit of $100.

Because different conditions can be easily examined by changing the numbers and the formulas in the spreadsheet, the club members could use the spreadsheet as a decision-making tool. For example, if the club members want to consider raising the price of washing a car to $7, they could see what happens to their expected income by changing the formula and having the computer recalculate the spreadsheet values in the income column. Or they could consider renting less costly space and recalculate the spreadsheet values in the expense column with a revised formula such as Expense = $2n + 15$. Example 1.1 further illustrates the use of a spreadsheet.

Example 1.1

Problem Solving: Golf Course Fees

A golf course charges a regular fee of $14 for an 18-hole round. It also offers a special summer package in which the first five rounds are $7 each, with each round after that costing $18 each. How many rounds would you have to play for the regular fee to be the best deal?

Working toward the Solution

Set up a computer spreadsheet using the data in the example. The cells in the first column should indicate the number of rounds to be played, and the cells in the second and third columns should contain the formulas for calculating the total cost for the number of rounds at the ordinary fee and at the special package fee. The values for the first seven rows are shown in the following sample computer spreadsheet.

	A	B	C
1	Rounds	Ordinary Fee ($)	Special Package ($)
2	1	14	7
3	2	28	14
4	3	42	21
5	4	56	28
6	5	70	35
7	6	84	53
8	7	98	71

Your Turn

Practice: Complete seven more rows of the spreadsheet to solve the problem. Use a paper-and-pencil copy of the spreadsheet and an ordinary calculator to fill in the values. Or use a computer spreadsheet, if a computer and software are available, to find the values.

Reflect: Explain the formula you used to find the values for the second and third columns. ■

Using a Calculator. Using a calculator allows routine mathematical tasks to be done quickly and easily. The efficiency that calculators provide frees the user to spend more time thinking about mathematical ideas, patterns, relationships, and problem solving and less time making calculations. Throughout the rest of this book, we provide numerous opportunities for you to try various tasks on different types of calculators. For now, we consider briefly some different types of calculators and their unique capabilities.

The simplest calculator is a *four-function calculator* that performs the basic operations of addition, subtraction, multiplication, and division. A second type of calculator is often called a *fraction calculator*. Its primary feature—in addition to the features of a four-function calculator—is its ability to add, subtract, multiply, and divide fractions and mixed numbers. Fraction calculators can also convert an improper fraction to a mixed number, simplify fractions, and convert a fraction to a decimal and vice versa. In this book, especially in our work with fractions in Chapter 6, we feature the Texas Instruments (TI) Math Explorer calculator, but other brands of fraction calculators may be used just as well.

A third type of calculator commonly used in mathematics is the *scientific calculator.* In addition to having all the features of a four-function calculator, most scientific calculators have the ability to carry out an entered program to complete a variety of combined calculations automatically. Other features of the scientific calculator include keys for most of the important algebraic and trigonometric functions and capabilities for calculating various statistical measures and producing random numbers.

A final type of calculator is the *graphing calculator.* In addition to having all the features of a scientific calculator, a graphing calculator can display graphs of data. For example, we could use the data from the car wash problem and a graphing calculator to plot graphs of expense and income for various numbers of cars, as shown on the following screen of a Texas Instruments TI-73 graphing calculator.

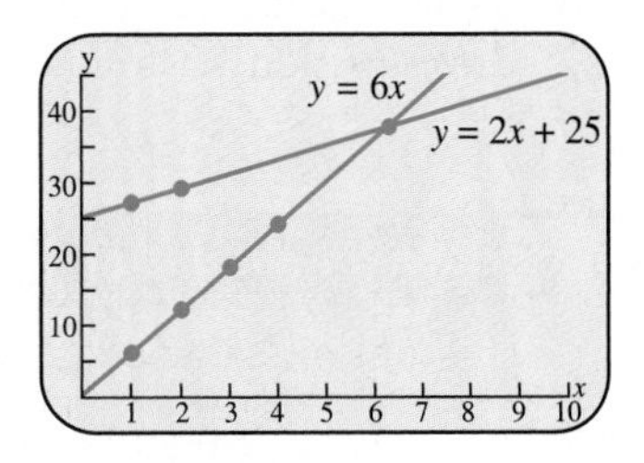

If we correctly set the scale on the (horizontal) x-axis and the (vertical) y-axis of the graph, the point where the two graphs intersect shows how many cars would have to be washed for the income to equal the expense. Appendix A contains more information on using graphing calculators.

Using Geometry Exploration Software. Another form of technology that can help in discovering and analyzing mathematical ideas is *geometry exploration software.* Appendix B contains information on using geometry exploration software, and in later chapters we explore its use in detail. But for now, let's consider briefly some of the things that geometry exploration software can do.

It allows the user to quickly produce and label a number of different examples of any geometric shape. Some types of this software even allow varying sizes and shapes dynamically in order to investigate the effect of changing them. Geometric exploration software also allows immediate measurement of the length of any segment, the size of any angle, and the area of any figure and the ability to compute with those measurements. The results of these computations are updated automatically when the data are changed.

To illustrate the value of geometry exploration software, suppose that we investigate what happens when we join the midpoints of the sides of a quadrilateral. We can instruct the computer to draw any type of quadrilateral we want, mark the midpoints of its sides, and connect them. We can also instruct the computer to calculate the area of the original large quadrilateral and the area of the smaller inner figure. After we have given these instructions to the computer, it can quickly

Area $ABCD = 1.084$	Area $ABCD = 1.036$	Area $ABCD = 1.204$
Area $EFGH = 0.542$	Area $EFGH = 0.518$	Area $EFGH = 0.602$

FIGURE 1.4 | Quadrilaterals produced by geometry exploration software.

produce additional random examples of this arrangement. Figure 1.4 shows three geometric figures that the geometry exploration software might produce. Mini-Investigation 1.3 follows up on this approach.

Draw some different quadrilaterals to illustrate your discovery. Then write a statement describing your discovery.

Technology Extension

Use geometry exploration software to produce several more examples of this relationship.

Mini-Investigation 1.3 **Finding a Pattern**

From the geometric figures and data shown in Figure 1.4, what can you discover about the following?

a. What geometric figure is formed when the midpoints of the sides of any quadrilateral are connected?

b. How does the area of the original quadrilateral compare to the area of the shaded quadrilateral?

Because geometry exploration software makes producing examples and measuring them so easy, it naturally leads to "what-if" questions. For example, you might be tempted to ask, "What if I were to join the midpoints of the sides of a triangle instead of a quadrilateral?" or "What if I were to join trisection points instead of midpoints?" In this book, especially in Chapters 10 and 11, we use geometry exploration software to explore patterns and relationships in geometric figures.

The **Technology Principle** in the NCTM *Principles and Standards* emphasizes that technology should be used to help all students learn mathematics. Every student should have access to a calculator and to a computer. Through the use of technology to solve problems and explore mathematical ideas students will be better prepared to live in a technologically oriented world.

Problems and Exercises | *for Section 1.1*

A. Reinforcing Concepts and Practicing Skills

1. Describe an experience in a previous mathematics class (any level) that you think typifies your experiences in learning mathematics. What do you especially remember about that experience?
2. List three events or people that contributed most to your view of mathematics and explain how each influenced you.
3. Suppose that you are directing a movie in which one of the leading characters is a mathematician. What movie star would you pick for the role? Identify the characteristic(s) of the movie star that match the characteristic(s) you associate with a mathematician.
4. What images of mathematics, if any, come to mind when you encounter technology, such as a computer?
5. In 1992, a company manufactured a Barbie doll that said the words, "Math is tough." Why do you think many people were upset with the manufacturer?
6. Write down directions for constructing a square that you would give a classmate over the telephone.
7. Suppose that *triangle* were not a word used in mathematics. How would you describe the shape that the word represents?
8. List three words that have special meanings in mathematics and explain their meanings.
9. List three symbols that you have used in mathematics and explain what they represent.
10. List three different ways of representing $\frac{3}{5}$.
11. List three different ways of representing 16.

SAFEST AGES ON THE ROAD

Rates of auto fatalities are lowest among drivers in mid-life, ages 45–59. Death rate per 100,000 licensed drivers[1]:

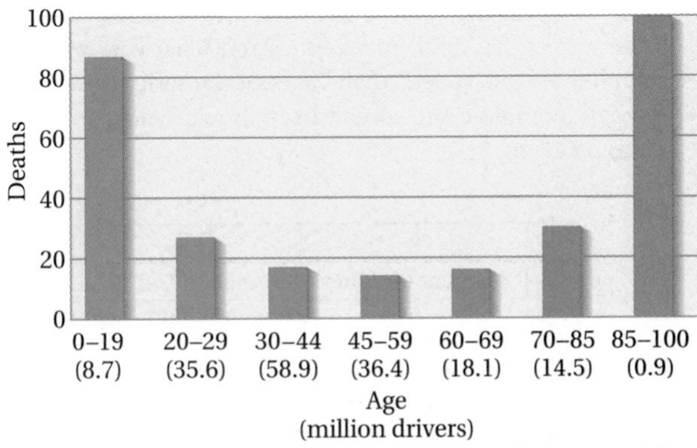

[1] 1993, latest available

Source of data: National Association of Independent Insurers. By Scott Boeck and Elys A. McLean, USA TODAY. *USA Today,* June 13, 1995, p. 1A. Copyright 1995, *USA Today.* Reprinted with permission.

12. What message is communicated by the previous graph?

Interpret the following circle graph to answer questions 13–15.

13. What general message does the graph convey to you?
14. What is the most frequent occurrence in terms of self-examinations?
15. What percentage of women perform self-examinations *at least* once a month?

REGULAR BREAST SELF-EXAMS

To aid early detection of breast cancer, the No. 2 cause of death for women, women are urged to perform a self-exam monthly. How often women actually perform them:

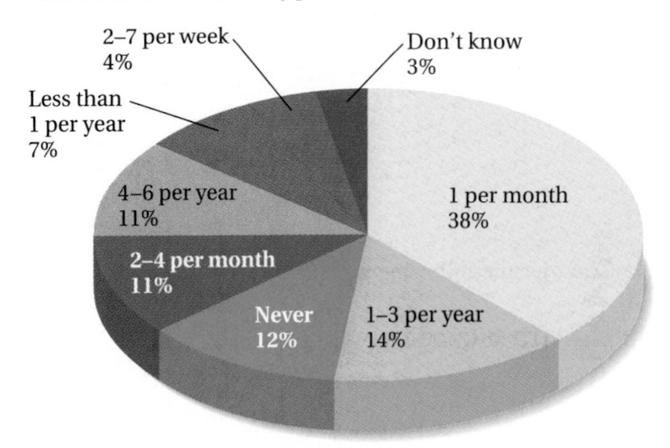

Source of data: Opinion Research Corp. survey for Sanus Health Plan. By Cindy Hall and Marcy E. Mullins, USA TODAY. *USA Today,* January 30, 1996, p. 1D. Copyright 1996, *USA Today.* Reprinted with permission.

16. Describe another way that the information in the above circle graph could be represented.
17. Consider the following computer spreadsheet.

	A	B	C
1	Cost of Item ($)	10% Discount ($)	Final Cost ($)
2	80	8	72
3	367	36.70	330.30

What formulas would you use to calculate the numbers in the following cells?

a. B2

b. C2

18. Which type of calculator would you use if you wanted to find $\frac{3}{8} + \frac{4}{5}$?
19. Which type of calculator would you use if you wanted to automatically find the value of F for several values of C in the formula $F = (9C \div 5) + 32$?
20. Which type of calculator would you use if you wanted to show a graph of the formula in Exercise 19?

21. Which features of geometry exploration software make it useful? Explain why.

B. Deepening Understanding

22. Which of the following do you think best communicates the cost of buying 1, 2, 3, . . . , 10 gallons of gasoline at $1.50 per gallon?

a. $C = 1.5n$

b.

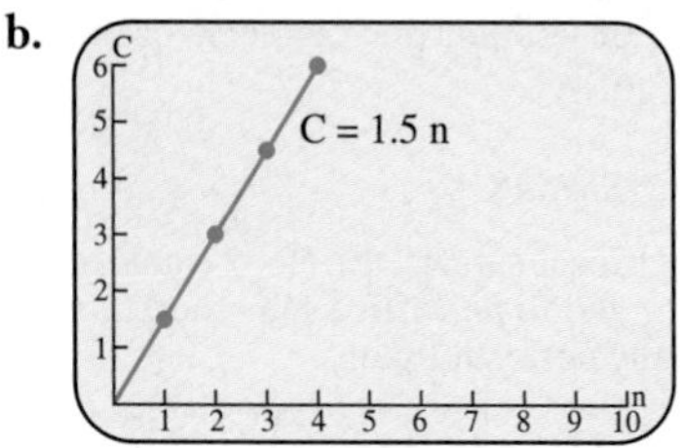

c.

Number of Gallons	Total Cost ($)
1	1.50
2	3.00
3	4.50
4	6.00
⋮	⋮

23. Consider the following graph.

COPING WITH MIGRAINES

How migraine sufferers say their families react when headaches keep them out of family activities:

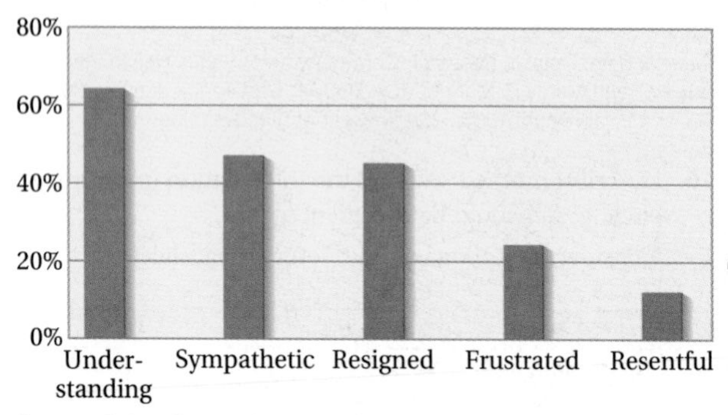

Source of data: Opinion Research Corp. for Glaxo Wellcome, Inc. By Cindy Hall and Elys A. McLean, USA TODAY. *USA Today,* February 1, 1996, p. 1A. Copyright 1996, *USA Today.* Reprinted with permission.

a. What questions could you answer with this graph?

b. What questions couldn't you answer with this graph?

c. Do you think that the title of the graph is appropriate? Explain.

d. Why is the sum of the percentages greater than 100 percent?

24. Consider the subjects of mathematics, art, and law. Which two do you think are the least alike? Explain.

25. Geometry exploration software was used to produce the following three examples of joining the midpoints C and

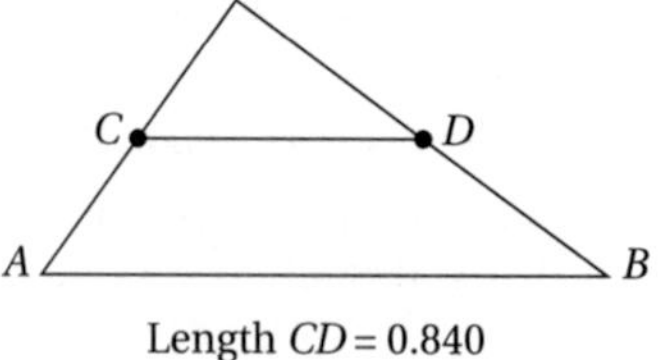

Length $CD = 0.840$

Length $AB = 1.680$

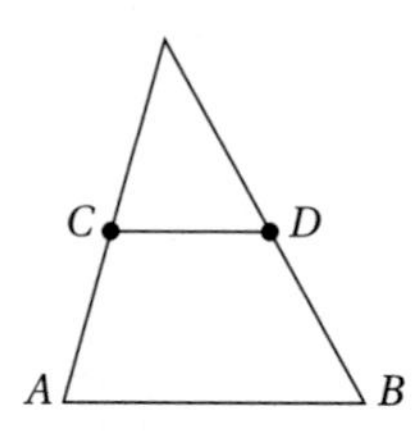

Length $CD = 0.520$

Length $AB = 1.040$

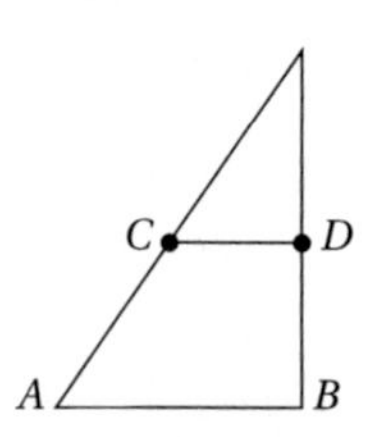

Length $CD = 0.460$

Length $AB = 0.920$

D of two sides of a triangle, along with the accompanying measurements. What do you discover when you compare the two lengths in each triangle? Write a description of your discovery.

C. Reasoning and Problem Solving

26. The Car Depreciation Problem. A newer car cost $20,000. Each year, the value of the car can be found by multiplying the previous year's value by 0.8. Thus after the first year the value is $16,000, after the second year the value is $12,800, and so on. After what year will the car be valued at less than $8,000? Describe how you would complete the spreadsheet shown below to solve the problem.

	A	B
1	Number of Years	Value of Car ($)
2	0	20,000
3	1	
4		
5		
6		

Use the following questions to help you obtain the answer.

a. What formula would you use for cell B3? cell B4? cell B5?

b. How many rows would you have in the spreadsheet? Explain.

27. Describe the idea illustrated in the following display and invent a word that could be used to communicate that idea.

All of these geometric figures illustrate the idea.

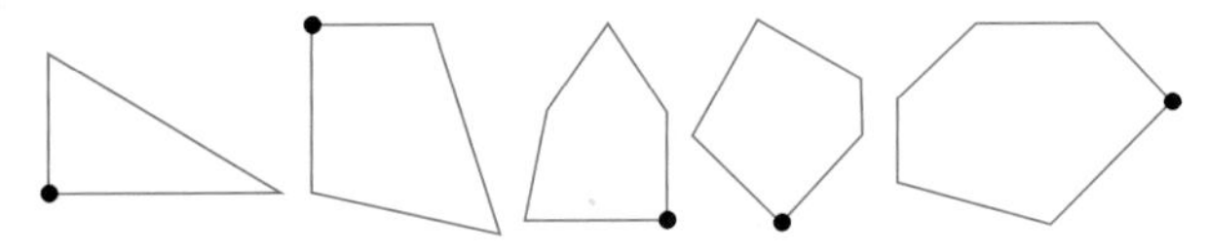

None of these geometric figures illustrate the idea.

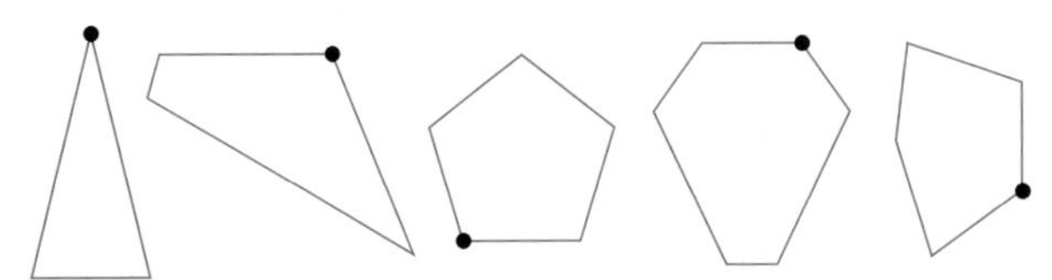

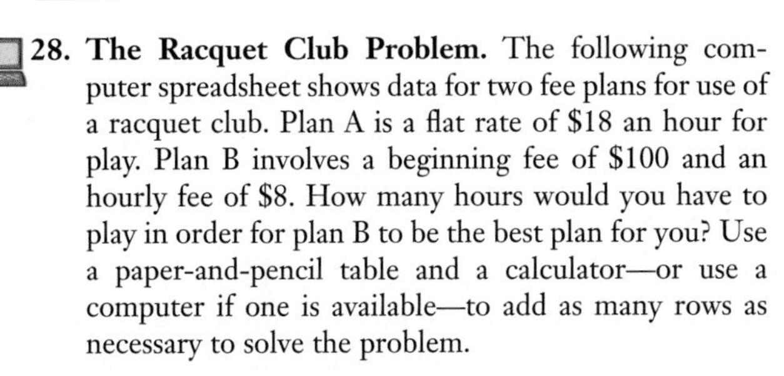

28. The Racquet Club Problem. The following computer spreadsheet shows data for two fee plans for use of a racquet club. Plan A is a flat rate of \$18 an hour for play. Plan B involves a beginning fee of \$100 and an hourly fee of \$8. How many hours would you have to play in order for plan B to be the best plan for you? Use a paper-and-pencil table and a calculator—or use a computer if one is available—to add as many rows as necessary to solve the problem.

	A	B	C
1	Number of Hours	Plan A (\$)	Plan B (\$)
2	1	18	108
3	2	36	116
4	3	54	124
5	4	72	132
6	5	90	140

Describe why the spreadsheet can be used effectively to solve the problem.

29. Geometry exploration software was used to produce the following figure (the dots are the midpoints of the sides of *ABCD*) and to measure the area of *ABCD* and *EFGH*. Estimate what part *EFGH* is of *ABCD*.

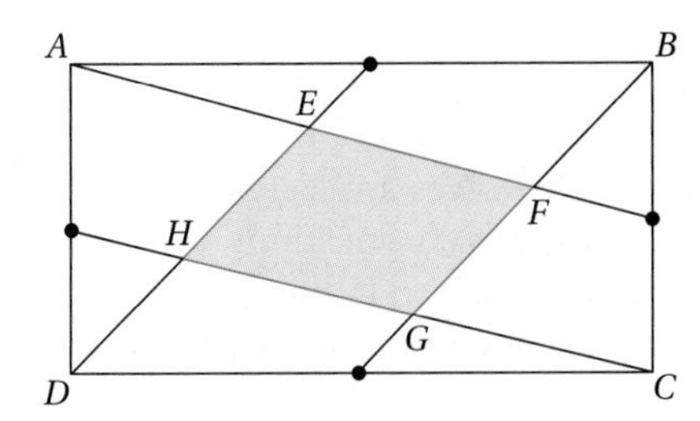

D. Communicating and Connecting Ideas

30. Work in a group of three, with each person assigned to one of three roles, to communicate information about the figure shown. The *teller* gives directions on how to draw the figure, without watching the drawer. The *drawer* draws the figure based on the teller's instructions, without looking at the figure. The *analyzer* watches the process, noting mathematical terms used or misused, and comments on the others' efforts.

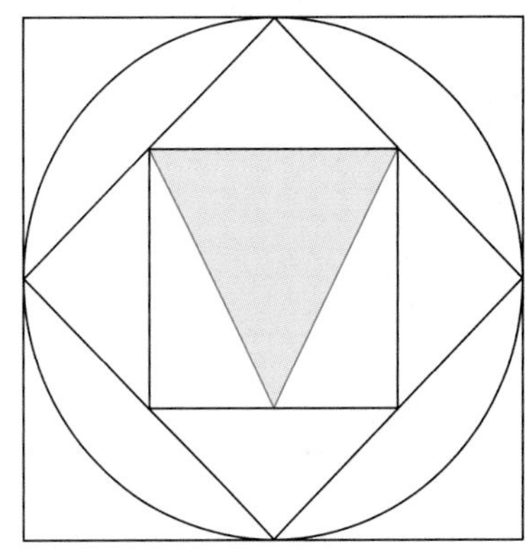

Write your observations about your role, discussing ways you could have been more effective in the communication process.

31. A researcher determined the amount of time (in hours) that students spent studying for a test, recorded their grades based on a score of 0 (low) to 6 (high), and compiled the following graph.

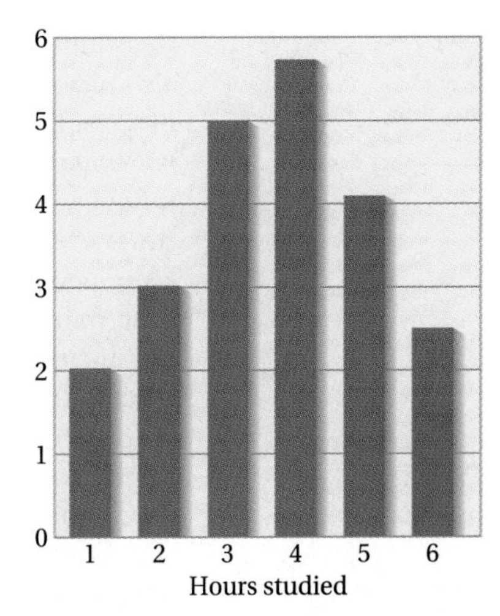

Write your interpretation of the data shown in the graph and compare your interpretation with those of others in your group.

32. **Making a Connection.** Consider the following problem: A farmer has hens and rabbits. They have a total of 50 heads and 140 feet. How many hens and how many rabbits does the farmer have?

    Make a table, or use a computer spreadsheet, to solve the problem. (*Hint:* Let the number of hens range from 50 to 0 and the number of rabbits range from 0 to 50. The number of feet can be determined as twice the number of hens plus four times the number of rabbits.)

33. **Historical Pathways.** Sometimes in mathematics a picture can communicate a lot of information. The Greek mathematician Thales (624–547 B.C.) chose points on a semicircle, connected them to the ends of the diameter of the circle, and measured the angle formed each time. He is said to have sacrificed a bull in joy over the discovery communicated by the following diagram. What do you think Thales discovered?

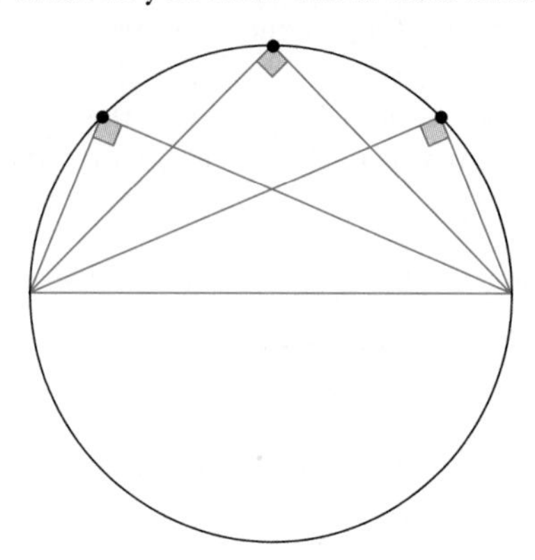

Section 1.2 MATHEMATICS AS REASONING

- Types of Reasoning
- Inductive Reasoning and Patterns
- Deductive Reasoning
- Proportional Reasoning
- Spatial Reasoning
- Importance of Reasoning Processes

In this section, we examine the view of mathematics as a process of reasoning. We consider four main types of reasoning utilized both in everyday life and in mathematics: inductive reasoning, deductive reasoning, proportional reasoning, and spatial reasoning.

Types of Reasoning

Reasoning not only plays an important role in making everyday decisions, but it also plays a central role in mathematics. Doing mathematics often requires the use of several different types of reasoning. For example, someone might use *inductive reasoning* to infer that an odd number multiplied by an even number is an even number. This inference might be based on looking at several examples of odds times evens and observing that in each case the result is an even number. Later the same person might use *deductive reasoning* to prove that an odd number times an even number is an even number, based on assumptions and the rules of logic.

In another situation, someone might use *proportional reasoning* to conclude that a flagpole is 100 feet tall, based on the person's height, the length of the person's shadow, and the length of the flagpole's shadow. In still another situation, someone might use *spatial reasoning* to draw conclusions about what its exposed surface would look like when a cube is sliced in certain ways.

As these examples suggest, reasoning involves a variety of mathematical entities, including numbers and geometric figures. If you think of mathematics as a process, you will realize that reasoning, as described in these examples, is the foundation of that process. Anytime you reach conclusions in these ways, you are using some type of mathematical reasoning. Let's now consider each type of reasoning in more detail, beginning with inductive reasoning.

Inductive Reasoning and Patterns

Inductive Reasoning. Mini-Investigation 1.4 sets the stage for understanding inductive reasoning.

Talk about situations in which you have used a similar type of reasoning.

Using Mathematical Reasoning

What is the main characteristic of the type of reasoning used in the cartoon?

"Water boils down to nothing . . . snow boils down to nothing . . . ice boils down to nothing . . . everything boils down to nothing."

Cartoon by Ed Fisher; © 1966, The Saturday Review, Inc. Reprinted by permission of Ed Fisher.

The preceding cartoon contains a somewhat humorous use of *inductive reasoning*, a process we more generally describe as follows.

Description of Inductive Reasoning

Inductive reasoning involves the use of information from specific examples to draw a general conclusion. The general conclusion drawn is called a **generalization.**

We can further describe inductive reasoning in the following way: After watching an event that gives the same results several times in succession, an observer detects a pattern or relationship and tentatively concludes that the event will always have the same outcome. For example, suppose you move into a new apartment and see a neighbor leave to walk her dog at 7 A.M. on Monday, again at 7 A.M. on Tuesday, and so on for the rest of the week. Based on your observations, you form a generalization and conclude, My neighbor always leaves at about 7 A.M. to walk her dog. We can use this example to illustrate the steps in the inductive reasoning process.

Procedure for Using the Inductive Reasoning Process

■ Check several examples of a possible relationship.	You observe your neighbor as she leaves to walk her dog each day for a week.
■ Observe that the relationship is true for every example you checked.	You observe that each day she leaves at 7 A.M. to walk her dog.
■ Conclude that the relationship is *probably* true for all other examples and state a generalization.	You conclude, "My neighbor always leaves at 7 A.M. to walk her dog."

To preview some ways in which we use the inductive reasoning process in this book, let's consider some mathematical situations. First, we double a number and add 1 each time. Then we use inductive reasoning to form a generalization.

Examples: $2 \times 3 + 1 = 7$; $2 \times 6 + 1 = 13$; $2 \times 7 + 1 = 15$; $2 \times 10 + 1 = 21$.

Generalization: When we double any number and add 1, the result is an odd number.

Next, let's triple a number and add 1 each time. Then we use inductive reasoning to form a generalization.

Examples: $3 \times 4 + 1 = 13$; $3 \times 6 + 1 = 19$; $3 \times 10 + 1 = 31$; $3 \times 12 + 1 = 37$.

Generalization: When we triple any number and add 1, the result is an odd number.

We can also do some calculations with the assistance of technology. Then we use inductive reasoning to form a general conclusion.

Examples: The screen shows several calculations carried out with a graphing calculator.

```
-2 × 7
                -14
-5 × 8
                -40
-1 × 11
                -11
```

Generalization: The product of a negative number and a positive number is a negative number.

Although the generalization that doubling and adding 1 always gives an odd number is true, we can produce another example, $3 \times 5 + 1 = 16$, which shows that the generalization that tripling and adding 1 always gives an odd number is false! Thus the use of inductive reasoning to form a generalization based on specific examples cannot ensure that the generalization will hold true for all possible cases. An example that disproves a generalization is of central importance to inductive reasoning and is called a *counterexample.*

Description of Counterexample

A **counterexample** is an example that shows a generalization to be false.

Example 1.2 further illustrates the need to look for a counterexample when you use inductive reasoning, and it presents some geometric situations that may surprise you!

Example 1.2

Finding a Counterexample

Use inductive reasoning to form a generalization from the following examples. Look for a counterexample that might prove the generalization false.

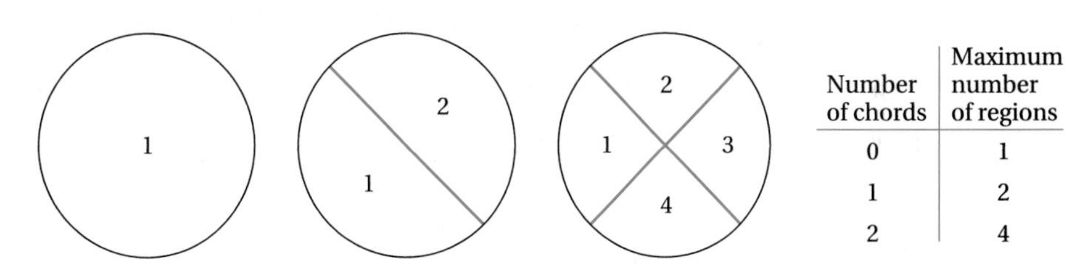

Number of chords	Maximum number of regions
0	1
1	2
2	4

Solution

Generalization: The number of regions formed by drawing chords in a circle doubles in each successive example.

Counterexamples: When we draw three chords in a circle, the maximum number of regions formed is seven, which is not the double of four regions.

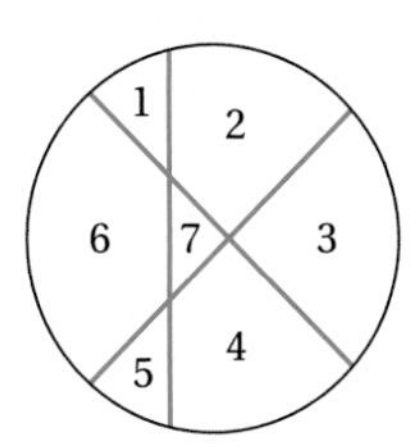

Your Turn

Practice: Use inductive reasoning to form a generalization from the following table. Then draw a large, 3-inch-diameter circle with six points on it to complete the last row of the table. Look for a counterexample that might prove the generalization false.

Reflect: Suppose you found that a generalization was correct for the first 100 examples. Does this prove the generalization true? Why or why not? ■

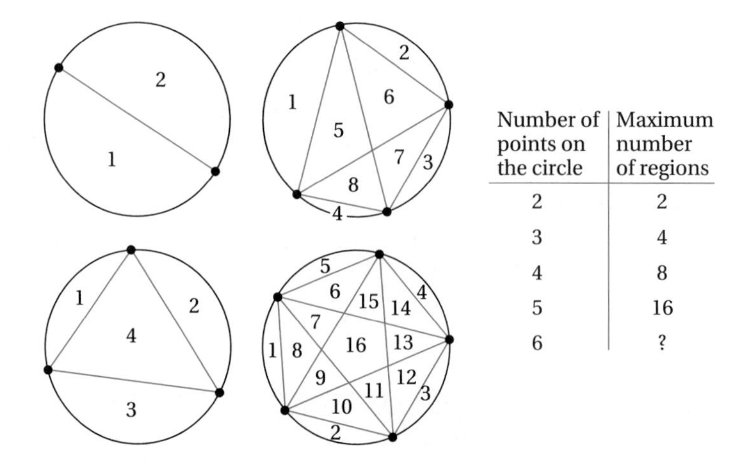

Number of points on the circle	Maximum number of regions
2	2
3	4
4	8
5	16
6	?

In Example 1.2, you could have formed the generalization that the maximum number of regions doubles each time and that the number of regions for six points would therefore be 32. The counterexample that can be found for this generalization shows that you must exercise care when using inductive reasoning. Clearly, you don't know whether a generalization formed by inductive reasoning is true or false. If you look at a lot of examples and can't find a counterexample for a generalization, you may conclude that the generalization probably is true, but you can't be sure until you have proved the generalization.

Patterns. Mathematics is sometimes defined as the science of studying patterns. When forming generalizations by inductive reasoning, you used patterns discovered by looking at several examples. Sometimes, as when you were looking for a pattern in the joining of midpoints of the sides of several quadrilaterals, the order of the examples didn't make any difference. At other times, as when you were looking at the number of regions formed by connecting points on a circle, as in Example 1.2, the order of the examples was crucial in helping you discover a pattern. In this section, we work primarily with ordered patterns.

Sequences. A pattern involving an ordered arrangement of numbers, geometric figures, letters, or other entities is called a **sequence.** The numbers, geometric figures, or letters that make up a sequence are called the **terms of the sequence.** The first four terms of a familiar numerical sequence are

$$1, 3, 5, 7, \underline{\ ?\ }, \underline{\ ?\ }, \underline{\ ?\ }.$$

The pattern is obvious, and we can easily extend the sequence by giving the next several terms. We can also look for patterns in sequences of geometric figures such as

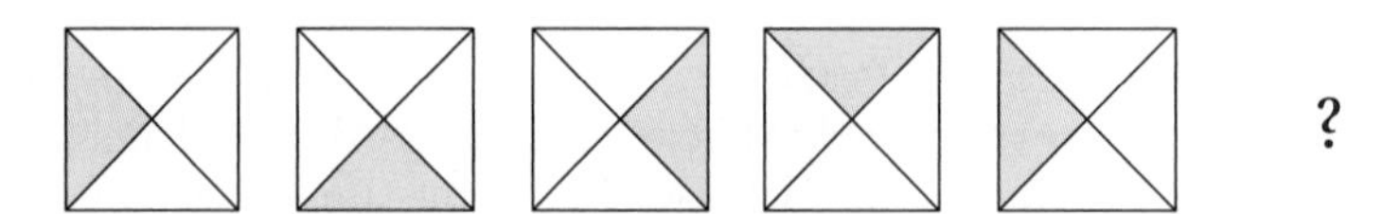

Discovering the pattern of rotation of the figure lets us anticipate the next several terms in the sequence.

Numerical sequences may be classified according to the methods used to find their terms. For example, a numerical sequence in which each term is obtained from the previous term by *adding* a fixed number is called an **arithmetic sequence.** The fixed number is called the **common difference** for the sequence. For example, the odd numbers form an arithmetic sequence with common difference 2; or 1, 3, 5, 7, 9, 11, ...

A numeric sequence in which each term is obtained from the previous term by *multiplying* by a fixed number is called a **geometric sequence.** The fixed number is called the **common ratio** for the sequence. An example of a geometric sequence with common ratio 2 is 1, 2, 4, 8, 16, 32, ...

The numbers of dots in the following triangle number arrays are an example of a sequence that is neither arithmetic nor geometric.

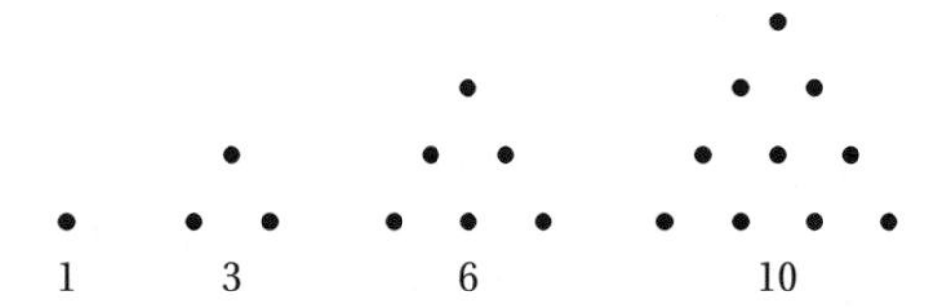

Note that for the dot array sequence, 2 is added to the first term to get the second; 3 is added to the second term to get the third; 4 is added to the third to get the fourth; and so on. You observe a pattern of increases, and inductively conclude that 5 would be added to the fourth term to get the fifth term. Example 1.3 extends the ideas of arithmetic and geometric sequences.

Example 1.3 Extending Sequences

Determine whether the sequence is arithmetic, geometric, or neither. If one exists, give the common difference or common ratio and then give the next three terms in the sequence.

a. 1, 3, 9, 27, 81, ...
b. 2, 5, 8, 11, 14, ...
c. 1, 4, 9, 16, 25, ...

Solution

a. We see a pattern: Each term is obtained by multiplying the term before it by 3, so the sequence is geometric, with common ratio 3. The next three terms are 243, 729, and 2,187.

b. We see a pattern: Each term is obtained by adding 3 to the term before it, so the sequence is arithmetic, with common difference 3. The next three terms are 17, 20, and 23.

c. We can't see a pattern: Succeeding terms aren't obtained by either adding or multiplying preceding terms, so the sequence is neither arithmetic nor geometric. The next three terms are the perfect squares 36, 49, and 64.

Your Turn

Practice: Determine whether the sequence is arithmetic, geometric, or neither. If one exists, give the common difference or common ratio and then give the next three terms in the sequence.

a. 10, 20, 30, 40, ...
b. 5, 5, 10, 10, 15, ...
c. 1, 4, 16, 64, ...

Reflect: Which type of sequence, arithmetic or geometric, increases faster? Explain. ■

To extend the ideas of arithmetic and geometric sequences, let's consider a general way to represent their terms. For example, suppose that a is the first term of an arithmetic sequence and that d is the common difference. We *add* the common difference to the first term to get the second term, so we can represent the second term as $a + d$. Adding d again, we obtain the third term, $a + d + d$, or $a + 2d$. Thus the first several terms of an arithmetic sequence may be written as

$$a, a + d, a + 2d, a + 3d, a + 4d, \ldots$$

Using inductive reasoning, we note that, because the second term is $a + 1d$, the third term is $a + 2d$, and the fourth term is $a + 3d$, the nth term of an arithmetic sequence is $a + (n - 1)d$.

We may represent a geometric sequence in a general way by using a similar line of reasoning. For example, suppose that a is the first term of the sequence and that r is the common ratio. We *multiply* the common ratio by the first term to get the second term, so we can represent the second term as ar. Multiplying by r again, we note that the third term is $a \times r \times r$, or ar^2. Thus the first several terms of a geometric sequence may be written as

$$a, ar, ar^2, ar^3, ar^4, \ldots$$

Using inductive reasoning, we note that as the second term is $a(1r)$, the third term is ar^2, and the fourth term is ar^3, the nth term of a geometric sequence is ar^{n-1}. In Exercise 55 on p. 35, you will use these ideas to find a specified term of a sequence when you are given some terms and the common difference or ratio.

Often, as in the B.C. cartoon in Figure 1.5 showing the effect of inflation on service charges, we associate a sequence of numbers with events, objects, or relationships between objects. In B.C.'s cartoon world, as inflation increases, will the next change be 6 clams, or will it be 8 clams? When you use inductive reasoning to discover a pattern in these numbers, you have discovered a regularity in a "real-world" situation.

FIGURE 1.5 | The beginning of a sequence.

By permission of Johnny Hart and Creators Syndicate, Inc.

Patterns in Real-World Data Tables. Real-world data are often recorded in a **table of values,** which shows relationships between different data categories. Often two types of patterns occur in such a table: *column extension* and *row relationship patterns.* Consider the following table of values created from this picture showing the way tree branches sometimes grow.

Level Number	Number of Branches
1	1
2	1
3	2
4	3
5	5
6	8

This table of values can be extended by discovering a **column extension pattern** in which each digit in the right-hand column is found by adding the two preceding digits in the column. Using this column extension pattern, we determine that the next digit in the column will be $5 + 8$, or 13. The sequence shown in the right-hand column of the table is known as a Fibonacci sequence. It was named after Leonardo Pisano (also called Fibonacci; 1170–1230), who discovered it. In a **Fibonacci sequence,** each term is the sum of the two preceding terms.

Now consider the data in the following table of values of predicted distances for travel on a bicycle at a predetermined rate of speed. This table illustrates how it is sometimes useful to also discover a **row relationship pattern;** in this case in each row the distance is always nine times the number of hours. Using this row relationship pattern, we determine that in the next row of the table the time will

Time in Hours (t)	Distance in Miles (d)
1	9
2	18
3	27
4	36
.	.
.	.
.	.
.	.
t	?

be 5 hours and the distance will be 45 miles. Note that this table can also be extended by discovering a column extension pattern. The numbers in the right-hand column form an arithmetic sequence with common difference 9. So, to extend the right-hand column, we simply add 9 and 36 to get 45.

Although column extension patterns are useful, using a row relationship pattern is often more efficient. For example, to find the distance for a 30-hour bike travel time, we would have to extend the right column in the table to include 26 more numbers. However, once we have found the row relationship pattern, we can simply multiply 30 by 9 to obtain the corresponding distance, 270 miles.

Example 1.4 gives further examples of real-world situations in which patterns may be used to discover a relationship between the numbers in a table of values.

Example 1.4

Problem Solving: Ramp Racer

A maker of miniature racing cars timed how far his prized car rolled down a ramp in various lengths of time and recorded the following information.

Length of Time in Seconds (s)	Distance in Feet (d)
1	1
2	4
3	9
4	16
.	.
.	.
.	.
s	?

Look for a pattern in the table and determine the distance the miniature car rolled down the ramp in 8 seconds.

Solution

Jeff's thinking: I looked at the right-hand column and noticed that the distances increased from 1 to 4, 4 to 9, and 9 to 16, or 3, 5, and 7 feet. So I continued this column pattern and increased 16 by 9 to get 25, increased 25 by 11 to get 36, increased 36 by 13 to get 49, and increased 49 by 15 to get 64. The car traveled 64 feet down the ramp in 8 seconds.

Grenada's thinking: I looked for a relationship between the numbers in each row. I noticed that 1 is 1 squared, 4 is 2 squared, 9 is 3 squared, and so on. So for 8 seconds, I used the same pattern to find that the distance would be 8 squared, or 64. The car traveled 64 feet down the ramp in 8 seconds.

Your Turn

Practice: A health spa gave out the following table to show the weight loss per week that a 166-pound person could expect from using the spa's facilities.

Number of Weeks	Weight (lbs.)
1	164
2	162
3	160
4	158

Look for a pattern in the table and determine the 166-pound person's weight after 8 weeks of workouts and dieting at the spa.

Reflect: Which pattern, the column extension or the row relationship, do you think is the most efficient way to solve the example problem? Why? ■

A graphing calculator can also provide visual information about the pattern of numbers in a table of values. For example, to solve the Ramp Racer problem in Example 1.4, we can show the points that represent the pairs of numbers from the table on a graph in which the x-axis represents the time (t) and the y-axis represents the distance (d). We scale the x-axis in 1-second intervals and the y-axis in 10-foot intervals and then enter the data to produce the screen in the margin.

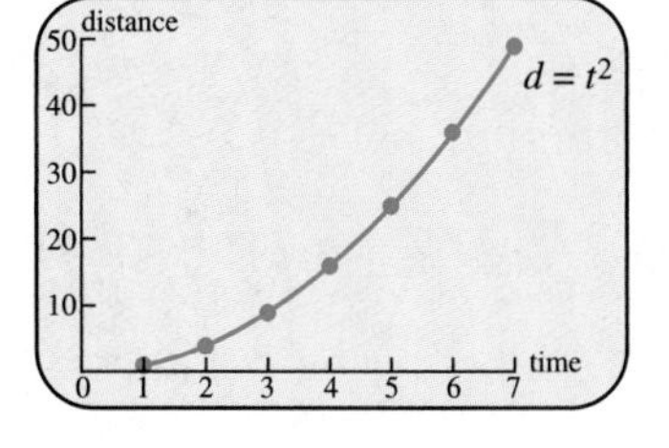

The graph does not show a pattern of constant change. That is, when the number of seconds increases by 1, the *distance* does not always increase by a constant amount. By magnifying portions of the graph on the screen, we can see that the *change in the change in distance* is constant and increases by two feet for each second of time increase.

Patterns in Sequences of Number Sentences. Not only can you discover patterns from sequences of numbers, you also can discover patterns from a sequence of number sentences. For example, consider the following sequence of number sentences involving sums of consecutive whole numbers:

$$1 + 2 = \frac{2 \times 3}{2}, \quad \text{or } 3;$$

$$1 + 2 + 3 = \frac{3 \times 4}{2}, \quad \text{or } 6;$$

$$1 + 2 + 3 + 4 = \frac{4 \times 5}{2}, \quad \text{or } 10;$$

$$1 + 2 + 3 + 4 + 5 = \frac{5 \times 6}{2}, \quad \text{or } 15;$$

$$1 + 2 + 3 + 4 + 5 + \cdots + n = ?$$

Study the pattern in these number sentences. Use inductive reasoning, observing that the last number to be added is the key to calculating the sum. If you multiply this last number by the number immediately following it and then divide this product by 2, you get the sum. Exercise 58, p. 35, asks you to look for a pattern in another sequence of number sentences.

Deductive Reasoning

We now consider a type of mathematical reasoning, called *deductive reasoning*, that is used in drawing logical conclusions and in presenting convincing arguments or proofs. We begin with two key processes that are involved in deductive reasoning:

- understanding if–then statements and deciding when they are true, and
- using rules of logic.

If–Then Statements. Mini-Investigation 1.5 illustrates the if–then type of statement often used in deductive reasoning.

Write an if–then statement about one of your favorite products.

Mini-Investigation 1.5 **Using Mathematical Reasoning**

What statement in the form **If ..., then....** could you write about the products in the following advertisement?

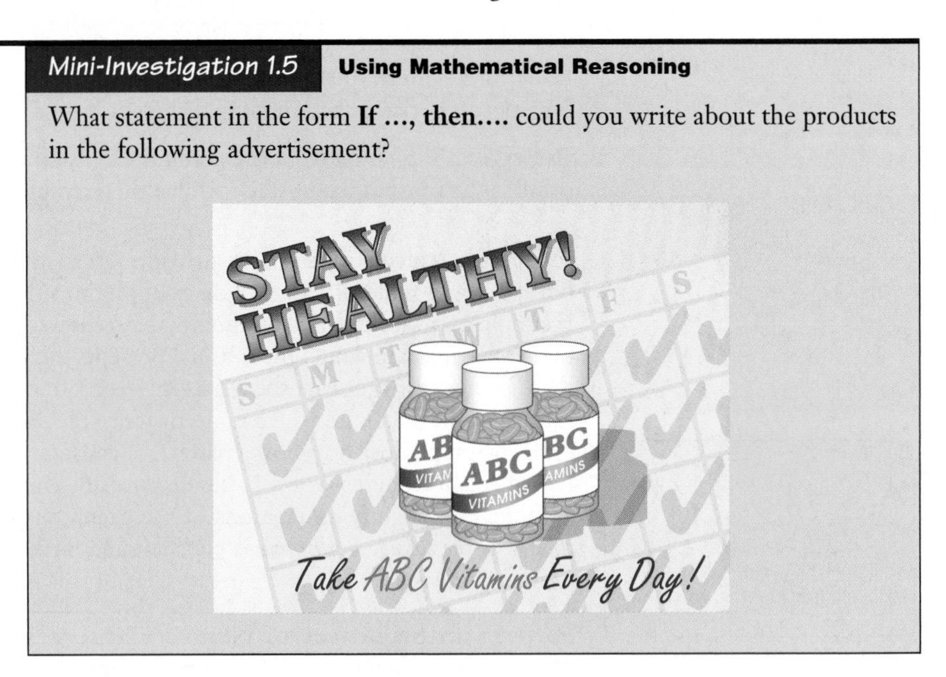

In mathematics, statements in *if–then* form are called **conditional statements.** The *if* part of a conditional statement is called the **hypothesis,** and the *then* part is called the **conclusion.** These ideas are illustrated in the following statement.

Hypothesis: If you wear Super Shoes, — Conclusion: then you will play like a champion.

If you wear Super Shoes, then you will play like a champion.

Example 1.5 shows how to decide whether a conditional statement is true or false.

Example 1.5 **Analyzing If–Then Statements**

Suppose that your basketball coach made the following if–then statement:

If you play well in practice, then you will start in tomorrow's game.

In which of the following cases would you feel that you were being treated unfairly and that the coach didn't tell the truth?

Case 1: You play well in practice (hypothesis true).
You start the game (conclusion true).
Case 2: You play well in practice (hypothesis true).
You do not start the game (conclusion false).
Case 3: You do not play well in practice (hypothesis false).
You start the game (conclusion true).
Case 4: You do not play well in practice (hypothesis false).
You do not start the game (conclusion false).

Solution

Case 2 is the only instance in which the coach did not tell the truth.

Your Turn

Practice: Suppose that a friend made the following if–then statement:

If I go on the trip, then I will bring you back a T-shirt.

Describe the conditions under which your friend would not have told the truth.

Reflect: Explain why the coach's statement in the example problem is true in cases 1, 3, and 4. ■

In Example 1.5, you may have discovered the following procedure for deciding whether a conditional statement is true or false.

Procedure for Deciding Whether a Conditional Statement Is True or False

1. First decide whether the hypothesis and the conclusion are true or false.
2. Then use the following to decide if the statement is true or false.
   a. When both the hypothesis and conclusion are true, the conditional statement is true.
   b. When both the hypothesis and conclusion are false, the conditional statement is true.
   c. When the hypothesis is true and the conclusion is false, the conditional statement is false.
   d. When the hypothesis is false and the conclusion is true, the conditional statement is true.

Truth Table for Conditional Statements. We can summarize the information above in a different way using what is called a *truth table*. Suppose we define conditions p and q in the following way.

p: US Women's soccer team scores at least two goals
q: US Women's soccer team wins the match

p	q	$p \rightarrow q$
T	T	T
T	F	F
F	T	T
F	F	T

Now consider the following conditional statement,

If the US Women's soccer team scores at least two goals, then they will win.

which can be represented symbolically by $p \rightarrow q$ (read "p implies q"). The truth table in the margin can be created.

The only time the statement $p \rightarrow q$ is considered false is when p is true and q is false. In our case this would mean that the soccer team scored at least 2 goals but did not win the match. Although it may seem strange, logicians have decided that if the hypothesis (p) of a statement is false, then the conditional statement $p \rightarrow q$ is considered true regardless of whether q is true or false.

Rules of Logic. We now consider two rules of logic that are frequently used in deductive reasoning. Let's first look at logic rule A, which is used when both an if–then statement and its hypothesis are true.

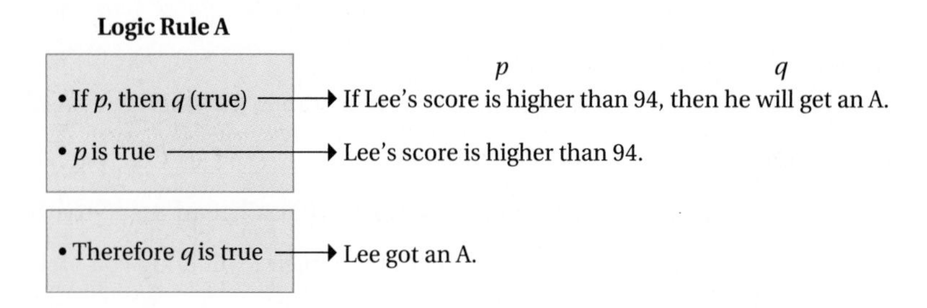

The letters p and q in the rule represent the hypothesis and conclusion, respectively. Logic rule A allows you to conclude that the conclusion is true.

Logic rule B is used when an if–then statement is true and its conclusion is false.

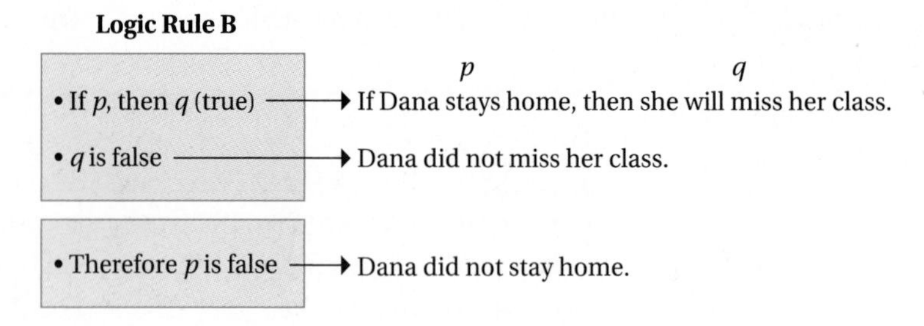

Logic rule B allows you to conclude that the hypothesis is false. Example 1.6 provides additional insight into the use of logic rules A and B.

Example 1.6 Using Rules of Logic

What conclusion can be drawn from the following conditional statements? Identify which rule of logic is used for each.

a. If today is Saturday, then we play the big game.
We do not play the big game.

b. If all sides of a quadrilateral are the same length, then the quadrilateral is a rhombus.
All sides of square $ABCD$ are the same length.

Solution

a. Using logic rule B, we can conclude that today is not Saturday.

b. Using logic rule A, we can conclude that square $ABCD$ is a rhombus.

Your Turn

Practice: Give your conclusion for each of the following conditional statements and indicate which rule of logic you used.

a. If two numbers are odd, the product of the numbers is odd.
The product of these two numbers is not odd.

b. If you send me a letter, then I will send you a letter.
You sent me a letter.

Reflect: Use your responses to parts (a) and (b) to explain how logic rule A differs from logic rule B. ■

Logic rule A is sometimes called **affirming the hypothesis** because it is used when the given hypothesis is true. Similarly, logic rule B is sometimes called **denying the conclusion** since it is used when the given conclusion is false. Other rules of logic, not described in this book, are also sometimes used in deductive reasoning.

Description of Deductive Reasoning

Deductive reasoning involves drawing conclusions from given true statements by using rules of logic.

Consider the following example of the use of deductive reasoning. Your counselor says, If you achieve at least a 3.0 grade average, then you will be admitted to the program. You work hard and get a 3.2 grade average. Using deductive reasoning, you conclude that you will be admitted to the program. If someone asks how you know that you will be admitted to the program, you support your conclusion by telling her about the true statement made by your counselor and your logical conclusion. We can use this example to illustrate the steps in the deductive reasoning process.

Procedure for Using the Deductive Reasoning Process

■ Start with a true statement, often in if–then form.	If you achieve at least a 3.0 grade average, then you will be admitted to the program.
■ Note given information about the truth or falsity of the hypothesis or conclusion.	You achieve a 3.2 grade average. The hypothesis is true.
■ Use a rule of logic to determine the truth or falsity of the hypothesis or conclusion.	You use a rule of logic to conclude that the conclusion is true. You will be admitted to the program.

In the deductive reasoning process, you might use two or more rules of logic in sequence to arrive at a final conclusion. Discussion of these more detailed uses of deductive reasoning is beyond the scope of this book. Mini-Investigation 1.6 will help you think more carefully about the process of deductive reasoning and identify some misuses of this process.

Talk about the reasons for your conclusions.

Mini-Investigation 1.6 Using Mathematical Reasoning

An instructor said that if a student gets an A on the final examination, then the student will get an A in the course. Kaya did not get an A on the final. Ted did get an A in the course. What can you conclude about Ted and Kaya?

You must be careful to use accepted rules of logic, such as logic rule A and logic rule B, when using deductive reasoning. Boxes C and D use an easily understood situation to show two invalid reasoning patterns that are often used.

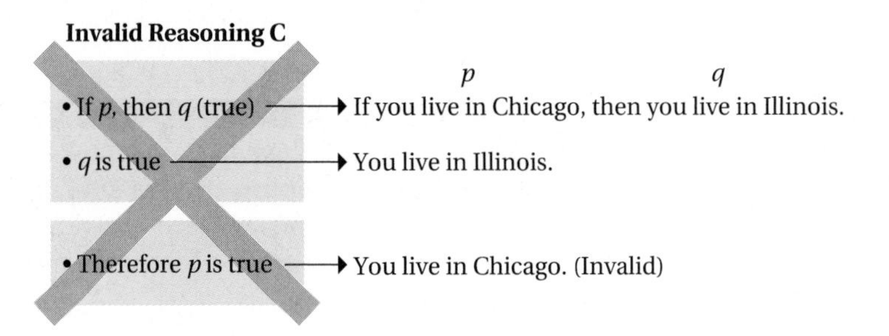

The invalid reasoning in box C is called **assuming the converse** and might have been used to draw the incorrect conclusion in Mini-Investigation 1.6 that Ted got an A on the final examination.

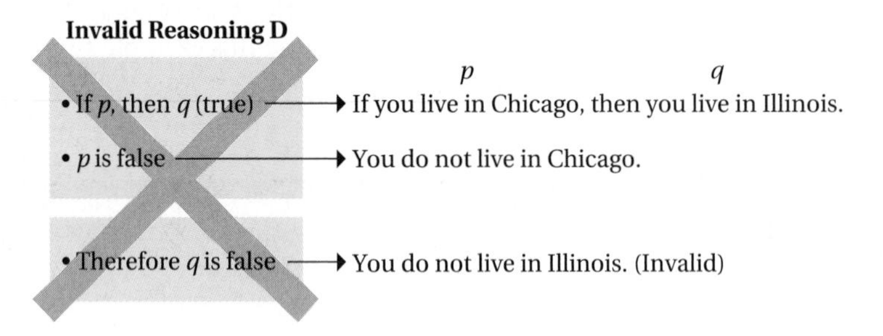

The invalid reasoning in box D is called **assuming the inverse** and might have been used to draw the incorrect conclusion in Mini-Investigation 1.6 that Kaya did not get an A in the course. Example 1.7 will help you learn to recognize invalid reasoning.

Example 1.7 Determining When Reasoning Is Valid or Invalid

Determine whether the reasoning is valid or invalid. If invalid, indicate whether it is based on assuming the converse or assuming the inverse.

a. Timothy's little sister knew that if a vehicle was a car, then it had four wheels. When given a toy wagon with four wheels, she concluded that it was a car.

b. Dan read that "if you play great basketball, then you wear Jumpup shoes." Cedric did not play great basketball, so Dan concluded that Cedric did not wear Jumpup shoes.

Solution

a. The reasoning is invalid (assuming the converse).

b. The reasoning is invalid (assuming the inverse).

Your Turn

Practice: Indicate whether the reasoning in each case is valid or invalid. If invalid tell whether it is based on assuming the converse or assuming the inverse.

a. Jordan's coach said, "If you win the most tryout games, then you will make the number one tennis team." Jordan did not make the number one team. So Jordan concluded that she had not won the most tryout games.

b. Jackie read that if a dog is of show quality, then it has registration papers. A friend's dog, Newton, had registration papers, so Jackie concluded that Newton was of show quality.

Reflect: Explain how you decided whether the reasoning was valid or invalid. ■

Proportional Reasoning

The following problem will help you think about another important type of reasoning.

> Two friends, Moose and Tiny, have pies of the same size. Moose cut his pie into 6 equal pieces and ate 2 of them. Tiny cut his pie into 12 equal pieces. How many pieces must Tiny eat in order to eat the same amount of pie as Moose?

You might reason that because Tiny cut his pie into twice as many pieces as Moose, he had to eat twice as many pieces. Or you might reason, 2 is to 6 as what number is to 12? and answer the question that way. Another possible method is to use ideas from your earlier study of algebra and solve the equation $\frac{2}{6} = \frac{n}{12}$. Each of these ways of thinking involves consideration of two equal ratios, $\frac{2}{6}$ and $\frac{n}{12}$. A statement asserting that two ratios are equal is called a **proportion.**

Description of Proportional Reasoning

Proportional reasoning involves drawing conclusions or solving problems with either the formal or informal use of proportions.

Even though describing it as a process with specific steps isn't very productive, proportional reasoning is a useful way to solve problems and draw conclusions about data. For example, it is used extensively in political and public issue polling. If 900 people in a carefully selected group of 1,500 people support a particular position on an issue, a pollster could use proportional reasoning to predict that 3000 people in a group of 5000 people might support that position. This prediction could involve informally thinking with proportions as follows:

900 out of 1,500 is 9 out of 15, or 3 out of 5. So I would predict that 3000 out of 5000 support the position.

Proportional reasoning can also help people make everyday decisions. For example, if a couple consistently spend $50 every two weeks for movies, they can use proportional reasoning to predict their movie expenditures for a year by solving the proportion $\frac{x}{50} = \frac{52}{2}$:

$$\begin{aligned} \frac{x}{50} &= \frac{52}{2} \\ 2x &= 50 \times 52 \\ &= 2600 \\ x &= 1300. \end{aligned}$$

Their movie expenditures for the year would be $1,300.

Example 1.8 shows some different ways that people might think when using proportional reasoning.

Example 1.8 Problem Solving: Meals to Go

For a canoe trip, the guide estimates that 9 pounds of food will be needed each day for every 4 people. If 28 people go on the trip, how many pounds of food will be needed for each day?

Solution

Reva's thinking: I thought that 4 is to 28 as 9 is to what? As 28 is 4 times 7, I multiplied 9 by 7 to get 63. My answer is that 63 pounds of food are needed.

Taylor's thinking: I just solved the proportion $\frac{28}{4} = \frac{x}{9}$ to get

$$\begin{aligned} \frac{28}{4} &= \frac{x}{9} \\ 4x &= 252 \\ x &= 63. \end{aligned}$$

So 63 pounds are needed.

Jeff's thinking: I decided that $\frac{9}{4}$ is $2\frac{1}{4}$, so I multiplied 28 times $2\frac{1}{4}$. The answer is 63 pounds.

Your Turn

Practice: The people on the canoe trip found that a 10-foot tent pole casts a shadow of 6 feet. How tall is a tent pole that casts a shadow of 24 feet?

Reflect: What is the proportion that Reva informally solved? ■

We consider proportions and proportional reasoning in more detail in Chapter 7.

Spatial Reasoning

Another important type of reasoning, used both in everyday situations and in mathematics when people make decisions about configurations in three dimensions, is *spatial reasoning.* For example, when you look at shapes like these and try to decide whether they could represent real-world objects, you are engaged in spatial reasoning.

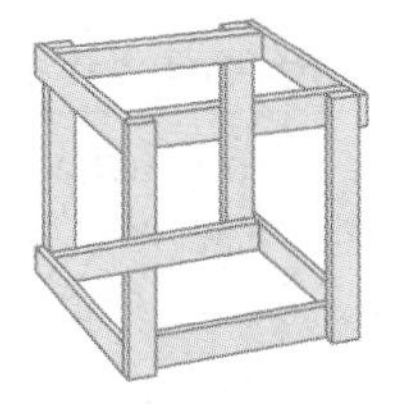
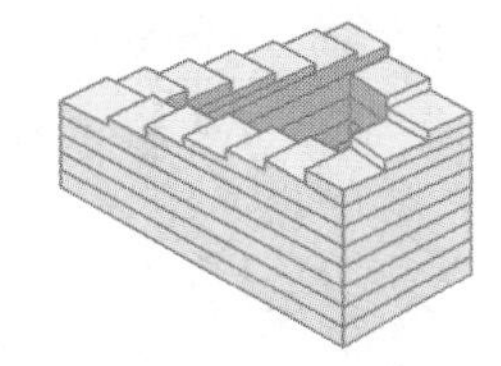

The preceding figure shows two clever 2-D drawings of apparent 3-D objects that contain contradictory visual information. Spatial reasoning also is involved in attempts to "see geometry" in natural objects, such as sand dollars or sunflowers. Its application can be dramatic, as in the use of geometric ideas to create real-world objects, such as the geodesic dome shown in Figure 1.6.

FIGURE 1.6 | A dramatic example of the use of spatial reasoning: A geodesic dome.

Spatial reasoning is used in identifying and analyzing the properties of models of 3-D geometric figures or objects and in analyzing what happens when we move those figures or objects to another location or alter them in some way. Sometimes spatial reasoning involves everyday objects as well as standard geometric figures. For example, spatial reasoning is required in arranging furniture in a room, creating an efficient kitchen, or determining where a fly ball will land when a batter hits the ball. Clearly, many types of situations require the use of spatial reasoning, which we describe as follows.

Description of Spatial Reasoning

Spatial reasoning involves visualizing three-dimensional (3-D) geometric figures or objects to draw conclusions about properties or relationships involving those figures or objects.

Example 1.9 further illustrates this important reasoning process.

Example 1.9 Problem Solving: Painted Blocks

A manufacturer of toy blocks first paints a large $3 \times 3 \times 3$ block and then cuts it as shown to form 27 smaller blocks. How many of the smaller blocks have no paint on them?

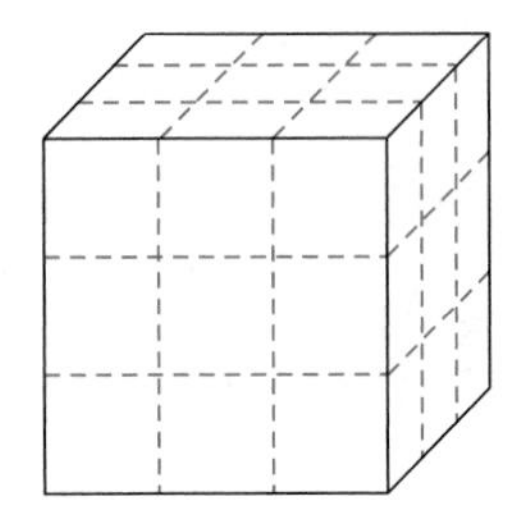

Solution

To find the number of smaller blocks that have no paint on them, we can first think about the horizontal layers in the large block. We can partially see the blocks in the top and bottom layers, all of which will have some paint on them. That leaves the middle horizontal layer to consider. We can visualize the nine smaller blocks in this layer and see that all of them will have paint on at least one side, except the one in the center. So only one smaller block in the large block has no paint on it.

Your Turn

Practice: Use the same large block and determine the number of smaller blocks that are

a. painted on only one face.

b. painted on 2 faces.

c. painted on 3 faces.

d. painted on more than 3 faces.

Reflect: How many unpainted cubes would you expect in a $4 \times 4 \times 4$ block that is cut into 64 smaller blocks? Explain your reasoning. ■

Mini-Investigation 1.7 extends the ideas presented in Example 1.9.

Make a table of your findings and use inductive reasoning to form a generalization about this situation.

Mini-Investigation 1.7 Finding a Pattern

What is the number of unpainted cubes in a $3 \times 3 \times 3$, a $4 \times 4 \times 4$, a $5 \times 5 \times 5$, and an $n \times n \times n$ cube?

One of the most important outcomes in teaching elementary school mathematics is for students to develop confidence in their ability to reason mathematically. Students should base their belief in mathematical statements on the reasoning process—not because someone says that something is true. Thus teachers need to emphasize reasoning to help them develop that skill.

Importance of Reasoning Processes

In this section, we reviewed four types of reasoning, which we can summarize as follows.

- *Inductive reasoning:* Reasoning based on generalizing from a pattern observed in several examples.

- *Deductive reasoning:* Reasoning based on using rules of logic to draw conclusions from given information.
- *Proportional reasoning:* Reasoning based on using ratios and proportions.
- *Spatial reasoning:* Reasoning based on visualizing properties and relationships in 3-D geometric figures or objects.

Each of these reasoning processes is important in mathematics. In fact, doing much in mathematics without these four kinds of reasoning would be very difficult.

For example, you might use spatial reasoning to analyze and compare the characteristics of different geometric figures. You could also use proportional reasoning for the same purpose, as in the study of similar triangles. After you have analyzed the geometric figures, you might use inductive reasoning to discover a possible generalization or theorem about them. Then you might try to find a counterexample to show that the theorem is false. If you don't find a counterexample, you could use deductive reasoning to prove that the discovered generalization or theorem is true. Thus many mathematical tasks require a combination of these types of reasoning. For example, the painted blocks problem in Example 1.9 requires both inductive and spatial reasoning.

The four types of reasoning described in this section occur over and over again in the development of mathematics and the solution to mathematical problems. As a body of knowledge, mathematics is changing dramatically. We may not always be able to understand what these changes are all about, but we can be assured that these changes are in large measure occurring because of the applications of these types of mathematical reasoning.

Problems and Exercises | *for Section 1.2*

A. Reinforcing Concepts and Practicing Skills

1. Use inductive reasoning and state a generalization regarding the following number sentences, which are true, about adding zero.

$$3 + 0 = 3; \quad 4 + 0 = 4;$$
$$376 + 0 = 376; \quad 8000 + 0 = 8000$$

Produce a counterexample for each of the generalizations in Exercises 2–5.

2. If a number is odd, then it ends in 3.

3. If a figure has four sides, then it is a square.

4. If a figure has two right angles, then it is a rectangle.

5. If two odd numbers are added, then the sum is an odd number.

6. For which type of sequence is each term found by
  a. adding a common difference to the preceding term?
  b. multiplying the preceding term by a common ratio?

Find the missing numbers in the following sequences; indicate whether the sequences are arithmetic, geometric, or neither; and give the common difference or ratio when appropriate.

7. 1, 4, 7, 10, 13, ___, ___, ___

8. 16, 8, 4, 2, 1, ___, ___, ___

9. 1, 4, 9, 16, 25, ___, ___, ___

10. 2, 4, 6, 10, 16, ___, ___, ___

11. 1, 3, 7, 15, 31, ___, ___, ___

12. 1, 3, 9, 27, ___, ___, ___

13. 1, 2, 4, 7, ___, ___, ___

14. 1, 5, 9, 13, ___, ___, ___

15. Consider the following example of inductive reasoning.

The numbers 4, 9, 16, 25, 36, and 49 are all the result of squaring a number. Therefore every number squared ends in 4, 5, 6, or 9.

Find several counterexamples that show this conclusion to be false.

16. Give a key characteristic of each type of reasoning.
  a. Inductive reasoning
  b. Deductive reasoning

Give the hypothesis and the conclusion for each of the statements in Exercises 17–20.

17. If an odd number is added to an even number, then the sum is an odd number.

18. A figure is a quadrilateral if it has four connected sides.

19. If it rains on Tuesday, then it will be nice on Wednesday.

20. A number is even if its square is even.

21. Julie's calculator screen is shown. (Recall that 2^3 represents 2^3.) Julie thinks that when she presses ENTER, a 1 will appear. What type of reasoning is Julie using? Explain.

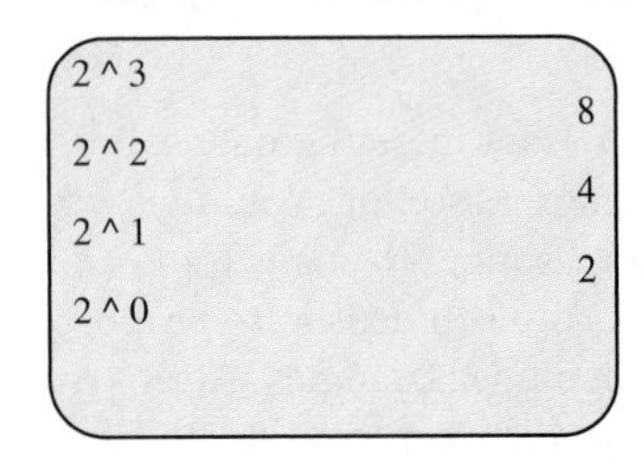

22. Suppose Jack says that, if the temperature is below 50°F, he won't play tennis. Determine which of the following cases, if any, indicate that Jack did not tell the truth.
   a. The temperature was below 50°, and he didn't play tennis.
   b. The temperature was below 50°, and he did play tennis.
   c. The temperature was above 50°, and he didn't play tennis.
   d. The temperature was above 50°, and he did play tennis.

23. Create a truth table using p, q, and $p \rightarrow q$ and describe the situation in which the following statement is false.

   If Serena wins the second set of her tennis match, she will win the match.

24. Describe two different situations in which the following statement could be true.

   Salaries will increase if inflation remains constant.

For Exercises 25–29, indicate whether the deductive reasoning used is an example of affirming the hypothesis or denying the conclusion.

25. If a number is a multiple of 10, then the square of the number is a multiple of 100. (Joe's age)^2 is not a multiple of 100. Therefore Joe's age is not a multiple of 10.

26. If a four-sided figure has four right angles, it is a rectangle. Shape $ABCD$ has four right angles. Therefore shape $ABCD$ is a rectangle.

27. If a woman lives in Atlanta, then she doesn't live in Chicago. Samantha lives in Chicago. Therefore Samantha doesn't live in Atlanta.

28. If a student is a preservice teacher, then he will study mathematics. Michael is a preservice teacher. Michael studies mathematics.

29. If a student is a preservice teacher, then she will study mathematics. Michelle doesn't study mathematics. Michelle isn't a preservice teacher.

Classify the following types of reasoning in Exercises 30–33 as deductive, inductive, proportional, spatial, or none of these.

30. If Katrina makes 7 of every 10 free throw shots, then she should make 28 of 40 free throws.

31. I've checked a lot of numbers and believe an even number is a number that ends in 0, 2, 4, 6, or 8.

32. All triangles have three sides. Geometric figure ABC is a triangle. Therefore $AB = BC = AC$.

33. The three triangles shown are equilateral. Each angle of each triangle measures 60°. Hence the angles of all equilateral triangles measure 60°.

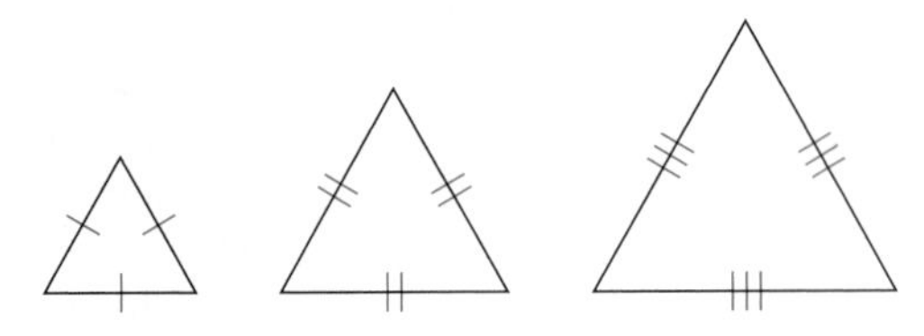

34. Suppose in a scale drawing that 1 in. = 10 ft.
   a. How many feet are represented by $2\frac{1}{2}$ in.?
   b. How many inches represent 15 ft.?
   c. What type of reasoning is used to solve parts (a) and (b)?

State whether the reasoning in Exercises 35–38 is valid or invalid. If invalid, indicate whether it is assuming the converse or assuming the inverse.

35. If a triangle is equilateral, then all its sides are congruent. Triangle ABC is equilateral. Therefore all sides of triangle ABC are congruent.

36. If Beth lives in Idaho, then Beth lives in the United States. Beth lives in the United States. Therefore Beth lives in Idaho.

37. If Tina shoots in the 80s in each practice round, then she will make the golf team. Tina did not shoot in the 80s in each practice round. Therefore Tina did not make the golf team.

38. If a number ends in 0, then it is an even number. Bill's age is an even number. Therefore Bill's age ends in 0.

39. Test your spatial reasoning. The pattern in part (a) can be folded to make an open-top box as shown. Which of the other patterns can be folded to make an open-top box?

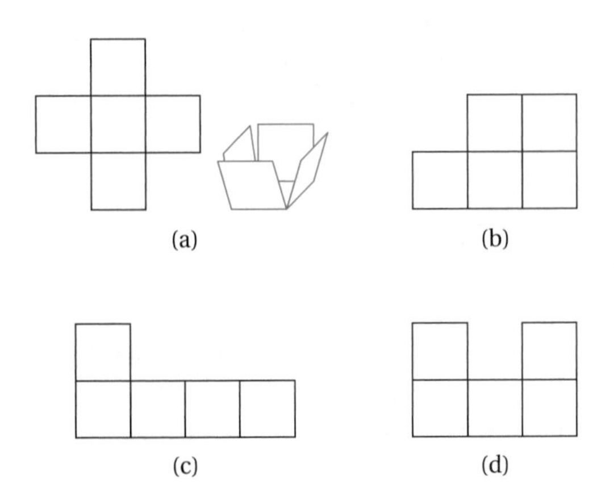

40. Determine a pattern in the following number sentences.

$$1 \times 12 = 12.$$
$$11 \times 12 = 132.$$
$$111 \times 12 = 1332.$$
$$1111 \times 12 = 13332.$$

Based on this pattern, indicate the product for 111111×12. Use a calculator to check your work.

B. Deepening Understanding

41. Create a mathematical situation involving deductive reasoning that leads to the following conclusion: The sides of geometric figure *ABCD* are all the same length.

42. Jennifer considered the numbers 4, 24, 44, and 64 and noticed that they all ended in 4 and that they can be divided evenly by 4, leaving no remainder. She concluded that all numbers that end in 4 can be divided evenly by 4. What type of reasoning did Jennifer use? Is her conclusion correct? Explain your answer.

43. Create a set of examples that suggests the generalization: The product of two numbers that end in 5 also ends in 5.

44. Use a calculator to create an arithmetic sequence or a geometric sequence. Write down the keystrokes you pressed to create the sequence.

45. In each case, write the conclusion that follows from the information given.

a. If a geometric figure is a square, then it is a rectangle. Figure *ABCD* is a square.

b. If a geometric figure is a triangle, then it is a polygon. Figure *ABC* is a triangle.

c. If a number is an odd number, then a number that is 2 greater than the number is also an odd number. Thirteen is an odd number.

46. How many more small blocks are needed to make a complete and solid larger block starting with the following configuration?

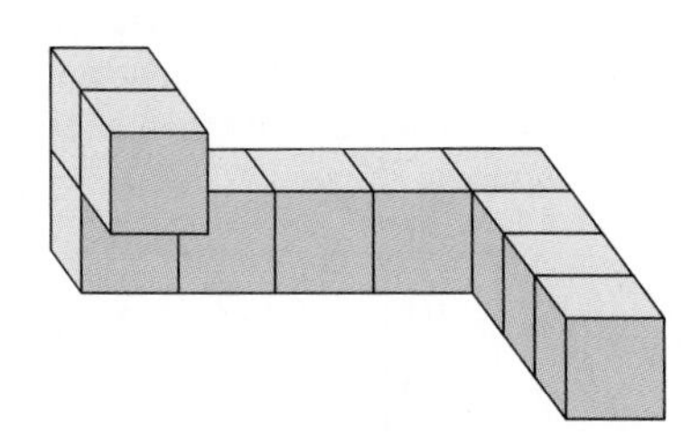

47. Use spatial reasoning to answer the following question: If a spider starts at one corner of a cube and walks only along edges, what is the greatest number of edges along which it could walk without walking along the same edge twice?

In exercises 48 and 49, consider the following examples and generalizations. Determine which, if either, of these two generalizations follows from the examples and whether they are true.

$$3^2 - 1 = 8;\quad 7^2 - 1 = 48;\quad 13^2 - 1 = 168;$$
$$17^2 - 1 = 288;\quad 23^2 - 1 = 528.$$

48. If 1 is subtracted from the square of an odd number, the result can be divided evenly by 8, leaving no remainder.

49. If 1 is subtracted from the square of an odd number, the result is a number that ends in 8.

Indicate whether Exercises 50–54 illustrate correctly the use of deductive reasoning. Explain your answers.

50. If Fred buys a new boat, then his friends will come to visit him. Fred's friends didn't come to visit. Therefore Fred didn't buy a new boat.

51. If a politician always tells the truth, then the politician will get elected. Politician Stretchit didn't always tell the truth. Politician Stretchit won't get elected.

52. If two odd numbers differ by 2, then their average is an even number. The average of two numbers that differ by 2 is odd. Therefore the two numbers are not odd.

53. If a geometric figure is a square, then its sides all have equal length. Figure *ABCD* is a square. Therefore Figure *ABCD* has all sides of equal length.

54. If you study a lot, then you will get good grades. Corine received good grades. Therefore Corine studied a lot.

C. Reasoning and Problem Solving

55. Ryan claims that two consecutive terms of a geometric sequence are 8 and 12 and that the next term is 16. Kevin says that the next term is 18. Fred maintains that the next term cannot be determined. Who is correct? Why?

56. Create an if–then statement that would make the underlined statement a valid conclusion.

Mick is a dog. Therefore Mick can fly.

57. Write as many if–then statements as you can that seem to be implied in the following advertisement from the Sleezy Company.

Come work for Sleezy and make a fortune. Work hard and you will get promoted fast. We train all of our employees—that's the reason they are so smart. You will like working for Sleezy.

58. Consider the following sums of numbers and how they are formed.

1 odd number

$$\overbrace{1} = 1$$

2 odd numbers

$$\overbrace{1 + 3} = 4 = 2^2$$

$$\overbrace{1 + 3 + 5}^{\text{3 odd numbers}} = 9 = 3^2$$

Predict the following sum and complete the generalization.

a. $1 + 3 + 5 + 7 + 9 + 11 = ?$

b. $1 + 3 + 5 + \cdots + (2n - 1) = ?$

59. The Vitamin Ad Problem. Frieda read a vitamin advertisement that stated, "If you want to feel your very best, take one Vigorous Vitamin each day. People who take Vigorous Vitamins care about their health." Frieda described the ad to her friend, telling her that the ad said that "if you take a Vigorous Vitamin each day, then you'll feel your very best," and that "if you don't take Vigorous Vitamins, then you don't care about your health." Did Frieda report the main ideas of the ad correctly to her friend? Analyze this situation and use what you learned about deductive reasoning in this section to support your answer.

60. As shown in the following figure, when a line segment is drawn from vertex A of triangle ABC to side $\overline{BC}$, how many triangles exist? Suppose that a second line is drawn from A to side $\overline{BC}$. How many triangles exist now? Suppose that a third line is drawn from A to $\overline{BC}$. Now how many triangles are formed?

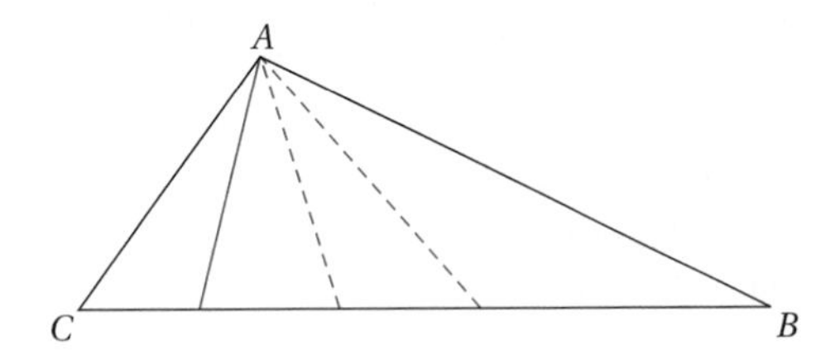

Write a statement about the number of lines drawn from a vertex and the number of triangles that are formed. What kind of reasoning did you use?

61. The Diet Problem. Bridget lost 3 pounds during the first week of her diet, another 1.5 pounds during the second week, and another 1.5 pounds during the third week. If she continues to lose 1.5 pounds per week, how much will she weigh in about 9 weeks if she weighed 140 pounds when she began dieting?

D. Communicating and Connecting Ideas

62. Work in a small group to design an advertisement for a magazine that communicates the following reasoning.

If you use Neato Hair Spray, then you will be a very popular person.

Exchange your group's ad with that of another group's ad. Analyze each other's ads to determine whether they accurately reflect the given statement.

"If we pull this off, we'll eat like kings."

63. Consider the Far Side cartoon. Create some sort of deductive argument that is based on the if–then statement in the cartoon. Compare your argument with those of others in the class.

64. Making Connections. Consider the Waltham advertisement on the next page. What if–then statements are the creators implicitly assuming? Analyze the ad in terms of the type of mathematical reasoning they are using.

65. Historical Pathways. A magic square is an array of numbers in which the sum of each row, column, and diagonal is always a certain magic number, say, 15. The patterns in magic squares were known to the ancient Egyptians and later to the Greek mathematician, Pythagoras (c. 582–507 B.C.). The sixteenth-century German artist Albert Dürer included a completed magic square, with the numbers shown, in his famous painting, *Melancholia*. Use mathematical reasoning to complete the magic square.

16		2	13
	10		8
		7	
4		14	1

If you don't see this sign on your pet food, there might be something missing.

It's the Waltham symbol. Waltham is the world's leading authority on pet care and nutrition.

At Waltham, hundreds of scientists, veterinarians, animal behaviorists and pet nutritionists study pets and their dietary needs on a day to day basis.

The end results are the exceptional foods that feed one third of the world's cats and dogs, keeping them healthy, happy and helping them recover when they're sick.

All the main meal pet foods developed by Waltham are highly palatable, 100% complete and perfectly balanced for your animal. The dog foods contain all the essential nutrients, including the all important amino acids. And for cats, Vitamin A and Taurine, which are crucial for sight, are part of every recipe.

So, to be sure your pets get all the good things they deserve, with nothing left out, look for the Waltham symbol on every can and bag of pet food you buy.

Call 1-800-WALTHAM, the Waltham Information Line with any questions on pet nutrition.

Reprinted with permission of Waltham.

Figure for Exercise 64

Section 1.3 MATHEMATICS AS PROBLEM SOLVING

- The Role of Problem Solving
- A Problem-Solving Model
- Problem-Solving Strategies
- Importance of Problem Solving

In this section, we explore the notion of mathematics as problem solving. We also look at the meaning of a problem, a model for the problem-solving process, and some useful problem-solving strategies.

The Role of Problem Solving

When people think of mathematics as solving problems, they view mathematics primarily as a thinking process and computation and other rule-oriented procedures as tools to be used when solving problems. Problem solving is central to the development and application of mathematics and is used extensively in all branches of mathematics. When someone is "doing mathematics," they are involved in the process of solving problems. Let's explore further the role of problem solving in mathematics and how to communicate with others effectively about problem solving. First, we need to look at the meaning of a problem and the problem-solving process.

The Meaning of a Problem. Although many possible answers exist to the question, "What is a problem?" the following description will serve to focus your attention on some key ideas.

Description of a Problem

A **problem** is a situation for which the following conditions exist.

a. It involves a question that represents a challenge for the individual.

b. The question cannot be answered immediately by some routine procedures known to the individual.

c. The individual accepts the challenge.

One key idea implied by this description is that what is a problem for one person may not be a problem for another person. For example, because of (a), (b), or (c) in the description, finding the product 10×12 is not a problem for you, but it probably would be a problem for the average second-grader.

Another key idea is that the word *situation* in the description may be interpreted very broadly and that many different kinds of problems exist. Clearly, a problem can be much more than just a standard word problem. For example, the following questions suggest some real-world situations that can present a challenge and are definitely problems.

How can I fix my car so it will run?

What is the most efficient way to travel by car to six different cities and return home?

What will be the estimated total cost for my junior year in college?

Other questions, such as "Can every map be colored with only four different colors if regions that have a border in common must be colored differently?" have been challenging mathematicians for centuries. This particular problem was solved recently, but several extremely important problems in mathematics remain unsolved. Even puzzle questions, such as "How can you cut a cake into eight pieces with three straight cuts?" legitimately qualify as problems. These examples, and others that you may generate, reinforce the need to take a broad view of the meaning of a problem. Mini-Investigation 1.8 encourages you to think about what constitutes a problem.

Write a paragraph describing your reaction and your solution, if any, to the problem.

Mini-Investigation 1.8 Solving a Problem

If you use the criteria for a problem presented in the description box, why would the following question be or not be a problem for you?

If the letters *a–z* are associated with the monetary values 1¢–26¢, what $1 word, if any, can you find?

The Meaning of Problem Solving. With the meaning of a problem in mind, let's now consider the meaning of problem solving.

Description of Problem Solving

Problem solving is a *process* by which an individual uses previously learned concepts, facts, and relationships, along with various reasoning skills and strategies, to answer a question or questions about a situation.

Thinking of problem solving as a process requires distinguishing it from a computational process, such as adding two 3-digit numbers. The process of adding numbers can be made into an *algorithm*, in which a sequence of prescribed steps always can be followed in a certain order to produce an answer. However, the problem-solving process cannot be made into an algorithm. Because of the many different varieties of problems, no single sequence of steps can always be used to solve a problem. Even though some helpful general guidelines and strategies are available, problem solving is a nonalgorithmic process that requires creativity, thought, and judgment.

Along with guidelines and strategies for exploring problems, many tools support the problem-solving process. Pencil and paper, compasses and protractors, rulers, random-number tables, calculators, geometry exploration software, and other computer programs are but a few of the many tools available. As you solve problems, you need to be aware of the situations in which these tools can help you achieve the desired results.

Note that, when discussing the problem-solving process, we distinguish between an **answer** to a problem, which is the final result, and a **solution** to a problem, which is the process used to find the answer. Thus, when asked to give your answer to a problem, you might give only the number 25. But when asked to give your solution to the problem, you might describe how you drew a picture, made a table, and extended a pattern to obtain the number 25.

When solving problems, you shouldn't look for a standard solution that others have developed. Rather, you should generate a solution based on your unique way of thinking. Example 1.10 illustrates that even a standard word problem can be solved in a variety of ways, each of which is appropriate.

Example 1.10

Problem Solving: Slices of Pizza

Gene, Tina, and Brenda have ordered a large pizza. Gene will eat three times as much pizza as Tina. Tina will eat twice as much pizza as Brenda. If the pizza is cut into 36 slices, how many slices will each person get?

Solution

Dede's thinking: When Gene gets 6 slices, Tina gets 2, and Brenda gets 1. But that totals only 9 slices, not 36. I want four times this number of slices so I use 24 slices for Gene instead of 6. Then Tina gets 8, and Brenda gets 4. I now have a total of 36 slices, which is the answer to the problem.

Megan's thinking: I fill in the columns of a table for Brenda and Tina first because Tina always gets twice as many slices as Brenda. Then I fill in the column for

Gene with the remainder of the 36 slices. I just add rows to the table until Gene has three times as many slices as Tina.

Brenda	Tina	Gene
1	2	33
2	4	

The numbers in the table are correct when Brenda gets 4 slices, Tina gets 8 slices, and Gene gets 24 slices.

Judd's thinking: I used what I learned in elementary algebra and wrote an equation to solve the problem. I let s equal the number of slices that Brenda ate, $2s$ equal the number of slices that Tina ate, and $6s$ equal the number of slices that Gene ate. Then, I solved the equation

$$s + 2s + 6s = 36$$
$$9s = 36$$
$$s = 4.$$

So Brenda ate 4 slices, Tina ate 8 slices, and Gene ate 24 slices.

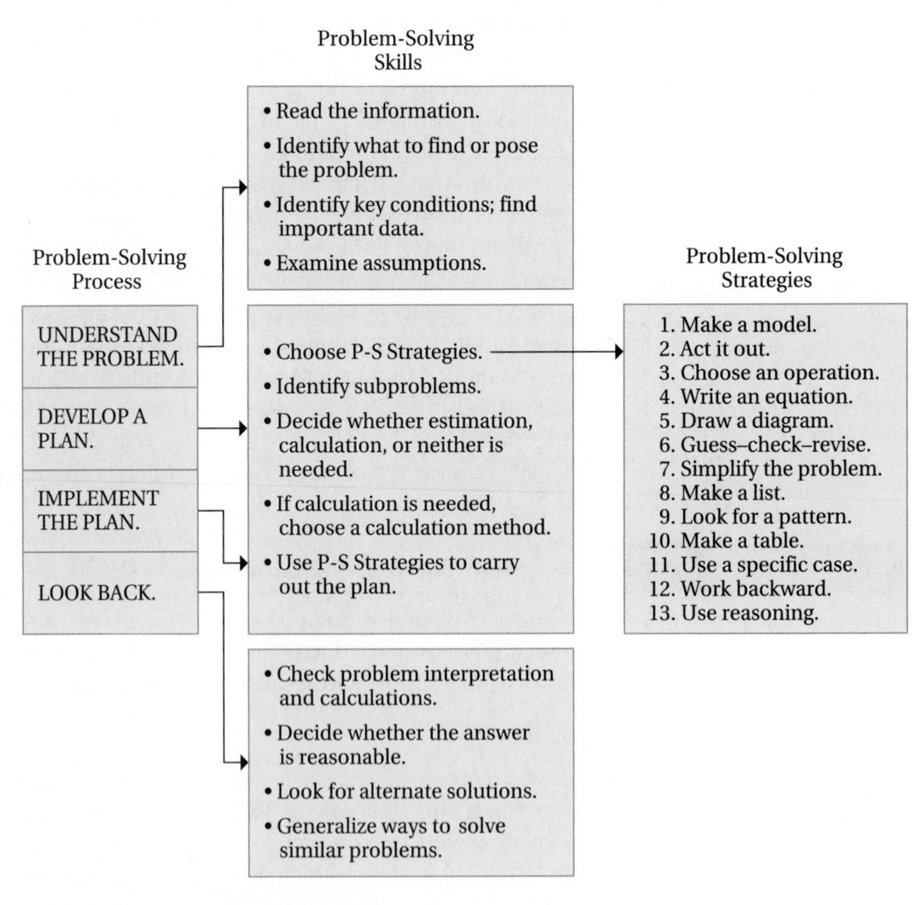

FIGURE 1.7 | Problem-solving model.

Your Turn

Practice: Find at least two different solutions to the following problem.

How many people can be seated at a large table formed by lining up eight square tables, each of which ordinarily seats four people, end to end?

Reflect: Which of the three solutions to the example problem do you think is "best"? Do you believe everyone who answers this question would agree with you? Why? ■

Exercise 30 on p. 47 provides an opportunity for you to use technology, if available, in the problem-solving process.

A Problem-Solving Model

Although no set rules exist for solving a problem, a general approach to this task has been developed. The following problem-solving model has four main phases. They are similar to those described by George Pólya (1945) in his classic book on problem solving, *How to Solve It.* Each of these phases may involve the use of some important problem-solving skills and problem-solving strategies, which are also shown in the model in Figure 1.7.

The process described in the model certainly is not a lock-step process, and flexibility across phases is important. The model contains sequential steps, but when actually solving a problem, you could legitimately deal with more than one phase at a time or jump back and forth among phases. The diagram in Figure 1.8 illustrates this idea and suggests the importance of correctly formulating problem statements.

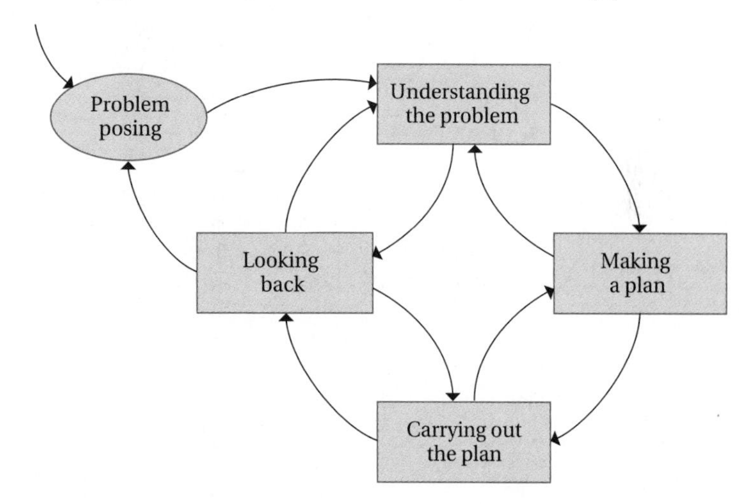

FIGURE 1.8 | The flexibility inherent in the problem-solving process.

Example 1.11 illustrates use of the model in Figure 1.7 in the problem-solving process.

Example 1.11

Problem Solving: Telephone Lines

A telephone company engineer needs to install phone lines for a political convention. Private lines are to connect eight committee chairpersons' desks with each other. How many phone lines will she need to install?

Working toward a Solution

Understand the problem	*What does the situation involve?*	Installing phone lines between desks.
	What has to be determined?	The number of lines to be installed.
	What are the key data and conditions?	Eight committee chairpersons' desks are to be connected by private lines.
	What are some assumptions?	A separate line is needed for each two-desk connection.
Develop a plan	*What strategies might be useful?*	Draw a diagram.
	Are there any subproblems?	Yes. How many lines are needed to connect each desk to all the other desks?
	Should the answer be estimated or calculated?	Calculate the exact numbers.
	What method of calculation should be used?	Use mental math.
Implement the plan	*How should the strategies be used?*	Draw a diagram showing 7 lines coming from each desk.

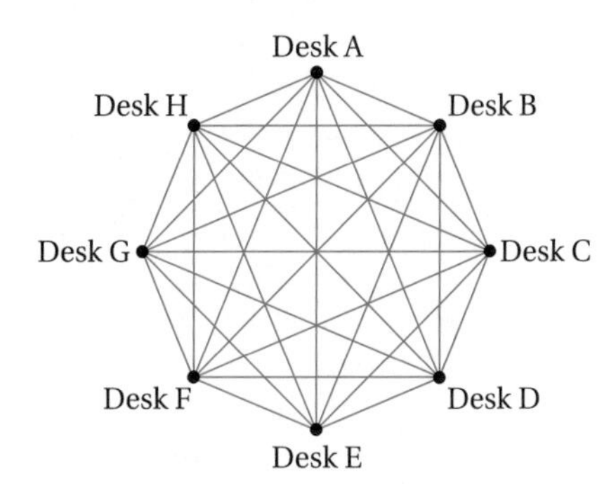

For 8 desks, $8 \times 7 = 56$. However, the line from desk A to desk B, for example, is the same as the line from desk B to desk A, so the number of lines needed is only $56 \div 2 = 28$.

	What is the answer?	The answer is that 28 telephone lines will be needed.
Look back	*Is the interpretation correct?*	*Check:* The diagram fits the data and conditions.
	Is the calculation correct?	*Check:* The mental math was correct.
	Is the answer reasonable?	Yes. It has to be much less than 56, as this number represents 7 lines from 8 desks without considering repetition.

Is there another way to solve the problem?	Make a table of values for 2, 3, and 4 desks and look for a pattern to help extend the table.

Desks	Lines
2	1
3	3
4	6
5	?
6	?
7	?
8	?

Your Turn

Practice: Solve the problem when 10 desks are to be connected.

Reflect: If you say that the number of lines needed is $\frac{1}{2}$ the product of the number of desks and 1 less than the number of desks, use examples to explain what you mean. ■

Estimation Application

The Dripping Faucet

Consider the following problem.

How much extra does it cost for the water lost in 1 year from a dripping faucet?

a. Estimate the data needed for the problem.

b. Assume that the cost of water is 4 gallons for 1¢ and estimate the answer to the question.

Determine how to check your estimate in (b).

Another important aspect of the problem-solving model is the use of estimation. **Estimation** is the process of determining an answer that is reasonably close to the exact answer. It can be used at different stages in problem solving. For example, you could use estimated data to help you understand a problem. Also, when developing and implementing a plan, you could choose the problem-solving strategy of guess–check–revise, which might involve estimation. You could also decide that an estimated answer is all that you need when finding a solution. Finally, when looking back, you could use estimation to help you decide whether the answer is reasonable. The Estimation Application features that appear throughout this book provide an opportunity for you to improve your estimating skills in problem solving.

Problem-Solving Strategies

When solving the pizza or telephone line problems in Examples 1.10 and 1.11, you may have guessed, checked, and revised your guess. Or you may have drawn a diagram, solved a simpler problem, made a table, solved an equation, or used other approaches. All these approaches are the problem-solving strategies shown in the model in Figure 1.7 and summarized in the following list.

Problem-Solving Strategies

- Make a model.
- Act it out.
- Choose an operation.
- Write an equation.
- Draw a diagram.
- Guess–check–revise.
- Simplify the problem.
- Make a list.
- Look for a pattern.
- Make a table.

- Use a specific case.
- Work backward.
- Use reasoning.

Example 1.12 illustrates the use of some of these strategies and provides practice in the use of others.

Example 1.12

Problem Solving: The Bicycle-Built-for-Two Rides

Each of six friends takes a bicycle-built-for-two ride with everyone else in the group. How many bike rides do they take?

Solution

Bill's thinking: Six of us can pair up and count the possibilities *(act out).*

Christine's thinking: We can use colored crayons as bike riders and count possible pairs *(make a model).*

Tom's thinking: Why not write down all possible pairs—Terry, Lynn; Terry, Phyllis; and so on *(make a list).*

Janie's thinking: First, find out how many rides if there were three friends. Once we figure out how to solve this simpler problem, we can consider the original problem *(solve a simpler problem).*

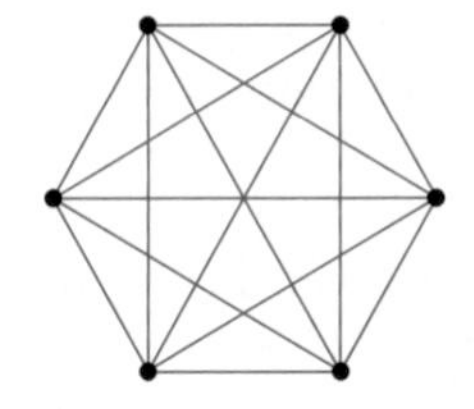

John's thinking: Drawing a picture is easier. Dots are friends and lines are rides *(draw a diagram).*

Ike's thinking: We could show simpler cases in a table *(make a table)* and then extend the table *(look for a pattern).*

Number of Friends	Number of Rides
1	0
2	1
3	3
4	6

Jerry's thinking: Each of six friends would ride with five other friends, making 6×5 rides. Randy riding with Peggy is the same as Peggy riding with Randy, so there are half that many rides *(use reasoning).*

Your Turn

Practice: Use the strategies listed to solve the following problem.

Donna put half of her baby-sitting earnings in a savings account. Of the remaining amount, she gave $7 to a Help the Hungry fund, leaving her with $2. How much money did she save originally?

a. *Work backward* and *choose operation(s).* (*Hint:* Start with $2 and work backward. Decide which operations you need to get back to the beginning.)

b. *Guess–check–revise.* (*Hint:* Guess $20 and check it. Revise your guess.)

c. *Use a specific case* and *write an equation.* (*Hint:* Choose a specific amount that Donna could have earned, and go through the operations used to get the final amount. Then use what you learned to write and solve an equation, letting n = the amount Donna earned and $\frac{1}{2}$ = the amount she put in the savings account.)

Reflect: Which strategy would you use to solve this problem? Why? ■

Problem-solving strategies play an important role in planning and carrying out solutions to problems. Choosing a useful strategy or several strategies is an important problem-solving skill. We further illustrate the use of these strategies as we solve problems in subsequent chapters.

All problem-solving strategies described are used extensively in PreK–8 classrooms. In the primary grades, act it out, make a model, and choose an operation are emphasized. Other strategies, such as draw a diagram, guess–check–revise, and look for a pattern, also may be introduced. As students progress through the middle grades, they are given many opportunities to use most of these strategies when they solve problems. Certain strategies, such as write an equation, work backward, use a specific case, and use reasoning, are emphasized in grades 7 and 8.

Importance of Problem Solving

When asked what they do, mathematicians will probably respond that a significant part of their activity involves solving problems. In fact, many practicing mathematicians may spend years working on a small number of important unsolved problems.

Not surprisingly, then, mathematics is used primarily to solve problems, both in mathematics and in the real world. Applications of mathematics in the physical sciences, social sciences, economics, engineering, architecture, computer science, and many other fields involve the use of mathematical models, ideas, and technology to help solve complex problems. On a more personal level, the real value of mathematics is that it helps individuals solve problems such as balancing a checkbook, planning a vacation, buying groceries, building a house, and playing a game, to name but a few.

When people think of why they learn facts, ideas, and procedures in mathematics, they may well conclude that they do so in order to solve problems. In its 1977 "Position Paper on Basic Skills," the National Council of Supervisors of Mathematics declared that "learning to solve problems is the principal reason for studying mathematics." [*Arithmetic Teacher* (October 1977), p. 20] This point of view has guided developments in the mathematics curriculum for schools for many years. Thus understanding and appreciating mathematics require acknowledgment that solving problems plays a central role.

Problems and Exercises | *for Section 1.3*

A. Reinforcing Concepts and Practicing Skills

1. Would you classify the following as a problem for you? Explain.

   Jenny found 24 shells on Monday and 36 shells on Tuesday. How many shells did she find?

2. Why is it impossible to give a prescribed sequence of steps that, if carefully followed, would produce the answer to any problem?

3. Show that the following problem may be solved in more than one way by using the methods in parts (a) and (b) below.

   A store clerk wanted to stack 21 boxes in a window display in a triangular arrangement like the one shown. How many boxes should the clerk place on the bottom row?

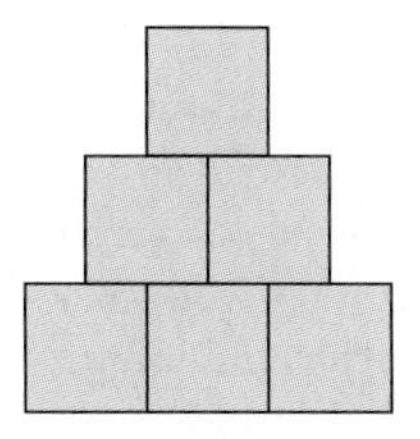

   a. Solve by extending a pattern.
   b. Solve by drawing a picture.

4. Show that some problems have more than one answer by giving at least two answers to the following problem.

   What change would a restaurant cashier give someone for $1 if he used no more than four of any coin and no coin smaller than a nickel or bigger than a quarter?

5. Consider the following problem.

   Tim said that he lost five coins, none of which was a half dollar, totaling 75¢. What coins might Tim have lost?

   How many solutions does the problem have?

   a. One **b.** More than one
   c. None

   Explain your answer.

6. What are four key stages in the problem-solving process?

In Exercises 7–12, indicate which problem-solving strategy or strategies might be used to solve the problem. You need not actually solve the problem.

7. Jill is 1 year older than Bill, and Bill is 1 year older than Phil. The sum of their ages is 60. How old is each person?
8. Bella is in a race with 20 other racers. Bella is fifth from last when the race begins. At the end of the race she is third. How many racers did she pass?
9. On the first day, Joe joined a secret club as its only member. Each day after that, one more member joined than on the previous day. What was the membership of the club after 40 days?
10. How many different arrangements of three letters can you make from the letters in the word *math*, if no repetition of letters is allowed?
11. Anton, Beth, Cal, and Diedre were hired as coaches for basketball, soccer, volleyball, and swimming. Anton's sister was among those hired, and she coaches soccer. Anton does not coach basketball. Beth coaches a water-sport. Which sport was each person hired to coach?
12. James opened a savings account on Monday, deposited \$35 on Tuesday, withdrew \$21 on Wednesday, withdrew \$17 on Thursday, and withdrew half of what was left in the account on Friday. If \$12 then remained in the account, how much money did James deposit on Monday?
13. Indicate whether an exact answer or an estimate is called for in each of the following questions. Justify your decision.
    a. How much paint do I need in order to paint my living room?
    b. How many seconds will a swimmer need to swim 100 meters freestyle in an Olympic Games race?
    c. How much would a vacation to Bermuda cost?
14. What information would you need in order to estimate the cost of buying and using a wireless phone in your car for a year?

B. Deepening Understanding

15. The distance around a rectangular room is 40 feet, and 96 tiles (each one is 1 square foot in size) are needed to cover the floor of the room. What are the room's dimensions? Describe the strategy you used to solve the problem.
16. Solve the following problem and describe the strategy that you used.

    Two 2-digit numbers have the same digits. Their sum is 77 and their difference is 27. What are the numbers?
17. A pilot and a copilot are to be chosen from a group of five pilots. How many different crews are possible? Describe your strategy for solving this problem.
18. What is the largest sum of money—all in coins but no silver dollars—that a person could have without being able to give change for a

    a. dollar. **b.** half dollar.
    c. quarter. **d.** dime.
    e. nickel.

    What strategy did you use to solve this problem?
19. Consider the following question.

    How many times between 4 A.M. and 2 P.M. will the minute hand of a clock pass the hour hand?

    Describe the strategy that you used to solve this problem.
20. Chu has to cut a lawn that is 60 feet by 60 feet with a lawn mower that cuts a 30-inch strip. He wants to figure out whether he should cut it back and forth in parallel strips or cut it in a spiral pattern in order to minimize the distance he walks with the mower.

Parallel pattern

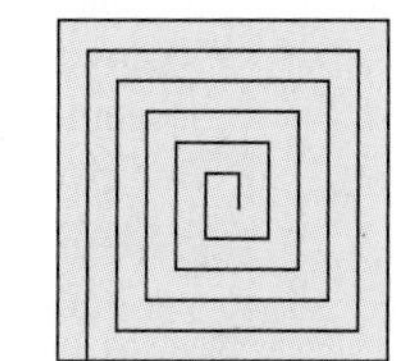

Spiral pattern

    What strategies would you suggest to Chu to help him solve the problem? Explain your selection. (You need not solve the problem.)
21. Place the digits 1, 2, 3, 4, 5, and 6 in the circles shown so that the sum of each adjacent pair of circles (both horizontally and vertically) is different. There are seven adjacent pairs of circles.

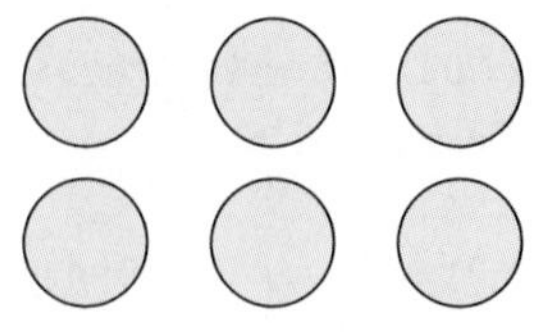

    What strategy did you use in solving this problem?
22. Suppose that you wanted to find the whole numbers represented by each of the letters in the following addition problem.

$$\begin{array}{r} SEND \\ \underline{MORE} \\ MONEY \end{array}$$

Which problem-solving strategy might be most helpful to you in solving this problem? (You need not solve the problem.)

C. Reasoning and Problem Solving

23. Create two problems based on the following situation.

A contractor is building a one-story brick ranch-style house, 30 feet by 60 feet. The height of the brick on the house will be 10 feet. Bricks cost $170 per 1,000. Covering each 100 square feet of the house takes 678 bricks.

24. Form a small group of classmates and exchange the problems you created in Exercise 23. Solve each others' problems.

25. Solve the problems stated in Exercises 7–12.

26. Write a paragraph or two telling which problem you solved in Exercise 25 was easiest and which was most difficult. Explain your reasons for each answer.

27. Consider the following problem.

There are five more children in Ms. Brown's room than in Mr. White's room. Ten children move from Ms. Brown's room to Mr. White's room. Now there are twice as many children in Mr. White's room as in Ms. Brown's room. How many children were in Ms. Brown's room before the move?

Suppose that Angela tries to solve the problem by making the table below. She is discouraged because this method doesn't seem to be generating a solution. What advice would you give her for solving the problem?

28. The Baby Problem. Use two different strategies to solve the following problem. Describe the strategies.

According to a book of records, a woman gave birth 40 times, bearing a total of 53 children. Eight of the births were twins or triplets. How many single babies, twins, and triplets were born?

29. The Balance Problem. Eight balls all look alike. One of them is heavier than the other seven, all of which weigh the same. How can you find the heavier ball if you use a balance scale only two times?

30. a. Use a computer spreadsheet to explore the pizza problem in Example 1.11. (*Hint:* Set up columns for Brenda, Tina, and Gene.)

b. Write a paragraph describing how you used spreadsheet formulas for the number of Brenda's slices in order to calculate the number of slices eaten by Tina and Gene.

D. Communicating and Connecting Ideas

31. Consider the following problem.

A rectangular garden 15 meters wide and 20 meters long is to be fenced. Posts should be set no more than 5 meters apart. How many posts are needed?

Create a series of questions based on the problem-solving model that you think would be helpful to someone trying to solve the problem.

32. Work with a partner to solve the following problem.

A beetle is at the bottom of a 10-foot well. Each day the beetle climbs up 5 feet. But at night when it sleeps, the beetle slips back 4 feet. On what day (fourth, fifth, and so on) will the beetle get out of the well?

Now compare your strategy for solving the problem with that of other pairs of students. In what way are the strategies the same? Different?

33. Making Connections. Consider the following two problems.

Five people come to a party and shake hands with each other. If each person shakes the hand of every other person, how many handshakes occur?

If each of five dots on a circle is connected to each of the other dots with a line, how many lines will be drawn?

a. Solve the two problems.

b. Write a short paragraph about the connection(s) that you see between the problems.

34. Historical Pathways. Henderson and Pingry (1953) published a well-known article, "Problem Solving in Mathematics," in which they describe the following three steps for the "pre-solution period of problem solving."

a. Orientation to the problem

b. Producing relevant ideas for solving the problem

c. Forming and testing hypotheses for solving problems

Describe the relationship between this model and the one presented in this section.

Table for Exercise 27

Brown	White	Brown	White	Brown	White	Brown	White
16	11	17	12	18	13	19	14
6	21	7	22	8	23	9	24

Section 1.4 A CONNECTED, BALANCED VIEW OF MATHEMATICS

- Importance of a Connected View of Mathematics
- Importance of a Balanced View of Mathematics
- Developing a Connected, Balanced View of Mathematics

In this section, we look briefly at some of the connections between different representations of an idea, between mathematical generalizations, and between mathematics and the real world. We describe how these types of connections—and a balanced view of mathematics—can help you understand what mathematics is all about.

Importance of a Connected View of Mathematics

A **connected view of mathematics** considers the various ideas in mathematics not as isolated, unrelated bits of knowledge, but as a body of ideas that are integrated in meaningful ways. Many different ways may be used to connect mathematical ideas and to gain a clearer view of how they fit together and how applications of these ideas fit together. We next look at some of these connections.

Connections between Different Representations of an Idea or Problem. An important connection in mathematics is between different ways to represent an idea, such as the idea of the fraction three-fourths illustrated in Figure 1.9.

Written or Verbal Representation

The fraction $\frac{3}{4}$ consists of the pair of numbers 3 and 4; 3 is called the numerator, and 4 is called the denominator. The denominator 4 states that fourths are involved, and the numerator 3 states the number of fourths being considered.

Visual or pictorial representation **Graphic representation**

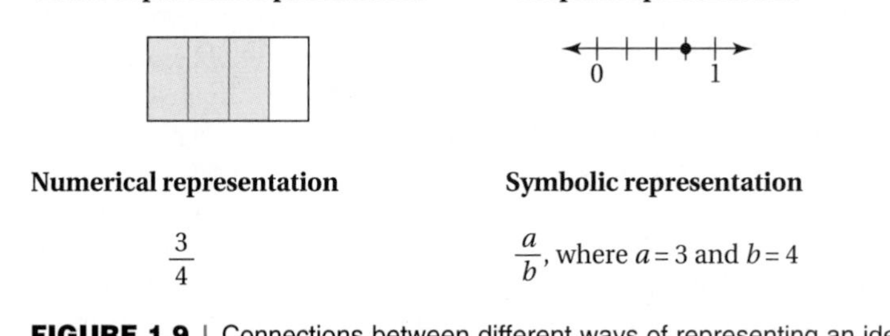

FIGURE 1.9 | Connections between different ways of representing an idea.

One or more of these representations—words, pictures, numbers, symbols, or graphs—may not be as familiar to you as the others. As you think back on the mathematical experiences that helped form your view of mathematics, you might ask questions such as, Was one type of representation emphasized more than others in your study of elementary school mathematics? What type of representation was used most often in algebra? Was a table of numbers ever used in geometry?

Knowing and using various representations for mathematical relationships and *making connections among several types of representations* are two important aspects of an ability to communicate about mathematics and mathematical ideas effectively. As you proceed through this book, you will find that we describe and explore various ways to represent and communicate mathematical relationships and ideas. Mini-Investigation 1.9 illustrates the connection between problems having a similar structure but which are represented in different ways.

Talk about the connections between the two problems.

Mini-Investigation 1.9 **Making a Connection**

How is the solution to the following handshake problem like the solution to the bicycle-built-for-two problem on p. 44?

How many different handshakes are there when six people each shakes everyone else's hand only once?

Connections between Mathematical Generalizations. Another aspect of making connections is to identify how different mathematical generalizations relate to each other. For example, recall that the length of a segment joining the midpoints of the sides of a trapezoid is the average of the lengths of the two bases, as shown in Figure 1.10. The same generalization applies to a triangle, if you think of it as a trapezoid with one base of length 0.

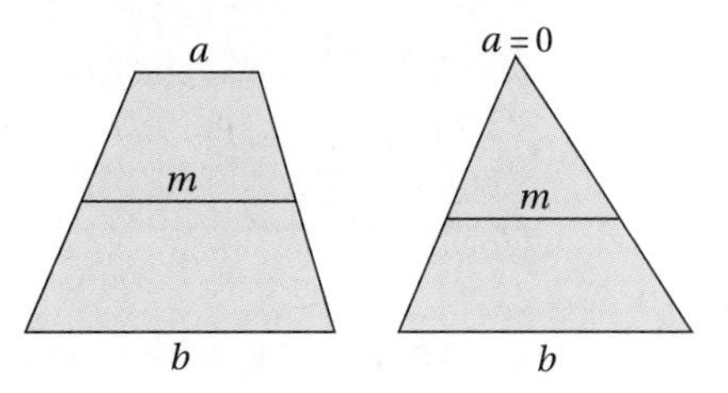

Trapezoid: $m = \frac{1}{2}(b + a)$ Triangle: $m = \frac{1}{2}(b + 0) = \frac{1}{2}b$

FIGURE 1.10 | Connecting generalizations about trapezoids and triangles.

To understand how two numerical generalizations connect, recall that in Exercise 58 on page 35, we discovered:

The sum of the first n odd numbers is n^2.

We can connect this outcome to finding the sum of the first n even numbers by observing that each even number is 1 less than its corresponding odd number, as shown for the first four odd and even numbers.

First four odd numbers: 1, 3, 5, 7
First four even numbers: 0, 2, 4, 6

From this observation, we conclude that the sum of the first four even numbers is 4 less than the sum of the first four odd numbers. Similarly, the sum of the first five even numbers is 5 less than the sum of the first five odd numbers, and so on for 6, 7, 8, ... even numbers.

From these examples we use inductive reasoning and hypothesize:

The sum of the first n even numbers is $n^2 - n$.

In later chapters, we explore a variety of connections between mathematical generalizations.

Connections between Mathematics and the Real World. Often, situations in the real world are modeled by using mathematics, and mathematical ideas are exemplified by using real-world objects.

Mini-Investigation 1.10 explores what might be meant by finding connections between mathematics and the real world.

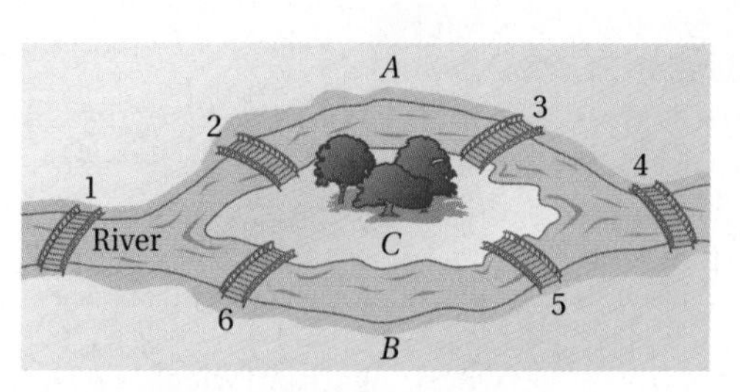

(a) Real-world situation: three land areas, *A*, *B*, and *C*, and six bridges, 1–6.

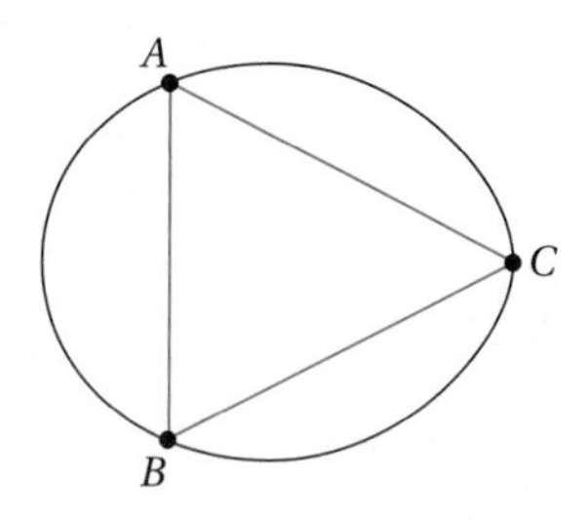

(b) Mathematical model: points represent land areas, and connecting lines represent bridges.

FIGURE 1.11 | Connecting real-world situations and mathematics.

Talk about your group's answer to the question and think of situations in which you connect mathematics and the real world.

Mini-Investigation 1.10 Making a Connection

Do you think that drawing a floor plan for a new house involves using a mathematical model to solve a real-world problem? Explain.

The idea of mathematics, real-world connections is further illustrated by considering a question about Figure 1.11(a)—can you start at land area *A*, cross each bridge just once, and end up at land area *B*? If we let points *A*, *B*, and *C* represent land areas *A*, *B*, and *C* and lines connecting the points represent the six bridges, we produce the model in Figure 1.11(b). The real-world question can now be translated into a question about the model: Can you start at point *A* and trace over each segment just once, ending at point *B*? We can easily answer this question (see the network section, 10.2) and thus solve the real-world problem.

The value of a connected view of mathematics is that the subject then makes sense logically, making it easier to understand and use. People who do not see the connections between various topics in mathematics often view it as a "bag of tricks," each of which must be learned in isolation and remembered. Those who see the connections in mathematics view it as a coherent, logical subject that makes sense because learning one topic builds a foundation for or a bridge to another topic.

Importance of a Balanced View of Mathematics

A **balanced view of mathematics** considers mathematics as an activity involving skills, concepts, relationships, and higher level processes. You should already be familiar with the importance of concepts, skills, and relationships as a result of your earlier study of mathematics. In this chapter, we have emphasized the key mathematical processes of communicating, using technology, reasoning, studying patterns to discover relationships, solving problems, and making connections. This balanced view of mathematics is an important foundation for the various activities that you will undertake throughout this book.

Why is a balanced perspective of mathematics important? One reason is that, although such a perspective recognizes the need for some routine skills, it stresses the higher level processes—reasoning, problem solving, and pattern finding—that are central to doing mathematics. A second reason is that it emphasizes the importance of communicating mathematical ideas. A third reason is that it focuses on connections within mathematics and facilitates learning. Envisioning mathematics as a series of interconnected ideas rather than a set of isolated rules gives the subject added meaning and makes it easier to learn. A fourth reason is that it recognizes the value of technology in the development, study, and use of mathematics.

Finally, the ability to connect mathematics with real-world situations not only gives meaning to the subject, but it also provides a basis for answering the often-asked question, "Why do I have to learn this?" That is, the ability to connect mathematics to the real world provides a basis for using mathematics to help make everyday decisions. For example, imagine the factors that you need to consider in making a decision about buying a car and how to finance it. They include the cost of the car itself and the amount required for a down payment, the monthly cost of financing the difference at a given rate of interest, the cost of insuring and maintaining the car, and the cost of driving the car—all of which need to be related to your income and other current expenses. Clearly, a balanced, usable understanding of mathematics can help you make this decision. To appreciate the importance of a balanced perspective on mathematics further, consider Example 1.13.

Example 1.13 Connecting Process, Concept, Skill, and Relationship

How much area is available for a circular walk if the outer part of the walk is determined by a circle with a radius of 40 feet and the inner part of the walk is determined by a circle with a radius of 36 feet? The circles are concentric. List a process, a concept, a skill, and a relationship needed to solve the problem.

Solution

Process: Problem solving

Concept: Concentric circles

Skill: Multiplying numbers, finding the area of a circle given its radius, and subtracting numbers

Relationship: The relationship between the area of a circle and its radius

Your Turn

Practice: Suppose that, for the circular walk described, the concrete is to be 3 inches thick and the question is, "How many cubic feet of concrete are needed to make the walk?" List a process, a concept, a skill, and a relationship needed to solve this problem.

Reflect: Why are all four aspects of mathematics (process, concept, skill, and relationship) essential to solving the problem correctly? ■

Developing a Connected, Balanced View of Mathematics

As you work through the various chapters in this book, you will be asked to investigate mathematical situations, sometimes individually and sometimes in small groups, and communicate your solutions and answers verbally, in writing, or with diagrams and graphs. You will be asked to engage in mathematical problem solving and reasoning, look for patterns, make connections between mathematical ideas, and relate mathematics to the real world. You will be encouraged to use technology whenever possible to help with these activities. Not all of these considerations apply equally to every section of every chapter, but on balance you will be challenged to engage in this process of doing mathematics and to consider the real nature of mathematical activity. In certain places, you will be asked to relate your view of mathematics to how to help children learn mathematics.

We hope that this book conveys to you a rich perspective of mathematics and the sense of excitement that doing mathematics can generate. Because of its consistency and connectedness, some have called mathematics the most beautiful testimony to human creation yet conceived. When mathematics is viewed as an art in creating, visualizing, reasoning, and solving problems, anyone can be a doer of mathematics. We hope that you will become a doer of mathematics.

Problems and Exercises | *for Section 1.4*

A. Reinforcing Concepts and Practicing Skills

1. How are the symbols $\frac{1}{2}$, 0.5, and 50% connected?
2. Give a verbal, visual, numerical, and graphic representation for the idea *one-half.*
3. Why might having several different types of connected representations for a single idea be useful, as in Exercise 2?
4. Solve the following two problems and state how they are connected.
   a. How many ways can you choose three letters from the four letters A, B, C, and D?
   b. How many different three-member committees can be formed from four people?
5. What connection, if any, do you see between the following geometric generalizations?
   a. The area of a square is equal to one-half the product of its distance around the square (*perimeter*) and the distance from the center of the square to one of the sides.
   b. The area of a circle is equal to one-half the product of the distance around the circle (*circumference*) and its radius.
6. What connection, if any, do you see between the following numerical generalizations?
   a. The sum of two odd numbers is an even number.
   b. The product of two negative numbers is a positive number.
7. What geometric model would you use to represent the real-world situation of a circular flower bed in the center of a square patio?
8. What geometric model would you use to represent two cities connected by three highways?
9. Describe two models that you could use to represent the different singles tennis matches that could be set up with four people.
10. Why is emphasizing the connections in mathematics useful?
11. Describe what is meant by a *balanced view of mathematics.*

B. Deepening Understanding

Provide the missing representations in Exercises 12–14.

	Visual	Numerical	Graphic
12.	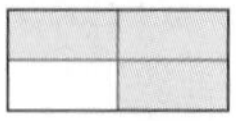	?	?
13.	?	$\frac{1}{2}$	?
14.	?	?	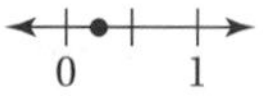

15. For what real-world situation(s) might the following figure be used as a model?

16. Consider the following mathematical activities involving decimals. Which of them do you consider to be the most important? Why?
    a. Multiplying and dividing decimals
    b. Recognizing the concept of place value in working with decimals, such as $0.231 = \frac{2}{10} + \frac{3}{100} + \frac{1}{1000}$
    c. Solving problems involving decimals, such as finding the average increases or decreases in the temperature for a given period of time
    d. Translating decimals and fractions—for example, $0.375 = \frac{3}{8}$
17. Does a balanced view of mathematics necessarily involve the use of technology? Explain.
18. To connect a visual representation to a numerical idea, consider the statement:

    The sum of two odd numbers is an even number.

    Use a representation like the following one to show that this statement is correct.

Odd number 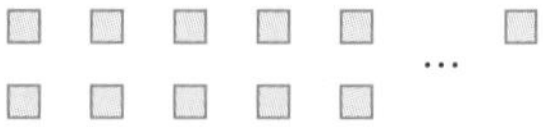

Even number 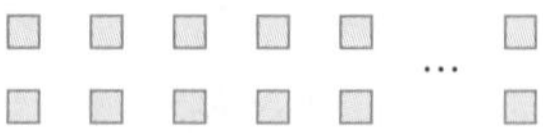

C. Reasoning and Problem Solving

19. Create a problem involving one of the following activities (or some other activity of your choice). Then create a second problem involving another activity that is connected to the first; that is, the same numbers and operations you used in solving the first problem can be used to solve the second problem.

a. Jogging
b. Playing a musical instrument
c. Playing tennis
d. Painting a picture
e. Designing a kitchen
f. Making a pizza
g. Playing basketball
h. Being a gymnast

20. If someone told you that a balanced approach to mathematics involved simply adding, subtracting, multiplying, and dividing numbers, how would you respond? Why?

21. **The House Plan Problem.** Teresa wants to create a house plan for a vacation house that measures 40 feet by 30 feet. If she wants to draw a model plan for her house on a regular $8\frac{1}{2}'' \times 11''$ sheet of paper, what is the greatest number of inches she can use to represent $10'$? (Use whole numbers only.)

22. **The Open House Problem.** Can you find a continuous path through the floor plan shown that will start in one room, go through each door only once, and end in another room? Let rooms = points and doors = segments and make a model for this real-world problem. Use the model to solve the problem. Describe your solution.

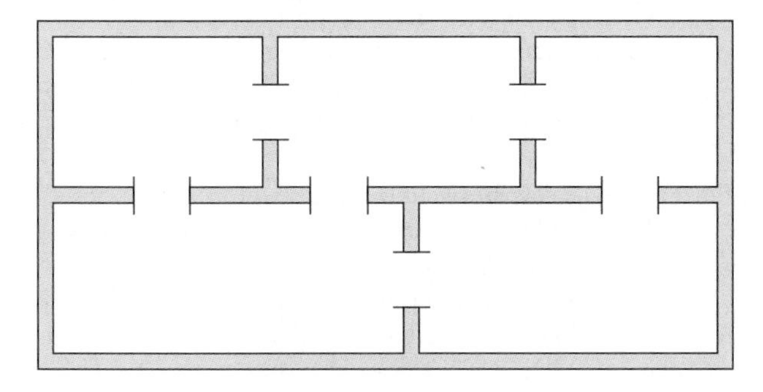

D. Communicating and Connecting Ideas

23. You have been asked to write a one-page memo to a vocal parent who maintains that teaching mathematics is a waste of the taxpayer's money because mathematics is nothing more than computing and calculators can do all that now. Write your memo and compare it with those written by others.

24. **Making Connections.** When you divide 1 by 7 and continue for several decimal places, the digits 142857 repeat over and over again. To shorten the answer you can say that $1 \div 7 \approx 0.142857$, where $\approx$ means *approximately equal to.*
   a. Use a calculator to help you write similar statements for
      i. $2 \div 7$. ii. $3 \div 7$.
      iii. $4 \div 7$. iv. $5 \div 7$.
      v. $6 \div 7$.
   b. What connections do you observe among the decimals in these statements?

25. **Historical Pathways.** Although many people strive to achieve a balanced view of mathematics, some people in the past have become famous for their amazing ability to do computational tasks. For example, Thomas Fuller was brought to Virginia as a slave in 1724, never learned to read or write, but could perform complex mental computations in seconds. Reportedly he could mentally find the product of two 9-digit numbers in seconds, probably using an extension of the following mental computation technique.

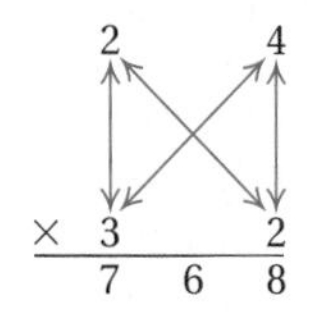

1. Multiply 4×2; write 8.
2. Multiply $2 \times 2 = 4$; $3 \times 4 = 12$. Add 4 and 12; write 6. Remember the 1.
3. Multiply 3×2; add 1; write 7.

Use this technique to find the product 32×43.

CHAPTER SUMMARY

Key Ideas: Questions and Answers

Section 1.1

- **What are some views and attitudes regarding mathematics?** (*p. 2*) People hold different views of mathematics. In general, they are formed by both in-school and out-of-school experiences. People have different attitudes toward mathematics. Some enjoy, but others fear, mathematics. These attitudes are developed through interactions with teachers, friends, and parents and perhaps by the way mathematics is conveyed in various media.
- **What role does communicating and representing play in mathematics?** (*pp. 3–5*) People communicate through the use of words, symbols, and graphs that represent various mathematical ideas in a variety of ways.
- **What role does technology play in communicating and representing mathematical ideas?** (*pp. 5–10*) Technology allows the exploration of mathematical topics that would otherwise be difficult. The use of computer spreadsheets, various types of calculators, and geometry exploration software are some of the technologies available for mathematical exploration and problem solving.

Section 1.2

- **What types of reasoning are used in mathematics?** (*pp. 14–15*) The four basic types of mathematical reasoning are inductive reasoning, deductive reasoning, proportional reasoning, and spatial reasoning.
- **What is inductive reasoning and how can it be used to find patterns?** (*pp. 15–24*) Inductive reasoning involves

the use of information from specific examples to draw a general conclusion, called a generalization. You can use inductive reasoning to find patterns involving examples from numerical or geometric sequences, tables of real-world data, or sequences of number sentences.

- **What is deductive reasoning?** (*pp. 24–29*) Deductive reasoning involves drawing conclusions from given true statements by using rules of logic.
- **What is proportional reasoning?** (*pp. 29–30*) Proportional reasoning involves drawing conclusions or solving problems with either the formal or informal use of proportions.
- **What is spatial reasoning?** (*pp. 30–32*) Spatial reasoning involves visualizing three-dimensional (3-D) geometric figures or objects to draw conclusions about properties or relationships involving those figures or objects.
- **Why are the reasoning processes important in mathematics?** (*pp. 32–33*) Almost all of mathematics has been developed by some combination of the four basic reasoning processes. They are important processes for doing mathematics.

Section 1.3

- **What is the role of problem solving in mathematics?** (*pp. 37–41*) Mathematics is a process of solving problems, which emphasizes the notion that mathematics is a method of thinking. Problem solving is the key to doing mathematics.
- **Is there a problem-solving model?** (*pp. 41–43*) A model for problem solving exists only in the sense that various strategies and skills for solving problems can be very helpful. Essentially, this approach involves understanding the problem, developing a plan, implementing the plan, and looking back. Within these four phases of problem solving, specific strategies and problem-solving skills can be identified.
- **How are problem-solving strategies used?** (*pp. 43–45*) Problem solving involves considering the range of possible strategies and then determining which strategy or strategies might be best for solving a particular problem.
- **Why is problem solving important in mathematics?** (*p. 45*) In one sense, the most important part of doing mathematics is solving problems. Through solving problems, people learn mathematics and how to relate mathematics to the real world.

Section 1.4

- **What is a connected view of mathematics and why is it important?** (*pp. 48–50*) A connected view of mathematics considers the various ideas of mathematics as interrelated. Such a view allows people to use different types of representations to communicate the same mathematical idea, make connections between mathematical ideas, and use mathematics to model real-world situations.
- **What is a balanced view of mathematics and why is it important?** (*pp. 50–51*) A balanced view of mathematics considers mathematics as an activity involving skills, concepts, relationships, and higher-level processes. Such a view of mathematics is necessary in order to do a broad range of mathematical tasks, effectively communicate mathematically, see connections both within mathematics and between mathematics and the real world, recognize the role of technology in mathematics, and understand the role of mathematics in personal decision making.
- **How will this book help you develop a connected, balanced view of mathematics?** (*pp. 51–52*) As you work through this book, you will be engaged in a variety of mathematical tasks and activities that will help you develop a connected and balanced view of mathematics.

Key Terms, Concepts, and Generalizations

Section 1.1

Views about mathematics (p. 2)
Attitudes toward mathematics (p. 2)
Mathematical communication (p. 3)
Technology (p. 5)
Spreadsheet (p. 6)
Calculators (p. 8)
Geometry exploration software (p. 9)

Section 1.2

Inductive reasoning (p. 15)
Generalization (p. 15)
Counterexample (p. 17)
Patterns (p. 18)
Sequence (p. 18)
Terms of the sequence (p. 18)
Arithmetic sequence (p. 19)
Common difference (p. 19)
Geometric sequence (p. 19)
Common ratio (p. 19)
Table of values (p. 21)
Column extension pattern (p. 21)
Fibonacci sequence (p. 21)
Row relationship pattern (p. 21)
Conditional statements (p. 24)
Hypothesis (p. 24)
Conclusion (p. 24)
Truth of conditional statement (p. 25)
Rules of logic (p. 26)
Affirming the hypothesis (p. 27)
Denying the conclusion (p. 27)
Deductive reasoning (p. 27)
Assuming the converse (p. 28)
Assuming the inverse (p. 28)
Proportion (p. 29)
Proportional reasoning (p. 29)
Spatial reasoning (p. 31)

Section 1.3
Problem (p. 38)
Problem solving (p. 39)
Answer (p. 39)
Solution (p. 39)
Problem-solving skills (p. 40)
Problem-solving model (p. 41)
Problem-solving strategies (p. 43)
Estimation (p. 43)

Section 1.4
Connected view of mathematics (p. 48)
Mathematical connections (p. 49)
Balanced view of mathematics (p. 50)

CHAPTER REVIEW

Concepts and Skills

1. Choose two symbols and two words that you have used in mathematics and explain what they represent.
2. What formulas would you have to enter to instruct the computer to calculate the numbers in cells C2 and C3 in the spreadsheet shown?

	A	B	C
1	Number of Items	Cost per Item (12)	Total Cost ($)
2	7	8	56
3	9	12	108

3. Produce a counterexample to each of the following conditional statements.
   a. If 10 is divided by another number, then the result will be less than 10.
   b. If a geometric figure is a rectangle, then it is a square.
   c. If a number is multiplied by 7, then the product is an odd number.
4. Give the hypothesis and the conclusion in each of the following statements.
   a. If the Bulls score 100 points, then they will win.
   b. Tom Hanks will win the Oscar, if he accepts the role in the proposed movie.

State the appropriate conclusion in Exercises 5 and 6 and whether the reasoning is affirming the hypothesis or denying the conclusion.

5. If a four-sided geometric figure has three right angles, then its fourth angle is a right angle. In figure *ABCD*, angles *A*, *B*, and *D* are right angles. Therefore ...
6. If you heat your cup of coffee in a microwave for 100 seconds, then it will be too hot to drink. Alice's heated coffee was not too hot to drink. Therefore ...
7. Use the following conditional statement. If a person lives in Austin, then the person lives in Texas.
   a. Create a valid reasoning situation in which affirming the hypothesis is used.
   b. Create a valid reasoning situation in which denying the conclusion is used.
   c. Create an invalid reasoning situation in which the converse is assumed.
   d. Create an invalid reasoning situation in which the inverse is assumed.
8. Create a truth table using p, q, and $p \rightarrow q$ and describe a situation in which each of the following statements is false.
   a. If a person lives in Texas, then the person lives in Austin.
   b. I can't go to the game if I spend my money for concert tickets.
9. What type of reasoning is involved in the following problem?

   Debra received 3 shares of new stock for every 2 shares she already owned. She owned 500 shares of stock. How many new shares did she receive?
10. Consider the following figure.

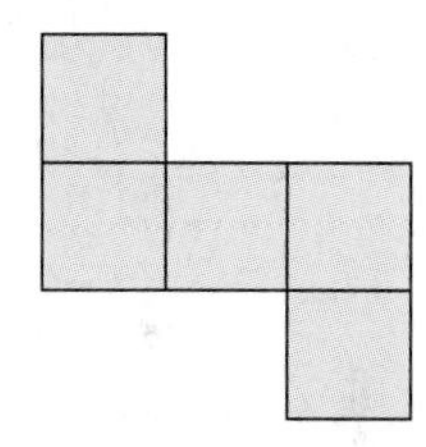

   a. Will this pattern fold to make an open-top box?
   b. What type of reasoning is involved in answering the question in part (a)?
11. Study the following sequences.
   i. 22, 33, 44, 55, 66, 77, 88, 99, ...
   ii. 1, 11, 12, 22, 23, 33, 34, 44, ...
   iii. 2, 6, 18, 54, 162, 486, 1458, ...
   a. Classify each sequence as arithmetic, geometric, or neither.
   b. Give the next three terms in each sequence.
12. Give the next statement in each sequence of statements.
   a. $1 \div 0.5 = 2$. $2 \div 0.5 = 4$. $3 \div 0.5 = 6$. $4 \div 0.5 = 8$. $5 \div 0.5 = 10$. $6 \div 0.5 = 12$. ?
   b. $5 \times 4 = 20$. $5 \times 14 = 70$. $5 \times 24 = 120$. $5 \times 34 = 170$. $5 \times 44 = 220$. $5 \times 54 = 270$. ?
13. Consider the following table of values.
   a. Look for a column extension pattern and write three more rows for the table.
   b. Look for a row relationship pattern and show the 25th row of the table. (*Hint:* Thinking about dou-

Time (weeks)	Growth of Plant (inches)
1	1
2	3
3	5
4	7
5	9
6	11

bling the number of items might help you determine a row relationship pattern.)

14. Name five problem-solving strategies that are frequently used.

15. Consider the following problem.

A basketball and a pair of basketball shoes cost a total of \$83. If the pair of shoes cost \$15 more than the ball, what did each cost?

a. What problem-solving strategy would you use to solve the problem?
b. Use the strategy to obtain the answer.

16. Give a verbal, visual, numerical, and graphic representation of the idea *one-third.*

17. What connection do you see between the following two problems?

a. How many different bike rides can be taken by four people if everyone rides only once with everyone else?
b. How many computer cables are needed to connect four desks so that each pair of desks is connected by only one cable?

18. Describe two different ways that you could represent the number 0.50.

19. What geometric model would you use to represent three highways that intersect pairwise in three different cities? Why?

Reasoning and Problem Solving

20. Write a short paragraph that communicates the message of the following bar graph.

Number of cars by make
owned by 100 one-car owners

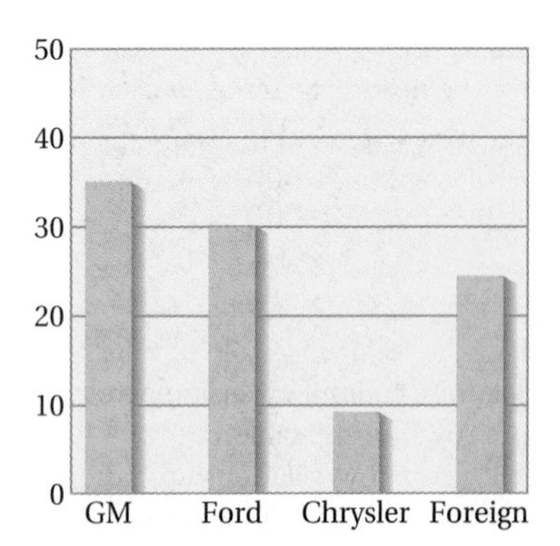

21. Create a graph different from the one above that could represent the same information about cars.

22. Consider the three following rectangles.

a. For each rectangle, determine the area (number of squares) and the perimeter (distance around).
b. What happens to the areas as you go from i to ii to iii?
c. What happens to the perimeters as you go from i to ii to iii?
d. Write a statement that describes the connection between the perimeters and areas of these three rectangles.

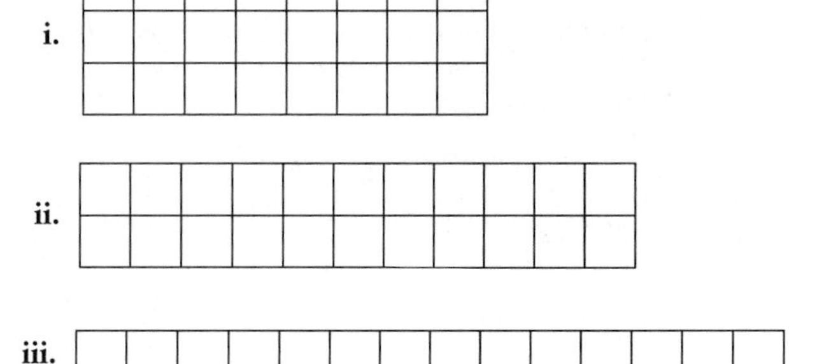

23. Consider the following ways to communicate directions for drawing a certain figure.

i. Draw a square. Draw a circle inside the square.
ii. Draw a square. Draw a circle inside the square that touches each side of the square.
iii. Draw a square. Draw a circle inside the square that touches the square.

a. In what way are the directions the same?
b. In what way might the directions result in the drawing of different figures? Show the resulting figures.

24. The Pizza Problem. Andy wants to buy three large pizzas. Rocky's Pizza Shop is having a special: Buy two large pizzas and get the third one at half price. A large pizza costs \$12.50, and Andy has \$32. Does he have enough money to buy the three pizzas? Explain your reasoning.

25. The Tennis Problem. A store that sells tennis clothing offers the following discount plan. If you pay \$100 up front, you may purchase any of the shirts or shorts for \$20. If the shirts or shorts all cost \$40, how many would you have to buy in order to break even? Explain your strategy.

26. Give a convincing argument that the advertisement encouraged the reader to engage in invalid reasoning.

Advertisement: If you want your skin to stay smooth, use Wrinkleflee cream morning and night. People who use Wrinkleflee treat their skin with loving care.

Reader's interpretation: If you use Wrinkleflee cream morning and night, then your skin will stay smooth, but if you don't use Wrinkleflee, then you don't treat your skin with loving care.

Alternative Assessment

27. Study the problem-solving model in Figure 1.7 *(p. 40)* and write a paragraph generally explaining the connection between the *problem-solving process, problem-solving skills,* and *problem-solving strategies.*
28. Work in a small group to decide the best way to explain to someone the difference between inductive and deductive reasoning.
29. Caitlin has solved the following problem (Exercise 8 in Section 1.3).

    Bella is in a race with 20 other racers. Bella is fifth from last when the race begins. At the end of the race she is third. How many racers did she pass?

    She obtained an answer of 14 racers, which she found out was the correct answer. However, Caitlin thinks that this answer depends on assumptions made about the race. For example, she claims that the number of times one runner can pass another and who finishes first and second make a difference. Consider her claim and discuss how these assumptions might affect the solution to the problem.
30. Make a table for the following problem.

    An underwater search expert took 14 days to find an old sunken treasure. The person who hired the expert agreed to pay him by putting $1 in a safe the first day, $2 in the safe the second day, $4 the third day, $8 the fourth day, and so on, doubling the amount put in the safe each day. When the expert opened the safe after 14 days, how much money did he get for the job?

    a. Use this problem as an example to help you write an explanation of what is meant by *column extension patterns* and *row relationship patterns.*

    b. In your written response to part (a), include a discussion of the advantages and disadvantages of the two types of table patterns for solving the problem.

2 Sets and Whole-Number Operations and Properties

Chapter Perspective

Whole numbers and operations involving them are essential to many activities and job-related tasks such as keeping records in sports, cooking, shopping, determining quantities, figuring costs, and calculating profits. In today's world of digital electronics, everything from sound to color can be expressed by whole numbers and thus be easily converted to discrete electrical impulses. Well-understood whole-number ideas, symbols, and the operations of addition, subtraction, multiplication, and division provide a mathematical means of communicating ideas and solving problems.

In this chapter we introduce *sets* to help you develop an understanding of whole numbers and give meaning to the symbols used to represent and communicate with whole numbers. We use models, set language, and mathematical definitions to describe the whole-number operations of addition, subtraction, multiplication, and division. Throughout the chapter, we focus on communication and problem solving with whole numbers.

Connection to the NCTM Principles and Standards

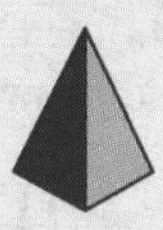

The NCTM *Principles and Standards for School Mathematics* (2000) state that instructional programs from prekindergarten through grade 12 should enable all students to

- understand numbers, ways of representing numbers, relationships among numbers, and number systems;
- understand meanings of operations and how they relate to one another;
- compute fluently and make reasonable estimates. *(p. 32)*

Connection to the PreK–8 Classroom

In grades PreK–2, students use a variety of models to connect number and operation words and symbols to situations involving addition and subtraction.

In grades 3–5, students use variations of these models to extend whole-number ideas to situations involving multiplication and division.

In grades 6–8, students use their understanding of whole numbers and whole-number operations to develop understanding of new number systems involving negative numbers and fractions.

Section 2.1 SETS AND WHOLE NUMBERS

- Connecting Sets to Whole-Number Ideas
- Using Sets to Define Whole Numbers
- Using Sets to Compare and Order Whole Numbers
- Important Subsets of Whole Numbers

In this section, we introduce the basic ideas of sets to convey the meaning of whole numbers, the ordering of whole numbers, and the relationships between different types of whole numbers. We use set ideas and set notation as a language to express what whole numbers are and how they are related. Mini-Investigation 2.1 will help you think about the meaning of whole numbers.

Write a paragraph or do a demonstration for the class describing your group's response to this question.

Mini-Investigation 2.1 **Communicating**

If intelligent beings were discovered on another planet, what methods might they use to teach their young about the whole number five, and what symbol might they use to represent this whole number?

Connecting Sets to Whole-Number Ideas

People use the numbers 0, 1, 2, 3, 4, . . . , called *whole numbers*, in a variety of ways every day and feel comfortable with what they are and how they are used. Mathematically, however, we need to ask, "What is a whole number?" We then need to look at the mathematical origins and meanings of whole numbers. We begin by exploring some important ideas about sets and the language associated with these ideas.

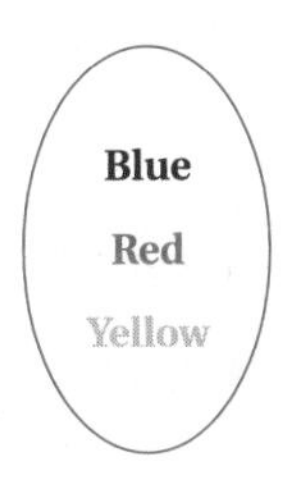

FIGURE 2.1 | Pictorial representation of the set of primary colors.

Sets and Their Elements. A **set** is any collection of objects or ideas that can be listed or described. For example, the set of primary colors contains the colors blue, red, and yellow. In mathematical shorthand, we list the objects in a set within braces: {blue, red, yellow}. As shown in Figure 2.1, we can also draw a loop to represent a set of objects pictorially.

A set is usually labeled with a capital letter. For example, if set A is the set of vowels in the English alphabet, we write

$$A = \{a, e, i, o, u\}.$$

Each individual object in a set is called an **element** of the set. Thus each of the vowels is considered an element of the set of vowels in the English alphabet. We symbolize the fact that i is an element of set A by writing $i \in A$. If an object such as w is *not* an element of set A, we write $w \notin A$.

A set with no elements, called the **empty,** or **null, set,** is denoted { } or $\varnothing$. An example of an empty set is: the set of living human beings more than 300 years old. A set with a limited number of elements is a **finite set.** A set with an unlimited number of elements is an **infinite set.** Later in this chapter, as we develop set language and whole-number ideas further, we extend our description of finite and infinite sets.

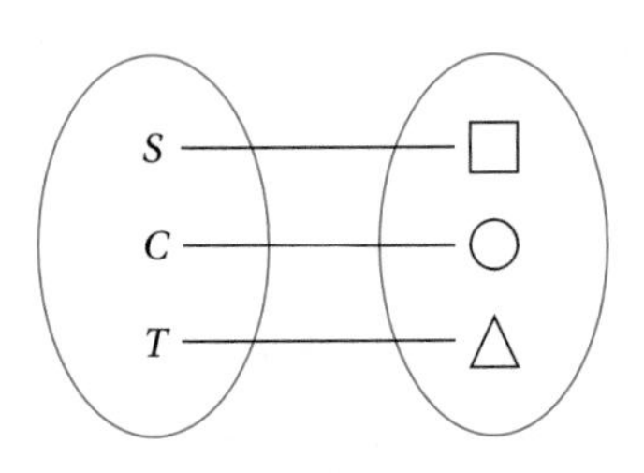

FIGURE 2.2 | One-to-One pairing (matching) of elements in two sets.

One-to-One Correspondence. Certain sets can have a common property in terms of their elements. That is, sometimes the elements of two sets can be paired or matched with each other with no extra elements left over, as illustrated in Figure 2.2. In the language of sets, this matching property is defined as follows.

Definition of One-to-One Correspondence

Sets A and B have **one-to-one correspondence** if and only if each element of A can be paired with exactly one element of B and each element of B can be paired with exactly one element of A.

The order of the elements in the sets makes no difference. The important point is that each element in one set can be matched with exactly one element in the other set with no extra elements in either set. For example, suppose that you are planning a conference and print exactly one name tag for each person who plans to attend. If each expected person does attend the conference, the order in which each arrives doesn't change the fact that you have a set of name tags and a set of people that can be matched or paired, with no extra people or name tags. Example 2.1 further illustrates the idea of one-to-one correspondence between sets.

Example 2.1 **Analyzing One-to-One Correspondence**

Consider the sets:

$$P = \{\text{one, two, three}\};$$
$$M = \{a, b, c, d, e\};$$
$$S = \{s, n, u, l, z\};$$
$$T = \{\text{blue, red, yellow}\};$$
$$V = \{\text{high, medium, low}\}.$$

Choose pairs of sets and decide if they have one-to-one correspondence.

Solution

Takota's thinking: I can draw lines for sets M and S to show that they have a one-to-one correspondence:

$$\begin{array}{c} M = \{a, b, c, d, e\}. \\ \;\;\;\;\;\;\;\;\; | \;\; | \;\; | \;\; | \;\; | \\ S = \{s, n, u, l, z\}. \end{array}$$

John's thinking: I can match set V with set P: high—one, medium—two, low—three; or low—one, medium—two, high—three. So sets V and P have one-to-one correspondence.

Alejandro's thinking: If I draw a picture and try to match sets P and M, every element in P has a matching element in M, but not every element in M has a matching element in P:

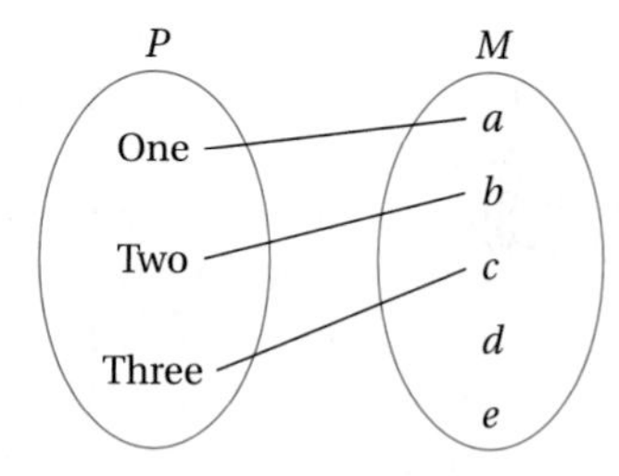

So there is no one-to-one correspondence between sets P and M.

Your Turn

Practice: Show two sets that have one-to-one correspondence.

Reflect: How can you verify the one-to-one correspondence without counting the elements in each set? ■

The idea of *if and only if* plays a crucial part in the definition of one-to-one correspondence. It ensures that the idea of one-to-one correspondence is based on the truth of the following two conditional statements.

1. If each element of set A can be paired with exactly one element of set B and each element of set B can be paired with exactly one element of set A, then sets A and B have a one-to-one correspondence.
2. If sets A and B have a one-to-one correspondence, then each element of A can be paired with exactly one element of B and each element of B can be paired with exactly one element of A.

Mini-Investigation 2.2 asks you to consider what would happen if these conditions were modified.

Draw a picture of two sets that would satisfy the definition if these words were omitted but that do not satisfy what we want one-to-one correspondence to mean.

Mini-Investigation 2.2 **Using Mathematical Reasoning**

How would the meaning of one-to-one correspondence change if the words *and each element of* B *can be paired with exactly one element of* A were omitted from the definition?

In early experiences with mathematics, young children need extensive practice in using one-to-one correspondence to prepare them for understanding whole numbers. The mathematics curriculum for the young child is built on everyday classroom experiences that exhibit one-to-one correspondence. For example, students record attendance on the attendance board by placing one token for each student present and use one-to-one correspondence to determine whether anyone is absent. The number of students working at an activity center in the classroom may be regulated by providing one workmat per student allowed at the center at one time. The idea of one-to-one correspondence determines whether there is room at the center or not because the presence of an unused workmat indicates that there is room for an additional student. In the construction of bar graphs on grid paper, students use one-to-one correspondence to match one piece of data to one unit of the height of a bar on the graph. By showing one-to-one correspondence, young children acquire the foundation needed for understanding whole numbers.

Equal and Equivalent Sets. The idea of one-to-one correspondence also provides the basis for describing relationships between two sets. Two sets that have a one-to-one correspondence are *equal* if they have *exactly* the same elements; that is, the sets are identical. For example, suppose that you collect children's books and have the complete works of Laura Ingalls Wilder, which you call your Wilder collection. You meet another person who also has the complete set of Wilder books, but she calls her collection her Little House set. Despite the different names, the two sets are equal because they contain exactly the same books.

Definition of Equal Sets

Sets A and B are **equal sets,** symbolized by $A = B$ (read "A is equal to B"), if and only if each element of A is also an element of B and each element of B is also an element of A.

By this definition, two sets that have one-to-one correspondence must also have *exactly the same elements* to be equal. The order of the elements within the sets isn't important. For example, $A = \{a, b, c\}$ and $B = \{c, b, a\}$ are equal sets.

When two sets that have *different elements* can be matched one-to-one, their *sameness* in the language of sets leads to the following definition.

Definition of Equivalent Sets

Sets A and B are **equivalent sets,** symbolized by $A \sim B$ (read "A is equivalent to B"), if and only if there is a one-to-one correspondence between A and B.

By this definition, two sets *with different types of elements* must have one-to-one correspondence in order to be equivalent. Thus the sets in Figure 2.2, page 60, and in Takota's thinking in Example 2.1, page 61, are equivalent sets. Example 2.2 gives further meaning to these ideas about the relationships between sets that have one-to-one correspondence.

Example 2.2 Identifying Equal and Equivalent Sets

If possible, give examples of sets A and B that are

a. equivalent, but not equal.
b. equal, but not equivalent.
c. neither equal nor equivalent.

Solution

a. $A = \{x, y, z\}$, and $B = \{\text{dog, cat, frog}\}$.
b. Impossible; all equal sets are equivalent.
c. $A = \{x, y, z, w\}$, and $B = \{1, 2, 3\}$.

Your Turn

Practice: Give different answers to parts (a) and (c).

Reflect: Under what conditions are two sets equivalent, but not equal? ■

Using Sets to Define Whole Numbers

We now have the ideas and language needed to describe mathematically the idea of a whole number. The cartoon in Figure 2.3 suggests at least one set of objects that might be helpful in doing so.

Although all Lucy's little brother sees now is fingers, he will soon notice many sets of objects that can be matched one-to-one with this particular set of fingers. The sets might be a set of marbles, a set of animals, a set of toys, a set of cookies, a set of letters, and so on. He will notice that each of these equivalent sets has a common property that adults describe with the word *three,* and later he will learn to recognize the symbol for it.

As illustrated in Figure 2.4, the number idea suggested by a collection of all sets equivalent to $\{a, b, c\}$ is the whole number three, which is represented by a whole-number symbol, 3. Each number idea embodied in such a collection of equivalent sets is a *whole number.* For the whole number four, think about all sets equivalent to

$$P = \{\text{one, two, three, four}\},$$

such as $T = \{\text{blue, red, yellow, green}\}$, $M = \{a, b, c, d\}$, $V = \{\text{high, medium, low, off}\}$, $W = \{\text{Tom, Jane, Bill, Dana}\}$, and so on. Further, the whole number four is represented by the symbol 4.

PEANUTS® by Charles M. Schulz

FIGURE 2.3 | PEANUTS reprinted by permission of United Feature Syndicate, Inc.

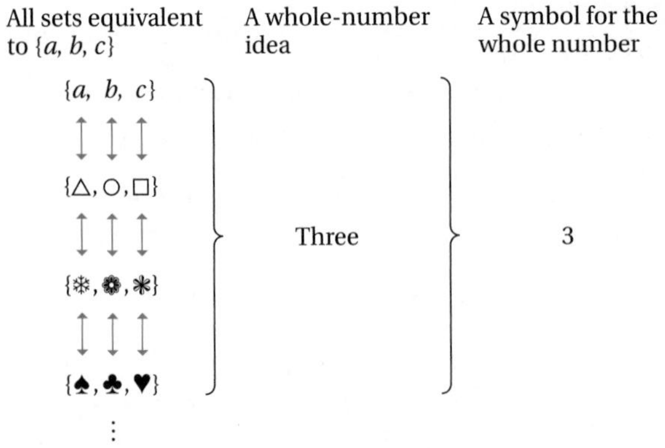

All sets equivalent to $\{a, b, c\}$	A whole-number idea	A symbol for the whole number
$\{a, b, c\}$ ↕ {△, ○, □} ↕ {❄, ✿, ✱} ↕ {♠, ♣, ♥} ⋮	Three	3

FIGURE 2.4 | A whole-number idea embodied in equivalent sets and represented by a whole-number symbol.

Definition of a Whole Number

A **whole number** is the unique characteristic embodied in each finite set and all the sets equivalent to it. The number of elements in set A is expressed as $n(A)$.

Estimation Application

Words, Words, Words

Choose a number from the set of whole numbers to estimate

a. the number of words on this page.

b. the number of words in this book.

When two finite sets A and B have the same number of elements, that is $n(A) = n(B)$, sets A and B are equivalent. Whenever any two finite sets A and B are equivalent, that is, $A \sim B$, they have the same number of elements, and $n(A) = n(B)$.

To complete the set of whole numbers embodied in all finite sets also requires an understanding of the meaning of the number zero, or 0. As the empty set contains no elements, the whole number represented by the empty set is 0, that is, $n(\varnothing) = n(\{\ \}) = 0$. The individual whole numbers, including zero, may be considered together as a single *infinite set* of numbers, called the **set of whole numbers,** denoted W:

$$W = \{0, 1, 2, 3, 4, 5, \ldots\}$$

According to research, some animals and birds can recognize the number of objects in a small set. People can usually determine which whole number represents the number of objects in a small set simply by looking at it. For larger sets, however, people often need some method to determine which whole number represents the objects in the set. **Counting** is the process that enables people systematically to associate a whole number with a set of objects. For example, to determine the number of objects in the set

{▲, ●, I, ■, ◆, ◗, ✖}

we use the counting process to set up a one-to-one correspondence between the number names and the objects in the set. That is, we say the number names in order and point at an object for each name. The last name said is the whole number of objects in the set.

{one, two, three, four, five, six, seven}
| | | | | | |
{▲, ●, I, ■, ◆, ◗, ✖}

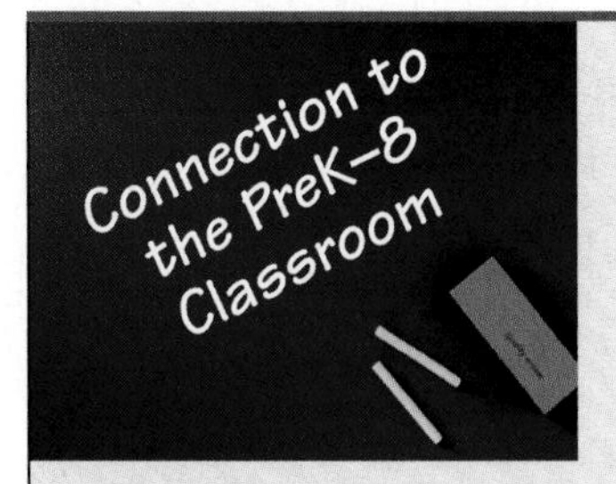

People sometimes think that young children understand counting with whole numbers when they have learned to say the number names in sequence as they point to objects. However, children often exhibit a lack of understanding of the need for one-to-one correspondence in counting by failing to assign a number name to each object or by assigning a number name to some objects twice. After they have overcome these difficulties, children must be able to realize that as they count each of several equivalent sets, they end with the same number name, and this word identifies the whole number of elements in each of the sets.

Using Sets to Compare and Order Whole Numbers

We have shown how the ideas and language of sets can help you understand whole numbers. We can use these same concepts to gain insight into how a pair of whole numbers compare and how comparison can help to order the whole numbers.

Procedure for Using One-to-One Correspondence to Compare Whole Numbers

When **comparing** two whole numbers, you can look at sets for each of the numbers. If a one-to-one correspondence cannot be made between the elements of two sets, the set with elements left over is said to have *more* elements than the other set and the whole number for that set is *greater than* that of the other set.

For example, when comparing the numbers 4 and 3, let's choose sets S and T with four and three elements, respectively. When we try to set up a one-to-one correspondence between the sets, we see that set S has an element left over:

$$
\begin{array}{l}
S = \{a, \quad b, \quad c, \quad d\}. \\
\qquad\;\; | \qquad\; | \qquad\; | \\
T = \{\text{blue, red, yellow}\}.
\end{array}
$$

Using the symbol 4 to represent $n(S)$ and the symbol 3 to represent $n(T)$, we write $4 > 3$ (read "four is *greater than* three"). The set that runs out of elements is said to have *fewer* elements than the other set, and its whole number is *less than* that of the other set. For sets T and S, we write $3 < 4$ (read "three is *less than* four"). Example 2.3 extends these ideas for comparing whole numbers.

Example 2.3 **Using One-to-One Correspondence to Compare Whole Numbers**

How can the following sets be used to compare the pairs of whole numbers 4 and 2 and 2 and 0?

$$A = \{\text{Rene, Harvey, Alicia, Pat}\}.$$
$$B = \{\ \}.$$
$$C = \{\text{basketball, baseball}\}.$$

Solution

Gerardo's thinking: I can draw a line from basketball to Rene and from baseball to Harvey, but then Alicia and Pat aren't matched with anything from set C:

$$A = \{\text{Rene, Harvey, Alicia, Pat}\}.$$
$$\diagdown \quad \diagdown$$
$$C = \{\text{basketball, baseball}\}.$$

So set A has more elements than set C, and $4 > 2$.

Twyla's thinking: There will always be elements left over when I try to match set C with set B, so set C has more elements than set B, and $2 > 0$.

Your Turn

Practice: Create sets of your own and draw matching lines to show the following comparisons.

a. $5 > 3$.

b. $4 < 6$.

c. $1 < 2$.

Reflect: Is there a greatest whole number? Why or why not? ■

Using Subsets to Describe Whole-Number Comparisons. Matching elements of their sets to decide whether one of two whole numbers is less than another quickly reveals that the elements of the set for the smaller number match one-to-one with only part of the elements of the set for the larger number. Some additional ideas and language about parts of sets can help in describing precisely the comparison when one whole number is less than another whole number.

Let's begin by returning to the earlier example of your complete collection of Laura Ingalls Wilder books and the person you met who also has a collection of Wilder books. Suppose that the other person isn't sure whether she has all of the Wilder books. You have a complete collection of books, so you could have more books than she has or she could have all the books that you have. In either case, in the language of sets, her set is a *subset* of your set, which is defined as follows.

Definition of Subset of a Set

For all sets A and B, **A is a subset of B,** symbolized as $A \subseteq B$, if and only if each element of A is also an element of B.

For example, when $A = \{a, b, c\}$ and $B = \{a, b, c, d, e\}$, A is a subset of B, written $A \subseteq B$. By the definition of subset, if $A = \{a, b, c, d, e\}$ and $B = \{a, b, c, d, e\}$, A is also a subset of B. Figure 2.5 illustrates these examples of a subset with a **Venn diagram,** which shows the relationships between two sets by enclosing the elements in each within a circle. These Venn diagrams show the subset relationships when B has some elements that are not in A and when A and B are equal.

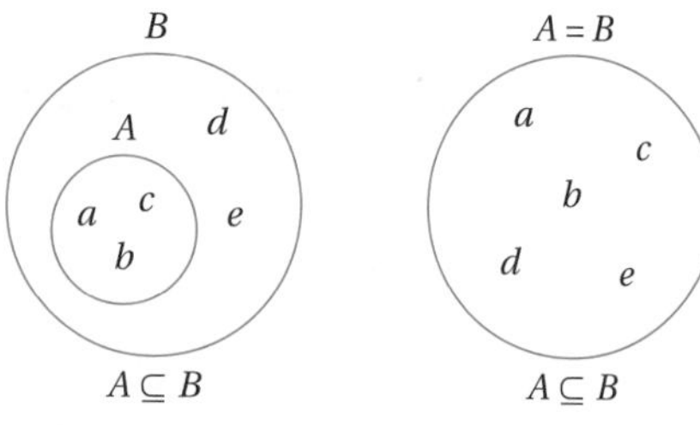

FIGURE 2.5 | Venn diagram illustrating subset relationships.

These diagrams clearly show that any set containing part or all of the elements of a set and *no other elements* is a subset of that set. That is, set A is not a subset of set B, written $A \not\subseteq B$, if and only if A contains elements that are not also in B. Note, however, that a set is a subset of itself, $A \subseteq A$, because every element in the first set is also in the second set.

Now let's suppose that the person you met with the Wilder book collection goes home to check her set and determines that she definitely is missing the Wilder book *Farmer Boy*. Hence your set contains more Wilder books than her set. In the language of sets, her set of books is a *proper subset* of yours, which we define as follows.

Definition of Proper Subset of a Set

For all sets A and B, **A is a proper subset of B,** symbolized by $A \subset B$, if and only if A is a subset of B and there is at least one element of B that is not an element of A.

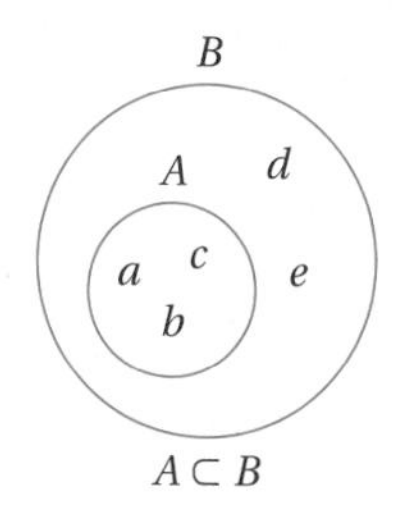

FIGURE 2.6 | Venn diagram illustrating a proper subset.

Thus the term *proper subset* identifies a subset that contains part, but not all, of the elements of a set. In other words, whenever A is a proper subset of B, B *always* contains elements that are not in A. The Venn diagram in Figure 2.6 illustrates this relationship for sets $A = \{a, b, c\}$ and $B = \{a, b, c, d, e\}$. Example 2.4 emphasizes the distinctions between a subset and a proper subset.

Example 2.4 Describing Relationships among Sets

Use subset and proper subset notation to describe the relationships among the following sets.

$$L = \{a, b, c, d, e\}, \quad V = \{a, e\}, \quad \text{and} \quad C = \{b, c, d\}.$$

Solution

$$V \subset L. \quad C \subset L.$$
$$V \subseteq L. \quad C \subseteq L.$$
$$L \subseteq L. \quad V \subseteq V. \quad C \subseteq C.$$

Your Turn

Practice: Create sets A, B, C, and D so that all three of the following relationships hold.

a. $A \subset B$.

b. $B \subseteq C$.

c. $C \subset D$.

Reflect: What has to be true for you to conclude that set A is not a subset of set B? ■

Let's now consider how subset relationships can be used to clearly describe whole-number comparisons. Note that, when one set has fewer elements than another, the set with fewer elements is equivalent to a proper subset of the set with more elements. For example, set A, with three elements, is equivalent to a proper subset, namely, $\{a, b, c\}$ of set B with five elements.

$$\begin{array}{l} A = \{p, q, r\}; \\ \quad\;\;\; | \;\; | \;\; | \\ B = \{a, b, c, d, e\}. \end{array}$$

Because each set represents a whole-number idea, this example suggests the following definition.

> **Definition of Less Than and Greater Than**
>
> For whole numbers a and b and sets A and B, where $n(A) = a$ and $n(B) = b$, a is **less than** b, symbolized as $a < b$, if and only if A is equivalent to a proper subset of B.
>
> Note that a is **greater than** b, written $a > b$, whenever $b < a$.

Ordering Whole Numbers. The specific order of whole numbers after 0 is indicated by the sequence of whole-number names used in counting: one, two, three, and so on. However, comparisons involving nonequivalent sets and the notion of *greater than*, which we define formally on p. 80, can be used to determine or verify this order. For example, when **ordering** the whole numbers, you probably think about the *next* whole number, which is always 1 greater than the number it follows. By showing a set for the whole number n and adding an additional element, you can produce a nonequivalent set for the next whole number, $n + 1$.

$$\begin{array}{ll} \text{A set with } n \text{ elements:} & \{e_1, e_2, e_3, e_4, \ldots, e_{n-1}, e_n\} \\ & \;\;\,|\quad\;|\quad\;|\quad\;|\qquad\quad\;\;|\quad\;\;| \\ \text{A set with } n+1 \text{ elements:} & \{e_1, e_2, e_3, e_4, \ldots, e_{n-1}, e_n, e_{n+1}\} \end{array}$$

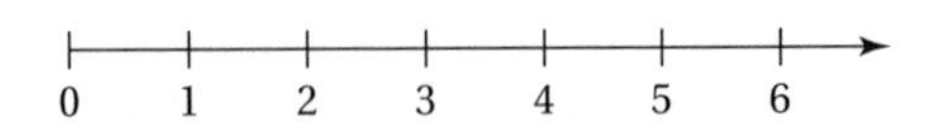

FIGURE 2.7 | Number line showing ordering of whole numbers.

Because you can always find a next number, you can convince yourself that the set of whole numbers continues indefinitely in a specific order and has an unlimited number of elements. Once you determine the order for the set of whole numbers, you can model the ordering geometrically on a number line, as shown in Figure 2.7. By convention, the order of numbers on the whole-number line is from left to right.

Important Subsets of Whole Numbers

The set of whole numbers, $W = \{0, 1, 2, 3, 4, 5, \ldots\}$, is an infinite set, with an unlimited number of elements. We can categorize the whole numbers in this set by looking at some of its subsets.

Some Special Subsets of the Set of Whole Numbers. Sometimes we want to be able to discuss the set of whole numbers except for zero. The infinite set, $N = \{1, 2, 3, 4, 5, \ldots\}$, is the **set of natural numbers** and is a proper subset of the whole numbers. It is also called the *set of counting numbers* because the numbers in the set are those that are used in counting.

If we form a proper subset of the set of whole numbers by choosing 0 and every second number following it, we produce the **set of even numbers,** $E = \{0, 2, 4, 6, 8, 10, \ldots\}$. The proper subset of the whole numbers that remains after the even numbers are removed is the **set of odd numbers,** $O = \{1, 3, 5, 7, 9, 11, \ldots\}$. In Chapter 4, we will discuss other properties of the whole numbers in these sets.

Sets W, N, E, and O are all infinite sets, and the elements of any of these sets can be matched in a one-to-one correspondence with the elements of any other of these sets. For example, we can create the following matching rule for setting up one-to-one correspondence between the set of whole numbers and the set of even numbers:

$$\begin{array}{l} W = \{0,\ 1,\ 2,\ 3,\ 4,\ \ldots,\ n\}; \\ \quad\quad\ \ |\ \ \ |\ \ \ |\ \ \ |\ \ \ |\ \qquad\ \ | \\ E = \{0,\ 2,\ 4,\ 6,\ 8,\ \ldots,\ 2n\} \end{array}$$

Unlike a finite set, an infinite set can have a one-to-one correspondence with one of its proper subsets. In fact, the definition of an *infinite set* is a set that can be put in one-to-one correspondence with a proper subset of itself. Mini-Investigation 2.3 extends these ideas about infinite sets.

Mini-Investigation 2.3 **Making a Connection**

Why is the set of odd numbers equivalent to the set of whole numbers?

Draw a diagram like the one in Figure 2.2, showing one-to-one correspondence between the set of odd numbers and the set of whole numbers, and use it to support your conclusion.

Finding All the Subsets of a Finite Set of Whole Numbers. To solve certain problems in mathematics, you may need to identify all the subsets of a finite set of whole numbers. For example, as the empty set is a subset of every set, the subsets of set $S = \{1, 2, 3\}$ are $\{\ \}$, $\{1\}$, $\{2\}$, $\{3\}$, $\{1, 2\}$, $\{1, 3\}$, $\{2, 3\}$, and $\{1, 2, 3\}$. Example 2.5 utilizes the problem-solving strategies of making a list and using logical reasoning to illustrate this idea.

Example 2.5 Problem Solving: The Board of Directors' Vote

A board of directors of a small corporation is preparing to vote on propositions 1, 2, 3, and 4 at its annual meeting. The CEO of the corporation believes that passage of any three or more of the propositions would be positive for the company. If half or less of the propositions are passed, she anticipates some future difficulties. How many possible outcomes of the voting would provide a positive situation, and how many would present difficulties?

Working toward a Solution

Understand the problem	*What does the situation involve?*	Voting on four propositions by the corporation's board of directors.
	What has to be determined?	The number of outcomes possible if three or more propositions pass and the number of outcomes possible if less than three pass.
	What are the key data and conditions?	There are four propositions.
	What are some assumptions?	That each board member will vote on all four propositions.
Develop a plan	*What strategies might be useful?*	Make a list and use logical reasoning.
	Are there any subproblems?	Find the number of possible subsets of {1, 2, 3, 4}.
	Should the answer be estimated or calculated?	No estimate or calculation is required.
	What method of calculation should be used?	No calculation is required.
Implement the plan	*How should the strategies be used?*	Let a subset of {1, 2, 3, 4}, such as {1, 2}, represent passage of selected propositions. List all the possible subsets of {1, 2, 3, 4}. Some of the subsets in the list are { }, {1}, {2}, and {1, 2}. After completing the list, draw conclusions from it to answer the question posed in the problem statement.
	What is the answer?	There are 16 possible outcomes: Five are favorable; nine present difficulties.
Look back	*Is the interpretation correct?*	Check: The approach fits the problem data.
	Is the calculation correct?	No calculation has been made.

Is the answer reasonable?	Check to determine whether the answer makes sense in terms of the data given.
Is there another way to solve the problem?	Look for generalizations that would make the solutions more efficient.

Your Turn

Practice: Solve a similar problem for five propositions to be voted on.

Reflect: How can you systematically list the subsets of a set so that you can be sure that you have listed them all? ■

Mini-Investigation 2.4 follows up on the ideas presented in Example 2.5. It involves a generalization about the number of subsets of a given set.

Write a formula for the number of subsets of a set with n elements.

Technology Extension: If the technology is available, use your formula to create a computer spreadsheet or graphing calculator table to produce a total of 10 rows in the table.

Mini-Investigation 2.4 **Finding a Pattern**

By completing this table, what can you discover about the total number of subsets for a given finite set?

Number of Elements in the Set	Number of Subsets of the Set
1	2
2	?
3	?
4	?
.	.
.	.
.	.
n	?

Problems and Exercises | *for Section 2.1*

A. Reinforcing Concepts and Practicing Skills

1. For which pairs of the following sets can you establish a one-to-one correspondence?

   $A = \{a, b, c, d, e\}$
   $B = \varnothing$
   $C = \{\ \}$
   $D = \{e, f, g, h, i\}$
   $E = \{\text{blue, green, yellow, red}\}$

2. Use the following sets to complete the given statements in as many ways as possible:

   $A = \{x, y, z\}$
   $B = \{a, b, c, d, e\}$
   $C = \{3, 1, 2\}$
   $D = \{1, 2, 3\}$

   a. Sets ______ and ______ are not equivalent and not equal.
   b. Sets ______ and ______ are not equivalent but are equal.
   c. Sets ______ and ______ are equivalent but not equal.
   d. Sets ______ and ______ are equivalent and equal.

3. Identify $n(S)$ for each of the given sets:
   a. $S = \{a, b, c, d, e\}$

b. $S = \{3\}$
c. $S = \varnothing$
d. $S = \{\varnothing\}$
e. $S = \{2, 4, 6, 8, 10, 12, \ldots\}$

4. Use sets to verify that $10 < 12$.

5. Use the following sets to complete the given statements in as many ways as possible:

$A = \{x, y, z\}$
$B = \{w, x, y, z\}$
$C = \{\ \}$

a. ______ is not a proper subset and not a subset of ______.
b. ______ is a proper subset but not a subset of ______.
c. ______ is a subset but not a proper subset of ______.
d. ______ is a proper subset and a subset of ______.

6. Predict how many subsets $S = \{c, d, e\}$ has. List all the subsets of S and compare the list to your prediction.

7. Create sets R, S, and T so that $S \subseteq T$, $R \subset S$, and $S \sim T$.

8. Create sets P and Q so that $P \subseteq Q$ and $Q \subseteq P$. What has to be true about sets P and Q?

9. a. Explain why $\{z\}$ is a subset of $\{w, x, y, z\}$.
b. Explain why $\{a\}$ is not a subset of $\{w, x, y, z\}$.

10. a. Explain why $\{\ \}$ is a subset of $\{w, x, y, z\}$.
b. Explain why $\{\ \}$ is a subset of every set.

11. a. Explain why $\{w, x, y, z\}$ is a subset of itself.
b. Explain why every set is a subset of itself.

12. For set $A = \{a, b, c\}$, generate all the proper subsets of set A. How do the whole numbers for the proper subsets of set A compare to the whole number for set A?

13. Use your understanding of sets and whole numbers to do the following.
a. Identify a finite subset of the set of whole numbers. Does it have a greatest element?
b. Identify an infinite subset of the set of whole numbers. Does it have a greatest element?
c. Can you identify an infinite subset of the set of whole numbers that has a greatest element? Why or why not?

14. Can you identify a subset (finite or infinite) of the whole numbers that does not have a least element? Why or why not?

15. List all the subsets of $\{a, b, c, d\}$.

16. Explain how the set of counting numbers differs from the set of whole numbers.

B. Deepening Understanding

17. How are the ideas of one-to-one correspondence and equivalent sets used in counting to identify whole numbers?

18. How are the ideas of subsets and proper subsets used in counting to identify relationships between whole numbers?

19. Create several sets equivalent to set C, where $C = \{a, b, c, d\}$.
a. Do these sets necessarily have *all* elements in common?
b. Do these sets necessarily have *any* elements in common?
c. What characteristic *must* these sets have in common?

20. You are ordering some frozen yogurt for dessert and have a choice of four toppings. You can get a combination of any or all of the toppings. What are your possible choices? How do these choices relate to finding all the possible subsets of a set?

21. a. List all subsets of the empty set.
b. List all proper subsets of the empty set.
c. How do the answers to parts (a) and (b) illustrate the relationship of 0 to every other whole number?

22. a. If two sets are equal, are they necessarily equivalent? Why or why not?
b. If two sets are equivalent, are they necessarily equal? Why or why not?
c. Draw a diagram of and write a statement describing the relationship between the ideas of two sets being equivalent and the idea of two sets being equal.

23. Decide whether the following statements are true or false and justify your answer.
a. If $n(A) = n(B)$, then $A = B$.
b. If $n(A) \neq n(B)$, then $A \neq B$
c. If $A = B$, then $n(A) = n(B)$.
d. If $A \neq B$, then $n(A) \neq n(B)$.

24. Decide whether the following statements are true or false and justify your answer.
a. If $n(A) < n(B)$, then $A \subset B$.
b. If $n(A) \leq n(B)$, then $A \subseteq B$.
c. If $A \subset B$, then $n(A) < n(B)$.
d. If $A \subseteq B$, then $n(A) \leq n(B)$.

25. Explain what a librarian means when he says, "Faculty may check out any *proper subset* of the newly acquired books in the library."

C. Reasoning and Problem Solving

26. Use the ideas about subsets presented in this section to answer the following problems and questions.
a. If $P = \{\ \}$, list all possible subsets of P. How many are there?
b. If $R = \{a\}$, list all possible subsets of R. How many are there?
c. If $S = \{a, b\}$, list all possible subsets of S. How many are there?
d. Show that the answers to parts (a)–(c) fit the generalization you discovered in Mini-Investigation 2.4 regarding the number of subsets of a finite set.

27. Verify that the set of whole numbers can be put in a one-to-one correspondence with the set of natural numbers.

$$W = \{0, 1, 2, 3, 4, 5, \ldots n\};$$
$$N = \{1, 2, 3, 4, 5, 6, \ldots ?\}.$$

28. Use the ideas of equal sets and subsets to verify that for every pair of whole numbers, a and b, only one of the following relationships exists: $a < b$, $a = b$, or $a > b$.

29. The Subcommittee Problem. A leader wants to determine how many different subcommittees she can form that include one or more of the following people: Andrews, Burton, Custer, Dunbar, and Ebel. She is willing to consider committees of 1, as well as larger committees. Determine how many and list the different possible committees.

D. Communicating and Connecting Ideas

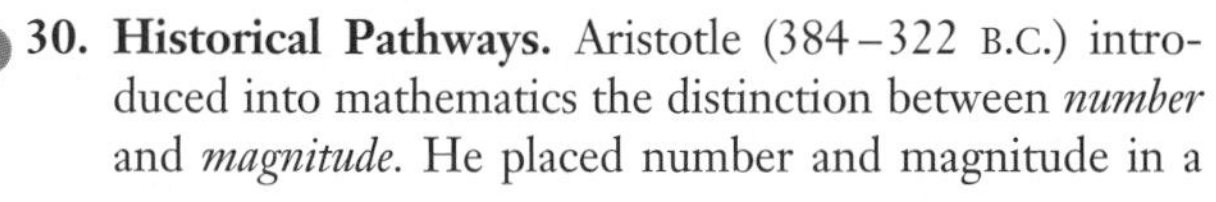

30. Historical Pathways. Aristotle (384–322 B.C.) introduced into mathematics the distinction between *number* and *magnitude*. He placed number and magnitude in a single category called *quantity* and divided this category into two cases: the discrete (number) and the continuous (magnitude). [Katz (1993)] The development of whole-number ideas through the use of sets of objects, as in this section, is an example of the discrete case. Lines, surfaces, volumes, and time are examples of the continuous case. With two or three other classmates, make a list of situations that are often described with whole numbers and categorize them as discrete situations or continuous situations.

31. Making Connections. With a partner, brainstorm a list of real-world situations that involve one-to-one correspondence. Identify some problems that would arise if one-to-one correspondence did not exist.

Section 2.2 | ADDITION AND SUBTRACTION OF WHOLE NUMBERS

- Using Models and Sets to Define Addition
- Basic Properties of Addition
- Modeling Subtraction
- Using Addition to Define Subtraction
- Comparing Subtraction to Addition

In this section, we use models to help explain addition and subtraction of whole numbers. We focus on the actions of joining and separating that embody these operations. We also examine the mathematical definitions and properties of the operations of addition and subtraction and investigate the relationship between them. Mini-Investigation 2.5 asks you to compare the operations of addition and subtraction.

Draw a picture to illustrate your comparison of addition and subtraction.

Mini-Investigation 2.5 | **Making a Connection**

How are addition and subtraction different, and how are they alike?

Using Models and Sets to Define Addition

Addition of whole numbers is often needed when **joining** two groups of discrete objects or two continuous lengths or regions. We use both discrete and continuous models as shown in Figure 2.8 and Figure 2.9 to describe the connection of addition to its real-world origins in the action of joining.

As an abstraction of the real-world situation shown in Figure 2.9, addition can be represented by joining two directed segments on the number line. If the endpoint of the directed segment representing a first number is placed at 0 and the endpoint of the directed segment representing the second number is joined to it with no gaps or overlaps, the sum of the two numbers can be read from the number

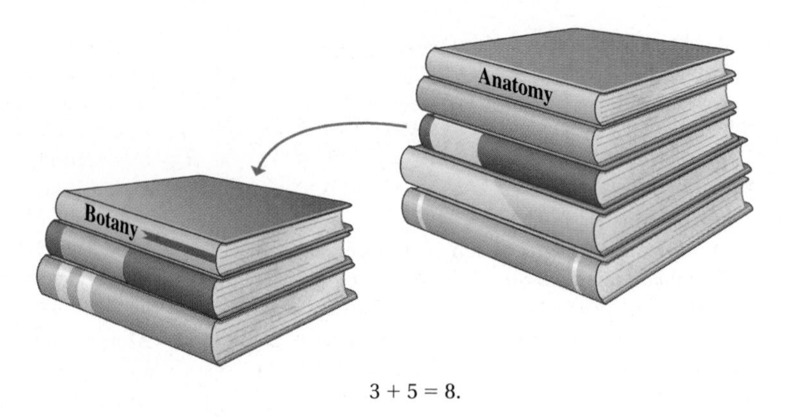

$3 + 5 = 8.$

FIGURE 2.8 | Addition as joining two groups.

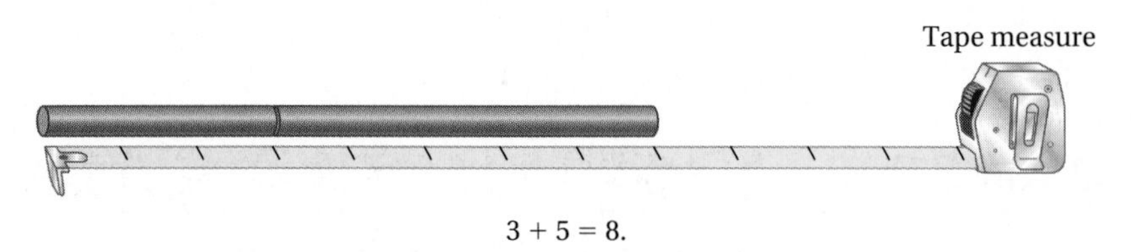

$3 + 5 = 8.$

FIGURE 2.9 | Addition as joining two lengths.

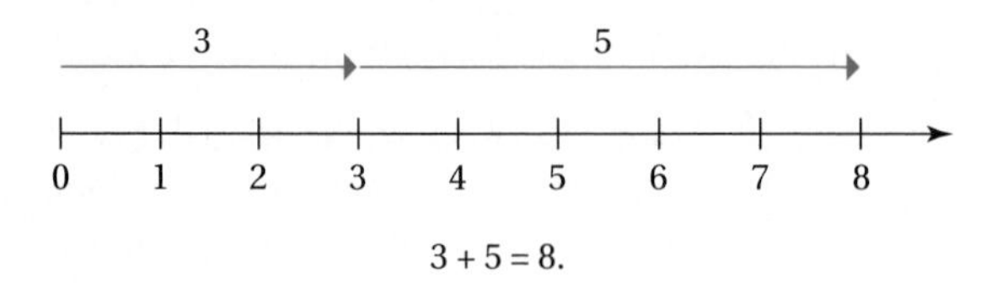

$3 + 5 = 8.$

FIGURE 2.10 | Number line model of addition.

line, as shown in Figure 2.10. These physical models provide an intuitive understanding of the operation of addition of whole numbers. However, in order to communicate clearly in mathematical situations, we need a proper definition of this basic operation. We develop the language used in the definition by exploring operations with sets.

Union of Two Sets. The following real-world example illustrates a useful idea about sets. If you have to read a set of five books in anatomy, $A = \{c, d, e, f, g\}$, and a set of three books in botany, $B = \{r, s, t\}$, the set of these books you have to read, $\{c, d, e, f, g, r, s, t\}$, is called the *union* of sets A and B.

Definition of Union of Two Sets

The **union of two sets** A and B is the set containing every element belonging to set A or set B and is written $A \cup B$ (read "the union of A and B").

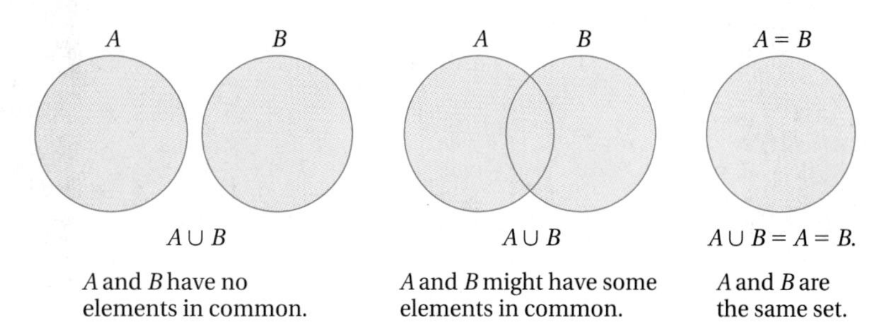

FIGURE 2.11 | Venn diagram illustrating the union of two sets.

The word *or* in the definition is the *inclusive or*, meaning "one or the other or both." Venn diagrams may be used to show the union of sets A and B, as indicated in Figure 2.11. The interiors of both sets are shaded to represent the union of sets A and B and to indicate that the union contains all elements in set A and all elements in set B. Example 2.6 extends the idea of the union of two sets.

Example 2.6 Finding the Union of Sets

If $D = \{a, b, c, d\}$, $E = \{b, c, d, e, f\}$, and $F = \{g, h, i\}$, find

a. $D \cup E$
b. $E \cup F$
c. $D \cup E \cup F$

Solution

a. $D \cup E = \{a, b, c, d, e, f\}$
b. $E \cup F = \{b, c, d, e, f, g, h, i\}$
c. $D \cup E \cup F = \{a, b, c, d, e, f, g, h, i\}$

Your Turn

Practice: Find the union of sets A and B, where $A = \{a, c, e\}$ and $B = \{b, d, f\}$.

Reflect: Make up a set G that is a subset of D. What is the union of set D and set G? How does $D \cup G$ compare to D? ■

Intersection of Two Sets. The following real-world example illustrates another useful idea about sets. If you have read a set of five books in political science, $P = \{a, b, c, d, e\}$ and a friend has read a set of six books in political science, $Q = \{a, b, c, f, g, h\}$, the set of books that you have both read, $\{a, b, c\}$, is called the *intersection* of sets P and Q.

Definition of Intersection of Two Sets

The **intersection of two sets** A and B is the set containing every element belonging to both set A and set B and is written $A \cap B$ (read "the intersection of A and B").

Two sets are said to be *disjoint* if and only if their intersection is the empty set; that is, the sets have no elements in common. Venn diagrams may also be used to show the intersection of two sets, as indicated in Figure 2.12. The

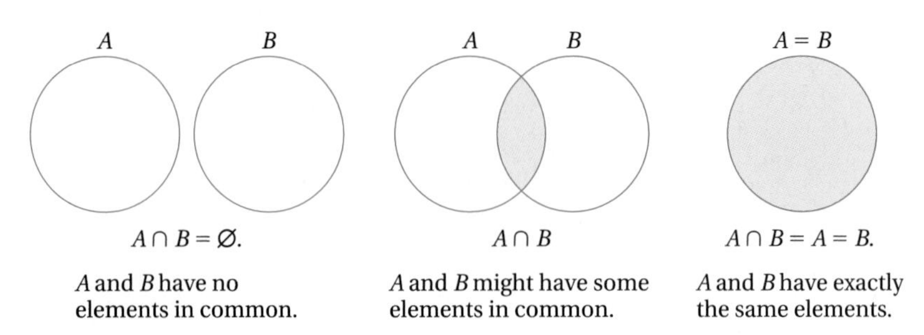

FIGURE 2.12 | Venn diagram illustrating the intersection of two sets.

shaded regions represent the intersection of the two sets and indicate the elements that are in both sets. Example 2.7 extends the idea of the intersection of two sets.

Example 2.7 Finding the Intersection of Sets

If $J = \{a, b, c, d\}$, $K = \{c, d, e, f\}$, and $L = \{f, g, h, i\}$, find

a. $J \cap K$
b. $K \cap L$
c. $J \cap L$

Solution

a. $J \cap K = \{c, d\}$
b. $K \cap L = \{f\}$
c. $J \cap L = \{\ \}$

Your Turn

Practice: Find the intersection of sets A and B, where $A = \{a, b, c\}$ and $B = \{a, b, c, d\}$.

Reflect: Make up a set M that is a subset of J. What is the intersection of set J and set M? How does $J \cap M$ compare to M? ■

Mini-Investigation 2.6 involves some reasoning about the intersection of sets.

Draw a picture to justify your answers.

Mini-Investigation 2.6 Using Mathematical Reasoning

If $n(A) = a$ and $n(B) = b$, what is the smallest number of elements that can be in $A \cap B$? The greatest number?

Addition of Whole Numbers. In Figure 2.8, we showed that the addition $3 + 5$ can be modeled by putting a set of 3 books together with a set of 5 books and counting the total number of books. One characteristic of this example is that the two sets of books were disjoint—that is, they had no book in common. This example can be generalized: For disjoint sets A and B and whole numbers a and b, if there are a elements in set A and b elements in set B, then $a + b$ is the number of elements in the union of set A and set B. We abbreviate this generalization as follows.

Definition of Addition of Whole Numbers

In the **addition of whole numbers,** if A and B are two disjoint sets, and $n(A) = a$ and $n(B) = b$, then $a + b = n(A \cup B)$. In the equation $a + b = c$, a and b are **addends,** and c is the **sum.**

Example 2.8 illustrates the importance of checking to determine whether two sets are disjoint before you use the operation of addition. The problem-solving strategies of choose an operation, draw a diagram, and use logical reasoning are helpful in solving this problem.

Example 2.8

Problem Solving: The Local Vote

The following votes came in from the local primaries.

	Democrat	Republican
County Commissioner	152	369
Justice of the Peace	200	225

How many people voted for County Commissioner? How many voted for Justice of the Peace? How many Democrats voted? How many Republicans voted?

Solution

The set of Democrats and the set of Republicans are disjoint, as shown by these diagrams, so we can add to find out how many people voted for County Commissioner and Justice of the Peace.

$152 + 369 = 521$ 521 people voted for County Commissioner.

$200 + 245 = 425$ 425 people voted for Justice of the Peace.

To find out how many Democrats voted, we cannot simply add the number of Democrats that voted for County Commissioner to the number of Democrats that voted for Justice of the Peace, as we do not know whether or not the two sets of Democrats are disjoint. Suppose we had further data indicating that the sets were not disjoint, and 100 Democrats voted for both positions, as shown in this diagram.

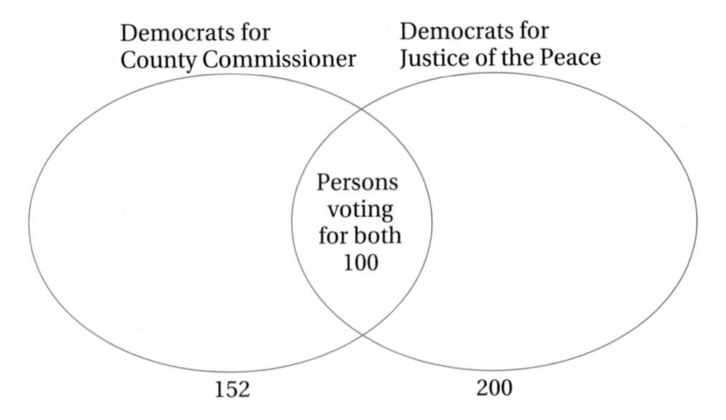

Then we could solve the problem by adding the pair of numbers for the two sets and subtracting the number that voted for both positions.

$152 + 200 - 100 = 252.$ 252 Democrats voted in the election.

If additional data were available, similar reasoning could be used to determine how many Republicans voted in the election.

Your Turn

Practice: How many people voted in the election?

Reflect: What characteristic of joined sets is important in the use of addition? ■

Basic Properties of Addition

Let's now identify and examine the basic properties of the operation of addition. The following "not-so-real-world" examples may cause you to think about some of these properties. Suppose that you are shopping for groceries in your favorite supermarket. When you reach the checkout counter, the checker makes the following comments.

1. Well, if you buy this item, I don't have any idea what your total will be; it might be \$200 or \$300, or the total might not even exist.
2. This box has nothing in it, but when I ring up its contents your total will change.
3. Would you like me to ring up the ice cream before or after the ice cream cones? The order will make a difference in your total.
4. Do you want me to ring up subtotals so you can keep track of what you are spending? Well, the way I choose to group the items in subtotals will affect your total cost. How would you like me to group them?

Each comment violates one of the properties of addition that are taken for granted in the everyday use of addition of whole numbers.

Closure Property of Addition. For each pair of whole numbers a and b, $a + b$ is a unique whole number. The **closure property of addition** guarantees that the addition of two whole numbers yields a whole-number sum—and a specific one, at that. We refer to this property by saying that the set of whole numbers is closed under addition. The word *closed* means that when any two numbers in a set are added, the result is in the same set. The word *unique* means that $a + b$ is related to exactly one whole number, although the whole number may be related to other pairs of addends. With reference to comment 1, this property assures you that the checker *will* be able to identify your total.

Identity Property of Addition. Zero is the unique whole number such that for each whole number a, $a + 0 = 0 + a = a$ The **identity property of addition** indicates that the behavior of a specific whole number is such that adding it to another whole number doesn't affect the number. Zero is called the *additive identity element.* With reference to comment 2, this property assures you that the contents of empty boxes will not change your total.

Commutative Property of Addition. For each pair of whole numbers a and b, $a + b = b + a$. According to the **commutative property of addition,** two whole numbers may be added in any order without changing the sum. With reference to comment 3, this property indicates that the checker can ring up your items in any order without changing your total.

Associative Property of Addition. For whole numbers a, b, and c, $(a + b) + c = a + (b + c)$. Addition can be done with only two numbers at a time, so adding three numbers involves first finding the sum of two of them. The **associative property of addition** ensures that the sum of three numbers will be the same regardless of which two are added together first. With reference to comment 4, this property indicates that how your grocery items are grouped for subtotals doesn't matter.

We summarize the properties of addition as follows.

Properties of Addition of Whole Numbers	
Closure property	For whole numbers a and b, $a + b$ is a unique whole number.
Identity property	There exists a unique whole number, 0, such that $0 + a = a + 0 = a$ for every whole number a. Zero is the *additive identity element.*
Commutative property	For whole numbers a and b, $a + b = b + a$.
Associative property	For whole numbers a, b, and c, $(a + b) + c = a + (b + c)$.

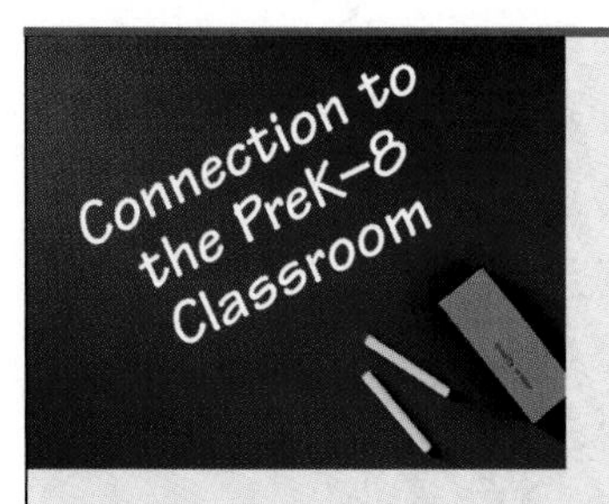

Although young students solve problems in mathematics without having memorized basic computational facts, the ability to recall sums of addends less than 10 becomes a useful tool for estimating and judging reasonableness of results. The identity, commutative, and associative properties of addition of whole numbers, taught informally, can provide students with a meaningful way to figure out sums, find patterns, and make connections that facilitate memory. For example, a student who understands "turn-around facts," or the commutative property of addition, is aware that knowing $7 + 5 = 12$ automatically tells her that $5 + 7 = 12$. If this sum proves difficult to remember, a student can figure it out by using "make-a-ten," or the associative property of addition, reasoning that $5 + 7$ is $5 + 5 + 2 = 10 + 2$, or 12.

We now briefly look at how the concepts of the operation of addition of whole numbers and the basic properties of this operation lead to a definition of *greater than* for whole numbers. In doing so, we extend the ideas presented in Section 2.1

where we used matching and subsets to compare whole numbers. To gain an initial understanding of the definition, let's consider a student's explanation of why 4800 is greater than 4756: If I enter 4756 on my calculator, I have to add to it to get 4800. So I know that 4800 is greater than 4756. The idea that a first number is greater than a second number if a whole number can be added to the second to get the first suggests the following.

Definition of Greater Than ($>$) and Less Than ($<$) for Whole Numbers

Given whole numbers a and b, a is **greater than** b, symbolized as $a > b$, if and only if there is a whole number $k > 0$ such that $a = b + k$. Also, b is less than a ($b < a$), whenever $a > b$.

We look further at this definition in Exercise 37 at the end of this section.

Modeling Subtraction

Subtraction of whole numbers is often represented by **taking away** a subset of a set, **separating** a set of discrete objects into two partial sets, or **comparing** two sets of discrete objects. We can use the model shown in Figure 2.13 to show take away. Figure 2.14 and Figure 2.15 help explain the meaning of subtraction as separating and comparing. As with addition, the representations for subtraction include continuous, as well as discrete, situations.

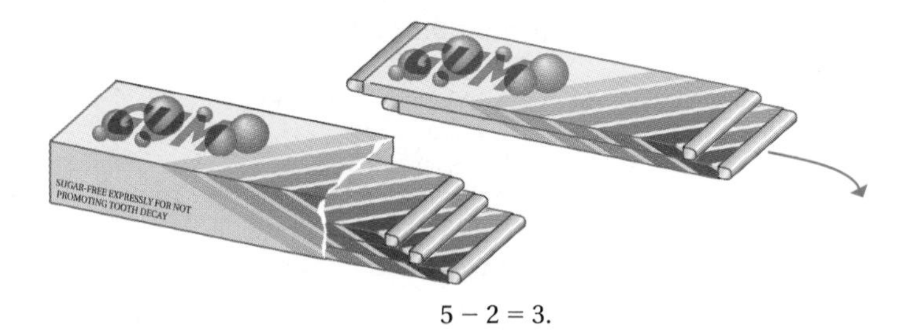

FIGURE 2.13 | Subtraction as taking away a subset of a set.

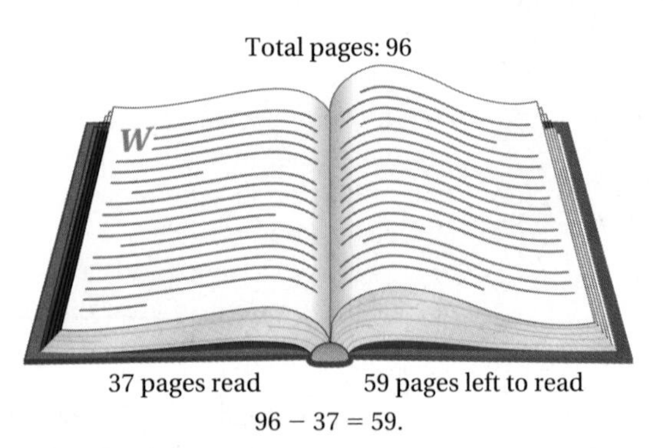

FIGURE 2.14 | Subtraction as separating a set into two disjoint sets.

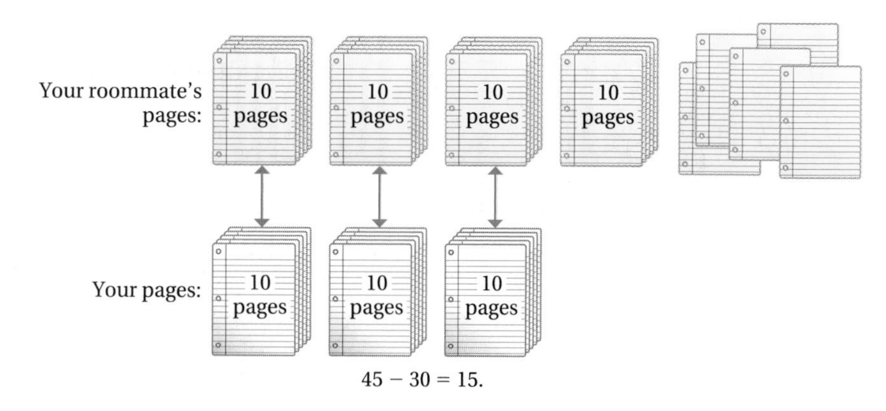

FIGURE 2.15 | Subtraction as comparing two sets.

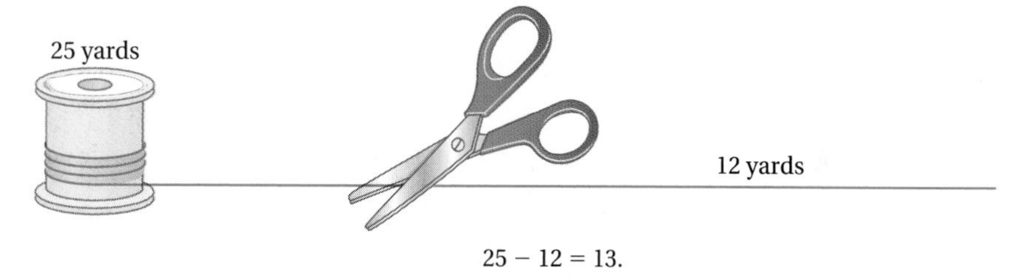

FIGURE 2.16 | Subtraction as taking away part of a length.

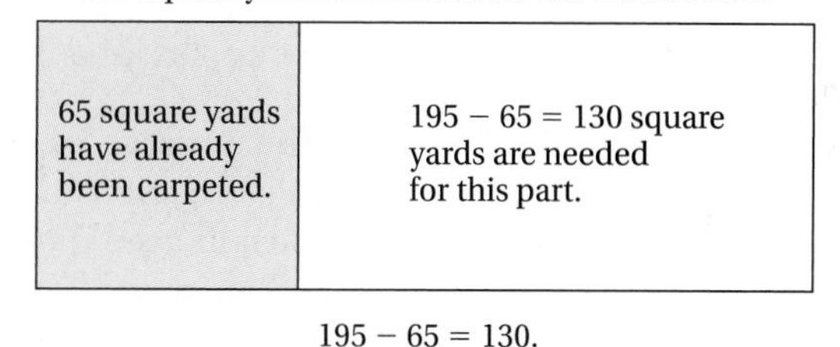

FIGURE 2.17 | Subtraction as separating an area into two parts.

Subtraction may be used to find how much string is left on a 25-yard spool if 12 yards have already been used (taken away), as illustrated in Figure 2.16. Subtraction may also be used to find how much carpeting is needed to finish carpeting a room when you know both the total amount and also the amount in the part of the room that is already carpeted (separation), as shown in Figure 2.17. Figure 2.18 shows that subtraction may also be used to find how much wider a living room window is than a bedroom window (comparing).

Subtraction may be represented geometrically by using two rays on the number line. The number line model of subtraction involves placing the endpoint of the directed segment representing a first number at 0 and joining the endpoint of

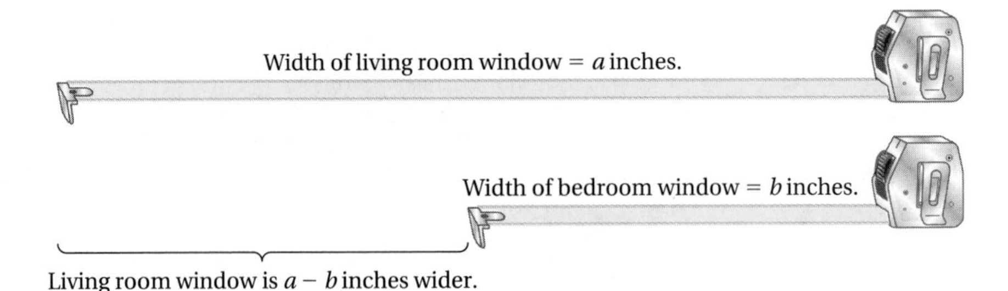

FIGURE 2.18 | Subtraction as comparing two lengths.

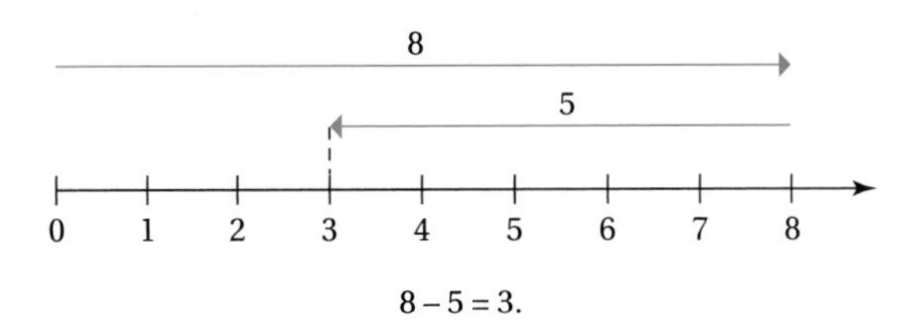

FIGURE 2.19 | Number line model of subtraction.

the directed segment representing the second number to it with no gaps or overlaps, but going in the *opposite* direction. The result of subtracting the second number from the first—the *difference* of the two numbers—is read from the number line, as indicated in Figure 2.19.

Using Addition to Define Subtraction

Subtraction as the Inverse of Addition. The following real-world situation illustrates the special relationship between addition and subtraction. How many more words do you still need to write if you have written 238 words of a required 350-word paper? You might naturally first think about the number sentence $? + 238 = 350$. However, to find the missing number of words, you might actually complete the subtraction $350 - 238 = ?$ Note that the answer to the subtraction in the second number sentence is the missing addend in the first number sentence. This situation illustrates the idea that addition, represented by joining two sets, and subtraction, represented by separating two sets, are **inverse operations.** That is, in doing addition, you combine two addends to find a sum, whereas in doing subtraction, you start with the sum and subtract an addend to find the other addend.

$$\begin{array}{ccccc} \text{Addend} & & \text{Addend} & & \text{Sum} \\ 112 & + & 238 & = & 350 \end{array}$$

$$\begin{array}{ccccc} \text{Sum} & & \text{Addend} & & \text{Addend} \\ 350 & - & 238 & = & 112 \end{array}$$

So the answer to a subtraction sentence is one of the addends in the related addition sentence. This outcome suggests the following definition.

Definition of Subtraction of Whole Numbers

In the **subtraction of the whole numbers** a and b, $a - b = c$ if and only if c is a unique whole number such that $c + b = a$. In the equation, $a - b = c$, a is the **minuend,** b is the **subtrahend,** and c is the **difference.**

Example 2.9 further examines the idea that addition and subtraction are inverse operations, as reflected in the definition of subtraction.

Example 2.9 Problem-Solving: Workout Time

How many minutes of a 40-minute exercise workout does a woman have left after riding the stationary bicycle for 15 minutes? Draw a diagram and use the inverse relationship of addition and subtraction to determine the answer.

Solution

Time is continuous, so we can use a number line to illustrate the problem.

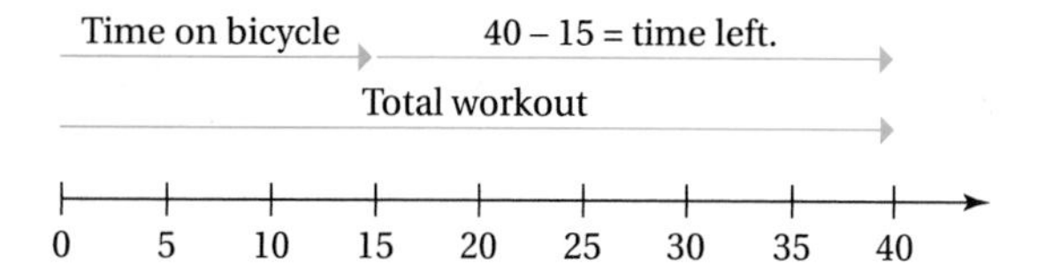

Using the information in the problem statement, we write the equation $n + 15 = 40$. By the definition of subtraction, $40 - 15 = n$ and $n = 25$. She has 25 minutes of exercise left.

Your Turn

Practice: Use the definition of subtraction to find a value of n when $52 - n = 51$.

Reflect: Does a whole number $c = a - b$ exist for every pair of whole numbers a and b? Why or why not? ■

Subtraction as Finding the Missing Addend. The inverse relationship between addition and subtraction forms the basis for a slightly different look at subtraction: finding the **missing addend.** A difference may be expressed in terms of a related addition problem involving finding the sum from two known addends. Then subtraction involves finding the missing addend from a known sum and one addend or

Sum		Addend		Addend
21	−	18	=	?

When asked to find the *difference* $21 - 18 = ?$, think of 21 as the sum and 18 as one of the addends. Then ask, What addend *adds* to 18 to give the sum 21?

Comparing Subtraction to Addition

Because the operation of subtraction is defined in terms of addition, subtraction of whole numbers should be examined for the existence of the closure, identity, commutative, and associative properties. This search requires looking for counterexamples for each of these properties in relation to subtraction, as in Example 2.10.

Elementary students can connect understanding of numbers to the operations of addition and subtraction by writing story problems to go with pictorial or concrete representations, as in the following example.

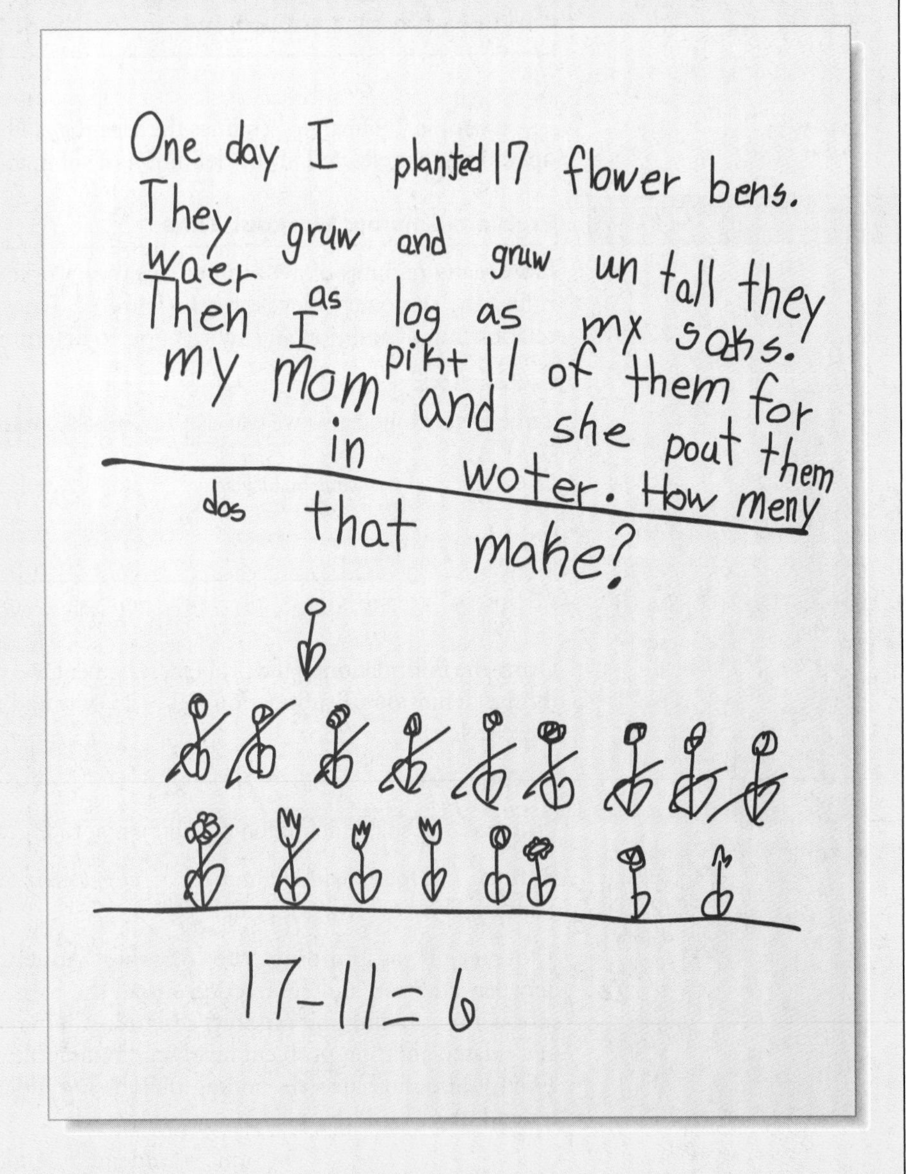

Exploration of the inverse relationship between addition and subtraction leads students to the understanding that there are no separate "subtraction facts" to learn. The mathematical sentence $17 - 11 = 6$ becomes just another way to represent the addition fact $6 + 11 = 17$.

Example 2.10 Comparing Addition and Subtraction Properties

Does whole-number subtraction have the closure property?

Solution

For subtraction to have the closure property, $a - b = c$ such that there is exactly one whole number c for every pair of whole numbers a and b. The counterexample $2 - 5$ proves that subtraction is *not* closed for the set of whole numbers since there is no whole number c such that $c + 5 = 2$.

Your Turn

Practice: Provide a counterexample showing that the commutative and associative properties do not hold for subtraction of whole numbers.

Reflect: Why is 0 *not* the identity element for whole number subtraction? ■

The counterexamples in Example 2.10 show that, even though subtraction of whole numbers is closely related to addition of whole numbers, the properties of addition do not hold for subtraction.

Problems and Exercises for Section 2.2

A. Reinforcing Concepts and Practicing Skills

1. Find the set that describes the union of each pair of sets.
   a. C = {people who are enrolled more than 20 years old}; D = {people who are enrolled in college}.
   b. $A = \{10, 20, 30\}$; $B = \{30, 40\}$.
   c. E = {the even whole numbers}; F = {the odd whole numbers less than 100}.
2. Find the set that describes the intersection of each pair of sets.
   a. C = {people who are more than 20 years old}; D = {people who are enrolled in college}.
   b. $A = \{10, 20, 30\}$, $B = \{30, 40\}$.
   c. E = {the even whole numbers}; F = {the odd whole numbers less than 100}.
3. If $R = \{s, e, n, d\}$, $S = \{m, o, r, e\}$, and $T = \{m, o, n, e, y\}$, find
   a. $R \cap S$
   b. $R \cup T$
   c. $S \cup T$
   d. $R \cap S \cap T$
4. Use sets to verify the following answers.
   a. $8 + 8 = 16$.
   b. $0 + 10 = 10$.
   c. $123 + 324 = 44$.
5. Use lengths on the number line to verify the following answers.
   a. $8 + 8 = 16$.
   b. $0 + 10 = 10$.
   c. $123 + 324 = 447$.
6. Create a word problem involving joining sets to go with the equation $2{,}000 + 356 = 2{,}356$.
7. Create a word problem involving joining lengths to go with the equation $2{,}000 + 356 = 2{,}356$.
8. Yesterday, 12 students brought their lunches from home. Seven of the students had peanut butter sandwiches and 5 of the students had apples. Did $7 + 5 = 12$ students have peanut butter sandwiches or apples in their lunches. Why or why not?
9. Which properties of addition verify the following conjectures? Model each situation with an appropriate equation or mathematical description.
   a. To figure out $2 + 8$, I don't have to start with 2 and count up 8 more. I can start with 8 and count up 2 more.
   b. It's easy to add with 0. The answer is the other addend.
   c. I can never remember what $8 + 7$ is. But I can remember $7 + 7$ is 14. So I just add on 1 more in my head to get 15.
   d. I almost made a mistake on my test because in one place I put that $188 + 59$ is 247 and in another place I wrote that it is 246. But I know the sum must be the same every time, so I checked them both to see which one was right.
10. Use the definition of *greater than* involving addition to verify the following inequalities.
    a. $3 < 5$
    b. $2500 < 3862$
    c. $4 \times 103 > 3 \times 103$
11. Use sets to verify the following answers.
    a. $16 - 8 = 8$
    b. $10 - 10 = 0$
    c. $447 - 324 = 123$

12. Use lengths on a number line to verify the following answers.
  a. $16 - 8 = 8$ **b.** $10 - 10 = 0$
  c. $447 - 324 = 123$

13. Use the definition of subtraction to rewrite the following subtraction equations as addition equations.
  a. $18 - 8 = n$ **b.** $25 - x = 15$
  c. $y - 83 = 129$ **d.** $a^2 - a = 30$

14. Use the definition of subtraction to determine the subtraction equations that are related to each of the following addition equations.
  a. $8 + 7 = n$ **b.** $14 + x = 25$
  c. $r + s = t$

For each situation in Exercises 15–19, create a word problem to go with the equation 12 − 5 = 7.

15. Separating a set into two sets

16. Separating a length into two lengths

17. Comparing two sets

18. Comparing two lengths

19. Finding the missing addend

B. Deepening Understanding

20. The practice problems at the bottom of the page appear in a mathematics study guide.
  a. What patterns could be used to generate the answers to these exercises?
  b. What mathematical characteristics of numeration, addition, and subtraction do these patterns illustrate?

21. Explore the use of the constant function available on many calculators to add the values of base-ten blocks by pressing + 100 === + 10 ==== + 1 == to represent 3 hundreds squares, 4 tens sticks, and 2 unit cubes. The display after each equals sign reads 100, 200, 300, 310, 320, 330, 340, 341, 342. What if you prefer to do the ones first, then add the tens, and then add the hundreds?
  a. Record the key sequence that you would use.
  b. Predict the display after each equals sign.
  c. Which properties of addition are involved?

22. Properties of addition may be investigated in relation to sets other than the set of whole numbers. Use the set of even numbers $\{0, 2, 4, 6, 8, 10, \ldots\}$ to answer the following questions. Justify your answers.
  a. Is this set closed under addition?
  b. Does this set have an additive identity element?
  c. Is addition with this set commutative?
  d. Is addition with this set associative?

23. If possible, identify a subset of the whole numbers having each of the following characteristics. Justify your choices.
  a. The subset is not closed under addition.
  b. The subset has no identity element under addition.
  c. The subset is not commutative under addition.
  d. The subset is not associative under addition.

24. A student points out, "If I have 20 marbles, lose 10 of them, then put 5 away, I have the same amount left as if I put 5 away first, then lose 10." Isn't that the commutative property of subtraction? What mathematics would you need to use in responding to this student?

25. Your study partner says that she figures out $9 + 6$ by thinking $(9 + 1) + 5 = 15$.
  a. What properties of place value and addition is she using?
  b. How could these properties be applied in learning other basic addition facts?

26. The basic facts table for addition of whole numbers is as follows.

+	0	1	2	3	4	5	6	7	8	9
0	0	1	2	3	4	5	6	7	8	9
1	1	2	3	4	5	6	7	8	9	10
2	2	3	4	5	6	7	8	9	10	11
3	3	4	5	6	7	8	9	10	11	12
4	4	5	6	7	8	9	10	11	12	13
5	5	6	7	8	9	10	11	12	13	14
6	6	7	8	9	10	11	12	13	14	15
7	7	8	9	10	11	12	13	14	15	16
8	8	9	10	11	12	13	14	15	16	17
9	9	10	11	12	13	14	15	16	17	18

  a. Describe how the table exhibits the closure property.
  b. Describe how the table exhibits the additive identity property.
  c. Describe how the table exhibits the commutative property.
  d. How are these properties important in learning basic addition facts?

27. Use the definition of *greater than* involving addition to verify that $24 < 30$.

Problems for Exercise 20:

110	120	130	140	150	160	170	180	190
+ 90	+ 80	+ 70	+ 60	+ 50	+ 40	+ 30	+ 20	+ 10

100	111	122	133	144	155	166	177	188
− 80	− 81	− 82	− 83	− 84	− 85	− 86	− 87	− 88

C. Reasoning and Problem Solving

28. **The Lunch Count Problem.** Use the following data.

	Lunch Count	
	Lunch from Home	School Lunch Ticket
Monday	11	12
Tuesday	6	13

a. Write four addition word problems and four subtraction word problems from this situation. At least two of the subtraction problems should represent comparing sets.
b. Write an equation that represents the solution to each of your problems. Could any be represented by more than one equation?

29. Subtraction can be used to answer questions about situations involving separating a set into two sets or comparing two sets.
a. How are these situations different?
b. How are they alike?

30. Why is it mathematically significant that the equation $14 - 8 = 6$ can represent several different types of physical situations?

31. Consider the equation $a + b - c + d = e$, which involves both addition and subtraction of whole numbers.
a. Predict what happens to e when a increases, when a decreases, when c increases or decreases, and when a increases and c decreases.
b. Test your predictions by using a computer spreadsheet, a programmable calculator, or a calculator with last-entry recall.
c. State your conclusions by using statements such as, "If a and c both increase, then e...."
d. Arrange your conclusions into two categories: e decreases if...; and e increases if....
e. Share your categories with other students and look for patterns.

32. Which of the following sets of whole numbers has (have) the closure property for addition?
a. $\{1, 3, 5, 7, 9, \ldots\}$ b. $\{0, 2, 4, 6, 8, \ldots\}$
c. $\{0, 3, 6, 9, 12, \ldots\}$ d. $\{1\}$ e. $\{0\}$

33. True or false: No finite set of whole numbers can have the closure property for addition. Support your answer.

34. A fictional, newly discovered number system appears to have used the four elements a, b, c, and d and the operation $*$ that behave as shown.

*	a	b	c	d
a	a	b	c	d
b	b	c	a	d
c	a	b	c	d
d	d	d	d	d

For example, $a * c = c$. Determine whether the operation $*$ has the following properties. Justify your decisions.
a. The closure property
b. The identity property
c. The commutative property
d. The associative property

35. Use the four elements a, b, c, and d and generate an operation, #, that has all four addition properties. Describe the operation in a table like that in Exercise 34.

36. Return to the picture you drew in Mini-Investigation 2.5 about how addition and subtraction are different and how addition and subtraction are alike. What changes would you make to this diagram now? Explain why.

37. Return to the definition of *greater than* for whole numbers.
a. Why is $k > 0$ an important part of the definition? What would happen if it were omitted?
b. How would a definition of *less than* based on whole-number operations and the properties of these operations differ from the definition of *greater than?*

D. Communicating and Connecting Ideas

38. Because subtraction is defined by using addition, you could argue that subtraction isn't an operation in its own right—that all joining, separating, and comparing situations could be described with an addition equation. With two or three other classmates, present an argument orally or in writing to support subtraction as an operation having its own definition and symbolism.

39. **Historical Pathways.** When recording on papyrus, the Egyptians used a set of mathematical symbols that included separate symbols for 1, 2, 3, 4, 5, 6, 7, 8, 9, 10, 20, 30, 40, 50, 60, 70, 80, 90, 100, and so on. For example, the symbol for 432 was written with the symbols for 400, 30, and 2 lined up next to each other. Addition with these symbols probably required the use of extensive addition tables showing the sums of each pair of symbols. How might the scribes have dealt with subtraction?

40. **Making Connections.** For many pairs of whole numbers, a and b, no whole number represents $a - b$. That is, whole numbers are not closed under the operation of subtraction. Based on your earlier experiences in mathematics, describe the kinds of numbers that, when included with whole numbers, bring closure to subtraction. What physical situations could these numbers model? How would these numbers be related to the set of whole numbers? How could these numbers be helpful when you use subtraction to compare two whole numbers?

Section 2.3 MULTIPLICATION AND DIVISION OF WHOLE NUMBERS

- Using Models and Sets to Define Multiplication
- Properties of Multiplication
- Modeling Division
- Using Multiplication to Define Division
- Comparing Division to Multiplication
- The Division Algorithm for Whole Numbers

In this section, we use models to help explain multiplication and division of whole numbers and the actions involved in these operations. We also examine the mathematical definitions and properties of the operations of multiplication and division and investigate the relationship between these operations. Mini-Investigation 2.7 asks you to compare the operations of multiplication and division.

Draw a picture to illustrate your comparison of multiplication and division.

Mini-Investigation 2.7 **Making a Connection**

How are multiplication and division different, and how are they alike?

Using Models and Sets to Define Multiplication

Models for multiplication are similar to models for addition. Addition can be modeled by joining two sets, and multiplication can be modeled by joining a certain number of equivalent sets. When these sets are arranged in equal rows and columns, as shown in Figure 2.20, this arrangement of the three equivalent sets is called a **rectangular array.** In any multiplication model involving discrete objects, one number represents the number of sets (rows) and one number represents the number of objects in each set (row). The idea behind these models and the languages of sets is used in the following description of multiplication.

> $a \times b$ is the number of elements in the union of a equivalent sets, each containing b elements.

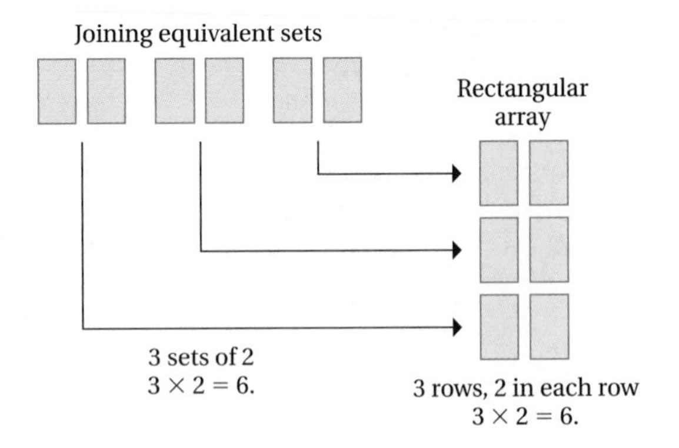

FIGURE 2.20 | Multiplication as joining equivalent sets and as a rectangular array.

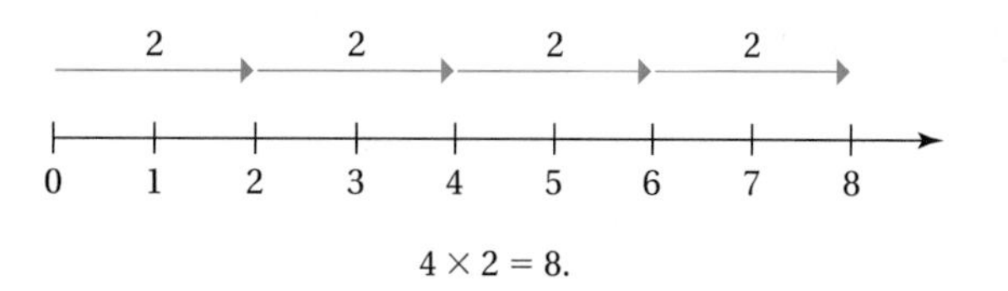

$4 \times 2 = 8.$

FIGURE 2.21 | Multiplication as joining segments of equal length on a number line.

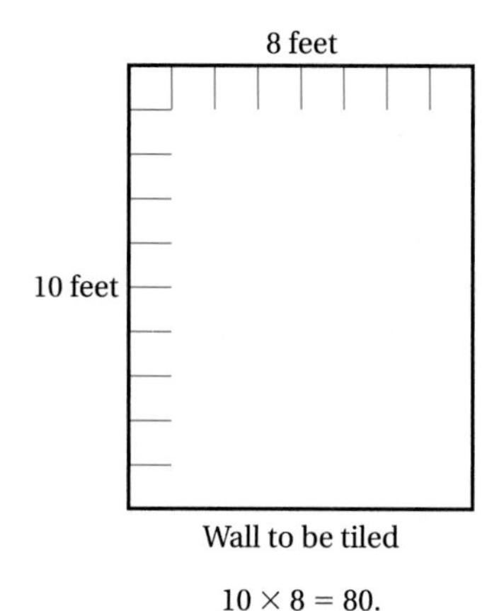

Wall to be tiled

$10 \times 8 = 80.$

FIGURE 2.22 | Multiplication as the area of a rectangle.

Addition also can be modeled by joining two directed segments along the number line, and multiplication can be modeled by joining a certain number of directed segments of equal length as illustrated in Figure 2.21.

In any multiplication model involving lengths, one number represents the number of segments being joined and one number represents the length of one segment.

Area Model. In an **area model** of whole-number multiplication, the two numbers being multiplied represent the dimensions of a rectangle. The area of the rectangle is the result of the multiplication. For example, if you are decorating 1-foot-square tiles to cover a rectangular wall that is 10 feet high and 8 feet wide, you naturally ask, "How many tiles do I need?" When you answer this question, you are demonstrating a fundamental meaning of multiplication, as illustrated in Figure 2.22.

The area model is similar to the array model because the area of a rectangle with whole-number length and width can be thought of as determining the total number of unit squares when the number in each row and the number of rows are known. However, the area and array models are different in that the rectangular dimensions aren't always whole numbers and in that the array model deals with discrete objects, whereas the area model deals with a continuous region.

Using Repeated Addition to Define Multiplication. The idea that 5×3, for example, can be found by putting together 5 sets with 3 objects in each set suggests that 5×3 can be interpreted as $3 + 3 + 3 + 3 + 3$. Thus we sometimes say that multiplication is **repeated addition.**

Using the Language of Sets to Define Multiplication. As described earlier, we can use set language to complete a description of $a \times b$ as the number of elements in the union of a disjoint equivalent sets, each containing b elements. Although this description provides a possible definition of multiplication of whole numbers, let's consider another set idea that can be used to define multiplication, called the *Cartesian product* of two sets. An example of the idea of an **ordered pair** used in the definition is (cat, dog), which is a different ordered pair than (dog, cat).

Definition of Cartesian Product

The **Cartesian product** of two sets A and B, $A \times B$ (read "A cross B") is the set of all ordered pairs (x, y) such that x is an element of A and y is an element of B.

For example, if $A = \{1, 2, 3\}$ and $B = \{a, b\}$, then $A \times B = \{(1, a), (1, b), (2, a), (2, b), (3, a), (3, b)\}$. Note that sets A and B can be equal. Example 2.11 illustrates the idea of a Cartesian product of a set with itself.

Example 2.11 Illustrating a Cartesian Product

In a particular game of chance, a player's turn consists of rolling a die twice. What are the possible results a player could get on a turn? How many results are there?

Solution

A player could roll a 1 first, then a 1, 2, 3, 4, 5, or 6 on the second roll. Similarly, if 2 came up on the first roll, the second roll could be 1, 2, 3, 4, 5, or 6. The entire set of possibilities may be represented in a diagram.

Number on second roll

Number on first roll	1	2	3	4	5	6
1	(1, 1)	(1, 2)	(1, 3)	(1, 4)	(1, 5)	(1, 6)
2	(2, 1)	(2, 2)	(2, 3)	(2, 4)	(2, 5)	(2, 6)
3	(3, 1)	(3, 2)	(3, 3)	(3, 4)	(3, 5)	(3, 6)
4	(4, 1)	(4, 2)	(4, 3)	(4, 4)	(4, 5)	(4, 6)
5	(5, 1)	(5, 2)	(5, 3)	(5, 4)	(5, 5)	(5, 6)
6	(6, 1)	(6, 2)	(6, 3)	(6, 4)	(6, 5)	(6, 6)

Each die can be modeled by a set of six numbers, $S = \{1, 2, 3, 4, 5, 6\}$. The 36 resulting pairs of numbers represent the Cartesian product, $S \times S$.

Your Turn

Practice: In another game, a player's turn consists of two spins of a spinner, with options 0, 1, 2, or 3 on the spinner. What are the possible results a player could get on a turn? How many results are there?

Reflect: What generalization can you discover from the two problems in this example? ■

We now use the idea of Cartesian product to give a commonly used definition of multiplication of whole numbers.

Definition of Multiplication of Whole Numbers

In the **multiplication of whole numbers,** if A and B are finite sets with $a = n(A)$ and $b = n(B)$, then $a \times b = n(A \times B)$. In the equation $a \times b = n(A \times B)$, a and b are called **factors** and $n(A \times B)$ is called the **product.**

The essential idea of the definition is that $a \times b$ is the number of elements in the Cartesian product of two sets, one set containing a elements and the other set containing b elements. Multiplication as the representation of a Cartesian product can occur in a variety of physical situations. As illustrated by Example 2.12, they include counting combinations and rectangular arrays. Useful problem-solving strategies include act it out or draw a diagram.

Example 2.12 Problem Solving: Color Combinations for Invitations

Suppose that you are using construction paper to make invitations for a club function. The construction paper comes in blue, green, red, and yellow, and you have gold, silver, or black ink. How many different color combinations of paper and ink do you have to choose from?

Working Toward a Solution

Understand the Problem	*What does the situation involve?*	Selecting colors of paper and colors of ink for invitations.
	What has to be determined?	The number of different combinations.
	What are the key data and conditions?	Four colors of paper; three colors of ink
	What are some assumptions?	Each invitation combines one color of paper and one color of ink.
Develop a Plan	*What strategies might be useful?*	Act it out or draw a diagram.
	Are there any subproblems?	Find the number of different invitations that can be made with blue paper.
	Should the answer be estimated or calculated?	Calculate the exact number.
	What method of calculation should be used?	Use the diagram to count.
Implement the plan	*How should the strategies be used?*	Use a choice, or tree, diagram to match each possible choice of paper with each choice of ink.

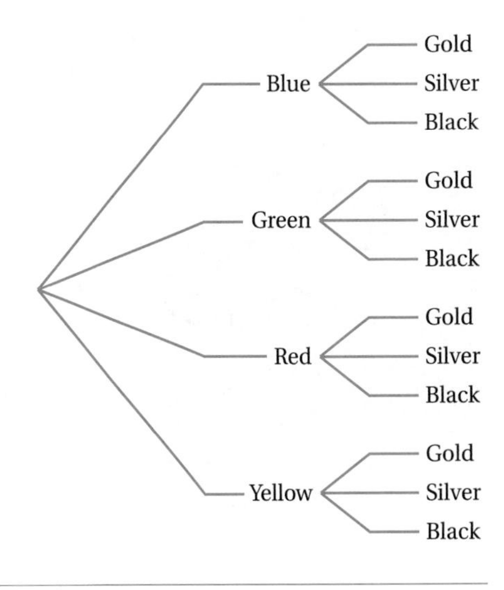

	What is the answer?	Four kinds of paper, each with three colors of ink, make 12 combinations.
Look back	*Is the interpretation correct?*	Check: The diagram fits the problem.
	Is the calculation correct?	Check: The count is correct.
	Is the answer reasonable?	Yes. Three choices of ink for each color of paper would triple the kinds of invitations made with one color of ink.
	Is there another way to solve the problem?	As shown, you can use an array of ordered pairs to list the kinds of invitations. Or you can use a multiplication equation: $4 \times 3 = 12$ combinations. In the array of ordered pairs, each color of paper is matched with each color of ink in an ordered pair that describes an invitation.

	Gold	**Silver**	**Black**
Blue	(B, G)	(B, S)	(B, Bk)
Green	(GR, G)	(GR, S)	(GR, Bk)
Red	(R, G)	(R, S)	(R, Bk)
Yellow	(Y, G)	(Y, S)	(Y, Bk)

Your Turn

Practice: From a committee, one woman's name will be drawn and one man's name will be drawn to see who serves on a community advisory board. If 9 women and 11 men are on the committee, how many different possible combinations of one woman and one man are there?

Reflect: What solution to the practice problem is suggested by the Cartesian product definition of multiplication? ■

Although multiplication is usually first introduced in the early grades as joining equivalent sets or as repeated addition, young students need to experience the other actions connected to multiplication. These experiences include looking for all possible pairs, finding the number in an array, and finding the number of tiles in a rectangle. They provide for a broader understanding of the concept of multiplication and its use in counting outcomes for sample spaces in probability and in determining areas and volumes of surfaces and solids.

Properties of Multiplication

As the pirate in the cartoon in Figure 2.23 seems to know, multiplication, like addition, is commutative. The properties that are important in addition—the closure property, the identity property, the commutative property, and the associative property—also hold for multiplication.

HERMAN®

FIGURE 2.23 | "Can you change two pieces of eight for eight pieces of two?"

We summarize the properties of multiplication as follows.

Properties of Multiplication of Whole Numbers	
Closure property	For whole numbers a and b, $a \times b$ is a unique whole number.
Identity property	There exists a unique whole number, 1, such that $1 \times a = a \times 1 = a$ for every whole number a. Thus 1 is the *multiplicative identity element.*
Commutative property	For whole numbers a and b, $a \times b = b \times a$.
Associative property	For whole numbers a, b, and c, $(a \times b) \times c = a \times (b \times c)$.
Zero property	For each whole number a, $a \times 0 = 0 \times a = 0$.
Distributive property of multiplication over addition	For whole numbers a, b, and c, $a \times (b + c) = (a \times b) + (a \times c)$.

The properties of multiplication different from the properties of addition include a different identity element for the identity property, the zero property, and the distributive property of multiplication over addition.

Identity Property of Multiplication. According to the **identity property of multiplication,** there is a specific whole number, *one*, that, when multiplied by a whole number, n, leaves n as the product:

> One is the unique whole number such that for each whole number a, $a \times 1 = 1 \times a = a$.

Thus 1 is called the *multiplicative identity element.*

Zero Property of Multiplication. The **zero property of multiplication** states that, whenever 0 is used as a factor, the product is always 0:

> For each whole number a, $a \times 0 = 0 \times a = 0$.

This property is important because 0 can often hide the value of another number in multiplication. For example, in the equation $a \times b = 0$, a could be any number if $b = 0$. The zero property of multiplication also leads to an important conclusion about division developed on page 98.

Distributive Property of Multiplication over Addition. The **distributive property of multiplication over addition** connects the operations of multiplication and addition. In general,

> For whole numbers a, b, and c, $a \times (b + c) = (a \times b) + (a \times c)$.

This property may be used to generate unknown multiplication facts from known ones and may be represented by looking at an array model from two different viewpoints, as shown in Figure 2.24.

The identity property, commutative property, zero property, and distributive property of multiplication may be used to help students learn multiplication facts. Of the 100 products with factors of 0–9, 19 have a factor of 0, and 17 more have a factor of 1. Seeing the patterns caused by the zero and identity properties automatically allows students to know 36 of the basic multiplication facts. Of the other 64 products, 8 are doubles ($a \times a$), leaving 56 products. The commutative property can be used to help students remember half of these products once the other half has been learned. For example, if a student knows 7×8, the student then knows that 8×7 is the same. The remaining 28 products can be connected to known products by using the associative or distributive properties. For example, 6×7 can be thought of as $(6 \times 6) + 6$, an application of the distributive property, and 5×4 can be thought of as $(5 \times 2) \times 2$, an application of the associative property.

$3 \times (5 + 1)$

3×5 | 3×1

$$3 \times (5 + 1) = (3 \times 5) + (3 \times 1).$$

FIGURE 2.24 | Array model of the distributive property of multiplication.

Modeling Division

Division of whole numbers may be modeled by separating a set of objects into equivalent subsets and asking appropriate questions. Division may also be represented with continuous models, but we focus primarily on the set models.

Finding How Many Subsets Model. One model for division involves separating a set into equivalent subsets of known size and asking the question, How many subsets? Figure 2.25 illustrates this model. To understand Figure 2.25, suppose that 300 invitations to a club function have been placed in envelopes that must be bundled for distribution. A stack of 25 envelopes seems to be convenient for bundling (and easy to count when combining bundles). When you bundle the set of 300 envelopes with 25 in each bundle and ask the question, "How many bundles are there?" you are demonstrating a fundamental meaning of division.

If you think about removing bundles of 25 to put in a mailbag, you could ask the question, "How many bundles (subsets) of 25 can be taken away from 300?" This question suggests a **repeated subtraction** interpretation of division. You could think about taking away 25 envelopes at a time and write $300 - 25 - 25 - 25 - \cdots = 0$. By counting the number of 25s subtracted to end up with 0 envelopes, you find the number of bundles that can be made and at the same time show that $300 \div 25 = 12$. As we are also determining how many bundles (units) of 25 are in 300, this view of division is sometimes called the **measurement interpretation of division.**

A total of 300 envelopes, with 25 in each bundle. How many bundles?

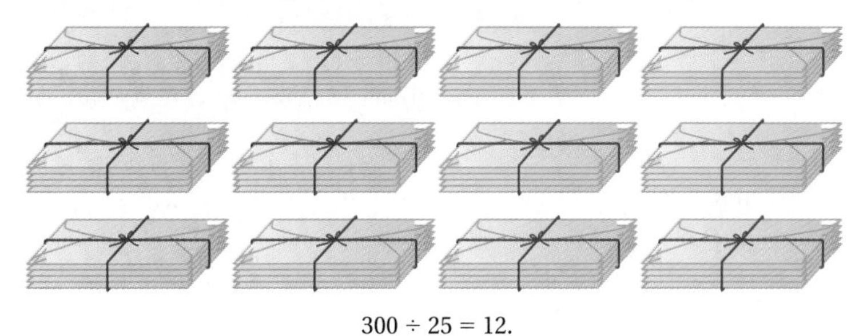

$300 \div 25 = 12.$

FIGURE 2.25 | Division as separating and finding how many subsets.

Finding How Many in Each Subset Model. A second model for division involves separating a set into a known number of equivalent subsets and asking the question, "How many are in each subset?" The following real-world situation illustrates this model. Suppose that the envelopes containing the 300 invitations to a club function need to be bundled for distribution and that 25 large rubber bands are available to wrap around the bundles. When you bundle the set of 300 envelopes into 25 bundles and ask the question, "How many invitations are in each bundle?" you are demonstrating a fundamental meaning of division, as illustrated in Figure 2.26.

If you think of starting with the 300 envelopes and 25 rubber bands, you could share envelopes with the rubber bands by placing one envelope in each rubber band in turn until all the envelopes are distributed. Then you could count the number held by each rubber band. This approach suggests a **sharing interpretation of division.**

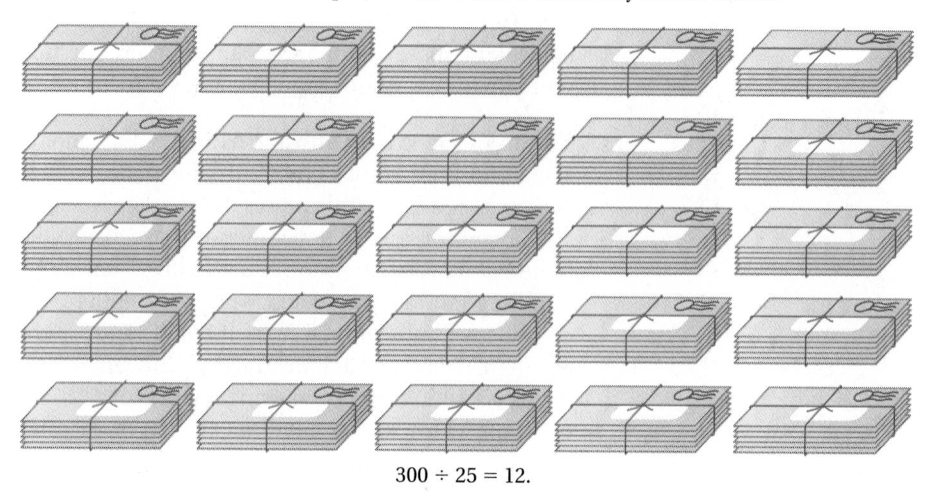

FIGURE 2.26 | Division as separating and finding how many in each subset.

Mini-Investigation 2.8 is intended to help you extend your understanding of these two interpretations of division of whole numbers.

Talk about which of the division models represents a process of one for Jim, one for Jan, one for Sue, and so on, that you might use to share some items with some friends.

Mini-Investigation 2.8 **Solving a Problem**

What two problems can you state that can be solved by using the division $24 \div 6$ and that illustrate the two interpretations of division described in Figures 2.25 and 2.26?

Example 2.13 provides a brief look at how the continuous counterparts of the set models might be used.

Example 2.13 **Problem Solving: Art Supplies**

Solve the problems in parts (a) and (b) and indicate which problem is a continuous counterpart of the *finding how many subsets* interpretation and which is a continuous counterpart of the *finding how many in each subset* interpretation.

a. Mrs. Chance has a roll of butcher paper 50 feet long. She wants to cut it into 2-foot lengths for her students to use in art class. Will she have enough for her class of 22 students?

b. Mrs. Chance has a roll of butcher paper 50 feet long. Only 10 students want to work with butcher paper today. If she wants to distribute the entire length of paper equally to the 10 students, how much paper should she give each of them?

Solution

a. Mrs. Chance wants to find the number of 2-foot pieces of paper in a 50-foot roll, so she divides: $50 \div 2 = 25$, or enough paper for 25 students. Because she knows the total length and the length of one piece and is trying to find the number of pieces, this problem is similar to *finding how many subsets.*

b. Mrs. Chance has 50 feet of paper and wants to find the length of a piece that she can cut for each of 10 students, so she divides: $50 \div 10 = 5$, or 5 feet of paper for each student. Because she is trying to find the length of each piece, this problem is similar to *finding how many in each subset.*

Your Turn

Practice: Solve problems (a) and (b). Which interpretation of division applies to each solution?

a. Rikki has 36 feet of ribbon to use in wrapping presents. If each present takes 4 feet, how many presents can she wrap with the ribbon?

b. Rikki has 36 feet of ribbon to use in wrapping presents. If she has four presents to wrap, what is the maximum amount of ribbon she could use for each present?

Reflect: In general, what question does division answer when you know the total length of a strip and are given the following information?

a. The length of each piece

b. The number of equal pieces to be cut ■

Using Multiplication to Define Division

Division as the Inverse of Multiplication. The following real-world situation helps explain the special relationship between multiplication and division. How many tables that seat 8 people will you need to provide banquet seating for 72 people? You might naturally first think about the number sentence $? \times 8 = 72$. However, to find the missing number of tables, you might actually complete the division $72 \div 8 = ?$ Note that the answer to the division in the second number sentence is the missing factor in the first number sentence. This situation illustrates the idea that multiplication, represented by putting several equivalent sets together, and division, represented by separating a set into several equivalent subsets, are inverse operations. That is, in doing multiplication, you multiply two factors to find a product, whereas in doing division, you divide the product by one factor to get the other factor.

Factor		Factor		Product
9	$\times$	8	=	72

Product		Factor		Factor
72	$\div$	8	=	9

So the answer to a division sentence is one of the factors in the related multiplication sentence. This outcome suggests the following definition.

Definition of Division

In the **division of whole numbers** a and b, $b \neq 0$, $a \div b = c$ if and only if c is a unique whole number such that $c \times b = a$. In the equation, $a \div b = c$, a is the **dividend,** b is the **divisor,** and c is the **quotient.** The operation $a \div b$ may also be written as $\frac{a}{b}$ or as $b\overline{)a}$.

Division as Finding the Missing Factor. The inverse relationship between multiplication and division is the basis for a slightly different view of division: finding the missing factor. That is, in multiplication you know two factors and find the product. In division, you know the product and one factor and find the other factor:

Product		Factor		Factor
36	÷	3	=	?

When asked to find the quotient $36 \div 3 = ?$, think of 36 as the product and 3 as one of the factors. Then ask, What factor multiplied by 3 gives the product 36?

The view of division as finding the missing factor helps you deal with the division when 0 is involved. Equations such as $8 \div 0 = ?$ have no whole number that when multiplied by 0 gives 8 or any number other than 0. For the equation $0 \div 0 = ?$ every number multiplied by 0 gives 0. Because equations such as $8 \div 0 = ?$ have no solution and the equation $0 \div 0 = ?$ has infinitely many solutions, 0 cannot be used as a divisor. However, dividing 0 by another number, say, $0 \div 8$ gives the quotient 0.

Comparing Division to Multiplication

Because the operation of division is defined in terms of multiplication, a natural question is whether the properties for multiplication of whole numbers hold for division. We examine this question in Example 2.14 by looking for counterexamples for each of the properties of multiplication as they relate to division of whole numbers.

Example 2.14 Comparing Multiplication and Division Properties

Does whole-number division have the closure property?

Solution

As with multiplication, division of whole numbers would have the closure property provided that, when we divide two whole numbers, the quotient is a unique whole number. As $11 \div 3$ doesn't have a whole-number answer, it serves as a counterexample proving that division is *not* closed for the set of whole numbers.

Your Turn

Practice: Provide counterexamples showing that the commutative, associative, and distributive properties for multiplication of whole numbers don't hold for division.

Reflect: Why is 1 *not* the identity element for whole-number division? ■

The counterexamples presented in Example 2.14 demonstrate that division doesn't have the same properties as multiplication.

Now that we have explored the operations of addition, subtraction, multiplication, and division, we need to take a brief look at how they relate to one another. As Figure 2.27 shows, the inverse relationship between addition and subtraction parallels the inverse relationship between multiplication and division. Also, the repeated addition relationship between addition and multiplication parallels the repeated subtraction relationship between subtraction and division.

The Division Algorithm for Whole Numbers

Although the mathematical definition for division of whole numbers restricts its use to numbers that "come out even," as in $24 \div 4 = 6$, whole-number division is often used to model physical situations that involve making groups of equal size

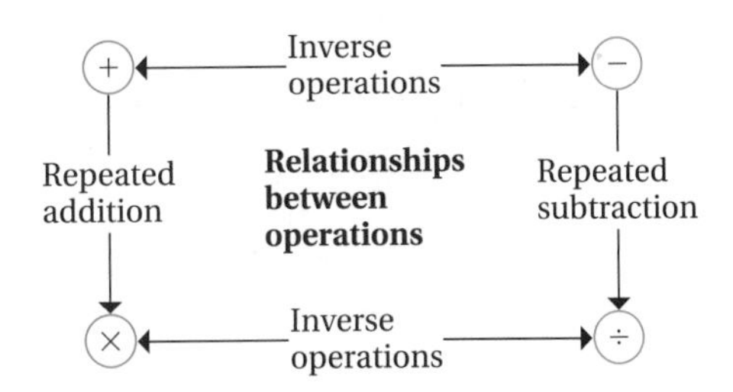

FIGURE 2.27 | Relationships between operations.

whereby part of a group is "left over." For example, if you are setting up a conference room for 25 people and want to have 4 people at each table, you might find $25 \div 4$ and note that you could have six full tables with 1 person left over. In such situations when we want to describe $a \div b$ $(b \neq 0)$ and no whole number c exists such that $c \times b = a$, we use the **division algorithm**. Essentially, it extends the definition of division to include the possibility of a remainder.

The Division Algorithm

For any two whole numbers a and b, $b \neq 0$, a division process for $a \div b$ can be used to find unique whole numbers q (quotient) and r (remainder) such that $a = bq + r$ and $0 \leq r < b$. For $a = 25$, $b = 4$, $q = 6$, and $r = 1$, $25 = (4 \times 6) + 1$.

In vertical form, we write

$$\begin{array}{r l} & 6 \text{ remainder } 1 \\ 4\overline{)25} & \\ -24 & (4 \times 6) \\ \hline 1 & \end{array}$$

We can also use a graphing calculator to illustrate the division algorithm. For example, using integer division on a graphing calculator to express $25 \div 4$ results in the following screen.

```
25 INT÷ 4
 Q=6
 R=1
```

In the symbolism of the division algorithm, $a = 25$ and $b = 4$. The graphing calculator produces the quotient, $Q = 6$, and the remainder, $R = 1$. From the calculator display, we can conclude that $25 = (4 \times 6) + 1$. Based on this example, you should be able to predict what will be displayed on the following screen when you press ENTER, and to express the data in a form described in the division algorithm.

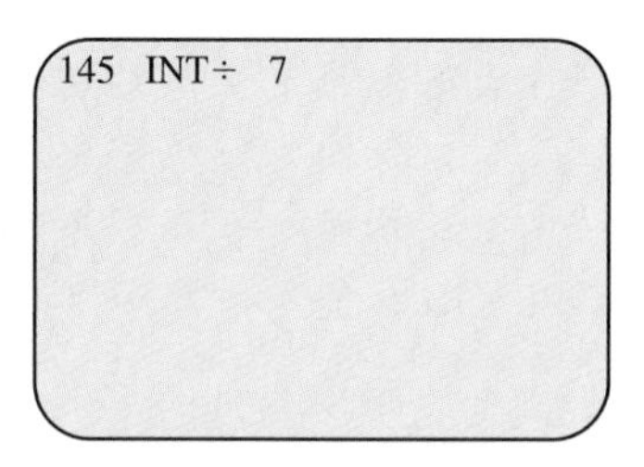

Mini-Investigation 2.9 extends the meaning of the division algorithm.

Talk about how the idea of the division algorithm would change if these words were omitted from the definition.

Mini-Investigation 2.9 **Using Mathematical Reasoning**

Why is the phrase "and $0 \le r < b$" included in the definition of the division algorithm for whole numbers?

For children, learning about and using operations involves more than just memorizing basic facts and properties. Using an operation also requires practical, common-sense reasoning. For example, if 27 students are going on a field trip and each car will hold 4 students, how many cars will be needed for the field trip? This situation is one of forming equivalent subsets, so the problem can be solved by using ideas from the division algorithm: $27 \div 4 = 6$ full cars and 3 students left over because $(4 \times 6) + 3 = 27$ and $0 < 3 < 4$. Students need to be able to interpret the results of the division algorithm in terms of the problem being solved. In this case, the answer to the problem is that seven cars will be needed for everyone to be able to go on the field trip.

Problems and Exercises *for Section 2.3*

A. Reinforcing Concepts and Practicing Skills

1. Use sets to verify the following products.
   a. $12 \times 3 = 36$. **b.** $14 \times 0 = 0$.
   c. $3 \times 100 = 300$.
2. Use lengths on a number line to verify the following products.
   a. $12 \times 3 = 36$. **b.** $14 \times 0 = 0$.
   c. $3 \times 100 = 300$.
3. Use a rectangular array to verify each of the following products.
   a. $12 \times 3 = 36$
   b. $14 \times 0 = 0$
   c. $3 \times 100 = 300$
4. Create a word problem involving joining equivalent sets to go with the equation $4 \times 30 = 120$.
5. Create a word problem involving joining segments of equal length to go with the equation $4 \times 30 = 120$.
6. Write a multiplication equation and find the area of each of the following rectangles.
   a. A rectangle 18 units high and 15 units wide
   b. A rectangle 1 unit high and 256 units wide
   c. A rectangle h units high and b units wide

Write a multiplication equation for and find the answer to each of the following questions in Exercises 7–9.

7. If a restaurant has five kinds of soup and eight kinds of salad, how many different kinds of soup and salad lunches can they make? (Each lunch consists of one soup and one salad.)
8. If a clothing manufacturer designs eight blouses, four skirts, and four vests that all mix and match, how many different outfits can be made with one blouse, one skirt, and one vest?
9. If a department store wants to advertise 72 different gift wrappings (based on a choice of one type of paper and

one color of bow), how many different types of wrapping paper and colors of bows should they plan to have on hand?

10. For the equation, $25 \times 4 = 100$, create an accompanying word problem involving
   a. a Cartesian product.
   b. an array.
   c. the area of a rectangle.

11. Which properties of multiplication verify the following conjectures? Model each situation with an appropriate equation or mathematical description.
   a. To remember 8×4, I just need to remember 4×8.
   b. It's easy to multiply with 0. It doesn't even matter what the other factor is.
   c. I can never remember what 8×6 is. But I *can* remember that 8×5 is 40. So I mentally add on one more 8 to get 48.
   d. I almost made a mistake on my worksheet because in one place I put that 9×3 is 27 and in another place I put that it is 29. But I know that it has the same answer every time, so I checked them both to see which one was right.
   e. I like to multiply by 10 or 100. The digits just move to different place-value positions.

12. Use the distributive property of multiplication over addition to find each product.
   a. $6 \times (5 + 4)$
   b. $(10 + 2) \times 10$
   c. $(2a + 3) \times (a + 4)$

For the equation $24 \div 6 = 4$, create an accompanying word problem for each situation in Exercises 13–17.

13. Separating a set into same size equivalent sets
14. Separating a segment into same length segments
15. Separating a set into a specific number of equivalent sets
16. Separating a segment into a specific number of segments of equal length
17. Finding the missing factor

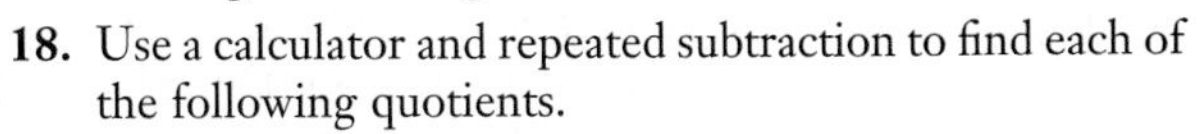

18. Use a calculator and repeated subtraction to find each of the following quotients.
   a. $135 \div 15$
   b. $98 \div 7$
   c. $1404 \div 26$

19. Use the definition of division to rewrite the following division equations as multiplication equations. If possible, find a whole-number value for each variable to make the statements true.
   a. $18 \div 6 = n$.
   b. $25 \div x = 5$.
   c. $y \div 42 = 126$.
   d. $0 \div b = c$.

20. Use the definition of division to determine the division equations that are related to each of the following multiplication equations.
   a. $15 \times 3 = n$.
   b. $9 \times y = 9$.
   c. $r \times s = t$.
   d. $8 \times 0 = 0$.

21. Use the division algorithm for whole numbers to find each quotient and remainder. Justify your answers.
   a. $19 \div 3$
   b. $256 \div 20$
   c. $2 \div 8$

B. Deepening Understanding

22. Study the following diagram.

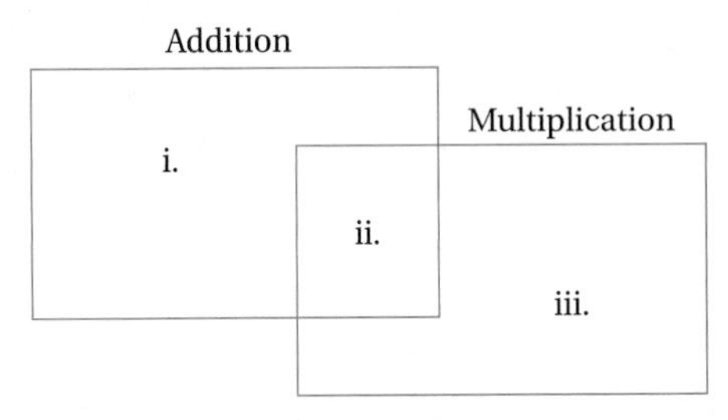

i. Characteristics of addition that are different from multiplication
ii. Characteristics that are common to both operations
iii. Characteristics of multiplication that are different from addition

   a. Complete the diagram to illustrate how the characteristics of multiplication and addition of whole numbers compare.
   b. Prepare a similar diagram for subtraction and division.

23. Predict the number of ordered pairs that will be generated from the Cartesian products in (a)–(c). Then find the Cartesian products and compare the number of ordered pairs in each one to your predictions. [*Hint:* Make a three-column (predicted $n(C)$, C, and actual $n(C)$), three-row (a, b, c) table and record your answers.]
   a. $\{a, b, c\} \times \{r\} = C$.
   b. $(1, 2, 3, 4\} \times \{a, b\} = C$.
   c. $\{\ \} \times \{3, 4, 5, 6, 7, 8\} = C$.
   d. Write the multiplication equation related to each of the Cartesian products.

24. Using the union of two sets to define addition requires that the sets be disjoint. Is that condition necessary for the two sets in the Cartesian product definition for multiplication? Why or why not? Give an example to support your response.

25. What role does multiplication play in our base-ten numeration system?

26. The product 12×6 can be found by thinking, $(10 + 2) \times 6 = (10 \times 6) + (2 \times 6) = 60 + 12 = 72$.
   a. What properties of multiplication are being used?
   b. Use these properties to verify the product 12×64.

27. Use the distributive property of multiplication over addition to rewrite each of the following sums as the product of two factors, where one of the factors is a sum.

a. $2a + 2b$
b. $14r + 18r^2$
c. $6c + 12d + 15e$

28. The basic facts table for multiplication of whole numbers is as follows.

×	0	1	2	3	4	5	6	7	8	9
0	0	0	0	0	0	0	0	0	0	0
1	0	1	2	3	4	5	6	7	8	9
2	0	2	4	6	8	10	12	14	16	18
3	0	3	6	9	12	15	18	21	24	27
4	0	4	8	12	16	20	24	28	32	36
5	0	5	10	15	20	25	30	35	40	45
6	0	6	12	18	24	30	36	42	48	54
7	0	7	14	21	28	35	42	49	56	63
8	0	8	16	24	32	40	48	56	64	72
9	0	9	18	27	36	45	54	63	72	81

a. Describe how the closure property of multiplication is reflected in the table.
b. Describe how the multiplicative identity property is reflected in the table.
c. Describe how the commutative property is reflected in the table.
d. Describe how the zero property is reflected in the table.
e. Describe how the distributive property of multiplication over addition can be illustrated by the table.
f. How are these properties important in learning basic multiplication facts?

29. How can the multiplication table be used to solve division equations?

30. Division can be used to model both separating a quantity into groups of a given size and separating a quantity into a given number of groups of equal size.
a. How are these situations different? (For example, what different questions are asked? What different actions are represented?)
b. How are these situations alike?

31. Consider the statement: $(q \times 15) + r = 95$.
a. Identify all possible pairs of whole numbers for q and r that make the sentence true.
b. Which pairs of values for q and r satisfy the division algorithm for whole numbers?
c. What could happen if the division algorithm didn't include the restriction that r is greater than or equal to 0 and less than the divisor?

32. The distributive property of division over addition has the form:

For whole numbers a, b, and c, with $c \neq 0$, $(a + b) \div c = (a \div c) + (b \div c)$, provided that whole number quotients exist for each division expression.

a. Identify four sets of three whole numbers that illustrate this distributive property of division.
b. What patterns do you see in these sets of whole numbers?
c. Why does the distributive property for division work only in particular formats?

C. Reasoning and Problem Solving

33. Describe as many patterns as you can find in the multiplication facts table in Exercise 28. Use the commutative, associative, and distributive properties to explain why these patterns occur.

34. Consider the data shown in the following picture graph and frequency chart.

Picture Graph

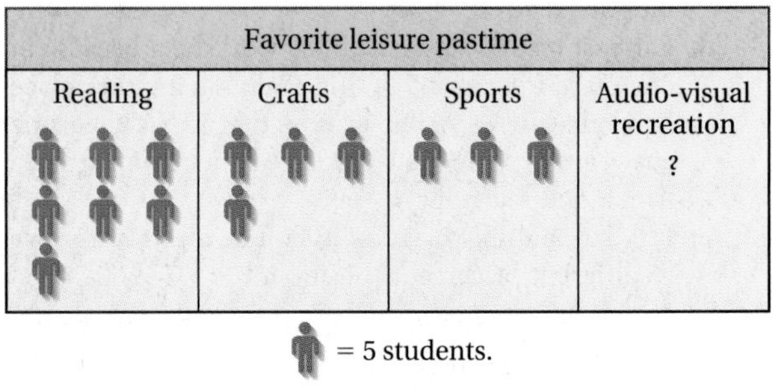

Favorite leisure pastime			
Reading	Crafts	Sports	Audio-visual recreation
			?

= 5 students.

Frequency Chart

Pastime	Reading	Crafts	Sports	Audio-visual recreation
Frequency	?	?	?	33

a. Write two multiplication word problems and two division word problems reflecting the relationship between the data in the graph and the chart.
b. Write an equation that represents the solution to each of your problems. Could either solution be represented by more than one equation?

35. The Candy Dish Problem. Eleanor made a deal with her mom. When she opened a bag of candy to put in her mom's four candy dishes, she would put an equal amount in each dish and then get to eat the ones left over. Last year, her mother received three new candy dishes as presents, and Eleanor believes that she has been getting to eat more leftover candy this year. Could that be true? Explain why or why not.

36. Consider an equation that involves both multiplication and division of whole numbers:

$$a \times b \div c = q \text{ (with remainder } 0 \leq r < c\text{)}.$$

a. Taking into consideration the use of the division algorithm to allow a remainder, make predictions about what happens to q and r when a increases, when a decreases, when c increases or decreases, when a increases and c decreases, and so on.

b. Test your predictions using a spreadsheet or graphing calculator.

c. State your conclusions in statements such as, If a and b both increase, then q . . . .

d. Arrange your conclusions into two categories: q decreases if . . . ; and q increases if . . . .

e. Share your categories with other students and look for patterns.

37. Which of the following sets of whole numbers has (have) the closure property for multiplication?

a. $\{1, 3, 5, 7, 9, \ldots\}$ **b.** $\{0, 2, 4, 6, 8, \ldots\}$
c. $(0, 3, 6, 9, 12, \ldots\}$ **d.** $\{1\}$
e. $\{0\}$

38. True or false: Every infinite subset of whole numbers has the closure property for multiplication. Support your answer.

39. Why can't 0 be used as a divisor? Use the definition of division to explain why the following statements are true.

a. $0 \div a = 0$. **b.** $a \div 0$ is undefined.
c. $0 \div 0$ is undefined.

40. Return to the picture you drew in Mini-Investigation 2.7 to show the similarities and differences between multiplication and division. What changes would you make to this picture now? Why?

D. Communicating and Connecting Ideas

41. The traditional sequence for teaching operations in the early grades is addition first, followed by subtraction, multiplication, and division. Some early arithmetic texts addressed the four operations in the order of addition, multiplication, subtraction, and division. With a team of classmates, present a debate, in written or oral form, presenting reasons for and against each approach.

42. Historical Pathways. Various definitions have been proposed throughout history for whole number multiplication and division. An early (1677) definition of multiplication stated, "Multiplication is performed by two numbers of like kind, for the production of a third, which shall have such reason [ratio] to the one, as the other hath to unite [unity, or the number 1]." [Smith (1953), p. 103] This definition could be restated as $a \times b = c$ means that $c \div a = b \div 1$. Compare this definition to the definitions of multiplication used in this section.

43. Making Connections. For many pairs of whole numbers, a and b, no whole number represents $a \div b$. What types of numbers are needed to bring the closure property to division? What physical situations could these numbers model? How would these numbers be related to the set of whole numbers? How could these numbers be helpful in using division to compare two whole numbers?

Section 2.4 NUMERATION

- Numeration Systems
- The Hindu-Arabic Place-Value Numeration System
- Other Early Numeration Systems
- Comparing Numeration Systems

In this section, we examine symbols used to represent whole numbers, with an emphasis on clarity and efficiency of communication. We consider some early methods of representing whole numbers and compare them with the system for representing whole numbers that is used today. Mini-Investigation 2.10 may help broaden your ability to communicate number ideas.

Write a paragraph evaluating the strengths and weaknesses of each of your methods.

Mini-Investigation 2.10 **Communicating**

Without using the number symbols 0–9 or the number words zero, one, two, . . . nine, devise three different ways to communicate to someone else the number of people in your class.

Numeration Systems

In Section 2.1, we used sets to develop an understanding of whole numbers and used the symbols 0, 1, 2, 3, and so on, to represent these numbers. To understand how these symbols are used to communicate whole numbers, recall that a symbol

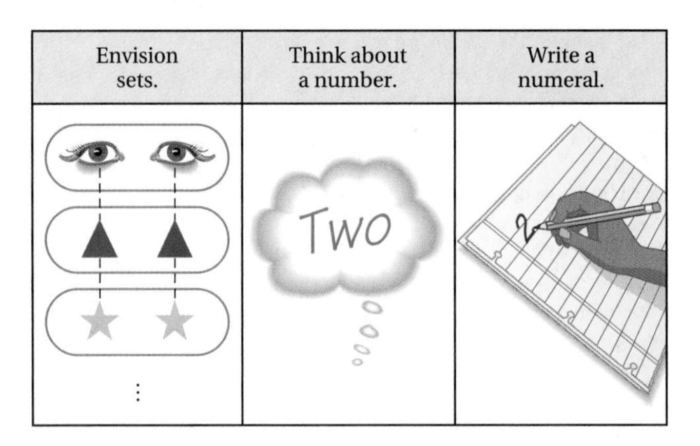

FIGURE 2.28 | Making connections between objects, whole-number ideas, and numerals.

is different from the number it represents. Just as the word *cat* is not itself an animal, the symbol 2 is not itself a number.

A written symbol, such as 2, that represents a number is called a **numeral.** Sometimes a numeral is referred to as a *name for a number.* For example, some familiar numerals (or names) for the number two are 2 and II. Figure 2.28 illustrates the connection between sets of two objects, the whole-number idea two, and the numeral 2.

Using numerals to communicate number ideas requires a method for representing all the whole numbers in some systematic way. That is, some type of *numeration system* is needed.

Definition of Numeration System

A **numeration system** is an accepted collection of properties and symbols that enables people to systematically write numerals to represent numbers.

The Hindu–Arabic Place-Value Numeration System

The **Hindu–Arabic numeration system,** developed by ancient Indian and Arabic cultures, is still in widespread use today and probably is the most familiar example of a numeration system. The ideas of grouping by tens and **place value** provide the cornerstones of the Hindu–Arabic numeration system. A look at grouping schemes other than grouping by tens can help you better understand the Hindu–Arabic system.

For example, if the grouping is based on threes, a set of three beans makes a group of the first size (3), a set of three of these groups makes a group of the next larger size (3×3), three of these groups make a group of the next larger size $(3 \times 3 \times 3)$, and so on. In Figure 2.29, a set of 21 cubes has been grouped by threes. The same set of cubes, grouped by fours, is shown in Figure 2.30.

The sizes and numbers of groups and leftovers change with each different grouping, but the number of cubes and the grouping procedure remain the same. The pattern used for writing and interpreting place-value numerals is based on this grouping procedure. The number of groups of each size is recorded in the numeral, with the rightmost digit representing the leftovers and the digits to the left representing increasingly larger groups. For example, in Figure 2.29, the grouping by

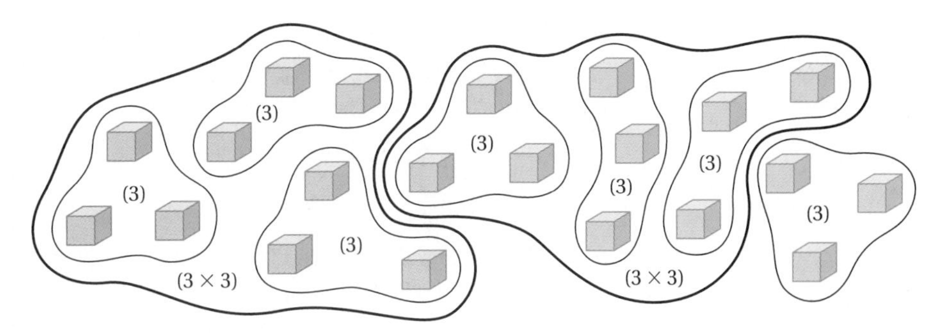

FIGURE 2.29 | Grouping by threes.

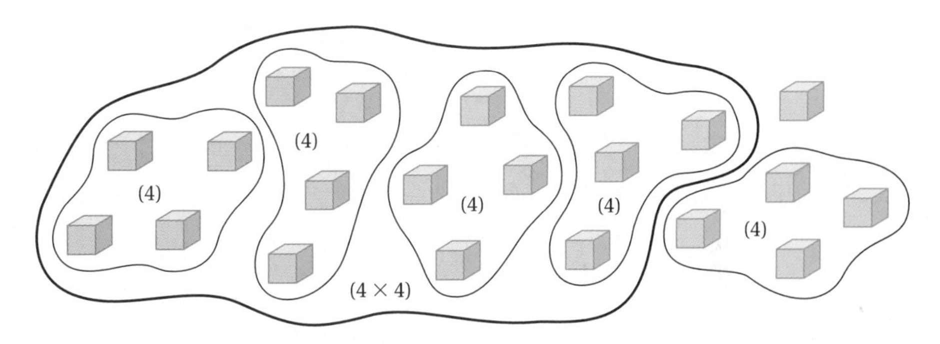

FIGURE 2.30 | Grouping by fours.

threes shown would be recorded as 210—that is, two groups of 3 × 3, one group of 3, and 0 left over. Similarly, in Figure 2.30, the grouping by fours shown would be recorded as 111—that is, one group of 4 × 4, one group of 4, and 1 left over.

To know what quantity is being represented by a place-value numeral, you must know the sizes of the groups on which the numeral is based. The group size used determines the **base** of the numeration system. For example, a base-two system results from making groups of 2, and a base-five system results from making groups of 5. A base-eight system uses groups of 8. Using groups of ten is the foundation of the Hindu–Arabic numeration system, and for this reason the system is often called the *base-ten place-value numeration system.* Since grouping by tens is the basis of the numeration system ordinarily used, only numerals in bases other than ten are written with a word indicating the base. For example, 120 indicates a base-ten numeral; 120_{five} indicates the base-five numeral. Example 2.15 helps clarify the role of different bases in a numeration system.

Example 2.15 Expressing Numerals with Different Bases

Explain why the quantity of tiles shown can be expressed as (a) 27 in base ten and (b) 102 in base five.

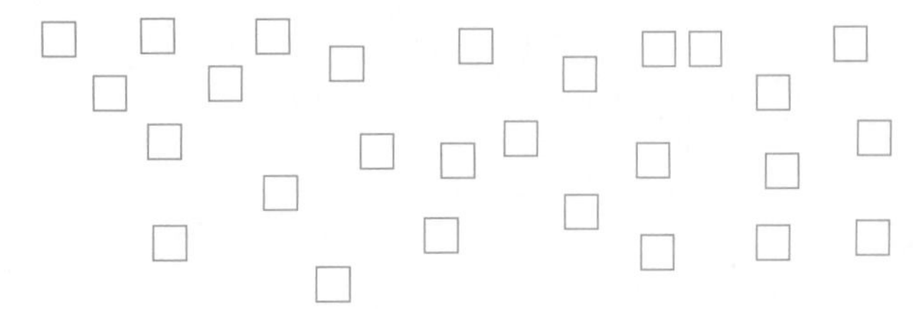

Solution

a. Using a base-ten system, we can group these tiles into two groups of ten with 7 tiles left over:

The numeral 27 represents the two groups of ten and the 7 ones.

b. Using a base-five system, we can group these tiles into groups of 5 and have enough of these groups of 5 to make one larger group of 5 fives, with 2 tiles left over:

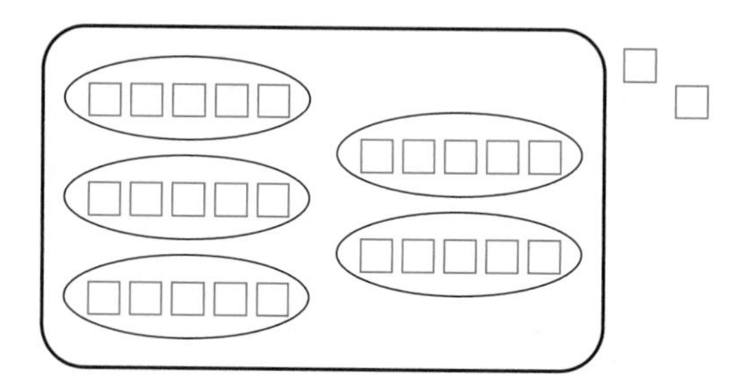

No group of 5 is left over, so we need to use a 0 in that position in the numeral. We record this quantity as 102_{five} so that everyone knows the group sizes used.

Your Turn

Practice: Draw pictures and explain why the number of tiles shown can be represented by both 38_{ten} and 123_{five}.

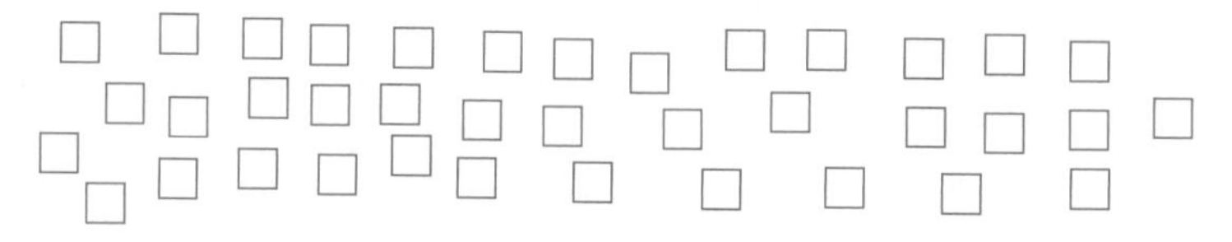

Reflect: What does the digit 1 in the numeral 123_{five} represent? ■

Not only do numeration systems with different bases have applications in computer systems, they also play an important role in helping you fully understand the characteristics of the base-ten numeration system. When you try to write a numeral in a different base, you're forced to think about the importance and use of both grouping and place value.

Models of Base-Ten Place Value. The fundamental idea of place value in the base-ten numeration system can be clarified with physical models. Different types of models enhance understanding at different levels of abstraction. **Proportional models** for place value actually exhibit the proportional differences in the values of the digits in a numeral.

Base-ten blocks are an example of a proportional model for place value. A set of base-ten blocks contains unit cubes (usually 1 cubic centimeter in size), a tens stick (10 unit cubes put together), a hundreds square (10 tens sticks put together),

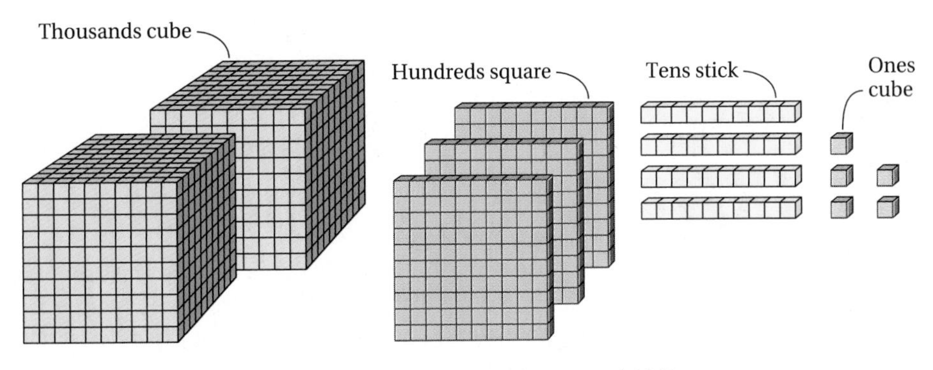

FIGURE 2.31 | Base-ten block model representing the numeral 2345.

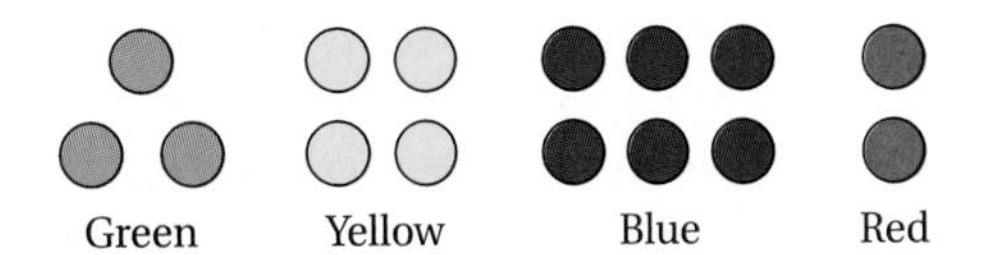

FIGURE 2.32 | Colored-chip model representing the number 3462.

a thousands cube (10 hundreds squares put together), and so on. Base-ten blocks are used in Figure 2.31 to model a 4-digit numeral.

In **nonproportional models** of place value the actual quantity expressed by the numeral isn't visible—usually just being represented by some object or set of objects. In a nonproportional colored-chip model, a red chip might represent one, a blue chip could represent ten ones or ten, a yellow chip might represent ten tens or one hundred, and a green chip might represent ten hundreds or one thousand. Based on these categories, the arrangement of chips shown in Figure 2.32 represents the number 3462.

Using Expanded Notation. Table 2.1 summarizes some of the important ideas of place value and suggests an expanded interpretation of a numeral. Note that in a base-five system, the value of each place in a numeral is 5 times that of the place to its right. In a base-eight system, the value of each place is 8 times that of the place to its right. In the base-ten system, the value of each place is ten times that of the place to its right. This pattern is generalized for a base-b system.

TABLE 2.1 | Place Values for Different Bases in the Numeration System

Base	**Place Values for a 4-Digit Numeral**			
Two	two × two × two	two × two	two	one
Five	five × five × five	five × five	five	one
Eight	eight × eight × eight	eight × eight	eight	one
Ten	ten × ten × ten	ten × ten	ten	one
⋮				
b	$b \times b \times b$	$b \times b$	b	one

Using the ideas in Table 2.1, we can show base-ten interpretations of numerals expressed in different bases:

$$\begin{aligned} 1324_{\text{five}} &= 1(125) = 3(25) + 2(5) + 4 = 229; \\ 1324_{\text{eight}} &= 1(512) + 3(64) + 2(8) + 4 = 724; \\ 1324_{\text{ten}} &= 1(1000) + 3(100) + 2(10) + 4 = 1324; \\ 1324_b &= 1(b \times b \times b) + 3(b \times b) + 2(b) + 4. \end{aligned}$$

When a numeral is written to show the sum of its digits times the value of each place, we say the numeral is written in **expanded notation.** For example, we can write the base-ten numeral 5,283 in expanded notation as

$$5283 = 5(10 \times 10 \times 10) + 2(10 \times 10) + 8(10) + 3,$$

or

$$5283 = 5(1000) + 2(100) + 8(10) + 3.$$

If we use exponents to show how many times 10 is used as a factor, the use of expanded notation becomes even more efficient:

$$5283 = (5 \times 10^3) + (2 \times 10^2) + (8 \times 10^1) + (3 \times 1).$$

Example 2.16 further illustrates the use of expanded notation.

Example 2.16 Expressing Numbers Using Expanded Notation

Express 2,150,389 in expanded notation.

Solution

Meei-Ling's thinking: According to the patterns in place value,

$$2{,}150{,}389 = (9 \times 1) + (8 \times 10) + (3 \times 100) + (5 \times 10{,}000) + (1 \times 100{,}000) + (2 \times 1{,}000{,}000).$$

Michael's thinking: Exponents indicate how far a digit is from the ones place, so in expanded notation this number is

$$(2 \times 10^6) + (1 \times 10^5) + (5 \times 10^4) + (0 \times 10^3) + (3 \times 10^2) + (8 \times 10^1) + (9 \times 1).$$

Your Turn

Practice: Use exponents to express 48,042 in expanded notation.

Reflect: How would you change the expanded notation for 48,042 to express a number that is 100 less? 10 less? 110 less? ■

Note that when you use exponents to express numbers in expanded notation, the thousands position is represented by the exponent 3, the hundreds by the exponent 2, and the tens by the exponent 1. Continuing this pattern leads to the conclusion that the ones position should be represented by the exponent 0. Using this notation means that 4378, for example, could be represented as

$$(4 \times 10^3) + (3 \times 10^2) + (7 \times 10^1) + (8 \times 10^0).$$

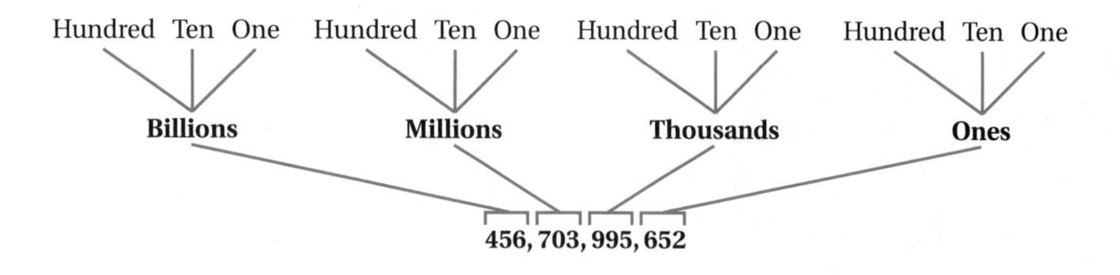

Read "456 billion, 703 million, 995 thousand, 652."

FIGURE 2.33 | Representing large numbers with numerals and periods.

Estimation Application

Length of Life

How old is someone who has lived 1 million hours? Estimate your answer by choosing one of (a)–(e).

a. Less than 1 year old
b. Between 1 and 10 years old
c. Between 10 and 50 years old
d. Between 50 and 100 years old
e. More than 100 years old

Determine a way to check your estimate.

Using Periods to Represent Large Numbers. The pattern that relates the value of a place in a numeral to the value of the place to its left allows the writing of numerals to express numbers far greater than those that can be shown with simple objects or drawings. To do so, we use successive groups of three digits, called **periods.** As illustrated in Figure 2.33, the three digits in each period are the number of hundreds, tens, and ones that indicate how many of the designated value of a period—ones, thousands, millions, billions, and so on—are involved. Some additional periods, to the left of billions, are trillions, quadrillions, quintillions, and so on.

Using Place Value to Compare Numbers. In Section 2.1, we compared two whole numbers by looking at sets for the numbers and matching elements of the sets to see which set had elements left over. In Section 2.2, we used addition to compare two whole numbers. By combining these set-matching and addition procedures, we can develop a procedure for comparing two numbers using the place-value relationships of the digits in the numerals. Example 2.17 illustrates the use of place-value ideas in comparing two numbers.

Example 2.17 Using Place Value to Compare Numbers

Which is greater, 4800 or 4756? Explain your reasoning.

Solution

Angela's thinking: If I made the numbers with base-ten blocks, each one would take 4 thousands cubes. But 4800 would need 8 hundreds squares, and 4756 would only need 7. The larger number of tens and ones pieces in 4756 wouldn't be enough to make up for the extra hundreds piece. So 4800 is greater than 4756.

Rodney's thinking: I lined the digits of the two numbers up according to their place-value positions. When I looked at the leftmost digits, I noticed that each number has the same number of thousands (4). Then I looked at the hundreds digits; 4800 has 8 hundreds but 4756 has only 7 hundreds. So 4800 is greater than 4756.

Your Turn

Practice: Which is smaller, 3402 or 2799?

Reflect: Can you choose digits for the blanks so that the number 2 _ _ _ will be larger than the number 3000? Explain. ■

Connection to the PreK–8 Classroom

The use of place value to compare whole numbers can be an important part of developing number sense as students analyze data or determine a winning strategy for a game, as in the following example.

A game for 2–6 players

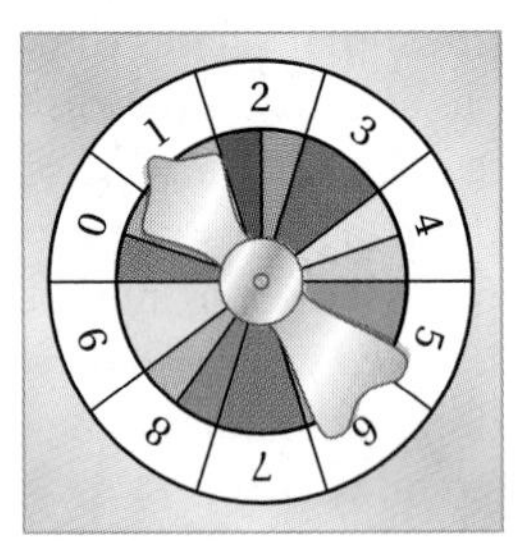

How

- Use a die or spinner with the digits 0 through 9. (Or draw playing cards Ace through 10, counting Ace as 1 and 10 as 0, replacing cards as you go.
- Players each draw three boxes on a piece of paper.

  □ □ □

- The leader spins or rolls a die and announces the number.
- Each person places that number in one of the three boxes.
- No changes are allowed after the digit is recorded.
- The leader spins or rolls two more times, and each time the players place the digit in any empty box.
- Each player reads his or her three-digit number. The player with the largest number wins.
- Repeat the game as many times as you wish, taking turns being leader.

Source: Stenmark, J. K., Thompson, V., and Cossey, R. *Family Math* (ISBN 0-912511-06-0), Berkeley: EQUALS Publications, Lawrence Hall of Science, Univ. of California, 1986, p.126.

Other Early Numeration Systems

Token Systems and Tally Systems. The actual history of how human beings moved from concrete objects to symbols to represent numbers has been lost for many cultures, but archaeological evidence indicates that in the area near present-day Iran, from 8000–3000 B.C., symbols in the form of clay tokens shaped like cones, spheres, disks, cylinders, and pyramids were used to represent measures of grain, oil, metal, and other goods in the economy. Thus one of the earliest known systems of numeration may be described as a **token system,** in which symbolic objects, or tokens, were used to represent quantities of actual objects.

In this system, each token represented one measure of a particular commodity, and sets of tokens were combined to represent larger quantities. Toward the end of the period, tokens also began to represent groups of measures (for example, a sphere meant 10 measures of grain), and sets of tokens began to appear enclosed in clay envelopes for ease of transporting and for storage in archives as long-term records. Because the quantity of tokens inside a clay envelope couldn't be determined without breaking it open, scribes began making imprints on the outside of the clay envelopes. A circular dent in the clay represented a spherical token, and a triangular dent, a cone-shaped token (Schmandt-Besserat, 1992). Figure 2.34 shows a sketch of these early tokens and an imprinted clay envelope.

The realization that the imprints made the tokens themselves unnecessary was an early step toward an actual written numeration system. A **tally system** is based on establishing a one-to-one correspondence between a single mark and a single object so that the marks represent the number of objects. Later, grouping was used as in representing seven hides with 𝍸 ||, to simplify numeration systems. Tally systems are still used sometimes to record one object at a time and later determine the total.

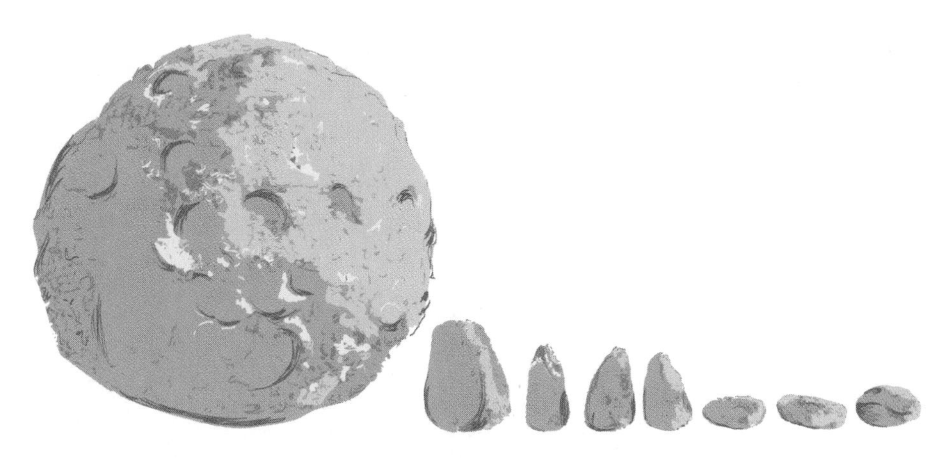

FIGURE 2.34 | Sketches of early tokens and clay envelopes imprinted to show tokens inside.

Egyptian Numeration System. The ancient Egyptians developed a type of tally system slightly more sophisticated than those based on strict one-to-one correspondence in about 3400 B.C. Found written on papyrus, wood, pieces of pottery, and stone, the **Egyptian numeration system** used picture symbols, called *hieroglyphics*, for 1, 10, 100, 1000, and so on, as depicted in Figure 2.35.

Egyptian symbol		Corresponding whole number
|	Reed	One
∩	Heel bone	Ten
𓍢	Coiled rope	One hundred
𓆼	Bent reed	One thousand
𓂭	Pointed finger	Ten thousand
𓆛	Burbot fish	One hundred thousand
𓁨	Astonished man	One million

FIGURE 2.35 | Egyptian hieroglyphs for whole numbers.

These symbols were combined and repeated as necessary to record whole numbers, such as ∩∩∩||||||| for 37. Example 2.18 compares the Egyptian numeration system with the Hindu–Arabic system.

Example 2.18 Writing Egyptian and Hindu–Arabic Numerals

a. Use Egyptian symbols to represent 5642.

b. Use Hindu–Arabic symbols to represent

Solution

a.

b. 240,318

Your Turn

Practice:

a. Use Egyptian symbols to represent 1,354,201.

b. Use Hindu–Arabic symbols to represent

Reflect: What advantages does the Hindu–Arabic numeration system have over the Egyptian system? ■

Babylonian Numeration System. In ancient Babylon, numbers were recorded on clay tablets, with a stylus or writing stick used to make imprints in the clay. The **Babylonian numeration system,** based on multiples of 60, utilized only two symbols. These symbols could be formed easily in the clay with a stylus but had multiple meanings:

1
One group of 60
One group of 60 sixties (60^2)

10
Ten groups of 60 ($10 \cdot 60$)
Ten groups of 60 sixties ($10 \cdot 60^2$)

The meaning of each symbol was determined by the context or the position in which it appeared. For example, the meaning of the single symbol that could mean 1, 60, or 60 sixties (3600) would be 60 rather than 1 if it appeared in a record of the number of eggs consumed in a 2-month period by a person who liked eggs.

Combinations of the two symbols, with spacing or size of symbol delineating different meanings of the same symbol, were used to represent whole numbers, as in the following.

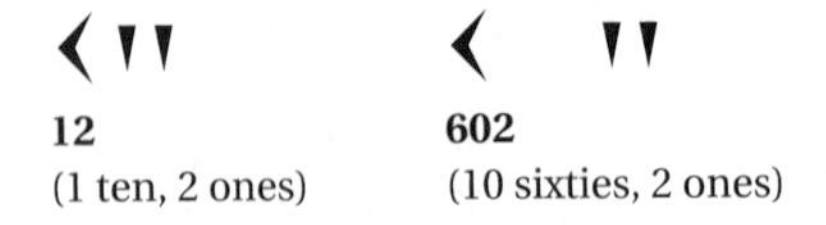

Much later, the Babylonians learned to help show different meanings by using a blank space (zero) represented by the symbol

as in the last line of Figure 2.36.

Babylonian numeral	Corresponding Hindu–Arabic numeral
	64
	$1(60^2) + 4(60) = 3840$
	$10(60^2) + 10(60) + 1 = 36601$
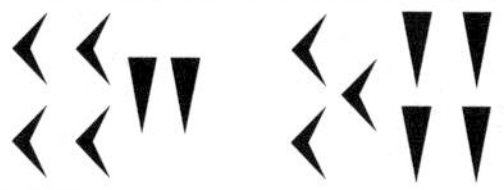	$42(60) + 34 = 2554$
	$1(60^2) + 0(60) + 1 = 3601$

FIGURE 2.36 | Examples of Babylonian numerals.

Figure 2.36 illustrates some Babylonian numerals and the equivalent Hindu–Arabic numerals, and Example 2.19 provides an opportunity to interpret a Babylonian numeral.

Example 2.19 Interpreting Babylonian Numerals

Suppose that a Babylonian had used a stylus to make the following numeral to show how many units of land were in a large valley.

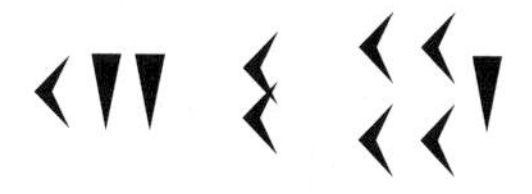

Write the Hindu–Arabic numeral that gives the same information.

Solution

The numeral represents $12(60^2) + 0(60) + 41$, or 43,241.

Your Turn

Practice: Write the Hindu–Arabic numeral that gives the same information as the Babylonian numeral

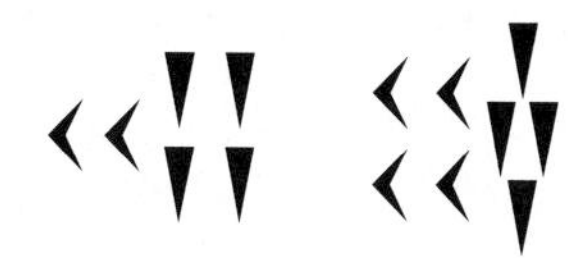

Reflect: What advantages does the Hindu–Arabic numeration system have over the Babylonian system? ■

Roman Numeration System. The numeration system developed between 500 B.C. and A.D. 100 by the Romans is still used today in certain contexts—for example, to indicate year dates on buildings and in credits on films, to number pages in the preface of a book, and to identify Super Bowl games. The **Roman numeration system** uses seven basic symbols to represent whole-number groupings:

Symbol	Whole Number
I	One
V	Five
X	Ten
L	Fifty
C	One hundred
D	Five hundred
M	One thousand

These seven symbols are combined and repeated as necessary to form Roman numerals. For efficiency, however, one of the conventions of the Roman numeration system is that no more than three of any one symbol are used in a numeral. Thus the quantity *four* is represented as IV, or one less than five, and *nine* is IX, or 1 less than 10. Very large numbers are represented by using a bar over a numeral to represent multiples of 1000, as in $\overline{\text{C}}$ for 100,000. Example 2.20 illustrates the use of symbols in Roman numerals.

Example 2.20 Interpreting Roman Numerals

Identify the numbers represented by the following Roman numerals.

a. MCCXLVII
b. CCCXXIV
c. CMXCIX
d. $\overline{\text{LVI}}$

Solution

Think about the symbols in groups.

a. M = 1000, CC = 200, XL = 10 less than 50, or 40, and VII = 7, so MCCXLVII = 1,247
b. CCC = 300, XX = 20, and IV = 4, so CCCXXIV = 324.
c. CM = 900, XC = 90, and IX = 9, so CMXCIX = 999.
d. LVI = 56, so $\overline{\text{LVI}}$ = 56,000.

Your Turn

Practice: Write the Roman numerals for the numbers between XXXV and LV.

Reflect: Does any pattern appear in the numbers (such as IV and IX) that are symbolized by subtractive representation? ■

Comparing Numeration Systems

The Hindu–Arabic system used today can be more clearly understood and appreciated by comparing and contrasting it with the other systems. In the following subsections, we compare systems based on grouping schemes, symbols, place value, and the use of zero. Using these criteria you will have an opportunity to judge the effectiveness of the different systems.

Grouping Scheme and Symbols. One way of comparing numeration systems is to look at the *grouping scheme* and *symbols* used. The Egyptian system grouped by tens, with a new symbol for ones, for tens, for ten tens, for ten ten tens, and so on. Theoretically, the Egyptian numeration system required infinitely many symbols. In contrast, the Babylonian system at first had only two symbols and grouped by sixties. The two symbols, representing the whole numbers one and ten, were used repeatedly and in different positions to write larger numbers.

The Roman system involves a modified scheme of grouping by fives and a combination of basic symbols, with bars as needed to represent larger numbers. More and more new symbols would be needed to represent larger and larger numbers. The Hindu–Arabic system involves grouping by tens and uses only 10 symbols: 0, 1, 2, 3, 4, 5, 6, 7, 8, 9.

Use of Place Value. Another way to compare numeration systems is in terms of use of **place value,** where the *position* of a symbol in a numeral determines the value it represents. The use of place value allows the writing of large numbers by imaginatively arranging old symbols without having to invent new symbols. The Babylonians probably were the first people to reduce the number of symbols needed to represent quantities by using place value, but their system had a somewhat confusing type of place value, in which the meaning of the symbol was often open to interpretation.

In contrast to the Babylonian system, the Egyptian system didn't utilize place value and required more and more symbols to represent larger and larger numbers. Like the Babylonian system, the Hindu–Arabic system is a place-value system, with each symbol having more than one possible value. For example, the symbol 2 can represent the whole numbers two or two hundred or two thousand, depending on its position in a numeral. For example,

in 52, the 2 represents the whole number two;
in 25, the 2 represents the whole number twenty; and
in 256, the 2 represents the whole number two hundred.

The Hindu–Arabic system makes the most efficient use of place value, for—unlike the Egyptian, Babylonian, and Roman systems—the Hindu–Arabic symbols are never used in the form of a tally. If a symbol is repeated in a numeral, each occurrence represents a distinct and different value, as in 2222. Note that in the Roman system the location, or place, of a numeral makes a difference, such as the location of the I in IV and VI. However, a single I without a bar over it, for example, always has the value 1, regardless of its place in a numeral.

Use of Zero. Another criterion for comparing numeration systems is the use of **zero**. The later Babylonian system and the Hindu–Arabic system are the only ones that have a symbol for zero. In these systems, the position of each digit in a numeral may be clearly indicated by the use of the symbol 0 when

TABLE 2.2 | Summary of Numeration System Characteristics

System	Grouping	Symbols	Place Value	Use of Zero
Egyptian	By tens	Infinitely many possibly needed	No	No
Babylonian	By sixties	Two	Yes	Not at first
Roman	Partially by fives	Infinitely many possibly needed	Position indicates when to add or subtract	No
Hindu–Arabic	By tens	Ten	Yes	Yes

necessary. For example, in each of the following numerals, the use of zeros indicates clearly that the digit 2 is in the third position from the right and represents 200:

10,204; 5280; and 200.

Table 2.2 summarizes the grouping schemes, symbols, and use of place value and zero of the numeration systems discussed in this section. Mini-Investigation 2.11 provides an opportunity to analyze the numeration systems further.

Write a paragraph supporting your conclusion about the first- and second-place winners.

Mini-Investigation 2.11 **Making a Connection**

If there were a "numeration system contest," which system in Table 2.2 do you think would win and which would come in second?

Problems and Exercises *for Section 2.4*

A. Reinforcing Concepts and Practicing Skills

1. Indicate whether each of the following statements is about numbers or is about numerals.
   a. If you cut vertically, half of 8 is 3.
   b. If you double 2 you get 4.
   c. **5** is larger than **8**.
2. Write the number of objects as a
   a. base-ten numeral
   b. base-two numeral.
   c. base-five numeral.

3. a. Express the number 408 using base-ten blocks.
   b. Express the number 3699 using expanded notation without exponents.
   c. Express the number 5,280,492 using expanded notation with exponents.
4. What quantity does the 3 represent in the following numerals?
   a. 4352 b. 324_{five}
   c. 311_b
5. Find the base-ten representation for each of the following numerals in other bases.
   a. 344_{five} b. 202_{three}
   c. 1010011_{two}
6. Find the representation of the number 179 in the following bases.
   a. Base six
   b. Base twelve
   c. Base two
7. a. $42_{five} = \_\_\__{six}$
   b. $32_{six} = \_\_\__{three}$
   c. $1010011_{two} = \_\_\__{four} = \_\_\__{eight}$
8. Use place value to order the following set of numbers from least to greatest.

   2245 962 980 1000 985 2222

9. Describe the basic characteristics of a tally system.

10. Write the Hindu–Arabic numerals for the numbers represented by the following numerals from the Egyptian systems.

a.

b.

c.

11. Write the Hindu–Arabic numerals for the numbers represented by the following numerals from the Babylonian system.

a.

b.

c.

12. Write the Hindu–Arabic numerals for the numbers represented by the following numerals from the Roman system.

a. CMIII
b. LIX
c. XCVII

13. Translate the Hindu–Arabic numerals 100, 61, 608, and 94 into the equivalent

a. Egyptian numerals.
b. Babylonian numerals.
c. Roman numerals.

14. Describe how you would determine whether a numeration system contains place value.

15. Describe the similarities and differences in the following numeration systems.

a. Egyptian and Hindu–Arabic
b. Babylonian and Hindu–Arabic
c. Roman and Hindu–Arabic

16. On a 10×10 hundreds chart, what patterns are direct results of the base-ten numeration system?

17. Use a 10×10 hundreds chart and describe a method for finding

a. 10 more than a number.
b. 10 less than a number.
c. 1 more than a number.
d. 1 less than a number.
e. 11 more than a number.
f. 11 less than a number.
g. 9 more than a number.
h. 9 less than a number.

18. Twenty-five in base-five notation is 100_{five}. (Verify with colored chips.)

a. What is the base-five numeral for 24?
b. What is the numeral for 1 less than 1000_{five}?
c. How and where does this pattern occur in a base-ten numeration system?

19. Explain the numeral *rstu* in base *b* where *r*, *s*, *t*, and *u* are single digits.

20. Use some or all of the digits 2, 5, 8, and 0 to form whole numbers in base-ten to complete each of the following sentences. Justify your choices.

a. When I drove home yesterday, traffic on the highway was moving about ______ miles per hour.
b. I spent about $______ on textbooks this semester, or $______ more than last semester.
c. I have ______ more days to complete my essay.
d. My roommate eats ______ times as much as I do.
e. The score of yesterday's basketball game was ______ to ______.
f. I have ______ checks left in my checkbook to last until next month.
g. We're expecting the fundraiser to clear about $______.
h. The campus post office sells approximately ______ stamps per week.

B. Deepening Understanding

21. Describe in words or with symbols the method you used to translate the

a. Egyptian numerals in Exercise 10 into Hindu–Arabic numerals.
b. Babylonian numerals in Exercise 11 into Hindu–Arabic numerals.
c. Roman numerals in Exercise 12 into Hindu–Arabic numerals.

22. Complete the sequence by writing the numbers in the numeration system shown, and describe the system's characteristic(s) you used in order to do so.

a. XXXVIII, ______, ______, ______
b. 98, ______, ______, ______
c. ______, ______, ______,
d. ______, ______, ______,
e. ______, ______, ______, CI
f. ______, ______, ______, 72

C. Reasoning and Problem Solving

23. Recall from the explanation of base-ten blocks that 1 usually is represented by a cube 1 cm × 1 cm × 1 cm. Ten is represented by a tens stick with dimensions 10 cm × 1 cm × 1 cm. One hundred is a hundreds square with dimensions 10 cm × 10 cm × 1 cm. One thousand is a larger cube, 10 cm × 10 cm × 10 cm. If you were to continue using base-ten blocks for the next place-value positions, what shape and dimensions would the millions block have? Explain your conclusion.

24. In a base-ten numeration system, the numerals for the first 10 counting numbers are 1, 2, 3, 4, 5, 6, 7, 8, 9, 10. In a base-five numeration system, the numerals for the first 10 counting numbers are 1_{five}, 2_{five}, 3_{five}, 4_{five}, 10_{five}, 11_{five}, 12_{five}, 13_{five}, 14_{five}, and 20_{five}. Record the base-five symbols for the first 100 counting numbers on a 10×10 hundreds chart. Compare the base-five numerals to the base-ten numerals in the same positions on another 10×10 chart. What generalizations can you form?

25. Record the base-five symbols for the numbers one through twenty-five on a 5×5 chart. What patterns do you see? Compare the patterns of the 5×5 chart to the base-ten numerals on the 10×10 chart.

26. What patterns would you expect to see on an 8×8 chart for base-eight numerals? Make a chart to test your conjectures.

27. Imagine the thoughts and actions of the scribe who discovered the possibility of communicating quantities with marks in a clay envelope rather than with actual tokens. Write a short description of this hypothetical scene.

28. Invent a new numeration system. Record the numerals for the whole numbers one through twenty. Give your list of new numerals to a classmate and ask that person to try to determine what the next 10 numerals would be. Record the comments the person makes while trying to figure out your numeration system. Critique your system based on the person's comments.

D. Communicating and Connecting Ideas

29. The Hamburger Stack. A famous fast-food chain placed signs at their restaurants, claiming: "Over 400 billion hamburgers have been sold." Think about how large a billion might be and estimate the height of this stack of hamburgers by choosing one of (a)–(f).

a. Higher than a Ferris wheel
b. Higher than the Empire State Building
c. Higher than a jet plane flying at an altitude of 7 miles
d. Higher than a 100-mile-high satellite in orbit
e. Higher than the distance to the moon
f. Higher than the distance to the sun

Determine a way to check your estimate.

30. Some numeration systems undoubtedly were developed around the idea of groups of ten because of human beings' 10 fingers. If only one hand's worth of fingers had been used, the Hindu–Arabic numeration system might have been based on groups of 5. With a partner or small group, brainstorm advantages and disadvantages to the base-ten and base-five numeration systems. Be prepared to defend either system as being "the best." Participate in a class debate.

31. Historical Pathways. For several hundred years throughout the region now known as Europe, the Roman numeration system was used to record whole numbers. Leonardo of Pisano, author of *Liber abbaci*, or *Book of Calculation*, published in 1202, is credited with the introduction of the Hindu–Arabic numeration system into European culture. What conjectures can you make about the acceptance of the Hindu–Arabic system when it was introduced in early Europe? [You might want to do some research in books by V. J. Katz (1993) or C. D. Boyer (1991) to test your conjectures.] With a small group of classmates, write a short script to show how you might react if a new numeration system were introduced in the United States.

32. Making Connections. A numeral for the quantity zero came fairly late in the development of mathematical symbolism. Make a list of situations in which using a symbol for zero would be desirable. What difficulties would you encounter if there were no symbol for zero?

33. Making Connections. The numeration system most commonly encountered in everyday experiences is based on groups of 10. However, important technological applications are based on numeration systems involving groups of 2, 8, and 16. For example, the two digits used in a base-two numeration system can be represented by an electrical circuit being off or on. Electrical circuits in a computer can then be combined to represent any whole number quantity in base-two. As 8 and 16 are powers of two, these bases might also be useful in computers. How might the base-ten number 421 be represented in (a) base-eight? (b) base-sixteen?

CHAPTER SUMMARY

Key Ideas: Questions and Answers

Section 2.1

- **What set ideas help with understanding whole numbers?** *(pp. 59–63)* A set is a collection of objects or ideas. It may contain from zero elements (the null or empty set) to infinitely many elements. Two sets that have a one-to-one correspondence are equivalent sets, and two equivalent sets that have exactly the same elements are equal sets.
- **What are whole numbers?** *(pp. 63–65)* A whole number is the number idea embodied in a collection of equivalent sets. Counting, or the establishment of a one-to-one correspondence between a set of number names and the objects in a set, is used to identify the whole number represented by a set.
- **How can set ideas help in comparing and ordering whole numbers?** *(pp. 65–69)* For whole numbers a and b and sets A and B, where $n(A) = a$ and $n(B) = b$, $a < b$ if and only if A is equivalent to a proper subset of B.

- **What are some important subsets of whole numbers?** *(pp. 69–71)* The set of whole numbers contains special subsets, such as the set of natural numbers, the set of even numbers, and the set of odd numbers. An infinite set is distinguished from a finite set by being shown to be equivalent to a proper subset of itself, as the set of whole numbers is equivalent to the set of even numbers.

Section 2.2

- **How can models and set language be used to define addition?** *(pp. 73–78)* Discrete models of sets and continuous models, including number lines, can be used to represent addition of whole numbers as a joining process. For two disjoint sets A and B, if $n(A) = a$ and $n(B) = b$, then $a + b = n(A \cup B)$. In the equation, $a + b = c$, a and b are addends and c is the sum.
- **What are some basic properties of addition?** *(pp. 78–80)* Addition of whole numbers has the closure property, the identity property, the commutative property, and the associative property. The identity element for addition is 0. For whole numbers a and b, $a > b$ if and only if there is a whole number $k > 0$ such that $b + k = a$.
- **How can models be used to explain subtraction of whole numbers?** *(pp. 80–82)* Discrete models of sets and continuous models including number lines, lengths, and areas can be used to represent subtraction of whole numbers as a taking away, separating, or comparing process.
- **How can addition be used to define subtraction?** *(pp. 82–83)* For whole numbers a and b, $a - b = c$ if and only if c is a unique whole number such that $c + b = a$. Thus subtraction may also be thought of as finding the missing addend. In the equation $a - b = c$, a is the minuend, b is the subtrahend, and c is the difference.
- **Does subtraction have the same properties as addition?** *(pp. 83–85)* No. Subtraction of whole numbers doesn't have the closure, identity, commutative, or associative properties.

Section 2.3

- **How can models and set language be used to define multiplication of whole numbers?** *(pp. 88–92)* Discrete models of sets and continuous models, including number lines, can be used to represent multiplication of whole numbers as a process of joining equivalent sets or equal lengths. Multiplication may be represented as repeated addition, counting elements in an array, finding the area of a rectangle whose dimensions are the numbers being multiplied, or finding the number of ordered pairs created in a Cartesian product of two sets. If A and B are finite sets with $n(A) = a$ and $n(B) = b$, then $a \times b = n(A \times B)$. In the equation, $a \times b = c$, a and b are factors and c is the product.
- **What are some properties of multiplication?** *(pp. 92–94)* Multiplication of whole numbers has the closure property, the identity property, the commutative property, the associative property, and the zero property. The identity element for multiplication is 1. Multiplication and addition are both involved in the distributive property of multiplication over addition.
- **How can models be used to explain division of whole numbers?** *(pp. 95–97)* Discrete models of sets can be used to represent the processes of division. The two division situations consist of finding how many equivalent subsets in a set (the repeated subtraction or measurement interpretation) and finding how many elements in each equivalent subset (the sharing interpretation). Continuous models, including lengths and areas, can also be used to show these two situations.
- **How can multiplication be used to define division?** *(pp. 97–98)* Looking at division as the inverse of multiplication leads to the use of multiplication to define division. For whole numbers a and b, $a \div b = c$ if and only if c is a unique whole number such that $c \times b = a$. In the equation $a \div b = c$, a is the dividend, b is the divisor, and c is the quotient. Division may also be thought of as finding the missing factor or as repeated subtraction.
- **Does division have the same properties as multiplication?** *(p. 98)* Division of whole numbers doesn't have the closure, identity, commutative, associative, zero, or distributive properties. The inverse relationship between addition and subtraction parallels the inverse relationship between multiplication and division. Also, the repeated addition interpretation of whole-number multiplication parallels the repeated subtraction interpretation of whole-number division.
- **What is the division algorithm for whole numbers?** *(pp. 98–100)* For whole numbers a and b, where $b \neq 0$, a division process for $a \div b$ can be used to find unique whole numbers q (quotient) and r (remainder) such that $(q \times b) + r = a$, and $0 \leq r < b$.

Section 2.4

- **What is a numeration system?** *(pp. 103–104)* A numeration system is an accepted collection of properties and symbols that allows systematic writing of numerals to represent numbers.
- **How can grouping help explain the Hindu–Arabic place-value numeration system?** *(pp. 104–110)* The meaning of a place-value numeral such as 123 can only be determined by knowing the grouping system on which it is based. For any base $b > 3$, 123 represents 1 group of b^2 plus 2 groups of b plus 3 ones. The Hindu–Arabic numeration system is a place-value system that uses a base of ten.
- **What were some other early numeration systems?** *(pp. 110–114)* Some early numeration systems involved tokens and tallying. The early Egyptian numeration system

used picture symbols based on groups of ten and tallying. The early Babylonian numeration system used two symbols based on groups of ten and place value based on groups of 60. The Roman numeration system used a combination of tallying and placement of seven basic symbols.

- **How is the Hindu–Arabic system like and unlike other systems?** *(pp. 115–116)* The Hindu–Arabic system uses ten symbols, including a symbol for zero, and place value based on groups of ten.

Key Terms, Concepts, and Generalizations

Section 2.1

Set (p. 60)
Element (p. 60)
Empty, or null, set (p. 60)
Finite set (p. 60)
Infinite set (p. 60)
One-to-one correspondence (p. 60)
Equal sets (p. 62)
Equivalent sets (p. 62)
Whole number (p. 64)
Set of whole numbers (p. 64)
Counting (p. 64)
Comparing (p. 65)
Subset of a set (p. 66)
Venn diagram (p. 66)
Proper subset of a set (p. 67)
Less than (p. 68)
Greater than (p. 68)
Ordering (p. 68)
Set of natural numbers (p. 69)
Set of even numbers (p. 69)
Set of odd numbers (p. 69)

Section 2.2

Joining (p. 73)
Union of two sets (p. 74)
Intersection of two sets (p. 75)
Addition of whole numbers (p. 77)
Addend (p. 77)
Sum (p. 77)
Closure property of addition (p. 78)
Identity property of addition (p. 79)
Commutative property of addition (p. 79)
Associative property of addition (p. 79)
Taking away (p. 80)
Separating (p. 80)
Comparing (p. 80)
Subtraction as the inverse of addition (p. 82)
Inverse operations (p. 82)
Subtraction of whole numbers (p. 83)
Minuend (p. 83)
Subtrahend (p. 83)
Difference (p. 83)
Missing addend (p. 83)

Section 2.3

Rectangular array (p. 88)
Area model (p. 89)
Repeated addition (p. 89)
Ordered pair (p. 89)
Cartesian product (p. 89)
Multiplication of whole numbers (p. 90)
Factor (p. 90)
Product (p. 90)
Closure property of multiplication (p. 93)
Identity property of multiplication (p. 93)
Commutative property of multiplication (p. 93)
Associative property of multiplication (p. 93)
Zero property of multiplication (p. 93)
Distributive property of multiplication over addition (p. 93)
Repeated subtraction (p. 95)
Measurement interpretation of division (p. 95)
Sharing interpretation of division (p. 95)
Division as the inverse of multiplication (p. 97)
Division of whole numbers (p. 97)
Dividend (p. 97)
Divisor (p. 97)
Quotient (p. 97)
Missing factor (p. 98)
Division algorithm (p. 99)

Section 2.4

Numeral (p. 104)
Numeration system (p. 104)
Hindu–Arabic numeration system (p. 104)
Base (p. 105)
Proportional models (p. 106)
Nonproportional models (p. 107)
Expanded notation (p. 108)
Periods (p. 109)
Token system (p. 110)
Tally system (p. 110)
Egyptian numeration system (p. 111)
Babylonian numeration system (p. 112)
Roman numeration system (p. 114)
Place value (p. 115)
Zero (p. 115)

CHAPTER REVIEW

Concepts and Skills

1. Use sets to show the following ideas.
   a. The meaning of 5
   b. $9 < 11$
   c. The meaning of 0
   d. $0 <$ all other whole numbers
2. Translate Hindu–Arabic numerals 200, 55, 398, and 120 into the equivalent Egyptian, Babylonian, and Roman numerals. (*Hint:* Make a table and record your answers in it.)
3. Describe the important mathematical characteristics of each system.

a. The Egyptian numeration system
b. The Babylonian numeration system
c. The Roman numeration system
d. The Hindu–Arabic numeration system

4. Express the quantity 156 as the equivalent numeral in each base given.
a. Base five b. Base two

5. Represent 1,208 with
a. base-ten blocks.
b. colored chips.
c. expanded notation without exponents.
d. expanded notation with exponents.

6. For sets $A = \{8, 9, 10, 12)$, $B = \{10, 11\}$, and $C = \{3, 6, 9, 12\}$, find
a. $A \cap B$ b. $A \cap C$
c. $A \cap B \cap C$ d. $(A \cap C) \cup B$

7. For each of the following actions, write a word problem that can be solved with the equation $85 + 62 = n$.
a. Joining lengths b. Joining sets

8. For each of the following actions, write a word problem that can be solved with the equation $85 - 62 = n$.
a. Taking away one length from another
b. Taking away a subset from a set
c. Separating a length into two lengths
d. Separating a set into two subsets
e. Comparing two lengths
f. Comparing two sets

9. For each of the following actions, write a word problem that can be solved with the equation $12 \times 25 = n$.
a. Joining equal lengths
b. Joining equivalent sets
c. Finding the area of a rectangle
d. Finding the number of ordered pairs

10. For each of the following actions, write a word problem that can be solved with the equation $125 \div 25 = n$.
a. Separating a set into equivalent subsets of a given size
b. Separating a length into equal lengths of a given size
c. Separating a set into a given number of equivalent subsets
d. Separating a length into a given number of equal lengths

Reasoning and Problem Solving

11. **Flower Arrangements.** The final assignment for 15 horticulture students is to design a flower arrangement by using, at most, four different types of flowers. Can the professor reasonably ask that no student use the same combination of flowers as another student uses? Why or why not?

12. Use expanded notation to describe the value of $1{,}045_b$.

13. Use the definition of subtraction to justify that $180 - 59 = 121$.

14. Use addition to justify that $123 > 85$.

15. Consider the set $S = \{0, 10, 20, 30, \ldots\}$. Support your answer to each of the following questions.
a. Is S closed under addition?
b. Does S have an additive identity?
c. Is S closed under multiplication?
d. Does S have a multiplicative identity?
e. How do the other properties of addition and multiplication compare between set S and the set of whole numbers?

16. Use the definition of division to justify that $120 \div 40 = 3$.

17. Use the definition of the division algorithm to justify that $125 \div 40 = 3$ with a remainder of 5.

18. **Which Answer Is Right?** Marcy and two friends had each written a division word problem that could be solved with the equation $18 \div 5 = \_\_\_$. One friend argued that the answer to her problem was 3 and the other argued that his answer was 4. Marcy believed that the answer to her problem was 3 with a remainder of 3. Write a paragraph explaining how all three could be right.

Alternative Assessment

19. Draw a diagram or concept map explaining the connections between the ideas of one-to-one correspondence, equivalent sets, equal sets, and whole numbers.

20. The ancient Incas used a *quipu*, an arrangement of colored and knotted cords tied to a base rope, as a way to record numbers. Design a numeration system based on this idea of knotted cords. Write a description of your system and compare its characteristics to those of the Hindu–Arabic system.

21. Choose three classmates and form a group. Each person is to prepare one of the following comparisons of two whole-number operations to present to the group: addition and subtraction, addition and multiplication, multiplication and division, or subtraction and division. Write a summary of the main ideas in your group's presentations.

22. Many students have difficulty memorizing basic arithmetic facts to the point of having immediate recall. Use the properties of addition and multiplication and outline a set of strategies that would be helpful for students as they learn basic arithmetic facts.

3 Estimation and Computation

Chapter Perspective

Mental computation and estimation techniques are important components in doing and using mathematics. For example, a law enforcement official who wants to determine how many people attended a rally in Washington, DC, could look at a photo like the one shown above. She could count the number of people in one square centimeter, and then count the number of square centimeters filled with people to reach her estimate. An estimate like this can be used to plan security for large gatherings.

Many mental calculation techniques are based on the properties of whole numbers and ideas of place value. Mental computation requires a solid understanding of numeration, a mastery of the basic facts, good number sense, and an ability to utilize mathematical reasoning. Learning techniques for estimating and doing calculations mentally gives a person more control over work in mathematics.

In this chapter, you will develop techniques for mental computation and estimation, and understand that flexibility in their application increases their usefulness. You will also explore paper-and-pencil procedures (called *algorithms*) for adding, subtracting, multiplying, and dividing whole numbers. You will learn why the steps in various algorithms make sense.

Connection to the NCTM Principles and Standards

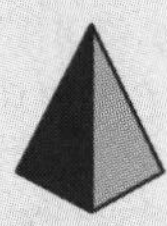

The NCTM *Principles and Standards for School Mathematics* (2000) recommend that the mathematics curriculum at grades PreK–8 include the study of estimation and computation techniques so students can

- develop and use strategies to estimate computations involving whole numbers, fractions, decimals, and integers;
- select appropriate methods and tools for computing with whole numbers, fractions, and decimals from among mental computation, estimation, calculators or computers, and paper and pencil, depending on the situation, and apply the selected methods;
- develop and analyze algorithms for computing with whole numbers, fractions, decimals, and integers and develop fluency in their use *(p. 214)*.

Connection to the PreK–8 Classroom

In grades PreK–2, children develop flexibility in working with numbers by composing and decomposing numbers in various contexts. They are introduced to mental calculation and estimation techniques with whole numbers and computational techniques for whole numbers.

In grades 3–5, mental calculation techniques, estimation techniques, and computational algorithms are extended to greater whole numbers and to fractions and decimals.

In grades 6–8, mental calculation techniques, estimation techniques, and computational algorithms are extended to integers. Choosing appropriate calculation techniques, including the selection of technology, is emphasized at all grades.

Section 3.1 STRATEGIES AND PROCEDURES FOR MENTAL COMPUTATION

- When to Use Mental Computation
- Mental Computation Techniques

In this section, we examine six techniques for mental computation. We also look at the role of the basic properties of whole numbers and place value in these techniques. We stress the importance of understanding numeration and mastering basic facts. Mini-Investigation 3.1 will help you think about the mental computation techniques you already use and should demonstrate that there is more than one way to do a mental computation.

***Write** an explanation of the thinking you used to arrive at each answer.*

Mini-Investigation 3.1 **Using Mathematical Reasoning**

Use mental computation to do the following calculations:

a. $1877 + 300$

b. 8×24

c. $6 \times 31 \times 5$

d. $775 - 38$

When to Use Mental Computation

Most of us have heard someone say, "I just did it in my head," when explaining how a calculation was done. This process of "doing it in your head," usually with the help of counting, numeration ideas, or basic properties, is what is meant by *mental computation.*

Description of Mental Computation

Mental computation is the process of finding an exact answer to a computation mentally, without pencil, paper, calculator, or any other computational aid.

Most people are aware of the paper-and-pencil procedures they use for doing calculations, but often aren't aware of the procedures they use to do calculations mentally. Moreover, as Mini-Investigation 3.1 may have shown, not everyone uses the same technique when doing a particular calculation or uses a specific technique in the same way.

Contrary to the cartoon shown in Figure 3.1, people are never *required* to use mental computation. That is, there are no set rules about *when* to use mental computation. Rather, individuals compute mentally when they feel confident that they can do so. Mental computation can be used to find exact answers and to estimate answers. Both situations call for examining the numbers involved and deciding whether the mental tasks can be carried out accurately.

"I asked you a question, buddy. ... What's the square root of 5, 248?"

FIGURE 3.1

Mental Computation Techniques

Understanding some specific mental computation techniques can help you efficiently and accurately carry them out. The techniques developed in this section are commonly used in mental computation.

Count On and Count Back. The **count on technique** is an efficient method for adding when one of the addends is 1, 2, or 3; 10, 20, or 30; or 100, 200, or 300; and so on. For example, in the calculation 45 + 30, you can start at 45 and count on by tens to get the sum: 45, 55, 65, 75. To count on, start by saying the larger addend and then count on to find the sum. The **count back technique** is an efficient method when subtracting 1, 2, or 3; 10, 20, or 30; and so on. For the calculation 871 − 2, you can start with the larger number, 871, and count back: 871, 870, 869. Most people can accurately count on or count back three numbers. Counting errors become common when people count on or back more than three numbers.

Procedure for Using the Count On and Count Back Techniques

When You Might Use These Techniques

Use this technique if one of the numbers to be added or subtracted is 1, 2, or 3; 10, 20, or 30; or 100, 200, or 300; and so on.

How to Use These Techniques

1. Begin by saying the larger number.
2. Count on to add or count back to subtract 1, 2, or 3; 10, 20, or 30; 100, 200, or 300; and so on.

Example 3.1 illustrates the counting on and counting back techniques.

Example 3.1 Debt and Expense Totals

a. A credit review shows that Sigmund owes $5800 on a car loan and has $3100 in credit card debt. What is Sigmund's total debt?

b. Expenses at a health fair last year were $1455. This year's committee was able to trim expenses by $200. What is the new cost for expenses?

Solution

a. Find 5800 + 3100. First, add the thousands. Start at 5800 and count on by 1000 three times: 5800, 6800, 7800, 8800. So 5800 + 3000 = 8800. Now start with 8800 and count on by 100: 8800, 8900. The total debt is $8900.

b. Find 1455 − 200. Start at 1455 and count back by 100 twice: 1455, 1355, 1255. So 1455 − 200 = 1255. The new cost for expenses is $1255.

Your Turn

Practice: Find the exact value of each expression by counting on or counting back. Explain the process you used.

a. 256 + 30

b. 12,200 + 2300

c. 962 − 3

Reflect: Why is counting on or back usually not effective with more than three numbers? ■

Choose Compatible Numbers. Some number combinations may be easy to add, such as 25 and 175, and others may be easy to multiply, such as 28×10. Numbers that are easy to compute mentally are called *compatible numbers.* The **choose compatible numbers technique** involves selecting pairs of compatible numbers to perform the computation, usually involving a basic fact. Most people can add and subtract multiples of 10 or 100 mentally—for example, $70 + 20 = 90$—and can multiply by multiples of 10 and 100—for example, $34 \times 100 = 3400$. However, beyond a few obvious cases, people must decide which numbers are compatible for them. Most people find that with some practice they have more compatible numbers at their disposal than they realized.

Procedure for Using the Choose Compatible Numbers Technique

When You Might Use This Technique

Use this technique if one or more pairs of numbers can be easily added, subtracted, multiplied, or divided,

or

Use this technique if numbers can be combined to produce multiples of 10, 100, or other numbers that make calculations easy.

How to Use This Technique

1. Look for pairs of numbers that are easy to calculate for the operation required. Do these calculations first.
2. Look for the other number combinations that can be calculated easily.

Example 3.2 demonstrates the use of the choose compatible numbers technique in a multiplication situation. Note how mastery of basic facts is needed in all of these calculations.

Example 3.2 Choosing Compatible Numbers in Multiplication

Look for compatible numbers to find the exact value for the computation $(2 \times 8) \times (5 \times 7)$.

Solution

Tony's thinking: I saw that 2 times 5 equals 10, and multiplying by 10 is easy. Then, 8 times 7 is 56 and 56 times 10 is 560. The product is 560.

Alba's thinking: I saw that 8 times 5 is 40. Then, 40 times 2 is 80 and 80 times 7 is 560. The product is 560.

Your Turn

Practice: Look for compatible numbers to find the exact value for the following expressions.

a. $(25 \times 9) \times (11 \times 4)$
b. $(5 \times 15) \times (20 \times 3)$

Reflect: What are some combinations of numbers that are compatible in multiplication? ■

Example 3.3 illustrates the use of the choose compatible numbers technique in addition.

Example 3.3

Problem Solving: Bike Costs

Suppose that you want to buy a new racing bike for next week's race. How much money would you need if your costs are

Bicycle	$715
Tax	67
Licenses and fees	15

Solution

You need to find $(715 + 67) + 15$. Start with 715 and 15, which are easy to add mentally and produce another number that is easy to use: $715 + 15 = 730$. Now 730 and 67 can be added mentally by counting on in the tens place:
$730 + 67 = 797$. You need $797.

Your Turn

Practice: Look for compatible numbers to find the exact value for each expression. Explain the process you used.

a. $4 \times 16 \times 25$
b. $63 + 18 + 27 + 12$
c. $120 + 385 + 115 + 280$

Reflect: Explain why compatible numbers are helpful in doing calculations. Give an example of a calculation in which it would be difficult to find compatible numbers. ■

The properties of whole numbers discussed in Chapter 2 provide the tools for proving that mental calculations associated with using compatible numbers work. For the bicycle cost in Example 3.3, the properties can be used to prove the mental manipulations.

Statements	**Reasons**
$(715 + 67) + 15 = (67 + 715) + 15$	Commutative property of addition
$= 67 + (715 + 15)$	Associative property of addition
$= 67 + 730$	Addition
$= 797.$	Addition

If we use a, b, and c instead of 715, 67, and 15, we can produce a general proof that $(a + b) + c = b + (a + c)$ and that choosing compatible numbers in this type of situation works. Mini-Investigation 3.2 provides an opportunity to use the properties of multiplication to verify that a mental computation technique works.

Write a proof showing that Tony's use of the choose compatible numbers technique in Example 3.2 works.

Mini-Investigation 3.2 **Using Mathematical Reasoning**

What basic properties of whole numbers would you use to prove that Tony's use of the choose compatible numbers technique in Example 3.2 works?

The formal justification of mental calculation techniques using basic properties is generally not part of the PreK–8 instructional program. Attempting to teach mental calculation techniques through the use of formal methods interferes in the learning of these techniques. Rather, mental calculation techniques should evolve from the natural ways that students operate with numbers. However, teachers need to be familiar with the properties of whole-number operations in order to be able to assess whether the techniques that students use are valid.

Break Apart Numbers. The **break apart numbers technique** involves breaking the numbers in a computation into manageable parts to permit the use of the basic commutative, associative, and distributive properties. This technique also draws on a firm understanding of numeration.

Some flexibility in thinking about large numbers is needed in order to break them apart for doing calculations mentally. For example, a number such as 134 can be thought about in many different ways, among which are

$$100 + 30 + 4,$$
$$100 + 34, \text{ and}$$
$$13 \text{ tens} + 4 \text{ ones}.$$

You can use your understanding of numeration and place value to break apart numbers in ways that make the computations easier. For example, think about the problem 345 + 130 as follows:

$$\begin{array}{r} 300 + 45 \\ +\ 100 + 30 \\ \hline 400 + 75 \end{array} = 475.$$

Note that the associative and commutative properties of addition allow the sums $(300 + 45) + (100 + 30)$ to be calculated in vertical form and give the same answer.

Another way to break apart 345 and 130 for adding is

$$\begin{array}{r} 300 + 40 + 5 \\ +\ 100 + 30 \\ \hline 400 + 70 + 5 \end{array} = 475.$$

Breaking apart numbers is also useful in subtraction involving a 0. For example, in the case of 305 − 24, think about 305 as 30 tens and 5 ones. Then, subtract 2 tens from the 30 tens and 4 ones from the 5 ones. In this situation, crossing out the 3 and writing 10 above the 0, as in the traditional paper-and-pencil procedure, shouldn't be necessary.

Another example of a break apart situation involves multiplication. For example, to multiply 3 × 42, you could break 42 into 40 + 2. You could then use the distributive property to justify finding the answer by adding the product 40 times 3 to the product 2 times 3.

Procedure for Using the Break Apart Numbers Technique

When You Might Use This Technique

Use this technique if simple calculations involving basic number facts result when the numbers are broken apart according to the place value of the digits.

How to Use This Technique

1. Think about each digit in a number according to its place value.
2. Do calculations with each ones value, tens value, hundreds value, or combinations of these values.
3. Recombine the parts to get the final answer.

Use Compensation. With the **use compensation technique,** first substitute a compatible number for one of the numbers so that you can do the calculation mentally. Then adjust the answer to *compensate* for the change made in the original calculation. Suppose, for example, that you wanted to buy two packages of computer disks that cost $10.95 each. Adding $11 twice is easy, $22, but that's too much. Adding 5¢ to $10.95 two times requires compensating by taking away 10¢. The exact answer to the original calculation is $21.90.

Procedure for Using the Use Compensation Technique

When You Might Use This Technique

Use this technique when a calculation can be chosen that is close to the original one and that is easy to do mentally.

How to Use This Technique

1. Change the original calculation to one that is easy to do mentally. Changing only one number usually makes the adjustment at the end easier.
2. Keep track of how you adjusted the original calculation.
3. Find the answer to the original calculation by compensating the answer to the adjusted calculation.

Example 3.4 illustrates the use of the compensation technique.

Example 3.4

Using Compensation in Multiplication and Addition

Find the exact value for each expression by using compensation. Explain the process you used.

a. 9×12

b. $65 + 38$

Solution

a. Multiplying by 10 is easy, so first find $10 \times 12 = 120$. But this calculation used 1 group of 12 too many: $120 - 12 = 108$. So $9 \times 12 = 108$.

b. *Nicole's thinking:* I'll add 65 and 40 to get 105. I then have to subtract 2 to get 103 because I added 2 too many.

Johan's thinking: I'll add $65 + 35$ to get 100. Then I have to add the remaining 3 to get 103.

Your Turn

Practice: Find the exact value for each expression by using compensation. Explain the process you used.

a. $234 + 68$

b. 19×6

c. $908 - 39$

Reflect: When should you compensate by subtracting? ■

The mental manipulations in Example 3.4(b) may also be verified with basic properties. We use the idea of substitution followed by applications of properties. Here is a proof of Nicole's thinking.

Statements	Reasons
$65 + 38 = 65 + (40 - 2)$	Substitute $(40 - 2)$ for 38
$= (65 + 40) - 2$	Associative property of addition
$= 105 - 2$	Adding
$= 103.$	Subtracting

Here is a proof of Johan's thinking.

Statements	Reasons
$65 + 38 = 65 + (35 + 3)$	Break apart 38
$= (65 + 35) + 3$	Associative property of addition
$= 100 + 3$	Adding
$= 103.$	Adding

Use Equal Additions. A technique called **use equal additions** is based on the idea that the difference between two numbers doesn't change if the same number is added to both of the original numbers. The key is to select the best number to add to the others.

For example, we know that $7 - 2 = 5$. Note that the difference remains the same if we add the same number to both 7 and 2.

$12 - 7 = 5.$	Adding 5 to each number
$17 - 12 = 5.$	Adding 10 to each number
$107 - 102 = 5.$	Adding 100 to each number

Thus we can use equal additions to find, say, 93 − 38. We know that the answer is the same if we add 2 to each number, so we obtain the answer by finding 95 − 40, or 55.

Procedure for Using the Equal Additions Technique

When You Might Use This Technique

Use this technique when one of the numbers in a subtraction calculation (usually the number being subtracted) can be changed so that it results in a computation that is easy to do mentally.

How to Use This Technique

1. Identify a number that can be added to one of the numbers in the original calculation to give a new computation that's easy to do mentally.
2. Add this number to both numbers in the original calculation. Compute.

Example 3.5 further illustrates the use of the equal additions technique.

Example 3.5

Using the Equal Additions Technique for Subtraction

Find the exact answer to each difference using the equal additions technique. Explain the process you used.

a. 64 − 28

b. 145 − 77

Solution

a. Subtracting a multiple of 10 is easy, so add 2 to both numbers. The result is 66 − 30 = 36. Thus 64 − 28 = 36.

b. Adding 23 to each number changes the expression to 168 − 100 = 68. Thus 145 − 77 = 68.

Your Turn

Practice: Find each difference by using the equal additions technique.

a. 83 − 29

b. 1456 − 397

Reflect: Can a kind of equal additions method be used to solve 78 + 26? (*Hint:* Think of using addition and subtraction.) ■

Procedure for Choosing a Mental Computation Technique. Example 3.6 illustrates a problem-solving situation that requires deciding which of the mental computation techniques are useful in solving a problem. The problem-solving strategy, *choose an operation*, is used in the example and relies on an understanding of the meanings of the operations. To solve the problem, we had to decide which operation or operations to use and in which order. We also had to decide whether some of the calculations could be completed mentally. Several steps are involved so we had to identify *subproblems* that needed to be solved to give answers needed in working toward the final solution.

Example 3.6 Problem Solving: The Jogging Calorie Burn

Marta and her husband Jay run together for 1 hour every other day. They run at an 8-minute-per-mile pace for the first 30 minutes and then slow to a 10-minute-per-mile pace for the last 30 minutes. Jay weighs 140 pounds and Marta weighs 110 pounds. Use the following chart to determine who will burn more calories during the hour. How many more?

Exercise Guidelines

Calories Burned by Running for 30 minutes

	A	B	C	D
1		Pace in Minutes per Mile		
2	Weight (lb)	10 minutes	9 minutes	8 minutes
3	100	240	260	290
4	110	265	290	325
5	120	290	315	345
6	130	315	345	390
7	140	345	370	415
8	150	365	405	440
9	160	385	430	480
10	170	410	440	500
11	180	480	500	530
12	190	540	550	580

Working toward a Solution

Understand the problem	*What does the situation involve?*	Two people exercising
	What has to be determined?	Who will burn more calories and how many more?
	What are the key data and conditions?	Marta and Jay run 30 minutes at an 8-minute-mile pace and then 30 minutes at a 10-minute-mile pace. They burn calories as shown in the above chart.
	What are some assumptions?	Assume that the data in the chart apply correctly to Marta and Jay.
Develop a plan	*What strategies might be useful?*	Choose an operation.
	Are there any subproblems?	Find the number of calories that Marta would burn during the hour. Find the number of calories Jay would burn during the hour. Then compare those answers.
	Should the answer be estimated or calculated?	The problem suggests that an exact answer is needed.

	What method of calculation should be used?	Most of the calculations seem to be ones that can be done mentally.
Implement the plan	*How should the strategies be used?*	Use the addition operation and mental computation techniques. *Marta:* 265 + 325; 200 + 300 is 500 by counting on; 65 and 25 are compatible numbers and 65 + 25 = 90. So 265 + 325 = 590. *Jay:* 345 + 415; 300 + 400 is 700 by counting on; 45 and 15 are compatible numbers and 45 + 15 = 60. So 345 + 415 = 760.
	What is the answer?	The answer is 760 − 590 = 170, or Jay burned 170 more calories.
Look back	*Is the interpretation correct?*	The correct data were used. The problem asked for two answers: Who burned more and how many more? Both answers are given.
	Is the calculation correct?	Yes. All calculations are correct.
	Is the answer reasonable?	To see whether 170 calories is reasonable, look at the data for a 10-minute mile. The difference is 80 for 30 minutes or 160 for an hour, so 170 seems reasonable.
	Is there another way to solve the problem?	Yes. We could subtract the calories burned for each half-hour, and then add the two differences.

Your Turn

Practice: Reading scores tend to drop as children watch more TV. The following chart shows the latest national reading assessment for eighth-graders.

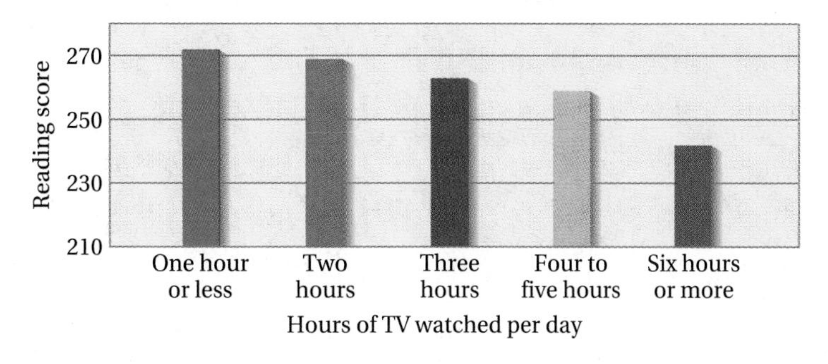

About how many fewer points in reading proficiency might an eighth-grade student who watches 4 to 5 hours of TV per day have than one who only watches 1 hour or less per day?

Reflect: What subproblems had to be solved in the example? Can you think of another way to determine that the answer seems reasonable? Explain. ■

Problems and Exercises | *for Section 3.1*

A. Reinforcing Concepts and Practicing Skills

Look for compatible numbers to find the exact answer mentally for each expression in Exercises 1–4.

1. $12 + 46 + 18 + 64$

2. $5 \times 18 \times 4$

3. $25 \times 28 \times 4$

4. $60 + 140 + 39 + 51$

Break apart numbers to find the exact answer mentally for each expression in Exercises 5–8.

5. $2579 - 372$

6. 132×3

7. 48×20

8. $367 + 622$

Count on or count back to find the exact answer mentally for each expression in Exercises 9–12.

9. $848 - 300$

10. $458 + 20$

11. $648 + 32$

12. $927 - 30$

Use compensation to find the exact answer mentally for each expression in Exercises 13–16.

13. $75 - 39$

14. $132 - 41$

15. 19×8

16. 29×50

Find the exact answer mentally for each expression by using a basic property. Identify the property you use in each case in Exercises 17–20.

17. $(85 + 18) + 12$

18. $(28 \times 5) \times 2$

19. $(18 \times 6) - (15 \times 6)$

20. $(186 + 67) + 33$

21. The owner of a large sporting goods store looks at the year's sales totals and wants to compare the four highest selling categories of goods and the four lowest selling categories (see table in next column). How much more is the total value of the highest four categories than the total value of the lowest four categories in the report?

22. Using data in the table in Exercise 21, how much less is spent on baseball and softball than the total spent on running footwear and apparel?

B. Deepening Understanding

Find the exact value of each expression in Exercises 23–32. Use any mental calculation technique you choose. Name the technique you use in each case.

23. $12 \times 40 \times 5$

24. $7 \times 14 + 3 \times 14$

25. $53 + 26$

26. 45×5

27. $24 + 39 + 76$

28. 9×17

29. $147 - 38$

30. $30 + 488$

31. $455 - 26$

32. $6 \times 22 \times 5$

Table For Exercise 21

	A	B
		$ Thousands
1	Basketball	100
2	Hunting—firearms	130
3	Golf	135
4	Archery	193
5	Scuba and skin diving	208
6	Bowling	244
7	Baseball and softball	275
8	Tennis	300
9	Snow skiing (alpine)	479
10	Running—footwear	700
11	Camping	850
12	Running—apparel	900
13	Exercise	930

33. Write an expression for which the exact sum might be found mentally by using compatible numbers.

34. Write an expression for which the exact sum might be found mentally by breaking apart numbers.

35. One way to compute 11×25 mentally is described in the following solution.

I thought of 11 as 10 + 1. I multiplied 10 by 25 which is 250 and then added 25 more for 275.

Give another way to find the exact value of this expression mentally.

36. Consider Carly's thinking:

To find 126 − 38, I'll first subtract 40 from 126, which is 86. Then I have to compensate by subtracting 2 more. The exact difference is 84.

Is her thinking correct? If so, use basic properties to prove her work. If not, explain the error in her thinking.

C. Reasoning and Problem Solving

37. The Mattress Problem. The advertisement on the next page shows the sale prices of different makes and sizes of mattresses. What is the most that you would pay altogether for one queen-sized 2-piece set and one twin-sized 2-piece set? Assume that there is no sales tax.

38. Use the data in the ad in Exercise 37 to determine how much you would save by buying one twin set and one

ROYALTY PREMIER

	EVERYDAY VALUE PRICE	SALE PRICE
Twin each piece	$199	$179
Full each piece	$279	$229
Queen 2 pc. set	$599	$499
King 3 pc. set	$799	$699

$179

queen set produced by W.S. Manufacturing rather than those produced by Royalty Premier.

39. Based on the data in the ad in Exercise 37, write a problem that can be solved by using mental calculation.

40. **Teen Spending Problem.** A survey of 1,000 students aged 12–19 showed that the average amount spent each week increases by age, as shown in the following table.

Age (years)	Amount ($)
12–14	36.25
15–17	56.90
18–19	81.24

How much more per week does an average 19-year-old spend than an average 12-year-old?

41. **Teen Spending Patterns.** In a certain year, female teenagers spent an average of $55.50 per week, and male teenagers spent an average of $48.50 per week. How much less per week did male teenagers spend, on average?

State whether an exact answer or an estimate is needed for each of the following situations in Exercises 42–47. Then identify the calculation method that you would use to find the answer. Explain your choices.

42. In 1994, a community's summer program fielded 28 soccer teams. Each team was filled to capacity with 21 players. Was the total number of players in the 1994 summer program more than the 500 players in the summer program of 1987?

43. A CD system that regularly costs $365.25 was on sale for $48.89 off the regular price. What was the sale price of the CD system, not including tax?

44. Attendance at student council meetings has been surprisingly high. Attendance at the last four meetings was 34, 46, 52, and 38. What was the total attendance for those four meetings?

45. A secretary ordered 30 boxes of pencils for school supplies. Each box contains 150 pencils. How many pencils did she order altogether?

46. A TV ad said that tires were on sale for $49.99 each, including tax. About how much would four new tires for your car cost?

47. A scale on a map said that 1 inch corresponds to 10 miles. On the map Kara measured three segments between cities and obtained $3\frac{1}{4}$ in., $2\frac{1}{2}$ in., and $2\frac{3}{4}$ in. What actual total distance corresponded to those that Kara measured on the map?

D. Communicating and Connecting Ideas

48. Create a real-world problem in which an exact answer is needed and the calculations might be done mentally.

49. **Making Connections.** Success with mental calculation depends greatly on a person's understanding of place value. Use the examples in this section to illustrate how a person with poor understanding of place value would have difficulty doing mental computations. Share your results with a small group of classmates.

50. Give examples of the types of calculations for which you use a calculator or computer. Give examples of the types of computational task that you do mentally. How do you decide which method to use?

Historical Pathways. *The Roman numeration system is essentially a base-ten system, although, as we showed in Chapter 2, it uses some other values: 5, 50, 500, and so on. For the following calculations, explain how the use of such values might make mental computation easier.*

51. 42 + 39

52. 2935 − 1495

Section 3.2 STRATEGIES AND PROCEDURES FOR ESTIMATION

- Computational Estimation Techniques
- Choosing an Estimation Technique

In this section, we examine four estimation techniques. They draw on an understanding of numeration and knowledge of basic facts. Like the techniques for mental calculation described in Section 3.1, they also involve decisions about whether an estimated answer is acceptable for a given situation and which technique should be used to arrive at the estimate. Mini-Investigation 3.3 will help you think about estimation techniques that you already use and should demonstrate that there is more than one way to estimate an answer.

Write an explanation of the thinking you used to arrive at each estimate.

Mini-Investigation 3.3 **Using Mathematical Reasoning**

Estimate the answer to the following calculations:

a. $478 + 223$
b. 8×26
c. $578 + 603 + 614 + 582$
d. $6563 - 2859$

There are three main types of estimation techniques.

Description of the Three Main Types of Estimation

1. Estimating a Quantity
Finding *how many* students, days, lunches, classes, and so on

2. Estimating a Measure
Finding *how much* length, area, volume, time, and so on

3. Estimating an Answer (Computational Estimation)
Finding a sum, difference, product, or quotient

In this section, we develop techniques for the third type, *computational estimation.*

Computational Estimation Techniques

Description of Computational Estimation

Computational estimation is a process for finding a number reasonably close to the exact answer for a calculation.

In this section, we look at four computational estimation techniques: rounding, substitution of compatible numbers, front-end estimation, and clustering. We focus on using these techniques in estimations that involve mental calculations. In most situations that call for an estimate, we can calculate the estimate mentally.

Rounding. Recall that the process of replacing a number or numbers in a calculation with the closest multiple of 10, 100, 1000, and so on, is called *rounding.* Using the **rounding technique,** we calculate—often mentally—with the rounded numbers to obtain an estimate. For example, to estimate $589 + 217$, we replace 589 with 600, replace 217 with 200, and mentally calculate $600 + 200$ to get the estimate 800.

Procedure for Using the Rounding Technique

When You Might Use This Technique

Use this technique when rounded numbers produce a calculation that can be done mentally.

How to Use This Technique

1. Find the digit in the place value to which you want to round. This is the key digit.
2. Identify the digit in the place value to the right of the key digit.
   - If that digit is less than 5, round down—that is, keep the key digit and replace all digits to its right with zeros.
   - If that digit is 5 or greater, round up—that is, add 1 to the key digit and replace all digits to its right with zeros.

Example 3.7 illustrates use of the rounding procedure to find estimates.

Example 3.7 Using Rounding for Estimating

Estimate the value of each expression using rounding and explain your thinking.

a. 16,942 − 7540
b. 23,562 ÷ 809

Solution

a. Rounding might be used in several ways for this task. We can round each number to the nearest thousand: 17,000 − 8000 = 9000. We can also round each number to the nearest hundred and get a closer estimate:

$$16{,}900 - 7500 = 9400.$$

b. The choice of the place value in which to round each number should be made so that an easy calculation results. Rounding each to the nearest hundred doesn't give an easy calculation. Rounding the first number to the nearest thousand and the second to the nearest hundred results in using a basic fact, 24 ÷ 8, to obtain the estimate: 24,000 ÷ 800 = 30. Thus the quotient is about 30.

Your Turn

Practice: Estimate the value of each expression by rounding.

a. 456 + 298
b. 78 × 145
c. 825 ÷ 79
d. 12,045 − 3907

Reflect: For which of the practice calculations was your estimate an overestimate of the exact answer? Can you decide without finding the exact answer? How? ■

Part (b) in Example 3.7 shows that both numbers do not have to be rounded to the same place value when rounding is used to estimate. Whenever we use the rounding technique, we should look at the numbers in the calculation and decide which numbers we can easily use to calculate mentally and round to the place values that produce those numbers.

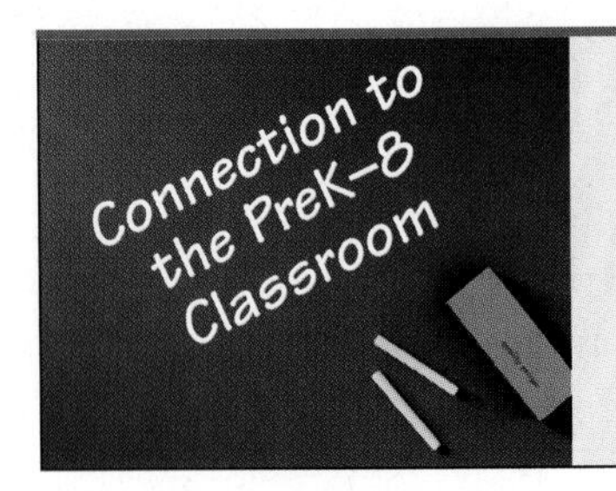

Rounding was the only estimation technique taught in elementary schools for many years. As in the case of mental calculations, estimation techniques are now part of most instructional programs. Estimation techniques are developed as part of the work with adding, subtracting, multiplying, and dividing whole numbers, fractions, and decimals. The PreK–8 instructional program should include explorations of a variety of estimation techniques.

Mini-Investigation 3.4 will deepen your understanding of rounding.

Mini-Investigation 3.4 **Solving a Problem**

What is the greatest whole number and what is the least whole number that, when rounded to the nearest 1000, round to 20,000?

Draw a number line to prove that your answers are correct.

Technology Extension: If you are using technology, determine the options, if any, that your calculator or computer software has for rounding numerical values. Share this information with your classmates.

Substitution of Compatible Numbers. In Section 3.1, we examined the mental calculation technique of looking for compatible numbers in a calculation and using them to calculate the answer mentally. In estimating, we use the **substitute compatible numbers technique,** which involves *replacing* some or all of the numbers in a computation with numbers that are easy to compute mentally—that is, compatible numbers—in order to obtain an estimate.

Suppose that a used car was selling for $3469, with tax and title fees of $334. To estimate the total cost for this car, we might substitute $3475 for the price and $325 for the tax and fees. The substituted numbers are close to the original numbers and easy to compute mentally. The estimated total price would be the sum of $3475 and $325, or $3800.

Procedure for Using the Substitute Compatible Numbers Technique

When You Might Use This Technique

Use this technique if numbers close to the numbers in the original calculation would make the estimate easy to do mentally.

How to Use This Technique

1. Identify the number or numbers in the original calculation that can be replaced by others to result in an estimate that is easy to do mentally.
2. Calculate with the new numbers to obtain the estimate.

Example 3.8 illustrates the use of the substitute compatible numbers estimation technique in subtraction. Example 3.9 demonstrates its use in multiplication.

Example 3.8

Problem Solving: The TV-Set Sale

A local department store is having a special one-day sale on all TV sets. The one that you've been wanting to buy costs \$358 at the regular price and is on sale for \$144. About how much will you save if you buy that TV set during the sale?

Solution

Multiples of 10 are easy to subtract, so one way to substitute compatible numbers is $350 - 150 = 200$. You will save about \$200.

Your Turn

Practice: Another TV set regularly costing \$483 can be purchased during the special sale for \$399. Use the substitute compatible numbers technique to estimate the savings on this TV set.

Reflect: Could you have used other compatible numbers in the practice problem? Explain. ■

Example 3.9

Substituting Compatible Numbers to Estimate

Estimate the value of 524×33 by substituting compatible numbers. Explain your thinking.

Solution

Multiples of 100 and 10 are easy to use, so one way to substitute compatible numbers is $500 \times 30 = 15{,}000$. The product is about 15,000.

Your Turn

Practice: Estimate the value of each expression by substituting compatible numbers. Explain your thinking.

a. $7243 + 815$

b. 326×47

c. $\$1.54 \times 12$

d. $815 - 149$

Reflect: What makes numbers compatible? Give an example of a calculation involving multiplication whose answer can be estimated by substituting compatible numbers. ■

Note that the approach used in Example 3.8 is not the same as rounding. That is, \$358 does not round to \$350 and \$144 does not round to \$150. In Example 3.9, the computation task that resulted from substituting compatible numbers, 500×30, is the same as rounding. Although the symbol manipulation looks identical, the thinking used is different.

Front-End Estimation. The simplest way to use the **front-end estimation technique** involves calculating with the leftmost, or front-end, digit of each number as if the remaining digits were all zeros. For example, suppose that we want to estimate the number of people who went to the local theater production on the opening weekend (Friday through Sunday nights). We know that the number of tickets sold for each of the three evenings were 225, 315, and 285. Using front-end estimation, we think 200 + 300 + 200, so our estimate is that about 700 people came to the play on the opening weekend.

In this situation, the exact answer is considerably higher than our estimate because 285 is quite a bit larger than 200, yet we used 200 in the calculation. We can adjust estimates obtained by using only the front-end digits to give an estimate closer to the exact answer. In this example, we might reason that the difference between the actual numbers and the numbers we used, 25 + 15 + 85, is about 100 and so adjust our estimate for the sum to 800.

Procedure for Using the Front-End Estimation Technique

When You Might Use This Technique

Use this technique when an estimate is needed quickly and a rough estimate is acceptable.

How to Use This Technique

1. Assume that all digits except the leading or front-end digit(s) in the numbers in a calculation are 0.
2. Do the calculation with the new numbers.
3. If you want a closer estimate, adjust the first estimate by using other digits or numbers for those assumed to be 0 and estimate again.

Using front-end estimation with adjustment, we can often obtain an estimate that is very close to the exact answer, as illustrated in Example 3.10.

Example 3.10

Using Front-End Estimation with Adjustment

Estimate the value of 563 + 325 by using front-end estimation with adjustment. Explain your thinking.

Solution

Adding the front-end digits, we get 500 + 300 is 800. The numbers that remain are 63 + 25, or about 85, and 500 + 85 = 585. Thus 563 + 325 is about 585.

Your Turn

Practice: Estimate the value of each expression by using front-end estimation with adjustment. Explain your thinking.

a. 809 − 367

b. 655 ÷ 388

c. 12,108 + 4589 + 23,547

Reflect: Will front-end estimation with adjustment always produce an underestimate of the exact answer? Explain. ■

Mini-Investigation 3.5 explores some of the difficulties involved with using front-end estimation with adjustment to estimate a product.

Talk about difficulties in making the adjustment in this case.

Mini-Investigation 3.5 **Using Mathematical Reasoning**

What is your estimate for 38×26 when you use front-end estimation with adjustment?

Clustering. In some addition calculations, the numbers tend to be approximately the same—that is, to cluster around a common number. When estimating in such situations, you can use the **clustering technique.** This technique involves looking for the number about which the addends cluster and then multiplying by the number of addends. For example, a salesperson recorded the number of customers that visited the store between 9 and 11 A.M. on weekdays. The numbers, 48, 55, 47, 52, and 53, all cluster around the same number, 50. The estimated total number of customers that came into the store between 9 and 11 A.M. during the week is $5 \times 50 = 250$.

Procedure for Using the Clustering Technique

When You Might Use This Technique

Use this technique to estimate sums when the addends in the calculation cluster around the same number. You can also use it in a similar manner for some products.

How to Use This Technique

1. Identify the number that each of the addends is close to and that is easy to compute with mentally.
2. Replace each addend with the same number.
3. Use multiplication to estimate the sum for the original addition calculation.

Example 3.11 further illustrates the clustering technique for estimation.

Example 3.11 **Problem Solving: Blood Donors**

The number of blood donors at a local hospital has been about the same for the first four months of the year. If the pattern continues, estimate the number of donors that might be expected for the year.

Month	Number of Donors
January	146
February	154
March	148
April	152

Solution

Each number is close to 150. We want to estimate the total for 12 months, so we think 12×150 is about 10 times 150, or 1500. The number of donors might be about 1500 for the year.

Your Turn

Practice: Suppose that the donors for other periods of time are as shown. Use clustering to estimate the indicated totals.

a. The donors for a four-month period are
September, 97 October, 120
November, 89 December, 106
About how many donors gave blood during these four fall months?

b. The number of donors for the first six months of a year are 126, 124, 125, 127, 129, and 123. About how many donors are there for the six-month period?

Reflect: Does clustering always give the same estimate as rounding? Use the example and part (a) of the practice, with numbers rounded to the nearest 10, to test your conclusion. ■

Clustering may also be used to estimate products. For example, to estimate the product $9 \times 13 \times 8 \times 12$, we observe that the numbers cluster around 10. We mentally find the product of $10 \times 10 \times 10 \times 10$, or 10,000, to arrive at the estimate.

Choosing an Estimation Technique

The problem in Example 3.12 requires making decisions about what calculations to estimate, what estimation technique to use, and whether giving the answer as a **range estimate,** where both a low and a high estimate are given to indicate a range, would be appropriate. The problem-solving strategy, *choose an operation*, and the use of mental calculation techniques also play a role in the solution.

Example 3.12 Problem Solving: Utility Costs

The sellers of a home told the potential buyers that their average monthly cost for gas was $65 and that their average monthly cost for electricity was $38. What might the buyers estimate for the total gas and electric costs for the year?

Working toward a Solution

Understand the problem	*What does the situation involve?*	The expected cost of gas and electricity in a new home.
	What has to be determined?	What is the total yearly estimated cost for both utilities?
	What are the key data and conditions?	The cost of gas averages $65 per month, and the average cost of electricity is $38 per month.
	What are some assumptions?	The buyers' expenditures for the utilities would be about the same as the sellers'.
Develop a plan	*What strategies might be useful?*	Choose an operation for the. calculations needed.
	Are there any subproblems?	Find the average yearly cost for gas and the average yearly cost for electricity. Then combine those costs to obtain the total.

	Should the answer be estimated or calculated?	An estimate is all that is called for.
	What method of calculation should be used?	Use multiplication and addition and try some different approaches by using mental computation and estimation techniques.
Implement the plan	*How should the strategies be used?*	Use the substitute compatible numbers technique to compute mentally the yearly cost of each utility. *Gas:* Substitute 70 for 68. $70 × 12 is 70 × 10 plus 70 × 2, so 700 + 140 = 840. *Electricity:* Substitute 40 for 38 $40 × 12 is 40 × 10 plus 40 × 2, so 400 + 80 = 480. Then use rounding to combine the costs mentally: 840 + 480 is about 840 + 500 = 1340. Try another approach by using front-end estimation. *Gas:* 60 × 12 is 60 × 10 plus 60 × 2, so 600 + 120 = 720. *Electricity:* 30 × 12 is 30 × 10 plus 30 × 2, so 300 + 60 = 360. The costs are easy to combine mentally: 720 + 360 = 1080.
	What is the answer?	The total yearly cost for both is between $1080 and $1340.
Look back	*Is the interpretation correct?*	Rereading the problem shows that the interpretation is correct.
	Is the calculation correct?	Yes. All mental calculations are correct.
	Is the answer reasonable?	The combined monthly cost rounds to about $100. As 100 × 12 is 1200, $1340 seems reasonable.
	Is there another way to solve the problem?	Yes. The combined monthly cost could have been obtained first by using the two estimation techniques and then multiplying.

Your Turn

Practice: Suppose that the average monthly cost for gas was $87 and the average cost of electricity was $43. What is a reasonable estimate for the total gas and electricity costs for a year?

Reflect: The example problem didn't call for a range estimate, but the answer was given as a range estimate. Why might a range estimate be useful in this type of situation? ■

Problems and Exercises | *for Section 3.2*

A. Reinforcing Concepts and Practicing Skills

Round each number in Exercises 1–6 to the place value of the bold digit.

1. 6783
2. 26.09
3. 209.8
4. 2995
5. 851
6. 749.9

Estimate each answer in Exercises 7–10 by rounding.

7. $4671 + 2509$
8. $26{,}897 - 19{,}456$
9. 864×22
10. $824 \div 18$

Estimate each answer in Exercises 11–14 by substituting compatible numbers.

11. $424 + 526$
12. $1195 - 195$
13. 23×8
14. $430 \div 6.8$

Use front-end estimation without adjustment to find the value of each expression in Exercises 15–18.

15. $433 + 126 + 678 + 400$
16. 119×54
17. $1286 \div 63$
18. $24{,}865 - 11{,}274$

Use front-end estimation with adjustment to find the value of each expression in Exercises 19–22.

19. $824 + 238$
20. $23{,}869 + 14{,}198$
21. $38 + 64 + 46 + 76 + 87$
22. $7653 - 2861$

Estimate each answer in Exercises 23–26 by using clustering.

23. $98 + 106 + 101 + 97 + 95$
24. $76 + 71 + 76 + 74$
25. $10.4 + 10.9 + 10.52 + 10.5$
26. $426 + 424 + 423 + 427$

Give a range estimate for each calculation in Exercises 27–30.

27. 54×38
28. $758 + 892$
29. $677 - 139$
30. 24.2×9.8

Estimate each answer in Exercises 31 and 32 using any technique you choose. Use reasoning to decide whether the exact answer is greater than or less than the estimate. Tell how you decided.

31. 325×42
32. $5678 + 2654$

B. Deepening Understanding

Estimate each of the following calculations in Exercises 33–40. Use any estimation techniques you want. Explain the techniques you use.

33. $765 + 824 + 799$
34. $5236 + 2810$
35. $8095 - 4877$
36. 328×72
37. $532 \div 92$
38. 15.8×9.28
39. $23{,}765 + 18{,}822 + 11{,}008 + 8320$
40. $24 \times 37 \times 4.8 \times 0$

Give an estimate for each expression in Exercises 41–46. Then state whether you believe that the exact answer is greater than or less than the estimate. Explain how you decided. Use a calculator to check the accuracy of your estimate.

41. $25{,}456 + 65{,}879$
42. 978×66
43. $788 + 808 + 766 + 803$
44. 54×23
45. $726 \div 67$
46. 12×26
47. An estimated sum is 900. One of the addends is 478. Give five possible numbers for the other addend.
48. An estimated product is 1500. One of the factors is 28. Give three possible numbers for the other factor.

Name the estimation technique or techniques used in Exercises 49 and 50.

49. $74 + 23 + 77$: I thought that 74 and 77 are about 75 and 2 times 75 is 150. Then 23 is about 20, so 150 plus 20 is 170. The sum is about 170.
50. $146 \div 68$: I thought that 14 and 7 are easy to divide, so I used $140 \div 70$, which is 2. The quotient is about 2.
51. Harvey looked at the expression $7653 - 2861$ and said, "I can tell immediately that the difference is less than 5000." Explain how Harvey could determine this result.
52. Estimate the value for the expression $328 + 144 + 478 + 530$. First, use front-end estimation with adjustment; then, use rounding. Which method do you think gives an estimate that is closer to the exact answer? Explain.
53. Write a numerical expression involving addition and another numerical expression involving multiplication, each with an estimated value of 500.

C. Reasoning and Problem Solving

54. **The Defective Bulb Problem.** Past experience has shown that approximately 3% of all bulbs of a certain type installed in a gymnasium have to be replaced in one month. A total of 3450 bulbs were installed. Use estimation to find two numbers between which the exact answer falls.

55. **The Garden Party Problem.** Victoria was planning a fund-raising garden party at her house for about 40 people. The cost per person from the caterer is $24.50. Should she overestimate or underestimate the total cost in planning for the party? Explain.

56. **The Turkey Cook-Time Problem.** Paul was estimating the time needed to cook a turkey. He knew the basic cooking time and the additional time needed per pound when cooking a turkey on the barbecue. No one wanted dry turkey meat. Should he overestimate or underestimate the time needed to cook the turkey? Explain.

57. **The Concrete Amount Problem.** Before ordering concrete for a construction project, a contractor has to estimate the number of cubic yards of concrete needed. The concrete company charges a substantial amount for each trip to the job site. Should the contractor underestimate or overestimate the amount of concrete to order? Explain.

58. **The Red Cross Problem.** The local Red Cross must base its budget entirely on the contributions received from people in the community. Its finance officer has examined the donation data from previous years. Should she underestimate or overestimate contributions during budget preparation? Explain.

59. **The Mustache Population Problem.** *The Unofficial U.S. Census* [Heymann (1991), p. 81] reported that one in four American men, 22,251,772, have mustaches. If this figure is correct, about how many men are in the United States?

60. **The Teacher Supply Problem.** In 1960, there were 991,000 elementary school teachers in the United States. In 1990, there were about 1,627,000 elementary school teachers. About how many more teachers were there in 1990 than in 1960?

61. **The Graduation Rate Problem.** In 1940, 75.5% of the people in the United States who were 25 years old and older did not have a high school diploma. In 1988, only 23.8% of the population 25 years old and older did not have a high school diploma. What is the approximate decrease in the percentage of people 25 years old and older who did not have a high school diploma?

62. **The Dog Data Problem.** Suppose that you were a newspaper reporter doing a story about the age of dogs in the United States. You want to use a headline that includes some numerical data from the following graph. Use the graph to estimate numerical information for your headline.

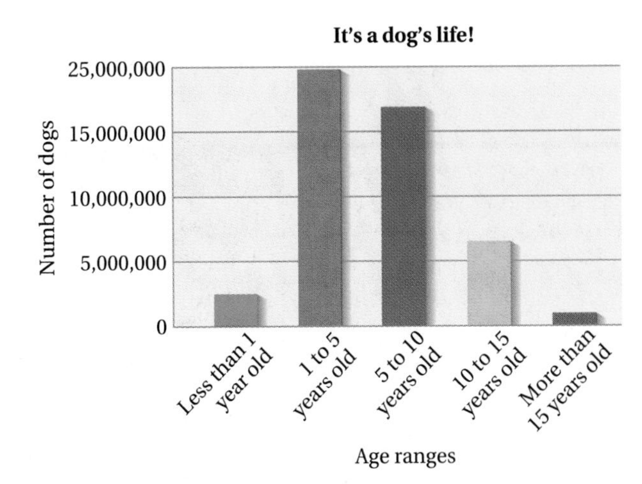

Source: Heymann, Tom. *The Unofficial U.S. Census.* New York: Fawcett Columbine, 1991, p. 144. Used with permission.

D. Communicating and Connecting Ideas

63. Interview some friends and ask them when they use estimating in doing calculations and how they do the estimating. Identify the technique or techniques discussed in this chapter that they use and share your findings with a small group of classmates.

64. **Making Connections.** Explain in writing how the idea of compatible numbers is involved in both estimation and mental calculations. Give examples to illustrate your points.

65. For a classmate who was absent from class, write a paragraph explaining the meaning of estimation and when it should be used. If appropriate, include examples to illustrate your ideas. Compare your paragraph with those written by two other people in your class and revise your work, if necessary, to make your explanation more complete.

66. **Historical Pathways.** A certain 1959 elementary school fourth-grade mathematics textbook contained no estimation or mental calculation lessons. Why do you think that estimation and mental math have become more important today than in the 1950s?

Section 3.3 ALGORITHMS FOR ADDITION AND SUBTRACTION

- Developing Algorithms for Addition
- Developing Algorithms for Subtraction
- Calculator Techniques for Addition and Subtraction

In this section, we examine the step-by-step procedures—the **algorithms**—for adding and subtracting whole numbers. We focus on using models and logic to make sense of the computational procedures for finding sums and differences, regardless of the algorithm used. Mini-Investigation 3.6 asks you to analyze the paper-and-pencil computational procedures you ordinarily use.

Write a detailed description of the procedure you used for finding the difference and compare it with procedures used by others.

Mini-Investigation 3.6 **Communicating**

How would you complete the following subtraction calculation, using paper and pencil?

$$\begin{array}{r} 2004 \\ -\ 1278 \\ \hline \end{array}$$

Developing Algorithms for Addition

A model is a useful tool for explaining an algorithm. For example, the use of base-ten blocks as a model to find the sum of two numbers involves actions with the blocks that can later help illustrate the procedures used in a paper-and-pencil algorithm for addition. In this subsection, we first look at an example that models addition. Then we develop the related paper-and-pencil algorithm. Finally, we use the properties of whole-number operations to verify that the steps in an addition algorithm are logical.

Using Models as a Foundation for Addition Algorithms. Example 3.13 shows how base-ten blocks can be used to find a sum and thus provide models that help explain the addition algorithms.

Example 3.13 **Using the Base-Ten Blocks Model for Addition**

The two numbers shown are modeled with the base-ten blocks. Use the base-ten blocks to find the sum of the two numbers and then write an equation to record the addition.

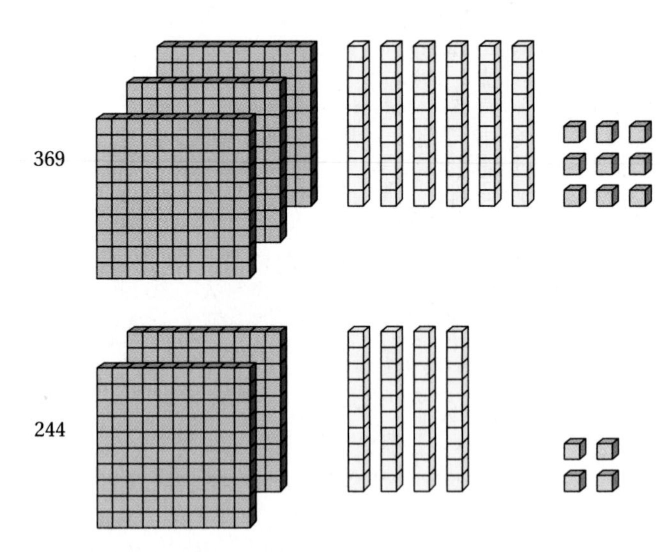

Solution

Linda's thinking: I started by putting together all blocks of the same type.

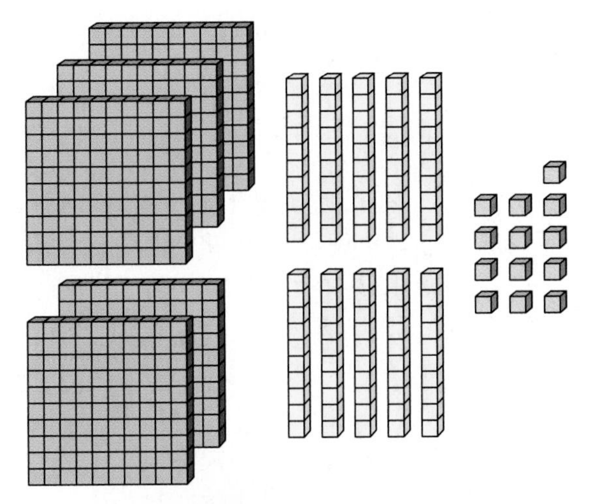

I then regrouped 10 tens to make 1 hundred.

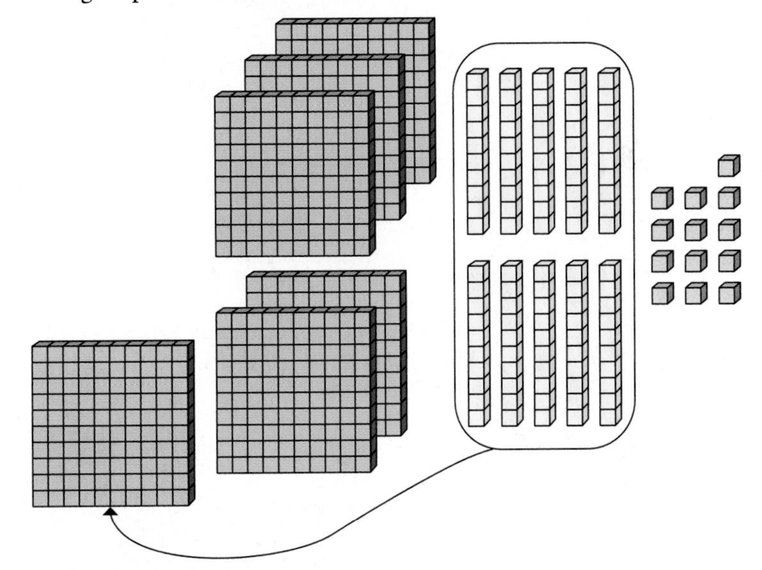

I then regrouped 10 ones to make 1 ten.

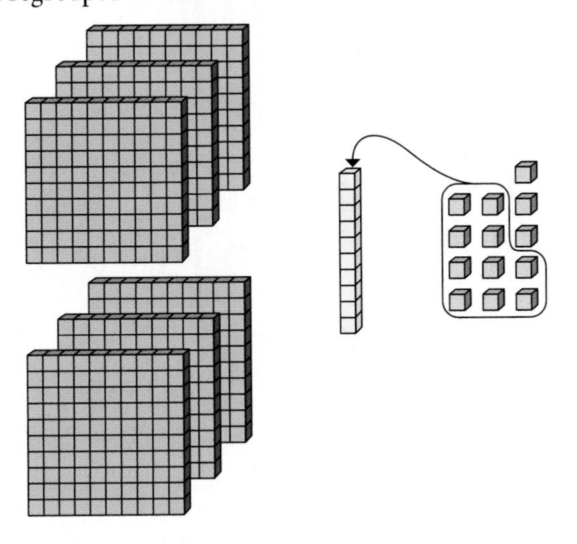

I found the sum, 613, and recorded the addition with a vertical equation:

$$\begin{array}{r} 369 \\ +\ 244 \\ \hline 613 \end{array}$$

Rosa's thinking: I began by putting together the ones. I then regrouped 10 ones to make 1 ten and was left with 3 ones.

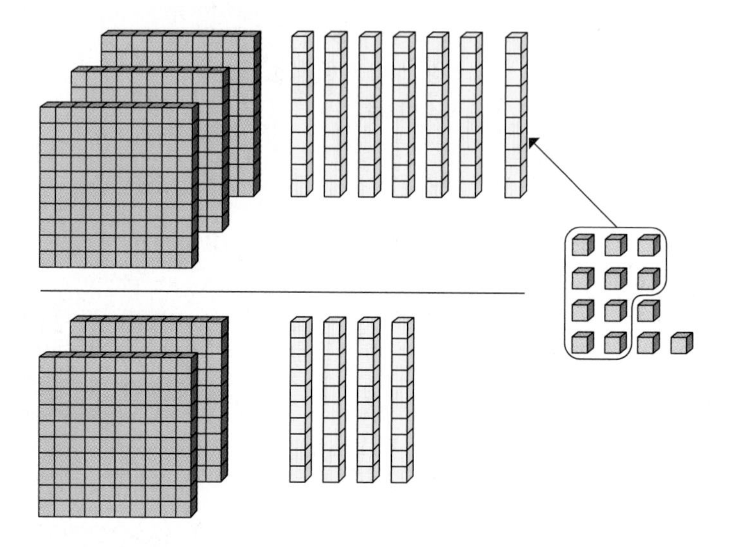

Next, I put together the tens. Then I regrouped 10 tens to make 1 hundred and was left with 1 ten.

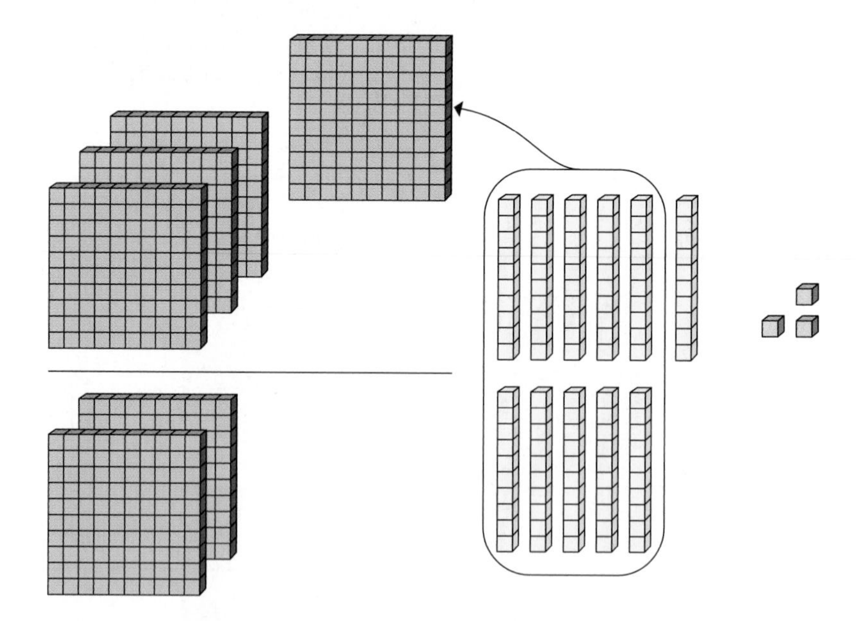

Finally, I put together the hundreds, which gave me 6 hundreds, 1 ten, and 3 ones.

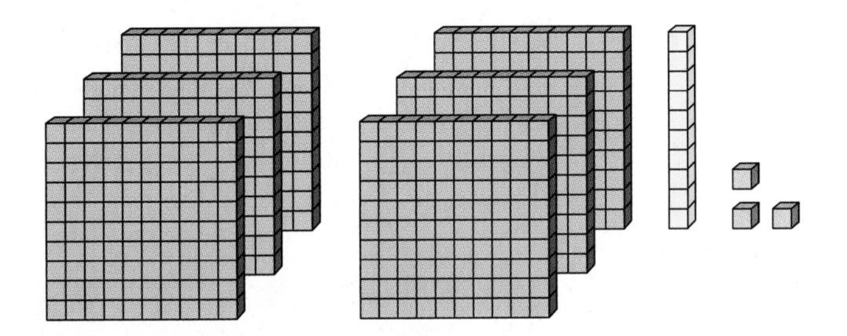

I observed the sum, 613, and recorded the addition with the equation
369 + 244 = 613.

Your Turn

Practice: Use base-ten blocks to show 374 + 128. Write an equation to record the addition.

Reflect: When you use base-ten blocks to add, does it matter whether you combine the ones, tens, or hundreds blocks first? Explain. ■

Example 3.13 demonstrated two different approaches to finding the sum of two numbers with base-ten blocks. In the first approach, the hundreds were combined, then the tens were combined, and finally the ones were combined. Following this combining, regrouping was done, starting with the tens. In the second approach, the ones were combined first and immediately regrouped from ones to tens. Then, the tens were combined and regrouped from tens to hundreds. Finally, the hundreds were combined. These two approaches suggest that the combining and regrouping can be done in different ways when you are adding two numbers.

Developing and Using Paper-and-Pencil Algorithms for Addition. Let's now look at two paper-and-pencil algorithms for addition that follow directly from the models in Example 3.13. We use the same calculation, 369 + 244, to show both of these algorithms. Along with the models in Example 3.13, these algorithms again emphasize that in mathematics even a routine task often may be done more than one way.

The first algorithm, which relates to Linda's work in Example 3.13, is the *expanded algorithm* in which the values of each place are added first and later combined.

Expanded Algorithm for Addition

Think			Write
			369
			+ 244
Add hundreds:	*300 + 200 = 500*	⟶	500
Add tens:	*60 + 40 = 100*	⟶	100
Add ones:	*9 + 4 = 13*	⟶	+ 13
Add the hundreds, tens, and ones:			613

In the expanded algorithm, the order in which the numbers with a given place value are added doesn't matter because all partial sums are recorded.

The second algorithm, called the *standard algorithm*, relates to Rosa's work in Example 3.13 and involves starting with the ones and proceeding to add, with regrouping, from right to left.

Standard Algorithm for Addition

Think	Write
	$\begin{array}{r} {}^{1\,1}\\ 369 \\ +\ 244 \\ \hline 613 \end{array}$
Add the ones and regroup:	3
Add the tens and regroup:	1
Add the hundreds:	6

Whenever we use the standard algorithm and there are 10 or more ones, we regroup 10 ones as 1 ten and then add the tens. If there are 10 or more tens, we regroup 10 tens to make 1 hundred and then add the hundreds, regrouping as needed. This process continues for as many digits as there are in the addends.

The word *carry* was not used in Example 3.13 or in the standard algorithm. Rather, the word *regroup* was used to convey the action with the blocks. As related to addition and subtraction, *carry* and *borrow* are no longer used in elementary school programs. The words *regroup* and *trade* are used instead to match more closely the physical actions done with manipulative models.

Example 3.14 illustrates the use of these two algorithms.

Example 3.14

Using the Expanded and Standard Algorithms for Addition

Use either the expanded or the standard algorithm to find the sum 562 + 783.

Solution

Leah's thinking: I add the ones, then the tens, and finally the hundreds. Each time I write the partial sum. Then I find the total of the partial sums.

$$\begin{array}{r} 562 \\ +\ 783 \\ \hline 5 \\ 140 \\ 1200 \\ \hline 1345 \end{array}$$

Rolando's thinking: I first add the ones. Then I add the tens and regroup. Finally, I add the hundreds. I know that 13 hundreds are 1 thousand and 3 hundreds when I write my answer.

$$\begin{array}{r} \overset{1}{5}62 \\ +\ 783 \\ \hline 1345 \end{array}$$

Your Turn

Practice: Use either the expanded or the standard algorithm to find the sum 674 + 598.

Reflect: Explain how Leah's thinking in the example differs from the thinking shown in the box explaining how to use the expanded algorithm. ■

Mini-Investigation 3.7 will help you review the role of models in explaining the algorithms for addition.

Talk about how the standard algorithm used by Rolando might be viewed as a shortcut for the expanded algorithm used by Leah.

Mini-Investigation 3.7 **Making a Connection**

How would you use base-ten blocks to model the algorithms used by Leah and Rolando in Example 3.14?

Using Properties of Whole Numbers to Verify Addition Algorithms. We can use mathematical reasoning to verify that the procedures used in the addition algorithms are logically correct. One such verification involves the use of the commutative, associative, and distributive properties of whole numbers, along with the properties of numeration.

Statements	Reasons
$369 + 244$	
$= [3(100) + 6(10) + 9(1)]$ $+ [2(100) + 4(10) + 4(1)]$	Writing a number in expanded form
$= [3(100) + 2(100)] + [6(10) + 4(10)]$ $+ [9(1) + 4(1)]$	Commutative and associative properties applied repeatedly
$= (3 + 2)100 + (6 + 4)10 + (9 + 4)1$	Distributive property
$= 5(100) + 10(10) + 13$	Basic addition fact
$= 5(100) + 10(10) + 1(10) + 3$	Writing a number in expanded form
$= 5(100) + 1(100) + 1(10) + 3$	Multiplying 10(10)
$= (5 + 1)(100) + 1(10) + 3$	Distributive property
$= 6(100) + 1(10) + 3$	Basic addition fact
$= 613.$	Writing in standard form

The statement and reason proof above can be interpreted as a verification of both the expanded algorithm and the standard algorithm. For example, on the one hand, line 3 suggests first adding the hundreds, tens, and ones, as in the expanded algorithm. Then line 7 verifies the addition of the partial sums in the expanded algorithm. On the other hand, lines 5 and 7, for example, verify the regrouping required in the standard algorithm.

Developing Algorithms for Subtraction

Models can be used to explain algorithms for subtraction in much the same way as they are used to explain addition algorithms. We first use models to illustrate the procedures for subtraction. Then, we use those procedures to develop pencil-and-paper subtraction algorithms. Finally, we use mathematical reasoning to justify the subtraction algorithm.

Using Models as a Foundation for Subtraction Algorithms. Our use of base-ten blocks in addition demonstrated that the procedures for finding a sum may be modeled in various ways. Similarly, using base-ten blocks to find differences demonstrates that many different procedures for modeling subtraction are also available. Example 3.15 shows how to use the take-away interpretation and base-ten blocks to model a procedure for subtracting.

Example 3.15 Modeling a Procedure for Subtraction

The larger number in the subtraction calculation shown is modeled with base-ten blocks. Find the difference by using the base-ten blocks and write an equation to record the subtraction.

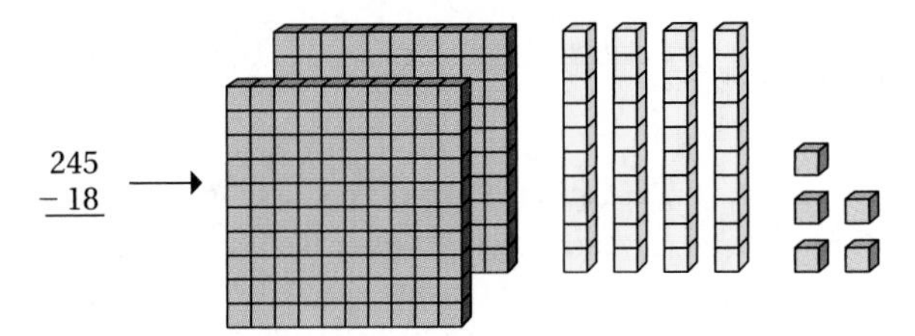

Solution

Winona's thinking: To have enough ones to take away 8, I begin by trading 1 ten for 10 ones. I then take away 8 ones from the 15 ones, leaving 7 ones.

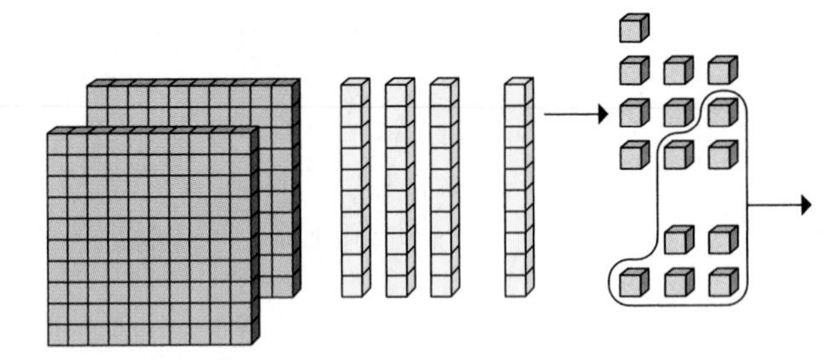

Next, I take away 1 ten from the 3 tens remaining and am left with 2 tens.

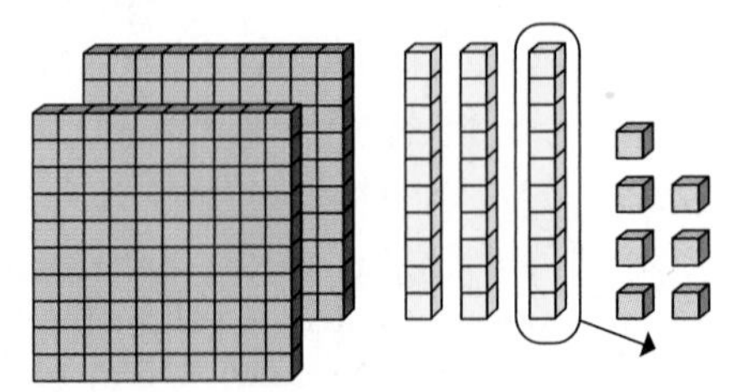

With no hundreds to take away, I now find the difference, 227, and record the subtraction:

$$\begin{array}{r} 245 \\ -\ 18 \\ \hline 227 \end{array}$$

Lyle's thinking: I start at the hundreds place and note that there are 0 hundreds to take away. I then take away 1 ten from the 4 tens, leaving 3 tens.

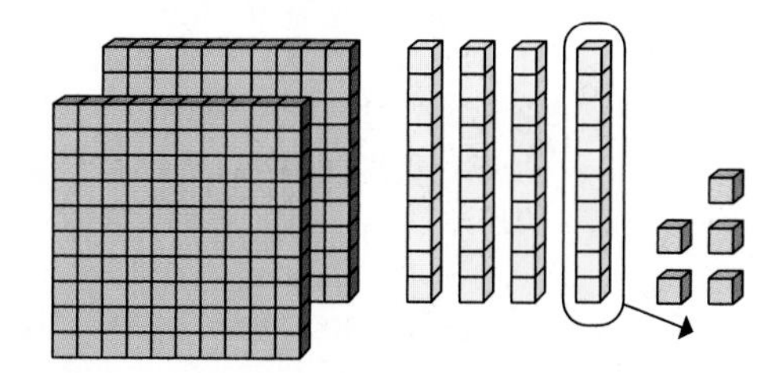

I now need to take away 8 ones but have only 5 ones. I take away the 5 ones, leaving 2 hundreds and 3 tens.

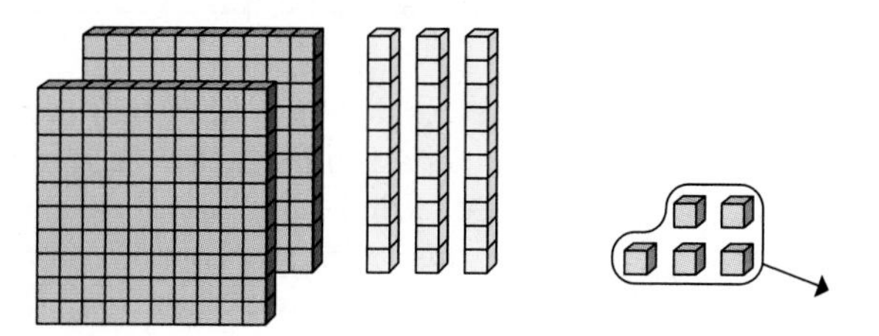

Now I trade 1 ten for 10 ones and take away 3 ones, leaving 2 hundreds, 2 tens, and 7 ones.

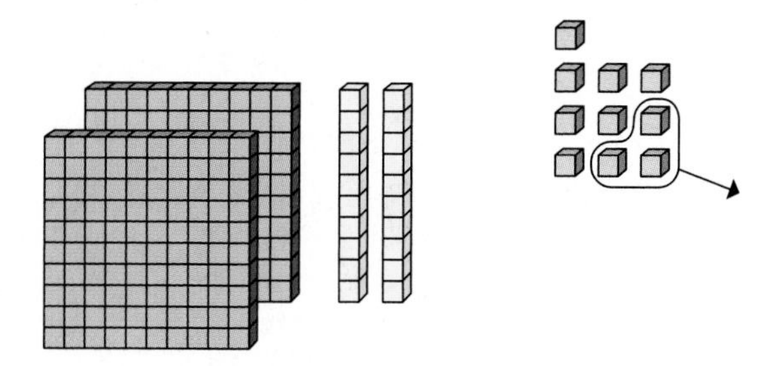

I find that the difference is 227 and record the subtraction with the equation $245 - 18 = 227$.

Your Turn

Practice: Find $137 - 48$ by using base-ten blocks. Show how you can use symbols to record your work.

Reflect: Explain how Winona's and Lyle's methods are alike and how they are different. ■

Example 3.15 demonstrates that subtraction can be done in different ways with base-ten blocks, which in turn suggests that different paper-and-pencil algorithms for subtraction exist.

Developing and Using Paper-and-Pencil Algorithms for Subtraction. Let's now look at two paper-and-pencil algorithms for subtraction. We use the calculation task modeled in Example 3.15 to develop these algorithms. The first algorithm is based on Lyle's work, whereby he subtracted the values of each place beginning on the left. In this algorithm, called the *expanded algorithm*, we start with the greatest number and repeatedly take away as much as is possible to do mentally before moving from left to right.

Expanded Algorithm for Subtraction

Think			Write
			245
There are no hundreds to subtract.			− 18
Subtract tens:	*245 − 10 = 235*	⟶	235
Take away 5 ones:	*235 − 5 = 230*	⟶	230
Take away 3 more ones:	*230 − 3 = 277*	⟶	227
			(the difference)

In the expanded algorithm, subtracting could begin at any place because the order of subtracting will not change the difference.

The second algorithm, based on Winona's work in Example 3.15, is called the *standard algorithm* and involves starting with the ones and proceeding to subtract, with regrouping, from right to left.

Standard Algorithm for Subtraction

Think	Write
	$\begin{array}{r} {}^{3\,1} \\ 2\not{4}5 \\ -\ 18 \\ \hline 227 \end{array}$
Regroup, subtract the ones.	(ones digit)
Subtract the tens.	(tens digit)
Subtract the hundreds.	(hundreds digit)

If not enough ones are available to subtract when we use the standard algorithm, we regroup 1 ten as 10 ones and then subtract the ones. If there are not enough tens to subtract, we regroup 1 hundred as 10 tens and subtract. We continue to regroup for as many digits as necessary in order to subtract.

Example 3.16 illustrates the use of these two algorithms. Vikki uses the expanded algorithm but records her work differently than Lyle did in Example 3.15.

Example 3.16 **Using the Expanded and Standard Algorithms for Subtraction**

Choose either the expanded or standard algorithm to find the difference 635 − 248.

Solution

Vikki's thinking:

$$
\begin{array}{rl}
635 & \\
-\ 200 & \text{I began by subtracting the hundreds} \\
\hline
435 & \\
-\ 30 & \text{Then I subtracted as many tens as I had.} \\
\hline
405 & \\
-\ 10 & \text{Then I subtracted 1 more ten.} \\
\hline
395 & \\
-\ 5 & \text{Next I subtracted as many ones as I had.} \\
\hline
390 & \\
-\ 3 & \text{Finally, I subtracted 3 more ones to get my answer.} \\
\hline
387 &
\end{array}
$$

Olav's thinking: I started with the ones and regrouped in each place as I went from right to left.

$$
\begin{array}{r}
{}^{5}\not{6}\,{}^{1}{}^{2}\not{3}\,{}^{1}5 \\
-\ 248 \\
\hline
387
\end{array}
$$

Your Turn

Practice: Use both the expanded and the standard algorithms to find the difference 1254 − 917.

Reflect: Explain each step in Olav's thinking. ■

The steps in the expanded and standard algorithms in Example 3.16 can be matched to the actions with the base-ten blocks to show that the steps make sense. Regardless of which algorithm is used for subtraction, it can be modeled with base-ten blocks. Mini-Investigation 3.8 reinforces the idea that various algorithms may be used for subtraction.

Talk about the procedure used in the algorithm and how it relates to the compensation method for mental calculations.

Mini-Investigation 3.8 **Making a Connection**

How would you explain the following algorithm, which someone used to complete the subtraction?

$$
\begin{array}{r}
425 \\
-\ 287 \\
\hline
225 \\
125 \\
+\ 13 \\
\hline
138
\end{array}
$$

(*Hint:* At one stage in the algorithm, the person subtracted 100, rather than 87.)

Using Properties of Whole Numbers to Verify Subtraction Algorithms. To verify that the algorithms presented for subtraction are logically correct, we use mathematical reasoning and the properties of whole numbers developed in Chapter 2. One such verification is given in Example 3.17.

Example 3.17 Verifying a Subtraction Algorithm

Subtract 245 − 18 by breaking apart numbers according to the place value of the digits and using properties of whole numbers. Give a reason for each step.

Solution

Statements	Reasons
$245 - 18$	
$= [2(100) + 4(10) + 5] - [1(10) + 8]$	Expanded notation
$= [2(100) + 4(10) + 5] + \{-[1(10) + 8]\}$	Definition of subtraction
$= [2(100) + 4(10) + 5] + [-1(10) + (-8)]$	Opposite of a sum is the sum of the opposites
$= 2(100) + [4(10) - 1(10)] + [5 + (-8)]$	Commutative and associative properties, definition of subtraction
$= 2(100) + (4 - 1)(10) + (5 - 8)$	Distributive property
$= 2(100) + 3(10) + (5 - 8)$	Basic fact
$= 2(100) + [(2 + 1)(10)] + (5 - 8)$	Substituting (2 + 1) for 3
$= 2(100) + [2(10) + 1(10)] + (5 - 8)$	Distributive property
$= 2(100) + [2(10) + 10] + (5 - 8)$	Basic fact
$= 2(100) + 2(10) + (10 + 5) - 8$	Associative property
$= 2(100) + 2(10) + 15 - 8$	Basic fact
$= 2(100) + 2(10) + 7$	Basic fact
$= 227.$	Standard form

Your Turn

Practice: Find the difference for 238 − 62 by breaking apart the number and using the properties of whole numbers. Give a reason for each step.

Reflect: What line in the above solution shows trading one ten for ten ones? Explain. ■

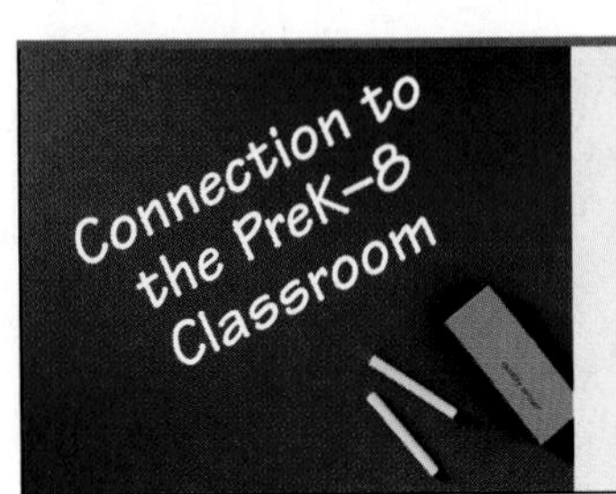

The elementary school mathematics program does not include proofs for addition and subtraction algorithms. However, teachers should understand that the properties of whole numbers provide the structure and proofs for the "shortcuts" invented in the elementary school classroom.

Calculator Techniques for Addition and Subtraction

Using a calculator to add or subtract when more than one step in a calculation is involved can sometimes pose a challenge. For example, consider the expression 172 − (63 − 39). This calculation can be completed in several ways, depending on the type of calculator used. We look at three of them.

When using a four-function calculator or a scientific calculator, many people fail to take advantage of the calculator's memory storage capability and instead complete the calculation by writing down numbers and reentering them. To use the memory storage, we first find $63 - 39$ because the parentheses indicate that the calculation within should be done first.

Key Sequence	**Display**
63 [−] 39 [=]	24

We then store the difference, 24, in the calculator's memory.

Key Sequence

[STO] [CE/C]

We are now ready to subtract 24 from 172 to complete the subtraction.

Key Sequence	**Display**
172 [−] [RCL] [=]	148

With a four-function calculator, we can use the parentheses keys to eliminate the need to store values.

Key Sequence	**Display**
172 [−]	172
[( ] 63 [−] 39 [ )]	24
[=]	148

With most graphing calculators, not only are parentheses keys available, but the screen display also shows the entire expression and the result of the calculation.

Key Sequence **Display**

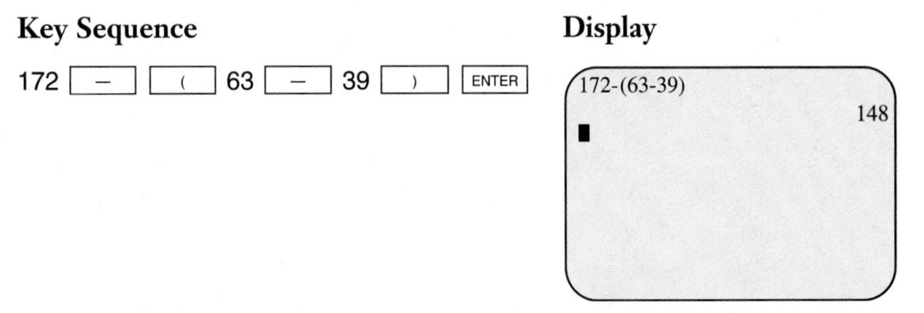

Example 3.18 further illustrates the use of these calculator techniques to add or subtract a combination of numbers.

Example 3.18 Problem Solving: Charity Funds

A charitable organization with an annual budget of \$150,000 recently paid out \$35,000 and \$28,000 to disaster victims. Use a calculator to determine how much money the organization has left for the rest of the year.

Solution

Find $150 - (35 + 28)$, where the numbers represent the number of thousands.

Alisha's method: I used a scientific calculator with a memory key.

Key Sequence	**Display**
35 [+] 28 [=]	63
[STO] [CE/C]	
150 [−] [RCL] [=]	87

The organization has $87,000 left to spend during the rest of the year.

Tranh's method: I used a graphing calculator.

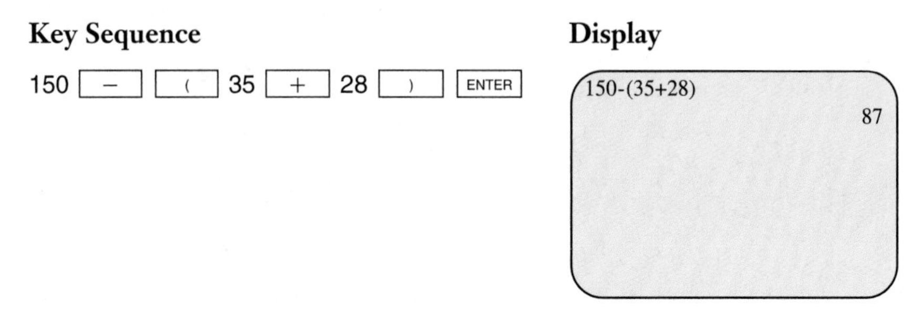

Thus $87,000 is still available for spending.

Elvira's method: I used a four-function calculator.

Key Sequence	**Display**
150 [−]	150
[(] 35 [+] 28 [)]	63
[=]	87

■ *Your Turn*

Practice: Use a calculator to solve the following problem.

One branch of a charitable organization paid out $28,000 of its $75,000 budget. The other branch has $36,000 left in its budget. How much money is now available in the organization's budget?

Reflect: What are the advantages and disadvantages of each of the three different types of calculators used in the example? ■

Example 3.19 illustrates a geometric problem that requires a decision about which numbers to add first in order to make efficient use of the calculator.

Example 3.19 Using a Calculator Efficiently

Use a four-function or scientific calculator to determine how much greater the perimeter of Figure A is than the perimeter of Figure B.

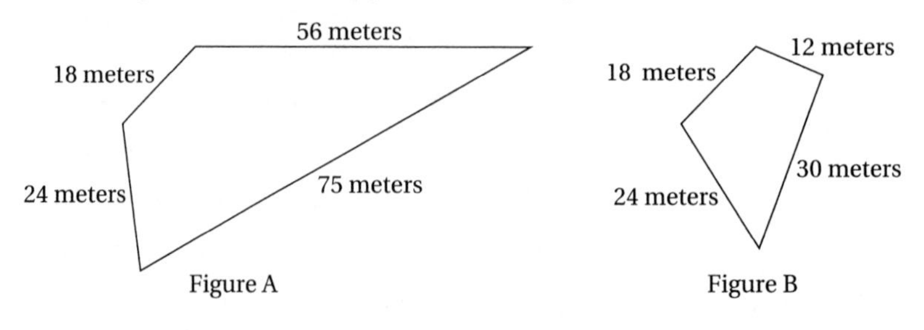

Solution

Calculate (24 + 18 + 56 + 75) − (18 + 12 + 30 + 24).

Key Sequence	**Display**
18 [+] 12 [+] 30 [+] 24 [=]	84
[STO] [CE/C]	0
24 [+] 18 [+] 56 [+] 75 [=]	173
[−] [RCL] [=]	89

The perimeter of Figure A is 89 m greater than the perimeter of Figure B.

Your Turn

Practice: Use a calculator to solve the following problem. How much greater is the perimeter of a triangle with sides 67 cm, 34 cm, and 158 cm than the perimeter of an isosceles triangle with base 27 cm and side 49 cm?

Reflect: Can another expression be used to calculate the answer in the example, and is using it on a calculator easier or more difficult than with paper and pencil? ■

When using a calculator to add and subtract a combination of numbers, we need to be aware of which operations the calculator does first and what grouping symbols, if any, the calculator uses. Mini-Investigation 3.9 encourages you to explore the *order of operation* characteristics of your calculator as they relate to addition and subtraction.

Write a short description of your calculator's interpretation of order of operations involving addition and subtraction. Do your classmates' calculators use the same interpretation?

Mini-Investigation 3.9 **The Technology Option**

How does your calculator interpret order of operations involving addition and subtraction? What grouping symbols are available on your calculator?

Solving a Multiple-Step Problem. The main value of computational algorithms—whether they are done mentally or with paper and pencil or whether they are built into a calculator or computer—is that they help solve problems efficiently. Addition and subtraction algorithms are used in solving problems when you must choose these operations in the process of obtaining a solution. Example 3.20 illustrates a problem of this type.

Example 3.20 **Problem Solving: Lunch Choices**

Select two different lunches from the menu on the following page that cost $5.00 or less.

Working toward a Solution

Understand the problem	*What does the situation involve?*	You want to select lunches from the menu board.
	What has to be determined?	What two different lunches will cost $5 or less?

Sandwiches and Other Meals	
Hot Dog	$1.75
Jumbo Hot Dog	$2.75
Hamburger	$2.50
Jumbo Hamburger	$3.50
Fries	$1.25
Onion rings	$1.50
Veggie plate	$2.75

Drinks	
Soft drink	
Small	$0.75
Medium	$1.25
Large	$1.50
Milk	$0.50
Juice	
Small	$1.25
Large	$1.75

	What are the key data and conditions?	The prices on the menu board are the data needed.
	What are some assumptions?	There are no restrictions on the number of items, but choosing a reasonable combination of items for lunch probably is best.
Develop a plan	*What strategies might be useful?*	Make a list and choose operations.
	Are there any subproblems?	First look for reasonable food combinations. Then find which cost $5 or less.
	Should the answer be estimated or calculated?	The cost for some combinations might be estimated, but others may need to be calculated to be sure that the total is $5 or less.
Implement the plan	*How should the strategies be used?*	List several combinations and their costs. Then use addition to find the total.

Jumbo hot dog: $2.75
Fries: $1.25
Large drink: $1.50
Total = $5.50

This combination costs too much, so try two more combinations.

Lunch 1:
Jumbo hot dog: $2.75
Fries: $1.25
Small drink: $0.75
Total = $4.75

		Lunch 2: Veggie plate: $2.75 Large drink: $1.50 Total = $4.25
	What is the answer?	Lunch 1 and Lunch 2 are two possible choices.
Look back	*Is the interpretation correct?*	Yes. The problem asks for any two choices that cost less than $5, so Lunch 1 and Lunch 2 work.
	Is the calculation correct?	Yes. The calculations are correct.
	Is the answer reasonable?	Yes. Both totals are less than $5 and both contain reasonable food combinations.
	Is there another way to solve the problem?	Yes. You could start at $5 and repeatedly subtract items.

Your Turn

Practice: What lunch combination comes close to $6.50?

Reflect: Did you use the solution to the example to help you solve the practice problem? Explain. ■

Problems and Exercises | *for Section 3.3*

A. Reinforcing Concepts and Practicing Skills

1. Draw a picture that shows how you would place base-ten blocks on your desk to start the task 328 + 567.
2. Draw a picture that shows how you would place base-ten blocks on your desk to start the task 484 − 195.
3. Name the property that justifies each step.

$[(2 \times 100) + (3 \times 10)] + [(1 \times 100) + (5 \times 10)]$

$= (2 \times 100) + [(3 \times 10) + (1 \times 100)] + (5 \times 10)$ _____ property

$= (2 \times 100) + [(1 \times 100) + (3 \times 10)] + (5 \times 10)$ _____ property

$= [(2 \times 100) + (1 \times 100)] + [(3 \times 10) + (5 \times 10)]$ _____ property

$= [(2 + 1) \times 100] + [(3 + 5) \times 10]$ _____ property

$= (3 \times 100) + (8 \times 10)$ Basic facts

$= 380.$ Expanded form to standard form

Tell whether each statement in Exercises 4–8 is correct or incorrect. If incorrect, rewrite it correctly.

4. $(75 \times 38) + (38 \times 75) = 2(75 \times 38)$.
5. $(145 - 67) - 28 = 145 + (67 - 28)$.
6. $6[2(3.5 + 8.6)] = 12(3.5) + 2(8.6)$.
7. $[(53 \times 32) - (32 \times 10)] = 32(53 - 10)$.
8. $5000 - (125 \times 8) \times 5 = [3(456 - 235)] \times 0$.

Identify and describe the error in each calculation in Exercises 9–12.

9. $\begin{array}{r} 84 \\ -\ 28 \\ \hline 64 \end{array}$

10. $\begin{array}{r} 167 \\ -\ 29 \\ \hline 148 \end{array}$

11. $\begin{array}{r} 2345 \\ -\ 238 \\ \hline 217 \end{array}$

12. $\begin{array}{r} 37 \\ +\ 75 \\ \hline 1012 \end{array}$

Evaluate each expression in Exercises 13–16 on a calculator.

13. $(28 + 75) + (134 - 12)$
14. $(910 - 635) - 129$
15. $789 - (23 + 45 + 345)$
16. $(546 - 232) - (124 - 76)$
17. Jorge solved Exercise 15 by using his graphing calculator, as shown on the next page. He knew that the result should be less than 500. What went wrong?

```
789 - 23 + 45 + 345
                  1156
▮
```

The following price list is for use in Exercises 18–21.

Tent	$179
Sleeping bags	85
Stove	68
Mats	23
Cooking utensils	47
Trail food	12
Lantern	115
Boots	137
Rain gear	89

18. What would the tent, boots, a sleeping bag, and a lantern cost?

19. How much more would a tent and a sleeping bag cost than boots and rain gear?

20. Using the prices given above, prepare two different lists of camping supplies whose total cost is close to but less than $200. Calculate the total for each list.

21. How much change would you get from a $1000 bill if you bought all the items listed above?

B. Deepening Understanding

22. Explain why the following approaches to subtraction make sense.

```
    9
  1 10 14        19 14
  2̸ 0̸ 4̸          2̸0̸4̸
 −   3 9       −  3 9
   1 6 5        1 6 5
```

23. Find two numbers whose sum is 973 and whose difference is 277.

Another way to develop the algorithm for addition is to write numbers in expanded form in Exercises 24–27.

174 = 1 hundred + 7 tens + 4 ones
+ 448 = 4 hundreds + 4 tens + 8 ones

= 5 hundreds + 11 tens + 12 ones
= 5 hundreds + 12 tens + 2 ones
(Regrouping 10 ones for 1 ten.)
= 6 hundreds + 2 tens + 2 ones
(Regrouping 10 tens for 1 hundred.)
= 622.

Use this method to complete each of the calculations in Exercises 24–27.

24. 234 + 125 **25.** 548 + 276

26. 838 + 627 **27.** 1256 + 867

An alternative format for using expanded notation is to use powers of 10, as in the example at the bottom of the page. Use this method to complete each of the calculations in Exercises 28–31.

28. 78 + 35 **29.** 136 + 123

30. 324 + 287 **31.** 1393 + 2736

To use the cashier's algorithm *for subtraction, start with the smaller number and add until you get the larger number. For example, to find the difference, $1300 − $625, start with $625 and count on.*

Say:	$625	$650	$700	$1200	$1300
Give:		25	50	500	100

The total is $675.

Use the cashier's algorithm to find each difference in Exercises 32–35.

32. $85 − $13 **33.** $135 − $84

34. $24 − $8 **35.** $55 − $32

The scratch algorithm, *or low-stress algorithm, for addition requires only adding 1-digit numbers. Whenever a 10 is reached, the units digit and the 1 ten are recorded separately. Start by adding down the ones column. Each time you get a sum of 10 or more, record a 1 for the 10 in the sum a half-space below and between the ones and tens columns and record the ones digit of the sum a half-space below and to the right of the ones column, as shown.*

Start in the ones column:

$\overset{2}{6}\ 8$

5_16_4 8 + 6 = 14. Record the 1 ten and the 4 ones.

$3\ 5_9$ 4 + 5 = 9. Record the 9 ones.

$+\ 7_16_5$ 9 + 6 = 15. Record the 1 ten and 5 ones.

5 Rewrite the 5 as the sum of the ones column and then add the tens made in the ones column, obtaining 2. Write 2 at the top of the tens column to show the 2 tens.

Example for Exercises 28–31

$174 = 1(10^2) + 7(10^1) + 4(10^0)$
$+\ 448 = 4(10^2) + 4(10^1) + 8(10^0)$

$= 5(10^2) + 11(10^1) + 12(10^0)$	
$= 5(10^2) + [(10 + 1)(10^1)] + [(10 + 2)(10^0)]$	Substitution
$= 5(10^2) + [(10(10^1) + (1)(10^1)] + [(10)(10^0) + (2)(10^0)]$	Distributive property
$= 5(10^2) + [(10^2 + 10^1) + 10^1 + (2)(10^0)]$	Simplifying
$= [5(10^2) + 10^2] + (10^1 + 10^1) + (2)(10^0)$	Associative property
$= 6(10^2) + (2)(10^1) + (2)(10^0)$	Distributive property and adding
$= 622.$	

Repeat the process in the tens place.

```
   2
  6₈8
 ₁5₃6
  3₆5
+₁7₃6
 2 3 5
```

Rewrite the 3 as the sum of the tens column, and then add the 1s written to the left of the tens column to obtain the tens that were regrouped to hundreds, which is 2.

With no hundreds column to be added, write 2 in the hundreds place in the sum.

Use the scratch method to find each sum in Exercises 36–39.

36. 34 + 56 + 88 + 94

37. 67 + 75 + 75 + 48

38. 234 + 654 + 768

39. 378 + 989 + 864 + 457

Explain each approach in Exercises 40 and 41 to finding 175 + 352. Can each approach be verified by using properties of whole numbers? Explain.

40.

```
  150        25
+ 350      + 2
  500        27    Sum: 527
```

41.

```
   75        127
+  52      + 400
  127        527
```

42. The following algorithm gives a correct answer. Explain why.

```
  4 8 8
+   6 5
  5¹5¹3
```

C. Reasoning and Problem Solving

43. The Cave-In Problem. A mining company is digging an 875-meter-long shaft. Each day workers are able to dig through 100 meters, but at night, owing to unstable ground, 50 meters of the shaft refills with rocks. At this rate, how many days will it take the workers to dig the 875-meter-long shaft?

44. Health Headlines. During a four-day winter weekend, the following numbers of students visited the university health center complaining of cold and flu symptoms.

Friday	245
Saturday	328
Sunday	197
Monday	276

Write a headline for the campus newspaper that tells approximately how many people visited the center during the four days.

Happy Trails Problem. *Use the following map to answer the questions in Exercises 45–47.*

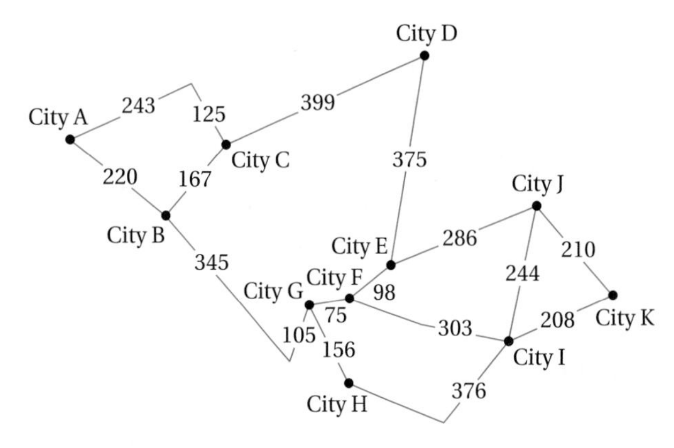

Note: All distances are in miles.

45. How many routes are possible from City A to City K with no backtracking? What is the mileage for each?

46. What route of travel gives a total distance of 1339 miles?

47. Write another problem for the same data.

48. Write a word problem for a real-world situation that can be solved by using the expression $(235 + 112) - 65$.

Planning a Purchase. *You want to buy a computer that costs $1349.*

49. How much more money do you need if you have saved $875?

50. If you can save $50 a month, in about how many more months will you be able to afford the computer?

D. Communicating and Connecting Ideas

51. Explain why the words *regrouping* and *trading* are used rather than *carrying* and *borrowing* in the paper-and-pencil algorithms for addition and subtraction.

52. Solve a problem by using an algorithm described in this section or one that you use for addition and subtraction, and provide a justification for the steps in the algorithm. Present your problem to another group to be solved and verified. Write a short paragraph comparing your group's work with that of the other group.

53. Historical Pathways. The *equal-additions algorithm* for subtraction was taught in many classrooms in the United States in the early part of the twentieth century. Here's how it works to find the difference, 204 − 139.

First add 10 to the top number and 10 to the bottom number, which will not change the difference.

```
  2 0¹4   Add 10 to the 4 ones in the top number.
- 1⁴3̸9   Add 10 to the 3 tens in the bottom number.
      5   Subtract the ones.
```

Then add 100 to the top number and 100 to the bottom number, which will not change the difference.

```
            Add 100 (10 tens) to the 0 in the tens
  2¹0¹4     column of the top number.
    2 4
- 1̸ 3̸9     Add 100 to the 1 in the hundreds place
-    65     in the bottom number.
            Subtract the 4 from the 10 tens.
```

Create three subtraction calculations and use the equal-addition algorithm to find each difference.

54. Making Connections. Recall that addition and subtraction are *inverse operations.* Addition undoes subtraction and subtraction undoes addition. Use examples to explain how to use this idea in checking subtraction.

Find the smallest and the largest solutions possible for the following problems.

55. Smallest

```
   99
   88
 + □□
  2□□
```

56. Largest

```
   99
   88
 + □□
  2□□
```

Section 3.4 ALGORITHMS FOR MULTIPLICATION AND DIVISION

- Developing Algorithms for Multiplication
- Calculator Techniques for Multiplication
- Developing Algorithms for Division
- Using a Calculator to Evaluate Multiplication and Division Expressions

In this section, we look at algorithms for multiplication and division of whole numbers. We begin by using models to help explain these algorithms and then use the properties of whole numbers to justify the algorithms.

Developing Algorithms for Multiplication

As with addition and subtraction algorithms, models provide a physical basis for explaining algorithms for multiplication. The models used here include base-ten blocks and pictorial models that represent multiplication as finding the area of a rectangle. Using the processes suggested by the models, we develop the related paper-and-pencil algorithms for multiplication. Finally, we use mathematical reasoning along with properties of whole numbers to verify that the steps in a multiplication algorithm are logically correct.

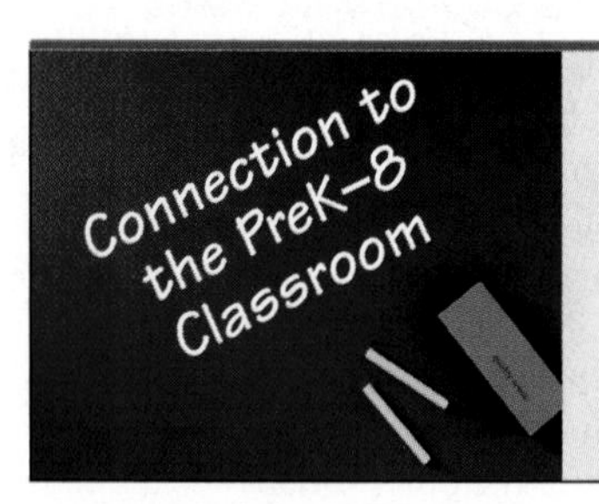

When working with children, teachers need to remember that the purpose of using base-ten blocks is not to teach an algorithm for using the blocks. Rather, the purpose is to use the blocks to help the child make sense of the steps to be done by using symbols—that is, to make sense of the paper-and-pencil algorithm. Thus the teacher needs to use the blocks to show how to find small products and move to other models or numbers to show how to find larger products.

One way of modeling the procedures for finding products is to use the *area of a rectangle interpretation* of multiplication developed in Chapter 2. Recall that the factors are the length and width of the rectangle and the product is the area of the rectangle. Example 3.21 shows how the product of two 2-digit numbers can be obtained by using base-ten blocks arranged to form a rectangle or by drawing a rectangle on graph paper.

Example 3.21 Using the Area Model for Multiplication

Use an area interpretation for multiplication to show 13×24.

Solution

We begin by using base-ten blocks to show a rectangle width and length of 13 and 24.

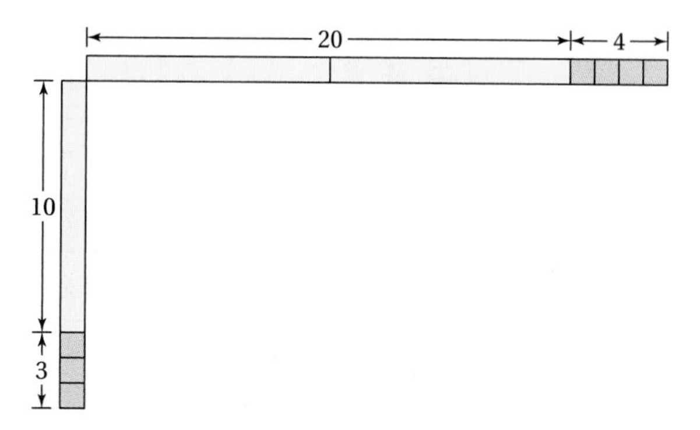

Next, we make the rectangle by using as few base-ten blocks as possible:

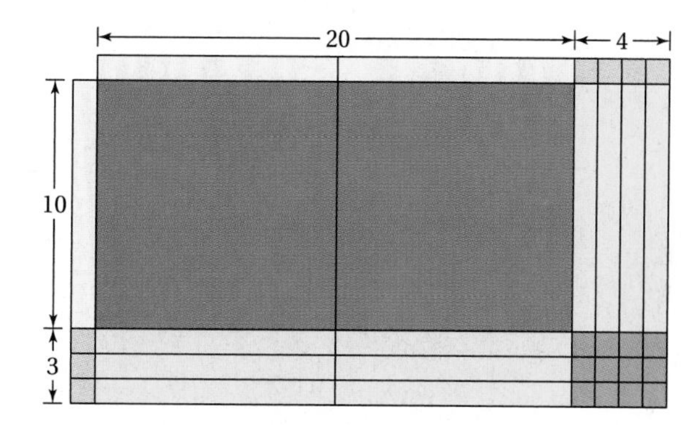

We now count the blocks to find the area of the rectangle, or the product. There are 2 hundreds, 10 tens, and 12 ones. Regrouping the ones and tens gives 3 hundreds, 1 ten, and 2 ones, so $13 \times 24 = 312$. The area is 312 square units.

Your Turn

Practice: Use base-ten blocks (or draw a picture) and an area interpretation for multiplication to obtain the product 12×21.

Reflect: How is the model used in the example different from using base-ten blocks to show 13 groups of 24 cubes each and regrouping to find the product 13×24? ■

In the area model for 13×24 in Example 3.21, the product 312 is made up of the areas 3×4, 3×20, 10×4, and 10×20. Because these products are combined to produce the final product, 312, we call them **partial products.** The idea of partial products is useful in developing paper-and-pencil algorithms for multiplication.

Developing and Using Paper-and-Pencil Algorithms for Multiplication. We now use the multiplication calculation modeled in Example 3.21 to examine two paper-and-pencil algorithms for multiplication. Partial products play an important role in each.

The first algorithm based on that model involves breaking apart the numbers according to the place value of each digit and multiplying each digit according to its place value to obtain the partial products. In this algorithm, called the *expanded algorithm*, the partial products are added to find the final product.

Expanded Algorithm for Multiplication

Think		Write
		24
		× 13
Multiply 3 × 4:	⟶	12
Multiply 3 × 20:	⟶	60
Multiply 10 × 4:	⟶	40
Multiply 10 × 20:	⟶	200
Add the partial products:	⟶	312

The second algorithm, called the *standard algorithm*, involves forming only two partial products.

Standard Algorithm for Multiplication

Think		Write
		1
		24
		× 13
Multiply 3 × 24:	⟶	72
Multiply 10 × 24:	⟶	240
Add the partial products:	⟶	312

In this case, the first factor is multiplied by the ones digit of the second factor and the numbers are regrouped to form the first partial product. Then the first factor is multiplied by the tens digit of the second factor.

Example 3.22 further illustrates the use of these two algorithms.

Example 3.22

Using the Expanded and Standard Algorithms for Multiplication

Choose either the expanded or standard algorithm to calculate the product 6 × 345.

Solution

Caleb's thinking: First, I multiplied the ones by 6, then the tens, and then the hundreds. Then, I added all the partial products.

$$\begin{array}{r} 345 \\ \times \quad 6 \\ \hline 30 \\ 240 \\ 1800 \\ \hline 2070 \end{array}$$

Makenzie's thinking: First, I multiplied the ones by 6 and regrouped. Then, I multiplied the tens by 6, added the extra tens, and regrouped. Finally, I multiplied the hundreds and added the extra hundreds.

$$\begin{array}{r} {}^{2\,3} \\ 345 \\ \times \quad 6 \\ \hline 2070 \end{array}$$

Your Turn

Practice: Calculate 34×26 by using the

a. expanded algorithm for multiplication.
b. standard algorithm.

Reflect: Explain how the standard algorithm is a shortcut for the expanded one. ■

Using Properties of Whole Numbers to Verify Multiplication Algorithms. Example 3.23 shows how the properties of whole numbers and properties of the base-ten numeration system can be used to verify a procedure for finding the product of two numbers. Example 3.23 uses the same multiplication task as in Example 3.21 and in the demonstrations of the expanded and standard algorithms. This common calculation shows the connections between the use of models, an algorithm for multiplying two numbers, and the reasoning involved in verifying the procedure.

Example 3.23 Connecting Model, Algorithm, and Reasoning.

Multiply 24×13 by breaking apart numbers according to the place value of the digits and by using properties of whole numbers. Give a reason for each step.

Solution

Statements	Reasons
$24 \times 13 = (20 + 4) \times (10 + 3)$	Expanded notation
$= (20)(10) + (20)(3) + (4)(10) + (4)(3)$	Distributive property, twice
$= 200 + 60 + 40 + 12$	Associative property, basic fact
$= 200 + 100 + 10 + 2$	Associative property, basic fact
$= 300 + 10 + 2$	Associative property, basic fact
$= 312$	Expanded to standard notation

Your Turn

Practice: Multiply 12×18 by breaking apart numbers according to the place value of the digits and by using properties of whole numbers. Give a reason for each step.

Reflect: How do you know that you would get the same product for 13×24? ■

Note that the products in the second step of the verification in Example 3.23—10×20, 10×4, 3×20, and 3×4—correspond to the partial products in the area model in Example 3.21 and in the box describing the expanded algorithm.

Calculator Techniques for Multiplication

The use of a calculator to multiply two numbers is straightforward and need not be reviewed here. Of more importance is the efficient use of a calculator. Mini-Investigation 3.10 asks you to explore the use of your calculator to determine products that involve the repeated use of one factor.

Write a brief description of the techniques you used and compare your techniques with those of your classmates.

Mini-Investigation 3.10 **Technology Option**

What techniques can you use with your calculator to multiply each of several numbers by one factor, without repeatedly reentering the factor?

Depending on the type of calculator, you have several options for completing the task described in Mini-Investigation 3.10. The following problem demonstrates techniques for multiplying by a factor repeatedly with three different types of calculator.

> Suppose that Tina earns $245 each week. How can she quickly determine her earnings for 4, 13, 26, and 52 weeks?

Estimation Application

The Best Deal

Estimate how much money you would save by purchasing a used car with cash rather than through the dealer's loan program.

Total Cost of the Deals

Cash: $6175

Loan: $198 per month for 36 months

Find a way to check your estimate.

To initiate repeated multiplication by 245 with a scientific calculator, begin by entering the constant multiplier 245 followed by the multiplication key. Then enter a number of weeks and press the equals key, repeating the number of weeks entry for as many different values as you want to use.

Key Sequence	Display
245 [×] 4 [=]	980
13 [=]	3185
26 [=]	6370
52 [=]	12740

Note: Do not clear the result before entering a new value for the number of weeks.

To use some four-function calculators, enter the first pair of factors with the constant factor 245 last and press the equals key. For additional values for the number of weeks, simply enter the number and press the equals key.

Key Sequence	Display
4 [×] 245 [=]	980
13 [=]	3185
26 [=]	6370
52 [=]	12740

To use a graphing calculator, use its list feature and enter the data for the number of weeks in list L1.

Key Sequence	**Display**
LIST	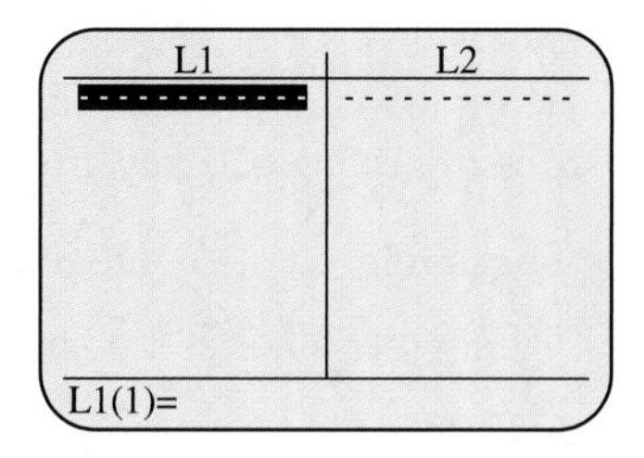
4 ENTER 13 ENTER 26 ENTER 52 ENTER	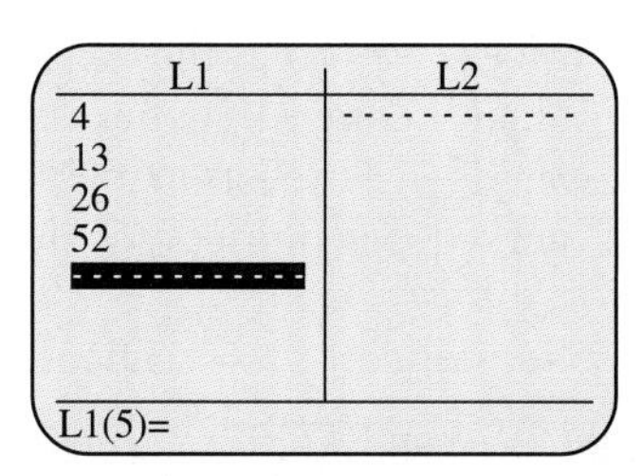

Move to the top of list L2, enter the formula L1 × 245, and press ENTER. The products are calculated and shown in list L2.

Key Sequence	**Display**
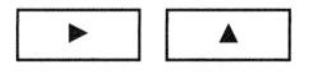	L1: 4, 13, 26, 52; L2: (empty); L2=
2nd [STAT] 1 ENTER × 245 ENTER	L1: 4, 13, 26, 52; L2: 980, 3185, 6370, 12740; L2(1)=980

Example 3.24 further illustrates the use of a calculator to multiply by a repeated factor and provides an opportunity to try it on your particular calculator.

Example 3.24 Problem Solving: Marked-Down Prices

Your favorite department store announces a sale in which the price of all merchandise will be marked down by 15%. Use a calculator to find 85% of the prices of three items you've been planning to buy whose regular prices are $2.95, $12.50, and $25.89.

Solution

Marion's thinking: I used a scientific calculator and began by entering the constant factor, .85, and pressing the multiplication key. Then I entered each price, pressing the equals key after each one.

Key Sequence	Display
.85 [×] 2.95 [=]	2.5075
12.5 [=]	10.625
25.89 [=]	22.0065

The displays showed the marked-down prices, which I can round to the nearest cent. The answers are $2.51, $10.63, and $22.01.

Tonya's thinking: I used a four-function calculator. To begin, I entered the first price, 2.95, times the constant factor, 0.85, and then pressed the equals key. Then, I entered the next two prices and pressed the equals key after each. I could have used as many prices as I wanted. The displays showed the marked-down prices for the three that I entered.

Key Sequence	Display
2.95 [×] .85 [=]	2.5075
12.50 [=]	10.625
25.89 [=]	22.0065

When I round the numbers in the displays to the nearest cent, the answers are $2.51, $10.63, and $22.01.

Carol Ann's thinking: Using a graphing calculator, I first entered the data in list L1. Then I moved to the very top of list L2, entered the formula L1 × .85, and pressed Enter. The display for list L2 showed the products for .85 times each price in list L1.

Key Sequence **Display**

[LIST]

2.95 [ENTER]

12.50 [ENTER]

25.89 [ENTER]

L1	L2
2.95	------------
12.5	
25.89	

L1(4)=

[►] [▲]

[2nd] [STAT] 1 [ENTER] [×] .85 [ENTER]

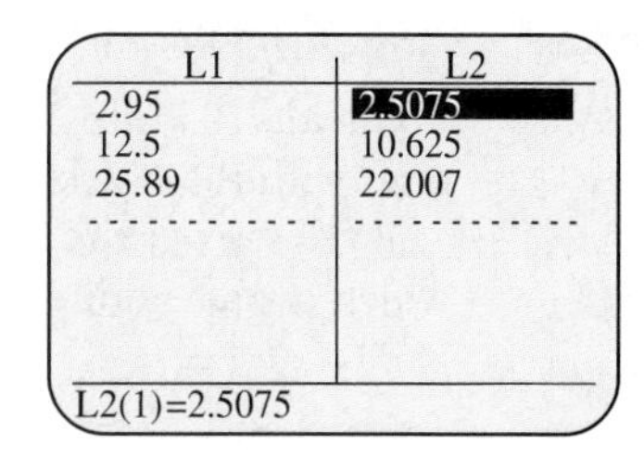

To the nearest cent, the marked-down prices are \$2.51, \$10.63, and \$22.01.

Your Turn

Practice: To find the new prices for items after a 10% price increase, multiply each item by 1.10 and round the results to the nearest cent. What would be the new prices for items now costing \$24.95, \$35.75, and \$120.49?

Reflect: How do the calculator techniques used in this example help save time? ■

Developing Algorithms for Division

In this section, we look at models of procedures for dividing whole numbers and then consider related paper-and-pencil algorithms. Although we used properties and mathematical reasoning to justify algorithms for addition, subtraction, and multiplication earlier in this chapter, justification of division is somewhat more complex and is not included in this book. Mini-Investigation 3.11 is intended to help you assess your understanding of the division algorithm.

Write a paragraph describing your group's response to the question.

Mini-Investigation 3.11 **Making a Connection**

What steps do you think are used in carrying out the following division task?

$$\begin{array}{r} 33 \text{ R } 4 \\ 8\overline{)268} \\ \underline{24} \\ 28 \\ \underline{24} \\ 4 \end{array}$$

Using Models as a Foundation for Division Algorithms. In Chapter 2 we examined several views of division. Two of those—the **repeated subtraction interpretation** and the **sharing interpretation**—can be used in modeling the procedures in a division algorithm. In the work with models in this section, we will focus only on the sharing interpretation for division.

In Example 3.25, we use base-ten blocks to model the sharing interpretation for division. Here we start with 105 blocks and 15 groups. We must divide, or share, the blocks evenly into these 15 groups.

Example 3.25 **Using the Base-Ten Blocks to Model Dividing**

Find 105 ÷ 15 by using base-ten blocks and a sharing interpretation of division. Trade when necessary to obtain needed blocks.

Solution

To divide 105 into 15 subsets, we again start by showing 1 hundred, 0 tens, and 5 ones with the blocks.

We start at the hundreds. With only 1 hundred to divide into 15 groups—which can't be done—we trade 1 hundred for 10 tens.

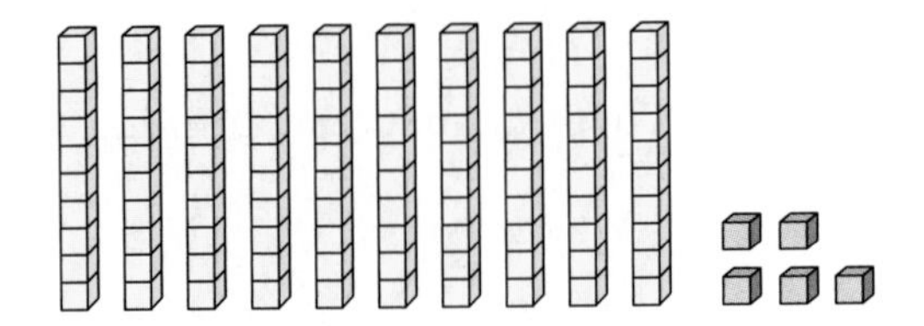

With no other tens, we still can't divide because we don't have enough tens (10) to divide them into 15 groups. So now we trade 10 tens for 100 ones. With the 5 other ones we already had, we now have 105 ones, which we can divide into 15 equal groups.

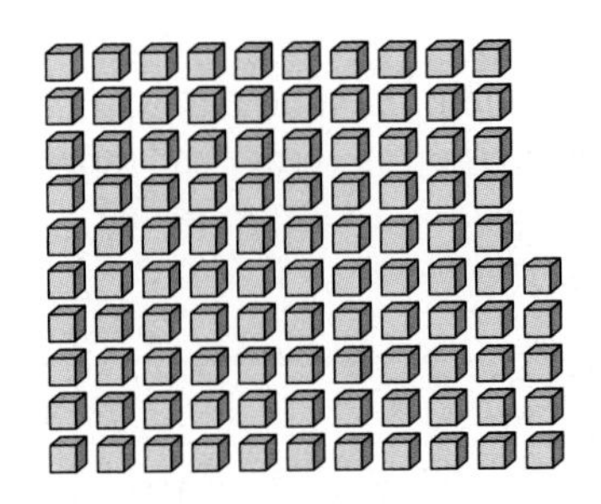

Seven ones can go into each of the groups, so $105 \div 15 = 7$.

Your Turn

Practice: Divide $647 \div 3$ by using base-ten blocks and a sharing interpretation of division. Trade when necessary to obtain needed blocks.

Reflect: What would happen in the practice problem if you began by dividing the ones, then tens, and so on? ■

The language that teachers use must be consistent with the interpretation of division being used. For the problem $126 \div 3$, the correct language to use when associating division with the repeated subtraction interpretation is, "How many 3s are in 126?" For the sharing interpretation, the correct language is, "If we divide (or share) 126 into three groups, how many should be in each group?" or "How can 126 be divided into three equal groups?"

Developing and Using Paper-and-Pencil Algorithms for Division. Let's now look at two paper-and-pencil algorithms for division.

Expanded Algorithm for Division

Think	Write	Think
	$15\overline{)105}$ with quotient 7	*The number of times the divisor is subtracted, 7, is the quotient.*
Subtract the divisor as many times as possible.	− 15	
	90	
	− 15	
	75	
	− 15	
	60	
	− 15	
	45	
	− 15	
	30	
	− 15	
	15	
	− 15	
	0	

Although this algorithm is simple to use, it can be quite inefficient. For example, imagine how many times you would have to subtract to divide 9,437 ÷ 13. One way to improve the efficiency of this algorithm is to subtract a greater number at least part of the time. For the 105 ÷ 15 in the expanded algorithm box, we know that two 15s make 30 and three 30s make 90, so we could first subtract 90 and then 15.

Example 3.26 further illustrates this idea.

Example 3.26 Using the Expanded Algorithm for Division

Use the expanded division algorithm to find 146 ÷ 18.

Solution

Bennie's thinking: I know that three groups of 18 is 54, so I subtract 54 as many times as I can, then I subtract two groups of 18, and have 2 left over.

$18\overline{)146}$	
− 54	three groups of 18
92	
− 54	three groups of 18
38	
− 36	two groups of 18
2	

Altogether I subtract 8 groups of 18, so my answer is 146 ÷ 18 = 8R2.

Luanne's thinking: I know that 10 × 18 = 180, so 10 is too large. I guess that I can take away six groups of 18 and then discover that I can take away two more groups for a total of eight groups.

$$\begin{array}{rl} 18\overline{)146} & \\ -\ 108 & \text{six groups of 18} \\ \hline 38 & \\ -\ 36 & \text{two groups of 18} \\ \hline 2 & \end{array}$$

Your Turn

Practice: Use the repeated subtraction approach to division with symbols only to find

a. 234 ÷ 17.

b. 4563 ÷ 36.

Reflect: Which student's thinking in the example problem is the most efficient? Why? ■

Each of the algorithms in Example 3.26 features the repeated subtraction approach to find the quotient, and each can produce correct results. Obviously, Luanne's solution uses the fewest steps. Also note how Bennie and Luanne used multiplication and estimation. Both tried to estimate, "How many 18s are in 146?" Their strategy was to find an estimate as close as possible to 146 without going over it.

The second algorithm for division—the *standard algorithm*—has several steps and is based on the sharing interpretation of division. Just as we used base-ten blocks to model the sharing interpretation of division in Example 3.25, we can show how they relate to symbols in each step of this box.

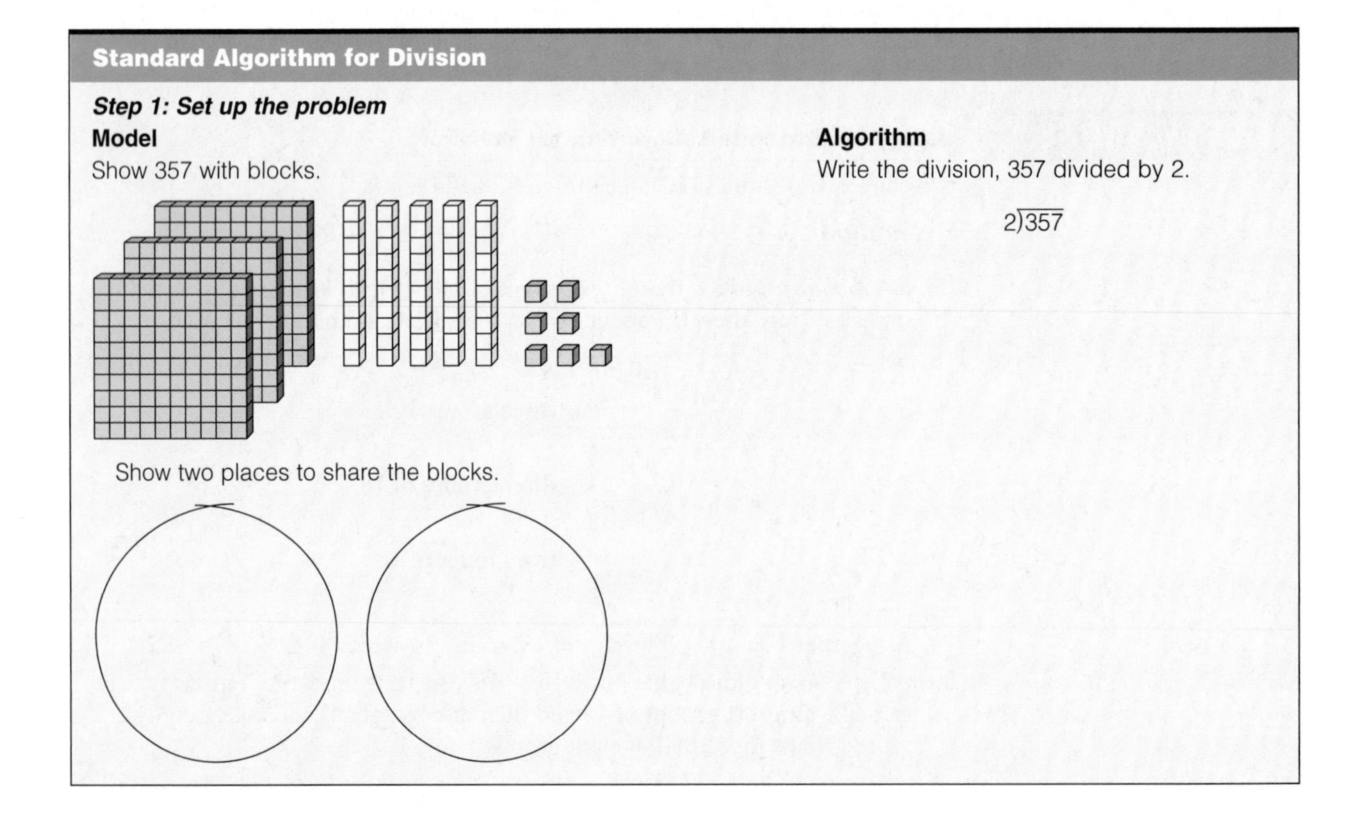

Step 2: Decide where to start

Model

The hundreds can be divided.

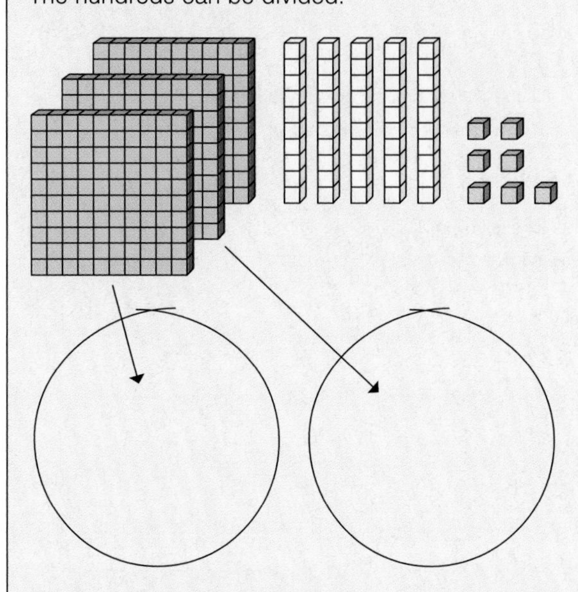

Algorithm

Start at the left, and choose the first place that can be divided. The 3 in the hundreds place can be divided by 2.

$2\overline{)357}$

Step 3: Divide the hundreds

Model

Divide the hundreds into two groups. One hundred group is left over, which cannot be divided into the two groups.

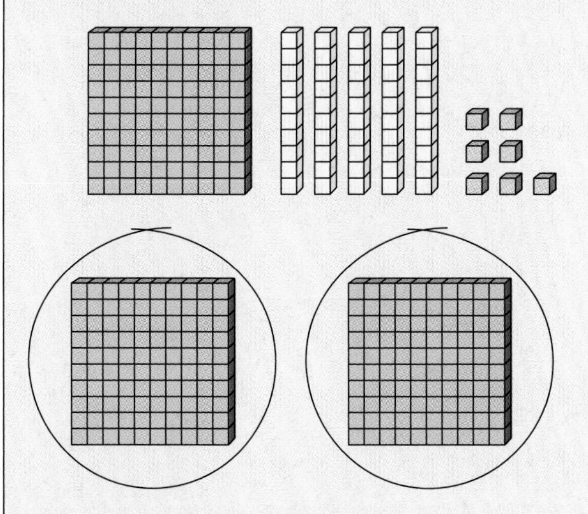

Algorithm

First divide in the hundreds place, writing the quotient digit above that place. Then multiply, subtract, and compare.

Number of hundreds in each group ↓

$$\begin{array}{r} 1 \\ 2\overline{)357} \\ -2 \\ \hline 1 \end{array}$$

Number of groups → 2; ← Number of hundreds shared (−2)

↑ Number of hundreds left over (should be less than the divisor 2)

Step 4: Divide the tens

Model

Trade the one leftover hundred for 10 tens to make 15 tens. Divide the tens, placing 7 in each group and 1 ten left over.

Algorithm

Divide in the tens place, writing the quotient digit above that place. Then multiply, subtract, and compare.

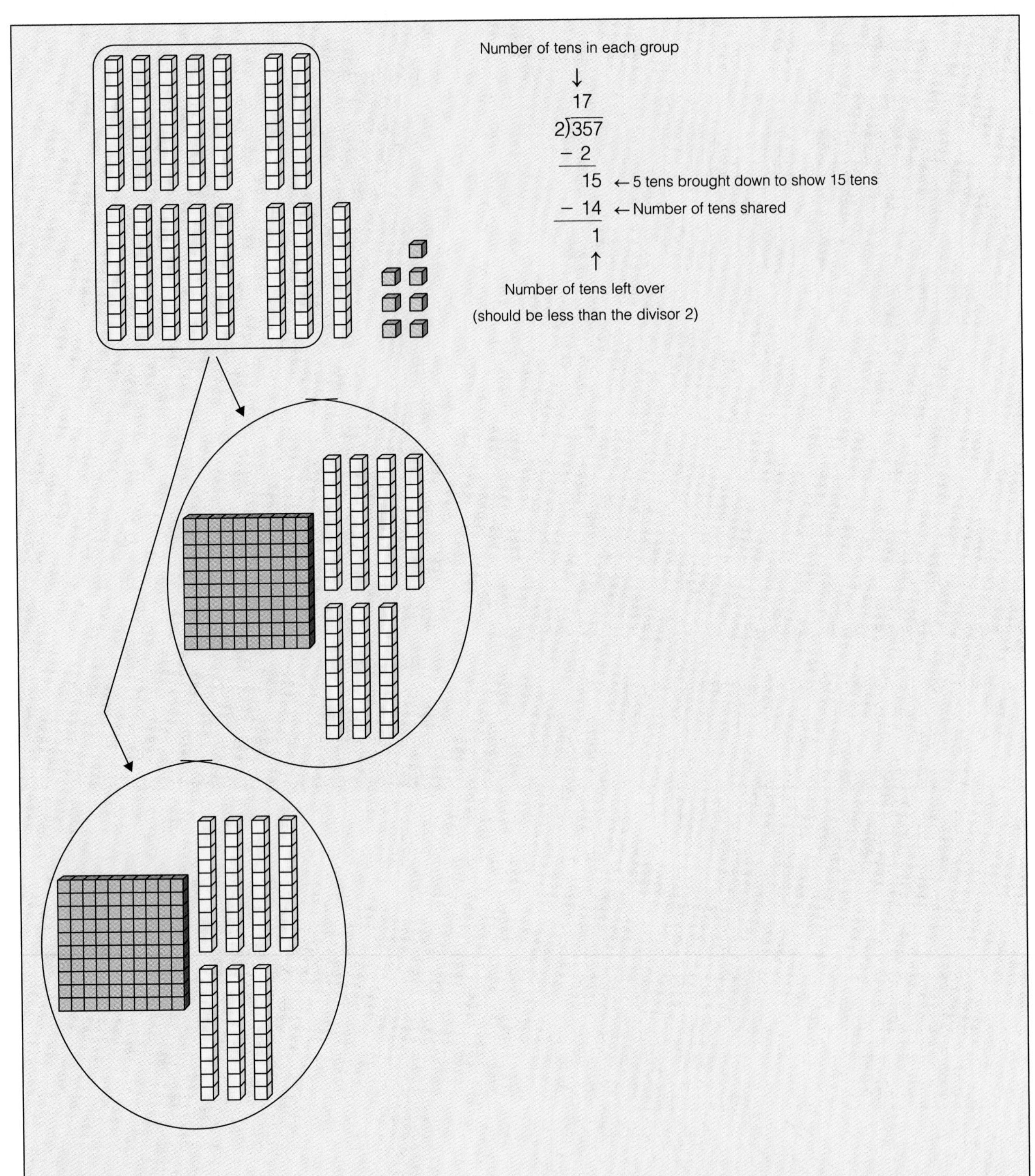

Step 5: Divide the ones

Model

Trade the leftover 1 ten for 10 ones to make 17 ones. Divide the ones, placing 8 in each group and 1 one left over.

Algorithm

Divide in the ones place, writing the quotient digit above that place. Then multiply, subtract, and compare.

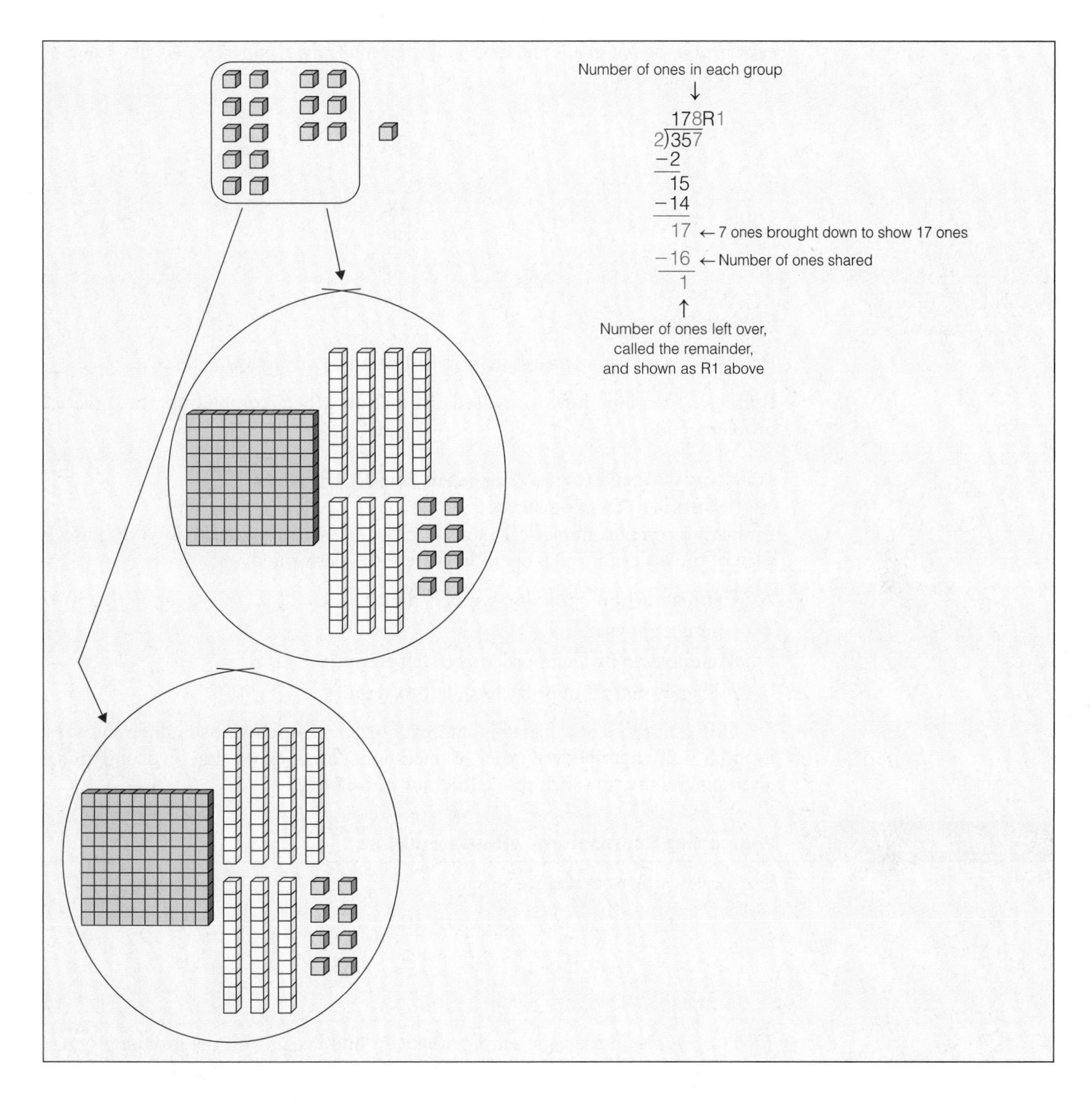

Example 3.27 further illustrates the use of the standard division algorithm.

Example 3.27 Using the Standard Division Algorithm

Use the standard division algorithm to calculate 408 ÷ 12.

Solution

We begin by deciding where to start. Four hundreds can't be divided into 12 groups, so we start with the tens and write the first digit in the quotient in the tens place. We estimate that 40 tens can be divided into 12 groups with three in each group. So we divide the tens, and then multiply, compare, and bring down to divide the ones.

$$\begin{array}{r} 34 \\ 12\overline{)408} \\ -36 \\ \hline 48 \\ -48 \\ \hline 0 \end{array}$$

Your Turn

Practice: Use the standard division algorithm to divide 897 ÷ 39.

Reflect: Explain how you used estimation when completing the practice problem. ■

Using a Calculator to Evaluate Multiplication and Division Expressions

Evaluating complex numerical expressions on a calculator usually involves knowledge of the order in which operations are to be performed.

- Compute within grouping symbols first.
- Compute powers.
- Multiply and divide in order from left to right.
- Add and subtract in order from left to right.

Different types of calculators interpret **order of operations** in different ways. Example 3.28 describes the order of operations for an expression involving multiplication and division with three different calculators.

Example 3.28 **Evaluating Expressions with Calculators**

Use a calculator to evaluate

$$\frac{32 + 18}{5} + (56 \times 2).$$

Solution

Jefferson's method: I used a scientific calculator and began with the grouping symbols. First, I evaluated the terms within parentheses and then the numerator of the fraction because the fraction bar is also a grouping symbol.

Key Sequence	Display
56 [×] 2 [=]	112
[STO] [CE/C]	
32 [+] 18 [=]	50
[÷] 5 [=]	10
[+] [RCL] [=]	122

Natasha's method: I used a four-function calculator and the following keystrokes.

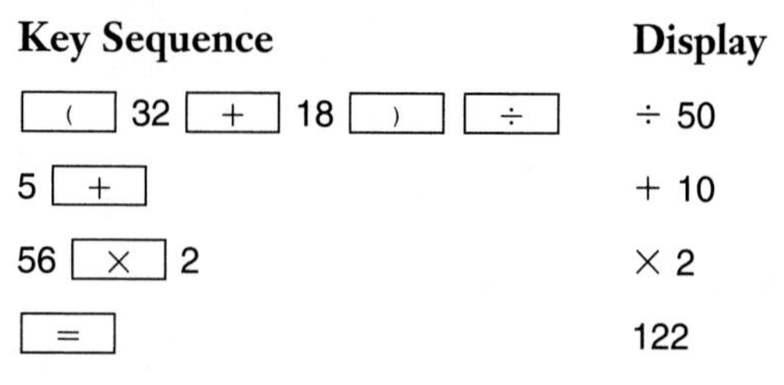

Cal's method: I used a graphing calculator.

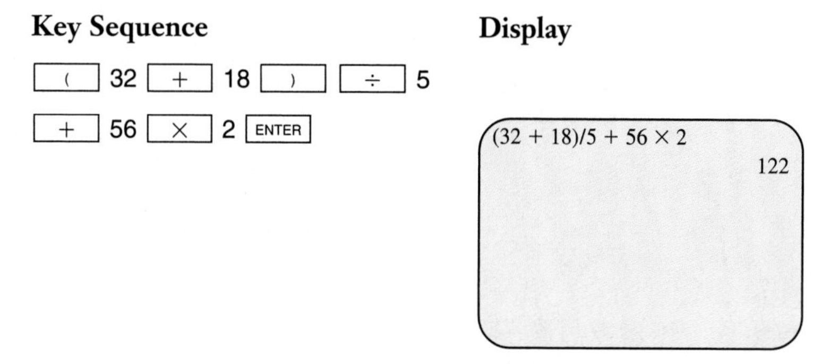

Your Turn

Practice: Use a calculator to evaluate

$$75 - \frac{124 - 64}{4}.$$

Reflect: Can other calculator techniques be used to evaluate the expression in the example? If so, show the keying sequence and the displayed results. ■

Some calculators are designed to give both the quotient and the remainder for a whole-number division. Mini-Investigation 3.12 challenges you to obtain the quotient and remainder without using this feature.

Write a paragraph describing the procedure you used to find the remainder.

Mini-Investigation 3.12 **Technology Option**

How can you use a calculator to obtain the quotient and remainder for $476 \div 9$ without using integer division commands?

Solving Problems Using Multiplication. Some problems involve both multiplication and division—as well as addition and subtraction—and so are solved by using the problem-solving strategy *choose the operations.* Once you have chosen the operations needed to solve a problem, you must then make two decisions: (1) whether an exact answer or an estimate is required and (2) the method of calculation—mental calculation, paper and pencil, or a calculator. Although the calculations needed in most situations that call for an estimate can be done mentally, sometimes they are sufficiently complex or so many calculations are called for that a calculator will come in handy. Example 3.29 uses multiplication and division and calls for an estimate. A calculator is used to solve several subproblems.

Example 3.29

Problem Solving: Time to Eat

About how many hours or days of your life to age 50 will you spend eating?

Working toward a Solution

Understand the problem	*What does the situation involve?*	Finding the time spent eating.
	What has to be determined?	The total number of hours spent eating in 50 years.
	What are the key data and conditions?	The total time is 50 years, and the problem statement asks for the number of hours and days.
	What are some assumptions?	You must assume a certain number of hours of eating each day and use that in all calculations.
Develop a plan	*What strategies might be useful?*	Choose an operation.
	Are there any subproblems?	Yes. First, find the time spent eating in 1 day. Then, extend that to 1 year and then to 50 years.
	Should the answer be estimated or calculated?	An estimate is sufficient because you are to find *about* how many hours per day.
	What method of calculation should be used?	A calculator probably makes sense with the large numbers likely to be involved.
Implement the plan	*How should the strategies be used?*	Estimate the amount of time spent eating in 1 day. Then use multiplication to find the total time for a year and convert it to hours. Then use multiplication to find the total for 50 years. Then find the number of days. An estimate for the time eating each day is 1 hour, 15 minutes, or 75 minutes. For 1 year, 75 min × 365 days = 27,375 min. Dividing by 60 minutes in an hour gives 456.25 hours in 1 year. For 50 years, 50 yr × 456.25 hr = 22,812.5 hr. Dividing by 24 hours in one day gives approximately 951 days.

	What is the answer?	You will spend about 22,813 hours, or 951 days, eating to age 50.
Look back	*Is the interpretation correct?*	Yes. The question wants to know the total for 50 years.
	Is the calculation correct?	Yes. The calculations are correct.
	Is the answer reasonable?	Yes. 951 days is about $2\frac{1}{2}$ years out of 50 years for a ratio of about 1 to 25; 75 minutes a day is about 1 hour out of 24 for a ratio of 1 to 24. These ratios are fairly close, so the answer seems reasonable.
	Is there another way to solve the problem?	Yes. You could have expressed the total time eating each day in hours and then multiplied by 365 to get the number of hours for 1 year. You also could have expressed the amount of time spent eating each day as a fractional part of a day and used that number as the basis for further calculations.

Your Turn

Practice: First estimate and then use your calculator to determine exactly how many seconds you have been alive.

Reflect: What subproblems did you solve in the practice problem, and how did they help you arrive at the final answer? ■

Problems and Exercises | *for Section 3.4*

A. Reinforcing Concepts and Practicing Skills

1. Draw a picture that shows how to model 3 × 43 using base-ten blocks. Give the product.
2. Draw a picture that shows how to model 2 × 36 by using base-ten blocks and the area of a rectangle interpretation for multiplication.
3. Consider the following calculation.

$$\begin{array}{r} 46 \\ \times\ 38 \\ \hline 368 \\ 1380 \\ \hline 1748 \end{array}$$

   a. Which numbers are the partial products?
   b. Why does the second partial product end in 0?
   c. Give a n estimate for the product.
4. Draw a picture of base-ten blocks and use the area interpretation for multiplication to show 34 × 28. Then complete the calculation by using a paper-and-pencil algorithm. Show the partial products in your picture.
5. What multiplication calculation is shown by the following figure? Write an equation representing what was done with the blocks.

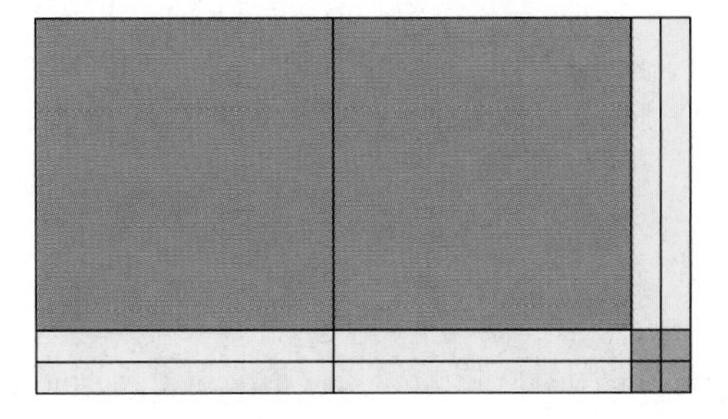

6. Explain in writing two ways to use addition to find 6 × 42.
7. Use the expanded algorithm for multiplication to find four partial products for 18 × 34. Then give the final product.
8. A teacher looked at the two calculations below and said, "When you show all partial products, you can multiply in any order. You just need to pay attention to place value." Use the calculations below to explain what this teacher means.

47	47
× 28	× 28
800	56
140	320
320	140
56	800
1316	1316

Use estimation in Exercises 9–12 to decide the interval in which the exact answer falls. For example, for 75 ÷ 4, think, How many 4s are in 75? Interval choices: 1–10, 11–20, 21–30, 31–40, 41–50, 51–60, 61–70, 71–80, 81–90, 91–100.

9. 65 ÷ 5 **10.** 76 ÷ 2 **11.** 86 ÷ 4 **12.** 125 ÷ 3

Use estimation in Exercises 13–16 to decide the interval in which the exact answer falls. For example, for 1275 ÷ 24, think, How many 24s are in 1275? Interval choices: 1–10, 11–20, 21–30, 31–40, 41–50, 51–60, 61–70, 71–80, 81–90, 91–100.

13. 2065 ÷ 32 **14.** 7563 ÷ 88

15. 4386 ÷ 74 **16.** 3089 ÷ 53

17. Use estimation to decide about how many groups of 32 are in 4357. Check on a calculator.

18. Use repeated subtraction and estimation to find how many groups of 15 are in 257.

Evaluate each numerical expression in Exercises 19–21 on a calculator. Round each answer to the nearest whole number.

19. 2208 ÷ (12 × 23) **20.** $\frac{34 \times 15}{5} \times \frac{364}{52}$

21. $124 + \frac{15 \times 12}{8}$

22. Juanita completed Exercise 19 on her graphing calculator, as shown. She knew that the result should be less than 2208. What went wrong?

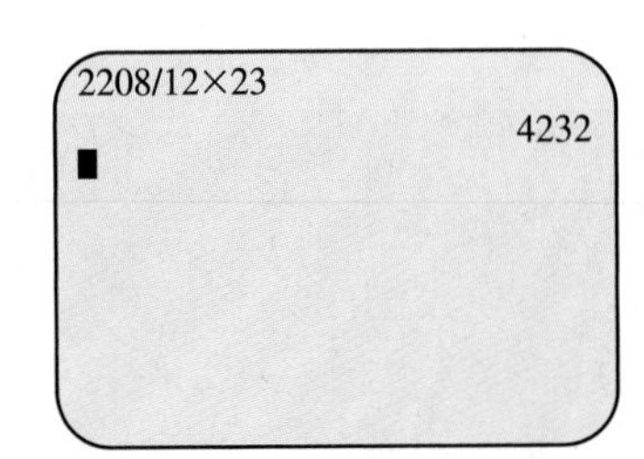

23. Suppose that you throw five darts at the game board shown.

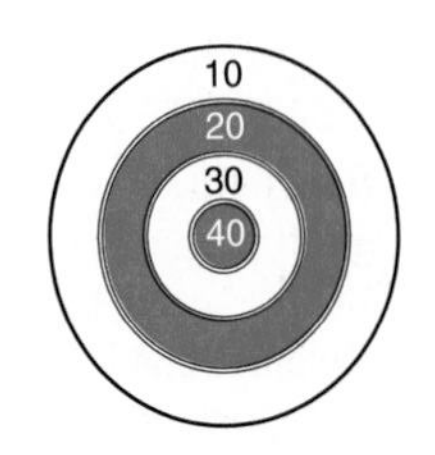

What is the highest possible score you can get? The lowest possible score?

B. Deepening Understanding

24. Write a division problem for the model depicted and the sharing interpretation of division.

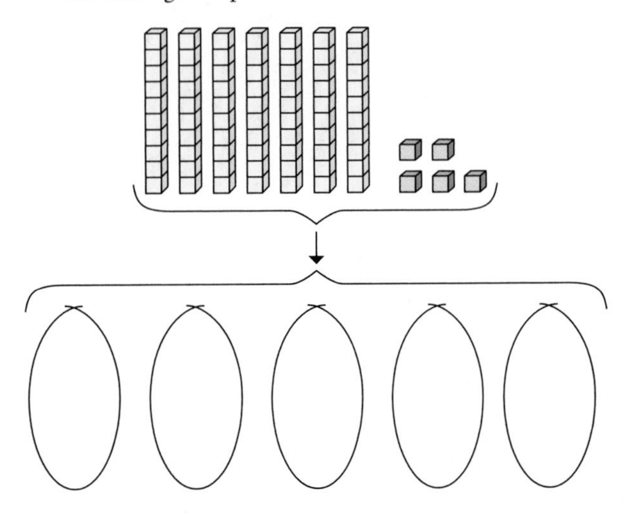

25. The following calculations are incorrect. Explain how estimation might help you discover the error in one case but not in the other.

74	48
× 23	× 32
212	96
148	124
1692	220

26. Use base-ten blocks and the sharing interpretation for subtraction to explain each step in the following algorithm. Write your explanation.

```
  126 R 2
6)758
 -6
  15
 -12
   38
  -36
    2
```

27. Which property of whole numbers indicates that the following equation is true?

$$3 \times 45 = (3 \times 40) + (3 \times 5).$$

Lattice multiplication *is an algorithm that reduces multidigit problems to single-digit multiplication tasks. To use this algorithm to multiply two 2-digit numbers, write one of the numbers above and the other to the right of a four-celled table. Multiply each digit at the top by each digit on the right. Doing so yields the products of the single digits. Write these products in the corresponding cells of the table, with the tens digits above the diagonals and the ones digits below the diagonals. Then start at the lower right, sum diagonally, and write the sums below and to the left of the table to show the final product. If a diagonal sum is a 2-digit number, regroup the tens digit to the next diagonal. As shown in the following diagram, we use the task 47 × 68 to illustrate the steps in this algorithm.*

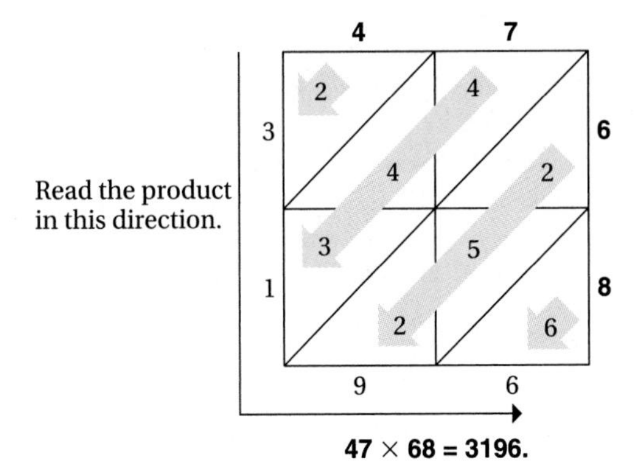

47 × 68 = 3196.

- First find the products $4 \times 6 = 24$ and $7 \times 6 = 42$. Write the products in the two top-row cells, separating each tens digit and ones digit in the product with a diagonal. Then find $4 \times 8 = 32$ and $7 \times 8 = 56$, placing these products in second-row cells.
- Starting at the lower right, find the sums of the diagonals, adding down the diagonals, and regrouping as needed. Write the sum outside the table beside the appropriate cell at the bottom or on the left.

$6 = 6$

$2 + 5 + 2 = 9$

$4 + 4 + 3 = 11$ (Record the 1 outside and regroup

$1 + 2 = 3$ the 1 ten to the next diagonal.)

- Read the final product starting at the upper left digit, 3, from the top down and to the right: 3196.

Use lattice multiplication to find each product in Exercises 28–31. (Hint: Remember to estimate first!)

28. 76 × 45 **29.** 81 × 90

30. 124 × 56 **31.** 65 × 92

32. Write an explanation for each of the boldface expressions in the division calculation:

$$\begin{array}{rl} 3\overline{)86} & \\ -\ 45 & \mathbf{15 \times 3} \\ \hline 41 & \\ -\ 36 & \mathbf{12 \times 3} \\ \hline 5 & \\ -\ 3 & \mathbf{1 \times 3} \\ \hline 2 & \mathbf{28 \times 3} \end{array}$$

33. Use multiplication to explain how you can find $32 \div 4$ when you know that $32 \div 8 = 4$. How can this relationship be useful in checking a division calculation?

C. Reasoning and Problem Solving

Annual Earnings. *A recent survey reported the following average annual earnings.*

Retail salesperson	$ 13,000
Taxi driver	$ 23,000
High school teacher	$ 38,000
President of the United States	$400,000

Use these data to solve Exercises 34–37 in any way you choose. Explain your solutions.

34. About how many years would a retail salesperson have to work to make what the president of the United States earns in 1 year?

35. How much more will a teacher make than a taxi driver in 20 years?

36. If a teacher puts 3 percent of the salary shown into a retirement fund each year, how many years will it take for the principal in the fund (excluding interest) to accumulate to the amount that a retail salesperson earns in 1 year?

37. If a taxi driver wants to triple her earnings in 5 years and for each of the first 4 years she makes $8,500 more than the preceding year, how much more than the preceding year does she need to earn in the fifth year?

38. Monthly Payments. About how much would each of the monthly payments be for a motorcycle costing $2375.99, including taxes and all charges, if the buyer makes equal payments over 3 years? Without computing the exact answer, indicate whether your estimate is over or under the exact answer and how you know.

Arrange the digits 1, 2, 3, 4, 5, and 6 in the boxes in Exercises 39 and 40 to make a product as close to the target number as possible.

39. Target = 50,000 **40.** Target = 125,000

□□□ □□□
× □□□ × □□□

Arrange the digits 1, 2, 3, 4, 5, and 6 in the boxes in Exercises 41 and 42 to make the largest and smallest products possible.

41. Largest Product **42.** Smallest Product

□□□ □□□
× □□□ × □□□

Retirement Dollars. *Suppose that, like most people, you want to have $1 million when you retire, and you decide to save one-fourth of your annual earnings to reach this goal. Use a calculator or computer to answer Exercises 43–45.*

43. If your earnings don't change over time, how much must you earn each year to save $1 million in 35 years?

44. If your spouse, who will retire before you do, earns $38,000 each year for the next 20 years and you decide to save one-third of your combined incomes each year, how much would you need to earn each year for the next 35 years to have jointly saved $1 million by the time you retire? Assume that your earnings are the same for each of the 35 years you work.

45. Suppose that you inherit $75,000, which you put into a retirement account. You decide to add $200 to the account at the end of the first year and, for each of the next 30 years, to double the amount you added in the previous year. How much will you have to put into your account by the time you retire in 30 years?

46. One side of a rectangle is 12 units long, and the rectangle has an area of 276 square units. Draw a picture to find the length of the other side of the rectangle. Explain in writing how this activity shows the inverse relationship between multiplication and division.

47. Use estimation to decide how many bottles can be put into 15 cartons if you have a total of 486 bottles and each carton holds the same number. Are there any extra bottles?

D. Communicating and Connecting Ideas

Making Connections. Suppose that the multiplication and division keys on your calculator were broken. Explain how you can calculate Exercises 48 and 49 by using other keys on the calculator.

48. 18×54 **49.** $207 \div 23$

50. Explain how the language used should be different when you describe the task $156 \div 12$ first as repeated subtraction and then as sharing.

Consider the expression 328 ÷ 6 in Exercises 51–53.

51. Write a real-world story problem that reflects a repeated subtraction interpretation for division and can be solved by using the expression.

52. Write another story problem for the same expression by using the sharing interpretation for division.

53. Exchange your group's problems with another group's and solve the problems. Show your work.

54. The following algorithm reflects what is called *short division.*

$$\begin{array}{r} 4\ 7\ 5 \\ 6\overline{)2 8^{4}5^{3}0} \end{array}$$

The reasoning processes used are the same as those for the algorithms developed in this section. Explain why this algorithm makes sense.

55. **Historical Pathways.** Elementary school mathematics textbooks in the 1960s and even into the 1970s taught a *stacking algorithm* as a transition to the standard algorithm. Here's one way the stacking algorithm might look for $526 \div 23$.

$$\begin{array}{r} \underline{22}\text{ R }20 \\ 2 \\ 10 \\ 10 \\ 23\overline{)526} \\ -\ 230 \\ \hline 296 \\ -\ 230 \\ \hline 66 \\ -\ 46 \\ \hline 20 \end{array}$$

Write a paragraph that describes how this algorithm might be used to explain the standard algorithm.

CHAPTER SUMMARY

Key Ideas: Questions and Answers

Section 3.1

- **What is mental computation and when should it be used?** *(p. 124)* Mental calculation is the process of finding an exact answer to a computation mentally, without paper and pencil, calculator, or any other computational aid. Use mental calculation when you believe that the calculations are sufficiently easy that you can do them without the aid of pencil/paper or a calculation device.

- **What are some mental computation techniques?** *(pp. 125–134) Count on and count back:* Add or subtract when one of the numbers to be added or subtracted is 1, 2, or 3; 10, 20, or 30; 100, 200, or 300; and so on.

Choose compatible numbers: Calculate with pairs of numbers that can easily be added, subtracted, multiplied, or divided or combined to produce multiples of 10, 100, or other numbers that make calculations easy.

Break apart numbers: Break apart numbers according to the place value of the digits to give simple calculations involving basic number facts that can be done mentally.

Use compensation: Use a calculation close to the original one that is easy to do mentally and compensate at the end to get the exact answer.

Use equal additions: Add a number to both numbers in a subtraction calculation so that the number being subtracted is easy to subtract from the other mentally.

Section 3.2

- **What are some computational estimation techniques?** *(pp. 136–142) Rounding:* Replace numbers with rounded ones that make calculations easy to do mentally.

  Substitute compatible numbers: Replace numbers with numbers close to the original ones to make the calculation easy to do mentally.

  Front-end: Compute using only the first digit of the numbers in the calculation. Adjustments may be made by using any estimation technique.

  Clustering: Replace addends in an addition calculation with one that is close to them all. Find the sum by completing the equivalent multiplication calculation mentally
- **How do you choose an estimation technique?** *(pp. 142–144)* The nature of the real-world situation and the kinds of numbers involved determine the specific estimation technique that makes most sense. Some situations call for a range estimate rather than one number. A range estimate gives a high estimate and a low estimate between which the exact answer falls.

Section 3.3

- **How can algorithms for addition be developed and used?** *(pp. 146–152)* Physical and pictorial models show that the steps in algorithms for addition make sense. Number properties can be used to explain steps in the standard algorithm. In the standard algorithm, addition is performed from right to left using place value and basic facts.
- **How can algorithms for subtraction be developed and used?** *(pp. 152–156)* Physical and pictorial models show that the steps in algorithms for subtraction make sense. Number properties can be used to explain steps in the standard algorithm. In the standard algorithm, subtraction is performed from right to left using place value and basic facts.
- **How can a calculator be used for multiple-step addition and subtraction calculations?** *(pp. 156–161)* The memory function on some calculators can store the results of an operation in one step until needed in another step. On other calculators, parentheses on the keypad are used to group operations and must be entered in a sequence specific to the calculation.

Section 3.4

- **How can algorithms for multiplication be developed and used?** *(pp. 164–168)* Physical and pictorial models show that the steps in algorithms for multiplication make sense. Number properties may be used to explain steps in the standard algorithm. In the standard algorithm, multiplication is performed from right to left by using place value and basic facts.
- **Are any calculator techniques useful in multiplication?** *(pp. 168–171)* Techniques that are specific to different types of calculators provide efficient ways to multiply by a repeated factor.
- **How can algorithms for division be developed and used?** *(pp. 171–178)* Physical and pictorial models show that the steps in algorithms for division make sense. Number properties can be used to explain steps in the standard algorithm. In the standard algorithm, division is performed from left to right using place value and basic facts. The steps in the standard division algorithm are explained best using sharing situations.
- **How can a calculator be used to evaluate multiplication and division expressions?** *(pp. 178–181)* An understanding of grouping symbols and order of operations, together with the specific features of different types of calculators, are necessary to evaluate expressions involving multiplication and division. The memory function on some calculators can store the results of a grouped operation until needed in another step. On other calculators, parentheses on the keypad are used to group operations and must be entered in a sequence specific to the calculation.

Key Terms, Concepts, and Generalizations

Section 3.1

Mental computation (p. 124)
Count on technique (p. 125)
Count back technique (p. 125)
Choose compatible numbers technique (p. 126)
Break apart numbers technique (p. 128)
Use compensation technique (p. 129)
Use equal additions technique (p. 130)

Section 3.2

Computational estimation (p. 136)
Rounding technique (p. 136)
Substitute compatible numbers technique (p. 138)
Front-end estimation technique (p. 139)
Clustering technique (p. 141)
Range estimate (p. 142)

Section 3.3

Algorithms (p. 146)
Expanded algorithm for addition (p. 149)
Standard algorithm for addition (p. 150)
Expanded algorithm for subtraction (p. 154)
Standard algorithm for subtraction (p. 154)

Section 3.4

Partial products (p. 165)
Expanded algorithm for multiplication (p. 166)
Standard algorithm for multiplication (p. 166)
Expanded algorithm for division (p. 173)
Standard algorithm for division (p. 174)
Order of operations (p. 178)

CHAPTER REVIEW

Concepts and Skills

Count on or back to compute the exact answer mentally in Exercises 1–4.

1. 54 + 30 **2.** 101 − 2
3. 648 − 200 **4.** 678 + 200

Look for compatible numbers to help you compute the exact answer mentally in Exercises 5–8.

5. (5 × 13) × (20 × 2) **6.** 85 + 27 + 15
7. 8 × 12 × 5 **8.** 32 + 27 + 18 + 23

Break apart numbers to compute the exact answer mentally in Exercises 9–12.

9. 548 + 261 **10.** 826 − 125
11. 431 × 2 **12.** 648 ÷ 2

Use compensation to compute the exact answer mentally in Exercises 13–16.

13. 78 × 2 **14.** 408 − 39
15. 253 + 58 **16.** 29 × 7

Use the equal additions method to compute the exact answer mentally in Exercises 17–20.

17. 48 − 27 **18.** 325 − 195
19. 210 − 88 **20.** 131 − 27

Round each number to the place value of the bold digit in Exercises 21–24.

21. 5748 **22.** 319.7
23. 6543 **24.** 603

Use rounding to estimate the answer in Exercises 25–28.

25. 665 + 243 **26.** 783 ÷ 4
27. 75 × 12 **28.** 704 − 337

Estimate by substituting compatible numbers in Exercises 29–32.

29. 3426 − 352 **30.** 128 × 9
31. 33 × 8 **32.** 774 + 122

Use front-end estimation with adjustment to estimate in Exercises 33–36.

33. 436 + 735 **34.** 1243 + 5488
35. 43 + 56 + 62 + 87 **36.** 43,650 + 45,436

Use clustering to estimate in Exercises 37 and 38.

37. 1243 + 1119 + 1228 + 1210
38. 544 + 556 + 550 + 547 + 553

Estimate and then find the exact answer in Exercises 39–42. Check your work.

39. 2546 + 4337 **40.** 56 × 34
41. 540 − 308 **42.** 765 ÷ 4

Reasoning and Problem Solving

Give two numbers between which the exact answer falls in Exercises 43–46.

43. 54 × 38 **44.** 567 + 327
45. 453 ÷ 3 **46.** 2345 − 1095

47. Use grid paper and an area model to show the partial products (192 and 240) in the following calculation.

$$\begin{array}{r} {}^{3}\\ 24 \\ \times\ 18 \\ \hline 192 \\ 240 \\ \hline 432 \end{array}$$

48. Write an explanation using base-ten blocks to explain each step in the following algorithm.

$$\begin{array}{r} 182\text{ R }2 \\ 3\overline{)548} \\ -3 \\ \hline 24 \\ -24 \\ \hline 8 \\ -6 \\ \hline 2 \end{array}$$

49. How are *compatible numbers* used in both mental calculation and estimation? Explain, using examples.

50. Copy and complete the following multiplication calculation. Use reasoning to find the missing digits.

$$\begin{array}{r} \square 4 \\ \times\ \square 7 \\ \hline 2\square 8 \\ \square 4 \\ \hline 5\square 8 \end{array}$$

51. Lottery, Lottery. A lottery winner can choose, without looking, two of four identical jars that contain the following amounts.

$154 K $25 K $750 K $3,500

The person's winnings are the sum of the amounts in the two jars. What are all possible amounts that can be won? What is the difference between the greatest and least amounts that can be won? Show your work.

52. **Tough Choices.** Another lottery winner has won $5,000 worth of prizes. She can select from the list of choices shown. Any money not spent on prizes has to be given back. She can buy more than one of each item.

Entertainment center: $1879
Ski outfits: $458
Tennis vacation: $2159
His-and-her golf clubs: $1875
Mountain bike: $656
Motor boat: $3799
Video camera equipment: $559

Select two different collections that she might choose and the total cost of each collection. How much money is left in each case?

53. What items should the lottery winner in Exercise 52 choose in order to return the least amount of money?

Alternative Assessment

54. Make up a 2-digit by 2-digit multiplication calculation. Write an explanation that you might give to a fifth-grader of what the steps in the traditional multiplication algorithm mean.

55. Students who do not have a strong understanding of place value usually have difficulty with the traditional algorithms for all four mathematical operations. Explain in writing why an understanding of place value is essential.

56. Other than using a calculator, how many ways can you use to find 12×24? Show each and explain in writing why each makes sense.

4 Number Theory

Chapter Perspective

Number theory is what we discover and verify about number properties, number patterns, and number relationships. The fascinating ideas of number theory have many practical and technological applications. For example, number theory is used in cryptology, the science of devising and decoding coded messages. The cryptology machine in the above photo was developed by the U.S. military many years ago to help in these processes. Today, much more sophisticated cryptology technology is available. Number theory, in addition to being an inherently interesting field of study, also forms the basis for the study of other important mathematical topics.

In this chapter, you can develop a greater sense of numbers and their relationships by exploring ideas associated with factors, multiples, divisibility, prime and composite numbers, and the greatest common factor and least common multiple properties.

Connection to the NCTM Principles and Standards

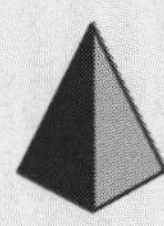

The NCTM *Principles and Standards for School Mathematics* (2000) indicate that the elementary school mathematics curriculum grades PreK–8 should include the study of number theory so that students can

- develop a sense of whole numbers and represent and use them in flexible ways, including relating, composing, and decomposing numbers *(p. 78)*;
- describe classes of numbers according to their characteristics such as the nature of their factors *(p. 148)*; and
- use factors, multiples, prime factorization, and relatively prime numbers to solve problems *(p. 214)*.

Connection to the PreK–8 Classroom

In grades PreK–2, students build number sense and readiness for number theory ideas. Expressing 6 as 2×3 helps them build the concept of factors. Skip counting by 5's helps them build the concept of multiples, and so on.

In grades 3–5 students build rectangular arrays with chips to investigate odd, even, and prime numbers, and use the idea of factors to classify numbers.

In grades 6–8 students utilize number theory ideas, such as prime numbers and divisibility properties, to discover patterns, reason about relationships, and solve problems.

Section 4.1 FACTORS AND DIVISIBILITY

- Connecting Factors and Multiples
- Defining Divisibility
- Techniques for Determining Divisibility
- Using Factors to Classify Natural Numbers

In this section we examine the idea of a factor of a number from several viewpoints. We also look at techniques for finding the factors of a number and use these techniques to develop shortcuts for deciding whether a number has a certain number as a factor. Finally, we classify natural numbers according to the number of factors they have and analyze these classifications.

Connecting Factors and Multiples

In general number theory is the study of the characteristics of and relationships involving the natural numbers. Many of the useful characterizations of the natural numbers are based on information about the *factors* and *multiples* of a number.

The labels on the following equations highlight the ideas of factors and multiples.

Factors of 12		A **multiple** of 3 A **multiple** of 4
↙ ↘		↓
3×4	$=$	$12.$

4

3 | 12 squares

The yellow rectangular array models the idea of factors and multiples described above. It shows visually that 3 and 4 are factors of 12, and that 12 is a multiple of 3 and 4.

Moving from the above examples to a general statement, we give the following definition.

Definition of *Factor* and *Multiple*

If a and b are whole numbers and $ab = c$, then a is a **factor** of c, b is a factor of c, and c is a **multiple** of both a and b.

The NCTM *Principles and Standards for School Mathematics* state that "Teachers should help students (primary grades) represent aspects of situations in mathematical terms, possibly using more than one representation" (p. 141). Representing skip counting as "jumps" on the number line can help children begin to understand the ideas of a factor and a multiple of a number. For example, each jump of three units on the number line below lands on a *multiple* of 3. The number line also shows vividly that 3 is a *factor* of 15, because $3 \times 5 = 15$.

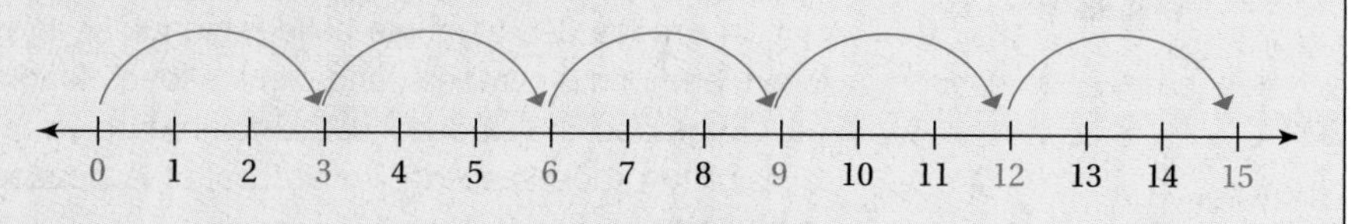

Finding Factors and Multiples. You can mentally produce *multiples* of small numbers. For example, counting by fives, 5, 10, 15, 20, … produces multiples of 5. Calculator displays also provide a way to multiply a factor to produce a list of multiples of the factor. Comparable keystrokes and displays from a four-function calculator and a graphing calculator that show multiples of 3 are as follows.

- A four-function calculator

Key Sequence	Display
[+] 3 [=]	3
[=]	6
[=]	9
[=]	12
[=]	15

- A graphing calculator

Key Sequence **Display**

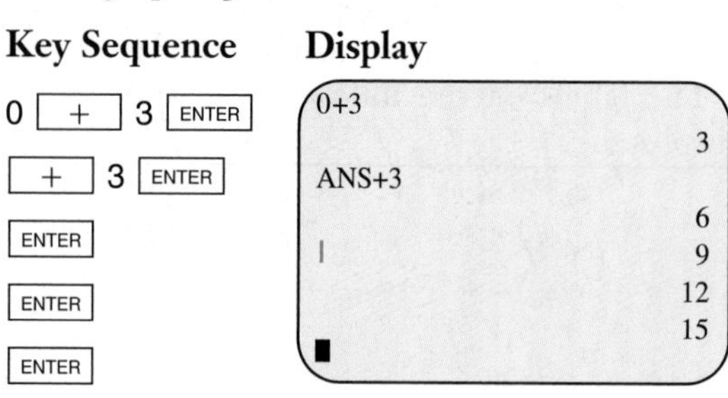

Using these calculators, we can extend the list of multiples as much as we want or easily produce a multiple of any number.

You can find the *factors* of a number by dividing. A simple way to decide whether a first number is a factor of a second number is to divide the second number by the first. If the quotient is a natural number with a remainder of 0, the divisor and quotient are each factors of the dividend.

With larger numbers, a calculator can help identify whether one number is a factor of another. Some calculators have what is called an *integer division feature*, [INT ÷], which gives the quotient and remainder for a division. We use the following keystrokes and display from a calculator that uses integer division to determine whether 24 is a factor of 5868.

Key Sequence	Display	
5868 [INT ÷] 24	I	24
[=]	244	12
	└ Q ┘	└ R ┘

Because the remainder is 12, not 0, 24 isn't a factor of 5868. Example 4.1 illustrates how making a decision about a factor of a number can be useful in solving a problem.

Example 4.1 Applying the Idea of a Factor of a Number

A movie rental store sold used movie video cassettes for $13 each, including tax. A clerk recorded a total income of $1009 for all the cassettes. If the total is possible, how would we know? If it isn't, what would be a possible total?

Solution

Using a calculator with an [INT ÷] function, we get the quotient 77 and remainder 8 when we divide 1009 by 13. The remainder isn't 0, so 13 isn't a factor of 1009, and the clerk's total is impossible. A possible total would be $1001.

Your Turn

Practice: Could computer systems that sell for $4649 each produce a total income of $1,111,117? Use a calculator with an [INT ÷] feature, if possible, to decide.

Reflect: How would you solve the practice problem if a calculator with an [INT ÷] feature is not available? ■

Finding All the Factors of a Number. Finding all the factors of a number is often useful. Rectangles cut from graph paper may be used to show all the factors of a number geometrically, as illustrated in Figure 4.1.

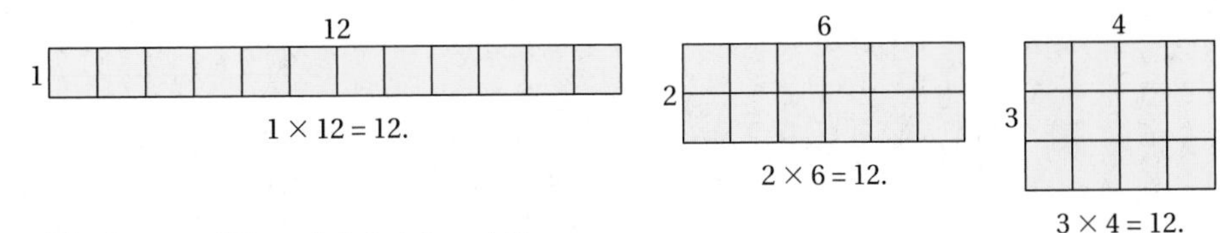

The factors of 12 are 1, 2, 3, 4, 6, and 12.

FIGURE 4.1 | Geometric interpretation of the factors of 12.

To find all the factors of a small number such as 12, you can divide 12 by each natural number in succession, beginning with 1. Dividing by 1, 2, and 3, you get the factor pairs (1, 12), (2,6), and (3, 4). Continued dividing will produce no new factors, so you find that the factors of 12 are 1, 2, 3, 4, 6, and 12. Using a calculator is a much more effective way of finding the factors of a larger number.

One way to use a scientific calculator to determine the factors of a number is to use its memory key. To do so, first enter [C], then enter the number, and then push [M+] to record it in the calculator's memory. Then try each divisor 2 (as shown), 3, 4, 5, and so on, in order.

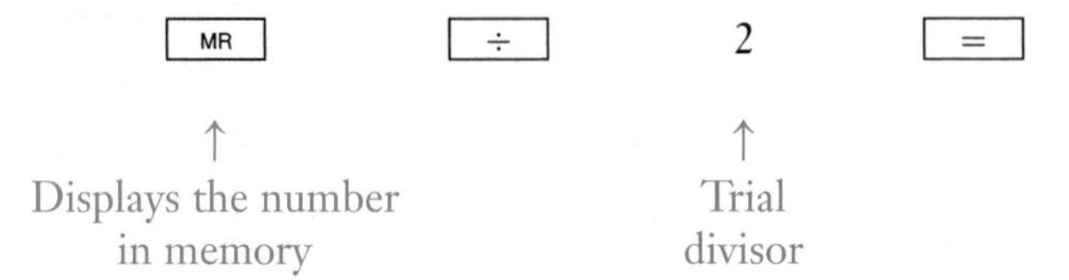

When the answer is a natural number, list the trial divisor and the natural number as factors of the number you put into the memory.

You may also use the [INT ÷] feature to find all the factors of a number. The following keystrokes and display for a graphing calculator show how to begin to find all the factors of 234.

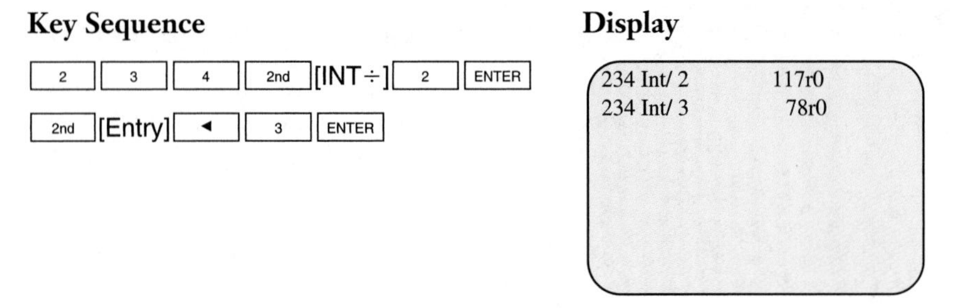

Repeat the command, checking divisors 4, 5, 6, and so on. If the value for the remainder, R, is 0, the divisor is a factor of 234. To find all the factors of 234, you could use a calculator to divide 234 by successive whole numbers and list the factors you find, as shown, along with 1 and 234.

Number	Divide by	Quotient	Factors
234	2	117	2, 117
234	3	78	3, 78
234	6	39	6, 39
234	13	18	13, 18
234	18	13	18, 13
234	39	6	39, 6
234	78	3	78, 3
234	117	2	117, 2

Thus the factors of 234 are 1, 2, 3, 6, 13, 18, 39, 78, 117, and 234. After many divisions you will have found all the factors of 234, but could you have stopped dividing earlier and had the same information? If so, *how do you know when to stop?* To decide, first note that the numbers in the *divide by* column are increasing and that the numbers in the quotient column are decreasing. The numbers "pass each other" between

13 and 18. A factor of 234 will appear in the divide by column first and reappear later in the quotient column, repeating earlier factors. As soon as the pass-by point is reached, you need check no further because no new factors will appear. When you come to a number already in your factor list—in this case 18—you can stop testing factors.

The pass-by point appears to be the place at which the factors are closest to being equal. Note that, as $\sqrt{234} \times \sqrt{234} = 234$, the pass-by point can't be greater than $\sqrt{234}$. This result suggests the following theorem.

Theorem: Factor Test

To find all the factors of a number n, test only those natural numbers that are no greater than the square root of the number, $\sqrt{n}$.

Example 4.2 provides an opportunity for you to apply the Factor Test Theorem.

Example 4.2

Applying the Factor Test Theorem

Find all the factors of 165.

Solution

Alison's thinking: I estimated $\sqrt{165}$ by thinking that it was between $\sqrt{144} = 12$ and $\sqrt{169} = 13$, so I only have to test the numbers through 12. I used my calculator and made a list.

Test 2: No	Test 8: No
Test 3: Factors 3, 55	Test 9: No
Test 4: No	Test 10: No
Test 5: Factors 5, 33	Test 11: 11, 15
Test 6: No	Test 12: No
Test 7: No	Test 13: No

The factors of 165 are 1, 3, 5, 11, 15, 33, 55, and 165.

Lee's thinking: I found the square root of 165 on my calculator. It's about 12.85, so I only have to test numbers through 12. I found that 2 wasn't a factor, so I knew that no even number could be a factor. I divided 165 by 1, 3, 5, 7, 9, and 11 and found the factors of 165: 1, 3, 5, 11, 15, 33, 55, and 165.

Your Turn

Practice: Find all the factors of 297.

Reflect: Did Alison have to test the number 13? Explain. ■

Defining Divisibility

Sometimes, as when deciding whether 144 people can be grouped by 3s, we need to make a quick check to see if one number is a factor of another. For example, we ask the question, "Is 3 a factor of 144?" The same question could be asked in other ways, such as, "Is 3 a *divisor* of 144?" or "Does 3 *divide* 144?" or "Is 144 *divisible* by 3?" The idea of divisibility is defined as follows.

Definition of Divisibility

For whole numbers a and b, $a \neq 0$, a **divides** b, written $\mathbf{a \mid b}$, if and only if there is a whole number x so that $ax = b$. Also, a is a **divisor** of b or b is **divisible** by a. Further, $\mathbf{a \nmid b}$ means that a does not divide b.

We illustrate the definition for 6 and 24 by noting that since there exists a whole number 4 such that $6 \times 4 = 24$, we can say that 6 is a factor of 24, 6 is a divisor of 24, and 24 is divisible by 6.

Techniques for Determining Divisibility

We naturally look for ways to discover and verify conclusions about whether one number is divisible by another. We first present a theorem on divisibility that is often useful in verifying what we have discovered. Consider the following examples. Let's use the facts that $7 \mid 35$ and $7 \mid 14$. Note also that $7 \mid (35 + 14)$, or 49. Similarly, $9 \mid 27$, $9 \mid 36$, and $9 \mid (27 + 36)$, or 63. These examples suggest the following divisibility theorem.

Theorem: Divisibility of Sums

For natural numbers a, b, and c, if $a \mid b$ and $a \mid c$, then $a \mid (b + c)$.

We can verify this theorem as follows. Suppose $a \mid b$ and $a \mid c$. Then from the definition of divisibility, we know that there exist whole numbers r and s such that $b = ar$ and $c = as$. By adding these equations we see that $b + c = ar + as$. It follows that $b + c = a(r + s)$, and since $r + s$ is a whole number, by the definition of divisibility, $a \mid (b + c)$.

Now let's try to discover a theorem about testing divisibility and see if we can use the Theorem on Divisibility of Sums to verify it. Table 4.1 shows how we might use patterns to discover divisibility tests for 2, 5, and 10.

TABLE 4.1 | Patterns for Discovering Divisibility Tests for 2, 5, and 10

Number divisible by 2	Last digit	Number divisible by 5	Last digit	Number divisible by 10	Last digit
0	0	0	0	0	0
2	2	5	5	10	0
4	4	10	0	20	0
6	6	15	5	30	0
8	8	20	0	40	0
10	0	25	5	50	0
12	2	30	0	60	0
14	4	35	5	70	0
16	6	40	0	80	0
18	8	45	5	⋮	⋮
⋮	⋮	⋮	⋮		

Three tests that are suggested by the patterns in Table 4.1 are stated as follows.

Theorem: Divisibility Tests for 2, 5, and 10

- A natural number n is divisible by 2 if and only if its units digit is 0, 2, 4, 6, or 8.
- A natural number n is divisible by 5 if and only if its units digit is 0 or 5.
- A natural number n is divisible by 10 if and only if its units digit is 0.

We verify the statement in the above theorem about divisibility by 2. The statements about divisibility by 5 and 10 can be verified in a similar manner. Let's first look at an example. Consider a number like 274. It can be written as $10(27) + 4$. We see that $2 \mid 10(27)$ and since we know that $2 \mid 4$, by the Theorem on Divisibility of Sums, we know that $2 \mid [10(27) + 4]$, or 274. So it follows that any number is divisible by 2 if its units digit is divisible by 2.

In general, we can express any number n in the form $n = 10q + r$, where q is the quotient when the number is divided by 10 and r is the remainder. Using the Theorem on Divisibility of Sums, we know that if $2 \mid 10q$ and $2 \mid r$, then $2 \mid (10q + r)$, or n. Since we know that $2 \mid 10q$, when we check and find that $2 \mid r$ (the units digit), we know that $2 \mid n$. The only units digits that 2 divides are 0, 2, 4, 6, and 8, so the first statement in the theorem is verified.

The idea of divisibility by 2 is used in the following definition.

Definition of Even and Odd Numbers

- A whole number is **even** if and only if it is divisible by 2.
- A whole number is **odd** if and only if it is not divisible by 2.

Now let's look at the patterns in Table 4.2 to discover another divisibility test, and see if we can use the Theorem on Divisibility of Sums to verify it.

TABLE 4.2 | Patterns for Discovering Divisibility Tests for 3 and 9

Number divisible by 3	Sum of the digits	Number divisible by 9	Sum of the digits
0	0	72	9
3	3	873	18
6	6	8883	27
9	9	77,886	36
12	3	99,999	45
15	6	⋮	⋮
18	9		
21	3		
24	6		
27	9		
⋮	⋮		

Two tests that are suggested by the patterns in Table 4.2 are stated as follows.

Theorem: Divisibility Tests for 3 and 9

- A natural number *n* is divisible by 3 if and only if the sum of its digits is divisible by 3.
- A natural number *n* is divisible by 9 if and only if the sum of its digits is divisible by 9.

We can use the Theorem on Divisibility of Sums to verify that the divisibility test stated above for 3 works. The divisibility test for 9 can be verified in a similar manner. Consider any 3-digit number, expressed in the form $a \times 100 + b \times 10 + c$. It can also be written as $a(99 + 1) + b(9 + 1) + c$, or as $(99a + 9b) + (a + b + c)$ we know that $99a + 9b$ is divisible by 3, and by the Theorem on Divisibility of Sums we can conclude that if $a + b + c$ is also divisible by 3, then the original 3-digit number is divisible by 3. And this last conclusion is the test for divisibility by 3, stated above.

We use some of the divisibility tests and the problem-solving strategy *use logical reasoning* in Example 4.3.

Example 4.3 Using Divisibility Tests

A manufacturer wants to put 3 items into each package and then 2 packages in each box. The packing supervisor wants to know whether the 67,986 items in stock will fill a whole number of boxes. Check the divisibility of 67,986 by 3 and by 2 to find out.

Solution

- The sum of the digits of 67,986 is $6 + 7 + 9 + 8 + 6 = 36$; 36 is divisible by 3, so 67,986 is divisible by 3.
- Because the last digit of 67,986 is the even number 6, 67,986 is divisible by 2.

Therefore 67,986 items will fill a whole number of boxes.

Your Turn

Practice: Suppose that the manufacturer puts 3 items in each package and then 5 packages in each box. Which of the following number of items will fill a whole number of boxes?

a. 1275 **b.** 1461
c. 2770 **d.** 2535

Reflect: Do you think that the 67,986 items in the example problem would fill exactly 6 boxes? Explain. ■

© Dennis O'Clair/Stone

A second important theorem that is useful in verifying certain divisibility tests is suggested by the following. We know that $2 \mid 20$ and that $5 \mid 20$. We can conclude that $2 \times 5(\text{or } 10) \mid 20$. We state this theorem as follows.

Theorem: Divisibility by Products

For natural numbers a, b, and c, if $a \mid c$ and $b \mid c$, and a and b have no common factors except 1, then $ab \mid c$.

We verify this theorem as follows. If $a \mid c$ and $b \mid c$, we know from the definition of divisibility that $ar = c$ and $bs = c$, where r and s are whole numbers. But since a and b share no factors other than 1, we know from the equations $ar = c$ and $bs = c$ that c contains all of a's factors and all of b's factors. Thus c is divisible by ab.

Now let's look at a divisibility test that can be verified using the above theorem. Numbers divisible by both 2 and 3, such as 6, 12, 18, 24, 30, 36, and so on, are also divisible by 6. And any multiple of 6 is also divisible by both 2 and 3. This relationship suggests the following divisibility test for 6.

Theorem: A Divisibility Test for 6

A natural number n is divisible by 6 if and only if it is divisible by both 2 and 3.

To verify the "only if" part of this theorem, we note that if a number n is divisible by 6, we can write $6k = n$, where k is a whole number. This can also be written as $2 \times 3 \times k = n$ or $2(3k) = n$, or $3(2k) = n$. Then by the definition of divisibility, we know that n is divisible by both 2 and 3.

To verify the "if" part of the theorem, we note that if $2 \mid n$ and $3 \mid n$, then using the Theorem on Divisibility by Products we can conclude that $2(3)$ (or 6) $\mid n$.

A similar approach could be used to verify the conclusion that a natural number is divisible by 10 if and only if it is divisible by both 2 and 5. We leave this verification for Exercise 58 on p. 204.

The patterns in Table 4.3 are helpful in discovering divisibility tests for 4 and 8.

TABLE 4.3 | Patterns for Discovering Divisibility Tests for 4 and 8

Number divisible by 4	Last two digits	Number divisible by 8	Last three digits
12	12	4160	160
124	24	7248	248
528	28	33,080	080
916	16	59,336	336
1732	32	97,328	328
9044	44	135,864	864
27,356	56	999,832	832
777,508	08		

Note that numbers formed by the last two digits of numbers divisible by 4 are also divisible by 4. Similarly, numbers formed by the last three digits of numbers divisible by 8 are also divisible by 8. These patterns suggest the "if" parts of the following tests.

Theorem: Divisibility Tests for 4 and 8

- A natural number n is divisible by 4 if and only if the number represented by its last two digits is divisible by 4.
- A natural number n is divisible by 8 if and only if the number represented by its last three digits is divisible by 8.

Exercises 59 and 60, pp. 204–205, ask questions to help you verify the above theorem.

The divisibility test for 4 involves mentally dividing a two-digit number by 4, and is fairly easy to use. For example, to decide if 5280 is divisible by 4, we check to see if the number represented by the last two digits of 5280, namely 80, is divisible by 4. Since $80 \div 4 = 20$, Remainder 0, we know that 5280 is divisible by 4.

The test for divisibility by 8 is somewhat more difficult to use, but usually we can mentally divide to use the test. If not, a calculator can be used to make the test slightly easier. For example, to decide if 467,863 is divisible by 8, we check to see if the number represented by the last three digits, namely 863, is divisible by 8. We can mentally divide to see that it is not, and that 467,863 is not divisible by 8. Alternatively, we could use a calculator to find that $863 \div 8$ is not a whole number.

The divisibility tests for 7 and 11 are interesting, but decisions about divisibility for these numbers are more efficiently made with the aid of a calculator. However, to satisfy any curiosity about whether tests exist for all the numbers 1–11, we state and discuss the tests for 7 and 11.

Theorem: Divisibility Tests for 7 and 11

- A natural number n is divisible by 7 if and only if the number formed by subtracting twice the last digit from the number formed by all digits but the last is divisible by 7.
- A natural number n is divisible by 11 if and only if the sum of the digits in the even-powered places minus the sum of the digits in the odd-powered places is divisible by 11.

For example, to test whether 3045 is divisible by 7, we check to see if $7 \mid [304 - 2(5)]$, or $7 \mid 294$. It does, so we conclude from the test that $7 \mid 3045$. Also, to test whether 1,234,607 is divisible by 11, we check to determine whether 11 divides the difference $(7 + 6 + 3 + 1) - (0 + 4 + 2)$, or 11. It does, so we conclude from the test that $11 \mid 1{,}234{,}607$. Example 4.4 further demonstrates the use of divisibility tests and the problem-solving strategy *write an equation.*

Example 4.4

Problem Solving: Boxing Baseballs

A sports shop sold baseballs to teams in boxes of 6 and in boxes of 12. The inventory manager reported that the shop had sold 9052 baseballs to teams during the past month. The owner asked the manager to check the inventory, saying that it was incorrect. How did he know?

Solution

If d is the number of boxes of 12 baseballs and h is the number of boxes of 6 baseballs, then according to the manager's report, $12d + 6h = 9052$. But both $12d$ and $6h$ are divisible by 6, so 9052 must be divisible by 6. The divisibility test for 6 shows that 6 doesn't divide 9052.

Your Turn

Practice: Suppose that the baseballs in the example problem came in boxes of 4 and boxes of 8. Could the shop have sold 9052 baseballs? Explain.

Reflect: Which theorem justified the shop owner's conclusion in the example problem that 9052 must be divisible by 6 for the inventory to be correct? ■

Other observations about divisibility suggest some additional useful divisibility theorems. For example, we notice that $6 \mid 36$ and $6 \mid 24$. It follows that $6 \mid (36 - 24)$ also. This suggests the first theorem in the following box, which is similar to the Divisibility of Sums theorem. Also, we note that $4 \mid 12$ and that $4 \mid 3(12)$. This suggests the second theorem in the following box. We also note that $3 \mid (27)$ or $(12 + 15)$ and that $3 \mid 12$. We conclude that $3 \mid 15$. This suggests the third theorem in the following box. The fourth theorem is similar to the third.

Theorems: Divisibility

For natural numbers a, b, and c,

- If $a \mid b$ and $a \mid c$, then $a \mid (b - c)$.
- If $a \mid b$ and c is any natural number, then $a \mid bc$.
- If $a \mid (b + c)$, and $a \mid b$, then $a \mid c$.
- If $a \mid (b - c)$, and $a \mid b$, then $a \mid c$.

You will be asked to verify and use one or more of the above theorems in the Problems and Exercises.

The *Principles and Standards for School Mathematics* affirm the role divisibility ideas can play in providing a setting for helping to develop children's reasoning abilities, in statements such as this: "Students can learn about reasoning through classroom discussion of claims that other students make. The statement, 'If a number is divisible by 6 and by 4, then it is divisible by 24' could be examined in various ways. Middle-grades students could find a counterexample—the number 12 is divisible by 6 and by 4 but not 24" (p. 58).

Using Factors to Classify Natural Numbers

Mini-Investigation 4.1 asks you to classify natural numbers according to the number and type of their factors. Such classifications play an important role in the development of some number theory ideas.

Make a table showing your data and another showing your classifications.

Mini-Investigation 4.1 **Finding a Pattern**

In what ways can you classify the numbers 1 through 30 by considering the factors of the numbers? (*Hint:* One classification might be all numbers that have 2 as a factor.)

One possible classification is that of numbers that have 2 as a factor, commonly called **even numbers,** or numbers that do not have 2 as a factor, called **odd numbers.** Another classification could involve the number of factors a number has. For example, the number 5 has exactly two factors, 1 and 5. Numbers with exactly two factors are important, and we consider them in the next section.

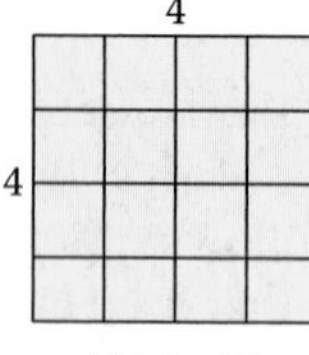

FIGURE 4.2 | Geometric model of the square number 16.

The class of numbers with an odd number of factors could be described as the **squares.** Any square number, such as 16, can be shown geometrically as a square (Figure 4.2). Consider the following classification. A number is called a **perfect number** when the sum of the factors of the number that are less than the number—its *proper factors*—equals the number. The sum of the proper factors of 6, $1 + 2 + 3$, equals 6, so 6 is a perfect number. When the sum of the proper factors of a number is less than the number, it is called a **deficient number.** When the sum of the proper factors of a number is greater than the number, it is called an **abundant number.** Two numbers are **amicable** if the sum of the proper factors of the first number equals the second number and if the sum of the proper factors of the second number equals the first number. Exercises 30–34, 48, and 49 at the end of this section give you the chance to answer questions about these classes of numbers.

The solution to the problem in Example 4.5 involves the use of a classification of natural numbers just described and the problem-solving strategies *solve a simpler problem*, *make a table*, and *look for a pattern*.

Example 4.5

Problem Solving: The Prize Winners

Prize winners for a giveaway program are to be determined as follows. Envelopes for the first 500 entries will be numbered 1 through 500 and laid out on a long table. The first official will walk along the table, opening all the envelopes. The second official will follow the first and close every even-numbered envelope. The third official will follow the second, reversing every third envelope by closing open envelopes and opening closed envelopes. The fourth official will reverse every fourth envelope, and so on until 500 officials have walked along the table and reversed envelopes. The envelopes that remain open after this procedure will be those of the contest winners! What numbers were on the winning envelopes?

Working toward a Solution

Understand the problem	*What does the situation involve?*	Choosing prize winners.
	What has to be determined?	The winning envelope numbers.
	What are the key data and conditions?	There are 500 envelopes. The first official opened all the envelopes, the second changed every second envelope, the third changed every third envelope, and so on, through the 500th official.
	What are some assumptions?	The officials walked along the table from the first envelope to the last envelope, following one another in numerical order.
Develop a plan	*What strategies might be useful?*	Solve a simpler problem, make a table, and look for a pattern.
	Are there any subproblems?	Determine which envelopes are open after the first, second, third, and so on, officials have walked along the table.
	Should the answer be estimated or calculated?	No calculation is needed.
	What method of calculation should be used?	Not applicable.

Implement the plan	*How should the strategies be used?*	Suppose that there were only 10 envelopes. Make a table and record the officials' actions.

Envelope Number

Official Number	1	2	3	4	5	6	7	8	9	10
1	O	O	O	O	O	O	O	O	O	O
2		C		C		C		C		C
3			C			O			C	
4				O				O		
5					C					O
6						C				
7							C			
8								C		
9									O	
10										C

	What is the answer?	Look for patterns in the table that will show which envelopes end up open. Extend the table to 20 envelopes, if necessary, to help solve the problem.
Look back	*Is the interpretation correct?*	Check the table for completeness and accuracy.
	Is the calculation correct?	Not applicable.
	Is the answer reasonable?	Look at the data and test the pattern to decide whether the answer is reasonable.
	Is there another way to solve the problem?	Logical reasoning could be used to solve the problem by looking at the number of factors for different types of numbers.

■ *Your Turn*

Practice: Complete the solution to the example problem for 1000 envelopes.

Reflect: How does this solution relate to the idea that square numbers have an odd number of factors? ■

Connection to the PreK–8 Classroom

The NCTM *Principles and Standards for School Mathematics* affirm the classroom role of the number theory ideas in Sections 4.1 and 4.2 with statements like these.

- Students (in grades 3–5) should recognize that different types of numbers have particular characteristics: for example, square numbers have an odd number of factors and prime numbers have only two factors (p. 151).
- For students (in grades 6–8) p Tasks p involving factors, multiples, prime numbers, and divisibility, can afford opportunities for problem solving and reasoning (p. 217).

Problems and Exercises | *for Section 4.1*

A. Reinforcing Concepts and Practicing Skills

1. How can you decide whether 7 is a factor of 87?

2. List the factors of 24.

3. To find the factors of 1225, what is the largest number you would have to test? What theorem assures you of that?

4. Use a calculator to find the factors of each number.
a. 78 **b.** 156
c. 252

5. List a factor and a multiple of 253 (other than 1 and 0).

6. Explain why 0 is a multiple of every whole number.

7. Show in two different ways that 13 is a factor of 299.

For Exercises 8–15, fill in the blank with "factor" or "multiple."

8. 8 is a ___ of 64.

9. 9 is a ___ of 3.

10. 0 is a ___ of every natural number.

11. 1 is a ___ of every natural number.

12. A ___ of a non-zero number is always less than or equal to the number.

13. A non-zero ___ of a number is always greater than or equal to the number.

14. A number $6n^2$ is a ___ of the number $2n$.

15. A number is a ___ of its cube.

16. List some multiples of the number $3n$.

17. How can you quickly decide whether a number is divisible by
a. 2? **b.** 3?
c. 4? **d.** 5?
e. 6?

18. Which of the following are divisible by 2? By 3? By 4? By 5? By 6?
a. 699 **b.** 1320
c. 2645 **d.** 5736

19. If a number n is divisible by 24, what are some other numbers n is divisible by?

In Exercises 20–22, fill in the missing digits so the number will be divisible by 3.

20. 4 _ 3 _

21. _ 7 _ _

22. 72 _ 94

In Exercises 23–25, fill in the missing digits so the number will be divisible by 6.

23. 5 _ 4 _

24. 24 _ 6 _

25. 82 _ 94

In Exercises 26–28, fill in the missing digits so the number will be divisible by 9.

26. 8 _ 3 _

27. 73 _ 5 _

28. 62 _ 95

29. Decide if the statement is true or false. Tell why.
a. $4 \mid 20$ **b.** 12 is a factor of 6
c. $24 \mid 24$ **d.** $0 \mid 8$
e. $4 \mid 0$ **f.** 60 is a multiple of 15
g. 24 is a divisor of 8
h. Every number is divisible by itself.

30. Show that 6 is a perfect number.

31. Show that 15 is a deficient number.

32. Show that 18 is an abundant number.

33. What is the smallest even, abundant number? Explain.

34. Give examples to show the following.
a. A square number has an odd number of factors.
b. There are both even and odd square numbers.
c. A number can be even, square, and abundant.

35. A chewing gum factory packages five sticks of gum into a small pack and three small packs into a large pack. A total of 43,860 sticks of gum are produced in a unit of time. Which divisibility rules can help you decide whether a whole number of full large packs of gum, with none left over, will be produced? Will a whole number of packages be produced?

36. How could you quickly decide whether 1516 people could be seated at a banquet 4 at a table with all full tables and no extra people? Can they?

37. Can 114 days be divided into a whole number of 7-day weeks?

38. Describe the different ways a designer could lay out parking lot spaces for 84 cars in rows, with an equal number of spaces in each row and no spaces left over.

39. Suppose that you were asked to introduce the number 28 to an audience. Write a paragraph telling as many things as possible about 28. Use as many of the terms defined in this chapter as you can.

40. A 73-year-old numerologist claimed that when she formed an 8-digit number by writing the year she was born twice in succession, it was divisible by her age. She was born in 1927. Use your calculator to verify that she was correct. Was this unusual, or could any other 73-year-old person say the same? Give examples to illustrate your answer.

41. To be a leap year, a year must be divisible by 4. However, some years that are divisible by 4, such as any year that is a multiple of 100, need to be checked further to see if they are leap years. Such a year will be a leap year only if it is also divisible by 400. For example, 1300 (a multiple of 100) is divisible by 4, but is not a leap year because it is not divisible by 400. Which of the following years are leap years? Why or why not?

a. 700 b. 1992 c. 1776
d. 2000 e. 2010

B. Deepening Understanding

42. **a.** If a number is divisible by 10, is it necessarily divisible by 5? **b.** If a number is divisible by 5, is it necessarily divisible by 10? Explain.

43. The following conjectures were made by students. Which do you think are false? Give a counterexample for each conjecture if possible.
  a. Only half of the non-zero even numbers up to 100 are divisible by 4.
  b. A number is divisible by 8 if it is divisible by 4 and divisible by 2.
  c. If 12 divides a number, 6 also divides the number.
  d. If a number is divisible by 4 and 6, it must be divisible by 24.

44. Describe the numbers that have exactly three factors by looking for a pattern.

45. Decide whether each statement is *sometimes*, *always*, or *never* true.
  a. A natural number has an unlimited number of multiples.
  b. A natural number has an even number of factors.
  c. With natural numbers other than 1, if a is a multiple of b, then a is a factor of b.
  d. An odd number is a factor of an even number.
  e. When n is a factor of a number, all factors of n are also factors of the number.

46. Try to find a counterexample for each, where n, a, and b are natural numbers. Give a numerical example for those that you think are true.
  a. If $n \mid ab$, then $n \mid a$ or $n \mid b$.
  b. If $n \mid (a + b)$, then $n \mid a$ and $n \mid b$.
  c. If $n \mid a$ and $n \mid b$, then $n \mid (a + b)$.
  d. If $n \mid a$ and $a \mid b$, then $n \mid b$.
  e. If $n \mid (a + b)$, then $n \mid a$ or $n \mid b$.
  f. If $n \mid a$ and $n \mid b$, and $a > b$, then $n \mid (a - b)$.

47. Show that 220 and 284 are amicable numbers.

48. The next perfect number greater than 28 is 496. Prove that it is a perfect number.

49 Which numbers between 30 and 40 are deficient? Which are abundant?

50. Make a table showing what happens when you add two even numbers, two odd numbers, and an odd and an even number. Repeat for multiplication.

51. A number that is the sum of some of its factors, such as $12 = 6 + 4 + 2$, has sometimes been called a *semiperfect* number. Find two more semiperfect numbers under 30 and prove that they are semiperfect.

52. A shopper had quarters and dimes and wanted to have the exact amount for a $4.56 purchase. Is that possible? Use divisibility to explain your answer.

53. A certain number has a remainder of 1 when divided by 2, 3, 4, and 5, but when divided by 7 the remainder is 0. What is the number? Is there more than one correct answer? Explain.

54. A sports shop sold tennis balls only in full boxes of 3 or in full boxes of 6. The shop's records showed that 5717 tennis balls had been sold during the past month. How did the owner know that the records were incorrect?

55. The House Numbers.
  a. Bill was generally superstitious and wondered whether his house number, 156, was deficient or abundant! The proper factors of 156 are 2, 3, 4, 6, 12, 13, 26, 39, 52, and 78. State how Bill can estimate the answer, without adding up all the factors.
  b. Janie's house number is 273, with proper factors of 3, 7, 13, 21, 39, and 91. How can Janie estimate, without adding all the numbers, whether her number is deficient or abundant?

C. Reasoning and Problem Solving

56. The following are some interesting theorems about number relationships. Select specific numbers and give an example for each theorem.
  a. Every 6-digit number with identical ones and thousands periods, such as 325,325, is divisible by 13.
  b. Every 4-digit palindrome is divisible by 11. (A palindrome number reads the same backward or forward, such as 2332.)
  c. Every even number greater than 46 is the sum of two abundant numbers.
  d. If the digits of any 2-digit number are reversed and the numbers subtracted, the difference is a multiple of 9.

57. Use numerical examples to illustrate that, if $n \mid (a + b)$ and $n \mid b$, then $n \mid a$.

58. A student generalized that "If a number is divisible by 2 and by 5, then it is divisible by 10." Choose the appropriate divisibility theorem, and use it to verify this generalization.

59. To verify the theorem about a divisibility test for 4, a teacher began her reasoning as follows:
  When you divide a number by 100, the remainder is the last two digits of the number. For example, $724 \div 100 = 7$,

Remainder 24. We can write this as $724 = 100(7) + 24$. So if n is any number with three digits or more, we can write $n = 100q + r$, where q is the quotient when n is divided by 100 and r is the number formed by the last two digits of n. Since $4 \mid 100q$, the number n will be divisible by 4 if it is also true that ______.

Complete this last statement, and give a divisibility theorem that supports your reasoning.

60. Use reasoning similar to that in Exercise 59 to verify the theorem on divisibility by 8.

61. Give a numerical example to illustrate the following theorem and then use the definition of divisibility to verify it.

For natural numbers a, b, and c, if $a \mid b$ and c is any natural number, then $a \mid bc$.

62. The fraction $\frac{16}{64}$ is interesting because you can use the *incorrect* procedure of crossing out a digit in the numerator and a digit in the denominator to get an equivalent fraction:

$$\frac{1\not{6}}{\not{6}4} = \frac{1}{4}.$$

Find three other fractions that have this characteristic. (*Hint:* The fractions have 2-digit numerators and denominators, with common factors of 13, 19, and 7, respectively.)

63. The Age-Guessing Contest Problem. A mother and daughter entered an age-guessing contest. The mother announced that she was about three times as old as her daughter and that their ages were alike in that each age left a remainder of 3 when divided by 4, a remainder of 2 when divided by 3, and a remainder of 1 when divided by 2. What were the ages of the daughter and the mother?

64. The License Plate Number Problem. An eccentric mathematics teacher ordered a 6-digit license plate with a number on it that contained all the digits 1 through 6 and met the following conditions. When read from left to right, its first two digits formed a number divisible by 2, its first three digits formed a number divisible by 3, and so on, ending with a complete 6-digit number divisible by 6. What was the teacher's license plate number?

65. The Football Score Problem. Bob's mother called her brother and reported that Bob's football team held the other team scoreless and made all its points on the touchdowns (6 points) and field goals (3 points), missing every point after the touchdown kick. She told him that the final score was 43–0. Her brother told her that this couldn't be the correct score. How did he know?

66. The House Number Problem. Cindy posed the following problem to her friend Roger.

Cindy: Can you figure out my house number? It's an even 3-digit number. Each digit represents the age of one of my three children. The product of the digits is 72 and the sum of the digits is 14.
Roger: Hmm ... I'm not sure.
Cindy: Oh, yes, my oldest child likes chocolate ice cream.
Roger: I have it!

What is Cindy's house number?

67. The Lost Receipt Problem. A secretary lost a receipt for 24 stenographer's notebooks that he had purchased for the office. He remembered that the middle two digits of the total 4-digit dollars and cents cost were 7 and 3. That made the total \$?7.3?, with the question marks representing the missing digits. He also remembered that the cost of each notebook was between \$2 and \$3. How much did each notebook cost?

D. Communicating and Connecting Ideas

68. Choose some numbers and work in a small group to figure out how to apply the tests for divisibility by 4, 6, 7, 8, 9, and 11 given on pp. 197, 198, and 199.

a. Use a calculator to select some numbers that are and some that aren't divisible by these numbers. Give an example to show that each test works.

b. Write an answer to the question, "Which tests, if any, are interesting but too complicated to be of much practical value in checking divisibility?" Present your conclusions and reasons for your choices.

69. Historical Pathways. Marin Mersenne (1588–1648) formulated the problem of finding *multiply perfect* numbers, in which the sum of the proper factors of a number is a multiple of the number itself. Show that 120 and 672 are multiply perfect numbers.

70. Making Connections. How do the ideas of divisibility and multiples relate to the study of fractions?

Section 4.2 PRIME AND COMPOSITE NUMBERS

- Defining Prime and Composite Numbers
- Techniques for Finding Prime Numbers
- The Role of Prime Numbers in Mathematics
- Greatest Common Factor and Least Common Multiple
- Prime and Composite Numbers and Relationships

In this section, we use the concept of factors of a number to develop the ideas of prime and composite numbers. We illustrate why prime numbers are considered *building blocks* for natural numbers and develop ways to find the prime factorization of any natural number. Prime factorization is then used to find the greatest common factor and least common multiple of two numbers. Finally, we explore some interesting patterns and relationships involving prime numbers. Mini-Investigation 4.2 asks you to consider members of a special set of numbers that will play a major role in this section.

Draw pictures on graph paper to illustrate that the numbers 8, 9, 10, and 12 can be shown by more than just a single-row rectangle.

Mini-Investigation 4.2 **Making a Connection**

Which numbers from 2 to 30 can be shown only by a single-row rectangle? Note, for example, that the number 4 can be shown by both a single-row rectangle and a double-row rectangle.

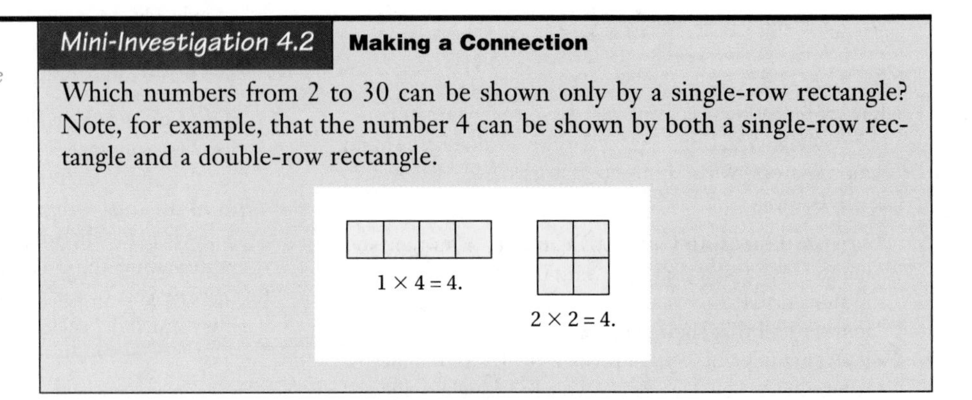

Defining Prime and Composite Numbers

In Mini-Investigation 4.2, you classified the numbers 2 through 30 into two sets, those that must be shown by a single-row or column rectangle (2, 3, 5, 7, 11, ...) and those that can be shown by a rectangle with more than one row of squares (4, 6, 8, 9, 10, 12, ...). The first five models for these numbers are shown in Figure 4.3.

Factors of the modeled numbers may be determined by looking at the length and the width of the rectangles. Note that the single-row numbers have exactly two factors, the number itself and 1. The multiple-row numbers have more than two factors. This classification suggests the following definition.

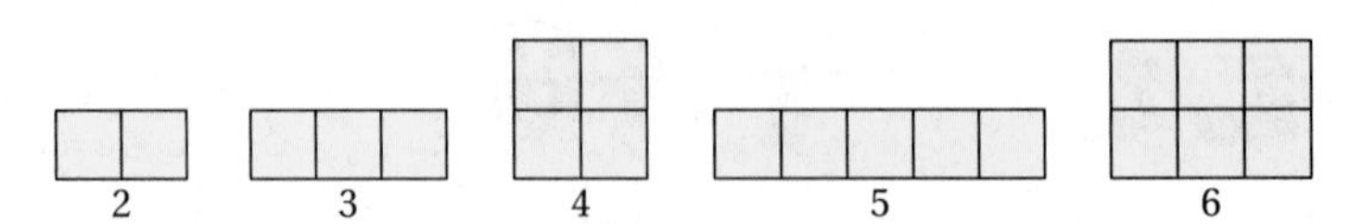

FIGURE 4.3 | Geometric models for five of the numbers 2–30.

Definition of Prime and Composite Numbers

A natural number that has exactly two distinct factors is called a **prime number.** A natural number that has more than two distinct factors is called a **composite number.**

Therefore the prime numbers are the numbers that may be shown only as a single-row rectangle and that have exactly two factors. The composite numbers may be shown by a rectangle that has more than one row of squares and thus have more than two factors. For example, the number 5 is a prime number because it has exactly two factors, 1 and 5. The number 6 is a composite number because it has more than two factors, 1, 2, 3, and 6.

PhotoDisc © 2001
Radio telescopes might send prime number sequences into space to communicate with intelligent beings in other galaxies.

The number 1 has only *one* distinct factor, so it is neither prime nor composite. This classification of the natural numbers greater than 1 into the two sets, prime and composite, gives us a useful language for discussing number properties and relationships. Exercise 45 at the end of this section will help you consider some common misconceptions about prime numbers.

Techniques for Finding Prime Numbers

Mini-Investigation 4.3 gives you an opportunity to use and analyze a method for identifying prime numbers that the Greek mathematician, Erastosthenes (200 B.C.), developed. It is called the **Sieve of Erastosthenes.**

A Sample Student Page: Representing Prime and Composite Numbers

The NCTM *Principles and Standards for School Mathematics* focus on the important role of representation in school mathematics learning. Some excerpts from the standards for grades 3–5 illustrate this focus. "Teachers can and should emphasize the importance of representing mathematical ideas in a variety of ways .... Many students need support in constructing pictures, graphs, tables, and other representations .... Organizing work in this way highlights patterns ... ." Study the page from *Scott-Foresman–Addison Wesley Math,* Book 5 (1998) shown in Figure 4.4 and answer the questions to see the importance of representations in developing number theory ideas.

- **What** number theory ideas are developed here?
- **What** different types of representations are used?
- **Comment** on the role of models and looking for patterns.

Chapter 4 Lesson 14

Exploring Prime and Composite Numbers

Problem Solving Connection
- Look for a Pattern
- Use Logical Reasoning

Materials
hundred chart

Vocabulary
prime number whole number with exactly two different factors
composite number whole number greater than 1 that is not prime

Explore

Eratosthenes (*ehr uh TAHS thuh neez*) was a mathematician from Cyrene (now known as Libya) who lived about 2,200 years ago. He studied **prime numbers**. The number 3 is an example of a prime number. It has only two factors, 1 and 3.

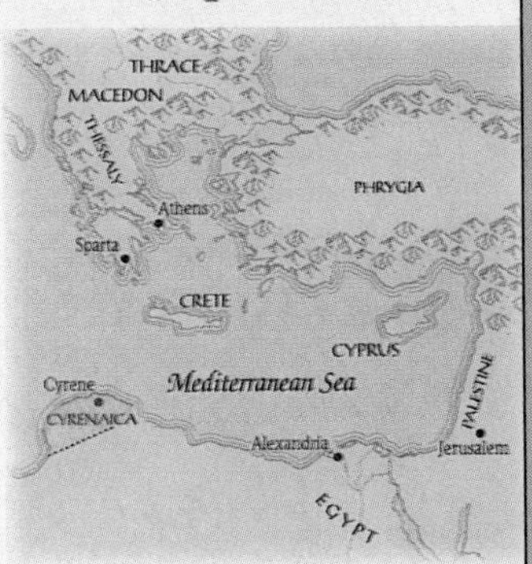

Cyrene was the capital of ancient Cyrenaica.

Composite numbers have more than two factors. An example of a composite number is 8. Its factors are 1, 2, 4, and 8.

1×8 ● ● ● ● ● ● ● ● 2×4 ● ● ● ● / ● ● ● ●

You tell whether a number is prime by testing possible factors. There are no other factors for 7 besides 7 and 1. So, 7 is a prime number.

1×7 ● ● ● ● ● ● ●

Eratosthenes' Sieve is a process that "strains" out composite numbers and leaves prime numbers behind. Use it to find the prime numbers between 1 and 100.

Did You Know?
A computer has been used to find a prime number that, if printed in a newspaper, would fill 12 pages.

Work Together

1. Use a hundred chart. Follow the directions to cross out composite numbers and circle prime numbers.
   a. Cross out 1. It has only 1 factor.
   b. Circle 2, the least prime number. Cross out all numbers divisible by 2.
   c. Repeat this step with 3, the next prime number.
2. Continue this process until you reach 100.
3. List all of the circled numbers. There should be 25.

1	2	3	4	5	6	7	8	9	10
11	12	13	14	15	16	17	18	19	20
21	22								

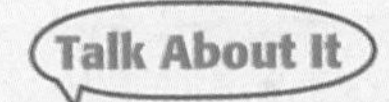

What pattern can you see in the list of prime numbers?

FIGURE 4.4 | Excerpt from *Scott Foresman–Addison Wesley Math.*

(*Source:* Page 204 from *Scott Foresman–Addison Wesley Math* Grade 5 by Randall I. Charles et al. Copyright © 1998 by Addison Wesley Longman, Inc., Reprinted by permission of Pearson Education, Inc.)

Write a description of what you know about the factors of the numbers crossed out and those not crossed out.

Mini-Investigation 4.3 **Finding a Pattern**

What do you discover about the numbers that remain in the following list after you cross out all the multiples of 2 other than 2 and do the same for 3, 5, and 7?

	2	3	4	5	6	7	8	9	10
11	12	13	14	15	16	17	18	19	20
21	22	23	24	25	26	27	28	29	30
31	32	33	34	35	36	37	38	39	40
41	42	43	44	45	46	47	48	49	50

The method presented in Mini-Investigation 4.3 identifies prime numbers in a list of consecutive natural numbers. Each time multiples of a number are crossed out, numbers that have that number as a factor are eliminated. Continuing this process eliminates all numbers that have more than two factors, leaving the prime numbers. This method identifies prime numbers from a consecutive list, but it isn't particularly useful in answering a question such as, "Is the number 323 prime?" To answer such a question, we would probably use a calculator to identify the factors of the number, as in Example 4.6.

Example 4.6 **Using a Calculator to Identify Prime Numbers**

Use a calculator to decide whether 323 is a prime number.

Solution

Janell's thinking: Here's how I used a scientific calculator. I entered 323 into memory.

Enter	Push
323	[M+]

I got the square root of 323. Because of the factor test theorem, I need test only factors through 17.

Push	Display	Push	Display
[MR]	323	[√]	17.972201

I tested all necessary factors and noted that 17 is a factor of 323.

Push	Display	Push	Enter	Push	Display
[MR]	323	[÷]	17	[=]	19

The number 323 isn't prime.

Tom's thinking: I used the table feature of a graphing calculator. First, I calculated the square root of 323 in order to determine the divisors to be checked, by entering the following.

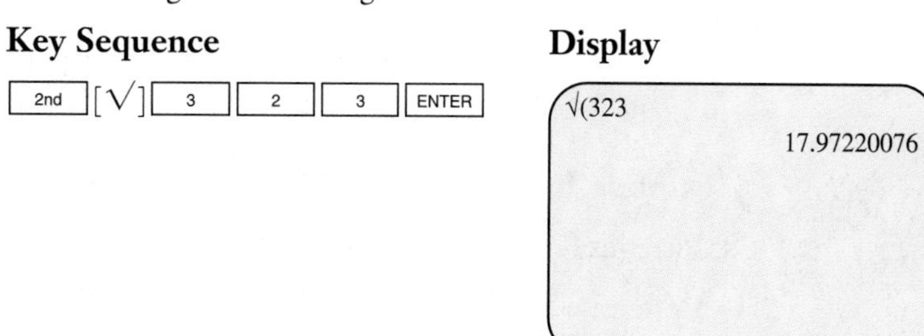

Next, I entered the function Y1 = $323/x$ under the [Y=] menu so that the calculator would divide 323 by successive values of x.

Key Sequence **Display**

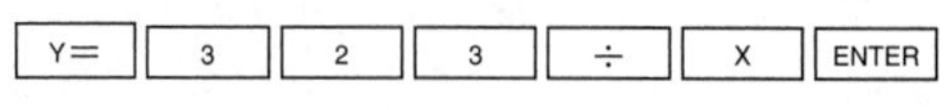

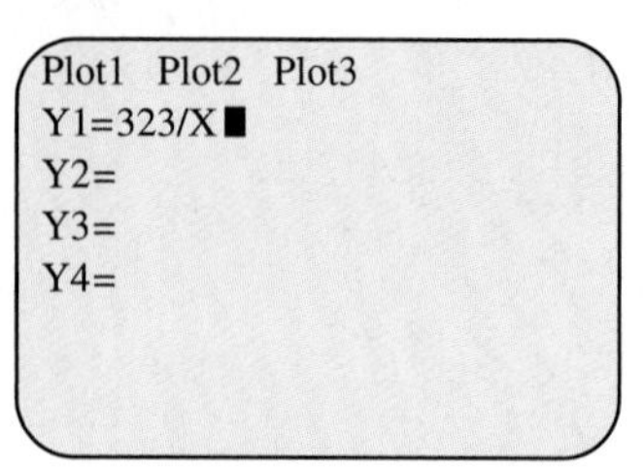

Then I used the table setup menu to set the calculator up to start with an initial value of 1 and to increase the size of x by 1 each time.

Key Sequence **Display**

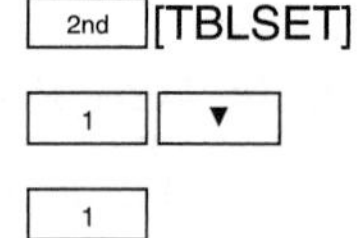

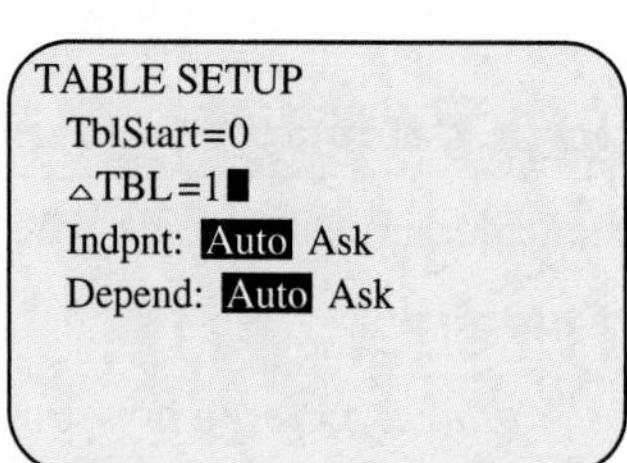

Finally, I displayed the table and used the arrow keys to scroll the Y1 column to search for whole-number results.

Key Sequence **Display**

X	Y1
1	323
2	161.5
3	107.67
4	80.75
5	64.6
6	53.833

X=1

X	Y1
12	26.917
13	24.846
14	23.071
15	21.533
16	20.188
17	19

Y1=19

When the divisor is 17, the quotient is exactly 19. Therefore 323 is not a prime number; it has factors other than 1 and itself.

Your Turn

Practice: Use a calculator to decide whether 431 is a prime number.

Reflect: After you tested the numbers 2, 3, and 5, did you need to test any multiples of these numbers? Explain. ■

Spotlight on Technology **Graphing Calculator Programs**

You have used a graphing calculator to do several different types of calculations. Now you will learn how to program graphing calculators to carry out repetitive calculations to complete many other tasks. In this feature, we focus on *programming a graphing calculator to determine whether a given number is prime.*

Follow the instructions provided to enter the program into one type of graphing calculator, a TI-73. With modest adaptation, the program can be used in many other types of graphing calculators.

- Press the keys [PRGM] [►] [►] [ENTER] to name your new program.
- Press [2nd] [TEXT] to display the keyboard for typing. Use the arrow keys and [ENTER] to select the letters from the alphabet table to type in your program name, for example, PRIMES.
- When you finish typing your program name, arrow to highlight [DONE] and press [ENTER] to finish typing. Press [ENTER] a second time to get into the program editor. Note that the screen shows your program name and a flashing cursor to the right of a colon (:). This display indicates you now can type in your program.
- Following these instructions, each line of the program PRIMES is shown, along with information on how to access the commands. *After you complete a line, always press* [ENTER] *to move to the next line. You do not need to type in a colon; each one is automatically placed after you hit* [ENTER].
- After entering the program PRIMES, run it. Press [PRGM] and then use the down-arrow key to highlight PRIMES. Press [ENTER] [ENTER] to start the program. The screen to the left shows that 1258 is not prime and the screen to the right shows that 56,989 is prime. Can you find a prime number larger than 56,989?

Program for Deciding Whether a Number Is Prime

`:INPUT "NUMBER=",N`	[PRGM] [►] [1] [2nd] [TEXT] and type "NUMBER=",N Finish typing by highlighting [DONE] and pressing [ENTER]
`:FOR (K,2,√(N),1)`	Press [PRGM] [4] to enter the **for(** command. Press [2nd] [TEXT] to enter K,2,√(N),1)
`:IF N/K = ROUND(N/K,0)`	Press [PRGM] [1] to enter **if,** [2nd] [TEXT] to enter N/K, and exit with [DONE] [ENTER]. Press [MATH] [►] [2] to enter **round(**. Press [2nd] [TEXT] to enter N/K,0) and exit the keyboard with [DONE] and [ENTER].
`:GOTO 1`	[PRGM] [0] [1]

Program line	Keystrokes
:END	PRGM 7
:CLRscreen	PRGM ► 8
:DISP N,"IS PRIME"	PRGM ► 3 , 2nd [TEXT] N,"IS PRIME" DONE ENTER
:GOTO 2	PRGM 0 2
:LBL 1	PRGM 9 1
:CLRscreen	PRGM ► 8
:DISP N,"IS NOT PRIME"	PRGM ► 3 2nd [TEXT] N,"IS NOT PRIME" DONE ENTER
:LBL 2	PRGM 9 2
:PAUSE	PRGM 8
:STOP	PRGM ▲ ▲ ▲ ENTER

The Role of Prime Numbers in Mathematics

Prime Numbers as Building Blocks. The prime numbers are building blocks for composite numbers, as you can see by breaking a composite number such as 30 into a product of prime factors. Factor trees, three of which are shown in Figure 4.5, graphically show this process. Even though three different factor trees are shown, the factors in the bottom row are the same, regardless of the order in which they appear. In each factor tree, the product of factors for 30 consists of 2, 3, and 5 multiplied in some order.

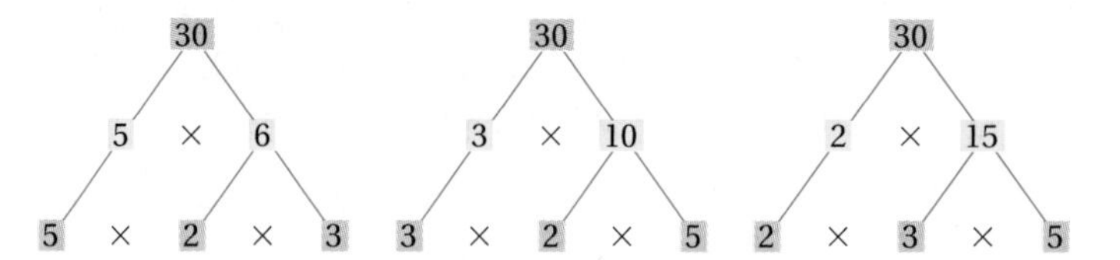

FIGURE 4.5 | Factor trees for the number 30.

The three factor trees shown suggest the following important theorem, which Euclid included in his book *Elements* in 320 B.C.

Fundamental Theorem of Arithmetic (Unique Factorization Theorem)

Each composite number can be expressed as the product of prime numbers in exactly one way, disregarding the order of the factors.

This theorem guarantees that each composite number is completely identified by a unique product of prime numbers, which might be considered its "fingerprint." Two different composite numbers cannot equal the same product of primes, and two different products of primes cannot equal the same composite number. The number 1 is defined as neither a prime number nor a composite number, because if 1 were one of these types of numbers, we could write $6 = 1 \times 2 \times 3$ and $6 = 1 \times 1 \times 2 \times 3$, and the prime factorization of 6 would not be unique.

Finding the Prime Factorization of a Number. When a number is expressed as a product of primes, the expression is called the **prime factorization** of the number. For example, $42 = 2 \times 3 \times 7$, so $2 \times 3 \times 7$ is the prime factorization of 42. When prime factors repeat in a prime factorization, we represent them with *exponents*. For $360 = 2 \times 2 \times 2 \times 3 \times 3 \times 5$, we write $360 = 2^3 \times 3^2 \times 5$.

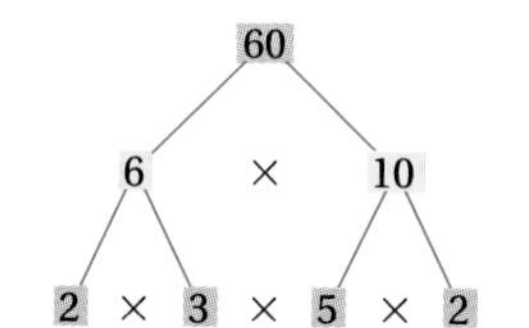

FIGURE 4.6 | The prime factorization of $60 = 2^2 \times 3 \times 5$.

Factor trees, such as those in Figure 4.5, can be used to record the process for finding the prime factorization of a number. For example, to find the prime factorization of 60, begin by breaking 60 into the product of two factors. Then break each of those factors into the product of two factors, and so on, until the numbers in the bottom row of the factor tree are prime and the prime factorization of 60 appears. This process for finding the prime factorization of 60 is depicted in Figure 4.6.

A *stacked division* procedure, based on the fundamental theorem of arithmetic, can also be used to find the prime factorization of 60. These two procedures show that $60 = 2^2 \times 3 \times 5$. It may also be used to find the prime factorization of a number.

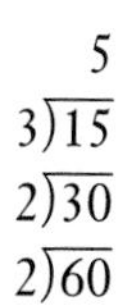

To use the division method for finding prime factorizations, choose the smallest prime that will divide the number. Divide by the prime and determine whether this prime will also divide the quotient. Repeat with successive quotients as long as possible. When this prime will not divide a quotient, use the next largest prime and continue the process. When you reach a quotient that is prime, you can see—in the spirit of the Frank and Ernest cartoon in Figure 4.7—a string of prime numbers, the quotient and divisors, that can be multiplied to produce the original number. The process essentially divides prime factors out of the number and uses these factors as the prime factorization. Example 4.7 illustrates how and then asks you to find the prime factorization of a number.

FRANK AND ERNEST® by Bob Thaves

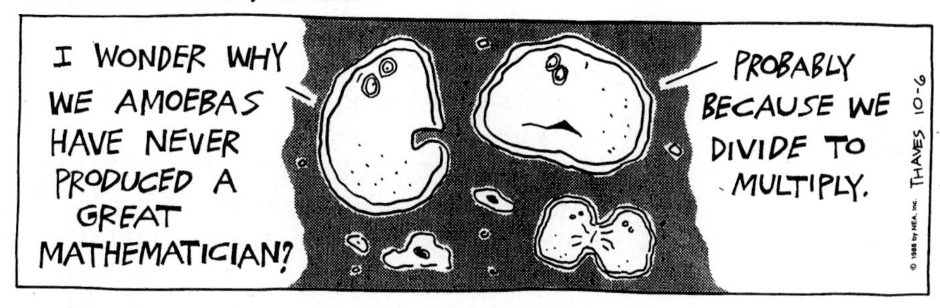

FIGURE 4.7 | Thaves/NEA, Inc. © 1988. Used with permission of Bob Thaves.

Example 4.7 Finding the Prime Factorization

Find the prime factorization of 126.

Solution

Dana's thinking: I just made a factor tree for 126. The bottom row of the factor tree is the prime factorization.

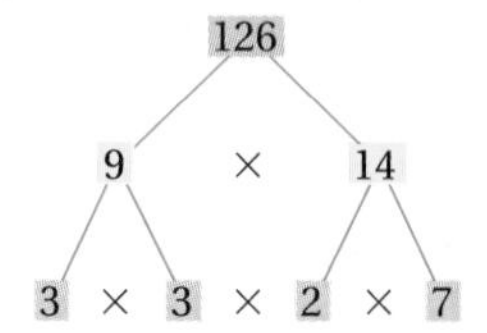

Jordan's thinking: I thought, 126 is 2 times 63, but 63 is 9 times 7 or 3 times 3 times 7. Then I wrote the equation $126 = 2 \times 3^2 \times 7$.

Wesley's thinking: I used the following division method.

$$\begin{array}{r} 7 \\ 3\overline{)21} \\ 3\overline{)63} \\ 2\overline{)126} \end{array}$$

The prime factorization of 126 is $2 \times 3^2 \times 7$.

Your Turn

Practice: Write the prime factorizations of 56, 150, and 252. Use exponents.

Reflect: How do you know that the answer to the example problem is the only prime factorization of 126? ■

Greatest Common Factor and Least Common Multiple

The ideas about sets you learned in Chapter 2 and the idea of the prime factorization of a number can be used to help find two important numbers—the greatest number that is a factor of two given numbers and the smallest number that is a multiple of two given numbers. These procedures will be developed in following subsections.

Greatest Common Factor. By listing all factors you could conclude that 16, for example, is the greatest number that is a factor of both 64 and 48. This conclusion suggests the following definition of a useful idea in elementary number theory.

Definition of Greatest Common Factor

The **greatest common factor (GCF)** of two natural numbers is the greatest natural number that is a factor of both numbers.

Note that the greatest common factor is sometimes called the greatest common divisor, and in that case is represented by GCD.

Using the Intersection of Sets to Find the Greatest Common Factor. To find the GCF when dealing with a pair of numbers that have a reasonably small number of factors, you can just list the factors of each number, identify the common factors, and select the greatest of these common factors. The Venn diagram in Figure 4.8 shows how this could be done.

This can be expressed using set notation as follows.

A = the set of factors of 24 = {1, 2, 3, 4, 6, 8, 12, 24}.
B = the set of factors of 30 = {1, 2, 3, 5, 6, 10, 15, 30}.
$A \cap B$ = the set of common factors of 24 and 30 = {1, 2, 3, 6}.

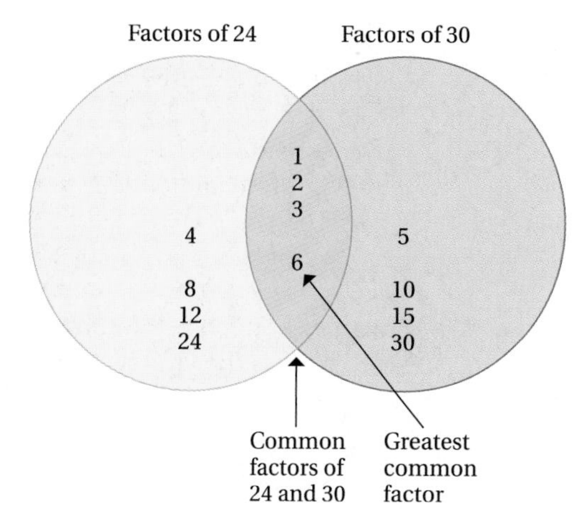

FIGURE 4.8 | Intersection of factor sets method of finding the GCF.

The greatest number in the set of common factors, 6, is the GCF of 24 and 30.

Using Prime Factorization to Find the GCF. A method for finding the GCF that is more efficient than listing factors, especially when dealing with larger numbers, involves writing the prime factorization of each number. First, write the prime factorization of each number.

$$24 = 2 \times 2 \times \underline{2 \times 3}$$
$$30 = \underline{2 \times 3} \times 5$$

Since 2×3 is the largest factor contained in both of the prime factorizations, 2×3, or 6, is the GCF of 24 and 30. In Example 4.8 you can use the GCF to solve a practical problem.

Example 4.8

Problem Solving: The Largest Square Tile

What is the largest size of square tile that could be used in making a mosaic that is 64 inches long and 48 inches wide?

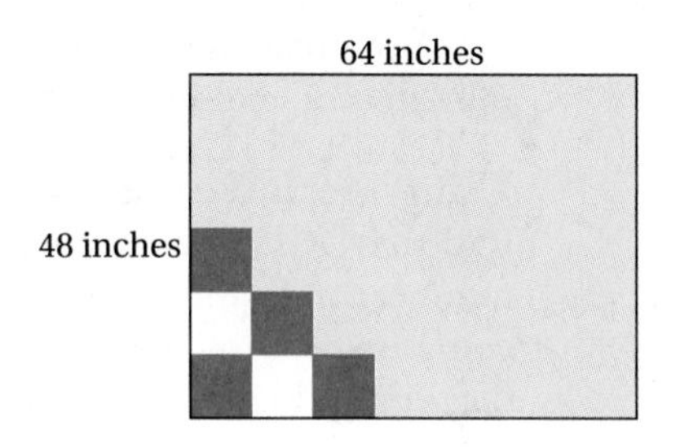

Solution

Kiree's thinking: To solve the problem, I need to find the GCF of 48 and 64. The factors of 48 are 1, 2, 3, 4, 6, 8, 12, 16, 24, and 48. The factors of 64 are 1, 2, 4, 8, 16, 32, and 64. Because 16 is the largest number in both lists, it's the GCF. The largest possible tile is 16 inches square.

Tavin's thinking: I used prime factorization and started with $48 = 2 \times 2 \times 2 \times 2 \times 3$, and $64 = 2 \times 2 \times 2 \times 2 \times 2 \times 2$. I discovered that $2 \times 2 \times 2 \times 2$ is contained in both prime factorizations, so the GCF of 48 and 64 is $2 \times 2 \times 2 \times 2$, or 16. The largest possible tile is 16 by 16.

Your Turn

Practice: Suppose the mosaic in the example is 60 inches long and 45 inches wide. What is the largest square tile that could be used?

Reflect: Which of the two methods in the example problem seems easier? Explain. ■

The **Euclidean algorithm** is the name given to a method that can be used with a calculator to find the GCF of a pair of larger numbers. The GCF of a pair of numbers is the same as the GCF of the smaller of the two numbers and the remainder when the larger is divided by the smaller. For example, GCF(30, 24) = 6, and GCF(24, 6) = 6. This idea is the basis for the Euclidean algorithm, as described by the following procedure.

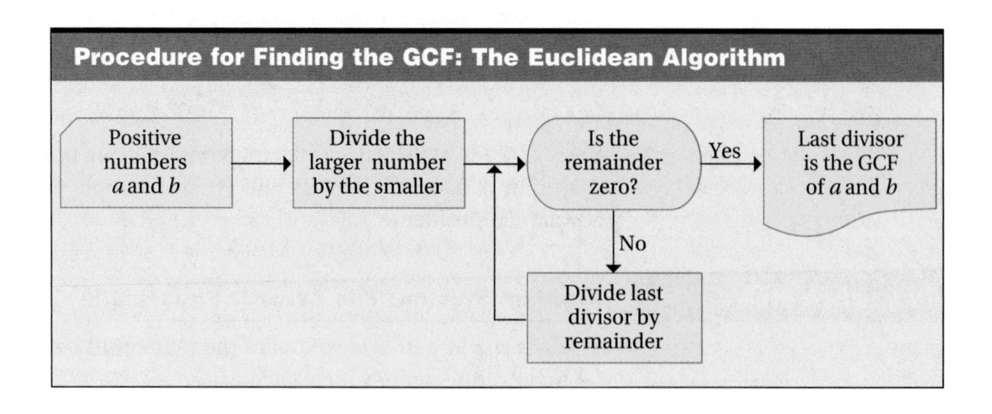

The Euclidean algorithm allows you to reduce the size of the numbers continually until you can easily find the GCF. Let's use the flow chart and the following divisions to help explain how to find the GCF of 84 and 308.

$$\text{a) } 84\overline{)308} = 3,\ 308 - 252 = 56 \qquad \text{b) } 56\overline{)84} = 1,\ 84 - 56 = 28 \qquad \text{c) } \textcircled{28}\overline{)56} = 2,\ 56 - 56 = 0$$

In Equation a, we divide the larger of the two numbers, 308, by the smaller, 84. The remainder, 56, isn't zero, so use it as the new divisor. In Equation b, we divide the last divisor, 84, by the remainder, 56. The remainder, 28, isn't zero, so use it as the new divisor. Finally, in Equation c, we divide the last divisor, 56, by the remainder, 28. The remainder is zero, so the last divisor, 28, is the GCF of 84 and 308.

Example 4.9 illustrates how the integer division feature on a calculator can be used along with the Euclidean algorithm to find the GCF of a pair of larger numbers.

Example 4.9 Using a Calculator to find GCF

Use a calculator that divides integers and shows remainders and the Euclidean algorithm method to find the GCF of 475 and 1501.

Solution

Key Sequence	Display
1501 [INT ÷] 475 [=]	3 76
475 [INT ÷] 76 [=]	6 19
76 [INT ÷] 19 [=]	4 0

The GCF of 1501 and 475 is 19.

Your Turn

Practice: Use a calculator with an [INT ÷] key to find the GCF of 2940 and 1260.

Reflect: How could you use a calculator *without* an [INT ÷] key to find the GCF of 2599 and 2825? ■

Least Common Multiple. It is useful to be able to find the smallest number that is a multiple of each of two given numbers. Suppose the two numbers are 9 and 12. By listing multiples of each, you could conclude that 36 is the smallest number that is a multiple of both 9 and 12. This conclusion suggests the following definition of a useful idea in elementary number theory.

Definition of Least Common Multiple

The **least common multiple (LCM)** of two natural numbers is the smallest natural number that is a multiple of both the natural numbers.

Two useful methods for finding the least common multiple of two numbers will be described in the following subsections.

Using Intersection of Sets to Find the Least Common Multiple. When dealing with a pair of small numbers, you can just list some of the multiples of each number, identify the common multiples, and select the smallest non-zero common multiple. The Venn diagram in Figure 4.9 shows this procedure.

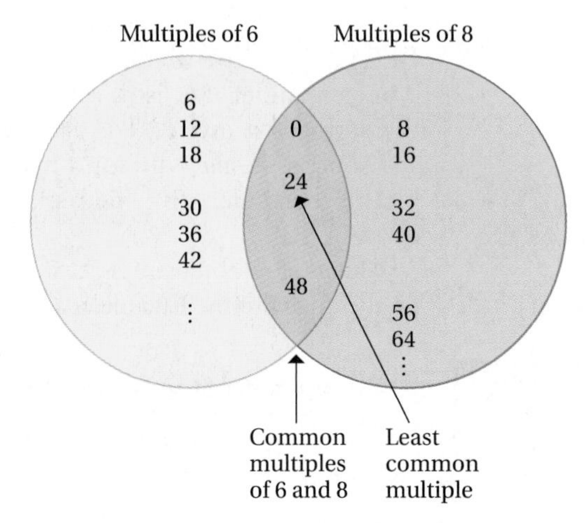

FIGURE 4.9 | Intersection of sets of multiples method for finding the LCM.

This can be expressed using set notation as follows.

C = the set of multiples of 6 = {0, 6, 12, 18, 24, 30, 36, 42, 48, ...}
D = the set of multiples of 8 = {0, 8, 16, 24, 32, 40, 48, 56, 64, ...}
$C \cap D$ = the set of multiples of 6 and 8 = {0, 24, 48, ...}

The smallest common multiple, 24, is the LCM of 6 and 8.

Using Prime Factorization to Find the Least Common Multiple. A method for finding the LCM that is more efficient than listing multiples, especially when dealing with larger numbers such as 28 and 30, involves writing the prime factorization of each number. First, write the prime factorization of each number.

$$28 = 2 \times 2 \times 7$$
$$30 = 2 \times 3 \times 5$$

Any non-zero multiple of 28 must at least have the factors 2, 2, and 7. Any non-zero multiple of 30 must at least have the factors 2, 3, and 5. Using these prime factors the largest number of times it appears in either of the prime factorizations, we find that the smallest multiple that has the necessary factors to be a multiple of both 28 and 30 is $2 \times 2 \times 3 \times 5 \times 7$, or 420. The LCM of 28 and 30 is 420.

Example 4.10 **Problem Solving: Orbiting Spacecrafts**

Two spacecrafts have elliptical orbits around the earth. Spacecraft Alpha makes one complete orbit in 90 minutes. Spacecraft Beta makes one complete orbit in 120 minutes. At this moment, the spacecrafts are beside each other in their orbits. In how many minutes will they be beside each other again?

Solution

Regina's thinking: The multiples of 120 are 120, 240, 360, etc. Multiples of 90 are 90, 180, 270, 360, etc., so the LCM of 120 and 90 is 360. The satellites will be in the same location after 360 minutes.

Vito's thinking: I used prime factorization and started with $120 = 2 \times 2 \times 2 \times 3 \times 5$, and $90 = 2 \times 3 \times 3 \times 5$. The LCM has to have three 2's, two 3's, and a 5 in it, so $2 \times 2 \times 2 \times 3 \times 3 \times 5$ works. The LCM of 120 and 90 is 360. It will take 360 minutes for the satellites to be in the same location.

Your Turn

Practice: Find the LCM of 9 and 12; 12 and 18; 28 and 42.

Reflect: Which of the students' thinking in the example solution makes the most sense to you? Explain why. ■

The GCF and LCM of two numbers can easily be found when using exponent notation in the prime factorization of the two numbers. For example $12 = 2^2 \times 3^1$, and $18 = 2^1 \times 3^2$. The GCF is found by using the *minimum* exponent for each prime power in the prime factorization of the numbers, so GCF (12, 18) = $2^1 \times 3^1$, or 6. The LCM, on the other hand, is found by using the *maximum* exponent for each prime power in the prime factorizations, so LCM (12, 18) = $2^2 \times 3^2$, or 36.

Exercise 50 at the end of this section gives another interesting procedure for finding the LCM of two numbers.

The NCTM *Principles and Standards for School Mathematics* state that "In grades 3–5 students should have learned to generate equivalent forms of fractions ..." (p. 215). "Middle grades students should continue to refine their understanding of addition, subtraction, multiplication, and division as they use these operations with fractions..." (p. 218). The ability to find the GCF of two numbers is used to find an equivalent, lowest-term representation for a fraction. Also, the LCM is used to find the least common denominator when adding fractions. Both ideas are used in algebra to simplify rational expressions.

Prime and Composite Numbers and Relationships

Patterns and relationships involving prime and composite numbers have always fascinated mathematicians and students alike. In this subsection, we present some of the interesting relationships involving LCM and GCF, as well as relationships involving prime numbers.

Relationships Involving GCF and LCM. As you complete Table 4.4, you might discover an answer to the simple question "How is the GCF related to the LCM?"

TABLE 4.4 | Discovering a Relationship between the GCF and the LCM

Pair of Numbers *a*, *b*	*a* × *b*	GCF (*a*, *b*)	LCM (*a*, *b*)	GCF (*a*, *b*) × LCM (*a*, *b*)
6, 8	48	2	24	48
6, 10	60	2	30	60
8, 9	72	1	72	72
9, 12	?	?	?	?
9, 15	?	?	?	?
12, 20	?	?	?	?

Using inductive reasoning and the table, you might generalize the following theorem.

Theorem: The GCF–LCM Product

The product of the GCF and the LCM of two numbers is the product of the two numbers.

The above theorem and the idea that numbers a and b are **relatively prime** if and only if GCF $(a, b) = 1$ can be used to answer other interesting relationship questions, such as the one in Exercise 42 at the end of this section. Example 4.11 applies the GCF–LCM theorem.

Example 4.11

Using the GCF–LCM Product Theorem

Find the GCF and LCM of 888 and 259.

Solution

Ryan's thinking: I first used the Euclidean algorithm and a calculator with the integer divide feature to find the GCF.

Key Sequence	**Display**
888 [INT ÷] 259 [=]	3 111
259 [INT ÷] 111 [=]	2 37
111 [INT ÷] 37 [=]	3 0

The GCF of 888 and 259 is 37. Using the GCF–LCM theorem, I then found the LCM.

$$\begin{aligned} \text{GCF } (888, 259) \times \text{LCM } (888, 259) &= 888 \times 259 = 229{,}992 \\ 37 \times \text{LCM } (888, 259) &= 229{,}992 \\ \text{LCM } (888, 259) &= 229{,}992 \div 37 = 6216 \end{aligned}$$

Lauren's thinking: I used a fraction calculator to find the GCF. Then I used the GCF–LCM theorem to find the LCM. I first entered the two numbers as a fraction into my calculator.

Key Sequence	**Display**
259 [/] 888	259/888

Now I use the [Simp] function to reduce the fraction.

Key Sequence	**Display**
[Simp=]	7/24

When I press [x⇄y], the common factor appears that was used to reduce the fraction.

Key Sequence	**Display**
[x⇄y]	37

(If the lower left corner of my calculator screen shows **N/D → n/d**, indicating that 37 isn't the GCF, I would need to use the [Simp] again and multiply common factors used for reducing to find the GCF.)

In this case, 37 is the GCF of 888 and 259. The LCM is $(888 \times 259) \div 37$, or 6216.

Your Turn

Practice: Find the GCF and LCM of

a. 1501 and 475.
b. 2599 and 2825.

Reflect: Explain in your own words how to find LCM (a, b) easily when you know GCF (a, b). ■

Relationships and Patterns Involving Prime Numbers. One interesting prime number topic is the frequency of occurrence of consecutive primes with a difference of 2, such as 3 and 5, which are called **twin primes.** Mini-Investigation 4.4 lets you search for some interesting prime number patterns, including those involving twin primes, in Table 4.5. To use this table, note that the 43rd prime, for example, appears at the intersection of row 4 and column 3; it is 191.

Make a graph to support your conclusion.

Mini-Investigation 4.4 **Finding a Pattern**

Do you think that the number of pairs of twin primes increases or decreases with each succeeding interval of 100 prime numbers?

The table of the first 199 primes raises several interesting questions.

- **Is there a largest prime number?** In answering *no* to this question, Euclid (300 B.C.) gave a clever proof. We present some of the ideas from Euclid's proof here. Euclid claimed that a largest prime number couldn't exist because if there were a largest prime, P, it would be contradicted by the fact that you could always produce a larger prime, N, by multiplying P by all the primes smaller than P and adding 1, namely, $N = (2 \times 3 \times 5 \times 7 \times 11 \times \ldots \times P) + 1$. To support this claim, Euclid had to prove two things. The first was that N is larger than P. That wasn't difficult to prove

TABLE 4.5 | The First 199 Prime Numbers

	0	1	2	3	4	5	6	7	8	9
0		2	3	5	7	11	13	17	19	23
1	29	31	37	41	43	47	53	59	61	67
2	71	73	79	83	89	97	101	103	107	109
3	113	127	131	137	139	149	151	157	163	167
4	173	179	181	191	193	197	199	211	223	227
5	229	233	239	241	251	257	263	269	271	277
6	281	283	293	307	311	313	317	331	337	347
7	349	353	359	367	373	379	383	389	397	401
8	409	419	421	431	433	439	443	449	457	461
9	463	467	479	487	491	499	503	509	521	523
10	541	547	557	563	569	571	577	587	593	599
11	601	607	613	617	619	631	641	643	647	653
12	659	661	673	677	683	691	701	709	719	727
13	733	739	743	751	757	761	769	773	787	797
14	809	811	821	823	827	829	839	853	857	859
15	863	877	881	883	887	907	911	919	929	937
16	941	947	953	967	971	977	983	991	997	1009
17	1013	1019	1021	1031	1033	1039	1049	1051	1061	1063
18	1069	1087	1091	1093	1097	1103	1109	1117	1123	1129
19	1151	1153	1163	1171	1181	1187	1193	1201	1213	1217

Estimation Application

A Big Number

Recently, the largest known prime number had 2,098,960 digits. Estimate the length of a strip of adding machine tape needed to hold this prime if it is typed on one line on the tape. Would it be as long as

a. less than half a football field?

b. one football field?

c. two football fields?

d. three football fields?

e. four or more football fields?

Make any necessary assumptions and calculate to check your estimate.

because $1 \times P + 1$ is greater than P, so $(2 \times 3 \times 5 \times 7 \times 11 \times \cdots \times P) + 1$ would be greater than $1 \times P + 1$ and hence also greater than P. The second thing that Euclid had to prove was that $N = (2 \times 3 \times 5 \times 7 \times 11 \times \cdots \times P) + 1$ is a prime number. That can also be proven because when N is divided by any prime, the remainder is always 1. A number, N, that isn't divisible by any other prime has only itself and 1 as factors and is, by definition, a prime number. Using this line of reasoning, Euclid was able to convince his mathematician friends, and others, that there is no largest prime number.

What is the largest prime number that has been found? The advent of the computer generated considerable activity in the search for larger and larger prime numbers. In 1978, two high school students, Laura Nickel and Curt Noll, of Hayward, California, showed that $2^{23,209} - 1$ was a prime. As this book went to press, the largest known prime was the number $2^{6,972,593} - 1$. Keeping up with the newest discoveries of largest prime numbers is extremely difficult, and a new largest prime may have been discovered since this book was printed. The estimation application gives a sense of the size of the current largest known prime number mentioned above.

Can two primes have an unlimited number of composites between them? To answer this question, first recall that $n!$, read "***n* factorial,**" represents the number $1 \times 2 \times 3 \times 4 \times 5 \times \cdots \times n$. For example, 5! represents

$1 \times 2 \times 3 \times 4 \times 5$, or 120. Then, if you want to find 100 consecutive composites, you can by using factorials and writing $(101! + 2)$, $(101! + 3)$, $(101! + 4)$, $(101! + 5)$, …, $(101! + 99)$, $(101! + 100)$, and $(101! + 101)$. The divisibility theorem, $a \mid b$ and $a \mid c \rightarrow a \mid (b + c)$, may be used to show that the first number in the list is divisible by 2, the second by 3, the third by 4, and so on. Therefore each of the 100 consecutive numbers is a composite number. By using any number you choose in place of 101, you can write as many consecutive composites as you want.

- **Are there still some unsolved problems in number theory?** This question can be answered by giving some interesting examples. Christian Goldbach (1690–1764) hypothesized that "every even number greater than 2 is the sum of two primes." *Goldbach's conjecture* may be illustrated by noting that $4 = 2 + 2$, $6 = 3 + 3$, $8 = 5 + 3$, $10 = 7 + 3$, and $12 = 7 + 5$. The proof of this conjecture is still one of the unsolved problems in mathematics. The *twin prime conjecture* that "there are an infinite number of pairs of primes whose difference is two" also has not been proven.

Another interesting unsolved problem is proof of the *odd perfect number conjecture* that "there is no odd perfect number." Although mathematicians have shown that even perfect numbers are all expressible in the form $2^{p-1}(2^p - 1)$, where $2^p - 1$ is prime, the total absence of an odd perfect number has never been verified. Related to this is the unsolved *Mersenne prime conjecture* (Marin Mersenne, 1588–1648) that "there are infinitely many Mersenne primes of the form $2^p - 1$" and the lack of verification of the conjecture that there are infinitely many even perfect numbers.

Pierre Fermat (1601–1665), aware of the Pythagorean theorem and the fact that there are numbers a, b, and c such that $a^2 + b^2 = c^2$, conjectured that $a^n + b^n = c^n$ can have no solutions in non-zero integers a, b, and c for $n > 2$. He wrote in the margin of a book, "For this I have discovered a wonderful proof, but the margin is too small to contain it." Until recently, this *Fermat conjecture* had been an unproved theorem for more than 350 years. However, in October 1994, the English mathematician Andrew Wiles presented a proof of the Fermat conjecture that has been accepted as correct by mathematicians specializing in number theory. His proof is often referred to as "the proof."

Talk about ways in which you think working on an unsolved number-theory problem might be like the situation(s) you chose from (a) through (d).

Mini-Investigation 4.5 **Making a Connection**

Which of the following do you think would be like working on an unsolved number-theory problem?

a. Solving a "whodunnit" mystery
b. Searching for genealogy connections
c. Discovering what's wrong with a car that won't run
d. Climbing the tallest mountain in the world

Problems and Exercises | *for Section 4.2*

A. Reinforcing Concepts and Practicing Skills

1. How many factors does a prime number have?
2. List the first 10 prime numbers.
3. How can you identify a composite number?
4. List the composite numbers between 30 and 40.
5. Which of the following are prime numbers? Why or why not?
   a. 8 **b.** 11
   c. 1 **d.** 51
   e. 221
6. Give a counterexample to show that the generalization, all odd numbers are prime, is false.
7. Use factor trees to write the prime factorization of each number.
   a. 18 **b.** 48
   c. 130 **d.** 51
8. Use the division method to find the prime factorization of each number.
   a. 504 **b.** 1176
   c. 2600 **d.** 3675
9. Use Table 4.5 on p. 222 to decide how many pairs of twin primes (primes that differ by only 2) there are in prime numbers less than 100.
10. Use a calculator to decide if 437 and 541 are prime numbers. Explain how you decided.
11. Use a calculator to help you write the prime factorization of 16,731.
12. List factors of the numbers to find the GCF of 42 and 28.
13. Use the intersection of sets method to find the GCF of 12 and 20. Draw a Venn diagram to show the procedure you used.
14. Use the intersection of sets method to find the GCF of 18 and 24. Use set notation to describe the procedure you used.
15. Use a method of your choice to find the GCF of 80 and 124. Explain why you chose to use this method.
16. Use the prime factorization method to find the GCF of each pair of numbers.
    a. 28, 42 **b.** 45, 60
    c. 36, 54
17. Use a calculator and the Euclidean algorithm method to find the GCF of each pair of numbers.
    a. 259, 888 **b.** 84, 308
    c. 1232, 7560
18. List the multiples of the numbers to find the LCM of 9 and 12.
19. Use the intersection of sets method to find the LCM of 4 and 18. Draw a Venn diagram to show the procedure you used.
20. Use the intersection of sets method to find the LCM of 6 and 21. Use set notation to show the procedure you used.
21. Use a method of your choice to find the LCM of 18 and 24. Explain why you chose to use this method.
22. Use the prime factorization method to find the LCM of each pair of numbers.
    a. 27, 36 **b.** 42, 60
    c. 28, 40
23. The LCM of a pair of numbers is 36. The product of the numbers is 108. What is the GCF of the numbers?
24. Find the GCF of each triple of numbers.
    a. 16, 28, 40 **b.** 30, 36, 48
25. Find the LCM of each triple of numbers.
    a. 9, 12, 15 **b.** 28, 40, 56
26. A karate instructor wanted to form small groups from 24 students on Monday and 42 students on Tuesday, with the same number in each of the small groups. What is the largest small-group size possible?
27. What is the shortest length of television cable that could be cut into either a whole number of 18-ft pieces or a whole number of 30-ft pieces?
28. A city siren is set to go off every 24 hours. A civil defense alarm is tested every 36 hours. If the siren and the alarm sound at the same time, after how many hours will they sound at the same time again?
29. A stamp collector has 280 stamps from North America and 264 stamps from South America. The collector wanted to place the same number of stamps on each page of a large stamp book displaying stamps from the Americas. What is the greatest number of stamps the collector can place on each page? How many pages would the book have?
30. Two motorcycles start around a race course at the same time. One cycle passes the starting point every 12 minutes and the other passes it every 15 minutes. How many minutes would elapse before both cycles pass the starting line together?
31. A field is 70 ft by 525 ft. It is to be divided into square garden plots, all the same size, and with sides having whole-number length. What is the largest size garden plot that could be chosen?

B. Deepening Understanding

32. Evaluate the formula $p = n^2 - n + 11$ by substituting each of the numbers 1–11 for n. For which value(s) of n is p not a prime?

33. The formula $p = n^2 - n + 41$ produces primes for $n = 1, 2, 3, \ldots, 40$, but $n = 41$ produces a composite number. How can you convince someone that p is composite when $n = 41$ by just looking at the formula?

34. List all the primes less than 100 that are of the form $n^2 + 1$.

35. Marin Mersenne conjectured that, if p is prime, then the number $2^p - 1$ is a prime. Show that this outcome holds for the first four primes. Show that it breaks down for the fifth prime, $p = 11$.

36. Do you think that the formula $p = 6n - 1$ will produce primes more than 50% of the time? Explain your reasoning (*Hint:* Table 4.5 on p. 222 may help.)

37. Finish arranging the numbers 2 through 100 in 6 columns.

	2	3	4	5	6
7	8	9	10	11	12
13	14	15	16	17	18
...					

Cross out the multiples of 2, 3, 5, and 7. The remaining numbers are prime. Look for and describe patterns.

38. Use Table 4.5 on p. 222 to answer the following questions.

a. What can you say about the ones digit of the prime numbers greater than 2?

b. What conjecture would you make about the existence and frequency of three consecutive odd number primes?

c. Prime numbers such as 13 and 31 are sometimes called *reversal primes.* Find four more pairs of reversal primes.

39. A company executive claimed that his license plate number, 5773, was a prime number. Do you agree? Justify your conclusion.

40. The following BASIC program determines whether an input integer greater than 1 is or is not prime. If the BASIC language is available on your computer, type the program into your computer and try it. Do you think that you could use the program to test any number, no matter how large? Explain.

```
10 INPUT N
20 FOR K = 2 TO SQR(N)
30 IF N/K = INT (N/K) THEN 50
40 GO TO 70
50 PRINT N; " IS NOT PRIME "
60 GO TO 90
70 NEXT K
80 PRINT N; "IS PRIME"
90 PRINT "IF YOU WANT TO CHECK ANOTHER NUM-
   BER, TYPE 1"
100 PRINT "IF NOT, TYPE 0"
110 INPUT C
120 IF C = 1 GO TO 10
130 END
```

41. Is the GCF of a pair of numbers always larger or smaller than the LCM of the numbers? Explain with an example.

42. If the GCF of two numbers is 1, what is the LCM of the two numbers? Why?

C. Reasoning and Problem Solving

43. Suppose that E is the set that contains 1 and the even natural numbers, or $E = \{1, 2, 4, 6, 8, 10, 12, 14, \ldots\}$ Answer these questions about E.

a. Why is 6 a prime in E?

b. What are the first five prime numbers in E?

c. Do you see a pattern for the prime numbers in E? If so, what is it?

d. Do you see a pattern for composite numbers in E? If so, what is it?

e. Does the formula $E = 2e - 2$ generate primes in E?

f. Are there infinitely many primes in E? Explain.

g. What formula generates all composite numbers in E?

h. If e is an element of E, can $2e$ be a prime number in E? Why?

i. Are there any twin primes in E? (*Hint:* Think about numbers generated from $2e - 2$ and from $2e$, where e is a natural number in E.)

j. How does Goldbach's conjecture relate to prime numbers in E?

44. The following are some interesting theorems about number relationships by the mathematician who conjectured or proved the theorem. Choose some specific numbers and give an example for each theorem.

a. Fermat: Every prime number of the form $4x + 1$ is the sum of two unique square numbers.

b. Chebyshev: Between every whole number greater than 1 and its double there is at least one prime.

c. Euclid: For a natural number p, if $2^p - 1$ is prime, then $2^{p-1}(2^p - 1)$ is a perfect number.

d. Ulam: If a natural number is even, divide it by 2. If it is odd, multiply it by 3 and add 1. If this process is applied repeatedly, you will arrive at 1. (*Hint:* Try the numbers 16 and 7.)

45. Give a convincing argument that 2 is the only possible even prime number.

46. Use variables to give a convincing argument that having three consecutive whole numbers that are all prime is impossible.

47. Give a convincing argument that no square number is a prime number.

48. Give two examples to illustrate the possible generalization that GCF (m, n) = GCF $((m + n)$, LCM $(m, n))$. Do the two examples prove that this generalization is true? Explain.

49. When you think about the sums of their proper factors greater than 1, the numbers 48 and 75 have a special relationship. What is this relationship? Show that the numbers 140 and 195 have this same special relationship.

50. Study the following procedure for finding the LCM of two or more numbers.

a. The procedure involves starting with the smallest prime and dividing one or the other of the two numbers by that prime as many times as possible before moving to the next prime. The procedure ends when the quotients are both 1. Explain why the procedure works.

2	90	24
2	45	12
2	45	6
3	45	3
3	15	1
5	5	1
	1	1

LCM (90, 24) = $2^3 3^2 5$.

b. Show how to use the procedure to find the LCM of 24, 28, and 45.

51. Show that this process for finding the number of factors a number has works for the number 24.

a. First, express the number as a product of powers of the prime numbers, 2, 3, 5, 7 ...;

b. Then, increase each exponent by 1 and calculate the product of the resulting numbers.

52. A student doing a curve-stitching art project knows that the number of different regular polygons he can construct with yarn on a 36-nail board is the number of factors that 36 has, minus 2. How many different polygons can he construct?

53. The Cheap Watch Problem. An enterprising watch dealer ordered several identical cheap watches. He kept the cost of each watch secret so that he could sell them to his friends for a lot more than he paid for them. His assistant, checking the invoice, was told that each watch cost a whole number of dollars, that the cost of a watch was more than the number of watches purchased, and that the total cost was $437. He quickly used his calculator to figure out how many watches were purchased and what they cost. How did he do it, and what were the number and cost?

54. The Chewing Gum Box Problem. A 2-digit number of large boxes contained a prime number of smaller boxes, each of which contained a prime number of packages, each of which contained a prime number of sticks of chewing gum. The total number of sticks of gum was 39,039. If no containers other than the large boxes held more than 15 of the next-smaller objects, and no two containers held the same number of next-size-smaller containers, how many large boxes were there?

55. The Area Code Problem. A person has a telephone area code that is a prime number, with no two digits the same. The ones digit is a prime, and when you mark it out, the remaining 2-digit number is also prime. When you mark out the ones and tens digits, the remaining 1-digit number is also a prime. What can you conclude about the state in which this person lives? (*Hint:* Use an area code map from a phone book and Table 4.5 on p. 222.)

56. Biorhythm Cycles Problem. Suppose that a baby's biorhythm cycles are together on the day of its birth. The physical cycle is 23 days long, the emotional cycle is 28 days long, and the intellectual cycle is 33 days long. How old will the baby be when the cycles all coincide again?

D. Communicating and Connecting Ideas

57. Make a flow chart that shows a procedure for determining whether a given number is prime.

58. Work with a group of your classmates to explore *lucky numbers*, which are the numbers found by the following process.

i. Write down the natural numbers to 100, as in the sieve of Erastosthenes, and mark out numbers as follows.

ii. Because 2 is the number following 1 in the list, mark out every second number, leaving the odd numbers.

iii. Because 2 has been used as a markout number, 3 is now the first unused number after 1 in the list. Mark out every third number from those remaining. (This step would remove 5, 11, 17, 23, ...)

iv. Because 7 is now the first unused number after 1 in the list, mark out every seventh number from those remaining. (This step would remove 19, 39, 61, ...)

v. Continue this process, using 9, 13, 15, and 21 as markout numbers. The remaining numbers are the lucky numbers.

a. How many lucky numbers are there that are less than 100? How many are prime? Composite?
b. How do the number of twin primes less than 100 compare to the number of twin lucky numbers?
c. How do the number of prime pairs differing by 4 compare with the number of lucky pairs differing by 4 for the lucky numbers and primes less than 100?
d. Does Goldbach's conjecture for primes seem to hold true for the lucky numbers?

59. If a graphing calculator is available, get help as needed and write a program for utilizing the Euclidean algorithm to find the GCF for any two numbers. (See Appendix A for more information on the graphing calculator.)

60. **Historical Pathways.** The French mathematician Pierre Fermat (1601–1665) is generally considered to be the founder of the theory of numbers. At first, he thought that the formula $P_n = 2^{2^n} + 1$ would always produce primes. Show that it does for $n = 0, 1, 2,$ and 3. It also does for $n = 4$, but for $n = 5$ the number produced is 641(6,700,417), which isn't prime.

61. **Making Connections.** Draw a Venn diagram to show how prime numbers, even numbers, odd numbers, composite numbers, and square numbers are related.

CHAPTER SUMMARY

Key Ideas: Questions and Answers

Section 4.1

- **How are the multiples and factors of a number found?** *(pp. 189–193)* Divide the number by successive whole numbers. If the remainder is 0, the divisor is a factor of the number. You need test only those whole numbers that are no greater than the square root of the number.
- **What is meant by divisibility?** *(pp. 193–194)* The whole number b is divisible by the number $a \neq 0$ if there is a whole number x so that $ax = b$.
- **Are there some easy ways for deciding whether one number is divisible by another?** *(pp. 194–200)* The most useful tests are the following. A number is divisible by (a) 2 if its ones digit is 0, 2, 4, 6, or 8; (b) 3 if the sum of its digits is divisible by 3; (c) 4 if the number represented by its last two digits is divisible by 4; (d) 5 if its ones digit is 0 or 5; (e) 6 if it is divisible by both 2 and 3; and (f) 10 if the last digit is 0.
- **How can information about factors be used to classify natural numbers?** *(pp. 200–202)* Some key classifications are: (a) an even number has 2 as a factor; (b) a square number has an odd number of factors; (c) a perfect number has a sum of its proper factors equal to itself; and (d) certain special numbers have exactly two factors.

Section 4.2

- **What are prime and composite numbers?** *(pp. 206–207)* A prime number is a natural number that has exactly two factors. A composite number has more than two factors.
- **What are some ways to find prime numbers?** *(pp. 207–212)* Use the sieve of Erastosthenes or use a calculator to test for factors.
- **What role do prime numbers play in mathematics?** *(pp. 212–214)* Every composite number may be expressed as the product of prime numbers in exactly one way, disregarding the order of the factors.
- **How can we find the greatest common factor and least common multiple?** *(pp. 214–219)* The greatest common factor (GCF) of two numbers is the greatest number that is a factor of both numbers. You can find the GCF by listing factors, by drawing Venn diagrams, by writing prime factorizations, or by using the Euclidean algorithm.

  The least common multiple (LCM) of two numbers is the smallest number that is a multiple of both numbers. You can find the LCM by listing factors, by drawing Venn diagrams, or by writing prime factorizations.
- **What are some patterns and relationships involving prime and composite numbers?** *(pp. 219–223)* Some examples are: (a) There is no largest prime number, but some very large primes have been found; (b) as many consecutive composite numbers as desired may be written; (c) every even number greater than 2 may be the sum of two primes.

Key Terms, Concepts, and Generalizations

Section 4.1

Factor (p. 190)
Multiple (p. 190)
Factor test theorem (p. 193)
Divides (p. 194)
Divisor (p. 194)
Divisible (p. 194)
Divisibility tests (p. 195)
Even numbers (p. 200)
Odd numbers (p. 200)
Squares (p. 200)
Perfect number (p. 200)

Deficient number (p. 200)
Abundant number (p. 201)
Amicable number (p. 201)

Section 4.2
Prime number (p. 207)
Composite number (p. 207)
Sieve of Erastosthenes (p. 207)
Fundamental theorem of arithmetic (p. 212)
Prime factorization (p. 213)
Greatest common factor (GCF) (p. 214)
Euclidean algorithm (p. 216)
Least common multiple (LCM) (p. 217)
The GCF–LCM theorem (p. 220)
Relatively prime (p. 220)
Twin primes (p. 221)
n factorial (p. 222)

CHAPTER REVIEW

Concepts and Skills

1. List all the factors of each number.
   a. 18 b. 32
   c. 48 d. 105
2. When using a calculator to find the factors of 625, what is the largest number you would have to test?
3. Use a calculator to find the factors of each number.
   a. 91 b. 143
   c. 663 d. 299
4. Describe how to decide if a number is divisible by
   a. 2. b. 3.
   c. 4. d. 5.
   e. 6.
5. Which of the numbers 987, 1436, 4674, and 5580 are divisible by
   a. 2? b. 3?
   c. 4? d. 5?
   e. 6?
6. Classify as true or false. If false, give a counterexample.
   a. If $3 \mid n$, then $9 \mid n$.
   b. If $10 \mid n$, then $5 \mid n$.
   c. If $2 \mid n$ and $4 \mid n$, then $8 \mid n$.
7. How do you know that 7 is a prime number but that 6 is a composite number?
8. Which of the following numbers are prime? How do you know?
   a. 89 b. 39
   c. 137 d. 217
9. Write the prime factorization of the numbers 90 and 420 by using
   a. a factor tree.
   b. the stacked division method.
10. Use the prime factorization of 36 to find how many factors 36 has.
11. Find the GCF of the numbers 48 and 108 by using
    a. the listing factors method.
    b. the prime factorization method.
    c. the Euclidean algorithm method.
12. Find the LCM of the numbers 24 and 32 by using
    a. the listing multiples method.
    b. the prime factorization method.
13. Is each pair of numbers relatively prime? Why or why not?
    a. 70, 231 b. 165, 182
14. The LCM of a pair of numbers is 72. The product of the numbers is 432. What is the GCF of the numbers? Explain how you know.

Reasoning and Problem Solving

15. Give a convincing argument that 2 is the only even prime number.
16. You know that a number is divisible by 6 if it is divisible by both 3 and 2. So why isn't a number divisible by 8 if it is divisible by both 4 and 2?
17. Produce the smallest number you can that is divisible by 2, 3, 4, 5, 6, and 7. Discuss whether you think a smaller such number exists and why.
18. What characteristic do the numbers 8, 10, 15, 26, and 33 have that the numbers 5, 9, 16, 18, and 24 don't have? (*Hint:* List the factors of the numbers.) Give two more numbers that have this characteristic.
19. **The Synchronized Sirens.** Test siren A was set to go off every 90 minutes. Test siren B was set to go off every 240 minutes. Today both sirens sounded at noon.
    a. In how many hours will they both sound at the same time again?
    b. How many times will each siren have sounded when they both sound at the same time again after today's noon sounding?
    c. How would you set a siren C so that it sounds at different times than A or B but sounds with A and B initially and later they all three sound together again?
20. Do you think that the formula $p = 6n + 1$, where n is a whole number, will produce a prime number more than 50% of the time? Give evidence to support your conclusion. (*Hint:* Table 4.5 on page 222 may help.)
21. **A Corny Experiment.** A corn field 560 meters by 528 meters is to be divided into small square research plots of the same size for experimental purposes.
    a. What are the dimensions of the largest research plot that could be chosen?

b. How many of the largest research plots could be made from the field?

c. If you wanted to have more smaller research plots, what other dimension might you use?

Alternative Assessment

22. Make a chart showing the different divisibility tests. Explain the connections between some of the tests. Discuss which tests you would encourage students to use and why.

23. Work with a small group to devise a good way to explain to someone how to find the prime factorization of a number. Give pros and cons of using factor trees, the division method, or a method that you have devised.

24. Analyze the prime factorization methods of finding the GCF and the LCM. Write a description in your own words of how to use these methods. At the end of your descriptions, indicate how these methods are alike and how they are different.

5

Understanding Integer Operations and Properties

Chapter Perspective

Positive and negative numbers, called integers, are universally used to give meaning to all sorts of real-world situations, from sports events to bank statements. For example, in the above photo, the integer -5 on the leader board shows that the score for one player was 5 strokes under par at that point in the golf match. Integers also help give logical meaning to number systems, and are used in the study of mathematics and technical applications of mathematics.

In this chapter we show how models and properties can help students understand how to add, subtract, multiply, divide, and order integers, as well as discover integer relationships and patterns.

Connection to the NCTM Principles and Standards

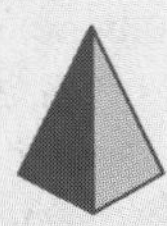

The NCTM *Principles and Standards for School Mathematics (2000)* indicate that the elementary school mathematics curriculum should include the study of number systems so that students can

- develop understanding of the relative position and magnitude of whole numbers and of ordinal and cardinal numbers and their connections *(p. 78)*;
- explore numbers less than 0 by extending the number line and through familiar applications *(p. 148)*; and
- develop meaning for integers and represent and compare quantities with them ... develop and analyze algorithms for computing with ... integers and develop fluency in their use *(p. 214)*.

Connection to the PreK–8 Classroom

In grades PreK–2, students develop readiness for integers by considering direction, such as in above-below and right-left and by playing "go in the hole by so many points" games.

In grades 3–5, students begin to work with integer ideas when they use below-zero numbers on thermometers, number line numbers to the left of 0, money owed, or lost yards in a football game.

In grades 6–8, students use models to understand operations with integers, solve equations with negative number solutions, and compare integers.

Section 5.1 ADDITION, SUBTRACTION, AND ORDER PROPERTIES OF INTEGERS

- Integer Uses and Basic Ideas
- Modeling Integer Addition
- Properties of Integer Addition
- Modeling Integer Subtraction
- Applications of Integer Addition and Subtraction
- Comparing and Ordering Integers.

In this section, we use models to examine some basic ideas about integers, present terminology and symbols for integers, and discuss some important uses of integers. We consider why the procedures for adding and subtracting integers work and how physical models help explain these procedures. Integer properties are used to prove that the procedures for adding and subtracting integers are mathematically correct. Procedures for comparing and ordering integers are introduced to provide another way of applying integers to real-world situations.

Integer Uses and Basic Ideas

Integers are numbers that provide a way to represent real-world situations involving both amount and direction. For example, the opposite quantities 32 degrees above zero and 32 degrees below zero may be represented by the integers $+32°$ and $-32°$.

Estimation Application

The National Debt

In a recent year, the national debt was $4.15 trillion, and the population of the United States was 260 million. First use an integer to represent the national debt. Then estimate how many years would be needed to pay off the national debt (ignore the interest) if each person in the United States contributed $100 a year for this purpose. Calculate to check your estimate.

Also, if $500 is withdrawn from a savings account and $200 deposited, the integers -500 and $+200$ could be used to represent these transactions. Integers are also used to show increases and decreases in stock prices, in many scientific endeavors, and even in sports. Without integers, people wouldn't be able to easily express information involving opposite quantities that include amount and direction.

From a mathematical perspective, some insight into integers is gained by considering the need to solve all types of equations. The whole numbers, 0, 1, 2, 3, ... are adequate for solving equations such as $x + 3 = 5$. But what about finding solutions to equations such as $x + 3 = 0$? If only the whole numbers existed, the equation $x + 3 = 0$ could not be solved. Negative integers are needed to solve such equations.

For every non-zero whole number n, then, a negative integer $-n$ may be created so that $n + -n = 0$. Thus the integer -3 is a solution to $x + 3 = 0$. We define the integers as follows.

Definition of Integers

The set of **integers,** I, consists of the **positive integers** (i.e., the non-zero whole numbers), the **negative integers,** and *zero.*

$$I = \{\ldots -4, -3, -2, -1, 0, 1, 2, 3, 4, \ldots\}$$

The **opposite** of an integer is the mirror image of the integer around 0 on the number line. Figure 5.1 shows the integers on a number line, and shows that the integers 3 and -3 are opposites.

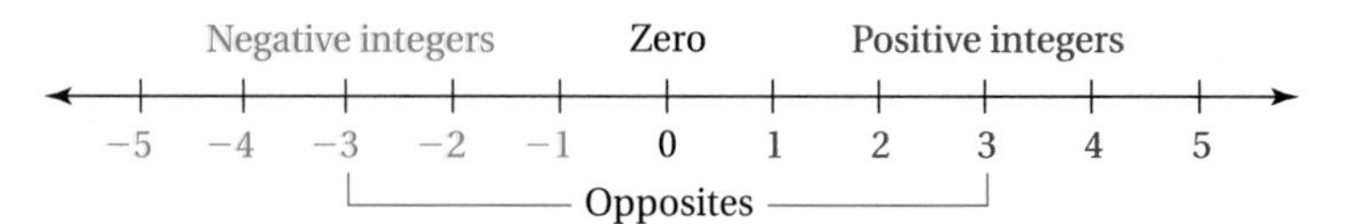

FIGURE 5.1 | Modeling integers on the number line.

The opposite of a *profit* is a *loss.*
The opposite of 3 is written as -3.
The opposite of a *loss* is a *profit.*
The opposite of -3 is written $-(-3)$, or 3.

The dash, −, is used in different ways when working with integers, but the place where it occurs makes it easy to understand its meaning. Here are some examples.

1. -5 represents the integer *negative* 5.
2. $-(-5)$ represents the *opposite* of the integer negative 5.
3. $8 - (-5)$ shows subtraction and represents 8 *minus* the integer negative 5.

The number line is also useful for modeling other integer ideas. For example, we might want to focus on *how far* an integer is from zero, but not the *direction* of the integer from zero. The idea of absolute value of an integer, defined as follows, allows us to do this.

Definition of Absolute Value

The **absolute value** of an integer is the number of units the integer is from 0 on the number line. The absolute value of an integer n is written $|n|$, and is positive for all $n \neq 0$.

A number line further illustrates the idea in the definition.

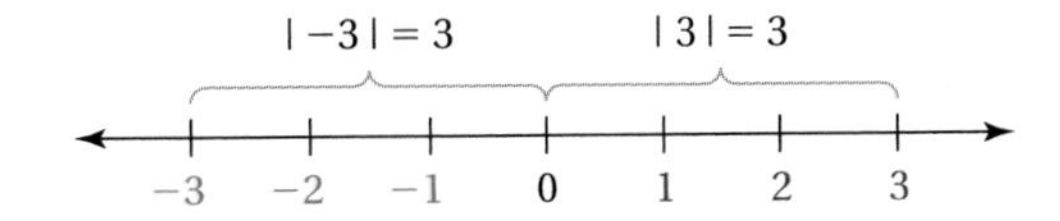

Both -3 and 3 are the same distance, 3 units, from 0 on the number line, so $|-3| = |3| = 3$. Example 5.1 further explains absolute value notation, and Exercise 43 on p. 252 provides an opportunity to interpret a more formal definition of absolute value.

Example 5.1

Using Absolute Value Notation

Use absolute value notation to write two symbols, each representing the number 5.

Solution

$|-5|$ and $|5|$

Your Turn

Practice: Use absolute value notation to write two symbols for each number.

a. 8 **b.** -4 **c.** $-(-6)$

Reflect: How can you decide whether $|-(5280 - 4998)|$ is positive or negative without doing the calculation? ■

Modeling Integer Addition

Models can be used to help formulate and verify rules for addition of integers. We consider two such models in the following subsections.

Using a Counters Model. To model integers, we can use the idea of "in the red" to show a loss and model the integer -1 with a red counter. We can use the idea of "in the black" to show a gain, and model the integer $+1$ with a black counter. Because a black counter shows a gain of 1 and a red counter shows a loss of 1, *a black counter and a red counter cancel each other.* This allows us to model a given integer in different ways, as shown in Figure 5.2.

Black and red counters may be used to give meaning to **addition of integers,** as illustrated in Figure 5.3.

Example 5.2 applies the use of counters to model a real-world problem.

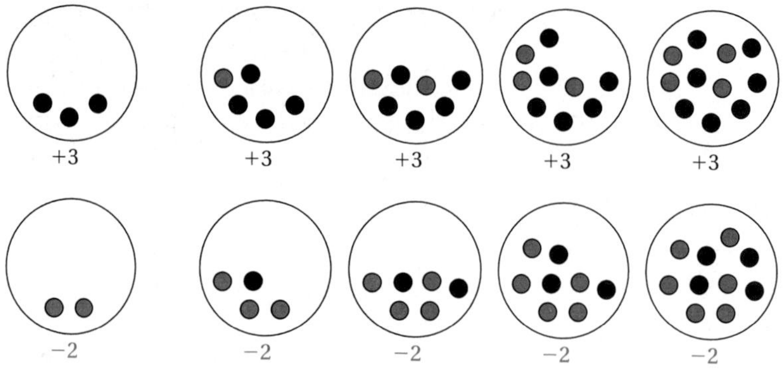

FIGURE 5.2 | Modeling integers with counters.

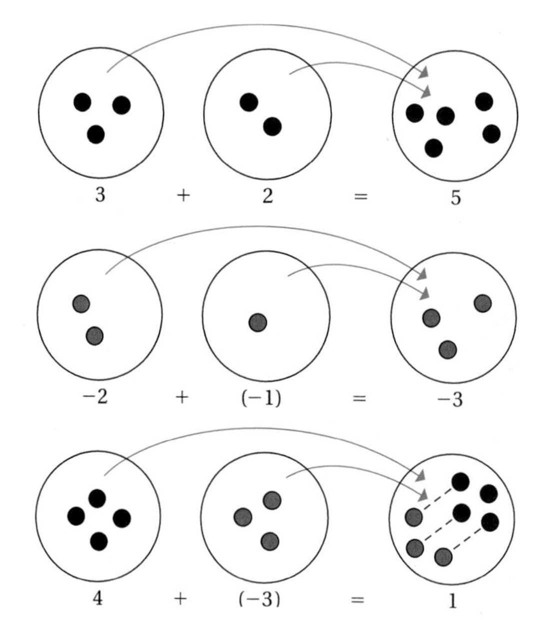

FIGURE 5.3 | Using counters to model integer addition.

Example 5.2 Using the Counters Model to Solve a Problem

How could counters showing integer addition be used to demonstrate the net result of a person receiving a check for $5 and paying a bill for $8?

Solution

Use 5 blacks (check dollars) and 8 reds (bill dollars) canceling black–red pairs when possible. Three reds are left. The person is short $3.

Your Turn

Practice: Use counters showing integer addition to determine the net effect of charges on 6 protons (+) and 11 electrons (−). Opposites cancel each other.

Reflect: Are there types of integer sums that can't be figured out by using counters? ■

Using a Charged Field Model. Another model for integer addition, similar to the counters model, is the charged field model. It is based on the idea that the same number of positive and negative charges produces a field with 0 charge, but an excess of positive charges, 4 for example, produces a $+4$ charged field and an excess of negative charges, 6 for example, produces a -6 charged field. Using this model, we start with a 0 charged field that can contain any number of $+$, $-$ charged pairs we choose. A positive number is modeled by putting positive charges into the field and a negative number is modeled by putting negative charges into the field. Addition is modeled by putting integral numbers of charges in the field in succession, and the sum is the charge of the resulting field. Figure 5.4 illustrates how to use this model to find the sum $4 + -6$.

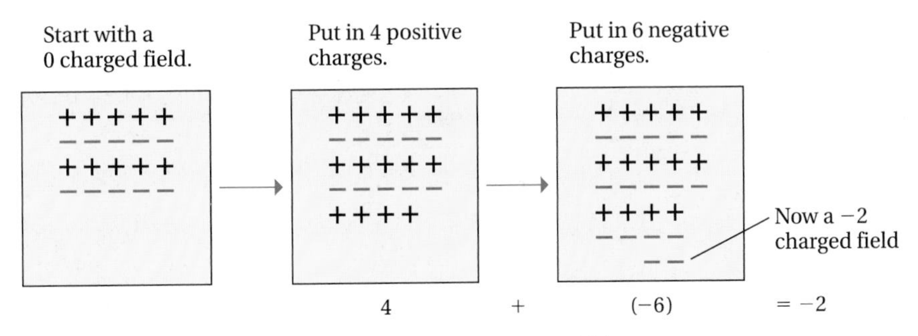

FIGURE 5.4 | Using a charged field to model integer addition

A charged field model can be used to model other types of integer addition, as in Example 5.3.

Example 5.3 Charged Field Modeling of Other Integer Sums

Use the charged field model to model $6 + -4$.

Solution

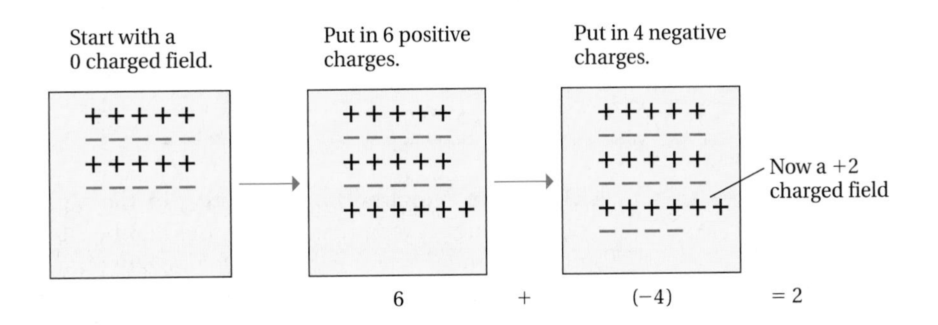

Your Turn

Practice: Use charged fields to model

a. $-6 + (-4)$ **b.** $2 + 5$

Reflect: How are the counters and charged field models for integer addition alike? How are they different? ■

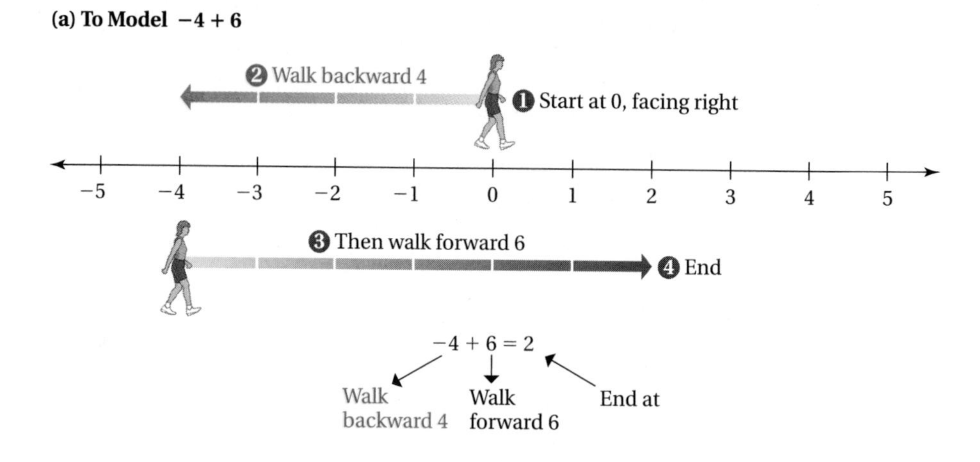

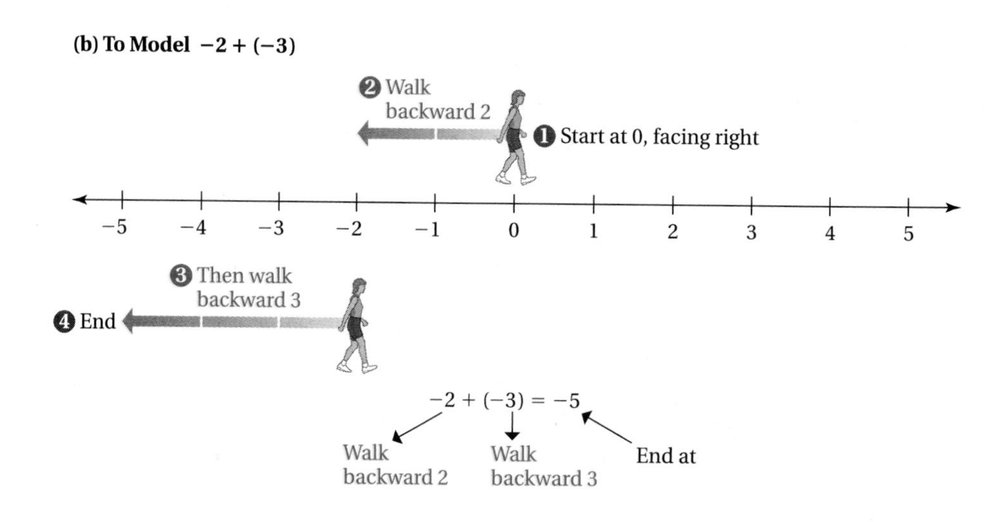

FIGURE 5.5 | Using the number line to model integer addition.

Using the Number Line Model. In the number line model in Figure 5.5, a person starts at 0 and "walks" on the number line. The person walks forward for positive integers and backward for negative integers. The addition symbol $(+)$ indicates combining the two walks. Follow steps 1–4 in Figure 5.5(a) and in Figure 5.5(b) to see how the model shows two types of integer addition.

Example 5.4 shows other types of integer sums that can be interpreted by using the number line model.

Example 5.4 Modeling Other Types of Integer Sums

Use the number line to model $5 + (-3)$.

Solution

a.

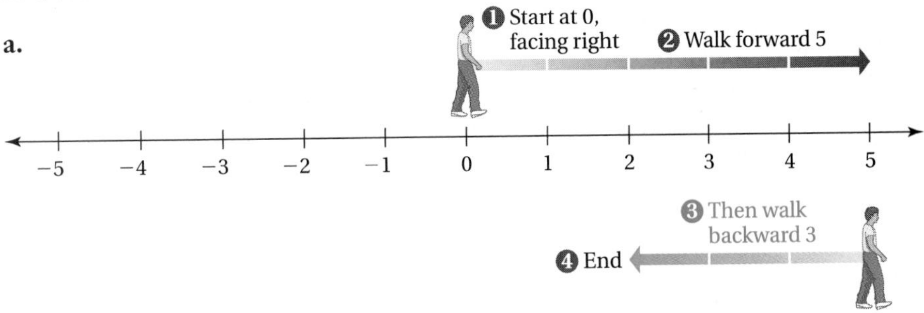

$5 + (-3) = 2$

Your Turn

Practice: Use the number line to model

a. $-5 + 3$.
b. $-3 + (-1)$.
c. $-5 + 7$.

Reflect: Explain how you could model addition on the number line without "walking backward." ■

Using a Calculator. Mini-Investigation 5.1 challenges you to use your calculator to explore patterns in adding positive and negative integers. Note that to enter a negative number such as −6 into some calculators, you push 6 followed by [+/−]. For other calculators, you push [−] and then the number 6.

Talk about how your conjectures compare with those of others and arrive at mutually acceptable ways to explain to someone how to add a positive and a negative integer.

Mini-Investigation 5.1 The Technology Option

After using your calculator to find several useful example sums, how would you complete these conjectures?

a. The sum of two negative integers is ...
b. The sum of a positive and a negative integer is ...

Formulating Procedures for Adding Integers. The counters model, the charged field model, the number line model, and a calculator certainly are useful in finding integer sums. However, a general procedure for finding any sum by considering only the numbers would be more efficient. Of the many different ways of describing how to add any pair of integers, the following procedures are among the most useful.

Procedures for Adding Integers

- *Adding two positive integers:* Add as with whole numbers.
- *Adding two negative integers:* Add absolute values. The sum is negative. That is, for negative integers a and b, $a + b = -(|a| + |b|)$.
- *Adding a positive and a negative integer:* Subtract the lesser of the absolute values of the integers from the greater. Give the answer the same sign as the integer with greater absolute value. That is, for positive integer a and negative integer b, with $|a| > |b|$, $a + b = (|a| - |b|)$; for positive integer a and negative integer b, with $|b| > |a|$, $a + b = -(|b| - |a|)$.

Connection to the PreK–8 Classroom

A Sample Student Page: Modeling Integer Addition

The NCTM *Principles and Standards for School Mathematics* (2000) state regarding grades 3–5 that "Negative numbers should be introduced at this level through the use of familiar models . . . .The number line is also an appropriate and helpful model" (p. 151). They also state regarding grades 6–8 that "Middle grade students . . . will learn to recognize, compare, and use an array of representational forms for . . . integers" (p. 280). A model for integer addition different from those in this chapter is shown in Figure 5.6. Study the page and answer the questions in the margin.

- **What** events in the model are represented by positive integers?
- **What** events in the model are represented by negative integers?
- **Using** the model, what postman story would you make up for $-6 + 4$?

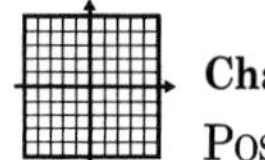

Chapter 12
POSTMAN STORIES

We would like to invent an arithmetic for numbers with signs.

To do this, we look at an example.

Suppose a postman brings you a check for \$3. We can represent this as ${}^{+}3$. If he brings you a **bill** for \$2, we can represent that as ${}^{-}2$.

(1) Suppose the postman brings you a check for \$5 and a bill for \$3. Are you richer or poorer? By how much?

Can you make up a postman story for each problem? What answer do you get for each problem?

(2) ${}^{+}2 + {}^{+}4 = ?$

(3) ${}^{+}5 + {}^{-}2 = ?$

(4) ${}^{-}2 + {}^{-}3 = ?$

(5) ${}^{+}5 + {}^{-}6 = ?$

(6) ${}^{-}7 + {}^{+}9 = ?$

(7) ${}^{-}5 + {}^{+}1 = ?$

(8) ${}^{-}3 + 0 = ?$

FIGURE 5.6 | Excerpt from the Madison Project

(*Source:* From *Explorations in Mathematics: The Madison Project,* Student Discussion Guide, by Davis, Robert B. © 1966, Addison-Wesley.)

The following uses common language to describe the procedures stated formally on page 237.

To Add Integers

Like Signs: Add the magnitudes and prefix the common sign

$$\begin{array}{r} +5 \\ + \ +3 \\ \hline +8 \end{array} \qquad \begin{array}{r} -5 \\ + \ -3 \\ \hline -8 \end{array}$$

Unlike Signs: Subtract the smaller magnitude from the larger. Prefix the sign of the larger magnitude.

$$\begin{array}{r} +5 \\ + \ -3 \\ \hline +2 \end{array} \qquad \begin{array}{r} -5 \\ + \ +3 \\ \hline -2 \end{array}$$

Properties of Integer Addition

The counters model, the charged field model, the number line, and the procedures for adding integers may be used to find several integer sums to help verify the following informally stated properties of integer addition.

- When you add any two integers, you always get another integer as the sum, so the set of integers is closed under the operation addition.
- Every integer has just one opposite that can be added to it to give zero.
- Zero is the only integer that doesn't increase or decrease an integer it's added to.
- You can add two integers in either order because you'll get the same sum either way.
- When adding three integers, you can start by adding the first two or the last two because you'll get the same sum either way.

In mathematics, variables help you express your ideas more efficiently. The basic properties of integers in the preceding list may be summarized by using variables as follows.

Basic Properties of Integer Addition

1. **Additive Inverse Property**
   For each integer a, there is a unique integer, $-a$, such that $a + (-a) = 0$.
2. **Closure Property**
   For all integers a and b, $a + b$ is a unique integer.
3. **Additive Identity Property**
   Zero is the unique integer such that for each integer a, $a + 0 = 0 + a = a$.
4. **Commutative Property**
   For all integers a and b, $a + b = b + a$.
5. **Associative Property**
   For all integers a, b, and c, $(a + b) + c = a + (b + c)$.

The properties of integer addition along with mathematical reasoning can be used to prove that the method we formulated for finding integer sums is correct. If our only goal were to devise a way to *find* sums, we could have met that goal by simply introducing the counters model. But there is more to mathematics than using models to understand ideas and *discover* procedures. Doing mathematics sometimes involves using basic properties and deductive reasoning to *prove* that the relationships we have discovered are true. For example, the following is a proof that $-4 + 7 = 3$. Each step in the proof is justified using a basic property or a whole number addition fact.

Statements	**Reasons**
$-4 + 7 = -4 + (4 + 3)$	Substitute $4 + 3$ for 7
$= (-4 + 4) + 3$	Associative property of addition
$= 0 + 3$	Additive inverse property
$= 3.$	Identity property

We can also use the properties to verify that $-5 + -3 = -8$. -8 is the only number that adds to 8 to give 0. But the following shows that $-5 + -3$ adds to 8 to give 0.

Statements	**Reasons**
$-5 + -3 + 8 = [-5 + (-3)] + (3 + 5)$	Whole number addition fact
$= [-5 + (-3 + 3)] + 5$	Associative property (applied twice)
$= (-5 + 0) + 5$	Additive inverse property
$= -5 + 5$	Additive identity property
$= 0.$	Additive inverse property

Since there is only one number that adds to 8 to give 0, $-5 + (-3)$ and -8 must represent the same number, and $-5 + (-3) = -8$.

Example 5.5 provides an opportunity to use the basic properties of integers in a simple proof. This use of deductive reasoning to verify procedures discovered with inductive reasoning illustrates one of the central activities in mathematics.

Example 5.5 Using Basic Properties of Integer Addition in a Proof

Use the properties and mathematical reasoning to prove $-8 + 3 = -5$.

Solution

Statements	**Reasons**
$-8 + 3 = [-5 + (-3)] + 3$	Replace -8 with $-5 + (-3)$
$= -5 + (-3 + 3)$	Associative property
$= -5 + 0$	Additive inverse property
$= -5.$	Identity property

So $-8 + 3 = -5$.

Your Turn

Practice: Prove that $8 + (-3) = 5$.

Reflect: Does the proof in this example prove that your procedure for adding integers is correct? Why or why not? ■

In Exercise 57 on page 253, you will have an opportunity to verify in general that the sum of two negative integers is a negative integer, that is, $-a + (-b) = -(a + b)$ where a and b are whole numbers.

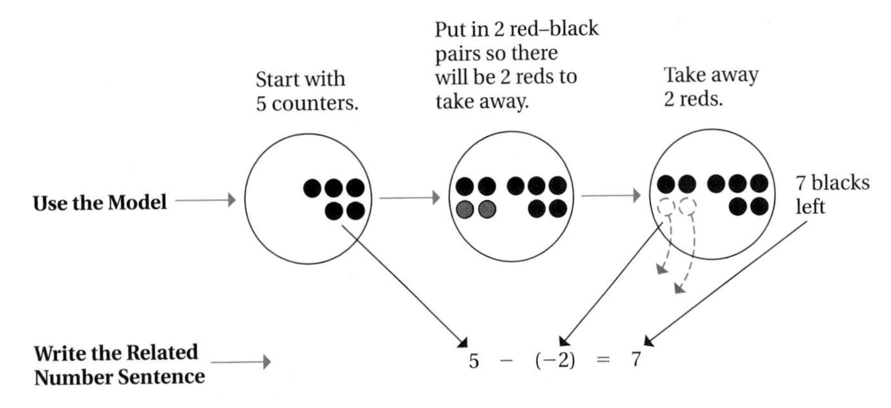

FIGURE 5.7 | Using counters to model integer subtraction.

Modeling Integer Subtraction

As with addition, models and the number line may be used to help explain integer subtraction. Mathematical patterns and relationships can then be used to generalize procedures for finding differences.

Using a Counters Model. A counters model may be used to find the difference of two integers. Recall that, because a black and a red counter cancel each other, you can include as many black–red pairs as you want when representing an integer, without changing its value. Figure 5.7 shows how this use of black–red pairs and a take-away interpretation of subtraction help in finding $5 - (-2)$.

Note in Figure 5.7 that taking away two red counters (subtracting -2) has the same effect as putting in two black counters (adding 2). This effect is a partial basis for a useful rule for subtracting integers that we present later. Example 5.6 applies the counters model to a game.

Example 5.6 Using the Subtraction Counters Model to Solve a Problem

In playing a money board game, a player had already taken \$4 from the bank. To purchase property, she borrowed another \$3. Use the subtraction counters model to find the number that represents her loan status.

Solution

The solution utilizes the problem-solving strategies *make a model* and *choose an operation.* Use a counters model and find $-4 - 3$ to answer the question. Since the problem involves subtracting 3, we need to take away 3 black counters. So we put in 3 red–black pairs so this will be possible.

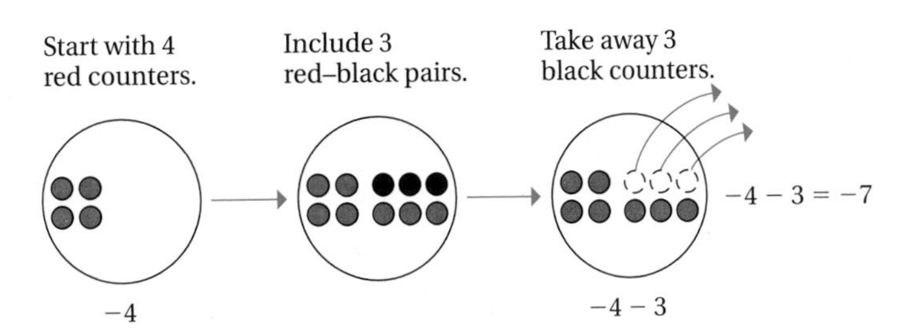

The player was \$7 in debt to the bank.

Your Turn

Practice: Suppose that the player had put $4 in the bank and then had borrowed $6. What number represents her loan status? Use a counters model to solve the problem.

Reflect: How would you complete the following? Explain, using the example. Taking away 3 black counters has the same effect as ? red counters. ■

You may have concluded that the problem in Example 5.6 could also be solved by finding the sum $-4 + (-3)$. This fact illustrates an important relationship between $-4 - 3$ and $-4 + (-3)$ that we discuss in more detail on page 246.

Using a Charged Field Model. The idea of a charged field, very similar to the counters model, can also be used to model integer subtraction. Just as with addition, a positive number is modeled by putting positive charges into the field and a negative number is modeled by putting negative charges into the field. Subtraction is modeled by putting integral numbers of charges in the field, followed by taking an integral number of charges from the field. The final difference is the charge of the resulting field. Figure 5.8 illustrates how to use this model to find the difference $-4 - (-6)$. Note that we can begin with a 0 charged field containing as many +, − pairs as we wish.

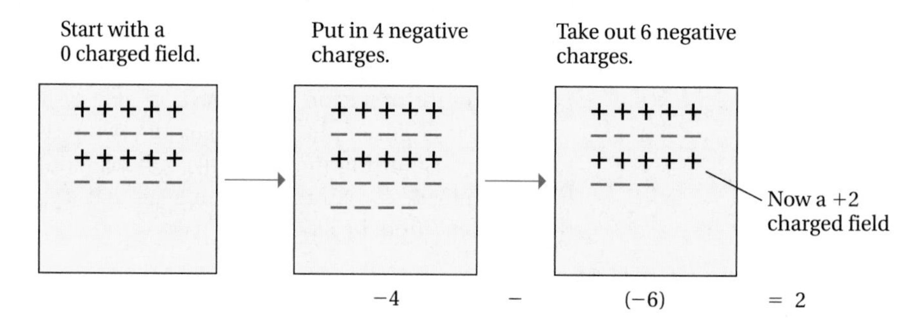

FIGURE 5.8 | Using a charged field to model integer subtraction.

A charged field model can be used to model other types of integer subtraction, as in Example 5.7.

Example 5.7 Charged Field Modeling of Other Integer Differences.

Use the charged field model to model $-4 - 6$.

Solution

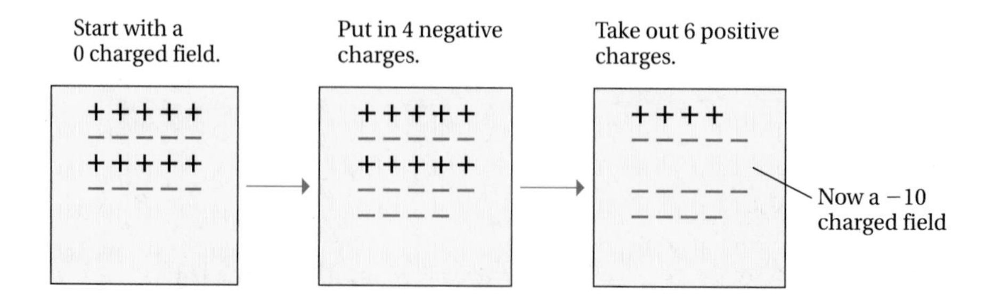

Your Turn

Practice: Use charged fields to model

a. $6 - (-4)$ **b.** $4 - 3$

Reflect: How are the counters and charged field models for integer subtraction alike? How are they different? ■

Using the Number Line. The number line may also be used to model subtraction of integers. As when modeling addition, the walker starts on the number line at 0, facing right. Also, positive integers are modeled by the walker walking forward and negative integers are modeled by the walker walking backward. The subtraction model differs from the addition model only in the modeling of the operation symbol. The subtraction symbol (−) is modeled by *changing the direction the walker is facing*. When modeling addition, the walker always faces right, but in subtraction, after the first walk, the walker turns to face left. Follow steps 1–5 in Figure 5.9 to see how to model three different subtraction situations.

(a) To Model 3 − 2

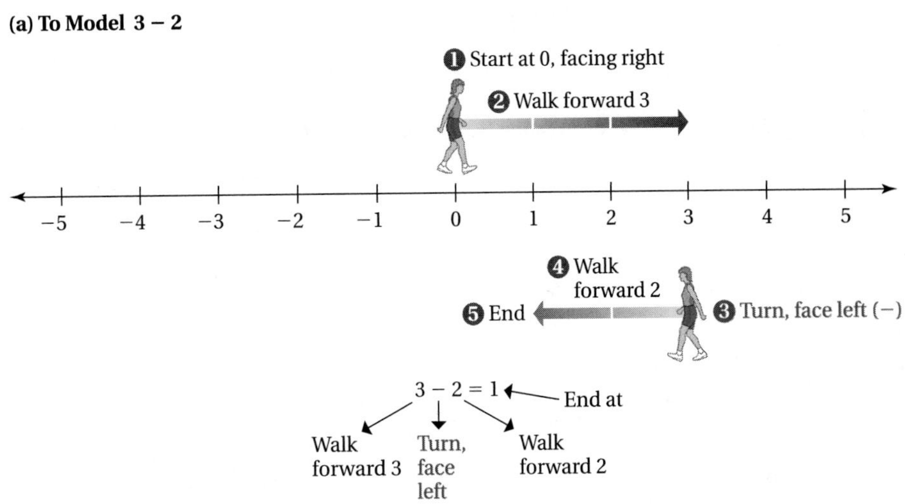

(b) To Model 3 − (−2)

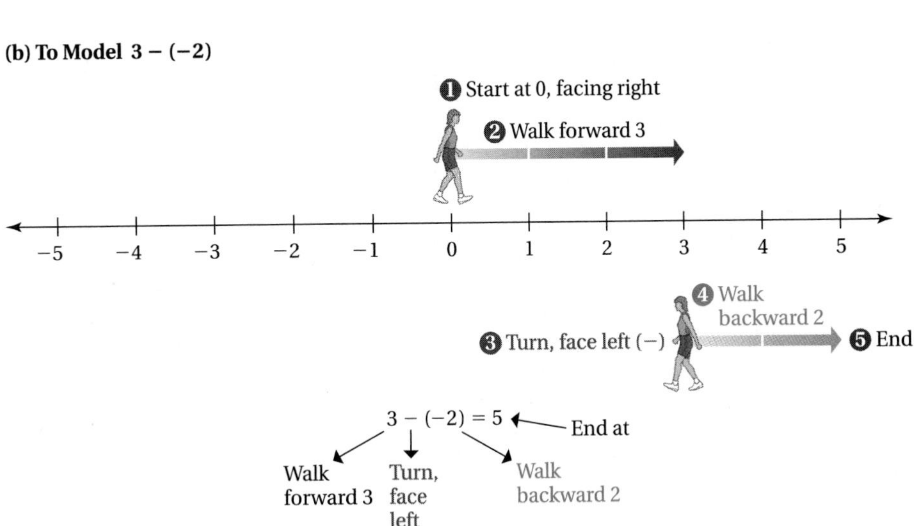

FIGURE 5.9 | Using the number line to model integer subtraction. *(Continued on next page.)*

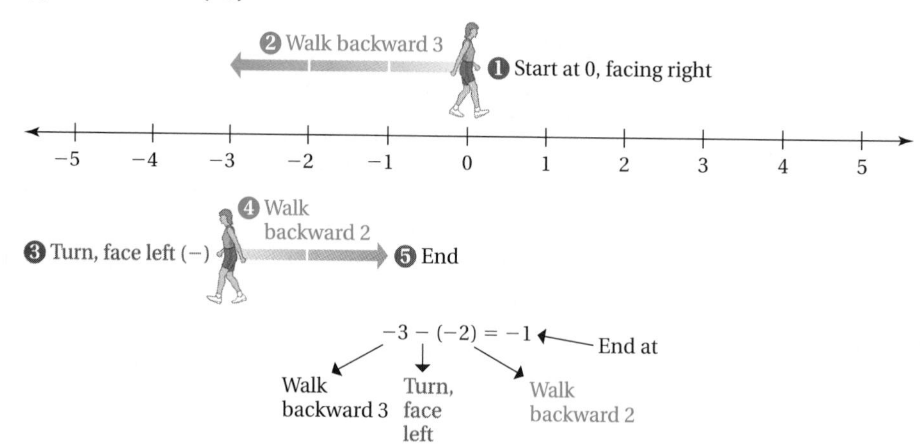

FIGURE 5.9 | *(cont.)* Using the number line to model integer subtraction.

Example 5.8 illustrates how to subtract a larger integer from a smaller integer.

Example 5.8 Using the Number Line Model for Integer Subtraction

Use the number line to model $4 - 6$.

Solution

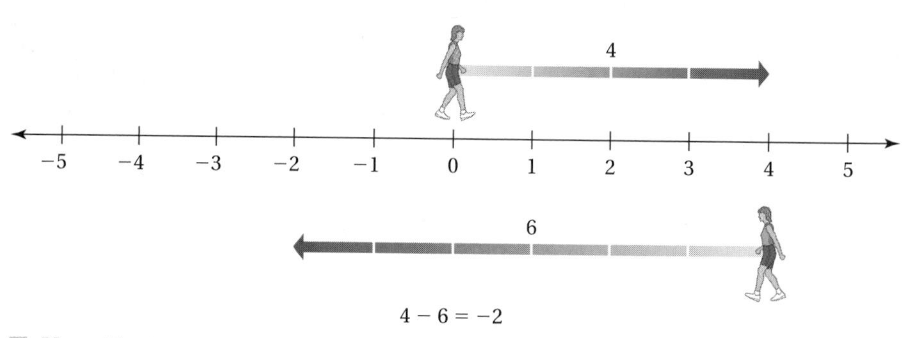

Your Turn

Practice: Use the number line to model

a. $2 - 3$. **b.** $-4 - (-3)$. **c.** $4 - (-1)$.

Reflect: Use this example to compare modeling subtraction with modeling addition on the number line. How are the two modeling procedures alike? Different? ■

Using Mathematical Relationships and Patterns. In Section 2.2, you thought about the idea, reviewed here, that subtracting a whole number can be thought of as *finding the missing addend.* This basic idea can also help explain subtraction of integers.

Addend		Addend		Sum		Sum		Addend		Addend
3	+	4	=	?	so	7	−	4	=	?

In addition you know two addends. You find the sum. In this case it is 7.

In subtraction you know the sum and one addend. You find the missing addend. In this case it is 3.

You can find the answer to the subtraction, $7 - 4 = ?$, by thinking: What number adds to 4 to give the sum 7? or If the sum is 7 and one addend is 4, what is the other addend? Looking at it another way, you might think of $7 - 4$ and 3 as each representing the number that adds to 4 to give 7. In any case, $7 - 4 = 3$.

This process works for all integers. We define it, using variables, as follows.

Definition of Integer Subtraction

For all integers a, b, and c, $a - b = c$ if and only if $c + b = a$.

The preceding ideas suggest a *missing addend* approach to subtraction of integers. We use this approach in Example 5.9.

Example 5.9 Using the Missing Addend Approach to Integer Subtraction

How could the idea of finding a missing addend be used to find $3 - (-2)$?

Solution

Jordan's thinking: I think, what number adds to -2 to give the sum 3? Since $5 + (-2) = 3$, I know that $3 - (-2) = 5$.

Dana's thinking: The sum is 3 and one addend is -2. The missing addend is 5 because $5 + (-2) = 3$.

Your Turn

Practice: Think about missing addends to find

a. $4 - 5$.
b. $-5 - (-3)$.
c. $-1 - 2$.

Reflect: A student said, "Subtracting by finding the missing addend is like what you do when you check subtraction." What do you think the student meant? ■

Mini-Investigation 5.2 challenges you to use a calculator to explore patterns to help you form generalizations about subtracting positive and negative integers.

Write a paragraph summarizing what you have discovered about subtracting integers.

Mini-Investigation 5.2 The Technology Option

After using your calculator to do several useful example subtractions, how would you complete these conjectures?

a. When a negative integer is subtracted from a positive integer, the difference is . . .

b. When a positive integer is subtracted from a negative integer, the difference is . . .

c. When a negative integer is subtracted from a negative integer, the difference is . . .

d. When a positive integer is subtracted from a positive integer, the difference is . . .

Modeling, thinking about missing addends, and using rules discovered from patterns of calculator-produced differences can always be used to subtract integers. However, there is an even easier method. When using the counters model to subtract integers, you have seen that taking away red counters has the same effect as adding black counters and that taking away black counters has the same effect as adding red counters. Using models or the calculator to find the following sums and differences reveals an interesting pattern, which is similar to the pattern obtained with red and black counters.

1. $4 - 2 = 2$ and $4 + (-2) = 2$.
2. $-3 - 2 = -5$ and $-3 + (-2) = -5$.
3. $2 - (-1) = 3$ and $2 + 1 = 3$.
4. $-5 - (-4) = -1$ and $-5 + 4 = -1$.

Note that, in the pairs of equations, *you can subtract an integer by adding its opposite.*

You have already discovered—using counters in Example 5.6 and patterns in the preceding list—that *you can subtract an integer by adding its opposite.* This generalization is summarized in the following theorem, which can be proved easily by using the basic properties of integer addition. This theorem makes subtracting any integer easier provided you know how to add integers.

Theorem: Subtracting an Integer by Adding the Opposite

For all integers a and b, $a - b = a + (-b)$. That is, to subtract an integer, add its opposite.

We use this theorem in Example 5.10 to subtract integers.

Example 5.10

Subtracting by Adding the Opposite

Evaluate $-7 - n$ for $n = -5$.

Solution

$-7 - (-5)$	Replace n with -5.
$= -7 + 5$	Use the theorem subtracting an integer by adding its opposite.
$= -2$.	

Your Turn

Practice: Evaluate

a. $-6 - n$ for $n = 9$.

b. $8 - n$ for $n = -4$.

Reflect: Do you think there is an easier way to find the difference $-7 - (-5)$ in the example? Explain your thinking. ■

We used both models and patterns to discover the theorem about subtracting an integer by adding the opposite. We now use the properties of integers to verify the theorem. We first show that the number represented by $a - b$ has the same unique properties as the number represented by $a + (-b)$, namely that they both add to b to get a. If we prove that $a - b$ and $a + (-b)$ have the same unique property, we will have proved that they are equal.

First, $(a - b) + b = a$ because of the definition of integer subtraction. Second,

$$\begin{aligned} [a + (-b)] + b &= a + [b + (-b)] && \text{Associative property} \\ &= a + 0 && \text{Additive inverse property} \\ &= a && \text{Additive identity property} \end{aligned}$$

So, from the first and second demonstrations above, we know that both $a - b$ and $a + (-b)$ have the unique property of adding to b to give the sum a, and hence $a - b = a + (-b)$.

We summarize the three procedures presented for subtracting integers as follows.

Procedures for Subtracting Integers

- *Take away:* To find $5 - (-2)$, take 2 red counters from a counter model for 5.

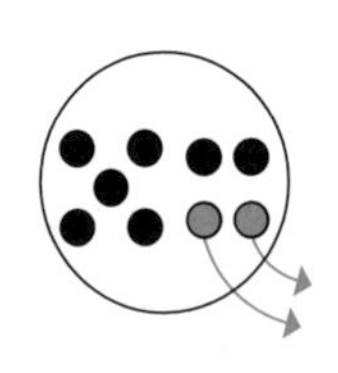

- *Missing addend:* To find $5 - (-2)$, think, "What number adds to -2 to give 5?"
- *Add the opposite:* To find $5 - (-2)$, find $5 + 2$.

Applications of Integer Addition and Subtraction

Everyday applications that involve negative number addition and subtraction occur less frequently than applications involving whole numbers. Often a real-world application, such as the following, is solved by adding or subtracting whole numbers rather than integers.

Problem: What is the temperature difference between a high temperature of 12°F and a low temperature of −9°F?

Solution with integers: $12 - (-9) = 21$. The difference in temperatures is 21°.

Solution with whole numbers: 12 (above) + 9 (below) = 21. The difference of the temperatures is 21°.

Example 5.11 illustrates a problem involving operations with integers. It can be solved by using the problem-solving strategies *make a list* and *choose an operation.*

Example 5.11

Problem Solving: The Bank Account Problem

A student had a balance of $125 in her checking account at the beginning of the month. Each time during the month the account becomes overdrawn, a $5 service charge is deducted from the account. The student deposited $85 in the account and then wrote checks for $160 and $135. Then she deposited $115 and later wrote a check for $45. Finally, she deposited $25. At the end of the month, what was the balance in the account?

Working toward a Solution

Understand the problem	*What does the situation involve?*	Transactions in a bank account
	What has to be determined?	The balance in the account at month's end
	What are the key data and conditions?	The initial balance is \$125; \$5 is deducted each time the account is overdrawn.
	What are some assumptions?	Assume that enough time elapsed after the account was overdrawn for the bank to deduct the service charges.
Develop a plan	*What strategies might be useful?*	Make a list and choose an operation
	Are there any subproblems?	Find the balance at key intervals. Calculate the final balance.
	Should the answer be estimated or calculated?	Calculate the exact numbers.
	What method of calculation should be used?	Use mental computation.
Implement the plan	*How should the strategies be used?*	***Transactions*** / ***Running Totals*** 125 85 / 210 −160 / 50 −135 / −85* 115 / 30 −45 / −15* 25 / 10 (*Account overdrawn)
	What is the answer?	$10 + (-10) = 0$. The balance at month's end was \$0.
Look back	*Is the interpretation correct?*	Yes. The calculations fit the problem data.
	Is the calculation correct?	Yes. The mental computation was correct.
	Is the answer reasonable?	Yes. The estimated sum of positive numbers is 250 and of negative numbers is −250.
	Is there another way to solve the problem?	First add all positives. Then add all negatives. Combine and check for times when the account is overdrawn.

Your Turn

Practice: Felipe had a balance of \$165 in his checking account at the beginning of the month. Each time during the month the account becomes overdrawn, a \$10 service charge is deducted from the account. Felipe wrote a check for \$190 and later deposited \$125. Then he withdrew \$115 and later made deposits of \$35 and \$140. Finally, he wrote a check for \$170. At the end of the month, what was the balance in the account?

Reflect: When did you add two or more negative numbers when solving the practice problem? Give the numbers and the sum you found. ■

Some applications of integer addition and subtraction that involve larger numbers are best solved by using a calculator. You may recall that one way to enter a negative number into some calculators is to first enter a whole number such as 5 and then push the change-sign key [+/−] to change the displayed number to its opposite. Example 5.12 involves a situation for which integer addition and subtraction with a calculator might be useful.

Example 5.12

Problem Solving: Another Bank Account Problem

A college student now has \$3426 in his checking account. He had started with a deficit of \$1798. What is the difference between what he has now and what he started with?

Solution

Use a calculator to find $3426 - (-1798)$.

Key Sequence	Display
3426 [−] 1798 [+/−] [=]	5224

or

Key Sequence	Display
3426 [−] [−] 1798 [=]	5224

or

Key Sequence	Display
3426 [−] [(−)] 1798 [=]	5224

Your Turn

Practice: Use negative numbers, whenever possible, and a calculator to solve the following problems.

a. A student started with a deficit of \$573 and after a deposit has \$389 in her account. How much did she deposit?

b. A student had a deficit of \$968 and then withdrew \$679. What was his balance after the withdrawal?

Reflect: Show another way you could use a calculator to find the difference in the example problem. ■

The NCTM *Principles and Standards for School Mathematics* (2000) state that in grades 6–8 "Students can use inductive reasoning to search for mathematical relationships through the study of patterns" (p. 262). Looking for a pattern after finding the differences in the following sequence on a calculator—even before they learn how to subtract integers—can help children discover how integer subtraction works. $3 - 3 = ?$ $3 - 2 = ?$ $3 - 1 = ?$ $3 - 0 = ?$ $3 - (-1) = ?$ $3 - (-2) = ?$ $3 - (-3) = ?$ $3 - (-4) = ?$

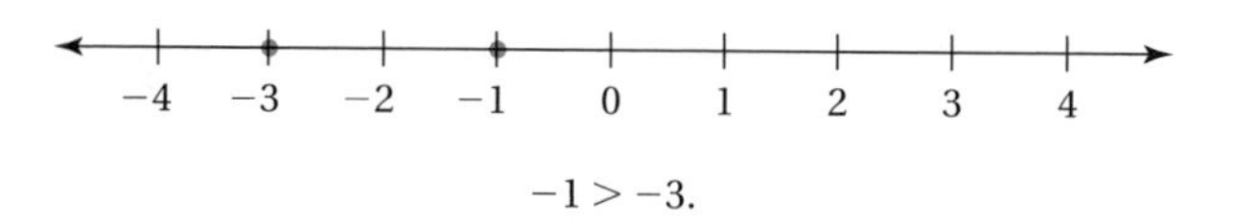

$-1 > -3.$

FIGURE 5.10 | Ordering integers with a number line model

Comparing and Ordering Integers

One everyday application of integers is deciding which of two directed quantities is greater. For example, which is warmer, a temperature of −18° or a temperature of +9°? Or which is the higher elevation, −1999 feet or −3001 feet? We present three methods for answering such questions.

Using the Number Line to Order Integers. Similar to whole numbers, as you go to the right on the number line, the integers get larger. For example, on the number line shown in Figure 5.10, −1 obviously *is greater than* −3, or $-1 > -3$, because −1 is to the right of −3. Similarly, −3 *is less than* −1, or $-3 < -1$.

The ideas of greater than and less than for integers can be used to order a set of integers. For example, the temperatures for five consecutive winter days, −1°, −3°, 5°, −6°, and 2°, can be ordered by listing them from lowest to highest. The ordering is −6°, −3°,−1°, 2°, and 5°.

Using Addition to Order Integers. If we are looking only for an intuitive way to decide which of two integers is greater, we could use the number line and go no further. However, to be able to read and understand mathematical writing and to prove mathematical generalizations, we need to use variables and symbols carefully when expressing our intuitive ideas.

For example, when thinking about a pair of whole numbers, say, 3 and 5, you know that you can always add to the lesser number, 3, to get the greater number, 5, as described in Chapter 2. The same idea holds for any pair of lesser and greater integers and can be expressed by using variables and symbols as follows.

Definition of Greater than (>) and Less Than (<) for Integers

$b > a$ if and only if there is a positive integer p such that $a + p = b$. Also, $a < b$ whenever $b > a$.

At first this definition may seem complicated, but when you think of it as simply a short, precise way to express carefully your intuitive ideas of greater than and less than, it makes sense. For example, you know that −1 is greater than −3 because you can add 2 to −3 (go two units right on the number line) to get −1.

Carefully expressed definitions, like the one above, make it easier to prove that something is true, as shown in Example 5.13.

Example 5.13 **Using the Definition of Integer Greater Than and Less Than in a Proof**

Prove that $2 > -6$.

Solution

We can *prove* that 2 is greater than −6 if we can find a positive integer that can be added to −6 to get 2. Because $8 + (-6) = 2$, by the definition of $>$, $2 > -6$.

Your Turn

Practice: Use the definition of $>$ and $<$ for integers to prove

a. $-13 > -15$.

b. $-4 < 1$.

c. $7 > -10$.

Reflect: How could you use subtraction to convince someone that the inequality in the example problem, $2 > -6$, is true? ■

Using a Calculator to Order Integers. Mini-Investigation 5.3 encourages you to use a calculator as yet another technique for comparing two integers and ordering a set of integers.

Write a few sentences to describe and justify your tecnhique and describe how you would use it to order three integers.

Mini-Investigation 5.3 **The Technology Option**

How can you use your calculator to order two integers?

Problems and Exercises *for Section 5.1*

A. Reinforcing Concepts and Practicing Skills

1. Give the integer that represents each situation. Then give the opposite of the situation and the integer that represents it.
  a. Lose 5 points
  b. Six seconds before blastoff
  c. A profit of $8500

Complete the sentence in Exercises 2–5.

2. The opposite of a positive integer is a ___ integer.

3. The opposite of a negative integer is a ___ integer.

4. The integer ___ is neither positive nor negative.

5. The absolute value of a non-zero integer is always a ___ integer.

6. Use counters to model these sums.
  a. $8 + (-3)$ **b.** $-7 + 4$
  c. $-2 + (-6)$

7. Use charged fields to model these sums.
  a. $4 + (-3)$ **b.** $(-3) + (-5)$
  c. $-5 + 2$

8. Use the number line to model these sums.
  a. $-5 + (-3)$ **b.** $6 + (-4)$
  c. $2 + (-6)$

9. Find the sums and describe the method you used.
  a. $11 + (-5)$ **b.** $-7 + 9$
  c. $-6 + 8$

10. State the additive inverse property for integer addition in your own words.

11. Give the missing reasons.

Statement	Reason
$9 + (-4)$	
$= (5 + 4) + (-4)$	Basic fact, $9 = 5 + 4$
$= 5 + [4 + (-4)]$	**a.** ?
$= 5 + 0$	**b.** ?
$= 5.$	**c.** ?

12. Use counters to model these differences.
  a. $-2 - 5$ **b.** $-8 - (-5)$
  c. $9 - (-3)$

13. Use charged fields to model these differences.
  a. $3 - 4$ **b.** $-4 - (-3)$
  c. $7 - (-5)$

14. Use the number line to model these differences.
  a. $-2 - (-5)$ **b.** $5 - (-3)$
  c. $3 - 7$

15. Find the difference and describe the method you used.
  a. $23 - (-4)$ **b.** $956 - (-437)$
  c. $6 - (-1)$

16. Find the difference by using the idea that $a - b = a + (-b)$.
  a. $-7 - (-8)$ **b.** $-9 - 5$
  c. $11 - (-4)$

17. Think about missing addends to find the difference.
  a. $-6 - (-2)$ **b.** $-8 - 1$ **c.** $7 - (-4)$

18. Evaluate for $n = -6$.
   a. $n - 9$ b. $-15 - n$
   c. $n - n$ d. $n + n$

Calculate the answers in Exercises 19–25.

19. a. $|-8|$ b. $-|8|$ c. $-|-8|$
   d. $|-(-8)|$ e. $|6 - 9|$
20. $-|-3 - (-8)|$
21. $|-8|-|10|$
22. $|-[-2 +(-5)]|$
23. $-|-6|-(-5)$
24. $|4 - |-6 - 8||$
25. $-|[2 + (-7)] - (-9)|$
26. Jackie bought a stock for \$34 a share. The daily changes for the next seven days after she bought the stock were $-5, -2, +1, +3, -6, +4, +8$. What was a share of the stock worth at the end of the seventh day?
27. The price of Allgain stock fell \$5 per share on Monday and rose \$9 per share on Tuesday. Write and solve an equation that shows the total change.
28. The yearly high temperature in a certain location was 75°. The yearly low was −9°. Write an integer difference that indicates the range in temperature and find this difference.
29. Order from largest to smallest.
   a. $-6, -15, 21, 0, -12, 8, -4, -9$
   b. $23, -18, 18, -15, 9, -2, -12, 17$
   c. $-13{,}570; -14{,}000; -13{,}999$
30. Use the definition of greater than to prove that $-5 < -2$.
31. Use a number line to show why $-13 > -19$.
32. The temperature in a walk-in refrigerator decreased by 23° to reach a level of −12°. Solve the equation $t - 23 = -12$ to find the original temperature.
33. After finding that he had been billed incorrectly for \$459, a store owner showed a profit for the week of \$9236. Solve the equation $a - (-459) = 9236$ to find the profit before the bill was dropped.
34. A football team lost 9 yards on a quarterback sack and then gained 12 yards on a pass play. Write an equation to find the result of these two plays and solve it. Use a negative integer in the equation.
35. Use a calculator to find the following sums and differences.
   a. $-789 + (-564)$ b. $4986 - (-5278)$
   c. $-876 + 1269$ d. $-7864 - (-9876)$

B. Deepening Understanding

36. If the statement is true for all integer values of p and q, explain why. If it isn't true, give a counterexample.
   a. $|p|$ is positive.
   b. $-q$ is negative.
   c. If p is the opposite of q, $p + q = 0$.
   d. The equation $px = q$ has no solution.
37. Find the next three numbers in each pattern.
   a. $-8, -4, 0, ?, ?, ?$ b. $15, 8, 1, ?, ?, ?$
   c. $5, -10, 10, -20, ?, ?, ?$ d. $1, -2, 2, -3, 3, ?, ?, ?$
38. The Chinese used red and black calculating rods as early as 300 B.C. How many years ago was this date?
39. Assume that the unit of time is a day, and 0 represents today. What number would represent
   a. a week from now? b. yesterday?
   c. a week ago? d. day after tomorrow?
   e. a month from now?
40. A student said that he could easily find the following sums mentally. How do you think he did it?
   a. $-48 + (-37) + 48 + 38 + 999$
   b. $|-24 + (-24)| + 75 + (-49) + (-74)$
41. If j and k represent integers, give the additive inverse of
   a. $-k$. b. $j + k$.
   c. j. d. $j - k$.
42. Does $|n + m| = |n| + |m|$ for all integers n and m? If so, give some examples. If not, give a counterexample.
43. The absolute value for an integer, for all integers n, can defined as follows.

$$|n| = -n \text{ if } n < 0 \text{ and } |n| = n \text{ if } n \geq 0.$$

   Choose numbers and illustrate the definition.
44. Do you think that the following properties hold for all integers a, b, c, and d? Substitute integers and find counterexamples, if possible.
   a. $a + (b - c) = (a + b) - (a + c)$.
   b. $a - b = b - a$.
   c. $|a - b| = |b - a|$.
   d. $-(a + b) = b - a$.
   e. $(a + b) - (c + d) = (a - c) + (b - d)$.
   f. $a - b = -(b - a)$.
   g. $(a - b) - c = a - (b - c)$.
   h. $|a - b| = |a| - |b|$.
45. Solve the equation. Use mental mathematics procedures or the guess, check, revise method.
   a. $x + (-6) = -5$. b. $n - 4 = -9$.
   c. $-8 + y = 12$. d. $z + (-4) = -4 + 12$.
   e. $t - (-9) = 2$. f. $7 + (-4) = r + (-4)$.
46. Write two different equations that can be used to solve this problem. A low of −4° is predicted for tomorrow. If the temperature is now 8°, by how much must it change?
47. Use the number line to show the answer to this question. A police car on patrol started at the station, traveled 2 miles east, and then traveled 5 miles west. How far and in what direction was the car from the station at that location?
48. Use integers and a calculator to solve the problems.

a. The element neon has the smallest difference between melting and boiling point (its liquid range) of all the elements. It melts at $-415.469°F$ and boils at $-410.894°F$. What is the liquid range of neon?

b. Of the metals, tungsten has the highest melting point at 6188°F, and mercury has the lowest melting point at $-37.892°F$. What is the range, in °F, of the melting points of all metals?

49. Use any two of the numbers shown for x and y to complete each number sentence.

$-21, 39, 45, -29, -49, 56, -61$

a. $x + y = -4.$ **b.** $x + y = -110.$
c. $x + y = 95.$ **d.** $x - y = 100.$
e. $x - y = 8.$ **f.** $x - y = -6.$

50. If you start with an integer, add -76, subtract -67, add 89, subtract 98, add -47, and subtract -74, the result is 0. With what integer did you start?

51. Estimate the following sums. State whether you used
i. rounding,
ii. substitute compatible numbers,
iii. front-end estimation,
iv. clustering, or
v. another method.
a. $715 + (-487)$
b. $-512 + (-498) + (-507) + (-476) + (-528)$
c. $-342 + 1256$
d. $-826{,}978 + 319{,}498$
e. $-428 + (-823) + 232 + 965 + (-697)$

C. Reasoning and Problem Solving

52. A student said that, unlike the whole numbers, integers allow us to solve any equation we want. Give a counterexample to disprove this generalization.

53. In a magic square, the sum of the numbers in each row, column, and main diagonal is the same, and is called the magic sum. Find the missing numbers in this magic square. The magic sum is -2.

?	?	−7	4
?	−3	?	−6
5	−8	2	?
−5	?	?	?

54. A 3×3 magic square uses the integers $-3, -2, -1, 0, 1, 2, 3, 4$, and 5. The number 1 is in the center and the magic number is 3. Construct the magic square.

55. Copy the figure and in each circle place a different one of the numbers $-3, -2, -1, 0, 1, 2$, or 3 so that the sum of the numbers on each side of the triangle is 2. Can you find another solution? If so, show it on another copy of the figure.

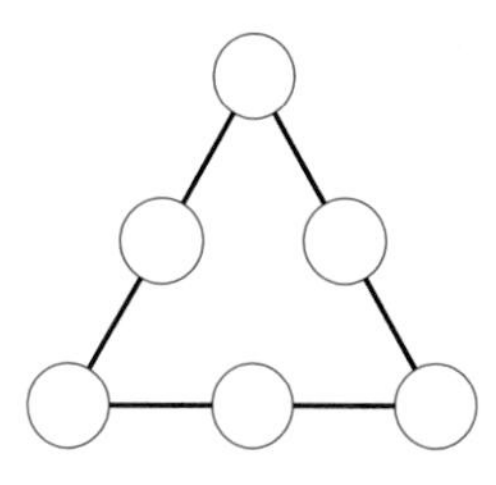

56. Explain how the statement in the paragraph above Example 5.6 on p. 241 relates to the Theorem on Subtracting an Integer by Adding the Opposite on p. 246.

57. Study the verifications on p. 240 and use the approach given with numbers to help you prove that $-a + (-b) = -(a + b)$, where a and b are any whole numbers.

58. The Budget Problem. A student made a 12-month budget. Each month he figured how much over or under the budget he was. He had been \$5 under in January and \$10 over in February, and had alternated being over and under each month. The amount over or under increased by \$5 each month through November.
a. How must he do in December to break even for the year?
b. Create a computer spreadsheet to represent this situation. What spreadsheet formula will you use to determine the solution to part (a)?

59. The Football Problem. A sports announcer watched the home team as it began play with a first and 10 on its own 20 yard line. She wrote the following on her notepad after watching several plays. +8, −3, +75, ExPt, K Off, +24, −8, +2, +12, Pnt. What is the net gain in yards so far by the home team? By the visiting team? Write a story about what might have happened on these plays.

60. The Pulley Problem. When pulley B goes 1 revolution, pulley A goes $\frac{3}{4}$ of a revolution.

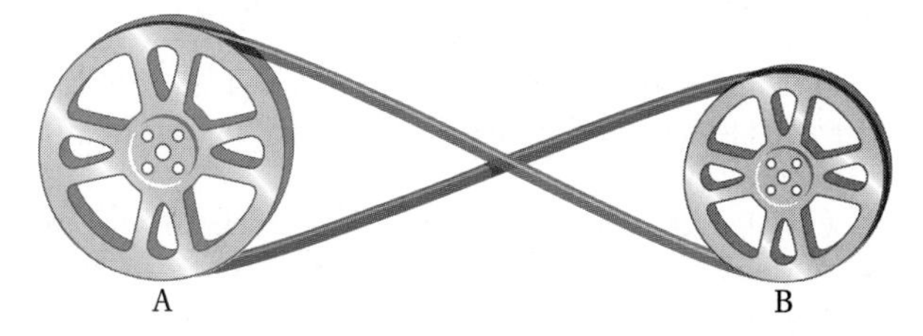

(*Hint:* Answer the questions in this problem using positive numbers for clockwise revolutions and negative numbers for counterclockwise revolutions.)
a. How would you describe the revolutions for pulley A when pulley B is turned +100 revolutions?
b. If pulley B moves at the rate of +60 revolutions per minute, at what rate does pulley A move?
c. If pulley A is turned −120 revolutions, how would you describe the revolutions for pulley B?

61. The Golf Tournament Problem. A golf tournament consisted of playing an 18-hole golf course three times (3 rounds). After 2 rounds, Paul Puttgood was listed on the tournament board as -8 (8 under par). David Drivefar was listed as -5. In the last round on the 7 holes where Drivefar made pars, Puttgood made seven birdies. (A birdie is a 1 under par score on a hole.) Their scores for each hole in the last round never differed by more than 1 stroke. Could Drivefar have beat Puttgood in the tournament? Explain your answer.

62. The Weathertalk Analysis Problem. Read the following story about the TV weatherperson and identify as many places as possible that the weatherperson could use addition or subtraction of integers to answer her questions. Write and solve the number sentences she could use.

Ms. Cloudcover, the TV weatherperson, noted that Sunday's temperature was 12°, much warmer than the $-7°$ on Friday and the $-2°$ on Saturday. She remembered that Thursday's temperature was 0° and checked to see how far Saturday's temperature had fallen from Thursday's temperature. She made a note to report the average temperature for the 3-day weekend and the temperature change from Friday to Saturday and from Saturday to Sunday. Then she began work on a report about how the high temperatures for each day of the past week had been over or under the normal high temperature for that day. She wrote, Sun: $-8°$, Mon: $-5°$, Tues: $-2°$, Wed: 6°, Thurs: 3°, Fri: $-4°$, Sat: 9°. She wondered whether the net effect of the week was more or less than normal. Before she left for the day, she noted that today's low temperature was 9° and figured the difference between it and the record low of $-16°$ for today's date.

D. Communicating and Connecting Ideas

63. Let three unit arrows to the right represent 3 and two unit arrows to the left represent -2.

$$3 \quad \rightarrow \quad \rightarrow \quad \rightarrow$$
$$-2 \quad \leftarrow \quad \leftarrow$$

When you combine the arrows, opposite arrows cancel each other, and you see that $3 + (-2) = 1$. How is this model for integer addition like one of the models in this chapter? Explain.

64. Find the age, to the nearest month, of the people in your class. Use the average age in months as your zero and positive and negative numbers to write the age of each person in relation to the average. Work in a group to find a good visual way to display this information. Discuss what is gained by communicating the data in this way and report your conclusions.

65. Write a convincing argument that subtracting an integer gives the same result as adding the opposite of the integer.

66. How could you use an elevator model to illustrate addition of integers? How is this model like one of the models you studied in this chapter? Explain.

67. Does the following procedure for adding integers give the correct sums?

To add integers with like signs, add the absolute values of the integers and give the sum the same sign as the integers. To add integers with unlike signs, subtract the lesser of the two absolute values of the integers from the greater. The sum has the same sign as the integer with the greater absolute value.

Explain your conclusion by using the sums $-7 + 4$ and $-3 + (-2)$.

68. Historical Pathways. Respond to the following remarks taken from historical references.

a. Antoine Arnauld (1612–1694) was a theologian and a mathematician. He argued that negative numbers were suspect because the proportion

$$\frac{-1}{1} = \frac{1}{-1}$$

is true and "how could a smaller be to a greater as a greater is to a smaller?" Explain in your own words what Arnauld meant. Do you think his reasoning was valid? Why or why not?

b. Bhaskara, a Hindu mathematician (1114–1185), when solving the equation $x^2 + 2x - 8 = 0$, commented: "The second value is in this case not to be taken, for it is inadequate...." What do you think he meant and why did he make this remark?

69. Historical Pathways. In Yoshio Mikami's book, *The Development of Mathematics in China and Japan* (1974), he states that the Chinese word *cheng* means positive and the word *fu* means negative. He also indicates that the following symbols were used by early Chinese mathematicians to represent positive (no diagonal slash) and negative (with a diagonal slash) numbers.

= −10,724

= −9

= 10,200

Write a paragraph describing the system of numeration you think the Chinese might have been using when writing these numerals.

70. Making Connections. Suppose that P = set of positive integers, N = set of negative integers, R = set containing 0, I = set of integers, E = empty set, and W = set of whole numbers. Use the symbols for set intersection, union, and subset of to write as many true statements as you can about P, N, R, I, and W.

Section 5.2

MULTIPLICATION, DIVISION, AND OTHER PROPERTIES OF INTEGERS

- Modeling Integer Multiplication
- Applications of Integer Multiplication
- Properties of Integer Multiplication
- Explaining Integer Division
- More Properties of Integer Multiplication and Division

In this section, we examine why rules for multiplying and dividing integers—such as a negative times a negative is a positive—work. We present and evaluate different physical models to explain integer multiplication. We then use the properties of integer multiplication and division to prove that some of the procedures that we developed are true.

Modeling Integer Multiplication

We first use physical models to understand **multiplication of integers**. Then pictorial representations, such as the number line, extend our insight.

Using a Counters Model. A *counters model* also may be used to understand integer multiplication. The setting of the model involves two handfuls of one-color counters, some in a bag and some in a pile. The counters are used with the actions indicated in Figure 5.11 to model four types of integer multiplication. The number of counters in the bag and pile at the beginning is your choice, but must be enough to perform the actions in the model. Example 5.14 provides an opportunity to use counters and model integer multiplication.

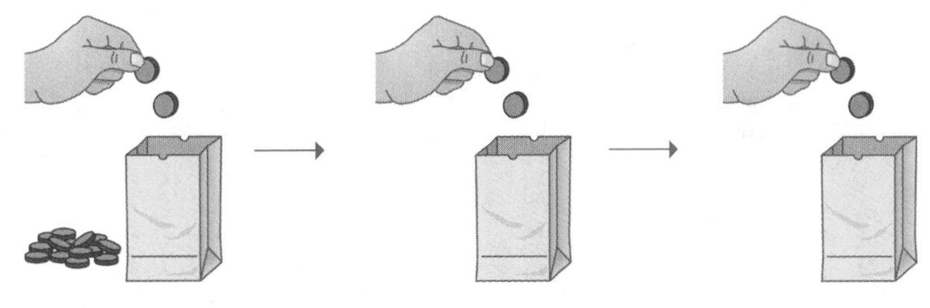

Action 1: Putting 2 Counters in the Bag (2)

Positive × Positive
Do action 1 three times. ⟶ $3(2) = \boxed{?}$ ⟵ Does the bag have more (+) or less (−) counters than before? How many more or less?

Negative × Positive
Do the opposite of action 1 three times. ⟶ $-3(2) = \boxed{?}$ ⟵ Does the bag have more (+) or less (−) counters than before? How many more or less?

Action 2: Taking 2 Counters out of the Bag (−2)

Positive × Negative
Do action 2 three times. ⟶ $3(-2) = \boxed{?}$ ⟵ Does the bag have more (+) or less (−) counters than before? How many more or less?

Negative × Negative
Do the opposite of action 2 three times. ⟶ $-3(-2) = \boxed{?}$ ⟵ Does the bag have more (+) or less (−) counters than before? How many more or less?

FIGURE 5.11 | Using counters to model integer multiplication.

Example 5.14 Using a Counters Model to Verify Integer Products

Show that $-4(-3) = 12$.

Solution

The second number is -3, so the action is taking 3 counters out of the bag. Because the first number is -4, do the opposite of the action 4 times. Putting 3 counters in the bag four times gives a total of 12 counters, or $-4(-3) = 12$.

Your Turn

Practice: Use a counters model to show that $-4(2) = -8$.

Reflect: Use the example and practice problems to explain the two important actions in the model. How do these actions represent negative and positive integers? ■

Using a Charged Field Model. The charged field model can also be used to model integer multiplication. Just as with addition and subtraction, a positive number is modeled by putting positive charges into the field and a negative number is modeled by putting negative charges into the field. Integer multiplication is modeled by putting in multiple same-size groups of charges when multiplying by a positive integer, and taking out multiple same-size groups of charges when multiplying by a negative integer. The product is the charge of the resulting field. Figure 5.12 illustrates how to use this model to find the products 2×-4 and -2×-4.

A charged field model can be used to model other types of integer multiplication, as shown in Example 5.15.

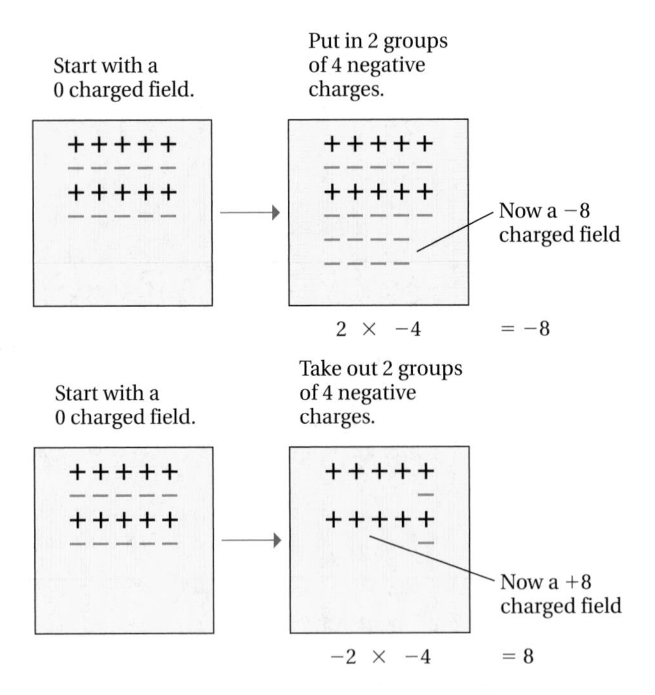

FIGURE 5.12 | Using a charged field to model integer multiplication.

Example 5.15 Charged Field Modeling of Other Integer Products

Use the charged field model to model -3×2.

Solution

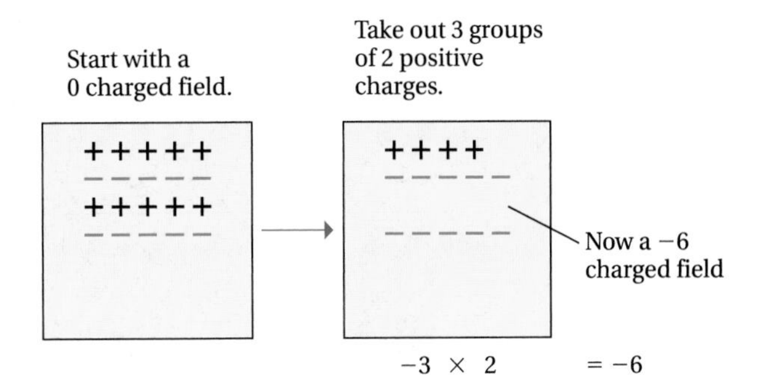

Your Turn

Practice: Use charged fields to model

a. -3×-2 **b.** 2×4

Reflect: How are the counters and charged field models for integer multiplication alike? How are they different? ■

Using the Number Line Model. The *number line model* also helps explain integer multiplication. Study the model shown in Figure 5.13 and complete the equations to see if it gives the same results as the other models.

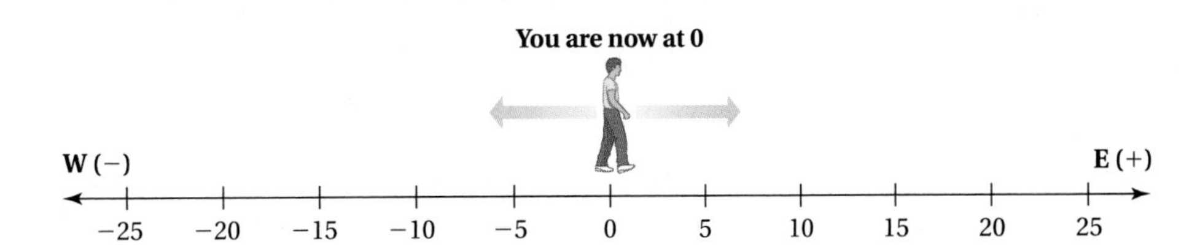

Action: Walking East at 5 Miles per Hour (+5)

Positive × Positive
Where will you be 3 hours from now? → $3(+5) =$ [?] ← Will you be east (+) or west (−) of 0? How far east or west?

Negative × Positive
Where were you 3 hours ago? → $-3(+5) =$ [?] ← Were you east (+) or west (−) of 0? How far east or west?

Action: Walking West at 5 Miles per Hour (−5)

Positive × Negative
Where will you be 3 hours from now? → $3(-5) =$ [?] ← Will you be east (+) or west (−) of 0? How far east or west?

Negative × Negative
Where were you 3 hours ago? → $-3(-5) =$ [?] ← Were you east (+) or west (−) of 0? How far east or west?

FIGURE 5.13 | Using a number line to model integer multiplication.

Another number line model that is sometimes used to help students understand integer multiplication is described in Figure 5.14. Note that in this model, the − sign is interpreted as "turn around" when in the first number, and "walk backward" when in the second number. The first magnitude is interpreted as the "number of steps" and the second magnitude is interpreted as the "length of each step."

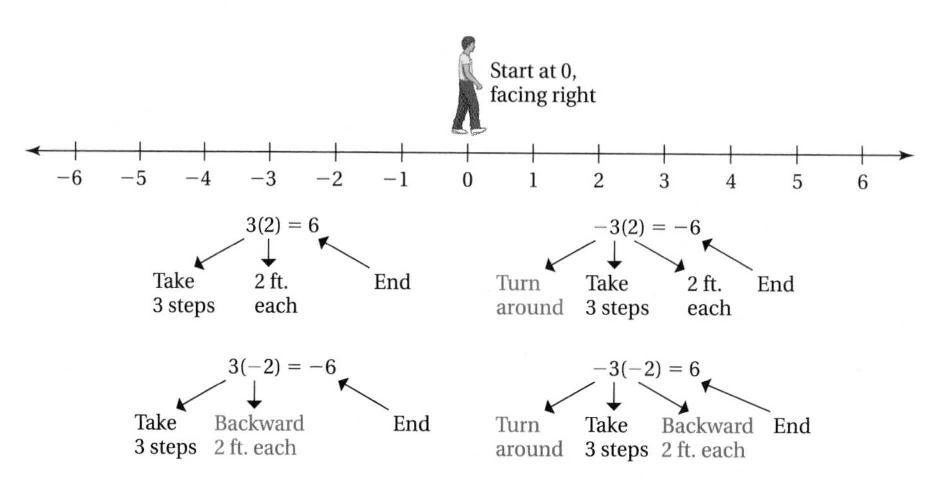

FIGURE 5.14 | Another number line model for integer multiplication.

Example 5.16 provides an opportunity to use these two number line models to understand multiplication of integers.

Example 5.16

Using the Number Line Model to Verify Integer Products

Show that $-3(-4) = 12$.

Solution

Stacy's thinking: The second number is -4, so the action is walking west at 4 miles per hour. Because the first number is -3, find where you were 3 hours ago if you are now at 0. Walking west at 4 miles per hour for 3 hours, you would have had to be at 12 then to be at 0 now, or $-3(-4) = 12$.

Eric's thinking: I think about starting at 0, facing right. Then I turn around, facing left, and imagine I'm a giant taking 3 backward steps of 4 ft. each. I end up at 12, so $-3(-4) = 12$.

Your Turn

Practice: Use the number line model of your choice to show that $5(-2) = -10$.

Reflect: Explain what the integers in $5(-2) = -10$ represent when you use the number line model for multiplying integers. ■

Using Mathematical Relationships, Patterns, and Reasoning. Number relationships also help explain integer multiplication. For example, the repeated addition interpretation of multiplication can be used to show that the product of a positive and a negative integer is a negative integer; that is, $3(-4) = (-4) + (-4) + (-4) = -12$.

Patterns also give an insight into integer multiplication. What does the extension of the following pattern suggest about the product of a negative integer times a negative integer?

$$\begin{aligned} 3(-4) &= -12 \\ 2(-4) &= -8 \\ 1(-4) &= -4 \\ 0(-4) &= 0 \\ -1(-4) &= ? \\ -2(-4) &= ? \\ -3(-4) &= ? \end{aligned}$$

The product increases by 4 each time. ↓

Mini-Investigation 5.4 challenges you to explore patterns in integer products you have found using your calculator.

Write a paragraph summarizing what you discovered about multiplying integers.

Mini-Investigation 5.4 **The Technology Option**

After using your calculator to find several useful example products, how would you complete these conjectures?

a. The product of a positive integer and a negative integer is ...
b. The product of a negative integer and a positive integer is ...
c. The product of two negative integers is ...
d. The product of two positive integers is ...

Mathematical reasoning can also help us draw some conclusions about multiplying integers. For example, we can use the known product $3(4) = 12$ to draw conclusions about $3(-4)$ and $-3(-4)$.

Because $3(4) = 12$, we know that $-3(4)$ would be the opposite of 12, or -12.
Because $3(-4) = -12$, we know that $-3(-4)$ would be the opposite of -12, or 12.

These conclusions suggest the generalization if $a(-b) = -(ab)$, then $-a(-b) = ab$, which supports our conjectures about multiplying integers.

The counters model, the number line, and the relationships, patterns, and reasoning covered in this section all support the following procedures for multiplying integers.

Procedures for Multiplying Integers

- *Multiplying two positive integers:* Multiply as with whole numbers.
- *Multiplying two negative integers:* Multiply absolute values. The product is positive. That is, for negative integers a and b, $a(b) = |a|\,|b|$.
- *Multiplying a positive and a negative integer:* Multiply absolute values. The product is negative. That is, for a positive integer a and a negative integer b, $a(b) = -|a|\,|b|$.

Applications of Integer Multiplication

Although many everyday applications involve multiplying positive integers, everyday applications that involve negative number multiplication occur less frequently. Often a real-world application, such as the following, can be solved either by multiplying integers or by multiplying whole numbers.

Problem: Over a 5-day period, the water level in the city swimming pool dropped 4 inches per day. What was the total change for the 5-day period?

Solution with integers: $5(-4) = -20$. The total change in water level was -20 inches.

Solution with whole numbers: $5(4) = 20$. The water level fell by 20 inches.

Example 5.17 features a *guess, check, and revise* problem-solving strategy to illustrate the use of integer multiplication to help solve problems.

Example 5.17

Problem Solving: The Weather Report

A TV meteorologist said that the evening temperature was triple the temperature that morning. Later he reported that the evening temperature was 6° less than the morning temperature. What was the morning temperature, or did the meteorologist make a mistake?

Working toward a Solution

Understand the problem	*What does the situation involve?*	Changes in temperature during a day
	What has to be determined?	The morning temperature
	What are the key data and conditions?	The evening temperature reading was 3 times the morning temperature reading but 6° less than it
	What are some assumptions?	Assume the meteorologist did not make a mistake and try to find a solution
Develop a plan	*What strategies might be useful?*	Guess, check, revise
	Are there any subproblems?	Triple the guessed temperature. Subtract 6 from the morning temperature to see if it matches the tripled temperature.
	Should the answer be estimated or calculated?	Calculate the exact numbers.
	What method of calculation should be used?	Use mental computation.
Implement the plan	*How should the strategies be used?*	Guess: 4° Check: $3(4°) = 12°$, which is not less than 4°.

		Guess: $-2°$ Check: $3(-2°) = -6°$, which is only $4°$ less. Guess: $-3°$ Check: $3(-3°) = -9°$, which is $6°$ less.
	What is the answer?	The morning temperature was $-3°$.
Look back	*Is the interpretation correct?*	Yes. The calculations fit the data.
	Is the calculation correct?	Yes. The mental computations were correct.
	Is the answer reasonable?	Yes. Three times a negative temperature would be a smaller negative temperature.
	Is there another way to solve the problem?	Write and solve an equation: $3t = t - 6$.

Your Turn

Practice: Bonnie's score in a math game was 2 more than double Reed's score but also 5 less than his score. What was Bonnie's score?

Reflect: In the example and practice problems, which is an easier strategy—guess, check, revise or write an equation? Explain your answer. ■

Although everyday applications involving the multiplication of integers are somewhat limited, multiplication of both positive and negative integers occurs often in science, economics, astronomy, and many other fields. Example 5.18 illustrates the use of integer multiplication to solve a problem in space flight science.

Example 5.18 Problem Solving: Satellite Life

The equipment aboard a satellite requires 10 watts of power to operate properly. Because this wattage is supplied by batteries, the operational life of a satellite, t (in days), was calculated by using a special formula. At one stage of evaluating the formula, the equation simplified to $-t = 25(-16)$. What was the operational life of the satellite?

Solution

For $-t = 25(-16)$, we multiply the integers and find that $-t = -400$. If the opposite of t is -400, then t must equal 400. The operational life of the satellite is 400 days.

Your Turn

Practice: Suppose that, at one stage of evaluating a formula for finding the half-life of a certain satellite power supply, the equation simplified to $-t = 25(-7)$. What would be the half-life of the power supply?

Reflect: What is another way you could have found the value of t in the example problem? ■

Properties of Integer Multiplication

The counters model, the number line, and the procedures for multiplying integers may be used to find several integer products to help make the following properties of integer multiplication seem plausible.

1. When you multiply any two integers, you always get another integer as the product.
2. One is the only integer that doesn't change an integer it's multiplied by.
3. You can multiply two integers in either order because you'll get the same product either way.
4. When multiplying three integers, you can start by multiplying the first two or the last two because you'll get the same product either way.

In mathematics, variables help you express your ideas more efficiently. The basic properties of integers in the preceding list may be summarized by using variables as follows.

Basic Properties of Integer Multiplication	
Closure property	For all integers a and b, ab is a unique integer.
Multiplicative identity property	1 is the unique integer such that for each integer a, $a \times 1 = 1 \times a = a$.
Commutative property	For all integers a and b, $ab = ba$.
Associative property	For all integers a, b, and c, $(ab)c = a(bc)$.

As with whole numbers, the distributive property *ties integer addition and multiplication together*. That is, when two numbers are first added and the sum is then multiplied by a third number, the result will be the same as when each number is first multiplied by the third number and the products are then added.

Distributive Property
For all integers a, b, and c, $a(b + c) = ab + ac$ and $(b + c)a = ba + ca$.

Another useful property, which can be proved by using other properties, is the zero property.

Zero Property for Multiplication
For all integers a, $a(0) = 0(a) = 0$.

We give a partial proof for the zero property in Exercise 55 at the end of this section and ask you to complete it.

As with integer addition, the properties of integers, accepted as basic postulates, help verify our techniques for multiplying integers. For example, even though we already know how to *find* products, the properties of integer multiplication, along with mathematical reasoning, can help us *prove* that our method of finding a product is correct. For example, we can use basic properties—and the fact $3(-4) = -12$, verified previously—to prove that $-3(-4) = 12$, justifying each step in the proof with one such property, as follows.

We know that 12 is the only number that adds to -12 to give 0. If we can show that $-3(-4)$ adds to -12 to give 0, it must equal 12.

Statement	**Reason**
$-12 + [-3(-4)] = 3(-4) + [-3(-4)]$	Substitute $3(-4)$ for -12
$= [3 + (-3)](-4)$	Distributive property
$= 0(-4)$	Additive inverse property
$= 0.$	Zero property for multiplication

So $-3(-4)$ must equal 12.

Example 5.19 illustrates use of the basic properties of integers to prove that the technique we use for finding the product of a positive and a negative integer gives the correct result.

Example 5.19 Using Basic Properties of Integer Multiplication in a Proof

Use the basic properties and reasoning to prove that $2(-3) = -6$.

Solution

The only number that adds to 6 to give 0 is -6. Thus if we can show that $2(-3)$ adds to 6 to give 0, it must equal -6.

Statement	**Reason**
$2(-3) + 6 = 2(-3) + 2(3)$	Substitute $2(3)$ for 6
$= 2(-3 + 3)$	Distributive property
$= 2(0)$	Additive inverse property
$= 0.$	Zero property for multiplication

But -6 is the number that adds to 6 to give 0 by the additive inverse property. Therefore $2(-3) = -6$.

Your Turn

Practice:

a. Prove that $-2(-3) = 6$.
b. Prove that $-3(4) = -12$.

Reflect: Do you believe that we proved

a. a negative integer times a negative integer is a positive integer?
b. a negative integer times a positive integer is a negative integer?

Why or why not? ■

We now use the same procedure used with numbers above to verify in general that the product of a positive integer and a negative integer is a negative integer. Because of the additive inverse property of integers, we know that $-(ab)$ is the only number that adds to ab to give 0. If we can verify that $a(-b)$ adds to ab to give 0, we will have verified that $a(-b)$ and $-(ab)$ are the same number, that is, $a(-b) = -(ab)$. We verify this as follows.

Statement	**Reason**
$a(-b) + ab = a(-b + b)$	Distributive property
$= a(0)$	Additive inverse property
$= 0$	Zero property for multiplication

So we conclude that $a(-b)$ and $-(ab)$ both add to ab to give 0, and since there is only one number with that property, $a(-b)$ and $-(ab)$ must represent the same number. Hence $a(-b) = -(ab)$, and if $a > 0$ and $b > 0$, this verifies that a positive integer times a negative integer is a positive integer.

In a similar manner, we can verify that the product of a negative integer times a negative integer is a positive integer. Because of the additive inverse property, we know that ab is the only number that adds to $-(ab)$ to give 0. If we can verity that $-a(-b)$ adds to $-(ab)$ to give 0, we will have verified that ab and $-a(-b)$ are the same number, that is, $-a(-b) = ab$. We verify this as follows.

Statement	**Reason**
$-a(-b) + -(ab) = -a(-b) + a(-b)$	$a(-b) = -(ab)$ was verified above.
$= (-a + a)(-b)$	Distributive property
$= 0(-b)$	Additive identity property
$= 0$	Zero property for multiplication

So we conclude that $-a(-b)$ and ab both add to $-(ab)$ to give 0, and since there is only one number with that property, $-a(-b)$ and ab must represent the same number. Hence $-a(-b) = ab$, and if $a > 0$ and $b > 0$ this verifies that a negative integer times a negative integer is a positive integer.

Explaining Integer Division

We can extend the relationship between whole-number multiplication and division, developed in Chapter 2, to help explain the relationship between integer multiplication and division. As with whole numbers, dividing by an integer can be thought of as *finding the missing factor.*

Factor		Factor		Product		Product		Factor		Factor
7	×	4	=	?	so	28	÷	4	=	?

When multiplying you know two factors. You find the product. In this case it is 28.

When dividing you know the product and one factor. You find the missing factor. In this case it is 7.

You can find the answer to the division, $28 \div 4 = ?$, by thinking, "What number multiplied by 4 gives the product 28?" or "When the product is 28 and one factor is 4, what is the missing factor?" Looking at it another way, you might think of $28 \div 4$ and 7 as each representing the number that multiplied by 4 gives 28. In any case, $28 \div 4 = 7$.

A Sample Student Page: Using Inductive and Deductive Reasoning

The *NCTM Principles and Standards for School Mathematics* (2000) state "In grades 6–8 students should extend their reasoning skills by ... using inductive and deductive reasoning to formulate mathematical arguments (p. 262). To see what this means, study the elementary school mathematics textbook page in Figure 5.15 and answer the questions in the margin.

- **Do** you think that Stacy proved the rules? Why or why not?
- **What** are some things that students can learn from this page?
- **How** does the page help students feel that mathematics makes logical sense and is something to think about?

Using Critical Thinking

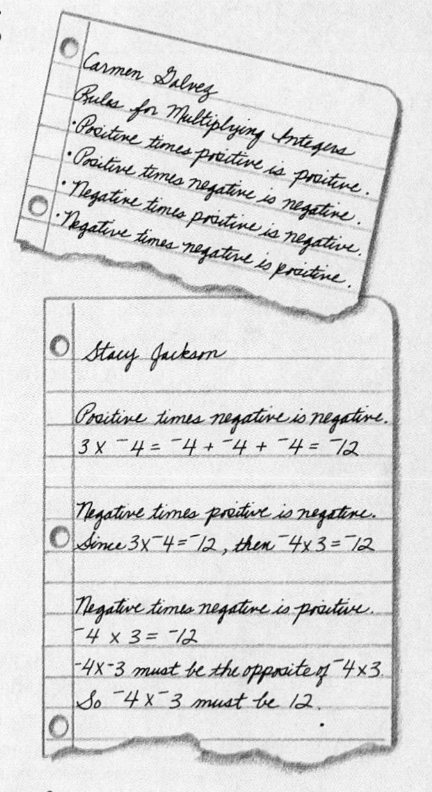

LEARN ABOUT IT

"These negative numbers are strange," Carmen said. "I can see that a positive times a positive is positive, but how could you ever *prove* the others, especially that a negative times a negative is a positive?"

"I can prove the rules using patterns," said Eric. "For example:

$$3 \cdot 2 = \mathbf{6}$$
$$3 \cdot 1 = \mathbf{3}$$
$$3 \cdot 0 = \mathbf{0}$$
$$\text{so } 3 \cdot {}^{-}1 = \mathbf{{}^{-}3}."$$

"Patterns are okay," Tanya agreed, "but you can't always be sure just by looking at a pattern."

"I think I've found a way to prove the last 3 rules," said Stacy as she wrote some equations on her paper.

TALK ABOUT IT

1. What were the students trying to do?
2. Why did Tanya discourage the use of patterns?
3. Explain Stacy's proof of the last 3 rules. Can you give reasons why each of her statements are true?
4. Do you think Stacy has proved the rules? Why or why not?

TRY IT OUT

Use Stacy's method or other methods, to convince someone that the following are true.

1. $3 \cdot {}^{-}2 = {}^{-}6$ **2.** ${}^{-}2 \cdot 3 = {}^{-}6$ **3.** ${}^{-}2 \cdot {}^{-}3 = 6$

FIGURE 5.15 | Excerpt from a middle school mathematics textbook.

(*Source:* From *Addison Wesley Mathematics, Grade 8* by R. E. Eicholz, P. G. O'Daffer, R. I. Charles, S. L. Young, and C. S. Barnett. © 1995 Pearson Education, Inc. publishing as Prentice Hall. Used by permission.)

This process works for all integers. We define it, using variables, as follows.

Definition of Integer Division

For all integers a, b, and c, $b \neq 0$, $a \div b = c$ if and only if $c \times b = a$.

Example 5.20 uses *missing factors* and the definition of integer division.

Example 5.20 Finding Integer Quotients

Find $-24 \div 6$.

Solution

Matt's thinking: What number multiplied by 6 gives the product -24? Because $-4(6) = -24$, I know that $-24 \div 6 = -4$.

Dave's thinking: The product is -24 and one factor is 6. The missing factor is -4 because $-4(6) = -24$.

Your Turn

Practice: Find the following quotients

a. $36 \div (-9)$.
b. $-54 \div (-6)$.

Reflect: How does finding the missing factor when dividing integers relate to the idea of multiplying to check division? ■

You can always use the definition of integer division to help you find quotients for integers. However, a general procedure to make finding quotients even easier would be helpful. Mini-Investigation 5.5 challenges you to use a calculator to explore patterns in dividing positive and negative integers to help in forming a generalization about finding quotients.

Write a paragraph summarizing what you discovered about dividing integers.

Mini-Investigation 5.5 The Technology Option

First use your calculator to find these quotients.

a. $782 \div (-23)$ **b.** $288 \div (-16)$
c. $-216 \div 9$ **d.** $-280 \div 8$
e. $-162 \div (-27)$ **f.** $-78 \div (-13)$
g. $234 \div 9$ **h.** $252 \div 28$

Then, use what you have discovered to complete the following conjectures.

1. The quotient of a positive integer divided by a negative integer is . . .
2. The quotient of a negative integer divided by a positive integer is . . .
3. The quotient of a negative integer divided by a negative integer is . . .
4. The quotient of a positive integer divided by a positive integer is . . .

The procedures for dividing integers are summarized as follows.

Procedures for Dividing Integers

- *Dividing two positive integers:* Divide as with whole numbers.
- *Dividing two negative integers:* Divide absolute values. The quotient is positive. That is, for negative integers a and b, $a \div b = |a| \div |b|$.
- *Dividing a positive and a negative integer:* Divide absolute values. The quotient is negative. That is, for a positive integer a and a negative integer b, $a \div b = -(|a| \div |b|)$.

Integer division can also be used to solve practical problems, as illustrated in Example 5.21.

Example 5.21

Problem Solving: The Average Temperature Change

Over a 6-month period, the total change in the average daily high temperature was -54°F. What was the average change per month?

Solution

We choose the operation division, which gives $-54 \div 6 = -9$. The average change per month was -9°F.

Your Turn

Practice: In a 4-month period, the total change in the average daily low temperature was -32°F. Find the average change per month.

Reflect: Do you think that the solution to the example problem suggests that the daily high temperature changed about the same amount each month? Explain. ■

More Properties of Integer Multiplication and Division

Integer multiplication and division properties aren't generally very useful for everyday use of mathematics at home, the gas station, or the store. However, mathematics is being used more and more in many trades and professions. Most people who want good jobs will need a solid foundation in mathematics. To understand mathematics, they will need to understand basic properties, like those that follow for integers, and build on them in later courses.

Some Integer Division Properties. In Chapter 2, we explored some of the properties of division for whole numbers. The definition of division and the 0 and 1 properties of multiplication can be used to verify quickly some similar properties of integer division. For example, for $a \neq 0$ the missing factor in $a \div a = ?$ obviously is 1 because $1 \times a = a$. Also, the missing factor in $a \div 1 = ?$ is a because $a \times 1 = a$. The missing factor in $0 \div a = ?$ is 0 because $0 \times a = 0$. When we look for a missing factor in $a \div 0 = ?$, we see that no integer would work unless $a = 0$. For $0 \div 0$, any integer would work and there is no unique quotient. For this reason, we agree that *integers can't be divided by 0.* The property $ab \div a = b (a \neq 0)$ is easily verified and is useful in solving equations. We summarize these division properties for integers as follows.

Properties of Integer Division

For all integers a and b, $a \neq 0$, $a \div a = 1$, $a \div 1 = a$, $0 \div a = 0$, and $ab \div a = b$. We do not divide an integer by 0, because no unique quotient exists.

Mini-Investigation 5.6 asks you to use the connection between division and multiplication.

Write an explanation of your verification.

Mini-Investigation 5.6 Making a Connection

How can you use the following division to verify that, for all integers a and b, $a \neq 0$, $ab \div a = b$?

$$a\overline{)ab}^{\;?}$$

Some Properties of Opposites for Integers. In working with models and the basic properties of integer multiplication, you may have discovered some ways of dealing with products that involve opposites of integers. Here are some ways to verify that some of the discovered generalizations are true.

For example, here is how you might reason to prove that $a(-b) = (-a)b = -(ab)$

1. Each of the expressions $a(-b)$, $(-a)b$, and $-(ab)$ can be added to ab to produce 0. To verify this for $(-a)b$, we see that $(-a)b + ab = (-a + a)b = 0b = 0$.
2. It can be verified in a similar manner that $a(-b)$, $-(ab)$ can also be added to ab to give the sum 0. (See Exercise 50, p. 271.)
3. From 1 and 2 above, we know that the three expressions $a(-b)$, $(-a)b$, and $-(ab)$ each have the same unique characteristic, that is, they add to ab to produce 0, and hence they are equal. So, $a(-b) = (-a)b = -(ab)$.

In a similar manner, we can verify that $a(-1) = -a$. Both $a(-1)$ and $-a$ can be added to a to produce 0. (Complete Exercise 49, p. 271 to verify this.) Because these two expressions both have the same unique characteristic, they are equal, and $a(-1) = -a$. Finally, we verified on page 264 that $-a(-b)$ and ab can be added to $-(ab)$ to give 0. Because these two expressions each have this same unique property, they are equal, and we write $-a(-b) = ab$. We summarize all these properties in the next box.

Properties of Opposites

For all integers a and b, $-(-a) = a$, $-a(-b) = ab$, $(-a)b = a(-b) = -(ab)$, and $a(-1) = (-1)a = -a$.

Let's examine how these properties might be used by considering the equation $-[-(x + 5)] + 8 = 13$. Note that $-[-(x + 5)]$ is of the form $-(-a)$, where a is $x + 5$. Because $-(-a) = a$, we conclude that $-[-(x + 5)] = x + 5$. From the original equation, we have $x + 5 + 8 = 13$, or $x + 13 = 13$. So by using a property of opposites, we see that we can simplify and solve the equation.

Some Distributive Properties for Integers. Some integer properties are very similar to the distributive property for multiplication over addition. For example, $a(b - c) = ab - ac$ appears to be true. We can view this as an application of the distributive property by thinking of the equivalent statement, $a[b + (-c)] = ab + a(-c)$. Also, $-(a + b) = -a + (-b)$ can be viewed as an application of the distributive property by thinking of the equivalent statement, $-1(a + b) = -1a + (-1b)$.

Distributive Property for Multiplication over Subtraction

For all integers a, b, and c, $a(b - c) = ab - ac$.

Distributive Property for Opposites over Addition

For all integers a, b, and c, $-(a + b) = -1(a + b) = -a + (-b) = -a - b$.

These properties aren't "rules" for doing something but rather are basic assumptions about the system of integers on which other relationships involving integers are based. For example, simplifying an expression in algebra may make finding a solution to a problem easier. You can apply general properties of integers to the simplification process and prove that it makes sense. These properties also apply to rational and real number systems, so understanding and using them is efficient. Example 5.22 illustrates how the distributive properties of integers can be used to verify simplification of an expression.

Example 5.22

Using the Distributive Properties of Integers

Find an equivalent expression for $-(n + 7)$ that does not have parentheses. Use properties to prove that the two expressions are equivalent.

Solution

$-(n + 7) = -n - 7$. The distributive property for opposites over addition supports this conclusion.

Your Turn

Practice: Find an equivalent expression that meets the conditions given. Then name or state a property or properties to prove that your expressions are equivalent.

a. $-4(n - 2)$, so there are no parentheses.

b. $5x \div 5$, so there are as few symbols as possible.

c. $-7(-n)$, to have as few $(-)$ signs as possible.

Reflect: How is the use of the distributive property for opposites over addition in the example problem like using the distributive property for multiplication over addition with $-1(n + 7)$? ■

Problems and Exercises | *for Section 5.2*

A. Reinforcing Concepts and Practicing Skills

1. Complete the following pattern to show that a negative number (-3) times a negative number (-5) is a positive number (15).

$$3(-5) = -15.$$
$$2(-5) = -10$$
$$1(-5) = -5.$$

   a. $0(-5) = ?$ b. $-1(-5) = ?$
   c. $-2(-5) = ?$ d. $-3(-5) = ?$

2. Use counters to model and find these products.
   a. $2(5)$ b. $2(-5)$
   c. $-2(-5)$

3. Use a charged field to model and find these products.
   a. $3(4)$ b. $3(-4)$
   c. $-3(-4)$

4. Use a number line to model and find these products.
   a. $2(3)$ b. $2(-3)$
   c. $-2(-3)$

5. Find the product. Describe the method you used.
   a. $9(-7)$ b. $-894(-567)$
   c. $-10(8)$

Calculate the answers in Exercises 6–12.

6. $12(-24)$
7. $|-8|(-13)$
8. $|-15||-16|$
9. $-|8|(-8)(-8)$
10. $[9 - (-7)]|-4|$
11. $-|-56|(14 + (-9))$
12. $[-2 + (-9)][4 - (-7)]$
13. Use a calculator to find each product.
   a. $95(-46)$ b. $-180(47)$
   c. $-34(-67)$

Calculate the answers in Exercises 14–17.

14. $(-5)(-4)(-3)$
15. $(-1)(-3)(-5)(-7)$
16. $(-2)(-3)(-4)(-3)(-2)$
17. $(-2)(-2)(-2)(-2)(-2)(-2)$

Use the results of Exercises 14–17 to discover how to complete the sentences in Exercises 18–19.

18. The product of an odd number of negative integers is a ____ integer.
19. The product of an even number of negative integers is a ____ integer.
20. Show that the distributive property holds true for integers a, b, and c by substituting these numbers in the statement of the property, p. 262.
   a. $a = -2, b = 3, c = -4$
   b. $a = 5, b = -3, c = -2$
   c. $a = -6, b = -5, c = -4$
21. Show that the associative property for multiplication holds true for integers, a, b, and c by substituting these numbers in the statement of the property, p. 262.
   a. $a = -2, b = 3, c = -4$
   b. $a = 5, b = -3, c = -2$
   c. $a = -6, b = -5, c = -4$
22. Simplify as indicated. State at least one property of multiplication or division that would prove your simplification process correct.
   a. $-(n + 4)$, to eliminate the parentheses.
   b. $6(-s)$, to eliminate the parentheses.
   c. $8(24 - n)$, to eliminate the parentheses.
   d. $12n \div 12$, to reduce the number of symbols.
23. Think about missing factors to find the quotient.
   a. $-56 \div 8$ b. $-48 \div -6$
   c. $36 \div -9$

Calculate the answers in Exercises 24–31.

24. $(-12 \div 3) + (-4)$
25. $-6[12 + (-7)]$
26. $[4 - (-8)] \div [-2 - (-6)]$
27. $-6(-54 \div -9)$
28. $[-8 + (-9)](8 + 9)$
29. $(-7)(-9) + 7(-9) + -9(7)$
30. $[-5 + (-4)][-15 - (-6)]$
31. $[53 + (-5)] \div [12 - (-4)]$
32. The temperature outside an airplane at a certain cruising altitude drops 7°C for each kilometer increase in altitude. Determine the change in outside temperature for an airplane that increases its altitude by 5 kilometers. Use positive and negative numbers.
33. Over a 4-month period, the total steady change in a corporation's sales income was $-\$12$ million. Determine the average change in sales income per month.
34. A stock's average change in price over a 5-month period was $-\$4$ per month. What was the total change in price during the period?

35. A 50-item health inventory gives the following scores for responses A, B, C, D, E, and F: A = 5, B = 4, C = 2, D = −2, E = −4, F = −5. A person's health score is calculated by totaling the scores for all responses. Will gave 16 A responses, 8 B responses, 8 C responses, 3 D responses, 11 E responses, and 6 F responses.
 a. What is Will's health score?
 b. What is Will's average negative response score?
 c. What is Will's average positive response score?

36. An error in recording bank charges caused the computer to mistakenly take $17 each month in charges from Alison's checking account for 9 months and $19 a month for the next 4 months. The error was corrected after 13 months. How much did the bank owe Alison? Write a numerical expression, involving some negative integers, that shows how you could solve the problem.

B. Deepening Understanding

37. How would you use a calculator in two different ways to verify that $81 \div (-27) = -3$? Write a description of how you would do so.

38. Copy and complete the multiplication magic square. The product in each row, column, and diagonal should be the same.

−2	?	12
−36	?	?
3	?	?

39. Use a calculator, start with 0, and subtract 12 ten times. What integer calculation does this show?

40. Evaluate each expression.
 a. $-4n$, for $n = -26$
 b. $-16rs$, for $r = -29$ and $s = 39$
 c. $-12x - 8y$, for $x = -24$ and $y = 16$
 d. $8p \div (-q)$, for $p = -130$ and $q = 65$

41. Use integer properties to explain why each is true.
 a. $-2(-y) = 2y$.
 b. $-(2x - 4) = -2x + 4$.
 c. $-1(ab) = -(ab)$.

42. Give a property of multiplication or division as a reason for each step.

$x(y - z)$
$= x[y + (-z)]$ a. ?
$= xy + x(-z)$ b. ?
$= xy + [-(xz)]$ c. ?
$= xy - xz.$ d. ?

43. Review the meaning of exponents and compute each of the following.
 a. 5^3
 b. $(-2)^2$
 c. $(-5)^3$
 d. $(-3)^4 \div (-3)^3$

44. Indicate *always positive* (P), *always negative* (N), or *can't tell* (C) for each expression in which n is a non-zero integer.
 a. n
 b. n^2
 c. n^3
 d. $-n^2$
 e. $-n$
 f. $(-n)^2$
 g. $-n^3$
 h. n^4
 i. n^5
 j. n^{100}

45. Copy and complete this table for multiplying integers.

×	Pos	Neg	0
Pos	?	?	?
Neg	?	?	?
0	?	?	?

Can you make a similar table for division? If so, make one. If not, explain why you can't.

46. Do you think each property holds for all integers a, b, c, and d? Substitute integers and find counterexamples, if possible.
 a. $(a + b) \div c = a \div c + b \div c$.
 b. $a \div (b + c) = a \div b + a \div c$.
 c. $(a + b)(a - b) = a^2 - b^2$.
 d. If $a < b$, then $a^2 < b^2$.
 e. If $a < b$, then $c - a < c - b$.

47. Find the pattern and give the next three integers in each sequence.
 a. 2, −4, 4, −8, 6, −12, 8, −16, ...
 b. 1, −2, 4, −8, 16, −32, ...
 c. 2, −6, 18, −54, 162, ...
 d. 2, −5, 9, −19, 37, −75, ...

48. Estimate the following products or quotients. Tell whether you used
 i. rounding,
 ii. substitute compatible numbers,
 iii. front-end estimation,
 iv. clustering, or
 v. another method.
 a. -97×48
 b. $-(19)(23)(-21)(-18)$
 c. $250 \div (-48)$
 d. $-689(-208)$
 e. $-3689 \div 438$

C. Reasoning and Problem Solving

49. Prove that $a(-1)$ and $-a$ both add to a to give 0 and so are equivalent.

50. Prove that $-ab$ and $-(ab)$ both add to ab to give 0 and so are equivalent.

51. Why is saying that $0 \div 0 = 5$ is incorrect, even though you can show that $5 \times 0 = 0$?

52. One student said, $3(-2)$ and $-3(-2)$ can't both be -6. I know that $3(-2)$ is -6 because $3(-2) = -2 + (-2) + (-2) = -6$. So $-3(-2)$ must equal 6. Another student said, I know that $3(-2)$ can be found by adding -2 three times, or $3(-2) = -2 + (-2) + (-2) = -6$. So $-3(-2)$ can be found by doing the opposite of adding -2 three times, which would be adding 2 three times. Therefore $-3(-2) = 2 + 2 + 2 = 6$.

    Do you believe these students' reasonings are valid? Why or why not?

53. What basic properties of integer multiplication do not hold for integer division? Give counterexamples to justify your choices.

54. **The Field Measuring Problem.** A farmer measured the length and width of a rectangular field and sent the results shown to the local agricultural agent. If the agricultural agent decides to use the farmer's data, what do you think would be the best way for the agent to record the area of the field?

55. State the missing reasons in the following proof that $a \times 0 = 0$.

Statements	Reasons
$a \times 1 + 0 = a \times 1.$	a.
$a \times 1 + a \times 0 = a(1 + 0)$	b.
$= a \times 1.$	c.

The proof above showed that

$$a \times 1 + 0 = a \times 1 \text{ and } a \times 1 + a \times 0 = a \times 1.$$

d. Since 0 is the only number that added to $a \times 1$ gives $a \times 1$, why must $a \times 0$ be 0?

56. **The Mile-Run Problem.** The record time for running a mile has decreased over the years, as shown.

Year	Min:Sec	Year	Min:Sec
1954	3:59.4	1964	3:54.1
1955	3:58.0	1965	3:53.6
1957	3:57.2	1966	3:51.3
1958	3:54.5	1967	3:51.1
1962	3:54.4	1975	3:49.4
1980	3:48.8	1993	3:44.3
1985	3:46.3		

The formula

$$s = \frac{-4y}{10} + 1020$$

was developed from these data to predict the record time in seconds, s, for a given year, y. How accurate do you believe this formula might be in predicting the next record time? (*Hint:* Use of a computer spreadsheet to create a scatterplot of the date/time data might be helpful.)

57. **The Barrel Weight Problem.** In the fifteenth century, European traders used integers to label barrels of grain or flour. A barrel labeled $+2$ was 2 pounds too heavy and one labeled -2 was 2 pounds too light. Suppose that a trader checked twenty-five 50-pound barrels and found the following number of labels: eight -3s, six -5s, nine $+4$s, and two -7s. What is the actual weight of all the barrels?

58. **The Cost of Smoking.** Estimate how much money a person who begins smoking at age 16 will spend in a lifetime for cigarettes. State the assumptions that you use in making the estimate.

D. Communicating and Connecting Ideas

59. How would you refute the argument that -50 is greater than -25 because a bill for \$50 is greater than a bill for \$25?

60. A student who used the number line model to show that $-4(-3)$ equals 12 claimed that he had *proved* that a negative number times a negative number is a positive number. Do you agree? Why or why not?

61. **Historical Pathways.** The Koreans used red and black bamboo rods to represent positive and negative numbers, respectively. How do you suppose they carried out the multiplication $(-3)4$?

62. Conduct an interview to find out how someone else responds to the question, "Why is a negative number times a negative number a positive number?" Write a report describing your findings.

63. **Making Connections.** Discuss in a small group and record the ideas expressed about how multiplication and division of integers are like addition and subtraction of integers and how they are different.

CHAPTER SUMMARY

Key Ideas: Questions and Answers

Section 5.1

- **What are integers and how are they used in everyday life?** *(pp. 231–233)* Integers are used to represent any real-world situation involving amount and direction. Some examples are temperature, debits and credits, stock price gains and losses, and golf scores. Mathematically, integers were created to provide solutions for equations such as $x + 1 = 0$. The set of integers consists of the positive integers, the negative integers, and zero.
- **What are some ways of explaining integer addition?** *(pp. 233–239)* Integer addition may be explained by using a counters model, in which black counters represent positive integers and red counters represent negative integers, with black and red counters canceling each other, or a charged field model. Trips forward and backward on a number line model may also be helpful. These models may be used to formulate procedures for adding integers.
- **What are some basic properties of integer addition and how are they used?** *(pp. 239–241)* Basic properties of integers are the additive inverse—for every a there is a $-a$ such that $a + (-a) = 0$; closure: $a + b$ is a unique integer; additive identity: for each a, $a + 0 = a$; commutative: $a + b = b + a$; and associative properties: $a + (b + c) = (a + b) + c$. The basic properties may be used to verify that the procedures used for addition are logically correct.
- **What are some ways of explaining integer subtraction?** *(pp. 241–247)* As with addition, integer subtraction may be explained by using counters, charged field, and number line models. Subtraction may be thought of as finding a missing addend; that is $a - b = c$, where $c + b = a$. Subtraction may also be thought of as adding the opposite; that is, $a - b = a + (-b)$.
- **What are some applications of integer addition and subtraction?** *(pp. 247–249)* Everyday applications that could involve integer addition and subtraction are often solved by using whole numbers. However, problems involving applications such as temperature and bank accounts can be solved by using integer operations, sometimes on a calculator.
- **How can two integers be compared and ordered?** *(pp. 250–251)* Two integers may be compared by recognizing that the integer farther to the left on the number line is the smaller integer. Two integers may also be compared by using the idea that the integer to which a positive integer is added to get the other is the smaller integer. A set of integers may be ordered by comparing them two at a time.

Section 5.2

- **What are some ways of explaining integer multiplication?** *(pp. 255–259)* Multiplication of integers may be explained by using the one-color counters model the charged field model, and the number line model. Ideas such as the following also help: (a) $3(-4) = -4 + (-4) + (-4) = -12$; and (b) because $3(-4) = -12$, then $-3(-4)$ should be the opposite of -12, or 12.
- **What are some applications of integer multiplication?** *(pp. 260–261)* Everyday applications of integer multiplication are limited, but multiplication of positive and negative integers occurs often in science, economics, astronomy, and other fields.
- **What are some basic properties of integer multiplication and how are they used?** *(pp. 262–264)* Basic properties of integer multiplication are closure: $a \times b$ is a unique integer; multiplicative identity: $a \times 1 = a$; commutative: $a \times b = b \times a$; and associative: $(a \times b) \times c = a \times (b \times c)$. The distributive property, $a \times (b + c) = a \times b + (a \times c)$, ties addition and multiplication together. The zero property, $a \times 0 = 0$, can be proved by using the other properties. The basic properties may be used to verify that the procedures used for multiplying integers are logically correct.
- **What mathematical relationship helps explain integer division?** *(pp. 264–267)* Division of integers may be thought of as finding the missing factor; that is, when $b \neq 0$, $a \div b = c$, where $c \times b = a$.
- **What important properties are related to integer multiplication and division?** *(pp. 267–270)* Important properties for integers a, b, and c are $a \div a = 1$, $a \div 1 = a$, $0 \div a = 0$, $ab \div a = b$, $-(-a) = a$, $-a(-b) = ab$, $(-a)b = a(-b)$, $a(-1) = -a$, $a(b - c) = ab - ac$, $-(a + b) = -1(a + b) = -a + (-b) = -a - b$.

Key Terms, Concepts, and Generalizations

CHAPTER REVIEW

Concepts and Skills

1. Classify each statement as true or false. If false, change it to make it true.
   a. The opposite of a negative integer is a positive integer.
   b. The sum of a positive and a negative integer is a negative integer.
   c. The absolute value of an integer is always a positive integer.
   d. The product of two negative integers is a negative integer.
   e. The integer 0 is sometimes positive and sometimes negative.
2. Write the integer that is represented by
   a. a collection of 5 red counters and 3 black counters.
   b. a collection of 1 red counter and 8 black counters.
   c. a point 9 units to the left of 0 on the number line.
   d. a point 3 units to the right of 0 on the number line.
   e. a point midway between -4 and 4 on the number line.
3. Evaluate
   a. $-|-7|$. b. $-|-4|$.
   c. $-[-(-8)]$. d. $|-|-9||$.
4. Show how to find the sum $6 + (-3)$ by using
   a. a counters model.
   b. a charged field model.
   c. a number line model.
   d. the rules for adding integers.
5. Show how to find the difference $2 - (-5)$ by using
   a. a counters model.
   b. a charged field model.
   c. a number line model.
   d. subtracting by adding the opposite.
6. Order from smallest to largest.
   a. $-3, 0, 2, -5, 8, 1, -12, 15$
   b. $34, -56, -17, 27, 46, -34, -60, 9$
7. How can the number line be used to compare two integers?
8. Use the definition of greater than to verify that
   a. $-7 < -5$. b. $1 > -5$.
   c. $0 > -3$.
9. Show how to find the product $-3 \times (-4)$ by using
   a. a counters model.
   b. a charged field model.
   c. a number line model.
   d. a pattern based on the idea that $3(-4) = -12$.
10. Show how to find the quotient $12 \div (-4)$ by using
    a. the idea of division as finding a missing factor.
    b. the procedures for integer division.
11. State a property of integers that verifies the following equalities.
    a. $3(x + 2) = 3x + 6$.
    b. $-3(4n) = (-3 \times 4)n$.
    c. $-n + n = 0$.
    d. $-(2 + x) = -2 + (-x)$.
    e. $3x \div 3 = x$.
    f. $4(-n) = -4n$.
12. Choose a non-zero integer for n to show that $-n$ can be evaluated as a positive number.

Reasoning and Problem Solving

13. Suppose that the unit of time is a month and that 0 represents June.
    a. What number would represent January? March? May? August? October? December?
    b. What month would the number -7 represent?
14. How would you convince someone that, for integers a and b, $a - b$ and $b - a$ are opposites of each other?
15. What basic principles support each statement in the verification that $-3 + 5 = 2$?

Statements	Reasons
$-3 + 5 = -3 + (3 + 2)$	Substitute 3 + 2 for 5.
$= (-3 + 3) + 2$	a. ?
$= 0 + 2$	b. ?
$= 2.$	c. ?

16. Give a counterexample to show that each integer property statement is false.
    a. Closure property for division
    b. Distributive property for division over addition
    c. Commutative property for division
    d. Associative property for division
    e. Associative property for subtraction
    f. Commutative property for subtraction

17. **Tricky Temperatures.** A crafty geography student observed that the difference between the record high and record low temperatures in Siberia is 192°F. He also observed that, if he lowered the high temperature by 4°, the result would be the opposite of the lower temperature. What are the record high and low temperatures in Siberia?

18. **Scoring in Golf.** A golfer played eight rounds on a tournament course, with the following scores (par is the expected score for a round; negative numbers represent the number of strokes under par for the round, and positive numbers represent the number of strokes over par for the round): $+5$, -2, $+3$, -1, -3, $+6$, $+2$, -7. The golfer played two more rounds, ending with an even par total for the ten rounds. Her score for the last round was 7 more than her score on the next-to-last round. What scores did the golfer have on the last two rounds?

19. **Quality Control.** A company that sells lawn fertilizer in 25-pound bags has a system of randomly checking the bags to determine whether they are the correct weight. The checker records the number of ounces the bag is over the desired weight with a positive integer on a tag and the number of ounces a bag is under the desired weight with a negative integer on a tag. In a sample of 50 bags, the following numbers of labels were attached: twelve -4s, sixteen -3s, nine $+7$s, ten -2s, and three 0s. What conclusions can you draw about this situation?

Alternative Assessment

20. Write a brief report summarizing the connections between the counters models for adding and subtracting integers and the more systematic methods that were formulated for adding and subtracting integers.

21. Make some small operation tables, similar to the one shown, to summarize the rules for integer addition, subtraction, multiplication, and division. Discuss the value and limitations of such tables for providing information about those operations.

+	Pos	Neg
Pos	Pos	S
Neg	?	?

S = depends on the situation.

22. Draw a diagram that explains how the relationship between integer addition and subtraction can be used to help in subtracting integers. Draw another diagram that explains how the relationship between integer multiplication and division can be used to help in dividing integers. Explain how these diagrams are alike and how they are different.

6 Rational Number Operations and Properties

Chapter Perspective

Both fractions and decimals describe a comparison between two quantities that is the basis for rational numbers. Measuring the worth of stocks requires describing quantities between whole number values. Stock prices, which have been reported in fraction form, are now commonly represented by decimals. For example, the price $58\frac{3}{8}$ is now reported as 58.375. An understanding of rational numbers and their related symbols provides an important foundation for extending the use of addition, subtraction, multiplication, and division to solve problems involving measurement, probability, and statistics.

In this chapter, you can develop a greater sense of the importance of the rational number system by exploring the symbolization used to represent rational numbers, the use of operations with rational numbers to solve problems, and the ordering and density properties of rational numbers.

Connection to the NCTM Principles and Standards

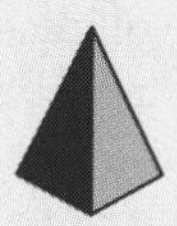

The NCTM *Principles and Standards for Mathematics* (2000) indicate that the elementary school mathematics curriculum grades K–8 should include the study of rational numbers so that students can

- understand and represent commonly used fractions, such as $\frac{1}{4}$, $\frac{1}{3}$, and $\frac{1}{2}$ *(p. 78);*
- develop understanding of fractions as parts of unit wholes, as parts of a collection, as locations on number lines, and as divisions of whole numbers *(p. 148);* and
- work flexibly with fractions, decimals, and percents to solve problems *(p. 214).*

Connection to the PreK–8 Classroom

In grades PreK–2, students use informal experiences to focus on dividing whole objects or sets of objects into equal parts and describing the equal parts with fraction language, such as "one out of three equal parts, or one-third."

In grades 3–5 students compare fractions to familiar benchmarks such as $\frac{1}{2}$ and use rational number sense to estimate sums and differences.

In grades 6–8 students learn to move flexibly among equivalent fractions, decimals, and percents in order to solve problems involving rational numbers.

Section 6.1 RATIONAL NUMBER IDEAS AND SYMBOLS

- Modeling Rational Numbers
- Defining Rational Numbers
- Using Fractions to Represent Rational Numbers
- Properties of Fractions
- Using Decimals to Represent Rational Numbers
- Connecting Rational Numbers to Whole Numbers, Integers, and Other Numbers

In this section, we examine the need for rational number ideas to describe quantities and answer questions that cannot be described or answered with integers. Mini-Investigation 6.1 asks you to reflect on ways of expressing rational number ideas.

Talk with other classmates about what they remember about learning to use these symbols and compare your experiences.

Mini-Investigation 6.1 **Communicating**

What symbols are used to describe quantities between 0 and 1?

Modeling Rational Numbers

People often need rational numbers to describe an amount greater than 0 but less than 1. For example, someone might want to describe what part of one cup of milk to use in a recipe, what part of one running lap has been completed, or what part of one class passed a test. Rational number ideas also describe amounts equal to or greater than 1, such as how much grapefruit there is when two halves are lying on a plate or how far two yard-long boards and a piece of another will reach.

Understanding the need for and use of rational numbers to describe a quantity between 0 and 1 first requires being able to (a) identify the **whole** represented by the integer 1, (b) separate that whole into **equal parts,** and (c) use an **ordered pair of numbers** to describe the portion of the whole under consideration. We use real-world situations that will later provide models for understanding rational numbers to look at each of these ideas.

Identifying the Whole and Separating It into Equal Parts. In the Mother Goose and Grimm cartoon, Mother Goose and the grocery checker are using different interpretations of 1 for each item. The cartoon reminds us that the numeral 1 can be used to describe a single apple or an entire set of apples. Also, the numeral 1 can be used to describe an entire quart of milk or an ounce of milk. If we use Mother Goose's interpretation of 1 as the whole bag of apples, there is no integer to describe a single apple. A new kind of number is needed. If 1 is used to represent the entire quart of milk, then a new kind of number is needed to describe any quantity of milk less than one quart.

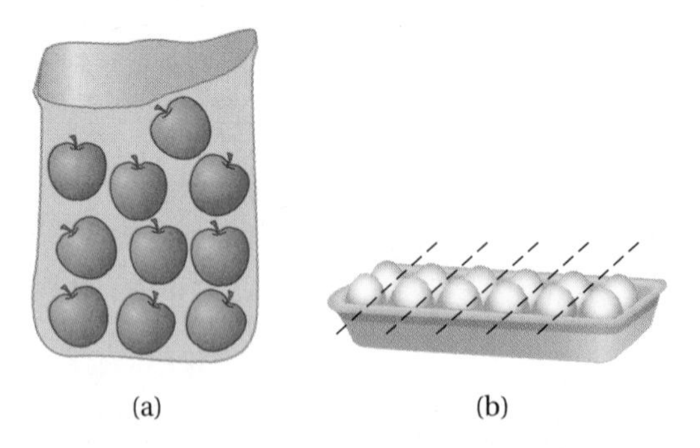

(a) (b)

FIGURE 6.1 | Models of (a) 10 equal parts; (b) 6 equal parts.

Once the whole is identified, we must be able to refer to equal parts of the whole. In the cartoon, the bag of apples could be separated into 10 equal parts. Even though the apples may not all be exactly the same size or shape, they are equal in the sense that they are each a single member of the one bag of apples, as shown in Figure 6.1(a). Similarly, the carton of eggs could be thought of as separated into 6 equal parts, with two eggs in each part, as shown in Figure 6.1(b).

Mini-Investigation 6.2 emphasizes the idea of separating a region into equal parts by asking you to demonstrate that there are various ways to do so.

Draw a picture of each of your ways and compare them to a classmate's.

Mini-Investigation 6.2 **Solving a Problem**

What are five different ways in which a square can be separated into four equal parts? What do these representations have to do with our understanding of fractions?

Using Two Integers to Describe Part of a Whole. Once you understand how to select a whole and divide it into equal parts, you need a way to describe, for example, what part of the bag of apples you eat if you eat three apples, or what part of the carton of eggs you need to bake a cake that uses four eggs. No single integer can give a complete description of these quantities, so language is needed that describes the part–whole relationship between the two integers—the number of pieces wanted or used and the number of equal pieces in the original whole. For example, in reference to the apples below, you could say, "I ate three of ten equal parts, or three-tenths, of the bag of apples."

In reference to the egg carton, you could say, "I used two of six equal parts, or two-sixths, of the carton of eggs to make the cake."

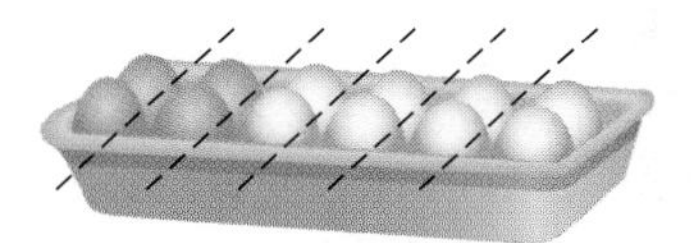

The idea of using a pair of numbers to describe a quantity can also be extended to amounts greater than or equal to one. If a carton of eggs was thought of as having three equal parts of four eggs each, four thirds, or one and one-third, cartons of eggs would be needed to make four cakes for the family reunion.

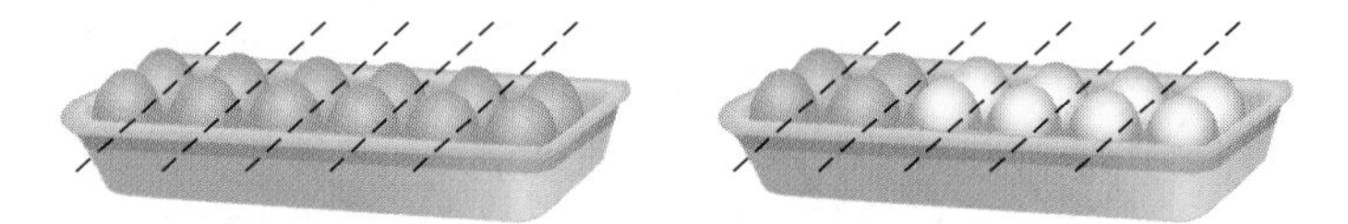

Example 6.1 extends the ideas about using language involving two integers to compare a part to a whole.

Example 6.1 Using Fraction Language

If puzzle piece B is 1 unit, use the idea of halves, thirds, fourths, and so on to describe puzzle pieces A and C.

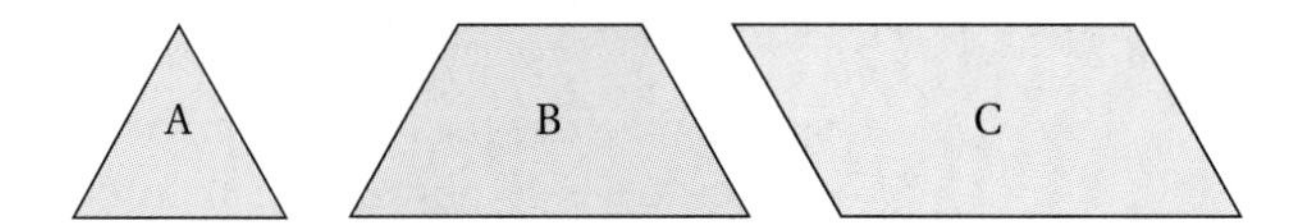

Solution

You can lay three of the triangles (piece A) on the trapezoid (piece B) and they just fit, so piece A is one-third of the piece B. You can also fit both piece B and piece A onto piece C, so piece C is four-thirds or one and one-third of piece B.

Your Turn

Practice: If the length of the line segment

represents 1 unit of length, use ideas of halves, thirds, and so on to describe the following lengths.

a.

b.

c.

Reflect: In each of your answers, identify the number you used to represent the part and the number you used to represent the whole. Is the number that represents the part necessarily less than the number that represents the whole? Explain. ■

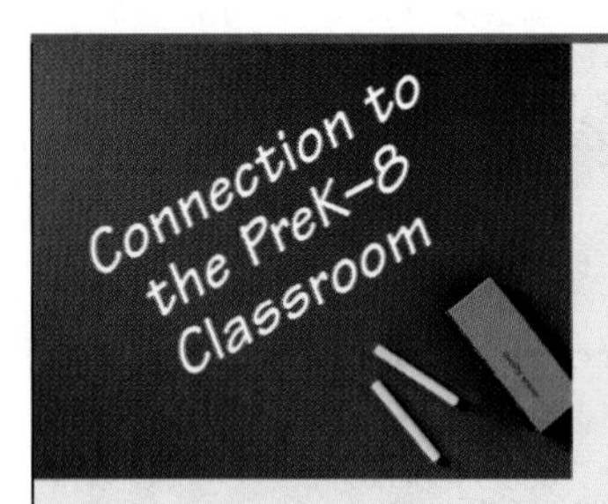

Students should experience using pairs of numbers to describe comparisons, as well as the language of halves, thirds, fourths, and so on, before focusing on more formal symbols for these ideas. Listening to and telling stories build a foundation for understanding word symbols. So too does talking about and listening to others talk about comparisons build a foundation for understanding number symbols. For example, students might be asked to find several situations, including sharing, that involve wholes and parts. They should then be asked to use language such as part, equal parts, whole, one of four parts, or one-fourth to talk about these situations.

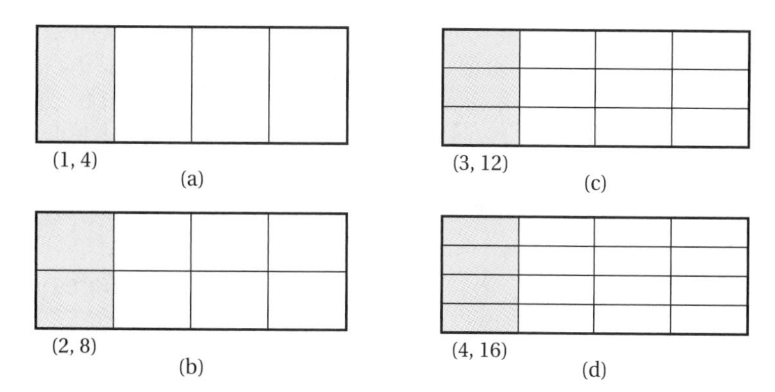

FIGURE 6.2 | Ordered pair representations of the same part of a whole and the same rational number.

Defining Rational Numbers

Rational numbers grow out of the need for descriptions of parts of wholes and are identified mathematically by the pairs of numbers that describe the relationships between parts and wholes. For example, consider four children sharing a candy bar. As indicated in Figure 6.2(a), with the candy bar divided into four equal parts, the one part of four equal parts received by each child can be described with the ordered pair $(1, 4)$. But suppose that the candy bar is already marked into eight equal sections, as many candy bars are. Then the portion received by each child could be described with the ordered pair $(2, 8)$, meaning two parts of eight equal parts [Figure 6.2(b)]. By extending this example [Figures 6.2(c) and (d)], we could describe the same amount of candy bar with many different ordered pairs, such as $(3, 12)$, $(4, 16)$, and so on. In other words, the relationships 1 of 4, 2 of 8, 3 of 12, and 4 of 16 all represent the same part of the whole candy bar—and the same rational number.

Similarly, in each of the cartons of eggs in Figure 6.3, the same number of eggs are brown, but are described with a different ordered pair of numbers. Therefore the ordered pairs $(1, 2)$, $(2, 4)$, $(3, 6)$, and $(6, 12)$ all represent the same portion of the carton and the same rational number.

These ideas suggest the following general description.

Description of a Rational Number

A **rational number** is the relationship represented by an infinite set of ordered pairs, each of which describes the same quantity.

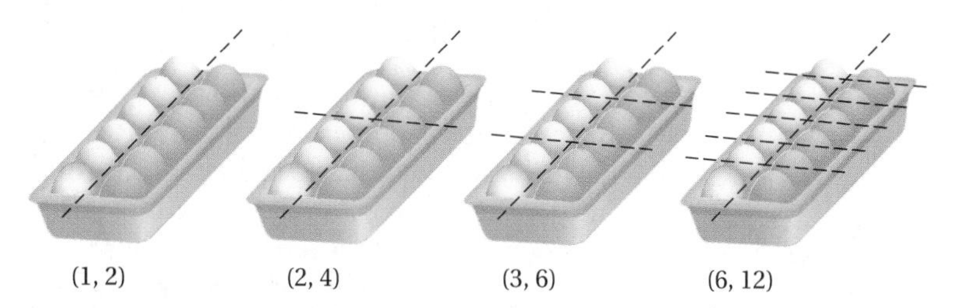

FIGURE 6.3 | Ordered pair representations of the same amount and the same rational number.

In the Family Circus cartoon, Dolly is concerned that PJ has more sandwiches than she does. When she understands the idea of rational numbers, she will see that 4 fourths (4 of 4 equal parts) is equal to 2 halves (2 of 2 equal parts), which is also equal to 1 out of 1 equal part, or 1 whole. These pairs of integers, $(1, 1)$ $(2, 2)$, and $(4, 4)$, all describe the same amount and thus the same rational number. In this case, the rational number is also an integer.

Reprinted with special permission of King Features Syndicate.

Using Fractions to Represent Rational Numbers

We first consider the meaning of a fraction and what it means for fractions to be equivalent.

Fractions and Equivalent Fractions. The following general definition of a fraction clarifies the symbols we use to represent rational numbers.

Definition of a Fraction

A **fraction** is a symbol, $\frac{a}{b}$, where a and b are numbers and $b \neq 0$. Here, a is the **numerator** of the fraction and b is the **denominator** of the fraction.

According to this definition, the numerator and denominator of a fraction can be integers or numbers other than integers. For example, $\frac{2}{3}$ is a fraction, but $\frac{\pi}{3}$ and $\frac{0.25}{0.47}$ are also fractions. Initially, we are concerned primarily with fractions having integer numerators and denominators because they are useful in beginning to discuss how to represent rational numbers.

For the four children sharing a candy bar, we used pairs of integers to describe the rational number for each piece of candy bar: 1 of 4 equal pieces, $(1, 4)$; 2 of 8 equal pieces $(2, 8)$; and so on. Because we use pairs of integers to express part–whole relationships, we can use fractions having integer numerators and denominators to repre-

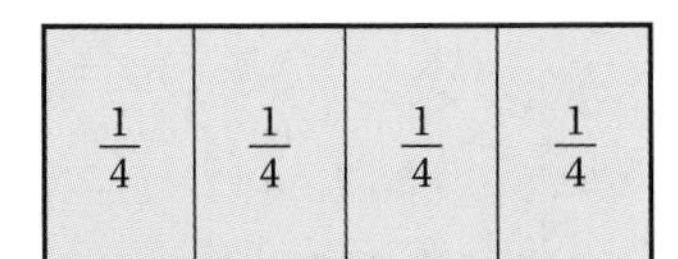

FIGURE 6.4 | The number of parts in the whole (denominator) and the number of parts being compared to the whole (numerator).

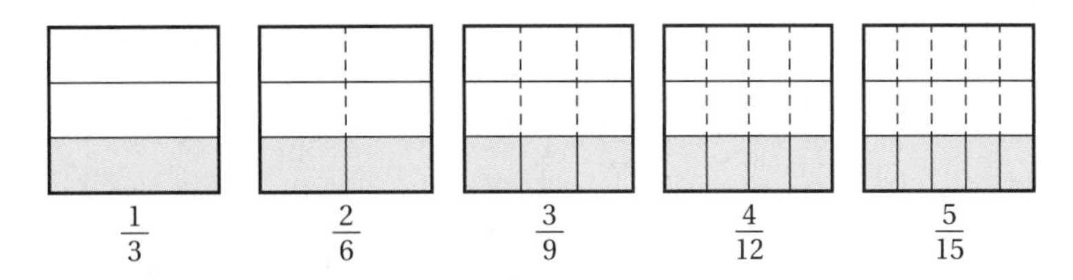

FIGURE 6.5 | Model for some equivalent fractions.

> **Estimation Application**
>
> **How You Spend Your Time**
>
> Estimate the fraction of your life so far that you've spent
>
> a. sleeping,
>
> b. eating,
>
> c. watching TV, and
>
> d. going to school.
>
> Describe how you made your estimates.

sent rational numbers. In this fraction notation, the denominator names how many parts are in the whole and the numerator counts how many parts are being compared to the whole. Figure 6.4 illustrates these ideas by using the candy bar example. The symbol $\frac{1}{4}$ (one-fourth) represents the 1 of 4 equal pieces that each child receives.

These models involving relationships between parts and wholes and ideas of rational numbers are also related to division. For example, dividing one candy bar among four children can be represented by a division sentence: $1 \div 4 = \underline{?}$. This sentence has no integer quotient, but a rational number can be used to complete it: One candy bar divided among four children gives $\frac{1}{4}$ candy bar per child, or $1 \div 4 = \frac{1}{4}$.

When the numerator of a fraction is less than the denominator, as in $\frac{1}{4}$, the fraction is called a **proper fraction.** When the numerator of a fraction is greater than or equal to the denominator, as in $\frac{9}{8}$, the fraction is called an **improper fraction.**

Just as the models presented earlier showed that many different number pairs can describe the same rational number idea, the models in Figure 6.5 show that many different fractions can represent the same rational number.

Each fraction describes the same shaded amount of the square, so we say that the fractions are equivalent. Note that for the fractions $\frac{2}{6}$ and $\frac{3}{9}$, $2 \times 9 = 6 \times 3$. This *cross-product* equality, which is true for all pairs of equivalent fractions, suggests the following definition.

> **Definition of Equivalent Fractions**
>
> Two fractions, $\frac{a}{b}$ and $\frac{c}{d}$, are **equivalent fractions** if and only if $ad = bc$.

We validate this idea of cross products in Section 6.4 and Chapter 7.

Using Fractions to Represent Rational Numbers. Because each rational number can be represented by infinitely many ordered pairs of integers that describe the same amount, it can also be represented by infinitely many equivalent fractions having integer numerators and denominators. This condition forms the basis for the diagram in Figure 6.6, which illustrates that, for each infinite set of equivalent fractions, there is exactly one rational number and one point on the number line. Clearly, any fraction from a set of equivalent fractions may be used as a symbol for the rational number represented by that set. Thus, whenever the numerator and denominator of a fraction are integers, the fraction represents a rational number and every rational number can be represented by a fraction having an integer numerator and denominator. Note that some fractions having noninteger numerators and denominators, such as $\frac{0.25}{0.50}$, can also represent rational numbers.

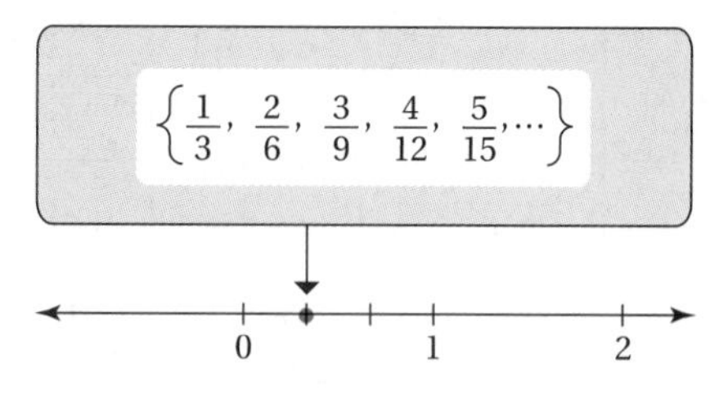

FIGURE 6.6 | Equivalent fractions and the rational number one-third.

Mini-Investigation 6.3 asks you to make connections between sets of equivalent fractions, rational numbers, and points on the number line.

Describe how you used the pattern to find the set of equivalent fractions and then list several members of the set.

Mini-Investigation 6.3 **Finding a Pattern**

What pattern can you discover in Figure 6.6 that would help you produce a set of equivalent fractions associated with the point on the number line that is one-fourth of the way from 1 to 2?

Properties of Fractions

The Fundamental Law of Fractions. Figure 6.5, which we used to illustrate the idea of equivalent fractions, suggests a useful property of fractions that you may have discovered in Mini-Investigation 6.3. Note that, when one dashed vertical line is drawn to divide the square into two parts, both the numerator and the denominator of the original fraction are doubled. Similarly, drawing vertical lines to divide the square into three parts triples the numerator and denominator, and so on. The fact that drawing lines has the effect of multiplying both the numerator and denominator by 2, 3, 4, 5, ..., while not changing the amount of the square that is shaded, suggests the following law.

Fundamental Law of Fractions

Given a fraction $\frac{a}{b}$ and a number $c \neq 0$, $\frac{a}{b} = \frac{ac}{bc}$.

The restriction that $c \neq 0$ in the fundamental law of fractions is an important restriction. Without it, when $c = 0$, the right-hand member of the equality would not be a fraction, for $bc = 0$. Beyond that, additional problems would arise if we let $c = 0$. Imagine for a moment that c could equal 0. Then

$$\frac{1}{2} = \frac{1 \times 0}{2 \times 0} = \frac{0}{0} \quad \text{and} \quad \frac{3}{4} = \frac{3 \times 0}{4 \times 0} = \frac{0}{0}$$

making it appear that $\frac{1}{2} = \frac{3}{4}$! Thus, in situations involving rational numbers, you must always be sure that the denominator is not allowed to equal 0.

Fractions in Simplest Form. In each set of equivalent fractions that represents a rational number, there is one fraction in which the denominator is positive and the numerator and denominator are integers with no common factors other than 1. This fraction is the one used when a rational number is expressed in *simplest form*.

Description of the Simplest Form of a Fraction

A fraction representing a rational number is in **simplest form** when the numerator and denominator are both integers that are relatively prime and the denominator is greater than zero.

Example 6.2 provides an opportunity to use these ideas about fractions.

Example 6.2 Finding Equivalent Fractions

Find two fractions equivalent to $\frac{10}{6}$, including its simplest form, and verify your choices.

Solution

Soon Li's thinking: First I used rectangles to show $\frac{10}{6}$. Then I drew bold lines to show $\frac{5}{3}$. Finally, I drew dashed vertical lines to show $\frac{20}{12}$.

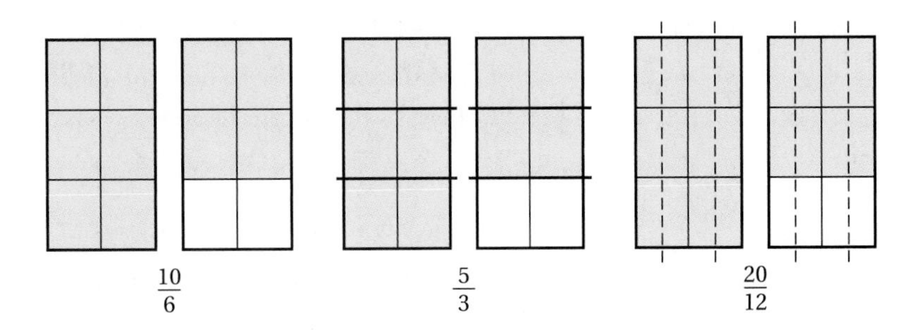

$\frac{20}{12}$ and $\frac{5}{3}$ are equivalent to $\frac{10}{6}$. The fraction $\frac{5}{3}$ is in simplest form.

Geraldo's thinking: If I think about $\frac{10}{6}$ on a number line, I see that $\frac{5}{3}$ and $\frac{15}{9}$ are also at that same place on the number line.

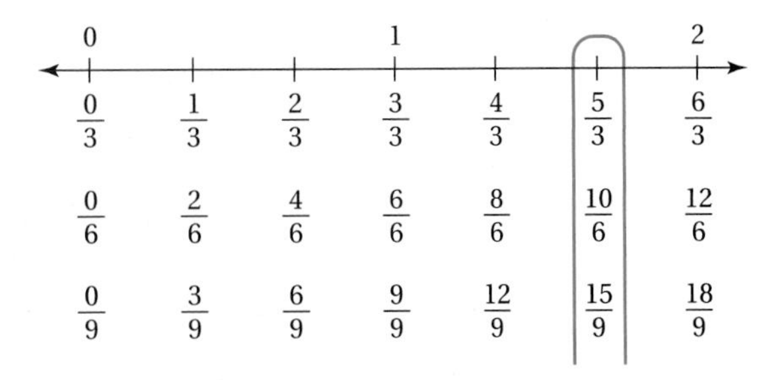

$\frac{5}{3}$ and $\frac{15}{9}$ are equivalent to $\frac{10}{6}$. The fraction $\frac{5}{3}$ is the one in simplest form because it has the smallest positive denominator in the pairs of integers that describe that point on the number line.

Myra's thinking: Since 10 and 6 have a common factor of 2, I think of $\frac{10}{6}$ as $\frac{(5 \times 2)}{(3 \times 2)}$ and by the fundamental law of fractions, $\frac{5}{3} = \frac{10}{6}$. As 5 and 3 have only 1 as a common factor, $\frac{5}{3}$ is the fraction in simplest form.

Eddie's thinking: I used a fraction calculator. First, I entered 10 ÷ 6 as a fraction.

Key Sequence	**Display**
10 [/] 6 [=]	N/D→n/d 10/6

The symbol N/D→n/d tells me that the fraction is not in simplest form, so I press [Simp] to reduce the fraction.

Key Sequence	**Display**
[Simp] [=]	5/3

The symbol N/D→n/d has disappeared, so I know that the fraction is now in simplest form. To generate another equivalent fraction, I multiply $\frac{5}{3}$ by a fraction equivalent to 1, such as $\frac{4}{4}$.

Key Sequence	**Display**
[X] 4 [/] 4 [=]	N/D→n/d 20/12

I've used the calculator to help me identify two fractions, $\frac{5}{3}$ and $\frac{20}{12}$, that are equivalent to the original fraction; $\frac{5}{3}$ is in simplest form.

Your Turn

Practice: Verify in three different ways that $-\frac{2}{3} = -\frac{4}{6} = -\frac{10}{15}$.

Reflect: Were your verifications any different with these negative rational numbers than they would have been if the rational numbers had been positive? Why or why not? ■

Using Decimals to Represent Rational Numbers

In this section, we look at other ways to express fractions in the Hindu–Arabic numeration system.

Decimals. The Latin prefix *decem* means "tenth," which suggests that a "decimal" is related to tenths, or fractions such as $\frac{1}{10}$, $\frac{9}{10}$, and so on. We can express a fraction based on tenths as a *decimal*, which we define as follows:

Description of a Decimal

A **decimal** is a symbol that uses a base-ten place-value system with tenths and multiples of tenths to represent a number.

The use of decimal notation to represent rational numbers is an extension of the use of place value to represent integers. If, as shown in Figure 6.7, we redefine the

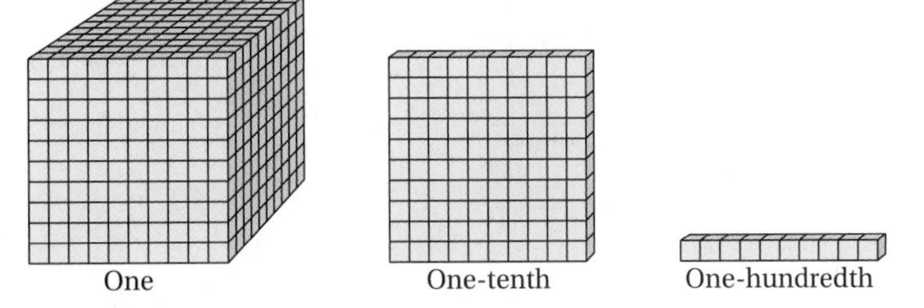

FIGURE 6.7 | Model for some basic decimals.

value of *one* for the base-ten blocks we used in modeling place value for whole numbers in Chapter 2, we have place-value models for some basic decimals.

To write these basic decimals, we extend the system for writing whole numbers by using a **decimal point** to identify the ones place. That indicates that the next digit to the right of the ones place shows tenths, the next hundredths, and the third digit to the right of the ones place shows thousandths. For example,

$$\frac{1}{10} = 0.1, \quad \frac{1}{100} = 0.01, \quad \text{and} \quad \frac{1}{1000} = 0.001.$$

Table 6.1 shows a place-value chart that extends the pattern for whole numbers to the right of the ones place. Note that the digits 5 and 2 in the decimal 0.52 represent 5 tenths and 2 hundredths. Even though there is no written denominator in the decimal 0.52, we understand that 5 tenths equal 50 hundredths and read 0.52 as "fifty-two hundredths." Similarly, the number 23.7 (2 tens, 3 ones, and 7 tenths) in the chart is read as "twenty-three and seven-tenths." The number written as 6.894 (6 ones, 8 tenths, 9 hundredths, and 4 thousandths) is read as "six and eight hundred ninety four thousandths."

Expanded Notation. Because it is based on place value, a decimal, like a whole number, can be expressed in **expanded notation.** To write 23.85 in expanded notation, we write

$$23.85 = (2 \times 10) + (3 \times 1) + \left(8 \times \frac{1}{10}\right) + \left(5 \times \frac{1}{100}\right).$$

Note the pattern in the second factors in each product—their value decreases by a factor of 10 for each position that you move to the right in the expanded product. Similarly, their value increases by a factor of 10 each time you move to the left in the expanded product.

TABLE 6.1 | Place Value Table Showing Some Decimals

Thousands	Hundreds	Tens	Ones		Tenths	Hundredths	Thousandths
1000	100	10	1	.	$\frac{1}{10}$	$\frac{1}{100}$	$\frac{1}{1000}$
			0	.	5	2	
		2	3	.	7		
			6	.	8	9	4

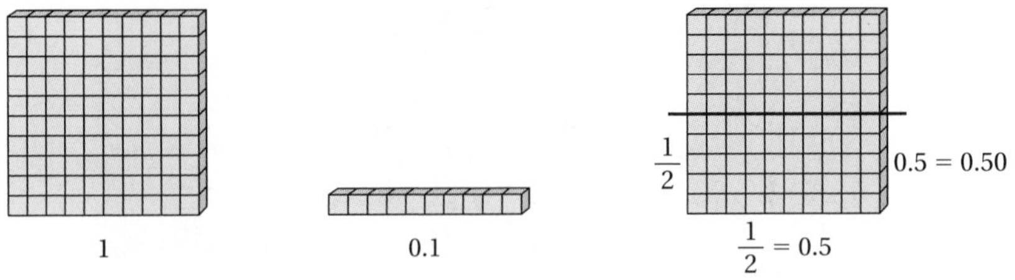

FIGURE 6.8 | Using base-ten blocks to show a rational number in fraction form as a decimal.

Writing a Decimal for a Fraction. Frequently, calculations involving rational numbers are easier to do in decimal form than in fraction form. For example, we can easily change $\frac{1}{2}$ to 0.5 and use base-ten blocks, where the hundreds square represents 1, to help us see that $\frac{1}{2} = 0.50$, as shown in Figure 6.8.

We can also use the fundamental law of fractions and write $\frac{1}{2} = \frac{(1 \times 5)}{(2 \times 5)} = \frac{5}{10} = 0.5$. We examine changing other fractions into decimal form in Example 6.3.

Example 6.3 Changing from Fraction to Decimal Form

Use the fundamental law of fractions to change each rational number into decimal form.

a. $\frac{2}{5}$

b. $\frac{7}{20}$

c. $\frac{7}{30}$

d. $\frac{3}{15}$

Solution

a. $\frac{2}{5} = \frac{2(2)}{5(2)} = \frac{4}{10} = 0.4$

b. $\frac{7}{20} = \frac{7(5)}{20(5)} = \frac{35}{100} = 0.35$

c. For $\frac{7}{30}$, there is no integer factor that when multiplied by 30 gives a power of 10. So there is no way to use the fundamental law of fractions to find an equivalent fraction with a numerator that is an integer and a denominator that is a power of 10. Some other way to change $\frac{7}{30}$ to decimal notation must be used.

d. For $\frac{3}{15}$, there is no integer factor that when multiplied by 15 gives a power of 10. However, $\frac{3}{15} = \frac{1}{5}$, so $\frac{3}{15} = 0.2$.

Your Turn

Practice: Use the fundamental law of fractions to change each rational number into decimal form.

a. $\frac{4}{25}$
b. $\frac{6}{75}$
c. $\frac{5}{12}$
d. $\frac{9}{12}$

Reflect: For which fractions were you able to use the fundamental law of fractions to find an equivalent fraction with a denominator that is a power of 10? What are some common characteristics of these fractions? ■

As indicated in Example 6.3, not all rational numbers can be expressed in fraction form with denominators that are powers of 10. Nevertheless, because a rational number represents a quotient, we can translate any rational number in fraction form into decimal form by dividing, as shown for the fractions $\frac{1}{4}$, $\frac{20}{3}$, and $\frac{4}{11}$.

$$\text{(a)}\quad \begin{array}{r} 0.25 \\ 4\overline{)1.00} \\ \underline{1\ 00} \\ 0 \end{array} \qquad \text{(b)}\quad \begin{array}{r} 6.666\ \ldots \\ 3\overline{)20.000\ \ldots} \\ \underline{18} \\ 2\ 0 \\ \underline{1\ 8} \\ 20 \end{array} \qquad \text{(c)}\quad \begin{array}{r} 0.3636\ \ldots \\ 11\overline{)4.0000\ \ldots} \\ \underline{3\ 3} \\ 70 \\ \underline{66} \\ 40 \\ \underline{33} \\ 70 \end{array}$$

Division (a) illustrates that for some fractions the remainder finally becomes 0. In this case, the resulting decimal in the quotient has a fixed number of places and is called a **terminating decimal.** Fractions that equal terminating decimals are also fractions that can be expressed as equivalent fractions whose numerators are integers and whose denominators are a power of 10. Division (b) shows a fraction that will never have a remainder of 0. Here, the remainder is 2, and hence the quotient digit 6 will repeat over and over again. A decimal in the quotient that has a digit or group of digits that repeat over and over is called a **repeating decimal.** Fractions that equal repeating decimals are also fractions that cannot be expressed as equivalent fractions whose numerators are integers and whose denominators are a power of 10. Division (c) shows a fraction that equals a repeating decimal in which a block of two digits repeats.

Mini-Investigation 6.4 focuses on a key idea that allows us to complete our characterization of decimals for rational numbers.

Write a convincing argument, using mathematical reasoning, to support your answer.

Mini-Investigation 6.4 **Using Mathematical Reasoning**

When you divide to find a decimal for a fraction, how do you know that every nonterminating decimal must repeat a certain set of digits over and over? (*Hint:* How many different remainders could you possibly have before getting a repeat? If a remainder repeats, what will happen with the quotient digits that occur after it?)

Divisions (a), (b), and (c) and Mini-Investigation 6.4 lead to the generalization that each rational number has some form of decimal representation.

Generalization about Decimals for Rational Numbers

Every rational number can be expressed as a terminating or repeating decimal.

Conversely, does a terminating or repeating decimal always represent a rational number? We consider this question later in this chapter. For now, we can use what we know about decimal notation and estimation to find a fraction that is approximately equal to a repeating decimal. For example, 1.555 ... is a little more than 1.5, so 1.555... is a little more than $\frac{3}{2}$. The decimal 0.232323 ... is close to 0.2, so a good fraction estimate is $\frac{2}{10}$, or $\frac{1}{5}$. However, 0.232323 ... is also close to 0.25, so $\frac{1}{4}$ is also a good estimate. The number 0.0290290... is about 0.03, so $\frac{3}{100}$ is a close estimate for it.

Scientific Notation. In sciences such as astronomy, physics, and biochemistry, many of the measurements made are extremely large or extremely small. These very large and very small rational numbers are usually expressed in a usable, efficient form called *scientific notation*, which we describe as follows.

Description of Scientific Notation

A rational number is expressed in **scientific notation** when it is written as a product where one factor is a decimal greater than or equal to 1 and less than 10 and the other factor is a power of 10.

For example, instead of writing 35 billion as 35,000,000,000, scientific notation may be used to write the number as the product of 3.5 and a power of 10:

$$35{,}000{,}000{,}000 = 3.5 \times 10^{10}.$$

Example 6.4 gives further meaning to the use of rational numbers in scientific notation.

Example 6.4

Expressing Measurements in Standard and Scientific Notation

At the Lawrence Livermore Laboratory in California, the Shiva laser reportedly concentrated 2.6×10^{13} watts into a pinhead-sized target for 9.5×10^{-11} seconds on May 18, 1978. Write these measurements in standard decimal notation.

Solution

If you imagine the numbers on a place-value chart, multiplying the number by 10 is like moving each digit of the number one space to the left. So the 2 in 2.6 moves 13 places to the left of the ones place when multiplied by 10^{13}.

10^{13}	10^{12}	10^{11}	10^{10}	10^{9}	10^{8}	10^{7}	10^{6}	10^{5}	10^{4}	10^{3}	10^{2}	10^{1}	10^{0}	.	10^{-1}
													2	.	6
2	6,	0	0	0,	0	0	0,	0	0	0,	0	0	0	.	

$$2.6 \times 10^{13} = 26{,}000{,}000{,}000{,}000.$$

Multiplying by 0.1, or 10^{-1}, is like moving each digit one place to the right, so the 9 in the 9.5 will move 11 places to the right of the ones place when multiplied by 10^{-11}.

10^0.	10^{-1}	10^{-2}	10^{-3}	10^{-4}	10^{-5}	10^{-6}	10^{-7}	10^{-8}	10^{-9}	10^{-10}	10^{-11}	10^{-12}
9.	5											
0.	0	0	0	0	0	0	0	0	0	0	9	5

$$9.5 \times 10^{-11} = 0.000000000095.$$

Your Turn

Practice: Write the following numbers in scientific notation.

a. 3,856,000

b. 0.00000098

Reflect: What patterns do you see in the place-value charts in the solution to the example? How can you use these patterns to judge whether a number reported in scientific notation is given correctly? ■

Connecting Rational Numbers to Whole Numbers, Integers, and Other Numbers

Previously, we noted that a whole candy bar could be described with the rational number $\frac{4}{4}$ (or $\frac{1}{1}$), as well as with the integer 1. Thus, for each integer a, $a = a \div 1 = \frac{a}{1}$. In other words, any integer a can be expressed by the fraction $\frac{a}{1}$, an element of the set of rational numbers denoted Q for quotients. Therefore every integer is also a rational number. In other words, I (the set of integers) is a subset of Q (the set of rational numbers). In fact, because some rational numbers are not integers, I is a proper subset of Q. Also, as the set of whole numbers, W, is a proper subset of I, then W is a proper subset of Q. The Venn diagram in Figure 6.9 illustrates these relationships.

The set of rational numbers itself is a proper subset of another set of numbers, the real numbers. In the set of **real numbers,** R, some numbers cannot be expressed as either repeating or terminating decimals. In other words, R contains numbers that cannot be described as comparisons between two integers. The numbers in R that are not rational numbers are called **irrational numbers** and include numbers such as pi (π) and the square root of 2 ($\sqrt{2}$). As a mat-

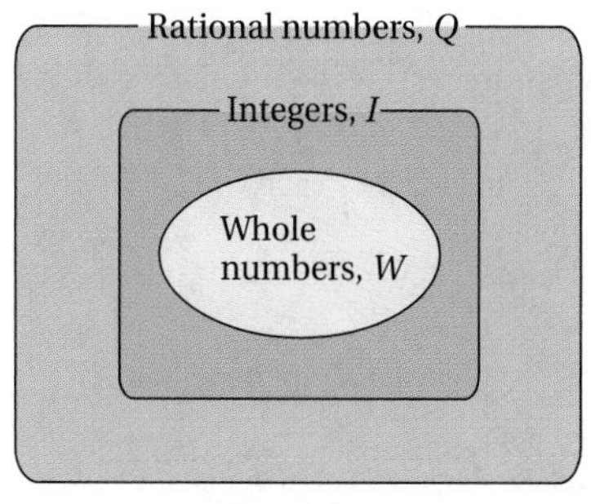

$W \subset I \subset Q$

FIGURE 6.9 | Relationship of rational numbers to integers and whole numbers.

ter of fact, the symbols π and $\sqrt{\ }$ were invented because symbols were needed for numbers that cannot be expressed with rational number notation but that are very important in mathematics. For example, in a square whose sides measure 1 unit, a diagonal of the square is $\sqrt{2}$ units long (as explained by the Pythagorean Theorem in Chapter 10).

Mini-Investigation 6.5 gives some hints to help you devise a proof that $\sqrt{2}$ is not a rational number.

Write a paragraph summarizing your argument for why $\sqrt{2}$ is not a rational number.

Mini-Investigation 6.5 **Using Mathematical Reasoning**

What reasons would you give to convince someone that each statement is true, given that a and b are nonzero integers?

a. If $\sqrt{2}$ *is* a rational number, $\sqrt{2} = \frac{a}{b}$.

b. $2b^2 = a^2$.

c. $2b^2 = a^2$ cannot possibly be true. (*Hint:* Show that the number of factors of 2 in a^2 is even, in $2b^2$, odd.)

d. So $\sqrt{2}$ *is not* a rational number.

Problems and Exercises *for Section 6.1*

A. Reinforcing Concepts and Practicing Skills

1. Use an **area model** to illustrate the quantity $\frac{3}{5}$. (Indicate clearly the corresponding representation for 1.)
2. Use a **linear model** to illustrate the quantity $\frac{3}{5}$. (Indicate clearly the corresponding representation for 1.)
3. Use a **discrete model** (a set of objects) to illustrate the quantity $\frac{3}{5}$. (Indicate clearly the corresponding representation for 1.)
4. Use an **area model** to illustrate the quantity 2.3. (Indicate clearly the corresponding representation for 1.)
5. Use a **linear model** to illustrate the quantity 2.3. (Indicate clearly the corresponding representation for 1.)
6. Use a **discrete model** (a set of objects) to illustrate the quantity 2.3. (Indicate clearly the corresponding representation for 1.)
7. Identify the value of each of the place-value blocks if the hundreds square represented *one*. Justify your answers.
8. Name a rational number that describes each of the following divisions.
   a. $-2 \div 3$ **b.** $18 \div 4$
   c. $0 \div 6$ **d.** $386 \div 1000$
9. Name the rational number that solves each of the following equations.
   a. $4 \times ? = 2$. **b.** $-18 \times ? = 1$.
   c. $? \times 6 = 3$. **d.** $? \times 1000 = 386$.
10. Give three equivalent fractions for each of the following rational numbers.
    a. $-\frac{6}{7}$ **b.** $\frac{0}{100}$
    c. $\frac{15}{25}$ **d.** $\frac{c}{d}, d \neq 0$
11. Complete the following table of symbols for the given rational numbers.

Fraction form	*Decimal form*
a. $\frac{160}{25}$	?
b. ?	0.386
c. ?	0.6666...
d. $\frac{4}{11}$	?

12. Express the following decimals in expanded notation.
    a. 385.192 **b.** 0.006
    c. 6.085
13. Explore with a calculator how the following numbers would be displayed in scientific notation. Record your findings.
    a. 3,480,000,000 **b.** 0.0000908
    c. 6.085

B. Deepening Understanding

14. For each of the following rational number models, show a corresponding model for 1.

a. $= \frac{1}{3}$ **b.** $= \frac{2}{3}$

c. 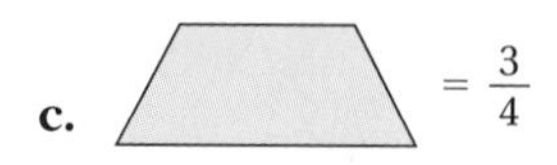 $= \frac{3}{4}$ **d.** 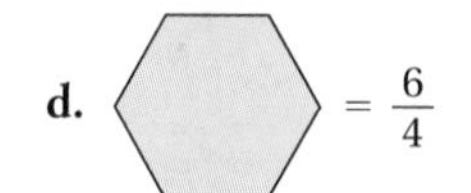 $= \frac{6}{4}$

e. 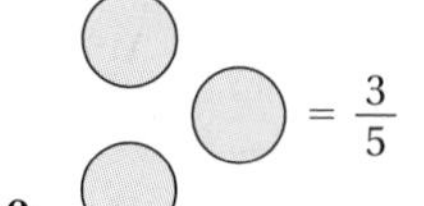 $= \frac{3}{5}$ **f.** 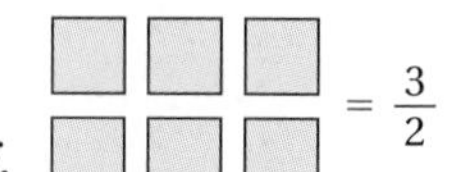 $= \frac{3}{2}$

15. Assume that each region represents 1. Copy each region and shade the appropriate part to show the approximate rational number quantity.

a. 0.28

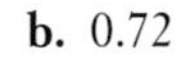

b. 0.72

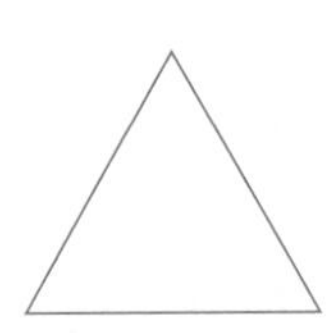

c. 0.45

d. 0.01

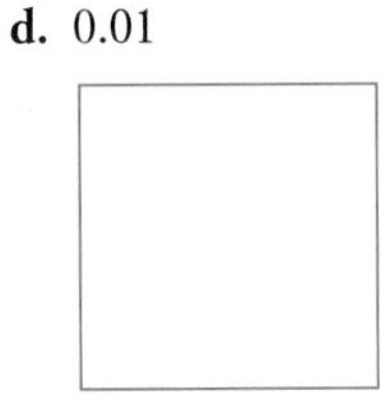

e. 1.18

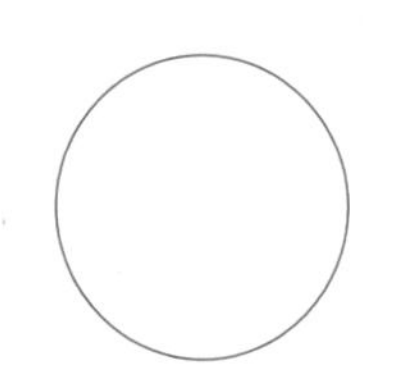

f. 0.005

16. In Exercise 15, describe the relationship between parts (d) and (f). How can you use the relationship to make your estimate?

17. Give fraction estimates of the following decimal numbers. Explain your reasoning.

a. 0.28 **b.** 3.125
c. 0.03939... **d.** 0.9999...

18. Give decimal estimates of the following fractions. Explain your reasoning.

a. $\frac{20}{30}$ **b.** $\frac{6}{99}$
c. $\frac{74}{7}$ **d.** $\frac{1}{300}$

19. Explain why $0.3 = 0.30$.

20. A *tangram* is a geometric puzzle made from seven pieces that form a square. Trace the pieces, cut them out, and answer the following questions.

a. If the area of the entire square equals 1 unit, label each piece with its rational number area.

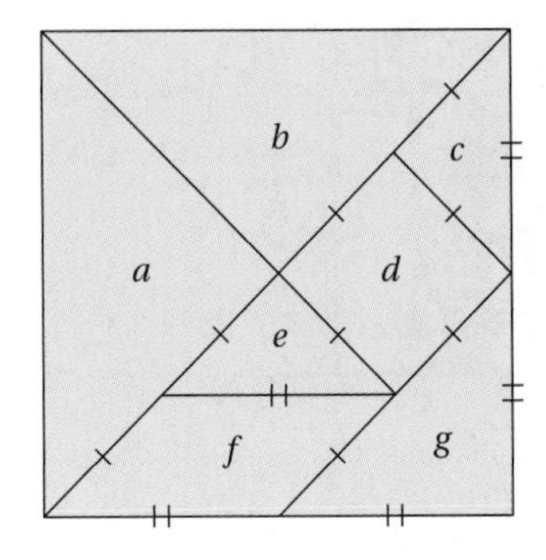

Segments marked (/) are all the same length. Segments marked (//) are all the same length.

b. If the area of piece (a) equals 1 unit, label each piece with its rational number area.

C. Reasoning and Problem Solving

21. The Service Project Problem. A group of local junior high school students meets once a month to plan a service project. Different numbers of students show up each month, and the projects planned require different amounts of time. Each month the group has to decide how much time each student will need to work during that month. Describe how they could use rational numbers to make these decisions.

22. Explain why the definition of a fraction restricts the denominator to being a nonzero integer.

23. Compare the use of 1 in understanding rational numbers to its use in understanding whole numbers. What is the same? What is different?

24. a. Draw a picture of a model for 1 that can be used to represent both $\frac{1}{3}$ and $\frac{1}{2}$.

b. Draw a picture of a model for 1 that can be used to represent both $\frac{3}{4}$ and $\frac{2}{3}$.

c. Draw a picture of a model for 1 that can be used to represent both $\frac{2}{5}$ and $\frac{7}{10}$.

d. Make a conjecture about the kind of model needed to be able to represent any pair of rational numbers.

25. The College Poll Problem. A polling agency conducted a survey on two different college campuses, asking women students if they would prefer to attend all-female mathematics classes. On the first campus, of 13,000 women students, 1,000 were polled and 48 said *yes.* The second college had 5,000 women enrolled, 1,000 were polled, and 48 said *yes.* Write a short paragraph and use rational number language to compare the results of the polls at the two campuses.

26. a. Separate the fractions $\frac{2}{6}$, $\frac{2}{5}$, $\frac{6}{13}$, $\frac{1}{25}$, $\frac{7}{8}$, and $\frac{9}{29}$ into two categories: those that can be written as a terminating decimal and those that cannot. Write an explanation of how you made your decisions.

b. Form a conjecture about which fractions can be expressed as terminating decimals.

c. Test your conjecture on the following fractions $\frac{6}{12}$, $\frac{6}{15}$, $\frac{28}{140}$, and $\frac{0}{7}$.

d. Use the ideas of equivalent fractions and common multiples to verify your conjecture.

27. Explain why any number of zeroes can be written to the right of the decimal point in a number symbol without changing the symbol's value.

28. Using a calculator with constant function capabilities and algebraic order of operations, enter [+] 1 [÷] 3 [=] [=] [=] [=] ... and record the display after entering each equals sign.

a. Do you expect $0.3333333 + 0.3333333$ to equal 0.6666667? Try it on your calculator. What happens? Why do you think that happens?

+ 1 ÷ 3	Display on Calculator
=	?
=	?
=	?
=	?
⋮	

b. Do you expect 0.3333333 added together 3 times to equal 1? Try it on your calculator. What happens? Why do you think that happens?

29. Use the division key or fraction keys on a calculator to change $\frac{1}{7}$, $\frac{2}{7}$, $\frac{3}{7}$, $\frac{4}{7}$, $\frac{5}{7}$, $\frac{6}{7}$, and $\frac{7}{7}$ to decimals. Record the results for each in a table like the one shown and answer the questions.

Fraction	Decimal Representation on Calculator
$\frac{1}{7}$	?
$\frac{2}{7}$	?
$\frac{3}{7}$	?
$\frac{4}{7}$	?
$\frac{5}{7}$	?
$\frac{6}{7}$	?
$\frac{7}{7}$	?

a. Look at the decimal you recorded for $\frac{2}{7}$. You would expect it to be twice as much as the decimal you recorded for $\frac{1}{7}$. Is it? Why or why not?

b. Look at the decimal you recorded for $\frac{5}{7}$. You would expect it to be five times as much as the decimal you recorded for $\frac{1}{7}$. Is it? Why or why not?

c. Look at the decimal you recorded for $\frac{7}{7}$. You would expect it to be seven times as much as the decimal you recorded for $\frac{1}{7}$. Is it? Why or why not?

d. Make a conjecture about the calculator's decimal representations for sevenths.

D. Communicating and Connecting Ideas

30. A certain telephone company commercial asserts that large discounts advertised by other companies may not be a better deal for the consumer. In the commercial, the statement is made that "They say you save $\frac{1}{5}$ of the cost, when really you save only $\frac{1}{20}$ of the cost." Describe a situation in which both fractions, in a sense, could give correct information.

31. Historical Pathways. Simon Stevin (1548–1620), a Dutch engineer and advisor in mathematically oriented fields such as finance and navigation, created a system for describing decimals. He proposed that a basic unit be identified as ⓪ and that whole-number increments of the unit be written as 1⓪, 2⓪, 3⓪, and so on. One-tenth of the unit would be written as ①, one-hundredth as ②, and so on. For example, 325 would be written 325⓪, and $\frac{325}{1000}$ would be written as 3①2②5③. Compare and contrast Stevin's notation with current decimal notation.

32. Making Connections. Form a group with three other classmates. See who can find the magazine or newspaper article that contains the greatest use of rational number ideas. With the rest of the class, discuss the kinds of articles you looked for and maybe some surprises that you found.

33. Making Connections. In the Sally Forth cartoon below, replace 25% with $\frac{1}{4}$ and 75% with $\frac{3}{4}$. Sally Forth considers that one option for equalizing salaries is for women to get a raise equal to $\frac{1}{4}$ of their current pay.

a. Discuss why the first option would *not* equalize men's and women's salaries for comparable work. (*Hint:* $\frac{1}{4}$ of what? $\frac{3}{4}$ of what?)

b. Would the second option, working only 6 hours instead of 8, equalize the rate of pay for comparable work? Why or why not?

c. What fraction of their salary would women need as a raise for the salaries to be equalized? Draw a diagram to justify your answer.

Reprinted with special permission of King Features Syndicate.

Section 6.2 | ADDING AND SUBTRACTING RATIONAL NUMBERS

- Modeling Addition and Subtraction of Rational Numbers
- Adding and Subtracting Rational Numbers in Fraction Form
- Properties of Rational Number Addition and Subtraction
- Adding and Subtracting Rational Numbers in Decimal Form
- Estimation Strategies

In this section, we extend the ideas of addition and subtraction from the set of integers to the set of rational numbers. We use models and mathematical definitions and properties to explain when and how to add and subtract rational numbers in their various forms. Mini-Investigation 6.6 asks you to explain an error commonly made in adding rational numbers.

Talk with other classmates and give them your convincing argument.

Mini-Investigation 6.6 **Using Mathematical Reasoning**

How could you convince someone that you don't find the sum $\frac{1}{2} + \frac{4}{5}$ by adding $1 + 4$ and $2 + 5$ to get $\frac{5}{7}$?

Modeling Addition and Subtraction of Rational Numbers

As with addition and subtraction of integers and whole numbers, addition of rational numbers is used to join quantities and subtraction is used to separate or compare quantities. We first consider models that help explain the ideas of adding and subtracting with fractions.

Modeling Adding and Subtracting: Fractions with Like Denominators. Consider the model shown in Figure 6.10 for finding a sum when the fraction denominators are alike. A recipe calls for $\frac{3}{4}$ cup of honey. How many cups are needed to double the recipe?

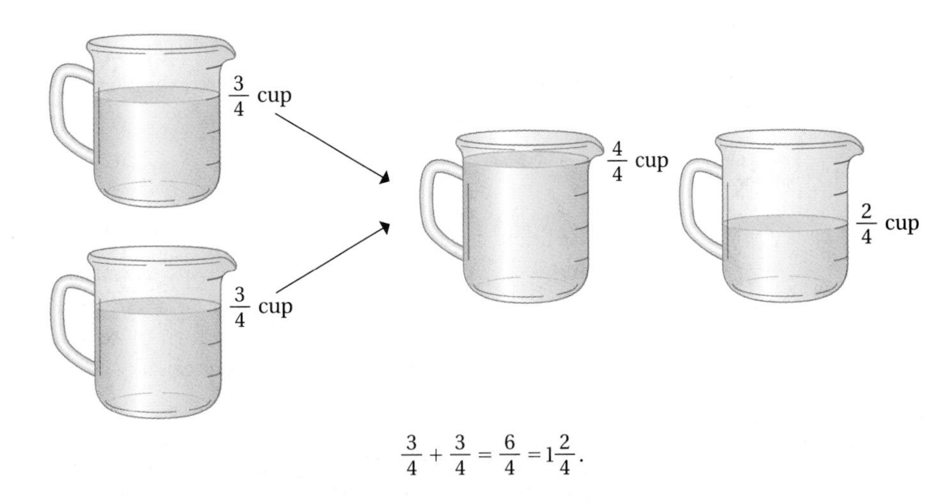

$$\frac{3}{4} + \frac{3}{4} = \frac{6}{4} = 1\frac{2}{4}.$$

FIGURE 6.10 | Model for adding rational numbers with like denominators.

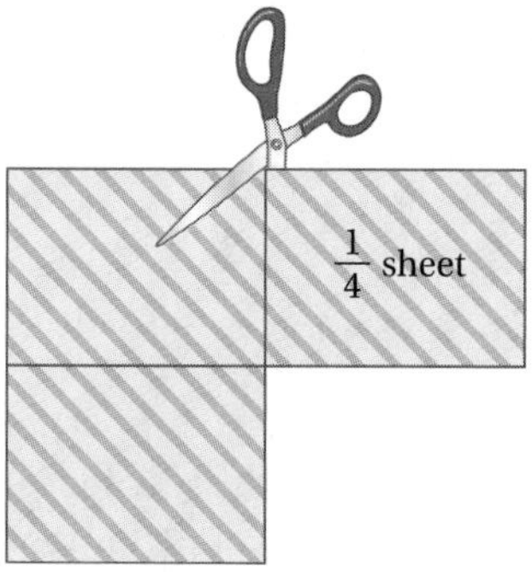

$\frac{1}{4}$ sheet

$\frac{3}{4} - \frac{1}{4} = \frac{2}{4}$.

FIGURE 6.11 | Model for subtracting rational numbers with like denominators.

Now consider the model shown in Figure 6.11 for finding a difference when the fraction denominators are alike. If you have $\frac{3}{4}$ sheet of wrapping paper, and it takes $\frac{1}{4}$ sheet to wrap a package, what part of a sheet of wrapping paper will you have left?

These models suggest the following generalization for adding and subtracting with like-denominator fractions.

Generalization about Rational Number Addition and Subtraction

In general, for rational numbers $\frac{a}{c}$ and $\frac{b}{c}$, $\frac{a}{c} + \frac{b}{c} = \frac{(a+b)}{c}$ and $\frac{a}{c} - \frac{b}{c} = \frac{(a-b)}{c}$.

This generalization is used in the next subsection when we develop algorithms to find sums and differences of rational numbers with unlike denominators. Addition of rational numbers can also be used to explain the connection between a **mixed number** ($I\frac{a}{b}$, where I is an integer and $\frac{a}{b}$ is a fraction with $b > a$) and an **improper fraction** ($\frac{c}{d}$, where $c \geq d$). Example 6.5 focuses on this connection.

Example 6.5

Working with Mixed Numbers and Improper Fractions

Use addition of rational numbers to explain why

a. $2\frac{1}{4}$ can be written as $\frac{9}{4}$.
b. $\frac{7}{3}$ can be written as $2\frac{1}{3}$.

Solution

a. The mixed number $2\frac{1}{4}$ means $2 + \frac{1}{4}$.

$$2 + \frac{1}{4} = 1 + 1 + \frac{1}{4} = \frac{4}{4} + \frac{4}{4} + \frac{1}{4} = \frac{4+4+1}{4} = \frac{9}{4}.$$

b. There are two 3s and 1 left over in 7, so $\frac{7}{3} = \frac{3}{3} + \frac{3}{3} + \frac{1}{3}$. That makes two wholes and $\frac{1}{3}$ left over. So $\frac{7}{3} = 2 + \frac{1}{3} = 2\frac{1}{3}$.

Your Turn

Practice: Use addition of rational numbers to show that

a. $3\frac{2}{3}$ can be written as $\frac{11}{3}$.

b. $\frac{7}{2}$ can be written as $3\frac{1}{2}$.

Reflect: Draw pictures to verify the relationships you found between the mixed numbers and improper fractions in the example. ■

Mini-Investigation 6.7 helps formulate a generalization about changing a mixed number to an improper fraction.

Write a description of the thinking you used to arrive at this generalization.

Mini-Investigation 6.7 **Using Mathematical Reasoning**

In general, the mixed number $A\frac{b}{c}$ is equal to what improper fraction?

Modeling Adding and Subtracting: Fractions with Unlike Denominators. We have shown that adding rational numbers with like denominators consists of adding the numerators just as integers are added while keeping the denominator the same. But what happens when the rational number addends do not have the same denominator? Mini-Investigation 6.8 helps you prepare to answer this question.

Draw a picture to illustrate your solution and compare it to the solutions of others in the class.

Mini-Investigation 6.8 Solving a Problem

How can you use square tiles of three colors (blue, yellow, and green) to make a rectangle that is $\frac{1}{2}$ green, $\frac{1}{3}$ blue, and $\frac{1}{6}$ yellow?

Your solution to Mini-Investigation 6.8 may have provided a model showing that the sum of the rational numbers $\frac{1}{2}$, $\frac{1}{3}$, and $\frac{1}{6}$ is $\frac{12}{12}$, or 1. Your response might also have indicated the benefit of choosing a model that uses equivalent fractions with like denominators. Consider the following situation. Suppose that a music teacher gets paid by the hour by the school district to give private lessons. She works $\frac{1}{2}$ hour with one student one day and $\frac{2}{3}$ hour with another student the next day. For how many hours should the music teacher expect to be paid in those two days? The model of this situation in Figure 6.12 shows the solution: The teacher gets paid for $1\frac{1}{6}$ hours.

First day: $\frac{1}{2}$ hour

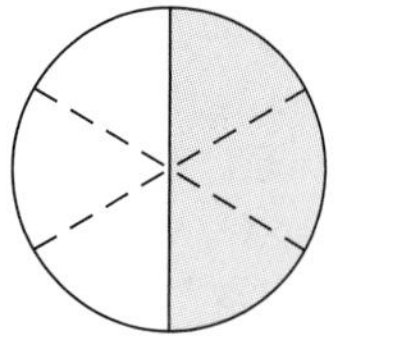

$\frac{1}{2} = \frac{3}{6}$ hour.

Second day: $\frac{2}{3}$ hour

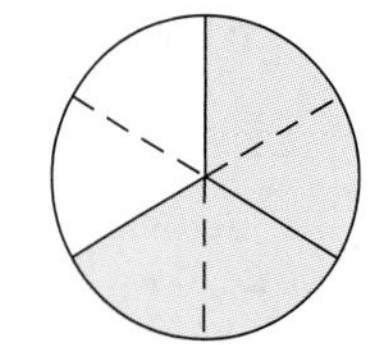

$\frac{2}{3} = \frac{4}{6}$ hour.

$$\frac{1}{2} + \frac{2}{3} = \frac{3}{6} + \frac{4}{6} = \frac{7}{6}, \text{ or } 1\frac{1}{6} \text{ hours.}$$

FIGURE 6.12 | Model for adding rational numbers with unlike denominators.

Figure 6.12 illustrates that models showing equivalent fractions with a common denominator can be used to find the sum of two rational numbers with unlike denominators. A similar approach can be used to find the difference of two rational numbers with unlike denominators. Example 6.6 further illustrates this idea.

Example 6.6 **Problem Solving: Lawn Mowing**

Tom and Dawn have a lawn service. On Thursday, Tom has enough time to mow $\frac{2}{3}$ of Mrs. Broach's large lawn, and Dawn has time to mow only $\frac{1}{4}$ of it. Did they get the entire lawn mowed on Thursday? How much more of the lawn did Tom mow than Dawn?

Solution

Tom mows $\frac{2}{3}$ of the lawn. Dawn mows $\frac{1}{4}$ of the lawn.

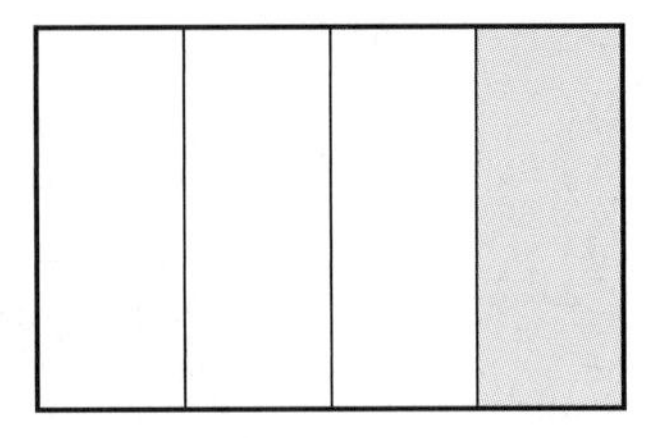

We can show thirds and fourths at the same time by separating the whole into twelfths (a multiple of 3 and of 4).

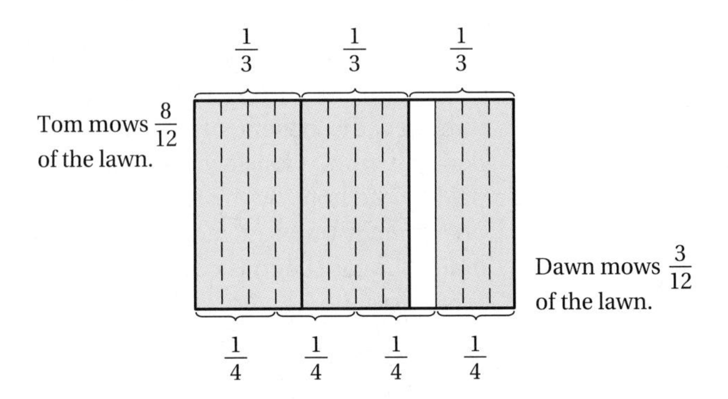

Then $\frac{2}{3} + \frac{1}{4} = \frac{8}{12} + \frac{3}{12} = \frac{11}{12}$ of the lawn will get mowed, so it is not quite finished. This answer seems reasonable because $\frac{2}{3} + \frac{1}{3} = 1$ and $\frac{1}{4}$ is a little less than $\frac{1}{3}$. (If one of them mowed a bit faster, they could finish the whole yard.) Since $\frac{2}{3} - \frac{1}{4} = \frac{8}{12} - \frac{3}{12} = \frac{5}{12}$, Tom mowed almost half of a lawn more than Dawn.

Your Turn

Practice: Use models and equivalent fractions to find the following sums and differences.

a. $\frac{5}{12} + \frac{1}{3}$

b. $\frac{5}{12} - \frac{1}{3}$

c. $4\frac{1}{3} - 2\frac{1}{4}$

d. $6\frac{1}{5} + 5\frac{2}{3}$

Reflect: What strategy did you use to choose the denominators of the sums and differences when drawing the models? ■

As with whole numbers and integers, computation with fractions can be done on many calculators. As a result, students need to be able to estimate in order to check their calculations. Having students connect addition and subtraction of rational numbers to an appropriate model is an important part of enhancing their abilities to estimate reasonable results. By thinking about the relative values of the rational numbers in a problem, students should be able to estimate whether a sum or difference should be less than 1, about 1, or greater than 1.

Adding and Subtracting Rational Numbers in Fraction Form

Rational number addition and subtraction, the fundamental law of fractions, and common multiples can be used to verify procedures for adding and subtracting rational numbers in fraction form. For example, after finding that the least common multiple of 3 and 4 is 12, we use 12 as the *common denominator* and verify the procedure for finding the sum of $\frac{2}{3}$ and $\frac{3}{4}$ as follows.

$$\frac{2}{3} + \frac{3}{4} = \frac{2(4)}{3(4)} + \frac{3(3)}{4(3)} \quad \text{Fundamental law of fractions}$$

$$= \frac{8}{12} + \frac{9}{12} \quad \text{Multiplication facts}$$

$$= \frac{17}{12}. \quad \text{Definition of rational number addition}$$

Similarly, two rational numbers in fraction form can also be subtracted by selecting a common denominator for the two fractions, using the fundamental law of fractions to find the appropriate equivalent fractions with the chosen common denominator, and then subtracting the new numerators. These ideas suggest the following procedure for adding and subtracting rational numbers with unlike denominators.

Procedure for Adding and Subtracting Rational Numbers Represented by Fractions

For rational numbers $\frac{a}{b}$ and $\frac{c}{d}$, $\frac{a}{b} + \frac{c}{d} = \frac{ad}{bd} + \frac{cb}{bd} = \frac{(ad + cb)}{bd}$, and $\frac{a}{b} - \frac{c}{d} = \frac{ad}{bd} - \frac{cb}{bd} = \frac{(ad - cb)}{bd}$.

Note that $\frac{a}{b} + \frac{c}{d} = \frac{(ad + cb)}{bd}$. This relation provides a shortcut that is often useful in quickly finding the sum of two rational numbers. For example, we could use the shortcut to find the sum $\frac{2}{5} + \frac{3}{4}$ by thinking that the numerator of the sum is $2(4) + 5(3)$, or 23, the denominator is 5(4), or 20, and the sum is $\frac{23}{20}$, or $1\frac{3}{20}$. The product bd is a common denominator for denominators b and d. Mini-Investigation 6.9 asks you to analyze further this common denominator.

Write a convincing argument supporting your conclusion.

Mini-Investigation 6.9 **Using Mathematical Reasoning**

In the algorithm described, is bd necessarily the *least* common multiple of b and d? Why or why not?

Example 6.7 illustrates the procedures for adding and subtracting rational numbers with unlike denominators.

Example 6.7 **Adding and Subtracting Rational Numbers**

Find the following sum and difference.

a. $2\frac{4}{5} + 1\frac{1}{3}$

b. $\frac{4}{7} - 2\frac{1}{2}$

Solution

a. $2\frac{4}{5} + 1\frac{1}{3} = 2\frac{12}{15} + 1\frac{5}{15} = 3\frac{17}{15} = 4\frac{2}{15}$.
b. $\frac{4}{7} - 2\frac{1}{2} = \frac{4}{7} - \frac{5}{2} = \frac{8}{14} - \frac{35}{14} = -\frac{27}{14} = -1\frac{13}{14}$.

Your Turn

Practice: Find the difference $4\frac{1}{6} - 2\frac{3}{4}$.

Reflect: What common denominator did you use to find the difference $4\frac{1}{6} - 2\frac{3}{4}$? Is it the denominator described by the boxed procedure? What other common denominators could you use? ■

Properties of Rational Number Addition and Subtraction

We can think of rational numbers as an extension of the integers, created not only as a way of describing a relationship between a whole and a part, but so that an equation such as $-2 \div 3 = n$ will have a solution. Because the integers are a subset of the set of rational numbers, checking to determine whether the properties of integers also hold for rational numbers is natural.

Addition and subtraction with rational numbers are used in the same types of situations as with integers. Therefore the inverse relationship between addition and subtraction in the system of rational numbers remains the same as for integers.

Definition of Rational Number Subtraction in Terms of Addition

For rational numbers $\frac{a}{b}$ and $\frac{c}{d}$, $\frac{a}{b} - \frac{c}{d} = \frac{e}{f}$ if and only if $\frac{e}{f}$ is the unique rational number such that $\frac{e}{f} + \frac{c}{d} = \frac{a}{b}$.

Because rational numbers are described by ordered pairs of integers, we can use the properties of addition for integers to show that the same properties hold for rational numbers.

Properties of Addition of Rational Numbers

Closure property	For rational numbers $\frac{a}{b}$ and $\frac{c}{d}$, $\frac{a}{b} + \frac{c}{d}$ is a unique rational number.
Identity property	A unique rational number, 0, exists such that $0 + \frac{a}{b} = \frac{a}{b} + 0 = \frac{a}{b}$ for every rational number $\frac{a}{b}$; 0 is the additive identity element.
Commutative property	For rational numbers $\frac{a}{b}$ and $\frac{c}{d}$, $\frac{a}{b} + \frac{c}{d} = \frac{c}{d} + \frac{a}{b}$.
Associative property	For rational numbers $\frac{a}{b}$, $\frac{c}{d}$, and $\frac{e}{f}$, $(\frac{a}{b} + \frac{c}{d}) + \frac{e}{f} = \frac{a}{b} + (\frac{c}{d} + \frac{e}{f})$.
Additive inverse property	For every rational number $\frac{a}{b}$, a unique rational number $-\frac{a}{b}$ exists such that $\frac{a}{b} + (-\frac{a}{b}) = -\frac{a}{b} + \frac{a}{b} = 0$.

Standard Procedure	**Verification with Fractions**
Align places and the decimal point. Add as with whole numbers.	Write in expanded notation. Add fractions and regroup. Use place value and basic properties. Write the answer as a decimal.

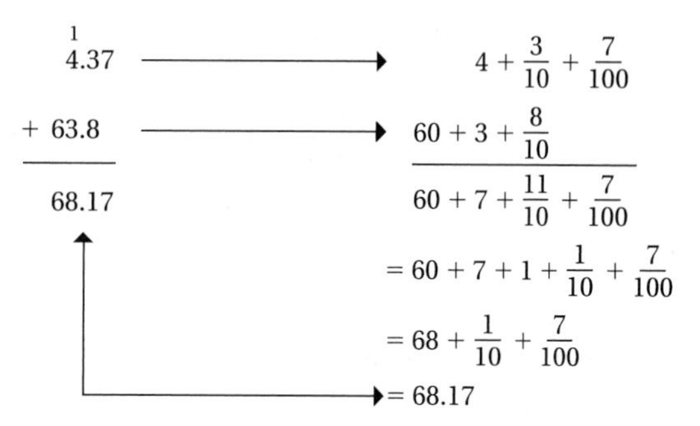

FIGURE 6.13 | Verification of the standard procedure for adding decimals.

Adding and Subtracting Rational Numbers in Decimal Form

The procedure for adding and subtracting with fractions could be applied to decimals by first writing fractions for the decimals. Alternatively, because decimals are based on place value, we can use algorithms similar to those used for adding and subtracting whole numbers to add with decimals. In this subsection we use fractions to verify the process used for adding or subtracting with decimals.

Consider the sum 4.37 + 63.8. Figure 6.13 shows a standard procedure for adding these decimals and verification of the procedure with fractions.

10^1	10^0	.	10^{-1}	10^{-2}
	$\overset{1}{4}$	.	3	7
+ 6	3	.	8	0
6	8	.	1	7

With the digits aligned according to their place-value positions, the decimal addends are just like fractions with common denominators—and the digits in the decimal notation become the numerators that are added. The table to the left emphasizes this relationship. In this case, the last place to the right is hundredths, so the denominator is one hundred. The numerators are the numbers formed by the digits in the problem, 437 and 6380. The sum would be $\frac{437}{100} + \frac{6380}{100} = \frac{6817}{100}$, or 68.17.

10^1	10^0	.	10^{-1}	10^{-2}	10^{-3}
4	$\overset{6}{\not 7}$	.	$\overset{15}{\not 6}$	$\overset{9}{\not 1}\!\not 0$	$\overset{1}{0}$
−	5	.	8	3	4
4	1	.	7	6	6

The approach presented for addition can also be used to explain subtraction of rational numbers in decimal form. The place farthest to the right in the numbers in the subtraction 47.6 − 5.834 is thousandths, so we can think of subtracting two fractions with denominator 1000. The numerators are the whole numbers 47,600 and 5834, as illustrated in the table to the left. The difference would be $\frac{47,600}{1000} - \frac{5834}{1000} = \frac{41,766}{1000}$, or 41.766.

We summarize these ideas with the following general procedure.

> **Procedure for Adding and Subtracting Rational Numbers Represented by Decimals**
>
> When adding and subtracting rational numbers in decimal form, align the place-value positions by aligning the decimal points and apply the algorithms for addition or subtraction of whole numbers.

Example 6.8 provides an opportunity to apply these ideas to a real-world problem.

Example 6.8

Problem Solving: The 400-Meter Relay Record

A world record for the women's 400-meter relay was set in 1985. Each of four women ran a 100-meter leg and recorded a total time of 41.37 seconds. Suppose that a team including Marcia and Alice wanted to beat the world record. Marcia's best time for a 100-meter leg is 9.80 seconds, and Alice's best time is 10.29 seconds. Determine the best times required for two other women who, with Marcia and Alice, would comprise a team that could beat the world record.

Solution

Marcia's and Alice's times total

$$\begin{array}{r} 9.80 \\ +\ 10.29 \\ \hline 20.09 \end{array} \text{ seconds,}$$

so the sum of the two other women's times must be less than

$$\begin{array}{r} 41.37 \\ -\ 20.09 \\ \hline 21.28 \end{array} \text{ seconds.}$$

If one had a best time of 10.50 seconds, the other could have a best time of less than

$$\begin{array}{r} 21.28 \\ -\ 10.50 \\ \hline 10.78 \end{array} \text{ seconds,}$$

or any two other times whose sum is less than 21.28 seconds.

Your Turn

Practice: A men's relay team wants to beat a local record of 42.35 seconds for the 400-meter relay. Bill's best time for a 100-meter leg is 10.47 seconds, and Jake's best time is 10.19 seconds. Lee and Wes have never beat 10.4 seconds. Calculate a set of reasonable times for Lee and Wes that would give the team of four a chance to beat the record.

Reflect: What is another way to solve the practice problem that requires fewer additions or subtractions? Explain your method. ■

Estimation Strategies

Sums and differences of rational numbers in either fraction or decimal form can be estimated by using the same techniques used to estimate sums and differences of whole numbers. The techniques discussed in Chapter 3 were rounding, front-end estimation, substitution of compatible numbers, and clustering. Because decimal notation is based on place value, these estimation techniques work just as they do with whole numbers. With a few adjustments, the techniques work equally well with numbers in fraction notation, demonstrated in Example 6.9.

Example 6.9 Estimating Rational Number Sums

Choose a technique and estimate the following sums.

a. $13\frac{3}{4} + 18\frac{2}{3} + 25\frac{1}{5}$

b. $1\frac{3}{4} + 2\frac{1}{3} + 1\frac{4}{5}$

Solution

a. Round each fraction to the nearest whole number: $13\frac{3}{4}$ is almost 14, $18\frac{2}{3}$ is almost 19, and $25\frac{1}{5}$ is close to 25; and $14 + 19 + 25$ can be rounded to $10 + 20 + 30 = 60$. Thus the sum is about 60.

Or use front-end estimation with adjustment: $13 + 18 + 25$ is close to $10 + 10 + 20 = 40$. Adjust for dropping the 3, 8, and 5 by adding 15 to make 55. Adjust for dropping the $\frac{3}{4}$, $\frac{2}{3}$, and $\frac{1}{5}$ by adding another 2 to obtain an estimate of 57.

b. Use substitution of compatible numbers: Change the sum to $1\frac{1}{2} + 2\frac{1}{2} + 2$. The $\frac{1}{2}$s are easy to add, and the estimate is 6.

Your Turn

Practice: Estimate the following differences.

a. $25.5 - 9.8$

b. $25\frac{1}{2} - 9\frac{4}{5}$

Reflect: What strategy did you use for each estimate, and why did you decide to use that particular strategy? ■

Problems and Exercises | *for Section 6.2*

A. Reinforcing Concepts and Practicing Skills

1. Make an illustration depicting area to find each sum or difference.
a. $\frac{3}{5} + \frac{1}{2}$ **b.** $\frac{3}{5} - \frac{1}{3}$
c. $0.6 + 0.5$ **d.** $2.45 - 1.6$

2. Make an illustration depicting the number line to find each sum or difference.
a. $\frac{3}{5} + \frac{1}{2}$ **b.** $\frac{3}{5} - \frac{1}{3}$
c. $0.6 + 0.5$ **d.** $2.45 - 1.6$

3. Change the following to improper fractions.
a. $6\frac{2}{3}$ **b.** $-4\frac{3}{4}$
c. $8\frac{1}{2}$ **d.** $10\frac{9}{10}$

4. Change the following to mixed numbers.
a. $-\frac{18}{5}$ **b.** $\frac{25}{8}$
c. $\frac{43}{10}$ **d.** $\frac{53}{7}$

5. Explain how you can use mental arithmetic to find the following sums.
a. $\frac{1}{3} + \frac{3}{4} + \frac{2}{3}$ **b.** $\frac{5}{6} + \frac{3}{8} + \frac{4}{6}$

6. Identify the properties of rational numbers that you used in your mental computation in Exercise 5.

7. Explain three different ways you can use mental arithmetic to find $2\frac{1}{3} - 1\frac{2}{3} + 3\frac{2}{3}$.

8. Explain three different ways you can use mental arithmetic to find $1\frac{3}{8} + 3\frac{3}{4} - 1\frac{1}{4}$.

9. Find each sum or difference.
a. $7\frac{5}{6} + 2\frac{2}{5}$ **b.** $\frac{3}{4} + \frac{4}{5} + \frac{5}{6}$
c. $9\frac{1}{8} - 6\frac{5}{6}$ **d.** $\frac{7}{8} - \frac{5}{6}$
e. $5.9 - 2.14 + 1.008$ **f.** $7.8 - 4.386$

10. Use the definition of subtraction for rational numbers to justify the solutions to the following equations.
a. $32.19 - 14.8 = ?$ because . . .
b. $\frac{15}{16} - \frac{1}{4} = ?$ because . . .

11. a. Find the sum $24.6 + 3.09$ by
i. changing these numbers into fraction form, applying the fraction algorithm, and changing the sum back into decimal form.
ii. applying the decimal algorithm to the numbers in decimal notation.
b. How do the two sums compare? How do the two procedures compare?

12. a. Find the difference $24.6 - 3.09$ by
i. changing these numbers into fraction form, applying the fraction algorithm, and changing the difference back into decimal form.

  ii. applying the decimal algorithm to the numbers in decimal notation.

b. How do the two differences compare? How do the two procedures compare?

B. Deepening Understanding

13. Without actually doing the calculations, classify each sum or difference as greater than 1, between $\frac{1}{2}$ and 1, or less than $\frac{1}{2}$. Explain your reasoning.

a. $\frac{7}{9} + \frac{9}{13}$ **b.** $\frac{4}{6} + \frac{6}{7}$

c. $\frac{4}{5} - \frac{3}{4}$ **d.** $\frac{3}{7} + \frac{2}{9}$

e. $4\frac{1}{5} - 3\frac{1}{20}$ **f.** $7\frac{3}{7} - 6\frac{2}{3}$

14. Without finding the actual sums and differences, arrange the following expressions in order from least to greatest value. Explain your reasoning.

$$\frac{3}{7} + \frac{4}{10},\ \frac{9}{11} - \frac{3}{4},\ 4\frac{4}{5} - 2\frac{1}{2},\ 6\frac{7}{8} + 3\frac{2}{7},\ \frac{5}{6} - \frac{1}{5}$$

15. Use the tangram pieces shown to make the figures shown in parts (a) and (b). If the value of the large triangle (either *a* or *b*) is 1, express the value of each figure with an addition equation and find the sum that represents the value.

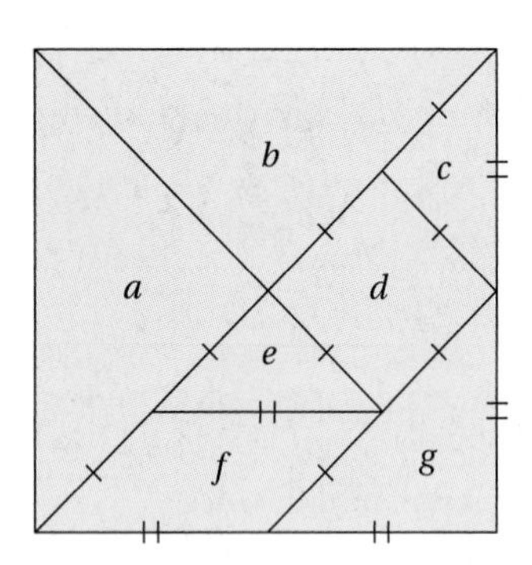

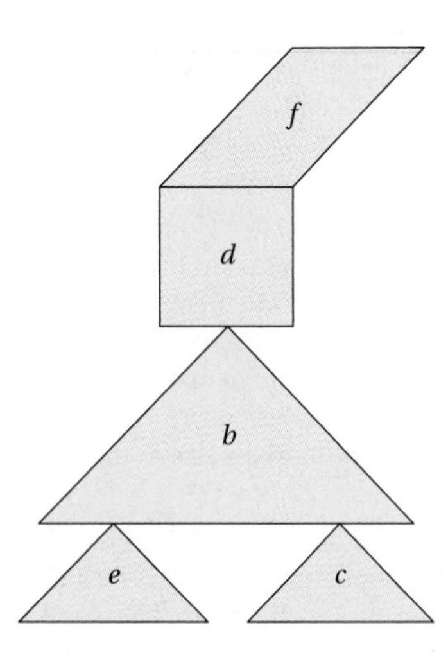

a.

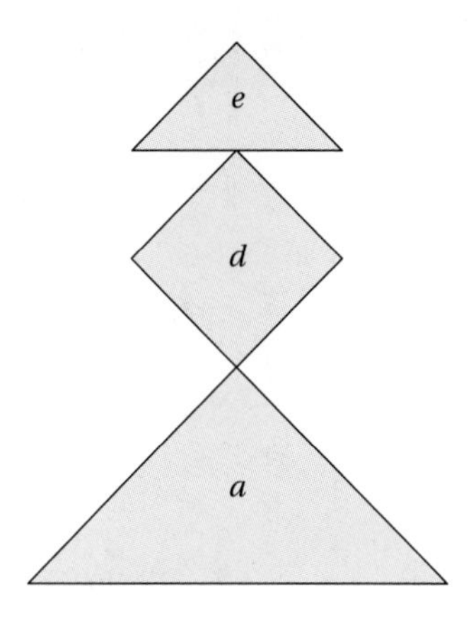

b.

c. How could you express the value of the shape in part (a) with a subtraction equation?

16. For the following equations, write a word problem, draw a picture, and find the solution to the equation.

a. $\frac{5}{12} - \frac{1}{3} = ?$ **b.** $9.6 - 0.81 = ?$

c. $\frac{1}{2} - \frac{2}{3} = ?$

17. Use counterexamples to show that, for rational numbers,

a. subtraction is not commutative.

b. subtraction is not associative.

18. Use the definition of additive inverse to find and verify the additive inverses of the following rational numbers.

a. 0.06   **b.** $\frac{3}{7}$

c. $-2\frac{5}{8}$ **d.** 0

19. a. Find three different pairs of decimals for which each pair has a sum of 75.03.

b. What strategies did you use to find the pairs of decimals in part (a)?

20. a. Find three different pairs of decimals for which each pair has a difference of 43.6.

b. What strategies did you use to find the pairs of decimals in part (a)?

21. The triangle of rational numbers shown is called the *harmonic triangle.*

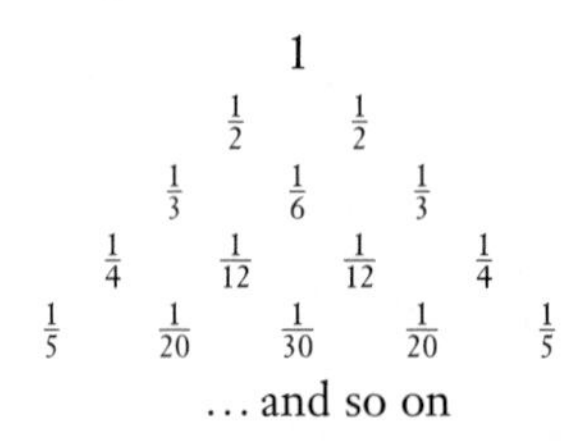

a. What are the next two rows of the triangle?

b. Using what you know about addition and subtraction of rational numbers, describe at least one pattern that appears in the harmonic triangle.

c. Rewrite the triangle in decimal form. Do the same patterns hold? Why or why not?

22. A student used her fraction calculator to examine the following pattern.

Key Sequence	Display
[+] 2 [/] 5 [=]	2/5
[=]	4/5
[=]	6/5
[=]	8/5
[=]	N/D→n/d 10/5

a. What will appear on the screen when she presses the equals sign again?
b. Including the five times indicated here, how many times must she press the equals sign until 42/5 appears on the display?

23. A student used his fraction calculator to explore the following pattern.

Key Sequence	Display
1 [/] 3 [+] 1 [/] 4 [=]	7/12
[=]	N/D→n/d 10/12
[=]	13/12
[=]	N/D→n/d 16/12

a. Why did the result 7/12 appear on the first display?
b. Explain what will appear on the screen when he presses the equals sign again.
c. Including the four times indicated here, how many times must he press the equals sign until N/D→n/d 34/12 appears in the display? How do you know?
d. Based on the pattern observed, will N/D→n/d 69/12 ever appear in the display? Explain.

24. Richie used his fraction calculator to examine the following pattern.

Key Sequence	Display
[−] 1 [/] 6 [=]	−1/6
[=]	N/D→n/d −2/6
[=]	N/D→n/d −3/6
[=]	N/D→n/d −4/6

a. What will appear on the screen when Richie presses the equals sign again?
b. Including the four times indicated here, how many times must Richie press the equals sign until −N/D→n/d −14/6 appears in the display?
c. Describe the set of rational numbers in Richie's pattern for which the symbol N/D→n/d will *not* appear. Explain.

25. Pamil used her fraction calculator to examine the following pattern.

Key Sequence	Display
1 [/] 5 [−] 2 [/] 9 [=]	−1/45
[=]	−11/45
[=]	N/D→n/d −21/45

a. Why did the result −1/45 appear in the first display?
b. Explain what will appear on the screen when Pamil presses the equals sign two more times.
c. Including the three times indicated here, how many times must Pamil press the equals sign until N/D→n/d −81/45 appears on the display? How do you know?
d. Describe the set of rational numbers in Pamil's pattern for which the symbol N/D→n/d *will* appear. Explain.

C. Reasoning and Problem Solving

26. The Salary Distribution Problem. Tatiana's goal for the year is to contribute $\frac{1}{10}$ of her earnings to charity and save $\frac{1}{10}$ for furthering her education. She pays approximately $\frac{1}{3}$ of her earnings in taxes each year. What fraction of her salary will she have left for living expenses?
a. Make an estimate of the answer and explain your estimate.
b. Draw a diagram to illustrate the problem.
c. Write an equation to describe the problem and solve the equation.

27. The Inheritance Problem. Ellis spent $\frac{1}{3}$ of his inheritance on outstanding loans, $\frac{1}{6}$ of it on entertainment, and $\frac{1}{4}$ of it on clothes. He has $600 left to put into savings. How much was his inheritance?
a. Make an estimate of the answer and explain your estimate.
b. Draw a diagram to illustrate the problem.
c. Write an equation to describe the problem and solve the equation.

28. The Travel Times Problem. Jolene used a combination of air and train travel to visit a friend. She spent $2\frac{1}{4}$ hours flying, $3\frac{3}{4}$ hours riding on the train, and $2\frac{3}{4}$ hours waiting between the plane and train. She traveled back home on a direct flight that took $3\frac{1}{4}$ hours. How much time did she save by taking the direct flight home?

29. Explain why one rule for addition and subtraction of decimals begins with: Line up the decimal points....

30. How is the algorithm for adding and subtracting decimals like the algorithm for adding and subtracting fractions? How are they different?

31. Use the commutative and associative properties of addition with rational numbers to find and justify a process for adding $2\frac{1}{4} + 3\frac{1}{3}$ without changing them to improper fractions.

32. **The Consecutive Raises Problem.** Suppose that you receive a raise in January that increases your salary by $\frac{1}{10}$ and another raise in June that increases your salary by $\frac{1}{20}$. When you add these fractions together, can you say that your salary has increased by $\frac{3}{20}$ for the year? Why or why not?

33. a. Place one of the integers 1, 2, 3, and 4 in each box in the addition sentence to make the greatest possible sum. (*Hint:* A fraction calculator will be helpful.)

$$\frac{\square}{\square} + \frac{\square}{\square} = ?$$

b. Where would you put the four integers 5, 6, 7, and 8 to make the greatest possible sum?
c. Where would you put 2, 5, 8, and 9?
d. Write a description of the pattern in the placement of the integers to generate the greatest sum.

34. Repeat the process in Exercise 33 to find and describe the pattern to make the least sum in each case.

35. a. Place one of the integers 1, 2, 3, and 4 in each box in the subtraction sentence to make the greatest possible difference. (*Hint:* A fraction calculator will be helpful.)

$$\frac{\square}{\square} - \frac{\square}{\square} = ?$$

b. Where would you put the four integers 5, 6, 7, and 8 to make the greatest possible difference?
c. Where would you put 2, 5, 8, and 9?
d. Write a description of the pattern in the placement of the integers to generate the greatest difference.

36. Repeat the process in Exercise 35 to find and describe the pattern to make the least difference in each case.

37. How would your patterns change in Exercises 33–36 if
a. one of the integers was 0?
b. one of the integers was negative?

38. Estimate the difference in the fraction of your income you pay in sales tax and the fraction you pay in income tax. What assumptions did you use in making this estimate?

D. Communicating and Connecting Ideas

39. With a partner, discuss the possible advantages and disadvantages of using *least* common denominators when adding and subtracting fractions. Write two paragraphs summarizing the two parts of your discussion.

40. **Historical Pathways.** The ancient Egyptians tried to avoid some of the computational difficulties of working with fractions by representing all rational numbers except $\frac{2}{3}$ as the sum of *unit fractions*, fractions with numerators of 1. For example, what we describe as $\frac{3}{4}$ would have been described by the ancient Egyptians as $\frac{1}{2} + \frac{1}{4}$.
a. Select a rational number and express it as the sum of unit fractions.
b. Express $\frac{2}{7}$ as the sum of two different unit fractions.
c. Find a rational number that could be written in more than one way as the sum of unit fractions. (*Hint:* A fraction calculator will be helpful.)

41. **Making Connections.** Use ideas of prime factorization and common multiples from Chapter 4 to describe the least common denominator of any two rational numbers, $\frac{a}{b}$ and $\frac{c}{d}$. Include consideration of the following conditions.
a. $b = d$.
b. b is a factor of d (or d is a factor of b).
c. b and d are not factors of one another, but have at least one common factor.
d. b and d are relatively prime (no common factors).

Section 6.3 MULTIPLYING RATIONAL NUMBERS

- Modeling Rational Number Multiplication
- Multiplying Rational Numbers in Fraction Form
- Properties of Rational Number Multiplication
- Multiplying Rational Numbers in Decimal Form
- Estimation Strategies

In this section, we extend the ideas of multiplication from the set of integers to the set of rational numbers. We present models of situations and actions, as well as mathematical definitions and properties, to help explain when and how to multiply rational numbers in their various forms. Mini-Investigation 6.10 asks you to make a connection between models used for whole-number multiplication and those used for rational number multiplication.

Talk about how multiplication of rational numbers might be similar to or different from multiplication of whole numbers.

Mini-Investigation 6.10 **Making a Connection**

If 3×2 is used to find the total amount of cheese in 3 boxes, each containing 2 pounds of cheese, what might $3\frac{1}{2} \times \frac{3}{4}$ be used to find?

Modeling Rational Number Multiplication

In Chapter 2, we used models to help explain the meaning of whole-number multiplication. These models included repeated addition, joining groups of equal size, area, and Cartesian products. In this subsection, we look at how some of these models can also be used with rational number multiplication and how to identify situations that call for multiplication of rational numbers.

Some types of rational number multiplication can be explained by using **repeated addition,** as in the following example.

$$3 \times \frac{4}{5} = \frac{4}{5} + \frac{4}{5} + \frac{4}{5} = \frac{12}{5}, \text{ or } 2\frac{2}{5}.$$

Multiplication of rational numbers, like multiplication of integers, can be modeled also by considering the **joining of equal-sized groups.** For example, consider the following problem.

If Doris picked $2\frac{1}{2}$ buckets of grapes and each bucket holds $\frac{1}{2}$ gallon, how many gallons of grapes did she pick?

If we apply the joining groups of equal size model of multiplication, $a \times b = a$ groups of b, the problem can be represented with the multiplication sentence, $2\frac{1}{2} \times \frac{1}{2} = ?$, as shown in Figure 6.14.

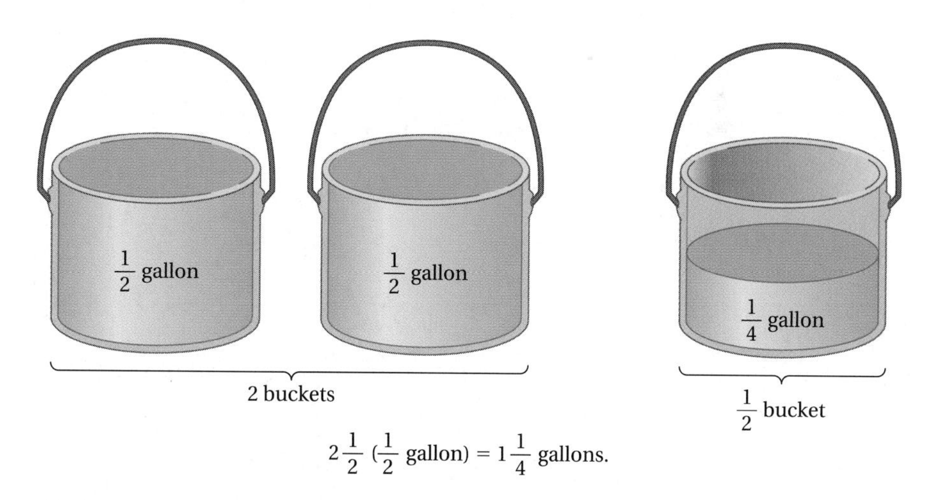

FIGURE 6.14 | Joining groups of equal size model of multiplication.

Next, we show that the **area model** of whole number multiplication can also be used to give meaning to rational number multiplication. Consider the following problem.

If a piece of glass needs to be $2\frac{1}{2}$ meters by $\frac{1}{2}$ meter and the glass is priced according to its area, what area would be used to determine the price?

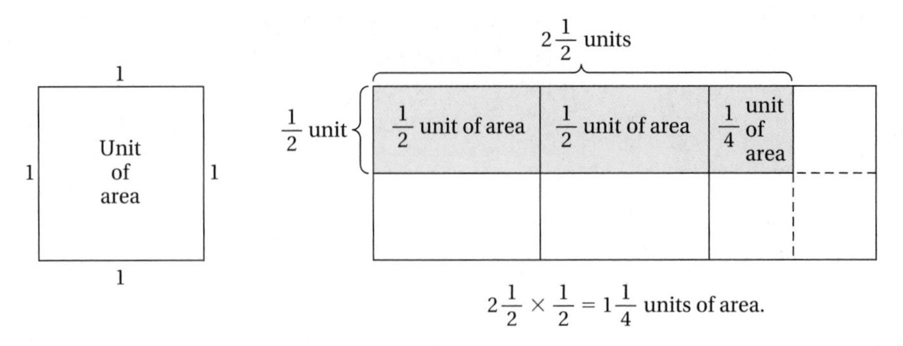

FIGURE 6.15 | Area model of multiplication.

In Figure 6.15, the product $2\frac{1}{2} \times \frac{1}{2}$ is represented by the area of the shaded rectangle with length $2\frac{1}{2}$ units and width $\frac{1}{2}$ unit. Taken together, the shaded parts of the rectangle represent the product.

In Example 6.10, we consider a situation that suggests the use of models to show rational number multiplication.

Example 6.10

Problem Solving: Partial Job Payments

John contracted to prepare and mail a set of fliers for a local club. He had completed $\frac{1}{3}$ of the job when a family emergency came up. He made an agreement with Gretchen to pay her for whatever part of the job she could complete while he was gone. She was able to complete $\frac{3}{4}$ of what he gave her to do. John finished the job when he returned. If the original contract was to pay John \$150 for the job, what payment should Gretchen receive?

Solution

Sharon's thinking: I drew a rectangle to represent the entire job. Then I separated the rectangle into sections to represent the parts done by John and the part done by Gretchen.

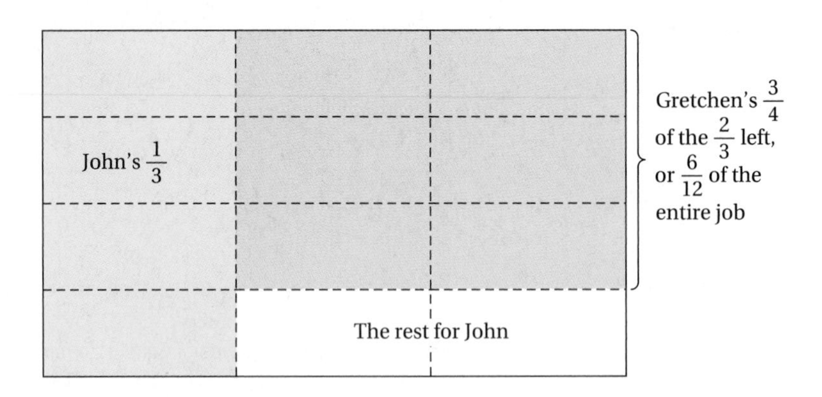

John left Gretchen $1 - \frac{1}{3}$, or $\frac{2}{3}$ of the job to do. She did $\frac{3}{4}$ of the $\frac{2}{3}$. The diagram shows the parts of the whole in terms of twelfths. Thus, Gretchen did $\frac{6}{12}$, or $\frac{1}{2}$, of the job and she should receive $\frac{1}{2} \times \$150 = \$75$.

Marco's thinking: I used the equation $\frac{3}{4}(1 - \frac{1}{3}) \times \$150 = \$75$.

Your Turn

Practice: Use a model to find how many cubic yards of concrete have been ordered in $4\frac{1}{3}$ loads that contain $3\frac{1}{4}$ cubic yards per load.

Reflect: What techniques could you use to estimate this product? What pattern do you see in the denominators obtained to solve these two problems? ■

According to the NCTM *Principles and Standards for School Mathematics* (2000), teachers need to be attentive to conceptual problems students may encounter as they transition from operations with whole numbers to operations with fractions. Students who have developed from their experiences with whole numbers a belief that "multiplication makes bigger" may have trouble making sense of multiplication with fractions, where a product may be smaller than either of the factors.

Multiplying Rational Numbers in Fraction Form

In the Peanuts cartoon, Sally is having trouble finding a product that involves rational numbers.

PEANUTS reprinted by permission of United Feature Syndicate, Inc.

What if Sally's big brother isn't around to tell her the answer? By using area models, we could help Sally discover a general method for multiplying rational numbers in fraction form. The sequence of unit square diagrams shown in Figure 6.16 can lead to such a discovery.

We call a fraction that has a numerator of 1 a **unit fraction.** The following generalization is suggested by Figure 6.16.

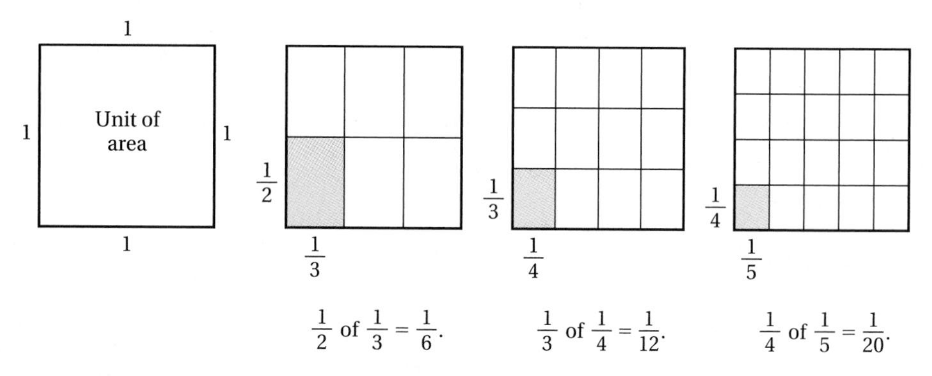

FIGURE 6.16 | Model of unit fraction (numerators of 1) multiplication.

Generalization about Multiplying Rational Numbers Represented by Unit Fractions

For rational numbers $\frac{1}{a}$ and $\frac{1}{b}$, $\frac{1}{a} \times \frac{1}{b} = \frac{1}{ab}$.

If we expand our models to include rational numbers with numerators other than 1, patterns begin to emerge in both the numerators and denominators of the products, as shown in Figure 6.17.

In general, the model depicted in Figure 6.17 shows that, *unlike* addition of rational numbers, the product of two rational numbers is the product of the numerators and the product of the denominators.

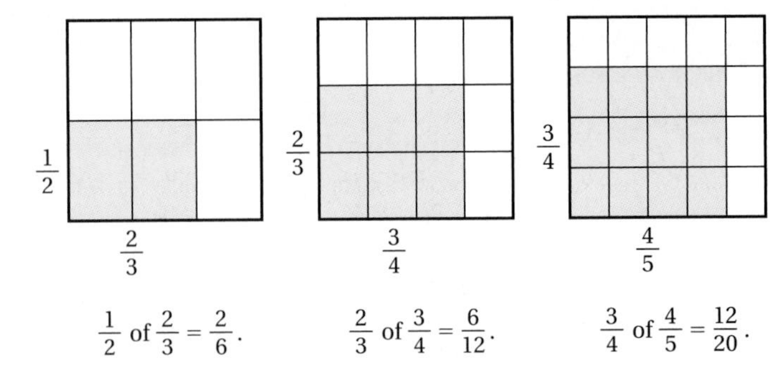

$\frac{1}{2}$ of $\frac{2}{3} = \frac{2}{6}$. $\frac{2}{3}$ of $\frac{3}{4} = \frac{6}{12}$. $\frac{3}{4}$ of $\frac{4}{5} = \frac{12}{20}$.

FIGURE 6.17 | Multiplication model of rational numbers with numerators other than 1.

The following procedure is suggested by Figure 6.17.

Procedure for Multiplying Rational Numbers in Fraction Form

For rational numbers $\frac{a}{b}$ and $\frac{c}{d}$, $\frac{a}{b} \times \frac{c}{d} = \frac{ac}{bd}$.

Example 6.11 is an application of the algorithm for multiplying rational numbers.

Example 6.11

Problem Solving: Voting Predictions

In a certain community, $\frac{2}{5}$ of the voters are Democrats, $\frac{3}{10}$ are Republicans, and $\frac{3}{10}$ are independents. Three-fourths of the voters who are Democrats and $\frac{1}{2}$ of the voters who are Republicans or independents are expected to vote for a proposed local school bond issue. Is the referendum expected to pass?

Solution

- $\frac{3}{4} \times \frac{2}{5} = \frac{(3 \times 2)}{(4 \times 5)} = \frac{6}{20}$ of the voters are Democrats expected to vote *Yes.*
- $\frac{1}{2} \times \frac{3}{10} = \frac{(1 \times 3)}{(2 \times 10)} = \frac{3}{20}$ of the voters are Republicans expected to vote *Yes.*
- $\frac{1}{2} \times \frac{3}{10} = \frac{(1 \times 3)}{(2 \times 10)} = \frac{3}{20}$ of the voters are independents expected to vote *Yes.*

$\frac{6}{20} + \frac{3}{20} + \frac{3}{20} = \frac{12}{20}$, or $\frac{3}{5}$, of the voters are expected to vote *Yes.*

Because $\frac{3}{5}$ is greater than $\frac{1}{2}$, the referendum is expected to pass.

Your Turn

Practice: Use the multiplication algorithm for fraction notation to find the following products.

a. $\frac{2}{3} \times \frac{2}{5}$
b. $1\frac{1}{2} \times \frac{2}{3}$
c. $-\frac{5}{2} \times \frac{1}{4}$
d. $\frac{3}{4} \times \frac{4}{3}$

Reflect: What estimation strategies can you use to determine whether your answers are reasonable? ■

Properties of Rational Number Multiplication

In order to use the properties of rational numbers to verify the multiplication procedure, we first consider which properties of integer multiplication hold for rational number multiplication. We then introduce one property that is unique to rational numbers.

Basic Properties of Rational Number Multiplication. The basic properties of multiplication for integers presented in Chapter 5 hold for rational numbers. However, a new basic property of multiplication appears for rational numbers, a property that didn't exist for integers.

This new basic property is analogous to the additive inverse property introduced with the extension of whole numbers to integers. It is the *multiplicative inverse property.* For every nonzero rational number, a unique rational number called its **multiplicative inverse** (or **reciprocal**) exists, such that the product of the number and its inverse equals the multiplicative identity of 1. Example 6.12 further illustrates this idea.

Example 6.12 Finding Reciprocals

Identify the multiplicative inverses for the following rational numbers. Verify your choices.

a. $\frac{1}{5}$
b. $\frac{2}{3}$
c. $-\frac{3}{2}$
d. 0

Solution

a. Because 5 fifths are needed to make 1, 5 is the reciprocal of $\frac{1}{5}$.
b. $\frac{2}{3} \times \frac{3}{2} = \frac{6}{6} = 1$, so $\frac{3}{2}$ is the reciprocal of $\frac{2}{3}$.
c. Because $\frac{3}{2} \times \frac{2}{3} = 1$, $-\frac{3}{2} \times -\frac{2}{3} = 1$. So$-\frac{2}{3}$ is the multiplicative inverse of $-\frac{3}{2}$.

d. Zero has no multiplicative inverse because every rational number times 0 is 0. We can never get 1 as a product with 0 as a factor.

Your Turn

Practice: Find the multiplicative inverses for the following rational numbers. Verify your choices.

a. -8

b. $1\frac{1}{4}$

c. 0.1

Reflect: What strategies did you use to find the multiplicative inverses? ■

Mini-Investigation 6.11 extends the idea of reciprocals.

Use the definition of multiplication of rational numbers and write a verification that the reciprocal of a nonzero rational number $\frac{a}{b}$ is $\frac{b}{a}$.

Mini-Investigation 6.11 **Finding a Pattern**

What pattern do you see between a rational number and its reciprocal, and can you verify that the pattern always holds?

The following is a summary of the basic properties of rational number multiplication.

Basic Properties for Multiplication of Rational Numbers

Closure property	For rational numbers $\frac{a}{b}$ and $\frac{c}{d}$, $\frac{a}{b} \times \frac{c}{d}$ is a unique rational number.
Identity property	A unique rational number, 1, exists such that $1 \times \frac{a}{b} = \frac{a}{b} \times 1 = \frac{a}{b}$ for every rational number $\frac{a}{b}$; 1 is the multiplicative identity element.
Zero property	For each rational number $\frac{a}{b}$, $0 \times \frac{a}{b} = \frac{a}{b} \times 0 = 0$.
Commutative property	For rational numbers $\frac{a}{b}$ and $\frac{c}{d}$, $\frac{a}{b} \times \frac{c}{d} = \frac{c}{d} \times \frac{a}{b}$.
Associative property	For rational numbers $\frac{a}{b}$, $\frac{c}{d}$, and $\frac{e}{f}$, $(\frac{a}{b} \times \frac{c}{d}) \times \frac{e}{f} = \frac{a}{b} \times (\frac{c}{d} \times \frac{e}{f})$.
Distributive property	For rational numbers $\frac{a}{b}$, $\frac{c}{d}$, and $\frac{e}{f}$, $\frac{a}{b} \times (\frac{c}{d} + \frac{e}{f}) = (\frac{a}{b} \times \frac{c}{d}) + (\frac{a}{b} \times \frac{e}{f})$.
Multiplicative inverse	For every *nonzero* rational number $\frac{a}{b}$, a unique rational number, $\frac{b}{a}$, exists such that $\frac{a}{b} \times \frac{b}{a} = \frac{b}{a} \times \frac{a}{b} = 1$.

We can formulate yet another useful property related to multiplication of integers by viewing multiplication as repeated addition. To find a rational number product when we multiply an integer a by a unit fraction $\frac{1}{b}$, consider that $a \times \frac{1}{b} = \frac{1}{b} + \frac{1}{b} + \cdots$ (a addends of $\frac{1}{b}$) $= \frac{a}{b}$. This property can be verified using the basic properties previously stated.

Property for Multiplying an Integer by a Unit Fraction

For any integer a and any unit fraction $\frac{1}{b}$, $a \times \frac{1}{b} = \frac{a}{b}$.

Using the Properties to Verify the Procedure for Multiplication. If we accept the properties and generalizations for multiplying with unit fractions, we can verify the procedure for multiplying rational numbers.

Statements	Reasons
$\frac{a}{b} \times \frac{c}{d} = (a \times \frac{1}{b})(c \times \frac{1}{d})$	Multiplying an integer by a unit fraction, $a \times \frac{1}{b} = \frac{a}{b}$
$= (a \times c)(\frac{1}{b} \times \frac{1}{d})$	Commutative and associative properties
$= (ac)(\frac{1}{bd})$	Multiplying with unit fractions, $\frac{1}{a} \times \frac{1}{b} = \frac{1}{ab}$
$= \frac{ac}{bd}$.	Multiplying an integer by a unit fraction, $a \times \frac{1}{b} = \frac{a}{b}$

We can also use the properties to verify that the following shortcut, sometimes used in multiplying fractions, is legitimate.

$$\frac{3}{\underset{2}{\cancel{4}}} \times \frac{\overset{1}{\cancel{2}}}{5} = \frac{3 \times 1}{2 \times 5} = \frac{3}{10}$$

The following numerical verification of the shortcut involves first applying the commutative and associative properties, and then applying the fundamental law of fractions to eliminate a factor of 2 from both the numerator and denominator.

$\frac{3}{4} \times \frac{2}{5} = \frac{3 \times 2}{4 \times 5}$	Multiplying rational numbers, basic facts
$= \frac{2 \times 3}{(2 \times 2) \times 5}$	Commutative property, basic facts
$= \frac{2 \times 3}{2 \times (2 \times 5)}$	Associative property
$= \frac{3}{2 \times 5}$	Fundamental law of fractions
$= \frac{3}{10}$.	Basic fact

Multiplying Rational Numbers in Decimal Form

We often encounter situations in which we want to multiply rational numbers that are in decimal form. For example, to find the area of a poster that measures 0.6 meter by 1.25 meters, we need to multiply 0.6×1.25. To find this product, we can use an extension of the standard procedure for multiplying whole numbers.

Mini-Investigation 6.12 lets you discover a generalization about multiplying decimals.

Write a paragraph to describe and justify your discovery. Are there any exceptions? Explain.

Mini-Investigation 6.12 **The Technology Option**

Use your calculator to compute 123×45, 12.3×45, 1.23×45, 123×4.5, 123×0.45, and 12.3×4.5. Based on the pattern you observe, what generalization can you make about the multiplication of decimals compared to the multiplication of whole numbers?

Figure 6.18 shows the standard procedure for multiplying with decimals and verification of the procedure with fractions.

Standard Procedure

Multiply as with whole numbers.
Put in the decimal point so there are as many places to the right of it as there are places to the right in the factors.

Verification with Fractions

Write a fraction for each decimal.
Multiply the fractions.
Write the answer as a decimal.

$$\begin{array}{r} {}^{1\,3}\\ 1.25 \\ \times\ \ 0.6 \\ \hline 0.750 \end{array} \qquad \frac{125}{100} \times \frac{6}{10} = \frac{125 \times 6}{100 \times 10} = \frac{750}{1000} = 0.750$$

FIGURE 6.18 | Verification of the standard procedure for multiplying decimals.

In a decimal, the number of decimal places is equal to the power of 10 in the denominator of the fraction that is equal to the decimal. For example, 0.6 has *1* decimal place and the power of 10 in the denominator of $\frac{6}{10}$ is *1*. Similarly, 1.25 has *2* decimal places and the power of 10 in the denominator of $\frac{125}{100}$, or $\left(\frac{125}{10^2}\right)$, is *2*. When we multiply 125×6 in the standard procedure, essentially we are multiplying the numerators of the fractions that are equal to the decimals. When we *add* the number of decimal places in the factors to get the number of decimal places for the product, essentially we are *adding* the exponents as we do when multiplying the denominators of the corresponding fractions, $\frac{125}{100}$ and $\frac{6}{10}$, as powers of 10. That is, to find $10^1 \times 10^2$, we add the exponents to get 10^3. The general idea used in this verification is expressed as follows.

Procedure for Multiplying Rational Numbers in Decimal Form

Multiply the factors as if they are whole numbers, then place the decimal point in the appropriate place in the product. The product of a factor with m places to the right of the ones place and a factor with n places to the right of the ones place will have $m + n$ decimal places to the right of the ones place (including 0s that are part of the product).

Mini-Investigation 6.13 focuses on the generalization about the number of decimal places in a product.

Talk with your classmates about other situations, such as writing numbers in sequence as part decimals and part fractions, that might hide or illuminate a pattern.

Mini-Investigation 6.13 Finding a Pattern

How would changing the fractions to simplest form at the beginning of the equation in Figure 6.18 have affected your search for a pattern to determine the number of decimal places in the product?

Example 6.13 elaborates on the procedure for multiplying decimals.

Example 6.13

Problem Solving: The Inventory Problem

Mary's job at the small local market is to record the inventory of the produce at the end of each day. Today she records 6.25 kilograms of tomatoes at \$0.89 per kilogram, 4.3 kilograms of oranges at \$0.49 per kilogram, and 5.75 kilograms of romaine lettuce at \$1.39 per kilogram. What is the total monetary value of the inventory at the end of the day?

Solution

$$\begin{array}{r} 6.25 \\ \times\ 0.89 \\ \hline 5625 \\ 5000 \\ \hline 5.5625 \end{array} \qquad \begin{array}{r} 4.3 \\ \times\ 0.49 \\ \hline 387 \\ 1\ 72 \\ \hline 2.107 \end{array} \qquad \begin{array}{r} 5.75 \\ \times\ 1.39 \\ \hline 5175 \\ 1\ 725 \\ 5\ 75 \\ \hline 7.9925 \end{array}$$

The market owner always rounds up, so the total monetary value is \$5.57 + \$2.11 + \$8.00 = \$15.68.

Your Turn

Practice: Use the algorithm for decimal notation to find the following products.

a. 4.26×1.5

b. -0.04×-0.01

c. 40.125×0.02

d. 0.8×1.25

Reflect: What estimation strategies can you use to determine whether your answers are reasonable? ■

Estimation Strategies

Products of rational numbers in either fraction or decimal form can be estimated by using the same techniques, based on place value, that are used to estimate products of whole numbers. Some of the techniques discussed in Chapter 3 were rounding, front-end estimation, and substituting compatible numbers. Because decimal notation is based on place value, these estimation techniques work just as they do with whole numbers. In fact, using estimation techniques with decimal notation can verify where the decimal point should be placed in the product, as shown in Example 6.14.

Example 6.14 Estimating the Decimal Point Placement in Products

Use estimation techniques to determine where to place the decimal in the following products.

a. $6.25 \times 0.89 = 55625$

b. $4.3 \times 0.49 = 2107$

c. $5.75 \times 1.39 = 79925$

Solution

a. If we use rounding to estimate the first product, 6.25×0.89 is about $6 \times 1 = 6$. Therefore, in the digits 55625 in the product, the decimal point must be placed after the first 5 in order to generate a reasonable answer of 5.5625.

b. Substituting compatible numbers to change 4.3×0.49 to 4×0.5 (or $4 \times \frac{1}{2}$) gives an estimate of 2 for the product and places the decimal point after the 2 in the digits 2107 to form a reasonable answer of 2.107.

c. Front-end estimation works well for the third product, changing 5.75×1.39 to 5×1 and indicating that the decimal point must be placed after the 7 in the digits 79925 to yield a reasonable answer of 7.9925.

Your Turn

Practice: Estimate where the missing decimal points should be located.

a. $6.19 \times 0.58 = 35902$

b. $0.096 \times 7.28 = 69888$

Reflect: What estimation strategy did you use for each product? What adaptations were needed, if any, to apply these strategies to fractions? ■

Problems and Exercises for Section 6.3

A. Reinforcing Concepts and Practicing Skills

1. Use an area model to show $\frac{2}{3} \times \frac{3}{5}$.
2. Use an area model to show $\frac{3}{2} \times \frac{3}{4}$.
3. Use an area model to show $\frac{2}{3} \times \frac{3}{2}$.
4. Use an area model to show $\frac{3}{4} \times \frac{3}{4}$.
5. What generalizations about products of fractions can you make from the area models you made in Exercises 1–4?
6. Identify the factors and the products represented by the areas below. (The large squares have sides of length 1 and represent 1 unit of area.)
7. Find the following products.
   a. $2\frac{1}{3} \times \frac{3}{7}$
   b. $(2\frac{3}{4} + 1\frac{1}{2})(\frac{2}{3})$
   c. 16.025×0.4
   d. $-\frac{3}{5} \times 3\frac{1}{2}$
   e. $\frac{ab}{c} \times \frac{c^2a}{b}$
   f. $\frac{3}{4} \times 4.8$

6a.

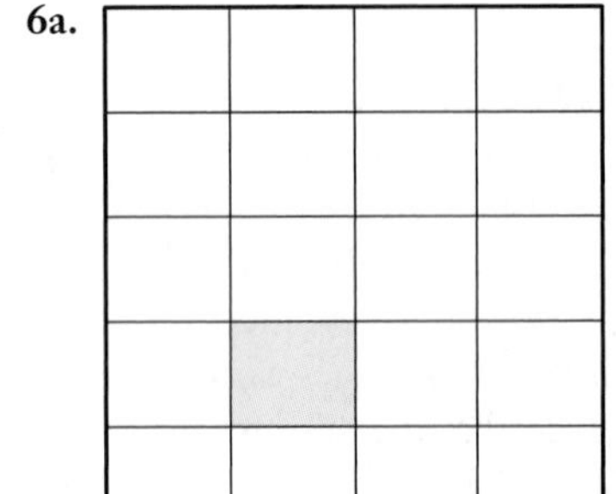

6b.

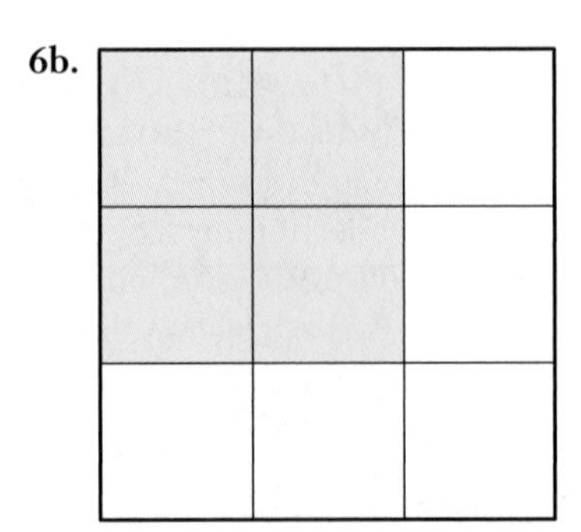

8. Identify which of the following situations can be represented by $\frac{1}{2} \times 10$. Explain your reasoning.
   a. Robert has 10 jawbreakers. He shares them with his sister. How much candy does he give her?
   b. Robert has $\frac{1}{2}$ as much candy as Jean. He has 10 pieces of candy. How much candy does Jean have?
   c. Samuel has twice as much candy today as he did yesterday. He has 10 pieces today. How much candy did he have yesterday?
   d. The 10 pieces of candy that Jennifer has left over from the party is $\frac{1}{2}$ of what she expected to have left over. How much candy did she expect to have left over?
9. Write a word problem that could be represented by $\frac{1}{3} \times \frac{1}{4}$.
10. Write a word problem that could be represented by 2.5×3.5.
11. Write a word problem that could be represented by $-\frac{3}{2} \times \frac{1}{3}$.
12. Write a word problem that could be represented by $\frac{1}{7} \times 7$.
13. Write a word problem that could be represented by 8×0.125.
14. a. Find the product of 24.6×3.09 by
      i. changing these numbers into fraction form, applying the fraction algorithm, and changing the product back into decimal form.
      ii. applying the decimal algorithm to the numbers in decimal notation.
   b. How do the two products compare? How do the two procedures compare?
15. Identify the multiplicative inverse for each rational number. Verify your choices.
   a. $\frac{5}{8}$ b. $-3\frac{3}{11}$
   c. 0.15 d. 9.18
   e. $(\frac{1}{2} + \frac{1}{3})(\frac{1}{2} + \frac{2}{3})$ f. $-x$

B. Deepening Understanding

16. Without actually doing the calculations, classify each product as greater than 1, between $\frac{1}{2}$ and 1, or less than $\frac{1}{2}$. Explain your reasoning.
   a. $\frac{14}{15} \times \frac{19}{20}$ b. 0.5×0.5
   c. $\frac{4}{5} \times 1.25$ d. $\frac{4}{5} \times 1.5$
   e. $\frac{9}{4} \times \frac{1}{3}$ f. $7 \times \frac{1}{6}$
17. Estimate the missing rational number factors for the following products. Explain your reasoning.
   a. $\frac{3}{4} \times ? = 16$. b. $? \times 0.25 = 0.5$.
   c. $-\frac{4}{3} \times ? = \frac{2}{3}$. d. $\frac{5}{6} \times ? = 1.5$.
18. Describe the connection between the multiplicative identity element for rational numbers and the fundamental law of fractions.
19. True or false: Every rational number has a multiplicative inverse. Support your answer.
20. Use the idea of multiplicative inverses to find the solutions to the following equations.
   a. $\frac{4}{3}x = 16$. b. $9y = 24$.
   c. $-\frac{p}{6} = \frac{92}{5}$. d. $\frac{4}{x} = \frac{12}{5}$.
21. What generalization is being applied in the following procedure for simplifying before multiplying rational numbers?
$$\frac{3}{4} \times \frac{2}{5} = \frac{3}{5} \times \frac{2}{4} = \frac{3}{5} \times \frac{1}{2} = \frac{3}{10}.$$
   Describe the generalization and then explain, with reasoning, whether it always holds, sometimes holds, or never holds.
22. Use estimation, rather than the decimal algorithm, to place the decimal points in the following products. Explain your reasoning.
   a. $31.8 \times 42 = 13356$. b. $0.318 \times 4.2 = 13356$.
   c. $318 \times 0.42 = 13356$. d. $3.18 \times 4.2 = 13356$.
23. Use a calculator to find three different rational number factors (in decimal or fraction form) whose product is 10.
   a. What strategies did you use to find the factors?
   b. What real-world situation could your set of three factors represent?
24. Marisha used her fraction calculator to examine the following pattern.

Key Sequence	Display
1 [×] 4 [/] 5 [=]	4/5
[=]	16/25
[=]	64/125

   a. What will appear on the screen when Marisha presses the equals sign again?
   b. If this pattern continues, how many times must Marisha press the equals sign until 4096/15625 appears on the display?
   c. The pattern cannot actually be extended to 4096/15625 with the fraction calculator. Try it and explain what does happen.
   d. What would have happened if Marisha had originally pressed [×] 4 [/] 5 [=]? Explain the result.
25. Jaime used his fraction calculator to examine the following pattern.

Key Sequence	Display
2 [/] 3 [×] 1 [/] 2 [=]	N/D→n/d 2/6
[=]	N/D→n/d 2/12
[=]	N/D→n/d 2/24

   a. Why did the result N/D→n/d 2/6 appear on the first display?
   b. Explain what will appear on the screen when Jaime presses the equals sign two more times.
   c. Including the three times indicated here, how many times must Jaime press the equals sign until N/D→n/d 2/768 appears on the display? How do you know?
   d. Based on Jaime's pattern, will N/D→n/d 2/246 ever appear on the display? Explain.

C. Reasoning and Problem Solving

26. **The Inventory Problem.** A spool of ribbon contained $25\frac{1}{3}$ yards of brocade ribbon. The ribbon was divided evenly among three stores. One store sold all of its ribbon. Another store sold $\frac{1}{2}$ of its ribbon. The third store was closed for inventory and sold none of its ribbon. How much of the original ribbon is left?
   a. Use rational numbers and operations to write an equation or set of equations to describe this situation.
   b. Can you think of a different equation or set of equations that could be used? If so, write it. If not, briefly explain why not.
   c. Use the equation or set of equations to answer the question.
   d. How did you determine whether your answer was reasonable?
27. **The Rental Depreciation Problem.** The owner of a rental house can depreciate its value over a period of $27\frac{1}{2}$ years, meaning that the value of the house declines at an even rate over that period of time until the value is $0.
   a. By what fraction does the value of the house depreciate the first year?
   b. If the house is judged to be worth $85,000, what is the value of the first year's depreciation?
28. **The Cellular Phone Service Problem.** The On-The-Air Cellular Phone company offers two packages for purchasing cellular phone service. The standard package has a monthly subscription charge of $24.95, with an additional charge of 35¢ per minute of use. The economy package has a monthly subscription charge of $30.00, which includes 30 minutes of local use. The usage charge after the first 30 minutes is 38¢ per minute during peak times and 15¢ per minute during offpeak times.

   To help make a decision about which service to use, Phyllis used a spreadsheet to compile the cost comparisons shown in the next column for various amounts of telephone time.
   a. What equation involving rational numbers did Phyllis use to compute the standard cost?
   b. What equation involving rational numbers did Phyllis use to compute the economy cost?
   c. What information about each of the packages do the numbers in the chart represent?
   d. Based on the information in the chart, which package is the better deal? Support your decision.
29. Respond to the conjecture, "Multiplication makes numbers greater."
   a. What observation might cause a person to form this conjecture?
   b. Under what circumstances is this conjecture true?
   c. When is this conjecture false?
30. Find the following rational number products and make a conjecture from the patterns you observe:

Comparison of Standard and Economy Rates for Telephone Use

Minutes	Standard Cost	Economy Cost*
5	$26.70	$30.00
10	28.45	30.00
15	30.20	30.00
20	31.95	30.00
25	33.70	30.00
30	35.45	30.00
35	37.20	31.90
40	38.95	33.80
45	40.70	35.70
50	42.45	37.60
55	44.20	39.50
60	45.95	41.40
65	47.70	43.30
70	49.45	45.20
75	51.20	47.10
80	52.95	49.00
85	54.70	50.90
90	56.45	52.80
95	58.20	54.70
100	59.95	56.60
105	61.70	58.50
110	63.45	60.40
120	66.95	64.20

*Economy cost is less than shown if some of the minutes are in "off peak" time.

   a. $\frac{3}{2} \times \frac{2}{7}$ b. $\frac{3}{3} \times \frac{3}{7}$
   c. $\frac{3}{4} \times \frac{4}{7}$ d. $\frac{3}{5} \times \frac{5}{7}$
31. Use the commutative, associative, and distributive properties of multiplication of rational numbers to justify a process for multiplying $2\frac{3}{5} \times 1\frac{1}{4}$ without changing the factors to improper fractions.
32. a. Use a different integer 1, 2, 3, or 4 for each of a, b, c, and d in the multiplication sentence, $\frac{a}{b} \times \frac{c}{d}$, to make the greatest possible product.
   b. Where would you put the four integers 5, 6, 7, and 8 to make the greatest possible product?
   c. Where would you put 2, 5, 8, and 9?
   d. Write a description of the pattern in the placement of the integers that generates the greatest product.
33. Repeat the process in Exercise 32 to find and describe the pattern that makes the least product in each case.
34. How would patterns change in Exercises 32 and 33 if
   a. one of the integers is 0?

b. one of the integers is negative?
c. more than one of the integers is negative?

D. Communicating and Connecting Ideas

35. a. Find the product of $\frac{1}{2}$ and $\frac{1}{4}$ in fraction form. Then write the fractions $\frac{1}{2}$ and $\frac{1}{4}$ in decimal form and find the product in decimal form. Which product is more accurate? Explain your answer.
b. Find the product of $\frac{1}{3}$ and $\frac{2}{3}$ in fraction form. Then write the fractions $\frac{1}{3}$ and $\frac{2}{3}$ in decimal form and find the product in decimal form. Which product is more accurate? Explain your answer.

36. **Historical Pathways.** Decimal notation was firmly established in the sixteenth century, largely owing to demands from new developments in astronomy, navigation, trade, engineering, and war that computations be performed more quickly and accurately. With a partner, discuss the advantages and disadvantages of multiplication in both fraction and decimal form. Write a brief paragraph summarizing your conclusions.

37. **Making Connections.** Multiplication of rational numbers is used to determine probabilities of multistage events, such as drawing a marble from box A and then drawing a marble from box B. If the probability of drawing a red marble from box A is $\frac{1}{3}$ and the probability of drawing a red marble from box B is $\frac{1}{4}$, the probability of drawing two red marbles, one from each box, is $\frac{1}{4}$ of $\frac{1}{3}$, or $\frac{1}{4} \times \frac{1}{3}$. Use multiplication of rational numbers to find the following probabilities.
a. The probability of rolling a 6 two times in a row if the probability of rolling a 6 on any single roll is $\frac{1}{6}$.
b. The probability of the lottery machine picking a 1 and then a 2 if the probability of a 1 on the first draw is $\frac{1}{50}$ and the probability of a 2 on the second draw is $\frac{1}{49}$.
c. The probability that you will guess someone's 4-digit PIN (personal identification number) for accessing that person's bank account if the probability of guessing each of the first, second, third, and fourth numbers is 0.1.

Section 6.4 DIVIDING, COMPARING, AND ORDERING RATIONAL NUMBERS

- Modeling Rational Number Division
- Defining Rational Number Division
- Dividing Rational Numbers in Fraction Form
- Dividing Rational Numbers in Decimal Form
- Estimation Strategies
- Comparing Rational Numbers
- Other Characteristics of Rational Numbers

In this section, we extend the ideas of division from the set of integers to the set of rational numbers. We use models of situations and actions, as well as mathematical definitions and properties, to develop understanding of when and how to divide rational numbers in their various forms. Mini-Investigation 6.14 asks you to give a real-world interpretation of rational number division.

Talk with other classmates and compare your stories to find out what questions you each have about division with rational numbers that are not also integers.

Mini-Investigation 6.14 Making a Connection

What story could you make up, using what you know about division of whole numbers and integers, that would go with the equation $\frac{3}{4} \div \frac{1}{2} = n$?

Modeling Rational Number Division

Division of rational numbers, like division of whole numbers and integers, is used to separate a quantity into groups of the same size. However, unlike the systems of whole numbers and integers, there are no remainders in division with rational numbers. Rather, a rational number describes the result.

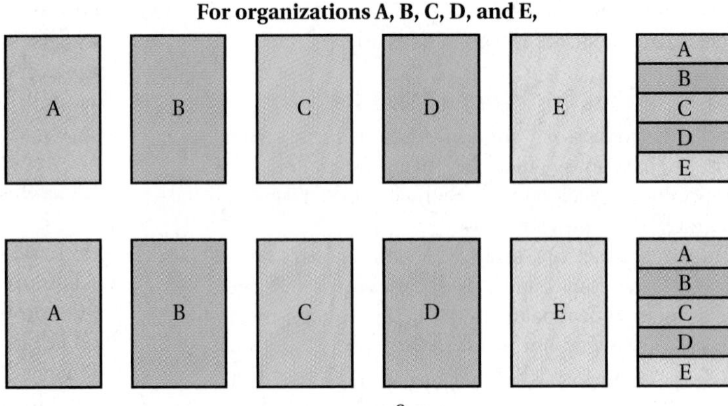

FIGURE 6.19 | Sharing model of rational number division: How many pages does each group get?

For example, the following situation involves the use of a sharing model to give meaning to rational number division. Suppose that 12 pages of space in the school newspaper are to be shared evenly by five student organizations. If we were using only integers, we would say that each organization gets 2 pages with 2 empty pages left over. However, as Figure 6.19 shows, if we are using rational numbers, we continue to divide the pages themselves into parts and say that each organization gets $2\frac{2}{5}$ pages.

We can consider this same situation in terms of a measurement model as shown in Figure 6.20. Suppose that space in the school newspaper is being allotted for sale in sets of 5 pages each and we want to know how many allotments we can get from the 12 pages. Instead of saying there are 2 whole allotments of printing space and 2 pages left over, as we would if working only with integers, we use rational numbers and describe the result of the division in terms of allotments only. The 2 pages left over are compared to the number of pages needed to complete

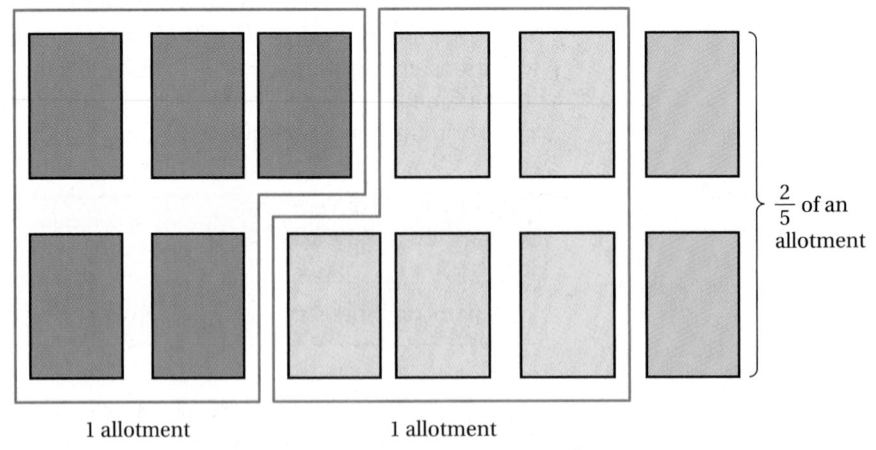

FIGURE 6.20 | Measurement model of rational number division: How many allotments of 5 pages each can be made from 12 pages?

A Sample Student Page: Dividing Whole Numbers by Fractions

Measurement models are often used to introduce division of fractions in the upper elementary grades, as shown in the lesson in Figure 6.21 from a sixth-grade textbook, where the question asked for $a \div b$ is: How many b's are in a? Teachers should also include questions leading to the other interpretation of division: $a \div b = c$, where b represents the number of groups and c is how many in one group. This sharing interpretation connects the operation of division to later work with ratio and proportion, where $2\frac{1}{2} \div \frac{3}{4} = c$ will be expressed as $\frac{2\frac{1}{2}}{\frac{3}{4}} = \frac{c}{1}$.

- ***Why*** *is this problem an example of the measurement interpretation of division?*
- ***Make*** *up another problem using fractions that involves the measurement interpretation of division.*
- ***What*** *might be the value of using models such as the one Peter used to divide fractions?*

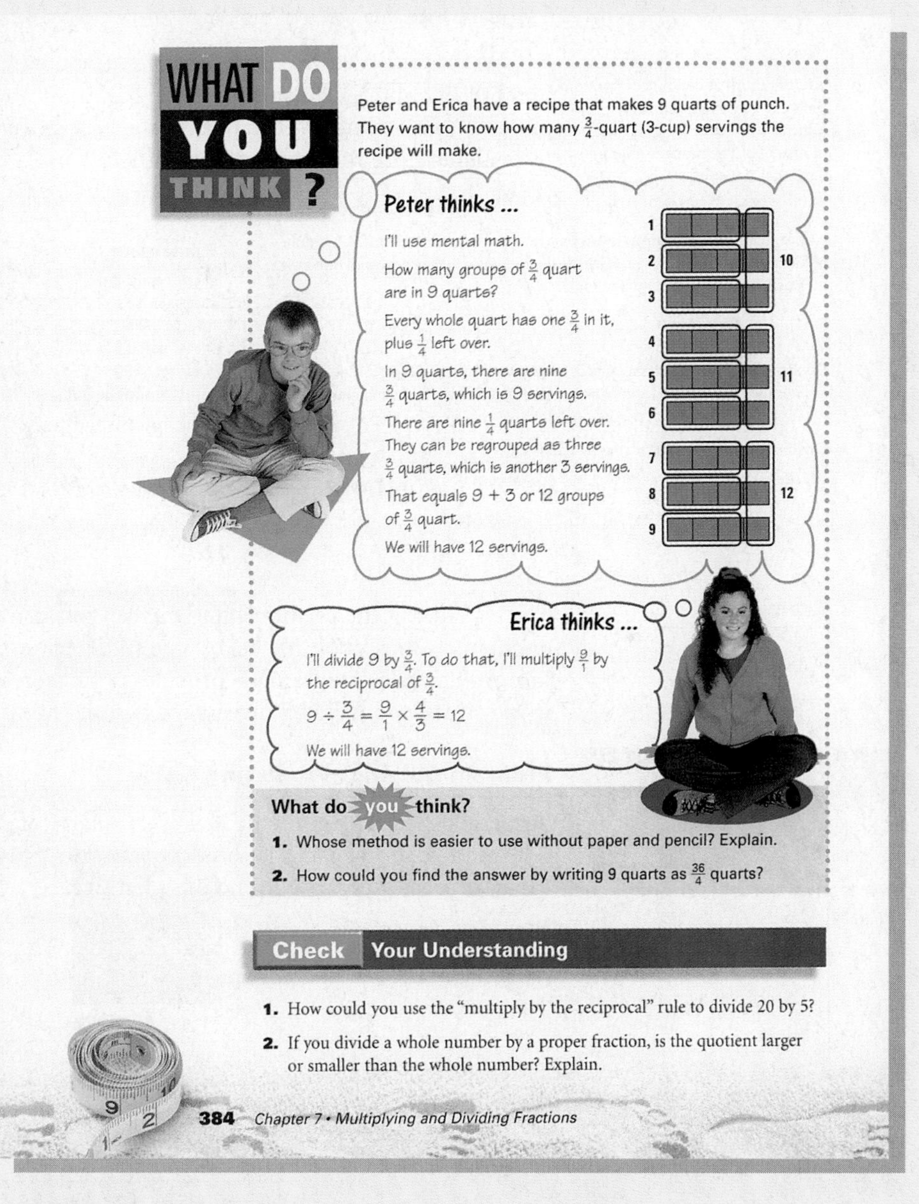

FIGURE 6.21 | Excerpt from a middle school mathematics textbook.

(*Source:* From *Scott Foresman–Addison Wesley Middle School Math, Course 1* by R. I. Charles, J. A. Dossey, S. J. Leinwand, C. J. Seeley, and C. B. Vonder Embse. © 1998 by Pearson Education, Inc. publishing as Prentice Hall. Used by permission.)

another printing space allotment. With 5 pages making a complete printing space allotment, the 2 pages left over can be described as 2 of the 5 pages needed, or $\frac{2}{5}$ of the allotment. So, $2\frac{2}{5}$ allotments are available.

An important feature of the measurement model not illustrated in Figure 6.21 is that the dividend doesn't always separate into a whole number of divisor groups. Often something is "left over" as a fraction of the divisor group. We illustrate this with the following.

> Fran needs $\frac{1}{2}$ meter of molding for the picture frame that she's working on. The kind of molding she wants to use comes in $\frac{3}{4}$-meter lengths. How many picture frames can she make with one piece of molding?

We use the measurement interpretation and ask, How many $\frac{1}{2}$s are in $\frac{3}{4}$? The number line model shown in Figure 6.22 indicates that there is 1 of the $\frac{1}{2}$-meter pieces in $\frac{3}{4}$ meter and enough of the $\frac{3}{4}$ meter left over to make half of another $\frac{1}{2}$-meter piece. Therefore there are one and a half $\frac{1}{2}$s in $\frac{3}{4}$.

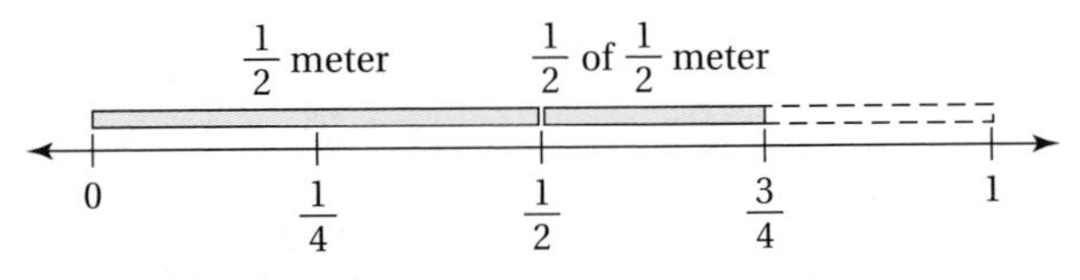

$$\frac{3}{4}\text{ meter} \div \frac{1}{2}\text{ meter per picture frame} = 1\frac{1}{2}\text{ picture frames.}$$

FIGURE 6.22 | Number line model for rational number division.

Visualizing the relationships between rational number dividends, divisors, and quotients is important for being able to determine reasonable answers when solving problems. Example 6.15 provides a further opportunity to interpret a real-world situation involving division of rational numbers.

Example 6.15 Problem Solving: Producing Salsa

Suzanne's special salsa recipe makes $4\frac{3}{4}$ quarts of salsa. The local market wants to sell it in decorative jars that hold $\frac{2}{3}$ quart. How many recipes would Suzanne have to make to get full jars with no salsa left over?

Working toward a Solution

Understand the problem	*What does the situation involve?*	Filling jars with salsa and trying to make the recipe amounts and the amount needed to fill the jars come out even.
	What has to be determined?	How many recipes must be made to produce a whole number of full jars?
	What are the key data and conditions?	The recipe makes $4\frac{3}{4}$ quarts. Each jar holds $\frac{2}{3}$ quart.
	What are some assumptions?	The recipes always come out with exactly the same amount; each jar holds exactly $\frac{2}{3}$ quart.

Develop a plan	*What strategies might be useful?*	Draw a diagram.
	Are there any subproblems?	How many $\frac{2}{3}$s are in $4\frac{3}{4}$?
	Should the answer be estimated or calculated?	Simple computation and counting, using estimation to check
	What method of calculation should be used?	Mental computation.
Implement the plan	*How should the strategies be used?*	Draw areas to represent the number of quarts made in a recipe. Because the dividend is expressed in fourths and the divisor in thirds, showing the quarts in twelfths makes comparing the two quantities easier. $4\frac{3}{4}$ quarts are shaded in the following:

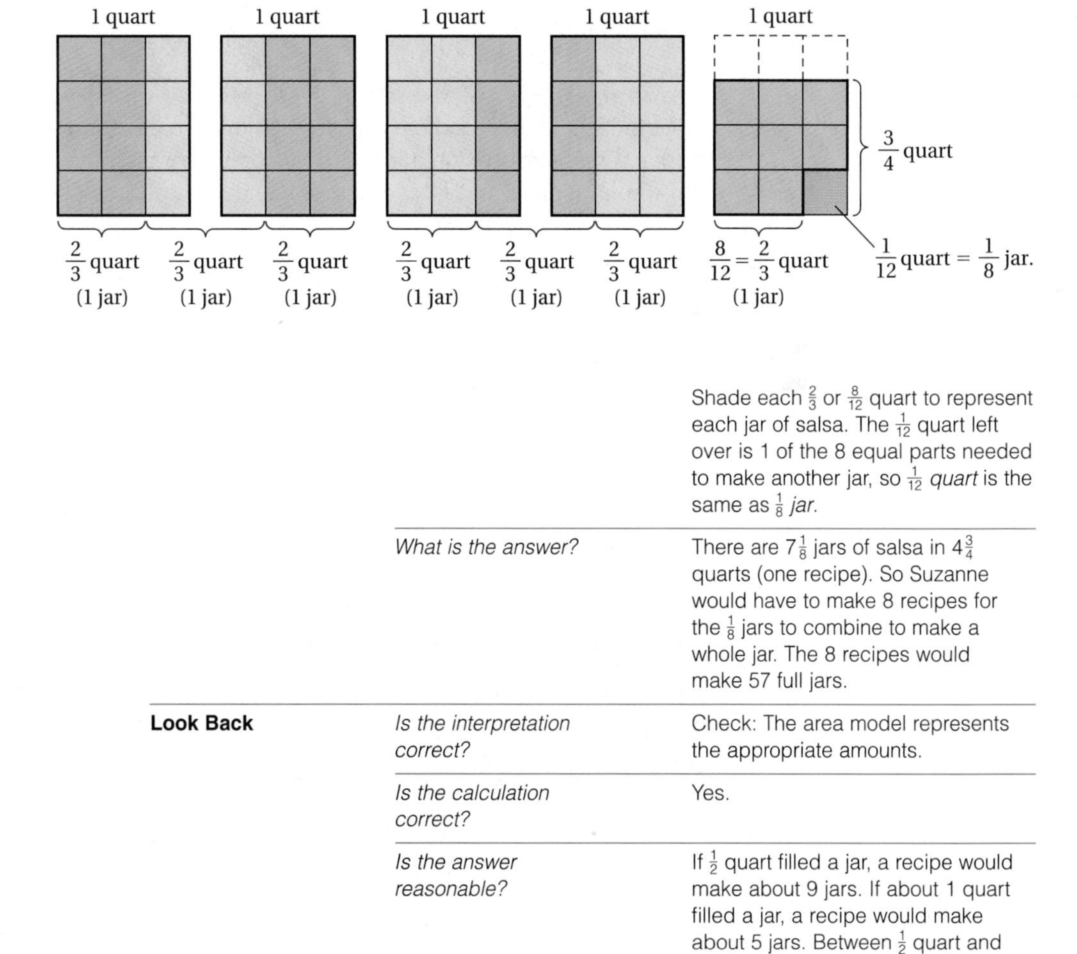

		Shade each $\frac{2}{3}$ or $\frac{8}{12}$ quart to represent each jar of salsa. The $\frac{1}{12}$ quart left over is 1 of the 8 equal parts needed to make another jar, so $\frac{1}{12}$ *quart* is the same as $\frac{1}{8}$ *jar*.
	What is the answer?	There are $7\frac{1}{8}$ jars of salsa in $4\frac{3}{4}$ quarts (one recipe). So Suzanne would have to make 8 recipes for the $\frac{1}{8}$ jars to combine to make a whole jar. The 8 recipes would make 57 full jars.
Look Back	*Is the interpretation correct?*	Check: The area model represents the appropriate amounts.
	Is the calculation correct?	Yes.
	Is the answer reasonable?	If $\frac{1}{2}$ quart filled a jar, a recipe would make about 9 jars. If about 1 quart filled a jar, a recipe would make about 5 jars. Between $\frac{1}{2}$ quart and 1 quart fills a jar, so the number of jars in a recipe should be between

	5 and 9, and $7\frac{1}{8}$ jars is a reasonable answer.
Is there another way to solve the problem?	The number of quarts could be represented on the number line and the number of $\frac{2}{3}$s in $4\frac{3}{4}$ marked. Or a fraction calculator could be used to find $4\frac{3}{4} \div \frac{2}{3}$.

Your Turn

Practice: If $4\frac{1}{2}$ yards of fabric are enough to cover $\frac{3}{4}$ of Allene's bedroom wall, how much fabric is needed to cover the whole wall? Draw a picture to model this problem and determine the answer.

Reflect: In the practice problem, which division question are you trying to answer: How many groups can be made? or What is the size of one group? Explain. ■

Mathematically, it makes no difference whether the task is to find how many groups (a measurement situation) or how many in one group (a sharing situation). Both involve division and can be answered by using a division algorithm for rational numbers.

Defining Rational Number Division

The models we used to help explain division of rational numbers clearly show that rational number division and multiplication have the same inverse relationship as do whole number and integer division and multiplication. That observation leads to the following definition.

Definition of Rational Number Division in Terms of Multiplication

For rational numbers $\frac{a}{b}$ and $\frac{c}{d}$, $c \neq 0$, $\frac{a}{b} \div \frac{c}{d} = \frac{e}{f}$ if and only if $\frac{e}{f}$ is a unique rational number such that $\frac{e}{f} \times \frac{c}{d} = \frac{a}{b}$.

Note that division of rational numbers has the same product ÷ factor = factor relationship as does division of integers. As integer division doesn't have the commutative, associative, or identity properties, we would suspect that these properties do not hold for rational number division. This idea is the focus of Mini-Investigation 6.15.

Talk with other classmates about their counterexamples and how they chose them.

Mini-Investigation 6.15 **Using Mathematical Reasoning**

Find counterexamples to show that division of rational numbers doesn't have the commutative, associative, or identity properties.

Rational numbers were designed to answer division questions, so the possibility of closure for division of rational numbers is worth considering. For each pair of rational numbers, with the exception of 0 as a divisor, a rational number quotient now exists. By restricting our discussion of properties of division to the set of rational numbers that are *not equal to* 0, so that choosing 0 as a divisor is not possible, we can say that division is closed for the set of nonzero rational numbers. This property is similar to closure for subtraction introduced by the negative integers.

Closure Property of Division for Nonzero Rational Numbers

For nonzero rational numbers $\frac{a}{b}$ and $\frac{c}{d}$, $\frac{a}{b} \div \frac{c}{d}$ is a unique nonzero rational number.

Other fairly evident properties of rational number division, $x \div x = 1$, $x \div 1 = x$, and $\frac{x}{a}(a) = x$ can be verified by using the definition of division and the identity properties for multiplication. These verifications are assigned as Exercise 32 at the end of this section.

Dividing Rational Numbers in Fraction Form

It has been humorously reported that in classrooms in the nineteenth century teachers often told their students that, to divide with fractions, "Thine is not to reason why—just invert and multiply!" This often-used procedure can be described as follows.

Procedure for Dividing Rational Numbers—Multiply by the Reciprocal Method

For rational numbers $\frac{a}{b}$ and $\frac{c}{d}$, where c, b, and $d \neq 0$, $\frac{a}{b} \div \frac{c}{d} = \frac{a}{b} \times \frac{d}{c}$.

We can use several different methods to verify that the rote procedure to divide fractions, just invert and multiply, works. Each such method presented in this subsection is based on properties of rational numbers. We show how each method can be used to divide fractions and how it validates dividing by multiplying by the reciprocal of the divisor.

The Common Denominator Method. This method involves first choosing a common denominator for each fraction and then proceeding to find the quotient. For example, to find $\frac{2}{3} \div \frac{1}{2}$, we can think of the inverse relationship between multiplication and division and the definition of multiplication of rational numbers:

$$\frac{2}{3} \div \frac{1}{2} = \frac{a}{b} \text{ if and only if } \frac{a}{b} \times \frac{1}{2} = \frac{2}{3}.$$

The answer to $\frac{2}{3} \div \frac{1}{2}$ is a fraction $\frac{a}{b}$ such that $a \times 1 = 2$ and $b \times 2 = 3$. Therefore $a = 2 \div 1$ and $b = 3 \div 2$. In other words, dividing the numerators of the fractions in the division problem produces the numerator in the quotient, and dividing the denominators of the fractions in the division problem produces the denominator of the quotient. When we then rewrite the fractions with common denominators, the quotient is simplified by having a denominator of 1.

$$\frac{2}{3} \div \frac{1}{2} = \frac{4}{6} \div \frac{3}{6} = \frac{4 \div 3}{6 \div 6} = \frac{4}{3} \div 1 = \frac{4}{3}$$

In general, the common denominator method for dividing rational numbers in fraction form is as follows.

Procedure for Dividing Rational Numbers—Common Denominator Method

For rational numbers $\frac{a}{b}$ and $\frac{c}{d}$, where $c \neq 0$, $\frac{a}{b} \div \frac{c}{d} = \frac{ad}{bd} \div \frac{bc}{bd} = \frac{ad}{bc}$.

The Complex Fraction Method. Another method that can be used to divide fractions involves the idea of rewriting the division as a fraction, then finding an equivalent fraction with a denominator of 1. When rewriting the division of two noninteger rational numbers, the fraction formed has a fraction numerator and a fraction denominator and is called a *complex fraction*. For example, we can use this idea to find $\frac{2}{3} \div \frac{1}{2}$.

$$\frac{2}{3} \div \frac{1}{2} = \frac{\frac{2}{3}}{\frac{1}{2}} = \frac{\frac{2}{3} \times \frac{2}{1}}{\frac{1}{2} \times \frac{2}{1}} = \frac{\frac{2}{3} \times \frac{2}{1}}{1} = \frac{2}{3} \times \frac{2}{1}, \text{ or } \frac{4}{3}.$$

In general, the complex fraction method for dividing rational numbers in fraction form is as follows.

Procedure for Dividing Rational Numbers—Complex Fractions Method

For rational numbers $\frac{a}{b}$ and $\frac{c}{d}$, where $c \neq 0$,

$$\frac{a}{b} \div \frac{c}{d} = \frac{\frac{a}{b}}{\frac{c}{d}} = \frac{\frac{a}{b} \times \frac{d}{c}}{\frac{c}{d} \times \frac{d}{c}} = \frac{\frac{a}{b} \times \frac{d}{c}}{1} = \frac{ad}{bc}.$$

The Missing Factor Method. Yet another method that can be used to divide rational numbers involves using the relationship between multiplication and division and the idea of the reciprocal of a fraction. For example, we can use this idea to find $\frac{2}{3} \div \frac{1}{2}$.

$\frac{2}{3} \div \frac{1}{2} = ?$	We are looking for a missing factor that, when multiplied by the factor $\frac{1}{2}$, gives the product $\frac{2}{3}$.
$\frac{2}{3} = \frac{1}{2} \times ?$	We use the definition of division to rewrite the equation as multiplication.
$\frac{2}{3} \times \frac{2}{1} = (\frac{2}{1} \times \frac{1}{2}) \times ? = ?$	We multiply by the reciprocal to solve for the missing factor.
$\frac{2}{3} \div \frac{1}{2} = \frac{2}{3} \times \frac{2}{1} = \frac{4}{3}$	We now know that both expressions are equal to the missing factor.

In general, the missing factor method for dividing rational numbers in fraction form is as follows.

Procedure for Dividing Rational Numbers—Missing Factor Method

For rational numbers $\frac{a}{b}$ and $\frac{c}{d}$, where b, d, $c \neq 0$,

$$\frac{a}{b} \div \frac{c}{d} = f, \text{ where } \frac{a}{b} = \frac{c}{d} \times f.$$

To find f,

$$\frac{a}{b} \times \frac{d}{c} = \frac{d}{c} \times \frac{c}{d} \times f = f.$$

So

$$\frac{a}{b} \div \frac{c}{d} = \frac{a}{b} \times \frac{d}{c} = \frac{ad}{bc}.$$

Example 6.16 demonstrates how to select and apply the different methods for dividing rational numbers in fraction form.

Example 6.16 Choosing Methods for Dividing with Fractions

Use a method of your choice other than invert and multiply to find $\frac{2}{3} \div \frac{5}{6}$.

Solution

Zach's thinking: I looked for the missing factor. I knew that it was $\frac{2}{3} \times \frac{6}{5}$, because the factor $(\frac{2}{3} \times \frac{6}{5})$ times the other factor $\frac{5}{6}$ gives the product $\frac{2}{3}$.

Dana's thinking: I just wrote equivalent fractions with common denominators. I divided numerators and denominators to find that $\frac{4}{6} \div \frac{5}{6} = \frac{(4 \div 5)}{(6 \div 6)} = \frac{4}{5}$.

Luke's thinking: I made a complex fraction,

$$\frac{\frac{2}{3}}{\frac{5}{6}}$$

and then multiplied its numerator and denominator by $\frac{6}{5}$ to make its denominator 1. Then I multiplied the fractions in the numerator to get $\frac{12}{15}$, or $\frac{4}{5}$.

Your Turn

Practice: Use each of the methods developed in this section to find the quotient $\frac{2}{5} \div \frac{3}{4}$. Show your solutions.

Reflect: Which of the methods for dividing fractions do you think is easiest to use? Why? ■

Dividing Rational Numbers in Decimal Form

The procedure for dividing rational numbers in decimal form is based on the fundamental law of fractions and on place value, as in the other operations involving decimal numbers. We now consider the rationale for that procedure. If we use the division interpretation of fractions, the division $12.45 \div 1.5$ can be written as a fraction:

$$12.45 \div 1.5 = \frac{12.45}{1.5}.$$

The fundamental law of fractions states that equivalent fractions—and therefore equal quotients—can be generated by multiplying both the numerator and denominator by the same nonzero number. Our goal is to make the division easier by creating a divisor (denominator) that is an integer. For example,

$$\frac{12.45}{1.5} = \frac{12.45(2)}{1.5(2)} = \frac{24.90}{3},$$

and

$$\frac{12.45}{1.5} = \frac{12.45(10)}{1.5(10)} = \frac{124.5}{15}.$$

Both of these equivalent fractions, $\frac{24.90}{3}$ and $\frac{124.5}{15}$, have an integer divisor and give us the correct quotient. However, the second procedure, in which we multiply by the necessary power of 10 to eliminate the decimal places to the right of the decimal point in the denominator, is easier to use mentally when dividing decimals. We use these ideas to complete the division, using a caret (∧) to show the change in the location of the decimal point after we multiply both the dividend and divisor by 10. We then write the decimal point in the quotient above the caret in the dividend and divide as with whole numbers.

```
        8.3
1.5∧)12.4∧5
     12 0
        45
        45
         0
```

The procedure developed for dividing decimals can be stated in the following way.

Procedure for Dividing Rational Numbers in Decimal Form

To find the quotient of two decimals, in which the divisor has n places to the right of the decimal point, multiply both the divisor and the dividend by 10^n to make the divisor an integer and to locate the decimal point in the quotient. Then use the whole-number division algorithm to find the digits in the quotient.

Example 6.17 further illustrates this process.

Example 6.17 Dividing with Decimals

Use the procedure for division of rational numbers in decimal form to find $15.3 \div 0.75$ and verify your answer.

Solution

By the fundamental law of fractions,

$$15.3 \div 0.75 = 15.3(100) \div 0.75(100) = 1530 \div 75.$$

So, when we find the quotient in (b), we have found the quotient for (a).

$$\text{(a) } 0.75\overline{)15.3} \qquad \text{(b) } \begin{array}{r} 20.4 \\ 75\overline{)1530.0} \\ \underline{150} \\ 30 \\ \underline{0} \\ 30\,0 \\ \underline{30\,0} \\ 0 \end{array}$$

This answer is correct because

$$\begin{array}{r} 20.4 \\ \times\ 0.75 \\ \hline 1020 \\ \underline{14280} \\ 15.300. \end{array}$$

Your Turn

Practice: Find the quotient $0.125 \div 0.05$ and verify your answer.

Reflect: What estimation strategies can you use to determine whether your answers to the example and practice problems are reasonable? ■

Estimation Strategies

Quotients of rational numbers in either fraction or decimal form may be estimated with the same techniques used to estimate quotients of whole numbers. Mini-Investigation 6.16 focuses on the importance of using estimation to determine whether an answer obtained by dividing decimals is reasonable.

Write a description of an instance in which you used an estimate of a decimal quotient and how you arrived at that estimate.

Mini-Investigation 6.16 **Communicating**

Why is the ability to estimate reasonable answers important when you are dividing decimals?

The estimation techniques discussed in Chapter 3 included rounding, front-end estimation, and substituting compatible numbers. Because decimal notation is based on place value, these estimation techniques work just as they do with whole numbers. The representation of division as a fraction can also be used to estimate quotients of decimals. Example 6.18 shows how estimating and finding decimal quotients are used in a practical situation.

Example 6.18 **Problem Solving: Price of Gasoline**

Jamil bought gasoline yesterday and is concerned that he might have been over-charged. His credit card receipt shows that he paid $23.96 for 16.3 gallons. What price per gallon did Jamil pay?

Solution

Robine's thinking: The price per gallon is \$23.96 ÷ 16.3 gallons. If I use front-end estimation and think of the division expression as the fraction $\frac{23}{16}$, the numerator is about half again greater than the denominator, so the quotient will be about 1.5. However, although estimation gives an idea of a reasonable quotient, it isn't accurate enough to determine whether Jamil was overcharged. Using the algorithm for dividing decimals gives

$$\frac{23.96}{16.3} = \begin{array}{r} 1.469\ldots \\ 163\overline{)239.60\ldots} \\ \underline{163} \\ 766 \\ \underline{652} \\ 1140 \\ \underline{978} \\ 1620 \end{array}$$

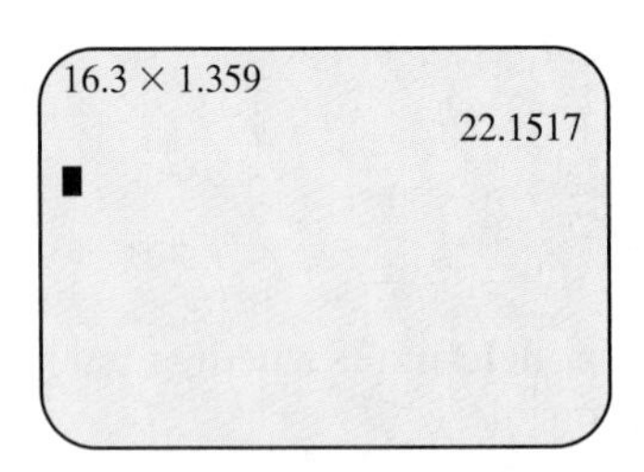

Thus Jamil paid \$1.47 per gallon for the gasoline, which was the price listed on the station's billboard.

Tiku's thinking: I bought gasoline last week for \$1.359, so I tried that as the price per gallon and multiplied by 16.3 gallons. At that price, Jamil would have paid \$22.15, but he paid \$23.96.

So I recalled my last entry and changed the price to \$1.509, but I was too high. I then tried \$1.409, which was too low. At \$1.469, 16.3 gallons cost \$23.94, which is very close to the amount on Jamil's receipt. That seems to be a reasonable price for the gasoline.

22.1517
16.3 × 1.509
24.5967
16.3 × 1.409
22.9667
16.3 × 1.469
23.9447

Your Turn

Practice: Use the procedure for dividing rational numbers in decimal form to find the following quotients.

a. 14 ÷ 0.05

b. −0.9 ÷ 4

c. 18.71 ÷ 3.18

Reflect: What estimation strategies would you use to determine whether your answers to the practice problems are reasonable? ■

Comparing Rational Numbers

The various models and symbols for rational numbers lead to a variety of ways to compare them. Models, common denominators, and place value combine to give a definition of *greater than* for rational numbers. We build up to that definition in the following subsections.

Using Models to Compare Rational Numbers. Just as models are important in explaining rational numbers, models are the basis for explaining how rational numbers compare. A *fraction wall* made from stacked unit rectangles (Figure 6.23) presents a useful way to compare fractions less than 1. The lengths of the bars in the wall allow easy comparison of fractions such as $\frac{3}{8}$ and $\frac{1}{3}$.

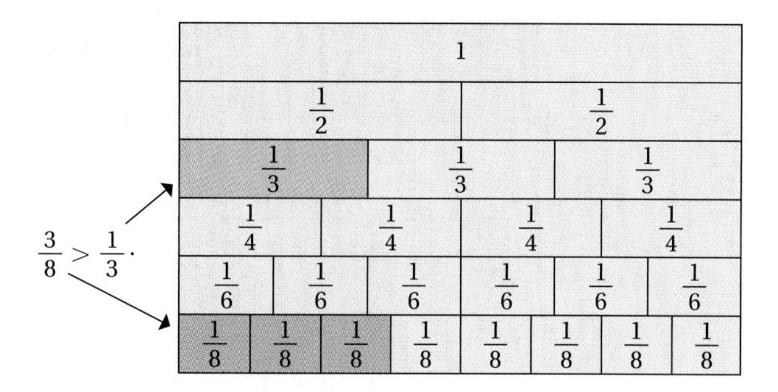

FIGURE 6.23 | Fraction wall model for comparing rational numbers.

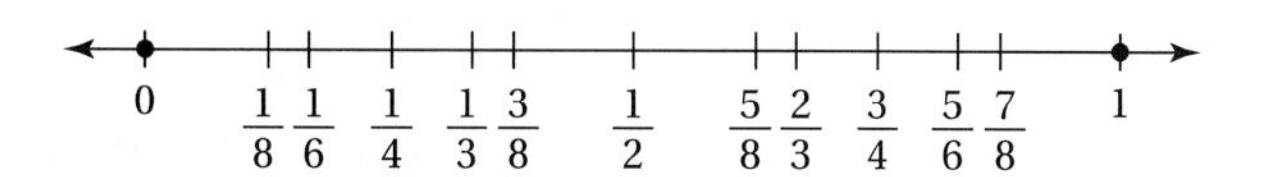

FIGURE 6.24 | Number line model for comparing rational numbers.

A fraction wall can be condensed into a number line, as shown in Figure 6.24. The positions of the rational numbers on the number line are used to compare and order rational numbers, as with whole numbers and integers.

Using Common Denominators to Compare Rational Numbers. As suggested by some of the models in the preceding subsection, we can also use the idea of common denominators to compare rational numbers. Just as the whole is the basis for comparison involving models, denominators can be the basis for comparison involving fractions. We can use models to verify that when the denominators of two fractions are the same, the one with the greater numerator represents the larger rational number. For example, $\frac{2}{5} > \frac{1}{5}$ and $\frac{4}{7} > \frac{2}{7}$, which leads to the following generalization.

Generalization about Comparing Rational Numbers That Have Like Denominators

For rational numbers $\frac{a}{b}$ and $\frac{c}{b}$, where $b > 0$, $\frac{a}{b} > \frac{c}{b}$ if and only if $a > c$.

The preceding generalization is the basis for using common denominators to compare rational numbers. To compare two rational numbers with different denominators, we use the fundamental law of fractions to write fractions equivalent to those being compared but that have common denominators. For example, we compare $\frac{3}{4}$ and $\frac{2}{3}$ as follows.

$$\frac{2}{3} = \frac{2(4)}{3(4)} = \frac{8}{12} \text{ and } \frac{3}{4} = \frac{3(3)}{4(3)} = \frac{9}{12};\ \frac{9}{12} > \frac{8}{12},\ \text{so } \frac{3}{4} > \frac{2}{3}.$$

By writing the pair of rational numbers in terms of a common denominator, we are comparing the same size parts of the whole. Once the rational numbers have common denominators, we need only look at the numerators to make the comparison. This approach leads to the following generalization.

Generalization about Comparing Rational Numbers That Have Unlike Denominators

For rational numbers $\frac{a}{b}$ and $\frac{c}{d}$, where $b > 0$ and $d > 0$, $\frac{a}{b} > \frac{c}{d}$ (or $\frac{ad}{bd} > \frac{bc}{bd}$) if and only if $ad > bc$.

If we apply this generalization for comparing fractions to a pair of fractions that are equal, then $ad = bc$, and we have a verification of the cross-product equality of equivalent fractions discussed earlier in Section 6.1. The cross-product simply generates the numerators of the fractions when represented with the product of the two denominators as a common denominator.

Using Place Value to Compare Rational Numbers. The general procedure for using place value to compare decimals is the same as the procedure for using place value to compare whole numbers, in that we start on the left with the place with the largest value and compare each place as we move to the right. Using this process to compare 0.375 to 0.333 ... is illustrated as follows.

```
0 . 3 (3) 3 ...
↕   ↕ (↕)
0 . 3 (7) 5        So 0.333 ... < 0.375.
```

The rationale for this process is based on the use of common denominators. By comparing corresponding place-value positions in two decimals, we are using common denominators that are powers of 10 to make the comparison. For example, $\frac{3}{10} = \frac{3}{10}$, but $\frac{37}{100} > \frac{33}{100}$, $\frac{375}{1000} > \frac{333}{1000}$, and so on.

Using a Definition of Greater Than to Compare Rational Numbers. As with whole numbers and integers, whenever a positive rational number is added to a first rational number to get a second rational number, the second number is greater than the first. For example, $\frac{2}{5} + \frac{1}{5} = \frac{3}{5}$, so we know that $\frac{3}{5} > \frac{2}{5}$. This relation suggests the following definition of *greater than* for rational numbers.

Definition of Greater Than for Rational Numbers

For rational numbers $\frac{a}{b}$ and $\frac{c}{d}$, $\frac{a}{b} > \frac{c}{d}$ if and only if a rational number $\frac{e}{f} > 0$ exists such that $\frac{a}{b} = \frac{c}{d} + \frac{e}{f}$.

The statement $\frac{c}{d} < \frac{a}{b}$ can also be read as $\frac{a}{b}$ is greater than $\frac{c}{d}$, so this definition also applies to *less than* relationships. The form (fraction or decimal) of the rational numbers often determines which process of comparison is the most appropriate. Example 6.19 gives further insight into comparing rational numbers.

Example 6.19

Problem Solving: Pencil Price and Quality

Wallece decided to trim the office budget by ordering from a company quoting cheaper prices. In a box of 2000 pencils, 243 were defective. In her previous order of 99 pencils from a company charging more, only 12 were defective. Should she continue to order from the company offering the cheaper pencils? Why or why not?

Solution

Walter's thinking: I need to compare $\frac{243}{2000}$ to $\frac{12}{99}$, which is almost the same as $\frac{12}{100}$. If I rename $\frac{12}{100}$ in terms of 2000ths, it becomes $\frac{12(20)}{100(20)} = \frac{240}{2000}$. I think that the two fractions are almost equal and that she should stick with the cheaper pencils.

Jaime's thinking: I can compare the two fractions $\frac{243}{2000}$ and $\frac{12}{99}$ with the products 243(99) and 2000(12). Because $243(99) = 24{,}057$ and $2000(12) = 24{,}000$, I know that $243(99) > 2000(12)$, so $\frac{243}{2000} > \frac{12}{99}$. She should order the more expensive pencils if they don't cost too much more.

Delania's thinking: I can use my calculator to change these fractions into decimals. The calculator display shows the following.

243 [/] 2000 [=] 0.1215, and 12 [/] 99 = 4/33 ≈ 0.1212121.

So $0.1215 > 0.1212\ldots$ But there is very little difference between the quality of the two products. The first company's product has a slight edge in terms of quality, but the second company's product is cheaper.

Your Turn

Practice: Verify that $\frac{4}{11} < \frac{5}{13}$ by using

a. estimation or models.

b. common denominators.

c. a calculator and place value.

d. the definition of greater than.

Reflect: Which method of comparison did you find most useful? Under what conditions might each method be more useful than the others? ■

Other Characteristics of Rational Numbers

As we have discussed, many of the properties of rational numbers and rational number operations are the same as those for integers. However, rational number properties that don't exist for integers are multiplicative inverses and closure for division in the set of nonzero rational numbers. In the following subsections, we look at some additional characteristics that are unique to the rational numbers.

Denseness of Rational Numbers. When someone speaks of a dense forest, they usually are referring to lots of trees very close together. Mini-Investigation 6.17 asks you to apply the forest illustration to an important characteristic of rational numbers.

Talk about how many rational numbers you think there are between $\frac{5}{12}$ and $\frac{6}{12}$.

Mini-Investigation 6.17 **Solving a Problem**

What three different rational numbers can you name that are between $\frac{5}{12}$ and $\frac{6}{12}$?

As you might suspect from Mini-Investigation 6.17, there are infinitely many rational numbers between 0 and 1, between 1 and 2, and so on. Moreover, there are infinitely many rational numbers between any pair of rational numbers, such as $\frac{5}{12}$ and $\frac{6}{12}$. This important characteristic of rational numbers may be described as follows.

Denseness Property for Rational Numbers

For any two rational numbers, $\frac{a}{b} < \frac{c}{d}$, at least one rational number $\frac{e}{f}$ exists such that $\frac{a}{b} < \frac{e}{f} < \frac{c}{d}$.

We can find rational numbers between any two rational numbers by using two of the same techniques we used in comparing rational numbers: common denominators and place value. The techniques for finding a rational number between any two rational numbers are illustrated in Example 6.20. A calculator might be useful in finding or verifying decimal representations for the fractions involved.

Example 6.20 **Finding a Rational Number between Two Other Rational Numbers**

Find three rational numbers between $\frac{2}{3}$ and 0.7272....

Solution

Doug's thinking: The decimal 0.7272... is a little more than $\frac{7}{10}$. If I get the common denominator for $\frac{2}{3}$ and $\frac{7}{10}$, I can find numerators for fractions between them.

$$\frac{2}{3} = \frac{20}{30} = \frac{40}{60} = \frac{80}{120}$$

$$\frac{81}{120}$$

$$\frac{82}{120}$$

$$\frac{83}{120}$$

$$\frac{7}{10} = \frac{21}{30} = \frac{42}{60} = \frac{84}{120}.$$

Because $\frac{7}{10}$ is less than 0.7272..., the fractions I have named are also less than 0.7272....

Geetha's thinking: In decimal form, $\frac{2}{3} = 0.6666\ldots$. If I compare the decimal places with the same values, by lining up the decimal points I can name numbers that are between the two decimal values.

$$\begin{array}{l} 0.6666\ldots \\ 0.67 \\ 0.675 \\ 0.7 \\ 0.7272\ldots \end{array}$$

Your Turn

Practice: Find three rational numbers between $0.555\ldots$ and $\frac{1}{2}$.

Reflect: What strategies did you use to find the rational numbers? ■

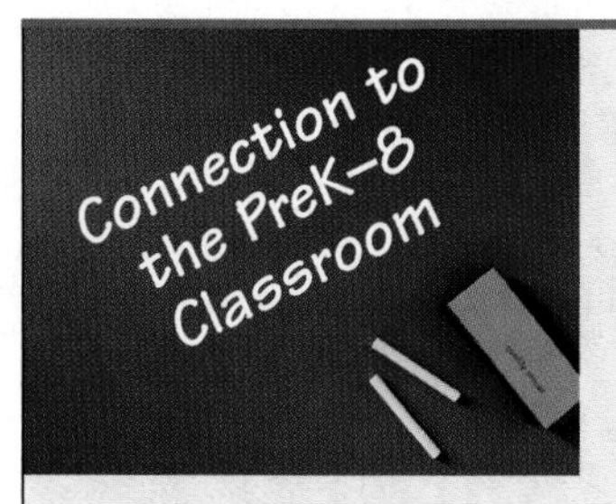

Although denseness of the rational numbers is not a topic usually identified in the elementary mathematics curriculum, it is essential to understanding the behavior of rational numbers. By using a variety of rational numbers between 0 and 1 and describing the relationships between rational numbers with appropriate language, teachers can help students understand that $\frac{3}{5}$ is not the *only* rational number between $\frac{2}{5}$ and $\frac{4}{5}$ and that 0.6 is not the number "next to" 0.5. With students' strong background in discrete counting with whole numbers, emphasizing the dense nature of the set of rational numbers is important.

Repeating Decimals and Fractions. One of the characteristics of rational numbers developed earlier was that every rational number in fraction form can be written as a terminating or repeating decimal. But can each terminating or repeating decimal be written as a fraction? Using the idea of place value, we can easily write a terminating decimal, such as 0.25, as the fraction $\frac{25}{100}$. In Example 6.20, when Doug was trying to use common denominators, being able to write a fraction for $0.7272\ldots$ would have been helpful. Fortunately, there is a way to find a fraction for any repeating decimal, r, which we demonstrate next.

- Let $r = 0.222\ldots$; then $10r = 2.222\ldots$. Subtracting yields

$$\begin{array}{r} 10r = 2.222\ldots \\ -\ r = 0.222\ldots \\ \hline 9r = 2 \end{array} \quad \text{so } r = \frac{2}{9}.$$

- Let $r = 0.3535\ldots$; then $100r = 35.3535\ldots$. Subtracting yields

$$\begin{array}{r} 100r = 35.3535\ldots \\ -\ r = 0.3535\ldots \\ \hline 99r = 35 \end{array} \quad \text{so } r = \frac{35}{99}.$$

- Let $r = 0.147147\ldots$; then $1000r = 147.147\ldots$. Subtracting yields

$$\begin{array}{r} 1000r = 147.147147\ldots \\ -\ r = 0.147147\ldots \\ \hline 999r = 147 \end{array} \quad \text{so } r = \frac{147}{999} = \frac{49}{333}.$$

As demonstrated, choosing an appropriate power of 10 and subtracting to eliminate the repeating numerals leads to the integer numerator and denominator of the fraction that is equal to the repeating decimal.

Problems and Exercises | for Section 6.4

A. Reinforcing Concepts and Practicing Skills

Use appropriate models to find the following quotients.

1. $2\frac{1}{3} \div \frac{2}{3}$
2. $\frac{3}{5} \div \frac{1}{2}$
3. $12 \div \frac{3}{4}$
4. $1 \div \frac{5}{6}$

Identify the division equation represented by each model.

5.

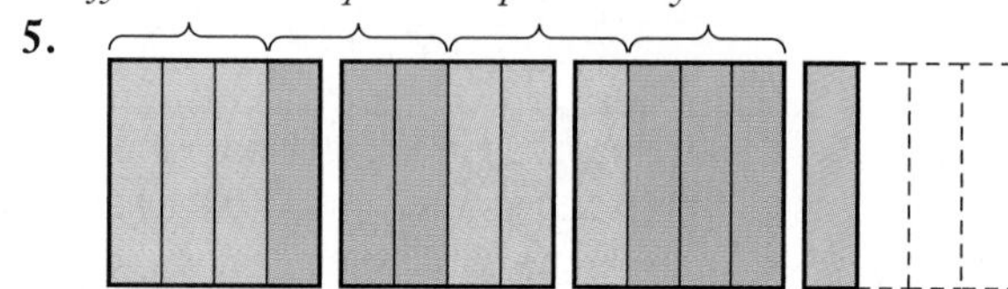

6.

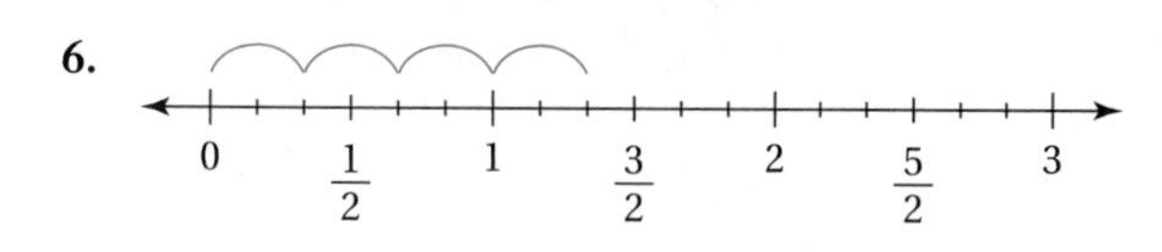

7. Identify which of the following situations can be represented by $4 \div \frac{3}{4}$. Explain your reasoning.
   a. Sue had $\frac{3}{4}$ of a room to paint in 4 hours. How much of the room must she paint each hour to finish on time?
   b. Kendra has 4 yards of electrical wiring to use in setting up her science fair experiment. She needs to cut it into lengths of $\frac{3}{4}$ yard. How many pieces will she be able to make from the 4 yards?
   c. A rotating videocamera in the parking garage goes through 4 complete rotations in $\frac{3}{4}$ hour. How many rotations will it make in 1 hour?
   d. In a local farmers' market, $\frac{3}{4}$ of the profits are cycled back into the local farming community. If you spend $4 there, how much of your money goes to local farmers?
8. Write a word problem that could be represented by $\frac{1}{2} \div \frac{2}{3}$.
9. Write a word problem that could be represented by $1 \div 1\frac{1}{2}$.
10. Write a word problem that could be represented by $-16 \div -\frac{3}{4}$.
11. Write a word problem that could be represented by $9.48 \div 0.96$.
12. Use the common denominator method to find the quotient $\frac{3}{4} \div \frac{2}{5}$.
13. Use the complex fraction method to find the quotient $\frac{3}{4} \div \frac{2}{5}$.
14. Use the missing factor method to find the quotient $\frac{3}{4} \div \frac{2}{5}$.
15. Use the invert and multiply method to find the quotient $\frac{3}{4} \div \frac{2}{5}$.
16. a. Find the quotient $0.9 \div 0.45$ by
      i. changing these numbers into fraction form, applying the fraction procedure, and changing the product back to decimal form.
      ii. applying the decimal procedure to the numbers in decimal form.
    b. How do the two quotients compare? How do the two procedures compare?
17. Find the following quotients by any method you choose.
    a. $\frac{9}{10} \div \frac{2}{3}$ b. $4\frac{1}{2} \div 5\frac{1}{4}$
    c. $-14.68 \div 1.9$ d. $1 \div \frac{4}{5}$
18. Choose a method to compare each pair of rational numbers and determine which one in each pair is greatest.
    a. $\frac{3}{8}, \frac{19}{24}$ b. $-\frac{8}{3}, -\frac{8}{5}$
    c. 0.3232…, 0.3232 d. 35.1, 35.01
19. Find three rational numbers between each pair of rational numbers.
    a. $\frac{5}{6}, \frac{7}{8}$ b. $\frac{13}{15}, \frac{14}{15}$
    c. $-0.2, -0.1$ d. $-\frac{1}{4}, \frac{1}{4}$
20. Change each repeating decimal to fraction form.
    a. 0.0303… b. 9.123123…
    c. 0.14545… d. 5.27111…

B. Deepening Understanding

21. Without actually doing the calculations, sort the following quotients into three groups: greater than 1, between $\frac{1}{2}$ and 1, and less than $\frac{1}{2}$. Explain your reasoning.
    a. $3\frac{2}{9} \div 5\frac{1}{7}$ b. $14 \div \frac{2}{3}$
    c. $5.6 \div 4.9$ d. $\frac{3}{4} \div \frac{2}{3}$
    e. $1 \div \frac{8}{3}$ f. $0.1 \div 0.2$
22. a. Use the properties of rational number operations to find a value for x that solves the equation $\frac{2}{7}x = \frac{3}{5}$.
    b. Use the properties of rational number operations to find a value for x that solves the equation $\frac{1}{6}x + \frac{5}{6} = \frac{5}{12}$.
23. Use estimation, rather than the decimal procedure, to place the decimal points in the following quotients. Explain your reasoning.
    a. $691.04 \div 5.6 = 0123400$
    b. $69.104 \div 56 = 0123400$
    c. $0.69104 \div 0.56 = 0123400$
    d. $691.04 \div 56 = 0123400$

24. Use a calculator to find 10 different pairs of rational numbers (in decimal or fraction form) for which each pair has a quotient of 10.

a. What strategies did you use to find the pairs of numbers?
b. What patterns do these pairs of numbers exhibit?
c. Did it matter which one of the numbers was used as the divisor? What does this tell you about division?

25. Aranxa used her fraction calculator to examine the following pattern.

Key Sequence	**Display**
1 [÷] 3 [/] 4 [=]	4/3
[=]	16/9
[=]	64/27

a. What will appear on the screen when Aranxa presses the equals key again?
b. If this pattern continues, how many times must Aranxa press the equals key until 4096/729 appears on the display?
c. What would have happened if Aranxa had originally pressed [÷] 3 [/] 4 [=]? Explain the result.

26. Hiroshi used his fraction calculator to examine the following pattern.

Key Sequence	**Display**
1 [/] 2 [÷] 2 [/] 3 [=]	3/4
[=]	9/8
[=]	27/16

a. Why did the result 3/4 appear on the first display?
b. Explain what will appear on the screen when Hiroshi presses the equals key two more times.
c. Including the three times indicated here, how many times must Hiroshi press the equals key until 2187/256 appears on the display? How do you know?
d. For Hiroshi's pattern, will any display include the symbol N/D → n/d? Explain.

27. Find a repeating decimal between each pair of terminating decimals.

a. 0.1, 0.2 **b.** −4.73, −4.74
c. $\frac{3}{5}, \frac{4}{5}$ **d.** $-\frac{1}{4}, \frac{1}{4}$

28. Use estimation to order the following pairs of rational number expressions from least to greatest.

a. $5 \div 2\frac{2}{3}, 1 \div \frac{1}{7}$ **b.** $2\frac{2}{3} \times \frac{1}{4}, 6 \div \frac{3}{4}$
c. $5.26 \times 0.23, 0.63 \times 2$ **d.** $-1.3 \times 0.5, -\frac{1}{2} \times \frac{4}{3}$

29. Describe the set of rational numbers whose reciprocals are between 0 and 1.

30. How could you use a fraction calculator to compare two rational numbers without converting the fractions to decimals? Write an explanation of your procedure, using specific fractions.

C. Reasoning and Problem Solving

31. Respond to the conjecture, "Division makes numbers smaller."

a. What observation might cause a person to form this conjecture?
b. Under what conditions is this conjecture true?
c. Under what conditions is this conjecture false?

32. Use the definition of division, the identity property, and other properties of multiplication to verify that for rational numbers $x \neq 0$

a. $x \div x = 1.$ **b.** $x \div 1 = x.$
c. $(x \div a)a = x.$

33. a. Find the following rational number quotients.

i. $1 \div \frac{3}{4}$ **ii.** $1 \div \frac{5}{6}$
iii. $1 \div \frac{3}{7}$ **iv.** $1 \div \frac{4}{9}$

b. Make a conjecture from the patterns you observe in this set of quotients.
c. Use properties of rational number division to prove your conjecture, or find a counterexample to disprove it.

34. The Olympic Race. At the 1988 Summer Olympics in Seoul, South Korea, Florence Griffith-Joyner ran the 100-meter dash in 10.49 seconds (a world record). What was her speed in miles per hour? (1 meter is approximately 39.37 inches.)

35. Square Footage. The following is a copy of the printout that a realtor gave Sally. It shows information about several houses on the market in a certain area.

	A	B	C	D	E
1	Homes for Sale: Lendorf Estates Subdivision				
2	Address	Asking Price	Bed/Bath	Price per Square Foot	Total Square Feet
3	4202 Autumn Circle	$ 82,900	3/2	$ 54.10	?
4	4107 Shawnee	66,500	3/2	55.50	?
5	3211 Willow Oaks	74,500	3/2	44.10	?
6	3902 Ravenwood	75,000	3/2	58.60	?
7	3613 Brighton	87,900	4/2	55.00	?
8	4026 Oak Valley	179,900	4/2	56.40	?

Sally wants to know the total square footage in each house, but that column is blank in the printout.

a. Explain how she can calculate the total square footage from the information on the printout.

b. Calculate the square footage for each house.

c. What spreadsheet formulas could the realtor use to calculate the total square footages in cells E3 through E8?

d. Suppose that Sally plans to offer \$2000 less than the asking price for a house.

i. Recalculate the price per square foot for each house.

ii. Include a column F in the spreadsheet and write spreadsheet formulas in that column to recalculate the price per square foot for each house.

e. What formula could be entered in cell B9 to determine the average (mean) asking price for these houses? Is the mean an appropriate value to use for the asking price of a typical house in this list? Explain.

36. Grocery Scanning. In the cartoon at the bottom of the page, Ditto has fun scanning a grocery item over and over again.

a. Assume that some of the groceries have already been scanned and write an equation describing \$2325.68 as the total of the prices of the previously scanned groceries, g, plus the product of the price of Ditto's item, D, and how many times he scanned it, s.

b. Use a calculator to find three sets of rational number values for g, D, and s that solve the equation.

c. What operations did you use in your equation?

d. What operations and strategies did you use to obtain the answers?

37. a. Place one of the integers 1, 2, 3, and 4 in each of the boxes in the division sentence to make the greatest possible quotient.

$$\frac{\square}{\square} \div \frac{\square}{\square} = ?$$

b. Where would you put the four integers 5, 6, 7, and 8 to make the greatest possible quotient?

c. Where would you put 2, 5, 8, and 9?

d. Write a description of the pattern in the placement of the integers that generates the greatest quotient.

38. Repeat the process in Exercise 37 to find and describe the pattern that makes the least quotient.

39. How would your patterns change in Exercises 37 and 38 if

a. one of the integers is 0?

b. one of the integers is negative?

c. more than one of the integers is negative?

40. The Mathematics Contest. The two junior high schools in town both participate in a national mathematics contest. In one school, 156 of 723 students participate. In the other school, 285 of 1208 students participate. Use comparison of rational numbers to determine which school has the higher rate of participation in the contest.

41. Granola Recipe. One recipe for granola makes 6 servings and requires $\frac{2}{3}$ cup of sugar. A similar recipe makes 8 servings and calls for $\frac{3}{4}$ cup of sugar. Use comparison of rational numbers to determine which recipe contains more sugar per serving.

42. Change the following repeating decimals to fractions and use the pattern to make a conjecture.

a. 0.222... 0.333... 0.444... 0.555..., and so on

b. 0.1212... 0.2323... 0.3434... 0.4545..., and so on

43. Show that 0.999 ... = 1 by

a. changing $\frac{1}{3}$ to decimal form and multiplying by 3.

b. changing $\frac{1}{9}$ to decimal form and multiplying by 9.

c. using the pattern from Exercise 42.

44. Write a short paragraph explaining why there is no rational number *right next to* 0.

45. Show why the arithmetic mean, or average, of two rational numbers $\frac{a}{b}$ and $\frac{c}{d}$ always lies between them. That is,

$$\frac{a}{b} < \frac{\frac{a}{b} + \frac{c}{d}}{2} < \frac{c}{d}$$

Illustrate your reasoning with an example for which the generalization holds.

Reprinted with special permission of King Features Syndiate.

46. Decide whether $\frac{n}{(n+1)}$, $n = 0, 1, 2, 3, 4, \ldots$, is an increasing sequence. (In an increasing sequence, each term is greater than the one before it.) Explain your reasoning.

47. The Batting Record. Laura has a *hit to at-bat* record described by the fraction

$$\frac{\text{number of hits}}{\text{number of at-bats}} \quad \text{or} \quad \frac{h}{b}$$

If she gets a hit each of her next three times at bat,

a. what will her hit to at-bat record be? (Write it in an expression, using h and b.)

b. will her hit to at-bat record increase, stay the same, or decrease? Explain your reasoning. (Pick some numbers to try, look for a pattern, make a conjecture, and then test and verify your conjecture.)

48. Which is greater, a rational number $\frac{a}{b}$ or its square $\frac{a^2}{b^2}$? Use rational number properties and operations to support your answer. (Pick some numbers to try, look for a pattern, make a conjecture, and then test and verify your conjecture.)

D. Communicating and Connecting Ideas

49. a. Find the quotient $\frac{2}{3} \div \frac{5}{6}$ in fraction form. Then write the fractions $\frac{2}{3}$ and $\frac{5}{6}$ in decimal form and find the quotient in decimal form. Which quotient is more accurate? Explain your answer.

b. Find the quotient $\frac{1}{2} \div \frac{3}{5}$ in fraction form. Then write the fractions $\frac{1}{2}$ and $\frac{3}{5}$ in decimal form and find the quotient in decimal form. Which quotient is more accurate? Explain your answer.

50. Historical Pathways. In Calandri's arithmetic, published in Florence, Italy, in 1491, the following example of an algorithm for division of rational numbers appears. [Newman (1988), p. 452]

$$\frac{2}{5} \div \frac{7}{9} = \frac{18}{5} \div 7 = \frac{18}{35}.$$

a. Verify with multiplication that this quotient is correct.

b. Describe this algorithm, in general, for division with any pair of rational number fractions that do not have divisor 0.

c. Try this algorithm with another pair of rational numbers.

d. Can you think of a pair of rational number fractions for which this algorithm would not work (a counterexample)? How could you use the procedures you know (common denominator or invert and multiply) to verify Calandri's algorithm?

51. Historical Pathways. Because of the denseness of rational numbers, early mathematicians believed that placing "all" the rational numbers along a line would name all the points on the line and therefore that every length could be described by a rational number. However, in about the fifth century B.C., a group of Greek mathematicians called the Pythagoreans showed that no rational number describes the length of a diagonal of a square with sides 1 unit in length. The length of the diagonal, according to the Pythagorean theorem, is the number that when multiplied by itself (squared) gives a product of 2, written as $\sqrt{2}$.

a. Estimate a rational number that you think would give a product of 2 when squared and test it by squaring it. Compare the product to 2. How close is it?

b. Based on the result of your first estimate, make another estimate, square it, and compare the product to 2.

c. Continue the process until your product is within $\frac{1}{100}$ of 2. What is your estimate of $\sqrt{2}$?

d. Explain why you should not expect to be able to record $\sqrt{2}$ as a terminating or repeating decimal.

52. Making Connections. Rational numbers are often used to compare certain characteristics. For example, changes in stock values are recorded and compared with rational numbers. Baseball players' batting averages are recorded in decimal form and compared in order to make draft and salary decisions. Make a list of situations that you can think of, or find examples in a newspaper or magazine, in which rational numbers are used not only for recording, but also for comparisons that lead to predictions or decision making. Combine your list with those of others in the class.

CHAPTER SUMMARY

Key Ideas: Questions and Answers

Section 6.1

- **What language and ideas are needed to explain rational numbers?** *(pp. 277–281)* After identifying a *whole* and separating the whole into equal parts, we use two integers to describe a portion of the whole. For example, three of four equal parts is $\frac{3}{4}$ of the whole.
- **What are rational numbers?** *(pp. 281–282)* A rational number may be thought of as the relationship represented by an infinite set of ordered pairs, each of which describes the same quantity.
- **What are fractions and how can they be used to represent rational numbers?** *(pp. 282–284)* A fraction is a symbol, $\frac{a}{b}$, with a called the numerator and b called the denominator. Two fractions are equivalent ($\frac{a}{b} = \frac{c}{d}$) if

and only if $ad = bc$. Each infinite set of equivalent fractions, all with integer numerators and denominators, describes a rational number. Therefore for each rational number there is an infinite set of equivalent fractions and one point on the number line. The fraction $\frac{a}{b}$ also represents $a \div b$.

- **What are some properties of fractions?** *(pp. 284–286)* The fundamental law of fractions states that $\frac{a}{b} = \frac{ac}{bc}$ for any fraction $\frac{a}{b}$ and number $c \neq 0$. A fraction is in simplest form when the numerator and denominator are both integers that are relatively prime and the denominator is greater than zero.
- **What are decimals and how can they be used to represent rational numbers?** *(pp. 286–291)* A decimal is a numeral that utilizes a base-ten place-value system consisting of tenths and multiples of tenths to represent a number. Every rational number can be expressed as a terminating or repeating decimal. Scientific notation is a special form of decimal notation.
- **How is the set of rational numbers related to whole numbers, integers, and other numbers?** *(pp. 291–292)* The set of whole numbers is a subset of the set of integers, which in turn is a subset of the set of rational numbers. The union of the set of rational numbers and the set of irrational numbers (two disjoint sets) is the set of real numbers.

Section 6.2

- **How can models be used to explain addition and subtraction of rational numbers?** *(pp. 295–298)* Discrete models of sets and continuous models of length, area, and volume can be used to represent addition of rational numbers as joining quantities and subtraction as separating or comparing quantities. In general, for rational numbers $\frac{a}{c}$ and $\frac{b}{c}$, $\frac{a}{c} + \frac{b}{c} = \frac{(a+b)}{c}$ and $\frac{a}{c} - \frac{b}{c} = \frac{(a-b)}{c}$.
- **How can rational numbers in fraction form be added and subtracted?** *(pp. 299–300)* Using the definition of rational number addition and subtraction, the fundamental law of fractions, and the idea of common multiples, we can verify the following procedure for adding and subtracting fractions $\frac{a}{b}$ and $\frac{c}{d}$: $\frac{a}{b} + \frac{c}{d} = \frac{ad}{bd} + \frac{cb}{bd} = \frac{(ad+cb)}{bd}$ and $\frac{a}{b} - \frac{c}{d} = \frac{ad}{bd} - \frac{cb}{bd} = \frac{(ad-cb)}{bd}$.
- **How do the properties of rational number addition and subtraction compare to the properties for integers?** *(pp. 300–301)* For rational numbers, as for integers, there is an inverse relationship between addition and subtraction. The properties of addition of rational numbers are the same as those of addition of integers: the closure, identity, commutative, associative, and additive inverse properties. Subtraction of rational numbers has the closure property as does subtraction of integers.
- **How can rational numbers in decimal form be added and subtracted?** *(pp. 301–302)* Decimal notation is based on place value, so the process for adding and subtracting rational numbers in decimal form consists of aligning the place-value positions by aligning the decimal points and applying the procedure for addition or subtraction of whole numbers.
- **What estimation strategies are appropriate for adding and subtracting rational numbers?** *(pp. 302–303)* Rounding and front-end estimation with adjustment are useful techniques for estimating sums and differences of rational numbers in both fraction and decimal form.

Section 6.3

- **How can models help explain rational number multiplication?** *(pp. 307–309)* Some models for whole-number multiplication can be extended to rational numbers to represent multiplication of rational numbers as joining groups of equal size, finding the area of a rectangle, and (in certain situations) repeated addition.
- **How can models be used to develop a procedure for multiplying rational numbers in fraction form?** *(pp. 309–311)* We can use area models to find patterns that lead to the conclusions that for rational numbers $\frac{1}{a}$ and $\frac{1}{b}$, $\frac{1}{a} \times \frac{1}{b} = \frac{1}{ab}$ and that for rational numbers $\frac{a}{b}$ and $\frac{c}{d}$, $\frac{a}{b} \times \frac{c}{d} = \frac{ac}{bd}$.
- **How can properties of rational number multiplication be used to verify the procedure for multiplication?** *(pp. 311–313)* The properties of multiplication of rational numbers include all the properties of addition of integers: the closure, identity, commutative, associative, distributive, and zero properties. A new property, the multiplicative inverse property, exists for multiplication of rational numbers. It is similar to the additive inverse property for integer addition. The use of the multiplicative inverse, or reciprocal, verifies (and can simplify) the multiplication procedure for rational numbers in fraction form.
- **How can rational numbers in decimal form be multiplied?** *(pp. 313–315)* Changing rational numbers from the decimal form to the fraction form clearly shows that part of the process is to multiply the numerators as integers. The next step is to locate the decimal point in the product. The patterns in the denominators reveal that the product of a factor with m places to the right of the decimal point and a factor with n places to the right of the decimal point will have $m + n$ decimal places to the right of the decimal point (including 0s that are part of the product).
- **What estimation strategies are appropriate for multiplying rational numbers?** *(pp. 315–316)* Rounding, front-end estimation, and substitution of compatible numbers are useful techniques for estimating products of rational numbers in both fraction and decimal form. These estimation strategies can be used to place the decimal points in products when decimals are multiplied.

Section 6.4

- **How can models help explain rational number division?** *(pp. 319–324)* Models of sets and continuous models of length, area, or volume can be used to represent division of rational numbers as finding how much in each group of equal size or how many groups of equal size.
- **How do the properties of rational number division compare to the properties for integers?** *(pp. 324–325)* For rational numbers, as for integers, an inverse relationship exists between multiplication and division. A new property, the closure property, exists for division of *nonzero* rational numbers. It is similar to the closure property for integer subtraction.
- **How can mathematical properties or models be used to verify a procedure for dividing rational numbers in fraction form?** *(pp. 325–327)* Common denominators, complex fractions, and missing factors can all be used to verify that, for rational numbers $\frac{a}{b}$ and $\frac{c}{d}$, $\frac{a}{b} \div \frac{c}{d} = \frac{a}{b} \times \frac{d}{c} = \frac{ad}{bc}$.
- **How can rational numbers in decimal form be divided?** *(pp. 327–329)* For a divisor with n places to the right of the decimal point, use the fundamental law of fractions to multiply both the divisor and dividend by 10^n to make the divisor an integer and to locate the decimal point in the quotient. Then use the whole-number division algorithm to find the digits in the quotient.
- **What estimation strategies are appropriate for dividing rational numbers?** *(pp. 329–330)* Rounding, front-end estimation, and substitution of compatible numbers are useful techniques for estimating quotients of rational numbers in both fraction and decimal form. These estimation strategies can be used to place the decimal points in quotients when decimals are divided.
- **How can rational numbers be compared?** *(pp. 330–333)* Common denominators can be used to compare rational numbers in fraction form, and place value can be used to compare rational numbers in decimal form. By definition, for rational numbers $\frac{a}{b}$ and $\frac{c}{d}$, $\frac{a}{b} > \frac{c}{d}$ if and only if a rational number $\frac{e}{f} > 0$ exists such that $\frac{a}{b} = \frac{c}{d} + \frac{e}{f}$.
- **What are some other characteristics of rational numbers?** *(pp. 333–336)* The denseness property for rational numbers states that for any two rational numbers, $\frac{a}{b} < \frac{c}{d}$, at least one rational number $\frac{e}{f}$ exists such that $\frac{a}{b} < \frac{e}{f} < \frac{c}{d}$. Every repeating decimal can be expressed in fraction form with integers in the numerator and denominator.

Key Terms, Concepts, and Generalizations

Section 6.1

Whole (p. 278)
Equal parts (p. 278)
Ordered pair (p. 278)
Rational number (p. 281)
Fraction (p. 282)
Numerator (p. 282)
Denominator (p. 282)
Proper fraction (p. 283)
Improper fraction (p. 283)
Equivalent fractions (p. 283)
Fundamental law of fractions (p. 284)
Simplest form (p. 285)
Decimal (p. 286)
Decimal point (p. 287)
Expanded notation (p. 287)
Terminating decimal (p. 289)
Repeating decimal (p. 289)
Scientific notation (p. 290)
Real numbers (p. 291)
Irrational numbers (p. 291)

Section 6.2

Rational number addition (p. 296)
Rational number subtraction (p. 296)
Mixed number (p. 296)
Improper fraction (p. 296)
Adding and subtracting in fraction form (p. 299)
Closure property of addition (p. 300)
Identity property of addition (p. 300)
Commutative property of addition (p. 300)
Associative property of addition (p. 300)
Additive inverse property (p. 300)
Adding and subtracting in decimal form (p. 301)
Estimation strategies for adding and subtracting rational numbers (p. 302)

Section 6.3

Repeated addition (p. 307)
Joining equal-sized groups (p. 307)
Area model (p. 307)
Unit fraction (p. 309)
Rational number multiplication (p. 310)
Multiplicative inverse or reciprocal (p. 311)
Closure property of multiplication (p. 312)
Identity property of multiplication (p. 312)
Zero property of multiplication (p. 312)
Commutative property of multiplication (p. 312)
Associative property of multiplication (p. 312)
Distributive property of multiplication over addition (p. 312)
Multiplicative inverse property (p. 312)
Multiplication in decimal form (p. 314)
Estimation strategies for multiplying rational numbers (p. 315)

Section 6.4

Sharing model of division (p. 320)
Measurement model of division (p. 320)
Rational number division (p. 324)
Closure property of division for nonzero rational numbers (p. 325)
Invert and multiply method for dividing rational numbers (p. 325)
Common denominator method for dividing rational numbers (p. 326)
Complex fraction method for dividing rational numbers (p. 326)

CHAPTER REVIEW

Concepts and Skills

1. Draw a diagram to show the meaning of each rational number.
   a. $\frac{5}{8}$ b. $2\frac{3}{5}$ c. 0.125
2. State the fundamental law of fractions and give an example of it.
3. Complete the following chart of symbols for the rational numbers shown.

Fraction form	*Decimal form*
$\frac{5}{12}$	?
?	0.125125 ...
?	0.56

4. Write the following rational numbers in scientific notation.
   a. 0.0000000001 b. 11 million
   c. 0.036754 d. 24,000,059
5. Name the additive and multiplicative inverses for each rational number.
   a. $\frac{2}{3}$ b. $-\frac{4}{3}$
   c. 0.25 d. 0
6. Find each sum or difference.
   a. 28.32 + 7.521 + 3.5 b. $-\frac{21}{5} + \frac{16}{3}$
   c. 53.007 − 25.89 d. $-2\frac{2}{3} - 5\frac{1}{8}$
7. Find each product.
   a. $2\frac{1}{10} \times -8\frac{1}{3}$ b. $7\frac{1}{8} \times 6\frac{1}{4}$
   c. 0.04 × 6.011 d. 2.04 × 3.25
8. Find each quotient.
   a. $\frac{2}{3} \div -9$ b. $\frac{4}{3} \div 3\frac{7}{8}$
   c. 36 ÷ 3.2 d. 5.301 ÷ 0.31
9. Write a word problem, draw a diagram, and solve the following rational number equations.
   a. $1\frac{3}{4} + 1\frac{1}{2} = s$. b. $1\frac{3}{4} - 1\frac{1}{2} = d$.
   c. $1\frac{3}{4} \times 1\frac{1}{2} = p$. d. $1\frac{3}{4} \div 1\frac{1}{2} = q$.
10. Order the following sets of rational numbers from least to greatest.
    a. $2\frac{1}{2}$; 2.333...; 2.51; $\frac{6000}{29}$
    b. 3.33; $\frac{577}{154}$; 3.121212...; $3\frac{5}{8}$
    c. 22; $\frac{32}{11}$; $\frac{477}{154}$; 2.7171...

Reasoning and Problem Solving

11. Is 2.357357... a rational number? Use the definition of rational number to justify your answer.
12. Describe the connections between rational numbers and division.
13. **The Grocery Shopping Problem.** Trish bought $\frac{1}{2}$ pound of cheese, a $6\frac{3}{4}$-pound ham, $2\frac{3}{4}$ pounds of roasting ears, $2\frac{1}{8}$ pounds of new potatoes, and 1 pound of butter. The grocery sackers at this store have been advised to limit the weight of the contents of each sack to 12 pounds or less.
    a. Could the sacker put all of Trish's purchases into one sack? Why or why not?
    b. If the sacker does decide to use two sacks, how could the contents be divided between the sacks to keep their weights fairly equal? Justify your answer.
14. Use the ideas of additive and multiplicative inverses of rational numbers to solve the following equations.
    a. $-5x + 13 = -12$. b. $\frac{1}{3}x - 7 = 25$.
    c. $3x - 4 = x + 12$. d. $x + \frac{11}{9} = \frac{2}{3}$.
15. **The Budget Cuts Problem.** Martha's employer told her that, owing to cuts in the budget, her salary starting next week would be $\frac{3}{4}$ of her present salary. Martha reported to her best friend that her salary had been reduced by $\frac{3}{4}$.
    a. Is Martha's statement true or false? Justify your answer.
    b. What fraction of her new salary would Martha need for an increase to return it to her previous salary?
16. **The Sewing Supplies Problem.** The seamstress purchased 2.4 meters of \$1.90-per-meter fabric; 5.4 meters at \$3.90 per meter; 7.5 meters at \$4.95 per meter; 25.20 meters of cording at \$0.25 per meter; and two spools of thread at \$0.70 per spool.
    a. How much did she spend on this purchase?
    b. If she charges \$10.50 per hour for sewing, how many hours must she work to pay for these supplies?
17. **Which Answer Is Right?** One of the jobs of student workers is to check the homework keys that are printed each day for the students. Charlie was using his calculator to check quickly through the division problems. For one problem with a divisor of 5, the quotient read

32 R2, but Charlie's calculator read 32.4. Which answer is correct? Justify your decision.

18. Let A be the set containing all rational numbers that are less than 5. Is there a rational number, q, in set A such that all other numbers in set A are less than q? Why or why not?

Alternative Assessment

19. Write a paragraph, or design a flow chart, to describe an efficient process for deciding whether any two rational numbers are equivalent. (Remember, the rational numbers might be given in either decimal or fraction form or both.)

20. When young students first encounter multiplication and division of rational numbers, they are often confused because their generalizations that multiplication makes numbers get bigger and division makes numbers get smaller don't hold in many situations. Identify some examples and some language that could help students correct their generalizations.

21. Find a partner to work with. Prepare a presentation to convince the rest of the class that, for representing rational numbers, decimal notation is better than fraction notation (or vice versa). Present your argument to the class.

22. In elementary mathematics textbooks or journals, find three games or activities based on rational numbers. Write a short paper critiquing the three games or activities according to their mathematical worth in terms of developing rational number concepts.

7 Proportional Reasoning

Chapter Perspective

In mathematics, the concept of ratio involves the comparison of two quantities, for example, 4 inches to 6 inches or one part oil to 50 parts gasoline. Perhaps the father shown above has prepared such a mixture of oil and gasoline for his lawnmower, a task that requires an understanding of ratio. The concept of ratio is directly related to the ideas of rational number, percent, and proportion. All of these concepts are fundamental to the development of mathematics and various applications in real-life situations.

In this chapter, we explore the concepts of ratio, proportion, and percent and consider ways of solving real-world problems that involve these concepts. We also consider a variety of situations that involve proportions and how to distinguish situations that involve proportions from those that don't.

Connection to the NCTM Principles and Standards

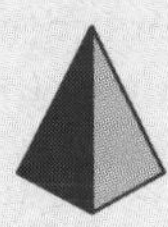

As noted in the NCTM *Principles and Standards,* proportional reasoning is fundamental to much of middle school mathematics. Proportional reasoning permeates the middle grades curriculum and connects a great variety of topics. The study of proportional reasoning in the middle grades should enable students to

- work flexibly with ... percents to solve problems;
- understand and use ratios and proportions to represent quantitative relationships; and
- develop, analyze, and explain methods for solving problems involving proportions, such as scaling and finding equivalent ratios. *(p. 214)*

Thus reasoning and solving problems involving proportions play a central role in the mathematical processes emphasized in the NCTM *Principles and Standards.* The concept of ratio is fundamental to being able to reason proportionally.

Connection to the PreK–8 Classroom

In grades PreK–2 children explore informally different figures that have the same shape to develop intuitive ideas about similar figures.

In grades 3–5 children develop the concept of ratio and compare different ratios to determine if they are equal. Corresponding sides of similar figures are compared to determine the relative sizes of the figures.

In grades 6–8 students investigate proportional reasoning and properties of proportions. They learn the concept of percent and its connection to ratio. They solve problems involving percents and different types of interest problems.

Section 7.1 THE CONCEPT OF RATIO

- Meaning of Ratio
- Uses of Ratio
- Ratios in Decimal Form

In this section, we consider various meanings associated with the word *ratio* and how these meanings arise in mathematics and in real-world situations. We also discuss the different ways of representing ratios. The concept of ratio is fundamental to understanding how two quantities vary proportionally and sets the stage for the development of this idea in Section 7.2. In Mini-Investigation 7.1, we ask you to contrast statements that reveal different interpretations of ratio.

Write a paragraph describing the similarities and differences in these two statements.

Mini-Investigation 7.1 **Making a Connection**

How are the following two statements involving the concept of ratio alike and how are they different?

a. At the regional medical center, there are six nurses for every two doctors.

b. A race car driver can drive about 6 kilometers in 2 minutes.

Meaning of Ratio

As illustrated in Mini-Investigation 7.1, the idea of a ratio has more than one interpretation or meaning. In this subsection, we look at two different interpretations, give a general definition of ratio, and clarify the meaning of equivalent ratios.

A Ratio as a Comparison. One interpretation of ratio is that it compares two like quantities. For example, statement (a) in Mini-Investigation 7.1 makes a comparison between two types of people, doctors and nurses. Statement (b) gives a comparison between two unlike quantities, distance and time. Each of these interpretations suggests the following definition.

Definition of a Ratio

A **ratio** is an ordered pair of numbers used to show a comparison between like or unlike quantities, written x to $y, \frac{x}{y}, x \div y$, or $x : y$ $(y \neq 0)$.

Thus we can think of a ratio as a fraction, say, $\frac{2}{3}$, or as a quotient, say, $2 \div 3$. In the ratio $\frac{x}{y}$ or $x : y$ $(y \neq 0)$, the numbers x and y are called the *terms* of the ratio. Although the second term of a comparison can be zero (compare the number of male U.S. presidents to the number of female U.S. presidents), for computational reasons the second term of a ratio cannot be zero because division by zero is undefined. That is, the ratio $x : y$ must satisfy the condition that $y \neq 0$.

Equivalent Ratios. Because a ratio can be expressed as a fraction, we can think of ratios as being equivalent just as we can think of fractions as being equivalent. You can apply the idea of equivalent ratios in Example 7.1.

Definition of Equivalent Ratios

Two ratios are **equivalent ratios** if their respective fractions are equivalent or if the quotients of the respective terms are the same.

Example 7.1

Problem Solving: Party Soft Drinks

Nikita is planning a party and has to buy soft drinks. She estimates that for every five people, three will drink diet cola and two will drink nondiet cola. If she decides to buy 40 cans of cola, how many of each kind should she buy?

Solution

The ratio of diet cola drinkers to nondiet cola drinkers is $3:2$. Other equivalent ratios are $6:4$, $9:6$, $12:8$, $15:10$, $18:12$, $21:14$, and $24:16$. This last ratio sums to 40 (the total number of cans Nikita will buy), so she should buy 24 cans of diet cola and 16 cans of nondiet cola.

Your Turn

Practice: Freddie thinks that there will be three times as many diet cola drinkers as nondiet cola drinkers. If Freddie is correct and Nikita buys 40 cans of cola, how many of each kind should she buy?

Reflect: If the ratio of diet cola drinkers to nondiet cola drinkers was 2 : 3 rather than 3 : 2, how would that change the answer to the practice problem? ■

Uses of Ratio

In this subsection, we explore each type of ratio—comparing like quantities or measures and comparing different quantities or measures. We also examine situations in which each type is used.

Using Ratios to Compare Like Quantities or Measures. A ratio involving two like quantities permits three types of comparisons: (a) part to part, (b) part to whole, and (c) whole to part. If we think of the union of the previously mentioned set of nurses and the set of doctors as being the set of medical staff, these three types of comparisons may be expressed in the following way.

(a)	**Part to part**	Nurses to doctors, $\frac{6}{2}$; or doctors to nurses, $\frac{2}{6}$
(b)	**Part to whole**	Nurses to staff, $\frac{6}{8}$; or doctors to staff, $\frac{2}{8}$
(c)	**Whole to part**	Staff to nurses, $\frac{8}{6}$, or staff to doctors, $\frac{8}{2}$

An opportunity to form these different types of ratios is provided in Example 7.2.

Example 7.2 Forming Different Types of Ratios

Marcie and Emmett went to a restaurant and noted that there were 36 seats for nonsmokers and 24 seats for smokers. Form the indicated ratios.

a. Part to part
b. Part to whole
c. Whole to part

■ *Solution*

Marcie's thinking:

a. The ratio of seats for nonsmokers to seats for smokers is 36 : 24, or 3 : 2.
b. The ratio of seats for smokers to total number of seats is 24 : 60, or 2 : 5.
c. The ratio of the total number of seats to seats for nonsmokers is 60 : 36, or 5 : 3.

Emmett's thinking:

a. The ratio of seats for smokers to seats for nonsmokers is 24 : 36, or 2 : 3.
b. The ratio of seats for nonsmokers to total number of seats is 36 : 60, or 3 : 5.
c. The ratio of the total number of seats to seats for smokers is 60 : 24, or 5 : 2.

■ *Your Turn*

Practice: Create a situation involving two like quantities in which (a) a part to part ratio is 2.5 : 1 and (b) a whole to part ratio is 3.5 : 1.

Reflect: In the example problem, if there had been 72 seats for nonsmokers and 48 seats for smokers, would the ratios in parts (a), (b), and (c) have changed? Explain. ■

Using Ratios to Show Rates Involving Time. When you think of a rate, one of the first things that probably comes to mind is that of rate of speed, in which a distance traveled is compared to the elapsed time for the trip, or the ratio of distance to time. You are familiar with the speedometer on a car, which indicates the rate of speed in *miles per hour*. A rate of 55 miles per hour, for example, indicates that the ratio of miles traveled to hours elapsed is 55 to 1. The rate of speed also is often given in *feet per second* or *meters per second*. A rate is a ratio comparing quantities that involve different units.

Using Ratios to Show Rates Involving Unit Price. At the grocery store, you might compare the cost of two kinds of cereal by calculating the cost for one ounce of each of the cereals. In this case, you would be using the notion of unit price, that is, the cost of a single unit of cereal. Unit pricing is based on the concept of ratio as a rate. For example, saying that six soft drinks cost $2 compares soft drinks and dollars. The ratio of 6 soft drinks to 2 dollars can be expressed as $\frac{6}{2}$. An equivalent ratio would be $\frac{3}{1}$. Because $\frac{3}{1} = \frac{1}{\frac{1}{3}}$ we can also say that each soft drink costs one-third of a dollar, or $0.33 per can. Example 7.3 involves a practical example of the use of unit pricing.

Example 7.3 **Problem Solving: Unit Pricing**

A 25-ounce box of cereal costs $4.00. What is the maximum amount that a 15-ounce box of cereal could cost and still be the better buy?

Solution

If the 25-ounce box costs $4.00, then 1 ounce costs $0.16. Thus a 15-ounce box should cost $2.40(15 × $0.16) if it is to be the same value. Hence the maximum the smaller box could cost would be $2.39 for it to be the better buy.

Your Turn

Practice: If a 25-ounce box of cereal costs $3.75, what is the maximum amount that a 12-ounce box of cereal could cost for the smaller box to be the better buy?

Reflect: The example problem was solved by first finding that 1 ounce costs $0.16. Solve the problem by first finding how many ounces can you get for $1.00. Do you get the same answer? ■

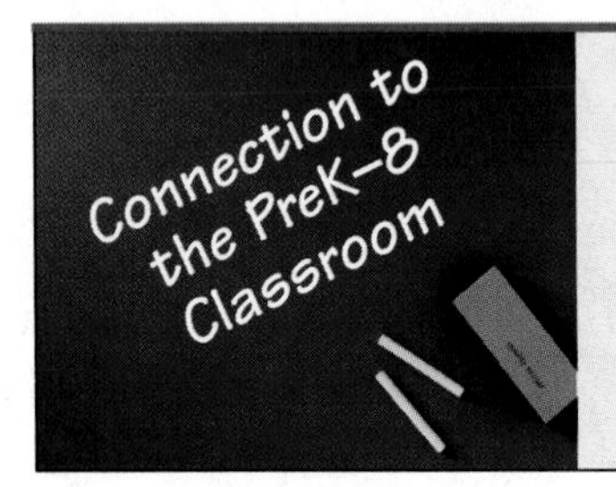

The use of the calculator to solve unit pricing problems allows students to solve more realistic problems without having to worry whether the unit price "comes out even" or whether the numbers are "too big." Students in the middle grades can carry out a project in which they try to buy the most groceries for the least cost. To do so, they can take a calculator to a store and compare the costs of items by finding unit prices.

Using Ratios to Show Comparisons of Other Unlike Quantities or Measures. Other types of rates include the comparison of different units. Different units of length are represented by scale drawings where 1 inch might represent 50 miles. Other forms of rates include the cost of building a house, expressed as dollars per square foot, and the cost of electricity, expressed as cents per kilowatt hour of use. The word *per* is often used in expressing ratios as rates, as when we express automobile fuel use in miles per gallon.

Ratios in Decimal Form

A decimal rather than a ratio is used to describe some real-world comparisons. For example, the cost factor for a car dealer for a new car might be .85. This can be expressed as the ratio .85 : 1, 85 : 100, or 8500 : 10,000. The last ratio suggests that a car that lists for $10,000 costs the dealer $8,500.

Example 7.4 demonstrates the use of a decimal to express a ratio.

Example 7.4 **Problem Solving: Batting Averages**

Pat claims that a softball player who is batting .111 can raise her batting average by more than 50 points (thousandths), from .111 to more than .161, if she gets a hit in her next time at bat. Ellen claims that one hit cannot change a batting average by 50 points. Who is correct and why?

Solution

Pat is correct. Suppose that the player had batted only 9 times and had 1 hit. Her batting average would be .111 because the ratio of hits to at bats would be 1 : 9. If she got a hit the next time at bat, the ratio would change to 2 : 10, giving her a new batting average of .200—an increase of more than 50 points.

Your Turn

Practice: Suppose that Teresa's free throw shooting average is .700. If she attempts 15 free throws in the next game, how many free throws must she make to increase her free throw shooting average?

Reflect: What might be another way to solve the practice problem? ■

Problems and Exercises | *for Section 7.1*

A. Reinforcing Concepts and Skills

1. The ratio of boys to girls in a class is 4 : 3. What is the ratio of girls to boys? Students to boys? Students to girls?

Express the ratios in Exercises 2–9 in simplest form.

2. 20 cm to 4 cm
3. $5 to 25 cents
4. 2 feet to 6 feet
5. 2 feet to 4 yards
6. 4 kg to 6 kg
7. 4 kg to 0.6 kg
8. 4 in. to 10 in.
9. 8 gal to 30 gal
10. Katrina makes 8 free throws every 10 attempts. At this rate, how many free throws can she expect to make in 15 attempts? 20 attempts? 75 attempts?
11. Who has the higher batting average: Ted, who has 12 hits in 40 at bats, or Pete, who has 16 hits in 50 at bats? Why?
12. The ratio of the circumference, c, of a circle to its diameter, d, is about 3 : 1. (Approximately equal to π). Use this information to complete the spreadsheet.

	A	B
1	*c*	*d*
2	6	?
3	?	3
4	10	?
5	?	9
6	21	?
7	?	4

13. Copy the following table and complete it so that the ratios $\frac{x}{y}$ will all be equivalent to $\frac{3}{8}$. (Some values may not be whole numbers.)

x	6	?	?	12	4.5	?
y	?	24	4	?	?	20

14. Copy the following table and complete it so that the ratio $\frac{x}{y}$ will be equivalent to $\frac{2}{5}$. (Some values may not be whole numbers.)

x	1	3	?	?	4	?
y	?	?	8	10	?	25

15. Identify an integer x for which $\frac{2x}{5}$ represents a ratio greater than 2 : 1 but less than 3 : 1.

16. At one point in a season, Reggie Miller had made 180 free throws in 210 attempts. Express his rate of making free throws as a decimal number to three places.

17. At one point in a season, Tom Glavine allowed an average of 2.5 runs per 9 innings pitched. If Glavine allowed 2 runs in 6 innings the next time he pitched, would this result be better or worse than his average? Why?

18. A recipe for biscuit mix calls for 8 cups of flour, $1\frac{1}{3}$ cups of nonfat dry milk, and 5 tablespoons of baking powder, along with other ingredients. If you had only 6 cups of flour, what would be the comparable amounts of nonfat dry milk and baking soda?

19. If a sewer line drops 1 foot every 40 feet, how many feet will it drop over a distance of 100 feet?

Express each ratio in Exercises 20–23 as a fraction with a denominator of 1.

20. 30 : 80 **21.** 60 : 30 **22.** 30 : 7.5 **23.** 50 : 240

24. Identify two rate pairs that represent the same ratio as a heart rate of 72 beats per 60 seconds.

25. San Francisco won 31 games and lost 26. What is their ratio of games won to games played?

Calculate the unit price for each item in Exercises 26–29.

26. $1.80 for 1.2 liters of soda

27. $1.20 for three grapefruits

28. two melons for 86 cents

29. $1.60 per 0.5 kilogram of peanut butter.

B. Deepening Understanding

30. A certain mix of grass seed calls for five parts blue grass and three parts clover seed. Make a table that maintains this ratio for 8, 16, 24, and 40 pounds of grass seed.

31. How could the following information be represented as a ratio?

Of 20 people interviewed, none drove a foreign car.

32. Alisha finds that she uses 12 gallons of gasoline for every 250 miles that she drives. Her friend Nicole uses about 15 gallons of gasoline for every 300 miles that she drives. Who gets the better gasoline mileage and why?

33. Suppose that Barry Bonds has 148 hits in 500 at bats. How many hits does he need in his next 10 at bats to raise his batting average to .300?

34. Marge says that if Terry has 7 hits in her next 20 at bats, she will increase her batting average, which is currently .305. Vickie says that Marge is wrong. Who is right and why?

35. Product A and product B have the same unit price of $1.20. Product A costs $10, and product B costs $8. What does that tell you about the ratio of the quantities of product A and product B?

36. Six bars of soap cost $1.70. At what price would four bars of soap be a better buy?

37. Bill and Juanita were comparing prices of ready-to-drink orange juice. The 128-ounce container of Sunshine State OJ sells for $3.19, and the 96-ounce container sells for $2.59. The following is the work that two students did to determine the better buy.

Bill: I found that 128 ÷ 3.19 is about 40.125 and that 96 ÷ 2.59 is about 37.066. I'll buy the 96-ounce container because 37.066 is smaller than 40.125.

Juanita: I did the same calculations, but because 40.125 is larger than 37.066, the 128-ounce container is the better buy.

a. Is either student correct? Explain.

b. Is there another way to solve this problem? If so, how?

c. If you wanted to help Bill and Juanita to understand their calculations better, what would you do?

38. The ratio of girls to boys in Mr. Maddox's class of 30 students is 3 : 2. The ratio of girls to boys in Ms. Wood's class of 27 students is 2 : 1. Which teacher has more girls? How many more girls?

39. Express $\frac{1}{3}$: 1 as a ratio whose terms are whole numbers.

40. Express $\frac{1}{3} : \frac{1}{2}$ as a ratio whose terms are whole numbers.

≈ **41.** A company markets toothpaste in two sizes of tubes. The large size is 7.8 ounces and sells for $2.00; the small size is 6.1 ounces and sells for $1.50. Estimate the cost of each size tube per ounce and decide which is the better buy. Use your calculator to check your estimate.

C. Reasoning and Problem Solving

42. Determine the ratio of miles driven to gallons used (miles per gallon) so that the ratios remain the same. Copy the graph and complete the two partial bars.

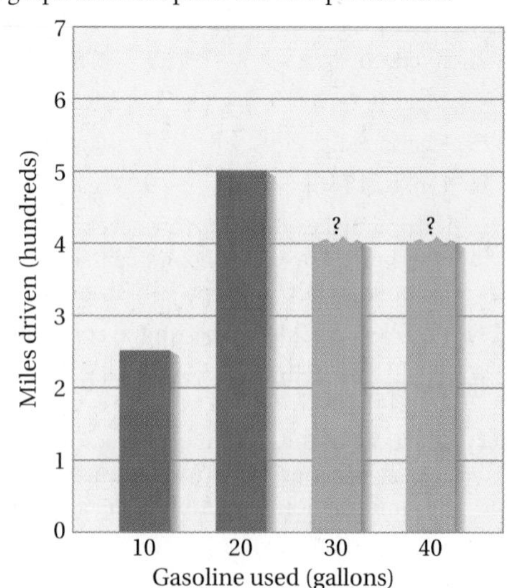

43. **Survey Problem.** Jackie conducted a survey on buying TV sets and found that for every 10 people surveyed the ratio of RCA owners to Sony owners was 3 : 5 and the ratio of Sony owners to Panasonic owners was 5 : 4. Her friend Wilma claims that these ratios aren't possible because 3 + 5 + 4 is greater than 10. Who is right and why?

44. **Swimming Problem.** At a local swimming club, Jim's best time in the 25-yard freestyle is 20 seconds. The coach decided to enter Jim in the 100-yard freestyle race at the next meet and gave his time as 80 seconds. In what sense is the coach's entry reasonable? In what sense is the coach's entry not reasonable?

D. Communicating and Connecting Ideas

45. **Making Connections.** Suppose that 50 million adults in the United States own a gun and that the average number of firearms per owner is shown in the following graph. Write a brief story (one or two paragraphs) that could appear in a local newspaper. The article should use the graph and include the concept of ratio.

Source of data: USA Today/CNN/Gallup Poll Survey. From "Gun Owners Don't Fit Stereotypes" by Cliff Vancure, USA TODAY. *USA Today,* December 30, 1993, p. 5A. Copyright 1993, *USA Today.* Used with permission.

46. Suppose that a reporter uses the graph in Exercise 45 and creates the following headline for her newspaper article: **Every Person in the United States Owns More Than Four Firearms.** Should her editor accept the headline? Explain your reasoning.

47. **Making Connections.** What ratio of sales to year is represented in the following graph? (Express the ratio in the form of sales per year.) Is the ratio the same for each year? Write a paragraph to explain this situation to another person. Use the term *ratio* in your paragraph. Write a caption for your graph that could be used as a newspaper headline.

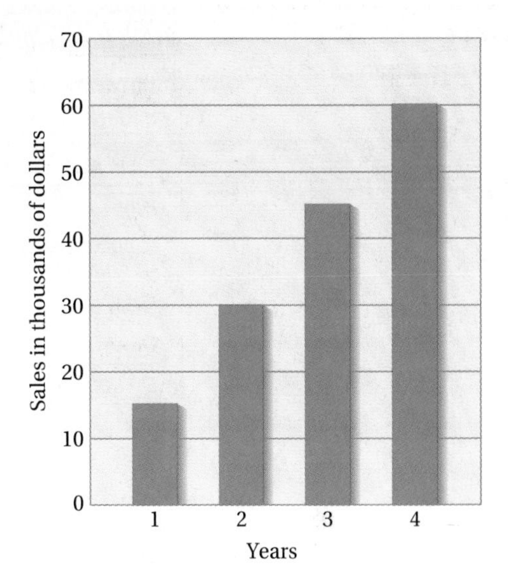

48. **Historical Pathways.** The problem of representing the ratio of two quantities when the ratio cannot be represented by two whole numbers was a significant problem for ancient mathematicians. Consider, for example, square $ABCD$, with each side 1 unit long. The length of the diagonal $\overline{AC}$ is $\sqrt{2}$. Thus the ratio $AC:AB$ is $\sqrt{2}:1$. But $\sqrt{2} = 1.414213562\ldots$. Using the numbers one through ten, what ratio of whole numbers would best approximate the ratio $\sqrt{2}:1$?

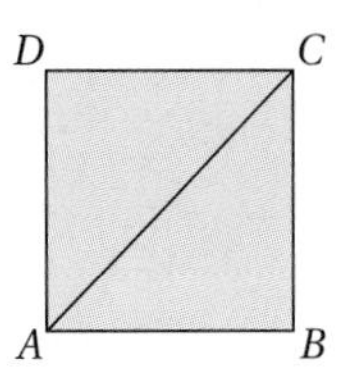

Section 7.2 PROPORTIONAL VARIATION AND SOLVING PROPORTIONS

- Recognizing Proportional Variation
- Characteristics of Quantities That Vary Proportionally
- Proportional Variation in Geometry
- Properties of Proportions
- Solving Proportional Problems

In this section, we consider a constant ratio of two quantities as the sizes of the two quantities vary. In doing so, we illustrate situations in both numerical and geometric contexts. After investigating proportional situations, we present several ways to solve proportions. Mini-Investigation 7.2 asks you to look at what happens to the ratio of two quantities as they vary.

Write about the differences between Tables A and B, using the language of ratio.

Mini-Investigation 7.2 **Using Mathematical Reasoning**

How are Tables A and B, which describe the costs of the same ad in two newspapers, alike and how are they different?

Table A

Day	1	2	3	4	5	6	7
Cumulative cost of ad ($)	2	4	6	8	10	12	14

Table B

Day	1	2	3	4	5	6	7
Cumulative cost of ad ($)	2	4	6	7	8	9	10

Recognizing Proportional Variation

To clarify what it means for two quantities to vary proportionally, we can use the tables in Mini-Investigation 7.2. In Table A, as the number of days increases, the cost also increases; the ratio of days to cost is *always the same.* That is, in Table A, $\frac{1}{2} = \frac{2}{4} = \frac{3}{6} = \frac{4}{8} = \frac{5}{10} = \frac{6}{12} = \frac{7}{14}$; note that these ratios are equivalent. In Table B, as the number of days increases, the cost also increases; however, the ratio of days to cost is *not always the same.* For example, in Table B the ratio $\frac{1}{2}$ is not equivalent to the ratio $\frac{4}{7}$. The important characteristic of the relationship in Table A leads to the following definition.

Definition of Quantities Varying Proportionally

Two quantities **vary proportionally** if and only if, as their corresponding values increase or decrease, the ratios of the two quantities are always equivalent.

Example 7.5 demonstrates how to complete tables for two quantities that vary proportionally.

Example 7.5

Completing Tables So That Quantities Vary Proportionally

Complete the table so that the two quantities vary proportionally.

x	2	3	4	5	6	7	8
y	6	?	?	?	?	?	?

Solution

The remaining *y*-row terms are chosen so that all the ratios of corresponding terms in the table are equivalent.

x	2	3	4	5	6	7	8
y	6	9	12	15	18	21	24

Your Turn

Practice: Create a table in which two quantities vary proportionally.

Reflect: What must be true if the corresponding values in the table vary proportionally? ■

Characteristics of Quantities That Vary Proportionally

A Multiplicative Relationship. As we have shown, when two quantities vary proportionally, the ratios of their corresponding values are always equivalent. A closely related property of quantities that vary proportionally is illustrated by the following table showing the relationship between the number of people and the total ticket cost. These quantities vary proportionally, with total ticket cost always three times the corresponding number of people. That is, a multiplicative relationship exists between the two quantities.

Number of people	1	2	3	4	5	6	7
Total ticket cost ($)	3	6	9	12	15	18	21

This and the preceding examples clearly show that, when two quantities vary proportionally, the value of one quantity is always the same multiple of the corresponding value of the other quantity. We state this relationship as follows.

Multiplicative Property of Quantities That Vary Proportionally

When quantities a and b vary proportionally, a nonzero number k exists such that $\frac{a}{b} = k$, or $a = b \cdot k$, for all corresponding values a and b.

This type of proportional variation is known as **direct proportionality.**

Example 7.6 further illustrates these ideas.

Example 7.6

Problem Solving: The Downpour

At Tawana's house a torrential downpour lasted for 7 hours. After 3 hours, her rain gauge indicated that 1.2 inches of rain had fallen. If the rain fell at a constant rate, how many inches of rain would have fallen by the time the downpour stopped?

Working toward a Solution

Understand the problem	*What does the situation involve?*	Accumulated rain in a storm
	What has to be determined?	The amount of rain after 7 hours
	What are the key data and conditions?	After 3 hours, 1.2 inches of rain had fallen.
	What are some assumptions?	The rate of rain remains constant.
Develop a plan	*What strategies might be useful?*	Make a table or a graph.

	Are there any subproblems?	Determine the amount of rain accumulated at the end of each hour.
	Should the answer be estimated or calculated?	The answer should be calculated.
	What method of calculation should be used?	Paper and pencil or mental computation
Implement the plan	*How should the strategies be used?*	Create a table such as the following.

Time (hr)	0	1	2	3	4	5	6	7
Rain (in.)	0	?	?	1.2	?	?	?	?

	What is the answer?	The rate of rain is a constant, so 1.2 inches of rain collects in 3 hours. Therefore the rate is 0.4 inch per hour. Thus 7×0.4, or 2.8, inches of rain fell during the 7-hour period. Check: Does $\frac{3}{1.2} = \frac{7}{2.8}$? Yes.
Look back	*Is the interpretation correct?*	Check: The table represents the problem correctly.
	Is the calculation correct?	Yes. All calculations give the same result when done a second time.
	Is the answer reasonable?	If less than 0.5 inch of rain fell per hour, it seems reasonable that less than 3.5 inches of rain would fall in 7 hours. The answer, 2.8 inches of rain, seems reasonable.
	Is there another way to solve the problem?	Multiply 1.2 inches of rain by $\frac{7}{3}$. This gives 2.8 inches of rain. You could also draw a graph and use it to estimate the amount of rain.

Your Turn

Practice: Solve the following problem.

If 2.8 inches of rain had fallen in 10 hours, how much would have accumulated at the end of 3 hours?

Reflect: What is there about the example problem that allows you to treat it as a proportional situation? In what way was the solution based on proportions? ■

A Constant Change Relationship. Another useful relationship can be observed when quantities vary proportionally. A unit change in any value of one quantity produces a constant change in the corresponding value of the other quantity. Consider the following table.

Hours walked	1	2	3	4
Calories burned	350	700	1050	1400

First, the quantities in the table vary proportionally; the ratios of the corresponding values are always the same. For example, $\frac{1}{350} = \frac{4}{1400}$. When the number of hours walked changes by 1, the corresponding change in the number of calories burned is constant, increasing by 350 each time.

This relationship suggests the following.

Constant Change Property of Quantities That Vary Proportionally

When quantities *a* and *b* vary proportionally and the ratio of *a* to *b* is 1 to *n*, a unit change in a value of *a* always evokes a constant change of *n* in the corresponding value of *b*.

The Idea of a Proportion. When two quantities vary proportionally, the ratios of the corresponding values of these quantities are equivalent. When solving problems, we have shown that choosing two of these equivalent ratios and writing an equation is often helpful. For example, suppose that the number of gallons of gasoline used varies proportionally with the number of miles traveled, as shown in the following table.

Number of gallons	1	2	3	4	5	6	7
Number of miles	20	40	60	80	100	120	140

The ratio of gallons to miles is always the same, so we can write the equation $\frac{1}{20} = \frac{5}{100}$ to verify that we can go 100 miles on 5 gallons of gasoline. We can also write and solve the equation $\frac{1}{20} = \frac{g}{500}$ to figure out how many gallons of gasoline would be needed to drive 500 miles. The usefulness of equations in indicating that two ratios are equal leads to the following definition.

Definition of a Proportion

A **proportion** is an equation stating that two ratios are equivalent.

Proportional Variation in Geometry

Although situations in which two quantities vary proportionally necessarily involve equivalent ratios and hence are numerical, many geometric situations give rise to proportional variation.

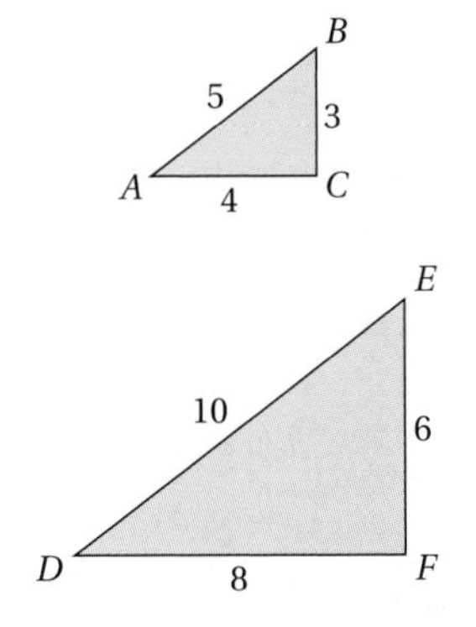

FIGURE 7.1 | Similar triangles.

Similar Figures and Proportional Variation. Recall that geometric figures that have the same shape are called **similar figures.** Measuring the lengths of the corresponding sides of similar figures gives lengths that vary proportionally, as shown in Figure 7.1. From them we can write the following proportions:

$$\frac{3}{6} = \frac{5}{10}, \quad \frac{4}{8} = \frac{3}{6}, \quad \text{and} \quad \frac{5}{10} = \frac{4}{8}.$$

The idea that the sides of similar figures vary proportionally can be used to solve many types of problems. One of its earliest mathematical applications was in determining the height of objects. For example, if a 6-foot-tall man casts a shadow 8 feet in length, how tall is a flagpole that casts a 20-foot shadow? Using the similar triangles formed and *h* to represent the height of the flagpole, we can represent the situation as a proportion involving ratios of height to shadow as follows: $\frac{h}{20} = \frac{6}{8}$. Solving the proportion leads to an answer of 15 feet for the height of the flagpole. Similarity

is also involved in using a blueprint. However, the Frank and Ernest cartoon suggests that even clear reasoning involving proportional variation isn't enough to produce the best finished product when the scale ratio for the blueprint isn't correct.

Some geometric figures are always similar; for example, all squares are similar and all equilateral triangles are similar. That is, the lengths of the corresponding sides of any two squares or equilateral triangles always vary proportionally. More generally, any two *regular* polygons of the same type have side lengths that vary proportionally. But as a rectangle is not a regular polygon, this property doesn't hold for any two rectangles. In Figure 7.2, rectangles A and B have the same shape; their sides are proportional; that is $\frac{2}{4} = \frac{3}{6}$. However, rectangle C doesn't have the same shape as either A or B. Note that the ratios of the corresponding sides ($\frac{4}{3}$ and $\frac{6}{5}$) are not equivalent and that the side lengths don't vary proportionally.

One application of this idea is being able to enlarge a 2-inch by 3-inch photo into a 4-inch by 6-inch photo without having to crop the photo. But a 3-inch by 5-inch photo couldn't be enlarged to fit a 4-inch by 6-inch frame without cropping the edges of the photo. Mini-Investigation 7.3 asks you to analyze further the idea of photo enlargement.

Write a convincing argument that justifies your conclusion.

Mini-Investigation 7.3 Solving a Problem

You have a photograph 4 inches by 6 inches. You want to enlarge it to fit into a frame that measures 8 inches by 10 inches. Can you do so without losing part of the picture?

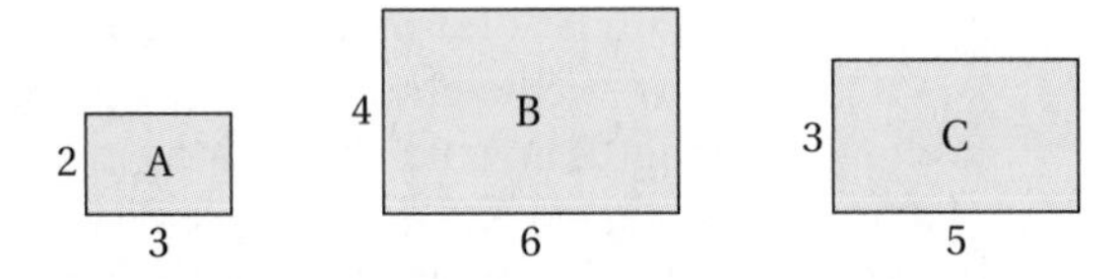

FIGURE 7.2 | Similar and nonsimilar rectangles.

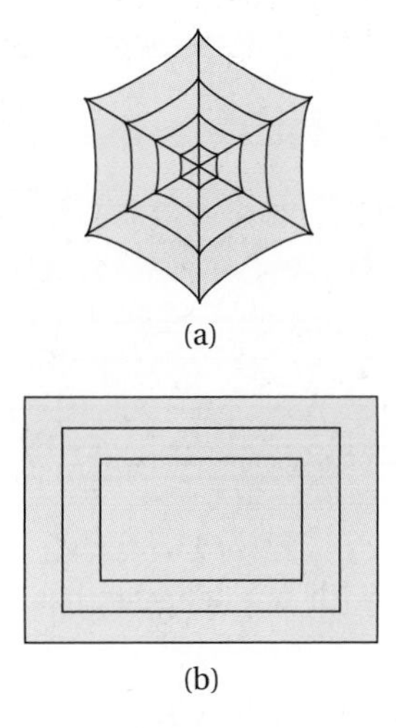

FIGURE 7.3 | Similar geometric figures

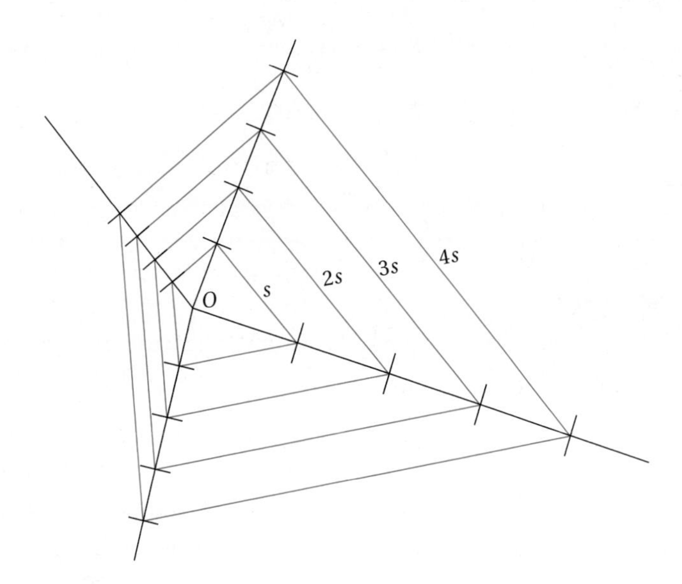

FIGURE 7.4 | Proportional line lengths.

Figure 7.3 also illustrates similar geometric figures that have the same shape but are different sizes. The side lengths of the figures in part (a), which could represent a spider's web, and the side lengths of the figures in part (b), which could represent stacked mats, vary proportionally.

Figure 7.4 depicts figures with the same shape that are formed by using the center O, of the figure and doubling, tripling, and quadrupling the distances from the center to the first marks. The lengths of the corresponding sides formed are similarly doubled, tripled, and quadrupled. A comparison of the lengths of the sides shows that they vary proportionally.

Example 7.7 demonstrates the use of this idea to create larger figures that are similar to a given figure.

Example 7.7 Creating Similar Figures

Copy the following figure and create two triangles, one whose sides are half as long as those of triangle ABC and one whose sides are twice as long as those of triangle ABC.

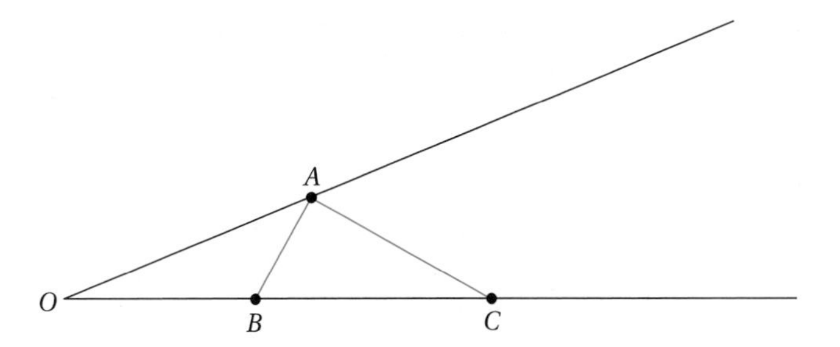

Solution

Find points D, E, and F so that $OD = \frac{1}{2}OA$, $OE = \frac{1}{2}OB$, and $OF = \frac{1}{2}OC$. Draw triangle DEF. Then find points G, H, and I so that $OG = 2OA$, $OH = 2OB$, and $OI = 2OC$. Draw triangle GHI.

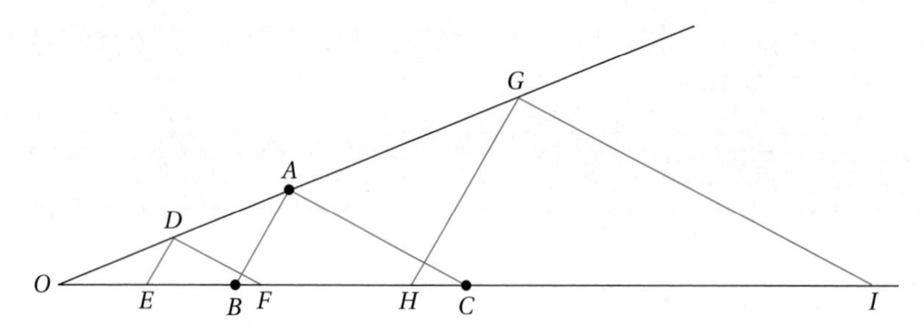

As a check, determine whether $DF = \frac{1}{2}AC$ and $GI = 2AC$.

Your Turn

Practice: Draw a triangle and create another triangle with side lengths that are three times as long as those of the original triangle.

Reflect: Are the corresponding side lengths of triangles *DEF* and *GHI* in the example problem proportional? Why or why not? ■

Properties of Proportions

Recall that a proportion is an equation whose members are equivalent ratios; for example, $\frac{2}{3} = \frac{4}{6}$. In this subsection, we develop four important properties that will be helpful in solving proportions.

If you look at a variety of different proportions, such as $\frac{3}{4} = \frac{6}{8}$, $\frac{1}{10} = \frac{10}{100}$, $\frac{3}{2} = \frac{6}{4}$, you can see that the two products found by multiplying a numerator of one fraction by the denominator of the other fraction are always equal. This result suggests the first property of proportions.

Cross-Product Property of Proportions

For integers a, b, c, and d ($b \cdot d \neq 0$), $\frac{a}{b} = \frac{c}{d}$ if and only if $ad = bc$.

The following statements illustrate the cross-product property of proportions.

If $\frac{3}{7} = \frac{12}{28}$, then $3 \cdot 28 = 7 \cdot 12$.
Conversely, if $3 \cdot 28 = 7 \cdot 12$, then $\frac{3}{7} = \frac{12}{28}$.

We can verify part of this property as follows. Suppose that $\frac{a}{b} = \frac{c}{d}$, $bd \neq 0$. Then

Statements	Reasons
$\frac{a}{b}(bd) = \frac{c}{d}(bd)$	Property of equations; multiply each side by bd.
$ad\left(\frac{b}{b}\right) = cb\left(\frac{d}{d}\right)$	Commutative and associative properties of multiplication, $r\left(\frac{s}{t}\right) = \frac{rs}{t}$
$ad = cb.$	Properties of fractions; identity property for multiplication

Conversely, we can verify the other part of this property as follows. Suppose that $ad = bc$, $b \cdot d \neq 0$. Then

Statements	Reasons
$\frac{ad}{bd} = \frac{bc}{bd}$	Property of equations; divide each side by bd
$\frac{a}{b} = \frac{c}{d}.$	Fundamental law of fractions

We now consider a second property of proportions. Suppose that the ratio of teaspoons of plant food to cups of water could be written as either $\frac{2}{5}$ or $\frac{4}{10}$, giving the proportion $\frac{2}{5} = \frac{4}{10}$. Alternatively, the ratio of cups of water to teaspoons of plant food could be written as either $\frac{5}{2}$ or $\frac{10}{4}$, giving the proportion $\frac{5}{2} = \frac{10}{4}$. This relationship suggests a second useful property of proportions.

Reciprocal Property of Proportions

For nonzero integers a, b, c, and d, $\frac{a}{b} = \frac{c}{d}$ if and only if $\frac{b}{a} = \frac{d}{c}$.

The following statements illustrate the reciprocal property of proportions.

If $\frac{2}{3} = \frac{4}{6}$, then $\frac{3}{2} = \frac{6}{4}$.
Conversely, if $\frac{3}{2} = \frac{6}{4}$, then $\frac{2}{3} = \frac{4}{6}$.

We can establish the reciprocal property by using the cross-product property as follows. Suppose that $\frac{a}{b} = \frac{c}{d}$ for nonzero integers a, b, c, and d. Then

Statements	Reasons
$ad = bc$	Cross-product property
$bc = ad$	Property of equality
$\frac{b}{a} = \frac{d}{c}$.	Cross-product property

A similar reasoning process can establish that, if $\frac{b}{a} = \frac{d}{c}$, then $\frac{a}{b} = \frac{c}{d}$.

Solving Proportional Problems

Proportions, like most equations, can be solved in a variety of ways. In this section, we consider four different methods for solving proportions. The methods are related but involve different procedures for finding the answer. Which method is the most efficient depends on the type of proportion being solved. Before we explore specific methods, Mini-Investigation 7.4 asks you to consider methods that you might normally use to solve a proportion using a calculator.

Talk about how your method compares with a method used by a classmate and what the two methods have in common.

Mini-Investigation 7.4 The Technology Option

How can you use a calculator or computer to solve for a missing value in a proportion such as $\frac{x}{528} = \frac{78}{117}$?

Using Properties of Equations to Solve Proportions. A standard way to solve proportions involves use of the property of equations from algebra that allows multiplying each side of an equation by the same number without changing the equality. It also uses the rational number multiplication generalization that $a(\frac{x}{a}) = x$. For example, suppose that a battery pack weighs 12 pounds on earth and 2 pounds on the moon. What would a generator that weighs 48 pounds on earth weigh on the moon? We could write the proportion $\frac{x}{48} = \frac{2}{12}$ and solve it to find the answer.

Statements	Reasons
$48(\frac{x}{48}) = 48(\frac{2}{12})$	Property of equations; multiply each side by 48
$x = 8.$	Rational number multiplication, $a(\frac{x}{a}) = x$

The generator weighs 8 pounds on the moon.

Example 7.8 applies these techniques for solving proportions.

Example 7.8 Using Proportions to Solve Problems

Rob has figured out that, for every \$100 he earns, he must hold back \$25 to pay his taxes. If he earns \$2660 one summer, how much does he need to hold back for taxes?

Solution

Let $x =$ the number of dollars Rob must save for taxes. We set up the following proportion to solve the problem:

$$\frac{x}{2660} = \frac{25}{100}.$$

We then multiply each side of the equation by 2660:

$$2660\left(\frac{x}{2660}\right) = 2660\left(\frac{25}{100}\right)$$
$$x = 665.$$

Rob must save \$665 for taxes.

Your Turn

Practice: Suppose that Rob's tax rate was \$30 for every \$100 earned and that he had earned \$4880 that summer. How much would he owe in taxes?

Reflect: Identify two other proportions that could be set up to solve the example problem correctly. ■

Using Cross Products to Solve Proportions. Another commonly used method for solving proportions is based on the cross-product property. It involves converting the proportion to an equation having a more familiar form. For example, on a camping trip, 3 pounds of potatoes are needed for every 2 campers. The leader needs to buy potatoes for 24 campers. To solve this problem, we write and solve the proportion

$$\frac{x}{24} = \frac{3}{2}$$

$2x = 72.$ Cross-product property

$x = 36.$ Property of equations; divide each side by 2.

For 24 campers, 36 pounds of potatoes are needed.

In Example 7.9 we consider another application of the cross-product method.

Example 7.9 Applying the Cross-Product Method

A TV network claims that for every 100 minutes of television, there are 23 minutes of commercials. If this same rate holds for all television shows, how many minutes of commercials would you expect for a 30-minute show?

Solution

Let $x =$ the number of minutes of commercials for a 30-minute show. If the same rate holds, we can set up the proportion

$$\begin{aligned} \frac{23}{100} &= \frac{x}{30} \\ 23 \cdot 30 &= 100x \quad \text{(Cross-product property)} \\ 100x &= 690 \\ \frac{100x}{100} &= \frac{690}{100} \\ x &= 6.9. \end{aligned}$$

A 30-minute show would have 7 minutes (rounded) of commercials.

Your Turn

Practice: Solve $\frac{4}{9} = \frac{14}{x}$.

Reflect: How can you use estimation to convince someone that the solution you obtained when solving $\frac{4}{9} = \frac{14}{x}$ is a reasonable answer? ■

Equating Numerators or Denominators to Solve Proportions. Solving the proportion $\frac{x}{7} = \frac{4}{7}$ is rather easy, as the answer, $x = 4$, can be determined by inspection. The equate numerators or denominators method for solving proportions involves changing the original proportion into one like $\frac{x}{7} = \frac{4}{7}$, in which either the numerators or the denominators are equal. The solution can then be found by equating the remaining denominators or numerators. Consider the following proportions.

$$\text{i. } \frac{9}{x} = \frac{27}{13}. \quad \text{ii. } \frac{5}{16} = \frac{x}{8}. \quad \text{iii. } \frac{x}{8} = \frac{5}{14}.$$

Proportion (i) can be solved by equating the numerators, as shown.

Statements	Reasons
$\frac{9}{x} = \frac{27}{13}$	
$\frac{3 \cdot 9}{3 \cdot x} = \frac{27}{13}$	Fundamental law of fractions
$\frac{27}{3x} = \frac{27}{13}$	Basic fact
$3x = 13$	Equating the denominators
$x = \frac{13}{3} = 4\frac{1}{3}$	Property of equations; divide each side by 3

Proportion (ii) can be solved by equating the denominators, as shown.

Statements	Reasons
$\frac{5}{16} = \frac{x}{8}$	
$\frac{5}{16} = \frac{2x}{2 \cdot 8}$	Fundamental law of fractions
$\frac{5}{16} = \frac{2x}{16}$	Basic fact
$5 = 2x$	Equating the numerators
$x = \frac{5}{2} = 2\frac{1}{2}$	Property of equations; divide each side by 2

Solving proportion (iii) by equating numerators or denominators is not convenient. Of course, it can be solved by multiplying the first ratio, $\frac{x}{8}$, by $\frac{14}{14}$ and multiplying the second ratio, $\frac{5}{14}$, by $\frac{8}{8}$ to produce the same denominator, $8 \cdot 14$. Solving proportion (iii) by the cross-product method probably would be easier. Example 7.10 illustrates that equating numerators or denominators to solve a proportion can sometimes be carried out with mental computation.

Example 7.10 Applying the Method of Equating Numerators or Denominators

Solve: If you can buy six oranges for \$2, how many can you buy for \$10?

Solution

Let x = the number of oranges you can buy for \$10. We set up the proportion

$$\frac{6}{2} = \frac{x}{10}.$$

Using the fundamental law of fractions mentally, we obtain

$$\frac{5 \cdot 6}{5 \cdot 2} = \frac{x}{10}, \quad \text{or} \quad \frac{30}{10} = \frac{x}{10}.$$

Thus $x = 30$ oranges.

Your Turn

Practice: A package of three grapefruits sells for \$1.80. How many grapefruits could you buy for \$9?

Reflect: Suppose that, in solving the example problem, the proportion $\frac{x}{6} = \frac{10}{2}$ had been set up. How would you equate the numerators or denominators in this proportion? Would you get the same answer? Why or why not? ■

Converting One Ratio to a Decimal to Solve Proportions. Another method for solving proportions, particularly useful when a calculator is available, involves converting one of the ratios to a decimal. For example, suppose that men, on average, have 2 pounds of muscle for every 5 pounds of body weight. How many pounds of the weight of a 150-pound man is muscle? We write the proportion $\frac{x}{150} = \frac{2}{5}$ and solve it:

$$\begin{aligned} \frac{x}{150} &= \frac{2}{5} \\ \frac{x}{150} &= 0.4 && \text{Convert } \tfrac{2}{5} \text{ to a decimal} \\ x &= 150(0.4) && \text{Property of equations;} \\ &= 60. && \text{multiply each side by 150} \end{aligned}$$

Therefore 60 pounds of a 150-pound man is muscle.

When you use this method with a calculator, the ratio sometimes converts to a terminating decimal. For example, if you were solving the proportion $\frac{x}{150} = \frac{5}{6}$, the calculator will round the fraction $\frac{5}{6}$ to the decimal .833333333. As a result, the final answer may be only an approximation. Mini-Investigation 7.5 asks you to identify conditions under which that will happen. Use a four-function calculator, a fraction calculator, or a graphing calculator in your investigation.

Write about how you might overcome situations in which an exact solution does not result.

Mini-Investigation 7.5 **The Technology Option**

When will the value of x found in solving the proportion $\frac{x}{150} = \frac{5}{6}$ be exact and when will it be an approximation?

As you may have discovered in Mini-Investigation 7.5, in some cases you can avoid inexact solutions by applying the reciprocal property before solving the proportion. For example, we could solve the proportion $\frac{14}{x} = \frac{5}{6}$ as follows:

$$\frac{14}{x} = \frac{5}{6}$$

$$\frac{x}{14} = \frac{6}{5} \quad \text{Reciprocal property}$$

$$\frac{x}{14} = 1.2 \quad \text{Convert } \tfrac{6}{5} \text{ to a decimal}$$

$$x = 14(1.2) \quad \text{Property of equations;}$$
$$= 16.8 \quad \text{multiply each side by 14.}$$

In other cases, you may want to choose a method that avoids approximation and produces an exact solution. Example 7.11 shows how to apply this method.

Example 7.11 Using Decimals to Solve Proportions

Solve the following proportions.

a. $\frac{2x}{5} = \frac{7}{8}$. **b.** $\frac{8}{3} = \frac{5}{x}$.

Solution

a.
$$\frac{2x}{5} = \frac{7}{8}$$
$$\frac{2x}{5} = 0.875$$
$$2x = 4.375$$
$$x = 2.1875.$$

b.
$$\frac{8}{3} = \frac{5}{x}$$
$$\frac{3}{8} = \frac{x}{5} \quad \text{Reciprocal property}$$
$$0.375 = \frac{x}{5}$$
$$x = 1.875.$$

Your Turn

Practice: Solve the following proportion by converting one ratio to a decimal.

$$\frac{2}{3} = \frac{7}{2x}.$$

Reflect: Explain how the reciprocal property can help you solve the practice problem. ■

Many real-world problems can be solved by setting up and solving a proportion. Example 7.12 demonstrates how to do so.

Example 7.12

Problem Solving: Gasoline Mileage

Josh plans to drive 400 miles before stopping for gasoline. He checks his gauge and estimates that he has used 5 gallons of gasoline from his 14-gallon tank. He plans to use only 12 gallons before stopping for gasoline, as he doesn't want to risk running out with no filling station nearby. His odometer indicates that he has traveled 164 miles of the 400 miles he plans to travel. At this rate, will he be able to travel 400 miles before stopping for gasoline?

Working toward a Solution

Understand the problem	*What does the situation involve?*	Finding the distance that can be traveled on a certain number of gallons of gasoline.
	What has to be determined?	Whether Josh can travel 400 miles on 12 gallons of gasoline.
	What are the key data and conditions?	Josh has traveled 164 miles on 5 gallons of gasoline. He plans to use 12 gallons of gasoline. He needs to travel 400 miles.
	What are some assumptions?	A constant rate of gasoline consumption
Develop a plan	*What strategies might be useful?*	Write a proportion. Use proportional reasoning.
	Are there any subproblems?	No.
	Should the answer be estimated or calculated?	Calculated
	What method of calculation should be used?	Set up a proportion to determine how far Josh can travel on 12 gallons of gasoline. Solve the proportion. Compare this result with 400 miles.
Implement the plan	*How should the strategies be used?*	Josh can drive 164 miles on 5 gallons of gasoline. We want to know how far he can travel on 12 gallons, so we set up the proportion: $\frac{164}{5} = \frac{x}{12}$. Converting $\frac{164}{5}$ to a decimal gives $32.8 = \frac{x}{12}$; or $x = 393.6$ miles.
	What is the answer?	Josh can expect to travel 393.6 miles on 12 gallons of gasoline. As 393.6 is less than 400, he will probably not be able to travel 400 miles on 12

		gallons of gasoline.
Look back	*Is the interpretation correct?*	We have the ratio of miles per gallon in each case (164 miles per 5 gallons and *x* miles per 12 gallons). The interpretation is correct.
	Is the calculation correct?	164 ÷ 5 equals 393.6 ÷ 12. The calculation is correct.
	Is the answer reasonable?	If Josh can travel 164 miles on 5 gallons, he can travel 328 miles on 10 gallons. As 393.6 miles is somewhat more than 328 miles, the answer seems reasonable.
	Is there another way to solve the problem?	We can also solve the problem in the following way. If Josh travels 164 miles on 5 gallons, he should be able to travel 328 miles on 10 gallons. Thus Josh's gas mileage is 32.8 miles per gallon. Two additional gallons would give him another 65.6 miles. The sum of 328 and 65.6 is less than 400. He will have to stop for gasoline before traveling 400 miles.

Your Turn

Practice: What is the minimum number of miles that Josh would have had to travel on 5 gallons of gasoline to be assured of traveling 400 miles on 12 gallons?

Reflect: If you were Josh, at what point on your gasoline gauge would you read the odometer to make the problem easier to solve? Why? ■

Problems and Exercises | *for Section 7.2*

A. Reinforcing Concepts and Practicing Skills

1. Complete the following table so that the two quantities vary proportionally.

x	1	1.5	2	2.5	3	3.5
y	3	?	?	?	?	?

In Exercises 2–5 identify the property of proportions that allows each conclusion to be drawn.

2. If $\frac{2}{5} = \frac{4}{10}$, then $2 \cdot 10 = 4 \cdot 5$.

3. If $\frac{2}{5} = \frac{4}{10}$, then $\frac{10}{4} = \frac{5}{2}$.

4. If $\frac{r}{m} = \frac{t}{s}$, then $\frac{m}{r} = \frac{s}{t}$.

5. If $\frac{c}{b} = \frac{m}{x}$, then $b \cdot m = c \cdot x$.

In Exercises 6–8, estimate the answer then solve the proportion. How close was your estimate?

6. $\frac{x}{20} = \frac{49}{100}$. **7.** $\frac{9}{40} = \frac{x}{60}$.

8. $\frac{11}{30} = \frac{x}{48}$.

9. Solve each proportion two different ways and state which method you think is easier.

a. $\frac{4}{15} = \frac{x}{5}$. **b.** $\frac{x}{7} = \frac{4}{25}$.

c. $\frac{9}{7} = \frac{27}{2x + 5}$.

In Exercises 10–29, solve the proportions by using whichever method you think is the easiest.

10. $\frac{2}{3} = \frac{x}{81}$. **11.** $\frac{27}{x} = \frac{3}{7}$.

12. $\frac{x}{5} = \frac{8}{15}$. **13.** $\frac{3}{x} = \frac{15}{16}$.

14. $\frac{3}{8} = \frac{x}{20}$. **15.** $\frac{4}{9} = \frac{x}{21}$.

16. $\frac{x}{36} = \frac{5}{6}$. **17.** $\frac{3}{8} = \frac{x}{30}$.

18. $\frac{2}{3} = \frac{6}{x - 1}$. **19.** $\frac{2}{7} = \frac{x + 1}{14}$.

20. $\frac{7}{20} = \frac{x + 2}{10}$. **21.** $\frac{3x}{10} = \frac{13}{40}$

22. $\frac{8}{x} = \frac{12}{9}$. **23.** $\frac{x}{5} = \frac{224}{720}$.

24. $\frac{6}{x} = \frac{9}{x + 18}$. **25.** $\frac{0.5}{8.5} = \frac{10}{x}$.

26. $\frac{344}{144} = \frac{2x}{3}$.

27. $\frac{12}{5x} = \frac{3}{4}$.

28. $\frac{3}{8} = \frac{x - 4}{4}$.

29. $\frac{20 + x}{x} = \frac{3}{7}$.

30. Explain why the lengths of the corresponding sides of rectangles *ABCD* and *EFGH* vary proportionally.

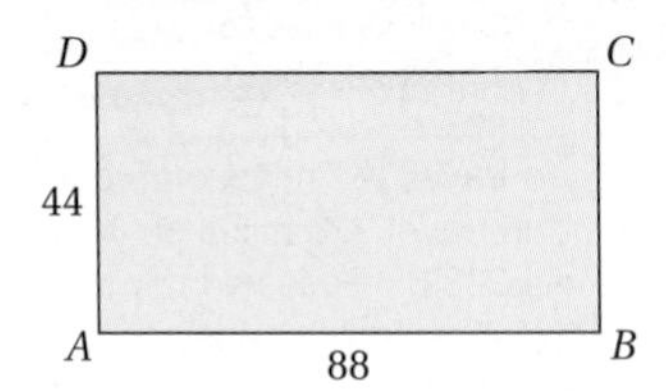

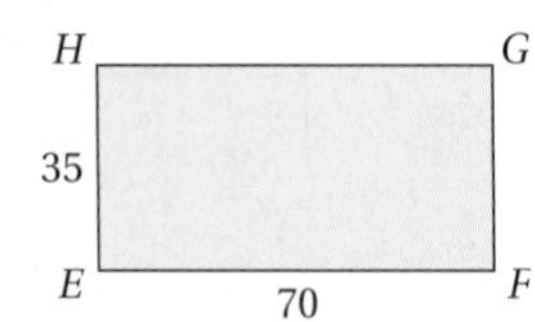

31. Explain why the lengths of the corresponding sides of rectangles *ABCD* and *EFGH* do not vary proportionally.

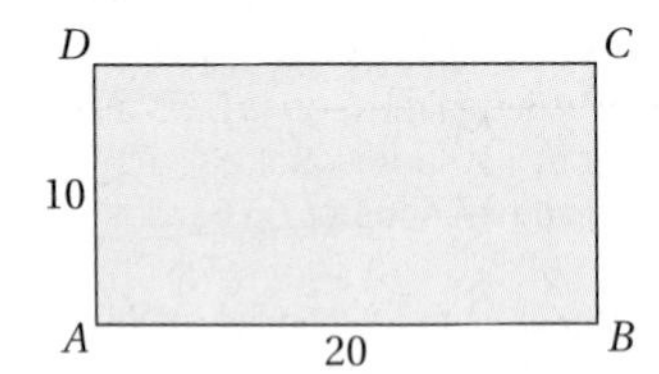

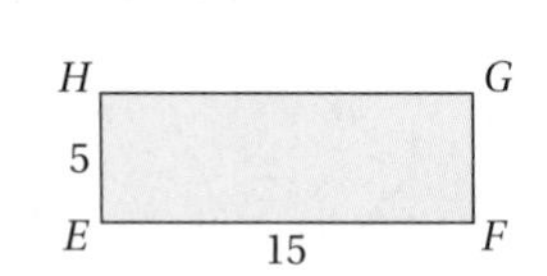

32. Draw a rectangle like *ABCD* and then draw two other rectangles, one larger and one smaller, whose side lengths vary proportionally to the corresponding side lengths of *ABCD*. Your rectangles should have sides on rays $\overrightarrow{AB}$ and $\overrightarrow{AD}$.

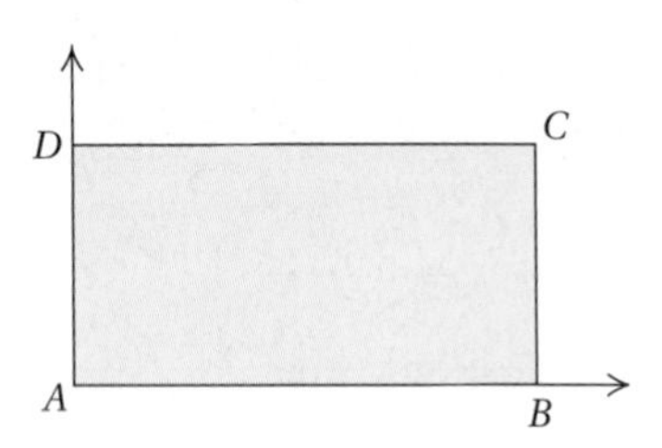

33. Create a table of values so that the two quantities vary proportionally but the second quantity is always negative. Determine the value of k so that $y = kx$ represents the relationship between the first quantity, x, and the second quantity, y.

34. Do the x- and y-values in each of the following tables represent two quantities that vary proportionally? Explain your reasoning.

a.

x	1	2	3	4	5	6
y	0	−1	−2	−3	−4	−5

b.

x	1	2	3	4	5	6
y	−1	−2	−3	−4	−5	−6

c.

x	100	90	80	70	60	50
y	0	10	20	30	40	50

35. The slope of a roof rises 7 vertical feet for every 12 horizontal feet. How many vertical feet will the roof rise if it runs 20 horizontal feet?

36. A parking lot can allocate spaces for 3 compact cars for every 2 spaces for full-sized cars. If part of the lot has space for 40 full-sized cars, how many compact cars could it hold if the parking lines were redrawn?

37. A U.S. nickel contains 3 ounces of copper for every ounce of nickel. How much nickel is required for 14 pounds of copper?

38. If 6 bars of soap cost $2.35, how much would 10 bars of soap cost?

39. Ricky went to the grocery store and found that it was having a sale on soft drinks: 6 cans for $1.80. How much would 24 cans cost? Which method of solving proportions did you use?

40. If a person spends $32 to drive 1000 miles, how much would she spend at the same rate to drive 15,000 miles?

41. A survey found that for every 30 minutes of television there were 9 minutes of commercials. If this is correct, how many minutes of commercials would you expect during 5 hours of television?

42. A tennis magazine averages about 150 pages per issue. There are seven ads for every 3 pages. How many ads would you expect in a typical issue?

B. Deepening Understanding

43. Suppose that the population of a type of insect doubles every day. Is the population growth varying proportionally with the number of days? Explain your reasoning.

44. If you can buy three raffle tickets for $5, how many tickets can you buy for $55? Discuss two different ways to solve this problem.

45. A student said that the table given in Exercise 34(b) cannot represent two quantities that vary proportionally because the y-value is decreasing. Do you agree or disagree? Explain your reasoning.

46. On a vacation to Germany, tourists found that when they converted their dollars to German marks they received 1.5 marks for every dollar, less a service charge of 2 marks. Create a table showing the number of marks they received when they converted $10, $20, $30, $40, $50, $60, $70, $80, $90, and $100 to marks. Argue that the number of dollars given and the number of marks received do or do not vary proportionally.

47. Bret and Barbara found out that for a given charity a family whose income is $30,000 gave $400 and a family whose income is $300,000 gave $4000. Bret says that the family with the higher income gave more than the family with the lower income in proportion to its earnings. Barbara disagrees. Who is correct and why?

48. Kay claims that if $\frac{a}{b} = \frac{c}{d}$, then $\frac{a}{a-b} = \frac{c}{c-d}$ will always be true. Teresa says that it doesn't always work. Who is right and why?

49. Consider the proportion $\frac{2x}{5} = \frac{8}{17}$.

a. Use your calculator to solve this problem by first changing $\frac{8}{17}$ to a decimal.

b. Use your calculator to solve the proportion but *after* you have applied the second property of proportions and then converted $\frac{17}{8}$ to a decimal.

c. Compare the answers that you obtained in parts (a) and (b). Were the answers the same or different? Is that what you would expect? Why or why not?

50. The Lawnmower Problem. Vivian needs a gasoline/oil mix for her lawnmower. The ratio of mix is 50 parts gasoline to 1 part oil. How many liquid ounces of oil (to the nearest whole number) does she need for a 2-gallon mix? (There are 128 ounces in a gallon.)

51. A student was trying to solve the proportion $\frac{3x}{4} = \frac{5}{7}$ with his calculator. Here is the keystroke sequence he used and the calculator display.

Key Sequence	**Display**
5 [÷] 7 [=]	.7142857
[÷] 4 [=]	.1785714
[×] 3 [=]	.5357143

The student was unsure of the solution. Is it correct? Explain.

52. An engineer set up the proportion $\frac{12}{t} = \frac{210}{50}$ to solve a problem. She used a graphing calculator to determine that $t = 2.857142857$, as shown below. Explain what she did to solve the proportion. Is the solution correct?

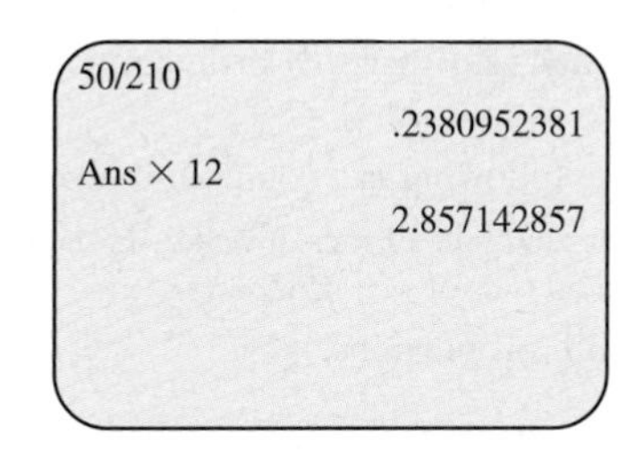

53. The Lawn Problem. Dan, Shelly, and Kirby worked together sodding a lawn. Together they earned \$4000. Dan worked 40 hours, Shelly worked 25 hours, and Kirby worked 15 hours. How should they divide the money?

54. The Photograph Problem. An aerial photograph has a scale of 3 inches equals 20 miles. If the photograph is 12 inches by 18 inches, how many square miles are represented?

55. The Party Problem. Liz needs to buy soft drinks for a party for 30 people. When she gets to the store, she finds that she can buy six cans for \$1.80. She assumes that each person will drink two cans. When she checks her wallet, she finds that she has only \$20. Does she have enough money to buy the soft drinks? Explain your reasoning.

56. The Potion Problem. A nurse has to make up a 30-ounce solution that is two parts glycerin and three parts water. Write a set of directions to the nurse for creating a solution with the proper mixture.

C. Reasoning and Problem Solving

57. Solve the proportion $\frac{x}{3} = \frac{7}{9}$. Create three other proportions that would give the same answer. Compare both the new proportions and the methods used to obtain them with those generated by others in the group.

58. The Picture Frame Problem. Yolanda has a picture that is 4 inches by 4 inches. If she has the picture enlarged so that it fits into a 10-inch by 10-inch frame, will some of the picture have to be cropped? If the picture is enlarged to fit an 8-inch by 10-inch frame, will some of the picture have to be cropped? Explain your reasoning in each case.

59. Suppose that Yolanda's picture is 4 inches by 6 inches. Give the sizes of two larger frames that she could use that would not require the picture to be cropped. Could her picture be put into a square frame without being cropped? Explain.

60. Create a table that compares the number of squares arranged as shown to the number of corners of the figure. (Make the table for one square, two squares, ..., five squares.) Argue that the table does or does not represent a proportional situation.

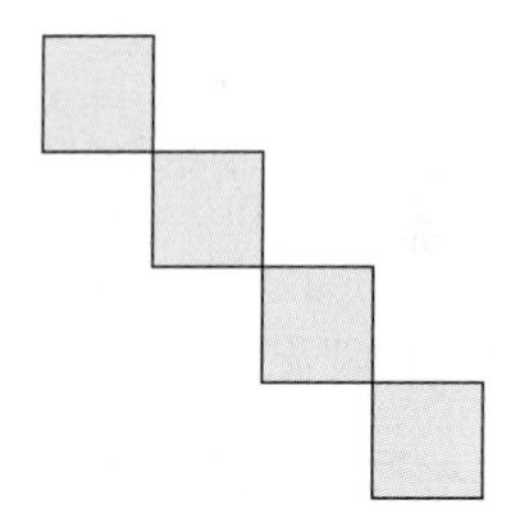

61. Sang says that she used her calculator in the following way to solve the proportion $\frac{6}{x} = \frac{14}{7}$. I divided 14 by 7, which gives 2. Then I hit the reciprocal key and got 0.5. I multiplied this answer by 6 to get the answer, 3. Sang claims that this method always works. Do you agree? Why or why not?

62. Use deductive reasoning and what you know about proportions to establish the following property of proportions.

$$\text{If } \frac{a}{b} = \frac{c}{d}, \text{ then } \frac{a+b}{b} = \frac{c+d}{d}.$$

63. The Physics Problem. Suppose the relationship between the force (F), mass (M), and velocity (V) of a moving object in a certain situation is determined by the formula $F = MV^2$ where M is the mass of the striking object and V is the velocity of the striking object. Is the force proportional to the mass? To the velocity? Explain your reasoning.

D. Communicating and Connecting Ideas

64. Explain why each of the following statements does or does not use the term *proportional* correctly.

a. The amount of sales tax paid is proportional to the cost of a purchase.

b. A person's weight is proportional to the person's height.

c. The amount of weight a person gains is proportional to the increase in calories eaten.

d. The value of a bar of gold is proportional to its weight.

e. The distance around a circle is proportional to the distance across the circle.

65. Consider the following students' methods of solving the proportion $\frac{8}{5} = \frac{20}{x}$ and determine whether the methods were correct. Explain your reasoning.

Richie: First I divided 20 by 8 and got $2\frac{1}{2}$. Then I multiplied $2\frac{1}{2}$ by 5 and got $12\frac{1}{2}$ for my answer.

Martha: I think of it this way. Every time a person takes 8 steps, another person takes 5 steps. So 16 steps correspond to 10 steps, and 24 steps correspond to 15 steps. As 20 is halfway between 16 and 24, I reasoned that the answer was halfway between 10 and 15. I got $12\frac{1}{2}$ for the answer.

66. A friend of yours is trying to solve the proportion

$$\frac{x + 3}{8} = \frac{2x - 4}{4}.$$

She is confused, and has asked you to help her. Write a specific set of directions that she can use to solve for x.

67. Write a paragraph in which you use the following terms appropriately: proportion, ratio, proportional variation, and rectangle. Share your paragraph with others. Did others use the terms in different contexts? If so, how did the contexts differ?

68. Making Connections. A survey conducted in 1972 found the marriage rate was about 10.9 marriages per 1000 people. A similar survey in 1991 found the marriage rate was about 9.4 per 1000 people. For a population of 1 million, how many more marriages would you expect in 1972 than in 1991? Why would you be wrong to conclude that for a particular city fewer marriages occurred in 1991 than in 1972?

69. Historical Pathways. Ancient mathematicians were challenged by ratios that can't be represented by whole numbers. They generally tried to create ratios with whole numbers that approximated a given ratio as closely as possible. For the following two similar triangles, set up a proportion involving x. Use your calculator to approximate $\sqrt{5}$ and find a value of x that approximates a solution. (You may want to use trial and error.) Let the numerator and denominator for x be any whole numbers from 1–99.

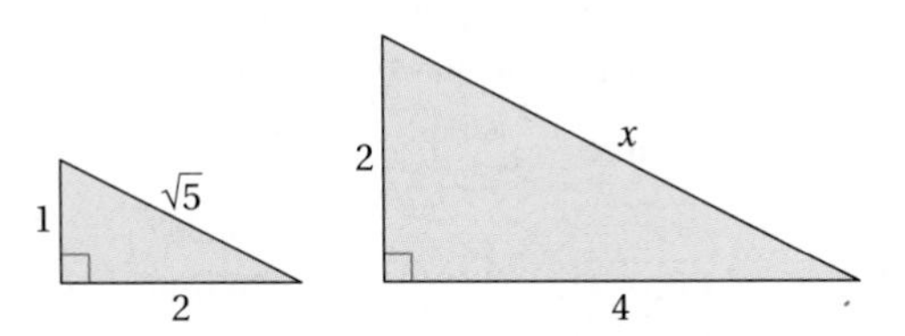

70. Historical Pathways. In a mathematics book published in the early 1900s [Harvey (1909), pp. 316–318], the author identified and illustrated how to solve two types of proportions: simple and complex. The following is a simple proportion.

If 10 men can build 50 rods of wall in a certain time, how many men can build 80 rods of wall in the same time?

The author said that this problem could be solved by using the proportion $\frac{50}{80} = \frac{10}{x}$.

The following is a compound proportion.

If 10 men can build 50 rods of wall in 12 days, how many men can build 80 rods of wall in 16 days?

Solve both proportions.

Section 7.3 SOLVING PERCENT PROBLEMS

- Definition of Percent
- Connecting Percents, Ratios, and Decimals
- Types of Percent Problems
- Solving Percent Problems
- Percent Increase or Decrease
- Simple and Compound Interest
- Estimating the Percent of a Number

In this section, we focus on percent, which is a special type of ratio. We define *percent* and show how to convert a percent to a fraction or a decimal and vice versa. We illustrate different types of percent problems, model solutions to percent problems, and consider various practical applications of percents, including finding the percent of increase or decrease associated with an event. Finally, we discuss simple and compound rates of interest.

Definition of Percent

In Chapter 6, we indicated that percent notation provides a concise way to represent a fraction with a denominator of 100. The word *percent* comes from the Latin *per centum* (meaning by the hundred or per hundred). Use of the words *per hundred* indicates that a percent can also be thought of as a ratio. For example, 6 percent means 6 of 100, 6 (of something) per 100 (of something), or that the ratio of one quantity to another quantity is $\frac{6}{100}$. This idea suggests the following definition.

Definition of Percent

A **percent** is a ratio with a denominator of 100.

Whether you think of percent as a fraction or a ratio, percent is represented by the symbol %; that is,

$$n\% = \frac{n}{100}.$$

Examples of this notation are

$$6\% = \frac{6}{100}, \quad 72\% = \frac{72}{100}, \quad 200\% = \frac{200}{100}, \quad 0.5\% = \frac{0.5}{100},$$

$$100\% = \frac{100}{100}, \quad \text{and} \quad 33\tfrac{1}{3}\% = \frac{33\frac{1}{3}}{100}.$$

The equation $100\% = \frac{100}{100}$ clearly shows that 100% of an amount means *all* of that amount. For example, 100% of 5 gallons of water means 5 gallons of water, and 100% of \$4589 means \$4589. Moreover, 200% of an amount means $\frac{200}{100}$ times the amount, or twice the amount. For example, 200% of \$15 means \$30. Similarly, 300% of 2 quarts means three times as much, or 6 quarts.

A pictorial representation of percent also is instructive. For example, in Figure 7.5, the 10×10 grid consists of 100 small squares, each representing 1% of the entire grid, which represents 100%. Grids (A) through (C) in Figure 7.5 represent 1%, 100%, and 72%, respectively. In (A), only one square is shaded. In (B), all the squares are shaded, and the diagram can be thought of as 1 whole unit (or grid). In (C), 72 of the squares are shaded.

Percents of less than 1 or more than 100 can be represented in a similar manner. If one small square represents 1%, then one-half of that square would represent $\frac{1}{2}$%, or 0.5%, as shown in (D) in Figure 7.5. A percent greater than 100 is represented by the number of full and partial grids required to depict the percent. In (E), two entire grids are shaded (to represent 200%) and 60 of the squares in the last grid are shaded, giving a total representation of 260%.

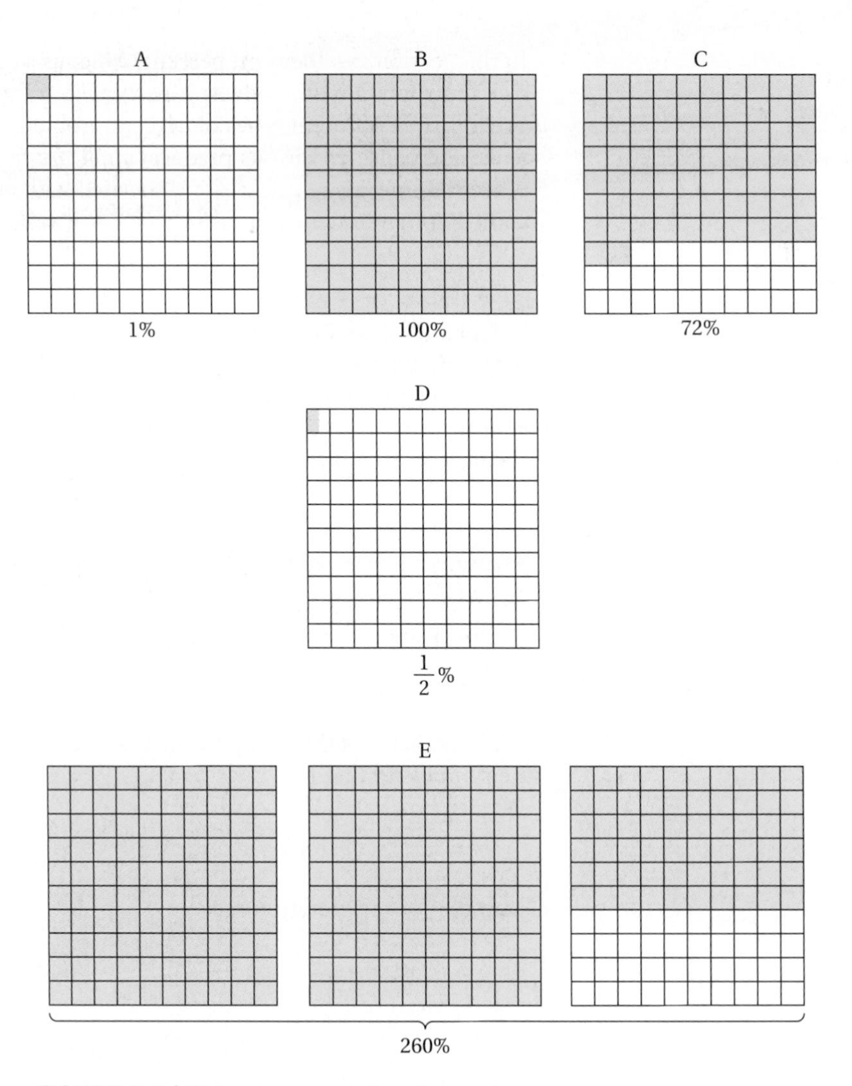

FIGURE 7.5 | Pictorial representation of percents.

Connecting Percents, Ratios, and Decimals

When dealing with everyday situations, we often need to write a percent as a fraction or decimal or to write a fraction or decimal as a percent. For example, if the ratio of the number of people attending a meeting to the total membership is 1 : 4, we may want to report that only 25% of the members attended. If 50% of the members were less than 30 years old, we might want to use $\frac{1}{2}$ to calculate the actual number of members under 30.

To *convert a percent to a decimal*, simply divide it by 100 and drop the % symbol. That is, 40% = 40 ÷ 100, or 0.40. Conversely, to convert from a decimal to a percent, multiply the decimal by 100 and insert the percent sign. That is, $0.36 = (100 \times 0.36)\% = 36\%$.

To *convert a percent to a fraction*, think of the basic meaning of percent; that is, $n\% = \frac{n}{100}$. Thus $80\% = \frac{80}{100}$ and $\frac{3}{4}\% = \frac{\frac{3}{4}}{100}$. Then simplify these fractions as you would any other fraction. In these cases, $80\% = \frac{4}{5}$ and $\frac{3}{4}\% = \frac{3}{400}$.

To *convert a fraction to a percent or decimal,* begin by writing an equivalent fraction with denominator 100. For example, to find a percent or decimal for $\frac{1}{20}$ think of $\frac{1}{20}$ as $\frac{5}{100}$, so $\frac{1}{20} = 0.05 = 5\%$

For fractions such as $\frac{1}{3}$, attempting to write an equivalent fraction with denominator 100 might not be productive. Instead, begin by using a calculator or pencil and paper to divide the numerator by the denominator:

$$\begin{array}{r} 0.33\ldots \\ 3\overline{)1.00} \\ \underline{9} \\ 10 \\ \underline{9} \\ 1 \end{array} \qquad \text{or} \qquad 0.33\frac{1}{3}$$

Then write $\frac{1}{3} = 0.33\ldots = 33\frac{1}{3}\%$.

Example 7.13 further illustrates these procedures.

Example 7.13

Converting Among Fractions, Percents, and Decimals

a. Change 125% to a fraction.

b. Change 0.005 to a percent.

Solution

a. 125% means $\frac{125}{100}$, which reduces to $\frac{5}{4}$.

b. $0.005 = \frac{5}{1000} = \frac{0.5}{100}$, or 0.5%.

Your Turn

Practice: Find the missing terms in the following table.

Percent	Fraction	Decimal
6%	?	?
?	$\frac{7}{5}$	?
?	?	0.0045

Reflect: What would have to be the relationship between the numerator and the denominator of a fraction if, when it is changed to a percent, the result is

a. less than 1%?

b. greater than 100% ■

Types of Percent Problems

Most percent problems are of three basic types. In this subsection, we show how to identify, formulate, and solve each type of problem. In Mini-Investigation 7.6, you are asked to begin considering how percents might be involved in real-world problems.

Talk about how you might estimate the missing percents in parts (a) and (b).

Technology Extension: What features on your calculator help you work with percents?

a. How does the % key found on some calculators work and how could you use it to solve the problem $56 \times 40\% = ?$

b. How does the way that the percent features of your calculator work, including the % key, compare to similar features on other calculators used by your classmates?

Mini-Investigation 7.6 Solving a Problem

What numbers would you use to calculate the following percents?

a. The credit card debt per account in 1984 was __?__% of the credit card debt in 1993.

b. The increase from 1984 to 1993 was __?__% of the debt in 1984.

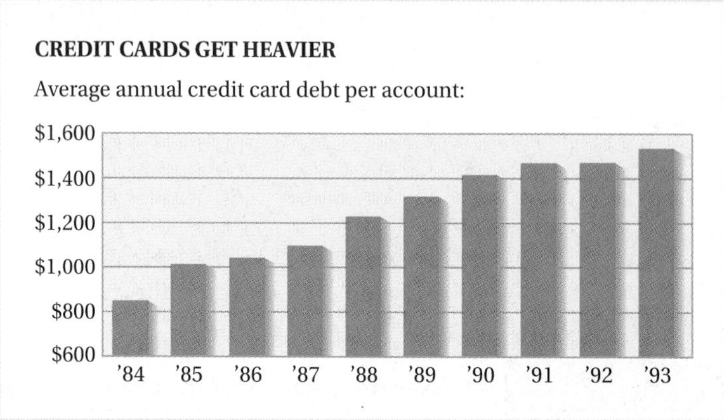

Source of data: RAM Research's Bankcard Update. By Cindy Hall and Sam Ward, USA TODAY. *USA Today,* June 1, 1994, p. A1. Copyright 1994, *USA Today.* Reprinted with permission.

In Mini-Investigation 7.6, you considered one type of percent problem that often occurs in the real world. Both parts (a) and (b) represent a type of percent problem in which two numbers are known and the problem involves finding the percent that one number is of the other. To take a further look at this and two other types of percent problems, consider the following.

a. Laura sees a dress in a store that regularly sells for $250 but is marked down 20%. How much will she save if she buys the dress on sale?

b. Barry determined that he had traveled 240 miles, which he estimated as being 40% of the trip. If his estimate is correct, how long is the trip?

c. Jere learned that 72 of her 150-member senior class went to college. What percent of her senior class went to college?

In (a), the whole, $250, and the percent, 20%, are given. This problem calls for *finding a percent of a number.* In (b), part of the whole, 240, and the percent, 40%, that this part represents of the whole are given. This problem involves *finding a number when a percent of it is known.* In (c), the whole, 150, and the number, 72, that represents part of that whole are given. In this case, as in Mini-Investigation 7.6, the problem calls for *finding the percent that one number is of another.* The following examples further illustrate these types of problems.

Finding a percent of a number

- Mr. Gates wants to buy a house for $80,000. He needs to pay 15% of the purchase price as a down payment. How much is the down payment?
- Three hundred students took the final examination in chemistry. Twelve percent of the students received an A. How many students received an A?

Finding a number when a percent of it is known

- Forty-five percent of the people who were interviewed approved of the new health plan. If 650 people approved of the plan, how many people were interviewed?
- A school survey determined that 640 students, or 60% of the student body, participated in various extracurricular activities. How many students are in the student body?

Finding the percent that one number is of another

- In a class of 22 students, 13 are girls. What percent of the class are girls?
- Of the 52 vehicles in a car dealer's lot, 21 are sedans. What percent of the vehicles are sedans?

Example 7.14 demonstrates how to formulate the three types of percent problems.

Example 7.14 Problem Solving: Movie Goers

Consider the following situation.

> A movie goer was interested in the types of movies that people like to see. The local theater was showing three films: a romance, a western, and a science fiction movie. On a Sunday afternoon, she interviewed each person who bought a ticket.

Based on this situation, create a problem for each of the three types of percent problems discussed in this section.

Solution

Karen's thinking:

Finding a percent of a number A total of 210 people bought tickets. Forty percent of these people bought a ticket to the romance movie. How many people bought tickets to this movie?

Finding a number when a percent of it is known Sixty percent of the people who saw the western said that they liked it. If 15 people said they liked it, how many saw the western?

Finding the percent that one number is of another Of the 210 interviewed, 60 said they were going to see the science fiction movie. What percent of the people went to see the science fiction movie?

Shannon's thinking:

Finding a percent of a number Katie interviewed 50 people who bought tickets. She found out that 15% of them were seeing their movie of choice for a second time. How many people who were interviewed were seeing the movie of their choice for a second time?

Finding a number when a percent of it is known Katie determined that the 50 people she interviewed comprised 40% of the total number of people who went to the movies that day. How many people went to the movies that day?

Finding the percent that one number is of another Katie found that 10 of the 50 people interviewed went to the western, 22 went to the romance, and 18 went to the science fiction movie. Of the 50 people interviewed, what percent of the people did not go to the western?

Your Turn

Practice: Create your own problems based on the situation given in the example problem.

Reflect: What pieces of information do you need in order to create a problem that fits the *finding a percent of a number* category? The *finding a number when a percent of it is known* category? The *finding the percent that one number is of another* category? ■

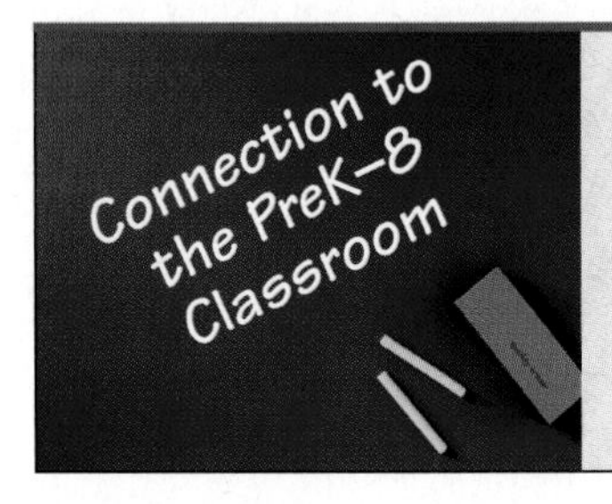

Elementary school students encounter many examples of percents. Many read about sales and the discounts that are given for items they're interested in. Often these discounts are expressed as percents. Elementary school students sometimes find their grades expressed as percent correct. These practical uses of percent can be important motivating factors for students as they begin to learn how to solve the three different types of percent problems.

Solving Percent Problems

Regardless of which of the three standard types of percent problems you may be trying to solve, a percent diagram can help you find the solution. Consider Figure 7.6, which represents 25% of 12 people. The top side of the bar represents percents—100% being the entire bar and 25% being the shaded part. The bottom side of the bar represents the number of people—12 people in all and 3 people for the shaded part.

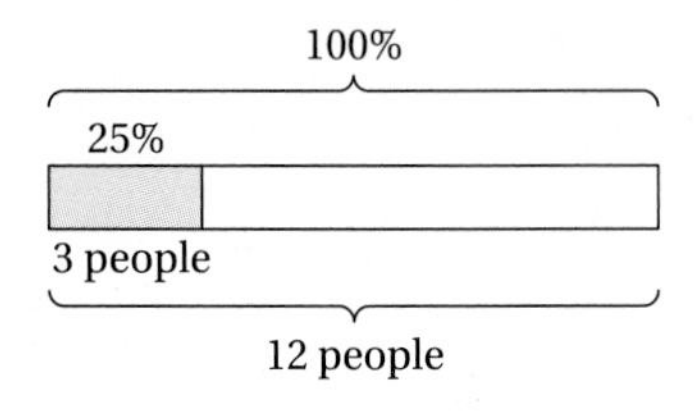

FIGURE 7.6 | A percent diagram

In the following subsections, we use such a percent diagram to represent visually and help set up a proportion to solve each of the three types of percent problems. A calculator or computer may be useful in carrying out the calculations in these and other percent problems.

Finding a Percent of a Number. Consider the following problem, which involves finding a percent of a number.

> Laura sees a dress in a store that regularly sells for $250 but is marked 20% off. How much will she save if she buys the dress on sale?

This type of percent problem is probably the easiest to solve. We do so by simply finding 20% of $250, or $20\% \times 250 = 0.20 \times 250 = 50$. Laura will save $50.

However, using a percent diagram to solve the same problem is instructive. First, above the bar in the diagram (Figure 7.7), we show the entire bar as 100% and the shaded part as 20%. Below the bar, we show $250 as the whole amount and $\$x$ as the shaded amount. Note that four amounts are represented, with three known and one unknown. Using this diagram, we set up a proportion, with one ratio comparing the percents and the other ratio comparing the dollars, and then solve it:

100%
20%
x dollars
$250

FIGURE 7.7 | Using a percent diagram to find a percent of a number.

$$\frac{20}{100} = \frac{x}{250}$$

$$x = \frac{250(20)}{100} = 50$$

Thus Laura will save \$50.

Alternatively, we could make a ratio of the parts and equate it to a ratio of the whole amounts. This approach would result in a proportion with ratios comparing percents to dollars, which we then solve producing the same answer:

$$\frac{20}{x} = \frac{100}{250}$$
$$x = \frac{250(20)}{100} = 50$$

Estimation Application

Increasing Income

A pay-per-view cable company charges \$4.94 per movie. It averages 79,536 customers per movie. The company estimates that, if it lowers its rate by 10%, it will experience a 20% increase in customers. Estimate the company's anticipated income under the proposed plan. Find a way to check your estimate.

Finding a Number When a Percent of It Is Known. Consider the following problem, which involves finding a number when a percent of it is known.

> Barry determined that he had traveled 240 miles, which he estimated as being 40% of the trip. If his estimate is correct, how long is the trip?

First, note that the whole trip is 100% and that the 240 miles is 40%. So the entire bar in the diagram (Figure 7.8) represents both 100% and x miles. The shaded part represents both 40% and 240 miles. Note that four amounts are represented, with three known and one unknown. Next, we use the percent diagram to set up a proportion for the problem and then solve it as before:

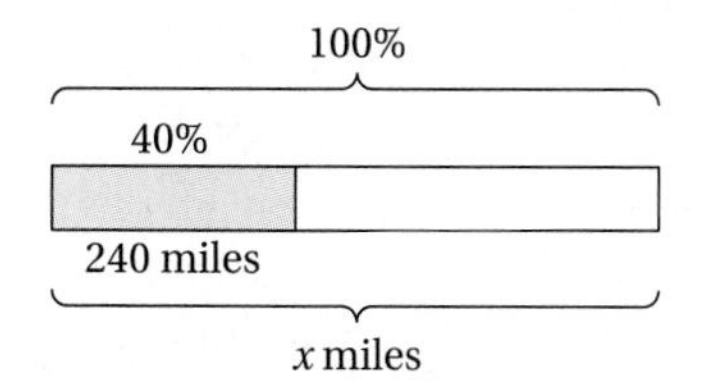

FIGURE 7.8 | Using a percent diagram to find a number when a percent of it is known.

$$\frac{40}{100} = \frac{240}{x}$$
$$x = \frac{240(100)}{40} = 600.$$

The total length of the trip is 600 miles.

Finding the Percent That One Number Is of Another. Consider the following problem, which involves finding the percent that one number is of another.

> Jere learned that 72 of her 150-member senior class went to college. What percent of her senior class went to college?

First, note that the whole senior class of 150 students is 100% and that the percent of the class amounting to 72 students, x, is unknown. So the entire bar in the diagram (Figure 7.9) represents both 100% and 150 students; the shaded part represents both x% and 72 students. Note that four amounts are represented, with three amounts known and one unknown. Next, we use the percent diagram to set up a proportion for the problem and then solve it:

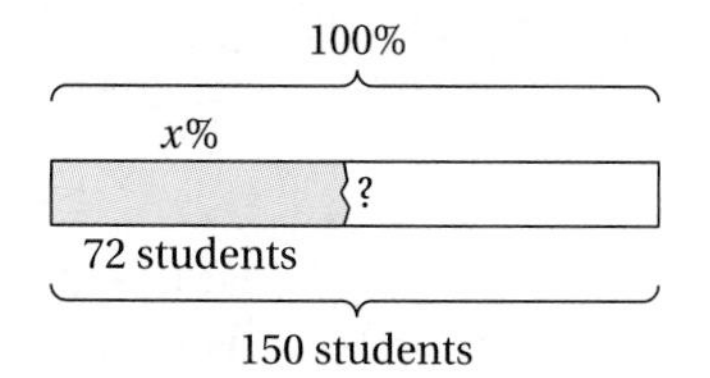

FIGURE 7.9 | Using a percent diagram to find the percent that one number is of another.

$$\frac{x}{100} = \frac{72}{150}$$
$$x = \frac{72(100)}{150} = 48.$$

Thus 48% of Jere's class went to college.

Example 7.15 presents a final illustration of solutions to standard types of percent problems.

Example 7.15 Using a Percent Diagram to Solve Problems

Carrie told a friend that last month she paid \$720 in federal income tax. If Carrie is in the 28% tax bracket, how much money did she earn last month?

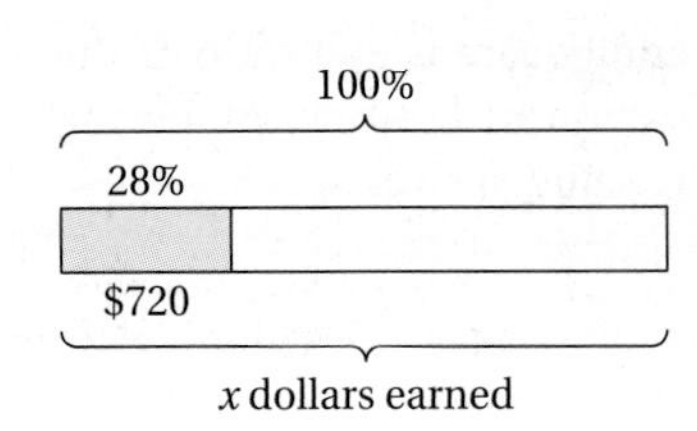

Solution

Let x represent the amount of money that Carrie earned last month. Then prepare a percent diagram.
Set up the proportion and solve it:

$$\frac{28}{100} = \frac{720}{x}$$
$$28x = 72{,}000$$
$$x = 2571.428571\ldots$$

Carrie earned $2571.43 (to the nearest cent).

Your Turn

Practice: Hank budgeted about 12% of his earnings for an investment program. If he had saved $3840, how much money had he earned?

Reflect: How would you estimate 28% of $2571.43? ■

Percent Increase or Decrease

The technique you learned for solving a standard percent problem that involves finding the percent that one number is of another may be used to find the percent of increase or decrease. Mini-Investigation 7.7 asks you to consider your interpretation of the idea of percent of increase or decrease.

Talk about the reasons for your conclusion.

Mini-Investigation 7.7 **Communicating**

Do you think that the editorial writer who wrote the following used the idea of percent of increase correctly?

> The number of violent crimes in our city increased from 720 last year at this time to 1400 this year. This represents nearly a 200% increase in violent crimes. It is time that the city council took serious action regarding criminal activity in our city.

To develop the idea of percent of increase or decrease, we first contrast it with the basic situation in which percent is used. A common use of percent is to talk about the percent of an amount, as we did when discussing the first of the three types of percent problems in the preceding subsection. For example, we might say that General Motors sells 35% of all motor vehicles sold or that a sales tax is 6% of a purchase. To find a percent of an amount, recall that you simply multiply the percent, expressed as a decimal number, by the amount. In contrast, the following statements illustrate the idea of percent increase or decrease.

> General Motors' stock went up 10%.
> The average attendance at the home football games this year is down 25%.

Both statements indicate an increase or decrease *relative to the original situation.* For example, if General Motors' stock went from $40 a share to $44 a share, it would have increased by $4. To find the percent increase, compare the increase with the original cost of a share. The ratio is $\frac{4}{40} = \frac{1}{10} = 0.10$, so the stock increased by 10%.

If the average attendance at football games decreased from 2000 to 1500, the decrease is 500 people. Compared to the original attendance, the drop is $\frac{500}{2000}$, or 25%. These examples suggest the following procedure for finding the percent increase or decrease.

Procedure for Finding the Percent Increase or Decrease

Step 1. Determine the *amount* of increase or decrease.

Step 2. Divide this amount by the *original* amount.

Step 3. Convert this fraction or decimal to a percent.

The following table illustrates these steps for an increase from 80 to 100 and a decrease from 150 to 90.

	Finding Percent Increase	**Finding Percent Decrease**
Original amount	80	150
New amount	100	90
Increase/decrease	20	60
Fraction/decimal number	$\frac{20}{80}$, or 0.25	$\frac{60}{150}$, or 0.40
Percent	25%	40%

Returning to Mini-Investigation 7.7, you can see that the writer erred in his use of percent. The increase in violent crimes was 680. To determine the percent of increase, he should have compared 680 to the original number of crimes and used the fraction $\frac{680}{720}$ to calculate the percent increase, which is approximately 94.4%. Thus a reasonable estimate is 100% rather than 200%. Example 7.16 applies the procedure for finding the percent increase or decrease.

Example 7.16 Finding the Percent of Increase or Decrease

Which represents a greater percent increase in the number of cars sold?

a. An increase from 3500 to 3850

b. An increase from 3800 to 4175

Solution

a. The amount of increase is 350 cars, which gives the fraction $\frac{350}{3500}$, or a 10% increase.

b. The amount of increase is 375 cars, which gives the fraction $\frac{375}{3800}$, or about a 9.9% increase.

The percent increase described in part (a) is greater than in part (b).

Your Turn

Practice: Which company has the greater percent decrease in its stock price?

a. The price of an IBM share drops from 48 dollars to 45 dollars.

b. The price of a General Electric share drops from 98 dollars to 94 dollars.

Reflect: How does the example problem show that the largest numerical increase doesn't necessarily have the largest percent increase? ■

Simple and Compound Interest

Banks and investment houses often advertise a rate of return on money invested with them in an attempt to attract customers. Also, information is readily available about the rates that must be paid for borrowing money or carrying a credit card balance. This rate of return or payment is called the **interest rate** and is expressed as a percent. The amount of money invested or borrowed is called the **principal**. The actual amount of money received from an investment or paid on a loan is called the **interest** earned. For example, you might invest your *principal*, \$1000, in a bank at an *interest rate* of 6%. The interest earned or paid can be calculated by applying a rate of simple interest or compound interest, whichever is appropriate, to the principal. We consider both types of interest.

Simple Interest. **Simple interest** is computed by multiplying the rate of interest (a percent) by the principal, that is, by solving a standard type of percent problem. For a principal amount of \$1000 and simple interest at 6%,

$$\text{Simple interest} = \$1000(0.06) = \$60.$$

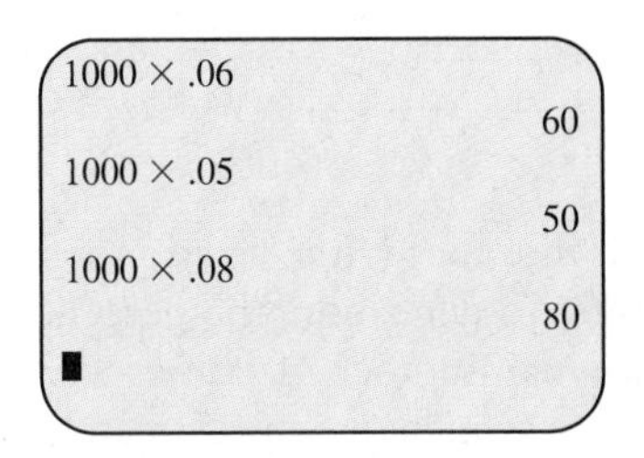

FIGURE 7.10 | Simple interest calculations on a calculator.

The calculator screen shown in Figure 7.10 demonstrates the calculation of simple interest for rates of 6%, 5%, and 8%.

We assume for our purposes that simple interest is paid *annually*. In the preceding example, the interest for the second year would be computed in the same way as for the first year and again would be \$60. Note that the simple interest for the second year is paid only on the original principal of \$1000, not on the principal plus interest earned, or \$1060. A simpler way to find the amount of simple interest for any period of time is to use a formula. The **formula used to calculate simple interest** is

$$i = prt,$$

where i is the amount of interest, p is the amount of principal, r is the annual rate of simple interest, and t is the number of years.

Example 7.17 demonstrates how to apply the formula for calculating simple interest.

Example 7.17 Calculating Simple Interest

a. Determine the amount of interest earned over a 2-year period for an investment of \$15,000 with a simple interest rate of 6.5% per year.

b. Determine the interest charges for the first month if the balance on the credit card is \$2000 and the annual rate is 18%.

Solution

a. $i = prt$

$= 15{,}000 \times 0.065 \times 2 = 1950.$

The interest earned would be \$1950 for the 2-year period.

b. *Carla's thinking:* The annual charge would be 2000×0.18, or \$360. Hence the month's charge is \$30.

Geoff's thinking: The annual charge is 18%, so the monthly charge is 1.5%. Since $2000 \times 0.015 = 30$, the month's charge is \$30.

Your Turn

Practice: Determine the amount of interest earned over a 6-month period if \$800 is invested at a simple rate of 7% per annum.

Reflect: What value would you use for t in the practice problem if you wanted to calculate the interest earned over 3 months? ■

Example 7.18 shows how to find simple interest in a real-world situation.

Example 7.18 Credit Card Charges

Suppose that a credit card company does not have an annual fee but charges an 18% annual rate of interest. A second credit card company charges \$30 per year but charges an annual rate of only 15%. What is the maximum amount a person could pay interest on and still be better off with the first credit card company than with the second company?

Working toward a Solution

Understand the problem	*What does the situation involve?*	Charges by credit card companies.
	What has to be determined?	The maximum amount for which the first credit card company offers the better deal.
	What are the key data and conditions?	The first company charges 18%, or \$18 per \$100, per annum. The second company charges 15%, or \$15 per \$100, per annum but also charges an initial fee of \$30.
	What are some assumptions?	That purchases are made with both credit cards.
Develop a plan	*What strategies might be useful?*	Make a table.
	Are there any subproblems?	Determine the charges by both companies.
	Should the answer be estimated or calculated?	Calculated.
	What method of calculation should be used?	A spreadsheet or a graphing calculator can be used to make a table indicating the interest charged for various amounts charged to each credit card. We can then calculate the answer by using the spreadsheet.
Implement the plan	*How should the strategies be used?*	The following table can be created. As a first approximation, we consider the values from \$500 to \$1500.

	A	B	C
1	$ Amount	Option I	Option II
2	500	90	105
3	600	108	120
4	700	126	135
5	800	144	150
6	900	162	165
7	1000	180	180
8	1100	198	195
9	1200	216	210
10	1300	234	225
11	1400	252	240
12	1500	270	255

For $1000 on which interest is paid, the cost associated with the two companies is the same.

	What is the answer?	Thus $999 is the maximum amount that can be charged and the cardholder pays less to the first company.
Look back	*Is the interpretation correct?*	We determined the amount of money for which the interest charged by the two companies is the same. The interpretation is correct.
	Is the calculation correct?	We can check our spreadsheet calculations by using a calculator. The calculations are correct.
	Is the answer reasonable?	Charges of $1000 on the cards result in the same cost, so a $999 maximum amount seems reasonable.
	Is there another way to solve the problem?	We can also think of the problem as the difference between the two companies. It is 3%, so for every $100 for which interest is charged, the first company charges $3 more. To make up a $30 difference would require ten $100 charges, or $1000. Thus, for any amount of charges less than $1000 per year, the first company offers the better deal.

Your Turn

Practice: What would have been the maximum amount if the second company had charged an annual fee of $20 per year?

Reflect: How can the example problem be solved without using a table? ■

Compound Interest. Recall that with simple interest the amount of interest earned is based on the principal only, not on any of the money previously earned as interest. We used the example of earning $60 each year on principal of $1000 at 6% simple interest. That is, with simple interest, the amount of interest does not change from year to year. In contrast, with **compound interest** an investor earns money each year not only on the principal but also on the amount of interest previously earned. For example, if interest is compounded annually, for the second year the person would earn 6% on the principal, \$1000, plus 6% on the \$60 interest earned the first year. Thus the interest earned the second year would be \$1060 × 0.06, or \$63.60.

Interest is not always compounded annually; sometimes it's done quarterly (4 times a year) or even daily (365 times a year). The differences in the amount of interest are usually small for small sums of money, but for larger sums of money or for larger rates of interest the differences in the amount of money accumulated can be substantial. The following table can be generated by using a spreadsheet. It contrasts the amounts earned on \$10,000 invested at 6% simple interest per year, 6% compounded quarterly, and 6% compounded monthly.

	Simple Interest	**Compounded Quarterly**	**Compounded Monthly**
Initial Investment	10,000.00	10,000.00	10,000.00
End of year 1	10,600.00	10,613.64	10,616.78
End of year 2	11,200.00	11,264.93	11,271.60
End of year 3	11,800.00	11,956.18	11,966.81
End of year 4	12,400.00	12,689.86	12,704.89
End of year 5	13,000.00	13,468.55	13,488.50
End of year 6	13,600.00	14,295.03	14,320.44
End of year 7	14,200.00	15,172.22	15,203.70
End of year 8	14,800.00	16,103.24	16,141.43
End of year 9	15,400.00	17,091.40	17,136.99
End of year 10	16,000.00	18,140.18	18,193.97

Thus, over a 10-year period, a person would accumulate about \$2194 more from a \$10,000 investment with interest of 6% compounded monthly than from simple interest at 6% annually.

Suppose a bank advertises an interest rate of 8% compounded quarterly, and you invest \$1000. The total amount you would have after the first quarter would be $\$1000 + \$1000(0.02) = \$1000(1.02) = \$1020$. At the end of the second quarter, you would have $\$1020 + \$1020(0.02) = \$1020(1.02) = \$1040.40$. At the end of the third quarter, you would have $\$1040.40 + \$1040.40(0.02) = \$1040.40(1.02) = \$1061.208$. At the end of the fourth quarter, you would have $\$1061.208 + \$1061.208(0.02) = \$1061.208(1.02) = \$1082.43$. We can simplify this process to show that the money accumulated for the four quarters is $1000(1.02)(1.02)(1.02)(1.02)$, or $1000(1.02)^4$. Using a calculator, we get $1000(1.02)^4 = 1082.43$, the amount calculated earlier. The **formula for calculating the amount of compound interest** is

$$a = p\left(1 + \frac{i}{n}\right)^{nt},$$

where a is the total amount of principal plus interest, p is the principal, i is the annual rate of interest, n is the number of times the interest is compounded per year, and t is the number of years. The product nt represents the number of times the investment is compounded over the life of the investment. For example, if the annual rate of interest is 9% and the interest is compounded monthly for 3 years, then $i = 0.09$, $n = 12$, $t = 3$, and $nt = 36$. Example 7.19 further illustrates use of the formula to calculate compound interest.

Example 7.19

Calculating Compound Interest

Determine the amount of money that would result from a \$1000 investment at an annual interest rate of 8% compounded daily for 5 years.

Solution

For interest compounded daily for 5 years, $n = 365$, $t = 5$, and $nt = 365(5)$, or 1825. The principal is 1000 and $i = 0.08$. Using the formula and a calculator, we obtain

$$a = 1000\left(1 + \frac{0.08}{365}\right)^{1825}$$
$$= \$1491.76 \text{ (to the nearest cent).}$$

Your Turn

Practice Use a calculator to determine the interest on an investment of \$5000 at a 10% annual rate for five years if the interest is compounded monthly.

Reflect: Without using the formula or a calculator, argue that the answer you obtained in the practice problem is a reasonable answer. ■

Estimating the Percent of a Number

At times estimating a percent of a number is helpful; for example, when you want to leave a tip in a restaurant or when you want to price a pair of slacks on sale. The strategies that you learned in Chapter 3 for estimating products of whole numbers may also be used for estimating percents. The strategies of rounding and front-end estimation are particularly useful. For example, rounding and using 30% or 40% give reasonable estimates of 37%. One of the most important estimation facts is that 10% of a number is one-tenth of that number. Consequently, 20% is two-tenths, 30% is three-tenths, and so on. Thus to estimate 30% of a number simply estimate 10% of the number and then triple this estimate. Similarly, to estimate 60% of a number find one-fifth of the number and then triple the estimate.

Certain fractions easily convert to percents and vice versa, as shown in the following table.

Percent	Fraction
10%	$\frac{1}{10}$
20%	$\frac{1}{5}$
25%	$\frac{1}{4}$
$33\frac{1}{3}$%	$\frac{1}{3}$
40%	$\frac{2}{5}$
50%	$\frac{1}{2}$
60%	$\frac{3}{5}$
$66\frac{2}{3}$%	$\frac{2}{3}$
75%	$\frac{3}{4}$
80%	$\frac{4}{5}$

We demonstrate how to use these conversions in estimating in Example 7.20.

Example 7.20 Estimating with Percents

Estimate 72% of 150.

Solution

Seventy-two percent is close to 75%, or $\frac{3}{4}$. One-fourth of 150 is about 40. Hence three-fourths is about 120. (The actual answer is 108.)

Your Turn

Practice: Estimate 45% of 184.

Reflect: Why was the estimate given in the example solution an overestimate? ■

Example 7.21 shows how to use well-known fraction–decimal conversion to estimate a solution to a real-world problem.

Example 7.21 Using Fraction–Decimal Conversions to Estimate Percents

Marsha is thinking of buying a dress that originally was priced at $75.95 and was being marked down 35%. Estimate the cost of the dress.

Solution

Thirty-five percent is close to $33\frac{1}{3}\%$, or $\frac{1}{3}$. One-third of 75 is $25. The new price should be about $50. (The actual sale price is $49.37.)

Your Turn

Practice: Estimate a 17% tip on a meal that costs $24.52.

Reflect: What would be another way to estimate the cost of the dress in the example problem? ■

Problems and Exercises *for Section 7.3*

A. Reinforcing Concepts and Practicing Skills

In Exercises 1–6 represent each percent on a 10 × 10 grid.

1. 25% **2.** 107%

3. 50.5% **4.** 210%

5. 0.6% **6.** 2.5%

Find the missing parts in Exercises 7–14.

	Percent	Fraction	Decimal
7.	35%	?	?
8.	?	?	0.42
9.	?	$\frac{7}{4}$	?
10.	?	?	2.45
11.	12.5%	?	?
12.	?	$\frac{1}{200}$	?
13.	?	?	0.0025
14.	120%	?	?

In Exercises 15–18, find the simple interest for the indicated length of time.

15. $4,000 at 5% interest for one year

16. $6,000 at 12% interest for two years

17. $10,000 at 6.5% interest for two years

18. $15,000 at 5.5% interest for six months

In Exercises 19–24, find the percent of increase or decrease.

19. From 1000 to 2000 **20.** From 2000 to 1000.

21. From 480 to 2400 **22.** From 125 to 150

23. From 600 to 200. 24. From 200 to 50.

25. At a large state university, 60% of the students who apply are accepted. Of those who are accepted, 40% actually enroll. If 20,000 students apply, how many actually enroll?

26. In a survey of 1500 people who owned the same make of car, 900 indicated that they would buy the same make of car again. What percent of these people were satisfied with their car?

27. Two percent of the readers of a local newspaper wrote to the editor complaining about a particular editorial. If 90 people wrote letters, how many readers did the newspaper have?

28. An investor earned $1400 interest on a $5000 investment over a period of 4 years. What rate of simple interest per year did she earn?

29. The list price of a car is $17,500. By purchasing a combination of luxury items, the cost can increase by as much as 25%. How much could the car cost with all the extras?

30. Luis, Felipe, and Maria had the following scores on their target practice from their different teachers.

Luis: $\frac{49}{59}$ Felipe: $\frac{58}{72}$ Maria: $\frac{74}{92}$

If these scores were represented as percents, who had the best record? the worst?

31. $25 can be saved by buying a jacket that is marked down 20%. What was the jacket's original price?

32. The number of students at a university increased from 6000 students in 1960 to 18,000 in 1968. What was the percent increase in enrollment from 1960 to 1968?

33. The number of students at another university increased from 14,000 students in 1971 to 30,000 in 1975. What was the percent increase in enrollment during this period?

34. The selling price of a house was dropped from $200,000 to $190,000 to make it sell faster. By what percent did the price drop?

35. Todd and Cassidy went on a diet together. Todd's weight dropped from 220 pounds to 190 pounds. Cassidy's weight dropped from 150 pounds to 135 pounds. Who had the greater percent weight loss?

36. Use a spreadsheet or a calculator with a TABLE feature to show the yearly balances for a 3-year, $1000 certificate of deposit that earns 6% interest compounded weekly. Show the formula(s) you used in your spreadsheet computations.

37. Use the percent key on a calculator to find the following products.
a. 48 × 40%
b. 288 × 56%
c. 125 × 0.5%
d. 462 × 125%

38. Estimate 63% of $495.

39. Estimate a 17% tip on a meal that costs $36.

40. Estimate the drop in the Dow Jones Industrial Average if it falls 5% from its value of 10,637.

41. One and one-half percent of the people who bowl score 200 or higher, on average. If 725 people bowled one evening, estimate the number of them that scored 200 or better.

42. Can a percent of increase be more than 100%? Why or why not?

43. Can a percent of decrease be less than 1%? Why or why not?

B. Deepening Understanding

44. Allen was earning $30,000 working for a small business. The business had a bad year, and Allen was asked to take a 10% cut in pay with the assurance that he would receive a 10% increase in salary the following year. If Allen agreed, what would be his salary in 2 years?

45. Suppose that a share of stock loses 15% in value and then gains 15%. Will its value be more or less than it was initially? Support your conclusion.

46. Kiisha is trying to estimate the cost of a reception that she will be hosting. She has invited 125 people and thinks that 60% of them will come. If the cost is $12 per person, estimate the cost of the reception. Explain how you obtained your estimate.

47. Suppose that a $10,000 investment is compounded quarterly and generates $4000 interest over a 5-year period. What simple interest would generate this same amount of money on an investment of the same size?

48. **The Discount Problem.** Renaldo bought a $600 television set that was on sale for 15% off. He then had to pay 6% tax on the discounted price. Renaldo wanted to pay the 6% tax on the $600 and then take the 15% discount on the original cost plus the tax. If the store would do it, would he save any money? Explain your reasoning.

49. For each of the following problems, create a percent bar. (Do not solve the problems.) For which problems are the percent bars the same except for the numbers?
a. Jack is hoping that his car insurance, which now costs $1500 a year, will go down 20% when he reaches his 25th birthday. How much of a decrease can he expect after he is 25 years old?
b. Ivie has saved $180 toward a summer trip that he is expecting to make. This amount is 30% of his goal. What is his goal?
c. Mr. Plumb lost 30 pounds from a high of 240 pounds. What percent weight loss did he achieve?
d. Grady determined that a 5% increase in salary would raise his salary $1600. What is Grady's salary?

50. Which is the best investment or are they all the same?
a. A growth of 5% followed by a growth of 10% followed by a growth of 15%
b. A growth of 15% followed by a growth of 10% followed by a growth of 5%

c. Three years of growth at 10% per year

51. Determine the worth of a $1000 investment at the end of the first year if it is compounded daily at an annual rate of 8%. What rate of simple interest would produce this same amount of money?

52. Would you rather receive simple interest at the rate of 15% per year or 1.5% per month? Explain.

53. Use a calculator to compute the value of a $1000 investment at the indicated interest rates (compounded annually) at the end of each year.

	6%	8%	12%
a. Year 6	?	?	?
b. Year 7	?	?	?
c. Year 8	?	?	?
d. Year 9	?	?	?
e. Year 10	?	?	?
f. Year 11	?	?	?
g. Year 12	?	?	?
h. Year 18	?	?	?

54. Consider the following rule of thumb for determining when an investment doubles for any given percent.

The number of years required for an investment to double may be approximated by dividing 72 by the whole number representing the percent.

Evaluate the accuracy of this rule based on the table you produced in Exercise 53.

55. The population of a city is 350,600. If 0.1% of the population earns more than $200,000, about how many people are in this category?

56. **The Car Problem.** A car that costs $20,000 depreciates about 10% per year. If the depreciation is compounded, what is the car worth after 5 years?

57. **The Inflation Problem.** If a person makes $30,000 in 2000 and the inflation rate is 4% annually, how much is this salary worth in the year 2004 (in terms of 2000 dollars)?

58. **The Retirement Problem.** Ever since he was a child, Rudy's parents had urged him to save money for retirement. When Rudy started working at age 20, he began saving $100 per month and did this for 10 years. At the end of 10 years this money had earned an additional $1000 interest. He then invested, using no new money, the savings plus the $1000 interest at 10%, and did this until age 60. How much money did he have at age 60?

C. Reasoning and Problem Solving

59. If a number is increased 5%, by what percent must it be decreased to obtain the original number?

60. If a number is decreased 20%, by what percent must it be increased to return to the original number?

61. If a number is increased 100%, by what percent must it be decreased to return to the original number?

62. **The Investment Problem.** Suppose that Rod invests $1000 at 6% compounded daily and Sheila invests $1000 at 7% (per year) simple interest. In how many years will Rod's investment be worth more than Sheila's investment? Complete the following table to answer your question. (The amounts for the first 2 years are given; you may not have to do all 10 years.)

Year	6% compounded daily	7% simple
1	$1061.83	$1070.00
2	$1127.49	$1140.00
3	?	?
4	?	?
5	?	?
6	?	?
7	?	?
8	?	?
9	?	?
10	?	?

63. **The Interest Rate Problem.** Mark claims that if interest rates double, a person makes twice as much money. Monica claims that if interest rates double a person makes more than twice as much money. Create a situation in which Mark is correct and another one in which Monica is correct.

64. **The Salary Increase Problem.** A worker makes $30,000 a year. After one year he gets a 5% raise. The next year he gets a 10% raise. If he had received no raise the first year, what percent increase would give him the same salary as consecutive raises of 5% and 10%?

65. Suppose that a person's consecutive salary increases are a% and b%. Would the person be better off, worse off, the same, or can't be determined if the two raises had each been $\frac{(a+b)}{2}$%? Explain your reasoning.

D. Communicating and Connecting Ideas

66. Suppose that you are trying to explain to someone how to determine a 15% tip at a restaurant without using paper and pencil. What directions would you give that person? Compare your directions with those of others in the class.

67. Write a one-page story on any topic of your choosing that uses the words *percent*, *ratio*, *interest*, and *decimal* in an appropriate way.

68. An employer wants her employees to be able to estimate mentally the 6% sales tax on the items in her store. What set of directions might she give the employees so that they could estimate a 6% sales tax?

69. **Connections.** Consider the survey result shown at the right. About how many couples were interviewed if 100 more men than women wanted to include the phrase "to honor and obey" in the wedding ceremony?

70. **Historical Pathways.** Italian manuscripts in the middle ages contained various expressions for percent. For example, Giorgio Chiarino (1481) used "xx.per.c" for 20 percent and "vii in x perceto" for 7 to 10 percent. Speculate on how you think 30%, 60%, and 5% might have been represented. [Amundson (1969), p. 147]

WEDDED TO TRADITION?

Prospective brides vs. grooms who say they will include the traditional phrase "to honor and obey" in the wedding ceremony:

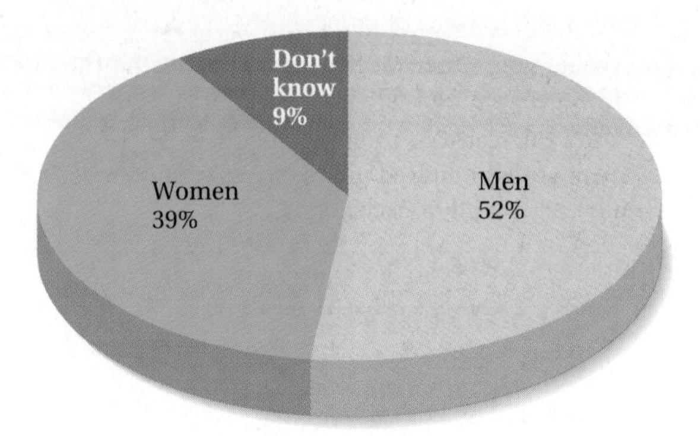

Source of data: A. Rudman and Associates and Freixenet survey. By Anne R. Carey and Marcia Staimer, USA TODAY. *USA Today,* June 13, 1995, p. D1. Copyright 1995, *USA Today.* Reprinted with permission.

CHAPTER SUMMARY

Key Ideas: Questions and Answers

Section 7.1

- **What are some ways to think about ratios?** *(pp. 346–347)* A ratio is an ordered pair of numbers used to compare like or unlike quantities. A ratio may be written x to y, $\frac{x}{y}$, $x \div y$, or $x:y$ $(y \neq 0)$. Two ratios are equivalent if their respective fractions are equivalent or if the quotients of the respective terms are the same.
- **What are some ways that ratios are used?** *(pp. 347–348)* Ratios can be used to compare like quantities or measures. A rate is a ratio that compares quantities that involve different units, such as rates involving time and rates involving unit pricing.
- **When might ratios be expressed as decimals?** *(p. 349)* Sometimes, using decimals as ratios is convenient; the decimal is assumed to have a denominator of 1.

Section 7.2

- **When do two quantities vary proportionally?** *(pp. 352-353)* Two quantities vary proportionally when, as their corresponding values increase or decrease, the ratios of the two quantities are always equivalent.
- **What are the characteristics of quantities that vary proportionally?** *(pp. 353–355)* When quantities a and b vary proportionally, a nonzero number k exists such that $\frac{a}{b} = k$, or $a = b \cdot k$ for all values a and b. When quantities a and b vary proportionally and the ratio of a to b is 1 to n, a unit change in a value of a always evokes a constant change of n in the corresponding value of b.]
- **How can proportional variation be recognized in geometric figures?** *(pp. 355–358)* The lengths of corresponding sides of similar figures vary proportionally.
- **What are some useful properties of proportions?** *(pp. 358–359)* Two useful properties of proportions are the cross-product property and the reciprocal property.
- **How can proportions be solved?** *(pp. 359–365)* Proportions can be solved by using properties of equations, by using the cross-product property, by equating numerators or denominators, and by converting one ratio to a decimal.

Section 7.3

- **What does percent mean?** *(pp. 369–370)* A percent is a ratio with a denominator of 100.
- **How can conversions among percents, ratios, and decimals be made?** *(pp. 370–371)* To convert a percent to a decimal, divide it by 100 and drop the % symbol. To convert a percent to a fraction, use the definition of percent and reduce the resulting fraction.
- **What are some types of percent problems?** *(pp. 371–374)* The three standard types of percent problems are (1) finding a percent of a number, (2) finding a number when a percent of it is known, and (3) finding the percent that one number is of another.
- **What are some ways to solve percent problems?** *(pp. 374–376)* Standard percent problems can be solved

by using percent bars and proportions to represent the problems.

- **How can solving a percent problem help to find percent increase or decrease?** *(pp. 376–378)* Percent increase or decrease can be found by (1) determining the amount of increase or decrease, (2) dividing this amount by the original amount, and (3) converting this fraction or decimal to a percent.
- **How can solving a percent problem help to find simple and compound interest?** *(pp. 378–382)* Simple interest is computed by multiplying the rate of interest (a percent) by the principal. The formula for calculating simple interest is $i = prt$. Compound interest pays interest on the interest accrued plus the original principal. The formula for calculating compound interest is $a = p(1 + \frac{i}{n})^{nt}$.
- **How can the percent of a number be estimated?** *(pp. 382–383)* The strategies for estimating products of whole numbers can also be used to estimate products involving percents. Certain types of percent are easy to compute mentally.

Key Terms, Concepts, and Generalizations

Section 7.1
Ratio (p. 346)
Equivalent ratios (p. 346)
Unit price (p. 348)

Section 7.2
Proportional variation (p. 352)
Multiplicative property of quantities that vary proportionally (p. 353)
Constant change property of quantities that vary proportionally (p. 355)
Proportion (p. 355)
Similar figures (p. 355)
Cross-product property of proportions (p. 358)
Reciprocal property of proportions (p. 359)
Solving proportional problems (p. 359)

Section 7.3
Percent (p. 369)
Types of percent problems (p. 371)
Solving percent problems (p. 374)
Percent increase or decrease (p. 376)
Interest rate (p. 378)
Principal (p. 378)
Interest (p. 378)
Simple interest (p. 378)
Formula for calculating simple interest (p. 378)
Compound interest (p. 380)
Formula for calculating compound interest (p. 381)
Estimating the percent of a number (p. 382)

CHAPTER REVIEW

Concepts and Skills

1. Use the following situation to create the ratios (a) whole to part, (b) part to part, and (c) part to whole.

   Twenty-five sixth-graders are taking a French class together. Eight have a parent who speaks French in the home at least occasionally. The remainder had never heard French spoken prior to the class.

2. Which car gives the better gasoline mileage in making the same trip: the car that goes 320 miles on 10 gallons of gasoline or the car that goes 500 miles on 15 gallons of gasoline?
3. Order the following prices from best to worst.
   a. 10 ounces for \$1.65 **b.** 9 ounces for \$1.50
   c. 6 ounces for \$0.96
4. If 6 cans of soft drinks cost \$1.68, estimate the cost of buying 18 cans of soft drinks.
5. Complete the following table so that p and q vary proportionally.

p	2	3	4	5	6
q	2.5	?	?	?	?

6. Give the dimensions of rectangles $MNOP$ and $XYTZ$ so that both are similar to rectangle $ABCD$.

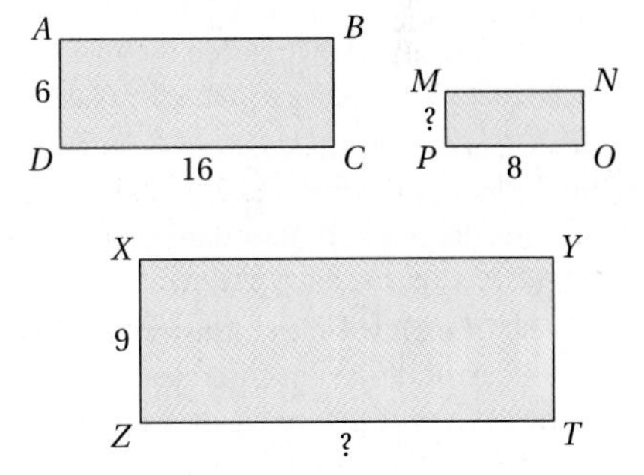

7. Answer true or false if r and s vary proportionally.
   a. $r = s \cdot k$ for some number k and for all values of r and s.
   b. $r = s + k$ for some number k and for all values of r and s.
   c. $\frac{r+2}{s+2} = \frac{r}{s}$.
   d. $\frac{2r}{2s} = \frac{r}{s}$.

8. If $\frac{m}{n} = \frac{p}{q}$, state two conclusions that follow, based on the properties of proportions.
9. Solve the following proportions.
   a. $\frac{x-2}{12} = \frac{5}{6}$.
   b. $\frac{0.4}{5} = \frac{x}{7.5}$.
   c. $\frac{15}{4} = \frac{45}{x}$.
   d. $\frac{4}{x} = \frac{21}{56}$.
10. If 3 sacks of chips cost $4.80, how much would you expect 10 sacks to cost?
11. Cindy plays on the golf team. She can buy a dozen golf balls for $28. But she wants only three balls. How much should she expect to pay if the price per ball remains the same?
12. How many squares on a 10 × 10 grid should be shaded to represent 3.5%?
13. Complete the following table.

Percent	Decimal	Fraction
$\frac{1}{4}$%	?	?
?	1.50	?
?	?	$\frac{19}{200}$

14. Kathy and Tommie put down 20%, or $25,500, to purchase a new house. What is the price of the house?
15. A survey of 1700 car drivers found that 87% of them used seat belts. How many indicated that they used seatbelts?
16. A car manufacturer claimed that less than 0.1% of seat belts fail in frontal accidents. If there were 100,000 frontal accidents, what would be the maximum number of seat belts that could be expected to fail?
17. Attendance for a major league baseball team averaged 29,500 for games in 1994 and 22,000 in 1995. What was the percent decrease in attendance from 1994 to 1995 (to the nearest tenth of a percent)?
18. In one year, the Dow Jones Industrial Average increased from 3945 to 4386. What percent increase does this represent to the nearest tenth of a percent?
19. A bank pays 5.5% simple interest on a certificate of deposit. If a person buys $2200 worth of the certificates, how much interest would be earned over a 2-year period?
20. How much more is an investment of $10,000 invested at 6% compounded daily worth than $10,000 invested at 6% simple interest after 5 years?
21. At Paradise Inn, there is a sales tax of 6% and a city tax of 12%. If the room rate is $85 per night, estimate the total cost of staying two nights.

Reasoning and Problem Solving

22. Sybil is planning a trip to Paris. She notes that $1 buys 5.5 French francs. If she doesn't want to pay more than $100 per night for a hotel room, what is the maximum number of francs she should pay?
23. **The Market Research Problem.** A company conducted market research on a product it is developing. The marketing department contacted 1000 people, half of whom were men and half of whom were women. The results indicated that one of every two men would use the product and one of every five women would use the product. Of the people contacted how many more men than women said that they would use the product?
24. **The Savings Problem.** Andy and Barney are debating about who is the better saver. Andy earned $30,000 and saved $1200. Barney earned $20,000 and saved $900. Who was the better saver? Justify your choice.
25. **The Survey Problem.** A company that specializes in surveys estimates that for a particular survey it is planning, 3 of every 5 people will respond to a questionnaire. If a second questionnaire is sent to those who didn't respond, 1 of 3 can be expected to respond. If a third questionnaire is sent to the remaining nonrespondents, 1 of 8 is expected to respond. If 600 people are surveyed initially, how many total responses can be expected if the three questionnaires are used?
26. **The Free Throw Problem.** A basketball player has made 135 free throws in 180 attempts. Of the next 50 free throws attempted, how many would she have to make to raise her percent of free throws made by 5%?
27. Write a fraction that represents 0.1% of a; that represents 200.5% of b.
28. **The Stock Problem.** A share of stock sold for $50 on Monday. On Tuesday it lost 5% of its Monday's value. On Wednesday it lost 10% of its Tuesday's value. On Thursday it lost 5% of its Wednesday's value. What percent increase would the share of stock have to have on Friday to achieve its Monday's value of $50?
29. **The Bank Problem.** A bank pays simple interest but claims that its simple interest is equivalent to another bank's interest rate of 5% compounded daily. What percent simple interest does the first bank pay (to the nearest hundredth of a percent)?

Alternative Assessment

30. A group of students are talking about the area of a rectangle, which equals the width times the length. They were wondering what it would take to increase the area

of a rectangle by 50%. The following students offered their opinions on how it could be done.

Rob: You should increase the longest side, the length, by 50%.

Jan: You should increase the shortest side, the width, by 50%.

Michael: You need to increase both the length and the width by 50%.

Michelle: You can't make a general statement. You need to know the exact size of the rectangle.

Bill: You can increase one dimension by 100% and decrease the other dimension by 25%.

With which student(s) do you agree? Justify why you agree with some and not with the others.

31. In a strange land, far far away, the following definition is given.

A percent is a ratio with a denominator of 10.

What model would you use to represent percent? How would you represent 0.5%? What would a 5% increase mean in this case? A 50% increase? Discuss the difficulties that this definition of percent presents.

8 Analyzing Data

Chapter Perspective

Everyday life is filled with statistics, from the stats for the latest ball game to discussions about voter polls and their influence on the direction of the stock market. As a result of this glut of data and interpretations, the average person can not be ignorant of the ways in which information is collected, organized, represented, and analyzed. The basketball game box score that you see in your local paper (like the one to the left) may not look simple, but it presents a remarkable amount of data in a concise format. With the data already collected and organized in this way, the reader is able to draw conclusions about the relative skills of each player. When individuals carefully develop the ability to analyze data-based presentations and are able to see the patterns imbedded in them, they have taken a significant step toward becoming better informed and more productive. In this chapter, you will explore the various ways in which data are displayed and how you can explore such displays for patterns without falling into some of the common error patterns caused by faulty or biased representations.

HOME 68, VISITOR 66

HOME	M	FG-A	FT-A	O-REB	A	PF	PTS
Williams	39	7-13	3-5	4-7	0	0	17
Sykora	34	6-15	3-5	2-4	0	2	17
Litkowski	23	3-9	0-0	1-6	2	4	6
Breen	37	3-9	0-0	0-2	2	1	8
Smith	40	6-15	3-3	1-7	5	2	18
Bailey	1	0-0	0-0	0-0	0	0	0
Alford	12	0-3	0-0	1-2	1	1	0
Lee	11	0-2	0-2	3-4	2	4	0
Sarma	3	0-1	0-0	0-0	0	2	0
Totals		**25-67**	**9-15**	**12-32**	**12**	**16**	**66**

Percentages: FG .373, FT .600. **Three-point goals:** 7-17, .412 (Smith 3-6, Sykora 2-4, Breen 2-5, Alford 0-2). **Team rebounds:** 9. **Blocked shots:** 4 (Litkowski 2, Williams, Sykora). **Turnovers:** 15 (Smith 6, Litkowski 2, Williams 2, Sykora 2, Lee, Breen, Alford). **Steals:** 10 (Smith 5, Litkowski, Lee, Williams, Breen, Sykora).

Connection to the NCTM Principles and Standards

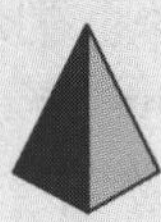

The NCTM *Principles and Standards for School Mathematics* (2000) indicate that the elementary school curriculum from prekindergarten through grade 8 should include the study of data analysis so that students can

- formulate questions that can be addressed with data and collect, organize, and display relevant data to answer them;
- select and use appropriate statistical methods to analyze data; and
- develop and evaluate inferences and predictions that are based on data (p. 48).

Connection to the PreK–8 Classroom

In grades PreK–2, students recognize the role played by data in answering questions and sort and classify objects and events based on their attributes and relation to questions being considered. Students are involved in creating representations based on concrete materials, pictures, and graphs and in answering questions based on such representations.

In grades 3–5, students design investigations to collect, represent, and analyze data related to a question of interest. They use averages and measures of spread to describe and differentiate between different data sets.

In grades 6–8, students are involved in comparing data about two populations or different characteristics of one population through one- and two-dimensional representations of relevant data. They draw conclusions based on the data they have, carefully avoiding common misconceptions in data usage and interpretation.

Section 8.1 TYPES OF DATA DISPLAYS

- Common Forms of Data Display
- Representing Frequency and Distribution of Data
- Representing Variation and Comparing Data
- Representing Part-to-Whole Relationships in Data

In this section we first focus on the common characteristics of data displays and then consider displays that show the frequency and distribution of data. We also introduce displays that show how data vary and that compare data. Finally, we demonstrate ways of displaying data that compare parts to the whole. Such data displays include frequency tables, dot plots, stem-and-leaf plots, and circle graphs. Mini-Investigation 8.1 begins your consideration of data displays by emphasizing the idea that every data display "tells a story."

Make a graph that someone could use to look for patterns in this set of data.

Mini-Investigation 8.1 **Communication**

What conclusions could you draw from the data in the following table about forest fires?

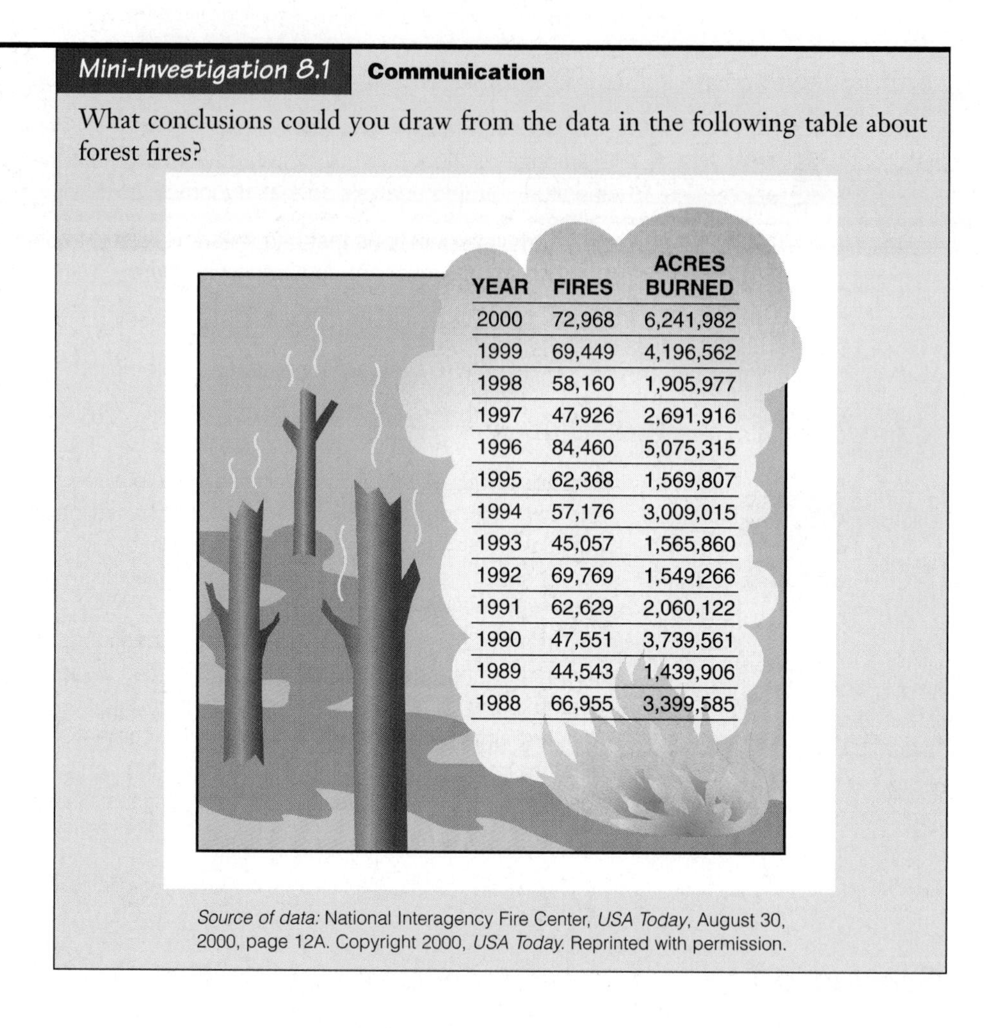

YEAR	FIRES	ACRES BURNED
2000	72,968	6,241,982
1999	69,449	4,196,562
1998	58,160	1,905,977
1997	47,926	2,691,916
1996	84,460	5,075,315
1995	62,368	1,569,807
1994	57,176	3,009,015
1993	45,057	1,565,860
1992	69,769	1,549,266
1991	62,629	2,060,122
1990	47,551	3,739,561
1989	44,543	1,439,906
1988	66,955	3,399,585

Source of data: National Interagency Fire Center, *USA Today*, August 30, 2000, page 12A. Copyright 2000, *USA Today*. Reprinted with permission.

Common Forms of Data Display

Since the beginning of human societies, people have recorded and interpreted numerical facts, or **data.** These activities have evolved into **statistics,** the science of collecting, classifying, and using data to interpret the significance of numerical information. To understand such information fully and to use it effectively require an ability to make **data displays,** select relevant data from such displays, and interpret the stories suggested by the displays.

The display of data in visual, or *graphic,* form can involve frequency tables, dot plots, stem-and-leaf plots, pictographs, histographs, bar graphs, or circle graphs. Each type of data display has special features that fit some data situations but not others. Many types of data displays may be created with spreadsheets or graphing calculators.

Technology aided or not, a graphic display of data should always have certain characteristics. Specifically, a graphic display should have a **title** explaining what the data represent and **labels** for the units or categories used in the display. The display itself should include **scales** that show the units and quantities of items portrayed. Such scales are usually shown along the number lines that form the axes of a graph. These features help the viewer interpret the data. That is, they make the story told by the data easier to understand and the related conclusions easier to grasp. Explore the technology you have available for this course—your scientific

TABLE 8.1 Coldest Reported Temperature (°F) at 70 U.S. Airports

Temperature	Tallies	Frequency
-50 to -41	II	2
-40 to -31	~~IIII~~ II	7
-30 to -21	~~IIII~~ ~~IIII~~ ~~IIII~~ III	18
-20 to -11	~~IIII~~ ~~IIII~~ ~~IIII~~ II	17
-10 to -1	~~IIII~~ ~~IIII~~ III	13
0 to 9	IIII	4
10 to 19	III	3
20 to 29	III	3
30 to 39	I	1
40 to 49		0
50 to 59	I	1
60 to 69	I	1

calculator, graphing calculator, computer software , or other aids to which you have access—and identify the options they give you for entering, organizing, scaling, summarizing, and displaying data.

Representing Frequency and Distribution of Data

Frequency Tables. Perhaps the most common form of displaying data is the **frequency table.** In this method of representing data, various data categories of interest are listed in a table. **Tally marks** are placed in each category to indicate the *frequency* with which observed data items fall into a given category.

For example, the data in Table 8.1 represent the coldest winter temperature observed at airport weather stations in 70 representative U.S. cities. Temperature categories based on 10° intervals were chosen, and a tally mark was made in one of these categories for each of the 70 temperatures in the set of data. Finally, the tally marks were counted to get the frequencies shown in the right-hand column.

The frequency table is a quick way to get a picture of the data. However, some of the specific data may be lost because data grouped into intervals rather than exact values are often shown. For example, in Table 8.1, one exact temperature might have been 26°. However, because only tallies for groups of temperatures were recorded, this specific temperature became only one tally in the 20–29° range. Nevertheless, the frequency table is a good way to get a quick look at the shape, or **distribution,** of the numerical data. Turn your book 90° to the left and examine the shape created by the tally marks. This shape represents the distribution of the temperatures. In this case, we quickly see two extreme temperature values. They represent Honolulu, Hawaii (53°), and San Juan, Puerto Rico (60°).

Dot Plots. One of the most elementary types of data displays is a **dot plot.** This form of display is an easy way to organize and view a set of data quickly. Dot plots are generally used with data sets of 40 numbers, or less. The numbers in the set are sometimes called **data points.** Consider the following readings that were found in measuring the depth of a river, to the nearest foot, at midstream at regular intervals along its course.

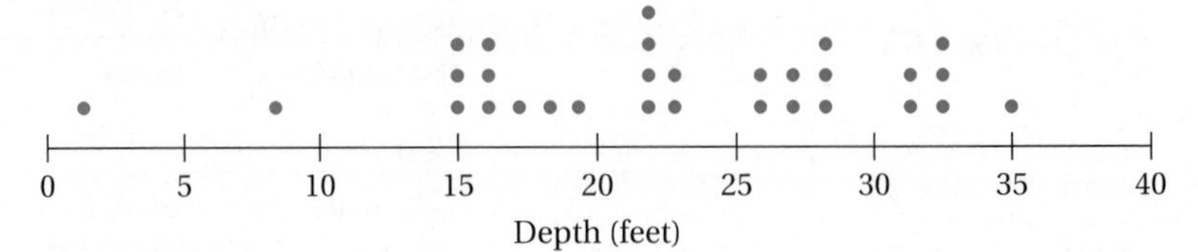

FIGURE 8.1 | A dot plot.

32	19	33	32	15	1	16	15	16	27
26	23	28	8	33	18	17	28	23	26
28	15	16	22	22	22	33	22	27	35

To make a dot plot of these data, we drew a horizontal number line with scale markings ranging from 0 to 40. Thus we established a scale starting with a mark for 0, the shallowest possible depth, and ending with a mark at 40, which slightly exceeds the maximum depth observed. To complete the dot plot we placed a dot above each of the depths observed, as shown in Figure 8.1, to indicate their frequencies.

An analysis of the dot plot indicates that the shallowest point measured in the river was 1 foot in depth and that the deepest point observed was 35 feet. These physical extremes in the variable being observed—in this case the depth of the river—set the lower and upper minimum boundaries for the scale on the horizontal number line, or **axis,** of the plot. The difference between the largest and smallest values in the data set is called the **range** of the data. In this case, the range is $35 - 1$, or 34 feet.

An examination of the graph shows that the depth readings of 1 foot and 8 feet are rather isolated and considerably shallower than the other depths observed. Such data points are often referred to as **outliers** because of their positions relative to other points. Outliers may occur at either end of a data set. Other groups of data points, such as those from 15 to 19, from 22 to 23, from 26 to 28, and from 32 to 35, are referred to as **clusters,** because they form identifiable subgroups of data points within the data set.

Dot plots make it easy to spot the extremes of the data set and to note both outliers and clusters of points. These extremes often signal differences or similarities in the data. Dot plots allow the viewer to see the location of the actual data values and to calculate the range. Both of these features are hidden in the categorized data of a frequency chart.

Example 8.1 demonstrates the use of data to make a dot plot.

Example 8.1 **Making a Dot Plot**

Use the following data to make a dot plot of the winning times for the men's 100-meter dash at the Olympic Games from 1900 to 2000.

11.0, 11.0, 10.8, 10.8, 10.8, 10.6, 10.8, 10.3, 10.3, 10.3, 10.4,
10.5, 10.2, 10.0, 9.95, 10.14, 10.06, 10.25, 9.99, 9.92, 9.95, 9.84, 9.87

Solution

Winning Times for Men's 100-Meter Olympic Dash

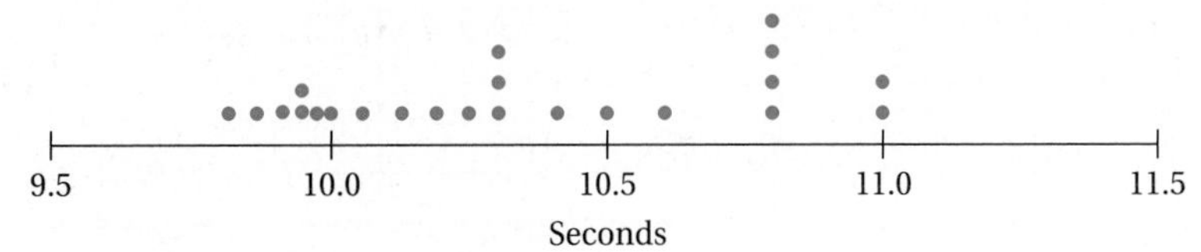

Source of data: http://www.trackandfieldnews.com.

Note that we might have spread the dot plot a bit more in order to alleviate the bunching of the last scores caused by the size of scale used.

Your Turn

Practice: Create a dot plot for the winning times for the men's 200-meter dash at the Olympic Games from 1900 to 2000, using the data given.

22.2, 21.6, 22.6, 21.7, 22.0, 21.6, 21.8, 21.2, 20.7, 21.1, 20.7, 20.6,
20.5, 20.3, 19.83, 20.00, 20.23, 20.19, 19.80, 19.75, 20.01, 19.32, 20.09

Reflect: Compare the nature of the data for the winning times of the two Olympic events shown in our dot plot and the one that you constructed. ■

Stem-and-Leaf Plots. Another common form of data display is the **stem-and-leaf plot.** This display is similar to the dot plot because it shows the actual data values, but it is usually shown in a vertical rather than horizontal format, more like the frequency chart. To make a stem-and-leaf plot, we begin by locating the extremes in the data set and finding the range for the data. For the river depth data, shown again below, we locate the extremes: 1 and 35 feet.

32	19	33	32	15	1	16	15	16	27
26	23	28	8	33	18	17	28	23	26
28	15	16	22	22	22	33	22	27	35

3
2
1
0

Then we separate each data value into its tens digit and its units digit. The tens digit serves as the *stem* and the units digit as the *leaf.* The stems are written to the left of a vertical line in either descending or ascending order, as shown.

Each of these stems has several leaves, or related units digits, associated with it. Filling them in with the appropriate stem to the right of the vertical bar results in an *unordered stem-and-leaf plot,* as follows, for the depth of river readings.

Depth of River Readings

Stem	Leaves	
3	232335	
2	7638836822227	2|7 represents 27 feet
1	956568756	
0	18	

If the leaves for a stem are ordered from smallest to largest, the resulting display is called an *ordered stem-and-leaf plot,* as shown.

Depth of River Readings

3	223335	
2	2222336677888	2 | 7 represents 27 feet
1	555666789	
0	18	

Other data sets might have 3-digit numbers such as 345, 457, 567, 382, 391, and 528. With 3-digit numbers, we might want to make the stems the hundreds digits and the leaves the remaining digits. In this case the leaves must be separated by a comma, as follows.

3	45, 82, 91	
4	57	3 | 45 represents 345
5	28, 67	

In other cases, the data might be numbers like 1.34, 1.53, 2.41, . . . Here the stem might be the whole number and tenths digit and the leaf the hundredths digit. This variation makes necessary the use of a **legend,** such as *2/7 represents 27 feet* in the graphs.

An ordered stem-and-leaf plot makes it easy to identify numbers that occur several times, like the 22 in the river depth data. Also, two stem-and-leaf plots can be combined back-to-back to compare better the data from two groups on the same measure. We illustrate this idea in Example 8.2.

Example 8.2 Making a Combined Stem-and-Leaf Plot

Make an ordered, combined stem-and-leaf plot for the winning men's and women's high jump heights (meters) for the modern Olympics, as follows:

Men:

1.81 1.90 1.80 1.78 1.91 1.93 1.93 1.98 1.94 1.97 2.03 1.98 2.04
2.12 2.16 2.18 2.25 2.23 2.25 2.36 2.35 2.38 2.34 2.39 2.35

Women:

1.59 1.66 1.60 1.68 1.67 1.76 1.85 1.90 1.82 1.92 1.93 1.97 2.02
2.03 2.02 2.05 2.01

Solution

We use a stem comprising the unit and tenths digits and a leaf comprising the hundredths digit, and show the combined stem-and-leaf plot as follows.

Winning Heights in Men's and Women's Olympic High Jump (meters)

Women's		Men's
	2.3	455689
	2.2	355
	2.1	268
53221	2.0	34
7320	1.9	01334788
52	1.8	01
6	1.7	8
8760	1.6	
9	1.5	

2.1 | 6 or 6 | 2.1 represents 2.16 meters

Source of data: http://www.trackandfieldnews.com.

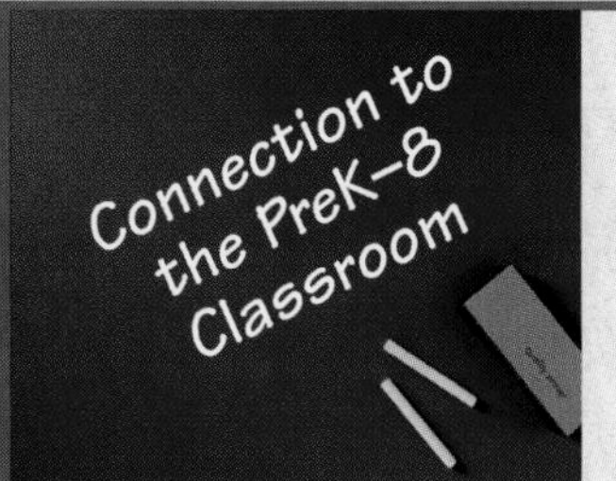

A Sample Student Page: Plotting Temperatures

The idea of a stem-and-leaf plot may have been a new form of data representation to you. However, it is widely used in industry and has already made its appearance in most elementary mathematics and science curricula as a quick way of searching for patterns in data. Study the page from *Scott Foresman–Addison Wesley Middle School Mathematics, Course 2* (2002), shown in Figure 8.2.

- *How are stems and leaves formed in this exercise? What other questions might be asked about the daily high temperatures?*
- *Like a dot plot, the stem-and-leaf plot makes it easy to find extreme values in a data set, to find the difference between the largest and smallest values, and to find the relative position of any piece of data to another. What are some questions about these ideas that can be answered with this stem-and-leaf plot?*

Example 2

Social Studies Link

Auckland is the largest city in New Zealand, but Wellington is the capital. New Zealand was the first nation to allow women to vote.

The table gives the high temperature in Auckland, New Zealand, on each day during the month of April. Make a stem-and-leaf diagram of the data.

Daily High Temperatures in Auckland, New Zealand (°F)					
75	67	83	90	79	74
70	71	72	78	76	67
66	80	77	77	84	74
64	76	79	82	76	85
71	81	69	83	75	84

Step 1 Order the data values from smallest to largest.

64, 66, 67, 67, 69, 70, 71, 71, 72, 74, 74, 75, 75, 76, 76, 76, 77, 77, 78, 79, 79, 80, 81, 82, 83, 83, 84, 84, 85, 90

Step 2 Separate each item into a stem and a leaf. Use the tens digit for the stem and the ones digit for the leaf.

Step 3 List the stems in a column from largest to smallest. List the leaves in order beside their stems.

Stem	Leaf
9	0
8	0 1 2 3 3 4 4 5
7	0 1 1 2 4 4 5 5 6 6 6 7 7 8 9 9
6	4 6 7 7 9

The diagram shows that most of the data values are in the 70s. The plot also clearly shows the highest temperature, 90°F, and the lowest temperature, 64°F.

Make an organized list.

Try It

a. Make a line plot of the data.
b. Make a stem-and-leaf diagram of the data.
c. Compare the two graphs.

Ages of Employees Surveyed					
21	27	30	33	17	20
15	23	21	30	42	24
30	17	21	16	22	23
16	21	30	17	23	28

Check Your Understanding

1. How would you put 589 on a stem-and-leaf diagram? What about 6?
2. How do you think the stem-and-leaf diagram got its name?
3. Explain how you could use a line plot to decide if there were any outliers in a data set.

18 *Chapter 1 • Making Sense of the World of Data*

FIGURE 8.2 | Excerpt from a middle school mathematics textbook.

(Source: From *Scott Foresman–Addison Wesley Middle School Math, Course 2* by R. I. Charles, J. A. Dossey, S. J. Leinwand, C. J. Seeley, and C. B. Vonder Embse. © 2002 by Pearson Education, Inc. publishing as Prentice Hall. Used by permission.)

Your Turn

Practice: Create a combined stem-and-leaf plot for the winning men's and women's discus throws, in meters, for the last twelve Olympic Games.

Women:

53.67	55.09	57.25	58.27	66.62	68.95
69.95	65.35	72.29	70.06	69.66	68.40

Men:

56.36	59.18	60.99	64.77	64.39	67.49
66.65	66.60	68.81	65.13	69.40	69.30

Reflect: What are the advantages of the combined stem-and-leaf display over a frequency table for describing the winning distances for the men's and women's discus throw at the modern Olympic games? ■

Representing Variation and Comparing Data

Pictographs, Histographs, and Bar Graphs. A graph commonly used in newspapers and popular magazines is the **pictograph.** In a pictograph, a small *icon*, or figure, is used to represent the data values. For example, to make the pictograph shown in Figure 8.3, first set up the age ranges and use the data to find the average household income for each range. Then choose an icon, such as a bundle of bills, to use in representing the average incomes. Based on the range of the average incomes and the manageable size of a graph, about 10 bundles seems about right, let each bundle represent $10,000 of income. Use a **key** (or *legend*) on the graph to indicate what the icon represents. Note that a fraction (one-half or one-fourth) of the icon may be used, as in the 15–24 age category, to represent a data value more accurately.

Pictographs have the advantage of quickly and visually showing differences in numerical values of different categories, but in doing so they lose precision of representation. For example, in Figure 8.3, estimating the actual value of a household income any closer than, perhaps, to the nearest $1000 for a given age group would be difficult.

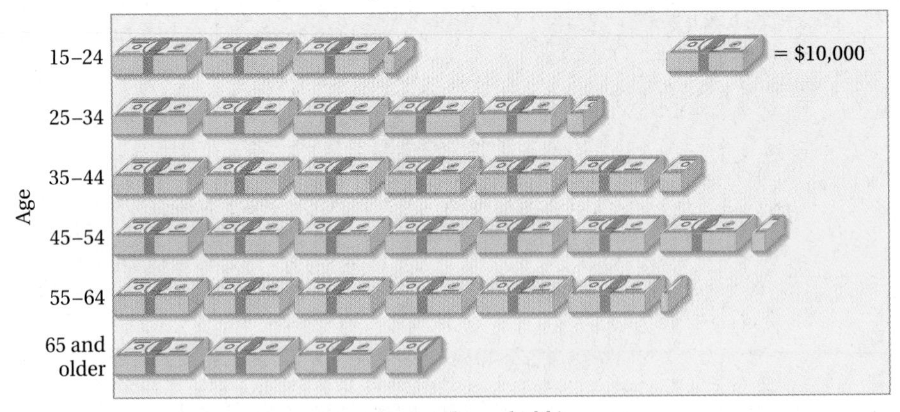

FIGURE 8.3 | A pictograph.

(*Source:* http://www.census.gov/hhes.)

Probably the most familiar forms of data display, often seen in the popular press, are the histograph and the bar graph. The **histograph** is similar to a stem-and-leaf plot, but the individual leaves are no longer distinguishable. The horizontal scale on a histograph runs from the smallest values on the left to the largest values on the right. Solid rectangular regions, whose heights are the same number of units as the number of leaves on a particular stem on the stem-and-leaf plot, replace the leaves.

Spotlight on Technology

Creating a Histograph Graphing calculators are useful not only in doing calculations of all types, but can also be used to display data graphically. In this feature we show you how to create a histograph to display student heights. The procedures for creating a scatterplot and a box-and-whisker plot are similar and are covered in Appendix A.

Step 1: Deselect functions that you don't want to be plotted.

Press [Y=]. If the equals sign of a function is highlighted, its graph will be displayed when you press [GRAPH]. For each function that you don't want to be displayed, move the cursor on to its equals sign and press [ENTER] to deselect that function's graph.

Step 2: Enter data values in a list.

See Entering Data in Appendix A to review data entry. Here, we used list L1 for the data set. The data represent heights of 15 female students enrolled in a summer mathematics program in California.

L1	L2
158	------------
176	
150	
165	
158	
158	

L1(6) = 158

Step 3: Turn on the stat plot, indicate plot type, and identify the data location.

a. Press [2nd] [PLOT]. Use the arrow keys to highlight 1:PLOT1 ... and press [ENTER].

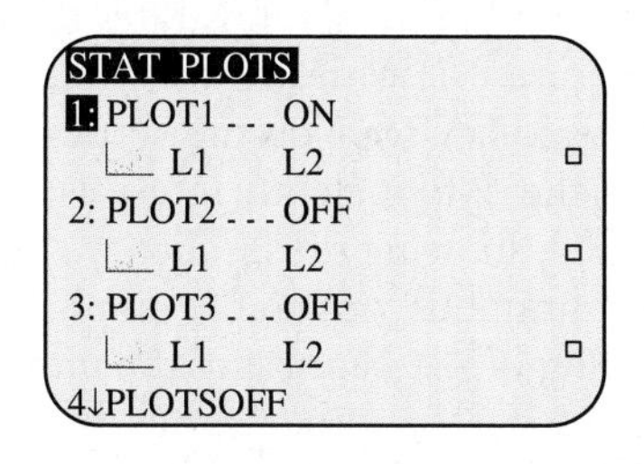

b. To turn on stat plot #1, highlight ON and press [ENTER].

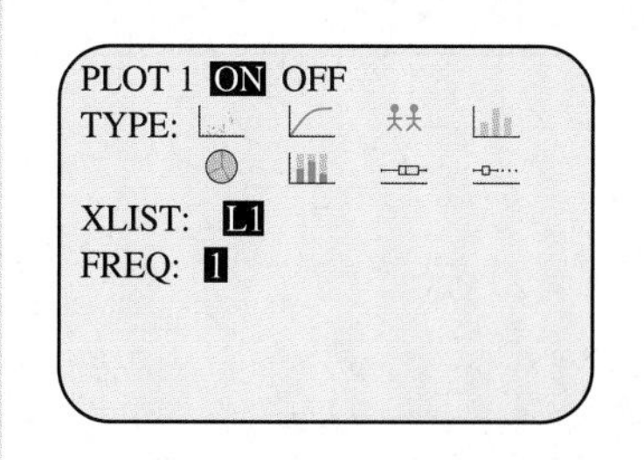

c. To indicate plot type, on the TYPE line highlight the histograph icon and press [ENTER].

d. To indicate location of the data, on the XL line highlight L1 and press [ENTER].

e. To indicate the frequency of occurrence of each data value, on the F line highlight 1 and press ENTER.

Step 4: Adjust window dimensions.

a. Press WINDOW.

b. Key in values for XMIN, XMAX, and XSCL. For XMIN and XMAX, refer to the smallest and largest values in the list. Here, we chose 150 and 180. The value for XSCL represents the size of each interval in the histograph. The value 5 indicates that each interval will span 5 units. Note that (XMAX − XMIN) ÷ XSCL determines the number of intervals to be shown in the histograph. Press ENTER to move to the next line.

```
WINDOW
 XMIN = 150
 XMAX = 180
 ΔX = .3191489361...
 XSCL = 5
 YMIN = 0
 YMAX = 6
 YSCL = 1
```

c. Key in values for YMIN, YMAX, and YSCL. The value YMAX should be larger than the greatest frequency in any interval of the histograph. The value YMIN is typically 0, the smallest possible number of occurrences in any interval. The value for YSCL indicates the distance between tic marks for the vertical scale of the histograph.

Step 5: Display the histograph.

Press GRAPH.

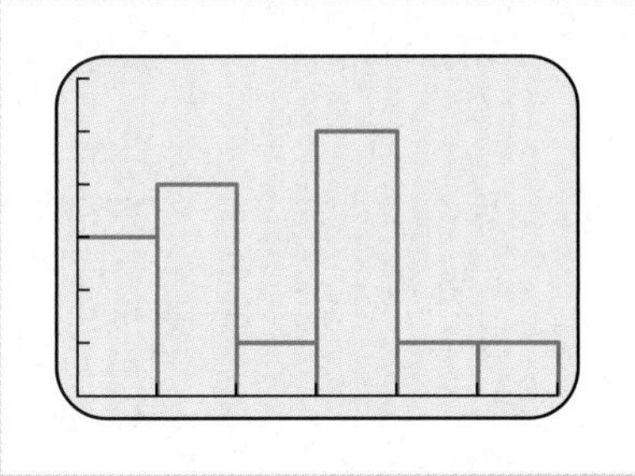

The graphing calculator allows you to create a histograph quickly and to use a visual representation of information in interpreting data relationships.

The histograph in Figure 8.4 illustrates the data on the depth readings for the river plotted in Figure 8.1, with depths shown in 5-foot intervals along the horizontal axis. The frequency scale on the vertical axis shows the number of readings falling in each of the 5-foot depth intervals. Note that the vertical rectangular regions in Figure 8.4, showing the frequency of depths in each interval, adjoin each other and give a continuous look to the graph. As with a frequency chart, the exact value of each reading is lost because of the intervals used.

Histographs, as well as other forms of graphs, can be easily created with graphing calculators. Sample instructions on the commands necessary to enter and edit data, select graphing options, and make relevant graphs are presented in Appendix A.

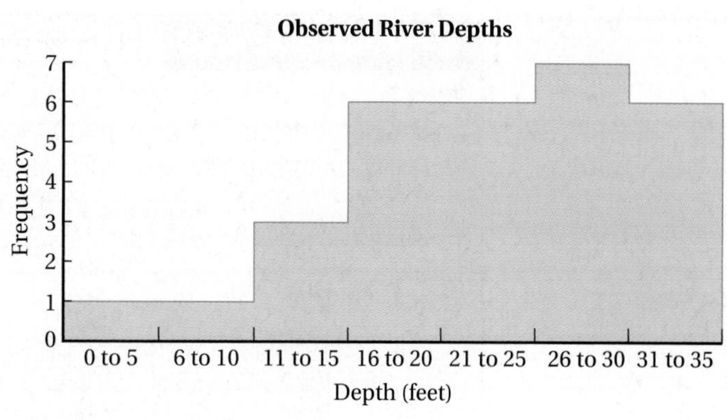

FIGURE 8.4 | A histograph.

A **bar graph** is used when the categories represent discrete groups, or at least nonadjacent intervals. This type of graph has space between the rectangular regions, unlike the histograph. For example, we chose the bar graph in Figure 8.5 to display information about public school expenditures because the data involve four discrete years from 1970 to 2000. These four years are shown along the horizontal axis, and an appropriately chosen range of dollars is shown along the vertical axis.

Histographs and bar graphs are often used to show variation over time or to compare items. However, they often hide exact values of data, as when intervals or joined categories are used, in which case they force the viewer to estimate the actual frequencies. Mini-Investigation 8.2 gives you the chance to select the most appropriate form of display for some data.

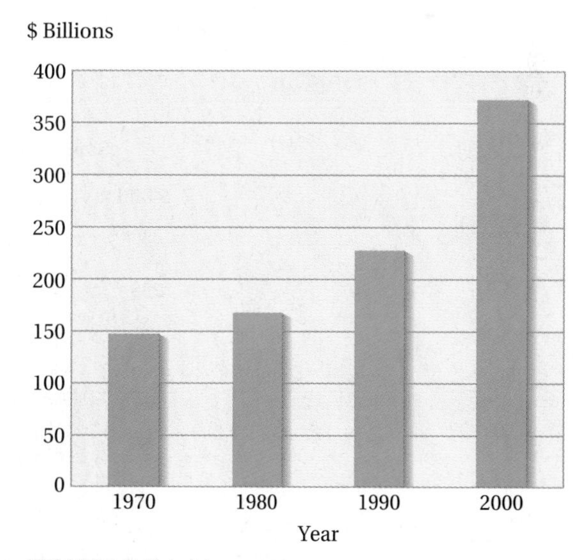

FIGURE 8.5 | A bar graph.

(*Source of data:* National Center for Education Statistics, *Digest of Educational Statistics,* 1999. Washington, D.C.: Department of Education, p. 34.)

Write a paragraph justifying your choice of graph.

Mini-Investigation 8.2 Using Mathematical Reasoning

Which form of display—pictograph, histograph, or bar graph—would be the most appropriate for displaying the following information on number of students attending public and private schools in the years indicated?

Year	Students in Public Schools (millions)	Students in Private Schools (millions)
1965	46	8
1970	52	8
1975	54	7
1980	50	8
1985	49	8
1990	52	8
2000	59	9

In certain bar graphs, the bars are divided to show sets of related data and trends in those data. Example 8.3 shows how to construct this type of bar graph.

Example 8.3 Creating a Divided Bar Graph

Create a divided bar graph for the following data on the source of funds for school expenditures (billions of 1990 dollars).

	Year		
Source	1970	1980	1990
Federal	25.2	30.9	29.2
State	74.6	105.2	132.7
Local	76.0	70.7	86.8

Solution

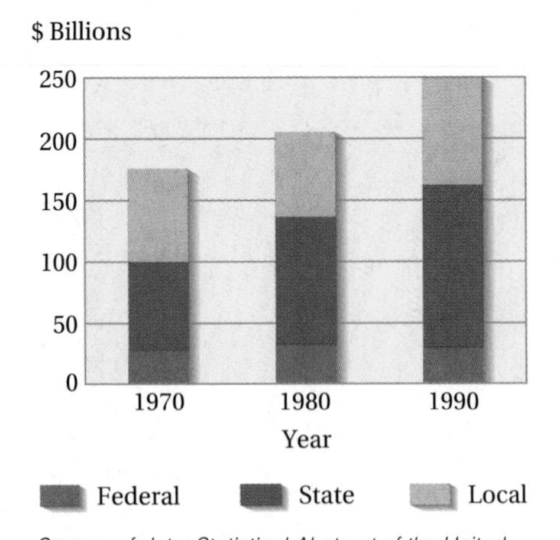

Source of data: Statistical Abstract of the United States, 1992. Washington, D.C.: U.S. Department of Commerce, Bureau of the Census, p. 141.

Note that the divided bar graph shows that, although the amount of support from local and federal funds has remained fairly constant, the amount of funding from state funds has steadily increased over the 20-year period studied.

Your Turn

Practice: Create a divided bar graph for the data on the racial and ethnic origins of the children enrolled in preprimary education (millions) in the United States for the years shown.

Group	Year 1975	Year 1991
White	3.4	5.1
Black	0.6	0.9
Hispanic	0.1	0.7

Reflect: Describe, in writing, the difference in the nature of the enrollments for the groups in the practice problem between 1975 and 1991. ■

Representing Part-to-Whole Relationships in Data

A **circle graph** or **pie graph** is often used to compare parts to a total in a data display, such as to compare the racial origins to the total population for the data given in Table 8.2. To show this comparison, a circular region is subdivided into a number of pie-shaped sectors, as in the circle graph in Figure 8.6. In this circle graph,

TABLE 8.2 | Racial Origin of U.S. Population in 2000

Racial Origin	Percent of Population
White	82.1
African-American	12.7
Native American/Eskimo	0.8
Asian or Pacific Islander	4.0
Other	0.4

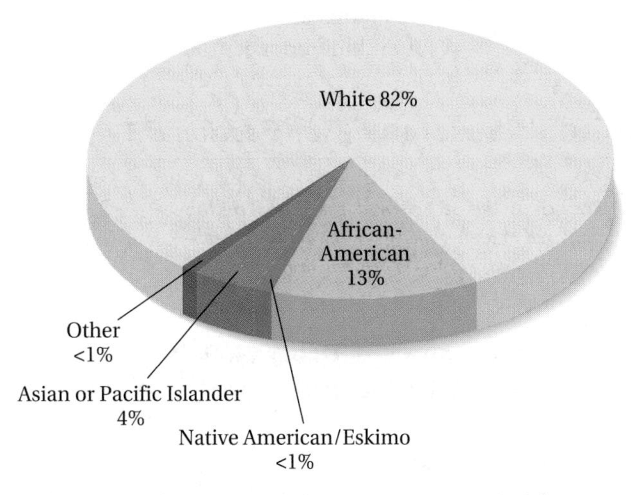

FIGURE 8.6 | Percent of U.S. population by racial origin.

(*Source of data:* http://www.census.gov/population.)

Education Level of Unemployed Workers

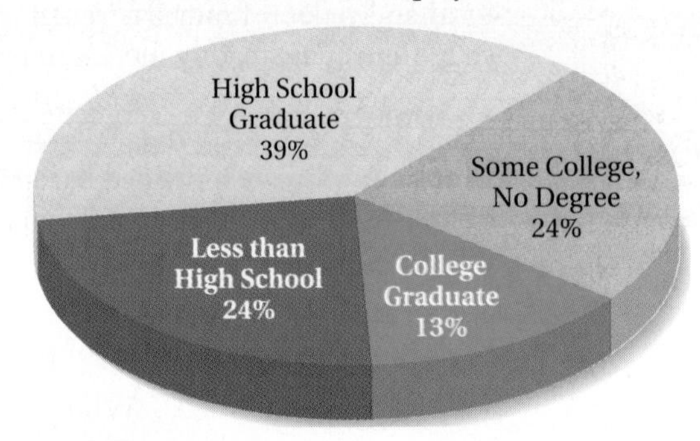

FIGURE 8.7

(*Source of data: Statistical Abstract of the United States, 1995.* Washington, D.C.: U.S. Department of Commerce, Bureau of the Census, p. 422.)

the area of the circle represents the total U.S. population in 2000. The area of the sectors of the circle represent the percent of the total population associated with each of five different racial origins.

To make the circle graph for these data, we first note that 82.1% of the area of the circle should be devoted to a sector representing white residents, 12.7% to African-American residents, and so on. Each sector of the circle representing a racial origin has a central angle that is the same percentage of the 360° central angle of the circle as the racial origin is of the U.S. population. For example, the sector representing African-American residents has a central angle of 46° because 12.7% of 360° is about 45°. The angles for the other sectors are found in a similar manner.

Circle graphs are often used in newspapers or magazines to show how different subgroups of the U.S. population feel about important issues. Other circle graphs show how leadership decisions have affected the government or a corporation. For example, the circle graph shown in Figure 8.7 shows how education and unemployment are related. Note that the information usually provided by a key or legend has been incorporated into the display. Although, like frequency charts, circle graphs are best used to show the relationship of parts to a whole, circle graphs often hide information because the exact values of parts aren't always given.

Problems and Exercises | *for Section 8.1*

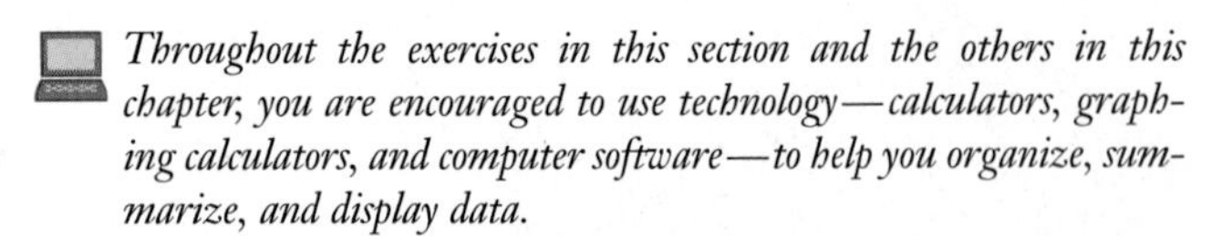

Throughout the exercises in this section and the others in this chapter, you are encouraged to use technology—calculators, graphing calculators, and computer software—to help you organize, summarize, and display data.

A. Reinforcing Concepts and Practicing Skills

1. Make a dot plot representing the grades for Ms. Johnson's geography students.

A C A C A A A C C A
B B B A D D F A B C
C B B A B B C B A D

2. Create a pictograph using automobile symbols to represent the world motor vehicle production information shown.

Year	Number of Cars
1970	29,300,000
1975	33,000,000
1979	43,600,000
1982	40,700,000
1984	42,000,000
1987	44,600,000

Table for Exercise 6

1.002	1.000	1.007	1.009	1.004	1.001	1.005	1.003	1.003	1.003
1.007	1.009	1.000	1.001	1.009	1.000	1.005	1.009	1.004	1.003
1.007	1.007	1.007	1.005	1.001	1.000	1.007	1.004	1.002	

3. Use the data in the grade distribution table to construct a circle graph showing the distribution of grade types in Ike's class.

Grade	Frequency
A	2
B	8
C	11
D	2
F	1

4. How would the graph change if there were twice as many of each grade?

5. Use the data in the graphic display to make a horizontal bar graph comparing the most popular sports for high school girls.

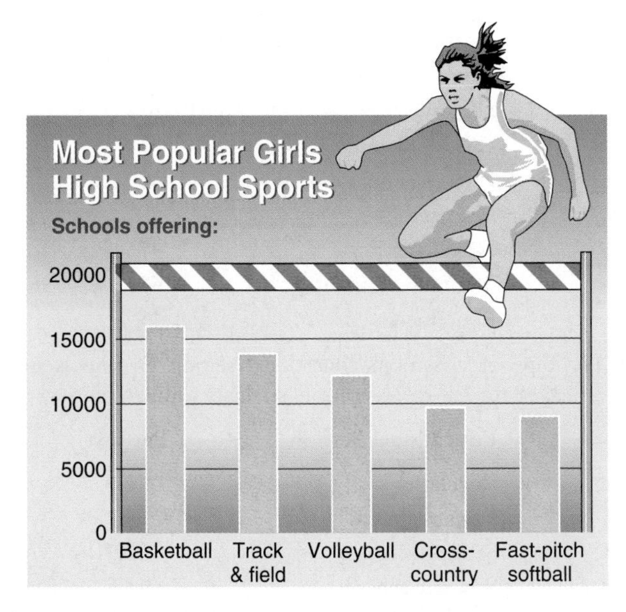

Source of data: National Federation of State High School Associations. By Elys A. McLean, USA TODAY. Copyright 1992, *USA Today.* Reprinted with permission.

6. Using the table at the top of the page, create a dot plot showing the widths of widgets manufactured at Acme, Inc., recorded during a quality control test.

7. Examine the data about the U.S. population in the following table.

Population by Region (millions)	Year 1970	Year 1990
Northeast	49.06	50.81
Midwest	56.59	59.67
South	62.81	85.45
West	34.84	52.79

a. Make a divided bar graph showing the relationship between the 1970 and 1990 populations.
b. Which region of the country grew most rapidly during the 20-year period?

8. Refer to the bar graph and write a short paragraph describing the backgrounds of people who voted (according to exit polls) for the newly elected mayor of Johnsville and the new mayor's likely political leanings.

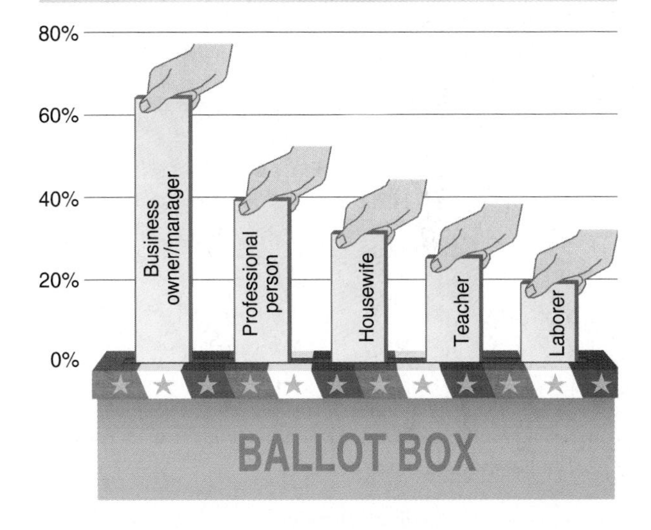

9. The table at the bottom of the page shows the distribution of the year of manufacture of the cars on Sam's Bold and Gold Used Car Lot.
a. Draw a stem-and-leaf plot of the years of manufacture.
b. What is the range of the data?
c. What model year is represented most frequently on the lot?

Table for Exercise 9

1976	1987	1989	1992	1992	1992	1983	1979	1978	1975
1986	1986	1985	1982	1990	1990	1991	1984	1979	1982

10. Write a short paragraph describing the information in the following graph.

Source of data: Land O'Lakes Holiday Bakeline poll. *USA Today,* November 30, 2000, p. 1A. Copyright 2000, *USA Today.* Reprinted with permission.

11. Write a short paragraph describing the distribution of grades shown in the following stem-and-leaf plot.

Grades on Geometry Examination

Stem	Leaf
10	000
9	1124599
8	135669
7	2333479
6	0123667

6 | 0 means 60

12. Write a short paragraph comparing the grades of the students in the first-hour geometry class with those in the last-hour geometry class.

Geometry Grades

First-Hour Class		Last-Hour Class
0	10	000
94443321111	9	1124599
9988777655543	8	135669
5531	7	2333479
	6	0123667
1	5	

6 | 0 means 60

B. Deepening Understanding

In Exercises 13–18, select what you consider to be the most appropriate form for displaying the data. For each set, develop and appropriately label a graph. Then write a short paragraph justifying your selection.

13. Annual pedestrian fatalities per 100,000 population.

Country	Fatalities
Hungary	6.6
Germany	6.4
Austria	5.9
Denmark	5.6
Finland	5.6
Poland	5.2
France	5.1
United Kingdom	4.2
United States	4.0
Italy	3.8

14. Stress units of various life experiences.

Experience	Stress Units
Death of spouse	100
Divorce	73
Marital separation	65
Jail term	63
Death of family member	63
Personal injury or illness	53
Marriage	50
Fired from job	47
Marital reconciliation	45
Retirement	45

15. Type of classroom educational setting in schools in 1988 for 18- to 21-year-old students with disabilities.

Educational Setting	Percent
Regular class	12.9
Resource room	35.2
Separate class	32.7
Separate school	14.7
Residential facility	2.8
Home/hospital	1.6

16. Prices (in dollars) for last year's automobiles, as listed in a sale ad in the automobile section of the newspaper.

12,997	14,891	18,666	21,799
17,495	17,266	21,517	14,414
14,257	15,634	15,734	27,825
28,015	19,514	14,182	19,135
19,555	12,333	7,700	9,898

17. Time (in years) for common objects to disintegrate if left as trash.

Object	Time
Traffic ticket	0.1
Cotton rag	0.5
Degradable plastic bag	0.3
Cotton rope	0.7
Wool stocking	1.0
Bamboo pole	2.0
Unpainted wooden stake	3.0
Painted wooden stake	13.0
Wooden telephone pole	36.0
Railroad tie	35.0

18. Product manufacturing levels for the 50 states in 1989 ($ millions).

23	25	12	145	19	89	441	203	228	212
105	256	182	94	94	53	100	11	11	31
49	15	99	26	136	28	130	60	130	227
66	92	68	38	37	79	52	340	13	16
11	66	25	65	28	28	96	52	697	20

19. Consider the data on AIDS deaths, by age for the period 1982–1995, as shown in the circle graph. Use the data in the graph to answer the following questions.

a. Which age group accounted for the highest percentage of AIDS deaths? The lowest?

b. What hypotheses might you make for the patterns noted in part (a)? Explain.

c. Does the percentage of deaths decrease as age increases? Why or why not?

d. Explain the pattern that led to your answer to part (c).

Distribution of AIDS Deaths, by Age: 1982 Through 1995

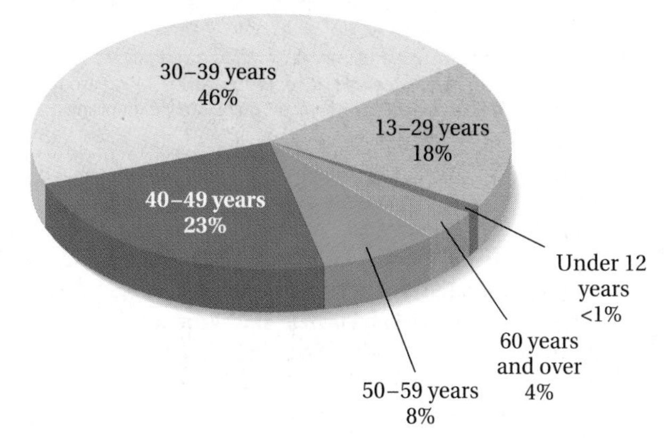

Source: Statistical Abstract of the United States, 1995. Washington, D.C.: U.S. Department of Commerce, Bureau of the Census, p. 97.

20. Consider the following set of ordered pairs denoting the number of inches of (November snowfall, December to May snowfall) for Minneapolis–St. Paul, Minnesota, for each of the winters from 1950–1951 through 1969–1970. Analyze these data and describe any patterns you see.

(5.6, 83.3), (10.8, 67.4), (10.1, 32.8), (1.9, 23.8), (6.4, 27.1), (6.0, 36.7), (6.8, 32.3), (10.3, 10.9), (3.3, 15.8), (6.9, 21.2), (2.4, 37.8), (2.5, 78.8), (5.6, 28.9), (0.0, 28.9), (4.3, 69.4), (1.6, 34.5), (3.4, 74.8), (0.8, 16.4), (4.9, 63.2), (3.8, 57.0)

C. Reasoning and Problem Solving

21. Another form of bar graph display is the back-to-back bar graph shown.

Suicide by Five Methods (A–E)

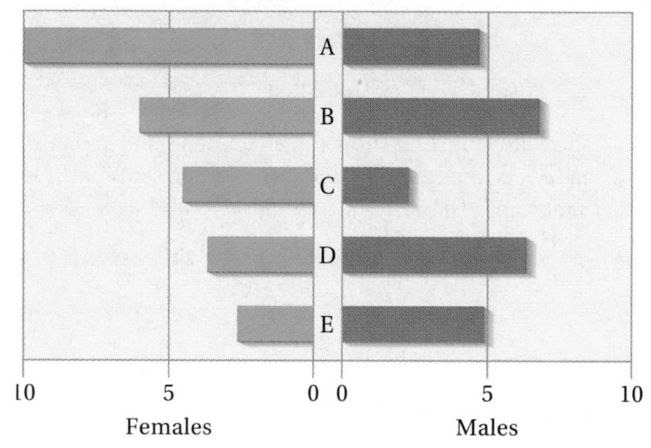

Source of data: Statistical Abstract of the United States, 1992. Washington, D.C.: U.S. Department of Commerce, Bureau of the Census, p. 90.

Use the following data about suicide deaths in the United States in 1989.

Age	Gender: Male	Gender: Female
Under 15	200	100
15–24	4100	800
25–34	5300	1200
35–44	4100	1200
45–54	2700	900
55–64	2500	800
65–74	2700	600
75–84	1900	400
85 and older	600	100

a. Make a back-to-back bar graph displaying this information.

b. Describe any patterns in the data.

c. Describe any differences by age and gender in suicide rates.

22. Bar graphs may have side-by-side columns for data related to the same person, as shown.

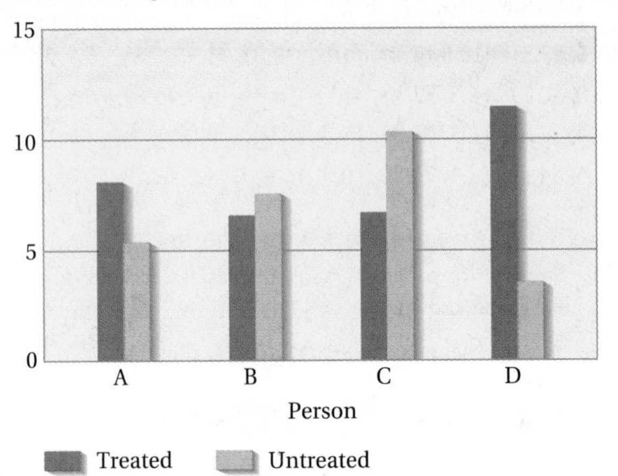

The following data show the effect of a hand cream on perceived roughness of men's hands. The six-point scale runs from 0 (smooth) to 6 (rough).

Person	Treated	Untreated
A	5	6
B	0	2
C	3	3
D	4	5
E	4	4
F	5	6
G	1	0
H	0	2
I	2	4
J	1	3

a. Make a side-by-side bar graph for this information.
b. Describe any patterns in the data.
c. What differences did you note in the softening of hands as a result of the use of the treatment?
d. As an advertiser of this product, what would you claim, based on these data?

23. An approach similar to that used in Exercise 12 may be used to create back-to-back stem-and-leaf displays. The stems are common and enclosed between a pair of lines, as we showed on page 396. Consider the two sets of test scores from Class A and Class B.

Class A	120	130	118	140	140	135	126	130	125
	127	119	115	128	129	131	132	141	135
Class B	128	131	127	132	141	137	118	132	130
	135	140	141	130	126	141	138	121	142

a. Make a back-to-back stem-and-leaf graph for these groups of data.
b. Write a short description of any differences in the patterns of performance in the two groups.
c. Identify any clusters or outliers in either group.

24. Describe a situation in which a stem-and-leaf display would be preferable to a circle graph for displaying data. Provide some specific information.

25. Describe a situation in which a dot plot would be preferable to a frequency table for examining some data. Give a specific example and illustrate your reasoning.

D. Communicating and Connecting Ideas

26. Write a short paper describing a time when you collected data and made a graphic display for a project or a report. What was your display about and how effective was it in explaining the case you wanted to make?

27. **Making Connections.** Study the following display, which contains information about John D. Rockefeller's oil empire. Write a paragraph interpreting this information and what it tells about the control that Rockefeller, who owned Standard Oil, had on the oil industry.

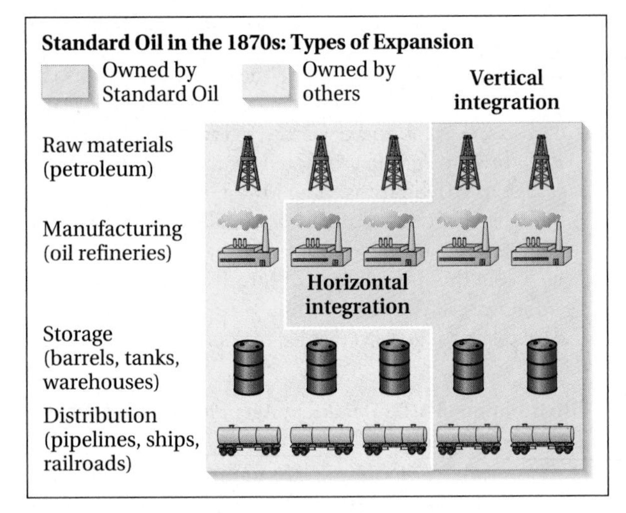

Source: From *The United States and Its People* by David C. King, Norman McRae, and Jaye Zola. © 1993 by Pearson Education, Inc. publishing as Prentice Hall. Used by permission.

28. **Historical Pathways.** William Playfair, a British mathematician of the late 1700s and early 1800s, is often considered the father of graphic displays of data. Do some research on Playfair and write a paper about his life, his use of statistics, and the problems confronting him that led Playfair to his methods.

Section 8.2 DATA DISPLAYS THAT SHOW RELATIONSHIPS

- Displaying Two-Variable Data
- Scatterplots and Trend Lines

In this section we examine two common data displays—line graphs and scatterplots—that are used to show a relationship between two variables. We use these displays to look for patterns in the data and to consider the possibility of cause and effect. Mini-Investigation 8.3 asks you to make a graphic display and look for patterns in it.

Talk about the patterns you see in the graph and whether women will eventually swim faster than men in the event.

Mini-Investigation 8.3 **Finding a Pattern**

How would you make a graphic display that shows both the men's and women's winning performances (in minutes) over time in the 1500-meter swimming freestyle event at the Olympic games for the years shown?

Year	Men's Time	Women's Time	Year	Men's Time	Women's Time
1948	4:41.0	5:17.8	1972	4:00.27	4:19.44
1952	4:30.7	5:12.1	1976	3:51.93	4:09.89
1956	4:27.3	4:54.6	1980	3:51.31	4:08.76
1960	4:18.3	4:50.6	1984	3:51.23	4:07.10
1964	4:12.2	4:43.3	1988	3:46.95	4:03.85
1968	4:09.0	4:31.8			

Displaying Two-Variable Data

The graph shown in Figure 8.8 illustrates a real-world situation in which the relationship between two variables, years of service and average annual paid days off, is of interest. Consider a slanted line suggested by the tops of the bars in the graph.

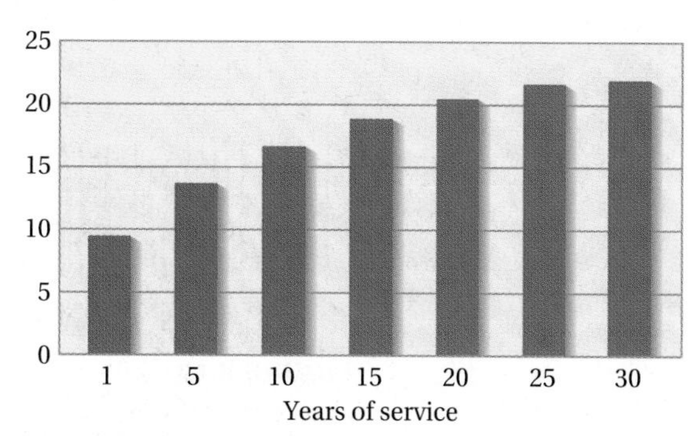

FIGURE 8.8

(*Source of data:* U.S. Bureau of Labor Statistics. By Anne R. Carey and Marcia Staimer, *USA Today. USA Today*, March 6, 1997, p. B–1. Copyright 1997, *USA Today*. Reprinted with permission.)

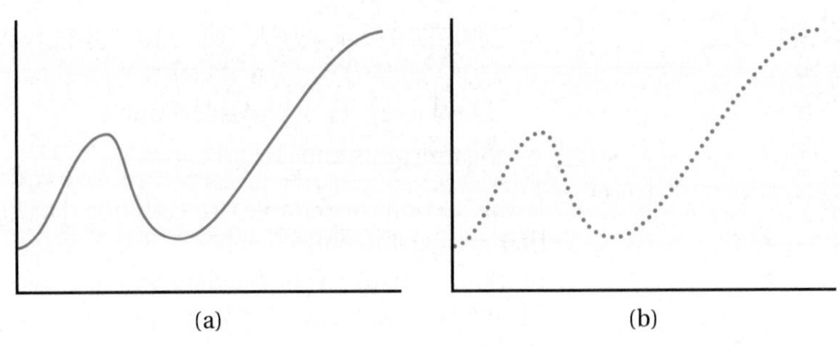

FIGURE 8.9 | (a) Graph of continuous data; (b) graph of discrete data.

Such a line would describe a trend in the data. In Mini-Investigation 8.3, you may have created a graph that showed a relationship between two variables, years and winning times, in the 1500-meter freestyle.

In this subsection we examine various types of data displays that are commonly used to illustrate relationships between two variables. The choice of a method usually depends on whether the data are **continuous data,** measurable at each point in time (such as your actual body temperature), or **discrete data,** measurable at only some finite number of points (such as the daily high temperatures observed over the past month). In the first case, the display should show the values for each point in time; in the second case, the display should show only a finite number of readings or data points. These different requirements result in two distinct types of graphs, as shown in Figure 8.9.

Line Graphs. Perhaps the most common method of showing the relationship between two variables is the **line graph.** This type of graph may be used to describe either discrete or continuous events. It is constructed by plotting the ordered pairs of data, say, time of day and temperature, on a rectangular coordinate grid. The axes on the grid should be scaled so that they represent accurately the nature of the changes. If the data are continuous, only a finite number of the data points can actually be graphed, but solid line segments can be drawn between data points, connecting them to show continuity. If the data are discrete, the data points can be plotted and then connected with dotted line segments.

The electroencephalograph shown in Figure 8.10 illustrates a relationship involving continuous data. It provides a line graph of electrical behavior of an individual's brain during a petit mal seizure. The axis scales are not shown.

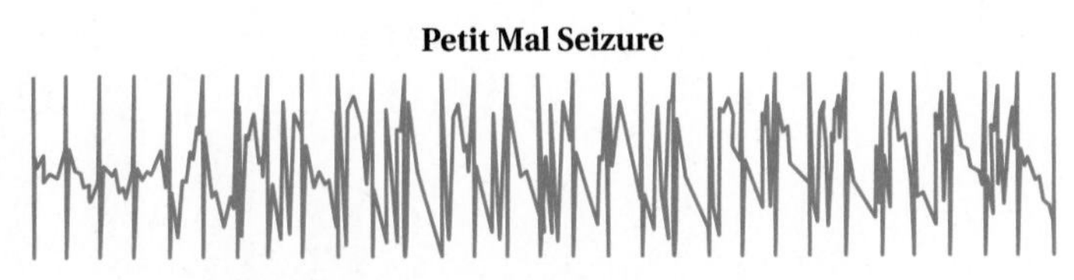

FIGURE 8.10

(*Source:* Miller, B. F., and Keane, C. B. *Encyclopedia and Dictionary of Medicine, Nursing, and Allied Health.* Philadelphia: Saunders, 1978, p. 326. Reprinted with permission.)

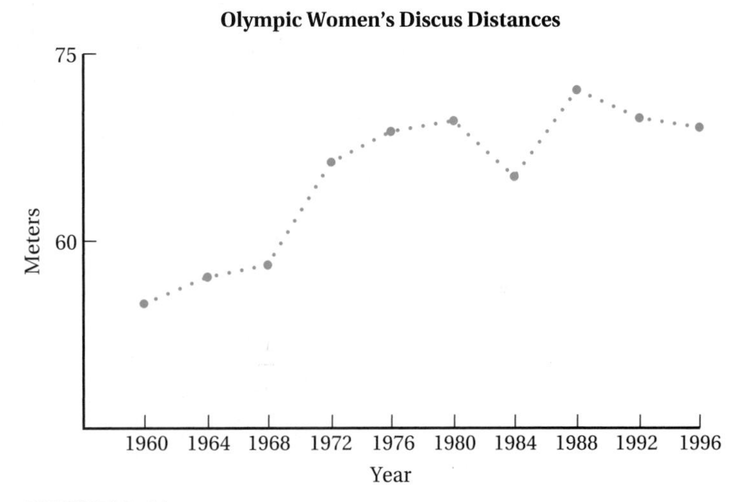

FIGURE 8.11

(*Source of data:* http://www.trackandfieldnews.com.)

In Figure 8.10, the actual information for the brain activity is produced for each point in time by a measuring instrument. The continuous graph is produced by the instrument itself, and no drawing of segments or arcs is needed. A similar type of graph is produced by seismic instruments that record and show the earth's movement during an earthquake.

Another example of a line graph is that shown in Figure 8.11, which shows the winning distances for the women's discus throw at each of the Olympic games from 1960 through 1996. Here the data are discrete, as there is no Olympic competition, and thus no winning distance, during the intervening years.

Even though the graph is discrete, we connected the points with dotted lines to help you look for patterns in the data. The estimation application encourages you to use the trends observed in the graph to make a prediction.

Line graphs provide an excellent way of studying changes, or trends, over time. However, when looking for patterns in line graphs, be sure to examine the scales and visual representation of the data. These features can be chosen in ways that give a particular slant or emphasis to the data. Line graphs have certain limitations, such as an inability to represent *exact* amounts of change in a certain quantity at various points in time. For example, a dot plot, a histograph, or a scatterplot (the next form of graph that we discuss), rather than a line graph, might be the best way to represent the actual enrollment at the end of the first 10 days of successive fall semesters at a college. Example 8.4 provides experience in interpreting a line graph.

Estimation Application

Extending the Throw

Estimate a projected winning throw for the women's discus for the 2000 Olympic Games from the data provided in the graph shown in Figure 8.11. How does your estimate compare with the actual winning throw of 68.40 meters in 2000?

Example 8.4 Interpreting a Line Graph

Interpret the story portrayed by the line graph for 2000 daily closing averages for the Dow Jones Industrial Average of the New York Stock Exchange. If you were writing an article on stocks, what would you say, based on the information in the graph?

Dow Jones—12 Month Daily Closes

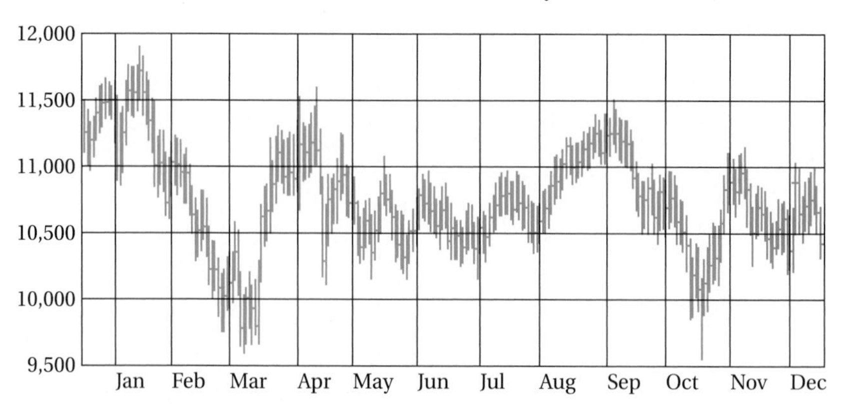

Source: Associated Press/Wide World Photos, 2001. Used with permission.

Solution

The Dow Jones Industrial Average varied a great deal during 2000. It began the year over 11,000 and climbed to nearly 12,000 in January. Then it went into a steep decline, dropping to nearly 9,500 points before recovering some of its value. During the middle of the year, the average varied about 10,500. It then rose briefly, but fell again to the year low in October. Moving into the 2000 election, it increased, but dropped back then again with the indecision following the election, ending the year with mixed action.

Your Turn

Practice: Write a paragraph describing the information presented in the following graph on the U.S. fishery products market during the past 30 years. The graph was drawn as a continuous line graph, although the data on which it is based are

Fishery Products—
Domestic Catch and Imports: 1960 to 1990

Live weight, billions of pounds

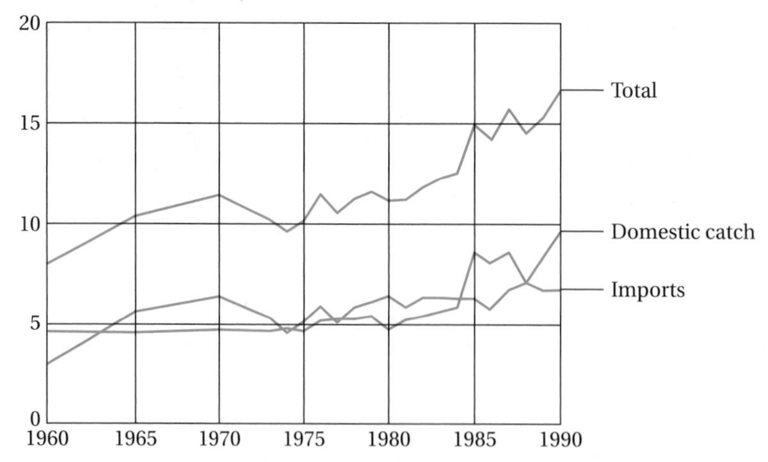

Source: Statistical Abstract of the United States, 1992. Washington, D.C.: U.S. Department of Commerce, Bureau of the Census, p. 668.

Relationship of Years of Operation and Employees

Years	Employees
1	4
2	6
3	6
4	7
5	8
6	8
7	10
8	12
9	15
10	18

FIGURE 8.12 | Scatterplot showing a relationship.

daily data. It appears to be continuous owing to the small fluctuation from one day to the next for the scale used.

Reflect: What advantages does such a multiple line graph have over individual line graphs in representing these types of data? ■

Scatterplots and Trend Lines

Scatterplots. For two-variable data that include a selected number of ordered pairs, the appropriate graph to use is the **scatterplot.** For example, you might want to examine the relationship between the number of years that a store has been open and the number of employees it had in each of those years. Here, the data consist of **ordered pairs** of the form (year, employees in year). The graph of these ordered pairs is called a scatterplot because it shows how ordered pairs of data are positioned relative to one another and to the values on the graph's axes. The scatterplot shown in Figure 8.12 consists of appropriately scaled and labeled axes and graphs for 10 ordered pairs. The pattern is one of constantly increasing numbers of employees over time, with a slightly more rapid increase in the later years.

Scatterplots can be easily created with graphing calculators. Sample instructions on the commands necessary to make a scatterplot are contained in Appendix A.

In some scatterplots, strong relationships appear. For example, consider the scatterplot shown in Figure 8.13, which shows the relationship between the overall "productivity" of 10 states, in terms of their gross state products (GSP), in dollars per student versus the amount of money spent by the state per student on education.

The general pattern in the graphs of the ordered pairs of data in this scatterplot is from the lower left to the upper right. This pattern represents a **positive correlation** between the values for the two variables. In other words, lower GSP values tend to be matched with lower support values and higher GSP values tend to be matched with higher support values for the states studied.

A second example of a scatterplot is shown in Figure 8.14. Here, the average percent of eighth-grade students in 10 states who watch 6 or more hours of TV daily is paired with the overall average mathematics proficiency of these students on the National Assessment Proficiency scale (0 to 500) for 1990. This pattern of data points falling from the upper left to the lower right represents a **negative correlation** between the values of the variables. That is, higher scores on one variable tend to be matched with lower scores on the other variable.

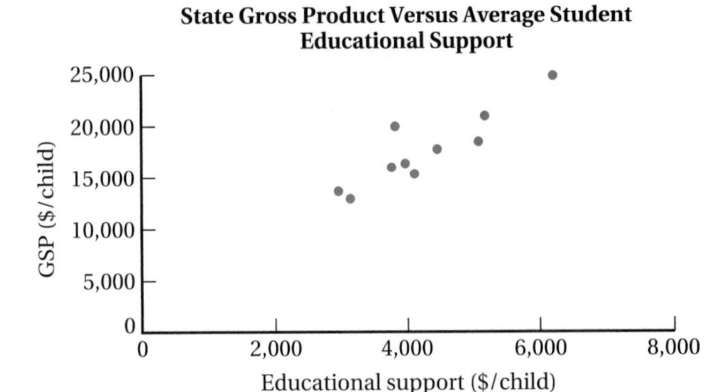

FIGURE 8.13

(*Source of data:* Mullis, I. V. S., et al. *The State of Mathematics Achievement.* Washington, D.C.: National Center for Education Statistics, 1991. Used with permission.)

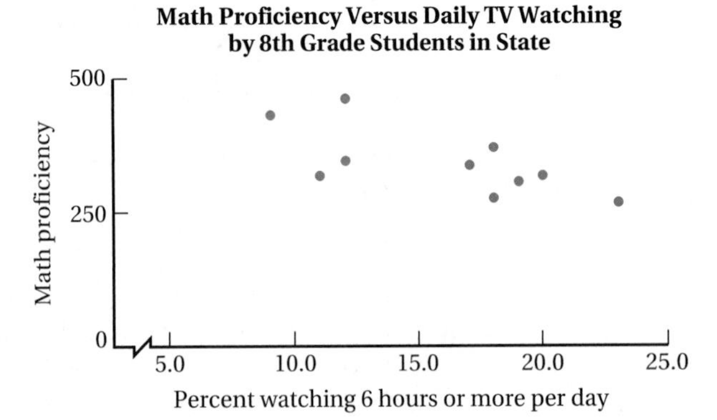

FIGURE 8.14

(*Source of data:* Mullis, I. V. S., et al. *The State of Mathematics Achievement.* Washington, D.C.: National Center for Education Statistics, 1991.)

Individuals interested in policy decisions often use scatterplots and try to interpret the relationships between variables by drawing **trend lines** to model the relationship between the variables in a scatterplot. The trend line shown in Figure 8.15 indicates an approximate relationship between the mathematics proficiency score and the percentage of eighth-grade students in a state watching 6 or more hours of TV daily. Later in this book we demonstrate how to find equations for such lines.

Caution: The existence of a visual relationship in a scatterplot does not mean that a high score on one variable *causes* a high score on the other, but merely that they tend to be related in some way. *No causation* is implied; only a report, or measure, of the pattern or relationship observed is shown. Sophisticated statistical analyses must be used to investigate possible causation. Example 8.5 provides an opportunity to analyze a scatterplot further.

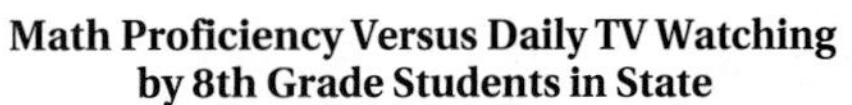

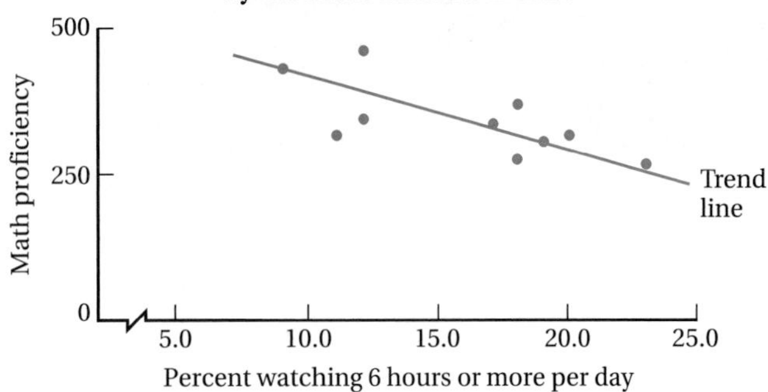

FIGURE 8.15

(*Source of data*: Mullis, I. V. S., et al. *The State of Mathematics Achievement.* Washington, D.C.: National Center for Education Statistics, 1991.)

Example 8.5 Analyzing a Scatterplot

Describe the pattern or relationship in the following scatterplot, which relates the number of grams of protein and the number of grams of fat in popular fast-food entrees.

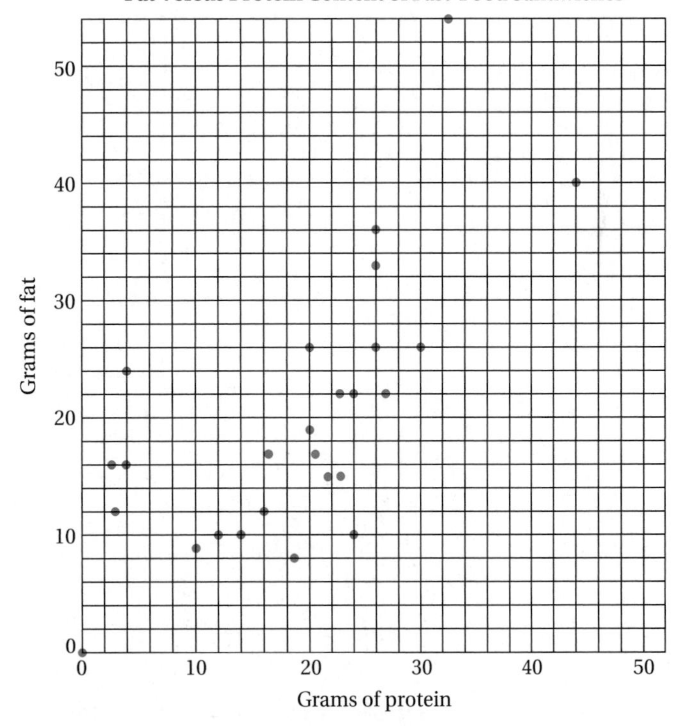

Source: From *Exploring Data* by James M. Landwehr and Ann E. White. © 1995 AT&T Corporation. Published by Dale Seymour Publications. Used by permission.

A Sample Student Page: Connecting Scatterplots to Health

Figure 8.16 contains a student page from *Scott Foresman-Addison Wesley Middle School Mathematics, Course 3* (2002), showing the use of a scatterplot to analyze the relationship between age and body temperature.

- **How** would you describe the pattern in the age–temperature scatterplot?
- **What** concepts are introduced in this lesson that may be new to the students?
- **How** would you answer the Try It and Check Your Understanding questions?

When creating a scatterplot, you must first decide which data category to show on the horizontal axis and which category to show on the vertical axis. Then you must be sure to scale each axis to include all values that occur in the data set.

Science Link

The temperature of the human body is normally 98.6° Fahrenheit.

Example 2

Create a scatterplot for the given data and determine a trend line between age and body temperature.

Use age to label the horizontal axis. Use body temperature to scale the vertical axis.

Plot the points.

There is no clear trend between age and body temperature.

Older and Warmer?

Body temperature(°F): 100, 99.5, 99, 98.5, 98, 97.5, 97

Age: 15, 17, 19, 21, 23, 25, 27

Age	Body Temperature (°F)	Time in Morning
18	98.2	7:00
18	99.2	9:00
19	97.7	8:00
19	98.0	12:00
20	98.2	8:00
20	97.0	11:00
21	99.0	10:00
21	98.4	7:00
22	98.8	12:00
23	98.0	11:00
24	97.4	8:00
25	97.8	9:00
25	97.4	10:00
27	97.5	7:00
27	98.8	8:00

Try It

Use the data above to create a scatterplot and draw a trend line to determine a relationship between body temperature and time of day.

a. Is there a trend?

b. What is the relationship between body temperature and time of day?

Check Your Understanding

1. In your own words, define scatterplot. How can you tell the difference between a scatterplot and a line graph? Explain.
2. Describe how you can tell whether a scatterplot indicates a trend. When can it be useful to notice a trend?
3. If a portion of a scatterplot shows a trend, and a portion of it does not, can this still be useful? Explain.

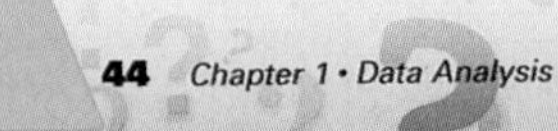

44 *Chapter 1 • Data Analysis*

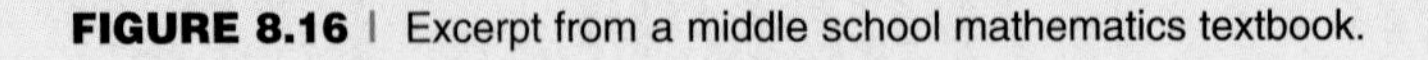

FIGURE 8.16 | Excerpt from a middle school mathematics textbook.

(*Source:* From *Scott Foresman–Addison Wesley Middle School Math, Course 3* by R. I. Charles, J. A. Dossey, S. J. Leinwand, C. J. Seeley, and C. B. Vonder Embse. © 2002 by Pearson Education, Inc. publishing as Prentice Hall. Used by permission.)

Solution

The data are positively correlated. As the amount of protein in a fast-food product increases, so does the amount of fat in the same product. On average, it appears that each additional gram of protein consumed is accompanied by about two-thirds to three-fourths of a gram of fat.

Your Turn

Practice: Use the graph below to describe the apparent relationship between favorite foods and the percent of fat in the foods. What type of correlation is represented by the data?

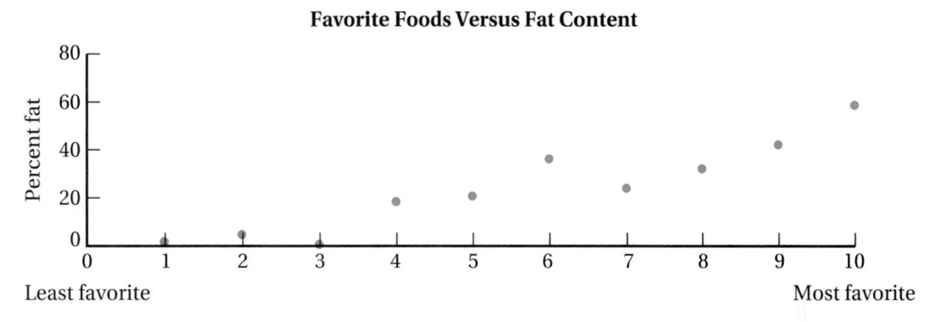

Reflect: Describe how a change of scale might affect the interpretation of the data in the graph. ■

Problems and Exercises | *for Section 8.2*

A. Reinforcing Concepts and Practicing Skills

1. What can you say about the pattern of percent of increase in the employer's healthcare costs, based on the graph shown?

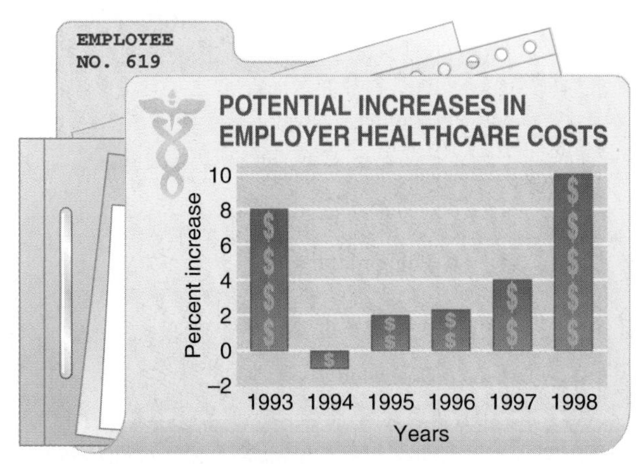

Source of data: Foster & Higgins, "Health-Care Inflation: It's BAAACK!" *Business Week*, March 17, 1997, p. 29.

2. Make a line graph displaying the data on fish lengths from a trial catch by forestry agents tallied in the frequency chart shown. Center the data in each interval on its midpoint on the horizontal axis of the plot.

Salmon Catch Lengths at Bonneville Dam

Length (centimeters)	Number
Less than 70	|||
70–79	𝍸 ||
80–89	𝍸 𝍸 |||
90–99	𝍸 |||
100–109	𝍸 𝍸 ||||
110–119	𝍸 𝍸 |||
Greater than 119	𝍸 ||

3. Make a multiple line graph, with dotted connecting trend segments, for the regional U.S. population data shown (millions). Graph each region's data with a separate dotted trend line.

	Year	
Region	**1970**	**1990**
Northeast	49.06	50.81
Midwest	56.59	59.67
South	62.81	85.45
West	34.84	52.79

4. Which of the following two graphs is more appropriate for displaying the number of individuals of differing ages at a picnic? Describe the flaw in reasoning promoted by the inappropriate graph. Would a scatterplot be more effective than either of these forms for the data?

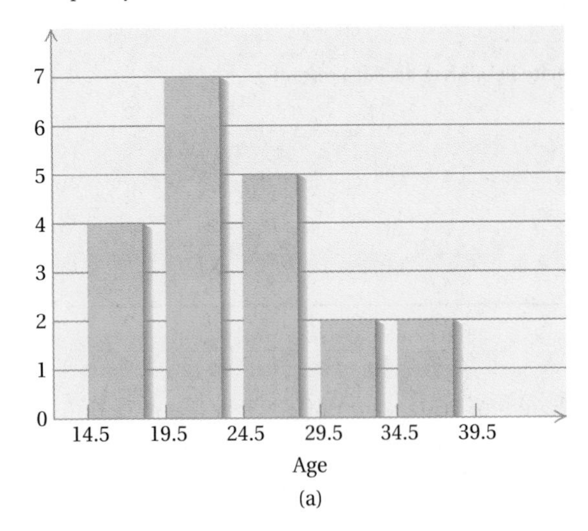

(a)

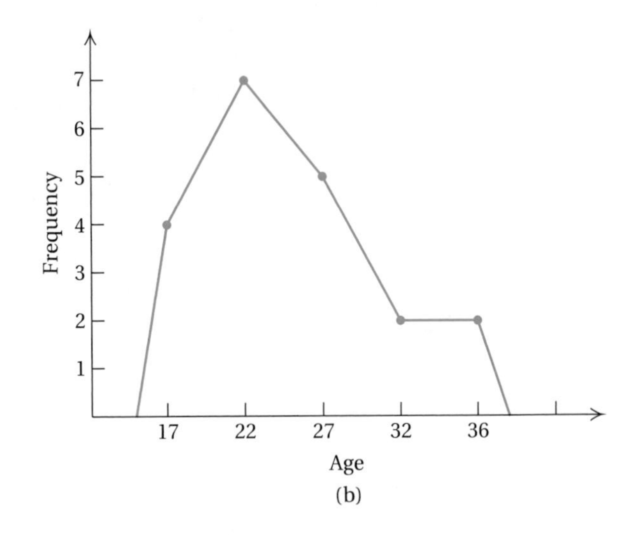

(b)

5. What conclusions can you draw from the U.S. government's analysis of the patterns of domestic consumption of timber products for the years indicated in the graph at the bottom of the page?

6. If the pattern between ACT scores and class averages is as indicated by the graph on the top of the next page, what class average would you predict for someone who has an ACT score of 25? What ACT score would you predict for someone with a class average of 60?

Graph for Exercise 5

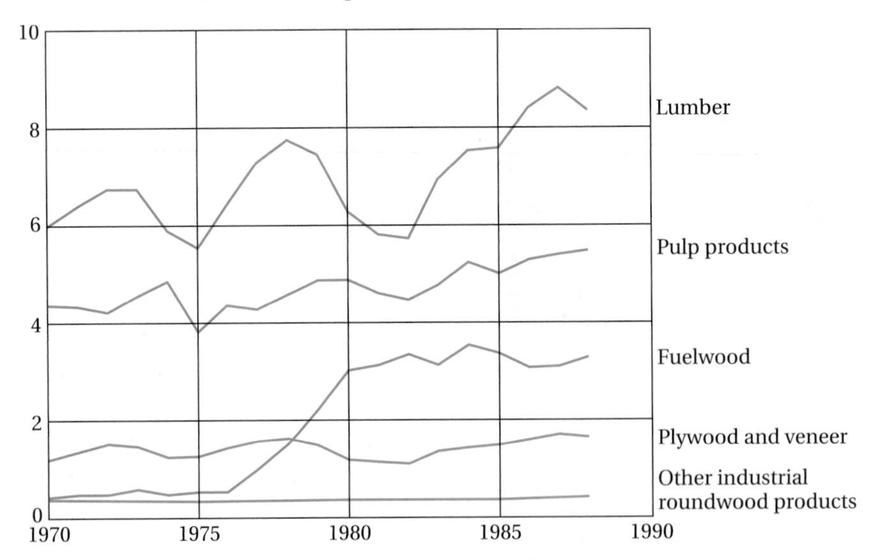

Source: Statistical Abstract of the United States, 1992. Washington, D.C.: U.S. Department of Commerce, Bureau of the Census, p. 668.

Graph for Exercise 6

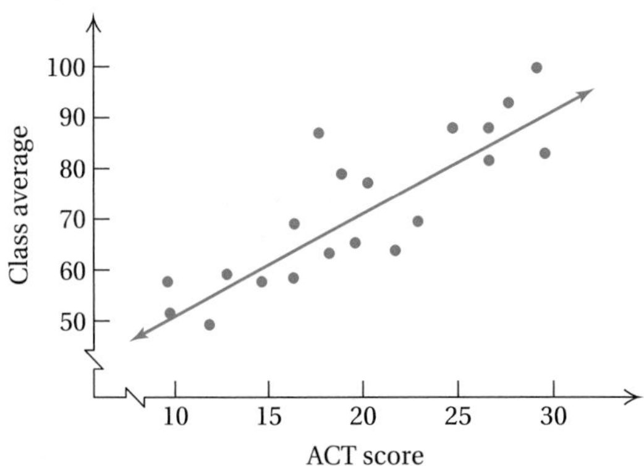

B. Deepening Understanding

7. For each of the graphs shown, indicate whether it reveals positive, negative, or zero correlation.

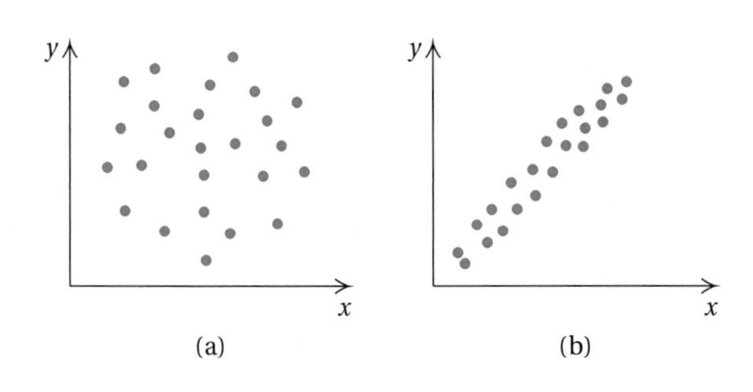

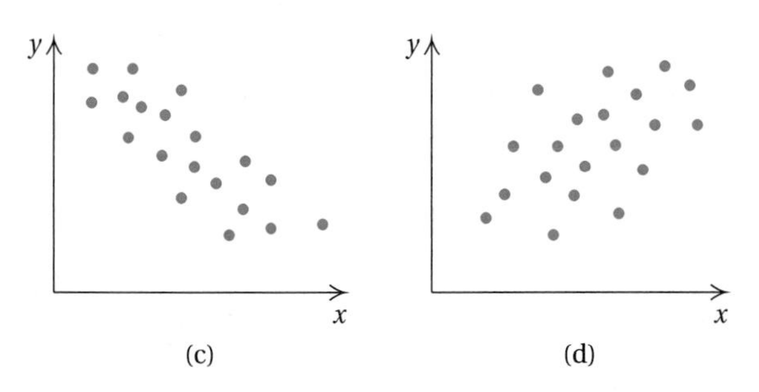

For Exercises 8–10, refer to the line graphs at the top of the next column.

8. Without knowing anything more about the nature of the variables involved, which graph(s) would you assume to be most representative? Why?
9. If you wanted to convince someone that the value of the variable related to the vertical axis responds sharply to small changes in the horizontal variable, which graph would you use?
10. What is the effect of lengthening the scale of the vertical axis in parts (b) and (d) or shortening the scale of the vertical axis in parts (a) and (c)? How might that influence your interpretations of the graphs?

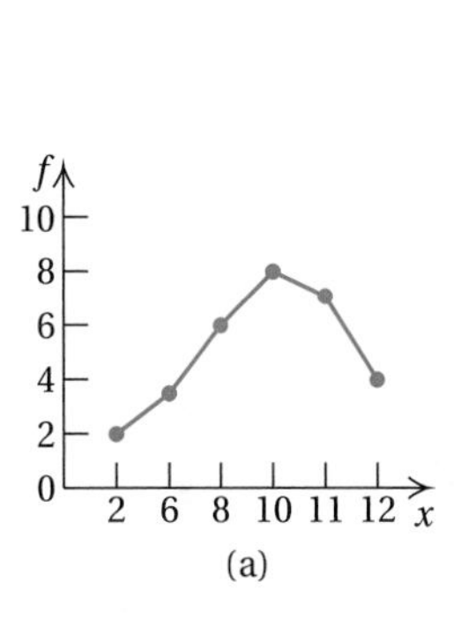

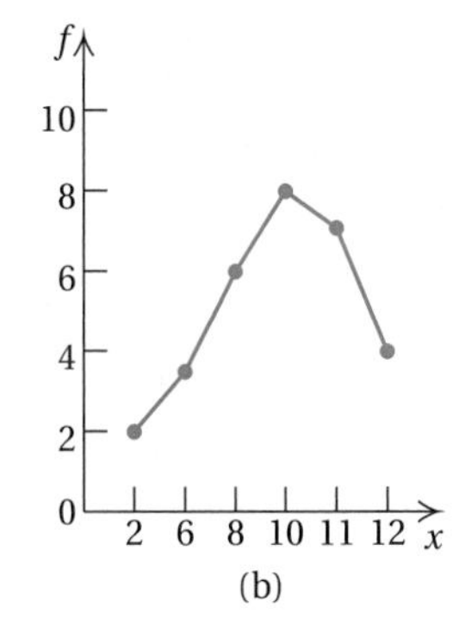

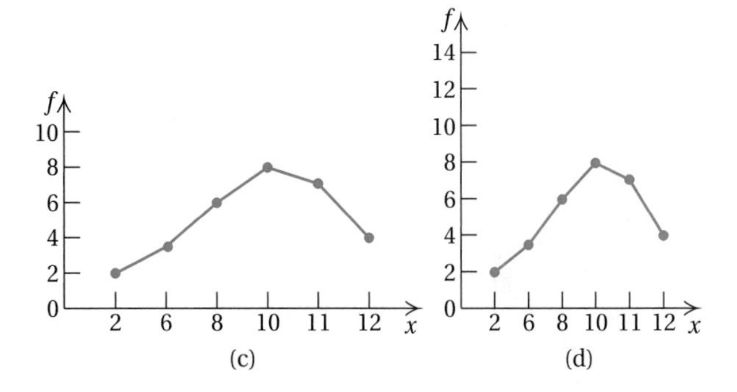

C. Reasoning and Problem Solving

11. Would it be appropriate to use the trend line shown on the graph in Exercise 6 to predict the class average of someone having an ACT score of 34? Why or why not?
12. The chart at the top of the next page is an example of a quality control chart. Samples of parts and products are drawn from an assembly line and checked against engineering specifications. Write a short report describing the various features of the chart and the performance of the machine relative to the stated limits over time, as if you were preparing a report for the company's chief executive officer (CEO).
13. The data display below was created with a graphing calculator. It shows a scatterplot of minutes played (horizontal axis) and points scored (vertical axis) for all players in one professional basketball game. The graphing window includes 0 to 60 minutes and 0 to 40 points.

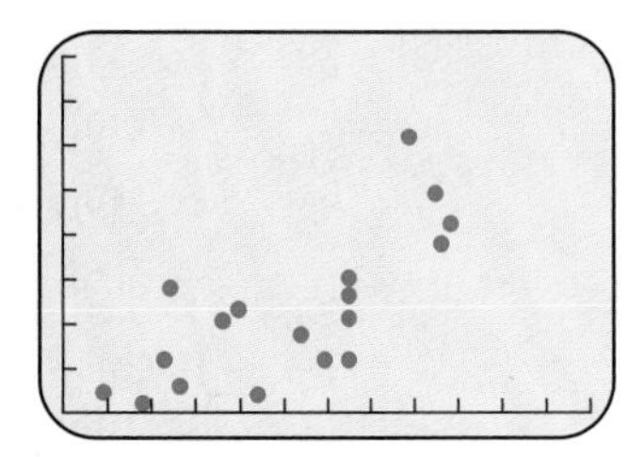

Chart for Exercise 12

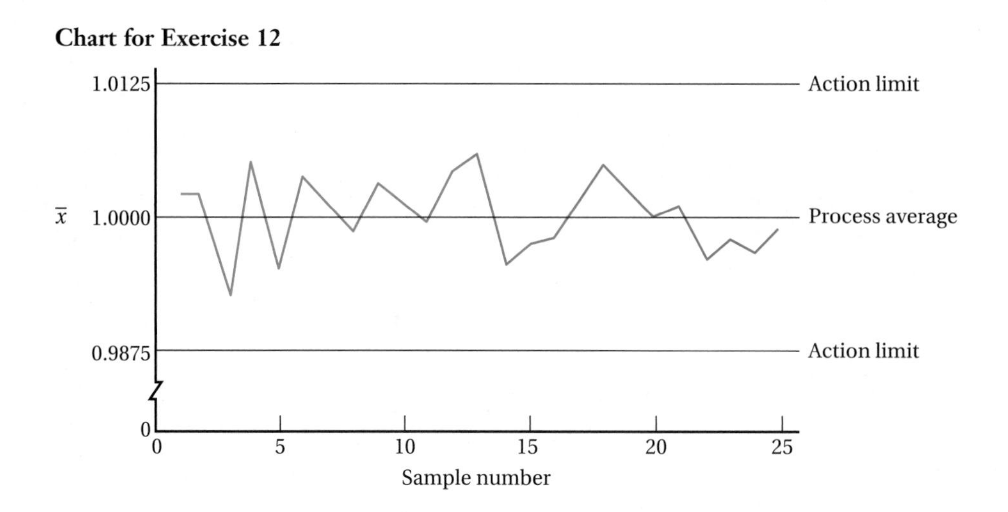

a. Is there a relationship between minutes played and points scored? Explain.

b. Without changing the data values, how could this data display be altered to enhance or downplay the relationship you identified in part (a)?

14. For the data in the following table, select a type of display and describe the steps that you would take in developing it to show the relationship between age and deaths/1000 people from car accidents.

Age	1–14	15–24	25–44	45–64	65–
Death rate/1000	4.6	6.2	16.6	8.8	15.7

15. Develop a graphic representation of the following U.S. population information. Describe how you decided on the form of display that you used.

D. Communicating and Connecting Ideas

16. Take a survey of your class members, recording their heights and their navel heights, in centimeters, from the floor. Prepare a scatterplot of these data. What trend do you see in the data? What is the ratio of mean navel height to total height? How does this ratio compare to the reciprocal of the value of the golden ratio ($\frac{1}{r} = 0.61803...$)?

17. **Historical Pathways.** The British scientist Sir Francis Galton published a paper in 1885 dealing with predicting the values of one variable from the values of another. His method was similar to what you do when you use a trend line to predict values from data. Look up Sir Francis Galton and the history of correlation. Write a short report detailing the history of this field and Galton's role in it.

Year	1800	1830	1860	1890	1920	1950	1980	1990
Population (millions)	5.3	12.9	31.4	62.9	105.7	150.7	226.5	248.7

Section 8.3 DESCRIBING THE AVERAGE AND SPREAD OF DATA

- Different Types of Averages
- Numerical Descriptions of the Spread and Distribution of Data
- Graphical Descriptions of the Spread and Distribution of Data
- Additional Ways to Describe the Spread and Distribution of Data

In this section we begin to analyze a set of data by looking at the idea of an *average* member of the set and how much *difference*, or *spread*, exists among the members of the data set. We demonstrate how to choose and apply different types of "averages," as well as the different ways to describe the spread in the data.

Different Types of Averages

Averages, or **measures of central tendency,** as statisticians refer to them, are numerical values used to describe the overall *clustering* of the members of a data set. By clustering, we mean the tendency of the members of the data set to be described by a single number. This idea of central tendency has led to the creation of four concepts of *average.*

Arithmetic Mean. The best known measure of central tendency of a set of data is the number called the **arithmetic mean,** most commonly referred to as the *average,* or **mean.** The value of the mean is found by summing all of the data values and dividing by the number of data points. For example, if the data values are 3, 5, 7, and 9, there are four data points and the mean of the set of data is $(3 + 5 + 7 + 9) \div 4 = 24 \div 4 = 6$. In the following definition we use variables to describe this average.

Definition of Arithmetic Mean

The arithmetic mean, or average, of n numbers $x_1, x_2, x_3, \ldots, x_n$ equals

$$\bar{x} = \frac{x_1 + x_2 + x_3 + \cdots + x_n}{n}$$

The symbol $\bar{x}$ denotes the arithmetic mean and is read "x bar."

Consider the data set 1, 1, 3, 3, 3, 4, 4, and 5. The mean, or average value, for this set is the sum of the values, 24, divided by 8, or 3. We can represent these data by using stacks of unit blocks as weights on a number line to illustrate the idea of the arithmetic mean as a clustering or balance point. On the number line shown in Figure 8.17, a unit block on 5, for example, represents the number 5 from the data set. Two unit blocks on 4 represent the two 4's from the data set, and so on. The fulcrum is at 3, the mean of the data set.

To demonstrate visually the clustering at a mean, we move blocks to the center while maintaining the balance by compensating moves of blocks on the left side to the right and moves of blocks on the right side to the left. Specifically, we move the two blocks at 1 each 2 units to the right, a total of 4 units to the right. To maintain the balance, we move the two blocks at 4 each a unit to the left, effecting a movement of 2 units to the left. We also move the block at 5, 2 units to the left, making a total of 4 units moved to the left. As we make the same number of moves

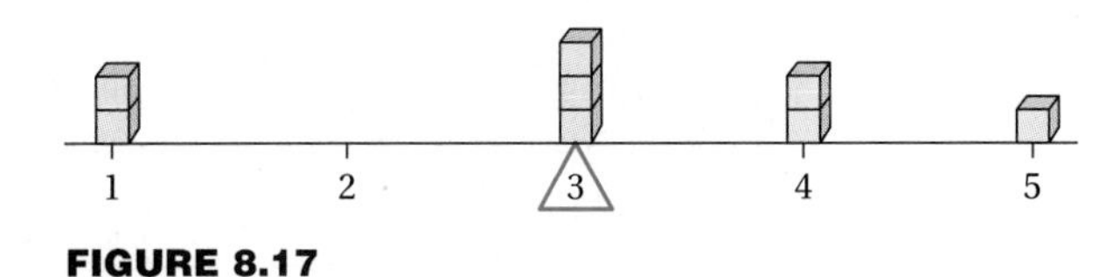

FIGURE 8.17

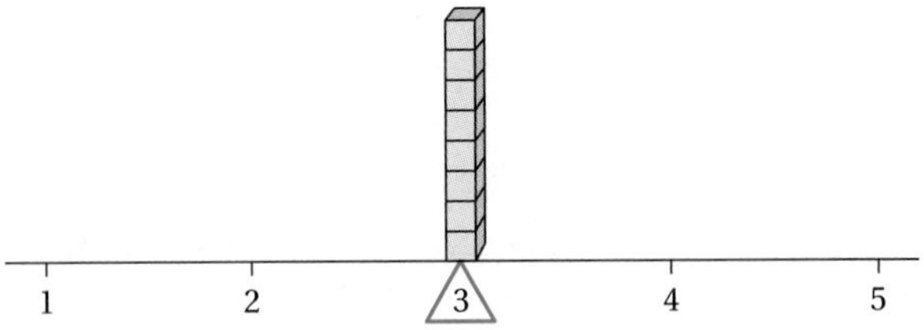

FIGURE 8.18

Estimation Application

Average Ages
Estimate the average (mean) of the ages of the McLean County Rowing Club if their ages are 32, 24, 17, 18, 26, and 33.

of unit blocks to the right as to the left, the number line remains in balance about its mean of 3, as shown in Figure 8.18. Thus the mean can be considered the "balance point" or "cluster point" for the numbers in the data set it represents.

Complete the estimation application by using the idea of the mean as a balance or cluster point.

Mini-Investigation 8.4 asks you to analyze further the idea of an arithmetic mean.

Write a short paragraph explaining how the arithmetic mean can relate to the numbers in the set that the mean represents.

Mini-Investigation 8.4 **Using Mathematical Reasoning**

Can the arithmetic mean for a set of numbers ever equal the largest number in the set? The smallest number in the set?

Median. The word *average* is also used to indicate "in the middle." Data analysts define the numerical value that describes the *midmost* value for an ordered data set to be the **median** for the list of numbers. For example, 3 is the median value for the ordered list of numbers 1, 2, 3, 6, and 9 because there are two numbers below 3 and two numbers above 3 in the set. If there are an even number of data points in the list, say, 1, 2, 3, 4, 6, and 9, the median is the mean of the two middle values, 3 and 4, or 3.5; that is, 3.5 is the median value for this data set. Note that in both cases we started with the data ordered from smallest to largest. Before you can find the median, the data must always be in either ascending or descending order. The following definition summarizes these ideas.

Definition of Median

The **median** of an ordered list of n numbers $x_1, x_2, x_3, \ldots, x_n$ is the numerical value that is (a) the middle number in the ordered list if n is odd or (b) the mean of the two numbers in the middle of the ordered list if n is even.

Another way of thinking about the median is that it is the value about which the data set is equally split or clustered. That is, equal numbers of the members of the ordered data set fall at or above the median and at or below the median. Note that the median itself doesn't have to be one of the values in the data set. Example 8.6 illustrates that, even though we can find the mean and the median, we may still have to interpret the data further in order to draw appropriate conclusions.

Example 8.6 Using the Mean and Median to Interpret Data

The following table contains information about how three judges rated the performance of three singers on a scale from 1 (poor) to 10 (excellent). Find the mean and median rating for each performer. Who should be declared the winner?

	Rating		
Judge	**Singer A**	**Singer B**	**Singer C**
1	1	4	3
2	8	5	4
3	6	9	8

Solution

The mean ratings for the performers are A, 5; B, 6; and C, 5. Their median ratings are A, 6; B, 5; and C, 4. Thus these two measures alone don't yield a clear-cut winner. Look at the overall rating given singer A and singer B by the judges: Judge 1 preferred B over A, judge 2 preferred A over B, and judge 3 preferred B over A. Based on this added evidence, we might select singer B as the winner.

Your Turn:

Practice: Suppose, however, that the ratings had been as follows. Who would you declare the winner?

	Rating		
Judge	**Singer A**	**Singer B**	**Singer C**
1	6	6	5
2	3	4	5
3	3	2	2

Reflect: Do any of the three singers in the practice problem have a legitimate complaint about your decision? Why or why not? ■

Mode. The third interpretation given to the word *average* is that it is the value in the data set that occurs most often (has the greatest frequency), called the **mode.** In the data set 1, 1, 3, 3, 3, 4, 4, and 5, the mode is 3 because 3 occurs more often than any other number. If the data set were 1, 1, 3, 3, 3, 4, 4, 5, 5, and 5 instead, the set would have two modes, 3 and 5. In this case, the data set is said to be *bimodal.* The mode can be thought of as the numerical value(s) where the greatest number(s) of members in the data set cluster. The following definition summarizes these ideas.

Definition of Mode

The **mode** of n numbers $x_1, x_2, x_3, \ldots, x_n$ is the numerical value, or values, that occurs most frequently in the set. A set may have more than one mode.

Midrange. A fourth way of interpreting the *average* of a set of n numbers is that it is the point midway between the largest and the smallest numbers in the set, called the **midrange.** Geometrically, you can think of the midrange as the numerical value associated with the midpoint of the segment connecting the largest and smallest numbers in the data set. For example, the midrange for the data set 3, 5, 8, 9, 12, and 13 is 8 because 8 is the mean of 3 and 13, the extreme values in the data set. The following definition summarizes these ideas.

Definition of Midrange

The **midrange** of n numbers $x_1, x_2, x_3, \ldots, x_n$ is the mean of the largest and smallest values in the set of data.

Example 8.7 demonstrates how to find each of the four measures of the "average" of a set of numbers.

Example 8.7

Using the Different Types of "Average"

Find the (a) mean, (b) median, (c) mode, and (d) midrange for the set of ages of people in Anne's quilting club: 15, 34, 23, 11, 34, 56, 38, 27, 18, and 29.

Solution

a. The sum of the data values is 285. When that sum is divided by the number of members, 10, the mean value is 28.5.

b. When the data set is ordered, 11, 15, 18, 23, 27, 29, 34, 34, 38, and 56, the median is the mean of the fifth and sixth values, as the number of data points is even. The median then is 28, or the average of 27 and 29. Five scores in the ordered set are less than 28 and five are greater than 28.

c. The mode is 34, as it occurs most often (twice).

d. The extremes are 11 and 56, so the midrange is $(11 + 56) \div 2$, or 33.5.

Your Turn

Practice: At State University, the heights of the five starting women's basketball players were 1.83, 1.83, 1.92, 1.85, and 1.97 meters. At State College, the five women starters were 1.84, 1.95, 1.99, 1.80, and 1.87 meters tall. Suppose that you were the sports information director at State College and wanted to find the mean, median, mode, and midrange of each set of data. Use the measure(s) of central tendency you believe to be most appropriate and write a brief press release about the comparative heights of the two teams.

Reflect: How might you interpret differently the comparative heights of the two teams? Which is the most defensible interpretation? ■

Sometimes you will need to compare two or more types of averages or even to decide which type of average is being used in a particular situation. For example, in 1992, the U.S. government reported that the average price of a one-family home was \$103,700. At the same time another source reported that the average price of a one-family home was \$130,016. Suppose that one of these prices was the mean and that the other was the median. How could you decide which was which? You could reasonably assume that the mean is pulled upward by the extreme values of expensive homes in the distribution but that the homes below the median wouldn't differ so much in value. Thus you could logically conclude that the higher price represents the mean and that the less expensive price represents the median. Exercises 7 and 8 at the end of this section ask you to decide which "average" is most appropriate for answering certain questions.

Numerical Descriptions of the Spread and Distribution of Data

In addition to using a measure of central tendency to help interpret the meaning of a set of data, we often also give meaning to a set of data by describing the way in which the values in the set of data are *spread* or *distributed.* In this subsection we look at both numerical and graphic ways to describe these characteristics of a set of data.

As we noted in Section 8.1, one of the measures commonly used to describe the spread of data values is the *range*, which is the distance between the largest and smallest values in a data set. For example, the range of the data set 1, 2, 3, 4, and 5 is $5 - 1$, or 4. You might think that, if the range of two sets of data is the same, the data sets will have the same spread. Although the range of the data set 1, 2, 3, 4, and 5 is equal to the range of the data set 1, 1, 3, 5, and 5, intuition indicates that the spread, or **dispersion,** of scores in the second set is somewhat greater because more scores fall at the extremes of the set. For this reason, data analysts have invented other measures of dispersion besides the range to describe the spread of the values in a data set.

Definition of Range

The **range** of a set of data is the difference between the largest and smallest values in the data set.

A second method of describing dispersion in data is the **interquartile range.** It might be thought of as the range of the middle half of the data. Consider the ordered data set of 12 bowling scores from a beginner's league:

21, 22, 34, 35, 36, 42, 44, 45, 49, 50, 52, 73.

The extreme scores are 21 and 73, giving a range of 52. The median score of 43 is the average of the two *midmost* scores. We mark the location of the median with a vertical bar:

21 22 34 35 36 42 | 44 45 49 50 52 73

Now let's consider the scores below and above the median as separate subsets of scores and find the median of each subset. The median of the six scores below 43 is 34.5, and the median of the scores above 43 is 49.5. These two values—and the overall median—divide the original set of scores into quarters:

	34.5		43		49.5	
21 22 34	|	35 36 42	|	44 45 49	|	50 52 73
	Q_1		Q_2		Q_3	

The points that divide the data set into quarters are known as *quartile points*, or **quartiles.** The first, 34.5, divides the first, or lower, quarter of scores from the second quarter. The median, 43, divides the second quarter of scores from the third quarter. The third, 49.5, divides the third quarter of scores from the fourth, or highest, quarter. The first and third of these points are given special names and symbols. The first is called the *first quartile point* and is denoted Q_1. The third point is called the *third quartile point* and is denoted Q_3. Sometimes the median is denoted Q_2. The interquartile range is the difference between Q_3 and Q_1. In this case, the interquartile range is $49.5 - 34.5$, or 15. This measure, like the median of a data set, relies not on the numerical values in the set, but on the position of data values in an ordered listing of the set.

Definition of Interquartile Range

The **interquartile range** (IQR) is the difference between the third quartile point (Q_3) and the first quartile point (Q_1) for a set of values.

Graphical Descriptions of the Spread and Distribution of Data

When working with measures of central tendency or dispersions you might think, wouldn't it be nice to be able to see in one glance the mean, median, range, and interquartile range of a set of data? Using a clever graphics idea that we introduce in the next subsection, you can do so.

Box-and-Whisker Plots. A good way to examine the spread of a set of data visually is with a graphic display called a **box-and-whisker plot.** To illustrate this type of display, let's consider a set of 12 test scores:

72, 75, 75, 76, 78, 83, 83, 86, 88, 88, 89, 91.

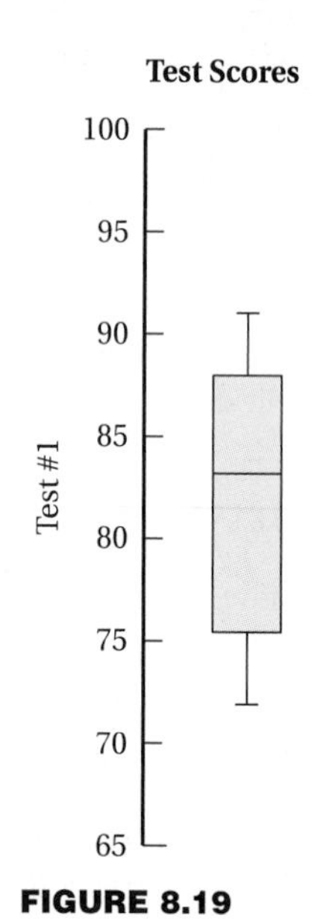

FIGURE 8.19

The range, or spread, of the test scores is represented in the plot in Figure 8.19 by the distance from the upper horizontal bar to the lower horizontal bar. The height of the box in the middle of the plot represents the interquartile range, or the distance between Q_3 (the top of the box) and Q_1 (the bottom of the box). In this data set, it is 12.5, or the difference between 88 and 75.5. Note that 50% of the scores in the set fall in the box in the plot, 25% fall above the box, and 25% fall below the box. The bar in the middle is the location of the median, 83. Sometimes the location of the mean for the data set is indicated by an asterisk.

The box-and-whisker plot also is an effective way to present and examine the relative features of two or more data sets. It provides information on measures of both central tendency and dispersion.

Box-and-whisker plot can be created easily with graphing calculators. Instructions for doing so are presented in Appendix A. Examine those instructions and make a box-and-whisker plot of the data set 72, 75, 75, 76, 78, 83, 83, 86, 88, 88, 89, and 91. Example 8.8 provides additional practice in making a box-and-whisker plot.

Example 8.8 **Making a Box-and-Whisker Plot**

Make a box-and-whisker plot for the weights, in pounds, of a litter of 15-day-old pigs raised on a special feed at Leman's pig farm:

7.2 9.8 12.3 13.1 13.3 8.4 10 11.6 9.9 11.5

A Sample Student Page: Representing Spread

Box-and-whisker plots are also found in the PreK–8 curriculum, as shown by the student page from *Addison Wesley Mathematics, Grade 8* (1995) in Figure 8.20. They, along with stem-and-leaf plots, are examples of statistical methods developed in the early 1970s that are already a part of the PreK–8 mathematics program.

- **What** special information does the box-and-whisker plot show about the members of the class?
- **What** does the box-and-whisker plot show about the spread or dispersion of the data?
- **What** information is available from the box-and-whisker plot that isn't available from the stem-and-leaf plot?

Box and Whisker Graphs

LEARN ABOUT IT

Graphic images help us interpret data. A **box and whisker graph** is one way to provide a picture of the central tendency of data.

EXPLORE Study the Table

You can draw a box and whisker graph to display the data given in the stem and leaf plot at the right.

stem	leaf
18	5
16	8
12	8,3
8	7,3,1
6	4,4
4	8
3	6,3

TALK ABOUT IT

1. How many items of data are listed in the stem and leaf plot?
2. List the data in order. Which item has the highest value? Which item has the lowest value?
3. What is the median of the data?

To complete a box and whisker graph of the above data follow the steps below.

- Find the median of the upper half of the data and label it Q_U to represent the **upper quartile.**
- Find the median of the lower half of the data and label it Q_L to represent the **lower quartile.**
- Mark an appropriate vertical scale and draw a box that connects the upper quartile to the lower quartile. A line across the box indicates the median. Label the median "MD." For this data, MD = 82.
- Draw lines, "whiskers," from the box to the **highest** (H) and **lowest** (L) data items.

200, 180, 160, 140, 120, 100, 80, 60, 40, 20, 10, 0

H = 185

$Q_U = 125$

MD = 82

$Q_L = 56$

L = 33

TRY IT OUT

Draw a box and whisker graph for each of these sets of data.

1. 12, 23, 24, 24, 28, 37, 49, 51, 53, 54, 54, 63, 65, 67, 92, 98
2. 23, 45, 46, 46, 49, 25, 72, 48, 63, 18, 29, 53

48

FIGURE 8.20 | Excerpt from a middle school mathematics textbook.

(*Source*: From *Addison Wesley Mathematics, Grade 8* by R. E. Eicholz, P. G. O'Daffer, R. I. Charles, S. L. Young, and C. S. Barnett. © 1995 by Pearson Education, Inc. publishing as Prentice Hall. Used by permission.)

Solution

When we order the pigs' weights, the range extends from 7.2 to 13.3 pounds. The interquartile range is 2.5 pounds, Q_1 is 9.8 pounds, and Q_3 is 12.3 pounds. These values determine the left and right ends of the box. The median for the data set is 10.75, as shown by the vertical line inside the box.

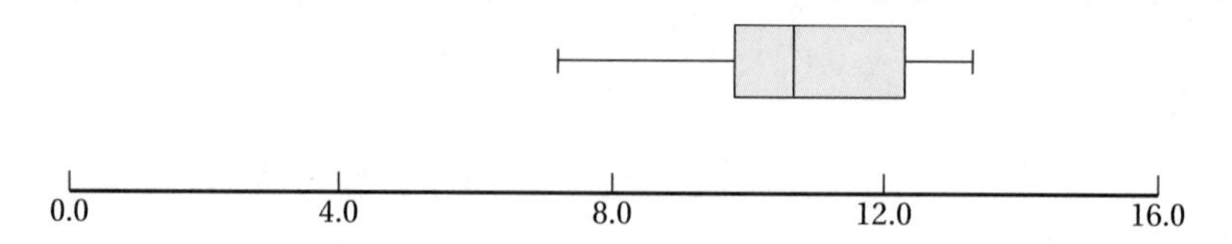

Your Turn

Practice: Suppose that a second litter of 15-day-old pigs, raised on a different feed, had the following weights.

6.8 8.5 9.5 9.2 10.1 11.3 11.7 7.9 7.5 13.3

Make a box-and-whisker plot for the weights of this litter of pigs.

Reflect: Compare your plot with the one shown in the example solution. What does this comparison show? ■

Mini-Investigation 8.5 gives you an opportunity to compare different box-and-whisker plots.

Write a paragraph explaining how you arrived at your decision.

Mini-Investigation 8.5 **Solving a Problem**

As a Little League coach, would you select team A, B, or C (all with identical range of batting averages)

a. if your main concern is winning?

b. if you wanted to have a challenging coaching experience?

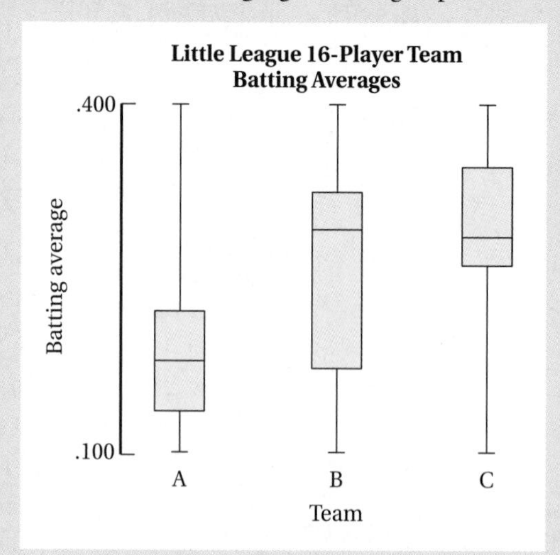

Source: Adapted from *Addison Wesley Mathematics, Grade 8* by R. I. Eicholz, P. G. O'Daffer, R. I. Charles, S. L. Young, and C. S. Barnett. © 1995 by Pearson Education, Inc. publishing as Prentice Hall. Used by permission.

Additional Ways to Describe the Spread and Distribution of Data

In research, two of the measures commonly used to describe the spread in a set of data are the closely related ideas of *variance* and *standard deviation*. Developed in the late 1800s by the English statistician Karl Pearson, these concepts today are the key to widely used statistical quality control methods in manufacturing. These concepts also provide the basis for statistical analysis and reporting of survey results, including the testing of hypotheses in experimental settings.

Variance. Karl Pearson decided that, if the mean of a data set can be thought of as a balancing point for the data set, a measure of dispersion for that data set should illustrate the way in which the values *deviated* from the mean. He suggested finding the distance that each point in the data set is from the mean of the set. For example, using the data set $A = \{1, 4, 7, 8, 10\}$ with mean 6, we calculate the difference, or **deviation,** of each value from the mean:

$$1 - 6 = -5,\ 4 - 6 = -2,\ 7 - 6 = 1,\ 8 - 6 = 2, \text{ and } 10 - 6 = 4.$$

Note that if we simply added the deviations, we would get 0. To ensure that we don't get 0, Pearson's method involves adding the *squares* of the deviations rather than the deviations and then dividing by the number of values to get what he called the *variance of the data set.* The variance of the data in set A, 10, is calculated as follows.

$$[(-5)^2 + (-2)^2 + (1)^2 + (2)^2 + (4)^2] \div 5 = 10.$$

To see how the variance of a data set describes the spread of values in the set, let's consider another five-element data set with mean 6, $B = \{4, 5, 6, 7, 8\}$, in which the values are closer together. The variance of the data in set B, 2, is calculated as follows.

$$[(4 - 6)^2 + (5 - 6)^2 + (6 - 6)^2 + (7 - 6)^2 + (8 - 6)^2] \div 5 = 2.$$

Note that the variance of the data in set A, $A = \{1, 4, 7, 8, 10\}$, is 10, whereas the variance of the data in set B, $B = \{4, 5, 6, 7, 8\}$, is 2. Thus, as represented by distance from the mean to the numbers in these sets, the variance does indeed give a measure of the spread of the values. The definition of variance is summarized as follows.

Definition of Variance

The **variance,** denoted σ^2, of a set of n data values $x_1, x_2, x_3, \dots, x_n$ having a mean m is the average squared distance of the data points in the set from m. Symbolically,

$$\sigma^2 = \frac{(x_1 - m)^2 + (x_2 - m)^2 + (x_3 - m)^2 + \cdots + (x_n - m)^2}{n}$$

The procedure for finding the variance of a set of numbers is summarized as follows.

Step 1: Find the mean m of the set of numbers.

Step 2: For each number x in the set of numbers, calculate the deviation (distance) of that number from the mean, $x - m$.

Step 3: Square all the deviation scores obtained in step (2), that is, $(x - m)^2$.

Step 4: Find the arithmetic mean of the squared deviation scores calculated in step (3).

These steps can often be performed directly on a spreadsheet. Some spreadsheets even have a variance formula in their library of formulas. Many calculators also have statistical function keys to assist in the calculation of such values. We illustrate finding the variance in Example 8.9.

Example 8.9

Finding the Variance

Find the variance of the set of scores 1, 2, 4, 5, and 6.

Solution

First, we calculate the mean score for the set, 3.6. We then find the deviations of the individual scores from 3.6 and square them. Finally, we find the mean of these squared deviation scores. In this case the variance is 17.20 ÷ 5, or 3.44.

Score	Deviation	(Deviation)^2
1	$(1 - 3.6) = -2.6$	6.76
2	$(2 - 3.6) = -1.6$	2.56
4	$(4 - 3.6) = 0.4$	0.16
5	$(5 - 3.6) = 1.4$	1.96
6	$(6 - 3.6) = 2.4$	5.76
Sum	0.0	17.20

Variance = 17.20 ÷ 5 = 3.44

Your Turn

Practice: Find the variance for the heights of the starting players for each of the women's basketball teams in the Your Turn part of Example 8.7 on page 424.

Reflect: How do these teams compare in terms of variability? What change of 5 centimeters in the height of one player would most dramatically change the variability of heights of either team? Why? ■

Standard Deviation. We have shown that the variance of a set of scores reflects the average *squared* distance of the scores from the mean. If we take the *square root* of the variance, we can create a measure of dispersion that is a little more reflective of the "average" distance of the scores from the mean, or the spread of the scores in a data set. This value, the square root of the variance for a set of values, is called the **standard deviation** for the values. Consider the data sets used to illustrate the variance, $A = \{1, 4, 7, 8, 10\}$ and $B = \{4, 5, 6, 7, 8\}$. The variance of A was 10, and the variance of B was 2. So the standard deviation of the data in set A is $\sqrt{10}$, or approximately 3.16, and the standard deviation of the data in set B is $\sqrt{2}$, or approximately 1.41. When we look at data set A, the "average" distance of a number in the set from the mean, 6, does seem to be a bit more than 3. Also, an "average" distance of a number in set B from the mean, 6, of a little less than 1.5 seems reasonable. The idea of standard deviation is summarized as follows.

Definition of Standard Deviation

The **standard deviation,** denoted σ, of a set of n data values $x_1, x_2, x_3, \ldots, x_n$ is the square root of the variance of those data values. Symbolically,

$$\sigma = \sqrt{\frac{(x_1 - m)^2 + (x_2 - m)^2 + (x_3 - m)^2 + \ldots + (x_n - m)^2}{n}}.$$

The ability to use variance and standard deviation as measures of dispersion determines, to a great degree, a person's ability as a data analyst. An experienced data analyst can construct an estimate of the shape of a display of data from these measures alone.

Problems and Exercises | *for Section 8.3*

A. Reinforcing Concepts and Practicing Skills

In Exercises 1–6, calculate the mean, median, mode, and midrange for the data set given.

1. $\{0, 0, 1, 2, 2, 4, 5, 10\}$
2. $\{6, 7, 8, 9, 10, 12, 15, 15, 20, 28\}$
3. $\{72, 80, 80, 82, 88, 90, 96\}$
4. $\{85, 61, 68, 73, 91, 68, 93\}$
5. $\{68, 74, 80, 82, 83, 85, 86, 88, 91, 93\}$
6. $\{-1, 7, 0, 14, -2, -15, 0, 8, 7\}$
7. Forty-nine employees in a factory earn a salary of \$25,000 a year and the manager earns a salary of \$125,000. What are the mean, median, and modal salaries for all 50 people? How do these statistics change if the manager's salary is excluded from the calculations?
8. Doug had scores of 80, 85, 75, and 80 on his first four exams in a course.
   a. Find the mean, median, and mode for these exam scores.
   b. Which "average" would Doug want the teacher to use in determining his grade?
   c. What score would Doug have to get on a fifth examination to raise his mean score to 84? Is it reasonable to expect Doug to achieve that score?

In Exercises 9–12, calculate the range, variance, standard deviation, and interquartile range for the data set given.

9. $\{8, 5, 9, 7, 10, 8, 6, 9, 8, 10, 7, 9\}$
10. $\{-1, 0, 1, 0, -1, 0, 1, 0, -1, 0, 1, 0\}$
11. $\{6, 7, 8, 9, 10, 12, 15, 15, 20, 28\}$
12. $\{72, 80, 80, 82, 88, 90, 96\}$

In Exercises 13–16, make a box-and-whisker plot for the given set of data.

13. Data in Exercise 9.
14. Data in Exercise 10.
15. Data in Exercise 11.
16. Data in Exercise 12.
17. The following box-and-whisker plot shows the network rating scores for 10 prime-time television shows for two different networks. What can you say about the relative standings of each network's set of shows?

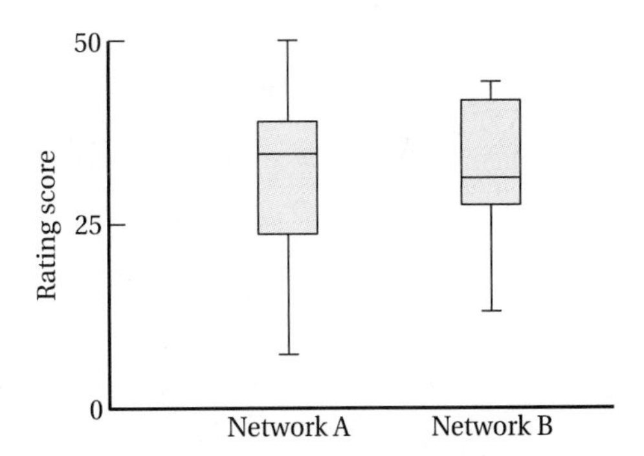

18. Data on the highest and lowest number of hits by three baseball teams within a specified period of time are described by the box-and-whisker plots shown on the next page.
    a. Which team had the largest range? The smallest range?
    b. Which team had the largest interquartile range? The smallest interquartile range?
    c. For which team is the mean nearly equal to the median? Why?
    d. For which team does the mean most likely exceed the median? Why?

Graph for Exercise 18

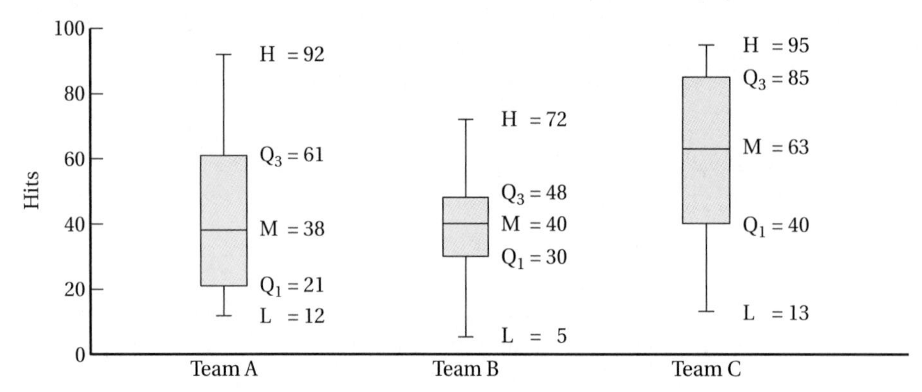

Source: Adapted from *Addison Wesley Mathematics, Grade 8* by R. E. Eicholz, P. G. O'Daffer, R. I. Charles, S. L. Young, and C. S. Barnett. © 1995 by Pearson Education, Inc. publishing as Prentice Hall. Used by permission.

B. Deepening Understanding

19. The average test score for four tests was a 72. What was the sum of the test scores?

20. If the oldest individual in an office is 57 and the range of employee ages is 29, what is the age of the youngest person in the office?

21. The mean score for 25 of 27 tests is 80. The other two scores are 30 and 35. What is the average of all 27 test scores?

22. Find the range for the set of scores {55, 67, 80, 85, 90, 92, and 98}. Then give a second set of seven scores that has the same mean as the first set but a smaller range.

23. A survey of potential injury severity ratings by percents for small, midsize, and large cars returned the following data.

Small	73, 83, 91, 95, 98, 100, 108
Midsize	56, 69, 76, 80, 83, 91, 93, 94, 94, 95, 100
Large	54, 59, 63, 67, 69, 69, 70.

a. Make a box-and-whisker plot for each type of car's injury potential by representing its range of values.

b. Compare the displays and write a short summary of your findings.

24. Suppose that Helen buys three items for 47¢ each, five items for 71¢ each, and two items for $1.21 each. What was the average cost of all the items she bought?

25. Steve drives at 30 miles per hour for the first 2 miles of a 4-mile trip. How fast will he need to drive for the remainder of the trip to average 60 miles per hour?

26. The following table shows the point totals for the five leading scorers on the Flyers and Bombers basketball teams.

a. How is the scoring of the two teams similar? How is it different?

Flyers

Player	Games Played	Total Points
Jackson	82	2580
Meyerson	82	2382
Koch	64	1817
Brooks	67	1849
Everson	81	2154

Bombers

Player	Games Played	Total Points
Anderson	82	2591
Watson	81	2414
Michaels	65	1833
Roberts	79	1840
Meeks	80	2112

b. Based on scoring data, who would win a game between the two teams? Explain.

27. Is it possible for the mean of a set of data to fall outside the box for its box-and-whisker plot? Explain your answer with an example.

C. Reasoning and Problem Solving

28. Is the standard deviation for a set of real number data points always less than or equal to the value of the variance for the same set of data? Explain.

29. Let's define the *average deviation* for a set of scores to be the arithmetic mean of the set of absolute values of the distances of the scores from the mean score for the set. What can you say about the relationship between the average deviation for a set of scores and its standard deviation? Why is *absolute value* needed in this definition?

30. If 10 is added to each score in a set of test scores, describe the effect of the addition on the mean, median, mode, range, and standard deviation of the set.

31. For a set of scores, will the interquartile range always be less than the range? Explain your answer with an example.

32. Add 5 points to the smallest and largest values in the data set {0, 12, 13, 15, 17, 17, 18, 20}. Then estimate and check the effect on the mean, median, mode, and midrange of the data set.

33. Can you show sets of eight numbers, each less than 10, that satisfy each of the following conditions?

a. The mode is to the left of the median and mean.

b. The mode is to the right of the median and mean.

c. The mode and the median are the same, and both are less than the mean.

34. Is each of the mean, median, mode, and midrange a member of the set of data that it describes? Why or why not.

35. The Appropriate Average Problem. The chief operating officer (COO) of Acme Widget Company makes $520,000 a year. Each of the other 99 employees of the company makes $20,000. Is the mean, median, or mode the most appropriate measure to use for giving the "average" salary for an Acme Widget employee? Explain.

D. Communicating and Connecting Ideas

36. Which of the measures of central tendency, in general, is located nearest to the "middle" of a data set? Write a short paper explaining your answer, stating any assumptions you made.

37. Using a calculator with a standard deviation key, test the hypothesis that the standard deviation for a set of data is about one-sixth the range of a set of data. Plan an experiment to test this conjecture and then carry it out. Write a description of your results.

38. Find two articles that discuss the variability of two groups or factors rather than their central tendencies. Summarize in writing the main points of the articles and be prepared to share them with the class.

39. Making Connections. The mean, variance, and standard deviation are called *parametric statistics* because they are influenced by the value of a computed parameter, the mean. The median, Q_1, Q_3, and interquartile range are nonparametric statistics, as they arise from position in the data set, not values. Compare in writing the effects of different types of data sets on these two types of statistics in describing the nature of a set of data points.

40. Historical Pathways. The concepts of variance and standard deviation are the work of Karl Pearson, a noted British statistician. Pearson first used the term *standard deviation* in a study of the outcomes from a roulette game. Explain how outcomes in games of chance are related to the concept of variance.

Section 8.4 DECISION MAKING WITH DATA

- Evaluating Data Collection Procedures
- Making Valid Conclusions from Data
- Misleading Conclusions from Graphs
- Evaluating a Data-Based Conclusion
- Connecting Data Analysis and Problem Solving

In this section we focus on some techniques for evaluating conclusions based on data. We present some ideas for analyzing the appropriateness of data collection, reporting, and graphic representation. We also discuss similarities between data analysis and problem solving. Mini-Investigation 8.6 encourages you to begin thinking about the validity of conclusions based on analysis of data.

Talk about situations that have been misrepresented by leading people to get others to draw false conclusions from data.

Mini-Investigation 8.6 **Communication**

According to the American author and humorist Mark Twain, British Prime Minister Benjamin Disraeli supposedly said:

> There are lies, damn lies—and statistics. [*Brewer's Quotations.* London: Cassell, 1994, p. 123.]

Why would someone in Disraeli's position make such a statement about statistics?

Many people believe that the use of statistics, or data, to support an argument is a strategy to cover up the truth. As evidence, consider the oft-heard statement that "liars figure and figures lie." Unfortunately, this assertion often is true, but it doesn't have to be. The inability of people generally to deal with data-based arguments is what makes it a possibility.

Clearly, the ability to draw valid conclusions from data and to reject invalid conclusions based on data are important skills for individuals in today's rapidly changing and technologically oriented world. When evaluating conclusions based on data, people need to be able to focus on the way the data have been collected, the logic of the conclusions based on them, and the manner in which the conclusions are represented graphically.

Evaluating Data Collection Procedures

Sometimes researchers and others draw conclusions from data, but don't explain how the data were collected. Unless you know that the collection procedures were legitimate, you shouldn't rely too much on the conclusion. Suppose that you read a "research study" that reported students studying from algebra book B improved 300 percent more than the students studying from algebra book A. Some of the questions that you should think of asking about how the study was conducted include the following:

- Was the **sample size** adequate?
- Were the students and teachers **randomly assigned** to the two groups?
- Was the test equally representative, or **valid,** for each of the algebra books?
- Was the test a **reliable instrument?** That is, would a student score about the same on it, or an equivalent test, if the student were to take it over and over again without prior knowledge of the exact questions?

The claim cited above actually was made in an advertisement for an algebra book. However, we have several concerns about the data collection procedures. First, only a small number of students were involved in the data set, and they weren't randomly assigned to the classes using the different algebra books. Thus one class might contain smarter students than the other class, or the students in one class might have had special educational or life experiences that the students in the other class hadn't had. There was also ample evidence that the test wasn't a valid instrument with which to compare the two texts, as it contained several items that only students using book B studied. The claim in the advertisement was based on data from a set of these items. Specifically, book B provided students with considerable instruction and practice on removing three levels of embedded grouping symbols, such as in the expression $\{[-3(2 - 4)] - [5 - 2(-4 - {}^{-}2)]\}$, whereas book A considered only two levels of grouping symbols and then only briefly. On a pretest, students in both groups answered correctly only 1 of the 10 questions on the examination dealing with simplifying these types of expressions. On the posttest, students using book B answered 4 of the 10 items correctly, but students using book A still answered only 1 of the 10 items correctly. Thus, in terms of this one isolated skill, students using book B improved 300% more than those using book A. However, the test was unfair to students using book A and was probably unreliable because of its short length.

This textbook comparison illustration suggests an important question about data collection that goes beyond the statistical questions already asked about sampling and testing:

> Have the most representative data been selected?

In this case, the answer is *no*. A test that contains 10 items on simplifying complicated expressions is not representative of the wide variety of important concepts and procedures in an algebra course. Clearly, anyone evaluating any conclusions drawn from those data must first carefully evaluate the data collection procedures.

Making Valid Conclusions from Data

Not only do you need to question data collection procedures, but you also need to evaluate carefully the logical basis for the conclusions drawn from the data. Some key signals can alert you to the possibility that conclusions about data may not be valid.

Beware of Vague or Undocumented Statements of Comparison. Often statements that purport to "prove" one point or another must be evaluated to determine whether the "proof" is valid. For example, return to the algebra book illustration discussed earlier. The broad conclusion that "students studying from algebra book B improved 300 percent more than students studying from book A" doesn't follow logically from the fact that students studying book B correctly answered 4 of 10 items involving a single procedure for simplifying complicated expressions, whereas students studying book A answered only 1 of 10.

Now, consider the statement:

> More people are killed in the average airplane accident than in the average automobile accident. Therefore driving is safer than flying.

Although the first statement may be true, the second doesn't necessarily follow logically from it. The actual average number of individuals killed in airplane crashes may be higher, but in general more individuals drive more miles than they fly. In fact, a fatality is 90 times more likely to happen per mile traveled by an individual in a car than by an individual traveling by airplane. Thus the risk to be considered should be the *exposure to risk* measured not in deaths per accident, but rather in deaths per mile traveled.

Another common advertising claim is that "95 doctors out of 100 surveyed preferred BUF-X for pain relief." Such claims also are often flawed. The data collectors may simply continue to sample groups of 100 doctors until they find a group in which 95 agree with the statement. To accept this type of a statement, you need to know that a reputable survey of 100 randomly chosen and knowledgeable doctors was conducted. You would also need to know what question was asked and what alternative responses the doctors could have made, if any, so that you could answer the question: preferred BUF-X to *what?*

Beware of Statements That Have Percents in Their Claims. Another type of common claim is that "50% of the females at Joe's Bar prefer Atomic Wings as their appetizer." Perhaps only two females have ever graced Joe's Bar, as it is such a dive, and only one of them stayed long enough to order an appetizer. She ordered Atomic Wings.

Beware of Conclusions Based on Correlations That Claim Causation. A third common misuse of data in drawing conclusions is to use data in which a strong positive correlation exists between two events to conclude that one event causes the other. Would you believe an argument that "older people have bigger feet"? Data collected only on individuals aged 1 to 12 would show a strong positive correlation between age and shoe size. However, that correlation quickly disappears if the range is 1 to 70 years of age. Similarly, the distance between telephone poles and the time between live births in the United States have a strong positive correlation. However, all that this correlation reports is the regularity of the two events; one doesn't cause the other.

Misleading Conclusions from Graphs

As with statements made about data, graphic displays of data can also misrepresent the facts involved. Figure 8.21 is a classic early example of how graphs can be misleading. The graph suggests that the commissions paid to travel agents by airlines declined in 1978. However, a careful reading of the graph shows that the information for 1978 covers only the first half of the year.

If the information for the first half of the year were doubled, in each case the commissions paid to travel agents would increase again for each airline in 1978. This display shows the misrepresentation of data that results from comparing *unlike data units.* If a significant change in a trend is apparent, you should always check to determine whether **like units** are being compared. Unlike units can quickly lead to bogus results and claims.

You also have to be very careful about checking *the scales* shown on the horizontal and vertical axes of a graph. Changes in scales can lead to significant differences in the way a graph appears. Consider two different representations for the

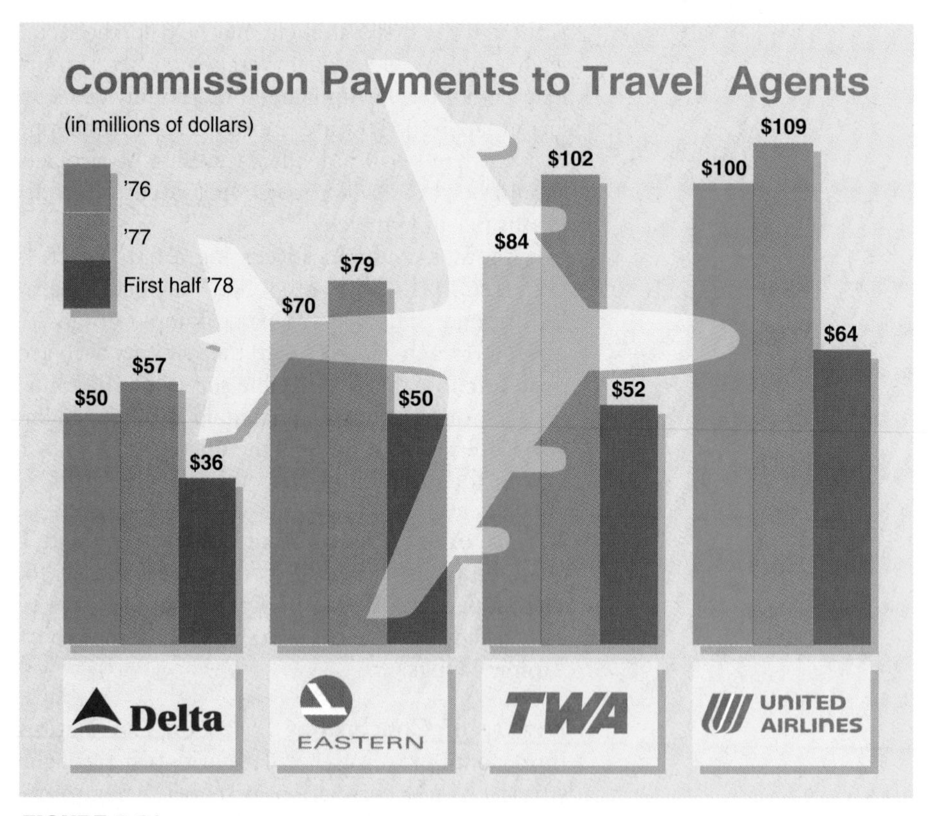

FIGURE 8.21

(*Source: New York Times Graphics*, August 8, 1978, p. D-1. Used with permission.)

A Sample Student Page: Evaluating Claims

An important goal of the PreK–8 curriculum is to prepare students to interpret data displays properly. Students need to be able to do so in order to shift data from a graphic representation to some other form for manipulation, to draw a trend line without fear of false conclusions, or to make estimates of some form of central tendency or measure of dispersion. Review the material from *Addison Wesley Mathematics, Grade 8* (1995) shown in Figure 8.22.

How could this material help students interpret data displays intelligently?

Using Critical Thinking

Mr. P. R. Smooth was describing the Prestigious Corporation.

"Our beginning salaries are very competitive," he said. "The average salary of the last 5 new employees was $20,000. Also, a recent sample showed that our average wage for hourly employees is $10.50. And on top of that, our profits are soaring!"

Mr. Smooth gave the information below to support his claims.

Last 5 New Employee Salaries	
Person A	$10,000
Person B	$10,000
Person C	$10,000
Person D	$10,000
Person E	$60,000
Average:	$20,000

Hourly Wage Sample	
Person A	Made $40 @ $4/hr
Person B	Made $40 @ $8/hr
Person C	Made $40 @ $10/hr
Person D	Made $40 @ $20/hr

Average hourly wage:
($4 + $8 + $10 + $20) ÷ 4
= $10.50

Prestigious Corp.

Profits
(in millions)

1.2
1.1
1.0
88 89 90 91

TALK ABOUT IT

1. What claims did Mr. Smooth make about Prestigious Corp.?
2. Do you think Mr. Smooth's claims about yearly salary, hourly wages, and profits are misleading? Why or why not?
3. What have you learned from this situation?

TRY IT OUT

1. Are the yearly salary data realistic? Explain.
2. How many hours did each person in the hourly wage sample work? What was the total number of hours worked? What were the total wages made? What is the average wage per hour using this data? Why is this different from Mr. Smooth's claim?
3. How could you change the scale and labeling so that it would look like the profits for the Prestigious Corporation weren't growing very fast?

FIGURE 8.22 | Excerpt from a middle school mathematics textbook.

(*Source:* From *Addison Wesley Mathematics, Grade 8* by R. E. Eicholz, P. G. O'Daffer, R. I. Charles, S. L. Young, and C. S. Barnett. © 1995 by Pearson Education, Inc. publishing as Prentice Hall. Used by permission.)

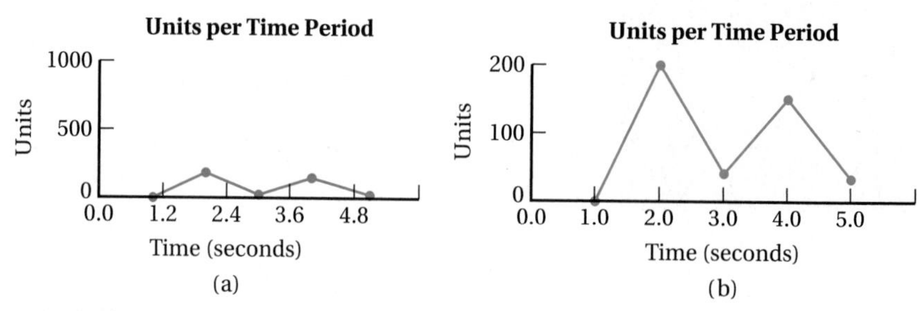

FIGURE 8.23 | (a) Small vertical scale; (b) large vertical scale.

same set of data shown in the line graphs in Figure 8.23. The graph in part (a) has a vertical axis of 1000 units, whereas the graph in part (b) has a vertical axis of 200 units. The larger vertical scale of the graph in part (b) appears to show greater variation in units per time period. In addition, the horizontal scales are different, and the horizontal scale in the graph in part (a) is scaled in multiples of 1.2 while the scale in part (b) is in units.

Mini-Investigation 8.7 lets you further analyze the idea of representing growth graphically.

Draw a graph that represents this data more accurately.

Mini-Investigation 8.7 **Making a Connection**

What point is the following graph trying to make, and would you consider making any changes in it if you could redesign it?

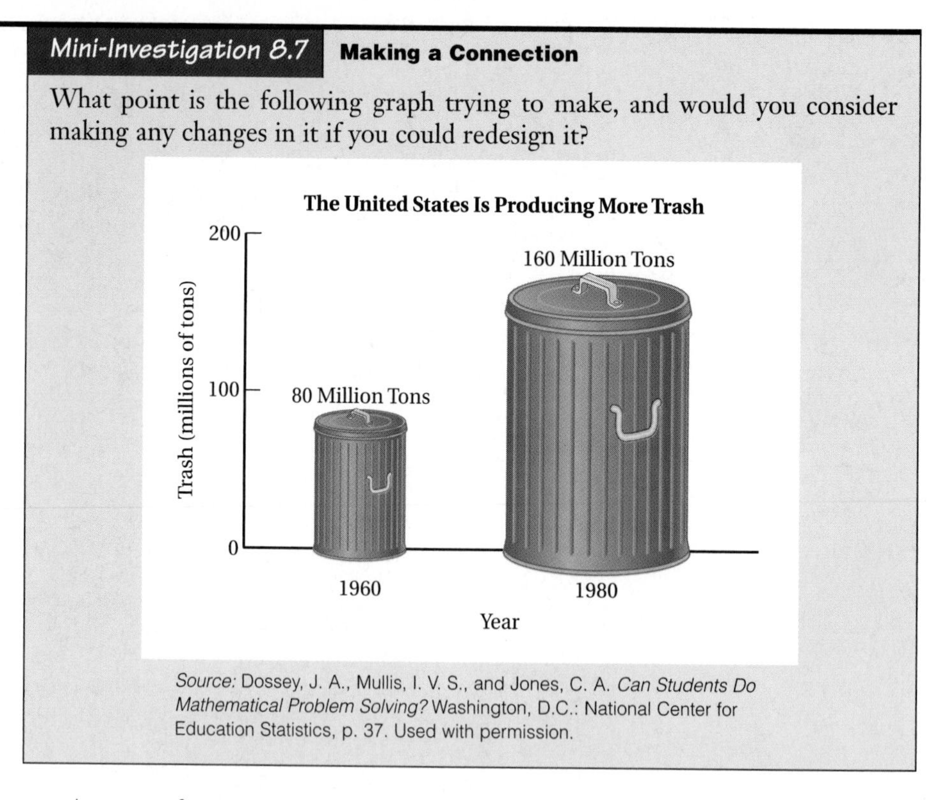

Source: Dossey, J. A., Mullis, I. V. S., and Jones, C. A. *Can Students Do Mathematical Problem Solving?* Washington, D.C.: National Center for Education Statistics, p. 37. Used with permission.

Any use of a measure in a graph—length, area, or volume—to describe growth must portray the nature of the growth with mathematical accuracy. If the growth shown is linear, the two line segments must accurately represent the actual size of that growth. If the growth is shown as area, the actual ratio of the areas must reflect the actual ratio of that growth. For example, if the original region in a display is

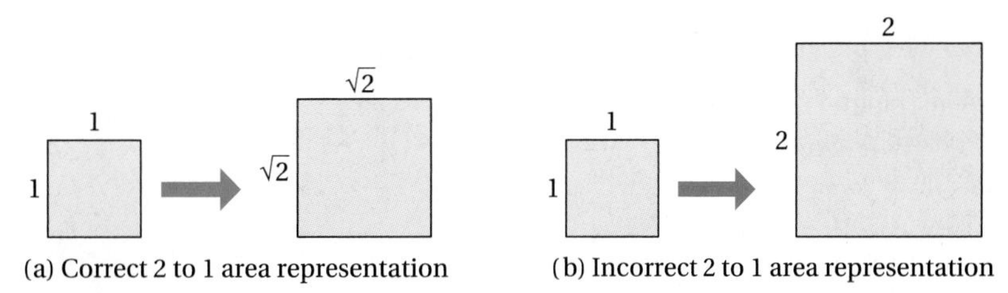

(a) Correct 2 to 1 area representation (b) Incorrect 2 to 1 area representation

FIGURE 8.24 | Visual area representation and misrepresentation of a 2 to 1 growth rate.

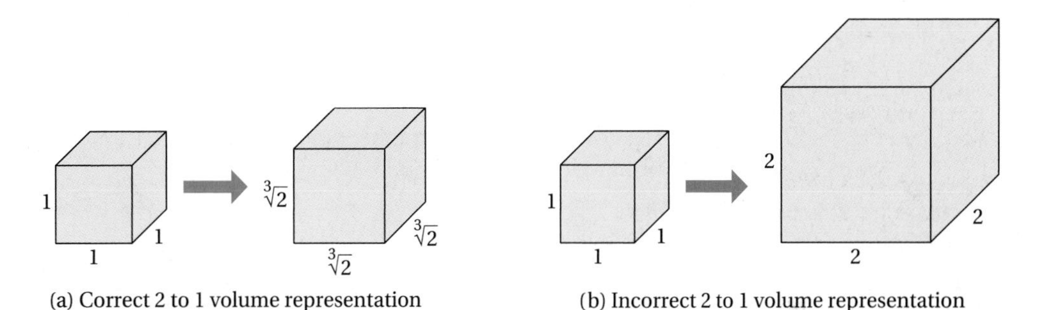

(a) Correct 2 to 1 volume representation (b) Incorrect 2 to 1 volume representation

FIGURE 8.25 | Visual volume representation and misrepresentation of a 2 to 1 growth rate.

shown as a 1×1 square and the reported growth rate is 2 to 1, the region for the new value should be a square with dimensions $\sqrt{2}$ and area 2, as shown in Figure 8.24(a). Often graphic displays purporting to show a 2 to 1 change show a larger region with dimensions 2×2 and area 4, erroneously suggesting a growth factor of 4, as shown in Figure 8.24(b). When evaluating the appropriateness of graphs, you should be aware that areas grow at the rate of the square of the linear growth rate.

If growth is shown as volume, the change in the linear dimensions of the new object must be the cube root, $\sqrt[3]{2}$, of the growth rate reported, as shown in Figure 8.25(a). Often graphic displays purporting to show a 2 to 1 change show a larger region with dimensions $2 \times 2 \times 2$ and volume 8, erroneously suggesting a growth factor of 8, as shown in Figure 8.25(b).

Recall the garbage-can graph in Mini-Investigation 8.7. The respective dimensions of the second garbage can should have been only 1.26 times as large, rather than twice as large, to reflect accurately the change in amounts of garbage. The Estimation Application provides an opportunity to apply the idea of volume representations of growth rate.

Example 8.10 further illustrates the idea of using area and volume graphics to show growth.

Example 8.10

Analyzing Graphs Showing Growth

The graph in part (a) shows the amount of sugar produced from Juan's sugar cane last year. Which of graphs (b) or (c) correctly represents his production this year if he doubled his output?

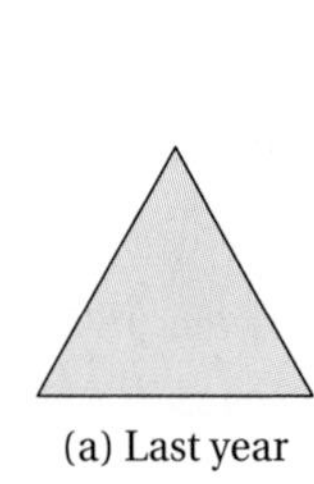
(a) Last year

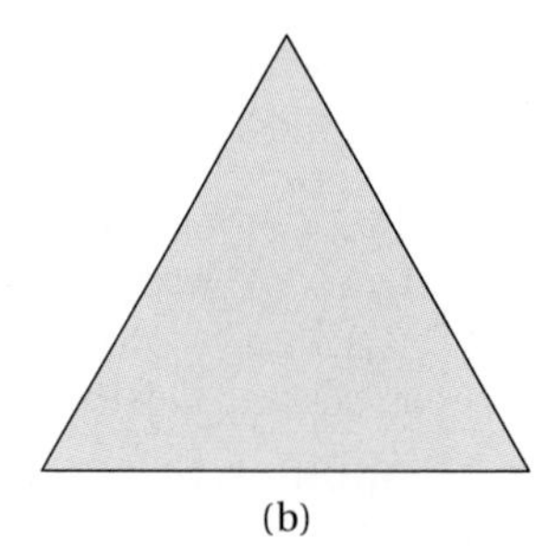
(b)

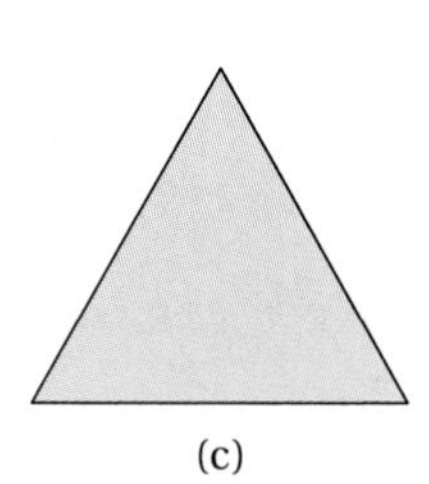
(c)

Solution

Representation (c) is the correct one because its area is double the area of the figure in part (a).

Your Turn

Practice: Create a graphic volume representation of a growth factor of 3 for oil production at well site #3.

Reflect: How would you construct the graphic if it were to be based on area rather than volume? ■

In summary, you can ask several questions to test a graphic display to ensure that it isn't misleading.

- Is the graph labeled appropriately?
- Are both axes labeled to identify the measures represented?
- Does the scale on each axis start at zero? If not, is an interruption in the scale clearly indicated?
- Are the numbers on the scales equally spaced and easily interpreted in relation to the data?
- If graphs are used to show growth or change, are the size ratios appropriate?

Evaluating a Data-Based Conclusion

In the preceding subsections we looked at several situations that might cause conclusions based on data to be misleading. To summarize the ideas presented, and to give some practical suggestions for evaluating data-based conclusions, we present the following questions.

- Does the conclusion or graphic seem reasonable?
- What is the source of the data? Is it representative of the population to which it is attributed?

Estimation Application

Growing Things

A graphic artist wants to make a graph based on volume that shows a growth of more than 300% in production of boxes in a box factory. She chooses a unit box to represent the initial production.

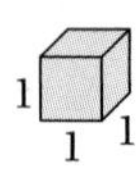

Estimate which box, (a), (b), or (c), the artist should select from her tech art file to represent the growth.

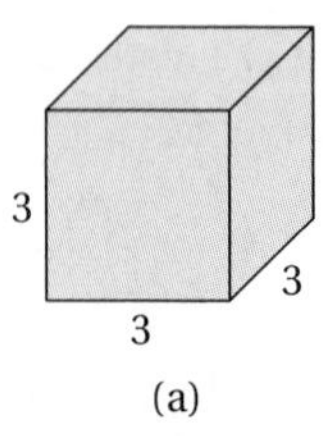

(a)

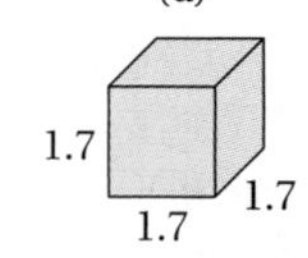

(b)

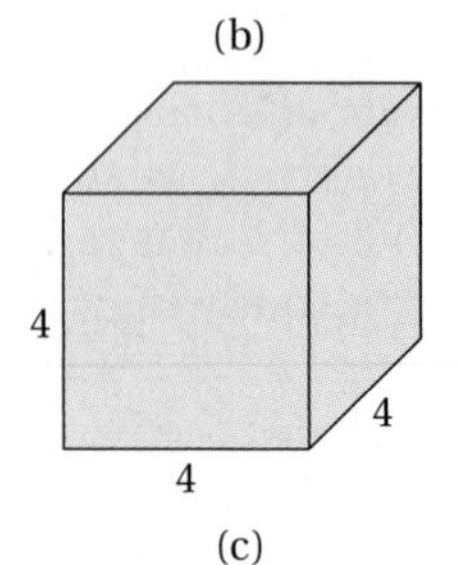

(c)

- Can the facts be verified or the study replicated? Or is it, perhaps, a one-time occurrence?
- Do the conclusions match your own observations?
- Is a causal relation erroneously assumed?
- Are comparisons or percents accurately based?
- Does the argument presented explain why the opposite claim doesn't hold?
- Are the scales and the units included in the graphic clear and not misleading?
- If the graph is showing change, are the units of measurement appropriately handled?

Example 8.11 utilizes the preceding questions to help evaluate data-based conclusions.

Example 8.11

Evaluating Data-Based Conclusions

Evaluate the following data-based conclusion and describe the misuse of data interpretation(s) involved in it:

> In 1991, citizens in Connecticut had the highest disposable personal income per person in the nation at $21,967 per capita. Therefore, to increase his or her disposable income, a person should move to Connecticut.

Solution

Living in Connecticut isn't the important factor here. An erroneous causal relationship is being assumed from the data. The reason for high per capita disposable income in Connecticut is due in part to the disproportionately large disposable incomes of the many corporate executives who live in southern Connecticut and to the fact that Connecticut is a state of average size. Moving to Connecticut will not guarantee that someone's disposable income will rise.

Your Turn

Practice: Consider the following data-based conclusions and describe the misuses of data involved in them.

a. Successful space launches are less frequent now than in the 1960s because of environmental concerns.

b. At Seawash High School, 98% of the students studied algebra from *Addison-Wesley Algebra* last year.

c. Over 70 percent of the individuals voting in the spring primary election reported voting for a Democratic candidate. Therefore the Democrats will win the general election in the fall.

Reflect: What words in each of the preceding data-based conclusions serve as clues that the conclusions are inappropriate? ■

Connecting Data Analysis and Problem Solving

The approach for analyzing a conclusion based on data is very similar to the process of problem solving discussed in earlier chapters. You should approach the analysis of a data-based conclusion by first *understanding the situation.* Next, *develop a plan* for examining the data and then *implement the plan* by studying the claims made. Finally, form your conclusions and examine them in the context of the data and other information by *looking back* to your original understandings. We illustrate this procedure in Example 8.12.

Example 8.12

Problem Solving: Fouling and Scoring

A student studying the following data table and scatterplot with a trend line for the Hornet basketball team drew the following conclusion: Players who score more points commit more personal fouls. Do the data support this conclusion?

Hornet Basketball Player Points and Fouls Data for Seven Games

Player	Total Points Scored	Total Fouls Committed
Hart	48	17
Ulrich	21	7
Schrock	18	9
Dossey	75	31
Kinder	6	6
Braman	42	24
Stafford	60	15
Walter	2	4
Garhke	3	0
Leman	37	16
Horcham	1	0
Charles	0	0
Steffe	2	0
Henderson	0	0

The following scatterplot of the data was graphed in a window [0, 100] by [0, 100] along with an accompanying trend line. See Appendix A for an explanation of window-size choice.

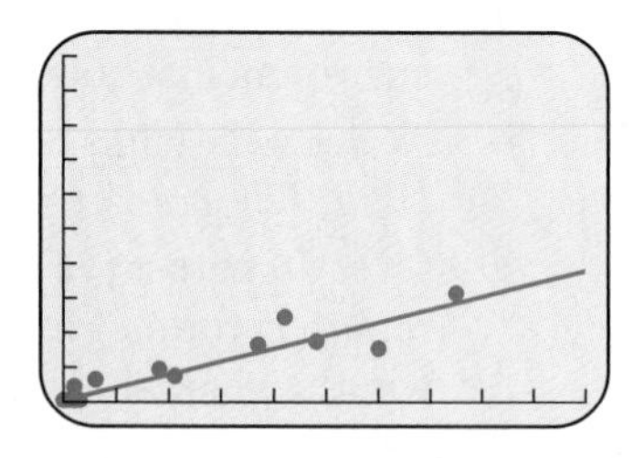

Working toward a Solution

Understand the problem	*What does the situation involve?*	Consideration of the relationship between the number of points scored and the number of fouls committed.
	What has to be determined?	Is the stated conclusion an appropriate one?

	What are the key data and conditions?	Data presented in the data table and scatterplot with trend line.
	What are some assumptions?	The data are valid.
Develop *a plan*	*What strategies might be useful?*	Look for a pattern, use reasoning, and draw a diagram.
	Are there any subproblems?	One might examine data for some subsets of players.
	Should the answer be estimated or calculated?	Little or no numerical calculation is required. At most an average or median might be needed.
	What method of calculation should be used?	A calculator could be used, if needed.
Implement the plan	*How should the strategies be used?*	Splitting the players into four groups, those scoring 0–10 points, 11–30 points, 31–50 points, and more than 50 points, we could find the median number of fouls per group.
	What is the answer?	Calculating the median number of fouls for each of the groups above, we get 0, 8, 17, and 23 fouls, respectively. Graphing these medians for each of the four groups of team members, we get the scatterplot shown. 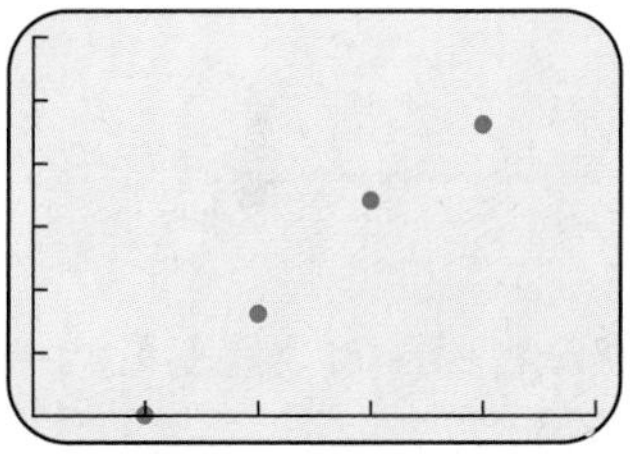 The data suggest that the number of fouls is positively related to the number of points scored for these four groups of players.
Look back	*Is the interpretation correct?*	Maybe. The data suggest a positive correlation exists, but do not prove the conclusion stated by the individual.
	Is the calculation correct?	Yes.
	Is the answer reasonable?	Yes. One might surmise that players scoring a large number of points might be playing more minutes and

	thus be exposed to the risk of making more fouls.
Is there another way to solve the problem?	Yes. One could collect data on other basketball teams, analyze patterns, and summarize findings over a larger number of teams and players.

Your Turn

Practice: Consider the following data about used-car prices obtained from Sunday newspaper advertisements and the conclusion drawn that older cars cost more. Then offer some possible interpretations.

Model Year	Price
1968	$20,185
1970	13,500
1972	11,750
1974	15,200
1985	500
1988	1,200
1989	7,500
1992	5,500
1994	8,450
1994	9,500
1995	11,000
1995	13,000
1996	18,000
1996	13,000
1996	17,000

Reflect: What factors other than age might enter into the pricing of used cars? ■

Problems and Exercises | *for Section 8.4*

A. Reinforcing Concepts and Practicing Skills

In Exercises 1–6, describe the misuse or misinterpretation of statistics that may be involved.

1. According to a survey, Miles Davis is the greatest American jazz musician.
2. Plato, St. Augustine, St. Thomas Aquinas, Hobbes, and Descartes were never married. Therefore anyone who wants to be a philosopher shouldn't get married.
3. Over 85 percent of the people at the baseball game said that hot dogs were their favorite snack. Therefore hot dogs must be the favorite American snack food.
4. Over 60 percent of the people who fly to Normal, Illinois, do so on American Airlines. Hence most people prefer American to any other airline.
5. A recent report said that the average professional baseball player is now making more than $1 million per year.
6. An ad for ARCBRITE claims that the detergent gets clothes twice as clean.
7. The classic graphic on the next page purports to show the effects of the Mideast oil crises in the 1970s. What difficulties do you see in this data display? Describe them and explain how you would correct them.
8. The display shown on the next page gives the percentages of a cattle operation owned by each of five partners. Does the 3-D style make it prone to any misinterpretation? If so, how could the data be more accurately represented?

B. Deepening Understanding

9. Babe Ruth is often considered the greatest home-run hitter

because of his long-standing single-season record of 60 home runs. Although Roger Maris later broke that record, with 61, many baseball purists rejected Maris's claim to be the home-run champion because Maris played in more games in his record-setting year than had Ruth. What do the career home-run data shown to the right indicate about who was the home-run champ?

Graphic for Exercise 7

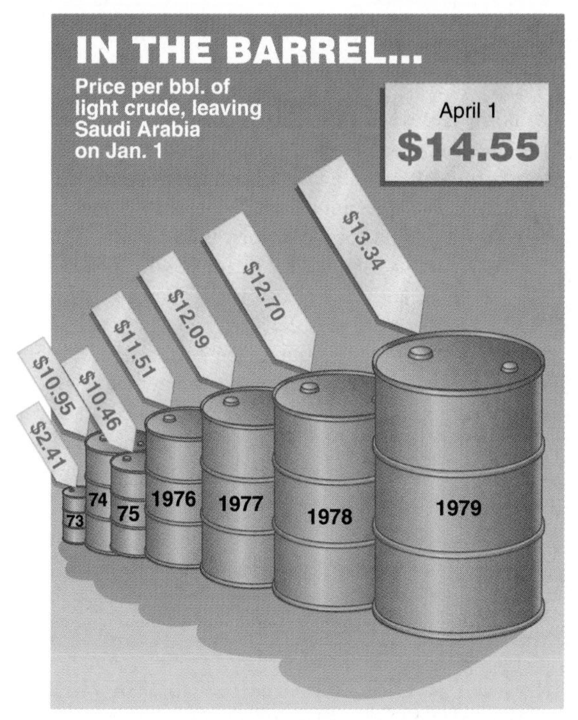

Source: Time, April 9, 1979, p. 57. © 1979 Time Inc. Reprinted by permission.

Graphic for Exercise 8

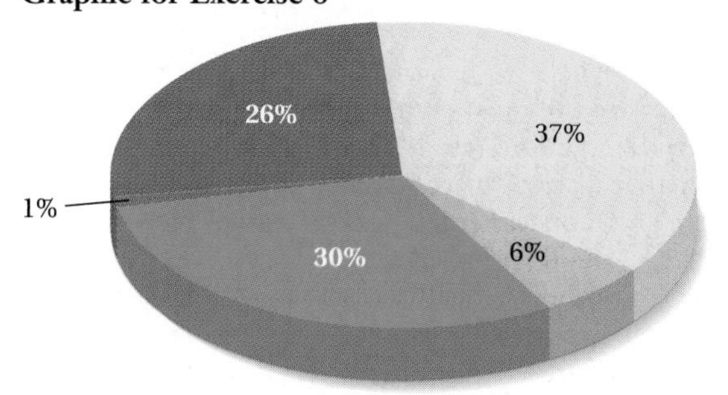

Cattle Operation Ownership

Graphic for Exercise 9

Home Runs per Season Played

Ruth		Maris
	0	8
	1	346
952	2	368
54	3	39
9766611	4	
944	5	
0	6	1

|6|1 means 61 homers

Source of data: Reichler, Joseph L. (ed.). *The Baseball Encyclopedia,* 7th ed. New York: Macmillan, 1998, pp. 1210–1211, 1419. Used with permission.

10. Consider the box-and-whisker graphs shown. They suggest that a U.S. Department of Agriculture study shows that "poultry dogs" have fewer calories per "dog" than either "beef dogs" or "meat dogs." Yet, one dietitian claims that some hot dogs made of poultry have 150% of the calories of some hot dogs made of beef. Can this claim be correct? Explain.

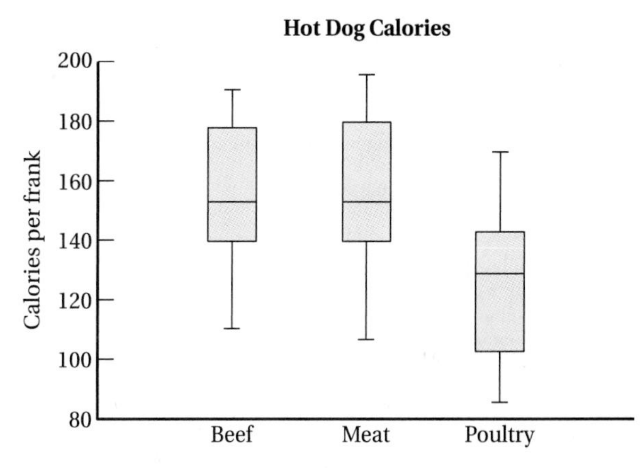

Source: Reprinted with permission from *On the Shoulders of Giants.* Copyright 1990 by the Academy of Sciences. Courtesy of the National Academy Press, Washington, D.C.

C. Reasoning and Problem Solving

11. Minority Opinion. Use the data at the bottom of the page to argue that college enrollment of minorities is increasing. Make your arguments in both numerical and graphic forms.

Data for Exercise 11

	Year				
	1970	**1975**	**1980**	**1985**	**1990**
Total enrollment (1000s)	9,095	10,880	12,524	11,387	13,621
Minority enrollment (1000s)	948	1,333	1,461	1,742	2,133

12. What patterns are apparent in the data display in Exercise 11? What generalizations can you make about the data?
13. Analysis of data collected from a classroom of students shows that the students who studied the most got the highest grades. Hence the study made the recommendation that all students study more so that they would get high grades. Is the conclusion a justifiable one? Why or why not?

D. Communicating and Connecting Ideas

14. Discuss the implication of the cartoon above. Is it a positive or negative image of data analysis? Explain.
15. **Making Connections.** Often statistics are used to justify decisions made in business and industry. How might the data be interpreted in the following situation to resolve the questions raised? Suppose that the current year's budget for a company is $1 billion. Suppose further that you are the chief financial officer (CFO) for the company and expect an overall increase of 5% in the budget for next year. Part of the board of directors is arguing for added resources for research, and others are calling for more investment. If you intend to give each department a 5% increase, explain how you could explain to each faction that they got "more" money this year while suggesting that the other group remained "fixed."

CHAPTER SUMMARY

Key Ideas: Questions and Answers

Section 8.1

- **What are some of the common forms of data display?** *(pp. 392–393)* Common forms of data display include the frequency table, dot plot, stem-and-leaf plot, pictograph, histograph, bar graph, and circle graph.
- **What are some ways to show the frequency and distribution of data?** *(pp. 393–398)* The frequency and distribution of data are often represented by tally marks, frequency tables, dot plots, and stem-and-leaf plots. These forms of data representation show the actual position and nature of the data.
- **What are some ways to show variation in or to compare data?** *(pp. 398–403)* The consideration of variation, both spread and range, in a data set or the comparison of two different data sets is often best accomplished by making a pictograph, a histograph, a bar graph, or a divided bar graph. These methods allow visual examination of the data set without strict reference to the actual numbers involved.
- **How can parts be compared to a total in a graphic display of data?** *(pp. 403–404)* Parts are best compared to the whole in displays of data by the use of a circle graph. The circle depicts the total data set, with sectors of the circle representing significant parts.

Section 8.2

- **What are the most common ways of displaying two-variable data?** *(pp. 409–413)* Two-variable data, whether discrete or continuous, may be displayed as a line graph or scatterplot. The line graph shows a trend clearly. Trend lines also may be superimposed on a scatterplot to show general relationships. In both cases the general trend shows the correlation between the variables involved.
- **What are trend lines?** *(pp. 413–417)* Trend lines are segments sketched on two-variable data plots to show the general nature of relations between the two variables involved.

Section 8.3

- **What are some different types of "averages"?** *(pp. 421–425)* The typical, or "average," member of a set of data may be described in several different ways. The most common measures of central tendency are the arithmetic mean, median, mode, and midrange.

- **How can the spread or distribution of data values be described numerically?** *(pp. 425–426)* The spread or distribution of data is most frequently described numerically by giving the range of the data. However, the interquartile range is often more revealing, as it shows the spread related to the middle 50% of the data set.
- **How can the spread and distribution of data values be described graphically?** *(pp. 426–428)* The graphic analysis of the distribution or spread of data is probably best accomplished with a box-and-whisker plot. It shows the range, the interquartile range, and the median, as well as the location of both Q1 and Q3 relative to the median and endpoints of the range, in the same display.
- **What are some more technical ways to describe how data values are spread or distributed?** *(pp. 429–431)* The most common way of numerically describing the dispersion of a data set is to give either the variance or standard deviation of the data. The standard deviation is the square root of the variance for a set of data.

Section 8.4

- **What are some questions that should be asked about data collection procedures?** *(pp. 434–435)* Questions normally should be asked about the nature of the sample (Was it randomly chosen? How large was the sample? Was it representative of the population?), the nature of the measure studied (Was it a valid measure? Was it reliable?), and the nature of the data set (Could it be replicated?).
- **What are some signals that conclusions based on data may not be valid?** *(pp. 435–436)* Points that may signal the need for a cautious approach to conclusions are vague and undocumented statements of comparison, heavy use of percents, and arguments that draw conclusions from correlations.
- **What are some ways that graphs may cause misleading conclusions?** *(pp. 436–440)* Graphs can mislead if unlike units, different scales, and inaccurate magnification of area and volume are used. In some cases, one graph may just be printed in a larger or smaller size than the comparison graph, even though all characteristics are correctly displayed.
- **What are some questions that should be asked about a data-based conclusion?** *(pp. 440–441)* You should always ask whether the information seems reasonable, is drawn from a reputable source, and can be verified or replicated. You should also ask whether the conclusion makes sense in terms of your own experience, you are being led to draw a conclusion based on correlations, the percents are accurately based, and the argument presented seems to explain why the opposite finding isn't valid. For graphs, you should always check their units, scales, and sizes.
- **What do data analysis and problem solving have in common?** *(pp. 441–444)* Like problem solving, data analysis calls for understanding the situation, developing a plan for examining it and the relevant data, implementing a plan to describe the situation, and looking back to see how the data and conclusions match the original questions raised.

Key Terms, Concepts, and Generalizations

Section 8.1

Data (p. 392)
Statistics (p. 392)
Data display (p. 392)
Title (p. 392)
Labels (p. 392)
Scales (p. 392)
Frequency table (p. 393)
Tally marks (p. 393)
Distribution (p. 393)
Dot plots (p. 393)
Data points (p. 393)
Axis (p. 394)
Range (p. 394)
Outlier (p. 394)
Clusters (p. 394)
Stem-and-leaf plot (p. 395)
Legend (p. 396)
Pictograph (p. 398)
Key (p. 398)
Histograph (p. 399)
Bar graph (p. 401)
Circle (pie) graph (p. 403)

Section 8.2

Continuous data (p. 410)
Discrete data (p. 410)
Line graph (p. 410)
Scatterplot (p. 413)
Ordered pairs (p. 413)
Positive correlation (p. 413)
Negative correlation (p. 413)
Trend line (p. 414)

Section 8.3

Measures of central tendency (p. 421)
Arithmetic mean (p. 421)
Mean (p. 421)
Median (p. 422)
Mode (p. 423)
Midrange (p. 423)
Dispersion (p. 425)
Range (p. 425)
Interquartile range (p. 425)
Quartiles (p. 426)

Box-and-whisker plot (p. 426)
Deviation (p. 429)
Variance (p. 429)
Standard deviation (p. 430)

Section 8.4
Sample size (p. 434)
Randomly assigned (p. 434)
Valid (p. 434)
Reliable instrument (p. 434)
Like units (p. 436)

CHAPTER REVIEW

Concepts and Skills

1. The data shown at the bottom of the page were developed by the English scientist Henry Cavendish in 1798 to describe the density of the earth in terms of its multiple of the density of water. Use these data to answer the following questions.
   a. What are the mean, median, mode, and range for the set of data?
   b. Make a dot plot, a stem-and-leaf plot, and a box-and-whisker plot for the data. Carefully label your axes and title your graphs. Are there any clusters or outliers in the data?
   c. What is the variance and standard deviation for the set of data?
   d. Based on your analysis of the data, what would be your estimate of the density of the earth in terms of the density of water?
2. The data in the table are the number and percent of U.S. high school students scoring in specific ranges on the Scholastic Assessment Test (SAT) in mathematics. Use these data to answer the following questions.

Score	Number	Percent
750–800	15,912	1
700–749	44,384	4
650–699	72,552	7
600–649	96,865	9
550–599	107,532	10
500–549	136,586	13
450–499	141,916	13
400–449	149,453	14
350–399	131,509	12
300–349	112,414	11
250–299	47,625	4
200–249	11,245	1

   a. Make a histograph and circle graph, showing the percent of students in each of the scoring intervals.
   b. Which graph in part (a) provides the better explanation for the distribution of scores on the SAT mathematics examination? Why?
3. The data in the table give the relative percents of male and female students scoring in various ranges on the SAT verbal examination in 1995. Use this information to answer the following questions.

Score	Percent Males	Percent Females
750–800	0	0
700–749	1	1
650–699	3	2
600–649	5	4
550–599	8	8
500–549	11	10
450–499	15	15
400–449	18	18
350–399	16	16
300–349	12	13
250–299	7	7
200–249	5	5

   a. Make a bar graph to compare the percents of males and females in each scoring range.
   b. Combine the scoring ranges into 100-point ranges and make another bar graph comparing males and females in each of the ranges. Does this graph show the information differently than the bar graph made in part (a)?
4. The table on the next page gives the stopping distance (feet) required for a car traveling at various speeds (miles per hour). Use these data to answer the following questions.
   a. Make a scatterplot of the data, carefully scaling your axes and labeling the plot.
   b. What type of correlation appears to hold for speed and stopping distance? What in the scatterplot indicates this relation to you?

Data for Exercise 1

5.50	5.57	5.42	5.61	5.53	5.47	5.65	5.39	5.58	5.27	5.85	
4.88	5.62	5.63	5.29	5.10	5.68	5.07	5.29	5.34	5.36	5.79	5.75
5.26	5.44	5.46	5.55	5.34	5.30						

Table for Exercise 4

Speed (mph)	Stop. Dist.
20	12.3
20	16.8
20	14.5
30	27.0
30	34.8
30	31.4
30	38.4
40	70.6
40	75.5
40	68.9
50	121.0
50	130.0
50	123.2
50	116.6
60	179.2
60	171.6
60	151.3
60	166.4

5. The points in the graph represent the pairings of confirmed sightings and telephone-reported sightings of tornadoes during the month of May over a 15-year period. Use the data presented in the graph to answer the following questions.

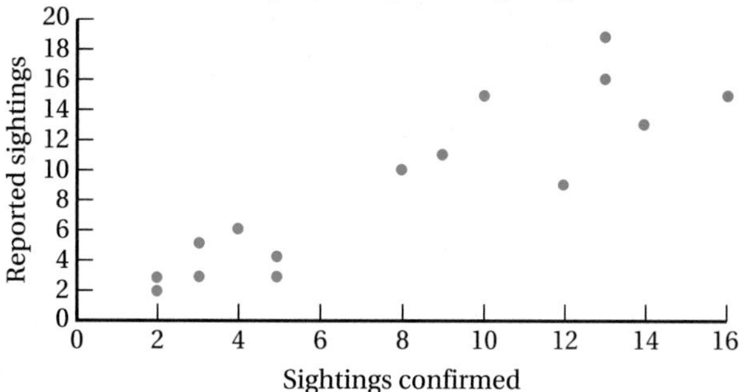

a. What are the median number of reported tornado sightings and the median number of sightings confirmed?

b. Are the number of reported sightings and number of sightings confirmed positively or negatively correlated? What suggests your answer?

c. Although the plot suggests a trend, why would a trend line not be appropriate for these data?

6. Suppose that a newspaper published the following in a boxed corner of the want ads section:

Should handgun control be abolished? You can call the shots if you participate in our poll tonight. If yes, call 1-900-For-Guns. If no, call 1-900-Ban-Guns. You will be charged 50¢ for the first minute.

Why will the results of this poll be biased?

7. To determine the best restaurant in a metropolitan area, a student conducted a survey at a snack bar at the city's professional baseball stadium on five randomly selected days. The first 50 individuals to purchase something at the snack bar were asked to state their preference of restaurant. Can the student draw a valid conclusion from the responses received? Why or why not?

8. Averageville is considering levying a flat tax on its citizens. To get an idea of the amount of revenue this tax would raise, the city council instructed the city staff to make a census of households to determine the "average" household income. The city accountant claims that they should use the median household income, rather than the mean. What do you think? Explain.

Reasoning and Problem Solving

9. **The Belting Bats Problem.** Michelle decides to test whether she hits the ball farther with an aluminum bat or a wooden bat. She uses a table of random numbers to decide the order in which she will use each bat. She then takes 15 swings (hitting the ball each time) with each bat, with the following results. Use both stem-and-leaf and box-and-whisker plots to compare the lengths of her hits. Develop an argument based on your data displays about her hitting performance with the two types of bat.

Aluminum bat distances	Wooden bat distances
105	86
115	117
123	121
130	123
139	127
142	127
147	139
149	142
151	147
156	151
160	153
167	154
167	156
174	159
177	163

10. **The Significant Stress Problem.** Engineering data revealed since the *Challenger* space shuttle disaster indicates that a postponement of the fatal launch was probably justified on the basis of the data and statistical analyses available at the time. The scatterplot indicates the number of incidents of problems with O-rings by temperature at the time of launch.

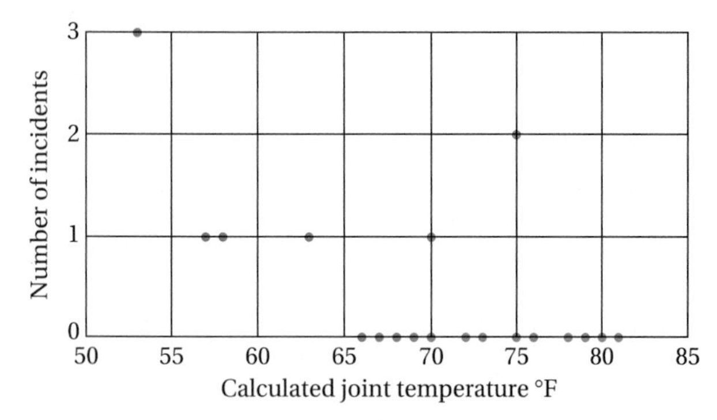

Source: Gordon, Gordon, Fusaro, Siegel, and Tucker, *Functioning in the Real World: A Precalculus Experience*, figure 3.3, p. 121. © 1997 Addison Wesley Longman Inc., Reprinted by permission of Pearson Education, Inc.

Use the data shown in the scatterplot to

a. discuss the overall correlation between lack of incidents and temperature;

b. identify any outliers in the data set; and

c. describe the pattern of the relationship between temperature and incidents when O-ring failure occurred (that is, the number of incidents is greater than 0). Recall that the temperature was 31°F at the time of the *Challenger* launch.

11. **The Rating Readers Problem.** The Right-On Reading Exam administered to 20 third-graders yielded the following scores: 40, 39, 26, 42, 35, 14, 15, 44, 38, 40, 31, 46, 45, 28, 14, 41, 51, 29, 33, and 35. Make a stem-and-leaf plot and histograph for these data to describe the students' performance. Which display do you prefer? Why?

12. The following table gives the number of international airline passengers (thousands) for each month of the years 1954–1956. Plot these data for each year on a line graph. Identify any major patterns you note and suggest a reason for each.

Month	1954	1955	1956
January	204	242	284
February	188	233	277
March	235	267	317
April	227	269	313
May	234	270	318
June	264	315	374
July	302	364	413
August	293	347	405
September	259	312	355
October	229	274	306
November	203	237	271
December	229	278	306

13. It is generally believed that individuals who are right-handed have a stronger grasp with their right hand than they do with their left hand. Design a simple experiment to determine whether this belief is true. Describe what you would use as your measure, how you would collect your data, and how you would use the information obtained to arrive at a conclusion.

14. **The Bulb Bargain Problem.** Suppose that you are the chief of maintenance for a large metropolitan hospital. You are trying to decide from what company to buy your next 5-year supply of light bulbs. Would you look to the company whose light bulbs have the greatest mean or the greatest median life? Explain your reasoning.

Alternative Assessment

15. Study the graphs on the next page. There are 20 students in Mr. Doug's class. On Tuesday *most* of the students in the class said that they had pockets in the clothes they were wearing. Which of the graphs most likely shows the number of pockets that each student's clothes had? Why?

Graphs for Exercise 15

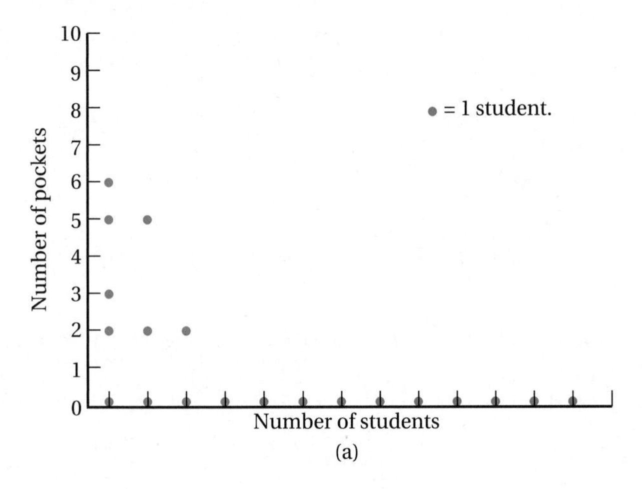

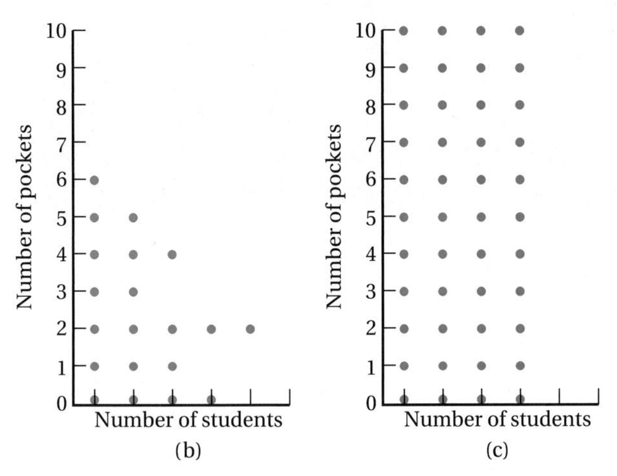

Source: Dossey, J. A., Mullis, I. V. S., and Jones, C. A. *Can Students Do Mathematical Problem Solving?* Washington, D.C.: National Center for Education Statistics, p. 106. Used with permission.

16. Work in a small group to design the best way of gathering information to determine which pizza shop is favored by students on your campus. Describe how you would collect, organize, interpret, display, and summarize the data and draw conclusions.

17. Marko Koers ran the mile 12 times during the track season. His times (seconds) for the races, in order of the races, were:

242, 239, 247, 236, 238, 239,
241, 244, 235, 248, 234, 231.

Describe this sequence of races, using descriptive statistics to make an argument that he ran better in his last six races of the season than he did in his first six.

9 Probability

Chapter Perspective

From the local weather forecast to the odds associated with the outcome of a Triple Crown horse race, probability has become part of people's daily lives. In addition, the theory of probability provides the basis for stocking inventories in shopping centers, exercising quality control in factories, scheduling television shows at various times during the week, and helping doctors choose medical treatments for patients. The recent surge in the use of probability has resulted from the availability of computer technology to process the large amounts of data needed to describe accurately the degree of "chance" associated with various events.

Connection to the NCTM Principles and Standards

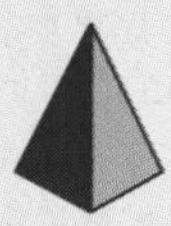

The NCTM *Principles and Standards for School Mathematics* (2000) indicate that the elementary school curriculum from prekindergarten through grade 8 should include the study of probability so that students can

- develop and evaluate inferences and predictions that are based on data *(p. 48)*; and
- understand and apply basic concepts of probability *(p. 48)*.

Connection to the PreK–8 Classroom

In grades PreK–2, students can describe various events associated with parts of a set of data and use that information to describe whether common events are likely or unlikely.

In grades 3–5, students can quantify the degree of likelihood associated with relatively common events, predict the probability of outcomes of simple experiments involving familiar situations, and test out the validity of their predictions via performing the experiment.

In grades 6–8, students can use the concepts of complementary and mutually exclusive events; can use proportions and probability concepts to make and test conjectures through performing simulations; and are able to compute probabilities for simple compound events through the use of lists, trees, and geometric models based on area.

Section 9.1 UNDERSTANDING PROBABILITY

- Defining Probability
- Identifying Sample Spaces
- The Addition Property of Probability

In this section we focus on the nature of chance, how it is quantified, and how you can communicate both the various outcomes associated with an experiment and their likelihood. In doing so, you will learn about the mathematical properties underlying probability.

Defining Probability

Probability, in its simplest form, deals with the use of numbers to describe the chance that something will happen. In the seventeenth century, Pascal, Fermat, D'Alembert, and others developed the first systematic approach to probability. However, the casting of lots and other games of chance date back to the beginning of recorded time.

A natural way to express chance is in the form of a ratio describing the number, or measure, of *favorable* occurrences to the number, or measure, of *total possible* occurrences of some specified result of an observable event. When someone says that the probability of an event occurring is 20%, that person is stating that the specified event is expected to occur 20% of the time, or, in general, about 1 of every 5 times.

The development of probability is based on the notion of an **experiment,** or observable situation, of interest. Associated with an experiment are several possible **outcomes** that occur as the result of performing the experiment. For example, the experiment of rolling a six-sided die results in the possible outcomes of rolling a 1, 2, 3, 4, 5, or 6. The set of all possible outcomes is the **sample space** associated with the experiment. In the experiment of rolling the die, the sample space is {1, 2, 3, 4, 5, 6}.

In some cases, these outcomes of an experiment have the same chance of happening. For example, each of the digits from 1 through 6 has an equal chance of occurring in the rolling of a fair six-sided die. In these cases, the outcomes are *equally likely,* and the sample space containing them is a **uniform sample space.** However, outcomes aren't always equally likely. For example, suppose that the sample space is {M, T, W, F, S}, or outcomes defined as the first letter in the English spelling of a day of the week. In this sample space, the outcomes aren't equally likely because the letters T and S are each the first letter of two different days of the week and thus twice as likely to occur as the letters M, W, and F, which are the first letters of only one day of the week. Because the outcomes aren't equally likely, the sample space is a **nonuniform sample space.**

Of interest sometimes is the chance that a subset of the outcomes in a sample space will occur. In such cases, an **event** describes any subset of the sample space. For example, suppose that you are interested in the chance that an even number comes up when a die is rolled. The event of rolling an even number is associated with any roll resulting in {2, 4, 6}, a subset of the sample space {1, 2, 3, 4, 5, 6} for rolling a die. The probability associated with the event of rolling an even number would be the ratio 3 to 6, or the ratio of favorable outcomes to total possible outcomes.

This situation involved a uniform sample space, as all of the outcomes had the same likelihood. In such settings, the probability, P, of an event A, written $P(A)$, is defined as follows.

Definition of Probability (Uniform Sample Space)

The probability, P, of an event A, written $P(A)$, is the ratio

$$P(A) = \frac{\text{Number of outcomes associated with the event } A}{\text{Number of outcomes in the sample space } S}$$

In other cases, when the outcomes have different probabilities associated with them (that is, in a nonuniform sample space), the definition of probability has to be altered slightly. In such settings, the probability, P, of an event A, written $P(A)$, is defined as follows.

Definition of Probability (Nonuniform Sample Space)

The probability, P, of an event A, written $P(A)$, is the ratio:

$$P(A) = \frac{\text{Measure of the outcomes associated with the event } A}{\text{Measure of all of the outcomes in the sample space } S}$$

A **random event** denotes an outcome of an experiment on which no external conditions were imposed. That is, outcomes of random events are likely to occur in relation to their natural probabilities of occurrence.

Finding the probability of an event requires being able to count or measure all possible outcomes associated with the experiment. Consider the event of flipping two coins and seeing whether they match, that is, whether both show *heads*, both show *tails*, or one shows *heads* and the other shows *tails*. We can express the sample space for this experiment as {S, D} for *same* and *different*. These outcomes are equally likely, as shown in the following table.

First Coin	Second Coin	Outcome
H	H	S
H	T	D
T	H	D
T	T	S

However, if the event is the number of heads showing after both coins are flipped, the sample space is {0, 1, 2}. These outcomes are not equally likely, because 0 can only occur in one way—(T, T), 1 can occur two ways—(H, T) and (T, H), and 2 can only occur in one way—(H, H). Thus the measures associated with 0, 1, and 2 are 1, 2, and 1, respectively, while the measure associated with the sample space is 4. This results in the following probabilities for the number of heads showing in the flipping of two coins:

$$P(0) = \frac{1}{4},\ P(1) = \frac{2}{4},\ \text{and } P(2) = \frac{1}{4}.$$

These probabilities may also be stated as 1 chance in 4, 1 chance in 2, and 1 chance in 4; as likely to occur 25%, 50%, and 25% of the time; or as 0.25, 0.50, and 0.25.

Example 9.1 further applies the idea of probability.

Example 9.1 Determining the Probability of an Event

A regular die is rolled. What is the probability of getting an outcome divisible by 3?

Solution

The set of possible outcomes is {1, 2, 3, 4, 5, 6}.

The set of favorable outcomes, or numbers divisible by 3, is {3, 6}.

$P(\text{a number divisible by } 3) = \frac{2}{6} = \frac{1}{3}$.

Your Turn

Practice: One ball is drawn from a bag containing five red balls, two green balls, and three blue balls. What is the probability that the ball is

a. red? **b.** blue?

c. yellow? **d.** a primary color?

Reflect: For the example problem, explain a way to find the probability of getting a number that isn't divisible by 3. ■

Probabilities like those in Example 9.1 are easy to describe because they deal with discrete settings in which outcomes are easily counted and compared. Many times, however, probabilities are used to describe the long-term behavior of continuous events of interest, such as weather conditions. The application of probability to weather forecasting is complex, as it is based on past records, weather conditions, and the forecaster's experience.

Identifying Sample Spaces

Some Basic Properties. Mathematicians called *probabilists* work to understand the theory of probability. In doing so, they have developed some very useful relations that make the practice of probabilistic reasoning easier. For example, if we know for sure that an event such as a dime landing heads or tails when tossed will occur, we say that the probability of the event is 1. If we know that an event such as a dime landing on its edge when tossed is impossible, we say that the probability of the event is 0. Thus the probabilities of other events that have some chance of occurring lie between 0 and 1. This reasoning suggests the following simple, basic properties of probability.

Basic Properties of Probability

The probability of an event A occurring, $P(A)$, is a real number between 0 and 1, inclusive. That is, $0 \le P(A) \le 1$.

If an event is certain, $P(A) = 1$. If an event is impossible, $P(A) = 0$.

Mini-Investigation 9.1 asks you to apply the basic properties of probability.

Talk about any difference in the probabilities that you and your classmates associated with the given statements. Try to come to an agreement on the probabilities.

Mini-Investigation 9.1 **Communication**

What probability would you associate with each of the following statements?

a. It is almost certain to rain today.
b. There is a slim chance that he will be on time.
c. Will we win? Maybe so, maybe not. It's hard to tell.
d. There's a chance that our house will sell, but it's far from a sure thing.
e. That horse has a good chance of winning, but there is some possibility that he won't.

Complementary Events. In some experiments the outcomes can easily be divided into two distinguishable sets. Flipping a coin is such an experiment: the person flipping the coin gets either heads, H, or tails, T. One of the two events must occur, so $P(\text{H}) + P(\text{T}) = 1$. Another way of thinking about the two events is heads and not heads, or $P(\text{H}) + P(\text{not H}) = 1$. We might also ask: If the probability of heads is 0.5, what is the probability of not heads? From the equation $P(\text{H}) + P(\text{not H}) = 1$, we conclude that $P(\text{not H}) = 1 - P(\text{H})$. These ideas suggest the definition on page 458.

Connection to the PreK–8 Classroom

A Sample Student Page: Considering the Likelihood of an Event

Figure 9.1 shows an activity for students just beginning to learn about probability. It's from a set of materials called the *Quantitative Literacy Series*, developed in a project sponsored by the American Statistical Association and the National Council of Teachers of Mathematics. Note how the questions are positioned to lead students to think about the difference between personal experience and long-term events, as well as toward seeing that the probability of certainty is 1 and impossibility is 0.

- ***What*** *is the purpose of such an exercise?*
- ***How*** *could this exercise be adapted for very young children?*
- ***How*** *would you explain the probability of the spinner ending up on a line?*

The Spinner

1. A spinner is divided into areas labeled *red* and *white*, as in the accompanying diagram.

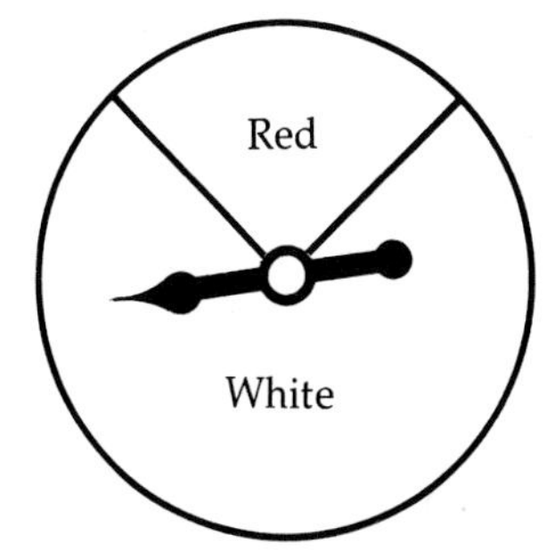

   a. If you were to spin the spinner, would you be just as likely to obtain red as white? If not, which color is more likely to occur? Why?

   b. Are you certain of getting at least one red in 100 spins?

   c. Is it very likely that you will not spin any reds in 100 spins?

   d. Is it possible never to spin a red in 100 spins?

2. Think about events that may occur in your life.

   a. List three events that are certain.

   b. List three events that are impossible—that is, they cannot occur.

   c. List three events that are highly likely.

   d. List three events that are unlikely.

FIGURE 9.1 | Excerpt on probability from a textbook.

(*Source:* From *Exploring Probability* by Claire M. Newman, Thomas E. Obremski, and Richard L. Scheaffer. © 1987 by Pearson Education, Inc., publishing as Dale Seymour. Used by permission.)

Definition of Complementary Events

Events A and B, which share no common outcomes but whose outcomes account for the total sample space, are known as **complementary events,** and their probabilities are related as $P(A) = 1 - P(B)$.

Example 9.2 features the properties of probability and the idea of complementary events.

Example 9.2

Using Complementary Events to Find Probabilities

A card is drawn at random from a standard deck of playing cards. What is the probability that the card drawn is

a. a spade?
b. not a spade?
c. a face card (jack, queen, king, or ace)?
d. a 2 through 10?

Solution

a. $P(\text{spade}) = 0.25$. As there are four suits of cards, each containing 13 cards, and 52 cards total, we have $P(\text{spade}) = \frac{13}{52}$, or $\frac{1}{4}$, or 0.25.
b. $P(\text{not spade}) = 1 - P(\text{spade}) = 1 - 0.25 = 0.75$. Or we could consider that three of the suits are not spades and account for three-fourths of the total cards.
c. $P(\text{face card}) = \frac{16}{52} = \frac{4}{13}$, or 0.31 (rounded to two places), because each suit has four face cards.
d. $P(2 \text{ through } 10) = 1 - P(\text{face}) = 1 - \frac{4}{13}$, or $\frac{9}{13}$, or 0.69 (rounded).

Your Turn

Practice: What is the probability of each of the following outcomes for the equally divided disk and spinner shown?

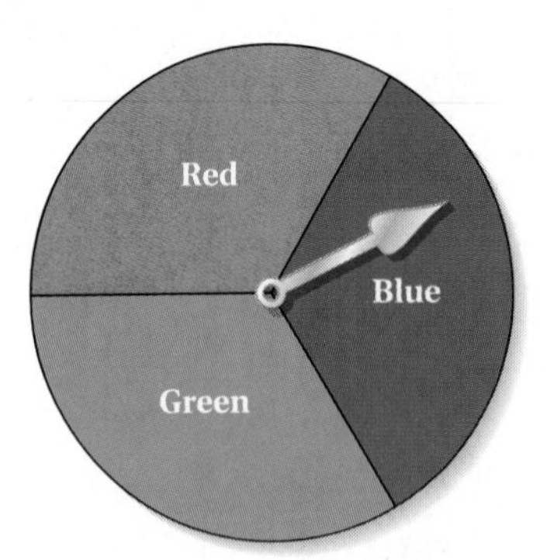

a. Red
b. Not blue
c. Not red
d. Not green

Reflect: What generalization can you make about the changes in the probabilities if the disk is changed to show four colors, each with equal area? ■

The Addition Property of Probability

Finding probabilities of more complex events requires the use of reasoning and some other key properties of probability. For example, suppose that an event is made up of two other events, A and B. Suppose further that we want to find $P(A$ or $B)$, that is, the probability that either event A or event B will occur. We consider two different cases of this problem in the following subsections.

Dealing with Mutually Exclusive Events. We can illustrate an event that is made up of two other events by finding the probability of getting a king or an ace, that is, $P(\text{king or ace})$, when drawing a card from an ordinary deck of 52 playing cards. The outcomes of the two events are as follows.

Outcomes involving a king: {K ♣, K ♦, K ♥, K ♠}
Outcomes involving an ace: {A ♣, A ♦, A ♥, A ♠}

We know that the $P(\text{king})$ is $\frac{4}{52}$, or $\frac{1}{13}$, and that $P(\text{ace})$ is the same. The probability of one or the other event happening is simply the sum of the individual probabilities. With 4 chances of getting a king and 4 chances of getting an ace, there are a total of 8 possible ways of successfully drawing a king or an ace from a 52-card deck. That is, $P(\text{king or ace}) = \frac{4}{52} + \frac{4}{52} = \frac{8}{52}$, or $\frac{2}{13}$. Because the events "drawing an ace" and "drawing a king" share no common outcomes, they are said to be **mutually exclusive events.** These ideas set the stage for an important property of probability.

The Addition Property of Probability

If A and B are mutually exclusive events, then $P(A \text{ or } B) = P(A) + P(B)$.

A Venn diagram illustrates how the addition law of probability works. Consider the sample space of an experiment being represented as the interior of a rectangle, as shown in Figure 9.2. The interior of the rectangle represents the total probability of 1, the set of all possible outcomes. The circle labeled A represents the probability of event A, and the circle labeled B, disjoint from A, represents the probability of event B, when A and B are subsets of the sample space.

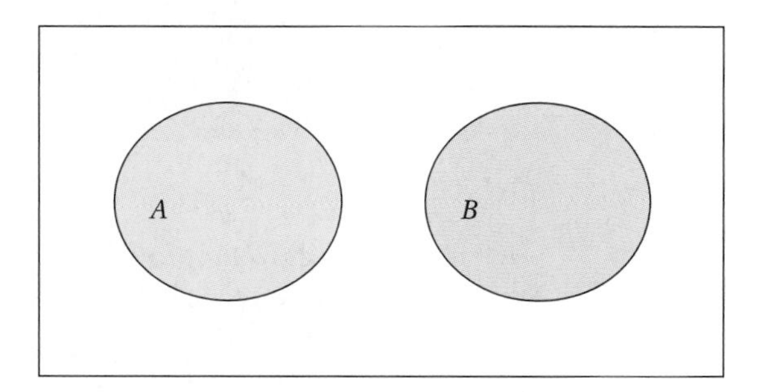

FIGURE 9.2 | Mutually exclusive events.

Thus if you want to know the probability of a result being either in A or in B, all that you need to do is add the probability associated with A to the probability associated with B. Doing so would give you $P(A \text{ or } B) = P(A) + P(B)$. Example 9.3 illustrates further the addition property of probability.

Example 9.3 Dealing with Compound Events

A key ring has eight similar keys on it. Three are for cars, two are for offices, one is for a boat, and the other two are for an apartment. Select one key in the dark and give the probability of selecting each different type of key. Then use these probabilities to find the probability of

a. selecting a car key or a boat key.
b. selecting an apartment key or an office key.
c. selecting a boat key or an apartment key.
d. selecting a boat key or an office key or an apartment key.

Solution

$P(\text{car key}) = \frac{3}{8}$.
$P(\text{office key}) = \frac{2}{8}$.
$P(\text{apartment key}) = \frac{2}{8}$, or $\frac{1}{4}$.
$P(\text{boat key}) = \frac{1}{8}$.

a. $P(\text{car or boat}) = P(\text{car}) + P(\text{boat}) = \frac{3}{8} + \frac{1}{8} = \frac{1}{2}$.
b. $P(\text{apartment or office}) = \frac{2}{8} + \frac{2}{8} = \frac{4}{8}$.
c. $P(\text{boat or apartment}) = \frac{1}{8} + \frac{2}{8} = \frac{3}{8}$.
d. $P(\text{boat or office or apartment}) = \frac{1}{8} + \frac{2}{8} + \frac{2}{8} = \frac{5}{8}$.

Your Turn

Practice: Suppose that you have a pair of disks with spinners like those shown, with the spinners equally likely to stop on any number.

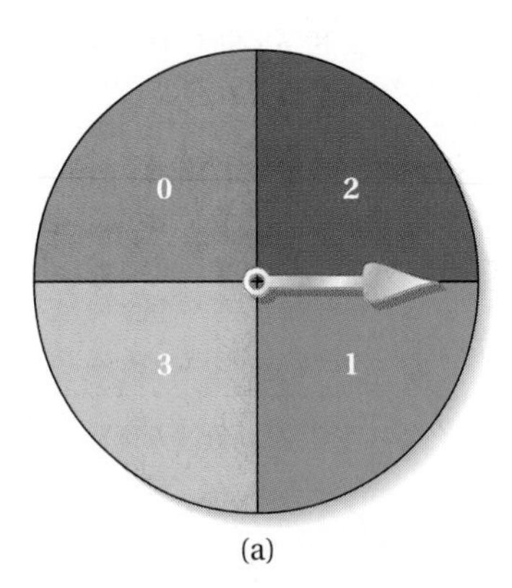

(a)

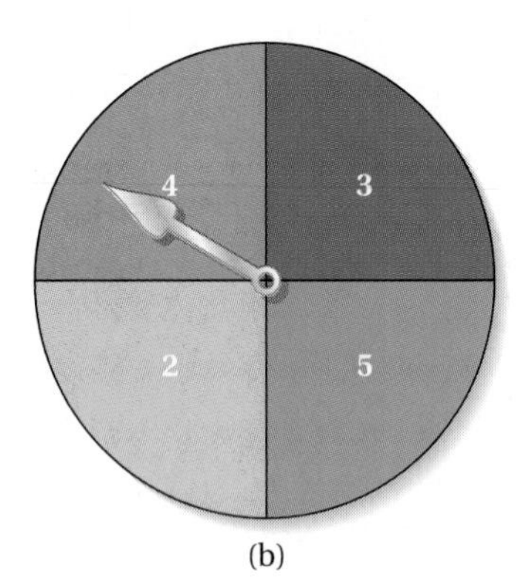

(b)

What are the following probabilities if each spinner is spun once and the numbers from each spin are recorded?

a. The sum of the digits is 4 or an odd number.
b. The numbers match.

c. The sum of the numbers is greater than 5 or even.

d. The sum of the numbers is 3 or even.

Reflect: Is the probability of the event (A or B) always greater than the probability of the event (A and B)? Explain. ■

Dealing with Events That Are Not Mutually Exclusive. Another application of the addition property of probability is illustrated by finding the probability of getting a king or a red card, that is, P(king or red card), when drawing a card from an ordinary deck of 52 playing cards. The events drawing a king and drawing a red card are not mutually exclusive because the draw of one card, such as the king of hearts, could be an outcome for both events. Let's consider the outcomes for these events.

Outcomes involving a king:

{K ♣, K ♦, K ♥, K ♠}

Outcomes involving a red card:

{2 ♥, 3 ♥, 4 ♥, 5 ♥, 6 ♥, 7 ♥, 8 ♥, 9 ♥ 10 ♥, J ♥, Q ♥ K ♥, A ♥,
2 ♦, 3 ♦, 4 ♦, 5 ♦, 6 ♦, 7 ♦, 8 ♦, 9 ♦, 10 ♦, J ♦, Q ♦, K ♦, A ♦}

Thus P(king) is $\frac{4}{52}$, or $\frac{1}{13}$, and P(red card) is $\frac{26}{52}$, or $\frac{1}{2}$. The events of drawing a king and drawing a red card are not mutually exclusive, so we can't find the P(king or red card) simply by adding the probabilities. However, we can add the probabilities and then eliminate the probability caused by the duplication in the sets of outcomes. The king of diamonds and the king of hearts are counted twice in the set of outcomes. Hence $P(\text{king and red card}) = \frac{2}{52}$, $P(\text{king or red card}) = \frac{4}{52} + \frac{26}{52} - \frac{2}{52} = \frac{28}{52}$, or $\frac{7}{13}$, or–about 0.54.

The Addition Property of Probability (for Non–Mutually Exclusive Events)

If A and B are non–mutually exclusive events, then $P(A \text{ or } B) = P(A) + P(B) - P(A \text{ and } B)$.

A Venn diagram may be used to illustrate this property. Events A and B overlap, as shown in Figure 9.3. The shaded area counts as part of $P(A)$ and again as part of the $P(B)$. Thus the $P(A \text{ and } B)$ has to be subtracted once to ensure that the

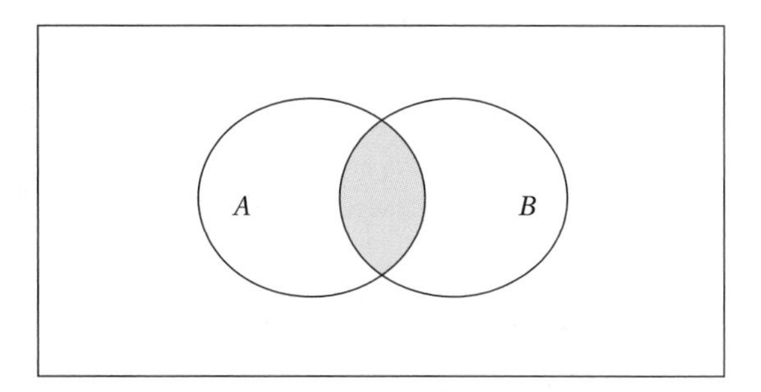

FIGURE 9.3 | Non–mutually exclusive events.

sum includes these outcomes, the probability of A and B occurring at the same time, only once in calculating the probability. Thus when the events are not mutually exclusive, $P(A \text{ or } B) = P(A) + P(B) - P(A \text{ and } B)$.

Example 9.4 applies the addition property of probability when the events are not mutually exclusive.

Example 9.4 Using the Addition Property of Probability

Suppose that each letter of the alphabet is written on a small slip of paper and placed in a box and a single slip is drawn at random. What is the probability of drawing a vowel, {a, e, i, o, u}, or a letter in the word *mathematics?*

Solution

In this case, $P(\text{vowel}) = \frac{5}{26}$, $P(\text{letter in } \textit{mathematics}) = \frac{8}{26}$, and $P(\text{vowel and letter in } \textit{mathematics})$ is $\frac{3}{26}$. Hence the probability of the event vowel or letter in *mathematics* is $\frac{5}{26} + \frac{8}{26} - \frac{3}{26} = \frac{10}{26}$, or $\frac{5}{13}$, or 0.38.

Your Turn

Practice: Consider the experiment of rolling a die. What is the probability of rolling a number greater than or equal to 3 or a number divisible by 2?

Reflect: How would you use a Venn diagram to illustrate the solution to the example problem? ■

Problems and Exercises for Section 9.1

A. Reinforcing Concepts and Practicing Skills

1. If a single letter is drawn from the set {m, a, t, h}, what is the probability that the letter drawn is a member of the set {m, a, t, h, e, i, c, s}?
2. If a single letter is drawn from the set {m, a, t, h, e, i, c, s}, what is the probability that the letter drawn is a member of the set {m, a, t, h}?

In Exercises 3–8, consider the experiment of rolling a single die.

3. What is the probability of rolling a number greater than or equal to 5?
4. What is the probability of rolling an odd number?
5. What is the probability of rolling a number such that when it is added to 2 gives a sum less than eight?
6. What is the probability of rolling a number that is divisible by 2 or 3?
7. What is the probability of rolling a prime number?
8. What is the probability of rolling a number that is divisible by 2 and 3?

In Exercises 9–16, consider the experiment of selecting a card from an ordinary deck of 52 playing cards and determine the probability of the stated event.

9. A face card is drawn.
10. A red card or a card showing a 5 is drawn.
11. A non-face card or a 7 is drawn.
12. A card that is not a king and not a spade is drawn.
13. A six is drawn.
14. A card with a prime number on it is drawn.
15. A card with a multiple of 4 on it is drawn.
16. A card with an even number and prime on it is drawn.
17. A disk with a spinner has three congruent regions of different colors. If you spin the spinner twice, what is the probability of landing on the same color with two successive spins?
18. A box contains three black balls and two gold balls. A ball is selected at random, its color is recorded, and it is then replaced. A second ball is then selected at random, and its color is recorded. The outcome associated with this type of selection is an ordered pair (first draw, second draw).
    a. List a sample space for this experiment.
    b. What is the probability that both balls are black?
    c. What is the probability that both balls are gold?
    d. Why doesn't the sum of the probabilities in parts (b) and (c) total 1?

19. An experiment consists of throwing a penny and a nickel. The respective values, head or tail, for each coin are recorded.
   a. List a sample space for this experiment.
   b. What is the probability that the penny lands heads?
   c. What is the probability that both coins land heads?
   d. What is the probability that the coins do not match?
20. Each capital letter of the alphabet is carefully printed on a piece of paper in block-letter form and placed in a bowl.

A B C D E F G H I J K L M
N O P Q R S T U V W X Y Z

An individual piece of paper is selected from the bowl. What is the probability that
   a. the letter is a consonant?
   b. the letter is formed from straight-line segments only?
   c. The letter has an enclosed region in its representation?
21. The following table shows data collected in the cafeteria at James O'Neill High School last week.

Class	Male	Female
Freshman	103	98
Sophomore	97	101
Junior	95	105
Senior	91	101

What is the probability that
   a. if a student were randomly selected, that student would be a male?
   b. if a student were randomly selected, that student would be a sophomore?
   c. if a sophomore were randomly selected, that student would be male?
   d. if a female were randomly selected, that student would be a junior or a senior?
22. Suppose that $P(A) = x$. The probability of the complementary event B is 0.7. What is the value of x?
23. Suppose that an experiment has five separate mutually exclusive outcomes: A, B, C, D, and E. If the sample space for the experiment is a uniform sample space, what is $P(A \text{ or } E)$?
24. An experiment has three possible outcomes: A, B, and C. If $P(A) = P(B)$ and $P(C) = 2P(A)$, what is the probability of each event?

B. Deepening Understanding

25. A basketball player has made 34 of her last 42 free throw attempts. What probability would you assign to the player's next free throw being successful? Explain your reasoning.
26. A counselor in a high school examined the records of 80 students who had participated in a special program during the past year. The data on their in-school suspensions are shown in the following table.

Number of in-school suspensions	Number of students
0	34
1	28
2	12
3	4
≥ 4	2

What is the probability of a randomly selected student having
   a. a prior in-school suspension?
   b. more than two in-school suspensions?
   c. For a randomly selected student, which is more likely: getting one without any previous in-school suspensions or one with at least one previous in-school suspension?
27. In a family with two children, what are the probabilities of the following outcomes, assuming that the birth of boys and girls is equally likely?
   a. Both are boys.
   b. The first is a girl and the second a boy.
   c. Neither is a girl.
   d. At least one is a girl.
28. Suppose that two cards are drawn from an ordinary deck of 52 playing cards without replacement. What is the probability that both cards are red?
29. What is the probability of flipping a coin two times and getting no heads?
30. If $P(A) = \frac{1}{2}$, $P(B) = \frac{1}{6}$, and $P(A \text{ or } B) = \frac{1}{2}$, what can you say about $P(A \text{ and } B)$?
31. If $P(A) = \frac{2}{3}$, $P(B) = \frac{1}{6}$, and $P(A \text{ and } B) = 0$, what can you say about $P(A \text{ or } B)$?
32. If $P(A) = \frac{4}{13}$, $P(B) = \frac{7}{13}$, and $P(C) = \frac{2}{13}$, what can you say about $P(A \text{ or } B \text{ or } C)$ when
   a. $P(A \text{ and } B \text{ and } C) = 0$?
   b. $P(A \text{ and } B) = 0$, $P(B \text{ and } C) = 0$, and $P(A \text{ and } C) = 0$?

C. Reasoning and Problem Solving

33. Three different gaming machines were in a line along a wall. The machines guarantee the following outputs with listed probabilities.

Machine A	Machine B	Machine C
$P(1) = \frac{2}{7}$	$P(0) = \frac{5}{9}$	$P(2) = 1$
$P(3) = \frac{3}{7}$	$P(6) = \frac{4}{9}$	
$P(5) = \frac{2}{7}$		

   a. Suppose that you want to select the game giving the highest average output over a long period of time. Which machine would you select? Why?

b. Suppose that you and two other players are starting to play at the three machines. Which machine would you select now, assuming that you get first choice? Why?

34. A box contains two black balls and three gold balls. Two balls are randomly drawn in succession from the box.
   a. If there is no replacement, what is the probability that both balls are black?
   b. If there is replacement before the second draw, what is the probability that both balls are black?

35. A 12-sided die can be made from a regular dodecahedron so that the digits 1 through 6 appear twice. Justify the claim that such a die can be used in any game in which a regular cubical (6-sided) die is used without any change in the outcome.

36. A die is thrown three successive times. What is the probability of obtaining a digit sum greater than or equal to 15?

37. What is the probability of rolling a regular die six times and getting each of the six possible numbers exactly once?

38. A recent survey of 150 students at a community college showed that 44% drank Cola, 66% drank Spirit, and 22% drank neither. What is the probability that a student drinks Spirit, if the student is known to drink Cola? In what way, if any, would this information be likely to help a soft drink company make decisions about advertising?

D. Communicating and Connecting Ideas

39. **Historical Pathways.** The French mathematician Georges-Louis Leclerc, Comte de Buffon (1707–1788), tossed a coin 4040 times and got 2048 heads. The English statistician Karl Pearson (1857–1936) tossed a coin 24,000 times, getting 12,012 heads. What probability would you associate with each of these experiments for $P(\text{H})$? Would a combination of their data give a better estimate of $P(\text{H})$? Why or why not?

40. **Making Connections.** The probability of a female birth is about 0.5. Over the period of a year, would there be more days when at least 60% of the babies born were girls in (select one)
   a. a large hospital?
   b. a small hospital?
   c. It makes no difference.

   Answer the question and then write a paragraph explaining your reasoning.

Section 9.2 CONNECTING PROBABILITY TO MODELS AND COUNTING

- Representing and Counting Experimental Outcomes
- Probabilities of Independent Events
- Conditional Probability

In this section we expand the ideas of probability by using models to represent outcomes and presenting a special property that helps in counting outcomes. We also look at ways to find probabilities of two-part events, when the outcome of the first part has nothing to do with the outcome of the second part. With geometric concepts, we also look at a situation in which additional information restricts the size of the sample space. Mini-Investigation 9.2 gives you an opportunity to focus on some of the ideas of this section by thinking through a real-world situation that involves representing and counting outcomes.

Talk about how you arrived at your answer and how your approach to the problem differs from that of a classmate.

Mini-Investigation 9.2 Making a Connection

What is the probability that the gorilla-sized pizza Viva ordered from Judi's Pizza Kitchen will satisfy a friend who hates deep-dish pizzas that contain pepperoni?

Judi's Gorilla Pizza Choices
Thin Crust or Deep-Dish
Sausage, Pepperoni, or Hamburger

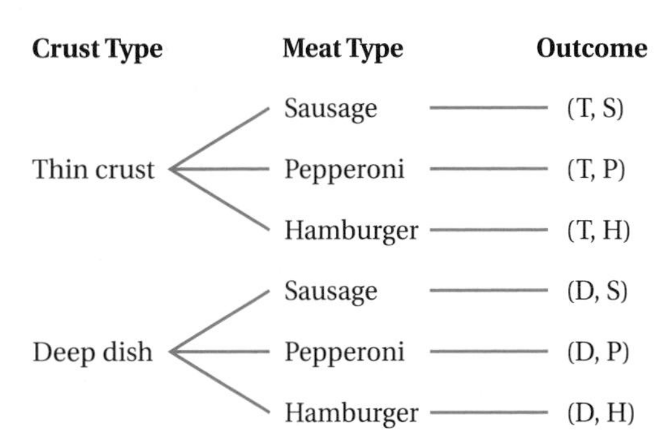

FIGURE 9.4 | Tree diagram for pizza topping combinations.

Representing and Counting Experimental Outcomes

Die	Coin	Outcome
1	H	(1, H)
	T	(1, T)
2	H	(2, H)
	T	(2, T)
3	H	(3, H)
	T	(3, T)
4	H	(4, H)
	T	(4, T)
5	H	(5, H)
	T	(5, T)
6	H	(6, H)
	T	(6, T)

FIGURE 9.5 | Tree diagram for die and coin experiment.

Using a Tree Diagram. The setting for some two-part probability problems can be modeled with a **tree diagram,** which depicts events and their outcomes. A tree diagram is an easy way to represent and count the outcomes of an event being considered. In Mini-Investigation 9.2, for example, you might have represented the six possible outcomes as depicted in Figure 9.4.

Now consider the experiment of rolling a die and then flipping a coin. The outcomes may be modeled with the tree diagram shown in Figure 9.5. The roll of the die can result in any one of 6 different outcomes. Each of the 6 outcomes can then be associated with 2 possible outcomes for the second part of the experiment. Thus there are 6×2, or 12, possible outcomes for the experiment. So the sample space has 12 members, as shown in the outcome column.

If we were interested in the event of rolling a 3 or flipping a head in the die–coin experiment, the outcomes associated with the event are $\{(1, H), (2, H), (3, H), (3, T), (4, H), (5, H), (6, H)\}$. Thus the probability of the event is $\frac{7}{12}$, or 0.58.

Structuring the possible outcomes with a tree diagram requires that you think through each of the possible outcomes for each individual part of the experiment. Then you have to show all the ways in which they might connect with each other during the experiment.

Second die

First die	1	2	3	4	5	6
1	2	3	4	5	6	7
2	3	4	5	6	7	8
3	4	5	6	7	8	9
4	5	6	7	8	9	10
5	6	7	8	9	10	11
6	7	8	9	10	11	12

FIGURE 9.6 | Box array for the sum from two dice.

Using a Box Array. Another way to display the sample space for an experiment is with a **box array,** which is a tabulation of all possible outcomes. Consider the experiment of rolling two dice and recording the sum of the numbers showing. The outcomes are represented by the box array shown in Figure 9.6. The digits in the leftmost column indicate the number showing on the first die, and the digits in the top row indicate the number showing on the second die. The number in each box (the intersection of row and column) in the table is the sum of the digits shown on the dice. As each of the first die's 6 outcomes could be associated with each of the second die's 6 outcomes, there were 6×6, or 36, possible outcomes associated with rolling two dice. When these outcomes are summed, only 11 *different* possible sums emerge, 2 through 12.

Using the Multiplication Property of Outcomes. Both the tree diagram and the box array may be used to illustrate an important property, the multiplication property of outcomes. It helps in determining the number of possible outcomes in two-part probability problems. For example, each of the 6 options in the die column of Figure 9.5 has 2 options in the coin column, giving a total of 6×2, or 12, outcomes for the experiment. In the box array of Figure 9.6, there are 6 rows (or options) for the first die and 6 options for the second die in each row. That gives a total of 6×6, or 36, outcomes for rolling two dice.

In general, the computation of probabilities for events similar to the two just illustrated involves first doing x things, each of which can be combined with y following things, giving x groups of y possibilities, or $x \times y$ choices in all. More formally, this multiplication property may be stated as follows.

The Multiplication Property of Outcomes

If event A has m outcomes and event B has n outcomes, then the experiment that has event A followed by event B has $m \times n$ outcomes. This property can be generalized to more than two events.

Example 9.5 applies the multiplication property to a real-world situation.

Example 9.5

Counting with the Multiplication Property

How many safety deposit boxes in a bank can be labeled differently by using one letter followed by a 2-digit number?

Solution

There are three spaces to fill. The first space can be filled in 26 ways (the number of letters in the alphabet). The second space can be filled in 9 ways (nine digits, as 0 isn't the first digit in a two-digit number). Finally, the third space can be filled in 10 ways. Using the multiplication property of outcomes yields $26 \times 9 \times 10 = 2340$ safety deposit boxes that could be labeled uniquely.

Your Turn

Practice: Suppose that a large bank uses identification numbers containing two letters followed by a 3-digit number for trust accounts. How many different accounts could the bank set up, with each having its own unique identification number?

Reflect: How would the answer for the example problem change if the use of 0 as the first digit in the 2-digit number were allowed? ■

Probabilities of Independent Events

When the outcome of one event has no influence on the outcome of a second event, the two events are said to be **independent events.** Let's consider three examples of such pairs of events. The two events have nothing to do with each other, and we are interested in the probability that the two events occur simultaneously.

a. A red die and a green die are tossed.
Event A: Red die shows even.
Event B: Green die shows odd.
We are interested in $P(A \text{ and } B)$.

b. Event A: A randomly chosen student in this class has blue eyes.
Event B: A randomly chosen student in this class is more than 20 years old.
We are interested in $P(A \text{ and } B)$.

c. A red ball is drawn from two boxes. Box I has two red balls and one blue ball, and Box II has one red ball and two blue balls.
Event A: A red ball is drawn from Box I.
Event B: A red ball is drawn from Box II.
We are interested in $P(A \text{ and } B)$.

We can use (c) to illustrate an important idea about finding probabilities involving independent events. Suppose that we want to find the probability of drawing a red ball from each of two boxes, I and II. Consider the box array shown in Figure 9.7, which represents the outcomes of the experiment.

Box I \ Box II	R	B	B
R	(R, R)	R, B	R, B
R	(R, R)	R, B	R, B
B	B, R	B, B	B, B

FIGURE 9.7 | Box array for drawing balls from two boxes.

Note that nine possible outcomes exist for drawing balls from each box, but that only two of them result in drawing a red ball from each box. So, using the events described in example (c), we find that $P(A \text{ and } B) = P(\text{red and red}) = \frac{2}{9}$. We also find that $P(A) = \frac{2}{3}$, $P(B) = \frac{1}{3}$, and $P(A) \times P(B) = \frac{2}{3} \times \frac{1}{3} = \frac{2}{9}$.

If we were to test other examples, we would find a pattern suggesting that the probability of two independent events occurring together equals the product of the probabilities of the two events. We express this property as follows.

Multiplication Property of Probabilities Involving Independent Events

If events A and B are independent, then $P(A \text{ and } B) = P(A) \times P(B)$.

In Example 9.6 we apply this process for finding the probability of the simultaneous occurrence of two independent events.

Example 9.6

Finding Probability for Two Independent Events

Suppose that a nickel and a penny are each flipped. What is the probability of getting at least one head? Both heads? Are the events of getting a head on the nickel and getting a head on the penny independent?

Solution

The sample space for this experiment may be represented by either a tree like the one shown in Figures 9.4 and 9.5 on page 465, or by an array.

Nickel/Penny	Heads	Tails
Heads	(H, H)	(H, T)
Tails	(T, H)	(T, T)

Examining the four outcomes (2×2) gives $P(\text{at least one head}) = \frac{3}{4}$ and $P(\text{both heads}) = \frac{1}{4}$. The outcomes of the two coin tosses are independent because the outcome of the flip of the penny isn't affected by the outcome of the flip of the nickel.

Your Turn

Practice: Suppose that an urn has two red balls and one white ball in it. Two balls are drawn from the urn in succession, with the first ball not being replaced prior to the drawing of the second ball. What is the probability that the two balls drawn are both red?

Reflect: What effect does not replacing the first ball drawn have on the probability of the event described? Does it increase or decrease the probability of drawing two red balls? Explain your reasoning in writing. ■

Conditional Probability

Sometimes in an experiment the question of stating the probability of a particular outcome is affected by additional information. For instance, consider the following events.

Event A: getting a 3 when rolling a die.

Event B: getting an odd number when rolling a die.

If we were asked what $P(A)$, that is, $P(3)$, is for rolling a single die, we would answer $\frac{1}{6}$. However, if we were asked the probability of getting a 3 on the roll of a die if we knew that the roll would result in an odd number, the situation would be quite different. In terms of the events A and B, this question is: What is the probability of A, given that B has occurred? As shown in Figure 9.8, the information that an odd number was rolled on the die shrank the sample space for the event from $\{1, 2, 3, 4, 5, 6\}$ to $\{1, 3, 5\}$.

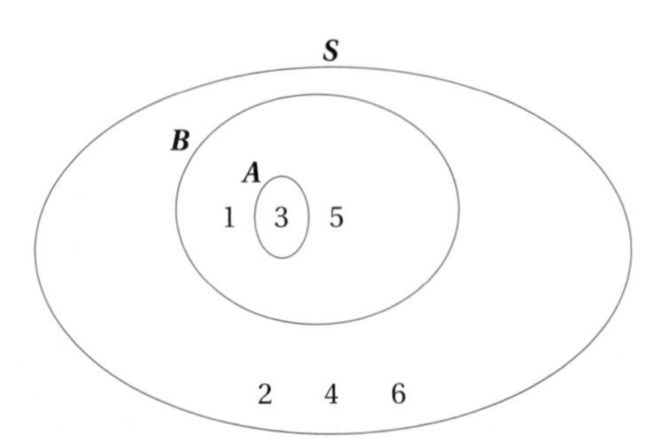

FIGURE 9.8 | Sample space for a conditional probability experiment.

The probability of getting a 3, given the added information that an odd number has been rolled, is $\frac{1}{3}$. We indicate the answer to the question, What is the probability of A, given that B has occurred, by writing $P(A|B) = \frac{1}{3}$, where the vertical bar, $|$, between A and B represents the word *given.*

Hence additional information about the outcome of an experiment reduces the size of the sample space from all possible outcomes in the original experiment to those outcomes that have the special characteristic disclosed by the added, or *conditioned,* information. When the probability of an event is conditioned by information about a related event that affects the nature of the sample space, we say that the probability quoted is a **conditional probability.** If we are considering the probability of an event A and we know that a related event B affecting the sample space has occurred, we must analyze the resulting probability carefully. Doing so usually requires shrinking the sample space related to A to accommodate what we know about the effects of B having occurred.

In Example 9.7 we further illustrate conditional probability. The method used is called **geometric probability** because it involves the use of geometric measurement principles to describe the event and sample spaces.

Example 9.7 Using Geometric Probability

What is the probability that a missing airplane is somewhere in the triangular region shown, given that from radio transmissions it is known to have landed somewhere in the smaller rectangular region?

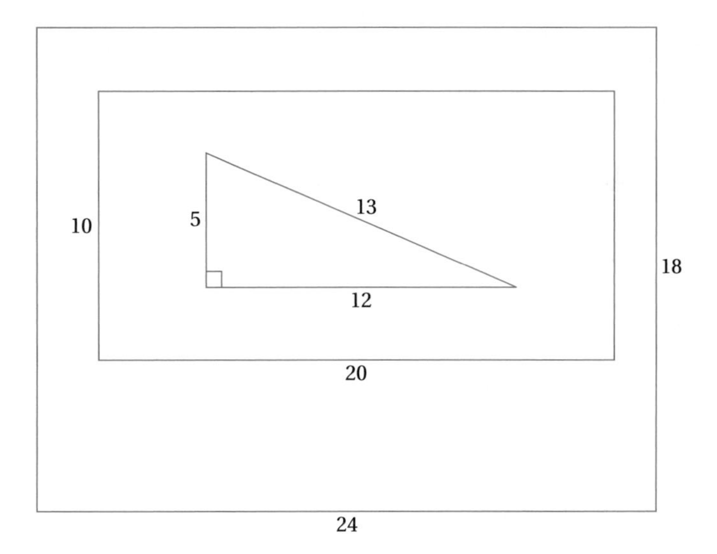

Solution

One could use the fact that the plane is in the smaller rectangular region and only consider the 10×20, or 200, square units of possible landing space. The triangle accounts for $\frac{1}{2} \times 5 \times 12$, or 30, square units of this space. So the probability of landing in the triangle, given the knowledge that the plane landed in the smaller rectangle, is $\frac{30}{200} = \frac{3}{20}$, or 0.15.

Your Turn

Practice: The land area of the earth is approximately 57.5 million square miles, whereas the oceans and lakes cover about 140 million square miles. What is the probability that a meteorite would land in the United States, given that the area of the United States is approximately 1.94 million square miles?

Reflect: Why is the practice problem called a geometric probability problem? How is geometry involved in its solution? ■

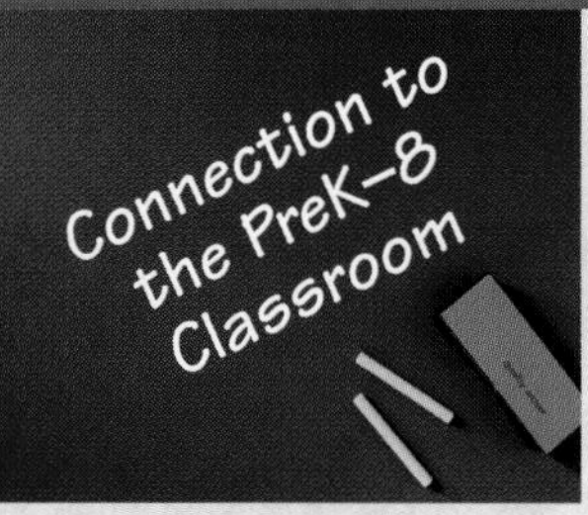

A Sample Student Page: Using Probability at Forks in the Road

Figure 9.9 is an example of a compound probability experiment contained in the Michigan State University Middle School Mathematics Project materials.

■ **How** would you solve this problem?

Mr. Icky Green is the dietitian at a middle school down the road. His favorite dish is Cheese and Rutabaga Surprise, which he serves twice a week in the school cafeteria. One day a group of students who were fed up with C and R Surprise approached Mr. Green with this proposition: The students would place a new cookbook in one of two rooms at the end of a maze. Mr. Green would walk the maze, randomly choosing a path at each intersection. If he enters the room with the cookbook, no more Cheese and Rutabaga Surprise. If he enters the empty room, C and R Surprise will appear with disgusting regularity for the rest of the year.

In which room should the students place the cookbook to have the best chance of being found by Mr. Green?

This is a map of the maze.

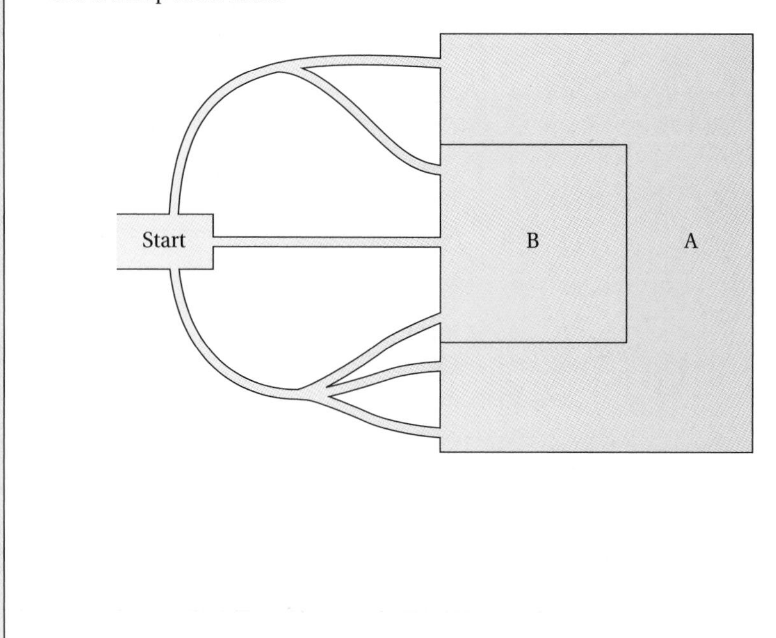

FIGURE 9.9 | Excerpt on probability from a middle grades textbook.

(*Source:* From *Middle Grades Mathematics Project Probability* by Elizabeth Phillips, Glenda Lappan, Mary Jean Winter, and William Fitzgerald. © 1986 by Pearson Education, Inc., publishing as Dale Seymour. Used by permission.)

Problems and Exercises | *for Section 9.2*

A. Reinforcing Concepts and Practicing Skills

1. How many 3-letter code symbols can be formed from the letters S, P, and Y without repetition?

2. How many 3-letter code symbols can be formed from the letters I, S, P, and Y without repetition?

3. How many 3-symbol codes can be made from the letters S, P, Y, and one digit from the set $\{0, 1, 2, \ldots, 9\}$ without repetition?

4. How many 3-symbol codes can be made from the letters S, P, Y, and two digits from the set $\{0, 1, 2, \ldots, 9\}$ without repetition?

5. How many ways can four cars be parked in a row of four parking spaces?

6. How many ways can three cars be parked in a row of four parking spaces?

7. If a child has a group of mix-and-match warm-up suit tops and bottoms, with six tops, four bottoms, and five pairs of shoes, what is the total number of possible outfits?

8. How many elements are in the sample space for an experiment that involves first flipping a coin, then rolling a die, and then spinning a spinner on a disk with three sectors on it?

9. Two cards are drawn without replacement from an ordinary deck of 52 playing cards. What is the probability that both cards are kings if the first card drawn was a king?

10. Two cards are drawn without replacement from an ordinary deck of 52 playing cards. What is the probability that both are spades if the first card drawn was a spade?

11. Three cards are drawn without replacement from an ordinary deck of 52 playing cards. What is the probability that the third card is a spade if the first two cards were not spades?

12. Three cards are drawn without replacement from an ordinary deck of 52 playing cards. What is the probability that the second and third cards are spades if the first card was not a spade?

13. Suppose that a natural number from 1 through 9 is selected at random. What is the probability that the number selected is odd if you know that it is prime?

14. Suppose that a natural number from 1 through 9 is selected at random. What is the probability that the number selected is even if you know that it is composite?

15. A bowl contains three blue balls and four white balls. One ball is drawn without replacement, and then a second ball is drawn. What is the probability that

a. both balls are blue?

b. both balls are blue, given that the first ball drawn was blue?

c. the balls are of different color, given that the first ball drawn was blue?

16. A pair of dice are rolled and the sum on their upturned faces is recorded. What is the probability that the sum showing is 8, given that one die is showing a 5?

17. In a subdivision, 67% of the homes have detached garages, 32% have a patio, and 13% have both. What is the probability that a house has a patio if you know that it has a detached garage?

B. Deepening Understanding

18. In the tests of a new pharmaceutical product, data were collected for use in the approval process required by the U.S. Food and Drug Administration (FDA). The data are shown in the following table. Some participants were given a placebo, an inert substance that looks like the drug; others were given the drug.

	No Help	Help
Drug	22	47
Placebo	31	20

What is the probability that

a. the participants perceived that their "medication" helped if they received the drug?

b. the participants perceived that their "medication" helped if they received the placebo?

c. the participants perceived that their "medication" helped?

19. At a quality control checkpoint on a manufacturing assembly line, 10% of the items failed check A, 12% failed check B, and 3% failed both checks A and B.

a. If a product failed check A, what is the probability that it also failed check B?

b. If a product failed check B, what is the probability that it also failed check A?

c. What is the probability that a product failed either check A or check B?

d. What is the probability that a product failed neither check A nor check B?

26. On a circular dart board with five concentric circles of increasing radii of 2, 4, 6, 8, and 10 inches, what is the probability of a single dart that hits the board hitting the innermost circle if you are sequentially given information that it is inside the circle with a radius of

a. 8 inches? **b.** 6 inches?
c. 4 inches? **d.** 2 inches?

21. In a game played on a rectangular gameboard of total dimensions 36 inches by 9 inches, where the top row is twice as high as the second row, contestants try to throw chips to land inside the sectors marked A. Given that a chip did not land in section B, what is the probability of a winning throw?

A	B
C	A

22. Jack is taking a 4-item true–false test. He has no knowledge about the subject of the test and decides to flip a coin to answer the items. What is the probability that he receives a perfect score? What is the probability on a 10-item test?

23. A parent's chance of passing a rare inherited disease on to a child is 0.15. What is the probability that, in a family of three children, none of the children inherit the disease from the parent?

24. Two cards are drawn from an ordinary deck of 52 playing cards with replacement. What is the probability that

a. both cards are of the same color?

b. both cards are from the same suit?

c. How would your answers to parts (a) and (b) change if the draws are made without replacement?

C. Reasoning and Problem Solving

25. In the maze shown, Bill picks his paths at random at each junction. What is the probability that,

a. with no knowledge of intermediate picks, Bill ends in room A?

b. if Bill selected the middle path at the first junction, he ends in room A?

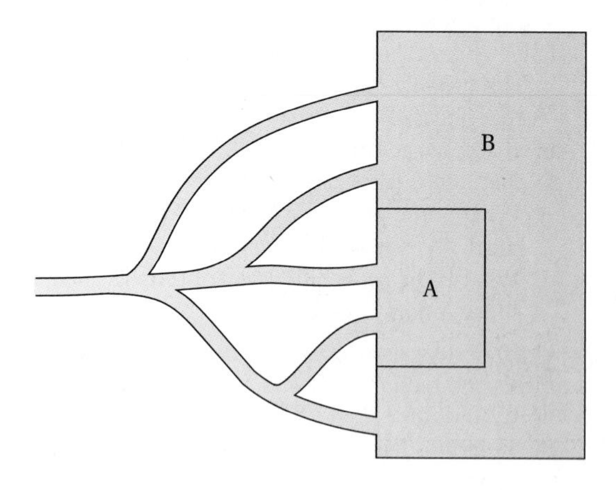

26. Twenty teams enter a single-elimination basketball tournament. How many games must be played to determine a winner?

27. How many different 3-digit numbers can be formed from the digits in the set {0, 1, 3, 5, 7, 9} if 0 isn't an acceptable first digit and repetitions are allowed if

a. the number must be divisible by 5?

b. the number must be divisible by 25?

c. the number must be divisible by 100?

28. What is the probability of throwing a sum of 20 on three throws of a dart at the board shown if the first throw was a 2?

3	5	3	5	3
5	7	9	7	5
3	9	2	9	3
5	7	9	7	5
3	5	3	5	3

29. Sue and Jane are playing in a "two out of three wins" checkers tournament. Assume that the two girls are equally matched going into the tournament and only games ending in a decision are counted; that is, draws are replayed.

a. Draw a tree diagram showing the possible outcomes for this tournament.

b. What is the probability that Jane wins in two games if she wins the first game?

c. What is the probability that Jane wins in three games if she wins the first game?

d. What is the probability that Jane wins in three games if she loses the first game?

e. What is the probability that the tournament goes three games?

30. Computer Game Problem. In playing a computer game, Viva has the option of throwing a number of "switches" to control a spaceship's movements. In an important situation, the probability that Viva throws switch A is $\frac{5}{9}$ and that she throws switch B is $\frac{2}{9}$. The probability that Viva throws neither switch A nor B is $\frac{3}{9}$. What is the probability that Viva throws either switch A or switch B?

31. Missile Targeting Problem. A battleship-launched cruise missile has a 95% chance of penetrating coastal defenses and an 80% chance of hitting its target if it penetrates the defenses. What should the admiral in charge of the mission be told about the chances that a missile will complete its task successfully?

32. Basketball Game Win Problem. A basketball player comes to the free throw line in an extremely hard-fought game to shoot a one-and-one free throw with the team behind by 1 point and no time left on the clock. The player normally shoots 50% from the free throw line. What is the probability that the team will win the game? What is the probability of a tie and overtime?

D. Communicating and Connecting Ideas

33. **Historical Pathways.** The French mathematician Jean D'Alembert argued that the calculation of a coin-flipping experiment for flipping a coin n times should not include the case of n heads or n tails in the calculation of the probabilities, as these outcomes couldn't possibly exist. State whether he was correct or incorrect and explain your reasoning.

34. **Making Connections.** In a two-player, coin-flipping game, the players agree that one will be "heads" and the other "tails." One point is awarded a player for each flip of the coin if the coin lands showing his designated side. The player who first scores 6 points wins the game. Suppose that the game is interrupted when the score is 5–4 in favor of player A. How should the $2 prize be divided between the players fairly if the game can't be completed?

Section 9.3 SIMULATIONS

- Approximating Probabilities with Simulations
- Creating Simulations with Technology

In this section we consider how simulation, or modeling, can be used to approximate probabilities that otherwise might be very difficult or impossible to determine. We show how random number tables and technology can be used to aid this process. We also demonstrate how simulation based on the use of technology is used to solve probability problems.

Approximating Probabilities with Simulations

Calculating an actual probability for an event may be too difficult or *experimenting* with the actual event may be too time-consuming. In such cases you can approximate a probability by "acting out" the events with a model. Using a model to act out an event is called **simulation.**

Procedure for Using Simulation to Approximate Probabilities

1. Clearly understand the event.
2. Select a model for the event.
3. Define a trial.
4. Conduct trials using the model.
5. Find the ratio of the number of trial successes to the total number of trials.

Estimation Application

How Often Does a Thumbtack Land Point Up?

Estimate the probability of a thumbtack landing on a tile floor with its point up. Develop a simulation to check your estimate.

The process of simulating an event by using a model to approximate probabilities is sometimes referred to as the **Monte Carlo approach.** The name comes from procedures used by casinos to study probabilities of winning casino games. This approach to the determination of probabilities was developed during World War II by the mathematician Stan Ulam.

The models used for simulation can vary, but one of the commonly used models involves the use of **random numbers,** generated by a calculator or computer. Such sets of random numbers are usually given as lists of digits 0 through 9, such as those shown on the next page.

33218	15401	97308	43001
27074	10242	39467	11178
94520	76707	86437	92553
90553	77985	70459	93855
31472	71631	09383	28369
...			

The order of digit occurrence in a random number list has no discernible pattern. Each digit appears in the list with equal frequency over a long period of time and thus is equally likely to occur in each location. Using a list of random numbers involves associating a digit (number) or groups of digits (numbers) with successful outcomes and other digits or groups of digits with failures. The situation is then simulated by selecting a string of digits and recording outcomes as successes or failures. For example, if *even* means success and *odd* means failure, 33216 means failure, failure, success, failure, success. *Repeating* this process over and over again and keeping records of successes and failures allows the building of a database that simulates what might happen with an actual process. The long-term ratio of successes to the total number of trials is then given as the approximated probability of the event occurring. Example 9.8 illustrates the process of using random numbers.

Example 9.8

Problem Solving: The Pressure Free Throws Problem

In a district playoff basketball game, a player is fouled in the act of shooting at the final buzzer with her team down by 1 point. The player has a 60% shooting average from the free throw line. What is the probability that she will make both free throws and win the game?

Working toward a Solution

Understand the problem	*What does the situation involve?*	Determining whether a basketball player with a 60% free throw average is likely to make two successive shots.
	What has to be determined?	The probability of the event of making two successive shots
	What are the key data and conditions?	The player makes 6 of 10 free throws over a period of time.
	What are some assumptions?	The pressure of the situation won't affect the player's average.
Develop a plan	*What strategies might be useful?*	Make a list; use logical reasoning
	Are there any subproblems?	Count successes and failures.
	Should the answer be estimated or calculated?	The probability could be simulated on a calculator.
	What method of calculation should be used?	Divide with a calculator. Simulate probability.
Implement the plan	*How should the strategies be used?*	Make a list of random numbers with the digits 0–9. Let 0–5

represent a made free throw, and 6–9 represent a miss. Consider each pair of digits in the 100 random digits shown as a model of one of the 50 repetitions of successive shots. How many result in two made free throws? Record the outcomes.

33218	15401
97308	43001
27074	10242
39467	11178
94520	76707
86427	92553
90553	77985
70459	93855
31472	71631
09383	28369

	What is the answer?	From the list of random numbers, 33 gives a success, 21 gives a success, 81 does not give a success, 54 gives a success, and so on. To solve the problem, complete the 50 trials to find the number of successes, and use this number to calculate the probability.
Look back	*Is the interpretation correct?*	Yes, based on 50 trials.
	Is the calculation correct?	Check the estimation of the probability.
	Is the answer reasonable?	Check it. The probability should be less than 60%, the probability of making one free throw.
	Is there another way to solve the problem?	Yes. We could multiply the single-shot probabilities, assuming independence, as we did in the simulation, giving a calculated probability of 0.6×0.6, or 36%. More trials in the simulation probably would have given an empirical answer closer to 36%.

Your Turn

Practice:

a. Complete the solution to the example problem.

b. What is the probability that a softball player makes two hits in five official at-bats if the player is currently batting .400? Use a simulation to give an estimate. The calculated probability is 0.346.

Reflect: In the practice problem, how do the results of a simulation differ from the actual calculated probability, if they differ at all, in interpreting the probability of the event? ■

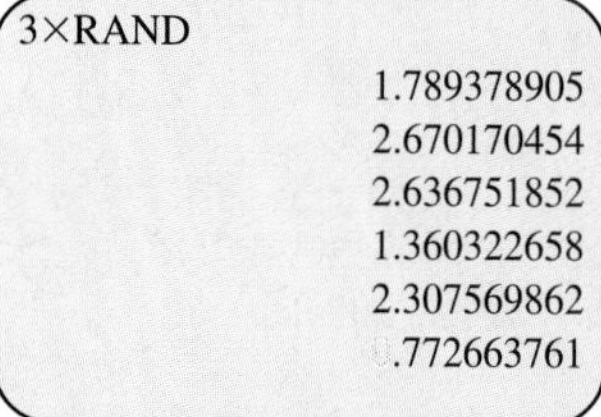

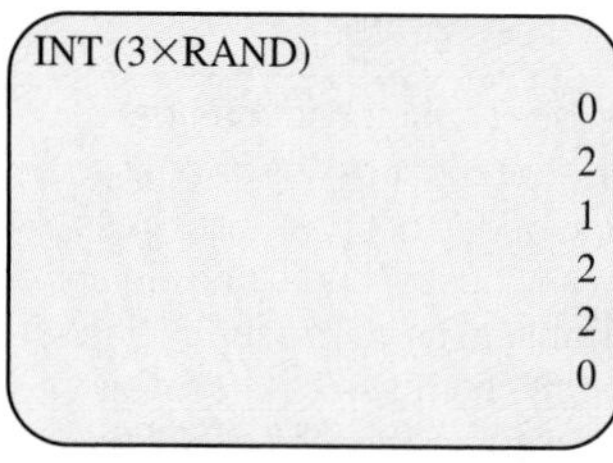

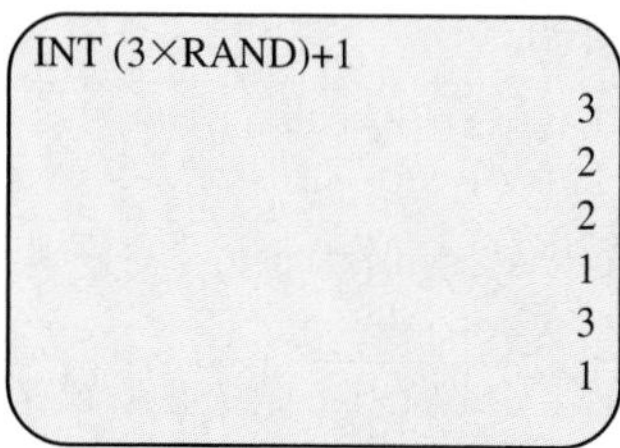

FIGURE 9.10 | Developing sequences of random numbers on a calculator.

Creating Simulations with Technology

A graphing calculator can be very helpful in producing a specified set of random digits to help you simulate experiments. For example, suppose that a box of cereal always contains one of three different small toys. You might wonder how many boxes of cereal, on average, you would have to purchase to be relatively confident of getting one of each of these three toys.

You can address this question by using a simulation. You need to model finding one of three toys in each box of cereal and can use the integers 1, 2, and 3 to model the situation. Use the commands ***Rand*** and ***Int*** built into some graphing calculators to generate random integers from that set.

The command ***Rand*** generates a random number between 0 and 1. If you multiply such a random number by 3 (***3 × Rand***), you get random numbers with values greater than 0 and less than 3. If you use just the integer portion of each of these random numbers, you have integers from the set 0, 1, and 2. The command ***Int*** on the graphing calculator gives you just the integer portion [***Int (3 × Rand)***]. Finally, to get whole numbers from the set 1, 2, and 3, add 1 to the integer portion. The graphing calculator command ***Int (3 × Rand) + 1*** thus provides a random number from the desired set of values. This procedure is illustrated in Figure 9.10.

Let the digit 1 represent the first toy, 2 represent the second toy, and 3 represent the third toy. Now simulate the event of buying cereal boxes until you have one of each toy. You can do so by producing a string of these random digits until the string has at least one 1, one 2, and one 3. At this point terminate the string and count the total number of "boxes" purchased to get all three toys by counting the number of digits in the string, as each digit represents the toy in a given box. Averaging the length of these strings, you can estimate the number of boxes you would have to purchase, on average, to be relatively confident of getting one each of the toys.

The screens in Figure 9.11 were created on a graphing calculator and represent the development of one such string of digits. Shown are the outcomes for seven repetitions of the simulation. Note that many calculators provide a single command that efficiently generates sets of random integers. Within the

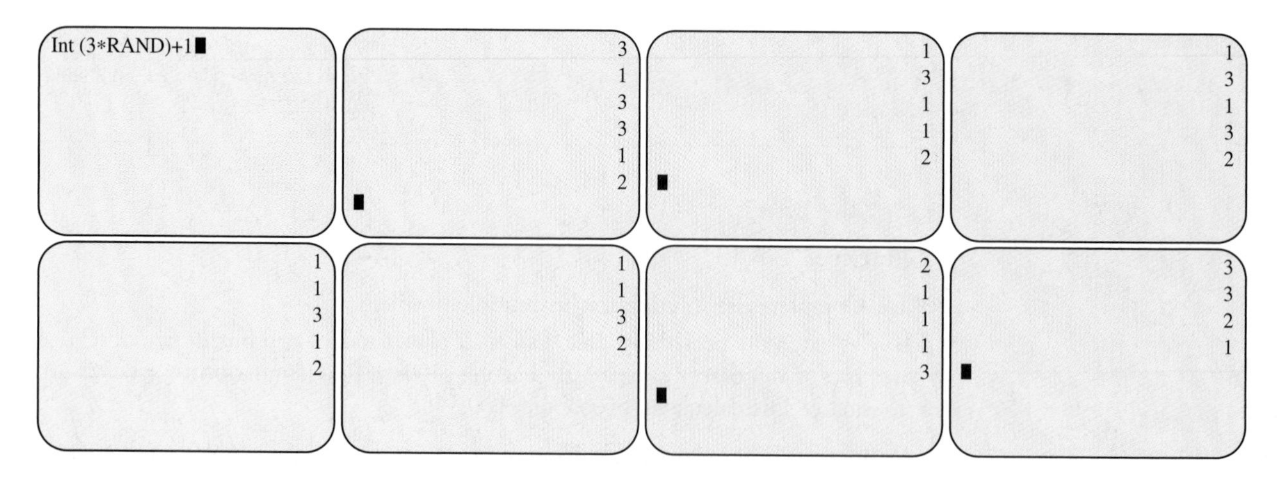

FIGURE 9.11 | Simulating the cereal box problem with random numbers.

A Sample Student Page: Finding Probabilities by Experimenting

Students need opportunities to see how mathematics is used in the world around them. Airlines, quality control experts, professional sports coaches, and many other professions use simulation in planning and carrying out their activities. Students need to understand that probability is an integral part of their daily lives whether or not they themselves are calculating the odds. Take the weather forecast, for example: The probability of rain tomorrow is 30%. What exactly does this forecast mean? The more that students understand, the better they can interpret their options or improve the choices they make. The exercise shown in Figure 9.12 is taken from a set of materials developed by the National Council of Teachers of Mathematics and the American Statistical Association to teach simulation to middle school students.

- ***How** many times would you need to repeat the experiment to begin to trust your data?*
- ***Would** you want to try this out with several mice to get an idea of what is "average"?*
- ***What** factors do you need to look for in planning such a simulation?*

Mouse Maze

Have you heard of psychologists doing experiments to find out how animals learn? Some of these experiments involve mice who are put in a maze, with food at an exit point of this maze. Suppose an experiment is run with the following maze. A mouse is dropped into the maze at point A, with an exit at the center of the maze at B. The mouse will reach the exit only if it makes a right turn. Suppose our mouse were to take the first right turn every time. Does that mean that we have a "smart" mouse? Or could it be that the mouse was making the turns at random and was just lucky?

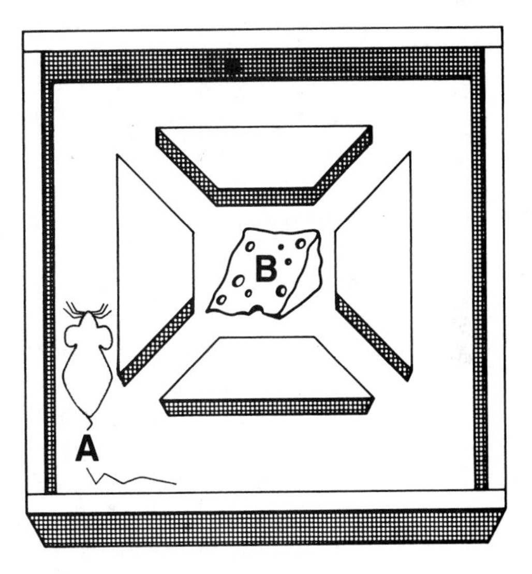

You can answer these questions by using simulation to find the probabilities that the mouse will reach the exit after passing 0, 1, 2, . . . turns. Toss a coin to simulate whether a mouse will make a right turn or keep going. Then record the number of tosses until the mouse reaches the exit. You should assume that the mouse cannot turn around in the maze.

FIGURE 9.12 | Excerpt on simulation from a middle grades textbook.

(*Source:* From *The Art and Techniques of Simulation* by Mrudulla Gnanadesikan, Richard L. Scheaffer, and Jim Swift. © 1987 by Pearson Education, Inc., publishing as Dale Seymour. Used by permission.)

probability menu is the command ***Randint()*** (press MATH and look under the **PRB:** menu for item 2). When you supply left and right endpoints, *a* and *b*, the command ***Randint(a,b)*** generates a random integer from the set $\{a, a + 1, \ldots, b - 1, b\}$. Thus you can use the command ***Randint(1,3)*** for the cereal box simulation.

When you examine these seven simulations, it appears that you could expect, on average, to get one each of the three toys in slightly less than five boxes purchased. This result is based on finding the average length, 4.86, of the number of boxes in a string before you have one each of the toys. However, the experiment was replicated only 7 times here. To really have some confidence in your data, you would probably want to replicate it *at least* 30 times. For a high degree of confidence, you might replicate the experiment 200 times. After that many times, the true nature of the situation would begin to emerge and the simulation would begin to approach the true value of the number of boxes needed.

Example 9.9 provides an opportunity to solve another problem by simulating an experiment.

Example 9.9 Simulating an Experiment

If someone were turning over cards from a shuffled ordinary deck of 52 playing cards, how many cards would have to be turned over until the first ace appeared?

Solution

Cleve's thinking: I just counted how many cards I had to draw each time until I turned over an ace. I repeated the process 50 times. It took quite a while to do it, but I found that the average number of cards I had to turn over was about 10.

Marietta's thinking: I used a graphing calculator and let the digits 1 through 13 simulate the cards ace (1) through king (13). I typed in "Int (13*Rand) + 1" or alternatively Randint(13,1), and used the program to select the digits 1 through 13 randomly by successively pressing the ENTER button. Each of the four horizontal rows of screens shown below the program screen represents a single experiment—until a 1 was found in the random digits provided.

I found that it took 27 tries in the first simulation (one 9 was deleted because that would have made five 9's in the deck); 3 tries on the second simulation; 19 tries on the third simulation; and 12 tries on the fourth simulation. When I continued this process a large number of times, the number of "turns" starts to level out between 10 and 11 cards, on average. So I feel pretty sure that I will turn over an ace, on average, by the 11th card.

Your Turn

Practice: How many cards, on average, would you expect to turn over before you had at least one of each face card (ace, jack, queen, and king)?

Reflect: Why is the answer to the practice problem not $\frac{4}{13}$? ■

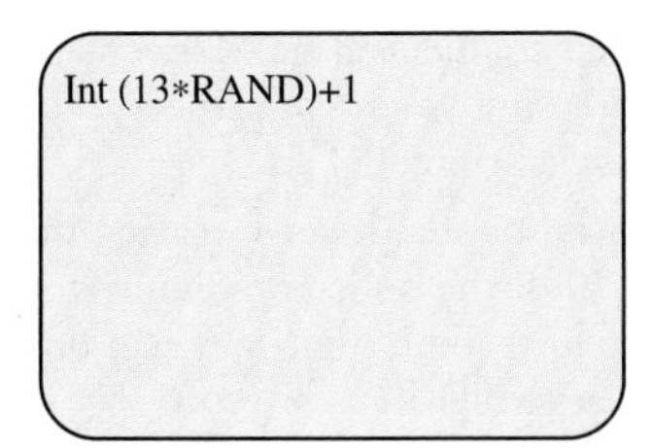

First Simulation

7	9	5	3
4	9	6	2
11	11	11	5
13	11	12	9
7	7	4	2
6	13	10	9
3	9	7	1

Second Simulation

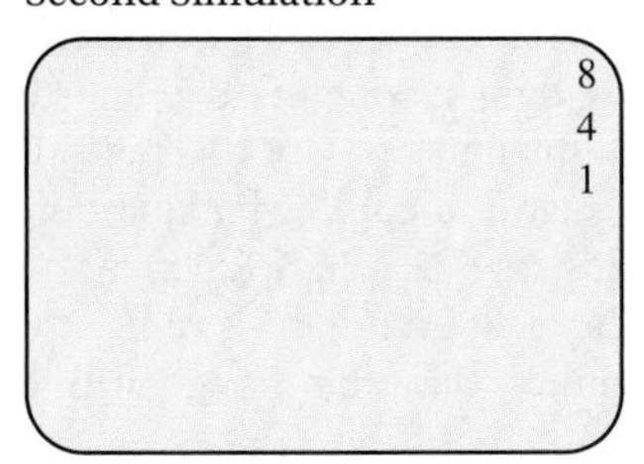

Third Simulation

Fourth Simulation

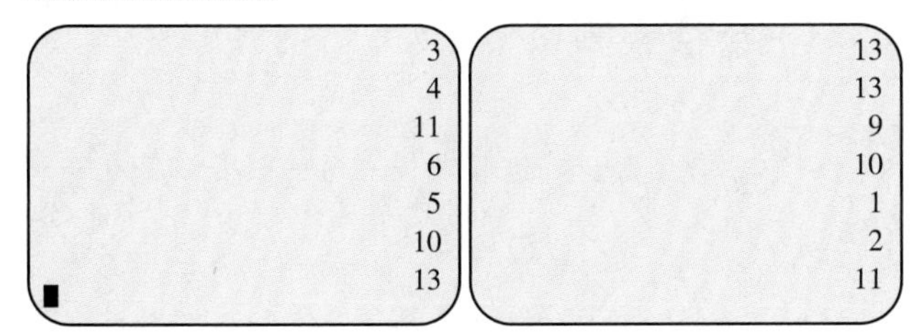

Problems and Exercises | *for Section 9.3*

A. Reinforcing Concepts and Practicing Skills

1. What is the probability that a family of four children has three boys and one girl?
2. Flip a coin 100 times. What empirical probability would you associate with the outcome *heads?*
3. A basketball player has a free throw average of 25%. What is the probability that the player will miss two free throws in a row?

4. Flip a coin and roll a die 72 times. Keep track of your outcomes. How closely do they match the theoretically expected values?
5. The manager of a quick-change oil shop estimates that, on average, the individual who makes one of the six reservations for a given hour won't show up. He therefore schedules seven cars for each hour. What is the probability that all seven of the cars with reservations will show up as scheduled?
6. In selecting a 2-digit number at random, what is the probability that the number selected is less than 45? How does your empirical result match the theoretical expectation?
7. Simulate the birth order, by gender, for a family of three children. How well does your simulation match the theoretical analysis if you assume that boy and girl births are equally likely?
8. A three-stage missile is launched with the following probabilities that each stage will work flawlessly: Stage I, 0.75; stage II, 0.80; and stage III, 0.95. What is the probability that the mission is launched flawlessly? Test the situation both theoretically and empirically. Run at least 40 trials of the simulation.
9. The graphing calculator ***Int(6 × Rand) + 1*** command was used to generate random integers for a simulation. What set of integers does this command make possible?
10. The graphing calculator ***Int(10 × Rand) + 11*** command was used to generate random integers for a simulation. What set of integers does this command make possible?
11. Supply the correct values for a and b in the graphing calculator ***Int(a × Rand) + b*** command to simulate accurately choosing a locker at random from the complete set of school lockers numbered 1 through 1000.
12. Supply the correct values for a and b in the graphing calculator ***Int(a × Rand) + b*** command to simulate accurately the number of crackers that Pat eats for lunch, based on the following information:

    Pat has soup and crackers for lunch every day. When the soup bowl arrives, Pat just empties the cracker barrel into the soup. The cracker barrel holds no more than 20 crackers, and it is refilled when it is empty.

B. Deepening Understanding

13. What is the probability that, if a fast-food restaurant is giving one of three different toys with a food purchase, you will get at least one of all three toys with the purchase of five meals?
14. Suppose that two teams enter the World Series evenly matched and the series is played at a neutral site. How many games of the best 4 of 7 series would you expect to be played before a winner is determined?
15. Suppose that two teams enter the World Series with one team expected to win 0.7 of the games between the two teams. The series is played at a neutral site. How many games of the best 4 of 7 series would you expect to be played before a winner is determined?
16. Police estimate the probability of a driver's wearing the shoulder belt is about 0.70. If the police randomly stop 20 motorists, what is the probability that they will find exactly 12 who aren't wearing belts?

C. Reasoning and Problem Solving

17. **Mixed-Up Keys Problem.** The valet at Big Bucks Inn dropped the tray containing the keys to the six cars in the lot. If the valet randomly puts the six keys back in the six slots, what is the probability that at least one of the keys is correctly reassigned to the car to which it belongs?
18. Suppose we assume the probability of observing a snipe on a walk is 0.13. What is the probability that one would see a snipe if one took seven walks?
19. In driving across town, the newspaper delivery truck driver must pass through four lights, I, II, III, and IV. The probability that any light is green is 0.4, and the probability that it is not green is 0.6. Use this information and a simulation to answer the following questions.
    a. What is the probability that each light is green?
    b. What is the probability that each light is red?
    c. What is the probability that the driver will be stopped by at least one red light?
    d. What is the probability that the driver will find the first light red and the last two lights green?

D. Communicating and Connecting Ideas

20. **Historical Pathways.** The French mathematician Georges Louis Leclerc, Comte de Buffon (1707–1788), developed a theory that predicted that the value of π could be estimated by dropping a needle of length L on a surface that was ruled with parallel lines a distance D apart, where $D < 2L$. A theoretical analysis shows that the probability of a needle touching or crossing one of these lines is $(2L)/(\pi \times D)$.

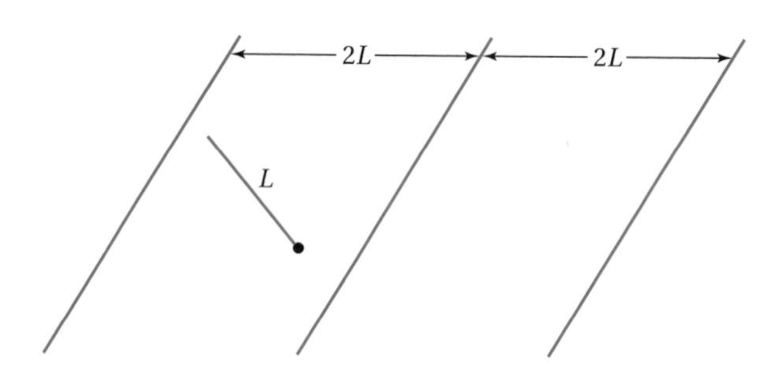

Carry out a simulation to find the probability of hitting or crossing a line. Then use your values for L and D to estimate the value of π.

21. **Making Connections.** Simulation may be used to estimate the probabilities associated with claims made by individuals who believe that they have extrasensory perception. Suppose that a person says that she can predict the order of cards in a suit from an ordinary deck of playing cards as she turns them over one by one. How many hits would you expect on a random basis in calling the order of these 13 cards?

Section 9.4 ODDS AND LONG-TERM BEHAVIOR

- Odds in Favor and Odds Against
- Expected Value
- Connecting Expected Value and Odds to Fairness

People often use probability to help decide what the *odds* are in favor of winning a bet or game. People also use probability to estimate expected results from participation in a situation involving chance over a long period of time. Mini-Investigation 9.3 provides an opportunity for you to consider whether you should participate in a situation involving chance.

Talk about methods that might be used to find how many people would have to be in a class to have "even odds" of having a birthday match. Check birth dates in your class.

Mini-Investigation 9.3 **Making a Connection**

A friend wants to bet that no 2 people in a class of 35 people have a birthday on the same day of the year. Should you bet with your friend?

Odds in Favor and Odds Against

In considering Mini-Investigation 9.3, you may have wondered what the odds are that two people in your class have a birthday on the same day. If someone who knew how to solve the problem mathematically told you that in a class of 23 the odds were 1:1 that 2 people would have a birthday on the same day, would you have been satisfied that you should take the bet? To decide, you need a clear idea of what is meant by the idea of *odds*. To illustrate this idea, let's consider the example of a horse running in a race. Suppose that the estimated probability that the horse would win is $\frac{5}{7}$. Then the probability that the horse would not win is $\frac{2}{7}$. The odds that the horse would win would be the ratio of the probability that it would win to the probability that it would not win, or $\frac{5}{7} \div \frac{2}{7} = \frac{5}{2}$. We summarize this idea more generally as follows.

Definitions of Odds

The **odds in favor** of an event A happening are given by the ratio of the probability that event A will happen to the probability that event A will not happen.

$$\text{Odds in favor of } A = \frac{P(A)}{1 - P(A)}$$

The **odds against** event A happening are the ratio of the probability that event A will not occur to the probability that event A will occur.

$$\text{Odds against } A = \frac{1 - P(A)}{P(A)}$$

Now consider the event of rolling double 6's, "boxcars," on a single roll of two dice. This event has special significance in several board games. The odds in favor of double sixes, 1 to 35, are given by

$$\frac{\frac{1}{36}}{1-\left(\frac{1}{36}\right)} = \frac{1}{35}.$$

Similarly, the odds against getting a double 6 are 35 to 1. In other words, the odds against an event may be expressed as the reciprocal ratio of the odds in favor. Sometimes these odds are written as 1::35, or 1:35.

In general, as an alternative to the equations in the preceding definitions, we could state the following.

$$\text{Odds in favor of } A = \frac{\text{Number of ways that } A \text{ could occur}}{\text{Number of ways that } A \text{ could not occur}}.$$

$$\text{Odds against } A = \frac{\text{Number of ways that } A \text{ could not occur}}{\text{Number of ways that } A \text{ could occur}}.$$

In Example 9.10 we apply these ideas.

Example 9.10 Finding Odds

Find the odds in favor of drawing a 2 or a red card from an ordinary deck of playing cards in a draw of a single card.

Solution

$P(2 \text{ or red}) = \frac{28}{52}$, and $P(\text{not } 2 \text{ and not red}) = \frac{24}{52}$. Hence the odds in favor of drawing a 2 or a red card is 28 to 24, or the ratio of the number of favorable outcomes to the number of unfavorable outcomes.

Your Turn

Practice: In the experiment of tossing three coins, what are the odds against the event of getting two heads and one tail?

Reflect: Use the practice problem to explain the difference, if any, between the odds in favor of getting two heads and one tail and the probability of getting two heads and one tail. ■

Expected Value

The mathematical expectation, or *expected value*, is the long-run average value over repeated plays of a payoff from a game or other event. Such an average value is weighted according to the probabilities involved. For example, consider playing a game of chance at a charity festival that involved the disk and spinner shown in Figure 9.13. The probabilities of the spinner's stopping on one of the four regions of the disk are $\frac{1}{2}$, $\frac{1}{3}$, $\frac{1}{12}$, and $\frac{1}{12}$. The payoffs associated with the regions are shown on the disk.

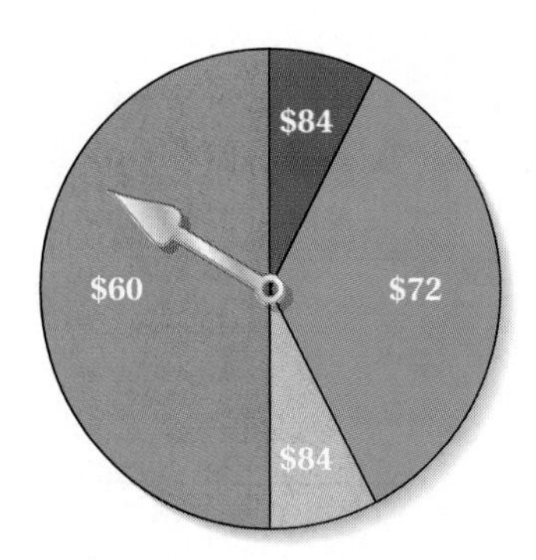

FIGURE 9.13 | Disk for expected value discussion.

The expected value of the payoff for such a game is the sum of the products of the payoffs associated with the various possible outcomes and their probabilities. For this wheel of fortune game the expected value of the payoff is

$$\text{Expected value} = \left(\frac{1}{2}\right)(\$60) + \left(\frac{1}{3}\right)(\$72) + \left(\frac{1}{12}\right)(\$84) + \left(\frac{1}{12}\right)(\$84) = \$68$$

The expected value indicates the average payoff from the game after repeated plays. We describe this idea more generally as follows.

Definition of Expected Value

The **expected value** of an experiment with outcomes $a_1, a_2, a_3, \ldots, a_n$, each with probability $P_1, P_2, P_3, \ldots, P_n$, is given by the sum of the products $a_1P_1 + a_2P_2 + a_3P_3 + \cdots + a_nP_n$.

Mini-Investigation 9.4 asks you to use the idea of expected value.

Write a brief paragraph explaining how you solved the problem to answer the question.

Mini-Investigation 9.4 **Problem Solving**

What would be the expected value of the charity game just described if the sector having the $60 payoff was halved and one of the sectors having an $84 payoff was increased by the same amount?

Connecting Expected Value and Odds to Fairness

To relate the idea of expected value to the fairness of a game or situation, let's suppose that the cost of playing the preceding charity festival game was $75. The expected value of $68 for a payoff in the game is less than the cost of playing the game, so we would expect to lose $7, on average, each time we played the game over a long period of time. Such a game, in which the long-term expected payoff is more or less than the cost of the game, is said to be an **unfair game.** If the expected value of the payoff were the same as the cost of playing the game, the game would be said to be a **fair game.** If the game isn't played for money, the outcomes associated with various events may be considered as points. Note that a *fair game* means that all players have an expectation of winning equal to their cost of playing. At a carnival, this expectation is that the contestant who has paid for playing the game and the booth owner who has paid for the prize awarded for a win have equal chances of winning.

Expected value is an important concept for anyone who likes to be involved in games or situations involving chance. Without an understanding of the concept and the ability to use it productively, someone can lose a lot of money. In Example 9.11 we analyze a simple game of chance.

Example 9.11 **Analyzing a Game of Chance**

Suppose that you and a friend play a game in which you each put 50¢ in the pot at the beginning and then each pitch a penny on the table. If the pennies are both heads, you win a dollar; if only one head appears, you win 50¢; and if no heads appear, you win nothing. Is the game fair?

Solution

The expected value of the amount won in one play of the game is

$$\text{Expected value} = \left(\frac{1}{4}\right)(\$1.00) + \left(\frac{1}{2}\right)(\$0.50) + \left(\frac{1}{4}\right)(\$0.00) = \$0.50.$$

The expected value is the same as the cost of playing, so the game is fair.

A Sample Student Page: Checking Games for Fairness

Expected value is an important statistical concept and is receiving increasing coverage in the school mathematics curriculum. The page shown in Figure 9.14 is from a seventh-grade textbook. It provides students with an exploration in which students are asked to consider a game and decide on its fairness. The building of such reasoning skills leads directly into the concepts of expected value.

- **What** happens if you change the game to involving the sum of the numbers rather than the product?
- **What** happens if you change the game to the sum being "prime" or "not prime"?
- **Why** is studying about the fairness of games an important school topic?

12-4 Odds and Fairness

You'll Learn ...

- to find the odds that an event happens

... How It's Used

Bird breeders need to know the odds of certain traits appearing in their chicks.

Vocabulary

experiment
event
odds
fair games

▶ Lesson Link In the last section, you found methods of counting all the ways that something can happen. Now you'll see how knowing all the possible outcomes can help you find the odds of an event. ◀

Explore Fairness

Materials: Number cubes

Multiplication Toss

You and a partner are about to play a game of chance called Multiplication Toss. Here are the rules of the game.

- Decide which player will be "even" and which will be "odd."
- Take turns rolling the cubes. For each roll, find the product of the numbers rolled. If the product is odd, the odd player gets a point; if it's even, the even player gets a point.
- Repeat until each player has rolled 10 times. The player with the most points wins.

1. Play Multiplication Toss several times. Switch from "even" to "odd" each time.
2. If you could choose to be the "even" player or the "odd" player, which would you prefer? Why?
3. Would you ever expect the "odd" player to win this game? Explain.
4. Is Multiplication Toss a fair game? Explain why or why not.

FIGURE 9.14 | Excerpt from a seventh-grade textbook exploring the fairness of a game.

(*Source:* From *Scott Foresman–Addison Wesley Middle School Math, Course 2* by R. I. Charles, J. A. Dossey, S. J. Leinwand, C. J. Seeley, and C. B. Vonder Embse. © 2002 by Pearson Education, Inc., publishing as Prentice Hall. Used by permission.)

Your Turn

Practice: Suppose that you and a friend play a game in which you each put 50¢ in the pot at the beginning and then pitch a total of three pennies on a table. If three heads show, you get \$1.00; if two heads show, you get 75¢; if one head shows, you get 50¢; and if no heads show, you get nothing. Is the game fair?

Reflect: In the practice problem, if the game isn't fair, how would you change it to make it fair? ■

We can also consider the idea of a fair game in terms of the odds of winning a game. Clearly, a game or situation in which the odds of winning are 1:1 would be considered a fair game because the probability of winning would be the same as the probability of losing. A game with any other odds for winning or losing would be considered an unfair game.

Problems and Exercises | *for Section 9.4*

A. Reinforcing Concepts and Practicing Skills

In Exercises 1–4, consider the experiment of drawing two cards without replacement from an ordinary deck of 52 playing cards.

1. What are the odds in favor of drawing two spades?
2. What are the odds in favor of drawing a spade and a heart?
3. What are the odds against drawing two hearts?
4. What are the odds against drawing two kings?
5. What are the odds against drawing a king from an ordinary deck of 52 playing cards?
6. What are the odds against getting two heads in three successive flips of a coin?
7. What are the odds against getting three heads in three successive flips of a coin?
8. What are the odds in favor of getting at least one head in three successive flips of a coin?

In Exercises 9–12, suppose that you are given the probability of an event A.

9. If $P(A) = \frac{5}{9}$, what are the odds in favor of A occurring?
10. If $P(A) = 0.333...$, what are the odds in favor of A occurring?
11. If $P(A) = \frac{6}{29}$, what are the odds in favor of A occurring.
12. If $P(A) = 0$, what are the odds in favor of A occurring?

In Exercises 13–16, suppose that you are given the probability of an event A.

13. If $P(A) = \frac{14}{37}$, what are the odds against A occurring?
14. If $P(A) = 0.2$, what are the odds against A occurring?
15. If $P(A) = \frac{1}{671}$, what are the odds against A occurring?
16. If $P(A) = .5$, what are the odds against A occurring?

In Exercises 17–21, assume that a die is thrown twice and the sum of the numbers showing is obtained.

17. What are the odds in favor of a sum of 6?
18. What are the odds in favor of a sum less than 4?
19. What are the odds in favor of a sum greater than 1?
20. What are the odds against a sum less than 9?
21. What are the odds against a sum of 7?
22. A lottery has a single grand prize of \$2000. What is the expected value associated with one ticket if
    a. 100 tickets are sold?
    b. 250 tickets are sold?
    c. 1,000,000 tickets are sold?
23. A card is drawn from two decks of cards. What are the odds against drawing a seven and a spade?
24. If the odds for Speedy Spider in the fourth race at Math Downs are 6 to 11, what is the probability that Speedy Spider will win the race?
25. A roulette wheel in Las Vegas has 18 red slots and 18 black slots numbered alternately with the numbers 1 through 36. Two green slots are numbered 0 and 00. The winning number is determined, or comes up, by the slot into which a single marble rolls. Players are not allowed to bet on green. The house wins all bets if 0 or 00 comes up.
    a. What are the odds of a green number winning?
    b. What are the odds of a prime number winning?
    c. What are the odds of a red number winning?
    d. What are the odds of an odd number winning?
26. Consider a disk and spinner with four equal sectors marked with the numerals 1, 2, 3, and 4. What is the expected value for a single spin? If the spinner is spun 3 times, what is the expected value for the sum of the 3 numbers spun?

B. Deepening Understanding

27. If the probability for snow today was forecast to be 40%, what are the odds against it snowing today?
28. On a TV game show, the contestant is asked to select a door and then is rewarded with the prize behind the door selected. If the doors can be selected with equal probability, what is the expected value of the selection if the three doors have behind them a $40,000 foreign car, a $3 silly straw, and a $50 mathematics textbook?
29. An insurance company sells a $50,000 insurance policy to a 25-year-old for $120 for the coming year. The mortality tables say that the probability that the individual will die during the year is 0.002. What is the insurance company's expected gain for the year on this policy?
30. Sam bought 1 of 250 tickets selling for $2 in a game with a grand prize of $400. Was $2 a fair price to pay for a ticket to play this game?
31. The odds that a woman over 60 will develop breast cancer are 1 to 25, and the odds that a woman over 60 will develop diabetes are 1 to 50. If these are independent events, what are the odds that a woman over 60 will develop both diseases? Is it reasonable to assume that these diseases are independent? Explain.
32. What are the odds against throwing either a sum of 3 or 8 in rolling a pair of regular dice?
33. The lottery payoff in a letter game at a church social is determined by the single letter the winner holds. There are four letters, A, B, C, and D. The probability that each will be drawn is $\frac{1}{10}$, $\frac{2}{10}$, $\frac{4}{10}$, and $\frac{3}{10}$, respectively. The payoffs are $2 for an A, $2 for a B, $2 for a C, and $2 for a D. Which letter has the best expected value? The worst?

C. Reasoning and Problem Solving

34. For each of the following events, find the odds in favor of the outcome presented. Then rank the four events from most likely to least likely.
    - **a.** Drawing a king of spades from the set of spades in an ordinary deck of 52 playing cards.
    - **b.** Getting an outcome of (2, H) in an experiment involving rolling a die and then flipping a coin.
    - **c.** Drawing a face card J, Q, K, A from an ordinary deck of 52 playing cards.
    - **d.** Picking up the June edition from a year's editions of a monthly magazine.
35. A $1 bet on a single roll of a roulette wheel (see Problem 25) wins $36 if the marble lands in the slot for the number selected. If that number doesn't come up, the person placing the bet loses the $1. What is the expected value of such a bet?
36. **Heads-Up Game Problem.** The game of *heads up* is played by flipping two coins. If they both land heads up, you win $0. If only one lands heads up, you win $10. If no coins land heads up, you win $0. If you pay $5 to play the game each time the coins are flipped, is the game a fair game?
37. The game of *dots* is played by rolling a fair die and receiving $1 for each dot showing on the top face of the die. What cost should be set for each roll if the game is to be considered a fair game?
38. What is the probability that in a group of 6 individuals at least 2 of them were born in the same month of the year? In a group of 9 individuals? In a group of 12 individuals? Assume that the probabilities of being born in any given month are equal.
39. **Stick Triangle Problem.** Consider a stick 1 unit long. What is the probability of breaking the stick into three pieces so that forming a triangle out of the three pieces is possible?
40. **Chip Game Problem.** Consider whether the following game is a fair game. Players 1 and 2 play by simultaneously flipping three colored chips having side color combinations of red–blue, red–white, and blue–white, respectively. Player 1 gets 1 point if two of the colors showing match. Player 2 gets 1 point if all the chips show a different color. The winner is the first player to score 15 points. Explain the reasoning that led to your conclusion about whether the game is fair.

D. Communicating and Connecting Ideas

41. **Historical Pathways.** The British mathematician William Playfair was the first person to establish mortality tables that gave the probability that a person of a given age would live to an established age. If the probability that a male of age 50 lives to the age 75 is 0.727272..., what are the odds in favor of the individual living to age 75?
42. **Making Connections.** We hear odds for horse races. They are the odds against the horse. If the odds for Speedy Spider were quoted at 1::10, Speedy is only expected to win 1 in 10 times. When you bet $1 on Speedy, the racetrack matches it with $9 and the winner takes all $10. Suppose that you bet $2 on horses with the racetrack odds for winning shown. How much do you win for each bet if the horse wins?

    a. 6::1 **b.** 8::7
    c. 2::1 **d.** 19::1

Section 9.5 PERMUTATIONS AND COMBINATIONS

- Listing Permutations
- Solving Permutations Problems
- Counting Combinations

In this section we describe a technique for determining the number of ways a set of objects can be arranged in a specified order. We also show how to count the number of ways a set of objects can be selected without regard to order. These techniques can help you more easily determine the number of possible outcomes for a probability experiment and help you connect to algebra and other areas of mathematics.

Mini-Investigation 9.5 gives you an opportunity to begin thinking about how to analyze situations in which the number of ways of doing something is the key question.

Write a description of your thought processes and list your answers to the questions.

Mini-Investigation 9.5 **Problem Solving**

Four candidates are running for office and only three can be elected.

a. In how many different ways can they be listed on the ballot?

b. In how many different ways can a group of three be elected?

Listing Permutations

Permutations. When considering question (a) in Mini-Investigation 9.5, you were concerned with the *order* of the candidates, that is, the first, second, third, or fourth candidate on the ballot. Many other practical situations require arranging a set of objects in order. For example, numbers and letters are arranged in order in making license plates, baseball players are arranged in order for batting purposes, and musical selections are arranged in order for a concert. When an order (first, second, third, and so on) is important in an arrangement, the arrangement has a special name.

Definition of a Permutation

A **permutation** of a set of elements is an arrangement of the elements in a *specified order.*

To further illustrate this idea, we note that, if candidates for an election were Allen, Barnes, Carr, and Dunn, each of the following ballot listings would also be a permutation of the candidates.

Allen, Dunn, Carr, Barnes
Dunn, Barnes, Allen, Carr
Carr, Allen, Dunn, Barnes

Finding the Number of Possible Permutations for a Set of Objects. Question (a) of Mini-Investigation 9.5 also asked: In *how many different ways* can the candidates be listed in order? In other words, how can we find the number of different permutations that are possible for a set of objects?

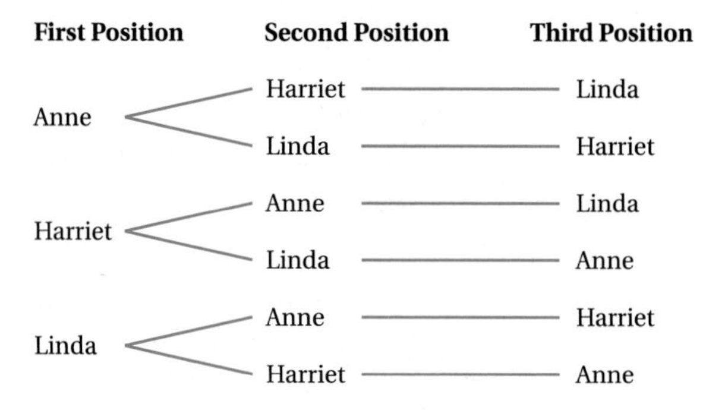

FIGURE 9.15 | Tree diagram for filling three positions on a tennis team.

To begin our search for a technique to find the number of permutations for a set of objects, let's consider the following situation. A coach wants to determine the number of different ways that he can play three tennis players of equal ability in the first three positions in an upcoming match. He begins by making a blank for each of the three positions:

________	________	________
First position	Second position	Third position

The tree diagram shown in Figure 9.15 can help the coach think more clearly about his choices.

Filling in the blank for each of the positions on the team can be thought of as a decision point for the coach. For example, the tree diagram indicates that he has three choices of names to write in the blank for the first position. For each of the three choices for the first blank, he has two choices of names to write in the second blank. Thus there are three groups of two, or $3 \times 2 = 6$, ways of filling in the first two blanks. For each way of filling in the first two blanks, there is just one way of filling in the third blank, so there are $3 \times 2 \times 1 = 6$ ways of positioning the three players. In other words, there are 6 possible permutations of positions for the three tennis players.

This process for finding the total number of permutations of the positions for three players uses the multiplication property presented in Section 9.2. Note that the multiplication property is based on the assumption that the *number* of ways to perform any step doesn't depend on the *particular* choice made in the previous step. As in the tennis coach's problem, no matter which of the three players is selected for the first position, there are still two possibilities for filling the second position.

We can use the same process to generalize a way for finding the total possible number of permutations for a set of n objects. To find the total number of permutations of three objects, we start with 3 and multiply by each consecutive digit from 3 down to 1.

If we represent the number of permutations of three objects as ${}_3P_3$ (read as "the number of permutations of three objects, taken all together"), we can write ${}_3P_3 = 3 \times 2 \times 1 = 6$. In a similar manner, we could show that ${}_4P_4 = 4 \times 3 \times 2 \times 1 = 24$. Recall that the products of all consecutive digits from n down to 1 is called n factorial and denoted $n!$. Hence we can write ${}_3P_3 = 3!$ and ${}_4P_4 = 4!$. These statements contain a pattern, from which we make the following generalization.

Procedure for Finding the Number of Possible Permutations of a Set of Objects

If ${}_nP_n$ represents the number of permutations of n objects, taken all together, ${}_nP_n = n!$.

Again, recall that, in order for later formulas involving permutations to work correctly, 0! is defined as 1. In Example 9.12 we use the generalization just stated to determine the number of permutations for sets of objects.

Example 9.12

Finding the Number of Permutations

How many different 6-digit security codes can be formed from the digits 3, 4, 5, 6, 7, and 8, if none of the digits can be repeated in a code?

Solution

Lena's thinking: I thought about how I could put the digits in a series of six blanks ____ ____ ____ ____ ____ ____. Any of the six digits could be put in the first blank. After it is used, there are five possible choices for the second blank, then four choices, and so on. So there are $6 \times 5 \times 4 \times 3 \times 2 \times 1$, or 720, ways to fill the blanks and thus 720 possible security codes.

Tom's thinking: The problem involves finding the different orders in which six digits can be arranged, so I used the procedure given earlier to find the number of permutations of six objects: ${}_6P_6 = 6 \times 5 \times 4 \times 3 \times 2 \times 1 = 720$. There are 720 possible security codes.

Your Turn

Practice: How many different possible 4-letter arrangements can you make from the letters in the word *math?*

Reflect: As the number of letters to be arranged increases, does the number of possible arrangements increase proportionally? Explain. ■

Permutations Where Some, but Not All, of the Objects in a Set Are Used. In certain cases the problem is to find how many different permutations are possible when only a proper subset of the objects in a set is used in each permutation. For example, suppose that a club wanted to use the digits 1, 2, 3, 4, and 5 to form a different 3-digit code number, with no digit repeating, for each member. What is the maximum number of members the club could have? The solution involves finding the number of 3-object permutations from a set of five objects. We can model each permutation by using three decision blanks.

______	______	______
First digit	Second digit	Third digit

The first blank may be filled by any one of five digits. The second may be filled by any one of the remaining four digits, and the third blank by any one of the three remaining digits. Using the multiplication property gives $5 \times 4 \times 3$, or 60, possible permutations. Hence the club could have code numbers for at most 60 different members.

We can use this example to generalize a process for finding how many different permutations are possible when only a proper subset of the objects in a set is used in each permutation. If we let ${}_5P_3$ represent the number of permutations of three objects that can be formed from a set of five objects, we can write ${}_5P_3 = 5 \times 4 \times 3 = 60$. The product for ${}_5P_3$ looks like 5!, but with 2×1, or 2!, canceled. We also note that 2 is $5 - 3$ and hypothesize that ${}_5P_3 = \frac{5!}{(5-3)!}$. To test this hypothesis, we use a process similar to the one we used to find the number of different 2-digit code numbers that can be formed from the digits 3, 4, 5, 6, 7, and 8. We find that ${}_6P_2 = 6 \times 5 = 30$. Again, the product for ${}_6P_2$ looks like 6! with $4 \times 3 \times 2 \times 1$, or 4!, canceled. As before, we note that $4 = 6 - 2$ and write ${}_6P_2 = \frac{6!}{(6-2)!}$. We use these ideas to form the following generalization, which can be verified by using algebra.

Procedure for Finding the Number of Permutations of *r* Objects Taken from a Set of *n* Objects

If ${}_nP_r$ represents the number of permutations of r objects that can be formed from a set o f n objects, then ${}_nP_r = \frac{n!}{(n-r)!}$.

Mini-Investigation 9.6 asks you to consider the technology you have available in order to find the number of permutations.

Talk about which of these devices will be easiest for you to use and why.

Mini-Investigation 9.6 **The Technology Option**

Explore the technology—scientific calculator, graphing calculator, or computer software—to which you have access. What options does it give you for calculating the number of permutations of r objects that can be formed from a set of n objects?

In Example 9.13, the formula for finding the number of permutations of r objects taken from a set of n objects is applied.

Example 9.13 Using a Formula to Find the Number of Permutations

If a manufacturer of combination locks wants to have 3 different numbers in a combination and there are 20 numbers on the dial, how many different possible 3-number combinations are there?

Solution

Amanda's thinking: I thought of it as having three blanks to fill. There are 20 choices for the first blank, 19 for the second, and 18 choices for the third blank. I used the multiplication principle to get $20 \times 19 \times 18$, or 6840, ways to fill the blanks. There are 6840 different 3-number lock combinations using distinct numbers.

Tan's thinking: I used the formula for the number of permutations of r objects taken from a set of n objects. In this case, $n = 20$ and $r = 3$, so ${}_{20}P_3 = \frac{20!}{(20-3)!} = 20 \times 19 \times 18$, or 6840 different 3-number lock combinations.

Your Turn

Practice: How many different 4-digit license plates can be formed from the digits 1, 2, 3, 4, 5, 6, 7, 8, and 9, digits not repeated?

Reflect: If you were to use the letters of the alphabet instead of the non-zero digits, would the number of 4-letter license plates be greater than the number of 4-digit license plates drawn from the nine non-zero digits? Why or why not? ■

Solving Permutations Problems

Permutations often play a vital role in solving certain types of probability problems. Such problems require finding the number of successes and the number of total possible outcomes when one or both of these numbers are determined by a permutation. A further complication may be a special condition placed on one or the other of the numbers (such as a particular number has to be in a designated position). In Example 9.14, we look at a problem that requires finding a number of permutations.

Example 9.14

Problem Solving: Code Choices

Three-digit code numbers are created by using the digits 2, 3, 4, 5, 6, and 7 without repetition. If one code number is selected at random, what is the probability that it begins with 5?

Working toward a Solution

Understand the problem	*What does the situation involve?*	Finding the likelihood of selecting a designated code number.
	What has to be determined?	Finding the probability that a code beginning with 5 is selected from the set of all possible 3-digit codes formed from the digits 2, 3, 4, 5, 6, and 7.
	What are the key data and conditions?	The number of digits in a code, three, and the number of digits from which we can select, six.
	What are some assumptions?	The selection of digits to the codes includes all possibilities, and the selection of the specific code to inspect is done randomly.
Develop a plan	*What strategies might be useful?*	Make a list; use logical reasoning; choose an operation.
	Are there any subproblems?	The actual calculation of the number of permutations is a subproblem of calculating the probability, or the ratio of successes to total possibilities.
	Should the answer be estimated or calculated?	Calculated.
	What method of calculation should be used?	Use mental multiplication.

Implement the plan	*How should the strategies be used?*	Make a list, using blanks to represent the numbers. For a code beginning with 5, the list might appear as 5 __ __; for any of the code numbers, __ __ __. Calculate the number of code numbers beginning with 5: $1 \times 5 \times 4$, or 20. Calculate the number of codes, $6 \times 5 \times 4$, or 120.
	What is the answer?	The ratio of $\frac{20}{120}$, or 0.166.... That is, the probability of drawing a code starting with 5 is approximately 0.17.
Look back	*Is the interpretation correct?*	Yes, we calculated the possible specific ways and the total possible ways. We then used these data to calculate the ratio of successes to total possibilities.
	Is the calculation correct?	Yes.
	Is the answer reasonable?	Yes. It seems reasonable that a code number starting with 5 would be selected more than one-sixth of the time.
	Is there another way to solve the problem?	Monte Carlo simulation could be used to approximate this probability.

Your Turn

Practice: A set of cards contains all possible 2-letter arrangements of letters of the alphabet without repetition. What is the probability of selecting a card containing two vowels from the set {a, e, i, o, u} when selecting at random from the total set of cards?

Reflect: Explain how permutations play a vital role in determining the probabilities in the example and practice problems. ■

Counting Combinations

Combinations. When considering question (b) in Mini-Investigation 9.5, you were *not concerned with the order* of the candidates. Rather, you simply wanted to look at selections of sets of three of the four candidates without regard for order. In many other practical situations subsets of a set are selected without regard for order.

For example, we might select committees of three from a board with seven members. A committee comprising Anderson, Brown, and Carver is the same as the committee comprising Brown, Carver, and Anderson because order is irrelevant. A second example is that of a basketball coach choosing a team of 5 from the 10 who tried out; the selection is not concerned with order. In this case the team consisting of Allie, Beth, Carol, Denise, and Effie is the same as a team indicated by arranging these same players in any other order. When we are interested in selecting a set of objects and order isn't important, the method of selection has a special name.

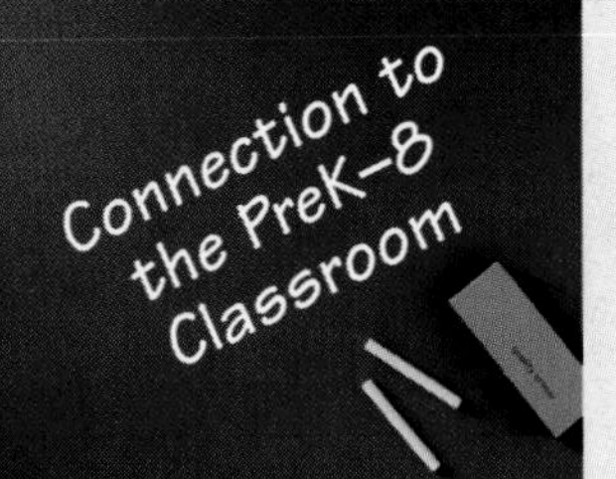

A Sample Student Page: Using Permutations to Plan a Strategy

In the middle grades, permutations serve as a basis for solving ordering problems in a variety of settings. The situation shown in Figure 9.16 shows how permutations might be employed in ordering the runners on an Olympic relay team.

- **How** would you explain in words that there is only one runner left for the third leg?
- **How** many orders would we have if the coach also decided to run Gail Devers as the first runner?
- **What** is the number of possible orders if the coach has no preconceptions of what she wants to do?

Example 1

In the women's 4 × 100 m relay, the U.S. team was made up of Chryste Gaines, Gail Devers, Inger Miller, and Gwen Torrence. The coach decided that Gwen Torrence would run the anchor leg. In how many different orders could the coach have Gaines, Devers, and Miller run the first 3 legs?

DID YOU KNOW?

In the 200 m butterfly final at Barcelona in 1992, the race was so fast that the first 7 finishers set personal records.

Chryste Gaines

Gail Devers

Inger Miller

Gwen Torrence

For the first runner, there are 3 possibilities. Once a selection is made for the first runner, there are only 2 runners left from which to choose for the second leg. After that person is chosen, there is only 1 choice for the third leg.

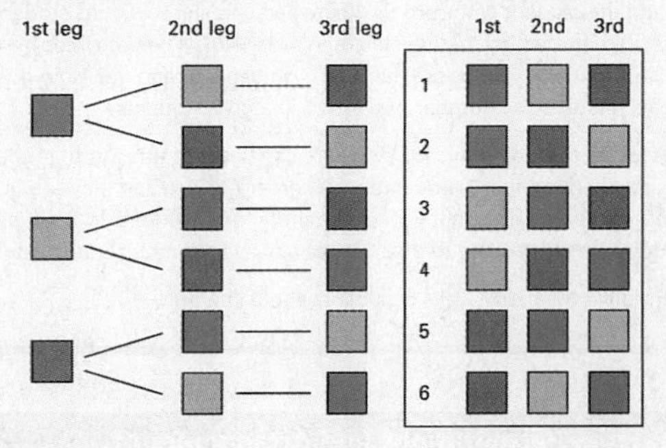

By the Counting Principle, the total choices are $3 \cdot 2 \cdot 1 = 6$ ways.

There are 6 possible ways that the coach could order the 3 runners.

Try It

There were 8 lanes in which swimmers could compete in the 200 m butterfly races. How many different ways could the swimmers be arranged in the lanes?

632 *Chapter 12 • Counting and Probability*

FIGURE 9.16 | Excerpt from an eighth-grade textbook exploring using permutations.

(*Source:* From *Scott Foresman–Addison Wesley Middle School Math, Course 3* by R. I. Charles, J. A. Dossey, S. J. Leinwand, C. J. Seeley, and C. B. Vonder Embse. © 2002 by Pearson Education, Inc. publishing as Prentice Hall. Used by permission.)

Definition of a Combination

A **combination** is a subset of a set of elements selected without regard to their order.

When forming combinations, we usually select proper subsets of a specified set. For example, if the specified set was the set of candidates for an election {Allen, Carr, Dunn, Barnes}, we could form combinations by selecting 3-element subsets of the 4-element set of candidates. One such combination would be {Allen, Carr, Dunn}. Note that the subset {Carr, Dunn, Allen} indicates the same combination.

To distinguish between a permutation and a combination, remember that, *for a permutation, order is involved*, but that *for a combination, order is not involved.* Mini-Investigation 9.7 will help you make the distinction between permutations and combinations.

Write a convincing argument, giving examples, to support your reasoning.

Mini-Investigation 9.7 **Making a Connection**

How would you try to convince someone that there are more permutations than combinations for subsets of three objects that can be formed from a set of five objects?

Finding the Number of Possible Combinations. Question (b) of Mini-Investigation 9.5 also asked: *How many different ways* are there to select a group of three from a set of four candidates? Using the language of combinations, we describe such tasks as finding the number of different combinations of four objects taken three at a time. A four-element set {A, B, C, D} has four different 3-element subsets:

{A, B, C} {A, B, D} {A, C, D} {B, C, D}

If we let ${}_4C_3$ represent the number of 3-element subsets that can be selected from a 4-element set, we can write ${}_4C_3 = 4$.

Now, let's consider the possible 3-member committees that can be selected from a group of five people. Using the first letters of last names, we can represent the five people as {D, G, M, R, T}. Note that one possible committee would be {D, M, G}. As a permutation, the subset {D, M, G} could be arranged 3! ways. However, as a combination, it is a single subset because with combinations we are only interested in the subset itself, not the order of its elements. This result suggests that we might find the number of combinations by first finding the number of permutations of five objects taken three at a time, or $5 \times 4 \times 3$, and then adjusting this by dividing out the number of different orderings of each set of three objects. We write the result as

$${}_5C_3 = \frac{5 \times 4 \times 3}{3 \times 2 \times 1} = 10.$$

These 10 combinations can be listed as

{D, G, M} {D, G, R} {D, G, T} {D, M, R} {D, M, T}
{D, R, T} {G, M, R} {G, M, T} {G, R, T} {M, R, T}

We can use this example to generalize a process for finding how many different combinations are possible for an r-element subset of an n-element set. To do so, we note that we can rewrite the preceding equation as

$$_5C_3 = \frac{_5P_3}{3!} = \frac{5 \times 4 \times 3}{3 \times 2 \times 1} = \frac{5 \times 4 \times 3 \times (5-3)!}{3 \times 2 \times 1 \times (5-3)!} = \frac{5!}{(5-3)!3!}.$$

Using this pattern of reasoning for a subset of r elements in a set of n elements, we write

$$_nC_r = \frac{_nP_r}{r!} = \frac{n!}{(n-r)!r!}.$$

We use these ideas to form the following generalization.

Procedure for Finding the Number of Combinations of *r* Objects Taken from a Set of *n* Objects

If $_nC_r$ represents the number of combinations of r objects that can be selected from a set of n objects, then $_nC_r = \frac{n!}{(n-r)!r!}$.

The number of combinations of n things taken r at a time, $_nC_r$, may also be written as

$$\binom{n}{r}$$

Various forms of technology can also assist in computing the values for combinations. Explore whether the calculators or spreadsheets you have access to are able to compute values for $_nC_r$ for various values of n and r. Do they give the same results as we obtained in the preceding illustrations above? Example 9.15 applies the techniques for finding combinations to jury selection.

Example 9.15 Finding Combinations

How many different 12-person juries could be selected from a potential juror pool of 20 individuals?

Solution

This situation calls for the value of $_{20}C_{12}$, which can be computed as

$$\frac{20!}{(20-12)!12!}.$$

Thus 125,970 different juries are possible. Note that order isn't important in selecting a jury—just which 12 people compose the jury is of interest.

Your Turn

Practice: How many different ways can a judge select a minijury of 6 individuals from a pool of 20?

Reflect: Why is the order in which the jurors are selected not important in selecting either a 12- or 6-person combination from the 20 potential jurors? ■

Problems and Exercises | *for Section 9.5*

A. Reinforcing Concepts and Practicing Skills

1. How many 4-digit license plates can be made by using the digits 0, 2, 4, 6, and 8 if repetitions are allowed? If repetitions are not allowed?
2. How many 5-digit license plates can be made by using two letters from the alphabet and three digits from 0, 1, 2, ..., 9 if repetitions are allowed? If repetitions are not allowed?
3. A teacher wants to write an ordered 5-question test from a pool of 14 questions. How many different forms of the test can the teacher write?
4. How many batting lineups of the nine players can be made for a baseball team if the catcher bats first, the shortstop second, and the pitcher last?
5. In a drug testing program at a prison, 4 prisoners must be selected from a pool of 10 prisoners who have volunteered. How many different ways can this selection of 4 prisoners be made?
6. How many games are played in a women's soccer conference if there are eight teams and all teams play each other twice?
7. How many games must be played in a single elimination tournament to determine a single winner if there are 20 teams in the tournament?
8. A class elects 2 officers, a president and a secretary/treasurer, from its 12 members. How many different ways can these two offices be filled from the members of the class?
9. A student has a penny, nickel, dime, and quarter distributed two coins apiece to his front two pants pockets. What is the probability that the front right pocket has more than 30¢ in it?
10. Five numbers are to be picked, without repetition, from 44 numbers to determine the winner of the Fortune Five game in the state lottery. If the order of the numbers is insignificant, how many different ways can a winning quintuple be selected? What is the probability of winning?
11. A bag contains a penny, a nickel, a dime, a quarter, and a half-dollar. How many different sums of money can be obtained by using these five coins?
12. A computer selects three numbers from 1 through 20, with replacement. What is the probability that all three numbers are less than or equal to 5?
13. How many different ways can a committee of 9 be selected from a larger organization of 20 people?
14. How many different license plates could be made if they have one or two letters followed by one, two, or three digits? Repetitions are allowed.
15. How many different 5-card hands can be dealt from an ordinary deck of 52 playing cards?
16. Ten horses are running in a race. How many different ways can first, second, and third place be awarded in the race? Assume no ties.

B. Deepening Understanding

17. How many 4-digit license plates with all different digits can be made by using the digits 0, 1, 2, 3, 4, and 5 if the number on the license plate must be even? Must be odd?
18. At a party, 55 handshakes took place as everyone shook hands with everyone else in attendance. How many people were at the party? Explain your reasoning.
19. Six people are running for four different offices on a school board. In how many different ways can those offices be filled?
20. A die is rolled seven times. What is the probability of rolling two 1s, three 2s, and two 6s?
21. Suppose that the U.S. Senate consisted of 57 Republican and 43 Democratic senators. How many different 12-person committees could be formed with 8 Republican senators and 4 Democratic senators?
22. A ship carries exactly 10 different signal flags. If each possible combination and ordering of 4 of these flags connotes a specific message, how many signals can be sent with these flags, taken four at a time?

C. Reasoning and Problem Solving

23. Consider a colored cube, with sides of colors red, orange, yellow, green, blue, and violet. How many different ways can the cube be placed on a surface in front of you? Specify the color of the face on the surface and the color of the face toward you each time.
24. A student asks, What's wrong with the argument that the probability of rolling a double 6 in two rolls of a die is $\frac{1}{3}$ because $\frac{1}{6} + \frac{1}{6} = \frac{1}{3}$? Write an explanation of your understanding of the student's misconception.
25. For all values of n and r, where $r \leq n$, does ${}_nC_r$ always equal ${}_nC_{n-r}$? Why or why not?
26. Consider the successive values of $(1 + 1)^n$, where n is a natural number. What pattern do you observe?
27. Five speakers are to be seated in a line at a speaker's table at a banquet. In how many ways can they be seated? Suppose, instead, that the table is circular. How does this affect the number of ways of seating the five individuals?
28. **Burger Ad Problem.** A fast-food chain advertises that it serves hamburgers in more than 1000 ways. The chain offers its burgers with various combinations of mustard, catsup, mayonnaise, Cajun spices, relish, pickles, lettuce, three cheeses, and three types of buns. Is the company telling the truth? Explain why or why not.

29. Estimate the number of personally constructed greeting cards possible at a machine if there are 12 designs, 30 messages, 18 closings, and 10 different paper stocks on which to print the card. Indicate how you made your estimate. How valid was your estimate?

30. In how many different ways can four individuals be seated around a circular table with four chairs?

D. Communicating and Connecting Ideas

31. Suppose that the combination lock (no pun intended) on a piggy bank has 10 numbers on it and the combination itself consists of 3 different numbers (all different). As a group, devise a method for generating and writing down combinations. How long will it take you to write down all combinations?

32. **Job Applicant Problem.** Fifteen applicants for a teaching position are to be interviewed to narrow the field to four candidates to be brought to the school for further interviews. How many possible groups of four could result if the candidates are ordered? If the candidates are not ordered? Is it reasonable, timewise, to do this?

33. **Historical Pathways.** The Dutch mathematician A. T. Vandermonde developed a related notation for factorial in 1772. He considered $[5]$ as 5! and $[5]^{-1}$ as

$$\frac{1}{5 \times 4 \times 3 \times 2 \times 1}, \text{ or } \frac{1}{5!}.$$

Describe $n!$ and $[n]^{-1}$ as n grows large.

34. **Making Connections.** A common application of counting involves counting the reconnections needed when a long-distance telephone cable is cut. If there were 120 smaller wires in the cable, how many different ways could the wires be reconnected in splicing the cable back together? Would placing the wires in 10 bundles of 12 help?

CHAPTER SUMMARY

Key Ideas: Questions and Answers

Section 9.1

- **What is probability?** *(pp. 453–456)* Probability is the study of chance. In most cases, probability may be viewed as the ratio of the number of favorable outcomes for an experiment to the number of all possible outcomes for the same experiment.
- **What are some useful ideas about sample spaces and the probability of events?** *(pp. 456–459)* Probability may be viewed as a decimal expressing the likelihood of an event taking place, where 1 indicates that it will certainly happen and 0 indicates that it can't possibly happen. The sum of the probabilities of an event occurring and not occurring must be 1, as it either happens or doesn't happen.
- **What is the addition property of probability?** *(pp. 459–462)* The probability of events A or B occurring can be found by adding their individual probabilities together when they share no common outcomes. However, should some outcomes be common to both events A and B, the probabilities associated with the common outcomes must be subtracted from the sum of the probabilities of A and B to get the probability of A or B occurring.

Section 9.2

- **What are some ways to represent and count experimental outcomes?** *(pp. 465–466)* The results of experiments may be represented or counted with tree diagrams and box arrays. These methods of depicting possibilities show why the multiplication property is a model for counting experimental outcomes, relating them to multiple additions or area models.
- **How can probabilities that involve independent events be determined?** *(pp. 466–468)* When independent events occur, the outcome to one event doesn't affect the outcome to the second event. A full tree diagram is a model for the event. Hence the probability of a pair of independent events is the product of the individual events' probabilities.
- **What is conditional probability?** *(pp. 468–470)* Conditional probability is the probability that an event A happens and we know that an event B, which influences event A, has already happened. It is analyzed by finding the ratio of the probability that both events A and B happen to the probability that event B happens. This analysis is often best done with a geometric figure.

Section 9.3

- **What is simulation, and how can it be used to approximate probabilities?** *(pp. 473–475)* A simulation is an experiment in which an event is "acted out" by means of a model reflecting assumptions of the probabilities of various parts of the desired event. Based on the model, samples representing the event are drawn, and a record is kept of the number of times the desired event occurs. Repeating this sampling process over and over leads to a value for the outcome based on the initial assumptions. If they were good, the simulation will give a good approximation of the probability of the desired event.

- **How can technology help in simulation?** *(pp. 476–479)* Simulation can approximate probabilities through the use of tables of random numbers, sets of marked chips or pieces of paper, or scientific calculators that have greatest integer and random number functions. Repeating a simulation over and over leads to an approximation of the value of the desired outcome.

Section 9.4

- **What is meant by *odds in favor* or *odds against?*** *(pp. 481–482)* Odds in favor of an event are the ratio of the probability of an event happening to the probability of it not happening. The odds against an event happening are the ratio of the probability of the event not happening to the probability of it happening.
- **What is meant by *expected value?*** *(pp. 482–483)* The expected value of an event is the long-run value that someone would expect to get from playing a game or observing some experiment. It is formed by summing the product of the outcomes with their associated probabilities for all possible outcomes to the experiment.
- **How are the ideas of expected value and odds related to the fairness of a game or situation?** *(pp. 483–485)* If the expected value of a game is equal to the cost of playing the game, the game is a fair game. That is, the odds in favor of winning are equal to the odds against winning.

Section 9.5

- **What is a permutation and how can the number of possible permutations for a set of objects be determined?** *(pp. 487–491)* A permutation is a particular arrangement of a set of objects. It also is the count of the number of arrangements, or orderings, of the set. In most cases it is a product of consecutive integers in reverse order, a product related to factorials. If a problem requires the count of objects ordered in a certain way, permutations are required in order to obtain the solution.
- **How can the techniques described be used to solve permutation problems?** *(pp. 491–492)* Many permutation problems can be solved through the use of tree diagrams, fill-in-the-blank techniques, or a formula for factorials.
- **What is a combination and how can the number of possible combinations in a set be determined?** *(pp. 492–495)* A combination is the occurrence of a given set of objects in an experiment without reference to the order of the events. We also refer to combinations as the count of different possible sets of objects, given some constraints, that can occur in a given situation. The number of possible combinations is determined by finding the number of possible permutations and then dividing this number by the number of ways individual collections of the same objects can be arranged differently to get to the actual number of individual groups, or combinations, of the objects.

Key Terms, Concepts, and Generalizations

Section 9.1

Probability (p. 453)
Experiment (p. 454)
Outcomes (p. 454)
Sample space (p. 454)
Uniform sample space (p. 454)
Nonuniform sample space (p. 454)
Event (p. 454)
Probability of an event, $P(A)$ (p. 454)
Random event (p. 455)
Complementary events (p. 458)
Mutually exclusive events (p. 459)
Addition property of probability (p. 459, 461)

Section 9.2

Tree diagram (p. 465)
Box array (p. 465)
Multiplication property of outcomes (p. 466)
Independent events (p. 466)
Multiplication property of probabilities involving independent events (p. 467)
Conditional probability (p. 468)
Geometric probability (p. 469)

Section 9.3

Simulation (p. 473)
Monte Carlo approach (p. 473)
Random numbers (p. 473)

Section 9.4

Odds in favor (p. 481)
Odds against (p. 481)
Expected value (p. 483)
Unfair game (p. 483)
Fair game (p. 483)

Section 9.5

Permutation (p. 487)
Combination (p. 494)

CHAPTER REVIEW

Concepts and Skills

1. An experiment consists of tossing three distinguishable coins. List
  a. the sample space.
  b. the elements of the sample space where tails occurs on the second coin.
  c. the event of a head on the third coin.
  d. the event of a head on the second coin and a tail on the third coin.

2. A jar contains four marbles, each a different color: red, blue, green, and yellow. If you draw two marbles from the jar, one after another, replacing the first before drawing the second, what is the probability of getting
  a. two red marbles?
  b. a red marble on the first draw and a green marble on the second draw?
  c. at least one red marble and one green marble?
  d. no yellow marbles?

3. What is the probability of drawing a card from an ordinary deck of playing cards and then rolling a die and getting
  a. the jack of diamonds and an even number?
  b. a red card and a number ≥ 3?
  c. a jack, queen, king, or ace and a 5?

4. Describe the numerical range of probabilities that you would expect to find for any experiment. Justify your answer.

5. If two dice are thrown, what is the probability of the following outcomes?
  a. The sum showing is 7 or greater.
  b. One die shows a 3, and the other a 5.
  c. Neither die shows an even number.
  d. One die is even or the sum is six.

6. The numbers 1–10 are placed on cards, and the cards are put in a hat. If two cards are drawn, in order, without replacement, what is the probability that
  a. both numbers are composite?
  b. one number is even and the other is odd?
  c. one number is even and the other a prime?

7. A jar contains four marbles, each a different color: red, blue, green, and yellow. If you draw two marbles from the jar, one after another without replacement, what is the probability of getting
  a. two red marbles?
  b. a red marble on the first draw and a green marble on the second draw?
  c. at least one red marble and one green marble?
  d. no yellow marbles?

8. A card is drawn at random from an ordinary deck of playing cards. Find the following odds.
  a. Odds in favor of drawing a 5, 6, or 7.
  b. Odds against drawing a king, queen, jack, or ace.
  c. Odds in favor of drawing the queen of spades.
  d. Odds against drawing an even-numbered (2, 4, 6, 8, 10) card.

9. According to a recent survey, 15% of all mathematics books break even in terms of profit, 35% lose $10,000, 20% lose $20,000, and 30% earn $55,000. What is the expected income for a newly published book in mathematics?

10. Use the disk and spinner shown to answer the following questions.

a. What is the probability that the spinner doesn't stop on yellow?
b. What is the probability that the spinner stops on red or blue?
c. What is the probability that the spinner stops on green or not orange?

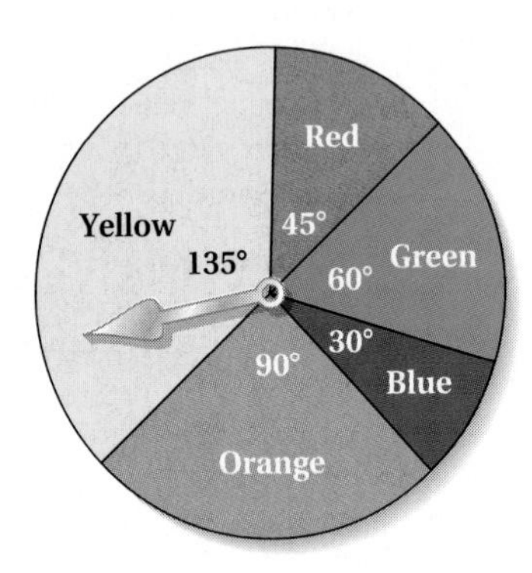

11. What is wrong with the following probabilistic statements?
    a. Because there are 50 states, the probability of a person's living in any one of them is 0.02.
    b. The probability that a person from a given high school attends a 4-year college or university is 0.5 and attends a 2-year college is 0.4. Therefore the probability that a person from that high school attends college is 0.9.
    c. The probability of making a free throw is 0.3 and of missing it is 0.6.
12. Suppose that pizzas can be ordered in four sizes (small, medium, large, and Illini-size), with three crust choices (thin, thick, and Chicago style), four choices of meat (sausage, pepperoni, hamburger, and none) and two types of cheese (regular or double). How many different styles of pizza can be ordered?
13. In how many ways can a teacher seat 15 students in a classroom having 15 desks?
14. If you know that 4 of the 10 items on a true–false test are false, how many different ways could you select four items to mark as being false?

Reasoning and Problem Solving

15. Describe the difference between two events being independent and being mutually exclusive.
16. Prizes are given at the end of the orchestra season to 6 of the 105 orchestra members. In how many different ways can the 6 winners' names be drawn from a hat if no one can receive more than one prize?
17. Mathman's Pizza offers choices of toppings, meat, and cheese. You can't see the number of choices in each category, but you do know that it makes 105 different pizzas. What do you think the number of cheeses might be? Why?
18. Consider the dart board with concentric rings shown. If the inner circle has a radius of 2 inches and each ring has a width of 2 inches, what is the probability of a person's hitting a bull's eye, knowing that the person hit the board?

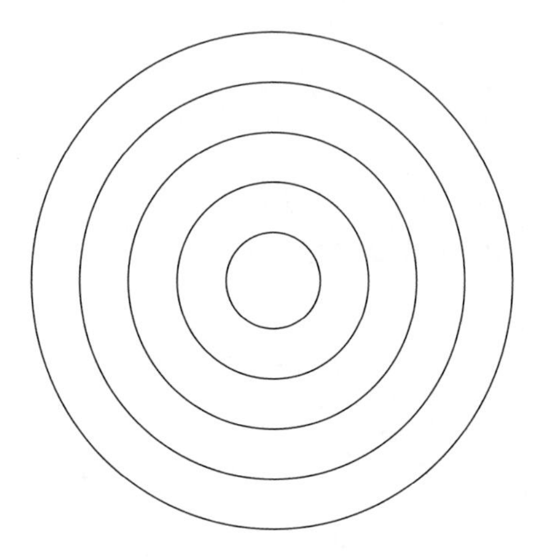

19. **Popping Good Problem.** A kernel of corn can fall into any one of five areas in the arrangement of silos shown. The radius of each circular portion of the silos is 20 feet. What is the probability that it falls in the shaded central section?

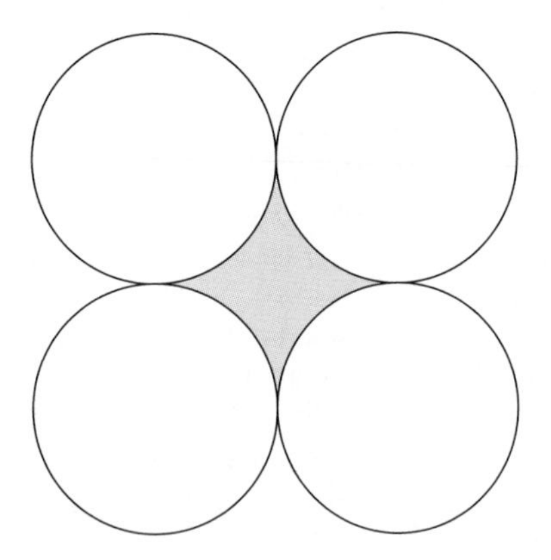

20. **Standby Problem.** A shuttle service regularly oversells the eight seats on its flight to the islands by two tickets. If the probability that any one passenger not

showing up is 0.1, what is the probability that at least one passenger won't have a seat on the shuttle on a given flight?

Alternative Assessment

21. A keymaker has three different types of blanks (thick, medium, thin) to make keys from. On each, she has the possibility of five different positions where metal can or cannot be removed. At each place where metal can be removed, there are two different cutting depths. Draw pictures of the possible keys and write a description of each, telling why you have identified all possible keys.
22. Working with a group of classmates, design an experiment to find the number of cards you would expect to turn over before finding the first even-numbered card in an ordinary deck of playing cards. Describe your experiment and then conduct it, giving the number of cards you expect to turn over.
23. Consider the dart board with concentric rings shown. If the inner circle has a radius of 2 inches and each ring has a width of 2 inches, what is the probability of a person's hitting a bull's eye, knowing that the person's dart hit one of the shaded areas? Explain how to answer this question without directly calculating any area.

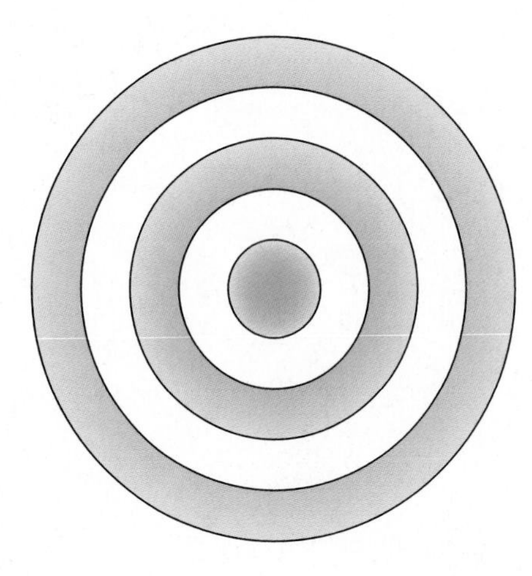

24. A box contains six red beads and four green beads. How many sets of four beads can be selected from the box so that two of the beads are red and two are green?

10 Introducing Geometry

Chapter Perspective

From road signs, carpentry, and architecture to interior design, advertising, art, and science, geometry plays a significant role in people's lives. For example, the functional use of a simple geometric figure in architecture is illustrated by the Pentagon building in Washington, DC, shown in the above photo. The building, almost 1 mile in perimeter, houses the U.S. Department of Defense. An understanding and appreciation of geometry is basic to understanding and appreciating mathematics, as well as solving mathematical problems. In this chapter we focus on basic ideas of geometry, and on visualizing two- and three-dimensional figures and relationships to solve problems. Our goal is to provide experiences that enable you to further develop your spatial sense.

Connection to the NCTM Principles and Standards

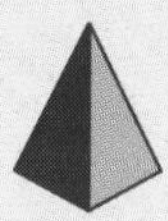

The NCTM *Principles and Standards for School Mathematics* (2000) indicate that PreK–8 students should study geometry so they can

- recognize, name, build, draw, compare, and sort two- and three-dimensional shapes *(p. 96);*
- classify two- and three-dimensional shapes according to their properties, and develop definitions of classes of shapes such as triangles and pyramids *(p. 164);* and
- Create and critique inductive and deductive arguments concerning geometric ideas and relationships, such as congruence . . . , and the Pythagorean relationship *(p. 232).*

Connection to the PreK–8 Classroom

The *Principles and Standards for School Mathematics* (2000) state that "The geometric and spatial knowledge children bring to school should be expanded by explorations, investigations, and discussions of shapes and structures in the classroom" (p. 97).

In grades PreK–2 children look at parts of solid shapes such as balls, cans, boxes, and party hats to develop concepts of plane figures, such as triangles, squares, rectangles, and circles.

In grades 3–5, children learn to classify and characterize basic geometric figures and the relationships among them by using drawings and cutouts.

In grades 6–8, students discover geometric relationships, for example, by drawing diagonals of different types of quadrilaterals and measuring to generalize how the diagonals are related.

Section 10.1 BASIC IDEAS OF GEOMETRY

- Seeing Geometry in the World
- Modeling and Defining Basic Geometric Ideas

In this section we help you recognize fascinating shapes and relationships in the real world and how they form the basis for geometry. As you make and analyze space figure models, you can sharpen your ability to visualize important geometric figures, patterns, and relationships.

Seeing Geometry in the World

Geometry in Nature. When you see natural objects such as the hexagonal tessellated honeycombs made by bees, or a snowflake, you realize that human beings have always been surrounded by geometric forms. Not only have people noticed them, but they also have appreciated and abstracted ideas from these forms to develop the body of knowledge we now call geometry.

The repetitive geometric shape of a honeycomb.
PhotoDisc © 2001

The geometric intricacy of a fully formed snowflake.
© Gerben Oppermans/Stone

These fascinating figures are just two of nature's mysteries, which include some amazingly intricate and interesting geometric relationships. Another such relationship is the ratio of the number of counterclockwise sunflower seed spirals to the number of clockwise spirals, which often is 55:34 or 34:21. These numbers occur in the **Fibonacci sequence** 1, 1, 2, 3, 5, 8, 13, 21, 34, 55..., discovered by Leonardo of Pisa, nicknamed Fibonacci, who lived from 1170 to 1230. The Fibonacci numbers also appear in other plants that have spiral growth patterns. For example, the spiral ratio in a pine cone is 8:5, in a pineapple, 13:8, and in a daisy head, 34:21. This fascinating sequence also occurs in the way branches grow on trees, as shown in Figure 10.1.

It has been discovered that the ratios of successive terms in the Fibonacci sequence approach the value 1.618. This ratio, called the **golden ratio**, was of great interest to the early Greeks, and we describe it in more detail in later chapters. The

Sunflower seed spirals.

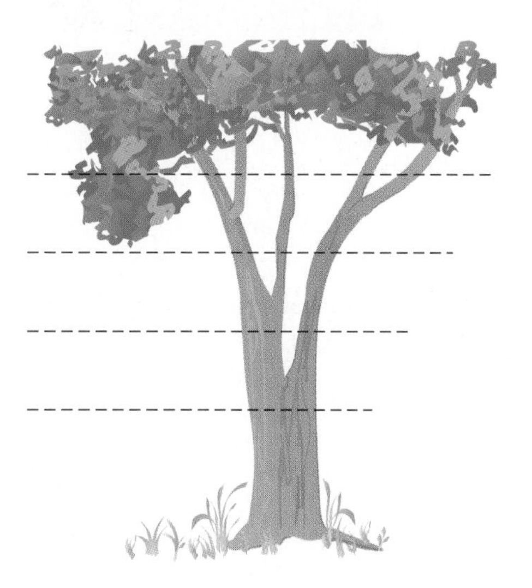

FIGURE 10.1 | The Fibonacci sequence in tree branch growth. Starting at the bottom and counting the number of branches cut by the dotted lines, we get 1, 2, 3, 5, and so on.

Source: From Mathematical Snapshots, 3rd ed., by H. Steinhaus. Copyright © 1969 by H. Steinhaus. Used by permission of Oxford University Press, Inc.

golden ratio is another of nature's amazing surprises and has been observed in spiral shells, starfish, and many other creatures. Whoever first discovered these intriguing geometric relationships in nature must have been very excited about the

The golden ratios can be used to describe the relationship between key parts of a starfish and of snail shell spirals.

discovery. In Mini-Investigation 10.1 you are asked to look further at nature's grand design and explore special ratios related to the human body.

Write *a description of your findings and how they relate to drawings of the human body.*

Technology Extension: *If available, use a calculator or computer to show a scatterplot of the data collected for each relationship considered in Mini-Investigation 10.1.*

Mini-Investigation 10.1 Finding a Pattern

Collect data and look for a pattern in the following ratios from the human body.

a. Height compared to navel height

b. Nose to chin compared to mouth to chin

c. Chin to hairline compared to chin to eyebrow

Geometry in Human Endeavors. As human beings progressed, they began to use ideas based on the geometric figures and relationships found in nature to serve their needs and to make their own beautiful objects. For example, the Egyptians used geometric techniques to re-mark land boundaries washed away by the Nile River each year. Thus the word *geometry*, which means "earth measure," came into use. The Egyptians also applied geometry to construction of the great pyramids and other edifices.

In the Great Pyramid of Gizeh the ratio of the slanted edge length to the distance from the ground center to the base edge is the golden ratio.

Modeling and Defining Basic Geometric Ideas

Points, Lines, Planes, and Space. As we explore the world around us, characteristics of the things we see suggest the abstract geometric ideas of *point*, *line*, *plane*, and *space*.

For example, as we look into the sky and see a tiny far-away pinpoint of a star, this real-world visual experience suggests the abstract geometric idea of a **point**, a location that has no length, width, or height.

Also, as we gaze into a starry sky and view the seemingly unlimited expanse of natural space, we can imagine the abstract geometric idea of **space**, the set of all points that has no boundaries.

When we see the trail of a jet plane clear across the sky, long after the plane is gone, we envision the abstract geometric idea of a **line**, a set of points in a straight, unlimited length with no thickness or endpoints.

And when we stand in a desert location where we can see for miles in every direction, the appearance of an unlimited flat place gives us an idea of the abstract geometric idea of a **plane**, a set of points in a flat surface that has no thickness and no edges.

Even simple everyday objects, such as the box in Figure 10.2, can suggest points, lines, planes, and space. In many formal developments of geometry, the ideas of point, line, plane, and space are so basic that they are accepted as "undefined terms" and are often explained with real-world models.

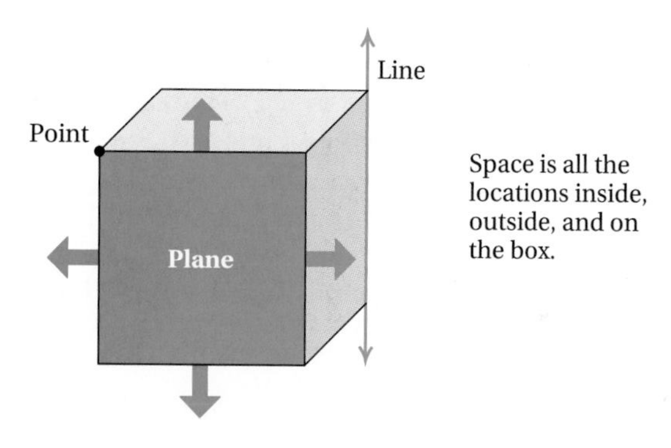

FIGURE 10.2 | Using a box to model points, lines, planes, and space.

To create a geometry that most closely matches our world of experience, we assume that two different points are contained in exactly one line and that three points that are not contained in one line are contained in exactly one plane. For example, in Figure 10.3(a) there is only one line that contains both points A and B (called line AB and written $\overleftrightarrow{AB}$), and in Figure 10.3(b) there is only one plane that contains points D, E, and F (called and written plane DEF).

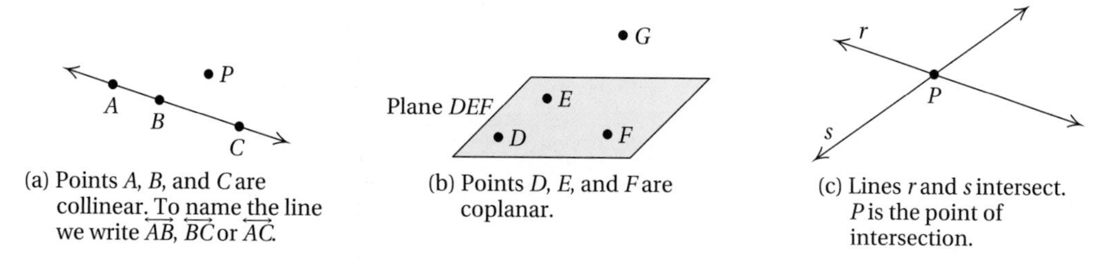

(a) Points A, B, and C are collinear. To name the line we write $\overleftrightarrow{AB}$, $\overleftrightarrow{BC}$ or $\overleftrightarrow{AC}$.

(b) Points D, E, and F are coplanar.

(c) Lines r and s intersect. P is the point of intersection.

FIGURE 10.3 | Relationships among points, lines, and planes. *(Continued on next page.)*

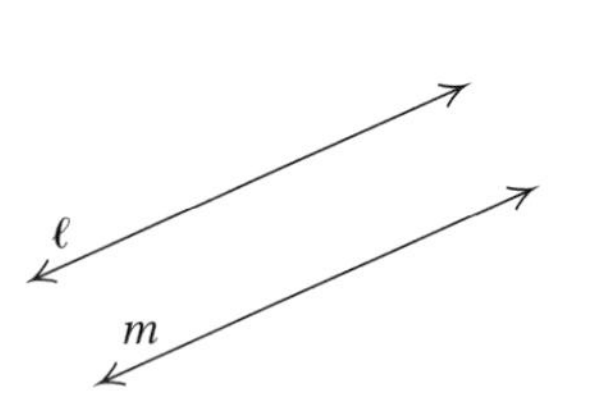

(d) Lines ℓ and m are parallel. We write $\ell \parallel m$.

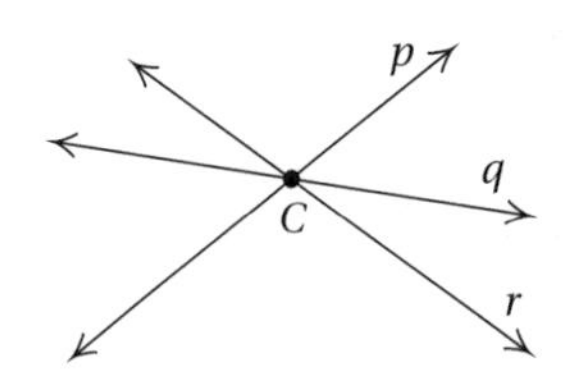

(e) Lines p, q and r are concurrent. C is the point of concurrency.

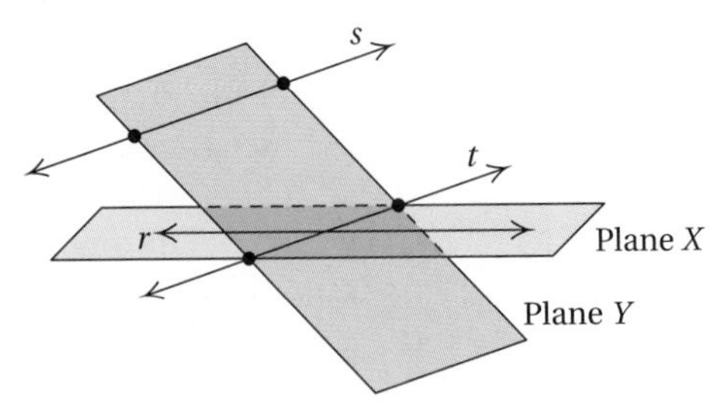

(f) Lines s and r are skew. Planes X and Y intersect in line t.

FIGURE 10.3 | *(cont.)* Relationships among points, lines, and planes.

Note that we sometimes denote a plane with one letter, such as Plane X in Figure 10.3(f). Points that can be contained in one line are called **collinear** (Figure 10.3(a)) and points that can be contained in one plane are called **coplanar** (Figure 10.3(b)). In Figure 10.3(a), points A, B, and P are noncollinear because they cannot be contained in one line, and points D, E, F, and G in Figure 10.3(b) are noncoplanar because they cannot be contained in one plane. Two different lines r and s can **intersect** in one point, called the **point of intersection**, as in Figure 10.3(c), or they can be **parallel** like lines ℓ and m in Figure 10.3(d). **Skew** lines r and s that cannot be contained together in any one plane are shown in Figure 10.3(f). Three different lines that intersect in one point, such as lines p, q, and r in Figure 10.3(e), are **concurrent**. Two different planes, X and Y, can *intersect* in a line t as in Figure 10.3(f). If two planes don't intersect, then they are *parallel*. In Example 10.1 we apply these point-line-plane relationships.

Example 10.1

Identifying Some Point-Line-Plane Relationships

Identify examples in the illustration of the relationships described in (a) through (f).

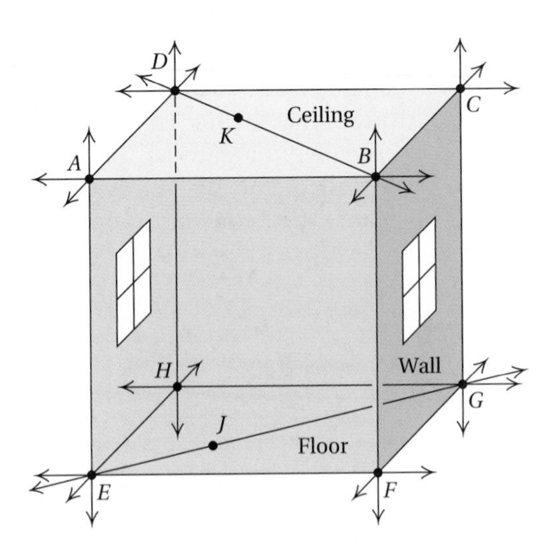

a. collinear points
b. coplanar points
c. intersecting lines
d. parallel lines
e. skew lines
f. concurrent lines
g. intersecting planes
h. parallel planes

Solution

a. Points D, K, and B
b. Points A, K, and B
c. $\overleftrightarrow{AB}$ and $\overleftrightarrow{BC}$
d. $\overleftrightarrow{AB}$ and $\overleftrightarrow{EF}$
e. $\overleftrightarrow{DB}$ and $\overleftrightarrow{EG}$
f. $\overleftrightarrow{AD}$, $\overleftrightarrow{BD}$, and $\overleftrightarrow{CD}$
g. Planes DAB and FBC
h. Planes ADB and EFG

Your Turn

Practice: Identify examples of the relationships described in (a) through (f) that are different from those given in the solution.

Reflect: How can you be sure that $\overleftrightarrow{DB}$ and $\overleftrightarrow{EG}$ are skew lines? Explain. ■

Segments, Rays, and Angles Real-world models can also be used to introduce other basic geometric figures that can then be defined using the geometric ideas of point, line, and plane. Table 10.1 gives meaning to three very basic figures: segments, rays, and angles.

To solve problems involving segments and angles, you may need to find measures of these geometric figures. A segment possesses a property of how far it extends, or how many unit segments must be laid next to each other to reach its

TABLE 10.1 | Models and Definitions for Segments, Rays, and Angles

Model	Description/Symbol	Definition
A, B, segment *AB*	**Segment** *AB*, written $\overline{AB}$	The set of points *A* and *B* and all the points between *A* and *B*.
A, B, ray *AB*	**Ray** *AB*, written $\overrightarrow{AB}$	The point *A* and all the points on $\overleftrightarrow{AB}$ on the same side of *A* as point *B*.
A, B, C, angle *ABC*	**Angle** *ABC*, written $\angle ABC$. *B* is the vertex.	The union of two noncollinear rays ($\overrightarrow{BA}$ and $\overrightarrow{BC}$) that have a common endpoint (*B*).

end. To describe the property of length, we assign a real number, called the **length measure,** to each segment. In Figure 10.4, the length measure of $\overline{AB}$ is 3.5 centimeters, so we write $AB = 3.5$ centimeters. Note that AB, without the overbar, denotes the length of $\overline{AB}$. So AB denotes distance while $\overline{AB}$ denotes a set of points.

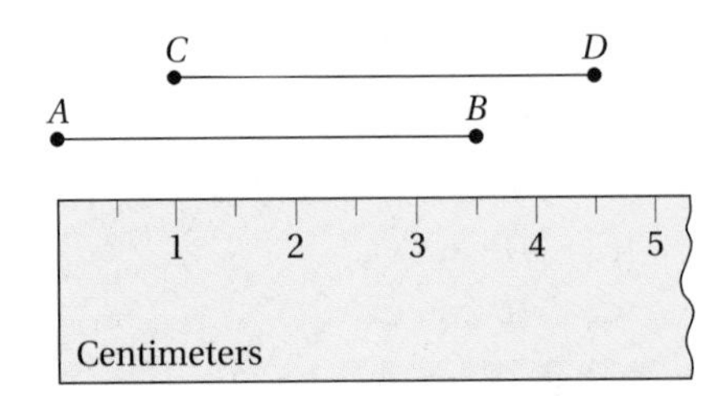

FIGURE 10.4 | Measuring segments.

Two segments are **congruent segments** if they have the same length. Since the length of $\overline{CD}$ in Figure 10.4 is also 3.5 centimeters, we know that the two segments are congruent and write $\overline{AB} \cong \overline{CD}$. We make immediate use of the idea of congruent segments by using it to define the **midpoint of segment** $\overline{EF}$ as point M on segment $\overline{EF}$ such that $\overline{EM} \cong \overline{MF}$, as shown in Figure 10.5.

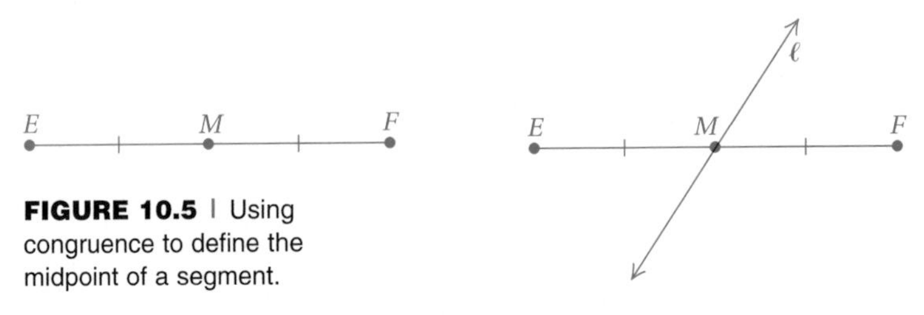

FIGURE 10.5 | Using congruence to define the midpoint of a segment.

FIGURE 10.6 | Line ℓ bisects $\overline{EF}$.

Any point, segment, ray, line, or plane that contains the midpoint of a segment is a **bisector of the segment,** as shown in Figure 10.6, where line ℓ is a bisector of $\overline{EF}$.

An angle possesses a property of size or amount of rotation. To describe this property, we assign a real number between 0 and 180 to each angle. This number is called the **degree measure** (m) of the angle. In Figure 10.7 the **protractor,** a device for measuring angles, shows that the degree (or angular) measure of $\angle ABC$ is 60. We write $m\angle ABC = 60°$. Two angles are congruent only if they have the same measure. Since it is also true that $m\angle DEF = 60°$, we write $\angle ABC \cong \angle DEF$.

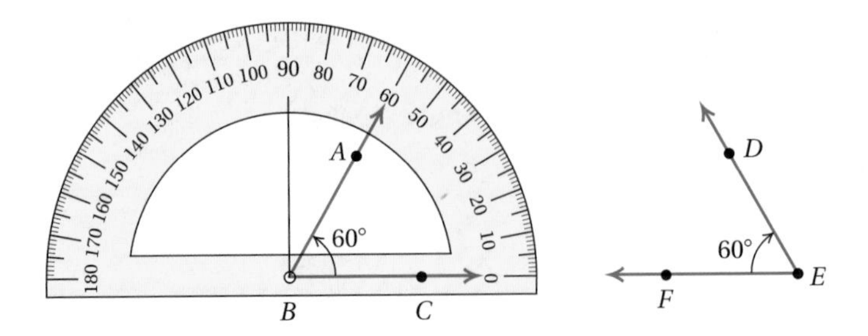

FIGURE 10.7 | Using a protractor.

In popular uses of angular measure, where the amount of rotation is of central concern, it is common to refer to a figure formed by rotating a side of an angle 180° as a **straight angle,** even though the resulting figure itself is really a line. Also, when the rotation goes beyond 180° but less than 360°, the figure is sometimes referred to as a **reflex angle,** even though as a figure it may be congruent to an angle with measure less than 180°. When there is no rotation, the figure is often referred to as a **zero angle,** even though, as a figure, it is really a ray. These "angles" are shown below.

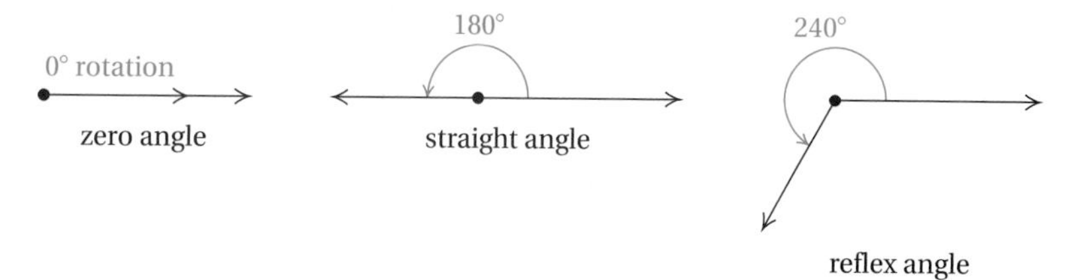

An angle partitions the plane into two regions, called the *interior* and the *exterior* of the angle. In Figure 10.8, if R = the set of all points on A's side of $\overleftrightarrow{BC}$ and S = the set of all points on C's side of $\overleftrightarrow{BA}$, then the **interior** of $\angle ABC$ is $R \cap S$. The set of points not in the interior or on the angle is the **exterior** of the angle.

The ray in the interior of an angle that forms two congruent angles is the **angle bisector** of the angle. For example, in Figure 10.9, $\overrightarrow{EG}$ is the angle bisector of $\angle DEF$.

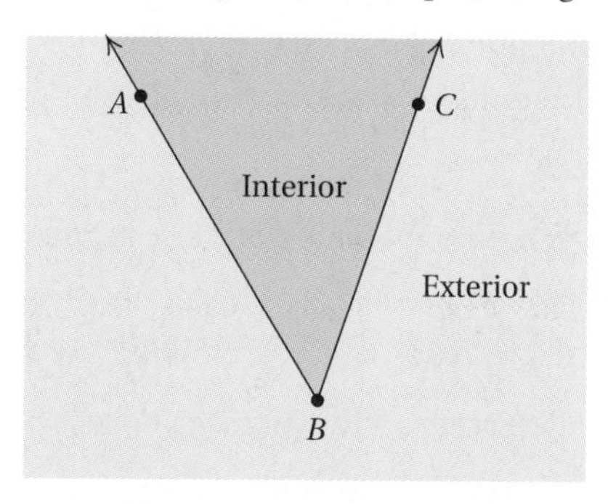

FIGURE 10.8 | Defining the interior of an angle.

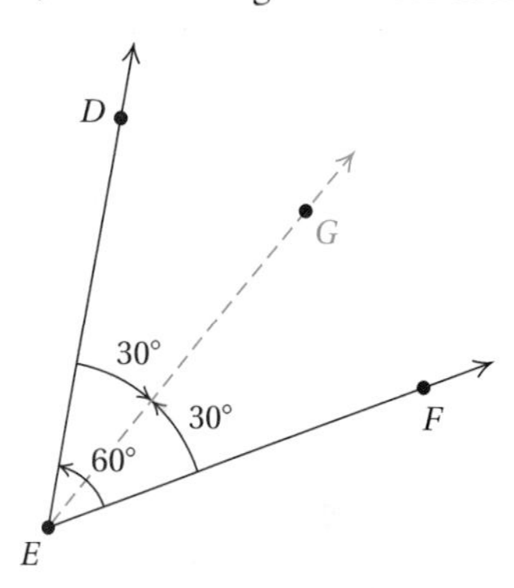

FIGURE 10.9 | Using congruence to define the angle bisector.

The cartoon below hints at the crucial role that Euclid played in early description of the properties of angles and other geometric figures. Example 10.2 involves finding the measures of segments and angles.

FRANK & ERNEST® by Bob Thaves

With permission of Bob Thaves.

Example 10.2 Measuring Segments and Angles

Refer to the pictures below and find the measures of the following segments and angles.

Segments: **a.** $\overline{AB}$ **b.** $\overline{BD}$ **c.** $\overline{CE}$
Angles: **d.** $\angle PAR$ **e.** $\angle PAU$ **f.** $\angle RAV$

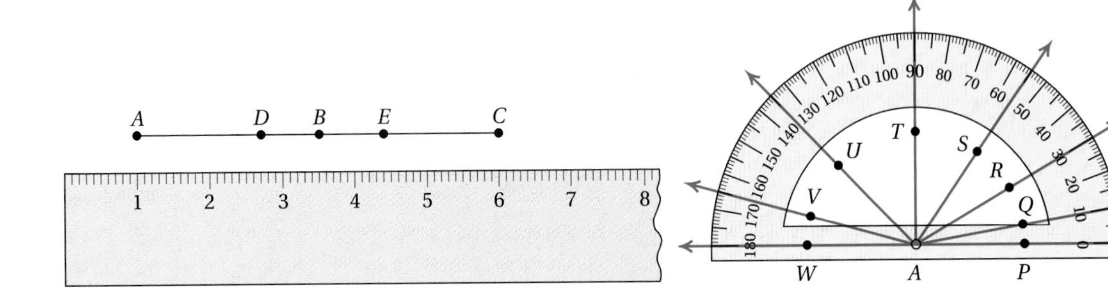

Solution

a. $AB = 2.5$ units **d.** $m(\angle PAR) = 30°$
b. $BD = 0.8$ units **e.** $m(\angle PAU) = 135°$
c. $CE = 1.6$ units **f.** $m(\angle RAV) = 135°$

Your Turn

Practice: Find the measures of these segments and angles.

Segments: **a.** $\overline{AE}$ **b.** $\overline{CB}$ **c.** $\overline{ED}$
Angles: **d.** $\angle PAT$ **e.** $\angle SAU$ **f.** $\angle SAW$ **g.** $\angle VAT$

Reflect: Referring to the picture above,

a. name two pairs of congruent angles.
b. name two pairs of congruent segments.
c. name an angle and its bisector. ■

Special Angles and Perpendicular Lines. Now that we have defined some basic ideas of geometry, we use them in Table 10.2 to describe and define some special types of angles and some useful angle and line relationships.

TABLE 10.2 | Defining Special Angles and Perpendicular Lines

Figure	Description/Symbol	Definition
A, B, C	$\angle ABC$ is a **right angle.**	Right angles are angles with a measure of 90°.

TABLE 10.2 | (continued)

Figure	Description/Symbol	Definition
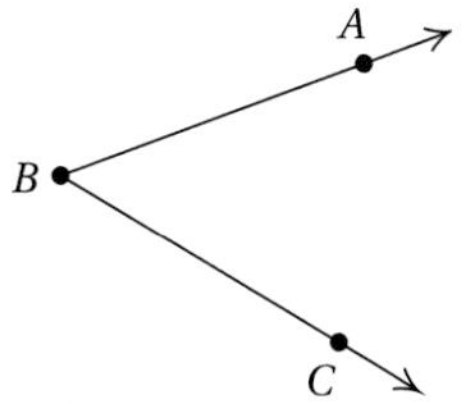 A, B, C	$\angle ABC$ is an **acute angle.**	Acute angles have measures between 0° and 90°.
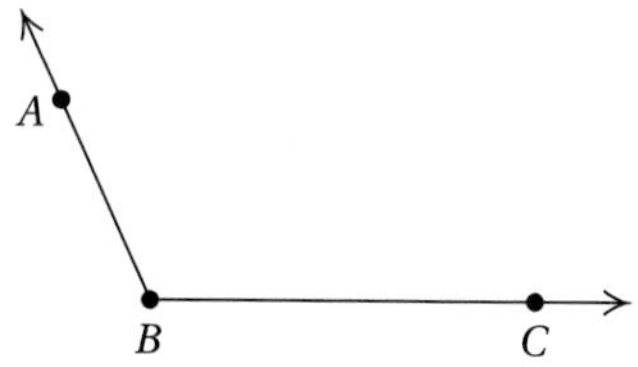 A, B, C	$\angle ABC$ is an **obtuse angle.**	Obtuse angles have measures between 90° and 180°.
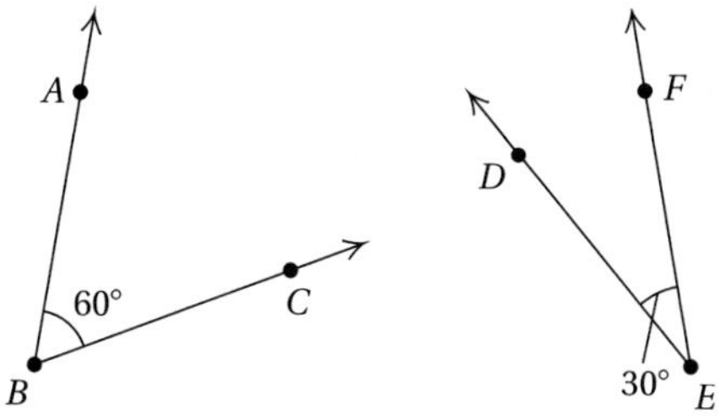 A, B, C, 60°; D, E, F, 30°	$\angle ABC$ and $\angle DEF$ are **complementary** to each other.	Complementary angles are a pair of angles whose measures have a sum of 90°.
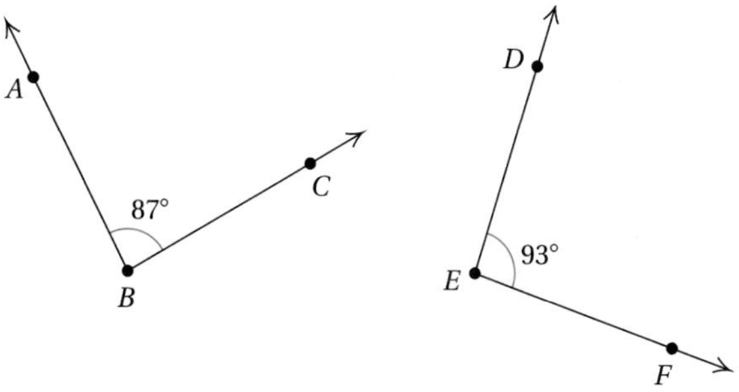 A, B, C, 87°; D, E, F, 93°	$\angle ABC$ and $\angle DEF$ are **supplementary** to each other.	Supplementary angles are a pair of angles whose measures have a sum of 180°.
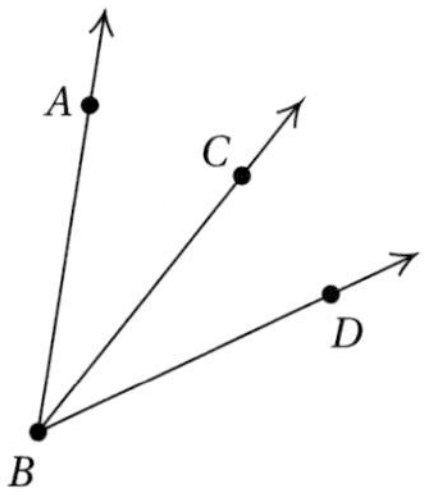 A, B, C, D	$\angle ABC$ is **adjacent** to $\angle CBD$	Adjacent angles are two angles that have the same vertex and a common side, but that have no common interior points.
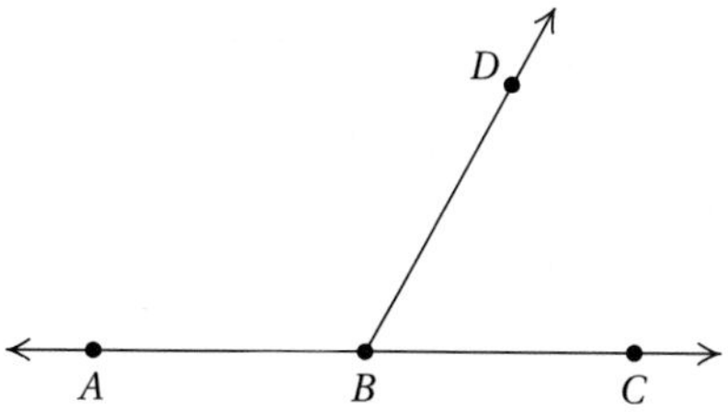 A, B, C, D	$\angle ABD$ and $\angle DBC$ are a **linear pair** of angles.	A linear pair of angles is a pair of adjacent angles with two non-common sides on the same line.

TABLE 10.2 | (continued)

Figure	Description/Symbol	Definition
	$\angle AED$ and $\angle BEC$ are a pair of **vertical angles.**	Vertical angles are a pair of angles that are formed by two intersecting lines and that are not a linear pair of angles.
	Line ℓ is **perpendicular** to line m. We write $\ell \perp m$.	Two lines are perpendicular if they intersect to form four right angles.
	Line ℓ is the **perpendicular bisector** of $\overline{CD}$.	The perpendicular bisector of a segment is a line that is perpendicular to the segment, and which divides it into two congruent segments.

The first three figures in Table 10.2 are single figures each with a special property. The remaining descriptions and definitions in the table are of *relationships* between two angles or between two lines. These figures and relationships in geometry are important and useful, and are applied in Example 10.3.

Example 10.3 Identifying Special Angles and Perpendicular Lines

In the illustration on the next page, identify examples of each of the figures and relationships described in Table 10.2.

Solution

a. $\angle GAC$ is a right angle.
b. $\angle DEG$ is an acute angle.
c. $\angle BEG$ is an obtuse angle.
d. $\angle DEA$ and $\angle BEA$ are complementary.
e. $\angle BEA$ and $\angle FEG$ are supplementary.
f. $\angle AEB$ and $\angle CEB$ are adjacent.
g. $\angle HEI$ and $\angle HEA$ are a linear pair.
h. $\angle BEC$ and $\angle GEH$ are vertical.
i. $\overleftrightarrow{AG} \perp \overleftrightarrow{GI}$.
j. $\overleftrightarrow{HB}$ is the perpendicular bisector of $\overline{AC}$.

Your Turn

Practice: Identify an example of each figure or relationship described in Table 10.2 that is different from those given in the solution.

Reflect: How are a pair of supplementary angles like a linear pair of angles? How are they different? ■

ACIG is a square. *ABED, BCFE, DEHG,* and *EFIH* are squares with all sides the same length.

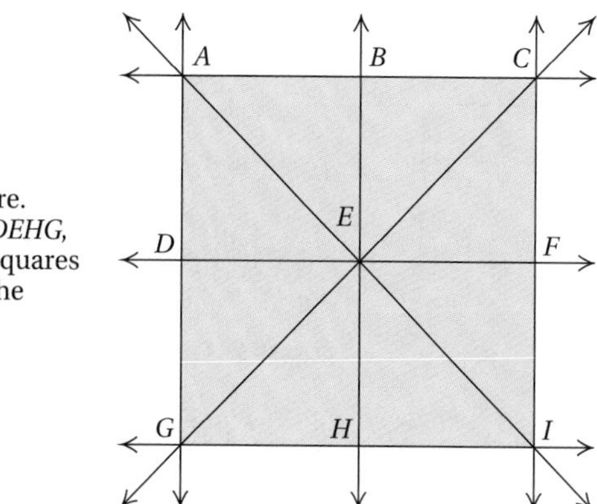

Circles and Polygons. Another type of geometric figure that we often find modeled in the real world is a closed curve. Imagine starting from home on a trip to run errands. As you travel from home to the library, to the gas station, to the grocery store, and back home, a **closed curve** made up of points in a plane can be used to represent your path. If your path has crossed itself at least once, then it is a **nonsimple closed curve,** as shown in Figure 10.10(b). If you have planned efficiently, and you have not had to retrace any points in your path except for the beginning and ending points, then it is a **simple closed curve,** as shown in Figure 10.10(c). A **circle** is a special type of simple closed curve in which all of the points in the curve are in one plane and are equidistant from a given point in the same plane, called the **center** of the circle. A segment whose endpoints are on the circle is called a **chord** of the circle, a chord that contains the center of the circle is called a **diameter** of the circle, and a segment with one endpoint at the center of the circle and one endpoint on the circle is called a **radius** of the circle, as shown in Figure 10.10(d).

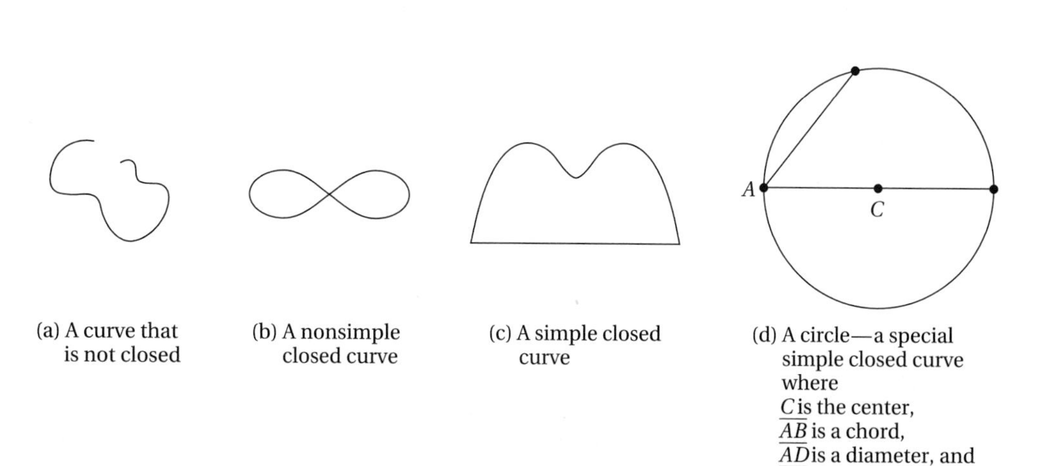

(a) A curve that is not closed

(b) A nonsimple closed curve

(c) A simple closed curve

(d) A circle—a special simple closed curve where C is the center, $\overline{AB}$ is a chord, $\overline{AD}$ is a diameter, and $\overline{AC}$ is a radius

FIGURE 10.10 | Examples of closed curves, including a circle.

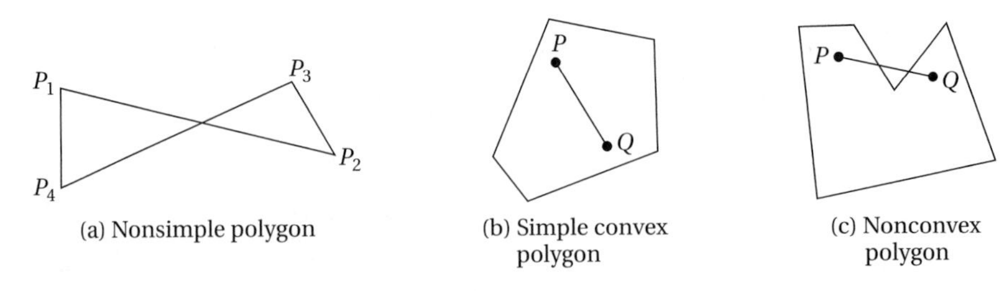

FIGURE 10.11 | Different types of polygons.

A **polygon** is a closed curve created from the union of segments meeting only at endpoints such that (1) at most, two segments meet at one point, and (2) each segment meets exactly two other segments at their endpoints. A **nonsimple polygon** has at least one pair of segments that intersect in a point other than their endpoints, as shown in Figure 10.11(a). A **simple polygon** has no such intersections. A **simple convex polygon,** shown in Figure 10.11(b), satisfies the property that for each pair of points P and Q inside the polygon, $\overline{PQ}$ lies inside the polygon. In a **simple nonconvex polygon,** at least one pair of points P and Q can be positioned inside the polygon so that part of $\overline{PQ}$ lies outside the polygon, as shown in Figure 10.11(c).

A polygon is named according to its number of sides, as indicated in Table 10.3. A polygon with 11 sides is called an 11-gon, and a polygon with 12 sides is called a dodecagon.

Names have been given for some polygons with more than 12 sides, but we prefer to use 13-gon, 14-gon, and so on. If the whole number n represents the number of sides of a polygon, we simply call it an ***n*-gon.**

Polygons in general can be classified by relationships of their sides and interior angles. An **interior angle** of a polygon is formed by two sides of the polygon that have a common vertex. A **regular polygon** is a simple polygon with all sides congruent and all interior angles congruent. Figure 10.12(a) shows two examples of polygons that are not regular, and Figure 10.12(b) shows an example of a regular polygon.

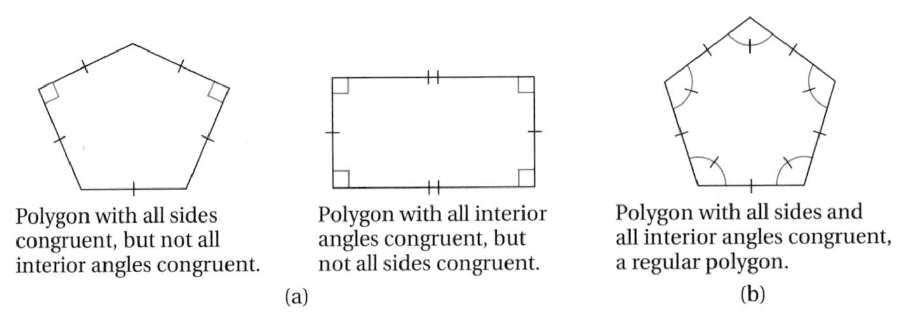

FIGURE 10.12 | Examples of (a) nonregular and (b) regular polygons.

TABLE 10.3 | Classification of Polygons by Number of Sides

Polygon	Number of Sides	Name
	3	**Triangle**
	4	**Quadrilateral**
	5	**Pentagon**
	6	**Hexagon**
	7	**Heptagon**
	8	**Octagon**
	9	**Nonagon**
	10	**Decagon**
	12	**Dodecagon**

Since a triangle is a special type of polygon that we will be studying in detail, we single it out and give the following definition.

Definition of a Triangle

A **triangle,** $\triangle ABC$, is the union of three segments determined by three noncollinear points A, B, and C. A, B, and C are the **vertices** of the triangle. $\overline{AB}$, $\overline{BC}$, and $\overline{AC}$ are the **sides** of the triangle.

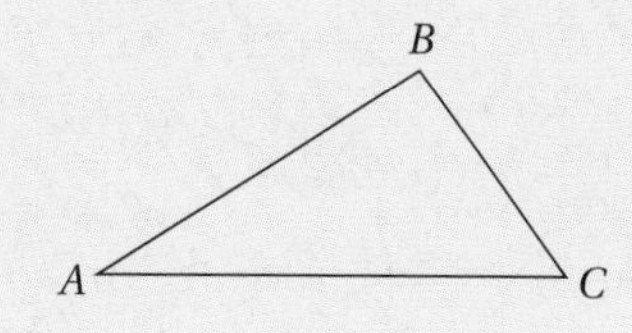

Two important segments associated with a triangle are defined in Figure 10.13. A **median** of a triangle is a segment from a vertex of a triangle to the midpoint of the side of the triangle opposite that vertex, as shown in Figure 10.13(a). An **altitude** of a triangle is a segment from a vertex of a triangle perpendicular to a line containing the side of the triangle opposite that vertex, as shown in Figure 10.13(b). We will explore the properties of medians and altitudes further in a later subsection.

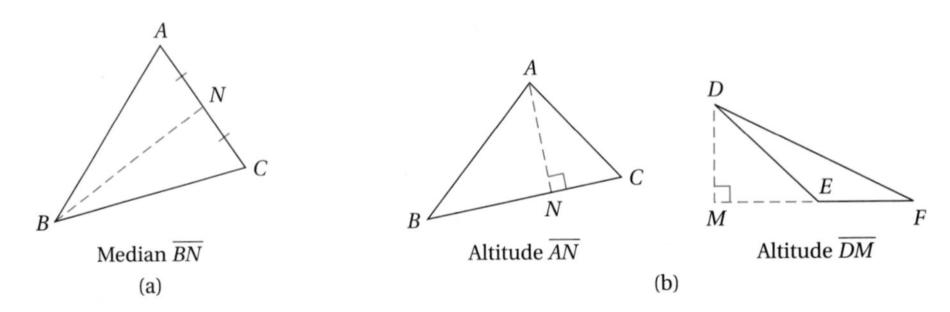

FIGURE 10.13 | A median and altitudes for some triangles.

It is useful to classify triangles according to the properties of their sides and interior angles, as in Table 10.4. Note that the definition of an isosceles triangle includes the equilateral triangle as a special case of an isosceles triangle.

TABLE 10.4 | Classification of Triangles

Triangle	Name	Properties
	Equilateral triangle	All sides congruent
	Isosceles triangle	At least one pair of congruent sides
	Scalene triangle	No congruent sides
	Right triangle	One right angle
	Acute triangle	All acute angles
	Obtuse triangle	One obtuse angle

Once triangles have been classified and defined according to the key characteristics in Table 10.4, we can explore them to discover additional properties. For example, we can discover that an equilateral triangle is also **equiangular,** that is, all the angles of an equilateral triangle are congruent. We can also discover and verify that the angles opposite the congruent sides in an isosceles triangle are congruent and thus an isosceles triangle also has two congruent angles. When we investigate the properties of a scalene triangle, we observe that it has no congruent angles. A right triangle also has several important additional properties, which we will consider in a later section.

Example 10.4 provides an opportunity to think about the different types of triangles, and identify examples of medians and altitudes.

Example 10.4 Identifying Triangles, Medians, and Altitudes

Identify each of the following in this figure. Write NP (not possible) if no such figure exists.

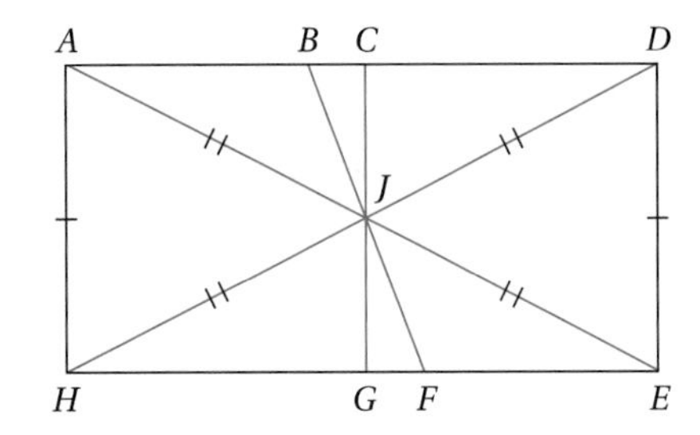

ADEH is a rectangle. *ACGH* and *CDEG* are squares with same length sides. $\overline{CG} \perp \overline{HE}$, $\overline{CG} \perp \overline{AD}$

a. equilateral triangle
b. isosceles triangle
c. scalene triangle
d. right triangle
e. acute triangle
f. obtuse triangle
g. median of a triangle
h. altitude of a triangle

Solution

a. NP
b. $\triangle AJD$
c. $\triangle HJF$
d. $\triangle AHE$
e. $\triangle DJE$
f. $\triangle JFE$
g. $\overline{JD}$ of $\triangle ADE$
h. $\overline{JC}$ of $\triangle DJB$

Your Turn

Practice: Identify an example of each of the figures in (a)–(h) that is different from those given in the solution.

Reflect: Can you select a triangle in the figure that fits into more than one of the triangle types listed in (a)–(f) above? Look for different ways to do this. ■

Quadrilaterals, like triangles, can be further classified according to properties of sides and interior angles, as shown in Table 10.5.

There are also relationships between the various types of quadrilaterals. Since a kite is defined to have *at least* two pairs of congruent adjacent sides, a square and a rhombus are both special types of kites.

A Sample Student Page: Properties of Quadrilaterals

In the *Principles and Standards for School Mathematics,* it is stated that "PreK–2 geometry begins with describing and naming shapes" (p. 97) and that "In grades 3–5, they should ... focus on identifying and describing a shape's properties and learning specialized vocabulary ..." (p. 165). Study the page from an elementary school mathematics textbook in Figure 10.14 and answer the questions.

- **How** does this page help students distinguish a rectangle from a parallelogram?
- **How** does this page implement the Principles and Standards statement about geometry?
- **Are** there opportunities for students to reason about geometry on this page? Explain.

Chapter 6
Lesson
4

Quadrilaterals

You Will Learn
how to classify quadrilaterals

Vocabulary
quadrilateral
a polygon with four sides

Types of Quadrilaterals
square
rectangle
parallelogram
rhombus
trapezoid

Learn

These Miao children, internationally known as Hmong, are from the Guizhou province in south central China. Their rectangular-shaped shoulder shawls are worn at one of the many festivals they celebrate each year.

Some of the shapes in Miao patterns are **quadrilaterals**, polygons with four sides.

There are about 8 million Hmong (mawng) worldwide.

Square
all sides the same length,
four right angles

Rectangle
opposite sides parallel
and the same length,
four right angles

Remember
Parallel lines do not meet. Right angles form square corners.

Some quadrilaterals do not have four right angles.

Parallelogram
two pairs of parallel sides

Rhombus
two pairs of parallel sides,
all sides the same length

Trapezoid
only one pair of
parallel sides

Talk About It

Is it true that a square is a rectangle and a rhombus? Explain.

FIGURE 10.14 | Excerpt from a grade 5 elementary school mathematics textbook.

(*Source:* Page 276 from *Scott Foresman–Addison Wesley Math Grade 5* by Randall I. Charles, et al. Copyright © 1998 by Addison Wesley Longman, Inc. Reprinted by permission of Pearson Education, Inc.)

TABLE 10.5 | Classification of Quadrilaterals

Quadrilateral	Name	Properties
	Trapezoid	Exactly one pair of opposite sides parallel
	Kite	At least two pairs of adjacent sides congruent, no side used twice in the pairs
	Parallelogram	Pairs of opposite sides parallel and the same length
	Rhombus	All sides the same length and opposite sides parallel
	Rectangle	Opposite sides parallel and the same length; all angles are right angles
	Square	All sides the same length; all angles are right angles

In Example 10.5, we explore quadrilateral relationships further.

Example 10.5 Describing Quadrilateral Relationships

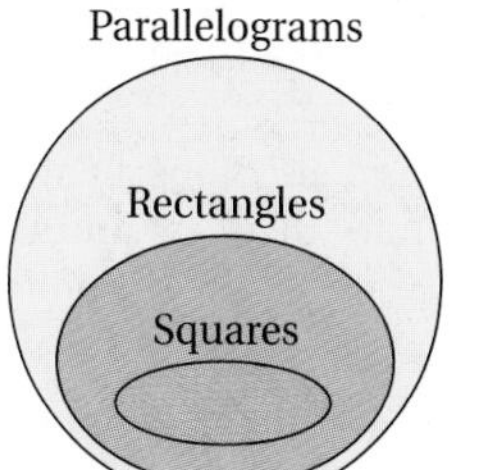

Explain the relationships among rectangles, parallelograms, and squares.

Solution

Mario's thinking: I thought about it like this.

a. Every square is a rectangle. There are rectangles that aren't squares.

b. Every rectangle is a parallelogram. There are parallelograms that aren't rectangles.

Cindy's thinking: I drew a Venn diagram to show the relationship.

Your Turn

Practice: Explain the relationships among rectangles, parallelograms, and rhombi.

Reflect: Why isn't every parallelogram also a rhombus? ■

Problems and Exercises | *for Section 10.1*

A. Reinforcing Concepts and Practicing Skills

1. What basic geometric ideas are suggested by these real-world situations?
   a. The tip of a pencil
   b. An ordinary drinking straw
   c. A beam of light from a spotlight
   d. Railroad tracks
   e. A long straight road, with no visible start or end
   f. A stop sign
   g. A bent elbow
   h. A bicycle wheel rim
2. Use the model of the box shown and give examples of
   a. intersecting lines
   b. parallel lines
   c. skew lines
   d. parallel planes
   e. intersecting planes
   f. concurrent lines
   g. coplanar points
   h. collinear points

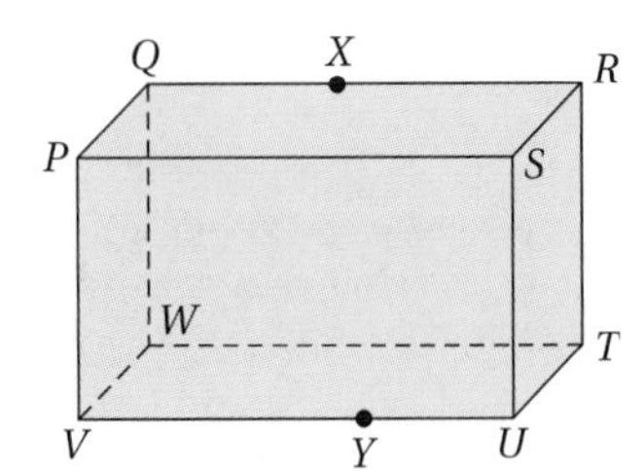

Draw and label the geometric figures in Exercises 3–8.

3. $\overleftrightarrow{JK}$
4. $\overline{RS}$
5. $\overrightarrow{XY}$
6. $\angle GHI$
7. $\triangle MNO$
8. Rectangle $PQRS$

Refer to the rectangular figure shown and name the figures described in Exercises 9–15.

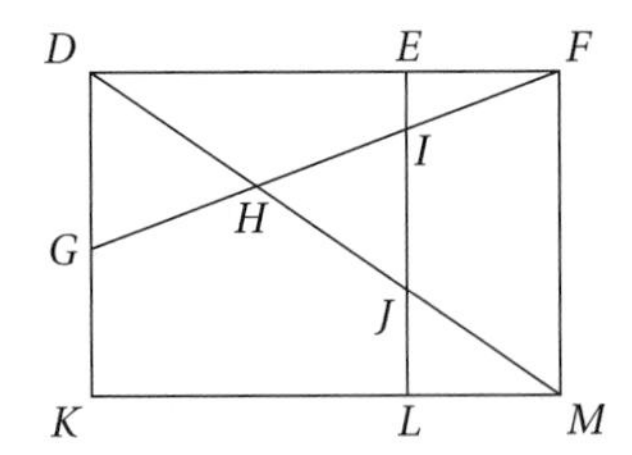

9. Three segments with the same endpoint
10. A pair of vertical angles
11. A pair of perpendicular lines
12. An obtuse angle
13. Three different types of triangles
14. Three different types of quadrilaterals
15. A polygon with more than four sides

Draw and label the figures described in Exercises 16–19.

16. A pair of angles that are complementary but that have no common vertex
17. A pair of angles that are supplementary but that have no common vertex
18. A linear pair of angles
19. A pair of vertical angles
20. If $\overrightarrow{BD}$ is a bisector of $\angle ABC$ and $m(\angle ABD) = 60°$, what can you conclude about $m(\angle CBD)$? Verify your conclusion.
21. Three lines r, s, and t are concurrent. How do you know that none of the lines are parallel to another of the lines?
22. Three points D, E, and F are collinear. Is there only one plane that contains these three points? Explain.
23. Can a pair of skew lines intersect? Explain why or why not.
24. If $\overrightarrow{AB} \cup \overrightarrow{AC}$ is a line, what is true about A, B, and C?
25. If $\overrightarrow{AB} \cap \overrightarrow{AC}$ is a ray, what is true about A, B, and C?
26. Draw a Venn diagram to show how parallelograms, squares, and rhombi relate to each other.
27. Write a description of how kites, squares, and rhombi relate to each other.
28. Explain why an equilateral triangle is also isosceles.
29. Draw a simple, nonconvex polygon that has four sides.
30. Draw a simple, nonregular pentagon with four congruent sides and with two of its interior angles right angles.
31. Is it possible for two obtuse angles to be a linear pair? for two acute angles? Explain.
32. Draw a segment that you think is 7 centimeters long. Then measure to find the length of the segments to the nearest millimeter.
33. Draw an angle that you think measures 65°. Then find the actual measure of the angle.

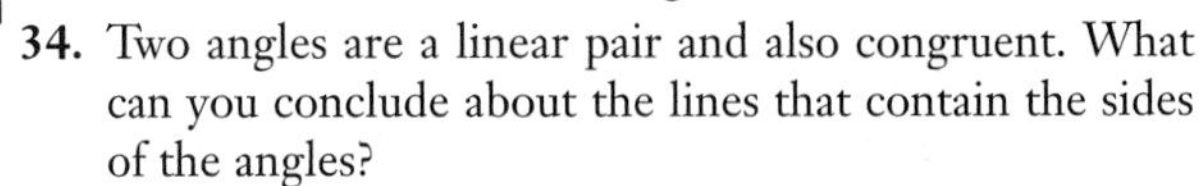

34. Two angles are a linear pair and also congruent. What can you conclude about the lines that contain the sides of the angles?
35. Is the ratio of the length to width of a 3-by-5-inch card or of a 5-by-8-inch card closest to the golden ratio?
36. Which is closer to a golden rectangle, a $4\frac{1}{2}$-ft × $7\frac{1}{4}$-ft window or a 4-ft × $6\frac{1}{2}$-ft window? (Use 1.618034 for the golden ratio.)

B. Deepening Understanding

Complete Exercises 37–42 about the Fibonacci sequence.

37. Write the first 15 terms of the Fibonacci sequence. What procedure did you use to do so?
38. Fibonacci concluded that the ratio of a term in his

sequence to the term before it gets closer and closer to the golden ratio (approximately 1.618) as he used larger and larger terms. Use a calculator or computer spreadsheet to show that this result holds for the terms you found in Exercise 37.

39. Square any term of the Fibonacci sequence. Then find the product of the term preceding and the term following that term. Repeat several times on a calculator or computer spreadsheet. What do you discover?

40. Estimate how many of the first 25 Fibonacci numbers are even. Extend the sequence to check your guess.

41. Use a calculator and a guess and revise strategy to find two decimals x and y such that $x - y = 1$ and $xy = 1$. What do you discover about the numbers?

42. The golden ratio can be transformed into its reciprocal by simply subtracting 1. Show that this outcome is true. How does this result relate to your findings in Exercise 41?

In Exercises 43–55, decide whether the statement describes something that will sometimes happen, always happen, *or* never happen. *Give a convincing argument to support your conclusions.*

43. Lines ℓ and m intersect to form four congruent angles that are not right angles.
44. The bisector of an angle forms two congruent angles.
45. A linear pair of angles are supplementary.
46. A pair of vertical angles include four pairs of supplementary angles.
47. Two supplementary angles have a common vertex.
48. A right triangle is an isosceles triangle.
49. A rhombus is a square.
50. An equilateral triangle is an acute triangle.
51. A trapezoid is a kite.
52. A right triangle is an obtuse triangle.
53. A parallelogram is neither a rhombus nor a rectangle.
54. An equilateral triangle is an isosceles triangle.
55. A parallelogram is a trapezoid.
56. If two parallel planes W and X intersect a third plane Y in two lines r and s, are lines r and s necessarily parallel? Explain your answer.
57. In the figure shown at the top of the next column, $\overrightarrow{PL}$ is perpendicular to $\overleftrightarrow{JN}$, $\overleftrightarrow{PK} \perp \overleftrightarrow{PM}$, and $m(JPK) = 70°$. Explain how to find the measures of $\angle KPL$, $\angle KPM$, $\angle JPM$, and $\angle NPM$. Verify that your measures are correct.

Study the techniques in Appendix D for using a ruler and compass for basic construction and construct the figures in Exercises 58–61.

58. Construct two adjacent right angles.
59. Draw an angle and construct its bisector.
60. Draw an angle and construct an isosceles triangle that contains two angles congruent to the angle you drew.

Figure for Exercise 57

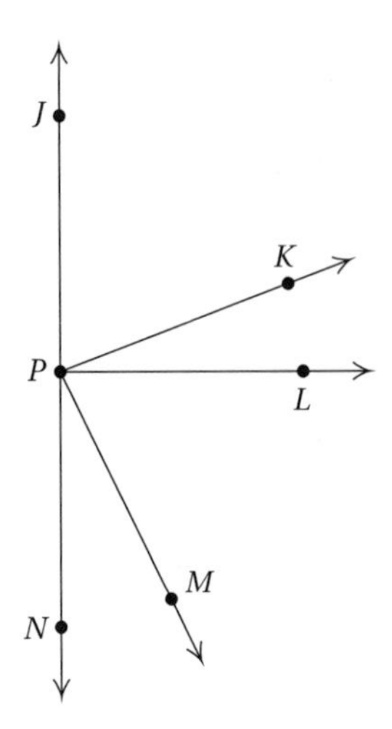

61. Draw a segment and construct an equilateral triangle with sides congruent to the segment you drew.

C. Reasoning and Problem Solving

62. Use a grid as shown here and mark three dots that, when connected, form a(n)

    a. equilateral triangle b. isosceles triangle
    c. scalene triangle d. right triangle

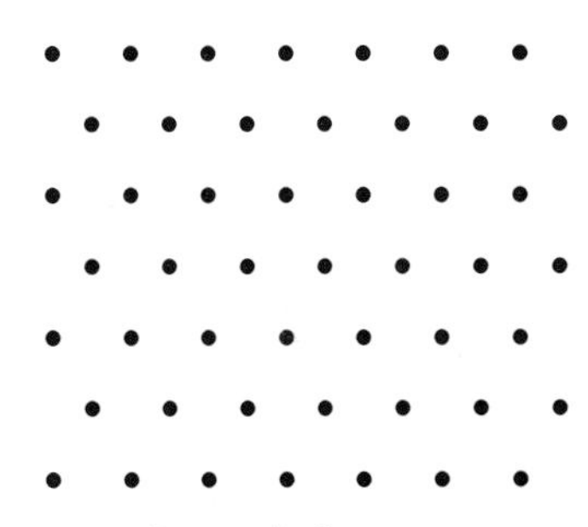

Isometric dot paper

63. Use a grid as shown here and mark four dots that, when connected, form a

    a. square b. rectangle
    c. rhombus d. parallelogram
    e. trapezoid f. kite

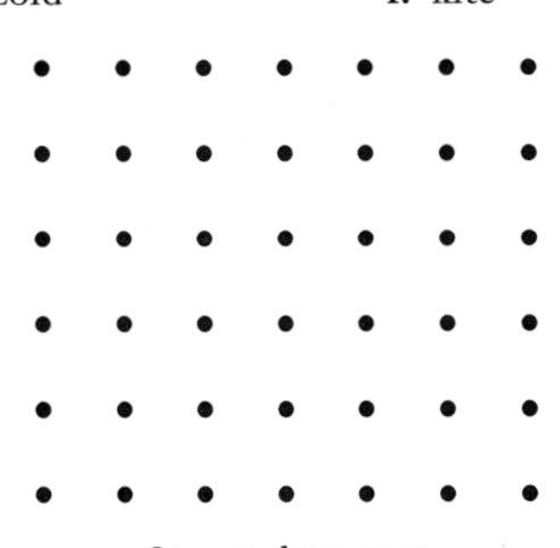

Square dot paper

64. a. Do you think that the following diagram correctly shows the relationships among the different types of quadrilaterals? Explain.

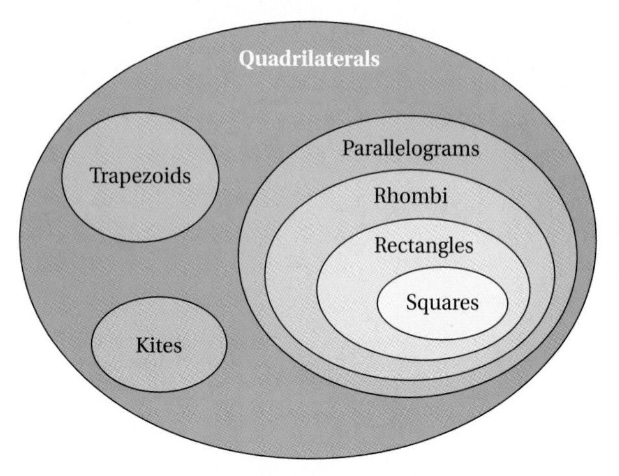

b. Draw a Venn diagram that correctly shows any part of the given diagram that is incorrect.

65. The Clock Problem. What is the measure of the angle between the two hands of a clock when the clock shows the time 3:36?

66. The Miniature Golf Problem. When a miniature golf ball is putted to hit a wall, it bounces off at an angle congruent to the angle at which it hit the wall, as shown in the figure below.

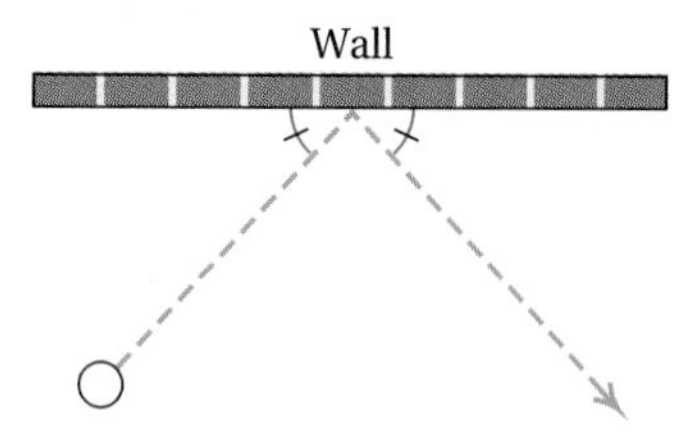

How would you hit the golf ball at the miniature golf course so that it would go in the hole as shown in the figure below? (*Hint:* You may wish to trace copies of this picture and use a protractor and a guess and check procedure.)

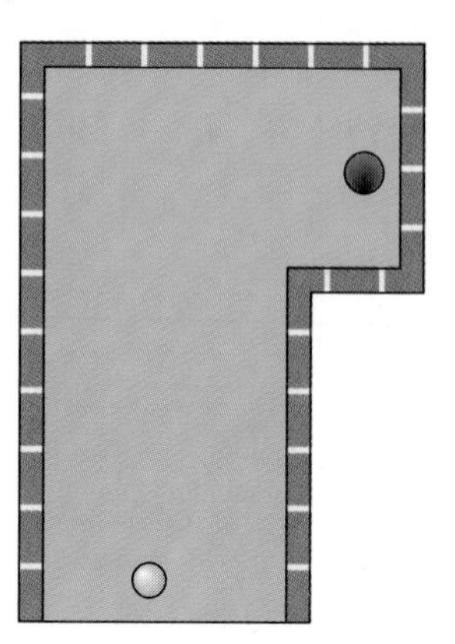

67. The Computer Network Problem. A creative office designer had eight programmer offices evenly spaced around the outside of a circular common work area. She wanted to run cables under the floor so that each pair of offices was connected by a cable. How many cables will she need to run? (*Hint:* You may wish to draw a geometric figure to model this problem.)

D. Communicating and Connecting Ideas

68. Work with a group of your classmates to make a set of 3-by-5-inch cards, each with a description of one of the figures or relationships described in this chapter. Make another colored set of cards, each with the name of one of the figures or relationships described on the other cards.

a. Try a group activity where people pin the figure or relationship cards on their backs and determine what their figure or relationship is by asking each other yes or no questions.

b. Devise and play a matching game with both sets of cards.

69. Geometry Exploration Software (GES) is described in Appendix B. It will be used to discover generalizations in geometry in Section 10.2. For now, look at Appendix B and complete the activities in parts B.1, B.2, and B.3 to learn how to use the software to create and measure basic figures.

70. Historical Pathways. When you cut a strip of paper 11 inches long and $1\frac{1}{2}$ inches wide and tape the ends together to make a loop, you notice that the loop has two sides and two edges. When you draw a line completely around the loop on the inside or on the outside and cut along the line, the loop separates into two distinct parts. The mid-nineteenth century German mathematician and astronomer Augustus Möbius discovered a way of producing a "loop" with only one side and only one edge by giving one end of the originally described strip a half twist before taping, as shown below. Such a loop is called a "Möbius strip."

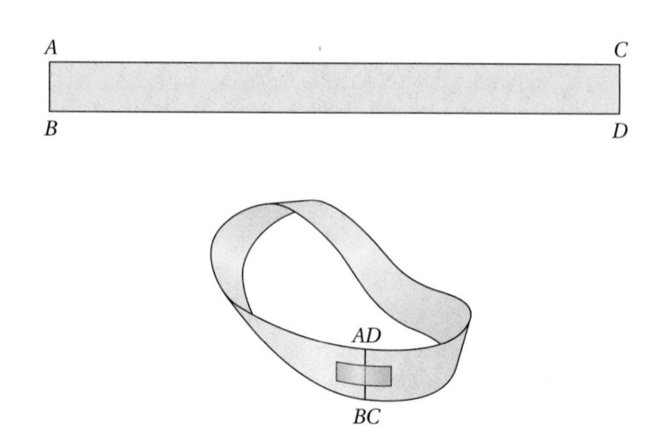

Make a Möbius strip and do the following experiments.

a. Color an edge of the band until you return to your starting point to show that the band has only one edge.

b. Show that the band has just one side by drawing a line down the center of the band, continuing until you arrive at your starting point. Then cut along this line. What happens?

71. Making Connections. Connect geometry and art by using a ruler and compass to make and color an artistic geometric design. First try this simple one, and then create one of your own.

1. Draw a circle with a radius at least 2 inches long.
2. With your compass opened as wide as the radius, make six points around the circle.
3. Connect each of the six points to each of the other points and color the resulting design in an interesting way.

72. The spreadsheet shown has the first two terms of the Fibonacci sequence in cells A1 and A2. What formula can be used in cell A3 to generate the third term? How can you generate the first 15 terms with the spreadsheet?

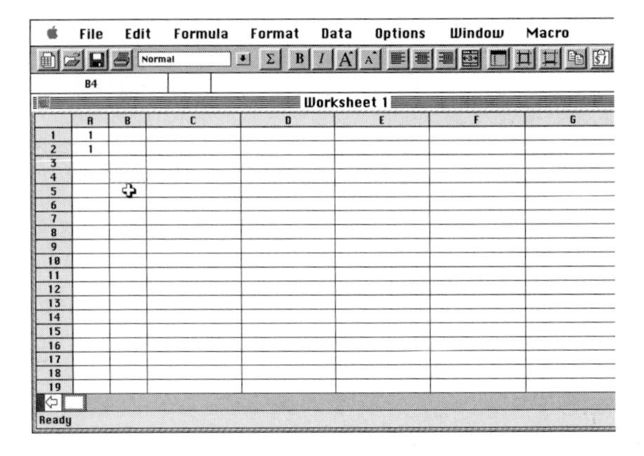

Section 10.2 SOLVING PROBLEMS IN GEOMETRY

- Point and Line Problems
- Segment, Angle, and Triangle Problems
- Quadrilateral, Polygon, and Circle Problems

In this section we help you use basic geometric ideas and relationships to solve and analyze real-world problems. As background, you need to look at parts of three-dimensional figures to visualize basic two-dimensional figures and relationships, classify them, and name them for ease in communication. Mini-Investigation 10.2 provides an opportunity for you to review your understanding of some basic geometric ideas introduced in Section 10.1.

Talk about meanings of terms you do not understand. See Appendix C for definitions.

Mini-Investigation 10.2 Communicating

Which letters in the figure can you use (along with any symbols needed) to give an example of each of the following terms?

a.	Point, line, plane	**b.**	Collinear points
c.	Coplanar points	**d.**	Intersecting lines
e.	Parallel lines	**f.**	Concurrent lines
g.	Perpendicular lines	**h.**	Skew lines
i.	Ray	**j.**	Segment
k.	Angle	**l.**	Triangle
m.	Quadrilateral	**n.**	Polygon
o.	Circle	**p.**	Chord
q.	Diameter	**r.**	Radius
s.	Parallel planes	**t.**	Perpendicular planes
u.	Intersecting planes		

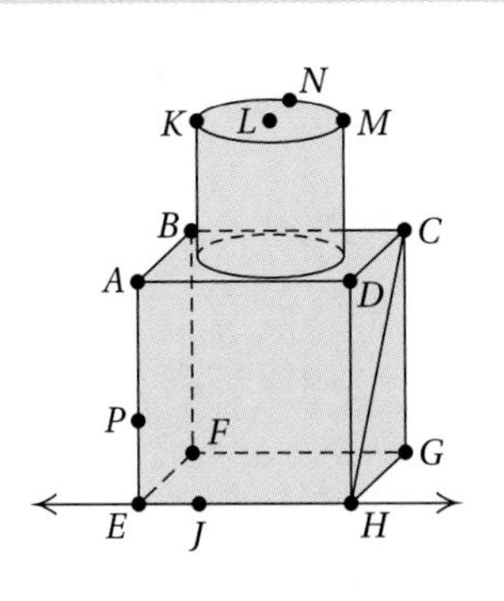

Point and Line Problems

Mini-Investigation 10.2 underscores the idea that points and lines are basic building blocks for other geometric ideas and relationships. A theory that uses points and lines, called *network (or graph) theory,* is very useful for solving some types of real-world problems. A familiar puzzle problem can be used to explain ideas associated with networks.

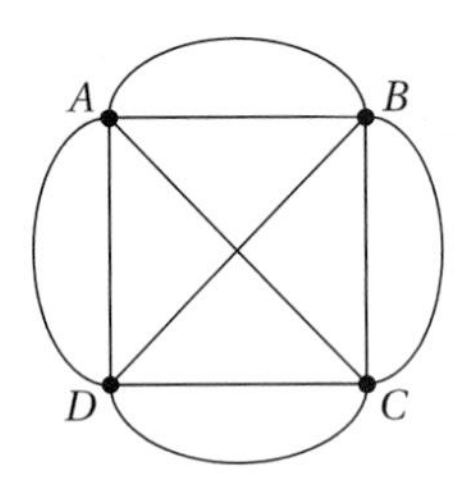

> Can you trace over a figure like the one shown without lifting your pencil from the paper and without going over any line twice?

This diagram is an example of a set of points *A, B, C,* and *D,* called **vertices,** and a set of segments or arcs, called **edges,** that join the vertices. The total configuration is called a **network.** If you can draw along the paths of a network without lifting your pencil from the paper so that each edge is traced only once, the network is a **traversable network.**

If the beginning and ending point of the traversable path is the same, the network is called **traversable type 1.** If the beginning and ending points do not coincide, the network is called **traversable type 2.** The number of edges meeting at a vertex influences whether a network is traversable, so some additional terminology is needed. When an odd number of edges meet at a vertex, as in the puzzle above, the vertex is called an **odd vertex.** When an even number of edges meet at the vertex, it is called an **even vertex.** The following theorem helps you decide about the traversability of a network.

Theorem: Network Traversability

1. If a network has all even vertices, it is traversable type 1, and the traversable path may begin at any vertex.
2. If a network has exactly two odd vertices, it is traversable type 2, and the traversable path must begin at one of the odd vertices and end at the other.
3. If a network has more than two odd vertices, it is not traversable.

The networks in Figure 10.15 illustrate the ideas in the above theorem.

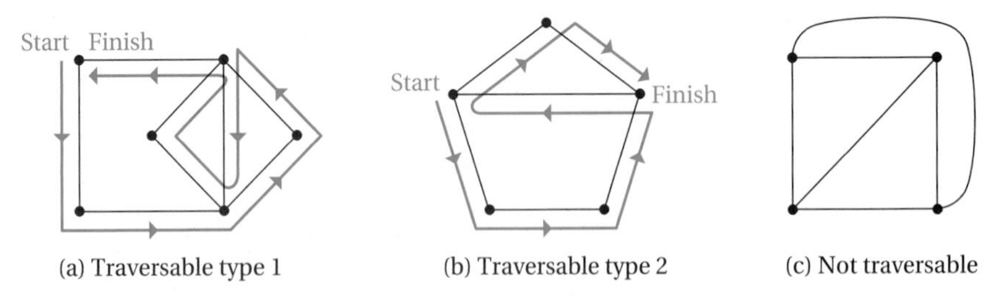

FIGURE 10.15 | Examples of traversable and nontraversable networks.

Example 10.6 shows how the Network Traversability theorem can help you solve a practical problem. The problem-solving strategies *draw a diagram* and *use logical reasoning* are useful in solving the problem.

Example 10.6 Problem Solving: The Highway Inspector's Route

A highway inspector lives at city A, as shown on the following map. She wants to inspect all the highways connecting cities A through F. Can she travel, starting at home, over each road once and only once and return to her home city A in making the inspection? If so, show how such a trip could be made.

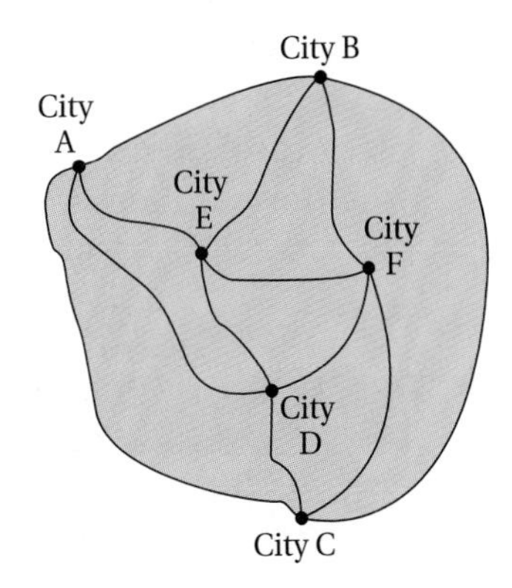

Solution

All the vertices are even, so the inspector knows that the network is traversable type 1. She can start at A, travel as shown, and return home without going over any road twice.

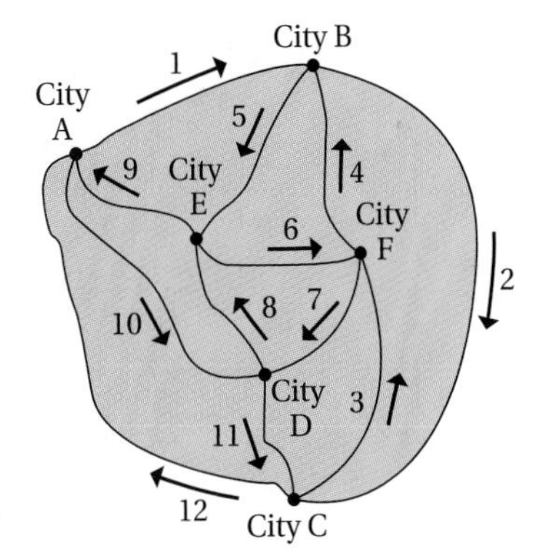

Your Turn

Practice: A salesperson lives in city P. Can he travel to each city shown on the following map and return to a friend's house in another city by going over each road once and only once? If so, where would his friend have to live?

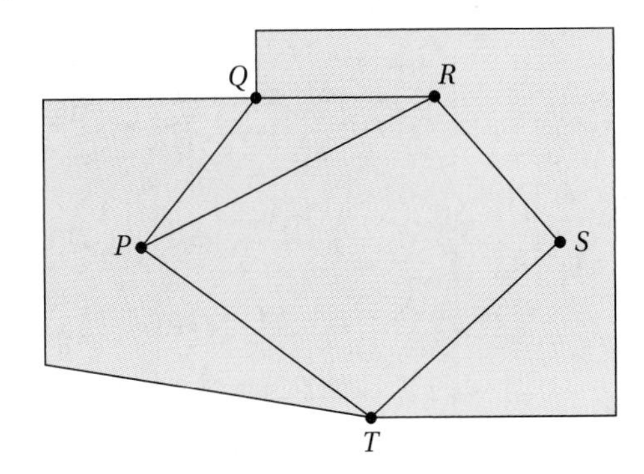

Reflect: What road could you add to the map in the practice problem so that the solution would be impossible? Explain. ■

As the preceding situations indicate, network theory involving points, lines, and their relationships is a useful tool for solving certain types of real-world problems. For another interesting problem of this type, see the Königsberg Bridge Problem in Exercise 48 at the end of this section.

Segment, Angle, and Triangle Problems

Mini-Investigation 10.3 lets you review ideas about segments, angles, and triangles to provide background for solving the problems that follow.

Talk about meanings of the terms that you don't understand. See Appendix C for definitions.

Mini-Investigation 10.3 Communicating

Which letters in the figure can you use (along with any symbols needed) to give an example of each of the following terms?

a. Obtuse angle
b. Acute angle
c. Right angle
d. Midpoint of a segment
e. Perpendicular bisector
f. Bisector of an angle
g. Vertical angles
h. Complementary angles
i. Supplementary angles
j. Linear pair of angles
k. Equilateral triangle
l. Isosceles, not equilateral, triangle
m. Scalene triangle
n. Acute triangle
o. Right triangle
p. Obtuse triangle
q. Altitude of a triangle
r. Median of a triangle

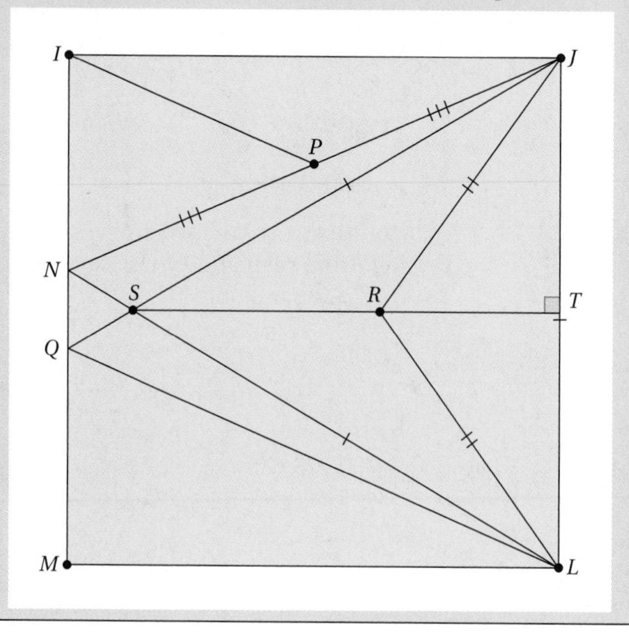

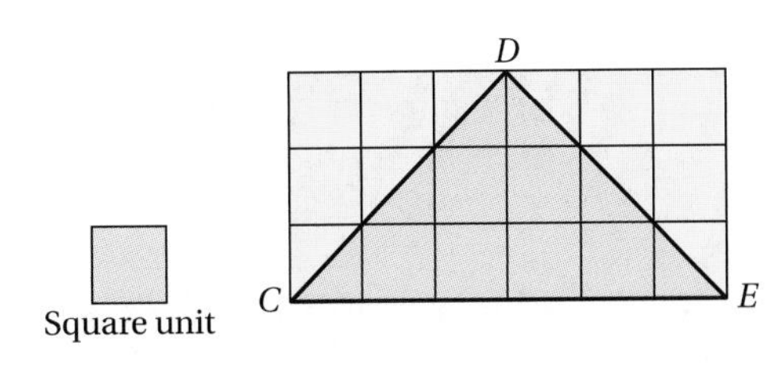

FIGURE 10.16 | One way to visualize the area of a triangle.

In Mini-Investigation 10.3, we thought about lengths of segments and sizes of angles to measure polygons. Another useful way to measure polygons is to use the idea of area.

A polygon possesses a property of the extent of the region enclosed or the number of square units needed to cover its inside. To describe this property, we assign a real number, called the **area measure** (A), to each polygon. In Figure 10.16, 9 square units are needed to cover the region inside the triangle, so the area measure of $\triangle CDE$ is 9. We write $A(\triangle CDE) = 9$ square units. Two polygons are congruent only if they have the same area and shape. Note that two differently shaped polygons can have the same area and not be congruent. We discuss congruent polygons more thoroughly in Section 10.4.

Also, we will carefully develop formulas for finding the area of some common polygons in Chapter 12. However, we review here some basic area formulas that you may recall from your earlier study of mathematics, and which may be used later in this chapter.

$$\text{area of a rectangle} = \text{length} \times \text{height}$$
$$\text{area of a parallelogram} = \text{base} \times \text{height}$$
$$\text{area of a triangle} = \frac{1}{2}\,\text{base} \times \text{height}$$

Estimation Application

Comparing Sizes of States

Which state—Illinois or Arkansas—has the larger area? How much larger?

a. If each one-fourth-inch square on the grid represents an area of 2000 square miles, estimate the areas of the two states to answer the questions posed.

b. Check your estimates by consulting an almanac or encyclopedia.

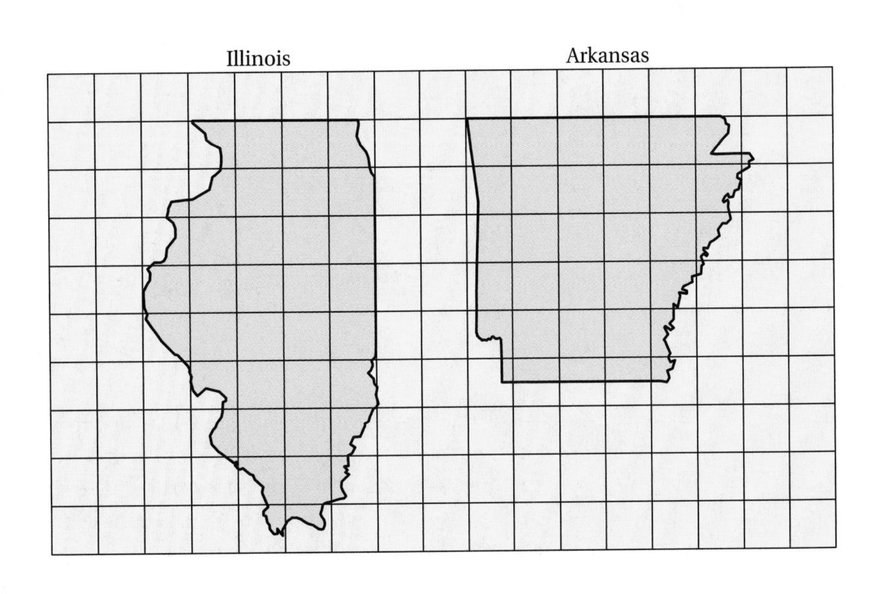

The computer can be a valuable tool in solving problems in geometry. In particular, computer software called **geometry exploration software** (GES) is quite useful. The Spotlight on Technology feature introduces you to this software.

Spotlight On Technology: Geometry Exploration Software (GES)

Judah Schwartz and Michal Yerushalmy developed the *Geometric Supposer* series as the first widely distributed geometry exploration software. The series opened the door to *what if*...? geometric explorations just as the electronic spreadsheet did for numerical explorations. The following GES printout shows that this software can automatically give the *length* of a segment, the *degree measure* of an angle, and the *area* of a polygon.

Length(Segment n) = 1.60 inches
Length(Segment p) = 1.60 inches
Length(Segment q) = 1.39 inches
Angle(FED) = 64 degrees
Angle(DEF) = 64 degrees
Angle(EDF) = 51 degrees
Area(Polygon 1) = 1.00 square inches

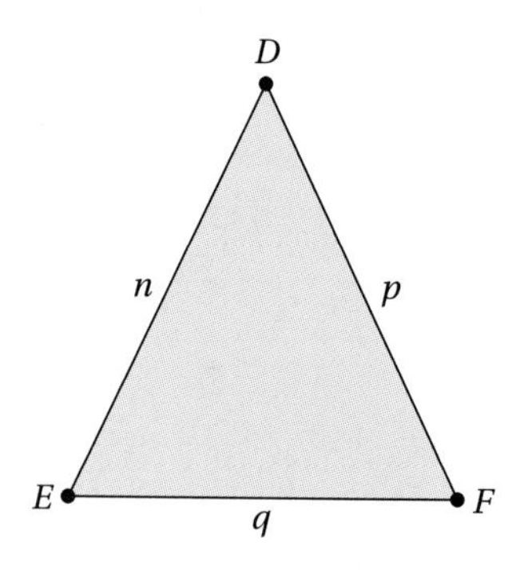

First-generation GES allowed accurate repetition of the construction of a figure and measurements associated with the figure. This initial exploration tool paved the way for dynamic geometry exploration software such as the *Geometer's SketchPad* and *Cabri.* With these software tools, you not only can generate and repeat geometric constructions, but you can also distort the resulting figures. This capability allows you to identify elements of the construction that change when distorted and those that don't. The following distortions of a construction of the three medians of a triangle illustrate this feature. What doesn't change?

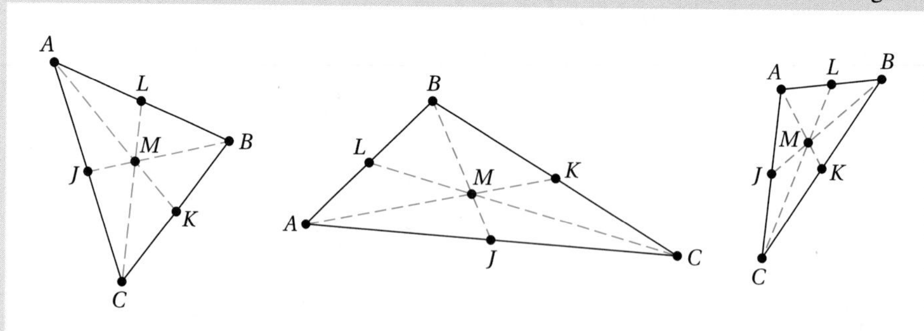

Exploring geometric properties through dynamic interaction with software tools actively engages you in generating conjectures and provides motivation for further investigation. The software's speed and efficiency allow you to examine many cases and consider extreme conditions in your search for relationships among geometric objects.

Many interesting and unexpected relationships in triangles have been discovered over the centuries, some of which we state here. Exercises 16, 17, and 18 at the end of this section ask you to use GES to illustrate those relationships.

Theorem: Concurrency Relationships in Triangles

- The three medians are concurrent in a point called the **centroid,** *G*, which is the balance point, or center of gravity, of a model of the triangle, and which is two-thirds the distance from each vertex to the opposite side.
- The three altitudes are concurrent in a point called the **orthocenter,** *H*.
- The three perpendicular bisectors of the sides are concurrent at a point called the **circumcenter,** *O*, which is the center of a circle containing the vertices. We say that this circle *circumscribes the* triangle and that the triangle is *inscribed in* the circle.
- The three angle bisectors are concurrent in a point called the **incenter,** *I*, which is the center of a circle that is tangent to each side of the triangle. We say that this circle is *inscribed in* the triangle.

Thus several special points are associated with every triangle. Because a triangle can be randomly chosen and three randomly chosen lines couldn't be expected to be concurrent, the preceding relationships are somewhat surprising. In Example 10.7 the problem can be solved by using one of the relationships about special points in a triangle. The problem-solving strategies *use logical reasoning* and *draw a picture* are helpful in solving the problem.

Example 10.7

Problem Solving: Football Stadium Location

Planners want to build a new football stadium at a location that is the same distance from each of three cities—Bay City, Hillview, and Prairie View. Where should the stadium be located?

Bay City

Stadium
?

Prairie View

Hillview

Working toward a Solution

The center of a circle that contains the three points representing the three cities will be the same distance from all three cities. Draw a triangle with the three cities as vertices. Using what we know about the special points associated with a triangle, the circumcenter of this triangle is the center of a circle that contains the vertices of the triangle. Use GES to draw the perpendicular bisector of each of the sides to locate the circumcenter. This point, where the perpendicular bisectors of the three sides of the triangle intersect, is the point where the stadium will be the same distance from each of the cities.

Your Turn

Practice:

a. Complete the solution to the example problem. Justify your solution.

b. Solve the following warehouse location problem.

A company wants to build a warehouse on a plot of land bordered by roads connecting three cities. Its goal is to build the warehouse so that it lies equidistant from the roads. Where should the warehouse be located? Justify your decision.

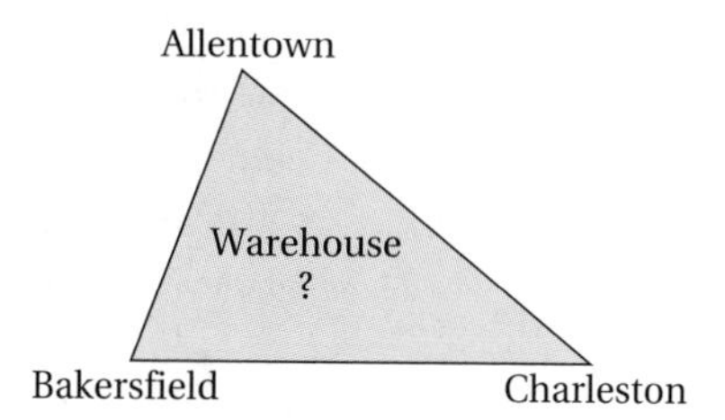

Reflect: How are the football stadium problem and the warehouse problem alike? How are they different? ■

The following relationship exists between the centroid, G, the orthocenter, H, and the circumcenter, O, as depicted in Figure 10.17. It is special because it is generally unexpected.

The line containing the centroid, orthocenter, and circumcenter is called the **Euler line** in honor of the Swiss mathematician Leonard Euler who discovered this relationship.

Theorem: The Euler Line

In every triangle, the centroid, the orthocenter, and the circumcenter are collinear.

In Example 10.8 we use GES to help solve another problem involving triangles, the lengths of segments, and the measures of angles. Construction and mea-

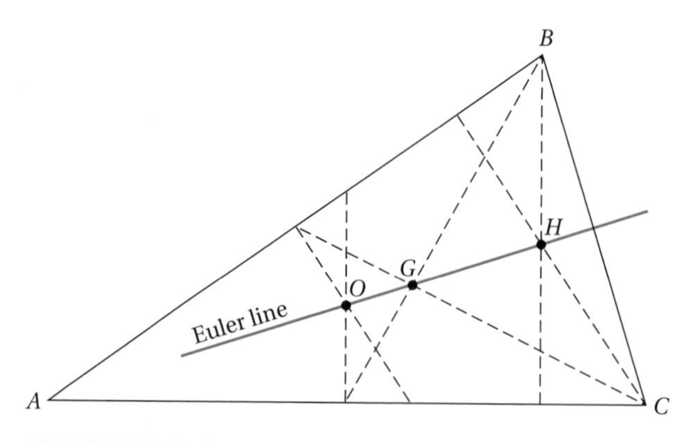

FIGURE 10.17 | The Euler line.

surement may be used to solve the problem if GES isn't available. The problem-solving strategies *use a specific case*, *draw a diagram*, and *guess, check, revise* are helpful in solving the problem.

Example 10.8

Problem Solving: The Three-City Water Supply Problem

A water pumping station was being cooperatively built by three cities that lie on the vertices of a scalene triangle. An engineer has been hired to recommend where the pumping station should be located so that the sum of the distances from the three cities, $a + b + c$, is the smallest in order to provide the most efficient piping system. The engineer is to present and justify a recommendation by specifying the location of the station in terms of angles.

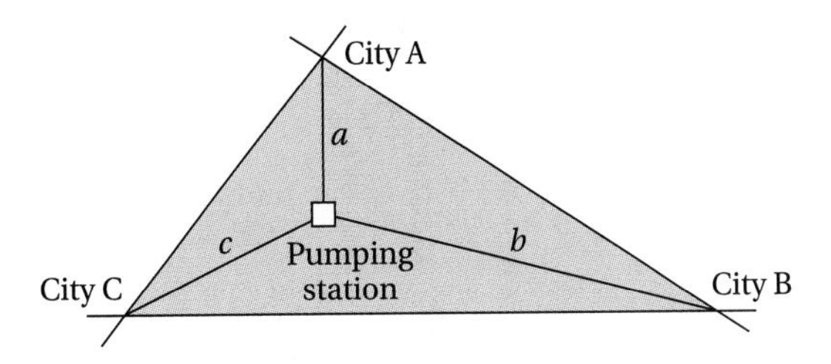

Working toward a Solution

Understand the problem	*What does the situation involve?*	Building a pumping station within a triangular area.
	What has to be determined?	The best location at which to build the station.
	What are the key data and conditions?	The sum of the distances from cities A, B, and C, $a + b + c$, must be as small as possible.
	What are some assumptions?	None.
Develop a plan	*What strategies might be useful?*	Use specific cases; draw a diagram; guess, check, revise.
	Are there any subproblems?	Setting up the problem with geometry exploration software.
	Should the answer be estimated or calculated?	Calculated.
	What method of calculation should be used?	Use the computer to find lengths and angle measures and do the calculations.
Implement the plan	*How should the strategies be used?*	Set the problem up on GES and choose a specific location for the pumping station, having the computer calculate distance sums and angles. Then move the pumping station to another location, calculating again. Repeat this guess, check, revise procedure until the sum $a + b + c$ is the smallest.

	What is the answer?	(Follow this procedure and obtain the answer on your computer, giving the angles between the lines from the pumping station to the cities.)
Look back	*Is the interpretation correct?*	Check: The GES setup fits the data.
	Is the calculation correct?	Check: The computer calculated correctly.
	Is the answer reasonable?	Yes. After seeing the distance sum at other locations, the solution seems reasonable.
	Is there another way to solve the problem?	Draw diagrams and measure with a ruler or make a model.

■ *Your Turn*

Practice: Use GES, if available, or construction and measurement to solve the following problem.

> An amusement park is to be built on a parcel of land having the shape of an equilateral triangle bordered by three highways. The owners want to build the park so that visitors would travel the least possible distance when coming from and going to the highways. Suppose that you were the company's engineer. Where would you recommend that the park be built? Support your choice.

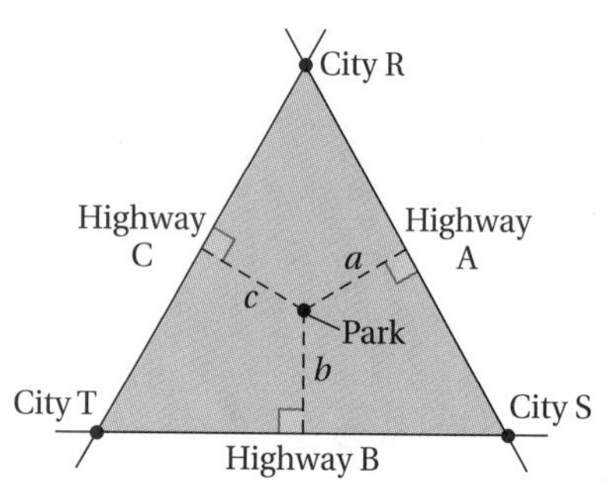

Reflect: How are the example and practice problems alike? How are they different? What geometric generalizations can you make, based on their solutions? ■

Quadrilateral, Polygon, and Circle Problems

In Example 10.9 we solve problems involving quadrilaterals and other polygons. The problem-solving strategies *make a model* and *use reasoning* are helpful in solving such problems.

Example 10.9 Problem Solving: Chinese Floor Tile

A wealthy Chinese businessman wanted to use convex (no dents) floor tiles to tile the floor of each of the seven great rooms of his mansion. He further stipulated that each of these basic tiles be triangles or quadrilaterals and constructable by using all seven pieces of the famous Chinese **tangram puzzle.** For example, the square shown, made of all seven tangram pieces, would be the basic tile shape for one of the rooms. There are six other possibilities. How can the tangram pieces be used to make these tiles?

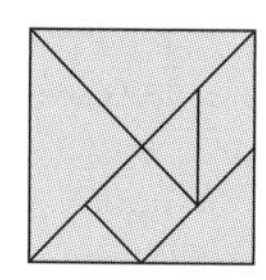
Tangram puzzle

Working Toward a Solution

Here is a start for each of the six remaining tiles.

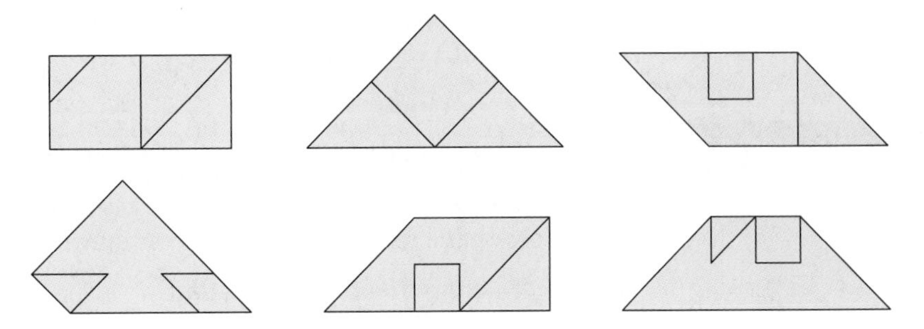

Your Turn

Practice:

a. Complete the solution to the example problem.

b. Solve the following Chinese tabletop problem.

The same wealthy Chinese businessman wants tables with convex tops, each made from all seven pieces of the Chinese tangram puzzle. He further stipulated that there were to be no triangular or quadrilateral tops. He had heard that the six possible shapes looked like those shown.

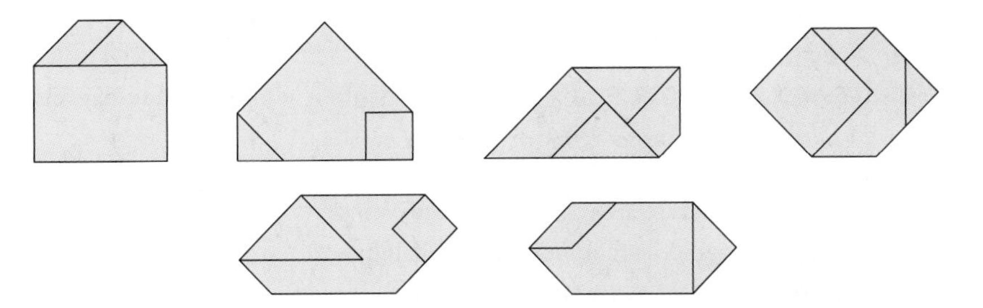

How can you use the tangram pieces to finish making the tabletops?

Reflect: How many different convex polygons do you think can be made if each contains all seven pieces of the tangram puzzle? Explain. ■

The *Principles and Standards for School Mathematics* state that children in grades 3–5 should "... draw and construct shapes, compare and discuss their attributes, classify them, and develop and consider definitions on the basis of the shapes' properties ..." (p. 165). Students in primary grades can do problems involving a few tangram pieces. In grades 3–5, the tangram puzzle provides a vehicle for constructing and analyzing shapes.

When solving problems related to tiling floors, designing wallpaper, carving wood, and other uses of geometric designs, you will often need to produce regular polygons accurately. You probably are already familiar with methods for constructing an equilateral triangle, a square, and a hexagon.

Although there are some relatively complicated ways to construct a regular pentagon, one simple way to produce one is to use a protractor and draw five radii of a circle that form angles of 72°. Also, GES or other computer software may be used to produce a regular pentagon. The problem in Example 10.10 requires formation of a regular pentagon.

Example 10.10 Problem Solving: Wood Carving Designs

A woodcarver wants to carve the design shown on the lid of a special jewelry box. How can he make an accurate larger drawing of the design?

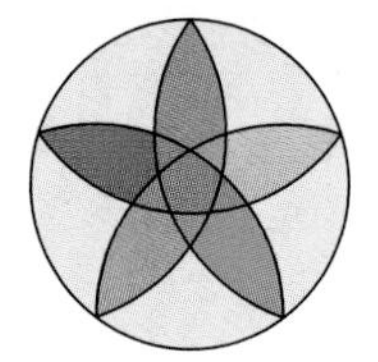

Working Toward a Solution

Draw a circle and measure angles to select five equally spaced points on it to form a regular pentagon. Use each point as a center and draw an arc passing through two other points.

Your Turn

Practice: Complete the solution to the problem.

Reflect: Do you think that it would be easier or more difficult to make a similar design with six petals instead of five? Explain. ■

Problems and Exercises for Section 10.2

A. Reinforcing Concepts and Practicing Skills

Decide if the statements in Exercises 1–10 are true, and explain why or why not.

1. Three collinear points are always also coplanar points.
2. Two concurrent lines are sometimes also parallel lines.
3. Two concurrent lines can never be perpendicular.
4. Two skew lines are also sometimes perpendicular lines.
5. According to definitions, a triangle never really contains a complete angle.
6. Two radii of a circle always make a diameter of a circle.
7. A right triangle cannot be an obtuse triangle.
8. An equiangular triangle doesn't have to be an equilateral triangle.
9. A perpendicular bisector, an angle bisector, and a midpoint of a segment all have a property in common.
10. Two of the angles of a triangle could be supplementary angles.
11. Use what you know about traversable networks to show that a hiker can walk all 21 miles of the trails shown on the map in the next column without walking over any trail twice. Where would the hiker start and where would the hiker end?
12. Show how to change the trails on the map in Exercise 11 so that you could start at the ranger station, traverse all the trails, and end at the pond.

Figure for Exercise 11

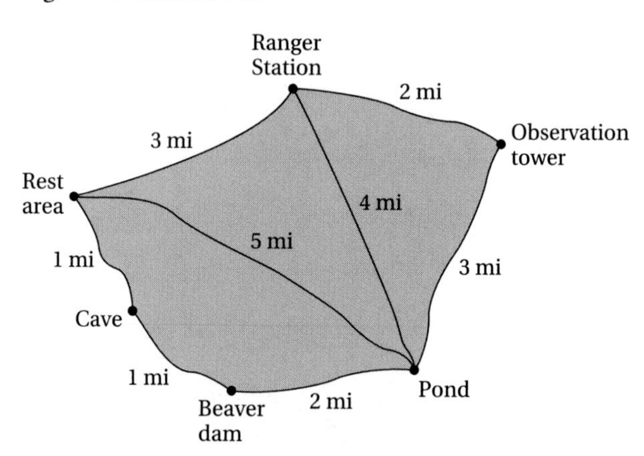

13. Draw a 16-mile trail map with five locations where a hiker could start at any location, walk all 16 miles, and return to the same location without walking over any trail twice.
14. Refer to the figure at the bottom of the next page. Can a salesman start at city A, traverse each road only once, and return to city A? How do you know?
15. Can you start in room A of the house pictured at the bottom of the next page, go through each door only once, and return to another room? How do you know?
16. Use GES and try several examples to illustrate that the three altitudes of a triangle are concurrent.

17. Use GES and try several examples to illustrate that the three perpendicular bisectors of the sides of a triangle are concurrent.

18. Use GES and try several examples to illustrate that the three angle bisectors of the angles of a triangle are concurrent.

19. What can you conclude about the centroid, orthocenter, and circumcenter of any triangle?

20. What can you conclude about the centroid, orthocenter, and circumcenter of an equilateral triangle?

21. By measuring the lengths only in the triangle in Figure 10.17, describe what might be the relationship between segments $\overline{OG}$ and $\overline{GH}$. How could GES help you test your idea?

22. The flat barge shown in the next column is 425 feet long. How long is the cruise ship? How do you know?

23. *AB* is 46 yards and *BC* is 247 yards. How far is it from *A* to *C* across the swamp? How do you know?

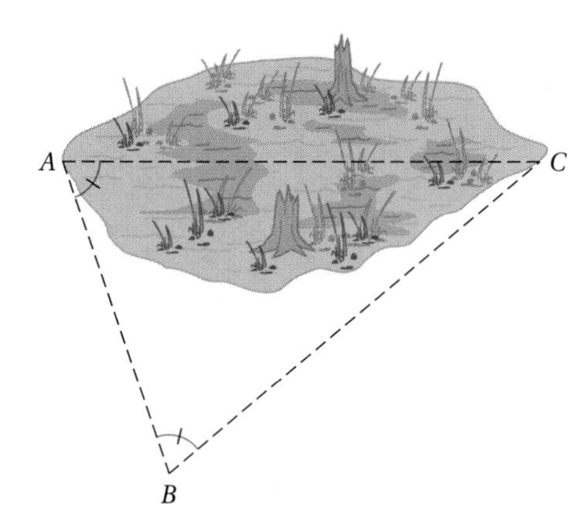

Figure for Exercise 14

Figure for Exercise 15

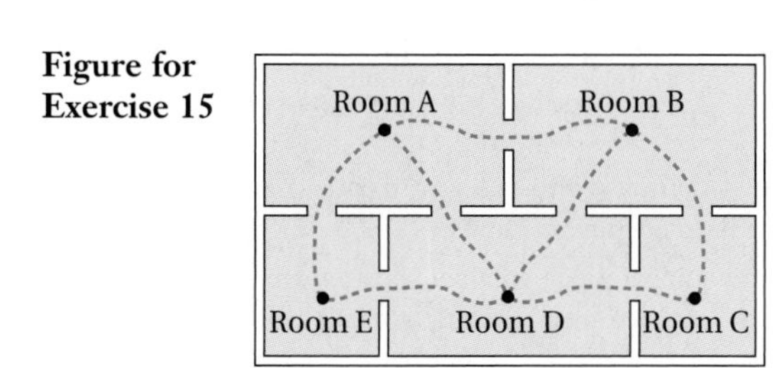

Figure for Exercise 22

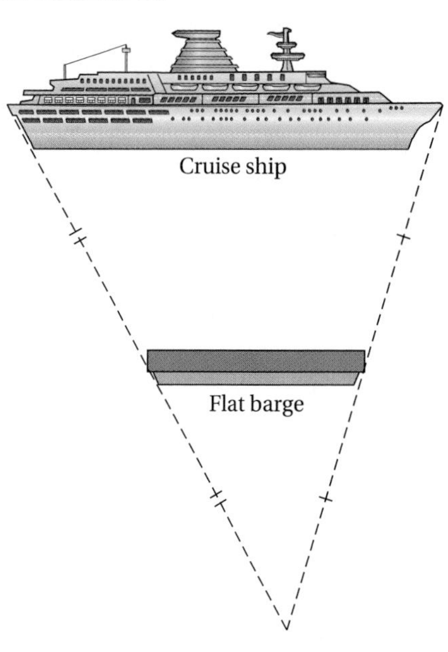

For Exercises 24–28, make a square like the following and cut along the dashed lines into five pieces.

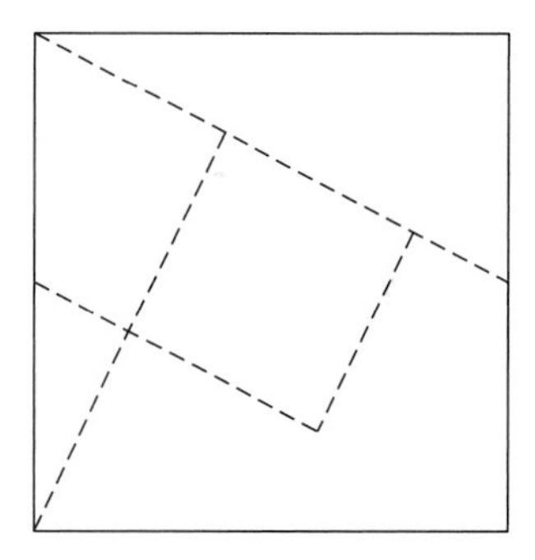

24. Arrange the pieces to form a rectangle. Here is a hint to get you started.

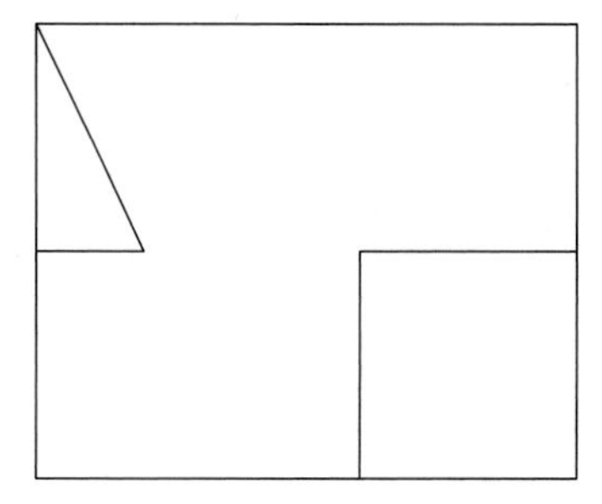

25. Arrange the pieces to form a triangle. Here is a hint to get you started.

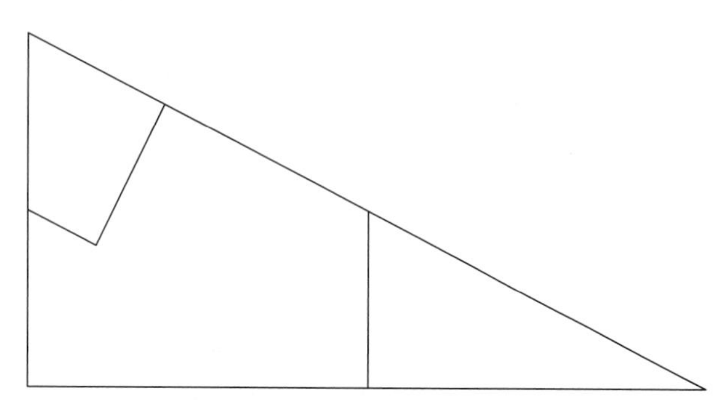

26. Rearrange the pieces to form a parallelogram. Here is a hint to get you started.

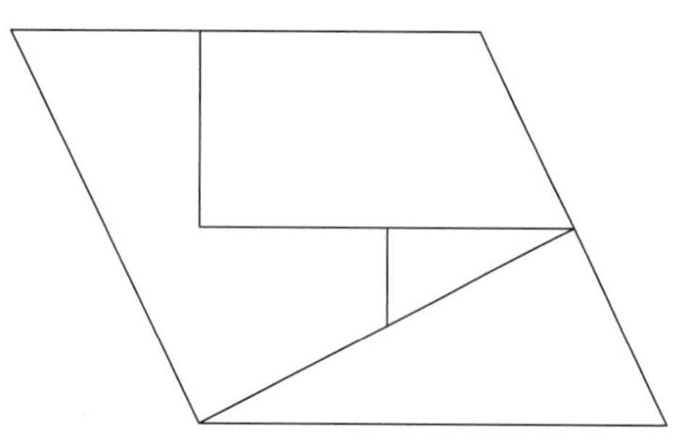

27. Rearrange the pieces to form a general quadrilateral. Here is a hint to get you started.

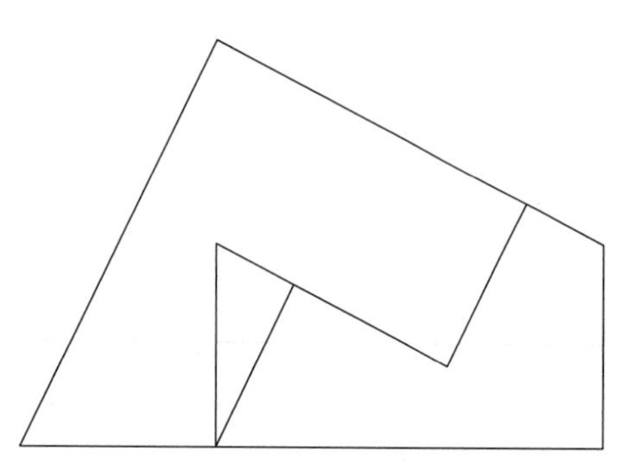

28. Rearrange the pieces to form a plus sign. Here is a hint to get you started. (See figure at top of the next column.)
29. Draw a regular pentagon inscribed in a circle and make a geometric design of your choice.
30. Draw a regular hexagon inscribed in a circle and make a geometric design of your choice.

B. Deepening Understanding

31. Use a ruler and compass (see Appendix D) to construct the following.
    - **a.** A segment congruent to another segment
    - **b.** An angle congruent to another angle

Figure for Problem 28

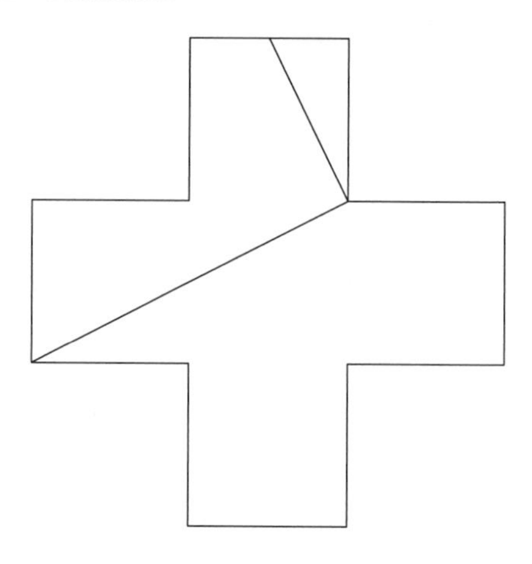

- **c.** The midpoint of a segment
- **d.** The bisector of an angle
- **e.** The perpendicular bisector of a segment
- **f.** A line perpendicular to a line from a point not on the line

32. Use GES and repeat parts (a)–(f) of Exercise 31.
33. Draw and label the following.
    - **a.** Two complementary angles with degree measures in the ratio of 2:1
    - **b.** An angle that is supplementary to an angle one-third its measure in degrees
34. Four lines can be drawn to intersect in only one point. If possible, show how four lines can be drawn to intersect in only two, three, four, five, and six points.
35. How many exterior angles does a triangle have? Draw a triangle and explain your reasoning.
36. Draw six lines that enclose the greatest number of regions possible.
37. Can you draw three different paths that start at point A on this network, traverse each edge only once, and return to point A? Show the paths if possible.

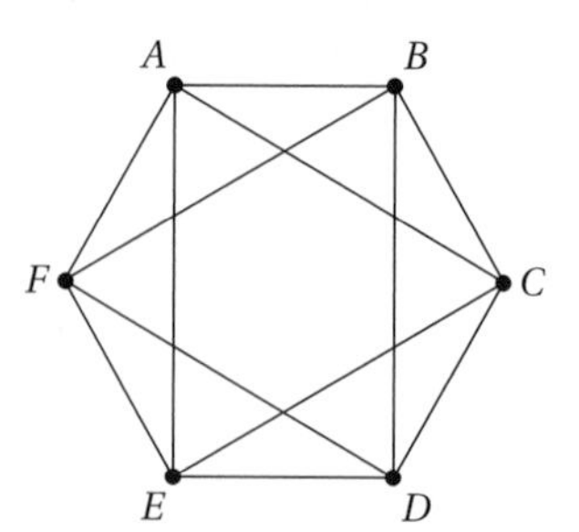

38. Use a compass and ruler to draw an interesting design that uses points and lines.
39. If the square piece in a set of tangram pieces is considered 1 square unit of area, what is the area measure of

the following pieces of the tangram? Explain your answers.

a. small triangle
b. medium triangle
c. large triangle
d. parallelogram

40. Using the square piece in a set of tangram pieces as 1 square unit of area, make a polygon with tangram pieces that has an area measure of the following:

a. 3.5 square units
b. 5 square units
c. 5.5 square units
d. 7 square units
e. 8 square units

C. Reasoning and Problem Solving

41. How many lines are determined by the array of points shown?

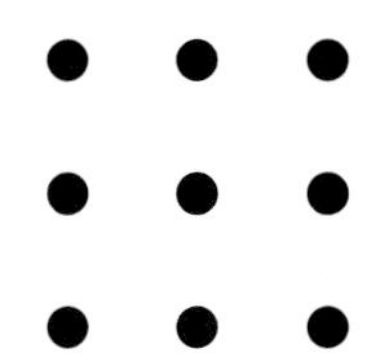

(*Hint:* You may first want to determine the number of 3-point lines and then use a separate sketch to determine the number of 2-point lines.)

42. On a geoboard with three nails on a side, find

a. two differently shaped right triangles.
b. three differently shaped isosceles triangles.
c. a scalene triangle.

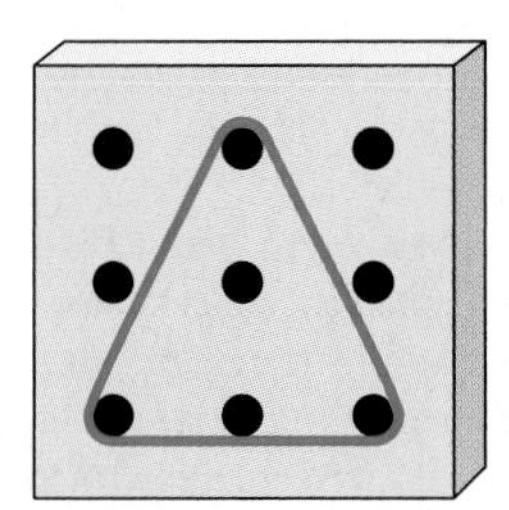

43. a. On a geoboard with five nails per side, determine the area of the *smallest square* you can create.

b. What is the area of the *largest square* you can create?

c. What squares can you create on this geoboard that have areas between the two extremes in parts (a) and (b)?

d. Can a square with an area of 6 units be constructed on this geoboard? Explain.

44. A segment joining the midpoints of two sides of a triangle is parallel to the third side and equal to $\frac{1}{2}$ of it. How does the length of the segments connecting the trisection points of two sides of a triangle compare to the length of the third side of the triangle? Use GES, tracing paper, or measurement to help you answer this question.

45. The Stained Glass Window Problem. A circular stained glass window is to be made with a center insert that has opposite sides parallel and all sides equal. The plan for the window shows the insert inside a dotted square, formed by four squares.

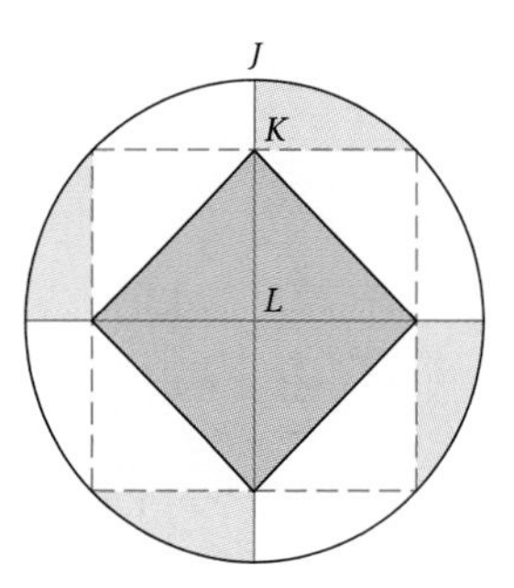

Lengths *JK* and *JL* are approximately 2 feet and 7 feet, respectively. As the window maker was trying to decide how to estimate accurately the length of a side of the center insert, her friend claimed that she had a close estimate of the length and hardly had to calculate at all. Explain how the friend might have easily estimated the length accurately.

46. Square 1 is a square of side length 10. Its midpoints are connected to create square 2. Square 3 is created from the midpoints of the sides of square 2. This pattern is continued. Use this information to complete three rows of a table of areas of successive squares. Look for a pattern and predict the areas for squares 4, 5, and 6.

47. The Ancient Wheel Problem. A part of an old wheel, as shown, was found in an archeological dig. A museum director would like to determine how large the complete wheel was. Decide how you could help the director. Can you find more than one way to do it? (*Hint:* The perpendicular bisector of a chord of a circle goes through the center of the circle.)

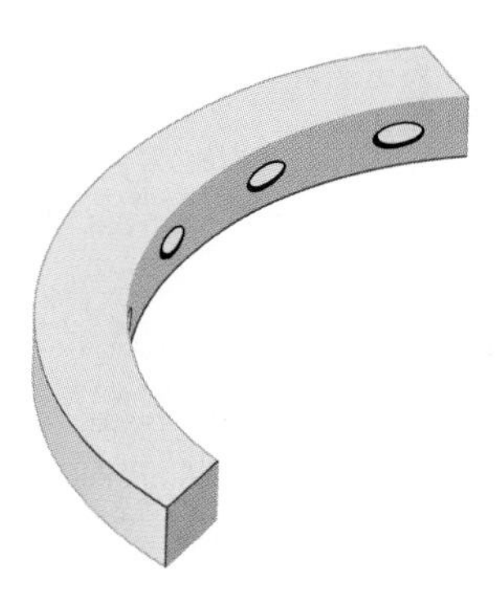

48. **The Königsberg Bridge Problem.** The city of Königsberg, formerly in East Prussia (now Kaliningrad in Russia) is located on the Pregel River. The regions of the city, partly on two islands, were connected by seven bridges, as shown. The townspeople were accustomed to asking visitors the following question: Is it possible to take a walk around town in such a way that, starting from home, I can return there after having crossed each bridge just once? Explore this question and give your answer.

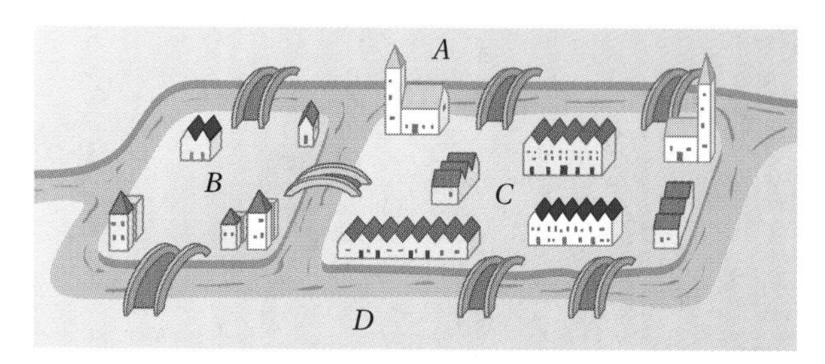

49. How can you make the following geometric figures with some or all of the tangram pieces? Sketch your solution.
    - **a.** An isosceles right triangle made from three pieces
    - **b.** A square made from four pieces
    - **c.** A square made from five pieces
    - **d.** A parallelogram made from the square in part (c) and two other pieces
    - **e.** A trapezoid made from the square in part (c) and two other pieces
    - **f.** A rectangle made from the square in part (c) and two other pieces
    - **g.** A pentagon made of four pieces

50. **The Strange Subdivision Problem.** An eccentric real estate developer always bought plots of land in the shape of scalene triangles and divided them into seven lots on which to build homes. His method was to divide each side of the plot into thirds and then connect each vertex with a division point, as shown. He offered to sell the entire plot for $999,999.

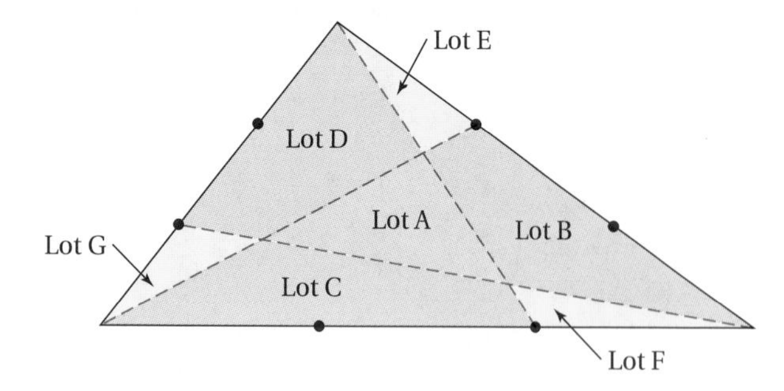

If the subdivided lots are priced in proportion to their areas, what would be a fair price for each lot? Predict first and then use GES to check your predictions and make generalizations.

51. Describe how you might solve the problem in Exercise 50 by
    - **a.** using a cardboard model of the triangle, cutting, and estimating.
    - **b.** using clear plastic graph paper with a small grid and counting.

D. Communicating and Connecting Ideas

52. The following concept card gives information to help you discover the defining characteristics of a geometric figure called a *triquad.* Work in a group and use the card to write a definition of a *triquad.* List some questions you would like to ask to ensure that your definition is correct.

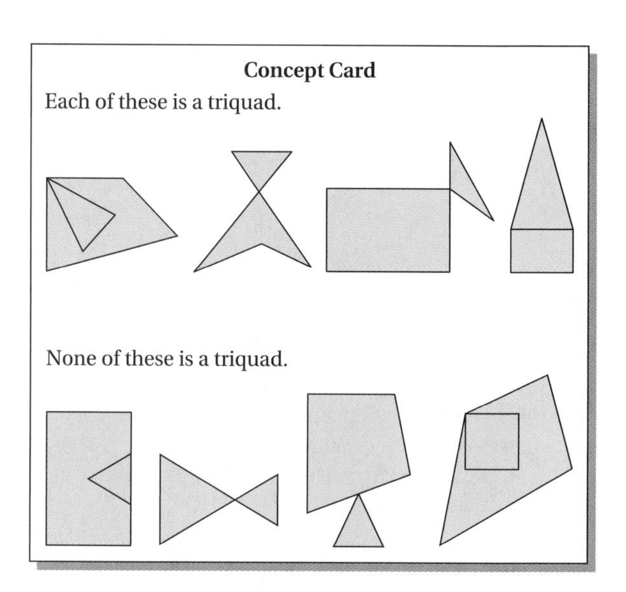

53. **Historical Pathways.** Sam Lloyd and H. E. Doudeney, two famous puzzle experts, have developed detailed legends about the tangram puzzle. However, no evidence exists that the puzzle actually was created by the ancient Chinese. According to Ronald Read (1965), a plausible theory is that in certain South Chinese dialects, the word *tang* means "Chinese." Sometime between 1847 and 1864, when the word *tangram* first appeared in Webster's dictionary, a Westerner may have combined *tang* with the familiar word *gram* to name the puzzle. Start with the three largest of the seven tangram triangle pieces, shown in part (a) of the following figure, and use the remaining four pieces to form the letter *C.* Then start with the three largest tangram triangle pieces, shown in part (b) of the figure, and use the remaining pieces to form the letter *T.* Sketch the results. This solution can remind you of *C*hinese *T*angrams. (See figure on the next page.)

Figure for Exercise 53

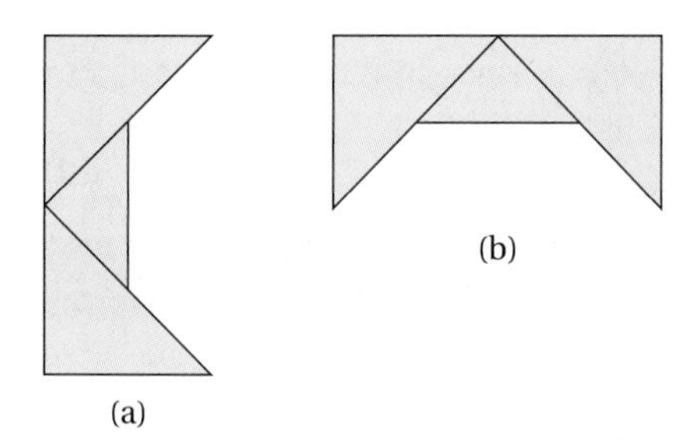

54. Making Connections. Connect geometry with the real world by making a list of geometric ideas and some objects that you see around you that suggest these ideas.

Section 10.3 | MORE ABOUT ANGLES

- Angles in Intersecting Lines
- Angles in Polygons
- Angles in Regular Polygons
- Angles in Circles

In this section we explore geometric relationships that occur among angles formed when various configurations of lines intersect. We also investigate relationships among angles associated with different types of polygons.

We will use the process of *inductive reasoning*, discussed in Chapter 1, to help us investigate these angle relationships. Inductive reasoning as used in geometry is a process in which specific examples of a geometric relationship or pattern are observed and used to form a generalization about the relationship or pattern. Recall that inductive reasoning is used to *discover* geometric generalizations but does not constitute a *proof*. The process of *deductive reasoning*, also described in Chapter 1, is used to prove that a discovered generalization is true. Sometimes a generalization thought to be true is later proved false by finding a *counterexample*.

Geometry Exploration Software (GES) or simply construction/drawing and then measuring will frequently be used to help in discovering generalizations.

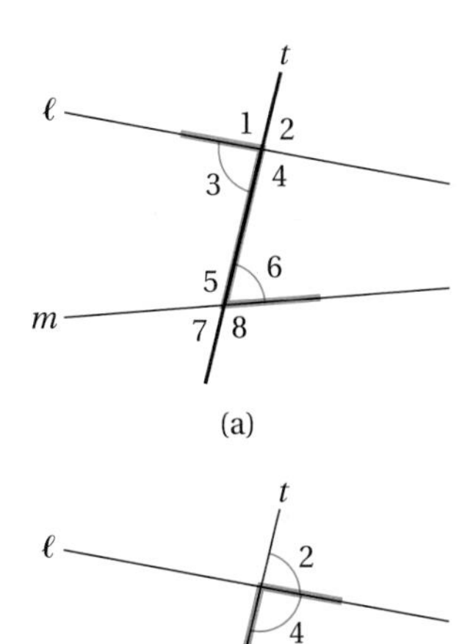

FIGURE 10.18 | Angles formed by intersecting lines.

Angles in Intersecting Lines

As you begin to look for generalizations in geometry, considering a basic idea such as a line, and the relationship between intersecting lines, is natural. In each part of Figure 10.18, line t, called a **transversal,** intersects two lines, ℓ and m, to form a total of eight angles.

Angles 3, 4, 5, and 6 are called *interior* angles, and angles 1, 2, 7, and 8 are called *exterior* angles. Some pairs of related angles have special names. For example, two nonadjacent interior angles on opposite sides of the transversal, such as angles 3 and 6 in Figure 10.18(a), form a **Z** and are called **alternate interior angles.** Two nonadjacent exterior angles on opposite sides of the transversal, such as angles 1 and 8 in Figure 10.18(a), form a double **V** and are called **alternate exterior angles.** Two nonadjacent angles on the same side of the transversal, one exterior and one interior, such as angles 4 and 8 in Figure 10.18(b), form an **F** and are called **corresponding angles.** Two interior angles on the same side of the transversal, such as angles 4 and 6 in Figure 10.18(b), form a **C** and are called

same-side interior angles. Angles 8 and 2 in Figure 10.18(b) are **same-side exterior angles.** When two lines, ℓ and m, are *parallel* and are intersected by the transversal, t, several pairs of angles are either congruent or supplementary or both, as indicated in the following generalizations.

Theorem: Angle Congruencies in Parallel Lines

Consider two lines cut by a transversal. If the lines are parallel,

a. the corresponding angles are congruent.

b. the alternate interior angles are congruent.

c. the alternate exterior angles are congruent

d. the same-side interior and same-side exterior angles are supplementary.

In Example 10.11 we apply the generalizations about the angles formed when parallel lines are cut by a transversal.

Example 10.11 Analyzing Angle Relationships

Lines r and s are parallel in the figure shown. Line t is the transversal.

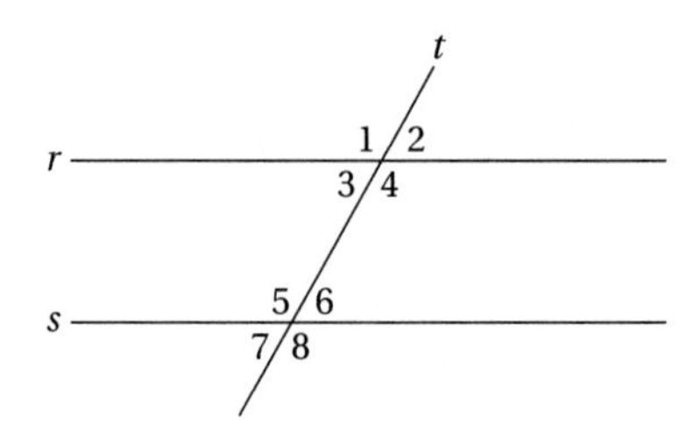

The measure of $\angle 3$ is 65°. Use $\angle 3$ to find the measure of each of the remaining angles. Support your answer.

Solution

$m\angle 1 = m\angle 4 = 115°$ ($\angle 1$ and $\angle 4$ are supplementary to $\angle 3$).
$m\angle 2 = 65°$ ($\angle 3$ and $\angle 2$ are vertical, congruent angles).
$m\angle 6 = 65°$ ($\angle 3$ and $\angle 6$ are congruent alternate interior angles).
$m\angle 7 = 65°$ ($\angle 3$ and $\angle 7$ are congruent corresponding angles).
$m\angle 5 = 115°$ ($\angle 3$ and $\angle 5$ are supplementary same-side interior angles).
$m\angle 8 = 115°$ ($\angle 3$ is supplementary to $\angle 1$, and $\angle 1$ and $\angle 8$ are congruent alternate exterior angles).

Your Turn

Practice: Suppose that $\angle 8$ in the figure has a measure of 108°. What is the measure of the other numbered angles?

Reflect: Explain why $\angle 3$ and $\angle 8$ are supplementary. ■

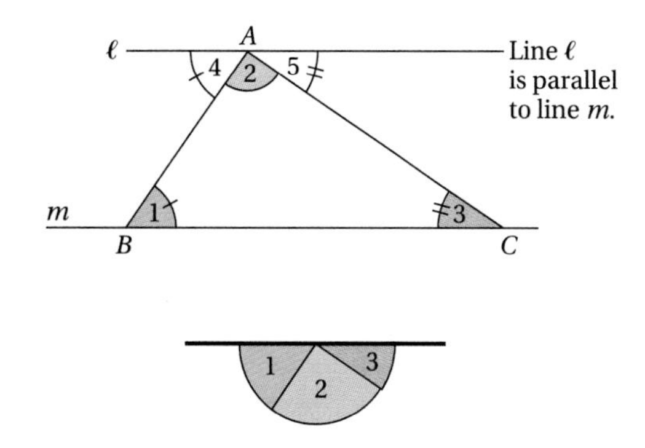

FIGURE 10.19 | Verifying a generalization about the sum of the angles of a triangle.

Angles in Polygons

The angles of polygons are the bases for many generalizations. In this subsection we look at the sum of angles of triangles and quadrilaterals.

The Sum of the Interior Angles of a Triangle. The angles formed by the sides of a polygon are called **interior angles of the polygon.** To discover or verify a generalization about the sum of the interior angles of a triangle, use a model, cut off the angles, and see if they fit along a straight line, as indicated in Figure 10.19.

Or you could use GES, measure angles A, B, and C, and calculate the sum of the measures. You could also use deductive reasoning and the idea that alternate interior angles formed by parallel lines and a transversal as shown in Figure 10.19 are congruent to conclude logically that $m\angle 1 + m\angle 2 + m\angle 3 = 180°$. These approaches help us verify the following theorem.

Theorem: Angle Sum for Triangles

The sum of the measures of the interior angles of a triangle is 180°.

The Sum of the Interior Angles of a Quadrilateral. Discovering that the sum of the measures of the interior angles of a triangle is 180° naturally leads to speculation about the sum of the measures of the interior angles of a quadrilateral. Example 10.12 illustrates ways to discover a generalization about the sum of the measures of the interior angles of a quadrilateral.

Example 10.12

Finding the Sum of Interior Angles of a Quadrilateral

Form a generalization about the sum of the measures of the interior angles of a convex quadrilateral.

Solution

Janelle's solution: When I cut off the corners of a quadrilateral and place them together, they fit exactly around a point, forming a complete circle. I think that the sum of the measures of the angles of a quadrilateral is 360°.

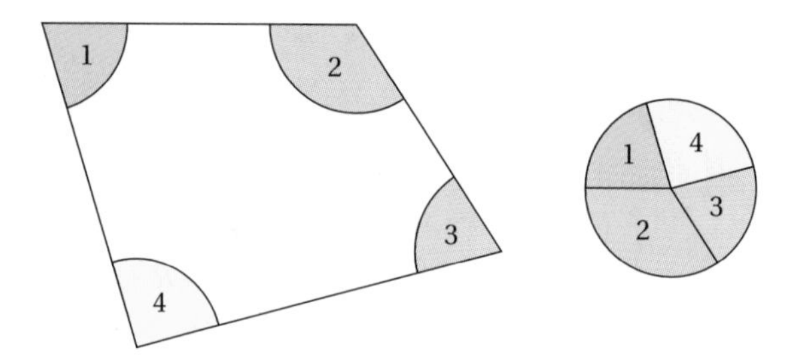

Paul's solution: I noticed that every quadrilateral can be separated into two triangles. The angles of the quadrilateral are made up of all the angles of the two triangles. So it looks like the sum of the measures of the angles of a quadrilateral is $2 \times 180°$, or $360°$.

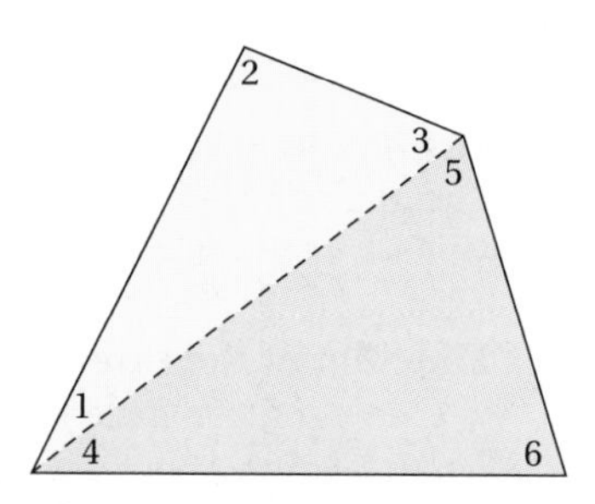

Sylvia's solution: I used GES both to create a quadrilateral and to calculate the sum of its four interior angles. I used the mouse to distort the quadrilateral and noticed that the sum of the angles didn't change. I think that for any quadrilateral, the interior angles sum to 360°.

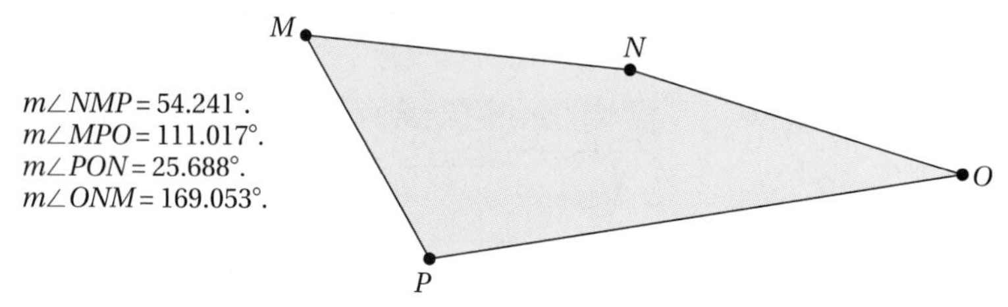

$m\angle ONM + m\angle PON + m\angle MPO + m\angle NMP = 360.000°$.

Your Turn

Practice: Use a cutout quadrilateral of a drastically different size than those shown and determine whether the students' hypotheses in the example problem hold.

Reflect: Which of the solutions in the example problem, if any, *prove* that the generalization made about the angles of triangles and quadrilaterals holds? Explain. ■

The ideas developed in Example 10.12 suggest the following theorem.

Theorem: Angle Sum for Quadrilaterals
The sum of the measures of the interior angles of a quadrilateral is 360°.

TABLE 10.6 | Determining the Interior Angles of Convex Polygons

Figure	Number of Sides	Number of Triangles That It Can Be Separated into from a Single Vertex	Sum of the Angle Measures
Triangle	3	1	1(180°), or 180°
Quadrilateral	4	2	2(180°), or 360°
Pentagon	5	3	3(180°), or 540°
Hexagon	6	4	4(180°), or 720°
n-gon	n	?	?

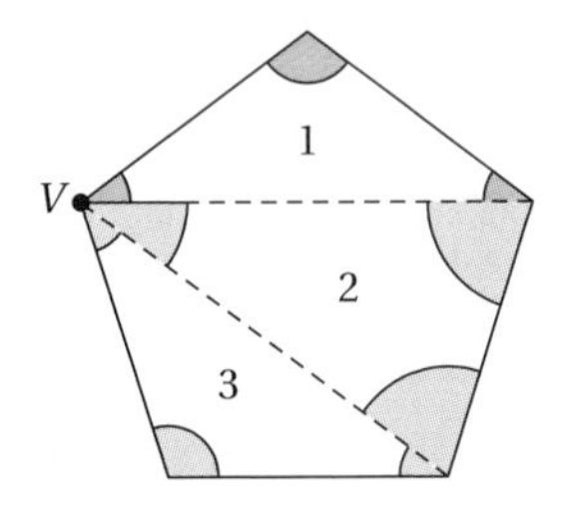

FIGURE 10.20 | Verifying a generalization about the sum of the interior angles of a pentagon.

The Sum of the Interior Angles of Any Polygon. Table 10.6 shows how specific examples can be used to look for a pattern and inductively discover a general formula for determining the sum of the interior angles of any convex polygon. Figure 10.20 illustrates the ideas presented in the table for a pentagon.

Table 10.6 suggests the following theorem, which can be discovered by generalizing the relationships in the table.

Theorem: Angle Sum for Any Polygon

The sum of the interior angles of an n-gon is $(n - 2)180°$.

In Example 10.13 we use the polygon angle sum theorem.

Example 10.13

Finding the Sum of Interior Angles of a Polygon

Find the sum of the measures of the interior angles of an octagon.

Solution

Wesley's thinking: I noticed that an octagon could be divided into six triangles, so I multiplied 6 by 180° to find the total number of degrees for the octagon. The sum of the measures of the interior angles of an octagon is 1080°.

Lee's thinking: I used the formula $I = (n - 2)180°$, or $I = (8 - 2)180° = 1080°$. The sum of the interior angle measures of an octagon is 1080°.

Your Turn

Practice: Find the sum of the measures of the interior angles of a decagon.

Reflect: Do you think that, as you choose polygons with greater and greater numbers of sides, there is a limit to the sum of the measures of the interior angles? Explain. ■

Sum of the Exterior Angles of a Polygon. An angle formed by a side of a polygon and extension of an adjacent side as shown in Figure 10.21(a) is an example of an **exterior angle of a polygon.** Two such angles can be drawn at each vertex for a triangle, but we usually consider only one exterior angle at each vertex.

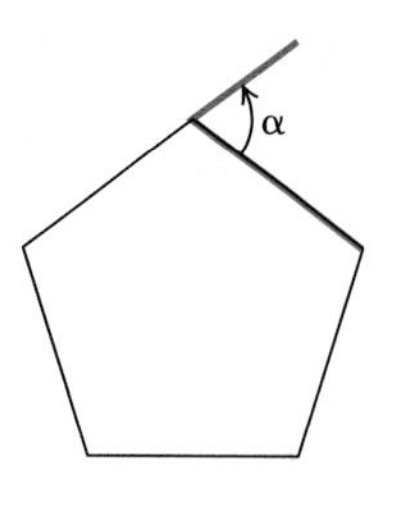

(a) An exterior angle of a polygon

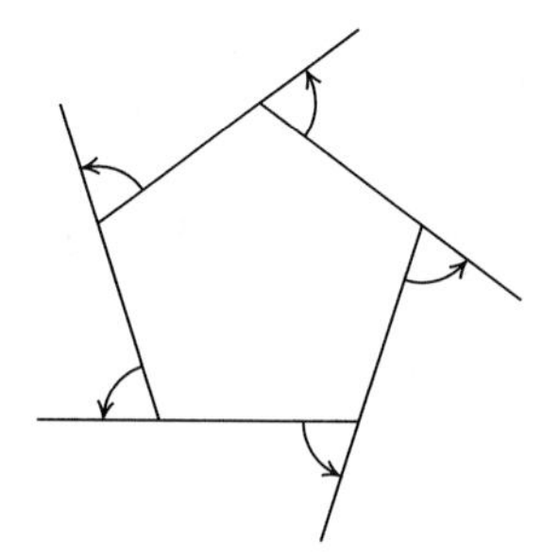

(b) One set of exterior angles of a polygon

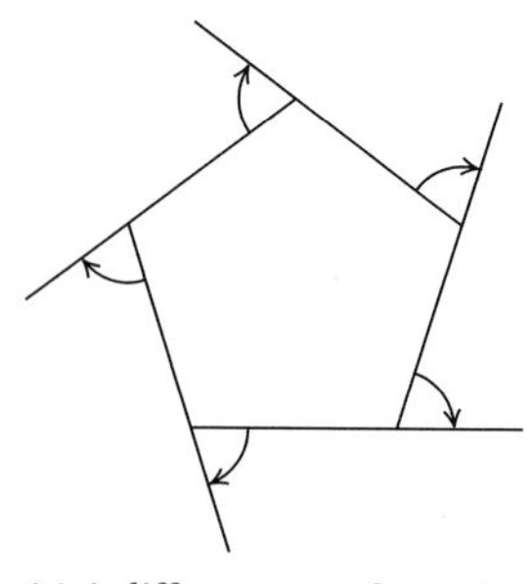

(c) A different set of exterior angles of the polygon

FIGURE 10.21

When we refer to the *sum of the measures of the exterior angles* of a triangle, we mean the sum of the measures of the set of angles shown in Figure 10.21(b) or 10.21(c), but not both.

Mini-Investigation 10.4 suggests an interesting way for you to use a model, along with inductive reasoning to investigate the sum of the measures of the exterior angles of any polygon.

Draw polygons and try this method with a quadrilateral, a pentagon, and a hexagon. Discuss what you find.

Mini-Investigation 10.4 **Using Mathematical Reasoning**

Through how many degrees do you turn the heavy color arrow when you slide it along each side of the triangle and turn it through each exterior angle,

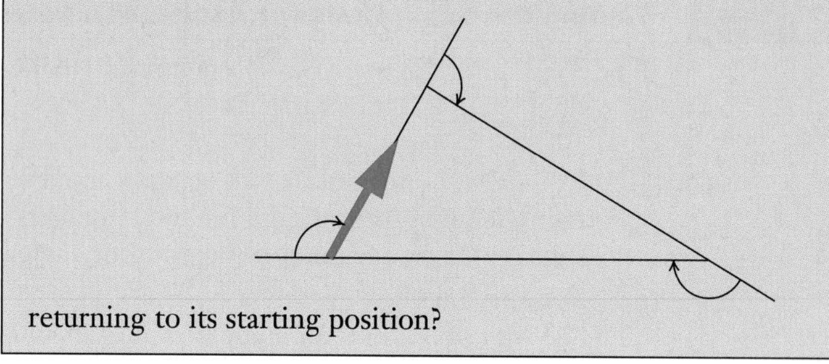

returning to its starting position?

In Mini-Investigation 10.4 you used a model to discover the following theorem for determining the sum of the exterior angles of any polygon.

Theorem: Exterior Angle Sum for Any Polygon

The sum of the exterior angles of any polygon is 360°.

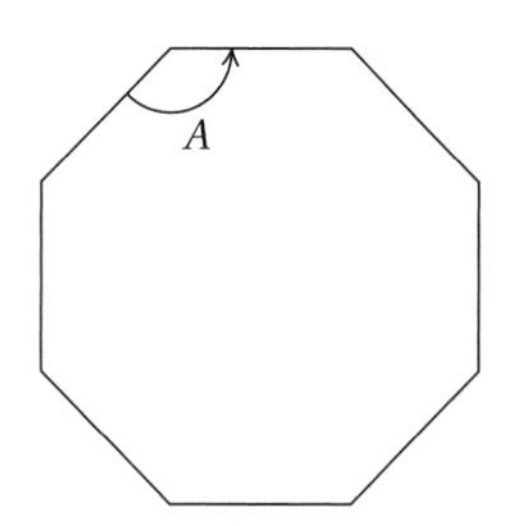

FIGURE 10.22

Angles in Regular Polygons

The theorems developed about the sum of the interior and exterior angles of an n-sided polygon also apply to regular polygons. The symmetric nature of regular polygons makes looking for other generalizations about their angles both possible and interesting. These generalizations are useful in the development of such topics as star polygons, which we consider in Section 11.3.

For example, information about the sum of all the interior angles of a regular polygon helps us find the measure of a single interior angle of the polygon. For the regular octagon shown in Figure 10.22, the sum of all the interior angles is $(8 - 2)180°$, or $1080°$. So the measure of interior angle A is $\frac{1080°}{8}$, or $135°$.

More generally, since the measures of *all* the interior angles of a regular n-gon is $(n - 2)180°$, and a regular n-gon has n congruent interior angles, we state the following theorem about the measure of a single interior angle of the n-gon.

Theorem: Interior Angle Measure for a Regular Polygon

The measure of each interior angle of a regular n-gon is $\frac{(n - 2)180°}{n}$.

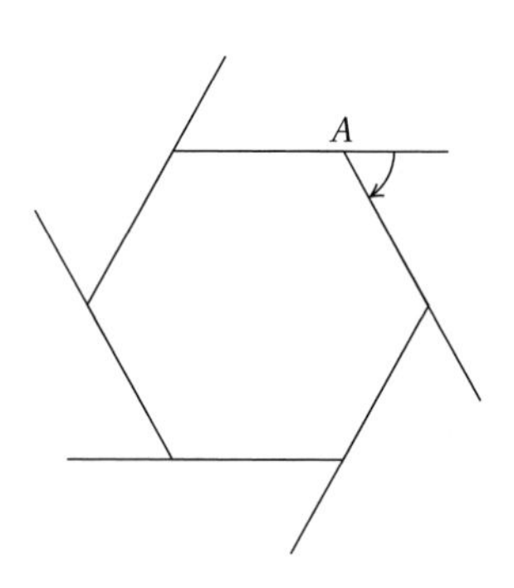

FIGURE 10.23

Similarly, you can use information about the sum of all the exterior angles of a regular polygon to find the measure of a single exterior angle of the polygon. For the regular hexagon shown in Figure 10.23, there are six congruent exterior angles with clockwise orientation. Because the sum of all these exterior angles is $360°$, the measure of one exterior angle is $\frac{360°}{6}$, or $60°$.

More generally, since the measures of all the exterior angles of a regular n-gon is $360°$, and an n-gon has n congruent exterior angles, we state the following theorem about the measure of a single exterior angle of the n-gon.

Theorem: Exterior Angle Measure for a Regular Polygon

The measure of an exterior angle of a regular n-gon is $\frac{360°}{n}$.

Interestingly, another expression for the measure of an interior angle of a polygon can be developed from the preceding theorem.

$$m(\text{interior angle}) = 180° - m(\text{exterior angle}), \text{ so}$$

$$m(\text{interior angle}) = 180° - \frac{360°}{n}$$

The theorem that the measure of an interior angle of an n-sided regular polygon is $\frac{(n-2)180}{n}$ can be used to verify algebraically that the measure of an exterior angle of the regular polygon is $\frac{360}{n}$. Since the exterior angle and the interior angle are supplementary, we can state that

$$m(\text{exterior angle}) = 180 - \left[\frac{(n - 2)180}{n}\right] = \frac{(180n - 180n + 360)}{n} = \frac{360}{n}$$

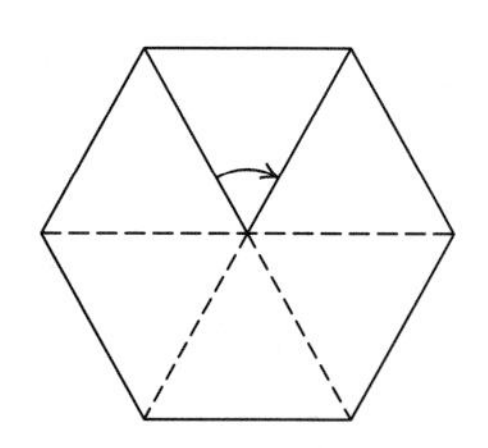

FIGURE 10.24

The number of degrees in a central angle of a regular hexagon, as shown in Figure 10.24, is found by noting that, for the six sides, there are six congruent central angles. The total degrees for all central angles is 360°, so the measure of one central angle is $\frac{360°}{6}$, or 60°.

More generally, because the n same-size central angles of a regular n-gon total 360°, we can state the following theorem.

Theorem: Central Angle Measure for a Regular Polygon

The measure of a central angle of a regular n-gon is $\frac{360°}{n}$.

In Example 10.14 we apply the theorem on finding the measures of an interior angle of a polygon.

Example 10.14

Problem Solving: The Stop Sign

PhotoDisc © 2001

A factory makes metal signs. Its machines have been set at an angle that allows them to cut out pentagonal direction signs. How much should this angle be increased or decreased to change the machines so they will cut out metal shapes used to make standard octagonal stop signs?

Solution

The interior angle of a polygon is found using the formula $\frac{(n-2)180}{n}$.

For the pentagon, $\frac{(5-2)180}{5} = \frac{3(180)}{5} = 108°$.
For the octagon, $\frac{(8-2)180}{8} = \frac{6(180)}{8} = 135°$.

The angle of the machine must be increased from 108° to 135°, or by 27°.

Your Turn

Practice: Suppose that the factory in the example problem was changing from a hexagonally shaped sign to a decagonally shaped sign. By how much would the angle have to be increased or decreased?

Reflect: Does the angle for the machine need to be increased less to go from a square to a pentagonal sign than from a pentagonal to a hexagonal sign? Explain your conclusion and then generalize beyond four-, five-, and six-sided figures. ■

Angles in Circles

Figure 10.25 shows some generalizations about angles associated with a circle. Think of point P starting on the circle, as in part (a), and moving outside (b) and then inside (c) the circle. Using GES to verify how the measure of the angle at P changes as point P is actually moved is instructive. When P is on the circle, the rays from P cut off a portion of the circle, called an **arc**, and $m\angle P$ is equal to $\frac{1}{2}m(\text{arc } s)$. When P moves outside the circle, $m\angle P$ is less than $\frac{1}{2}m(\text{arc } s)$ by the amount of $\frac{1}{2}m(\text{arc } r)$. When P moves inside the circle, $m\angle P$ is greater than $\frac{1}{2}m(\text{arc } s)$ by the amount of $\frac{1}{2}m(\text{arc } r)$.

If the sides of angle P in Figure 10.25(b) were one or the other or both tangent to the circle, instead of intersecting the circle in two points, the formula for finding the measure of angle P would be the same.

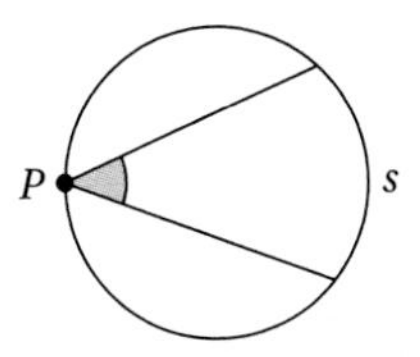

(a) Point P on the circle:
$m\angle P = \frac{1}{2}\, m(\text{arc } s)$.

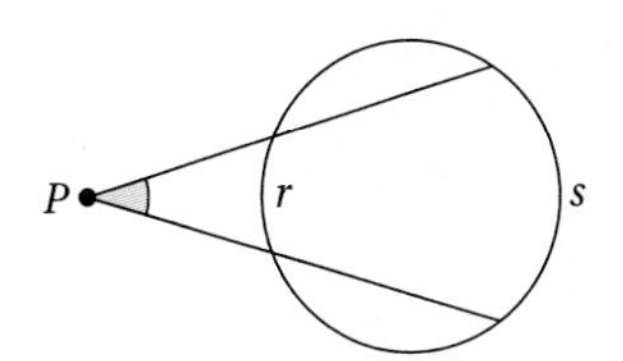

(b) Point P outside the circle:
$m\angle P = \frac{1}{2}\,[m(\text{arc } s) - m(\text{arc } r)]$

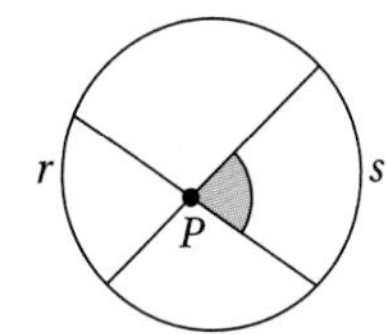

(c) Point P inside the circle:
$m\angle P = \frac{1}{2}\,[m(\text{arc } s) + m(\text{arc } r)]$.

FIGURE 10.25

In Example 10.15 we investigate further the relationships described in Figure 10.25.

Example 10.15 Relating Arc Measure to Angle Measure

Find the measure of angle P in Figure 10.25(a)–(c) if $m(\text{arc } s) = 80°$ and $m(\text{arc } r) = 36°$.

Solution

In part (a), $m\angle P = \frac{1}{2}(80° - 36°) = 22°$.
In part (b), $m\angle P = \frac{1}{2}(80°) = 40°$.
In part (c), $m\angle P = \frac{1}{2}(80° + 36°) = 58°$.

Your Turn

Practice: By how many degrees does $m\angle P$ differ from $m\angle Q$ in the following figure?

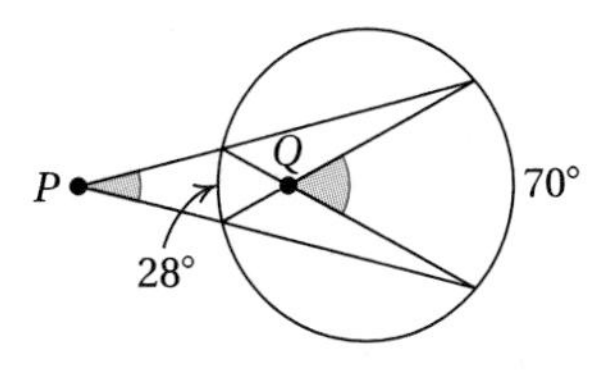

Reflect: What generalization can you make about $m\angle P$ in Figure 10.25(a) if it is inscribed in a semicircle? ■

On the next page we see how inductive reasoning in geometry is used in the PreK–8 classroom.

Connection to the PreK–8 Classroom

A Sample Student Page: Developing Inductive Reasoning

The *Principles and Standards for School Mathematics* state that "Geometry provides a rich context for the development of mathematical reasoning, including inductive and deductive reasoning, making and validating conjectures, and classifying and defining geometrical objects" (p. 233). The page from a seventh-grade textbook in Figure 10.26 illustrates the use of inductive reasoning in geometry. Study the page and answer the questions.

- **How** is the model in the Explore activity used to help with inductive reasoning?
- **What** generalization is made in the example to help students extend the pattern?

Inductive Reasoning
Discovering Geometric Patterns

LEARN ABOUT IT

EXPLORE **Make a Model**
Use isometric dot paper to make layers of triangles. Shade the inner 6—they make up the first layer. Put a dot in each triangle in the second layer—it has 18 triangles. Put a check in each triangle in the 3rd layer. Be sure you have 30 checks.

TALK ABOUT IT

1. How many triangles are in the 4th layer? in the 5th layer?
2. How could you find the number of triangles in the 25th layer—it's too hard to draw!

Inductive reasoning is often used to discover geometric relationships. You consider simple cases like the problem given, organize your work, and look for a pattern. When you discover a pattern, you test it on a few more simple cases. If the pattern holds true, you can give the answer to the original problem.

Example How many different-sized squares are there on an 8×8 checkerboard?

Solution: Use simple cases as shown to the right. The pattern for the first 3 cases says the number of squares in the 4×4 should be $1^2 + 2^2 + 3^2 + 4^2$. We count and find it.

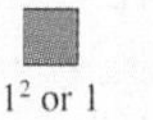
1^2 or 1

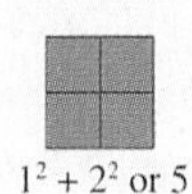
$1^2 + 2^2$ or 5

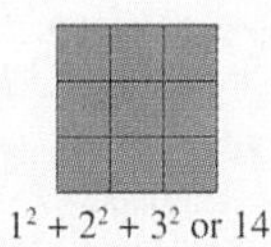
$1^2 + 2^2 + 3^2$ or 14

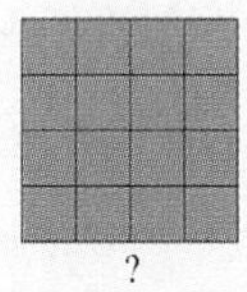
?

A checkerboard has $1^2 + 2^2 + 3^2 + 4^2 + 5^2 + 6^2 + 7^2 + 8^2$, or 204 squares!

TRY IT OUT

1. Find the number of vertices (corner points) on an 8×8 checkerboard without drawing it.

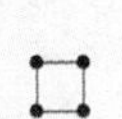
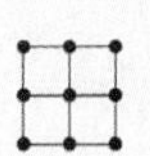
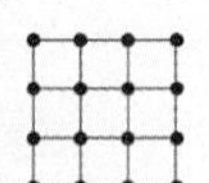
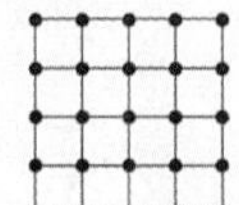

FIGURE 10.26 | Excerpt from a middle grades textbook.

Problems and Exercises | *for Section 10.3*

A. Reinforcing Concepts and Practicing Skills

1. Draw a pair of parallel lines cut by a transversal and label the angles formed.
   a. Name two pairs of alternate interior angles.
   b. Name four pairs of corresponding angles.
   c. Name two pairs of alternate exterior angles.
   d. Name two pairs of interior angles on the same side of the transversal.
   e. Name three pairs of different types of angles that are congruent.
   f. Name two pairs of different types of angles that are supplementary.
2. Complete the following statements about the measures of angles in the figure shown, where $r \parallel s$, and state the generalization that supports the statement.

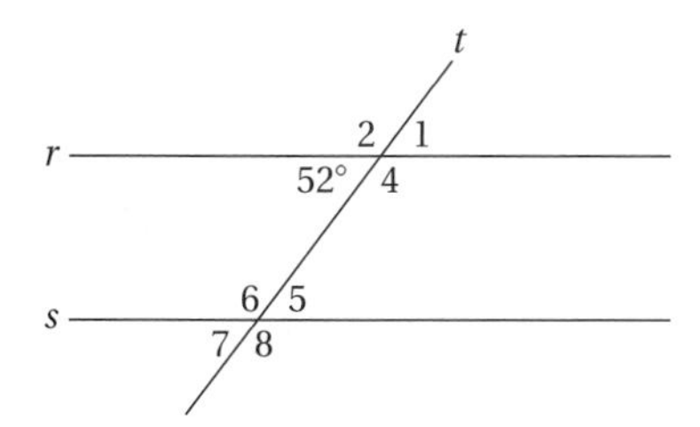

   a. $m\angle 1 = ?$ b. $m\angle 2 = ?$
   c. $m\angle 4 = ?$ d. $m\angle 5 = ?$
   e. $m\angle 6 = ?$ f. $m\angle 7 = ?$
   g. $m\angle 8 = ?$

Use the figure in Exercise 2 to answer Exercises 3 and 4. For each exercise, give a property of angles that helped you solve the problem.

3. If $m\angle 5 = 2x + 4$ and $m\angle 4 = 3x - 14$, what are the measures of angles 4 and 5?
4. If $m\angle 2 = 3x - 15$ and $m\angle 8 = 2x + 30$, what are the measures of angles 2 and 8?
5. A student concluded that, in the figure shown, since $m\angle 4 + m\angle 2 + m\angle 5 = 180°$, it was true that $m\angle 1 + m\angle 2 + m\angle 3 = 180°$. What property of angles formed when parallel lines are cut by a transversal would support this conclusion?

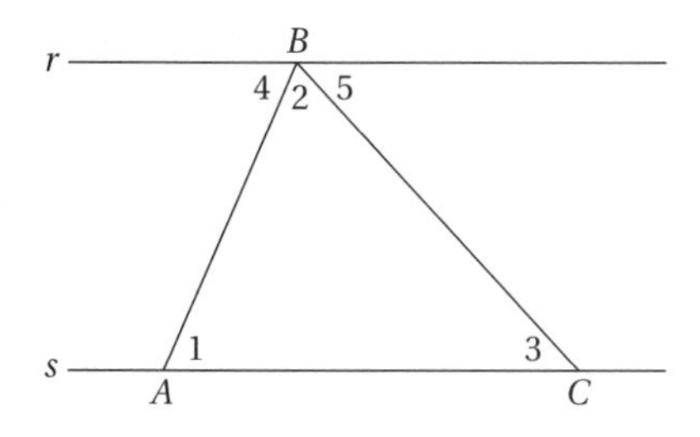

6. Use the data given by the figure and verify numerically that $m\angle 3 = m\angle 1 + m\angle 2$.

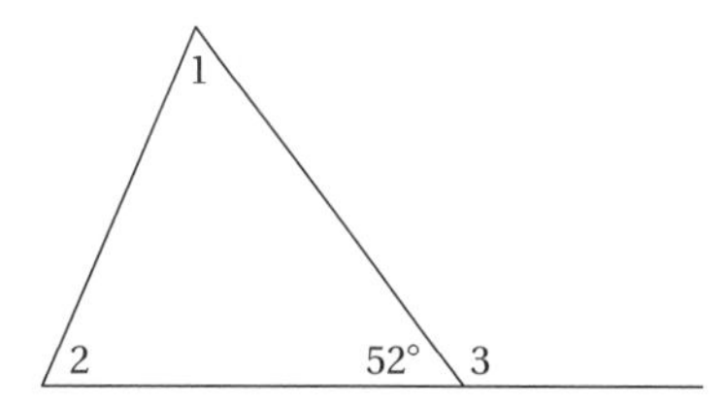

7. $ABCD$ is a square. Find the measures of angles 1–6. Explain your thinking.

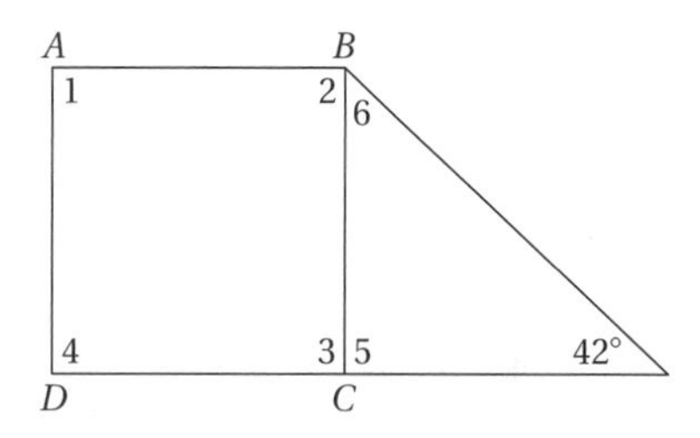

8. A student said "When I want to find the sum of all the interior angles of any polygon, I just subtract two from the number of its sides and multiply that number by 180." What formula, using n for the number of sides, describes what the student does?
9. Another student said "to find the measure of an interior angle of a regular polygon, I just divide the sum of all its interior angles by the number of its sides." What formula, using n for the number of sides, describes what the student does?

For Exercises 10–14, find the number of degrees in an interior angle, a central angle, and an exterior angle of the regular polygon named.

10. pentagon
11. hexagon
12. decagon
13. dodecagon
14. nonagon
15. What is the relationship between the measure of a central angle of a polygon and the measures of an interior and an exterior angle of the polygon?
16. As the number of sides of regular polygons increases, what happens to the measures of the interior angles of the polygons? The measures of the central angles?
17. What is the sum of the interior and of the exterior angles of a 100-gon?
18. Angles A and B of triangle ABC have measures 37° and 59°, respectively. What is the measure of $\angle C$?

19. Three angles of a quadrilateral each measures 87°. What is the measure of the fourth angle?

20. Home plate on a baseball field has three right angles and has the shape shown. What are the measures of the other angles?

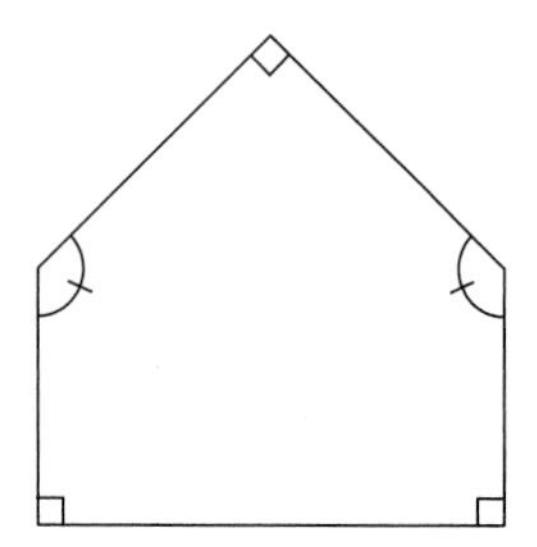

21. If $\overline{PR}$ is the diameter of the circle, explain how you would convince someone that quadrilateral $PQRS$ has at least two right angles.

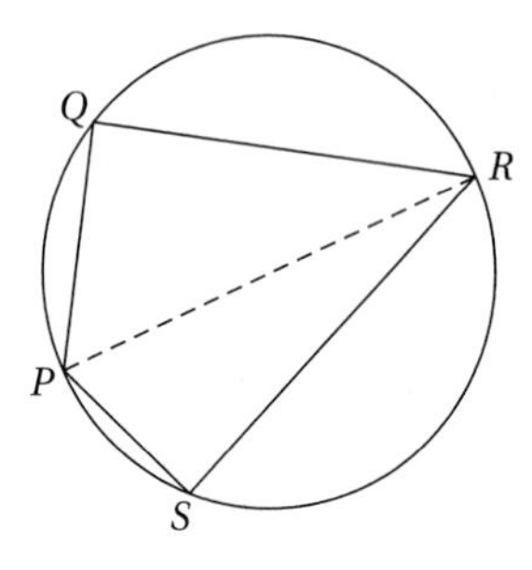

Find the measure of ∠P in Exercises 22–26.

22.

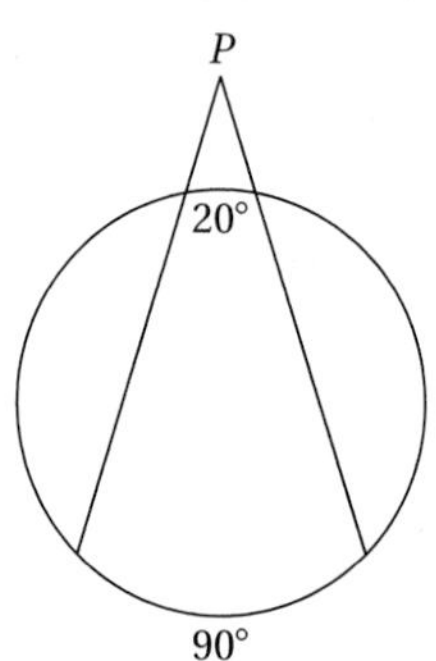

23.

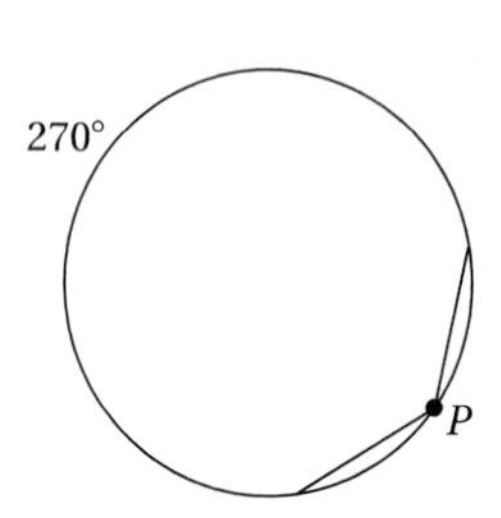

24.

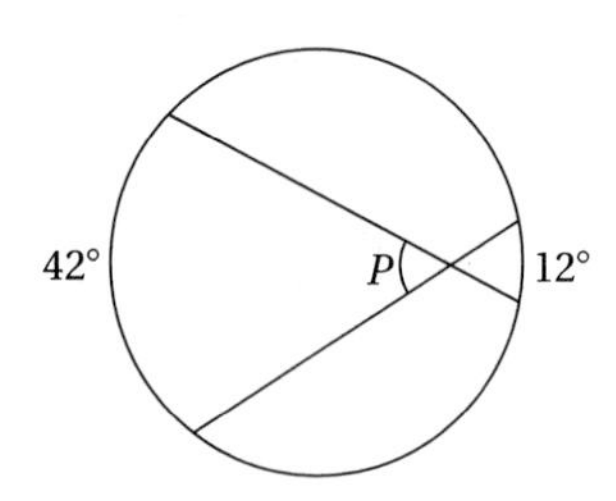

25.

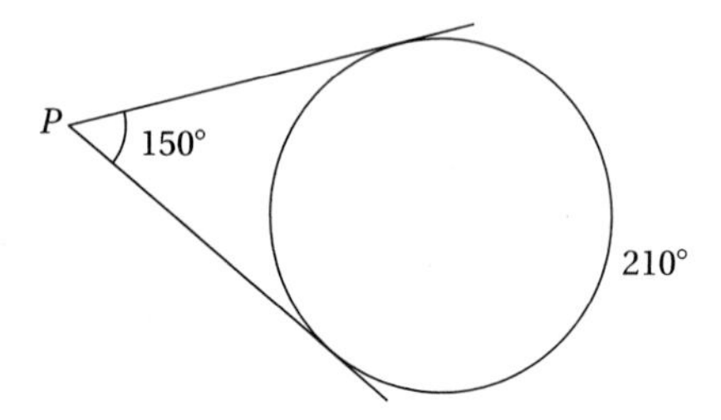

26.

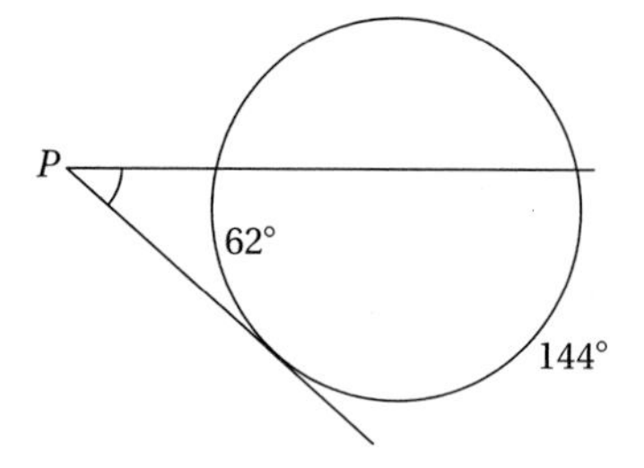

27. Find the degree measure of angles 1 and 2. What conclusion can you draw about angles that intercept the same arc on a circle?

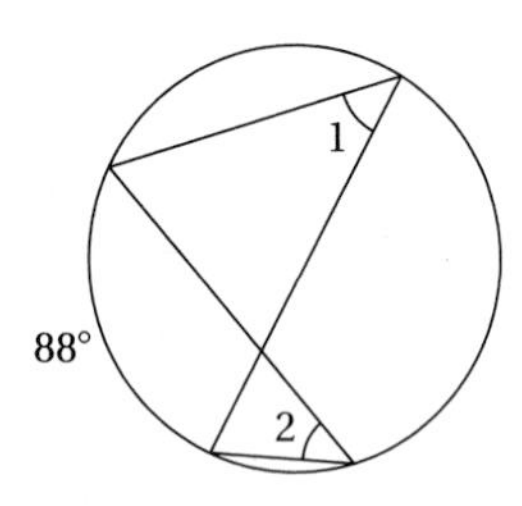

28. A design specialist drew the pattern on the next page, using regular pentagon $ABCDE$ inscribed in a circle. To make the mosaic pieces for the design, it would help her to be able to find the measure of $\angle DAC$. How could this measure be easily determined?

29. The skipper of a boat knows that the water is too shallow to bring his boat inside a "danger semicircle" that goes from lighthouse to lighthouse. How can the angle between his sightings of the two lighthouses help him decide when he is entering the danger zone? (See figure on next page.)

30. A manufacturer of stop signs wanted to set up a machine that would automatically cut out a metal sheet the correct size and shape for a stop sign. How should the angles on the cutter be set? Explain.

B. Deepening Understanding

31. If $r \parallel s$, use generalizations from this section to give a convincing argument for each of the following.
   a. $\angle 1$ is supplementary to $\angle 7$.
   b. $\angle 1$ is congruent to $\angle 8$.

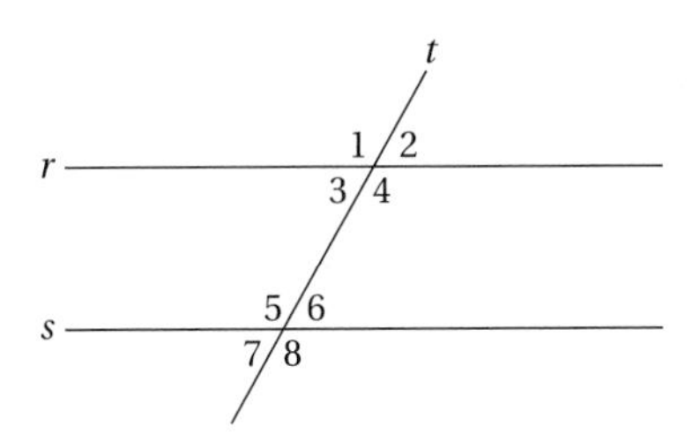

Figure for Exercise 28

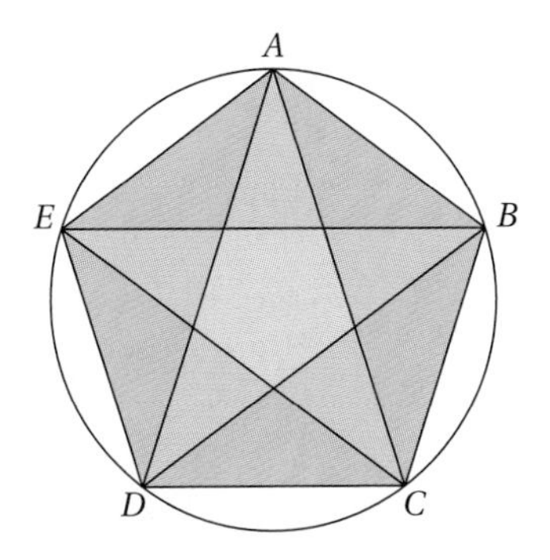

Figure for Exercise 29

32. Explain why a triangle cannot have
   a. two obtuse angles.
   b. an obtuse angle and a right angle.
33. What conclusion can you draw about the sum of the acute angles of a right triangle?
34. Describe a way to find the measure of $\angle B$ in the following figure.

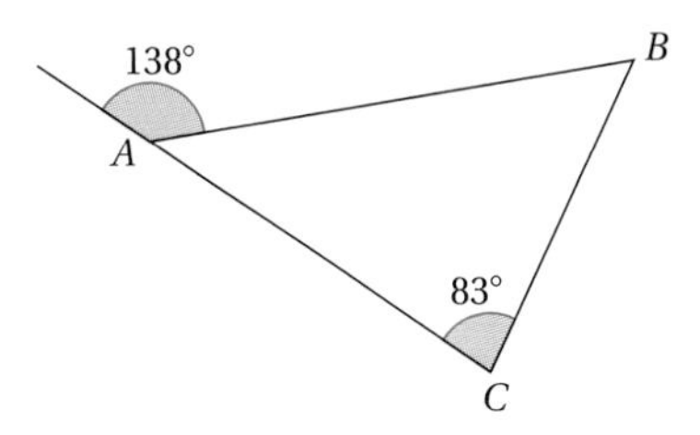

35. Refer to the figure shown and give a convincing argument that the sum of the interior angle measures of a hexagon is 720°.

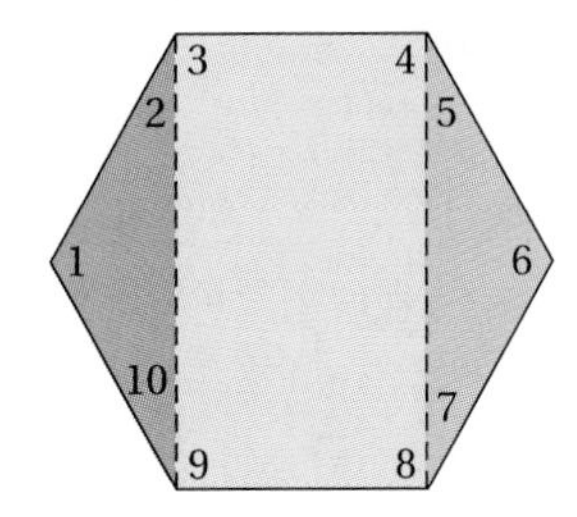

36. A wallpaper manufacturer wants to use a pattern of regular pentagons that fit around a point with no gaps or overlaps. Explain why this pattern would be impossible to achieve.
37. Will two regular pentagons and a regular decagon fit around a point with no gaps and no overlaps? Explain.
38. Prove that an equilateral triangle, a regular 7-gon, and a regular 42-gon will fit around a point with no gaps and no overlaps.
39. The measure of each interior angle of a regular polygon is eight times that of an exterior angle of the polygon. How many sides does the polygon have?
40. How many sides does a regular polygon have if each exterior angle is 40°?
41. How many sides does a regular polygon have if each interior angle is $157\frac{1}{2}°$?

Use GES or construction (see Appendix B for GES and Appendix D for constructions) for Exercises 42–48.

42. Begin with any triangle ABC. Bisect the external angles at A and C, and the internal angle at B. Extend these bisectors.
   a. What do you discover?

b. Change the shape of the triangle several times. Does the discovery hold true? What theorem could you state about this?

c. Consult a reference and find out what an *excenter* of a triangle is. How does it apply to this situation?

43. Begin with any parallelogram $ABCD$. On each side of the parallelogram construct squares lying external to the parallelogram. The centers of these squares, A_1, B_1, C_1, D_1, are vertices of what type of figure?

44. Draw a convex quadrilateral and its diagonals. Are the diagonals of the quadrilateral

a. always perpendicular?

b. never perpendicular?

c. sometimes perpendicular?

45. In the following figure, how does $\angle ABC$ relate to $\angle BCD$ and $\angle CDB$? Use triangles of different sizes and shapes to test your conjecture.

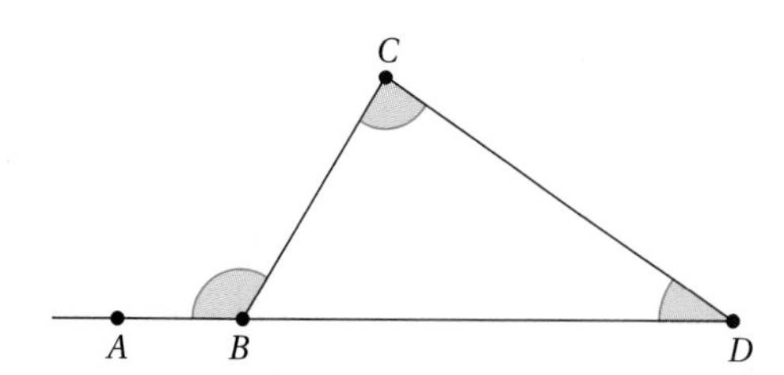

46. What generalization can you make about $\angle H$ and $\angle I$ in the figure shown? Use right triangles of different sizes and shapes to test your conjecture.

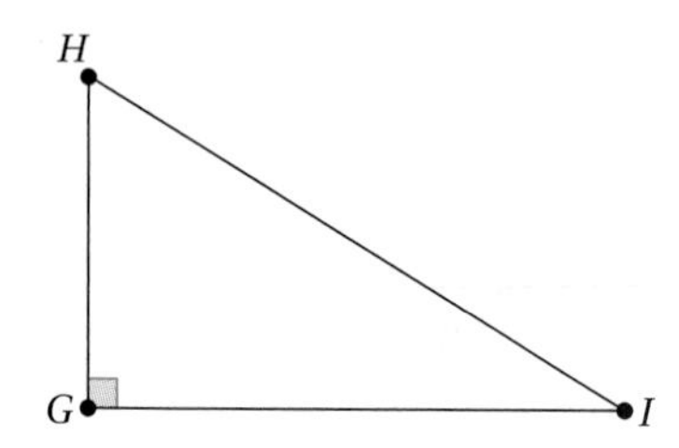

47. Begin with any convex quadrilateral. On its sides, construct equilateral triangles alternately exterior and interior to the quadrilateral. What type of figure is formed by the vertices of the triangles that aren't on the quadrilateral?

48. What generalization can you make about the sum of the measures of opposite angles, $m\angle P + m\angle R$ and $m\angle S + m\angle Q$, when quadrilateral $PQRS$ has vertices on a circle, as shown in the figure at the top of the next column? Plot points P, Q, R, and S in several different locations to test your conjecture.

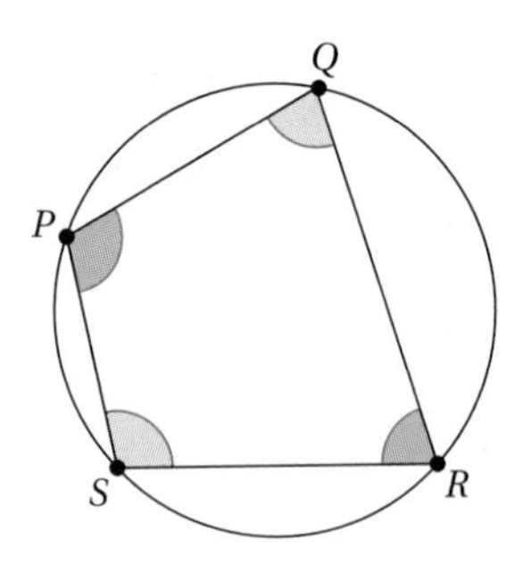

C. Reasoning and Problem Solving

49. The following figure shows the maximum number of regions that can be formed when two, three, and four points on a circle are connected by line segments.

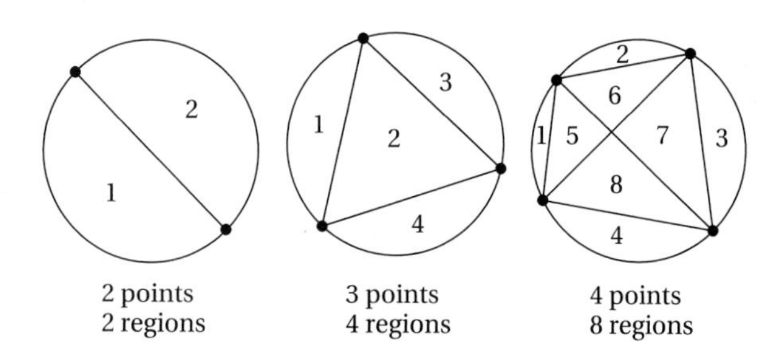

a. Draw a circle and find how many regions there are for five points.

b. Make a table showing the data for two, three, four, and five points. Then generalize how many regions there will be for six points.

c. Check your generalization for correctness. Find a counterexample if possible. Explain your findings.

50. Find the number of sides of a regular polygon if the sum of the measures of its interior angles is twice the sum of the measures of its exterior angles.

51. The Mosaic Problem. An artist who creates mosaics wants to have a reference list of possible combinations of three regular polygons that will fit around a point. She found one arrangement with three identical polygons, one in which two of the polygons were pentagons, three arrangements that had a square with a 5-, 6-, or 8-gon and another polygon, and five arrangements that had an equilateral triangle with a 7-, 8-, 9-, 10-, or 12-gon and another polygon. What are the 10 possible arrangements that were on her list?

52. The Small Tabletop Problem. A carpenter wants to find the center of a circular wooden disc that is to be used as the top of a small table in order to attach a support leg, as shown in the figure at the top of the next page. How can he use a carpenter's square to find the center?

53. An old book stated, "To find the measure of an interior

Figure for Exercise 52

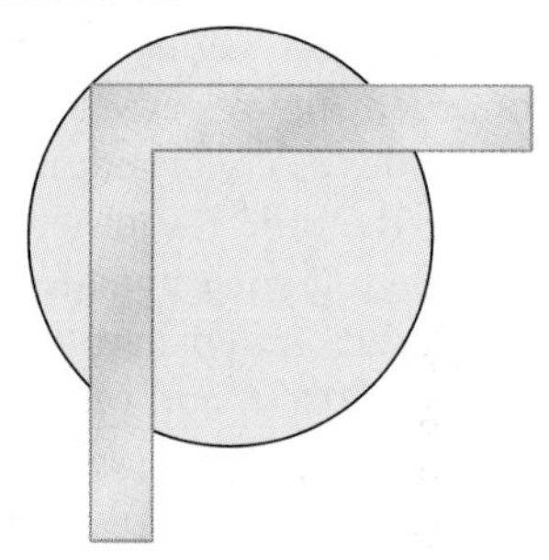

angle of a regular polygon, just subtract the measure of the central angle from 180°." Does this method work? If so, can you explain why? If not, how could you change the method so that it would work?

54. As the number of sides of a regular polygon increases, why does the sum of the measures of the exterior angles remain constant even though the sum of the measures of the interior angles increases?

D. Communicating and Connecting Ideas

55. Work in a small group to complete the following experiment.

Lay a straw between a pair of hinged mirrors, as shown. For which angles, θ, between the mirrors can a regular polygon be seen? Record your findings in a table or tables and state any generalizations you discover.

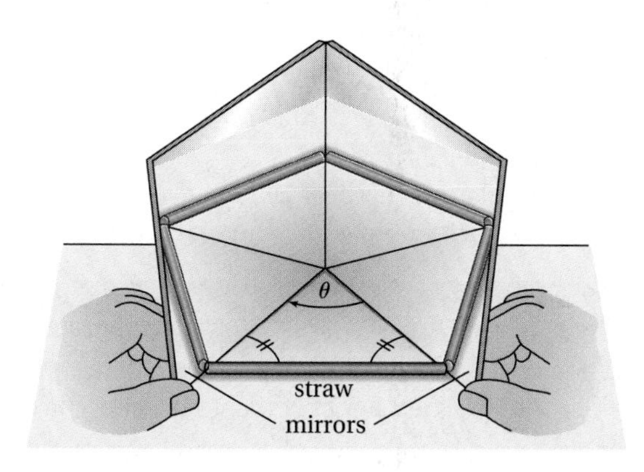

56. Historical Pathways. Eratosthenes (275 B.C.) used an ingenious method to calculate the distance around the earth. He assumed the sun's rays to be parallel and found that, when the sun was directly overhead in Syene, its rays made an angle of $7\frac{1}{5}°$ with a vertical pole 500 miles away in Alexandria. He also noted that the angle $a°$ compares to 500 miles as an angle of 360° compares to the earth's circumference. Use Eratosthenes's method to calculate the circumference of the earth. Check your answer with the value given in an encyclopedia and explain any discrepancy.

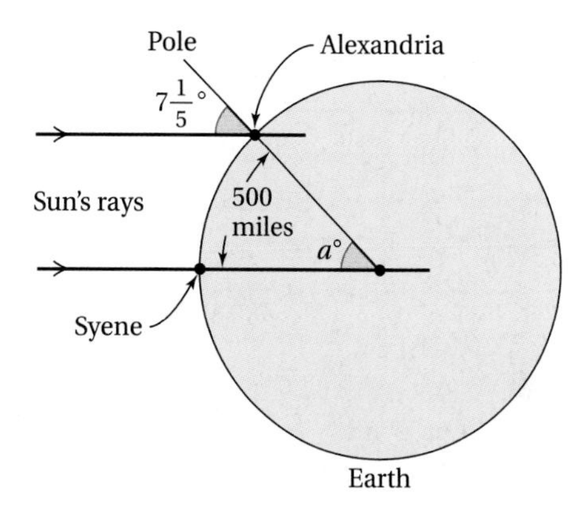

57. Making Connections. Show the connection between two different verifications of the generalization that the sum of the interior angles of an n-sided polygon is $(n - 2)180°$. Use the figure shown, write a convincing argument, and compare your argument to the verification given on p. 545.

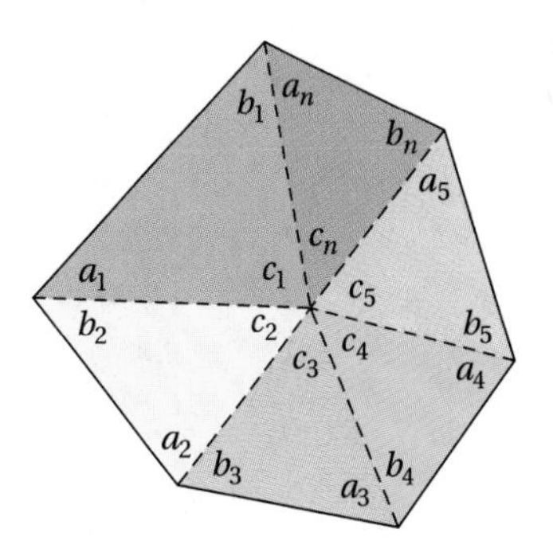

Section 10.4 | MORE ABOUT TRIANGLES

- Congruent Triangles
- The Pythagorean Theorem
- Special Right Triangles

In this section we examine geometric relationships that occur in and between triangles. We will frequently employ Geometry Exploration Software (GES) or drawing/constructing and measurement to help us in our use of inductive reasoning to discover these relationships.

Congruent Triangles

In Section 10.1, we discussed the ideas of congruent segments and congruent angles, and you may have concluded that congruent figures are the same size and shape. Similarly, we intuitively think of two triangles as congruent if they have the same size and shape, or if one will "fit on top" of the other. Another way of thinking about this idea is that two triangles are congruent if you can match up their vertices in such a way that matching sides and angles are congruent. For example, in Figure 10.27, if we match vertices A to F, vertices B to E, and vertices C to D, we see that $\angle A \cong \angle F$, $\angle B \cong \angle E$, $\angle C \cong \angle D$, $\overline{AB} \cong \overline{FE}$, $\overline{BC} \cong \overline{ED}$, and $\overline{AC} \cong \overline{FD}$, and that $\triangle ABC \cong \triangle FED$.

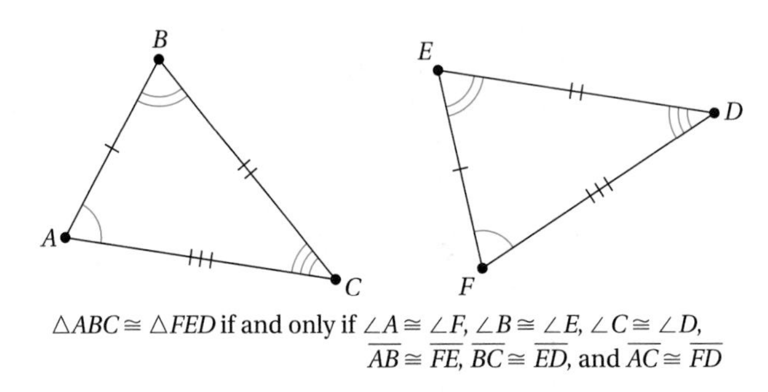

$\triangle ABC \cong \triangle FED$ if and only if $\angle A \cong \angle F$, $\angle B \cong \angle E$, $\angle C \cong \angle D$, $\overline{AB} \cong \overline{FE}$, $\overline{BC} \cong \overline{ED}$, and $\overline{AC} \cong \overline{FD}$

FIGURE 10.27 | A pair of congruent triangles.

The order in which the vertices are matched, as shown below, is significant in that it indicates which parts of the two triangles are congruent to each other. Not only do the vertices of the congruent angles show up in the matching, but if the three possible pairs of points are correspondingly considered, the matching segments are visible too.

$$\begin{matrix} A & B & C \\ \updownarrow & \updownarrow & \updownarrow \\ F & E & D \end{matrix}$$

The following definition describes the relationship between congruent triangles.

Definition of Congruent Triangles

Two triangles are **congruent** if and only if, for some correspondence between the two triangles, each pair of corresponding sides are congruent and each pair of corresponding angles are congruent.

Fortunately, it is not necessary to check all six of the congruent pairs in the definition to determine whether or not two triangles are congruent. Using GES, you may wish to explore the minimum conditions that determine the size and shape of a triangle. For example, can more than one size and shape triangle be formed when the lengths of the three sides, or the measures of two sides and the included angle, or the measures of two angles and the included side are held constant? Such an exploration might lead to the following generalizations about congruent triangles. Generalizations such as the ones that follow are so basic that they are often accepted as true without proof and are called *postulates.*

Triangle Congruence Postulates

- The Side-Side-Side (SSS) *Postulate*

Given a correspondence between two triangles, if three sides of one triangle are congruent to the corresponding three sides of the second triangle, then the two triangles are congruent.

- The Side-Angle-Side (SAS) *Postulate*

Given a correspondence between two triangles, it two sides and the included angle of one triangle are congruent to the corresponding two sides and the included angle of the second triangle, then the two triangles are congruent.

- The Angle-Side-Angle (ASA) *Postulate*

Given a correspondence between two triangles, if two angles and the included side of one triangle are congruent to the corresponding two angles and the included side of the second triangle, then the two triangles are congruent.

The following are three additional generalizations about congruent triangles that can be proved using the postulates, definitions, and theorems of geometry.

Triangle Congruence Theorems

- The Angle-Angle-Side (AAS) *Theorem*

Given a correspondence between two triangles, if two angles and a side opposite one of them in one triangle are congruent to the corresponding parts of the second triangle, then the two triangles are congruent.

- The Hypotenuse–Acute Angle (HA) *Theorem*

Given a correspondence between two right triangles, if the hypotenuse and an acute angle of one triangle are congruent to the corresponding hypotenuse and acute angle of the second triangle, then the two triangles are congruent.

- The Hypotenuse-Leg (HL) *Theorem*

Given a correspondence between two right triangles, if the hypotenuse and a leg of one triangle are congruent to the corresponding hypotenuse and leg of the second triangle, then the two triangles are congruent.

In Figure 10.28 we give four key statements that can be used in the proof of the AAS Triangle Congruence theorem. Think about why each of the statements is true, and how they could be used to help prove the theorem.

In Example 10.16, we apply some of the above generalizations to identify pairs of congruent triangles. We will study congruent triangles further in Chapter 11.

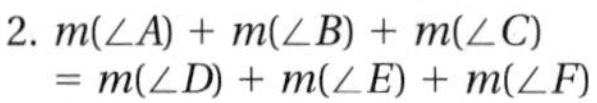

1. $\angle A \cong \angle D$, $\angle B \cong \angle E$, and $\overline{AC} \cong \overline{DF}$ (given data)
2. $m(\angle A) + m(\angle B) + m(\angle C) = m(\angle D) + m(\angle E) + m(\angle F)$
3. $\angle C \cong \angle F$
4. $\triangle ABC \cong \triangle DEF$

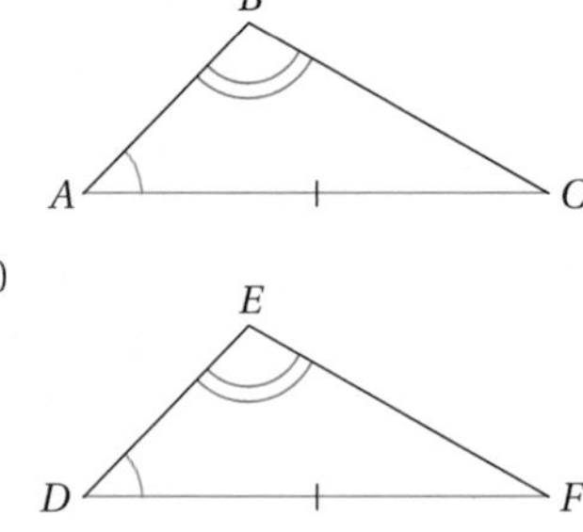

FIGURE 10.28 | Key ideas for proving the AAS Triangle Congruence Theorem.

Example 10.16 Identifying Congruent Triangles

In the figure below, use one of the properties of congruent triangles to identify a pair of congruent triangles.

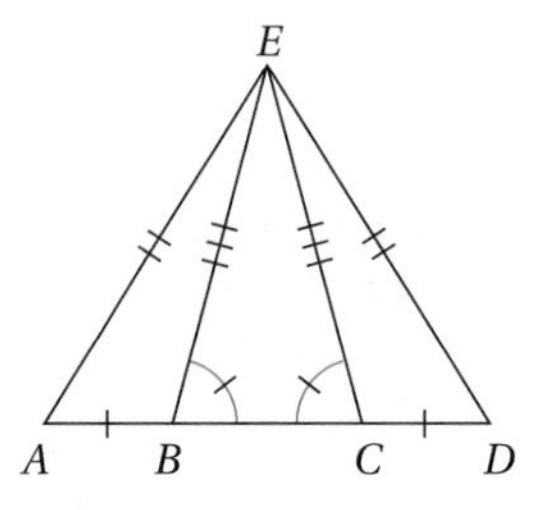

Solution

Hassid's solution: I can use the Side-Angle-Side property of congruent triangles to show that $\triangle ABE$ is congruent to $\triangle DCE$ because $\overline{AE} \cong \overline{DE}$, $\angle A \cong \angle D$, and $\overline{AB} \cong \overline{DC}$.

Loretta's solution: Since $\overline{AB} \cong \overline{DC}$, then $\overline{AC} \cong \overline{DB}$. Also $\overline{EC} \cong \overline{EB}$ and $\overline{AE} \cong \overline{DE}$. So, $\triangle ACE \cong \triangle DBE$ by the Side-Side-Side property of congruent triangles.

Your Turn

Practice: What is another way to prove that $\triangle ACE \cong \triangle DBE$ in the figure above?

Reflect: Suppose $\triangle EBD$ and $\triangle ECA$ are right triangles. What triangle congruence theorems or postulates could be used to prove that they are congruent? Explain. ■

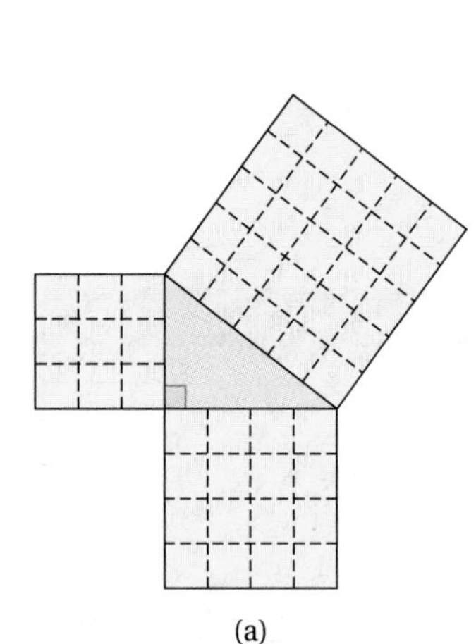

(a)

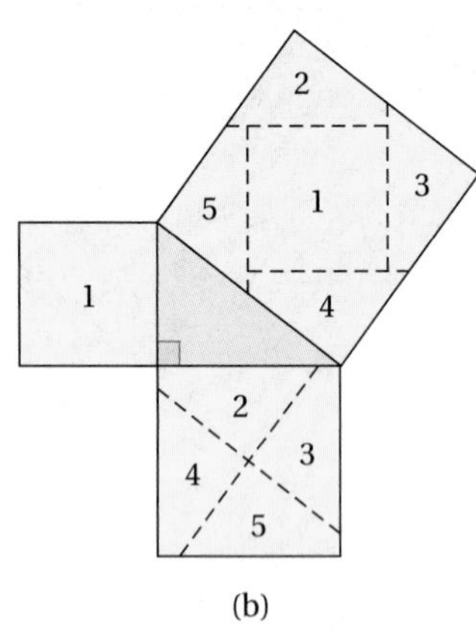

(b)

FIGURE 10.29 | Models for discovering the Pythagorean theorem.

The Pythagorean Theorem

A Dissection Model for the Pythagorean Theorem. The **Pythagorean theorem,** dealing with relationships in a right triangle, is one of the most famous and useful theorems in geometry. The models shown in Figure 10.29 suggest ways to discover the basic idea of this theorem, namely, that the sum of the areas of the squares on the **legs** of a right triangle is equal to the area of the square on the longer side, or **hypotenuse.** The cartoon on the next page reveals the humor in this terminology.

Note in Figure 10.29(a) that the ability to count the squares and discover the theorem depends on having a right triangle with side and hypotenuse lengths that are whole numbers. In Figure 10.29(b), however, the method of

FRANK AND ERNEST ® by Bob Thaves

With permission of Bob Thaves.

dissecting the squares doesn't depend on the lengths of the sides of the right triangle. To dissect the larger square on a leg, first find the center of the square. Then construct a line through the center perpendicular to the hypotenuse. Construct a second line through the center that is perpendicular to the first line. Pieces 1–5 can then be traced, cut out, and repositioned to fill exactly the square on the hypotenuse.

The following statement of the Pythagorean theorem emphasizes the relationship between the lengths of the legs of a right triangle and its hypotenuse. It is equivalent to the statement of the Pythagorean theorem given in the first paragraph in this subsection.

The Pythagorean Theorem

In a right triangle, the square of the length of the hypotenuse equals the sum of the squares of the lengths of the legs. When the lengths of the legs are a and b, and the length of the hypotenuse is c, $a^2 + b^2 = c^2$.

Because Pythagoras (580–496 B.C.) was the first to *prove* this important theorem, it is called the *Pythagorean* theorem. However, the Babylonians used the converse of the theorem more than 1300 years before Pythagoras, so it's possible that they were also aware of the theorem itself.

A Motion Model for the Pythagorean Theorem. We don't always verify a generalization in geometry by looking at static figures. As the sequence in Figure 10.30(a)–(e) shows, we can also verify generalizations by using a series of motions, in which a figure is dynamically changed to establish a result.

Connection to the PreK–8 Classroom

The NCTM *Principles and Standards for School Mathematics* include the following comment about the role of the Pythagorean theorem in the schools. "Visual demonstrations can help students analyze and explain mathematical relationships. Eighth graders should be familiar with one of the many visual demonstrations of the Pythagorean relationship—the diagram showing three squares attached to the sides of a right triangle" (p. 238).

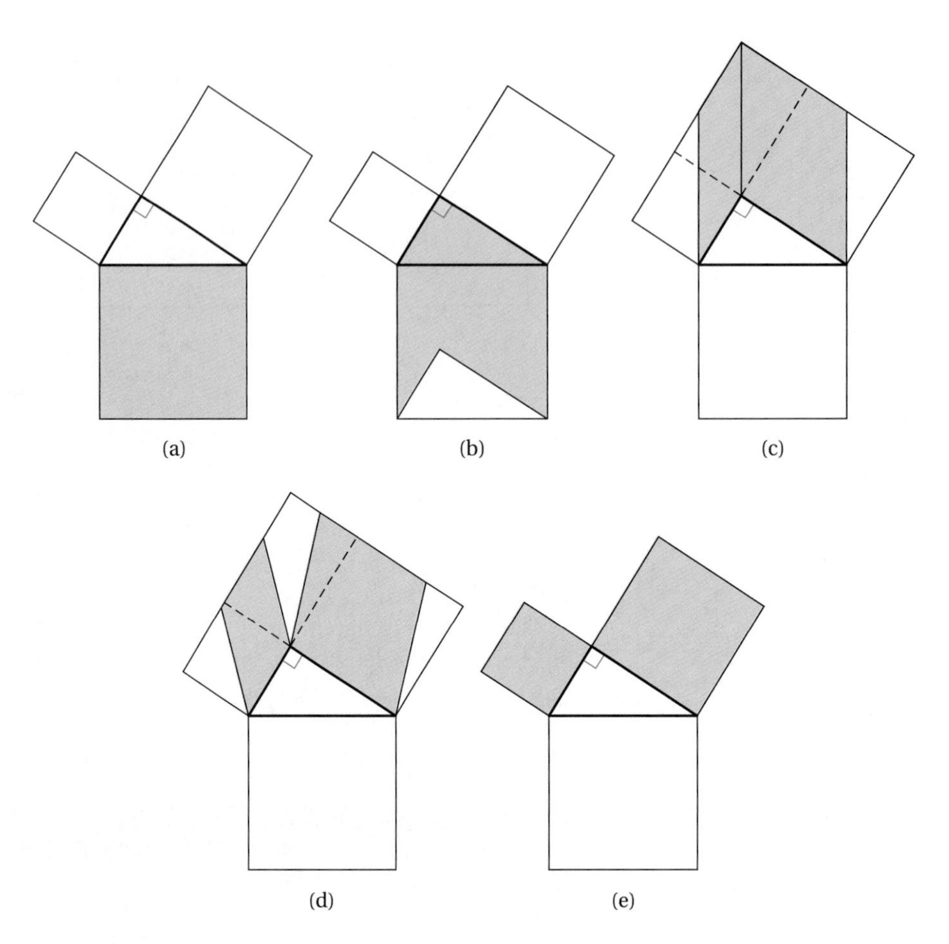

FIGURE 10.30 | Using motions to verify the Pythagorean theorem.

Note that when the shaded part in (c) is split as shown in (d) its area doesn't change, since the bases and heights of the two parallelograms have not changed, and we find the area of a parallelogram by multiplying the base times the height.

The Pythagorean theorem can be used to find the length of any side of a right triangle when the lengths of the other two sides are known, as illustrated in Example 10.17.

Example 10.17

Problem Solving: Packing Your Fishing Rod

Suppose that you can take your fishing rod on a trip only if you can carry it in your suitcase. The rod collapses to a length of 80 centimeters. You want to order a new suitcase from a catalog for the trip. The one you like best has interior dimensions of 66 centimeters (length) × 18 centimeters (width) × 46 centimeters (height). Discuss some ways to decide whether you can take the fishing rod in the new suitcase. Show your solution.

Working toward a Solution

Draw a picture of the suitcase, such as the one shown, and reason that the maximum length for the fishing rod would be from A to B. Note that the Pythagorean theorem can be used twice to find this length.

Your Turn

Practice:

a. Complete the solution to the example problem. Then check your work.

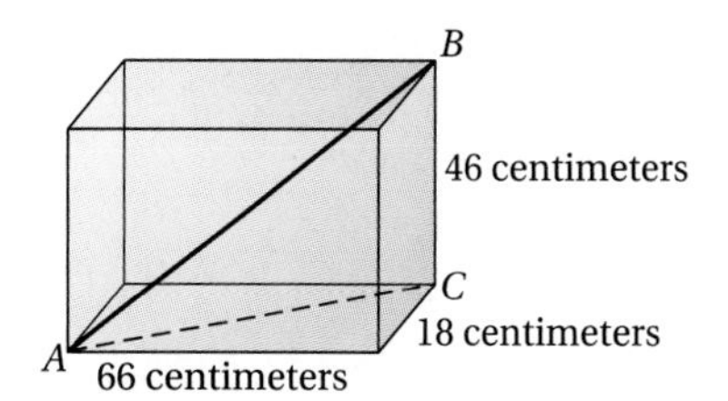

b. Solve the following problem.

> What is the longest piece of plumbing pipe that can be stored in an empty cabinet without shelves that is 6 feet high, 4 feet wide, and 2 feet deep?

Reflect: What subproblems helped you solve the example problem? ■

In Mini-Investigation 10.5, you explore an application of the Pythagorean theorem.

Write an explanation of your reasoning.

Mini-Investigation 10.5 Using Mathematical Reasoning

Why is it correct to say that the Babylonians and the Egyptians were assuming the *converse* of the Pythagorean theorem when they used a rope with knots 3, 4, and 5 units apart to lay out square corners when surveying fields or building pyramids, as suggested in these figures?

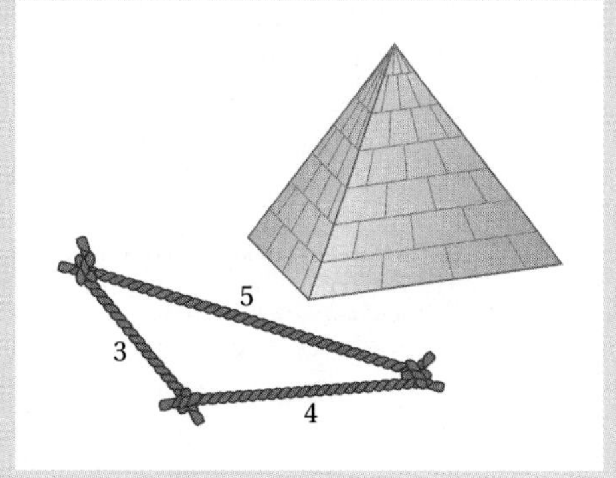

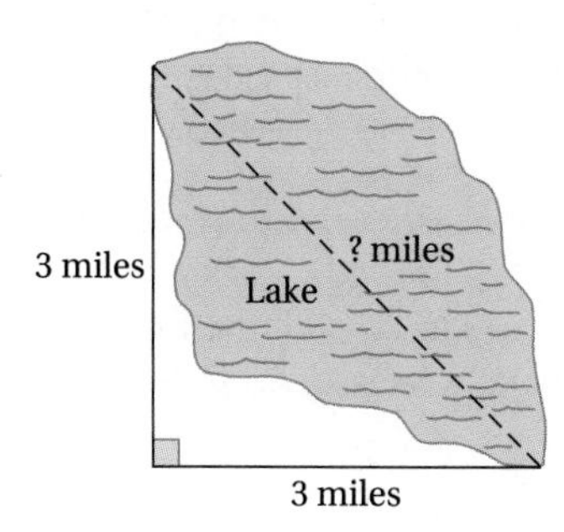

FIGURE 10.31

Special Right Triangles

Certain special right triangles are involved in applications of mathematics. For example, someone who wants to find quickly how far it is across a lake, as illustrated in Figure 10.31, might find the generalization suggested by looking for a pattern in the sequence of 45°–45°–90° triangles shown in Figure 10.32 on the next page quite useful. We chose legs of successive lengths, 1–5 units, as examples and used the

FIGURE 10.32 | A sequence of 45°–45°–90° triangles.

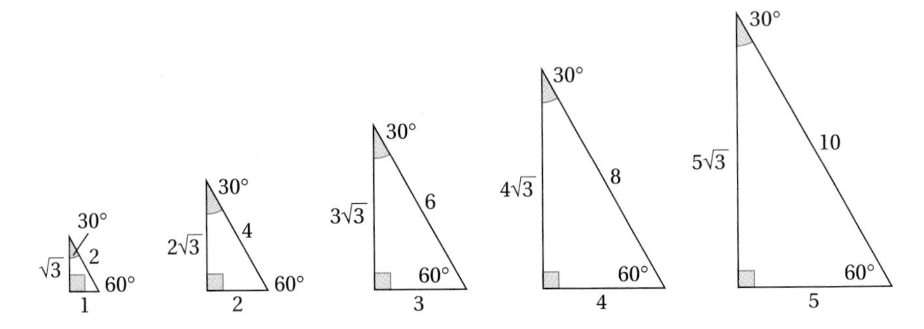

FIGURE 10.33 | A sequence of 30°–60°–90° triangles.

Pythagorean theorem to calculate the length of the hypotenuse of each triangle.

Theorem: 45°–45°–90° Triangle Relationships

The length of the hypotenuse of any 45°–45°–90° triangle is the length of a leg times $\sqrt{2}$.

We can use algebra to verify the above theorem as follows. Let a represent the length of each of the legs of a 45°–45°–90° triangle, and h represent the length of the hypotenuse. Then $h^2 = a^2 + a^2 = 2a^2$. It follows that $h = \sqrt{2a^2} = a\sqrt{2}$, which was to be verified.

The sequence of right triangles shown in Figure 10.33 suggests another useful generalization. Hypotenuse lengths were found using the Pythagorean Theorem.

Theorem: 30°–60°–90° Triangle Relationships

The length of the hypotenuse of any 30°–60°–90° triangle is 2 times the length of the shorter leg. The length of the longer leg is the length of the shorter leg times $\sqrt{3}$.

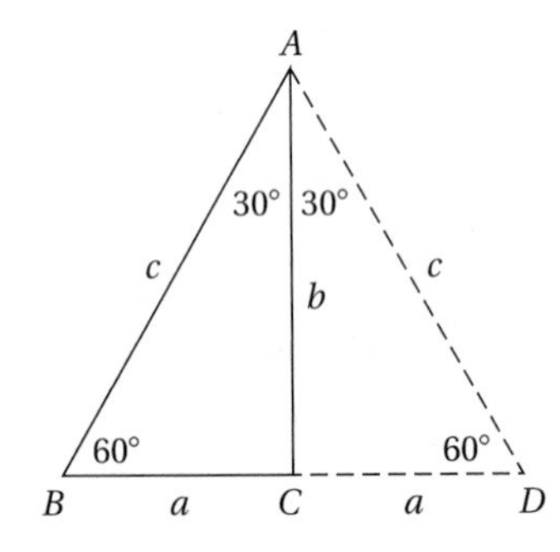

To verify this theorem, we begin with a 30°–60°–90° $\triangle ABC$ and draw $\triangle ADC$ so that $\triangle ABC \cong \triangle ADC$, as shown to the left.

Because $\triangle ABD$ is equiangular, it is also equilateral. So $c = 2a$, and the first part of the theorem has been verified. To verify the second part, we reason as follows.

$$b^2 + a^2 = (2a)^2$$
$$b^2 = 4a^2 - a^2$$
$$b^2 = 3a^2$$
$$b = a\sqrt{3}$$

In Example 10.18 we apply the special triangle generalizations.

Example 10.18

Problem Solving: The Stained Glass Window

The stained glass window shown in the figure is a regular hexagon and is to be placed in an opening in a wall.

a. How high must the opening be?

b. How wide must the opening be?

Solution

a. In the hexagon below, $\overline{AB}$ is the longer leg of a 30°–60°–90° triangle and has length $\sqrt{3}$. The height of the opening needed is $2\sqrt{3}$.

b. $\overline{AC}$ in the hexagon below is the hypotenuse of a 30°–60°–90° triangle and has length 2, which is half the length of any diagonal of the hexagon. The width of the opening needed is 2×2, or 4.

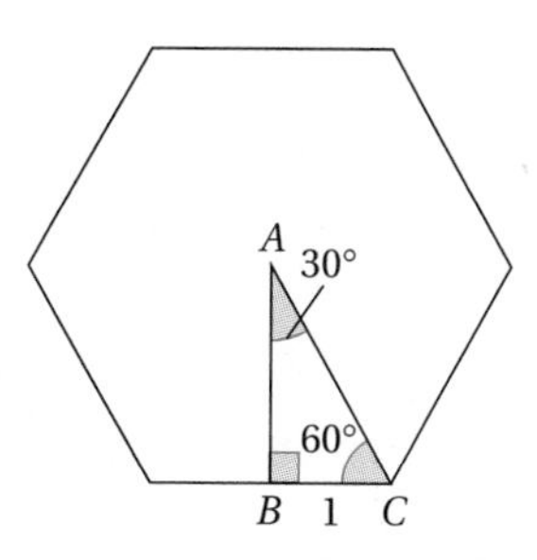

Your Turn

Practice: A carpenter removed a square-shaped window shown in part (a) of the figure. She wondered whether the height h of the regular hexagon-shaped window shown in part (b) of the figure would be more or less than that of the square window. What do you think? Explain.

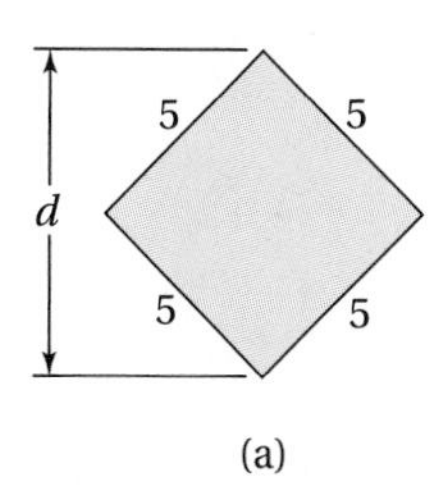

(a)

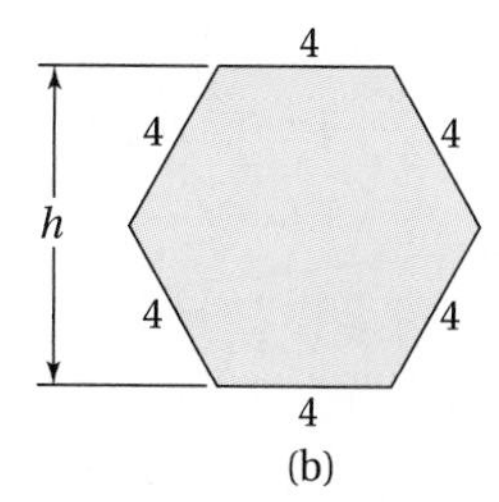

(b)

Reflect: A student claimed that the distance across a hexagon from vertex to vertex is twice the length of the side. Is this claim true for the example problem? Is it always true? If you think that it is, verify it. ■

Problems and Exercises | *for Section 10.4*

A. Reinforcing Concepts and Practicing Skills

For Exercises 1–6, if the triangles are congruent, write a congruence statement and give a reason why they are congruent.

1.

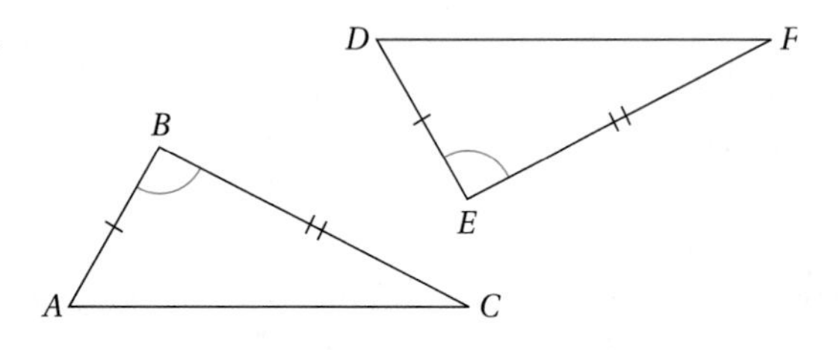

2.

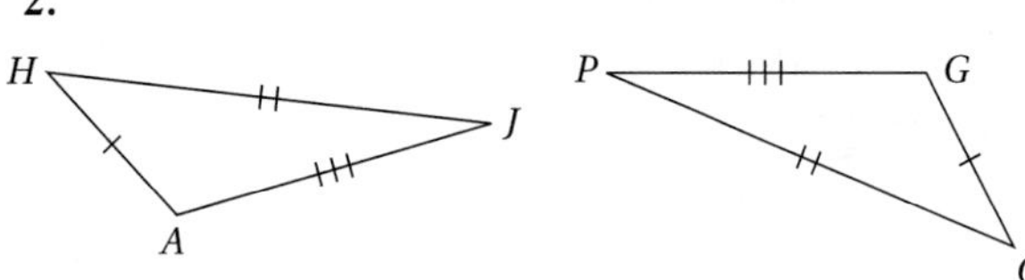

3.

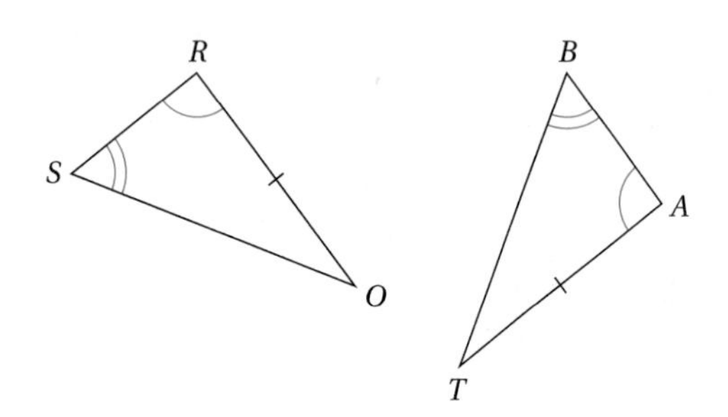

4.

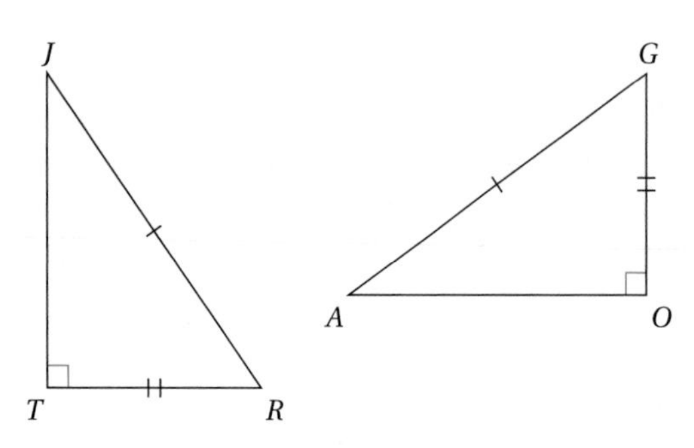

5.

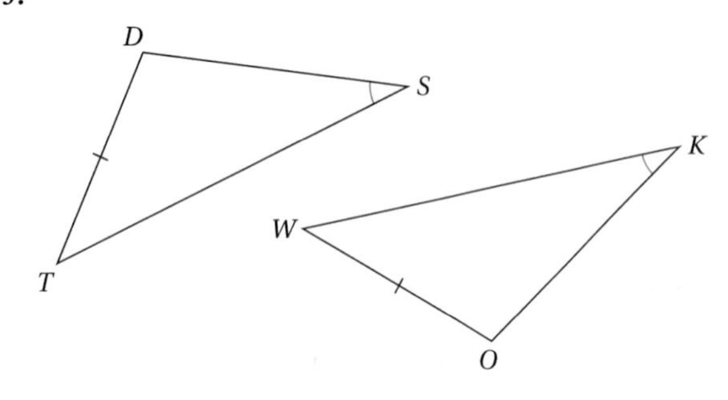

6.

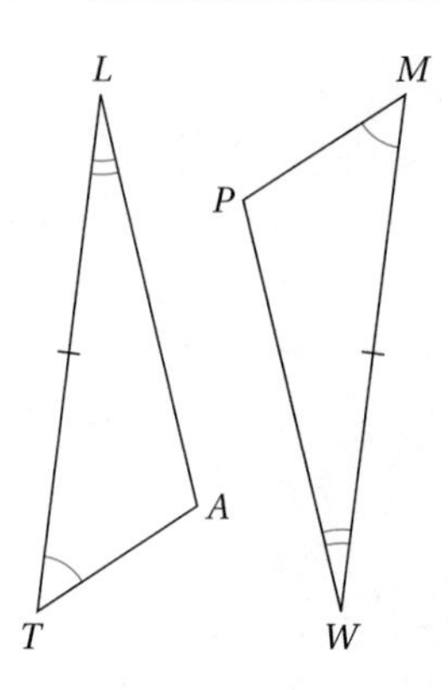

7. Give a counterexample to the following false generalization: If three angles of one triangle are congruent to the corresponding three angles of another triangle, then the triangles are congruent.
8. Give a counterexample to the following false generalization: If two sides and an angle of one triangle are congruent to the corresponding two sides and an angle of another triangle, then the two triangles are congruent.

In Exercises 9–14, find a triangle in the figure that is congruent to the triangle given. Make a statement about this congruence, and give a reason why the two triangles are congruent.

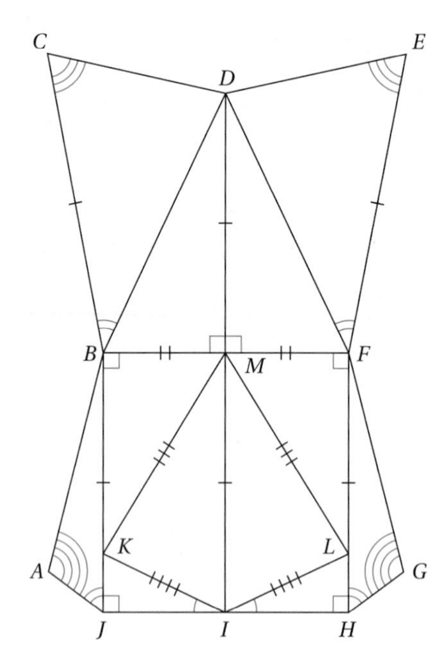

9. $\triangle BDM$	**10.** $\triangle MKI$
11. $\triangle BAJ$	**12.** $\triangle CDB$
13. $\triangle KBM$	**14.** $\triangle KJI$

15. Are all equilateral triangles congruent? What property justifies your conclusion?

Use the Pythagorean theorem to find the distance d *in each of the figures in Exercises 16–20.*

16.

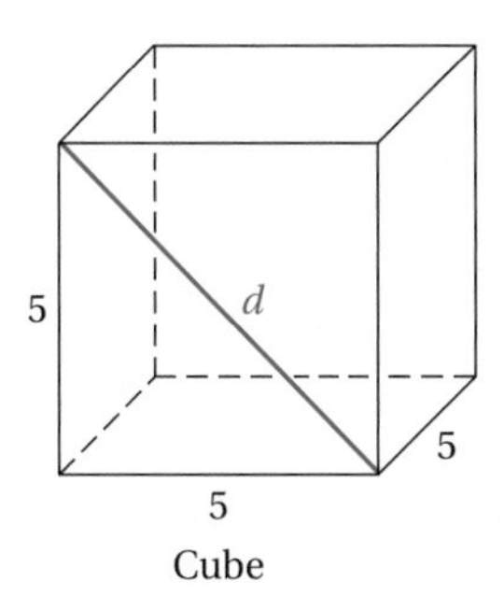

Cube

17.

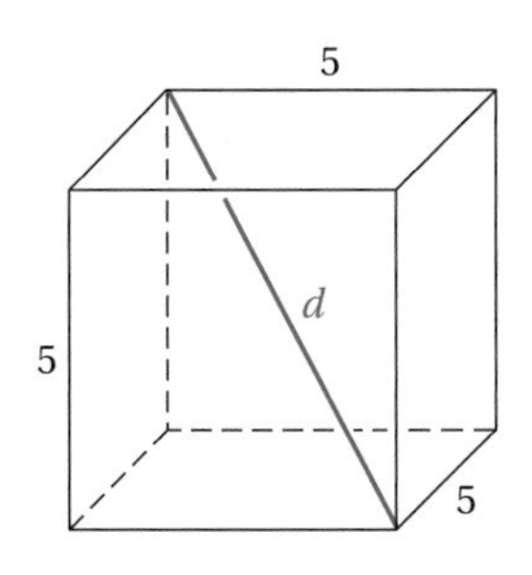

Cube

18.

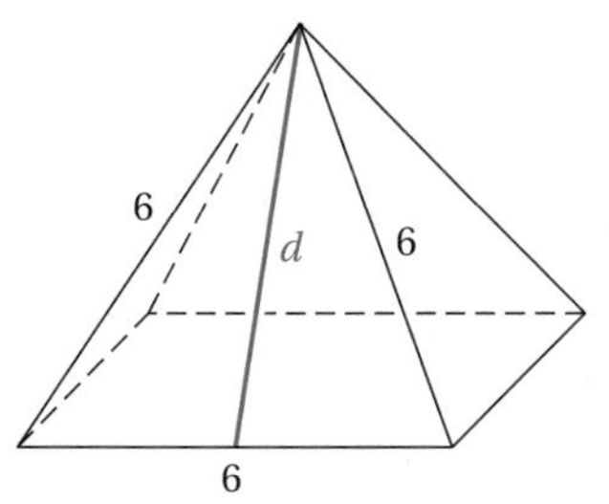

Right square pyramid with equilateral triangular faces

19.

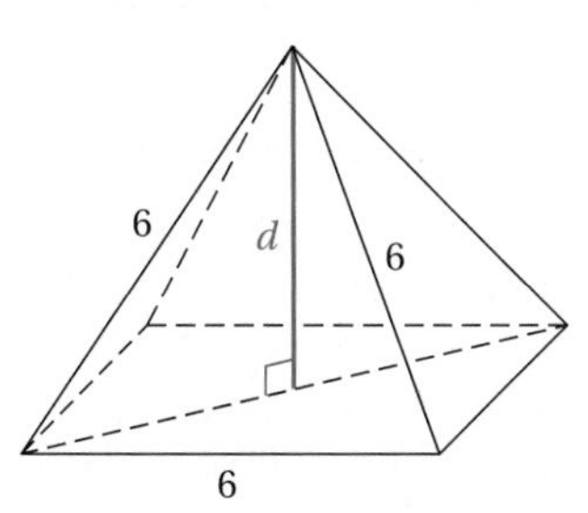

Right square pyramid with equilateral triangular faces

20.

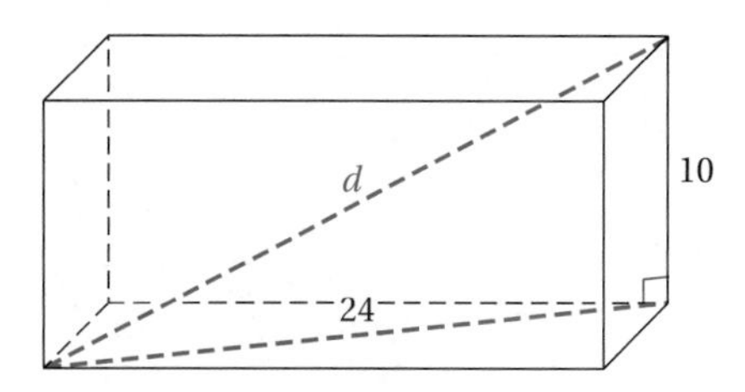

21. Use variables to state the Pythagorean theorem. State in your own words the main idea of this theorem.

22. Use variables to state the converse of the Pythagorean theorem. How does this theorem help you decide if a triangle with sides 5, 12, and 13 is a right triangle?

23. A construction foreman wants to find the length of the diagonal of a square lot that was 120 feet on each side. Explain how she can calculate this distance rather than measure it.

24. The shorter leg of a 30°–60°–90° triangle is 2. What are the lengths of the longer leg and the hypotenuse?

25. A television screen measures approximately 15.5 inches high and 19.5 inches wide. A television is advertised by giving the approximate length of the diagonal of its screen. How should the size of this television be advertised?

26. A TV store owner wanted to make a shelf to display television sets. How long should he make the support board, *b*?

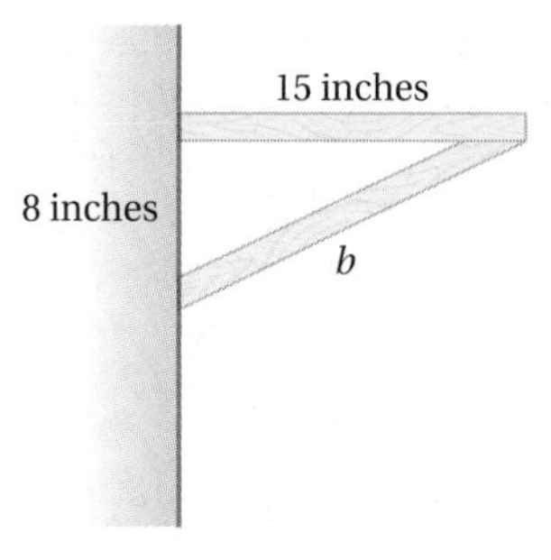

27. A 6-foot board is placed against a wall with its base 2 feet from the wall. How high above the ground is the top of the board?

28. Which set of numbers shows distances between knots in a rope that could be used to lay out a right angle for the corner of a building?

a. 5, 12, 13
b. 4, 5, 6
c. 8, 15, 17
d. 7, 11, 13
e. 7, 24, 25

29. A door is 6 feet 6 inches tall and 36 inches wide. Can a 7-foot square piece of plywood be carried through the door? Why or why not?

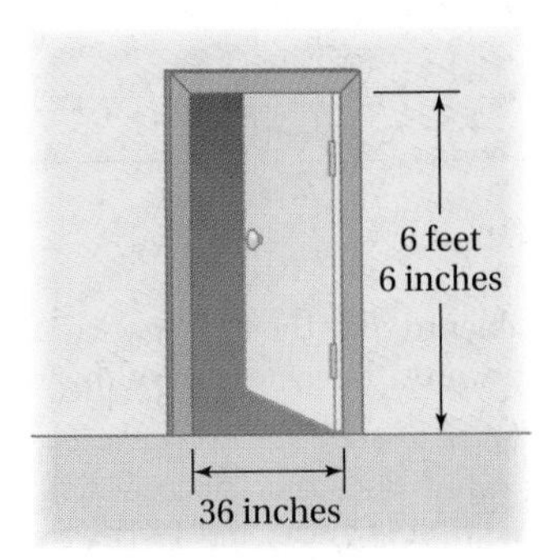

30. A grounds crew laying out a baseball diamond measured the distance from home plate to second base to be sure that the base path angle at first base was a right angle. What should the distance be?

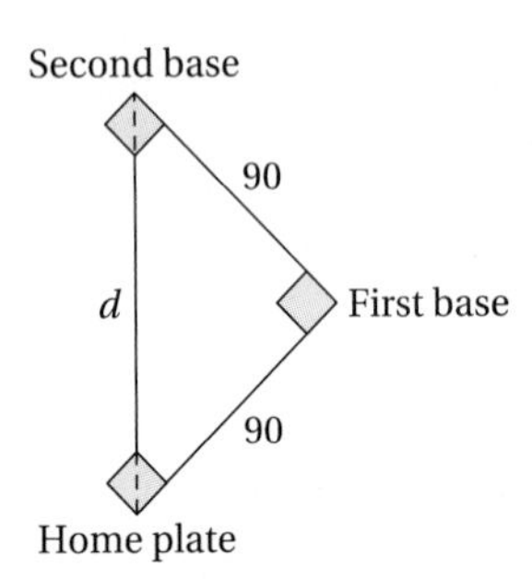

31. To determine how much hedge to plant around an irregularly shaped lot, a landscape architect drew the lot to scale on a grid. If 1 unit represents 25 feet, how many feet of hedge would be needed to outline the lot?

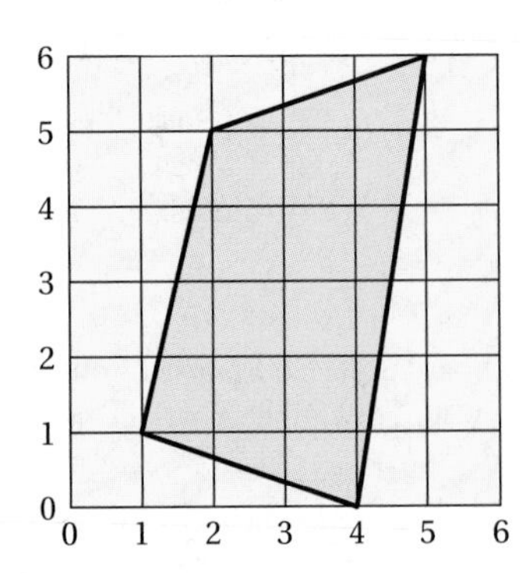

32. About how much rope would be required to make three support lines for a tree if the lines were tied to the tree 12 feet above the ground and to stakes 5 feet from the base of the tree?

33. Complete each statement.

a. To find the length of the hypotenuse of a 45°–45°–90° triangle, multiply the length of one of the legs by ___.

b. To find the length of the hypotenuse of a 30°–60°–90° triangle, multiply the length of the shorter leg by ___.

c. To find the length of the longer leg of a 30°–60°–90° triangle, multiply the length of the shorter leg by ___.

d. To find the length of the shorter leg of a 30°–60°–90° triangle, multiply the length of the longer leg by ___.

Find the value of the variables in each triangle in Exercises 34–36.

34.

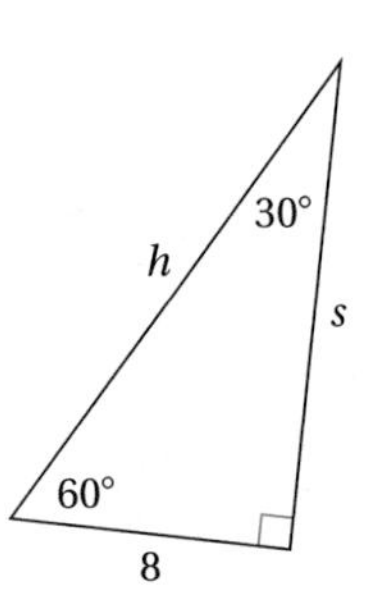

35.

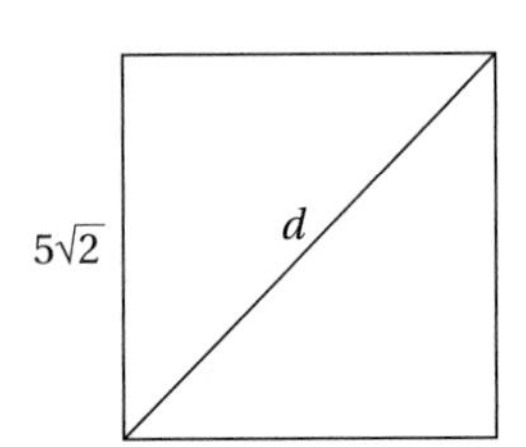

36.

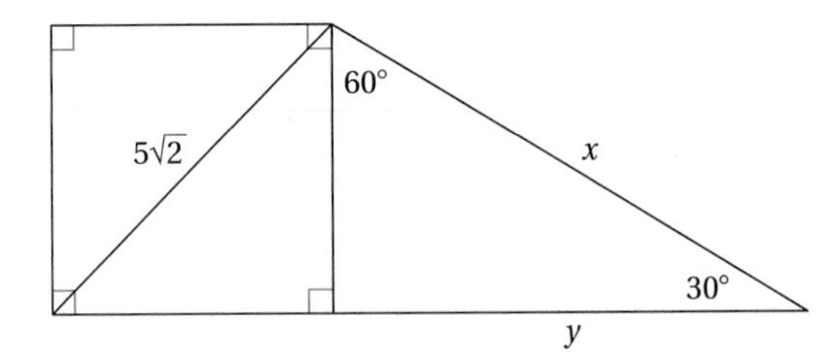

B. Deepening Understanding

Complete each statement in Exercises 37–39.

37. The length of a leg of a 45°–45°–90° triangle is ___ times the length of the hypotenuse.

38. The length of the hypotenuse of a 30°–60°–90° triangle is ___ times the length of the longer leg.

39. The length of the shorter leg of a 30°–60°–90° triangle is ___ times the length of the hypotenuse.

For Exercises 40–42, give the congruence property (SSS, SAS, ASA, or AAS) that justifies the conclusion.

40. If two legs of a right triangle are congruent to two legs of another right triangle, then the two right triangles are congruent.

41. If a leg and an acute angle of one right triangle are congruent to the corresponding leg and acute angle of another right triangle, then the two right triangles are congruent.

42. If the hypotenuse and an acute angle of one right triangle are congruent to the hypotenuse and corresponding acute angle of another right triangle, then the two right triangles are congruent.

43. A sculptor wants to make a pattern for a metal cone 10 inches high with a base diameter of 8 inches. A paper cone model of the sculpture is shown. How long should length s in the pattern be?

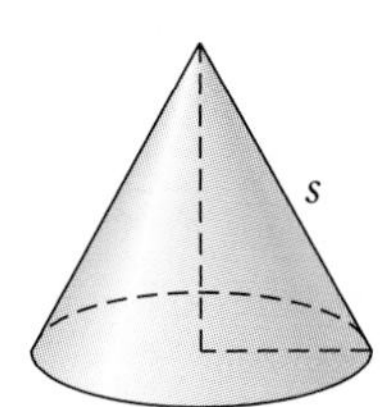

Paper cone model

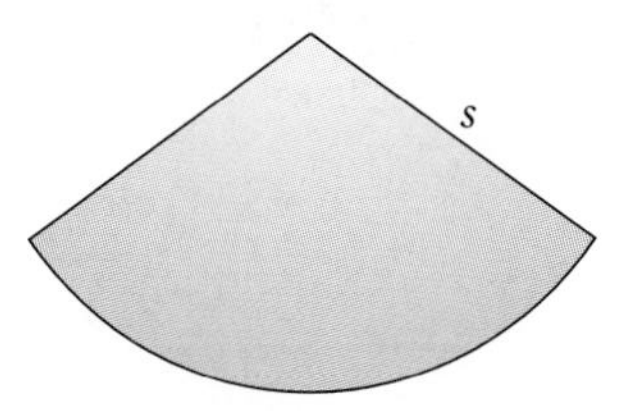

44. The Pythagorean theorem "always," "sometimes," or "never"

a. states a relationship among the sides of an obtuse triangle.

b. states a relationship among the sides of an isosceles triangle.

c. states a relationship among the sides of a scalene triangle.

d. states a relationship among the sides of a right triangle.

Explain how you decided in each case.

45. Which of the following triples of numbers can be the sides of a right triangle? Explain why.

a. $\sqrt{3}, \sqrt{4}, \sqrt{7}$ **b.** 0.3, 0.4, 0.5

c. 10, 24, 26 **d.** 2, 3, 4

46. How far is it from point $F(4, 3)$ to point $G(8, 5)$ on a coordinate grid? Give a convincing argument that your method gives the correct distance.

47. In the figure shown, find the lengths of the diagonals for $s = 2, 3, 4$, and 5. Look for a pattern and make a conjecture about a formula that a carpenter could use for finding the diagonal of any square with side s. Use the Pythagorean theorem to show that your method is correct.

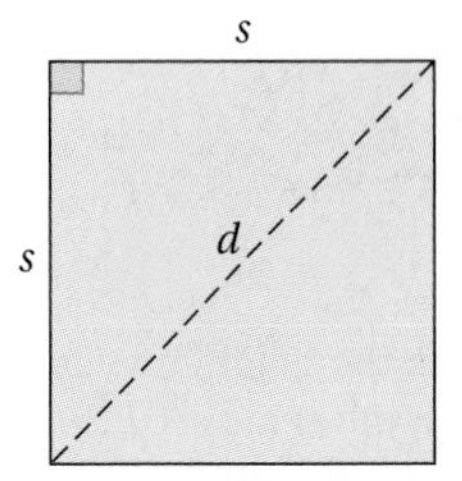

C. Reasoning and Problem Solving

48. The Curved Patio Problem. A landscape engineer is planning to add a patio to a home as part of a landscaping project. Her crew uses a thin flexible strip of wood as a border for the circular portion of the patio. The strip of wood is 16 feet long and is bent in the arc of a circle. Two radii, from the center of the circle to the ends of the arc, form a right angle. What is the approximate distance from one end of the wooden arc to the other?

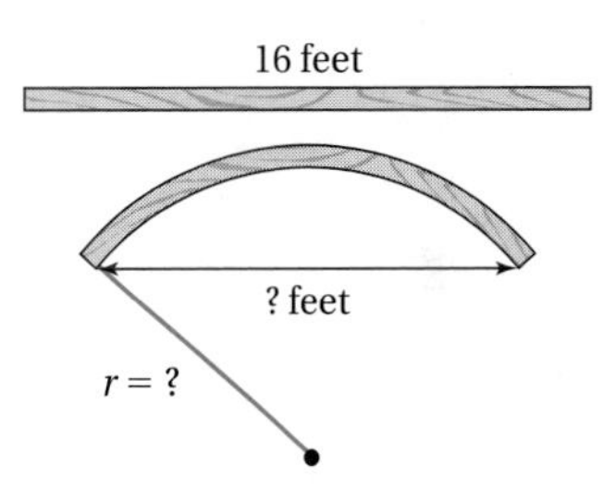

Use a computer, if available and helpful, for Exercises 49–56.

49. Construct external equilateral triangles to each side of a triangle. What figure is formed when the centroids of these equilateral triangles are connected?

50. Use a protractor or GES to draw the trisectors of each angle of a triangle. What figure is formed when you connect the three points of intersection of adjacent trisectors?

51. Begin with any triangle. Construct the following nine points: the midpoints of the three sides, the feet of the three altitudes, and the midpoints of the segments from the three vertices to the orthocenter. How are these nine points related? Is there any relationship between your discovery and the Euler line?

52. Begin with any triangle *ABC*, and construct the midpoints of the sides, *P*, *Q*, and *R*, and connect them as shown in the figure.

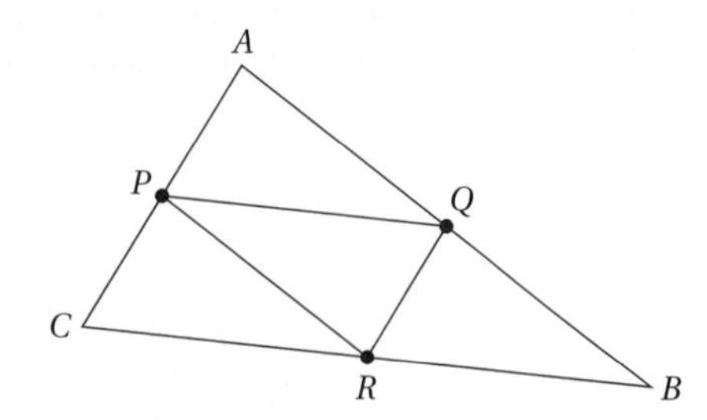

a. How does the length of segment $\overline{PQ}$ compare with the length of segment $\overline{CB}$?
b. How does triangle *PQR* compare with triangles *QPA*, *RCP*, and *BRQ*?

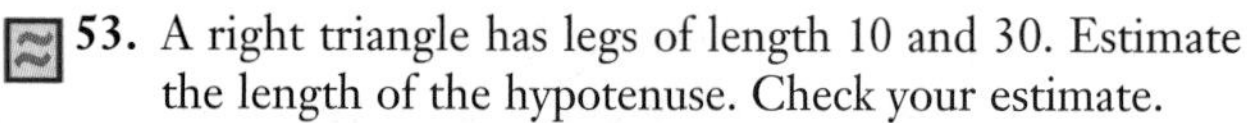

53. A right triangle has legs of length 10 and 30. Estimate the length of the hypotenuse. Check your estimate.

54. Give three or four key statements that could be used to help prove the HA Congruence theorem (see p. 557). Give reasons that support the truth of each statement.

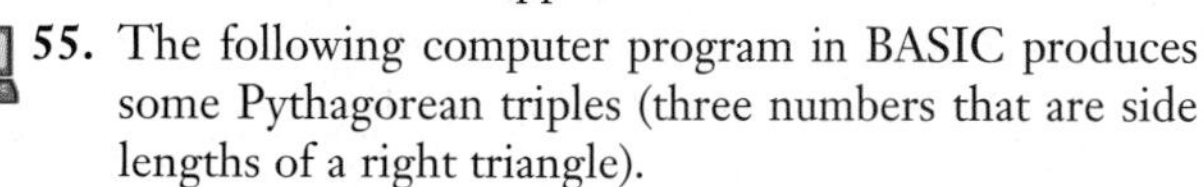

55. The following computer program in BASIC produces some Pythagorean triples (three numbers that are side lengths of a right triangle).

```
10 FOR M=2 TO 5
20 FOR N=1 TO M-1
30 LET A=M*M-N*N
40 LET B=2*M*N
50 LET C=M*M+N*N
60 PRINT A;",";B;",";C
70 NEXT N
80 NEXT M
90 END
```

a. What formula is used to produce the Pythagorean triples? How does M relate to N?
b. Run the program. Can you change the program so that it will produce more triples?
c. In *primitive* Pythagorean triples, the three numbers have no common factor other than 1. Give three primitive triples.
d. If n is a positive integer, then $(2n + 1, 2n^2 + 2n, 2n^2 + 2n + 1)$ is a Pythagorean triple. Change the program so that it uses this formula to produce triples. Do the two programs produce the same triples in the first 10 tries? Describe any differences.

56. A special cutter for use in milling, like the one shaded in the figure, is made by cutting seven right triangles from a seven-sided regular polygon (septagon). If the marked segment is cut the same length for each tooth, why are the sharp points of the cutter all the same size angle?

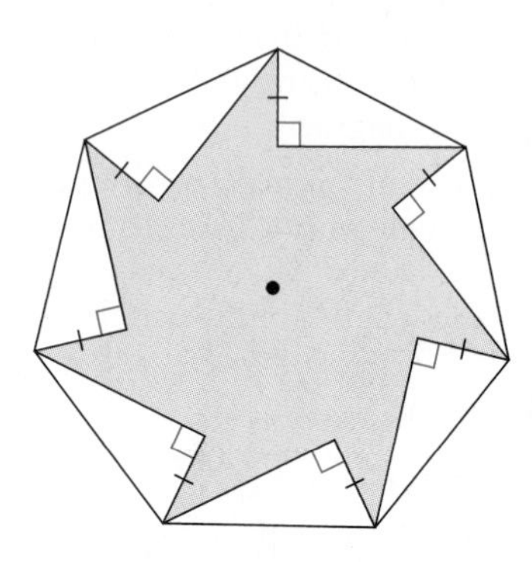

57. Use algebra and the following diagram to verify the Pythagorean theorem.

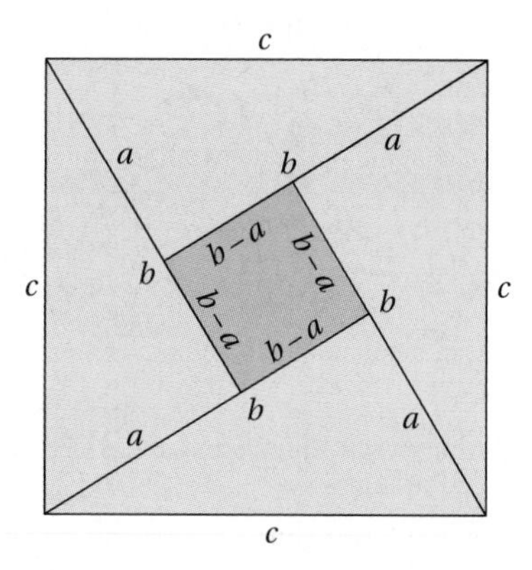

58. Use the diagrams shown below to write a convincing argument that $a^2 + b^2 = c^2$. No algebra, please!

Figure for Exercise 58

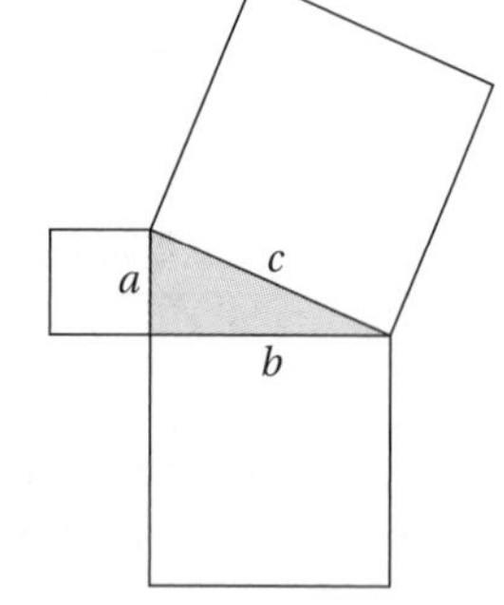

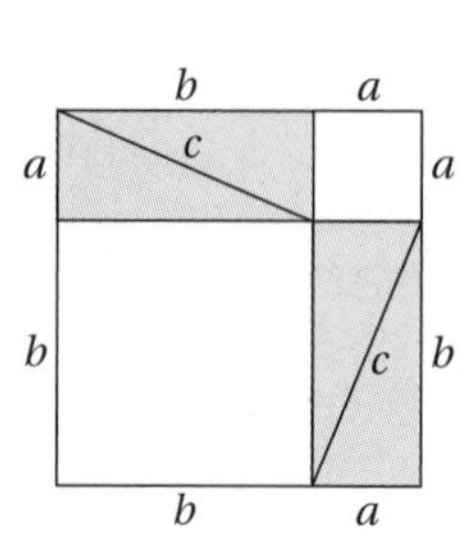

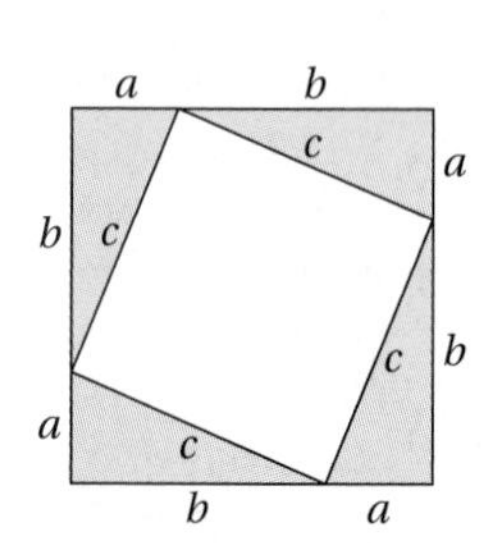

59. A student studied a right triangle and said:

> The square of a leg can be found by multiplying the sum of the other leg and the hypotenuse by the difference of that leg and the hypotenuse.

Do you agree? Use algebra to support your conclusion.

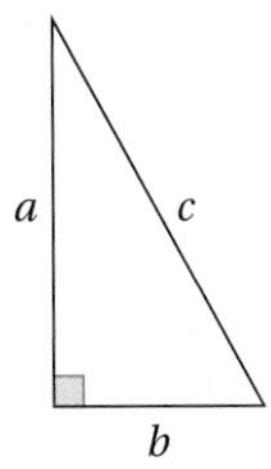

60. a. If you begin with a right triangle with sides a, b, and c and decrease the right angle so it is an acute angle, how does $a^2 + b^2$ compare with c^2?

b. Answer this question for the situation where you increase the right angle so that it is obtuse.

61. A person claimed: When the hypotenuse and a leg differ by 1, you can easily find the other leg. For example, just add 24 and 25; b is the square root of this sum.

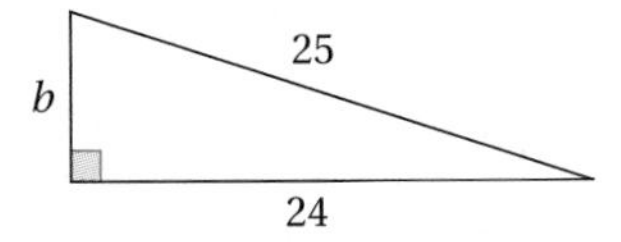

Do you agree? How could you decide? Give the missing side in the following triples.

a. ?, 7.5, 8.5

b. ?, 9, 10

c. ?, 40,41

d. ?, 60, 61

62. Which of the following do you think are true? Give reasons to support your decision.

a. When the sides of a triangle are m, $(\frac{m^2}{4}) - 1$, and $(\frac{m^2}{4}) + 1$, where m is an even number, the triangle is a right triangle.

b. If n is an integer and $a = 2n + 1$, $b = 2n^2 + 2n$, and $c = 2n^2 + 2n + 1$, then a, b, and c are the sides of a right triangle.

c. The 3, 4, 5 right triangle is one of several right triangles with sides that are consecutive integers. (*Hint:* Use x, $x + 1$, and $x + 2$ to represent consecutive integers.)

63. The Box Kite Problem. A kite maker wants to cut a wooden piece that would extend from A to B to strengthen a box kite. How long must the piece be?

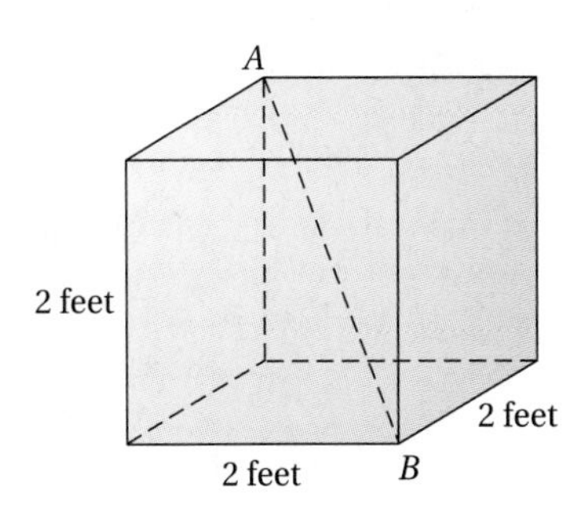

D. Communicating and Connecting Ideas

64. Work in a small group to discuss the following: The Pythagorean theorem states that "the sum of the areas of the squares on the legs of a right triangle is equal to the area of the square on the hypotenuse." What if the word *squares* were replaced by the word(s)

a. *semicircles?*

b. *equilateral triangles?*

Would the statement still be true? Devise ways to support your conclusions.

65. Historical Pathways. Throughout recorded history, people in various walks of life have had a recreational interest in mathematics. For example, Representative James A. Garfield discovered a proof of the Pythagorean theorem after a mathematical discussion with some members of Congress in about 1876. (He later became the twentieth president of the United States.)

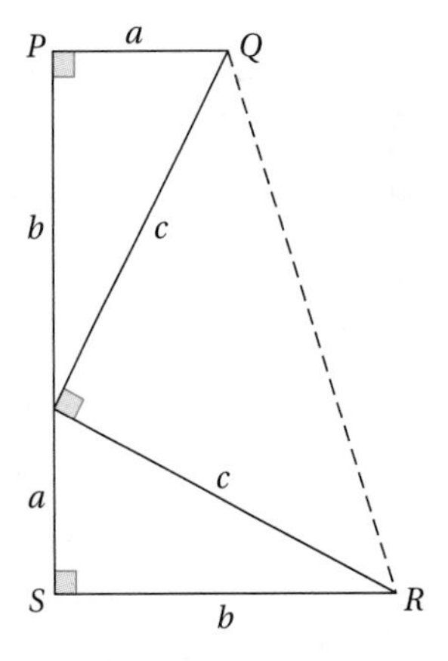

The key to the proof is to express the area of trapezoid $PQRS$ in two different ways and then to simplify the equation showing the equal expressions. Try to write the proof.

66. Making Connections. Illustrate the connection between algebra and geometry by using algebra and the figures in Exercise 58 to make a convincing argument that $a^2 + b^2 = c^2$. Discuss how your geometric and algebraic arguments are alike.

Section 10.5 MORE ABOUT QUADRILATERALS

- Properties of Quadrilaterals
- Quadrilaterals and Geometric Constructions
- What-If Questions about Quadrilaterals

In Section 10.1, we described a quadrilateral as a polygon with four sides. In this section, with the help of GES or other tools, we investigate properties of some basic quadrilaterals, including parallelograms, rectangles, rhombi, and squares. Mini-Investigation 10.6 provides a systematic way to do this.

Talk about how the information in the table relates to the diagram of the relationships between quadrilaterals that you investigated in Exercise 64 in Section 10.1.

Mini-Investigation 10.6 The Technology Option

Use geometry exploration software or other tools to draw several examples of each quadrilateral listed in the column headings of the table below. Use your examples to decide which characteristics hold for each type of quadrilateral, make a table like the one shown, and record your decisions by filling in the appropriate boxes in your table.

Attribute	Parallelogram	Rectangle	Rhombus	Square
Opposite sides parallel	?	?	?	?
Opposite sides congruent	?	?	?	?
Opposite angles congruent	?	?	?	?
Adjacent sides perpendicular	?	?	?	?
Diagonals congruent	?	?	?	?
Diagonals bisect each other	?	?	?	?
Diagonals perpendicular	?	?	?	?

Properties of Quadrilaterals

A **parallelogram** is defined as a quadrilateral with two pairs of parallel sides. It is natural to ask if there are other properties of a quadrilateral, other than two pairs of parallel sides, that would ensure that the quadrilateral is a parallelogram. For example, we might use GES to draw several quadrilaterals with opposite sides congruent, and measure to see if the quadrilaterals were parallelograms. Our conjecture might be: If the opposite sides of a quadrilateral are congruent, then the quadrilateral is a parallelogram. We can then use geometry theorems we have studied to verify our conjecture. Five key statements needed for this verification are shown in Figure 10.34. Think about the given information and theorems of geometry to see why these statements are true.

1. In quadrilateral $ABCD$, $\overline{AB} \cong \overline{CD}$ and $\overline{BC} \cong \overline{AD}$ (given data)
2. $\triangle ABC \cong \triangle CDA$
3. $\angle 1 \cong \angle 2$ and $\angle 3 \cong \angle 4$
4. $\overline{AB} \parallel \overline{CD}$ and $\overline{BC} \parallel \overline{AD}$
5. $ABCD$ is a parallelogram

B C 2 3 4 1 A D

FIGURE 10.34 | Key ideas for proving that a quadrilateral with opposite sides congruent is a parallelogram.

Other characteristics of a quadrilateral, such as information about angles or diagonals, might also ensure that the quadrilateral is a parallelogram. Other generalizations about parallelograms that might be discovered using GES are given as follows. They too can be verified using logical reasoning.

Properties of Parallelograms

A quadrilateral is a parallelogram if and only if

- opposite sides are parallel; or
- opposite sides are congruent; or
- one pair of opposite sides are both congruent and parallel; or
- opposite angles are congruent; or
- consecutive angles are supplementary; or
- its diagonals bisect each other.

A **rectangle** is defined as a quadrilateral with four right angles. It is also natural to ask if there are other properties of a parallelogram, other than four right angles, that would ensure that the parallelogram is a rectangle. For example, we might use GES to draw several parallelograms with congruent diagonals, and measure to see if the parallelograms were rectangles. Our conjecture might be: If the diagonals of a parallelogram are congruent, then the parallelogram is a rectangle. We can then use geometry theorems we have studied to verify our conjecture. Five key statements needed for this verification are shown in Figure 10.35. Think about the given information and theorems of geometry to see why these statements are true.

A summary of the characteristics of parallelograms that ensure that they are rectangles are given as follows. They can be discovered using GES and verified using logical reasoning.

Properties of Rectangles

A parallelogram is a rectangle if and only if

- it has at least one right angle, or
- its diagonals are congruent.

1. In parallelogram *ABCD*, $\overline{AC} \cong \overline{DB}$ (given data)
2. $\triangle ABC \cong \triangle CDA \cong \triangle BAD \cong \triangle DCB$
3. $\angle A \cong \angle B \cong \angle C \cong \angle D$
4. $\angle A$, $\angle B$, $\angle C$, and $\angle D$ are right angles
5. *ABCD* is a rectangle

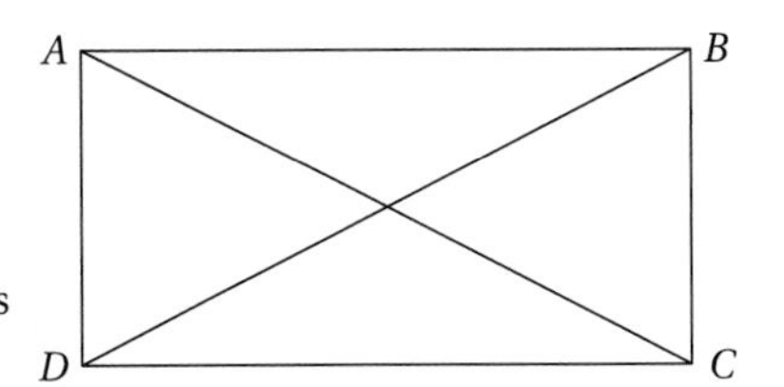

FIGURE 10.35 | Key ideas for proving that a parallelogram with congruent diagonals is a rectangle.

1. In parallelogram *ABCD*, $\overline{AC}$ and $\overline{DB}$ are perpendicular bisectors of each other (given data)
2. $\triangle AED \cong \triangle AEB \cong \triangle CEB \cong \triangle CED$
3. $\overline{AD} \cong \overline{AB} \cong \overline{BC} \cong \overline{CD}$
4. *ABCD* is a rhombus

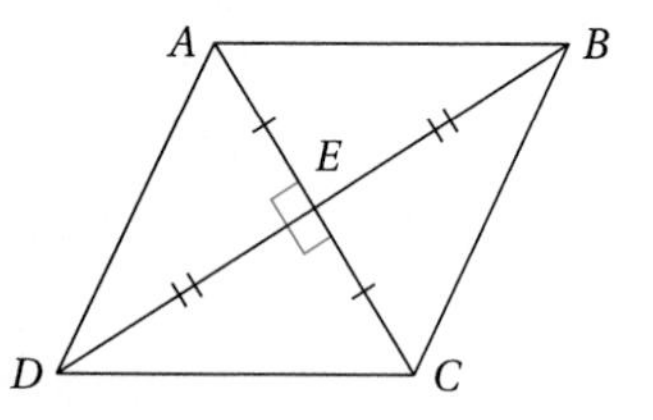

FIGURE 10.36 | Key ideas for proving that a parallelogram with diagonals that are perpendicular bisectors of each other is a rhombus.

A **rhombus** is usually defined as a quadrilateral with four congruent sides. A **square** is usually defined as a quadrilateral with four right angles and four congruent sides.

Let's explore to see if there are other properties of a parallelogram, other than four congruent sides, that would ensure that the parallelogram is a rhombus. For example, we might use GES, draw several parallelograms in which the diagonals are perpendicular bisectors of each other, and measure to see if the parallelograms were rhombi. Once we have conjectured that if the diagonals of a parallelogram are perpendicular bisectors of each other then the parallelogram is a rhombus, we can use geometry theorems we have studied to verify our conjecture. Four key statements needed for this verification are shown in Figure 10.36. Think about the given information and theorems of geometry to see why these statements are true. A summary of the characteristics of parallelograms that ensure that they are rhombi or squares, that might be discovered using GES, are given as follows. They can be verified using logical reasoning.

Properties of Other Special Parallelograms

A parallelogram is a rhombus if and only if

- it has four congruent sides; or
- its diagonals bisect the angles; or
- its diagonals are perpendicular bisectors of each other.

A parallelogram is a square if and only if

- it is a rectangle with four congruent sides; or
- it is a rhombus with a right angle; or
- its diagonals are congruent and perpendicular bisectors of each other.

Two other quadrilaterals of interest are the **kite,** defined as a quadrilateral with at least two pairs of adjacent sides congruent, no side used twice in the pairs, and the **trapezoid,** defined as a quadrilateral with exactly one pair of parallel sides. In Exercises 14, 22, and 51 at the end of this section, we will consider some generalizations about kites and trapezoids. In Example 10.19 you are asked to apply the generalizations stated above about quadrilaterals, parallelograms, rhombi, and squares.

Example 10.19

Drawing Conclusions about Different Types of Quadrilaterals

ABCD is a quadrilateral with diagonals $\overline{AC}$ and $\overline{BD}$. Draw a picture, and make three true statements about *ABCD* if *ABCD* is

a. a parallelogram **b.** a rectangle **c.** a rhombus **d.** a square

Solution

a. $\overline{AB} \cong \overline{DC}$, $\overline{AB} \parallel \overline{DC}$, $\overline{AC}$ bisects $\overline{DB}$
b. $\overline{AD} \cong \overline{BC}$, $\angle A$ is a right angle, $\overline{AB} \parallel \overline{DC}$
c. $\overline{AC}$ bisects $\angle A$, $\overline{AC}$ is the perpendicular bisector of $\overline{DB}$, $\overline{AD} \cong \overline{AB}$
d. *ABCD* is a rectangle, *ABCD* has four congruent sides, $\overline{AC} \cong \overline{DB}$

Your Turn

Practice: Complete the original task, giving statements different than those in the solution.

Reflect: Give names of quadrilaterals to complete this statement in several different ways. A ___ is also a ___. ■

Quadrilaterals and Geometric Constructions

Properties of quadrilaterals, described in the previous subsection, can be used to justify the procedures described in Appendix D for using only a straightedge and compass to construct basic geometric figures and relationships. For example, consider the procedure in Appendix D for constructing the bisector of an angle, with auxiliary lines drawn as shown in Figure 10.37.

In this figure, a compass is used with the same opening to mark arcs at points A, C, and E, so that $\overline{AB} \cong \overline{BC} \cong \overline{CE} \cong \overline{EA}$. So *ABCE* is a rhombus, and in a rhombus a diagonal bisects the angle. This verifies that $\overrightarrow{BE}$ is the bisector of $\angle B$.

The procedure for constructing the perpendicular bisector of a line segment (explained in Appendix D, in the section titled "Construct the Perpendicular Bisector of a Line Segment") is justified by using the property that the diagonals of a rhombus are perpendicular, as illustrated in Example 10.20.

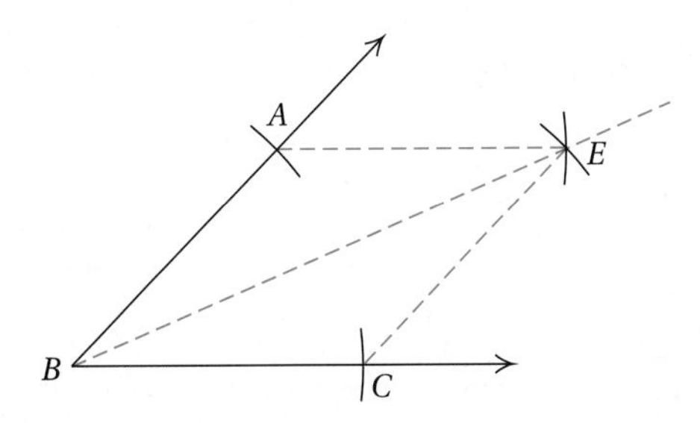

FIGURE 10.37 | Using the properties of a rhombus to justify the construction procedure for bisecting an angle.

Example 10.20 Justifying the Procedure for Constructing Perpendicular Bisectors

Examine the process for constructing the perpendicular bisector of a line segment (in Appendix D). How can you use the characteristics of a rhombus to explain why this construction works?

Solution

The intersection points C and D of the two arcs as shown in the diagram below (taken from Figure D.6d in Appendix D) are constructed so that $\overline{AC}$, $\overline{CB}$, $\overline{BD}$, and $\overline{DA}$, are all congruent to one another. Therefore, since quadrilateral $ACBD$ has four congruent sides, it is a rhombus.

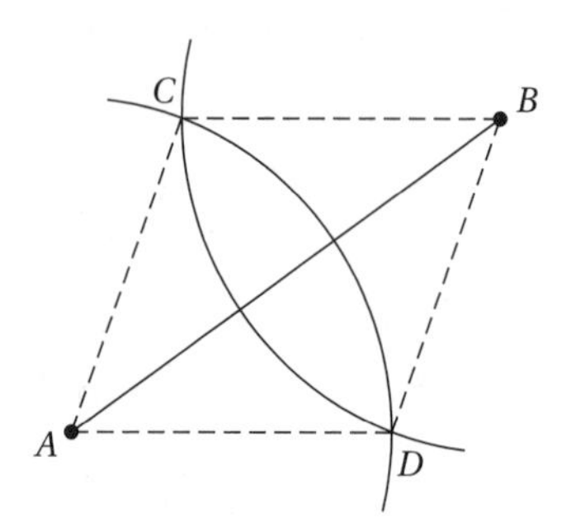

Since $\overline{AB}$ and $\overline{CD}$ are the diagonals of the rhombus, they are perpendicular bisectors of each other, and the construction process is verified.

Your Turn

Practice: How can the properties of a rhombus be used to verify the process for constructing perpendicular lines? (See "Construct a Perpendicular to a Line Through a Given Point on the Line" in Appendix D.)

Reflect: A quadrilateral is a kite if and only if its diagonals are perpendicular bisectors of each other (and a rhombus is a special type of kite). How could you change the process of constructing perpendicular lines to use kites that don't have to be rhombi? ■

Estimation Application

The Size of a State

On a map the State of Nevada appears roughly as a quadrilateral and has the dimensions shown. The map scale is 1 inch = 72 miles.

1. Estimate the area of Nevada in square miles.
2. Explain how you arrived at your estimate. Use (a) an encyclopedia or reference book or (b) GES to check your estimate.

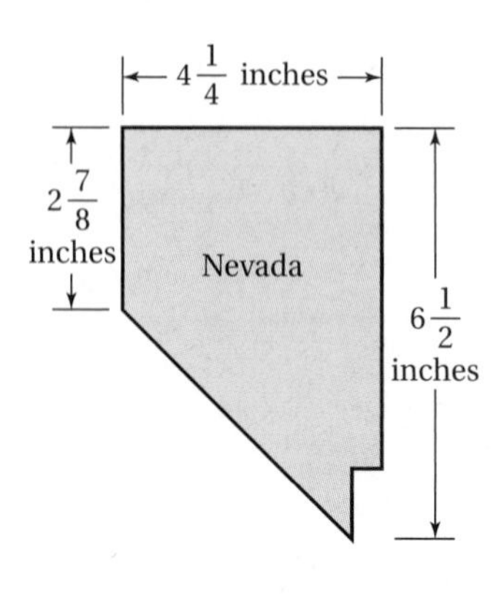

What-If Questions about Quadrilaterals

In a clever little book called *Grooks 2*, written by Piet Hien, there appears the following.

> We shall have to evolve
> problem-solvers galore—
> since each problem they solve
> creates ten problems more. (p. 32, Doubleday and Co., 1969)

Many new, interesting problems or generalizations are often sparked by asking What-if? questions about a problem we're solving. In fact, cultivating the habit of asking "What if...?"about problems or generalizations and varying the conditions can pay huge dividends in discovering new patterns and relationships in geometry.

For example, suppose you were working on the generalization that the segment joining the midpoints of two sides of a triangle is parallel to and has length $\frac{1}{2}$ that of the third side, as shown in Figure 10.38(a). You could ask the question, "What if the figure were a trapezoid, as in Figure 10.38(b), instead of a triangle?",

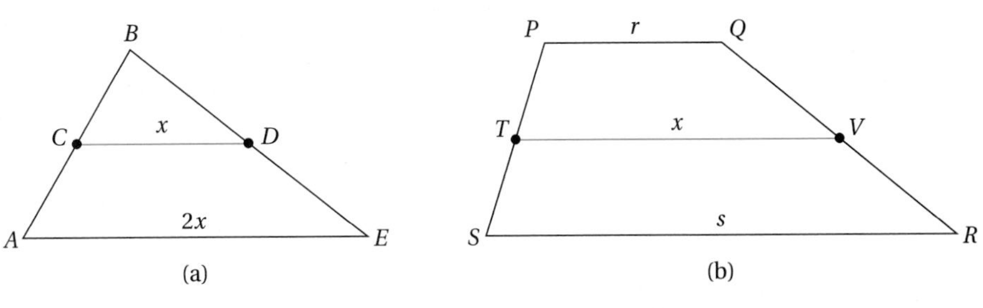

FIGURE 10.38 | Asking a what-if question about a generalization.

and a new generalization is waiting to be discovered. Exercise 51 at the end of this section asks you to discover this generalization.

A second example of the value of asking what-if questions involves the Strange Subdivision Problem that you solved in Exercise 50 at the end of Section 10.2. This problem illustrated the very interesting generalization that when segments are drawn to connect the vertices of a triangle to trisection points on the sides of the triangle, the area of the region formed in the center of the triangle is *one-seventh* the area of the original triangle. You could ask, "What if we begin with a quadrilateral instead of a triangle?" This question suggests the problem in Example 10.21.

Example 10.21

Problem Solving: The Strange Subdivision Revisited

An eccentric real estate developer always bought plots of land in the shape of quadrilaterals and subdivided them into nine lots. He used the following method: Find the trisection points of each side of the quadrilateral; then connect each vertex with a trisection point, as shown. He offered to sell the complete quadrilateral plot for $1,000,000. If the subdivided lots are priced in proportion to their areas, what would be a fair price for the shaded interior lot? Predict first and then solve and justify your solution.

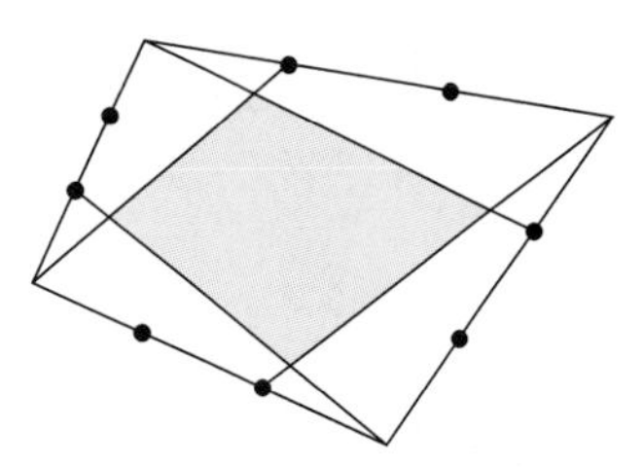

Working toward a Solution

Use GES, if available. Construct a large quadrilateral and draw the appropriate segments from its vertices to the trisection points of the sides. Now calculate the areas of the large quadrilateral and the shaded interior quadrilateral, and calculate the ratio of the two.

If GES isn't available, use construction and measurement.

Your Turn

Practice: Complete the solution to the example problem. State a generalization, if possible.

Reflect: Would your answers to the problem be the same regardless of the shape of the original quadrilateral? Why or why not? ■

In Example 10.21, what if we use a parallelogram instead of a general quadrilateral? What if we connect the vertices to midpoints rather than trisection points? What if we connect vertices to midpoints going in both directions around the parallelogram? These questions suggest the problem in Exercise 50 at the end of this section. When working with such a problem, it helps to use large figures and to use at least three different types of figures in each situation, to help ensure that the generalization discovered doesn't depend on the type, size, or shape of the figure. Asking what-if questions can be very interesting!

The *Principles and Standards for School Mathematics,* in discussing problem solving and geometry in grades 6–8, states that "An important aspect of a problem-solving orientation toward mathematics is making and examining conjectures raised by solving a problem and posing follow-up questions .... A teacher might orchestrate a discussion in which students pose a variety of What if questions. ... Such conjectures can easily be examined by using interactive geometry software which can also facilitate students' search for a counterexample to disprove a conjecture" (p. 261). For example, when students are studying the SSS property of congruent triangles, they might ask, "What if we applied this idea to quadrilaterals?" and potential generalizations can be discussed.

Problems and Exercises for Section 10.5

A. Reinforcing Concepts and Practicing Skills

For Exercises 1–3, give as many true statements about figure ABCD *as you can.*

1. Assume $ABCD$ is a parallelogram.
2. Assume $ABCD$ is a rectangle.
3. Assume $ABCD$ is a rhombus.

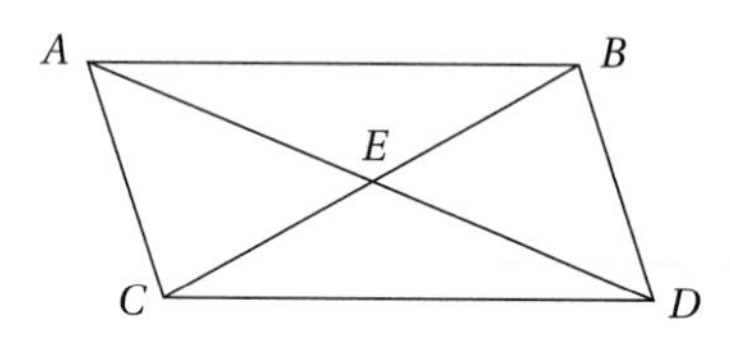

For Exercises 4–10, choose the word "parallelogram," "rectangle," "square," "rhombus," or "kite" that could be used in the blank to make the statement true. If none of the words can be used, write "none" and explain why not.

4. A quadrilateral is a ___ if and only if all of its angles are right angles.
5. A quadrilateral is a ___ if and only if all of its sides are congruent.
6. A quadrilateral is a ___ if and only if its diagonals are perpendicular bisectors of each other.
7. A quadrilateral is a ___ if and only if its sides are congruent and its diagonals are congruent.
8. A quadrilateral is a ___ if and only if one pair of opposite sides are parallel and the other pair of opposite sides are congruent.
9. A quadrilateral is a ___ if and only if a pair of adjacent sides are congruent.
10. A quadrilateral is a ___ if and only if its diagonals bisect its angles.

Draw the figure named with the conditions given in Exercises 11–17, or indicate "not possible."

11. parallelogram: diagonals that bisect each other
12. rhombus: all angles congruent
13. parallelogram: congruent and perpendicular diagonals
14. kite: no right angles, diagonals congruent
15. rectangle: all sides congruent, with diagonals that are not perpendicular
16. parallelogram: not all angles congruent, diagonals not congruent nor perpendicular
17. rhombus: not all angles congruent, diagonals not perpendicular

Answer Exercises 18–22 for each figure (a)–(f) on the next page, given only the conditions shown by the congruent angle and segment marks.

18. Must the figure be a parallelogram?
19. Must the figure be a rectangle?
20. Must the figure be a rhombus?

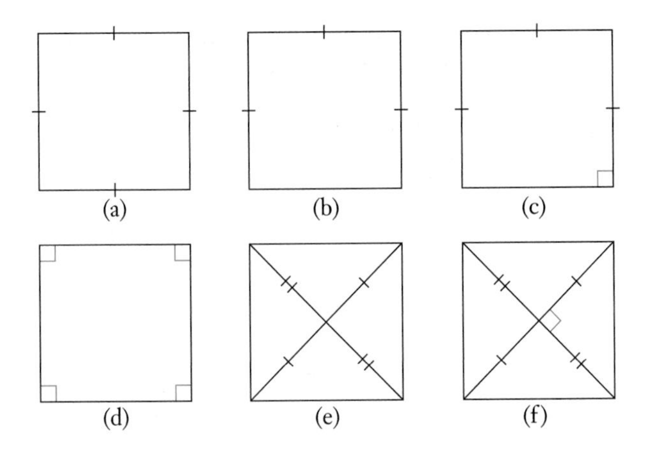

21. Must the figure be a square?
22. Must the figure be a kite?
23. The roads around a city subdivision form a parallelogram. Give two other facts about the roads around the subdivision.
24. A vegetable gardener's plot of land is in the shape of a rectangle. Give two other facts about the plot.
25. A builder used twine to determine whether a foundation was layed out in a true rectangular shape. He stretched the twine from one corner to its opposite corner. Then he checked to see if the same length of twine would stretch from the adjacent corner to its opposite corner. Which generalization about rectangles did he use to determine whether the foundation was rectangular?

26. The legs of an ironing board are equal in length, and bisect each other at the point where they cross. What generalization about parallelograms ensures that the ironing board will always be parallel to the floor, regardless of the height of the board above the floor?
27. The chains $\overline{AC}$ and $\overline{BD}$ on this swing are the same length and the distances between the chains, $\overline{AB}$ and $\overline{CD}$, are equal. What generalization about parallelograms assures that the seat of the swing will always be parallel to the bar $\overline{AB}$ at the top of the swing?
28. Give two ways in which a square differs from a nonsquare rectangle.
29. Give two ways in which a rhombus differs from a nonrhombic parallelogram.
30. Use the procedure for constructing the perpendicular bisector of a segment (Appendix D) to construct a rhombus. Explain why it is a rhombus.

B. Deepening Understanding

In Exercises 31–40, which of the conditions ensure that a quadrilateral is a parallelogram? For those conditions that do not characterize a parallelogram, give a counterexample.

31. Both pairs of opposite sides are parallel.
32. Both pairs of opposite sides are congruent.
33. A pair of congruent sides and a pair of parallel sides exist.
34. A pair of opposite sides are both congruent and parallel.
35. Both pairs of opposite angles are congruent.
36. A pair of opposite angles are congruent.
37. The diagonals bisect each other.
38. A pair of opposite sides are congruent.
39. A pair of consecutive angles are supplementary.
40. A pair of opposite sides are parallel.

Figure for Exercise 26

Figure for Exercise 27

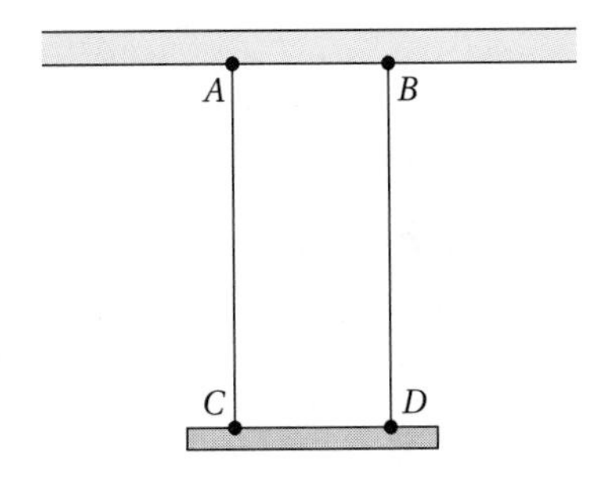

In Exercises 41–44, give a convincing argument that each statement is true.

41. A parallelogram with at least one right angle is a rectangle.

42. A parallelogram with adjacent sides congruent is a rhombus.

43. A parallelogram with a right angle and two adjacent sides congruent is a square.

44. Two parallel lines are equidistant at all points. (*Hint:* Draw two parallel lines, r and s. Then draw lines through any two points A and B on r, perpendicular to s. Consider the resulting quadrilateral.)

45. Would any generalization about a parallelogram, a rectangle, and a rhombus also hold true for a square? Explain.

46. State a generalization about a rectangle that isn't true for a parallelogram.

47. State a generalization about a rhombus that isn't true for a parallelogram.

48. Which quadrilaterals have

a. perpendicular diagonals?
b. opposite angles congruent?
c. two pairs of parallel sides?
d. a pair of congruent, adjacent sides?

49. Use GES and try several examples as needed to form generalizations about the following constructions.

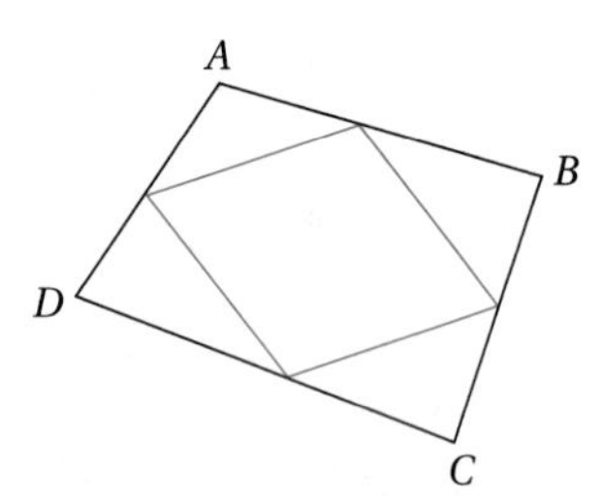

What type of figure is formed by joining the midpoints of the sides of the following shapes? How does the area of the figure formed compare to the area of the original figure?

a. a square
b. a rectangle
c. a parallelogram
d. a rhombus
e. a kite
f. a trapezoid
g. a general quadrilateral

C. Reasoning and Problem Solving

50. In the following parallelogram, the dashed lines show segments joining vertices and the midpoints of the sides going around in one direction. The gray lines show segments joining vertices and the midpoints of the sides going around in the opposite direction. How does the area of the shaded octagon produced, which is common to both center regions, compare to the area of the original parallelogram? Predict; then try examples using GES or constructions to form a generalization.

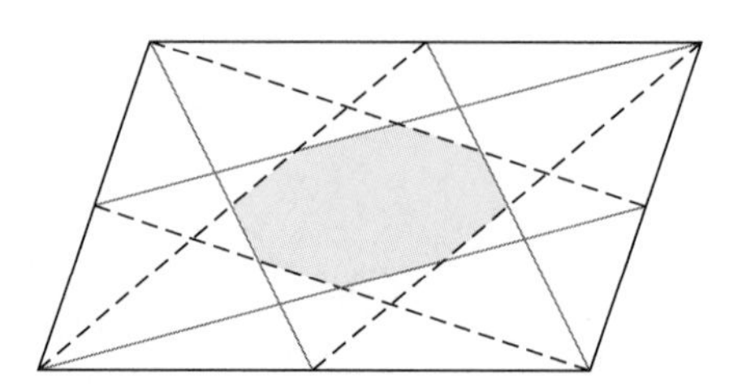

51. Connect the midpoints of the nonparallel sides of a trapezoid. How does the length of this segment relate to the measures of the two bases? Use GES to help you form a generalization.

52. Use GES to draw a conclusion about the bisectors of consecutive angles of a parallelogram.

53. Use the properties of a rhombus or a kite to verify the construction process titled "Construct a Perpendicular to a Line Through a Given Point Not on the Line" in Appendix D.

54. Construct external squares to each side of a parallelogram. What figure is formed when the centroids of these squares are connected?

55. Suppose you were given the information that the opposite angles of a parallelogram were congruent. How could you use this information and the idea that the sum of the angles of a quadrilateral is 360° to convince someone that the consecutive angles of the parallelogram are supplementary?

56. The Four-City Factory Problem. Four cities, A, B, C, and D, are located at the corners of a rectangle. A builder who was formerly a mathematics teacher wants to build a factory at point F inside the rectangle so that the relationship $a^2 + b^2 = c^2 + d^2$ between the distances from the factory to the cities holds.

a. Where should the factory be built to meet these conditions?

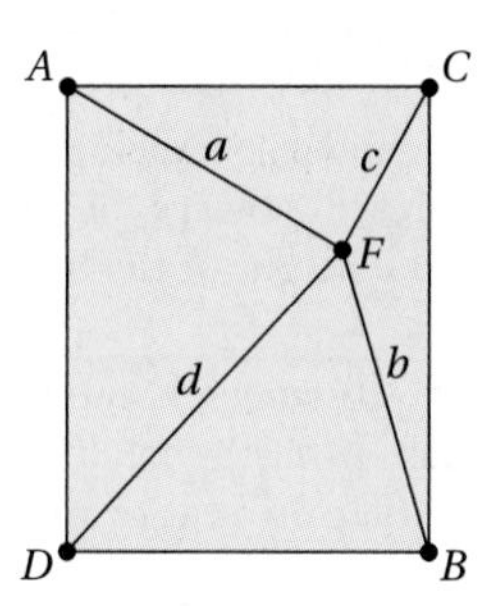

b. Make a conjecture about point F in the figure and use the Pythagorean theorem to give a convincing argument to support your conclusion. (*Hint:* Draw a line through F, perpendicular to $\overline{AC}$ and $\overline{DB}$, and consider the four right triangles formed.)

57. The vertices of a quadrilateral $ABCD$ are connected clockwise to the midpoints of the sides of the quadrilateral.

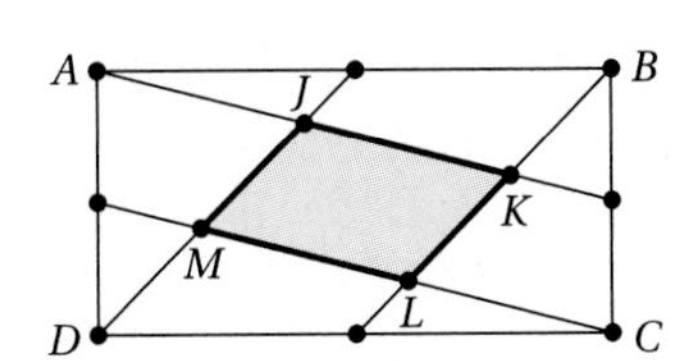

a. Use GES or other tools and look for a possible relationship involving the ratio of the area of the original quadrilateral to the area of the shaded interior quadrilateral, $JKLM$.
b. Ask a what-if question about this situation, describe the question, and discuss any generalizations you find.

D. Communicating and Connecting Ideas

58. Work in a small group to discuss how to systematically decide how many different parallelograms with $\overline{AB}$ as one side can be formed by connecting points of a 5 by 5 lattice.

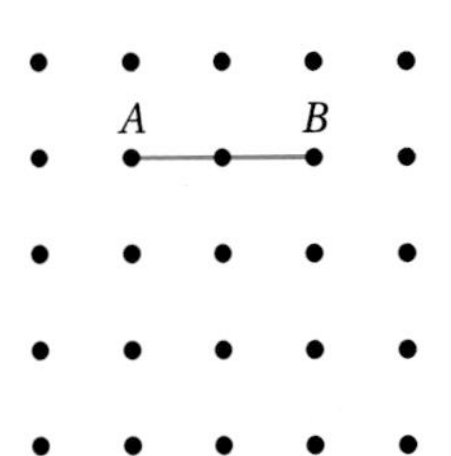

59. **Historical Pathways.** A mathematician named Pierre Varignon (1654–1722) was the first to publish the discovery (part of which you may have discovered in Exercise 49) that the figure formed when the midpoints of the sides of any quadrilateral are connected is a parallelogram (called the Varignon parallelogram of the quadrilateral), with area half that of the original quadrilateral. Use GES to discover a relationship between the perimeter of a Varignon parallelogram and the diagonals of the original quadrilateral.

60. **Making Connections.** Make a connection between geometry and architecture by identifying each type of quadrilateral discussed in this section in a human-made building, bridge, or other edifice in your community. Sketch and describe the occurrence of each figure.

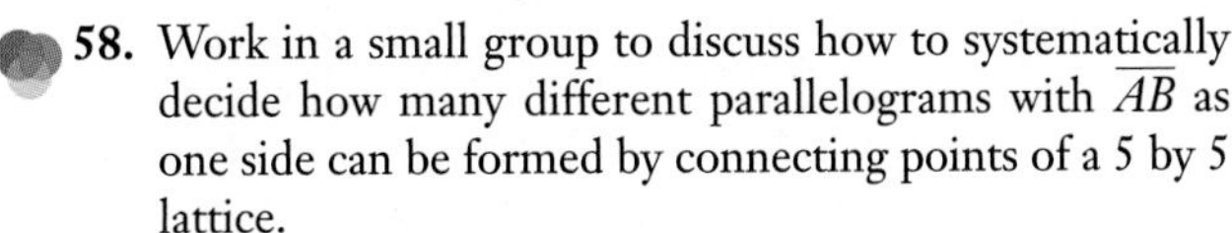

CHAPTER SUMMARY

Key Ideas: Questions and Answers

Section 10.1

- **What geometry can be seen in the world around you?** *(pp. 503–506)* Nature displays numerous geometric shapes, such as hexagons in honeycombs, pentagons in starfish, and the Fibonacci sequence in tree growth. Human-made edifices, such as pyramids of ancient Egypt and modern skyscrapers, also contain many different geometric shapes.
- **What are some key basic geometric figures and relationships?** *(pp. 507–521)* Geometry is based on the ideas of points, lines, planes, and space. All geometric figures, such as segments, rays, angles, triangles, polygons, and circles are essentially sets of points. Some basic relationships between lines are parallel lines, skew lines, and perpendicular lines. Some basic relationships between angles are supplementary angles, complementary angles, and vertical angles. It is important to classify different types of triangles, such as right, isosceles, and equilateral, and quadrilaterals such as parallelogram, rectangle, square, and rhombus.

Section 10.2

- **How can ideas about points and lines be used to solve problems?** *(pp. 526–528)* Problems such as mapping out a route so that a highway inspector would travel each highway only once to either arrive at another location or arrive back where he started utilize generalizations about traversibility of networks. A network is made up of points called vertices and lines called edges. Even vertices have an even number of edges emanating from them and odd vertices have an odd number of edges. A network that has all even vertices can be traversed starting at a vertex and ending at the same vertex (traversable type 1). A network that has exactly two odd vertices can be traversed by starting at one odd vertex and ending at the other (traversable type 2).

- **How can segments, angles, and triangles by used to solve problems?** *(pp. 528–534)* The area of a polygon is the number of square units needed to cover its inside. In a triangle, the medians, altitudes, perpendicular bisectors of the sides, and angle bisectors intersect respectively in points called the centroid, orthocenter, circumcenter, and incenter. These and other basic ideas of segments, angles, and triangles are used along with Geometry Exploration Software (GES) to solve problems such as locating a pumping station for a three-city water supply.
- **How can quadrilaterals, other polygons, and circles be used to solve problems?** *(pp. 534–536)* Ideas about polygons and GES can be used to solve problems such as those involving floor tiling and designing tabletop patterns. The seven-piece tangram puzzle is useful in modeling different types of polygons.

Section 10.3

- **What is true about angles formed by intersecting lines?** *(pp. 541–543)* If two parallel lines are cut by a transversal, the alternate interior, alternate exterior, and corresponding angles are congruent. The same-side interior angles and same-side exterior angles are supplementary.
- **What is true about the interior and exterior angles of any polygon?** *(pp. 543–546)* The sum of the measures of the interior angles of an n-sided polygon is $(n - 2)180°$. The sum of the measures of the exterior angles of any polygon is $360°$.
- **What is true about the angles of regular polygons?** *(pp. 547–548)* The measure of each interior angle of a regular n-gon is $\frac{(n-2)180°}{n}$. The measure of an exterior angle of a regular n-gon is $\frac{360°}{n}$. The measure of a central angle of a regular n-gon is $\frac{360°}{n}$.
- **Can any interesting generalizations be made about angles in circles?** *(pp. 548–550)* When lines from a point P are drawn to intersect a circle cutting arcs of length r and s, the following relationships hold.
  a. If P is on the circle, $m\angle P = \frac{1}{2}m(\text{arc } s)$.
  b. If P is outside the circle,
  $m\angle P = \frac{1}{2}[m(\text{arc } s) - m(\text{arc } r)]$.
  c. If P is inside the circle,
  $m\angle P = \frac{1}{2}[m(\text{arc } s) - m(\text{arc } r)]$.

Section 10.4

- **What generalizations can be made about congruent triangles?** *(pp. 556–558)* Two triangles are congruent if you can match up their vertices in such a way that matching sides and angles are congruent. It is not necessary to check the congruence of all six parts of a pair of triangles to determine if they are congruent. The SSS, SAS, and ASA Postulates, along with the AAS, HA, and HL Theorems give generalizations about some minimum conditions that ensure congruence.
- **What is the Pythagorean theorem?** *(pp. 558–561)* When the lengths of the legs of a right triangle are a and b and the length of the hypotenuse is c, then $a^2 + b^2 = c^2$, and conversely. This is the Pythagorean theorem and its converse.
- **What are some triangle ideas that are related to the Pythagorean theorem?** *(pp. 561–563)* The length of the hypotenuse of any 45°–45°–90° triangle is the length of a leg times $\sqrt{2}$. The length of the hypotenuse of any 30°–60°–90° triangle is 2 times the length of the shorter leg. The length of the longer leg is the length of the shorter leg times $\sqrt{3}$.

Section 10.5

- **What are some generalizations that can be made about quadrilaterals?** *(pp. 570–573)* Generalizations about quadrilaterals, summarized in Appendix C, include the fact that a quadrilateral is a parallelogram if both pairs of opposite sides are parallel or congruent, both pairs of opposite angles are congruent, consecutive angles are supplementary, or the diagonals bisect each other. A parallelogram is a rectangle if and only if it has at least one right angle or its diagonals are congruent. Also, a parallelogram is a rhombus if and only if it has four congruent sides, or its diagonals bisect the angles, or its diagonals are perpendicular bisectors of each other. Finally, a parallelogram is a square if and only if it is a rectangle with four congruent sides, or it is a rhombus with a right angle, or its diagonals are congruent and perpendicular bisectors of each other.
- **How can quadrilateral generalizations be used to justify geometric constructions?** *(pp. 573–574)* The generalization that the diagonal of a rhombus bisects an angle of the rhombus can be used to justify the geometric construction for bisecting an angle. The generalization that the diagonals of a rhombus are perpendicular can be used to justify the geometric construction for the perpendicular bisector of a line segment. Other quadrilateral generalizations can be used to justify other geometric constructions described in Appendix D.
- **What is the value of what-if questions about quadrilaterals?** *(pp. 574–576)* What-if questions about a problem we have solved or a generalization we have discovered are valuable in creating new and interesting problems and generalizations. We should develop the habit of asking what-if questions about our discoveries in geometry.

Key Terms, Concepts, and Generalizations

Section 10.1

Fibonacci sequence (p. 504)
golden ratio (p. 504)
point (p. 507)
space (p. 507)
line (p. 507)
plane (p. 507)
collinear (p. 508)
coplanar (p. 508)
intersect (p. 508)
point of intersection (p. 508)
parallel lines (p. 508)
skew lines (p. 508)
concurrent lines (p. 508)
segment (p. 509)
ray (p. 509)
angle (p. 509)
length measure (p. 510)
congruent segments (p. 510)
midpoint of a segment (p. 510)
bisector of a segment (p. 510)
degree measure (p. 510)
protractor (p. 510)
straight angle (p. 511)
reflex angle (p. 511)
zero angle (p. 511)
interior of an angle (p. 511)
exterior of an angle (p. 511)
angle bisector (p. 511)
right angle (p. 512)
acute angle (p. 513)
obtuse angle (p. 513)
complementary angles (p. 513)
supplementary angles (p. 513)
adjacent angles (p. 513)
linear pair of angles (p. 513)
vertical angles (p. 514)
perpendicular lines (p. 514)
perpendicular bisector (p. 514)
closed curve (p. 515)
nonsimple closed curve (p. 515)
simple closed curve (p. 515)
circle (p. 515)
center of a circle (p. 515)
chord of a circle (p. 515)
diameter of a circle (p. 515)
radius of a circle (p. 515)
polygon (p. 516)
nonsimple polygon (p. 516)
simple polygon (p. 516)
simple convex polygon (p. 516)
simple nonconvex polygon (p. 516)
n-gon (p. 516)
interior angle of a polygon (p. 516)
regular polygon (p. 516)
triangle (p. 517)
vertices (p. 517)
sides (p. 517)
quadrilateral (p. 517)
pentagon (p. 517)
hexagon (p. 517)
heptagon (p. 517)
octagon (p. 517)
nonagon (p. 517)
decagon (p. 517)
dodecagon (p. 517)
median (p. 518)
altitude (p. 518)
equilateral triangle (p. 518)
isosceles triangle (p. 518)
scalene triangle (p. 518)
right triangle (p. 518)
acute triangle (p. 518)
obtuse triangle (p. 518)
equiangular triangle (p. 519)
trapezoid (p. 521)
kite (p. 521)
parallelogram (p. 521)
rhombus (p. 521)
rectangle (p. 521)
square (p. 521)

Section 10.2

vertices (p. 526)
edges (p. 526)
network (p. 526)
traversable network (p. 526)
traversable type 1 (p. 526)
traversable type 2 (p. 526)
odd vertex (p. 526)
even vertex (p. 526)
area measure (p. 529)
geometry exploration software (p. 530)
Concurrency Relationships theorem (p. 531)
centroid (p. 531)
orthocenter (p. 531)
circumcenter (p. 531)
incenter (p. 531)
Euler line (p. 532)
Euler Line theorem (p. 532)
tangram puzzle (p. 534)

Section 10.3

transversal (p. 541)
alternate interior angles (p. 541)
alternate exterior angles (p. 541)
corresponding angles (p. 541)
same-side interior angles (p. 542)
same-side exterior angles (p. 542)
Angle Congruencies in Parallel Lines theorem (p. 542)
interior angles of a polygon (p. 543)
Angle Sum for Triangles theorem (p. 543)
Angle Sum for Quadrilaterals theorem (p. 544)
Angle Sum for Any Polygon theorem (p. 545)
exterior angle of a polygon (p. 545)
Exterior Angle Sum for Any Polygon theorem (p. 546)
Interior Angle Measure for a Regular Polygon theorem (p. 547)
Exterior Angle Measure for a Regular Polygon theorem (p. 547)
Central Angle Measure for a Regular Polygon theorem (p. 548)
arc of a circle (p. 548)
formulas for measures of angles associated with a circle (p. 549)

Section 10.4

congruent triangles (p. 556)
SSS postulate for congruent triangles (p. 557)
SAS postulate for congruent triangles (p. 557)
ASA postulate for congruent triangles (p. 557)
AAS theorem for congruent triangles (p. 557)
HA theorem for congruent triangles (p. 557)
HL theorem for congruent triangles (p. 557)
Pythagorean theorem (p. 558)
legs of a right triangle (p. 558)
hypotenuse of a right triangle (p. 558)
45°–45°–90° Triangle Relationship theorem (p. 562)
30°–60°–90° Triangle Relationship theorem (p. 562)

Section 10.5

parallelogram (p. 570)
properties of parallelograms (p. 571)
rectangle (p. 571)
properties of rectangles (p. 571)
rhombus (p. 572)
square (p. 572)
properties of rhombi (p. 572)
properties of squares (p. 572)
kite (p. 572)
trapezoid (p. 572)

CHAPTER REVIEW

Concepts and Skills

1. Write the first five terms of the Fibonacci sequence and give an example of how this sequence relates to geometry.
2. Give the dimensions of a window that is very close to being a golden rectangle.
3. Draw an example of each of the following. Label and write a symbol if appropriate.
   a. parallel lines
   b. perpendicular lines
   c. an obtuse angle
   d. skew lines
   e. vertical angles
   f. three different types of triangles
   g. three different types of quadrilaterals
   h. a pentagon
   i. medians of a triangle
   j. complementary angles
4. Explain the difference between each pair of geometric figures.
   a. complementary angles and supplementary angles
   b. a perpendicular bisector and an angle bisector
   c. an isosceles triangle and an equilateral triangle
   d. a kite and a rhombus
   e. a parallelogram and a rhombus
   f. a rectangle and a square
   g. a trapezoid and a parallelogram
   h. a right triangle and an obtuse triangle
   i. a simple closed curve and a nonsimple closed curve
   j. a convex polygon and a nonconvex polygon
5. If r and s are parallel, name
   a. four pairs of nonvertical angles that are congruent.
   b. four pairs of nonadjacent angles that are supplementary.
   c. two pairs of alternate interior angles.
   d. two pairs of alternate exterior angles.
   e. four pairs of corresponding angles.

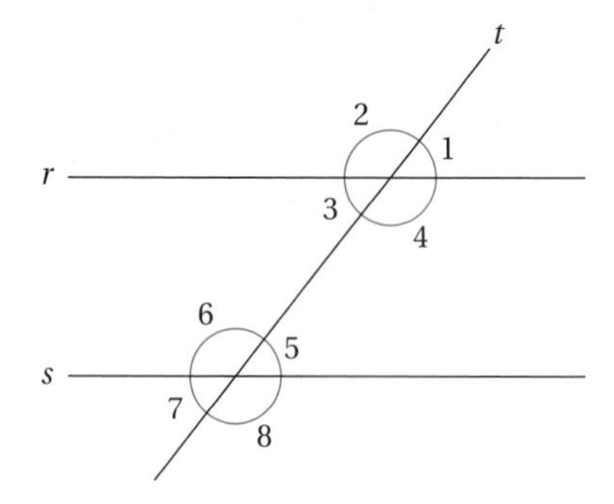

6. Can you start at P, traverse each trail only once, and end at Q? How do you know?

Figure for Exercise 6

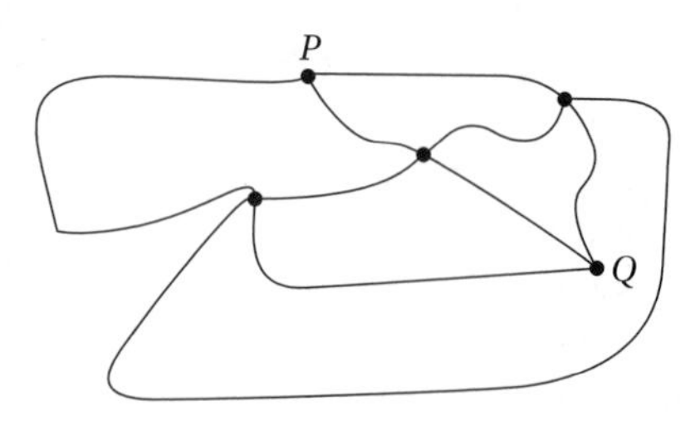

7. Find the number of degrees in an interior angle, a central angle, and an exterior angle of a regular pentagon.
8. The measure of $\angle A$ of triangle ABC is 54°, and the measure of $\angle B$ is 78°. What is the measure of $\angle C$? How do you know?
9. What is the sum of the interior angles of a pentagon shaped like a house?
10. What is the measure of an angle formed by two edges of a stop sign?
11. By how many degrees does the $m\angle P$ differ from the $m\angle Q$ in the following figure?

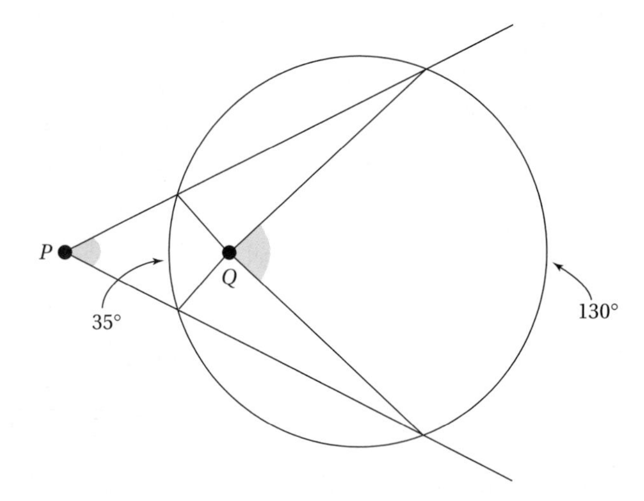

12. Give a fact about each of the following points associated with a triangle.
    a. incenter
    b. circumcenter
    c. orthocenter
    d. centroid
13. What is the Euler line for a triangle?
14. What theorem helps you find the length of one side of a right triangle when you know the length of the other two sides? Use variables to state this theorem.
15. A 13-foot ladder is placed against a wall with its bottom 5 feet from the wall. How high above the ground is the top of the ladder?

16. List two relationships that are true about pairs of sides or angles or diagonals in
 a. a parallelogram. b. a rectangle.
 c. a rhombus.

17. Find each of the following and state a generalization that could be used to do so.
 a. The length of the diagonal of a square with sides 16 centimeters long.
 b. The length of an altitude of an equilateral triangle with a side 12 centimeters long.

Reasoning and Problem Solving

18. Can a quadrilateral have three of its angles greater than 120°? Why or why not?

19. As the number of sides of a regular polygon increases, what change would you observe in the measure of an exterior angle of the polygon? Explain.

20. How can you use the generalization, the sum of the exterior angles of a regular n-gon is 360°, to develop a formula for finding the measure of a vertex, or interior, angle of a regular n-gon? Explain your reasoning.

21. Should the nonsimple polygon in the next column be classified as a quadrilateral? Why or why not?

22. **The Sliding Ladder Problem.** A painter put the top of a ladder at the bottom of a large sign on a wall. The bottom of the sign was 24 feet above the ground. When he slid the ladder so that its bottom was half the distance to the base of the wall, the top of the ladder was then 9 inches above the bottom of the sign. How long was the ladder, to the nearest foot?

Alternative Assessment

23. Make models to demonstrate that
 a. the sum of the measures of the angles of a triangle is 180°.
 b. the sum of the measures of the angles of a quadrilateral is 360°.

24. One approach used in this chapter to verify the Pythagorean theorem was to utilize a dissection model. Write a paragraph describing the different approaches used either in the development or in the exercises to verify this theorem.

Figure for Exercise 21

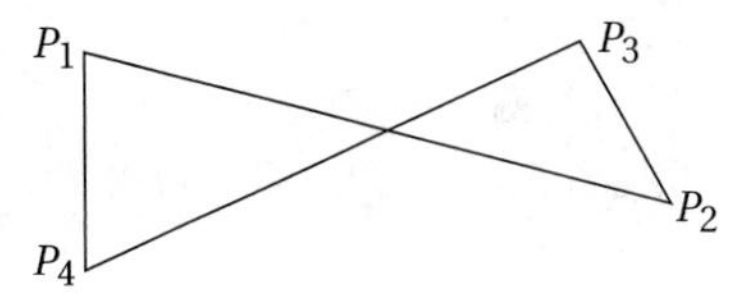

Extending Geometry

Chapter Perspective

The beauty of mathematics is accentuated through the lens of geometry. Geometric designs in buildings and other man-made structures reveal symmetry and give insight into this important mathematical idea. The intriguing world of patterns, designs, and tessellations brings to life some fascinating relationships among mathematics, nature, and art. In this chapter we focus on figures, relationships, and patterns in space, including geometric descriptions of motion, tessellations, and special polygons to extend the development of your spatial sense.

Connection to the NCTM Principles and Standards

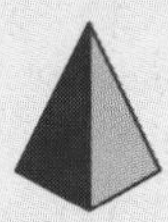

The NCTM *Principles and Standards for School Mathematics* (2000) indicate that the mathematics curriculum in geometry for grades PreK–8 should prepare students to

- apply transformations and use symmetry to analyze mathematical situations;
- use visualization, spatial reasoning, and geometric modeling to solve problems (p. 41).

Connection to the PreK–8 Classroom

In grades PreK–2, students use their own physical experiences with shapes, such as fitting pieces into a puzzle, to learn about slides, turns, flips, and symmetry.

In grades 3–5, students are ready to mentally manipulate shapes to make predictions, learn the mathematical language to describe their predictions, then verify their predictions physically.

In grades 6–8, students further develop spatial sense by analyzing motions of geometric figures and by putting polygons together to form patterns such as tessellations.

Section 11.1 TRANSFORMATIONS

- Translations, Rotations, and Reflections
- Connecting Transformations and Symmetry
- Transformations That Change Size
- Transformations That Change Both Size and Shape

In this section, we investigate the geometric transformations that do and do not affect a figure's shape or size. We use mathematical ideas and symbols to describe, analyze, and compare the different types of transformation. You will have opportunities to use special computer software to create and move the geometric figures. In Mini-Investigation 11.1, you are asked to use some motions that are related to geometric transformations.

Draw a picture to illustrate each motion that you used.

***Technology Extension:** Use geometry exploration software (GES) to create a triangle or quadrilateral. Explore the GES options for sliding, turning, and flipping the figure (see Appendix B). Describe how GES may be helpful in studying different ways to move figures.*

Mini-Investigation 11.1 Solving a Problem

In what ways (slide, turn, flip, or combinations) can you move pentomino (a) to test whether it matches pentominos (b), (c), (d), and (e)? (Use a tracing if one will help.)

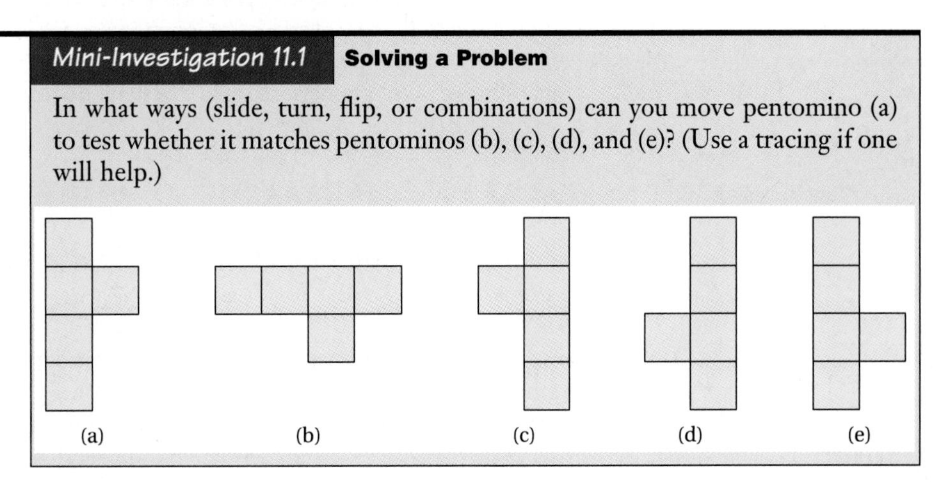

Translations, Rotations and Reflections

In Mini-Investigation 11.1, you may have observed that an object may be slid, turned, or flipped without changing its shape or size. With these three simple motions, or combinations of these motions, we can move a figure anywhere in space to match another figure that is congruent to it. We use these motions to describe transformations in the following subsections.

Slides or Translations. You encounter the motion of slides often in everyday life: a notebook sliding across a desk, a window frame sliding up and down, or a drawer sliding in and out. In Figure 11.1, the blue pentomino shape is the **image** of the yellow pentomino shape after all the points in the plane have been slid in the direction and distance indicated by the **slide arrow** (or **directed segment**) from point A to point B.

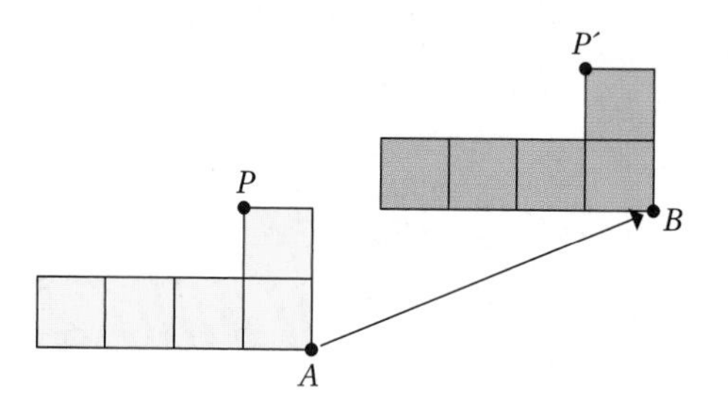

FIGURE 11.1 | A figure and its translation image.

To visualize the slide, trace the yellow pentomino outline and the slide arrow. Then slide point A on the tracing paper along the arrow to point B. (If you trace the entire arrow, you will not accidentally turn the tracing paper.) The pentomino shape you traced on your paper should now coincide with the blue pentomino shape, which is the image of the slide. The physical motion of sliding, mathematically called a *translation*, is defined as follows.

Definition of a Translation

If each point P in the plane corresponds to a unique point in the plane, P', such that directed segment PP' is congruent and parallel to the directed segment AB, then the correspondence is called the **translation** associated with the directed segment AB and is written T_{AB}.

To describe further the effect of a translation, we write

$$T_{AB}(P) = P'$$

to say that P' is the image of P under the translation associated with the directed segment AB. When considering the translation T_{AB}, point B becomes the image of point A. If we want to indicate a slide from B to A, we write T_{BA}.

Turns or Rotations. Turns in the real world include food rotating on a turntable in a microwave oven, doorknobs turning, and a Ferris wheel turning around its large axis as the seats turn on their individual axes. The blue pentomino shape shown in Figure 11.2 is the image of the yellow pentomino shape after all the points in the plane have been turned around point C in the direction and angle measure indicated by the turn angle, $\angle PCP'$.

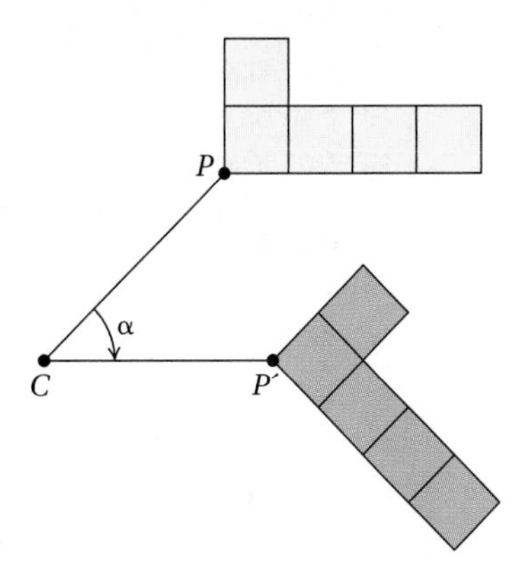

FIGURE 11.2 | A figure and its rotation image.

To visualize the turn, trace the yellow pentomino outline and $\overline{PC}$ of the turn angle. Place the point of your pencil on point C and hold point C still while you turn the tracing paper in the direction and angle measure of the turn angle until the $\overline{PC}$ drawn on your tracing paper coincides with $\overline{P'C}$ on the page. The pentomino shape you traced on your paper should now coincide with the blue pentomino shape, the image of the turn.

In mathematics a turn is called a *rotation*, and is defined as follows.

Definition of a Rotation

Consider a point C and an angle with measure α from $-180°$ to $180°$. If each point P in the plane corresponds to a unique point in the plane, P', such that $m\angle PCP' = \alpha$ and $PC = P'C$, then the correspondence is called the **rotation,** with center C and angle α, and is written $R_{C,\alpha}$.

When considering the rotation $R_{C,\alpha}$, we write $R_{C,\alpha}(P) = P'$ to say that P' is the image of P under the rotation associated with the center C and angle α.

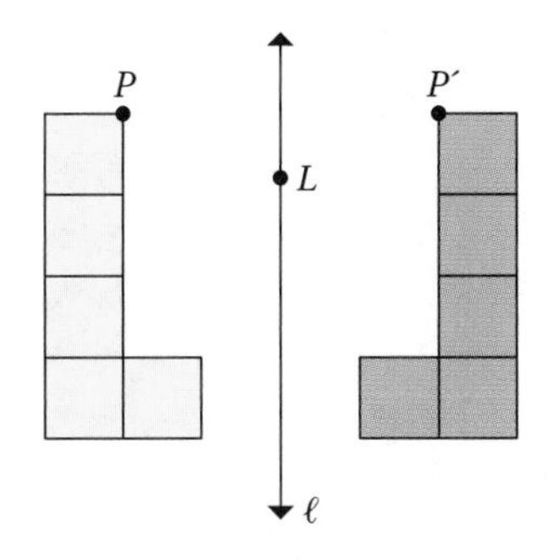

FIGURE 11.3 | A figure and its reflection image.

Flips or Reflections. Flipped images in the real world appear when you look in mirrors, turn transparencies over on the overhead projector, or press an inked stamp onto a piece of paper. The blue pentomino shape in Figure 11.3 is the image of the yellow pentomino shape after all the points in the plane have been flipped over line ℓ.

To visualize the flip, trace the yellow pentomino outline and line ℓ, marking point L on line ℓ. Lift your tracing paper, flip it over, and replace it on the page by laying the traced line ℓ over line ℓ on the page with point L on the tracing paper matching point L on the page. The pentomino shape that you traced should now coincide with the blue pentomino shape, the image of the flip. The physical motion of flipping (mathematically called a *reflection*) can be defined as follows.

Definition of a Reflection

If each point on line ℓ corresponds to itself, and each other point P in the plane corresponds to a unique point P' in the plane, such that ℓ is the perpendicular bisector of $\overline{PP'}$, then the correspondence is called the **reflection** in line ℓ, and is written M_ℓ.

With this notation, we write

$$M_\ell(P) = P'$$

to say that P' is the image of P under the reflection in ℓ. In Example 11.1, you are asked to identify the motion that produced the given image.

Example 11.1

Identifying Transformations

Identify which transformation (translation, rotation, or reflection), if possible, that would change each yellow figure shown to the corresponding blue image. Justify your answers and use correct notation to show the effect of the transformation on point A.

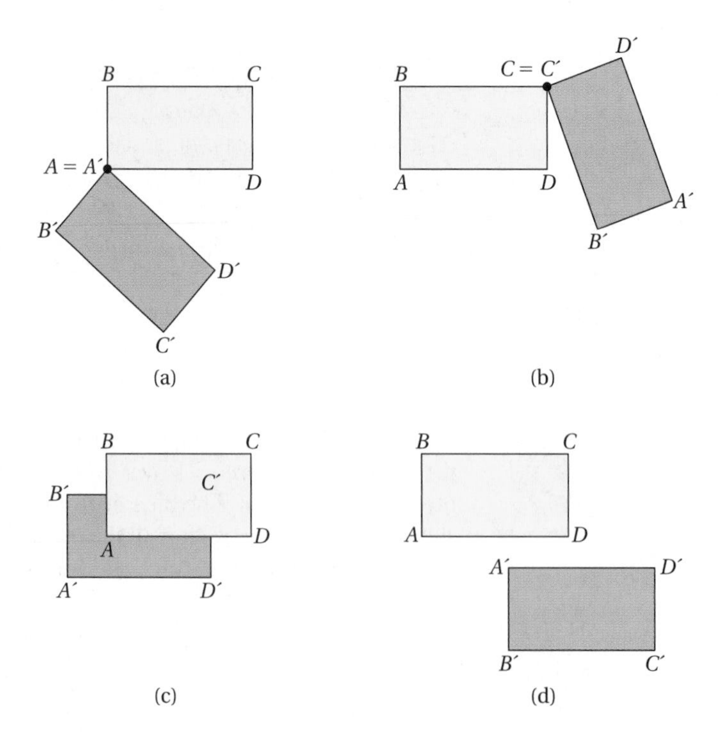

Solution

In part (a) the transformation is a reflection because the orientation of the figure has changed. In the original rectangle, reading from A clockwise around the figure gives $ABCD$. In the blue image, reading clockwise from A gives $A'D'C'B'$. We write $M_{\text{line of reflection}}(A) = A'$.

In part (b) the transformation is a rotation. Its orientation hasn't changed, but the segments that were horizontal now aren't. We write $R_{C,\text{ angle of rotation}}(A) = A'$.

In part (c) the transformation is a translation. If A is connected to A', B to B', C to C' and D to D', the segments formed are parallel and congruent. All the points moved the same distance in the same direction. We write $T_{BB'}(A) = A'$.

In part (d) the blue image of the yellow figure can't be produced with a single transformation. The orientation of the figure has changed, so it must have been reflected. But the image doesn't appear where it would if a mirror had been placed between the figures. Thus it has been translated *and* reflected.

Your Turn

Practice: Identify the type of transformation that would transform each yellow figure to the corresponding blue image. Use correct notation to show the effect of the transformation on point A.

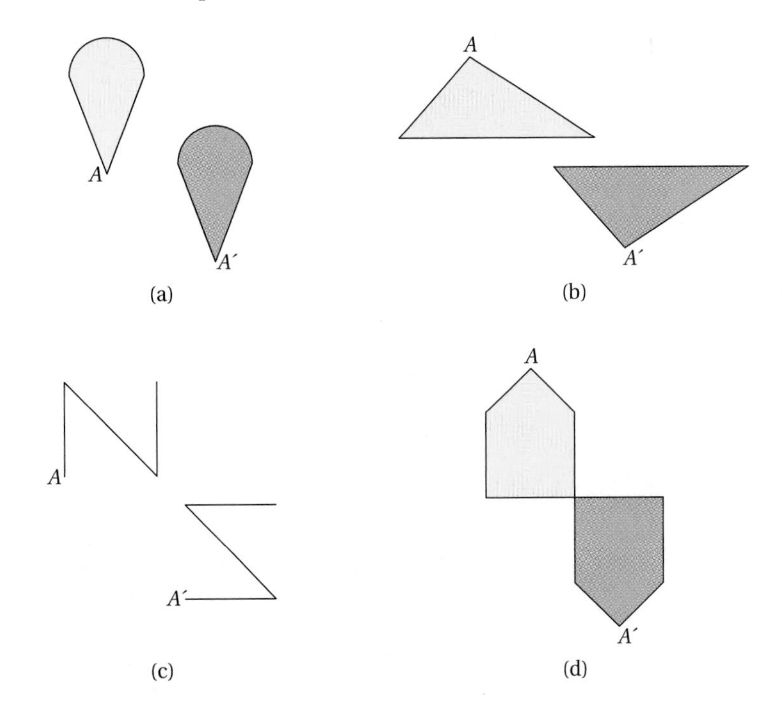

(a) (b) (c) (d)

Reflect: Which of the transformations in the practice problem changed the orientation of the points in the figure? ■

Combinations of Motions. As indicated in the solution to Example 11.1, no single translation, rotation, or reflection transformed figure $ABCD$ to figure $A'B'C'D'$ in part (d): It required a translation and then a reflection over the line of translation. This combination motion is called a **glide–reflection**. In Mini–Investigation 11.2 you are asked to use GES, if it is available, to explore the results of combining other motions.

Talk about which of those combinations of motions you could have completed in just one motion. Explain how.

Mini-Investigation 11.2 **Technology Option**

Use GES or other tools to describe the result of the following combination of motions on the figure shown.

a. Reflect ABC over line m and then reflect the resulting image over line n.

b. Rotate ABC 90° counterclockwise around point P and then reflect the resulting image over line n.

c. Translate ABC on line PQ from C to P and then rotate the resulting image 180° clockwise around point Q.

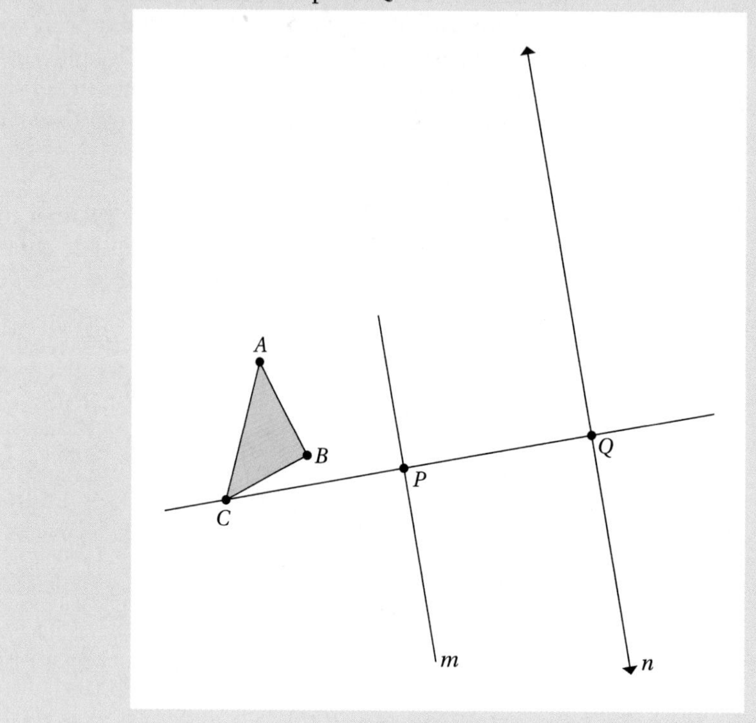

All two-motion combinations of rotations, reflections, or translations other than the translation followed by a reflection can be replaced by a single rotation, reflection, or translation. You will have an opportunity to explore combinations of motions in Exercises 32, 33, 38, and 40 at the end of this section.

Transformations and Congruence. Experiences with motions in the real world suggest that translations, rotations, reflections, and glide–reflections don't change the size or shape of a figure because the points in the plane maintain their relative distances from one another. In other words, distance is preserved. Mathematicians call a transformation that preserves distance and other characteristics a **congruence transformation,** or an **isometry**.

The properties of polygons and transformations suggest that translations, rotations, reflections, and glide–reflections are isometries. Thus, in an isometry, not only distance, but also "betweenness," angle measure, and size and shape are preserved. That is, when one of these properties is determined in a figure, it is the same in the transformed image of the figure. Thus a figure can be repositioned anywhere in space without changing its size or shape through the use of no more than the four types of transformation that we have discussed. As a result we can state the following transformation definition of **congruence.**

> **Definition of Congruence, Using Transformations**
>
> Two figures are **congruent** if and only if there exists a translation, rotation, reflection, or glide-reflection that sets up a correspondence of one figure as the image of the other.

An interesting illustration of how transformations can be used to solve problems is the crossword puzzle problem in Exercise 47 at the end of this section.

Connecting Transformations and Symmetry

Symmetry in the Plane. The idea of symmetry plays a central role in art, interior design, landscaping, and architecture, as well as in mathematics. When a plane figure can be traced and folded so that one half coincides with the other half, as shown in Figure 11.4, we have modeled the congruence transformation we have referred to as a reflection, in which each point in one half of the figure corresponds in a special way to a unique point in the other half of the figure and each point on the line of reflection corresponds to itself. In this case, we say that the figure has **reflectional symmetry**. The fold line is called the **line of symmetry**.

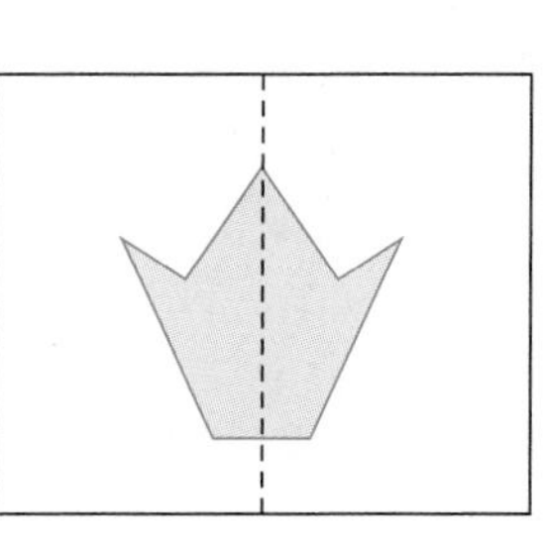

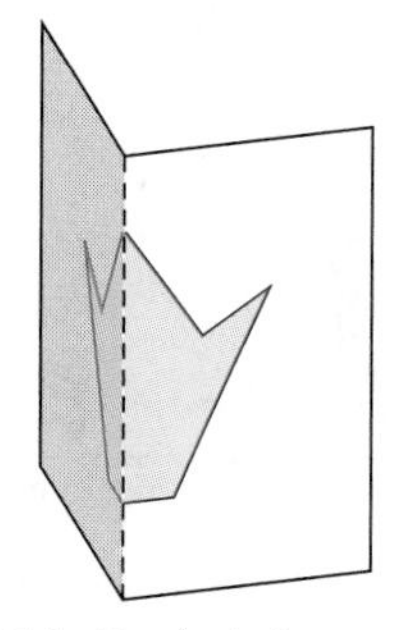

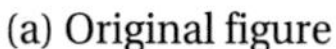

(a) Original figure (b) Fold—Do the halves match?

FIGURE 11.4 | Folding paper to illustrate reflectional symmetry.

A commercially produced piece of plexiglass, shown in Figure 11.5, can also be used to identify reflectional symmetry. If a plastic reflector can be placed on a figure so that the reflection of half the figure fits exactly on the other half, the line of the reflector is a line of reflectional symmetry.

When a tracing of a figure can be rotated $n°$ (with $n < 360$) about a fixed point to fit back on the figure, as shown in Figure 11.6, we have modeled the congruence transformation we have referred to as a rotation, in which each point of the figure corresponds in a special way to another point of the figure, except for the fixed point which corresponds to itself. In this case, we say that the figure has **$n°$ rotational symmetry.** The figure in Figure 11.6 has 90° rotational symmetry. The fixed point is called the **center of rotational symmetry**.

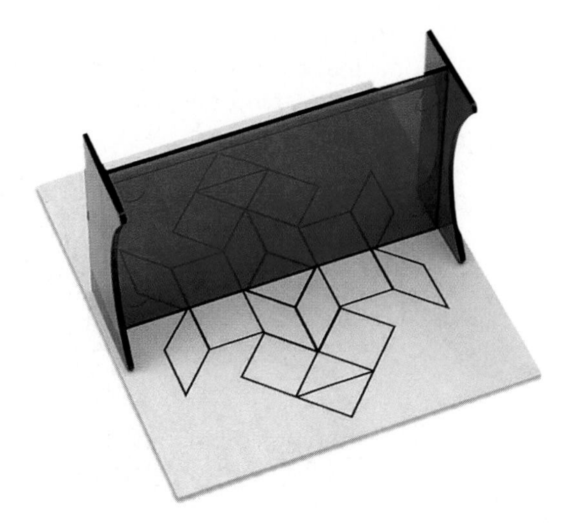

FIGURE 11.5 | Testing for reflectional symmetry with a plastic reflector. (*Source:* Courtesy of Learning Resources®)

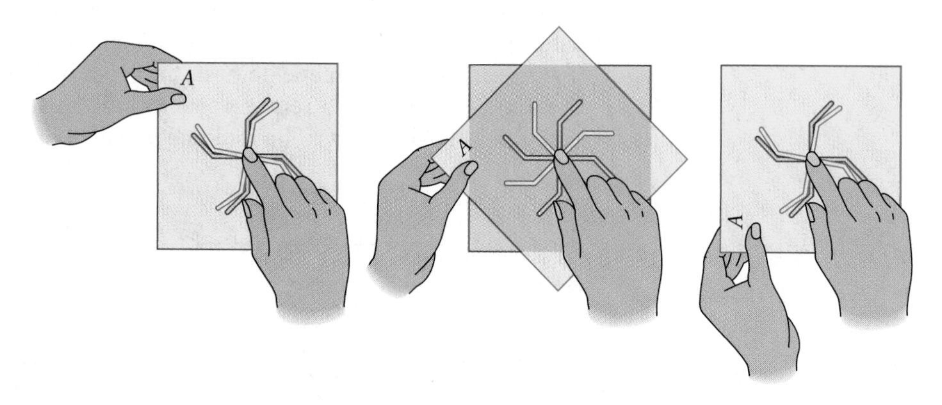

FIGURE 11.6 | Using tracing paper to test for rotational symmetry.

We apply these ideas about symmetry in Example 11.2.

Example 11.2 Describing Symmetry Properties

Describe the symmetry properties of the figure on the left.

Solution

a. The star shown has reflectional symmetry. There are five lines of symmetry, as shown at the top left of the next page.

b. A tracing can be made to coincide with the star after it is rotated 72°, 144°, 216°, or 288° about its center, so the star has rotational symmetry of these numbers of degrees.

Your Turn

Practice: Describe the symmetry properties of the figures shown on the next page.

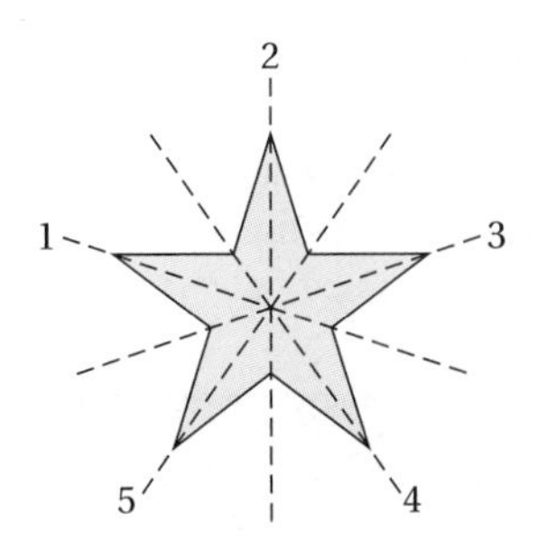

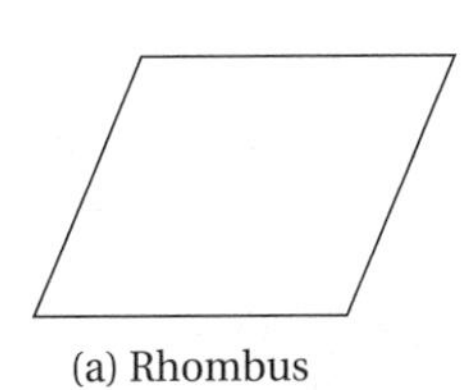
(a) Rhombus

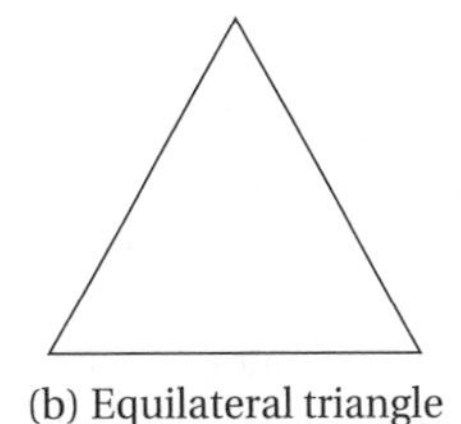
(b) Equilateral triangle

(c) Pinwheel

Reflect: Explain how the symmetry properties of a rhombus and a parallelogram differ. ■

In Chapter 10, we learned that triangles and quadrilaterals could be classified according to side and angle properties. Classes of triangles or quadrilaterals thus formed were given special names, such as equilateral triangles and parallelograms. Triangles and quadrilaterals also may be classified according to symmetry properties, but it isn't common practice to give special names to the classes of figures formed. In Exercise 49 at the end of this section you will be asked to classify triangles and quadrilaterals according to their symmetry properties.

Just as symmetry is basic to an understanding of the universe, so are ideas of symmetry necessary to an understanding of mathematics. Students in the primary grades fold and cut paper to make hearts, pumpkins, and other figures having lines of symmetry. Ideas of symmetry are gradually expanded as students progress through elementary school.

Transformations That Change Size

Size Transformations. The idea of *same shape* but not necessarily *same size* plays an important role in everyday life. For example, people enlarge photographs and blueprints and draw large circuit diagrams on a wall and shrink them to the size of a microchip.

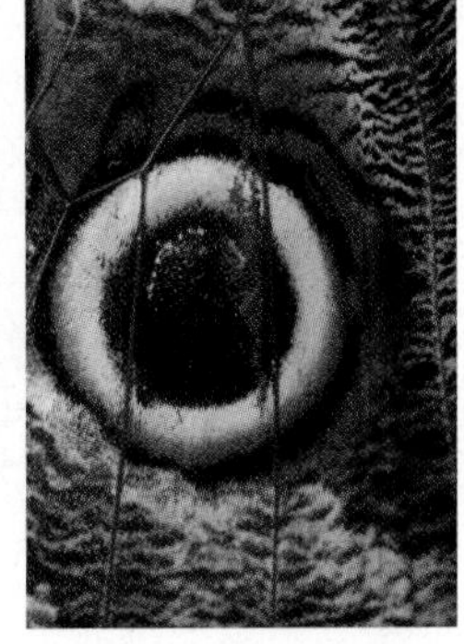

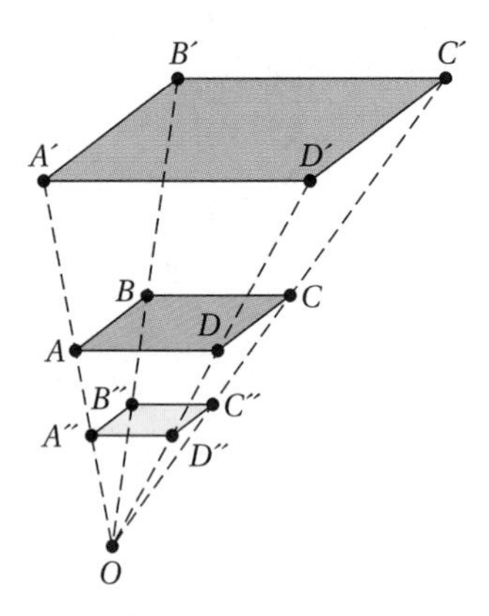

FIGURE 11.7 | Using a size-changing transformation to enlarge and shrink a figure.

What guarantees that these larger and smaller images retain the essential characteristics of the originals? One way is to use a special transformation that enlarges or shrinks the size of a figure along specified lines by multiplying distances by a given factor. The physical action of enlarging or shrinking, mathematically called a *size transformation*, is defined as follows (with reference to Figure 11.7).

Definition of a Size Transformation

If point O corresponds to itself, and each other point P in the plane corresponds to a unique point on $\overrightarrow{OP}$ such that $OP' = r(OP)$ for $r > 0$, then the correspondence is called the **size transformation** associated with center O and scale factor r, and can be written $S_{O,r}$.

With this notation, we can write $S_{O,r}(P) = P'$ to say that P' is the image of P under the size transformation with center O and scale factor r. The enlarging and shrinking lines are determined by a point called the **center of the size transformation**. The multiplier that enlarges or shrinks the lengths is called the **scale factor**. For example, the blue figure, $A'B'C'D'$, in Figure 11.7 is the image of parallelogram $ABCD$ when it is enlarged by using point O as the center and 2 as the scale factor. Note that OA' is two times OA, OB' is two times OB, OC' is two times OC, and OD' is two times OD. The yellow figure, $A''B''C''D''$, is the shrunken image of $ABCD$ when $\frac{1}{2}$ is the scale factor instead of 2. In this case, OA'' is one-half OA, OB'' is one-half OB, OC'' is one-half OC, and OD'' is one-half OD. Size transformations can also be carried out with a computer drawing program, as illustrated in Figure 11.8.

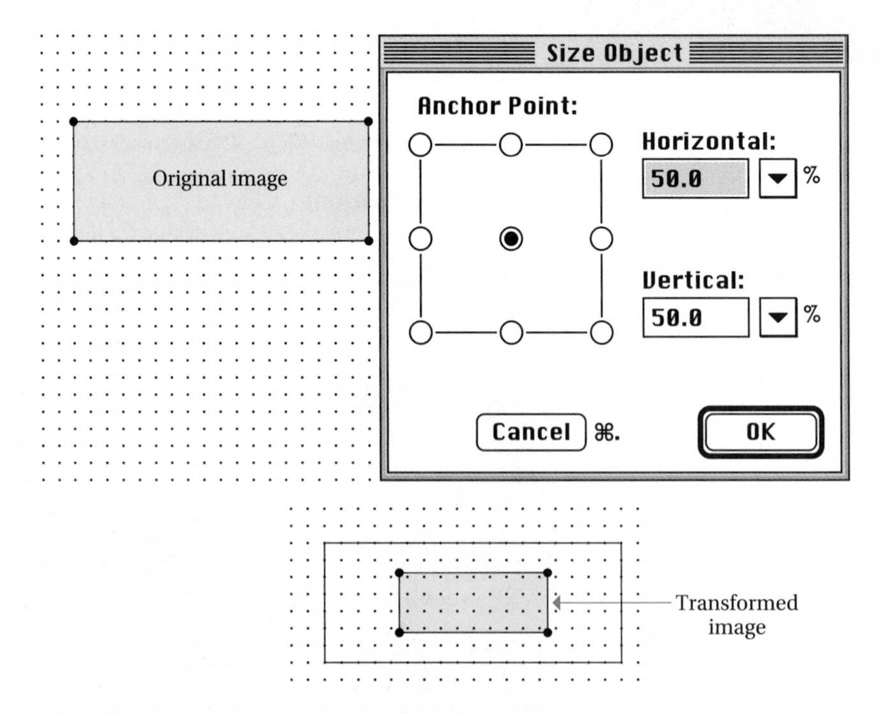

FIGURE 11.8 | Effecting a size transformation by using a computer drawing program.

(*Source:* Screen shot from Corel® WordPerfect 3.5® for the Macintosh. © 1996 Corel Corporation Ltd. Reprinted by permission.)

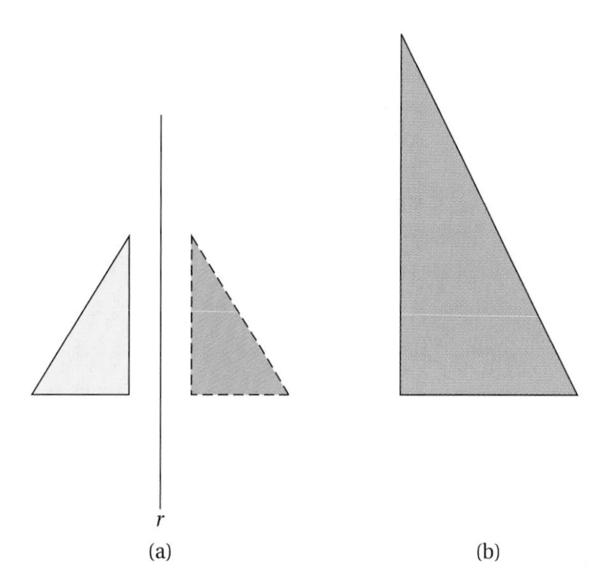

FIGURE 11.9 | Transforming a triangle into another triangle that is the same shape.

Size Transformations and Similarity. To relate size transformations to **similarity,** we first consider some of the size properties of transformations. Measuring the sides of the parallelograms in Figure 11.7 reveals that $\frac{BC}{AB} = \frac{B'C'}{A'B'}$. This result suggests that the ratios of lengths of sides of a figure and the ratios of the lengths of corresponding sides in its image are equal. Because of this property and the fact that size transformations also preserve betweenness and angle measure, size transformations also preserve shape. Sometimes a figure can be made to correspond to a figure with the same shape by a single size transformation—but not always. In Figure 11.9, the original triangle—on the left in part (a)—was first reflected in line *r*; then this reflected image was subjected to a size transformation to produce the blue image of the triangle shown in part (b).

This possible need for a combination of transformations to transform a figure into a figure of different size with the same shape suggests the following definition.

Definition of Similarity, Using Transformations

Two figures are **similar** if and only if there exists a combination of an isometry and a size transformation that generates one figure as the image of the other.

In Example 11.3, a size transformation along with the problem-solving strategies of *draw a picture* and *write an equation* are used to solve a practical problem.

Example 11.3

Problem Solving: The Poster Problem

The service organization at Lonnie's school is planning a fund-raiser with stars as the theme. Lonnie can only find a pattern for a star that is 8 centimeters high, which is too small for the posters. How can Lonnie enlarge the star pattern so that it is 20 centimeters high?

Solution

Trace the center O of a size transformation and the 8-centimeter star pattern in the lower left corner of the paper. The height needs to be changed from 8 to 20 centimeters, so the scale factor is $20 \div 8$, or 2.5. Draw the rays from O through the tips of the small star pattern and measure 2.5 times the distance from O to a small star tip along these rays and mark the tips of the larger star. Connect every other point to form the 20-centimeter star pattern.

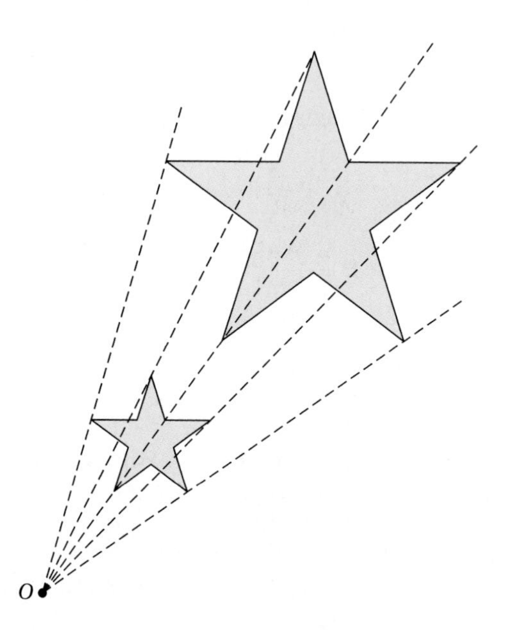

Your Turn

Practice: Use a size transformation to find an image of the pentagon that is three-fourths as wide as the one shown.

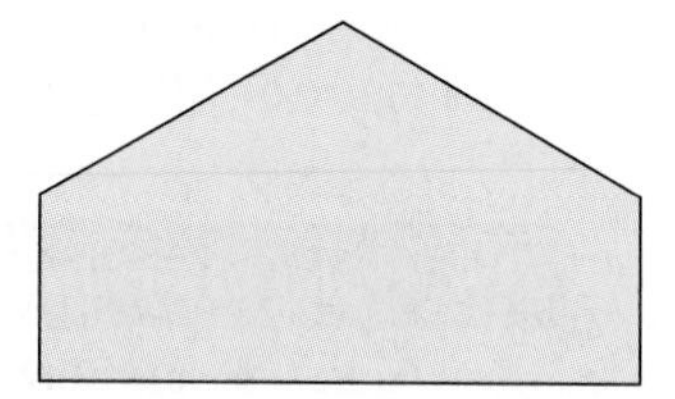

Reflect: Could you use a size transformation to find an image that is congruent to the original shape? Explain your answer and use mathematical notation to symbolize the transformation. ■

Transformations That Change Both Size and Shape

Topological Transformations. Transformations that represent shrinking, stretching, or bending a curve or surface without tearing it or joining points are called **topological transformations**. These transformations can change both the size and shape of a figure, as shown in Figure 11.10.

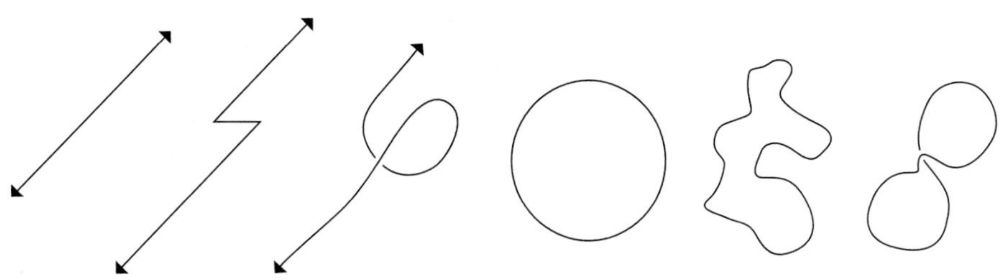

(a) Topological transformation of a line. (b) Topological transformation of a circle.

FIGURE 11.10 | Topological transformation of a line and a circle.

Note that the images have no holes or breaks that were not also in the originals. The fact that "connectedness" doesn't change in a topological transformation leads to other important characteristics that remain constant, such as the characteristic of having no ends. For example, squares, triangles, and all other polygons can be transformed by a topological transformation into a circle.

If one figure can be transformed to another by a topological transformation, the two figures are **topologically equivalent**. For example, a rubber band is topologically equivalent to a compact disk, and a bagel is topologically equivalent to a coffee mug. Also, a schematic diagram of a circuit is topologically equivalent to a diagram to scale of the actual circuitry; how the components and wires are connected and their relative locations are what is preserved.

However, a picture frame and a grocery sack are not topologically equivalent. If the picture frame were made of modeling clay, it could never be formed into the shape of a grocery sack by just stretching or squeezing the clay. Some parts would have to be joined or the clay broken apart, neither of which is allowed in a topological transformation.

In Mini-Investigation 11.3 you are asked to analyze topological transformations further.

Make a chart showing your classification. Talk with classmates and compare your charts.

Mini-Investigation 11.3 **Using Mathematical Reasoning**

How would you classify the digits 0–9 into topologically equivalent figures?

In Exercise 45 at the end of this section you are asked to use a topological transformation to solve an interesting problem, the castle court problem.

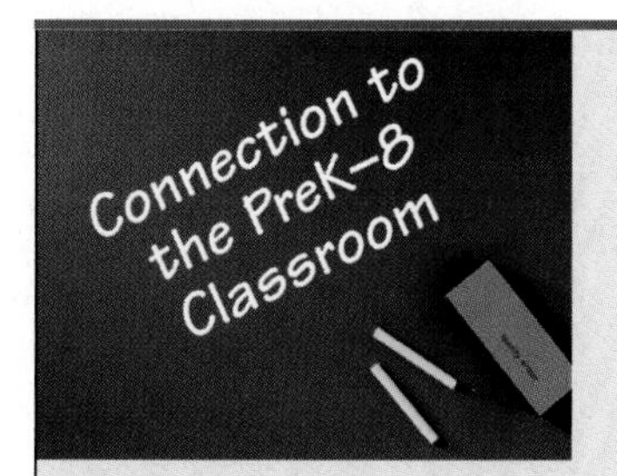

Many young children can distinguish topological properties such as connectedness, separation, inside, and outside before they can distinguish characteristics of rigid geometry such as length, angle measure, or relative position. Elementary school students extend their spatial visualization abilities by drawing on balloons and watching what changes and what doesn't change as the balloon is inflated and deflated. The topological concepts of a figure being connected and separating a plane are used later in classifying shapes and creating formal definitions for polygons and other geometric figures.

Problems and Exercises | *for Section 11.1*

A. Reinforcing Concepts and Practicing Skills

1. What information is necessary for identifying a specific slide?
2. How is the information in Exercise 1 related to the mathematical notation for a translation?
3. What information is necessary for identifying a specific turn?
4. How is the information in Exercise 3 related to the mathematical notation for a rotation?
5. What information is necessary for identifying a specific flip?
6. How is the information in Exercise 5 related to the mathematical notation for a reflection?

In each diagram in Exercises 7–10, identify the motion or combination of motions that would produce the image. Justify your choices.

7. 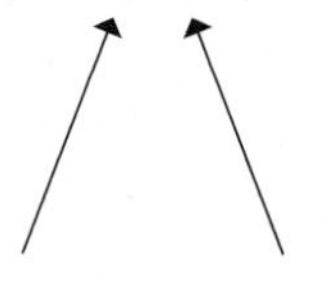

8. 

9. 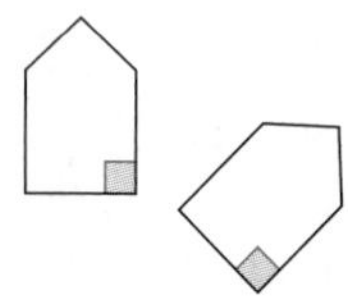

10. 

For Exercises 11–22, use tracing paper, graph paper, a geoboard, or geometry exploration software to find the image of the quadrilateral obtained from each transformation.

11. T_{MN}	12. T_{DE}	13. T_{RS}	14. T_{CB}
15. $R_{O,\alpha}$	16. $R_{D,45°}$	17. $R_{P,-90°}$	18. $R_{O,180°}$
19. M_l	20. M_r	21. M_s	22. M_t

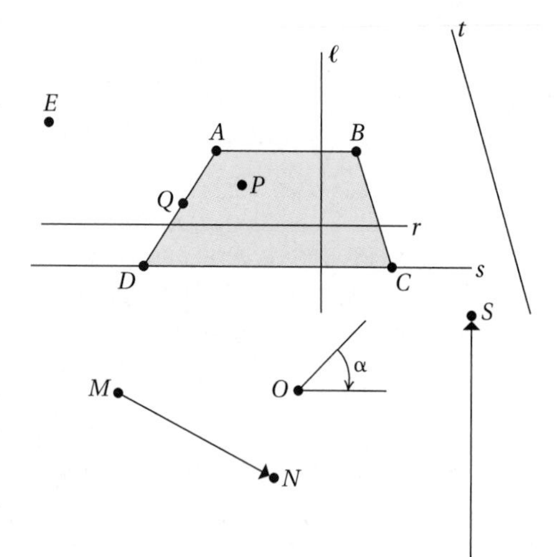

23. Describe the reflectional and rotational symmetry properties of each object. Use a plastic reflector or paper folding, or GES when needed.

   a. Propeller

   b. Flag

   c. Advertising logo

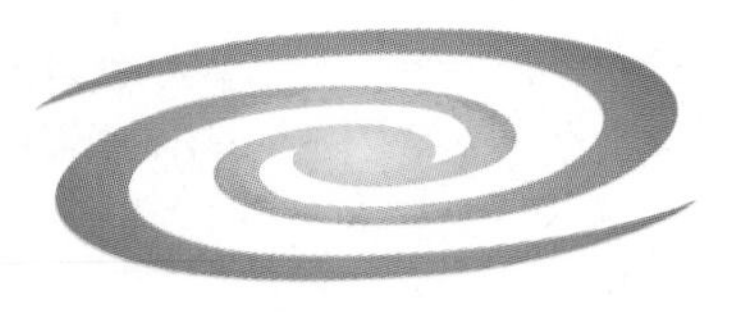

24. Decide whether the pair of objects in each part of the figure are topologically equivalent. Support your answer.

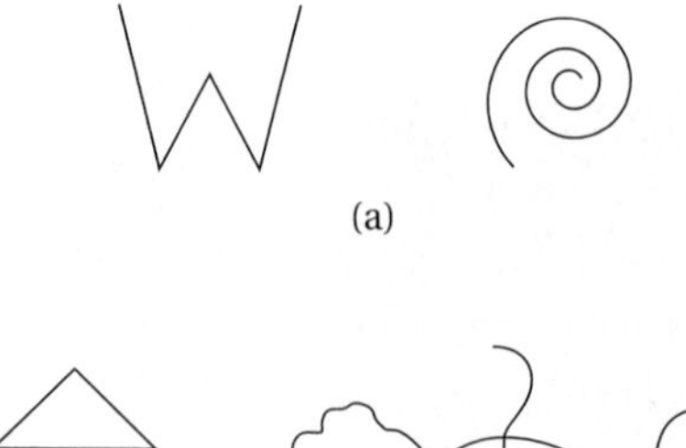

(a)

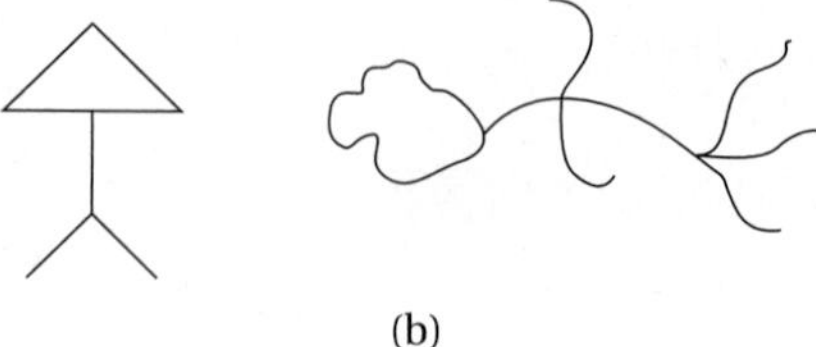

(b)

25. Identify two physical objects that are topologically equivalent to a
  a. ball
  b. donut.
  c. sack with two handles.

26. Classify the letters of the alphabet into groups of topologically equivalent figures.

27. Use the figure shown to find the images and ratios specified and verify your answers.

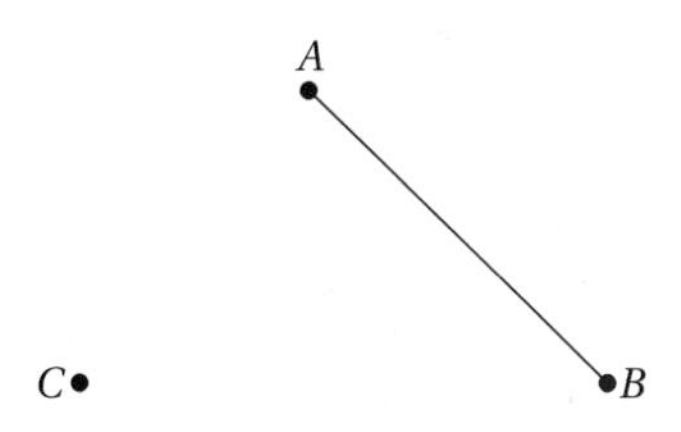

  a. Find the image of $\overline{AB}$ under the size transformation $S_{C,3}$

  b. $\dfrac{CA'}{CA}$ **c.** $\dfrac{CB'}{CB}$ **d.** $\dfrac{A'B'}{AB}$

  e. Find the image of $\overline{AB}$ under the size transformation $S_{C,0.5}$.

  f. $\dfrac{CA'}{CA}$ **g.** $\dfrac{CB'}{CB}$ **h.** $\dfrac{A'B'}{AB}$

28. Explain how to use the transformational definition of similarity to test whether each pair of figures is similar.

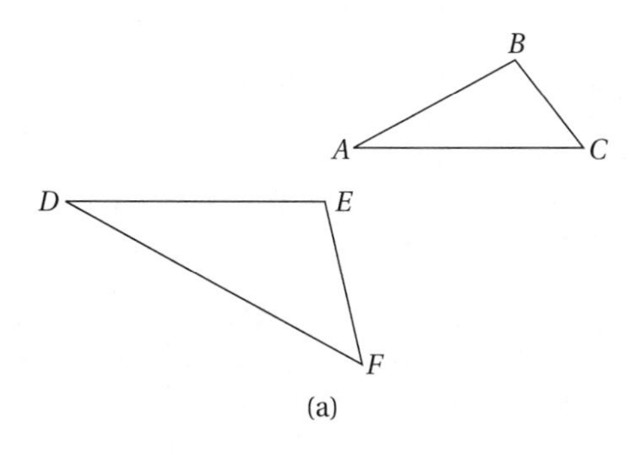

(a)

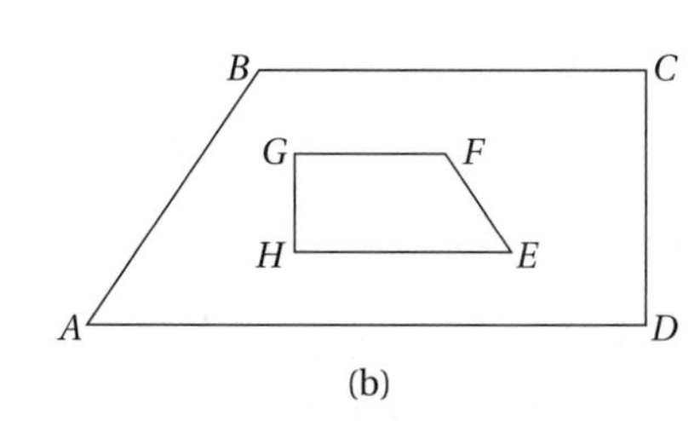

(b)

B. Deepening Understanding

29. Describe the symmetry properties of the following geometric shapes or figures.

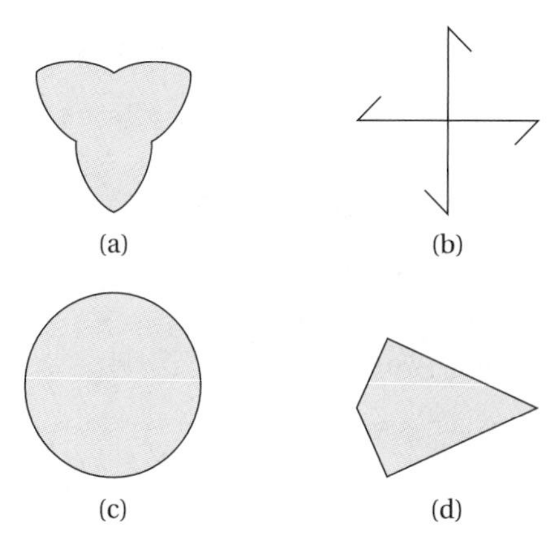

(a) (b) (c) (d)

30. Make a chart showing how the isometries are alike and how they are different. Consider the following characteristics.
  a. Reverse orientation
  b. Points that map into themselves
  c. Mathematical notation
  d. Role of segments, angles, and lines

31. Figures that correspond under a translation, rotation, or reflection are said to be *translation congruent*, *rotation congruent*, or *reflection congruent.* Position two congruent figures on a piece of paper to satisfy each combination of congruences listed. Use tracing paper, mirrors, or GES to justify your choices.
  a. All three congruences
  b. Translation and rotation congruences only
  c. Translation and reflection congruences only
  d. Rotation and reflection congruences only
  e. Translation congruence only
  f. Rotation congruence only
  g. Reflection congruence only
  h. None of the three congruences

32. Find the image of the quadrilateral when it is subjected to the glide–reflection T_{RS} followed by $M_{\overleftrightarrow{RS}}$. Does it matter whether you translate first and then reflect, or reflect first and then translate? Justify your answer.

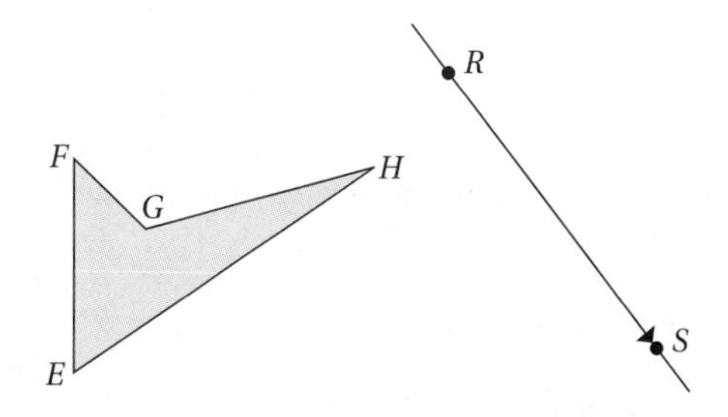

33. Use tracing paper, geoboards, or GES to find the images obtained from the following combinations of motions, where, for example, $T_{AB}{}^{*}T_{CD}$ indicates a translation from A to B followed by a translation from C to D.

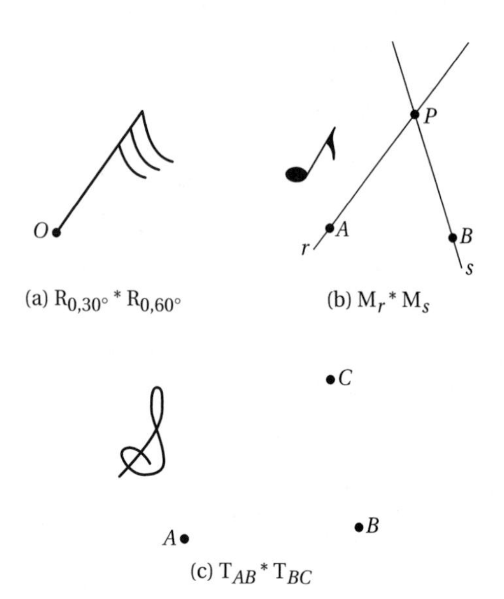

(a) $R_{0,30^\circ} * R_{0,60^\circ}$ (b) $M_r * M_s$

(c) $T_{AB} * T_{BC}$

34. What different types of symmetry, if any, do you see in these automobile manufacturers' logos?

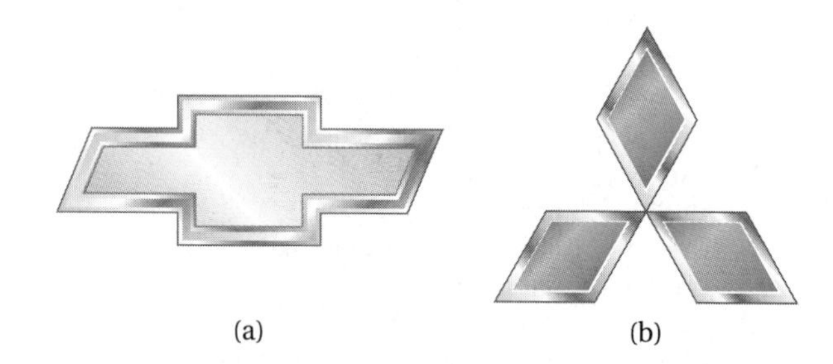

(a) (b)

(c) (d)

35. Create three logos for an imaginary corporation.

a. One is to have rotational symmetry but no reflectional symmetry.

b. One is to have reflectional symmetry but not rotational symmetry.

c. One is to have both rotational and reflectional symmetry.

36. Draw, if possible, a quadrilateral that has

a. a line of symmetry but no rotational symmetry.

b. rotational symmetry but no line of symmetry.

c. both reflectional and rotational symmetry.

C. Reasoning and Problem Solving

37. Copy the following figures on dot paper and find the original shape if the shaded shape is its image under the given transformation:

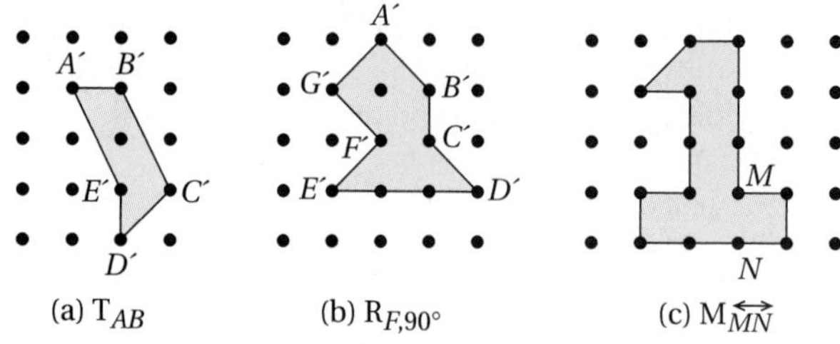

(a) T_{AB} (b) $R_{F,90^\circ}$ (c) $M_{\overleftrightarrow{MN}}$

38. Gina showed Maria the following GES sketch. Gina claimed that she created figure *XYZ* by using a combination of two motions applied to the original figure *ABC*. The dashed lines show the result of the first motion.

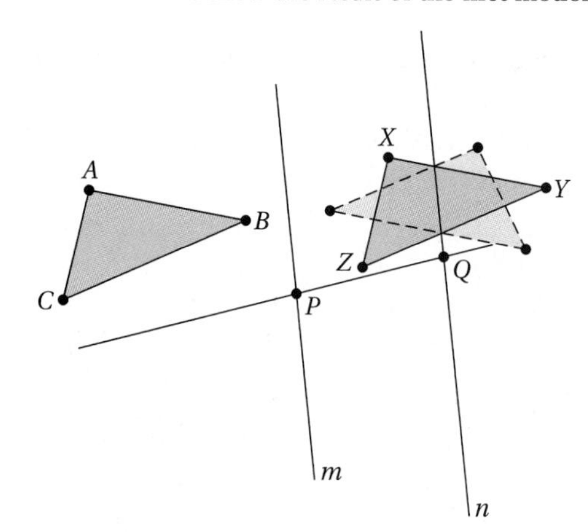

Describe the combination of motions that Gina could have used. Explain how *XYZ* could result from *ABC* in just one motion.

39. How can GES be used to show whether a transformation preserves distance? Create various figures to illustrate your techniques.

40. Matisto told his classmates that he obtained figure *K* from a figure *A* by using two GES transformations: $T_{XY}(A) = D$ and $R_{(0,0),90^\circ}(D) = K$.

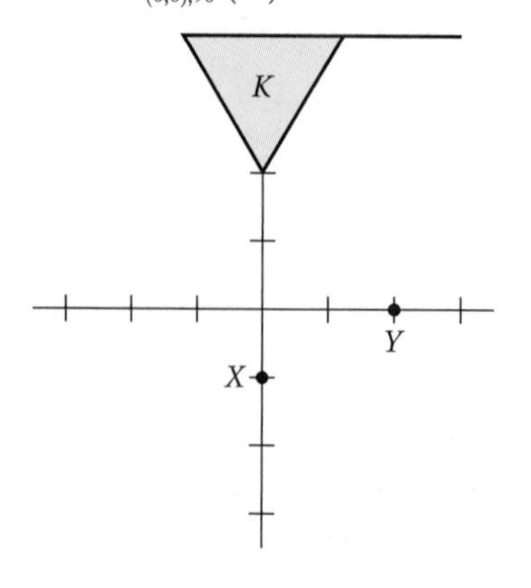

a. Copy the grid and use this information to locate figure A on it. Note that X is located at $(0, -1)$ and that Y is located at $(2, 0)$.

b. Is A congruent to K?

41. Lana told her classmates that she got figure P, shown here, from a figure W by using the size transformation $S_{(0,0),2}(W) = P$.

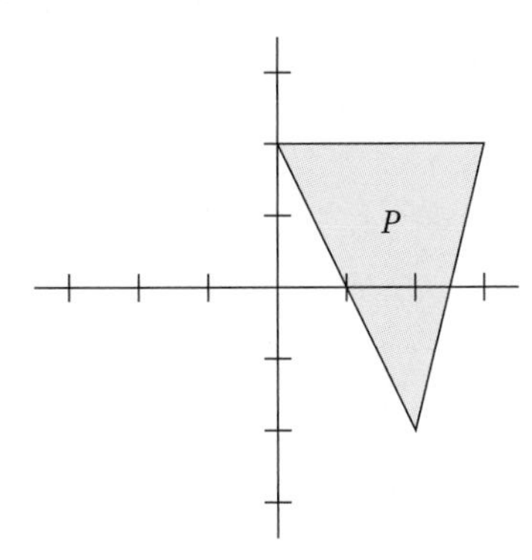

Copy the sketch and use this information to recreate figure W on it. Are P and W congruent? Explain.

42. The Picture Frame Problem. Larry tries to make a rectangular picture frame the same shape but larger than one he has by adding 5 units to both the length and the width. Are the picture frame rectangles similar? Explain and support your conclusion.

43. The Mirror Problem. What is the smallest size of mirror, hung with the top of the mirror even with the top of her head, that would allow Julie to see her total height? Explain with a diagram.

44. The Lot Purchase Problem. Carlos owns a house on a lot that is a quadrilateral with dimensions w, x, y, and z. The lot that he wants to buy is a quadrilateral with dimensions $4w$, $4x$, $4y$, and $4z$. Should he expect the lots to be the same shape? Explain.

45. The Castle Court Problem. When the gatekeeper at a castle was asked how to get to the inner chamber, he said: After you enter, always stick to the right-hand wall. Use the topological transformation of the map of the inner court in the next column to decide whether the gatekeeper is correct. In following this path, does the visitor go through all the halls of the inner court? If the gatekeeper had said, stick to the left-hand wall, would the visitor have reached the inner chamber?

46. Draw a topological transformation of a map of the main floor of a house with which you are familiar. Would following the gatekeeper's instructions in Exercise 45 take you through each room of the house and return you to the front door?

47. The Crossword Puzzle Problem. Crossword puzzles are designed in such a way that, if C is the center of the square, the transformation $R_{C,180°}$, generates an image of the colored squares in the blank puzzle that is exactly the same as those in the original blank puzzle. Copy the following crossword puzzle template and color in at least 16 squares to design a crossword puzzle having that characteristic.

Figure for Exercise 45

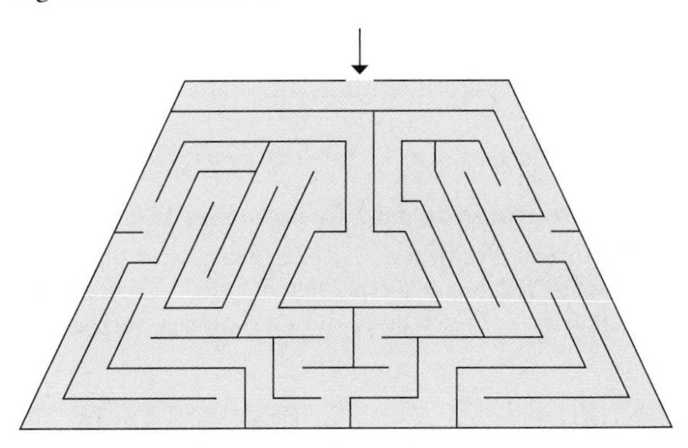

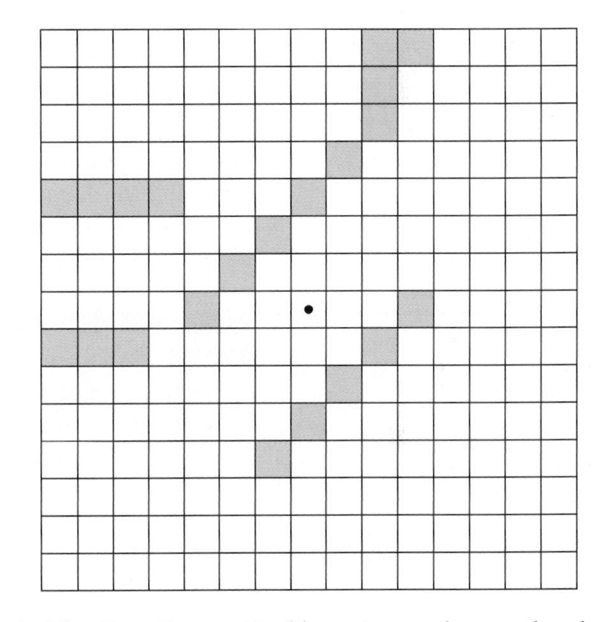

48. The Box Pattern Problem. A manufacturer found that the machines in his factory could make boxes most efficiently if the six-square box pattern had four squares in a row and had rotational symmetry. He asked a staff engineer to analyze all 35 possible box patterns and present a report on the patterns that had the needed characteristics. He asked the engineer to give reasons to assure him that all such patterns had been included. Prepare the engineer's report for the manufacturer.

49. How would you classify different types of triangles and quadrilaterals according to their symmetry properties? Make a chart to show your classifications.

50. How many other pentominos, like the one shown on the next page, have one line of symmetry? Draw a picture of each pentomino.

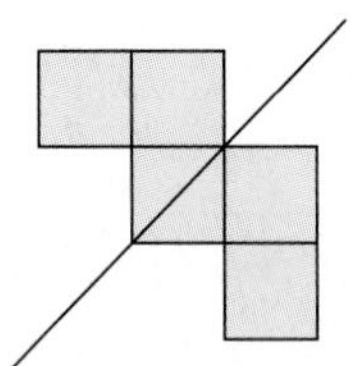

D. Communicating and Connecting Ideas

51. **Historical Pathways.** In 1872, Felix Klein published *Erlanger Programm*, a paper that described a new view of geometry as the study of those properties, such as the commutative property of transformation combination, that are preserved by certain types of transformations. In your group, answer the following questions and devise a way to support your answer.
    a. Does $T_{AB}*T_{PQ} = T_{PQ}*T_{AB}$
    b. Does $M_r*M_s = M_s*M_r$?
    c. Does $R_{A,180°}*R_{B,180°} = R_{B,180°}*R_{A,180°}$?
    d. Does $R_{C,\alpha}*R_{C,\beta} = R_{C,\beta}*R_{C,\alpha}$?
52. **Making Connections.** Write conjectures about the connections between **a.** reflections and line symmetry, and **b.** rotations and turn symmetry.

Section 11.2 GEOMETRIC PATTERNS

- Geometric Patterns and Tessellations
- Polygons That Tessellate
- Combinations of Regular Polygons That Tessellate
- Other Ways to Generate Tessellations
- Tessellations with Irregular or Curved Sides

In this section we explore tiling patterns that consist of geometric figures that fit without gaps or overlaps to cover a plane. We use charactreristics and properties of geometric figures to decide which polygon shapes can be used to tile a floor by themselves and which combinations of polygon shapes can be used to tile a floor. We also explore tiling with irregular figures. Mini-Investigation 11.4 gives you an opportunity to look at the general idea of tiling a floor and which polygons can be used to do so.

Draw tilings by using tracing paper to convince someone that your solution is correct. Verify your solution in another way and describe tilings you have seen in real-world situations.

Mini-Investigation 11.4 **Solving a Problem**

Which of the tiles shown can a manufacturer advertise correctly as tiles that can be used by themselves to tile a rectangular floor? (Note: Partial tiles may be used at the sides to exactly fill the rectangle.)

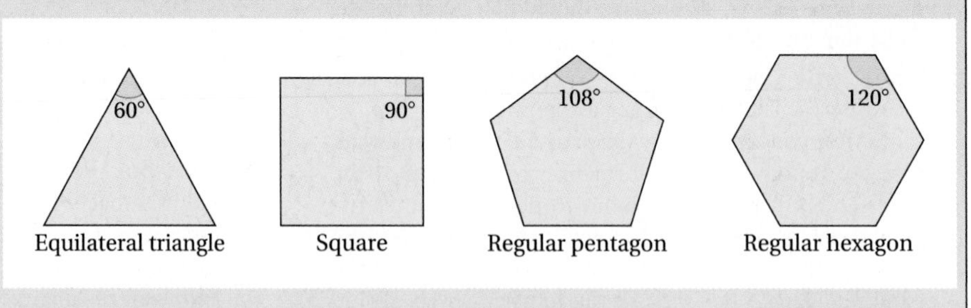

Geometric Patterns and Tessellations

From floor tiling to Escher art, the world is full of geometric patterns. The Greek wall pattern shown in Figure 11.11(a) is an example of a design repeating some basic element in a systematic manner, commonly known as a **pattern**. Another example is the interesting Mexican strip pattern shown in Figure 11.11(b). Note the translations in both patterns: the reflectional symmetry in Figure 11.11(a), and the rotational symmetry in Figure 11.11(b). We now define a special type of pattern.

(a)

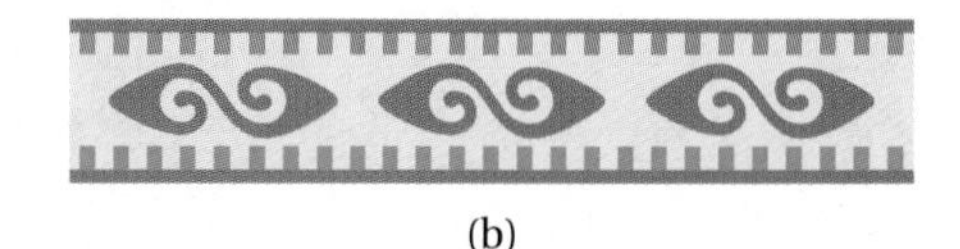

(b)

FIGURE 11.11 | Interesting geometric patterns.

(*Source:* (a) Adapted from Jones/Owen, *The Grammar of Ornament,* p. xvi, Bernard Quarterly, London, 1868; found in Weyl, Herman, *Symmetry.* Copyright © 1952 by Princeton University Press. Reprinted by permission of Princeton University Press; (b) Adapted from Grünbaum, Branko, and Shephard, G. C. *Tilings and Patterns—An Introduction.* Copyright © 1987 by W. H. Freeman and Company, p. 5. Used with permission.)

Definition of a Tessellation

A **tessellation** is a special type of pattern that consists of geometric figures that fit without gaps or overlaps to cover the plane.

Figure 11.12 shows an example of a tessellation. Because this pattern involves the use of regular octagons and squares, we say that a combination of these figures will *tessellate the plane* or, more simply, will *tessellate.* Sometimes the word *tiling* is used to mean "tessellation," and the word *tile* is used to mean "tessellate." In Mini-Investigation 11.4, you may have found that an equilateral triangle will tessellate the plane. As indicated by the tessellations in Figure 11.13, the figures in a tessellation can be curved and do not need to be polygons.

FIGURE 11.12 | Example of a tessellation.

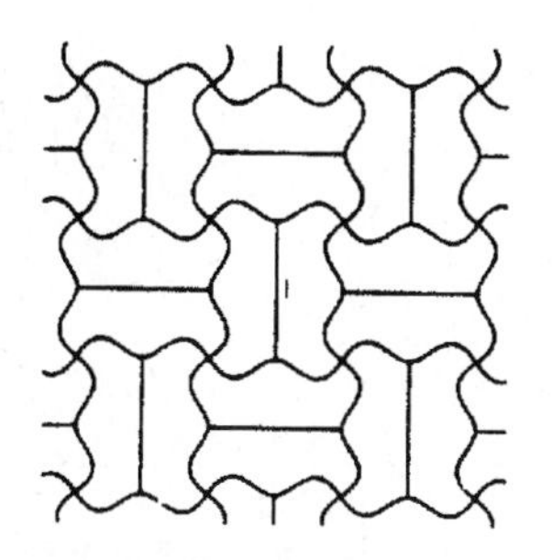

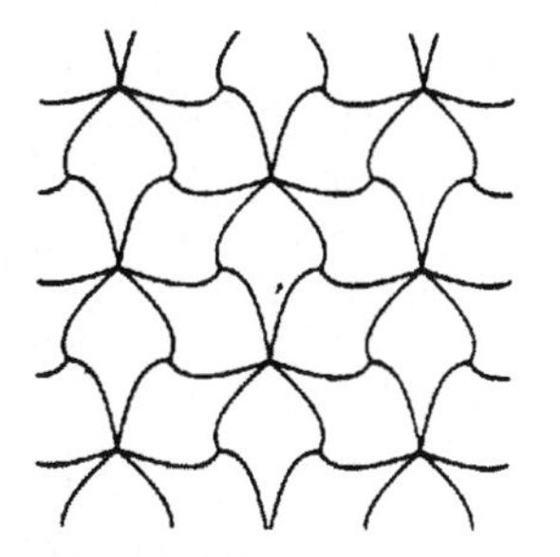

FIGURE 11.13 | Curved tessellations.

(*Source: Tilings and Patterns,* by Branko Grünbaum and G. C. Shephard. © 1987 by W. H. Freeman and Company. Used with permission.)

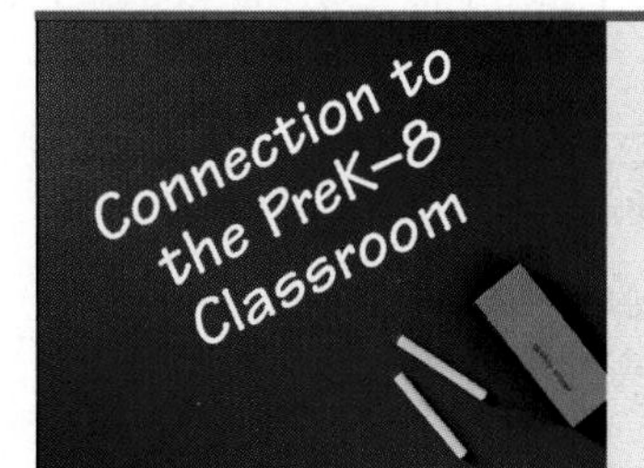

Analyzing tessellations enhances students' understanding of the various properties of different polygons, including angle measures, congruences, and symmetries. Creating tessellations provides students with an interesting context in which to apply their knowledge of translations, rotations, and reflections.

Polygons That Tessellate

The process of finding which polygons will tessellate the plane is an example of what makes mathematics challenging and interesting. It also enables us to broaden our ideas about polygons and transformations by applying them in analyzing tessellations. Let's take a brief look at part of this process.

In Mini-Investigation 11.4, you may have drawn the correct conclusion that an equilateral triangle, a square, and a regular hexagon can each be used to form a tessellation. Such a tessellation, made up of congruent regular polygons of one type, all meeting edge to edge and vertex to vertex, is called a **regular tessellation**.

Now let's consider tessellations involving nonregular polygons. Because the sum of the measures of the angles of a triangle is 180° and the sum of the measures of the angles of a quadrilateral is 360°, we can show that these types of figures can be made to fit around their vertices to tessellate the plane. We can verify that *any triangle or quadrilateral will tessellate the plane.*

A natural extension of that idea is to investigate whether pentagons and hexagons tessellate the plane. You may have found in Mini-Investigation 11.4 that *a regular pentagon does not tessellate the plane but that a regular hexagon will.* Some nonregular pentagons and hexagons, such as regions A and B in Figure 11.14, will also tessellate the plane. No convex polygon with more than six sides will tessellate.

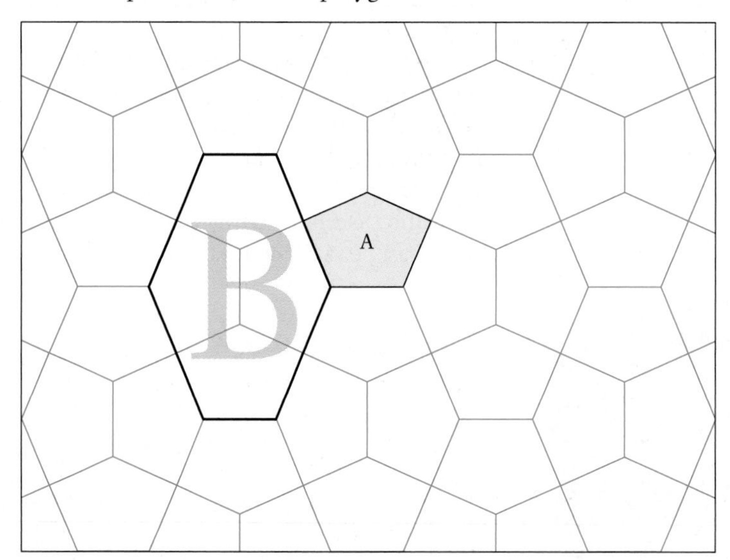

FIGURE 11.14 | Nonregular pentagons and hexagons that tessalate.
(*Source:* O'Daffer/Clemens/Libeskind, *Geometry: An Investigative Approach,* Second Edition, figure 4.15 from p. 95, © 1992. Reprinted by permission of Pearson Education, Inc.)

In Mini-Investigation 11.5 you are asked to verify a generalization about which regular polygons will not tessellate the plane.

Write a convincing argument in support of your conclusion.

Mini-Investigation 11.5 **Using Mathematical Reasoning**

How could you use the fact that the measure of an interior angle of a regular hexagon is 120° to convince someone that no regular polygon with more than six sides will tessellate the plane?

A Sample Student Page: Tessellations from Transformations

Experiences with tessellations enable elementary school students to integrate art with mathematics, as illustrated in Figure 11.15. They provide an interesting environment in which students can apply what they have learned about polygons and their angles. Geometric design fosters creative thinking and helps students further develop their spatial visualization abilities.

- *Which of the six shapes shown will tessellate the plane and which will not? How do you know?*
- *Would these tessellations be regular or not? Explain.*
- *Which shapes could be combined to create a semiregular tessellation? How do you know?*

Slides and Tessellations

8-10

▶ Lesson Link You know what happens to a figure when you flip it or turn it. Now you'll see what happens when you slide a figure to a new position. ◀

You'll Learn ...

■ to identify translations of figures and tessellations

... How It's Used

Graphic designers use translations and tessellations when designing logos and graphic artwork.

Vocabulary

translation

tessellation

Explore Slides and Tessellations

I've Got You Covered!

Materials: Tracing paper, Unlined paper

These shapes represent different tiles for sale. Which of the shapes could be used to cover the floor without having any space in between tiles?

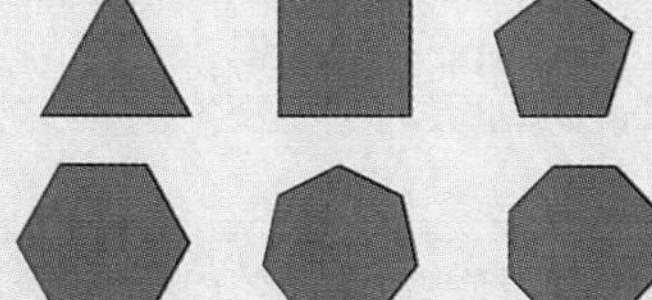

1. Copy each shape onto tracing paper several times. Copy the shapes as closely together as possible without overlaps. Then state whether each shape can or cannot be put together with copies of itself, leaving no spaces in between.
2. Draw a shape that does not appear above but that could be used to entirely cover a floor without any spaces in between.
3. Look at all of the shapes that can fit together without spaces in between. What patterns do you see that could help you determine if a shape would work without using tracing paper?

Learn Slides and Tessellations

When a figure is slid to a new position without flipping or turning, the new image is called a slide, or a **translation**.

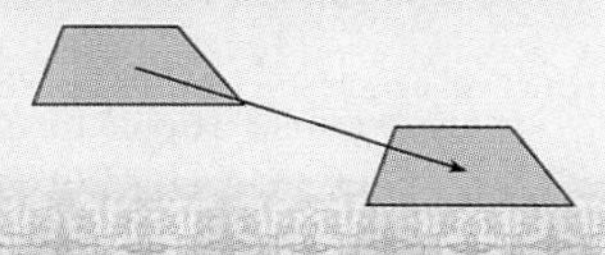

8-10 • Slides and Tessellations **453**

FIGURE 11.15 | Excerpt from a middle school mathematics textbook.

(*Source:* From *Scott Foresman–Addison Wesley Middle School Math, Course 1* by R. I. Charles, J. A. Dossey, S. J. Leinwand, C. J. Seeley, and C. B. Vonder Embse. © 1998 by Pearson Education, Inc. publishing as Prentice Hall. Used by permission.)

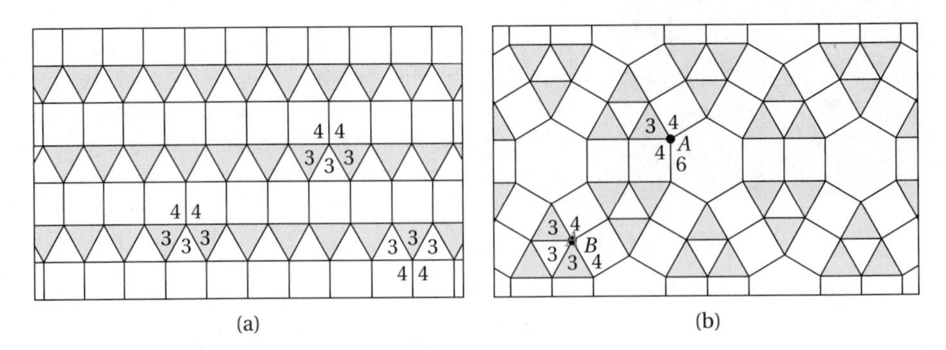

FIGURE 11.16 | Two types of tessellations involving the combination of regular polygons.

Combinations of Regular Polygons That Tessellate

Piet Hein has stated: "We will have to evolve problem solvers galore, for each problem they solve creates ten problems more." [Hein (1969), p. 32] In this spirit, the question about whether combinations of regular polygons can tessellate the plain naturally arises from the question about single-polygon regular tessellations that we answered earlier. Figure 11.16 depicts two combinations of regular polygons that do tessellate the plane.

The pattern of polygons in Figure 11.16(a) is a special kind of tessellation, defined as follows.

Definition of a Semiregular Tessellation

A tessellation formed by two or more regular polygons with the arrangement of polygons at each vertex the same is called a **semiregular tessellation.**

The tessellation in Figure 11.16(b) has two different arrangements at vertices and is not a semiregular tessellation. Because a semiregular tessellation is a special kind of tessellation, another natural question arises: *How many different semiregular tessellations are there?* This question is explored more fully in O'Daffer and Clemens (1992), in which they established that there are only 21 ways to arrange regular polygons around a point, with no gaps or overlaps. Of these 21 ways, only 8 can be extended to form a semiregular tessellation. Thus we conclude that *only eight semiregular tessellations are possible.* These tessellations are shown in Figure 11.17.

The symbol 4, 8, 8 or $4, 8^2$ can be used to describe tessellation (h) in Figure 11.17 because the arrangement around each vertex is square, octagon, octagon. The other tessellations can be symbolized in a similar manner.

The problem in Example 11.4 can be solved using ideas about semiregular tessellations. The problem-solving strategies *make a model*, *draw a diagram*, and *use reasoning* are helpful when solving the problem.

Example 11.4 **Problem Solving: The Ancient Temple Floor**

A floor restoration specialist was called upon to restore the floor tile in an ancient temple. He knew that the floor had been tiled only with equilateral triangles and regular hexagons and was a semiregular tessellation. The only remaining section of

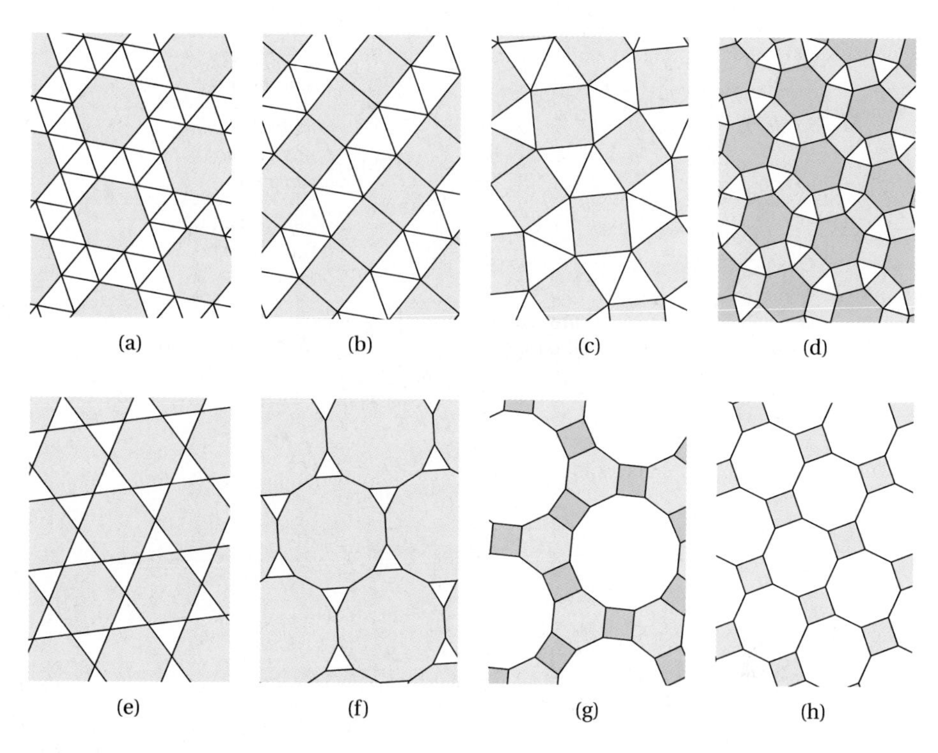

FIGURE 11.17 | The eight possible semiregular tessellations.

original tile is shown. Can the specialist use this information to restore the floor to its original appearance?

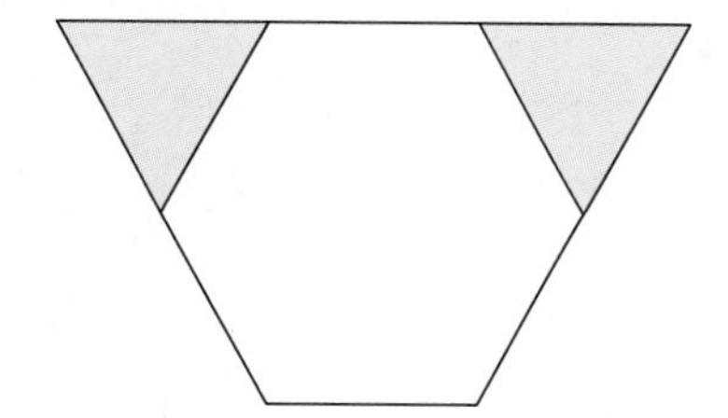

Working toward a Solution

Understand the problem	*What does the situation involve?*	Restoring a tiled floor in an ancient temple.
	What has to be determined?	What the original tiling looked like.
	What are the key data and conditions?	The only remaining section of tile is pictured. Only equilateral triangles and hexagons were used in the semiregular tiling.
	What are some assumptions?	Assume that the tiling must be formed by extending the available section.
Develop a plan	*What strategies might be useful?*	Use a model; draw a diagram; use reasoning.

	Are there any subproblems?	Deciding what arrangements will fit around a point.
	Should the answer be estimated or calculated?	No estimation or calculation is needed.
	What method of calculation should be used?	Not applicable.
Implement the plan	*How should the strategies be used?*	Use a model with cutout triangles and hexagons, or a diagram to extend the arrangement—for example, the following. Then use reasoning to decide whether this arrangement is a semiregular tessellation.
	What is the answer?	(Continue this process in Your Turn to look for other tessellations and solve the problem.)
Look back	*Is the interpretation correct?*	Yes. The diagram meets the conditions.
	Is the calculation correct?	Not applicable.
	Is the answer reasonable?	Look for all possibilities and decide.
	Is there another way to solve the problem?	We could find the number of degrees in the angles of the polygons and use these measures to look for different ways to arrange them around a point.

Your Turn

Practice: Complete the solution to the example problem.

Reflect: What if the example problem didn't specify that the floor tiling was a semiregular tessellation? Would that change the solution? Explain. ■

Other Ways to Generate Tessellations

Using a Kaleidoscope. A three-mirror kaleidoscope provides a model that uses reflections to create tessellations of the plane. To make a three-mirror kaleidoscope, you can fasten three mirrors of the same size together to form the vertical faces of an equilateral triangular prism. When you place an equilateral triangle piece of

FIGURE 11.18 | Viewing tessellations through a three-mirror kaleidoscope.

Source: Farnsworth Kaleidoscope Images

construction paper with a pattern on it in the base of the kaleidoscope and peer over the edge as in Figure 11.18, an interesting geometric pattern appears.

If we investigate the semiregular tessellations further, we find that only those with lines of symmetry that form a superimposed tessellation of equilateral triangles can be produced with a three-mirror kaleidoscope. The other tessellations can be produced with other types of kaleidoscopes, based on their lines of symmetry.

Using Computer Software. The following LOGO procedures will create a portion of the tessellation of hexagons shown here.

```
TO HEXAGON :SIDE
REPEAT 6[FD :SIDE RIGHT 60]
END

TO TESSHEX
PENUP BACK 80 LEFT 90 PENDOWN
REPEAT 4 [HEXAGON 30 RIGHT 120 FD 30
          LEFT 60 HEXAGON 30 FD 30 LEFT 60]
END
```

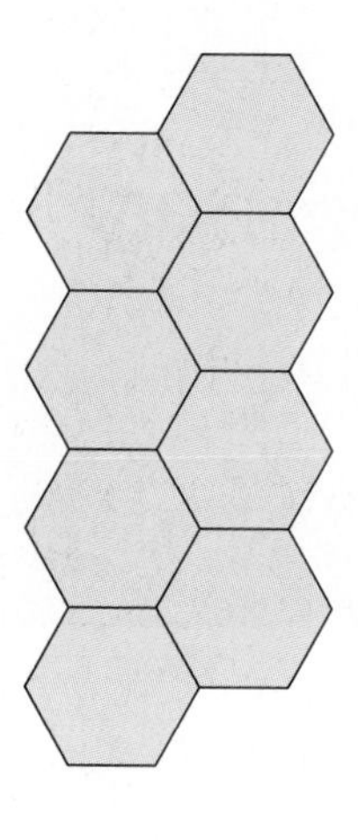

Similar procedures can be created easily to produce a portion of a regular tessellation of triangles or squares. In fact, writing LOGO procedures to produce the semiregular tessellations is both possible and interesting. Exercise 36 at the end of this section asks you to do so.

Tessellations with Irregular or Curved Sides

The Moors, who occupied Spain from 711 until 1492, were forbidden by their religion to draw living objects. To compensate, they mastered creative design, as indicated by the drawings of two of the simpler designs with which they decorated the walls of the Alhambra in Granada shown in Figure 11.19.

FIGURE 11.19 | Drawings from the Alhambra.
(*Source: M. C. Escher's Drawings from the Alhambra* © 2001 Cordon Art B.V. Baarn, Holland. All rights reserved.)

M. C. Escher, a Dutch artist whose drawings, including the one shown in Figure 11.20, have long been a special source of enjoyment for people interested in mathematics, made the following remarks about the Moors' influence on his art.

> This is the richest source of inspiration I have ever struck: nor has it yet dried up . . . . [A] surface can be regularly divided into, or filled up with, similar-shaped figures (congruent) which are contiguous to one another, without leaving any open spaces. The Moors were past masters of this. [Escher (1960), p. 11]

Interestingly, techniques are readily accessible for creating curved tessellation art. The basic approach is to begin with a shape that will tessellate and alter it, often using transformations, to produce the effect wanted. In the following subsections we describe some ways to do so.

Replacing Polygon Edges with Curved Lines. The basic figure for the tessellation in Figure 11.21 was created by starting at the midpoint of each edge of

FIGURE 11.20 | A classic drawing by M. C. Escher.
(*Source: M. C. Escher's Day and Night* © 2001 Cordon Art B.V. Baarn, Holland. All rights reserved.)

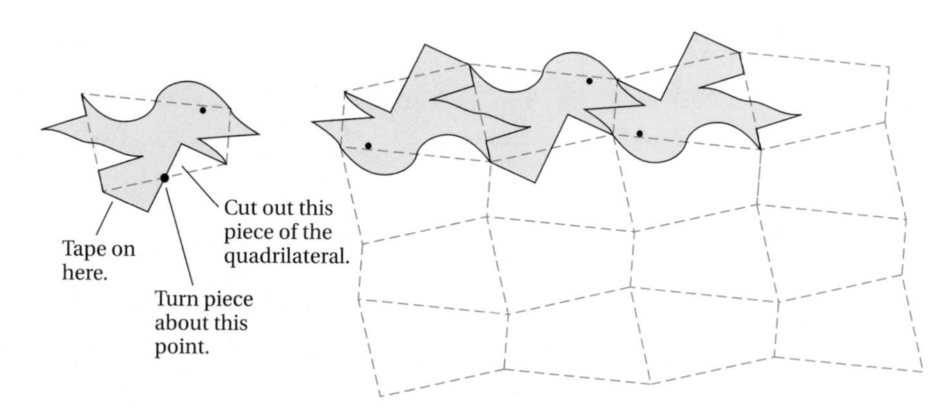

FIGURE 11.21 | The cut and turn method.

the basic quadrilateral, cutting out a piece, rotating it 180° about the midpoint, and taping it back on the original figure—called a *cut and turn* method.

A similar method uses only tessellating polygons with congruent adjacent sides to make a basic figure for a curved tessellation. As shown in Figure 11.22, it involves starting with a tessellating polygon, cutting out a piece that contains a complete side, rotating it about a corner, and taping it along the adjacent side.

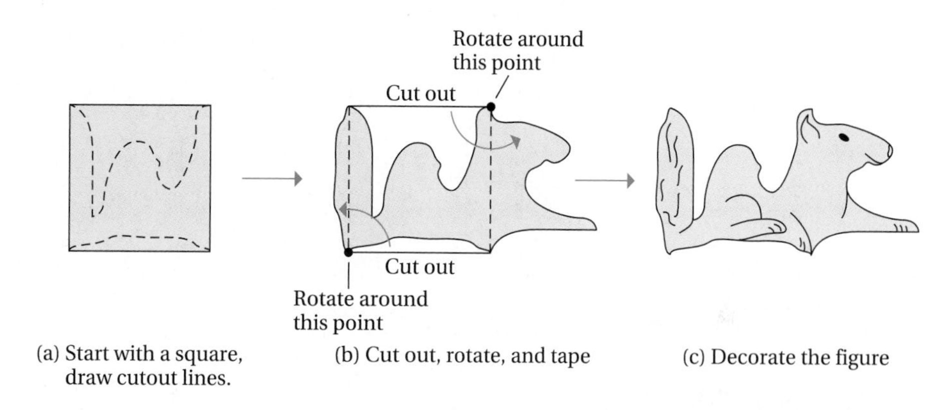

FIGURE 11.22 | The cut and turn method with special polygons.

Another way to replace each edge of a square or rectangle with a curve is shown in Figure 11.23. In this procedure, the basic figure of the tessellation was created by cutting out a piece that includes a side of the basic rectangle and sliding the piece so that the side on the cutout piece coincides with its opposite side—called a *cut and slide* method. Notice how the transformation we have called a translation is involved in this procedure.

Making a Design in the Polygon Interior. An appropriate design drawn in the interior of a polygon can produce interesting curved tessellations, as shown in Figure 11.24 on the next page. When creating such a tessellation, it is helpful to begin with a rough sketch of the tessellation you want to produce, and work backward to develop the interior design.

Example 11.5 demonstrates the creation of a curved tessellation and then gives you an opportunity to use one of the methods just described.

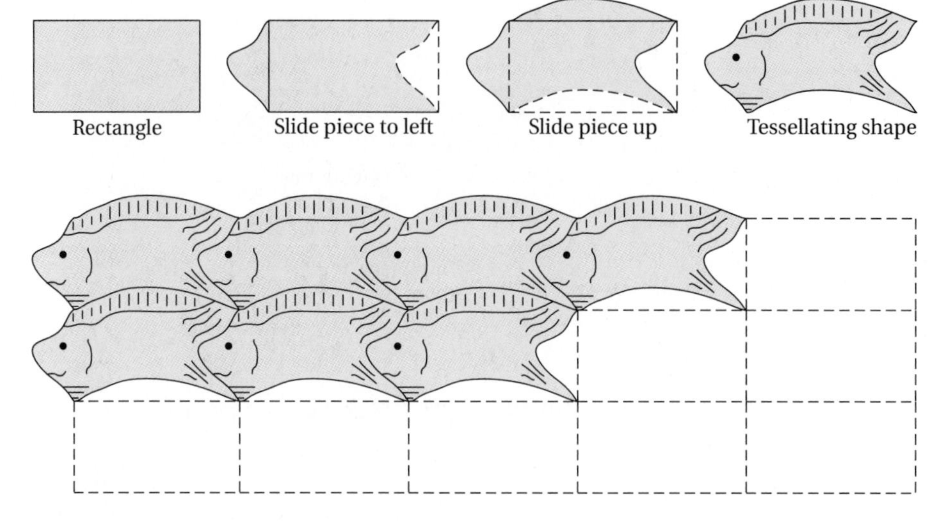

FIGURE 11.23 | The cut and slide method.

FIGURE 11.24 | The interior design method.

Example 11.5 Making a Curved Tessellation

Use one of the methods described to make a curved tessellation from a tessellation of equilateral triangles.

Solution

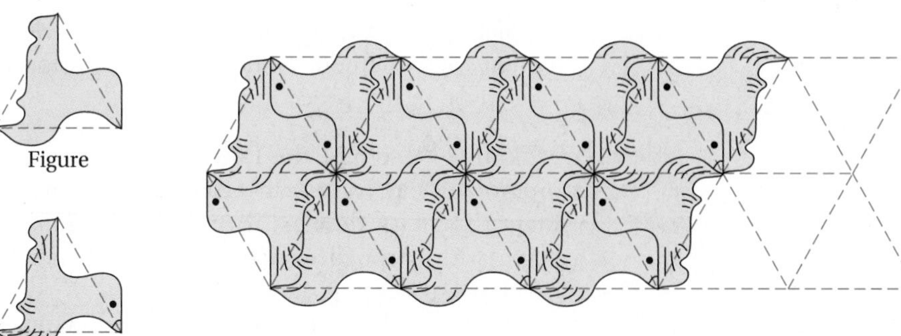

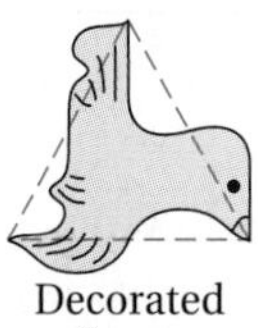

Your Turn

Practice: Use a method not used in the example problem to make a curved tessellation from a tessellation of equilateral triangles.

Reflect: What type of curved tessellation can you make with one method you could have used that you can't make with another? ■

Problems and Exercises | *for Section 11.2*

A. Reinforcing Concepts and Practicing Skills

1. Describe or draw an arrangement of polygons that has a characteristic that prohibits it from being a tessellation.
2. Describe or draw an arrangement of polygons that has a characteristic, different from the one selected in Exercise 1, that prohibits it from being a tessellation.
3. How many equilateral triangles will fit about a point when tessellating the plane? How do you know?
4. How many quadrilaterals will fit about a point when tessellating the plane? How do you know?
5. Use the idea that all parallelograms tessellate to show that the following triangle tessellates the plane.

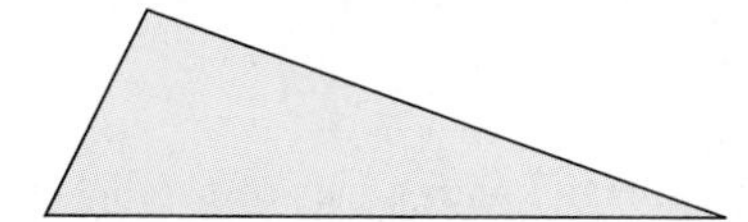

6. How many different types of regular tessellations are there? Describe them.
7. Give an example of a tessellation made of squares that is not a regular tessellation.
8. Give an example of a tessellation made of equilateral triangles that is not a regular tessellation.
9. Why does a regular pentagon not tessellate the plane?
10. Are the curved tessellations in Figure 11.13 regular tessellations? Explain.
11. Will a kite-shaped figure tessellate the plane? How do you know?
12. Will the nonconvex quadrilateral shown below tessellate the plane? Devise a way to convince someone that your answer is correct.

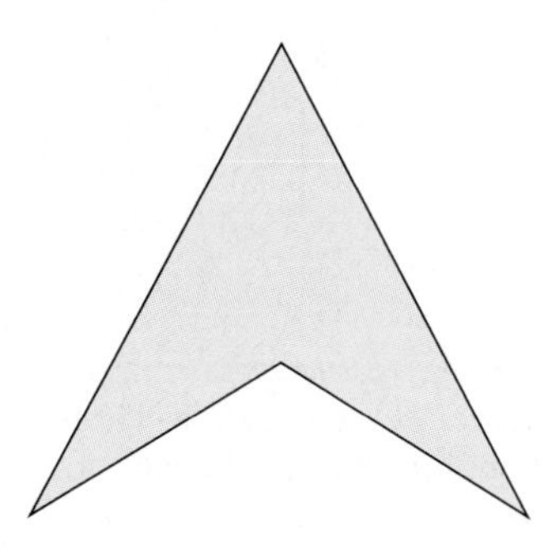

13. Why is it impossible for a regular polygon with more than six sides to tessellate the plane?
14. If this pattern were placed in a three-mirror kaleidoscope, what tessellation do you think would be produced? Describe the tessellation as accurately as possible and draw a picture to explain your thinking.

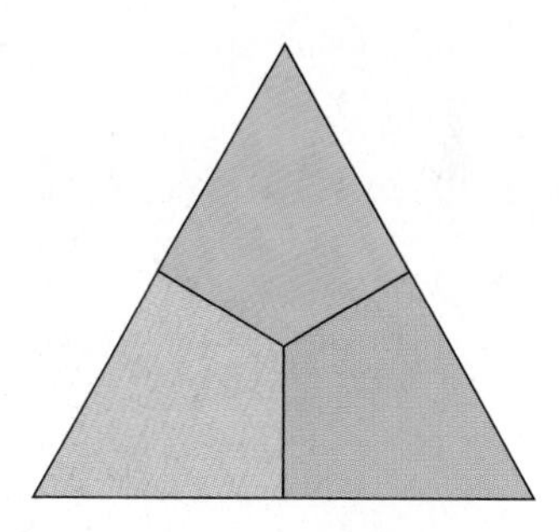

15. Give an example of a regular polygon, other than a pentagon, that won't tessellate the plane.
16. Could you tile a floor by using only pentagons such as the ones shown? Use tracing paper and show enough of the tessellation to convince someone of your conclusion.

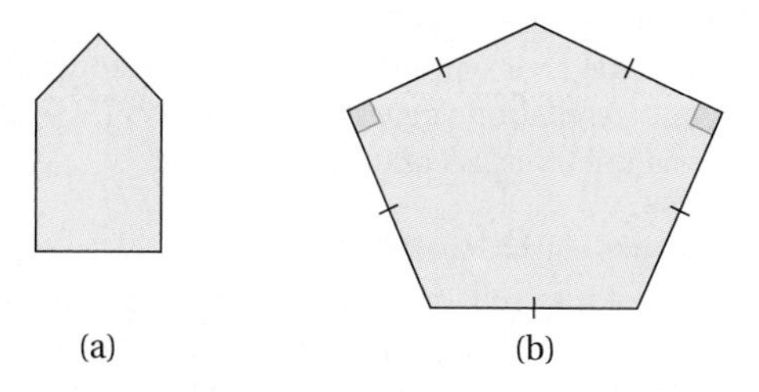

17. Why won't a regular octagon tessellate the plane by itself? Describe a combination of a regular octagon and another regular polygon that will tessellate the plane.
18. Use symbols to represent semiregular tessellations (b), (d), and (g) in Figure 11.17.
19. How many different semiregular tessellations are there?
20. Why is the tessellation shown on the next page not a regular tessellation?

Figure for Exercise 20

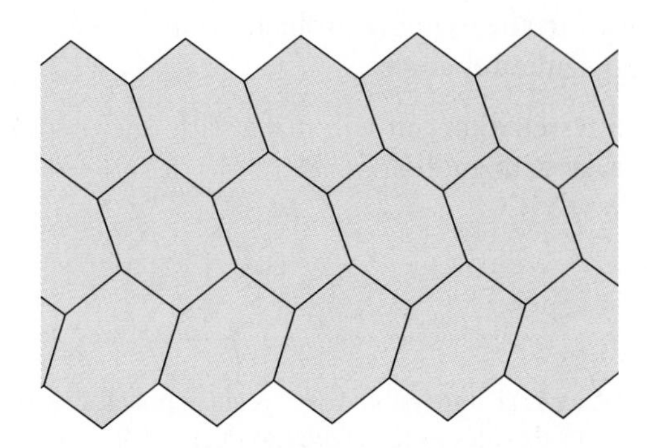

21. Why is the tessellation shown not a semiregular tessellation?

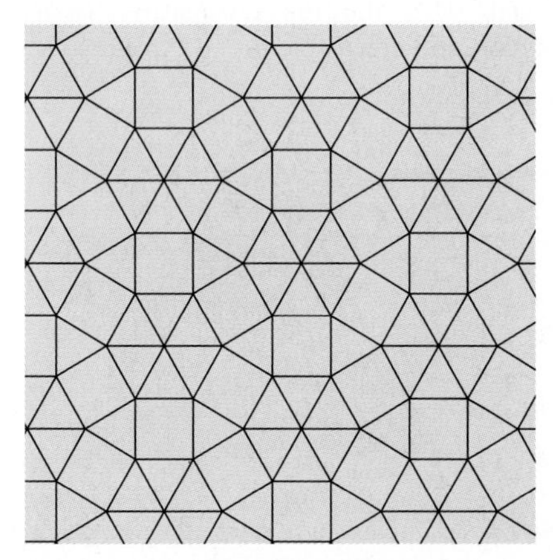

22. Draw the arrangement of regular polygons around a vertex described by 6, 6, 3, 3. Explain why it can't be extended to tessellate the plane.

23. What type(s) of symmetry, if any, is present in the completed tessellation suggested in Figure 11.21? Describe the symmetry as accurately as possible.

24. What type(s) of symmetry, if any, is present in the completed tessellation suggested in Figure 11.23? Describe the symmetry as accurately as possible.

B. Deepening Understanding

25. Trace each tessellation shown in the next column, recognizing that it actually covers the entire plane. In how many ways can you slide, turn, and flip each tessellation to match the original tessellation? Devise ways to describe these slides, turns, and flips.

26. The *dual* of a tessellation is a tessellation created by connecting the centers of neighboring polygons in the tessellation.

a. Draw the dual of the tessellation of regular hexagons.

b. Draw the dual of the tessellation of equilateral triangles.

c. Describe any relationships that you see between the results of parts (a) and (b).

Figures for Exercise 25

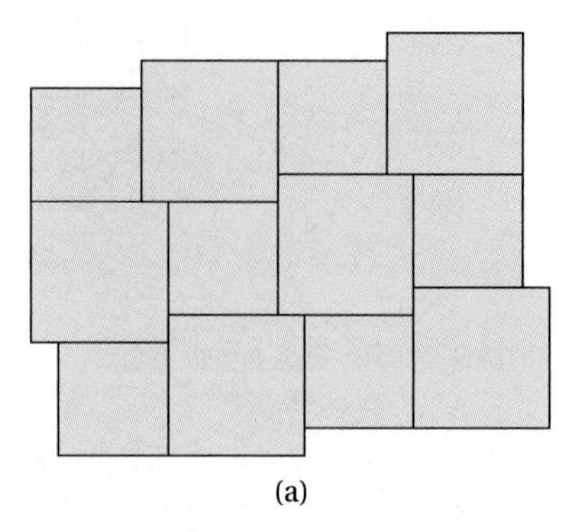

(a)

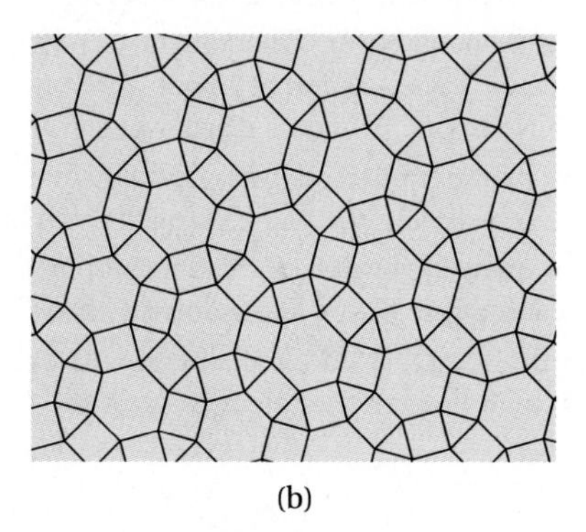

(b)

27. Every pentagon with a pair of parallel sides and every hexagon with three pairs of equal parallel sides will tessellate the plane. Draw an example of each of these types of tessellating figures.

28. Make a curved tessellation from one of the following geometric figures.

a. A tessellation of equilateral triangles

b. A tessellation of squares

c. A tessellation of general quadrilaterals

29. A *vertex figure* is formed by connecting the midpoints of the edges that form the vertex of a tessellation, as shown.

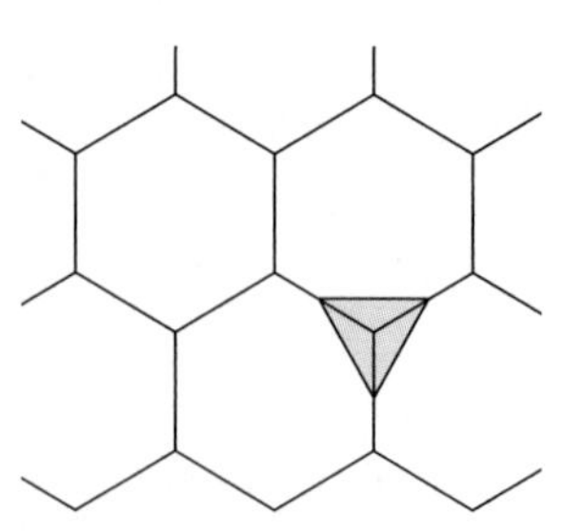

A *vertex figures tessellation* is created by drawing all the vertex figures of a tessellation. Draw the vertex figures tessellation for the tessellation of regular hexagons. Describe this tessellation.

30. Adapt the LOGO procedure on p. 609 to produce a portion of a tessellation of
  a. squares.
  b. equilateral triangles.

31. Write a LOGO procedure that will produce a figure consisting of an equilateral triangle and a square with a common side. Such a configuration might be used as a part of a procedure to produce the semiregular tessellation $3^3, 4^2$.

C. Reasoning and Problem Solving

32. Of the 12 types of pentomino, find at least 2 that will tessellate the plane and show a portion of the tessellation.

33. Of the 35 hexominos, find at least 2 that will tessellate the plane and show a portion of the tessellation.

34. A heptiamond is a figure formed by seven connected, nonoverlapping, and congruent equilateral triangles, each touching others only along a complete side. Of the 24 heptiamonds, all but 1 will tessellate the plane. Choose a heptiamond and show how to tessellate the plane with it.

35. Which tessellations should the following design make when placed in a three-mirror kaleidoscope? Explain.

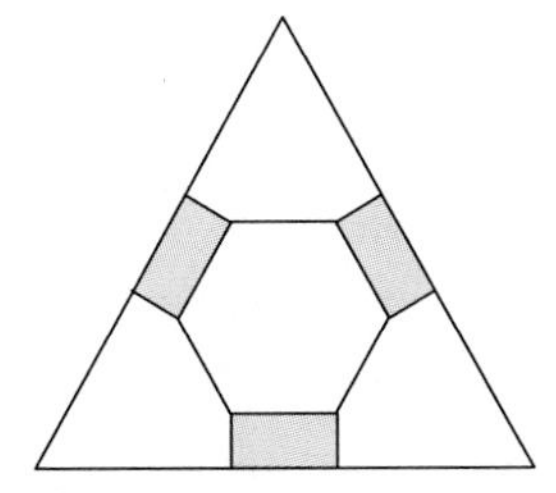

36. How would you instruct the LOGO turtle to have it produce semiregular tessellation (b) in Figure 11.17?
  a. Write a paragraph describing the instructions that you would give the turtle.
  b. Use GES or a computer drawing program to create a tessellation. What functions of the technology did you find most useful in completing your tessellation?

37. The Roman Bath House Tiling Problem. A floor restoration specialist was called upon to restore the floor tile in a Roman bath house. She knew that the floor had been tiled only with equilateral triangles and squares and that it was a semiregular tessellation. The only remaining pieces of the original tile are shown.

Can you help the specialist by drawing a diagram of what the original tiled floor looked like? Is there more than one possibility? Explain.

38. The Kaleidoscope Problem. A geometer made a kaleidoscope out of 3 mirrors put together to form a 45°–45°–90° right triangle. He asserted that he could put a pattern in the kaleidoscope that would produce a semiregular tessellation. His colleague disagreed. Who was correct, the geometer or his colleague? Justify your conclusion.

39. The famous tessellation shown was created by Johannes Kepler. Describe the types of figures that make up the tessellation.

D. Communicating and Connecting Ideas

40. The following pattern shows what appears to be a tessellation of squares, regular pentagons, regular hexagons, regular heptagons, and regular octagons. How would you convince someone that it is a fake tessellation of regular polygons?

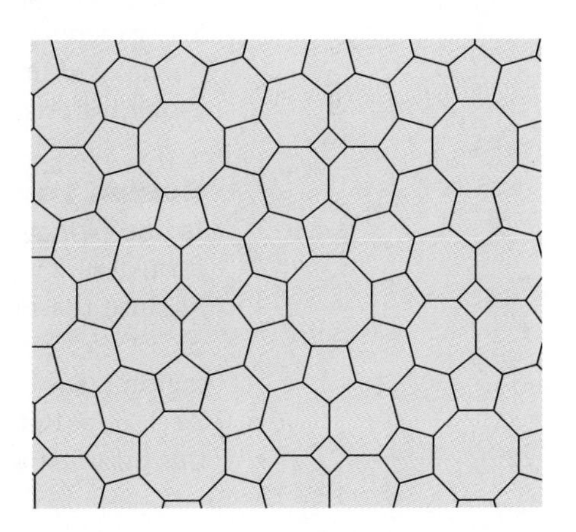

41. Discuss what is communicated to you by a tessellation such as the one shown in Exercise 39 or by an Escher design such as the one shown in Figure 11.20. How do you feel when you see it? What message, if any, does it convey to you about mathematics?

42. **Historical Pathways.** The pattern shown is a marble design found in a thirteenth century Roman church.

a. Is it a tessellation? Why or why not?

b. Describe at least two lines of symmetry of the design.

43. **Making Connections.** Illustrate the connection between tessellations and the Pythagorean theorem by explaining how the dissection of the following tessellation by the vertical and horizontal lines proves the Pythagorean theorem. Assume that the squares that make up the tessellation are the squares on the legs of a right triangle and that the squares formed by the vertical and horizontal lines are squares on the hypotenuse of the right triangle. How does this relate to the puzzle figure in Figure 10.29(b)?

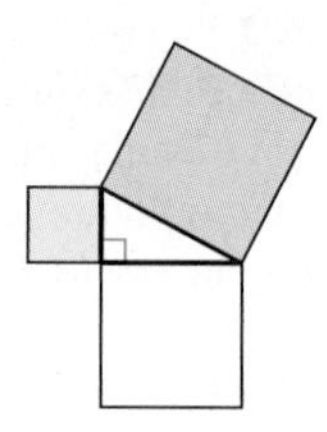

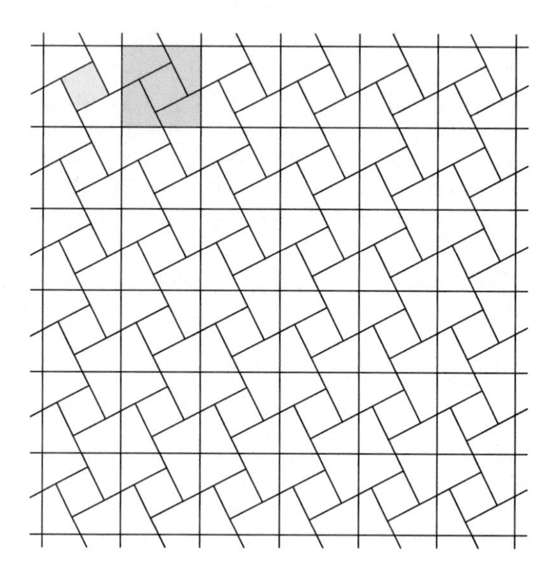

Section 11.3 SPECIAL POLYGONS

- Golden Triangles and Rectangles
- Star Polygons
- Star-Shaped Polygons

In this section we examine some special topics that involve polygons. We explore and generalize some properties of the pentagram, the golden rectangle, and star polygons. Mini-Investigation 11.6 gives you an opportunity to make some interesting historical connections.

Golden Triangles and Rectangles

In the pentagram shown in Mini-Investigation 11.6, an isosceles triangle that forms one of the points of the star, such as $\triangle BFG$, is called a **golden triangle** because the ratio of its longer side to its shorter side is the golden ratio, $\phi = (1 + \sqrt{5})/2 \approx 1.618$. We know that the measure of an interior angle of the pentagram, such as $\angle ABC$, is 108°, so we can conclude from the equations $2x + \alpha = 108$, $4x + \alpha = 180$, and $\beta = 2x$, that $x = \alpha = 36°$ and $\beta = 72°$. Using this information, try to identify 10 other golden triangles in the pentagram.

Talk about why the ancient Greeks, who chose the pentagram as a sacred symbol because of its special beauty, called $\triangle BFG$ a golden triangle.

Mini-Investigation 11.6 **Making a Connection**

How many different pairs of segments can you find in the following **pentagram** with a ratio of lengths equal to the golden ratio, $\phi \approx 1.618$?

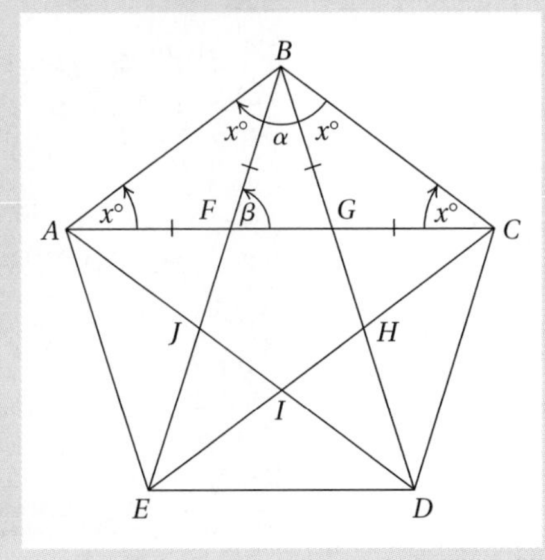

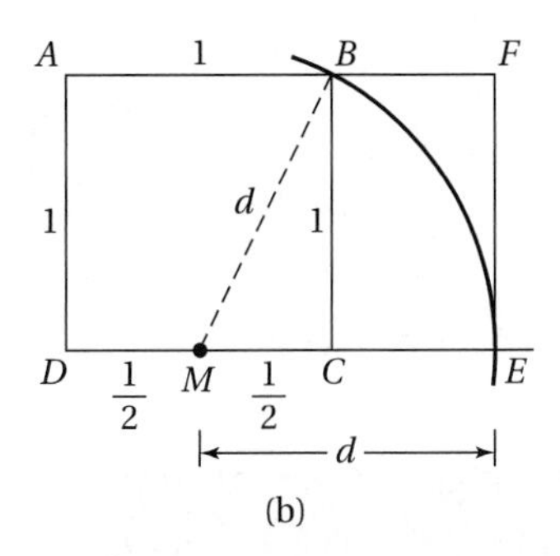

FIGURE 11.25 | Construction of a golden rectangle.

The golden triangle is interesting but hasn't received as much attention as the often discussed and used **golden rectangle.** Mathematicians, artists, architects, and others have long considered the golden rectangle an especially pleasing form—in nature, in buildings, and in works of art. To construct a golden rectangle, we start with a unit square, as in Figure 11.25(a), and find the midpoint M of one side. Next, we consider $\overline{MB}$. Then, as shown in Figure 11.25(b), we use M as the center and $\overline{MB}$ as the radius and strike an arc on the extension of $\overline{DC}$ at E. Finally, we construct $\overline{EF} \perp \overline{DE}$ and complete the golden rectangle $AFED$.

To verify that the ratio of the length to the width of rectangle $AFED$ is the golden ratio, and that $AFED$ is a golden rectangle, we use the Pythagorean theorem and show that $d = \frac{\sqrt{5}}{2}$. It then follows that $DE = \frac{1}{2} + \frac{\sqrt{5}}{2} = \frac{(1 + \sqrt{5})}{2}$. This length of $\overline{DE}$ is approximately $\frac{(1 + 2.236)}{2}$, or 1.618, and $AFED$ is a golden rectangle.

Star Polygons

Using Circles to Produce Star Polygons. The pentagram shown in Mini-Investigation 11.6 leads to consideration of another interesting type of polygon called a **star polygon.**

Star polygons are frequently used in advertising logos, artistic designs, quilts, and other decorative situations.

The star polygon shown in Figure 11.26 is constructed from five equally spaced points on a circle. Beginning at V_1, we go in one direction and draw a segment to every second point, returning to V_1. The segments drawn to form the star polygon are $\overline{V_1V_3}$, $\overline{V_3V_5}$, $\overline{V_5V_2}$, $\overline{V_2V_4}$, and $\overline{V_4V_1}$. Because the star polygon shown in Figure 11.26 was constructed from *five* points with segments connecting every *second* point, it is denoted $\{\frac{5}{2}\}$. This star polygon is a nonsimple polygon having five vertices with angles of equal measure and five congruent sides and is called a **regular star polygon**.

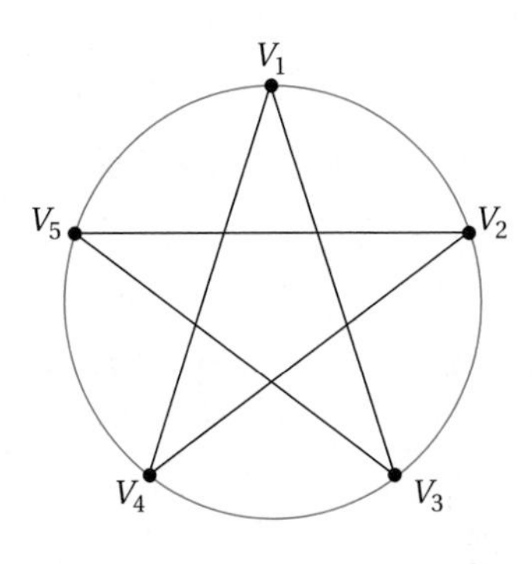

FIGURE 11.26 | A regular star polygon.

Mini-Investigation 11.7 lets you explore the properties of star polygons and make some generalizations.

Talk about which are regular polygons and which are the same star polygon—and why.
Technology Extension: If you have access to LOGO, use it to construct the star polygons with eight points and formulate possible generalizations.

Mini-Investigation 11.7 **Finding a Pattern**

What possible relationships can you discover by constructing the following star polygons?

a. $\left\{\frac{8}{4}\right\}$
b. $\left\{\frac{8}{6}\right\}$
c. $\left\{\frac{6}{2}\right\}$
d. $\left\{\frac{6}{7}\right\}$

Mini-Investigation 11.7 and the analysis of the star polygons with eight points suggest the following generalizations.

Generalizations About Star Polygons

n = the number of equally spaced points on the circle.
d = the dth point that segments are drawn to.

- The star polygon $\left\{\frac{n}{d}\right\}$ is the same as the star polygon $\left\{\frac{n}{n-d}\right\}$.
- The polygons $\left\{\frac{n}{1}\right\}$ and $\left\{\frac{n}{n-1}\right\}$ are regular polygons.
- The n-sided star polygon $\left\{\frac{n}{d}\right\}$ exists if and only if $d \neq 1$, $d \neq n - 1$, and n and d are relatively prime.

Exercise 40 at the end of this section asks you to consider another generalization about star polygons that might be true. In Example 11.6 we apply some of the ideas about star polygons.

Example 11.6 **Identifying Star Polygons**

Describe the star polygons having seven vertices and verify that all of them have been identified.

Solution

There are two star polygons with seven vertices, namely, $\left\{\frac{7}{2}\right\}$ and $\left\{\frac{7}{3}\right\}$, as shown.

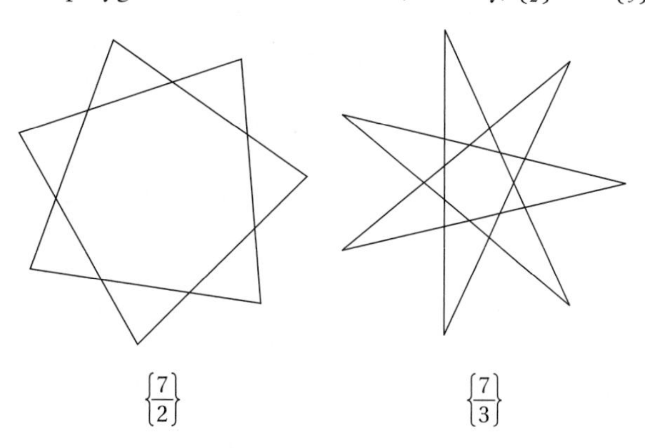

Note that $\left\{\frac{7}{1}\right\}$ and $\left\{\frac{7}{6}\right\}$ form regular polygons, not star polygons; $\left\{\frac{7}{5}\right\}$ and $\left\{\frac{7}{4}\right\}$ are the same star polygons as those shown here. So there are only two star polygons with seven vertices.

Your Turn

Practice: Describe the star polygons with nine vertices and verify that you have identified all of them.

Reflect: Explain how the process for producing star polygon $\left\{\frac{7}{2}\right\}$ is different from the process for producing the star polygon $\left\{\frac{7}{5}\right\}$ even though the final pictures are the same. ■

Star-Shaped Polygons

If we stop after only considering star polygons that can be produced by sequentially connecting points on a circle with line segments, we miss a lot of very interesting star-shaped figures. For example, a quiltmaker might want to use star-shaped figures like those shown in Figure 11.27.

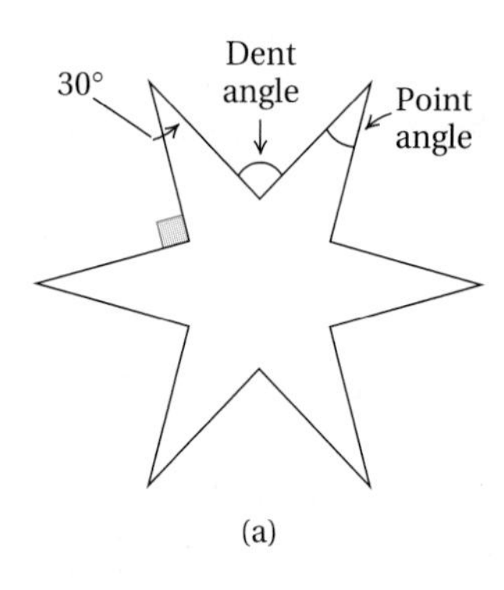

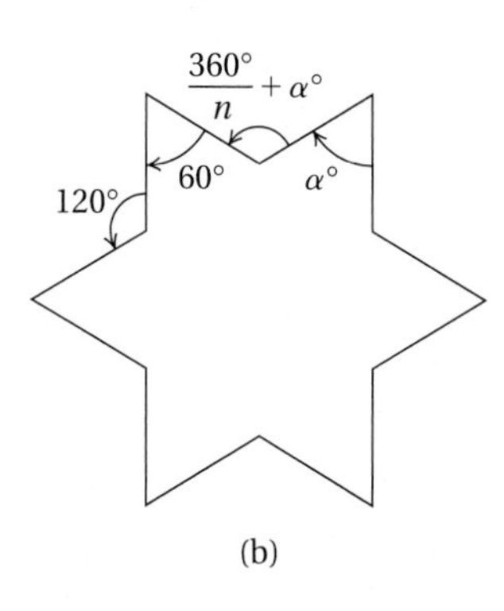

FIGURE 11.27 | Examples of star-shaped polygons.

To compare these six-pointed figures with six-pointed star polygons, we observe that $\left\{\frac{6}{1}\right\}$ and $\left\{\frac{6}{5}\right\}$ produce regular hexagons, $\left\{\frac{6}{2}\right\}$ and $\left\{\frac{6}{4}\right\}$ produce triangles, and $\left\{\frac{6}{3}\right\}$ produces a straight line. Thus the polygons shown in Figure 11.27 are not star polygons, nor can they be produced by connecting points on a circle with line segments and then erasing the interior parts of the segments. However, we can use the term **star-shaped polygon** to describe a nonconvex symmetric figure like the one shown in Figure 11.27(a) or (b) that isn't a star polygon. Star-shaped polygons have n star-tip points, $2n$ congruent sides, n congruent *point angles* with measure α, and n congruent *dent angles* β such that $\beta = \left(\frac{360}{n}\right) + \alpha$, or $\alpha = \beta - \left(\frac{360}{n}\right)$. Exercise 47 at the end of this section asks you to verify this relationship. A six-pointed, star-shaped polygon with point angle 30° is denoted by $6_{30°}$.

Figure 11.28 compares a *star polygon* and *star-shaped polygon*. It illustrates the importance of precise definitions, which are required to differentiate two different but closely related ideas. Although the basic shape of the five-pointed stars shown are the same here, that isn't always the case when an n-pointed star polygon and an n-pointed star-shaped polygon are compared.

This information about the relationship between dent angles and point angles is useful in constructing star-shaped polygons that meet our specifications. For example, if you want to produce a star-shaped polygon with a specific point angle

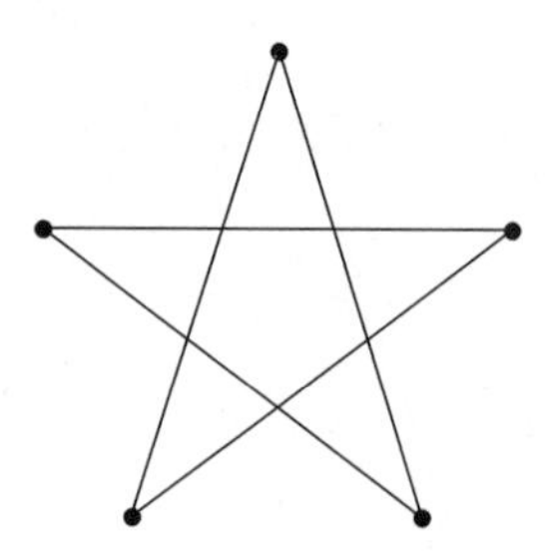

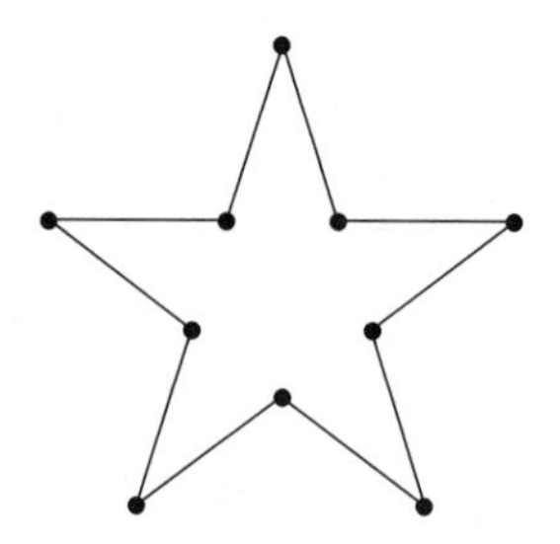

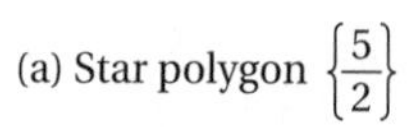

(a) Star polygon $\left\{\frac{5}{2}\right\}$

Nonsimple, nonconvex
5 vertices
5 sides
5 point angles

(b) Star-shaped polygon $5_{36°}$

Simple, nonconvex
10 vertices
10 sides
5 point angles

FIGURE 11.28 | Comparison of a star polygon and a star-shaped polygon.

for a special quilt, you can quickly calculate the measure of the dent angle for that figure. Conversely, if you want to make a star-shaped polygon with a specific dent angle for tessellation or other design purposes, you can quickly calculate the measure of the point angle for that figure. Example 11.7 demonstrates how to construct a desired star-shaped polygon.

Example 11.7

Problem Solving: The Star Design

An artist wants to make a painting that includes a five-pointed star-shaped polygon with a fairly thin point angle of 18°. How could the artist accurately construct the star-shaped polygon $5_{18°}$?

Solution

The measure of the point angle is 18°, so the dent angle is $\left(\frac{360°}{5}\right) + 18° = 90°$. We construct the star-shaped polygon as follows

a. Plot five equally spaced points on a circle.

b. Connect two of the points, A and B, and construct 45° angles at those two points. The intersection of the sides of these angles is point D, where the 90° dent angle is formed.

c. Use a compass to find the other dent angle points.

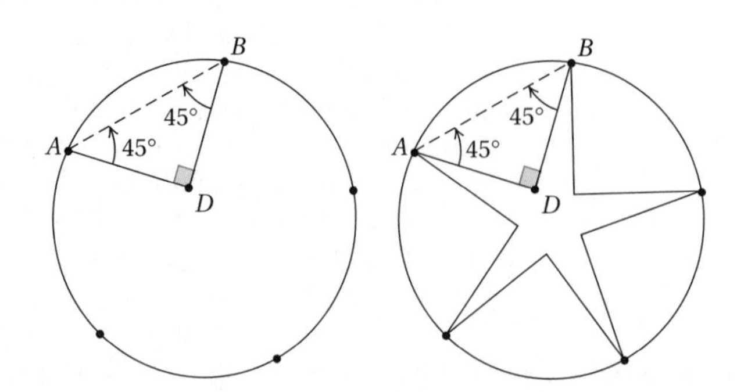

Your Turn

Practice: Construct a six-pointed star-shaped polygon with a point angle of 30°.

Reflect: If you want a dent angle of 120°, how will you perform the construction to find the dent angle vertex D? ■

The results in Example 11.8 allow us to investigate how star-shaped polygons can be used in tessellations. The problem-solving strategies *draw a picture* and *use logical reasoning* are helpful here.

Example 11.8

Problem Solving: The Quilt

A quiltmaker wanted to make a quilt from six-pointed star-shaped polygons and either squares or equilateral triangles. Is this combination possible? If so, what would the quilt pattern look like?

Working toward a Solution

The quiltmaker needs to decide which type of six-pointed star-shaped polygon to use. Drawing a picture and reasoning indicate that the dent angles of the star-shaped polygon must be either 90° (to fit a square), as shown in part (a) of the figure below, or 120° (to fit two equilateral triangles), as shown in part (b). Then the quiltmaker can calculate the measure of the point angles, if needed.

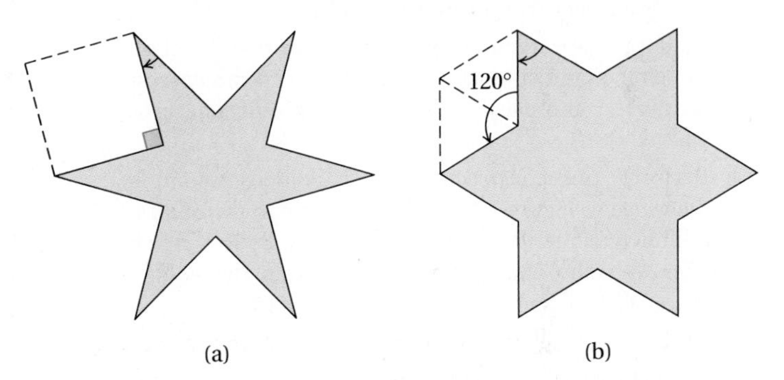

The following quilt pattern meets the conditions of the problem for the star-shaped polygon shown in part (a) of the preceding figure. The measure of the point angle is 30°.

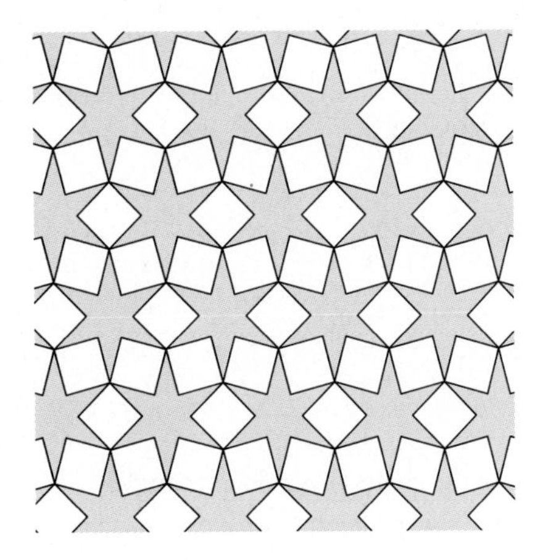

Your Turn

Practice: Draw the quilt pattern for the polygon shown in part (b) of the figure.

Reflect: Why can't you use a dent angle of 60°, which would hold a single equilateral triangle, to produce a third quilt pattern solution to the problem? ■

Making geometric designs has benefits for elementary school students. It provides an opportunity to learn and extend ideas of geometry. It also provides a path to success for students who may not be as successful in other areas of mathematics. It promotes active involvement and is an excellent activity for integrating mathematics and art.

Problems and Exercises for Section 11.3

A. Reinforcing Concepts and Practicing Skills

1. Is a 3-inch × 5-inch card shaped like a golden rectangle? Is a 5-inch × 8-inch card? Explain.
2. Find a picture frame or poster that you think is close to being a golden rectangle. Measure it to the nearest centimeter and check. What is the ratio of length to width?
3. Perform the construction of a golden rectangle described in Figure 11.25. Then measure to the nearest millimeter and calculate the ratio *AF*/*FE*. By how much does the ratio you calculated differ from the golden ratio, which is approximately equal to 1.618?
4. What are the defining characteristics of a golden triangle?
5. Name two pairs of segments in the following pentagram for which the ratio of their measures is the golden ratio.

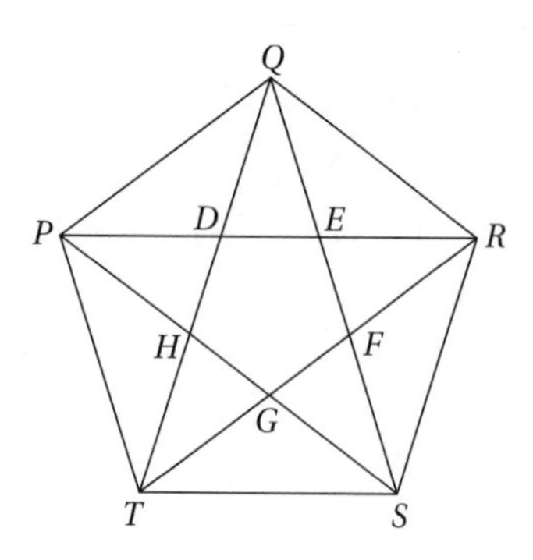

6. Describe the measures of the angles of a golden triangle.
7. How many sides does the star polygon $\left\{\frac{5}{2}\right\}$ have?
8. Why is the polygon that would be constructed by interpretation of $\left\{\frac{6}{4}\right\}$ not a star polygon?
9. Describe the polygons produced by interpretation of the following symbols.
   a. $\left\{\frac{8}{4}\right\}$ b. $\left\{\frac{8}{6}\right\}$
   c. $\left\{\frac{6}{2}\right\}$ d. $\left\{\frac{6}{7}\right\}$
10. How are the processes for constructing star polygons $\left\{\frac{5}{2}\right\}$ and $\left\{\frac{5}{3}\right\}$ alike? How are they different? What can you conclude about the resulting star polygons?
11. What is the number of degrees in the point angle of star polygon
    a. $\left\{\frac{5}{2}\right\}$? b. $\left\{\frac{8}{3}\right\}$?
    c. $\left\{\frac{9}{2}\right\}$?

For Exercises 12–15, describe the polygons produced by interpreting the symbols given.

12. $\left\{\frac{6}{3}\right\}$
13. $\left\{\frac{8}{2}\right\}$
14. $\left\{\frac{8}{7}\right\}$
15. $\left\{\frac{6}{5}\right\}$
16. Give an example to illustrate that the star polygon $\left\{\frac{n}{d}\right\}$ is the same as the star polygon $\left\{\frac{n}{n-d}\right\}$, where n is the number of equally spaced points on the circle, and d is the dth point to which the segments are drawn to form the polygon.
17. Give an example to illustrate that the star polygons $\left\{\frac{n}{1}\right\}$ and $\left\{\frac{n}{n-1}\right\}$, where n is the number of equally spaced points on the circle, are regular polygons.

For Exercises 18–20, find the number of degrees in the point angle of the star polygon.

18. $\left\{\frac{5}{3}\right\}$
19. $\left\{\frac{8}{5}\right\}$

20. $\left\{\frac{9}{7}\right\}$

21. How are star polygons and star-shaped polygons alike? How are they different?

22. How does the number of sides of the star polygon $\left\{\frac{9}{4}\right\}$ differ from the number of sides of the star-shaped polygon $9_{20°}$?

For Exercises 23–26, calculate the measures of the point angles of the star-shaped polygons shown.

23.

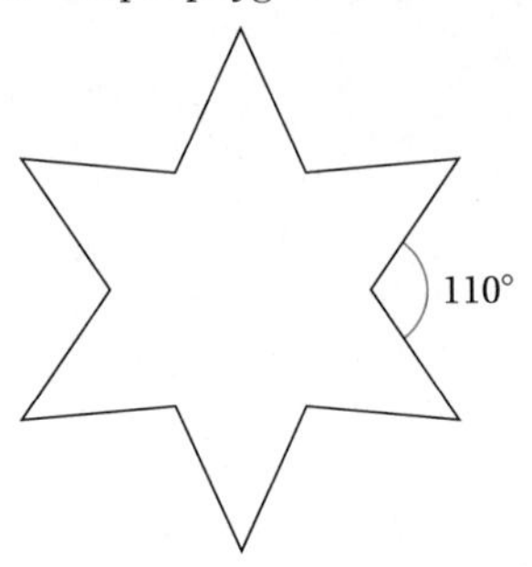

24.

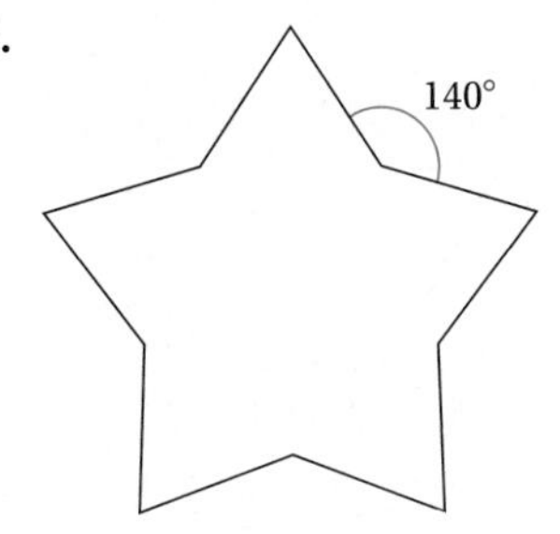

25.

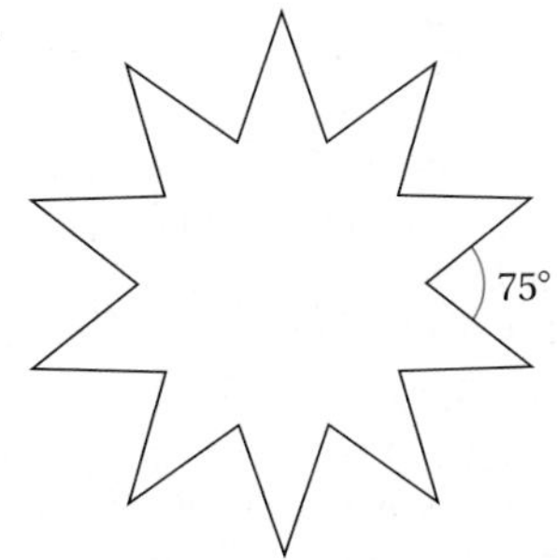

26.

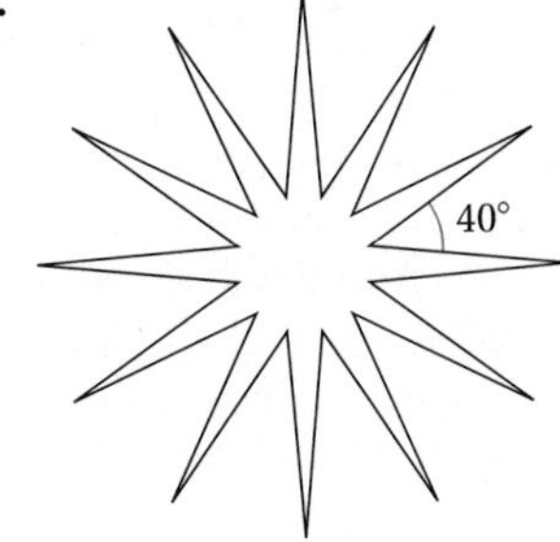

For Exercises 27–30, calculate the measures of the dent angles of the star-shaped polygons shown.

27.

28.

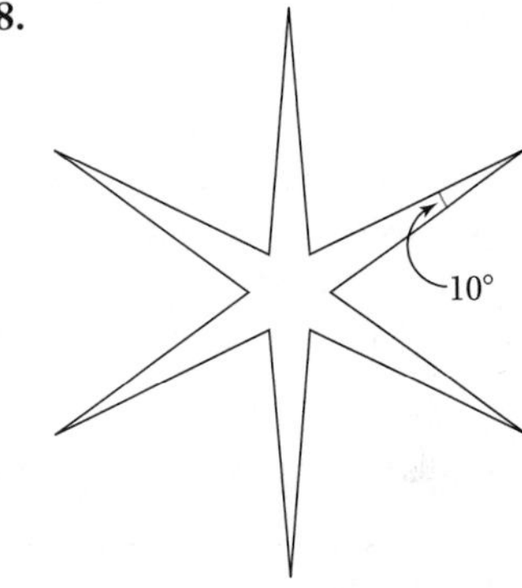

29.

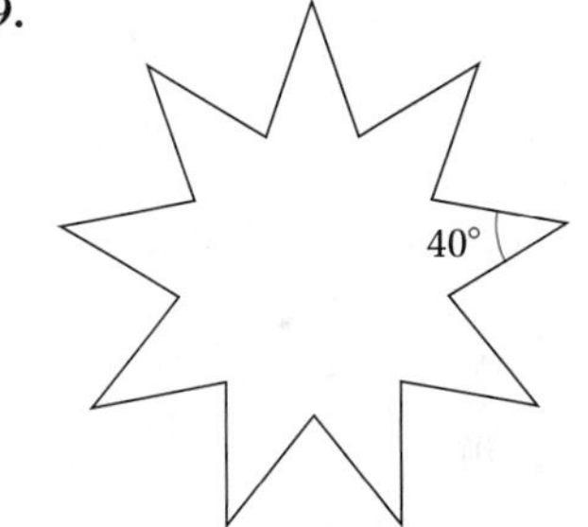

30.

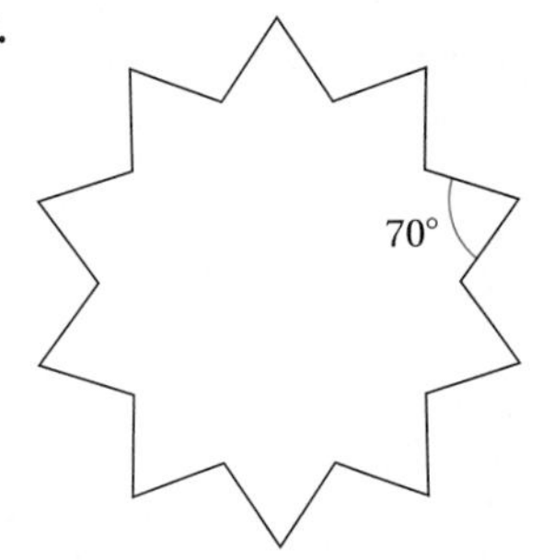

31. A wallpaper designer wants to use a star-shaped polygon with eight points and a point angle of 15°. What should be the dent angle of this polygon?

B. Deepening Understanding

32. Show that there are only six star polygons with less than 10 sides.

33. A quiltmaker thought that a six-pointed star-shaped polygon with a point angle of 30° would go together with squares to make a quilt. Was she correct? Explain.

34. How many star polygons are there with 12 sides? Give symbols for these polygons and verify your answer.

35. How many star polygons are there with 10 sides? Give symbols for these polygons and draw a picture of each.

36. Use GES and inductive reasoning to make a generalization about the sum of the measures of all the point angles in any nonsymmetric five-pointed star.

37. Use GES and try several different examples to check the generalization about the relationship between the point angle and the dent angle in a star-shaped polygon.

38. For the pentagram shown in the figure in Mini-Investigation 11.6, it has been said that the ratio of the length of the side of the large outer pentagon to the length of the side of the smaller inner pentagon is ϕ^2. Do you agree? Support your decision.

39. Construct a star-shaped polygon with eight points and a point angle measure of 45°.

C. Reasoning and Problem Solving

40. The number of numbers less than n and relatively prime to n is found by using the formula

$$n\left[1-\left(\frac{1}{p_1}\right)\right]\times\left[1-\left(\frac{1}{p_2}\right)\right]\ldots,$$

where $p_1, p_2, \ldots$ are prime factors of n. Show how to use this formula to help find the number of star polygons there are with 36 sides.

41. Describe three different six-pointed star-shaped polygons. If the area of the smallest is 1, estimate the areas of the others. How might you check your estimates?

42. Use the figure shown and the ideas about the sum of the exterior angles of a polygon to prove that what you discovered in Exercise 36 is true.

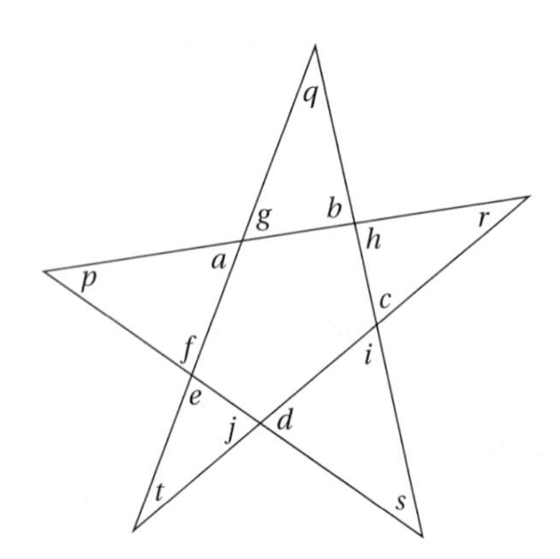

43. Analyze the following tessellations.

(a)

(b)

i. What regular polygons are used?

ii. What are the measures of the point and dent angles of the star-shaped polygons that are used?

iii. Are all vertices surrounded alike? If not, what different vertex arrangements are there?

44. The Wallpaper Design Problem. A wallpaper designer wants to use the tessellation shown on the next page as the base for some decorative wallpaper. To make the tessellation, he needs to make accurate templates for the 9-gon and the star-shaped polygon. What directions (angle measures, side lengths, and the like) would you give him for making the templates?

45. If the ratio of a to b in the square shown on the next page is the golden ratio, prove that the shaded rectangle is a golden rectangle. (*Hint:* Use properties of similar triangles.)

46. The ancient Greeks found that the proportion $\frac{l}{w} = \frac{(l+w)}{l}$ holds only for the length and width of a golden rectangle. If the width in the proportion is 1, show that the length is $\frac{(1+\sqrt{5})}{2}$.

Figure for Exercise 44

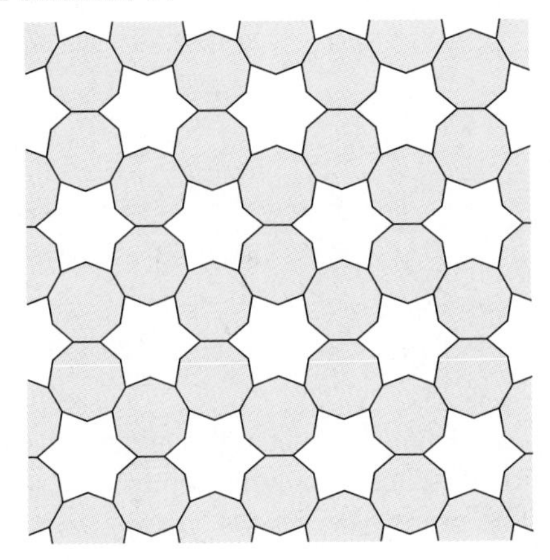

Figure for Exercise 45

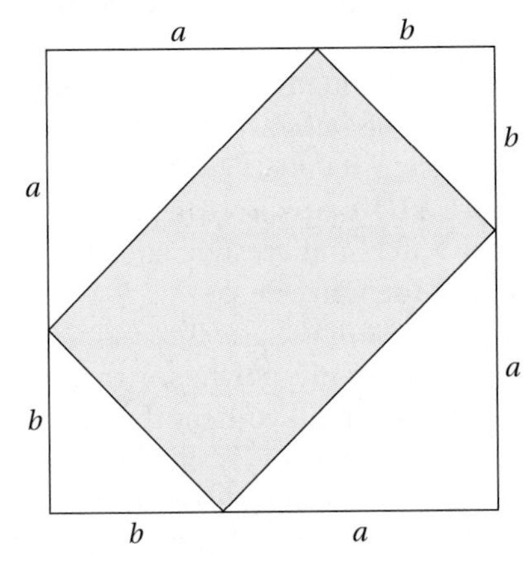

47. How would you use the following partial picture of a star-shaped polygon inscribed in a regular n-sided polygon to verify that $\beta = (\frac{360}{n}) + \alpha$, where β is the measure of the dent angle and α is the measure of the point angle?

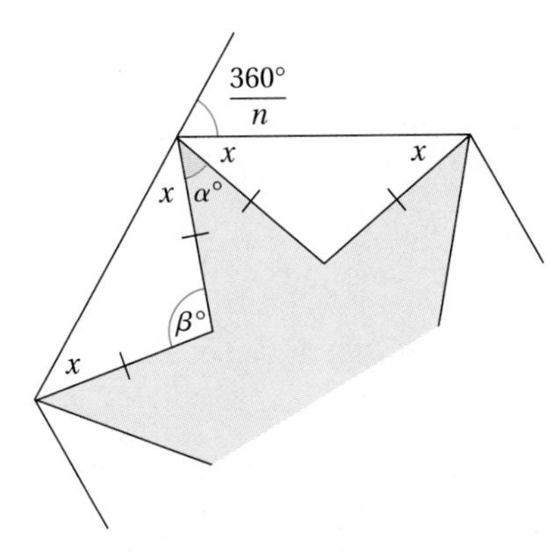

D. Communicating and Connecting Ideas

48. Discuss how you would produce an accurate drawing of a six-pointed star-shaped polygon with a point angle of 60°. Describe your procedure in detail.

49. Work in a small group and devise a procedure to test whether people really do think that the golden rectangle is the most aesthetically pleasing rectangle. Write a paragraph explaining your procedure and the results.

50. Historical Pathways. The problem of constructing tessellations that utilized star-shaped polygons challenged early Islamic artists. The following design is adapted from Islamic art found in a Russian mosque. Describe the different geometric figures used in this tessellation. If you assume that the five-pointed star is a pentagram star, draw some conclusions about the angle measures of the other figures.

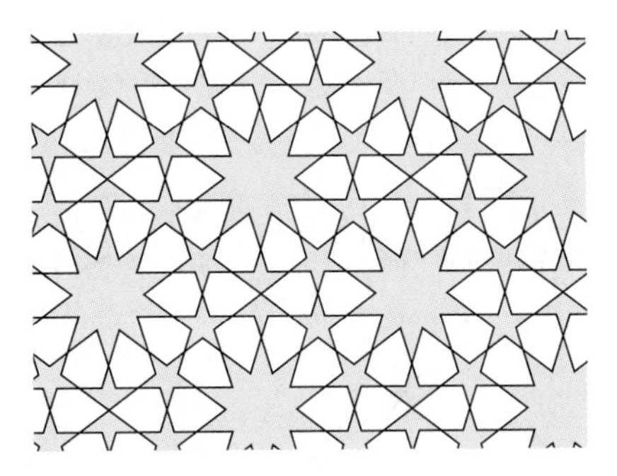

51. Making Connections. Show a connection between geometry and art by making and coloring a tessellation that uses the polygons given.

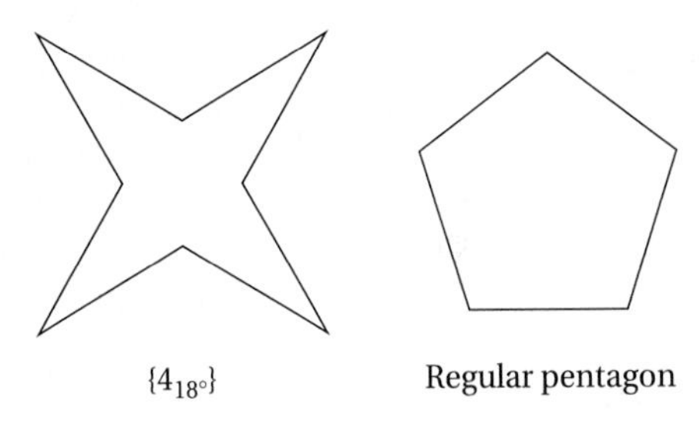

52. Making Connections. Connect algebra and geometry by finding an algebraic expression for the ratio of the length to width of golden rectangle $AHCD$. Do not use approximate values for square roots.

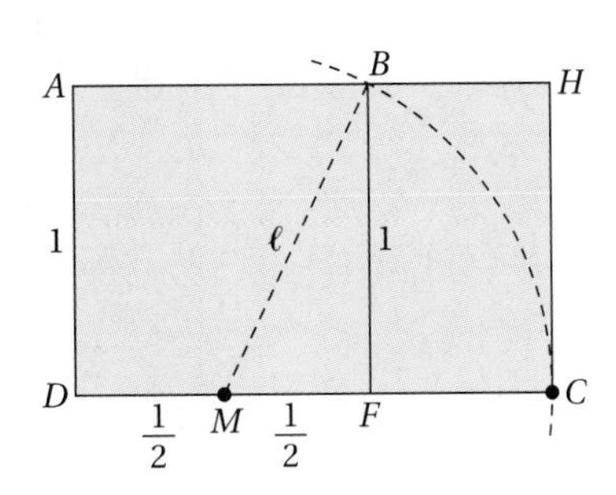

Section 11.4 THREE-DIMENSIONAL FIGURES

- The Regular Polyhedra
- Prisms and Pyramids
- Cylinders, Cones, and Spheres
- Symmetry in Three Dimensions
- Visualizing Three-Dimensional Figures

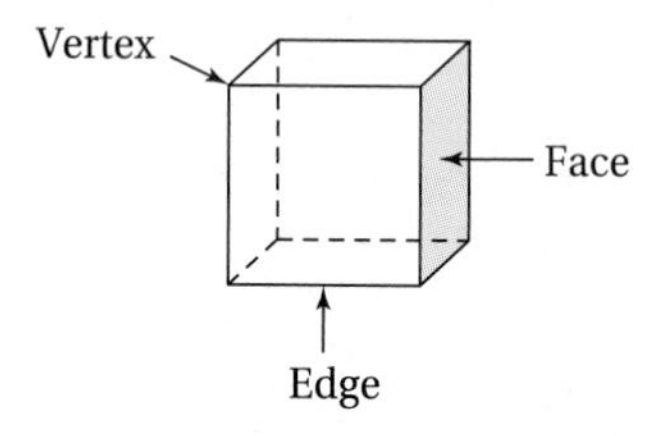

FIGURE 11.29 | Parts of a polyhedron.

The Regular Polyhedra

A **polyhedron** is a collection of *polygons* (triangles, rectangles, pentagons, hexagons, and the like) joined to enclose a region of space. A polyhedron has faces, edges, and vertices as shown in Figure 11.29. The prefix *poly-* means "many" or "several" and the suffix *-hedron* indicates "surfaces" or "faces."

The five polyhedra shown in Figure 11.30 can be modeled from cardboard, Styrofoam, or connected drinking straws and are called **regular polyhedra**. The ancient Greeks used prefixes that indicated the number of faces to name these polyhedra. *Tetra-* means "four," *octa-* means "eight," *dodeca-* means "twelve," and *icosa-* means "twenty." Because *hexa-*" means "six," a cube may be called a *hexahedron*. The regular polyhedra shown in Figure 11.30 are also called the Platonic solids because Plato associated earth, air, fire, water, and creative energy with these five solids and used them in his description of the universe.

These regular polyhedra might be thought of as basic three-dimensional building blocks of geometry because they are the only polyhedra that have edges of equal length and the arrangement of polygons at all vertices the same. A nonregular polyhedron with two different vertex arrangements is shown in Figure 11.31.

Evolving scientific theories of the universe long ago replaced Plato's description, but nature hasn't ignored the Platonic solids. Skeletons of tiny sea creatures called *radiolarians*, made of silica and measuring only a fraction of a millimeter in diameter, have the form of the octahedron, icosahedron, and dodecahedron. Also, several mineral crystals and some viruses take the form of these and other polyhedra. Mini-Investigation 11.8 gives you an opportunity to explore further the regular polyhedra.

Write an equation that expresses the relationships you found.

Mini-Investigation 11.8 **Finding a Pattern**

When you complete the following table and look for a pattern, what relationships between vertices, faces, and edges do you find?

Polyhedron	V	F	$V + F$	E
Tetrahedron	?	?	?	?
Cube	?	?	?	?
Octahedron	6	8	14	12
Dodecahedron	20	12	32	30
Icosahedron	12	20	32	30

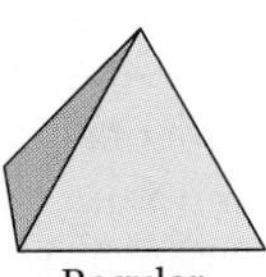

Regular tetrahedron

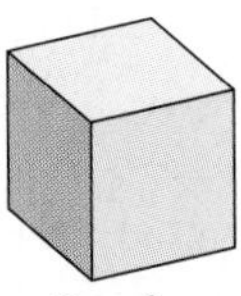

Regular hexahedron or cube

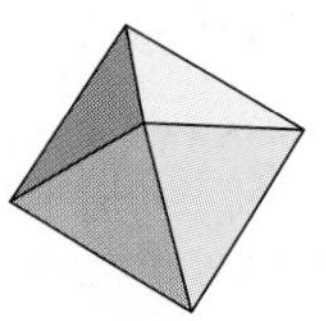

Regular octahedron

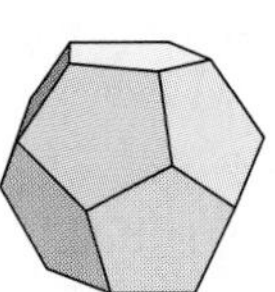

Regular dodecahedron

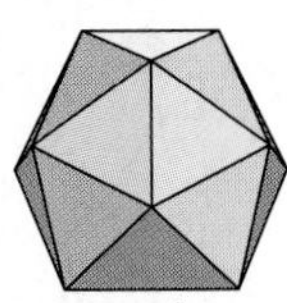

Regular icosahedron

FIGURE 11.30 | Regular polyhedra.

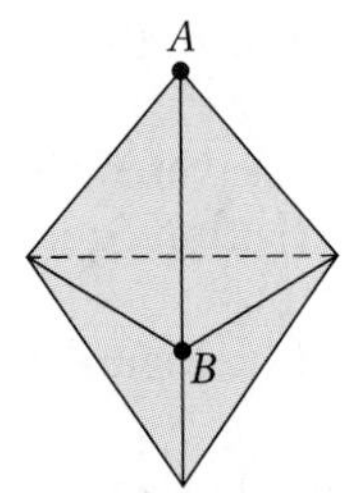

Three triangles surround vertex *A*.
Four triangles surround vertex *B*.

FIGURE 11.31 | A nonregular polyhedron.

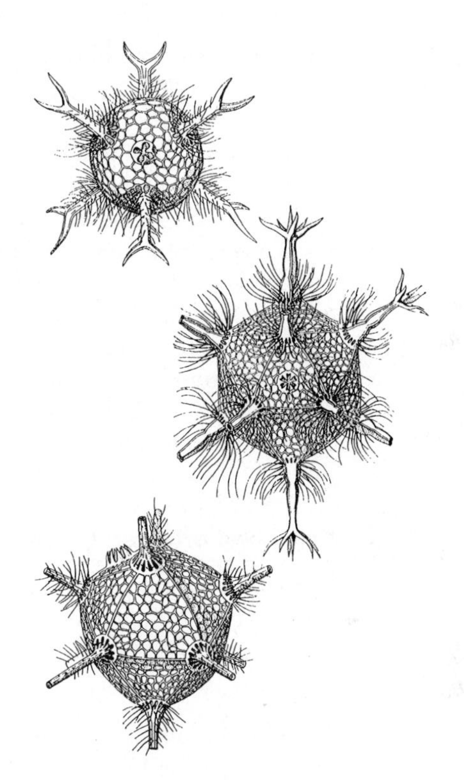

Geometric forms in the *radiolarian* skeleton.

Source: D'arcy W. Thompson, *On Growth and Form*. In J. T. Bonner (ed.), abridged edition, Cambridge University Press, England, 1971, p. 168. Reprinted with permission of Cambridge University Press.

The formula you discovered in Mini-Investigation 11.8 is called **Euler's formula**, named after the Swiss mathematician, Leonhard Euler (1707–1783). Example 11.9 involves the use of this formula.

Example 11.9 Using Euler's Formula

a. Is the polyhedron shown on the top of the next page a regular polyhedron? Why or why not?

b. Does Euler's formula hold for it? Explain.

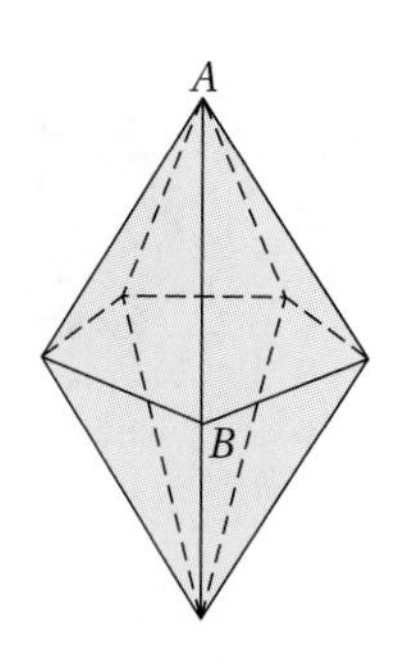

Solution

a. It isn't a regular polyhedron because, although the faces might all be congruent equilateral triangles, the arrangements of polygons at all vertices aren't the same: Five sides meet at vertex A, and four sides meet at vertex B.

b. In the polyhedron, $V = 7$, $F = 10$, and $E = 15$; $7 + 10 = 15 + 2$, so $V + F = E + 2$ and Euler's formula holds.

Your Turn

Practice: Is the polyhedron shown below on the left a regular polyhedron? Why or why not? Does Euler's formula hold for it? Explain.

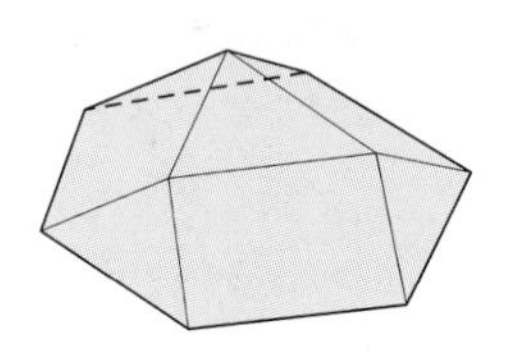

Reflect: Do you think that Euler's Formula holds for all polyhedra? Explain. ■

Many different polyhedra can be created simply by slicing off parts of one of the five regular polyhedra. For example, Figure 11.32 shows some of the results of slicing off the corners of a cube and an octahedron.

The idea of slicing or taking a cross section of a solid has many practical applications. For example, X-ray tomography is the use of computer graphic techniques to create a three-dimensional object solely from data about planar slices of the object. Contour maps, temperature analyses of materials, and biological analyses are other important uses of slicing.

Example 11.10 involves analyzing the polyhedron that results when the faces of a regular polyhedron are sliced in a certain way.

Example 11.10 Analyzing a Sliced Polyhedron

Describe the number and type of faces of the polyhedron formed when every corner of a tetrahedron is sliced as shown to the left. Show that Euler's formula holds.

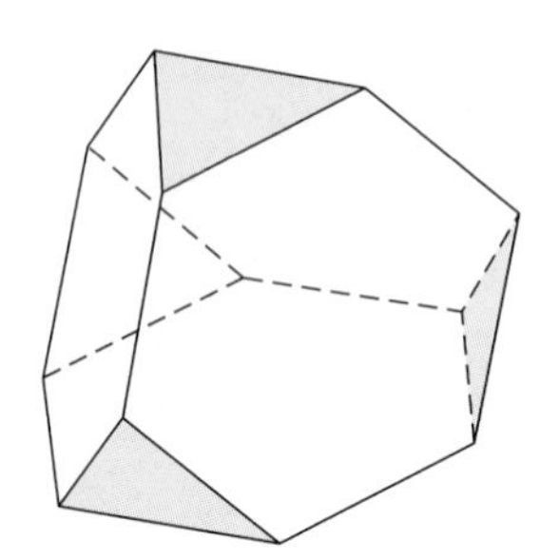

Solution

The polyhedron has 4 hexagonal and 4 triangular faces. It has 12 vertices, 8 faces, and 18 edges. As $12 + 8 = 18 + 2$, Euler's formula, $V + F = E + 2$, holds.

Your Turn

Practice: Complete the example problem by slicing off the corners of a cube.

Reflect: Do these examples prove that Euler's formula holds for any polyhedron formed by cutting off the corners of another polyhedron? ■

Prisms and Pyramids

Figure 11.33 shows other three-dimensional building blocks for geometry. A **prism** is a polyhedron with a pair of congruent faces (same size and shape), called *bases*, that lie in parallel planes. The vertices of the bases are joined to form the *lateral* faces of a prism. Prisms are named according to the shapes of their bases. For example, prisms with triangular bases are called *triangular prisms*.

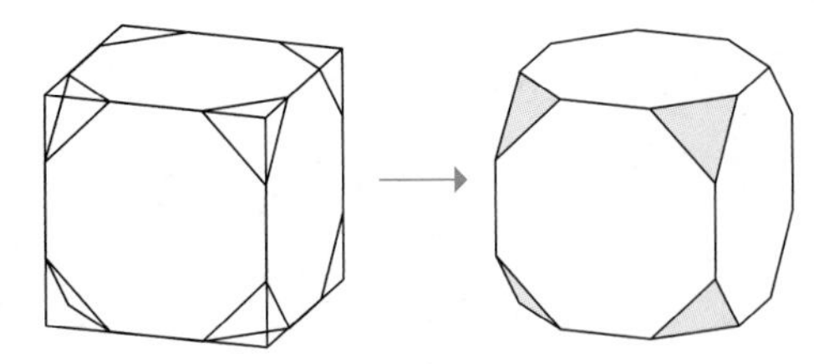

(a) Truncated cube

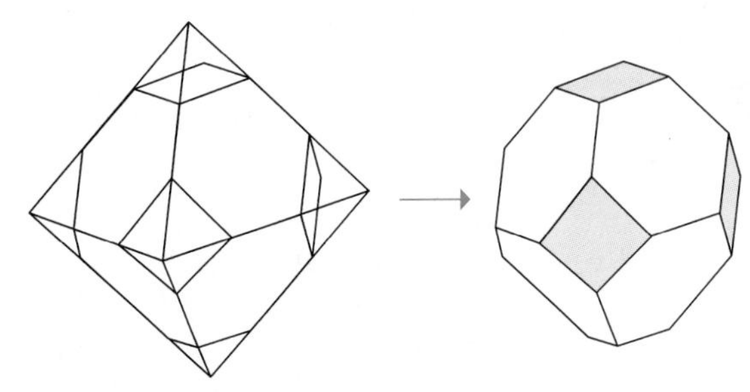

(b) Truncated octahedron

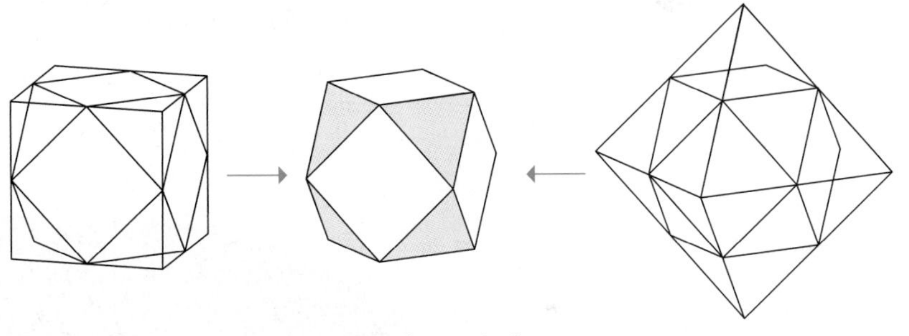

(c) Cube octahedron

FIGURE 11.32 | (a) Slicing off the corners of a cube can change the original square faces to regular octagons; (b) slicing off the corners of a regular octahedron can change the original triangular faces to regular hexagons; (c) slicing off the corners of a cube or an octahedron at the midpoint of each edge changes each of these polyhedra to a cube octahedron.

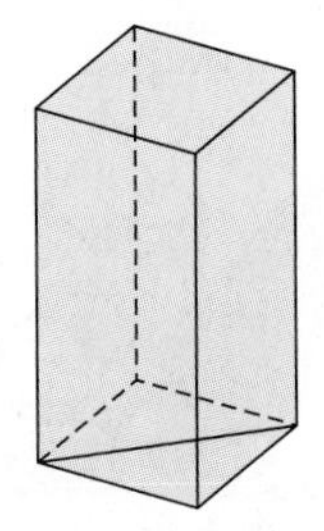

(a) Square, or rectangular, prism

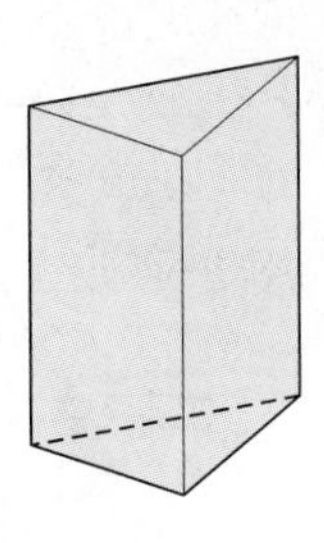

(b) Triangular prism

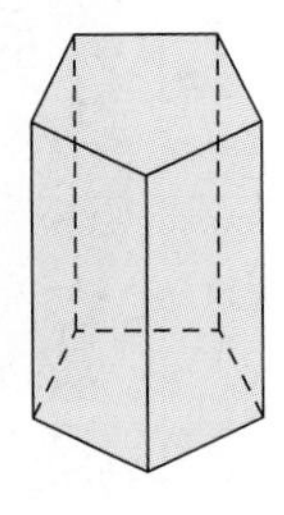

(c) Pentagonal prism

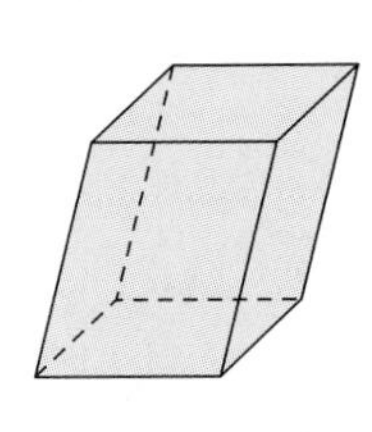

(d) Oblique square prism

FIGURE 11.33 | Prisms.

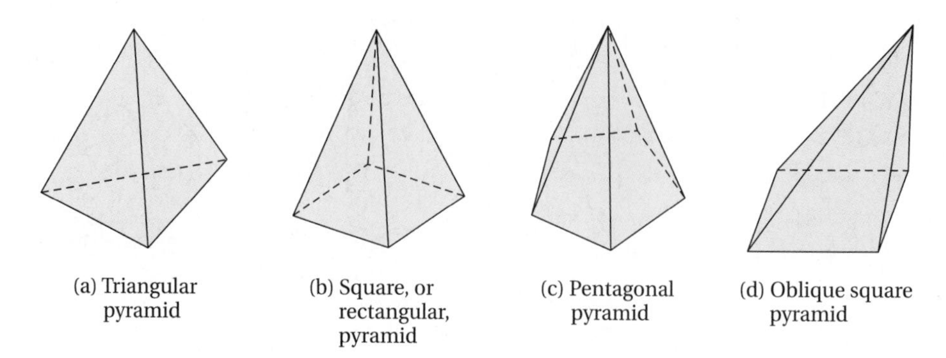

FIGURE 11.34 | Pyramids.

A **pyramid** is a polyhedron formed by connecting the vertices of a polygon, called the *base*, to a point not in the plane of the polygon, called an *apex*. As shown in Figure 11.34, pyramids also are named according to the shapes of their bases.

In contrast to the pyramid shown in the cartoon, the pyramids built by the ancient Egyptians were spectacular.

Mini-Investigation 11.9 suggests that you broaden your perspective of the applicability of Euler's formula.

"Actually, what I had in mind was something a little larger."

Draw a square pyramid and show that Euler's formula holds for it.

Mini-Investigation 11.9 **Using Mathematical Reasoning**

Do you think that Euler's formula holds for prisms and pyramids? Copy and complete a chart like the one below to help you investigate this question and support your conclusion.

Name of Polyhedron	Number of Base Edges	Number of Vertices = V	Number of Faces = F	Number of Edges = E	$V + F = E + 2$? (Yes or No)
Triangular prism	3	?	?	?	?
Square prism	?	?	?	?	?
n-gon prism	n	?	?	?	?
Triangular pyramid	3	?	?	?	?
Square pyramid	?	?	?	?	?
n-gon pyramid	n	?	?	?	?

Cylinders, Cones, and Spheres

If you imagine prisms, pyramids, and polyhedra with thousands of faces, these solid figures come very close to the solids with curved surfaces shown in Figure 11.35. In the **cylinder** and **cone**, the shaded circles are called *bases*. The *radius*, r, of the cylinder or cone is the radius of its base. The *height*, h, of a cylinder is the perpendicular distance from one base to the other. The line through the centers of the bases of a cylinder is called the *axis* of the cylinder. The *height*, h, of the cone is the perpendicular distance from the vertex, or apex, V, to the base. The line through the center of the base and the vertex is called the axis of the cone. Point O is the *center* of the **sphere**, and r is the *radius* of the sphere.

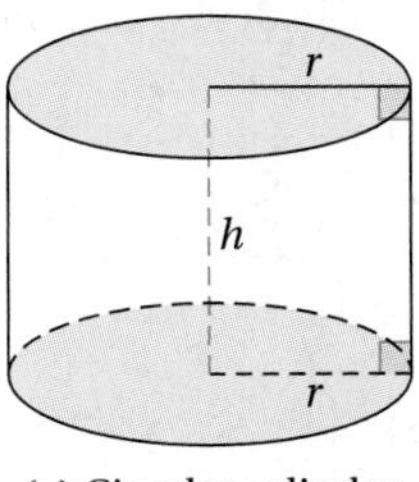

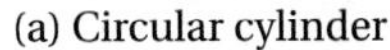

(a) Circular cylinder

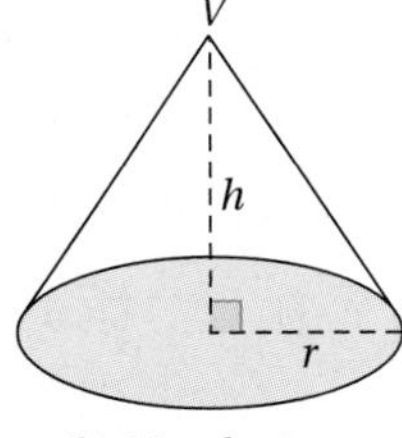

(b) Circular cone

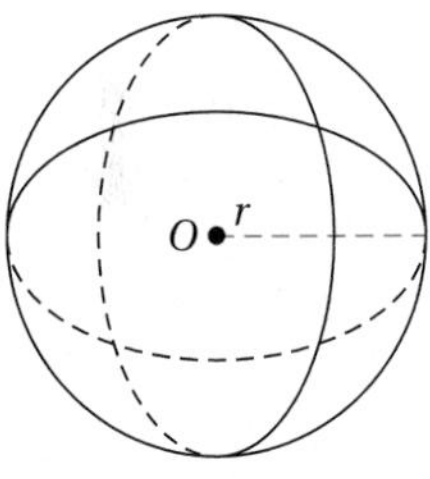

(c) Sphere

FIGURE 11.35 | Cylinders, cones, and spheres.

From ice cream cones to cans to architecture to the earth and moon, the everyday importance of these building blocks of geometry with curved surfaces is apparent.

Along with prisms, pyramids, and other polyhedra, cylinders, cones, and spheres are used to model real objects. Example 11.11 asks you to compare the characteristics of different three-dimensional figures.

Example 11.11 **Analyzing Cylinders, Cones, and Spheres**

a. How are a cylinder and a prism alike?

b. How are a cylinder and a prism different?

Solution

a. They both have pairs of congruent, parallel bases. Each has a dimension called *height*. They are both three-dimensional figures.

b. The bases of a prism are polygons; the bases of a cylinder are not polygons. A prism has lateral faces that are polygons; a cylinder has a lateral surface that is not made up of polygons.

Your Turn

Practice: How are a cone and a pyramid alike? How are they different?

Reflect: Which polyhedra are most like a sphere? Support your answer. ■

Symmetry in Three Dimensions

In Section 10.1, we discussed symmetry of two-dimensional figures. Symmetry is also an important characteristic of some three-dimensional figures. The apple shown in Figure 11.36, in its ideal form, has **reflectional symmetry.** The vertical slice that separates the apple into two symmetric parts goes through a **plane of symmetry** of the apple. You might think of this plane as a mirror. If half the apple were held against the mirror, it together with its reflection would appear to be a whole apple.

Slicing an ideal apple horizontally provides an example of **rotational symmetry.** The part of the apple showing a star can be rotated 72° about an axis through its center to appear to be in the same position. Thus we say that the piece of apple has 72° rotational symmetry. The apple piece can be rotated through 72° five times before returning to its original position. Thus we say that the apple piece has rotational symmetry of order 5. The axis through the center represents the **axis of rotational symmetry.**

A cube has several planes of reflectional symmetry. One plane of symmetry is parallel to a pair of faces of the cube and intersects edges at only one point, as shown in Figure 11.37(a). Because a cube has three different pairs of faces, there are

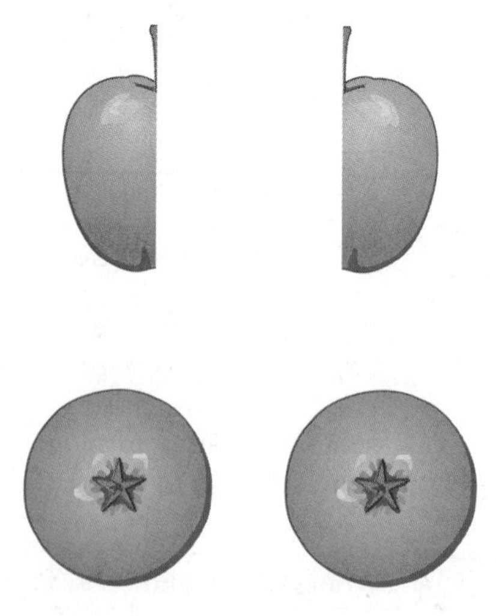

FIGURE 11.36 | Reflectional and rotational symmetry in an ideal apple.

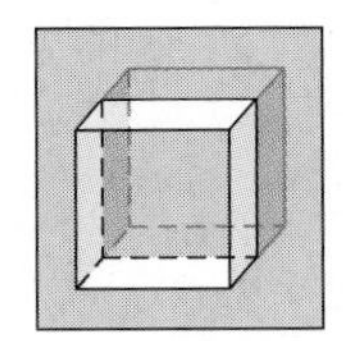

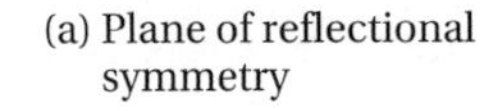

(a) Plane of reflectional symmetry

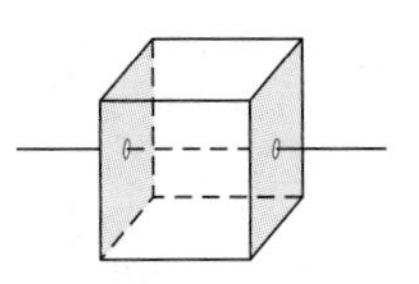

(b) Axis of rotational symmetry

FIGURE 11.37 | Symmetry of a cube.

three of this type of plane of symmetry. Another plane of symmetry contains a pair of edges of the cube. As there are six different pairs of edges, there are six of this type of plane of symmetry. A cube also has several axes of rotational symmetry. One type of axis goes through the center of opposite faces, as shown in Figure 11.37(b). there are three of this type of axis. Another type of axis goes through the midpoints of a pair of opposite edges of the cube; there are six of this type of axis. A third type of axis goes through opposite vertices of the cube; there are four pairs of opposite vertices, so there are four of this type of axis. Visualizing reflectional and rotational symmetry properties of the cube is a starting point for analyzing symmetry properties of other solid figures in order to solve practical problems.

Example 11.12 shows how to solve a practical problem by applying three-dimensional symmetry ideas. The problem-solving strategies *draw a diagram* or *make a model* (with drinking straws) or both are helpful when solving this problem.

Example 11.12

Problem Solving: Planning a Sculpture

An artist wants to build a sculpture that includes an object of tetrahedron shape with rods welded to its exterior. She decided to weld the rods so that they would look like extensions of all the axes of symmetry of the tetrahedron. What instructions would help her decide where to weld the rods?

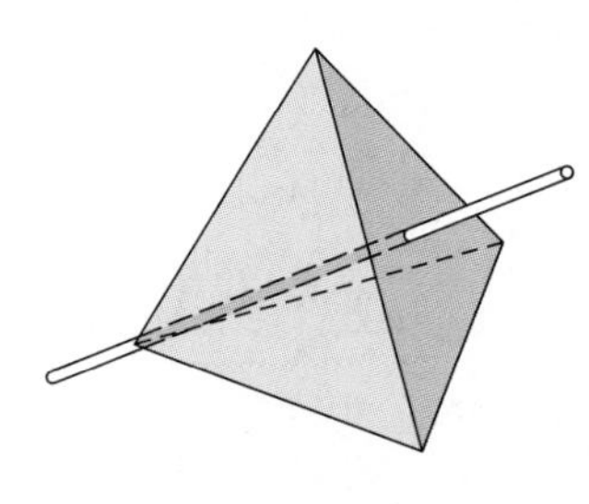

Working toward a Solution

One type of axis of rotational symmetry passes through a vertex and the center of a face opposite this vertex, as shown on the left, indicating where some rods should be welded.

Your Turn

Practice:

a. Give additional instructions to tell the artist where rods should be welded.

b. Suppose that the artist wanted to weld sheets of metal to show extensions of all planes of symmetry of the tetrahedron. Use the figure on the left to help give her instructions about where to weld them.

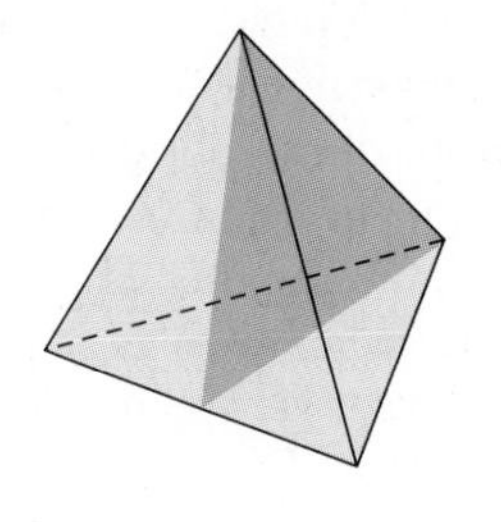

Reflect: If every plane of symmetry of a tetrahedron contains an edge and a midpoint of an opposite edge, explain how you could decide how many different such planes there are. ■

Visualizing Three-Dimensional Figures

One important aspect of spatial sense is the ability to represent a three-dimensional figure in a two-dimensional picture. There are several different types of two-dimensional representations.

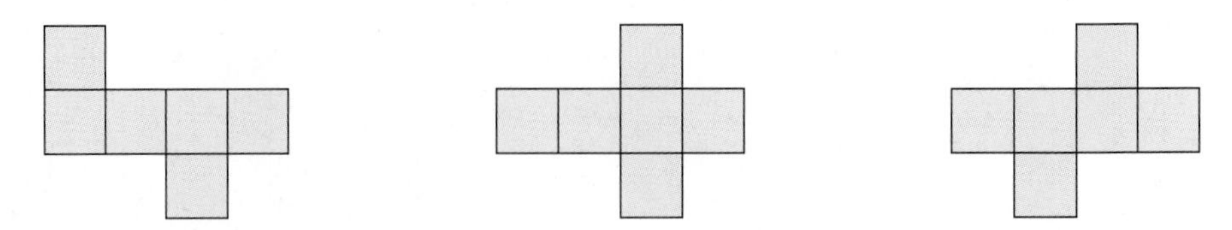

FIGURE 11.38 | Hexomino patterns for a cube.

Visualizing Polyhedra from Their Patterns. A **pattern** or **planar net** for a polyhedron is an arrangement of polygons that can be folded to form the polyhedron. For example, each of the patterns shown in Figure 11.38 can be folded to make a cube. They are three of 35 possible hexominos, which are made from six squares that are always connected by at least one common side. It is instructive to look at pictures of the patterns and try to visualize how they could be folded to make cubes. In Exercise 38 at the end of this section, you are asked to explore whether all hexominos can be folded to make a cube.

Example 11.13 shows how to visualize other polyhedra from patterns.

Example 11.13 Visualizing Polyhedra from Patterns

What polyhedron could be made with the pattern shown on the left?

Solution

If we visualize folding the pattern, we see that a tetrahedron could be formed.

Your Turn

Practice: What polyhedra could be made with these patterns?

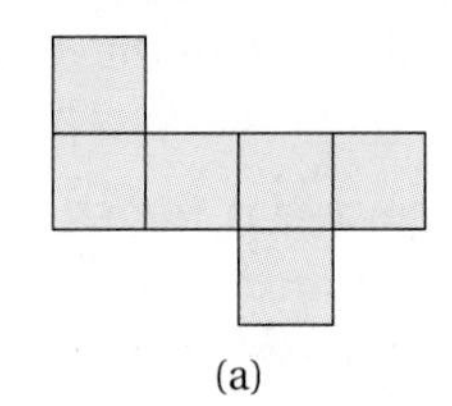

(a)

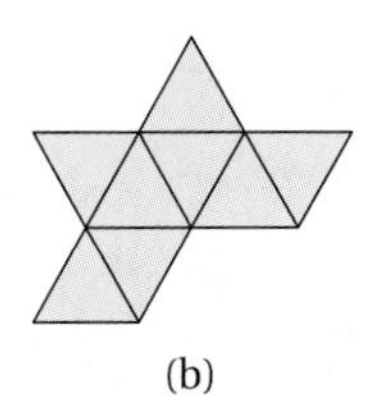

(b)

Reflect: Do you think that other patterns can be folded to make a tetrahedron? Explain, using an example if possible. ■

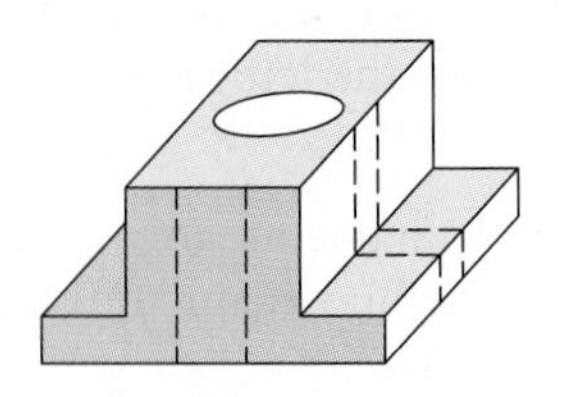

Visualizing Three-Dimensional Figures from Different Views. Architects, engineers, and others often need to represent a three-dimensional real-world object on a two-dimensional piece of paper. To give a total picture of the object as they visualize it they turn it to show an end view, a side view, and a top view and draw a picture of each view. Figure 11.39 illustrates this approach. Computer programs called computer-aided design (CAD) are now used to create such drawings.

In Example 11.14, we look at an object from three different viewpoints.

Example 11.14 Drawing End, Side, and Top Views

Draw the end view, side view, and top views of the iron casting shown above.

FIGURE 11.39 | End, top, and side views of an object.

Solution

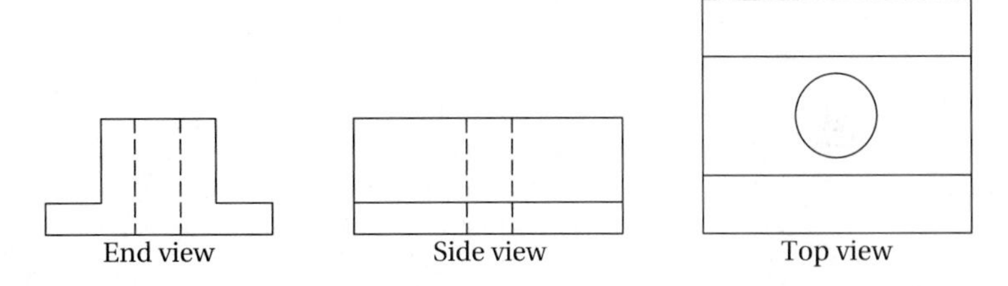

Your Turn

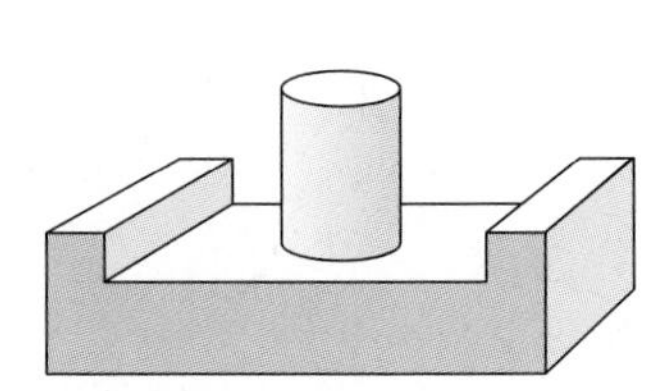

Practice: Draw the end, top, and side view of the object on the left.

Reflect: Can you draw a figure accurately after seeing only the side view and the end view? Explain. ■

Visualizing Three-Dimensional Figures and Their Shadows. In the transformations discussed in Section 11.1, three-dimensional objects remained three-dimensional and two-dimensional objects remained two-dimensional. Transforming a three-dimensional figure into a two-dimensional shadow involves the use of **shadow geometry**. In such transformations, straightness is preserved, but length may be changed, as Trixie has discovered with her sunbeam in the cartoon.

Reprinted with special permission of King Features Syndicate.

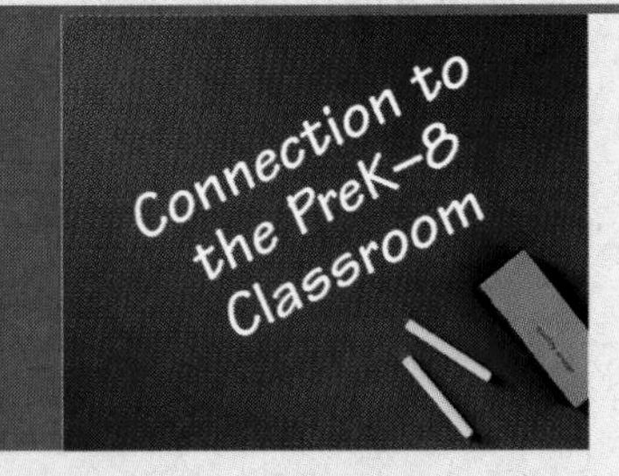

A Sample Student Page: Different Views of Three-Dimensional Figures

To develop space perception skills, students need to learn to look at actual two-dimensional objects, or drawings of objects on a page, and visualize what the object would look like from different points of view. Study the page from a middle school mathematics textbook shown in Figure 11.40 and answer the questions.

- **Do** you need to see all three views (front, side, and top) in order to determine the original figure? Why or why not?
- **Can** two figures have the same front, side, and top views? Explain.
- **When** would these views of a three-dimensional object be useful?

Learn Three-Dimensional Figures

Recall that a solid is a three-dimensional figure. Solids are often drawn in perspective to show that they are three-dimensional.

Solids can also be drawn using a flat view. Flat drawings show the solid from one view only. In order to record what the solid looks like, you usually need to show three views: front, side, and top.

Front Side Top

Language Link

The flat drawings that show the front, side, and top views of a solid are also known as *orthographic projections*. A three-dimensional picture of a solid is an *isometric projection*.

Example 1

Draw front, side, and top views of the solid. There are no hidden cubes.

Front Side Top

Try It

Draw front, side, and top views of the solid. There are no hidden cubes.

A solid also has other views, such as the back view and the bottom view. Because these views are mirror images of the front, side, and top views, they are not necessary when drawing a solid.

Check Your Understanding

1. Describe a solid with the same front, side, and top views.
2. Could you use the front, side, and top views of a solid to build it? Explain.

11-5 • Three-Dimensional Figures 603

FIGURE 11.40 | Excerpt from a middle school mathematics textbook.

(*Source:* From *Scott Foresman–Addison Wesley Middle School Math, Course 1* by R. I. Charles, J. A. Dossey, S. J. Leinwand, C. J. Seeley, and C. B. Vonder Embse. © 1998 by Pearson Education, Inc. publishing as Prentice Hall. Used by permission.)

Example 11.15 asks you to visualize the different shadows that could be made with a given three-dimensional object.

Example 11.15

Connecting Shadows to Three-Dimensional Objects

What are some shadow figures that you can make with a cylinder with a circular base?

Solution

Some possible answers are as follows:

a. A circle, with a light source on the line through the center of the circle.

b. A rectangle, with a light source on a line perpendicular to and bisecting the segment joining the centers of the bases.

c. A nonrectangular parallelogram, if the light source creates a beam that is bisecting but not perpendicular to the segment joining the centers of the bases.

Your Turn

Practice: What are some shadow figures you can make with a circular cone?

Reflect: What shapes of shadows can you *not* make with a cylinder? Why? ■

Visualizing Three-Dimensional Figures from Perspective Drawings. From daily observation, you have probably noticed that an object nearer to you appears larger than an object of the same size that is farther away. Not until late in the thirteenth century, however, did artists, many of whom were mathematicians, begin to explore this idea of *perspective* and develop ways to attain a two-dimensional representation of the actual appearance of three-dimensional objects.

One technique for depicting three dimensions on a flat surface is to base the orientation of the picture on a **vanishing point**. In a drawing in which the front surface of the object or scene is in a plane parallel to the plane containing the picture, lines that move away from and should appear parallel to the viewer are drawn to meet at the vanishing point. Such representations are called **perspective drawings** and are illustrated in Figure 11.41.

Example 11.16 gives you an opportunity to practice using a vanishing point to make a perspective drawing

(a) Draw the front of the box parallel to the plane of the paper.

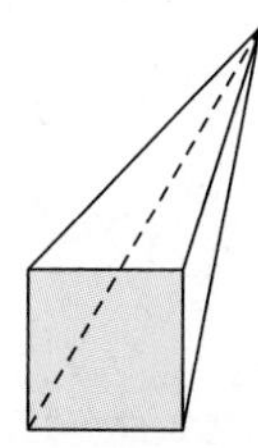

(b) Choose a vanishing point and use a ruler to connect the vertices of the box to the vanishing point.

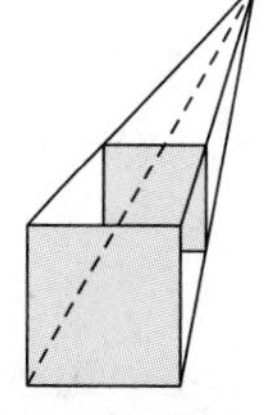

(c) Draw the back of the box with edges parallel to the corresponding edges of the front of the box.

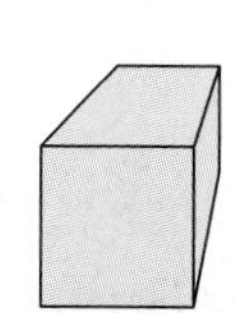

(d) Erase the lines that are no longer necessary.

FIGURE 11.41 | A rectangular box drawn with one-point perspective.

Example 11.16 Using a Vanishing Point to Make a Perspective Drawing

Use the technique shown in Figure 11.41 to make a perspective drawing of a hexagonal prism.

Solution

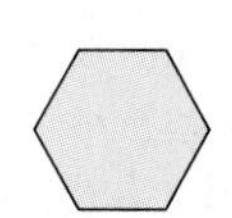

(a) Draw the hexagonal base parallel to the plane of the paper.

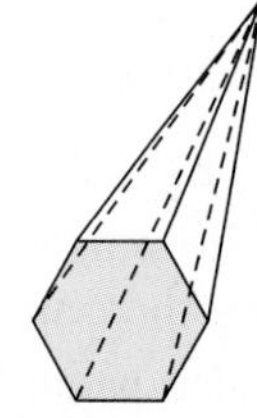

(b) Choose a vanishing point and connect it to each vertex of the hexagon.

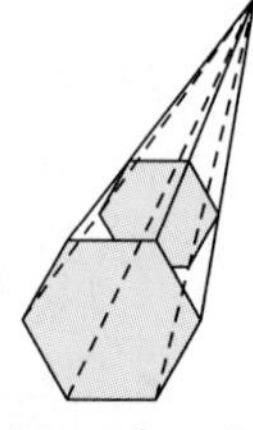

(c) Draw the other base of the prism with appropriate parallel edges.

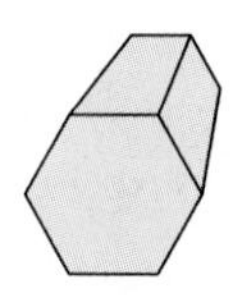

(d) Erase unnecessary lines.

Your Turn

Practice: Draw the prism from a different perspective by putting the vanishing point in a different place on the paper.

Reflect: How would you use this technique to draw a hexagonal pyramid? ■

Another technique for creating perspective drawings is the use of isometric dot paper, as shown in Figure 11.42. With isometric dot paper, the front surface of the object is not parallel to the plane containing the picture.

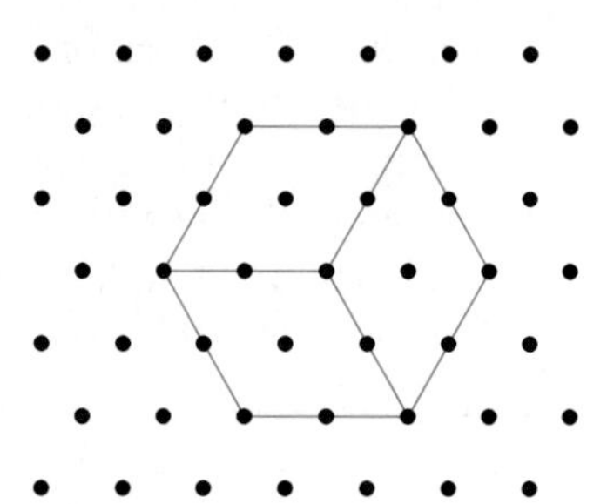

FIGURE 11.42 | A cube represented by a perspective drawing on isometric dot paper.

Problems and Exercises for Section 11.4

A. Reinforcing Concepts and Practicing Skills

1. a. What is the difference between a regular octahedron and a nonregular octahedron?
   b. Describe or sketch a nonregular octahedron.

2. Name the polyhedron that can be made from each of the patterns shown on the next page. Give the number of faces, vertices, and edges for each.

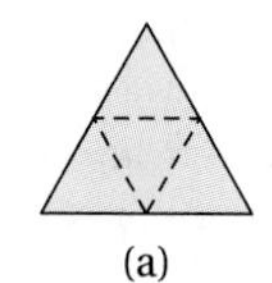

(a)

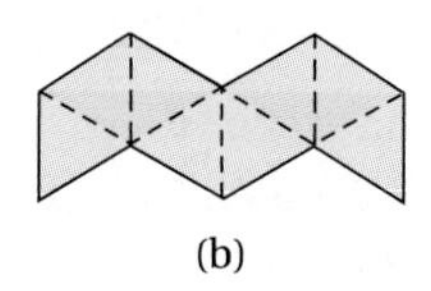

(b)

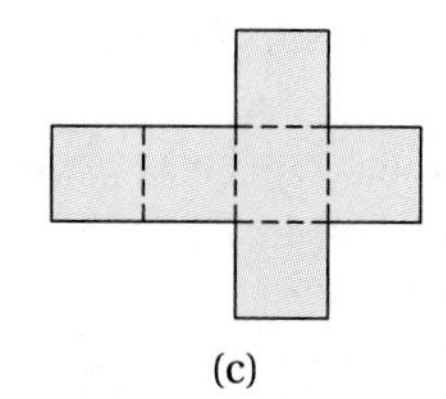

(c)

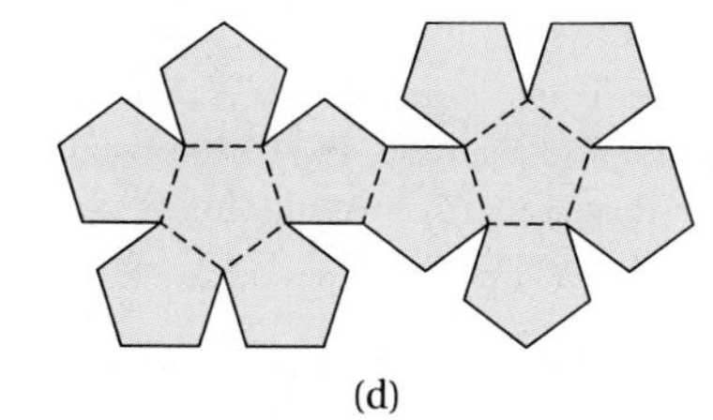

(d)

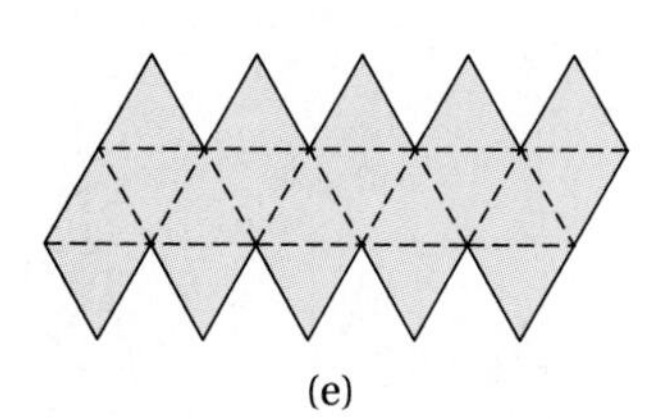

(e)

3. Name some real-world objects that can be modeled by the following geometric figures.
   a. a cube
   b. a rectangular prism other than a cube
   c. a triangular prism
   d. a pyramid
   e. a cylinder
   f. a cone
   g. a sphere
4. Name or describe some other hexahedrons besides a cube.
5. **a.** How many pairs of bases does a right rectangular prism have?
   b. Is there any other prism with more than one pair of faces that can be identified as bases?
6. Make a sketch of the following.
   a. a circular cylinder with radius of approximately 3 cm and height of approximately 20 cm
   b. a circular cylinder with radius of approximately 10 cm and height of approximately 3 cm
   c. a circular cone with radius of approximately 3 cm and height of approximately 20 cm
   d. a circular cone with radius of approximately 10 cm and height of approximately 3 cm
7. Show that Euler's formula holds for each of the polyhedra shown.

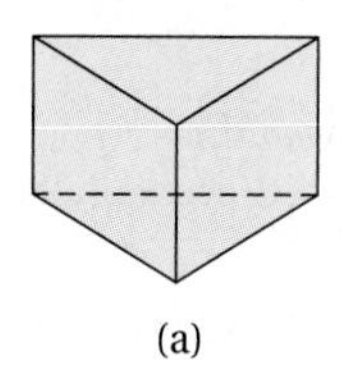

(a)

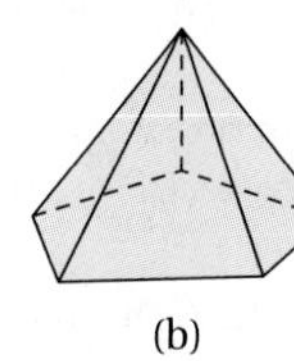

(b)

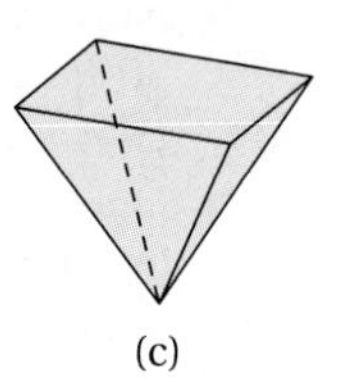

(c)

Copy and complete the following table.

Polyhedron	F	V	$F + V$	E
a.	?	?	?	?
b.	?	?	?	?
c.	?	?	?	?

8. Show that Euler's formula holds for the truncated octahedron and cube octahedron in Figure 11.32(b) and (c).
9. Sketch a square prism. How many planes of symmetry and axes of rotational symmetry does it have?
10. **a.** Describe the different types of planes of symmetry and axes of symmetry of a regular octahedron.
    b. How many planes and axes of symmetry does a regular octahedron have?
    c. Make a sketch and compare the symmetry properties of a regular octahedron and a cube.
11. Draw or describe a pyramid that has three planes of symmetry.
12. Describe the characteristics that a pyramid must have for it to have an axis of symmetry.
13. Which of the patterns shown can be folded to make an open-top box?

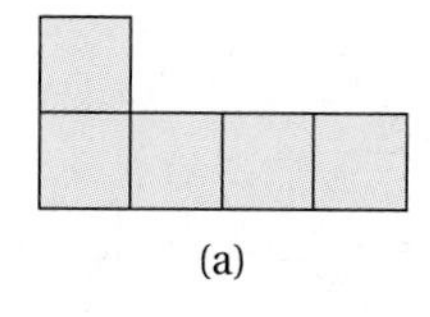

(a)

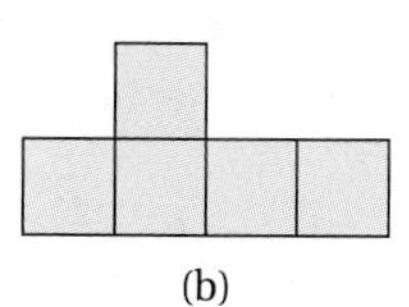

(b)

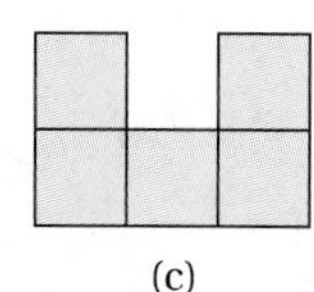

(c)

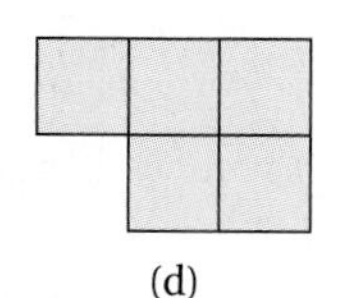

(d)

14. A pentomino is a figure formed by five connected, nonoverlapping congruent squares, each touching others only along a complete side. Of the 12 different pentominos, how many can be folded to make an open-top box?

15. a. Draw a pattern for a rectangular prism that is not a cube.
    b. Draw another pattern for the same rectangular prism.

16. a. Draw a pattern for a rectangular pyramid.
    b. Draw another pattern for the same rectangular pyramid.

17. Each solid figure shown is sliced by a plane. Identify the cross sections formed.

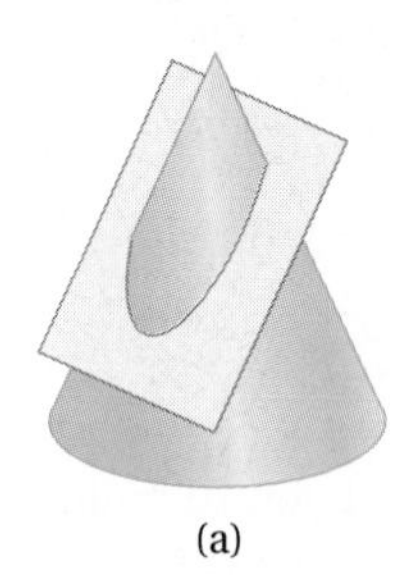

(a)

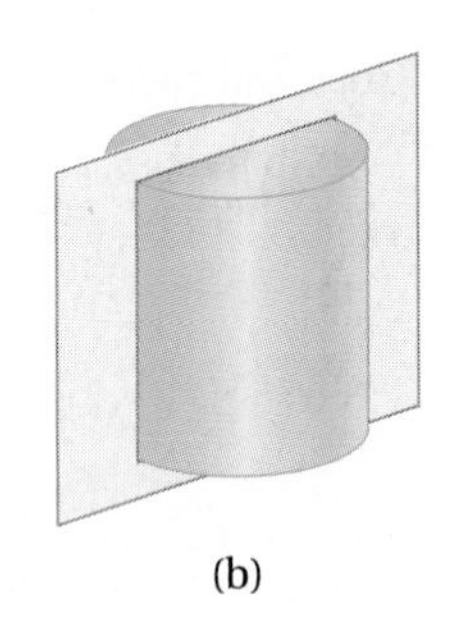

(b)

18. Sue built a figure with five cubes. The top and side views of the figure look as follows.

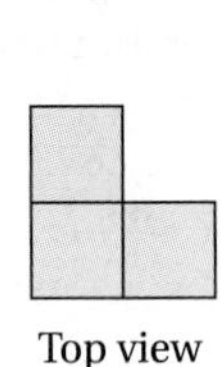

Top view

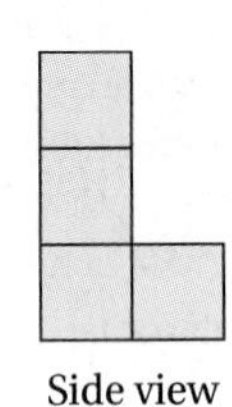

Side view

Draw or describe the figure she made.

19. Draw or describe all the figures you could make by joining five cubes that would have the following top view.

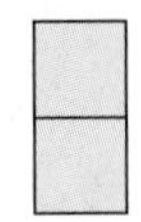

20. Draw the end, side, and top views of the following object.

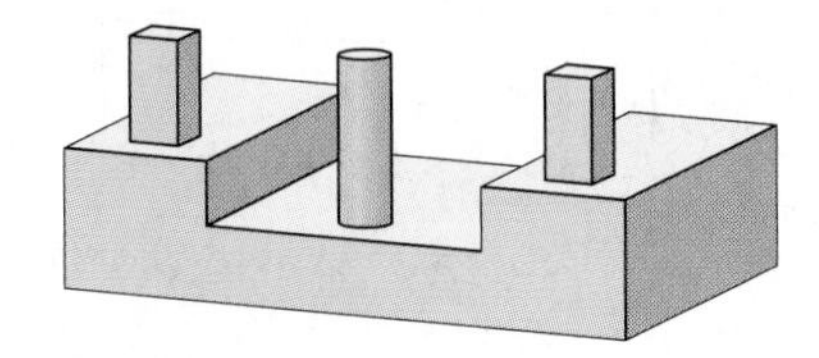

21. Describe at least four different shapes of shadows that you can make with a cube.

22. Make a perspective drawing of an isosceles triangular prism. Let the triangular bases be parallel to the plane of the paper.

23. Use isometric dot paper to draw a perspective representation of a shape that could be made with three cubes.

B. Deepening Understanding

24. What is the resulting polyhedron if
    a. the center points of the faces of an octahedron are connected?
    b. the center points of the faces of a tetrahedron are connected?

25. Why are the regular polyhedra good models for dice?

26. Compare the patterns in Exercises 15(a) and 16(a).
    a. How are they alike?
    b. How are they different?
    c. How does each illustrate the definition of prism and of pyramid?

27. An *antiprism* is like a prism except that the lateral faces are triangles and the bases have been rotated.
    a. Show that Euler's formula holds for the square antiprism shown.

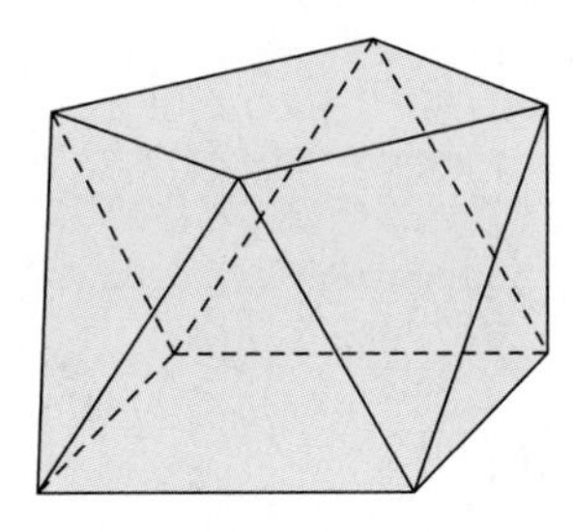

    b. How many faces, edges, and vertices would a triangular antiprism have? Does Euler's formula hold?

28. a. Place a vanishing point in about the middle of your paper to make a perspective drawing of a rectangular box.
    b. Keeping the front face of the box in the same position, move the vanishing point closer to the upper left corner of your paper and make a perspective drawing.
    c. Repeat with the vanishing point near the upper right corner of the paper, then near the lower right corner, and finally near the lower left corner.
    d. Compare the drawings and make a conjecture about the effect that the position of the vanishing point has on the perspective.

29. How could you show a cube from two different perspectives on the isometric dot paper on the next page?

Figure for Exercise 29

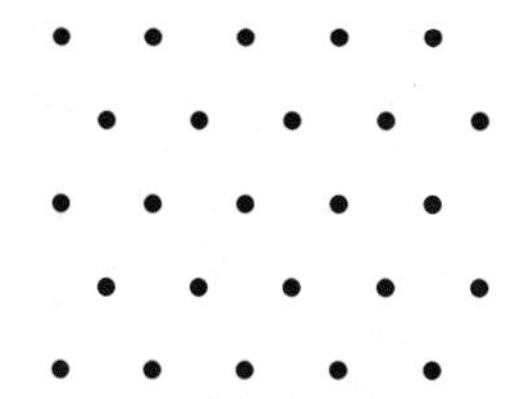

C. Reasoning and Problem Solving

30. Answer the following questions to determine why Euler's formula continues to hold for the polyhedron formed by cutting corners off of an octahedron, as in Figure 11.32(b).

a. For an octahedron, $V =$ ___, $F =$ ___, and $E =$ ___.

b. When you slice off one corner of the octahedron, you (gain or lose) ___ vertices, (gain or lose) ___ faces, and (gain or lose) ___ edges.

c. Therefore the total change in V is ___; the total change in F is ___; and the total change in $V + F$ is ___.

d. The total change in E is ___.

e. What does the comparison of the total change in $V + F$ to the total change in E tell you?

31. Use the above reasoning to determine why Euler's formula would hold for the polyhedra formed by slicing off the corners of any prism.

32. Show how to slice a cylindrical piece of cheese into eight congruent pieces with three slices.

33. What do you think the new exposed surfaces will look like when the following cube is separated into two parts by a slice that goes through the marked midpoints of its edges? Justify your answer.

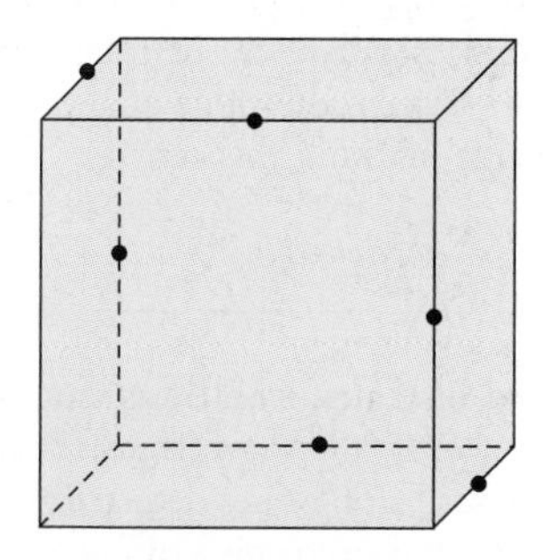

34. The Toy Blocks Problem. A toy factory manager has eight hundred 8-inch cubic blocks that have been painted red. She wants to produce small blocks by cutting each 8-inch cube into sixty-four 2-inch blocks. To utilize her machines as efficiently as possible, she wants to make the fewest number of cuts possible. She needs unpainted blocks to package in a block set and blocks painted on exactly two faces to package in another set. Prepare a report for the manager indicating

a. the fewest cuts needed to produce 64 cubes from one block.

b. the total number of unpainted blocks and blocks painted on two faces that can be produced.

Justify your conclusions in the report.

35. The Polyhedra Sculpture Problem. A sculptor found the following table of information about the dihedral angles (angles formed by adjacent faces) of the regular polyhedra. He knows that polyhedra that will fit all the way around an edge with no gaps can be used to make a sculpture he is planning.

Regular polyhedron	Degree Measure of Dihedral Angle
Cube	90°
Tetrahedron	70°32′
Octahedron	109°28′
Dodecahedron	116°34′
Icosahedron	138°11′

a. Use the table to determine which regular polyhedra could be used in his sculpture.

b. Could he use two tetrahedra and two octahedra in combination for his sculpture? Why or why not?

36. Construct a tetrahedron from a sealed envelope, as follows.

a. Find point C so that $\triangle ABC$ is equilateral.

b. Make a cut DE parallel to AB. Discard the other part of the envelope.

c. Fold AC and BC in both directions.

d. Pinch the envelope to join D and E.

e. Tape the opening to make a tetahedron.

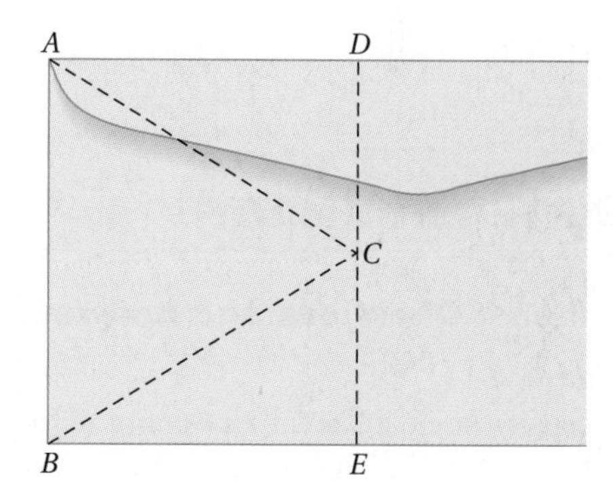

37. A deltahedron is a polyhedron with equilateral triangle faces.

a. Which deltahedra are also regular polyhedra?

b. Draw a picture of a deltahedron with six faces that is not a regular polyhedron.

38. Can every hexomino be folded to make a cube? Explain.

39. Could the two figures shown describe actual objects? Explain your conclusions.

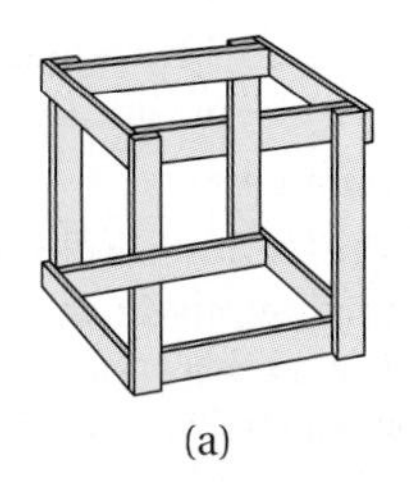

(a)

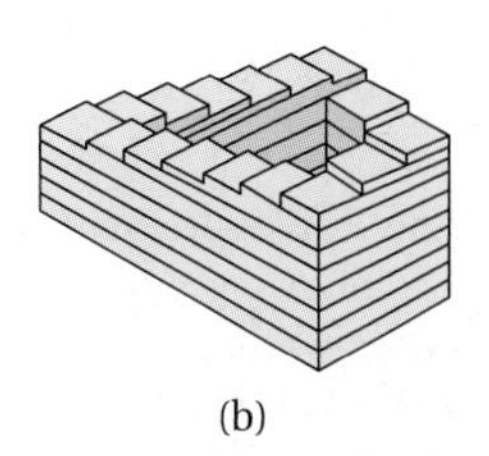

(b)

40. The Dollar Bill Problem. A magician claimed that he could make a tetrahedron out of a $1 bill. He began to fold the bill along the solid lines as shown. Do you believe that he could really do it? Explain your conclusion.

41. Without counting all the faces, edges, and vertices, how would you convince someone that Euler's formula still holds for the figure produced when one corner of a cube is cut off? Make a sketch.

D. Communicating and Connecting Ideas

42. In a small group, discuss differences, if any, between mathematical and everyday meanings of each term.

a. Sphere
b. Cube
c. Pyramid
d. Prism

Why is it important to have precise meanings for terms in mathematics?

43. Historical Pathways. The Greek philosopher Plato (430–347 B.C.), in his book *Timaeus*, associated the cube with earth, the tetrahedron with fire, the octahedron with air, and the icosahedron with water. He associated the dodecahedron with what was used to create the universe. To develop this cosmology, Plato used the relationships between these polyhedra. Follow the instructions given in order to view two of those relationships.

a. Draw a cube as follows.

- Draw square $ADHE$ in front and a partially dashed square $BCGF$ in back of it.

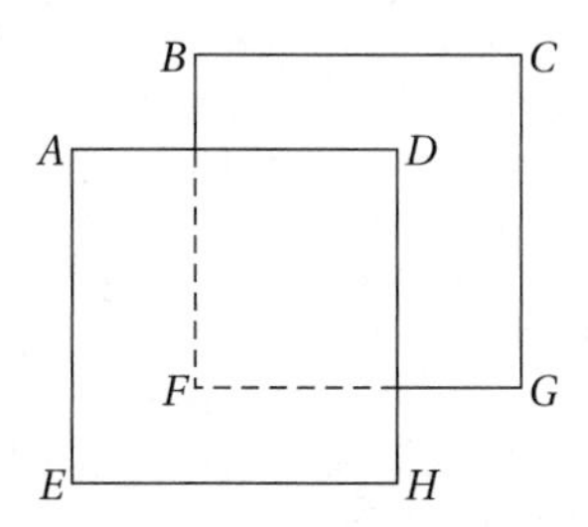

- Join vertices A and B, D and C, E and F, and H and G.

b. Now use a colored pencil to join all midpoints of adjacent faces of the cube. What polyhedron is formed?

c. Now draw another cube and use a colored pencil to join vertices B, D, E, and G with all possible segments. What polyhedron is formed?

44. Making Connections. Put a familiar object in a bag and have another person visualize it only by feeling, using names of polyhedra or other geometric ideas to describe it. Discuss the value of appropriate mathematical language in doing this activity.

CHAPTER SUMMARY

Key Ideas: Questions and Answers

Section 11.1

- **What are some different types of transformations?** *(pp. 586–591)* An object can be mapped onto any other figure congruent to it through the use of one or more of three types of transformations called isometries: translations, rotations, and reflections.
- **How can we use congruence transformations, or isometries, to describe symmetry?** *(pp. 591–593)* When a figure can be traced and folded so that one-half exactly coincides with the other half, it has reflectional symmetry about the line of symmetry. When it can be turned less than 360° about a fixed point called a center of rotational symmetry to fit back on itself, it has rotational symmetry.
- **What kind of transformations can change a figure's size?** *(pp. 593–596)* A size transformation enlarges or shrinks a figure along lines determined by a point called the center of the size transformation and according to a multiplier called the scale factor. A combination of one or more isometries and a size transformation can map a figure onto any other similar figure.
- **What kind of transformations can change both a figure's size and shape?** *(pp. 596–597)* Topological transformations preserve the connectedness of curves and surfaces but can change both size and shape.

Section 11.2

- **What is meant by geometric patterns and tessellations?** *(pp. 602–603)* A design repeating some basic element in a systematic manner is called a pattern. A tessellation is a special type of pattern that consists of geometric figures that fit without gaps or overlaps to cover the plane.

- **Which polygons will tessellate the plane by themselves?** *(pp. 604–605)* Equilateral triangles, squares, and regular hexagons are the only regular polygons that tessellate by themselves. Every triangle and quadrilateral will tessellate. Several nonregular pentagons and hexagons will tessellate. No convex polygon with more than six sides will tessellate, but some nonconvex polygons with more than six sides will tessellate.
- **Can combinations of regular polygons tessellate the plane?** *(pp. 606–608)* Many different tessellations can be formed with a combination of regular polygons. Of interest are semiregular tessellations, formed by two or more polygons with the arrangement of polygons at each vertex the same. There are eight different semiregular tessellations of the plane.
- **What other interesting ways are there to generate tessellations?** *(pp. 608–609)* A three-mirror kaleidoscope or computer software such as GES or LOGO can be used to generate tessellations using translations, rotations, and reflections.
- **How can tessellations with irregular or curved sides be formed?** *(pp. 609–613)* Techniques include replacing the edges of a tessellating polygon with curved lines (using rotation in the cut and turn or translation in the cut and slide methods) or making a design in the interior of a tessellating polygon.

Section 11.3

- **What are some characteristics of golden triangles and rectangles?** *(pp. 616–617)* In both a golden triangle and a golden rectangle, the ratio of its longer side to its shorter side is the golden ratio, $\phi = \frac{(1 + \sqrt{5})}{2} \approx 1.618$.
- **What are some generalizations that can be made about star polygons?** *(pp. 617–619)* If n = the number of equally spaced points on the circle and d = the dth point that segments are drawn to, (a) the star polygon $\{\frac{n}{d}\}$ is the same as the star polygon $\{\frac{n}{n-d}\}$; (b) the polygons $\{\frac{n}{1}\}$ and $\{\frac{n}{n-1}\}$ are regular polygons; and (c) the n-sided star polygon $\{\frac{n}{d}\}$ exists only if $d \neq 1$, $d \neq n - 1$, and n and d are relatively prime.
- **What are some characteristics of star-shaped polygons?** *(pp. 619–622)* Star-shaped polygons have n star-tip points, $2n$ congruent sides, n congruent point angles with measure α, and n congruent dent angles β such that $\beta = (\frac{360}{n}) + \alpha$. They differ from star polygons in that they are simple polygons, whereas star polygons are nonsimple.

Section 11.4

- **What are regular polyhedra?** *(pp. 626–628)* A polyhedron is a collection of polygons joined to enclose a region of space, forming a figure with faces, edges, and vertices. A regular polyhedron's faces are congruent regular polygons, and the arrangement of these polygons is the same at all vertices of the polyhedron. The five regular polyhedra include the regular tetrahedron, the cube, the regular octahedron, the regular dodecahedron, and the regular icosahedron.
- **What are some other important polyhedra?** *(pp. 628–631)* Prisms and pyramids are important types of polyhedra. A prism has two congruent, parallel bases joined by quadrilateral lateral faces. A pyramid has one base and triangular lateral faces joined at a common vertex called the apex.
- **What important three-dimensional figures are there other than polyhedra?** *(pp. 631–632)* Cylinders, cones, and spheres are three-dimensional figures with curved surfaces, rather than polygonal faces as in polyhedra. Cylinders are like prisms, in that they have two congruent, parallel bases. Cones are like pyramids, in that they have one base and an apex. A sphere might be considered like a regular polyhedron with a very large number of faces.
- **How can we apply the idea of symmetry to three-dimensional figures?** *(pp. 632–633)* A three-dimensional figure that can be sliced into two congruent parts along a plane of symmetry is said to have reflectional symmetry. A three-dimensional figure that can be rotated less than 360° about an axis of rotational symmetry until it matches itself is said to have rotational symmetry.
- **How can three-dimensional figures be visualized?** *(pp. 633–638)* Three-dimensional figures can be represented in two dimensions in a variety of ways. A pattern, or planar net, can be used to show an arrangement of polygons that can be folded to form a polyhedron. A three-dimensional object can be turned to show only its end view, its side view, or its top view. Shadows can also produce two-dimensional representations of three-dimensional objects. Finally, perspective drawings based on a vanishing point can be created to represent three-dimensional figures.

Key Terms, Concepts, and Generalizations

Section 11.1

Image (p. 586)
Slide arrow (p. 586)
Directed segment (p. 586)
Translation (p. 586)
Rotation (p. 587)
Reflection (p. 588)
Glide–reflection (p. 589)
Congruence transformation (p. 590)
Isometry (p. 590)
Congruence (p. 591)
Reflectional symmetry (p. 591)

Line of symmetry (p. 591)
$n°$ rotational symmetry (p. 591)
Center of rotational symmetry (p. 591)
Size transformation (p. 594)
Center of size transformation (p. 594)
Scale factor (p. 594)
Similarity (p. 595)
Topological transformation (p. 596)
Topologically equivalent (p. 597)

Section 11.2
Pattern (p. 602)
Tessellation (p. 603)
Regular tessellation (p. 604)
Semiregular tessellation (p. 606)

Section 11.3
Golden triangle (p. 616)
Pentagram (p. 617)
Golden rectangle (p. 617)
Star polygon (p. 617)
Regular star polygon (p. 617)
Star-shaped polygon (p. 619)

Section 11.4
Polyhedron (p. 626)
Regular polyhedra (p. 626)
Euler's formula (p. 627)
Prism (p. 628)
Pyramid (p. 630)
Cylinder (p. 631)
Cone (p. 631)
Sphere (p. 631)
Reflectional symmetry (p. 632)
Plane of symmetry (p. 632)
Rotational symmetry (p. 632)
Axis of rotational symmetry (p. 632)
Pattern or planar net (p. 634)
Shadow geometry (p. 635)
Vanishing point (p. 637)
Perspective drawings (p. 637)

CHAPTER REVIEW

Concepts and Skills

1. Describe the five regular polyhedra and explain how they are named.
2. Give the dimensions of a window that is a golden rectangle.
3. Show that Euler's formula holds for a square pyramid.
4. Draw a pattern of squares that could be used to form a cube.
5. Draw the front, side, and top views of the following object.

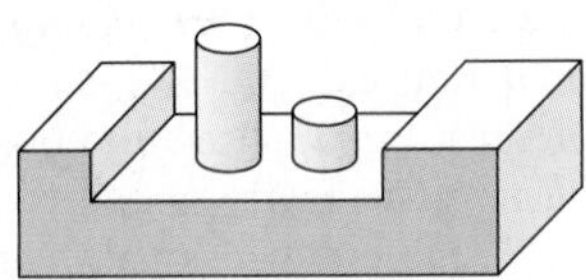

6. Draw a cube in perspective.
7. Draw a sketch of the star polygon $\{\frac{7}{3}\}$. Give another symbol for this same star polygon. Is it a regular polygon?
8. A six-pointed star-shaped polygon has a point angle of 45°. What is the measure of its dent angle?
9. Describe the symmetry properties of the following figures.

(a)

(b)

10. How many planes and axes of symmetry does a right circular cylinder have?
11. Name three regular polygons that will tessellate the plane.
12. Draw a small portion of a semiregular tessellation.
    a. Tell why it is semiregular.
    b. What is true about a tessellation that utilizes a combination of regular polygons but isn't semiregular?
13. Use tracing paper to find the image of the quadrilateral shown for the following transformations.

a. T_{AB}
b. $R_{A,90}$
c. $M_{\overleftrightarrow{AB}}$
d. $S_{E,2}$
e. $T_{AB} * R_{B,180°}$

14. Consider the figure:

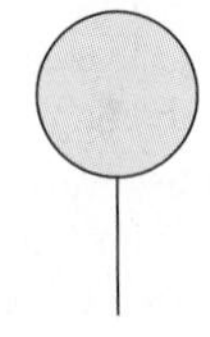

a. If it is drawn on a balloon that can be stretched, shrunk, and twisted, draw two different shapes that could be produced.
b. If it is made of wire, draw two different shapes that its shadow could take.

Reasoning and Problem Solving

15. If you marked the midpoints of the edges of a cube and sliced off all its corners through the midpoints of its edges, how many and what type of faces would the truncated figure have?

16. Do you believe the following generalization about the measure of a point angle of a star polygon is true?

The measure of a point angle of the star polygon $\left\{\frac{n}{d}\right\}$ is $\frac{(|n - 2d|180°)}{n}$.

Support your belief in writing, giving evidence.

17. How do the axes of rotational symmetry of an octahedron compare to the axes of rotational symmetry of a cube?

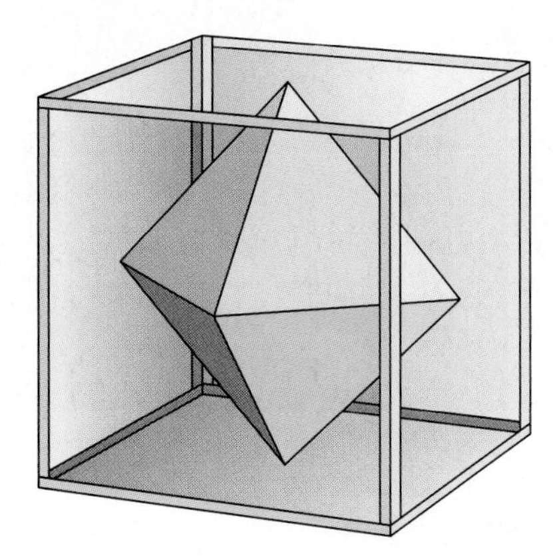

18. The Fancy Quilt Problem. Stacy wants to make a quilt that uses a 12-pointed star-shaped polygon that tessellates with equilateral triangles. To make the star, she needs to know the measure in degrees of the point angles and the dent angles of the star-shaped polygon. If possible, supply this information and sketch the quilt pattern. If not, explain what additional information you need in order to do so.

19. a. How are congruence transformations and size transformations alike?
b. How are they different?

20. Draw two topologically equivalent figures that look as different as you can make them. Justify why they are topologically equivalent.

21. Measuring Trees Problem. Jared wants an estimate of the height of some of the trees near his lake house. Use ideas of size transformations to design a plan for him to use his height to estimate accurately a tree's height without directly measuring the tree.

22. The Open-Top Box Pattern Problem. A manufacturer found that the machines in his factory could make open-top boxes most efficiently if the five-square box pattern had no more than two squares in a row and had reflectional symmetry. He could also manufacture the pattern more easily if it would tessellate the plane. Assume that you are the engineer given the task of analyzing all 12 possible open-top box patterns, find a pattern or patterns that meet the conditions, and verify that you have found the correct pattern or patterns.

Alternative Assessment

23. Use paper cutouts or GES to make a tessellation based on
a. slide images.
b. turn images.
c. flip images.

Write a paragraph for each one, describing your procedure.

24. Choose five different types of familiar rectangular objects that you estimate to be in the shape of a golden rectangle and analyze them to see how close they are to being a golden rectangle. Devise a technique to rate the closeness of rectangles to golden rectangles.

12 Measurement

Chapter Perspective

Measurement is an important part of the way people communicate their ideas of size to others. Questions such as "How long is it?" and "How big is it?" can be answered using concepts from measurement. For example, the surveyor pictured above needs to determine certain lengths as he surveys a site. Measurement involves approximation, so it's necessary to be aware of how numbers can be interpreted to indicate the ranges within which measurements are accurate. Also, the properties of geometric figures that you have learned provide the basis for understanding the formulas that are used to find the distances around planar figures, to find the areas of those figures, and to find the volumes of three-dimensional figures.

In this chapter we consider what it means to measure something, we examine various measurement systems, and we develop formulas that can be used in the measurement process. These formulas involve length, area, and volume of various objects. We also consider the effect of increasing a particular dimension of an object on the object's area and volume.

Connection to the NCTM Standards

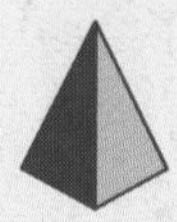

Measurement is a key concept in mathematics that appears throughout the school curriculum. As stated in the NCTM *Principles and Standards for School Mathematics* (2000), instructional programs from prekindergarten through grade 12 should enable all students to

- understand measurable attributes of objects and the units, systems, and processes of measurement;
- apply appropriate techniques, tools, and formulas to determine measurements *(p. 44)*.

As pointed out in the *Principles and Standards,* "Measurement helps connect ideas within areas of mathematics and between mathematics and other disciplines" *(p. 171)*. The *Principles and Standards* are very clear that measurement is a fundamental concept for children of all ages.

Connection to the PreK–8 Classroom

In grades PreK–2 children naturally raise questions that involve measurement, such as "How much did I grow last year?" or "How many jelly beans are in the jar?" Children's fascination about the size of real-world objects provides a basis for investigating many measurement concepts.

In grades 3–5 children study the length, area, weight, and volume of various geometric figures and solids. They encounter the standard units in the English and metric systems.

In grades 6–8 students continue to explore measurement concepts involving the perimeter, area, and volume of various figures and solids. They understand relationships among units and convert from one unit to another unit within the same system. They also study the effect on the area and volume of a figure when the dimensions are changed.

Section 12.1 THE CONCEPT OF MEASUREMENT

- The Measurement Process
- Types of Measures and Units Used
- Nonstandard Units of Measure
- Standard Units of Measure
- Measuring Time and Temperature

In this section we consider what it means to measure something and what is involved in the measurement process. We apply this process to determine length, area, volume, mass, temperature, and time. We address the reasons for having standardized units of measure. Finally, we consider the English and metric systems of measurement. In Mini-Investigation 12.1 you are asked to consider the various ways in which you commonly use measurement.

Talk about those situations with others in your group.

Mini-Investigation 12.1 **Making a Connection**

What are three different situations you have encountered that required measuring something?

The Measurement Process

The **measurement process** has three basic components. First, there has to be an entity to be measured. Second, a unit of measure must be chosen to apply to the entity being measured. Third, the number of units of measure that the entity contains must be determined. Determination of the number of units is called the *measurement of the object.* The following schematic diagram indicates how this measurement process works.

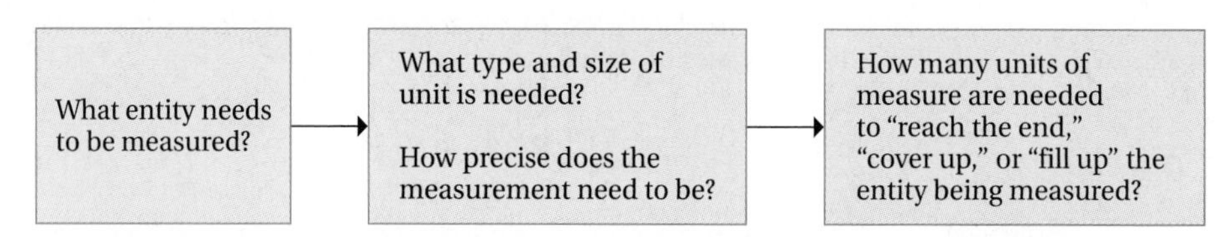

Measurement is the assignment of a number, called a *measure,* to the entity being measured. For example, suppose that the entity to be measured is the length of $\overline{AB}$.

A ———————————————————— B

Choose a unit segment such as ___ and decide how many of these units can be laid end to end to reach from A to B.

A |__|__|__|__|__|__|__|__|__|__|__|__|__|__|__| B

In this case, you can count to determine that $\overline{AB}$ contains 15 units. So the number 15 is assigned as the measure of the length of $\overline{AB}$, using the given unit.

Types of Measures and Units Used

When you measured $\overline{AB}$, you were measuring **length,** or the amount of distance between two points or objects. When you measure a rectangle by finding how many square units it takes to cover it, you are measuring **area.** When you measure a box by finding how many cubic units it takes to fill it, you are measuring **volume.**

Other types of units that you commonly encounter are **mass,** or how heavy or light something is; **temperature,** or how warm or cold something is; and **time,** or how long it takes for something to happen. Among the many other units of measure are those used in science: coulomb (amount of electrical charge), joule (amount of energy), and hertz (number of cycles per second of a wave-form). In this chapter, we focus primarily on the units of length, area, and volume.

The nature of the entity to be measured determines the type of unit that is needed, that is, a unit of length, area, volume, mass, temperature, or time. Table 12.1 gives some characteristics of objects to be measured and the types of units needed to measure them.

TABLE 12.1

Object	Type of Unit
Distance between two towns	Length
A person's height	Length
Carpet needed to cover a floor	Area
The amount of land owned by a farmer	Area
Amount of water to fill a swimming pool	Volume
Amount of liquid medication for a child	Volume
Amount of weight in an elevator	Mass
How warm it is outside	Temperature

Once you have determined the type of unit, you need to determine the size of unit that you want to use. If you wanted to measure the distance between two cities, you would use a longer unit of length rather than a shorter one. Example 12.1 illustrates choosing an appropriate type of measure and unit.

Example 12.1

Choosing the Type of Measure and Unit

In each of the following cases, indicate which type and size of unit would be most appropriate. Choose among length, area, and volume for the type of unit and small or large for the size of the unit.

a. The width of a finger
b. The amount of land on a ranch
c. The amount of perfume in a container
d. The size of a postage stamp

Solution

a. Length, small
b. Area, large
c. Volume, small
d. Area, small

Your Turn

Practice:

a. Identify an object that you would measure by using a short unit of length.
b. Identify an object that you would measure by using a large unit of volume.
c. Identify an object that you would measure by using a large unit of area.

Reflect: Why is the size of a unit an important consideration when you are making a measurement? ■

The length, area, or volume of an object doesn't change when different units of measure are used. What does change is the measure, that is, the number of units required to cover or fill the object. However, as the size of the unit increases, the measure of the object decreases. For example, if unit B is twice the size of unit A, an object with a measure of 12 units, using unit A, would have a measure of 6 units, using unit B. This result makes sense because a larger unit of measure would take fewer to cover or fill an object than a smaller unit of measure.

Nonstandard Units of Measure

The cartoon suggests that choosing widely accepted, standard units for measurement isn't always necessary. In fact, many useful measurements are made with handy, individually created units.

For Better or For Worse® **by Lynn Johnston**

In Mini-Investigation 12.2 you are asked to find the area of a figure, using a nonstandard unit of area.

Talk about how your answer would change if a legal-sized or note pad–sized sheet of paper were used.

Mini-Investigation 12.2 **Solving a Problem**

If the unit of area is a regular sheet of typing paper, about what is the area of one side of a typical door?

As suggested in Mini-Investigation 12.2, any planar figure can serve as a unit of area. Also, any object having length can be used as a unit of length, and any object that fills space can serve as a unit of volume.

Throughout history different nonstandard units of measure have been used. In 1 Samuel 17:4 it is written that "And there went out a champion out of the camp of the Philistines, named Goliath of Gath, whose height was six cubits and a span." The units cubit and span are depicted in Figure 12.1.

According to the Old Testament account, Goliath was tall but we can't be sure just how tall because of the ambiguity of the lengths of a cubit and a span. For the average man today, a cubit is a little more than 1.5 feet and a span would be less than one-half that. Applying those measures would mean that Goliath was more than 9 feet tall. However, a cubit in ancient times wasn't as long as a cubit is today because people were much smaller then. So the actual height of Goliath is unknown.

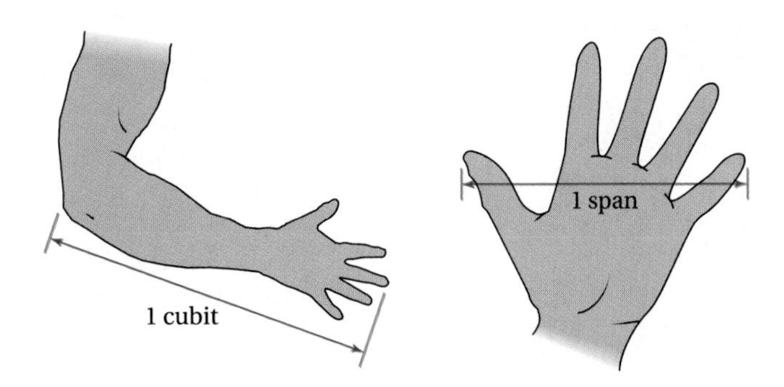

FIGURE 12.1 | Nonstandard units of length.

Although nonstandard units are useful sometimes, their widespread use results in miscommunication, frustration, and potential deceit. As human beings developed various types of systems, they also gradually developed standard units for measuring entities in those systems.

Standard Units of Measure

Because measurement is such a basic part of our lives when we buy something or when we order a product of certain dimensions, communication regarding the sizes of objects is of utmost importance. In order to ensure that that communication is clear and not ambiguous, measurements must be standard, that is, measures must mean the same thing to everybody. Two standard measurement systems that have been developed and agreed upon are the English system and the metric system. We consider each of these systems in the following subsections.

The English System. In the **English system** measures of length include the inch, foot, yard, and mile. A yard was originally introduced as the distance from the tip of the king's nose to the end of his extended arm. A foot originally represented the length of the king's foot. These definitions provided people with the same units of measure, but they didn't standardize the length of the units. Consequently, a standardized unit of length was later defined in terms of the length of a meter from the metric system. A meter is defined as the length of a platinum bar at 0° C that represents a small part of the distance along the prime meridian. In 1866 an Act of Congress defined a yard as $\frac{3600}{3973}$ of a meter. Although the English system isn't as efficient to use as the metric system, it is the system most familiar to people living in the United States. The most frequently used units of length and area in this country are the following.

Length		Area	
Unit	**Equivalents**	**Unit**	**Equivalents**
1 yard (yd)	3 feet (ft)	1 square foot (sq ft, or ft^2)	144 square inches (sq in., or in.^{2})
1 foot (ft)	12 inches (in.)	1 sq yd (or yd^2)	9 ft^2
1 mile (mi)	1760 yards (yd), or 5280 feet (ft)	1 acre	43,560 ft^2
		1 square mile (sq mi, or mi^2)	3,097,600 yd^2, or 27,878,400 ft^2

Estimation Application

The Milk Drinkers

If the average cow gives 2 gal of milk a day, estimate how many cows are needed to produce the milk drunk in the United States in a single day. Show the data you used to make the estimate, the source of the data, and how you made the estimate.

The units of length can also be used to create units to measure volume, such as cubic foot (ft^3), cubic yard (yd^3), and cubic mile (mi^3); however, this last unit of volume is rarely used. Typical everyday measures of volume are referred to as measures of capacity and include cup, pint, quart, and gallon. Equivalent measures of these quantities are given in the following table.

Capacity

Unit	Equivalents
1 pint (pt)	2 cups
1 quart (qt)	2 pt
1 gallon (gal)	4 qt

Weight

Unit	Equivalents
1 pound (lb)	16 ounces (oz)

Example 12.2 involves the conversion between units of measurement.

Example 12.2 Converting One Unit to Another

Henry Sandstone is selling a rectangularly shaped lot that measures 120 ft × 150 ft. Elizabeth Johnson says that the lot contains 6000 sq yd of land. Monica Rembert says, No way—and claims that the lot has only 2000 sq yd of land. Who is correct and why?

Solution

Ms. Rembert is correct. We can convert each measure to yards, and then the area is 40 yd × 50 yd, or 2000 sq yd. Alternatively, we could calculate the area in square feet, obtaining 18,000 sq ft. If we divide by 9, we convert this quantity to square yards, obtaining 2000 sq yd. Ms. Johnson divided by 3, not by 9.

Your Turn

Practice: How many square inches are in a square yard?

Reflect: If you were going to determine the area of a typical living room, would you use square inches, square feet, or square yards? Why? ■

The Metric System. Most countries other than the United States—and most manufacturers in the United States—use the Systeme International d'Unites (SI), or **metric system.**

One reason that the metric system is easier to use than the English system is its close relation to the decimal system of numeration. For example, in the metric system the prefixes given in the following table are used along with three basic units—the meter, m (measure of length), the liter, l (measure of capacity), and the gram, g (measure of mass)—for most common measurements. Note that the value of each prefix is $\frac{1}{10}$, or 0.1, as large as the one before it, much like the place values in the decimal system.

Basic Metric Units		Equivalent Units
kilo-	(k)	1000 basic units
hecto-	(h)	100 basic units
deka-	(da)	10 basic units
deci-	(d)	0.1 basic unit
centi-	(c)	0.01 basic unit
milli-	(m)	0.001 basic unit

For example, 1 cm is 0.01 m, and 1 km is 1000 m. The metric system is efficient because the prefixes allow units of different sizes to be created from one basic unit. The most commonly used prefixes are kilo, centi, and milli, although the others are used as appropriate. Example 12.3 demonstrates how metric prefixes can be used to compare a measure in one metric unit with a measure in another unit.

Example 12.3 Comparing Metric Measures

Fill in the missing amounts.

a. 3 km = ___ m
b. 0.5 g = ___ mg
c. 15 cl = ___ l

Solution

a. 3×1000, or 3000 m
b. 0.5×1000, or 500 mg
c. 15×0.01, or 0.15 l

Your Turn

Practice:

a. How many kg are in 250 g?
b. How many mm are in 2 km?

Reflect: Why are comparisons between different measures of length in the metric system easier to make than comparisons between different measures of length in the English system? ■

Mini-Investigation 12.3 lets you extend your understanding of metric units by considering the relationship between two important units of area.

Write a statement on how to convert the number of square meters in a housing development to the number of square kilometers in the development.

Mini-Investigation 12.3 Using Mathematical Reasoning

What fraction of a square kilometer is contained in a square meter?

You can get a feel for the size of the basic metric units by comparing them to familiar English system units, or to familiar everyday objects.

Meter—a little longer than a yard
Kilometer—about $\frac{3}{5}$ of a mile
Centimeter—about the width of a fingernail on the little finger
Millimeter—the thickness of a dime
Liter—a little more than a quart
Milliliter—about a teaspoon of water
Gram—about the same mass as two raisins or a paper clip
Kilogram—about 2.2 pounds
Milligram—about the mass of one human hair

Centimeter

Square centimeter

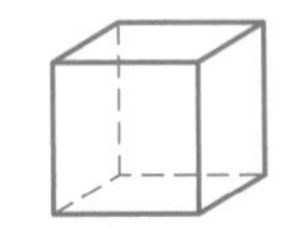

Cubic centimeter

FIGURE 12.2 | Metric units related to the centimeter.

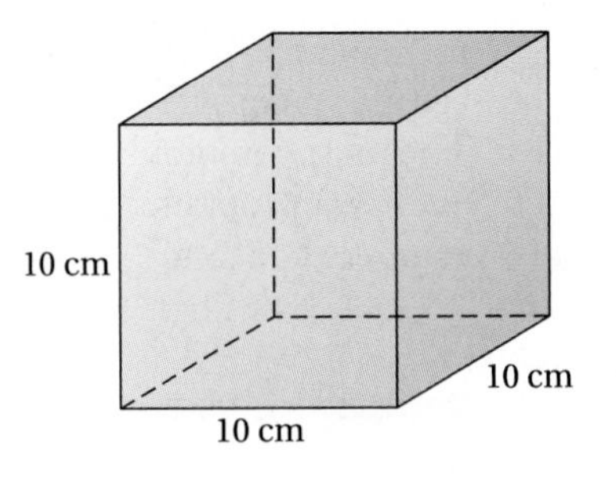

$1\ dm^3 = 1\ l.$

FIGURE 12.3 | Metric units: Relating the liter and the cubic decimeter.

Connections in the Metric System. The metric system was created to connect explicitly the measures of length, area, and volume. For example, the centimeter, a unit of length, may be used to define a unit of area and a unit of volume as shown in Figure 12.2. You can use the same means of creating units with other units of length, such as meter, square meter, and cubic meter as units of length, area, and volume.

There are also connections between measures of volume, measures of capacity, and measures of mass in the metric system. For example, a cubic decimeter has the same volume as a liter. That is, a cubic container 10 cm (1 dm) on a side would hold 1 l of water, as illustrated in Figure 12.3.

Furthermore, 1 l of pure water has a mass of 1 kg. This connection between volume (cubic decimeters), capacity (liters), and mass (kilograms) is a central feature of the metric system. In Example 12.4 we explore how the idea of capacity and volume relate to each other.

Example 12.4 Connecting Capacity and Volume

Explain why a 1-cm^3 container would hold 1 ml of water.

Solution

$1\ dm^3$ is 10 cm on a side and so contains $10 \times 10 \times 10$, or $1000\ cm^3$. It also holds 1 l, as previously shown. So, if a 1000-cm container holds 1 l, $\frac{1}{1000}$ of the container, or $1\ cm^3$, would hold $\frac{1}{1000}$ of a liter, or 1 ml.

Your Turn

Practice: Explain why water in a half-liter bottle would fit into a 500-cm^3 container.

Reflect: What two general types of measures does the connection in the example problem tie together? ■

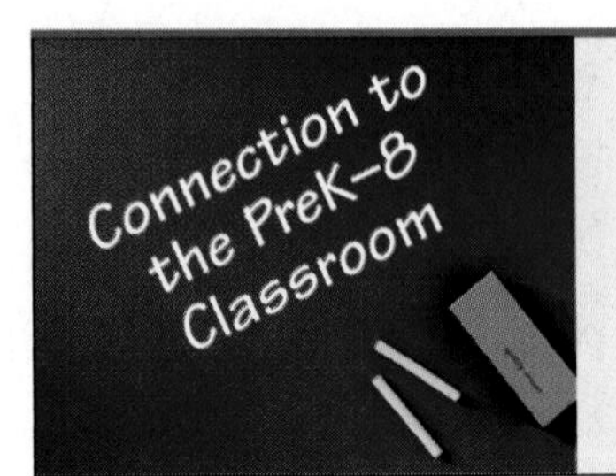

If we expect children to become proficient in using any system of measurement, they must be provided many opportunities for estimating and measuring objects. Children often are interested in body measurements, including height and weight, head size, and arm length. This interest provides a natural setting for creating a variety of measurement activities in either the English or metric system.

Measuring Time and Temperature

Two of the most frequently asked questions are "What time is it?" and "What is the temperature?" The units used most frequently to measure time are the hour (hr), minute (min), and second (s), with 1 hr equaling 60 min and 1 min equaling 60 s. In the United States, the time of day is based on the hours, minutes, and seconds from midnight to noon (A.M.) and from noon to midnight (P.M.). Thus, when you say that you will meet someone at 8 o'clock, you have to specify whether you mean A.M. or P.M., unless the context is clear. In Europe and in the military, time is measured on the basis of 24 hr in a day. Thus 8 A.M. would be 0800 (or just 8 in some places), and 8 P.M. would be 2000 (or just 20 in some places). This method makes the A.M. and P.M. suffixes unnecessary, although they are still used in some contexts.

Other units of time consist of the day, week (wk), month (mo), and year (yr). However, for really long periods of time, still other units are used, such as the decade (10 yr) and the century (100 yr). Astronomers, who are interested in how long it takes light (which travels approximately 180,000 mi/s) to travel from a star to earth, use the unit *light-year*, which is the distance light travels in a year. Thus a star that is 10 light-years away is $10 \times 180{,}000 \times 60 \times 60 \times 24 \times 365$, or approximately 5.7×10^{13} mi from earth! Keep in mind that a light-year is a measure of distance, based on time.

The two main scales used to measure temperature are the **Celsius scale** (metric) and the **Fahrenheit scale** (English). The Celsius scale is named after the Swedish astronomer Anders Celsius (1701–1744). It originally was called the centigrade scale because it is divided into 100 units (degrees or °). The Fahrenheit scale is named after Gabriel Fahrenheit (1686–1736), a German physicist. Both scales measure how warm or cold something is, so there is a connection between the measures of these two scales. Figure 12.4 demonstrates this

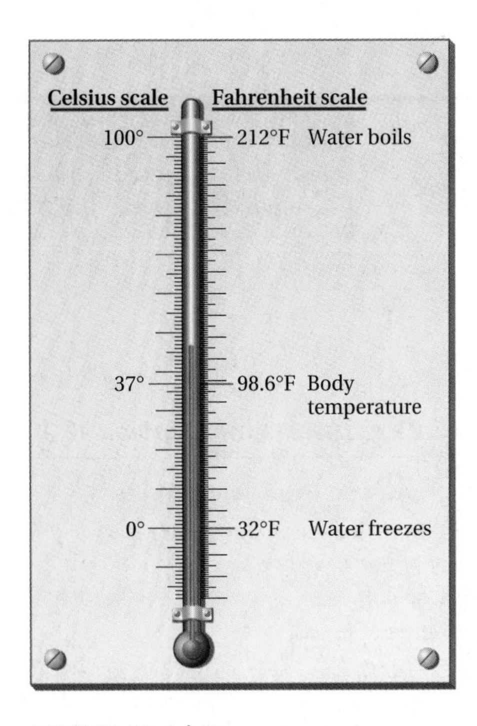

FIGURE 12.4 | Temperature scales.

relationship. The relationship between these two measurement scales may be expressed in the following ways.

$$F = \frac{9}{5}C + 32, \quad \text{and} \quad C = \frac{5}{9}(F - 32)$$

The expression on the left is used to change Celsius to Fahrenheit; the one on the right is used to change Fahrenheit to Celsius. Note that if $C = 100°$, then $F = (\frac{9}{5})100 + 32$, or 212°. Similarly, if $F = 32°$, then $C = (\frac{5}{9})(32 - 32)$, or 0°. Example 12.5 shows more about how to use these formulas.

Example 12.5 Connecting Celsius and Fahrenheit Formulas

a. How many degrees Celsius would be equivalent to 50° F?
b. How many degrees Fahrenheit would be equivalent to 30° C?

Solution

a. Use the formula $C = \frac{5}{9}(F - 32)$. Substituting gives $C = \frac{5}{9}(50 - 32)$, or 10°.
b. Use the formula $F = \frac{9}{5}C + 32$. Substituting gives $F = \frac{9}{5}(30) + 32$, or 86°.

Your Turn

Practice: What would be a reasonable Celsius temperature inside a house?

Reflect: If the temperature reading displayed on a bank on a summer day is given in both Celsius and Fahrenheit, which reading will have the greater number? ■

Mini-Investigation 12.4 helps you see how the relationship between the Celsius and the Fahrenheit scales can be used to estimate temperatures.

Write a description of how someone who knew that 0° C = 32° F and the relationship previously described could estimate the temperature in Fahrenheit, given the temperature in Celsius.

Mini-Investigation 12.4 Using Mathematical Reasoning

For each 5° increase in Celsius, what is the corresponding increase in Fahrenheit degrees?

Problems and Exercises for Section 12.1

A. Reinforcing Concepts and Practicing Skills

Measure the following objects by first using an average-sized paper clip and then using the length of your index finger.

1. Width of a paperback book
2. Length of a standard sheet of paper

Indicate whether the mass of the following objects could best be measured in grams or kilograms.

3. A safety pin
4. An elementary school student
5. A paperback novel
6. A grown collie
7. A 19-in. color television set

Complete the following statements by using milligram (mg), gram (g), or kilogram (kg).

8. A penny has a mass of about 3....
9. A young child has a mass of 30....

10. A bucket of water has a mass of 10....

11. An aspirin has a mass of 5....

12. An average-sized automobile has a mass of 1000....

13. A quarter has a mass of about 4....

Complete each of the following.

14. 6 qt = ___ pt

15. 10 pt = ___ gal

16. 4 sq ft = ___ sq in.

17. 12 sq ft = ___ sq yd

18. 150 mm = ___ cm

19. 25 l = ___ ml

20. 5 kg 200 g = ___ g

21. 1500 cm = ___ km

22. 2 m 300 mm = ___ cm

23. 0.012 l = ___ ml

24. 0.21 km = ___ m

25. 2000 mg = ___ kg

26. 600 cm^3 = ___ l

27. 10 cm^2 = ___ mm^2

Identify an object or mark off a distance that you think would have the following measures. Measure the objects or distance to determine how good your selection or estimate was.

28. 10 cm

29. 3 m

30. 10 g

31. 1 lb

32. 3 in.

33. 6 yd

34. 5 kg

35. 500 ml

Make the following conversions to the nearest degree.

36. 30° C = ___ ° F

37. −20° C = ___ ° F

38. 95° C = ___ ° F

39. 0° C = ___ ° F

40. 120° C = ___ ° F

41. 20° C = ___ ° F

42. 0° F = ___ ° C

43. −24° F = ___ ° C

44. 70° F = ___ ° C

45. 40° F = ___ ° C

46. 120° F = ___ ° C

47. 86° F = ___ ° C

Estimate the measure of the following objects by using the indicated units.

48. Length of a new pencil (in cm)

49. Length of a four-door sedan (in m)

50. Amount of liquid in a coffee mug (in ml)

51. The mass of a gallon of water (in kg)

52. The mass of a paperback book (in g)

Indicate whether you think that each of the statements in Exercises 53–57 is reasonable.

53. The water temperature is 30° C, which is too chilly to go in.

54. The temperature is 10° C, which means that the ice will be melting.

55. Unless the temperature gets up to 40° C, I don't think it's warm enough to play tennis.

56. The weather is beautiful—sunny and warm at 25° C.

57. At −5° C, the ice shouldn't melt.

58. Determine the normal body temperature (98.6° F) in degrees Celsius.

B. Deepening Understanding

59. Determine the area of an 8-in. × 10-in. rectangle by using the following units of area.

a. a 1-in. × 1-in. square

b. a 1-in. × 2-in. rectangle

c. a right triangle with legs 1 in. × 1 in.

60. If the unit of area is the equilateral triangle, determine the area of the parallelogram.

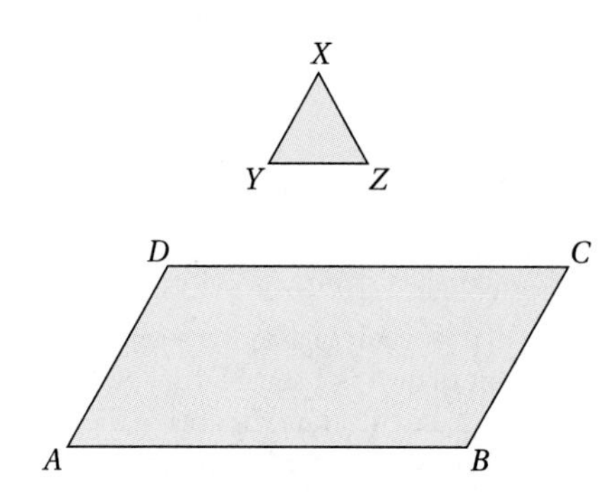

61. If the unit of area is the right trapezoid, determine the area of the trapezoid.

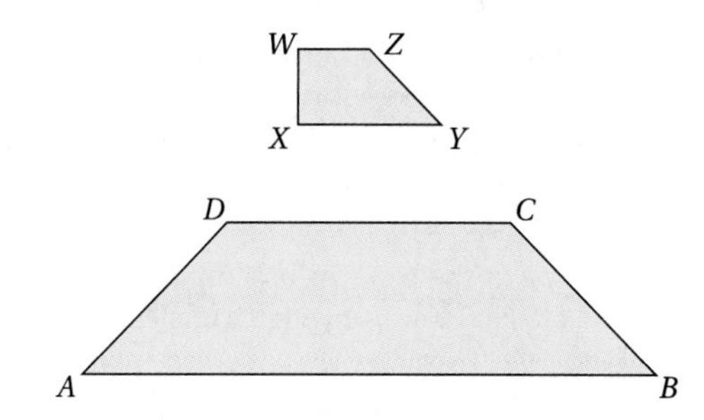

62. Assume that one square equals 1 unit of area. Determine the area of the figure shown on the geoboard.

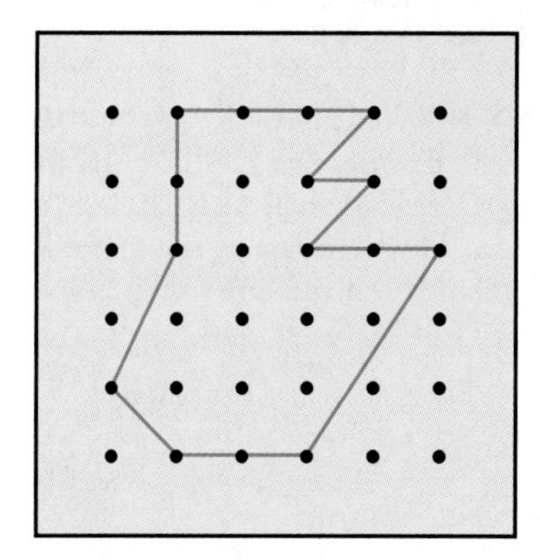

Indicate whether you think each statement in Exercises 63–67 is likely to be true. If not, provide a more reasonable measure.

63. A professional basketball player could be 4 m tall.

64. A young boy could have a mass of 25 kg.

65. An average-sized house contains about 200 m^2 of area.

66. The average person walking briskly can walk 7000 yd in about 30 min.

67. The height of a three-story building is 15 m.

68. Estimate the area of the following figure in mm^2.

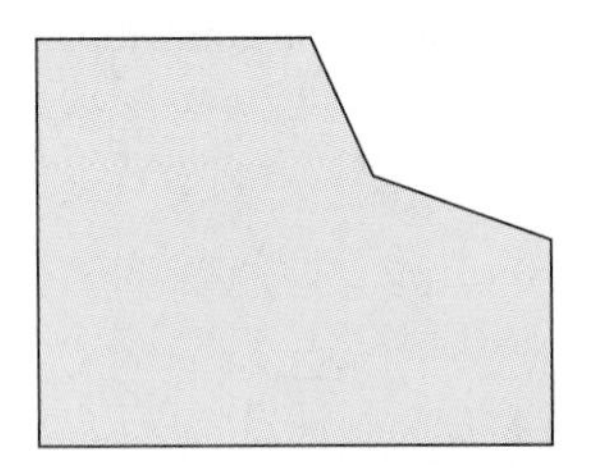

69. Which is colder, $-10°$ C or $-10°$ F? Why?

70. Suppose that the unit of area is a square 2 in. × 2 in. and another unit of area is a square 1 in. × 1 in. If these two units were used to measure the area of the same rectangle, how would you expect the two measures to compare?

71. Discuss what happens to the measure of an object if the unit of length continues to double.

C. Reasoning and Problem Solving

72. If the temperature rises by 10° on the Celsius scale, by how many degrees does it rise on the Fahrenheit scale?

73. If the price of gasoline in Germany is DM 1.4/l, what is the equivalent price of gasoline in dollars per gallon? (Assume that US \$1 = DM 2.0.)

74. Jackie goes to the store to buy some meat. The meat costs \$6.20/kg. How many grams of meat could she buy for \$10?

75. If you were buying nuts, would you rather pay \$8/kg or \$3/400 g? Explain.

76. If a sheet of paper has a mass of about 10 g and there are 500 sheets of paper in a ream, what is the mass of a ream of paper in kilograms?

D. Communicating and Connecting Ideas

77. **Making Connections.** The graph of the line $F = (\frac{9}{5})C + 32$ is shown. (Each hash mark represents 20 units.) At what point on the line will C = F? What interpretation can you give to this point?

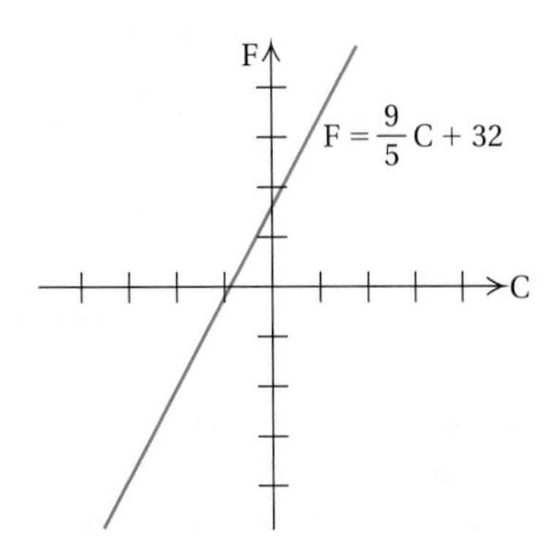

78. Consider the following sign. Write a rule that could give a driver an easy way to convert from kilometers to miles while driving.

79. Terry says that if the length of a unit is doubled, the object being measured will be half as long. How would you explain to Terry that he is wrong but that his thinking isn't totally incorrect?

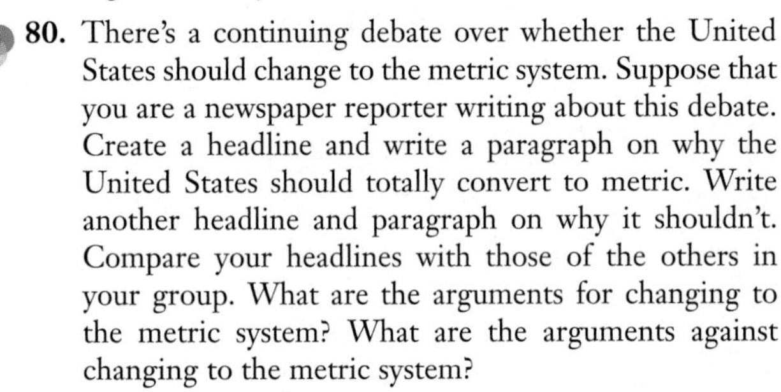

80. There's a continuing debate over whether the United States should change to the metric system. Suppose that you are a newspaper reporter writing about this debate. Create a headline and write a paragraph on why the United States should totally convert to metric. Write another headline and paragraph on why it shouldn't. Compare your headlines with those of the others in your group. What are the arguments for changing to the metric system? What are the arguments against changing to the metric system?

81. **Historical Pathways.** Two units of measure used in ancient times were the digit and the hand, shown below. Write a paragraph that describes misunderstandings that would arise if a king defined the kingdom's units of length as the digit and the hand of his 10-year-old son.

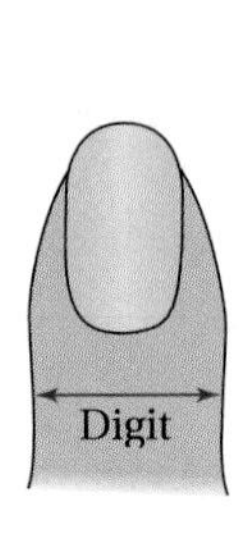

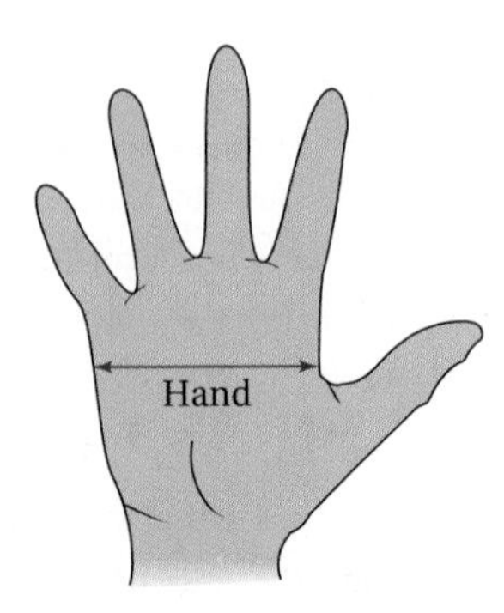

Section 12.2 MEASURING THE PERIMETER AND AREA OF POLYGONS

- Perimeters of Common Polygons
- Areas of Common Polygons
- Circumference and Area of Circles
- Comparing Perimeters and Areas of Similar Polygons

In this section we present formulas that are used to find the perimeter and area of rectangles, squares, parallelograms, triangles, trapezoids, and regular polygons. We also give formulas for finding the circumference and area of circles. We describe how to find the area of irregular polygons on the geoboard. Finally, we consider the relationship between the lengths of corresponding sides of similar polygons and the relationship between the perimeter and area of similar polygons.

Perimeters of Common Polygons

There are times when you will need to solve problems that involve finding the distance around a figure. For example, you would need to know the distance around a yard in order to buy enough materials to fence the yard. This idea suggests the following definition.

Definition of Perimeter of a Polygon

The **perimeter,** *P*, of a polygon is the sum of the lengths of each side of the polygon.

The most common planar (two-dimensional) figures for which perimeters are found are the rectangle, square, parallelogram, triangle, and some regular polygons. We consider how to find the perimeter for each of these types of figures and for other general polygons.

Perimeters of a Rectangle, Square, Parallelogram, and Triangle. We can find the perimeter of a square, rectangle, parallelogram, or triangle (Figure 12.5) by summing the lengths of its sides. In the case of the square, all four sides are equal, so the perimeter can be found by multiplying the length of one side by 4. In the case of a rectangle and a parallelogram the opposite sides are equal, thus making the perimeter twice the sum of the lengths of the two adjacent sides. For a triangle the perimeter is simply the sum of the lengths of the sides.

Recall that a regular polygon is a polygon whose angles are all congruent and whose sides are all congruent. Hence, the perimeter of a regular polygon is found by multiplying the length of one side by the number of sides. Thus we have the following formulas.

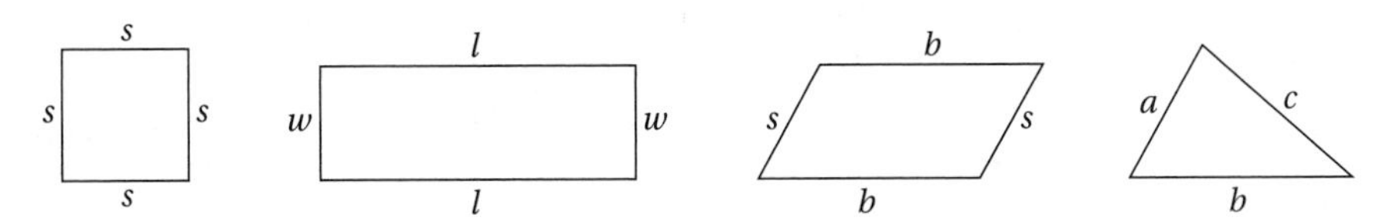

FIGURE 12.5

Formulas for the Perimeters of Selected Polygons

- **Square:** $P = 4s$, where s is the length of each side of the square.
- **Rectangle:** $P = 2(l + w)$, or $P = 2l + 2w$, where l and w are the length and the width of the rectangle.
- **Parallelogram:** $P = 2(b + s)$, or $P = 2b + 2s$, where b and s are the lengths of the consecutive sides of the parallelogram.
- **Triangle:** $P = a + b + c$, where a, b, and c are the lengths of the three sides of the triangle. If the triangle is equilateral, $P = 3s$, where s is the length of each side of the equilateral triangle.
- **Regular polygon:** $P = ns$, where n is the number of sides and s is the length of each side of the polygon.

Example 12.6 provides an opportunity to apply the perimeter formulas.

Example 12.6 Applying Perimeter Formulas

Find the perimeters of the following figures.

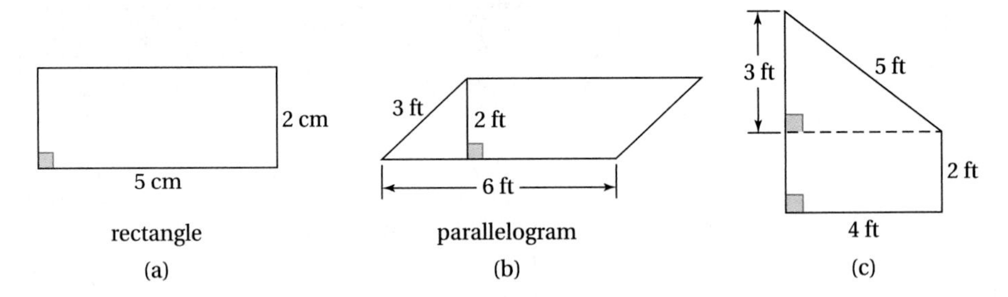

Solution

a. $P = 2(5 + 2) = 14$ cm
b. $P = 2(6 + 3) = 18$ ft
c. $P = (3 + 4 + 5) + (2 + 2) = 16$ ft

Your Turn

Practice: Find the perimeters of the following figures.

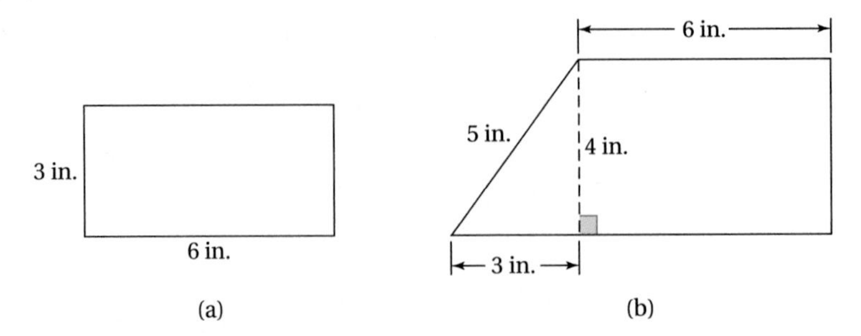

Reflect: Is any irrelevant information presented in the example and practice problems? Explain. ■

Areas of Common Polygons

In solving real-world or other types of mathematics problems, you will need to be able to find the number of units required to cover something—that is, its area. For example, if you wanted to know the amount of carpet needed to cover the floor of a room, you would need to find the area of the floor. Rather than counting units to find the area, you can use formulas that allow you to find the area more efficiently.

The most common polygons for which areas must be found are the rectangle, square, parallelogram, triangle, and trapezoid. We first develop formulas for the areas of these common polygons and then develop a formula for finding the area of regular polygons. We begin with the rectangle and the square.

Area of a Rectangle and Square. As you learned in Section 12.1, finding the area of a rectangle is a matter of determining the number of unit squares needed to cover the rectangle. In the simplest situation, think of counting the number of unit squares contained in the following rectangle.

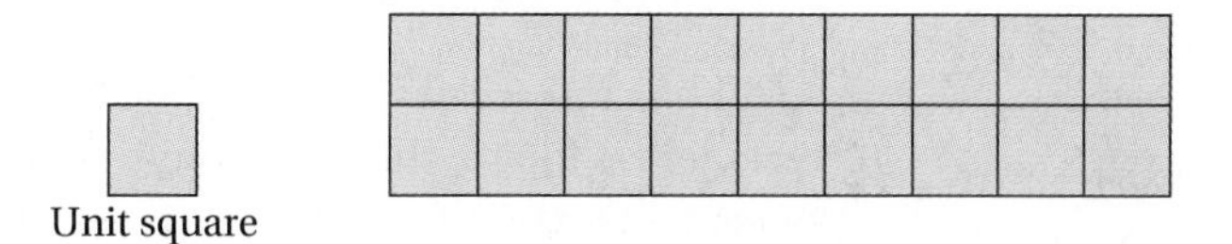

The area of the rectangle is 18 because 18 unit squares are required to cover the rectangle. But instead of counting you can multiply the length of the base (9 units) times the length of the width (2 units), or its height, to obtain the area. In general, you can find the area of any rectangle by multiplying its length by its width. In the case of the square, that involves multiplying the length of a side by itself. Thus we have the following formulas for finding the areas of rectangles and squares.

Formulas for the Areas of Rectangles and Squares

- **Rectangle:** $A = lw$, where l is the length and w is the width.
- **Square:** $A = s^2$, where s is the length of each side.

Example 12.7 shows how to apply these formulas.

Example 12.7

Finding the Area of a Rectangle

Find the area of a rectangle if the base is 2 cm longer than the height and the perimeter is 28 cm.

Solution

If the perimeter is 28, $l + w = 14$. Because $l - w = 2$, $l = 2 + w$. Hence $(2 + w) + w = 14$, or $2w + 2 = 14$ and $2w = 12$, by substitution and simplification. Thus $w = 6$, $l = 8$, and $A = 6 \times 8$, or 48 cm^2.

Your Turn

Practice: Find the area of a rectangle if its base is twice as long as its height and its perimeter is 36 ft.

Reflect: Can a square and a rectangle have the same area but different perimeters? Explain. ■

Area of a Parallelogram. The relationship between the area of a parallelogram and the area of a rectangle is shown in Figure 12.6, where we "cut off" the left corner of the parallelogram and placed it on the right side of the parallelogram to form a rectangle.

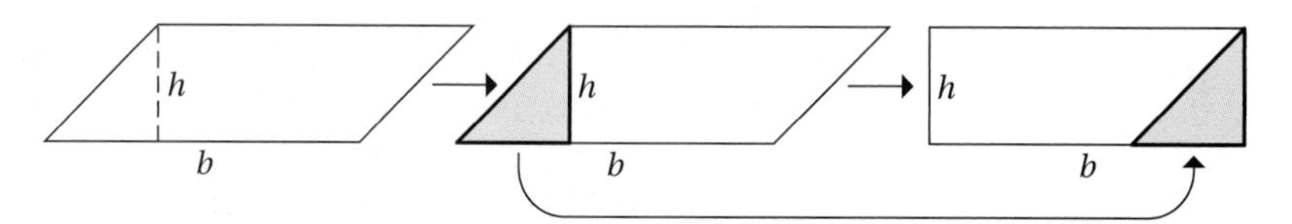

FIGURE 12.6 | Area of a parallelogram.

Thus the area of a parallelogram is found in the same way that the area of a rectangle is found, that is, by multiplying the length of the base, b, by the height of the parallelogram to that base. This approach gives the following formula.

Formula for the Area of a Parallelogram

$A = bh$, where b is the length of the base and h is the height to the base.

Example 12.8 demonstrates how to apply the formula for the area of a parallelogram.

Example 12.8 Finding the Area of a Parallelogram

Find the area of parallelogram $ABCD$.

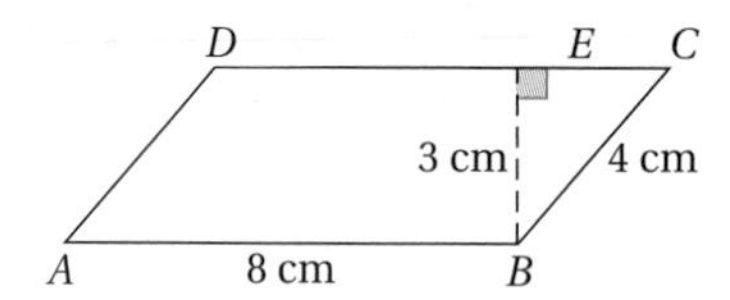

Solution

$$A = 8 \times 3, \text{ or } 24 \text{ cm}^2$$

Your Turn

Practice: Find the area of the parallelogram $ABCD$ if $AB = 10$ cm, $AD = 6$ cm, and $BE = 4$ cm.

Reflect: What information in the example problem isn't needed for finding the area of $ABCD$? Why? ■

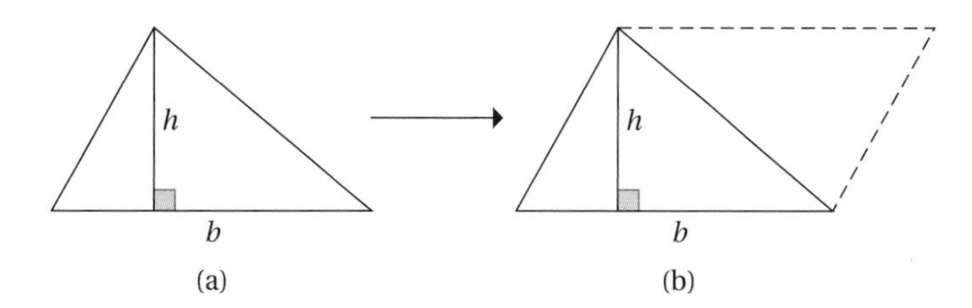

FIGURE 12.7 | Area of a triangle: Method 1.

Area of a Triangle. We can relate the area of a triangle to the area of a parallelogram or the area of a rectangle. Let's consider the parallelogram case first, with the help of Figure 12.7. We can use two copies of the triangle shown in Figure 12.7(a) to make a parallelogram, as shown in Figure 12.7(b). The area of the parallelogram is $A = bh$, so the area of the triangle is half that amount.

The diagrams shown in Figure 12.8 indicate how to arrive at the same conclusion by using a rectangle model. If we make a fold dividing the height of the triangle shown in part (a) in half (along the dotted line), we obtain the trapezoid shown in part (b). If we then make folds shown in part (c) (along the dotted lines), we obtain the rectangle shown in part (d).

Note that the base of the rectangle shown in Figure 12.8(d) is one-half that of the original triangle and that its height is one-half that of the original triangle (observe the way that the bases and the heights are folded). Thus the area of this rectangle is $\frac{1}{4}bh$. But, because of the nature of the folds, two of these rectangles would be needed to make the original triangle. Thus we need to multiply $\frac{1}{4}bh$ by 2, arriving at $\frac{1}{2}bh$ as the expression for the area of the triangle.

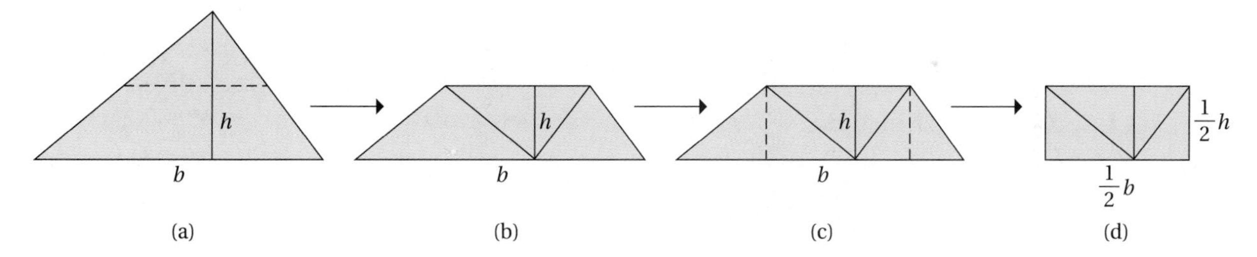

FIGURE 12.8 | Area of a triangle: Using a model rectangle.

An equilateral triangle has a special formula for finding its area. If each side of the triangle is s, we can use the Pythagorean theorem to express its height, and this height can be used to develop a general formula for finding its area. Consider the diagram and formula for its height, shown in Figure 12.9.

Using the formula $A = \frac{1}{2}bh$ and substituting for b and h, we get

$$A = \frac{1}{2}(s)\left(\frac{\sqrt{3}s}{2}\right).$$

Simplifying, we produce a formula for finding the area of an equilateral triangle. Thus we have the following formulas for finding the areas of triangles.

$$\left(\frac{1}{2}s\right)^2 + h^2 = s^2$$

$$\frac{s^2}{4} + h^2 = s^2$$

$$h^2 = \frac{3s^2}{4}$$

$$h = \frac{\sqrt{3}\,s}{2}.$$

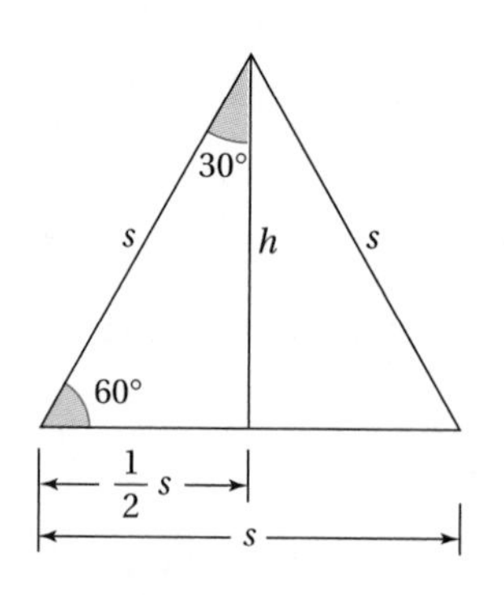

FIGURE 12.9 | Finding the height of an equilateral triangle.

Formulas for the Areas of Triangles

- **General triangle:** $A = \frac{1}{2}bh$, where b is the base and h is the height.
- **Equilateral triangle:** $A = \frac{\sqrt{3}s^2}{4}$, where s is the length of each side.

Any side of a triangle can serve as its base. The height is always the perpendicular to that base from the opposite vertex. Study the bases and heights of triangles shown in Figure 12.10. Note that in part (c) the height, or altitude, lies outside the triangle.

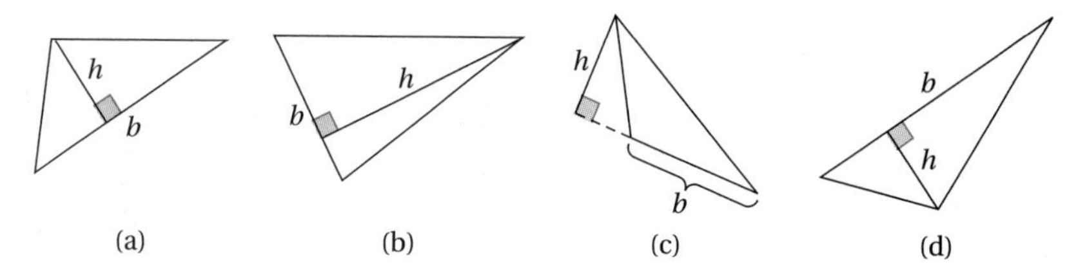

FIGURE 12.10

Example 12.9 illustrates how to apply the formulas for the area of a triangle.

Example 12.9 Finding the Area of a Triangle

Find the area of an equilateral triangle if each side is 4 cm.

Solution

The area of an equilateral triangle is given by the formula $A = \frac{\sqrt{3}s^2}{4}$, where s is the length of each side. For $s = 4$, $A = \frac{16\sqrt{3}}{4}$, or $A = 4\sqrt{3}$ cm^2.

Your Turn

Practice: Find the area of an equilateral triangle if each side is 6 in.

Reflect: If you know the perimeter of an equilateral triangle can you find its area? Explain. ■

In Mini-Investigation 12.5 you are asked to look for a relationship between the area of a triangle and a parallelogram that share a common base.

Write a statement about the heights of a triangle and a parallelogram that share a common base and have the same area.

Mini-Investigation 12.5 **Using Mathematical Reasoning**

If a triangle and a parallelogram share the same base and the triangle has twice as much area as the parallelogram, what can you conclude about their heights?

Area of a Trapezoid. The formula for the area of a trapezoid can be based on the area of a parallelogram. Consider Figure 12.11(a), in which trapezoid $ABCD$ with bases $\overline{AB}$ and $\overline{CD}$ is rotated to form trapezoid $C'D'A'B'$ that is congruent to $ABCD$ but is positioned differently. The two trapezoids are then placed together to make the large parallelogram shown in Figure 12.11(b).

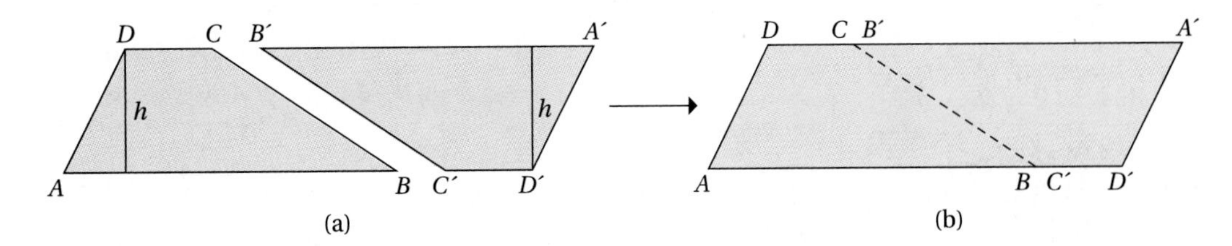

FIGURE 12.11 | Area of a trapezoid.

The area of the large parallelogram is given by the general formula $A = bh$, or

$$A(ADA'D') = (AB + C'D')h.$$

But $C'D' = CD$, so we have

$$A(ADA'D') = (AB + CD)h.$$

This formula represents the area of the two congruent trapezoids. Thus we need to divide $(AB + CD)h$ by 2, which gives the following formula for the area of a trapezoid. Note that the formula could be paraphrased as: The area of a trapezoid is the average of the lengths of the two bases times the height.

Formula for the Area of a Trapezoid

$A = \frac{1}{2}(a + b)h$, where a and b are the lengths of the two bases of the trapezoid and h is the height.

Many students become confused as to which sides of a trapezoid are the bases. Recall that the bases of a trapezoid are the parallel sides of the trapezoid. These parallel sides (bases) can be positioned in various ways, as shown in Figure 12.12.

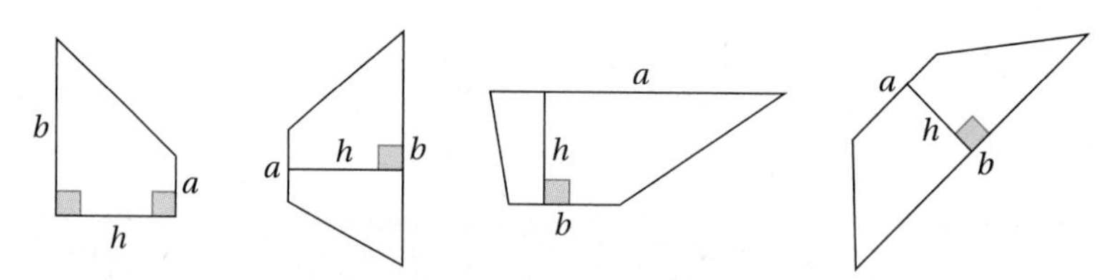

FIGURE 12.12

Example 12.10 shows how to apply the formula for the area of a trapezoid.

Example 12.10 Finding the Area of a Trapezoid

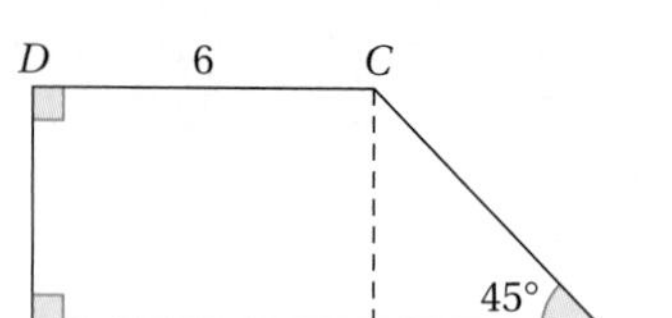

Find the area of the trapezoid shown on the left.

Solution

The bases are 10 and 6. We need to find the height. Triangle CEB is a $45°-45°-90°$ isosceles triangle. Because $EB = 4$, CE (the height) $= 4$. Hence

$$A(ADCB) = \frac{1}{2}(10 + 6)(4), \text{ or } 32 \text{ sq units.}$$

Your Turn

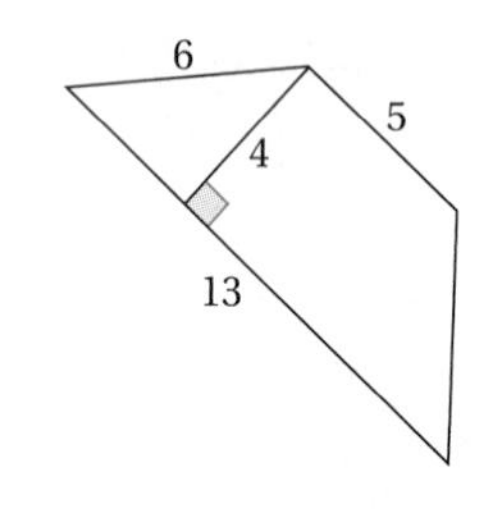

Practice: Find the area of the trapezoid shown on the left.

Reflect: Is any information shown on the diagram not needed for finding the area of the trapezoid? Explain. ■

In Mini-Investigation 12.6 you are asked to solve a real-world problem by using the formula for the area of a trapezoid.

Write about what would affect the accuracy of the answer if you used the formula for the area of a trapezoid to find the area of the lot.

Mini-Investigation 12.6 Solving a Problem

The Johnsons own a lot that has a creek running behind it. How could they find the area of the lot?

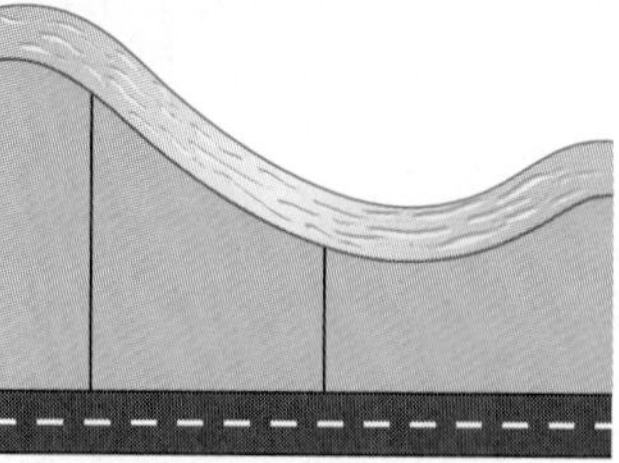

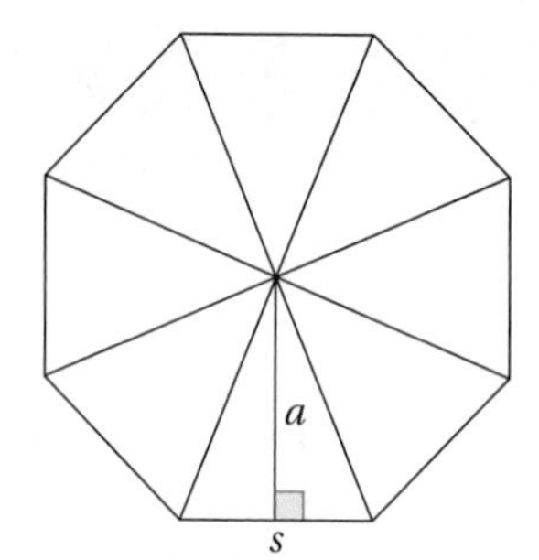

FIGURE 12.13

Area of a Regular Polygon. To develop a formula for finding the area of a regular polygon, let's consider the diagram of a regular octagon shown in Figure 12.13. The area of a regular polygon can be found by multiplying the number of triangles formed times the area of one such triangle. For example, the octagon can be divided into eight congruent triangles. To find the area of one of the triangles, we need to know its height (called the *apothem* and denoted a) and the length of side s. For each triangle, $A = \frac{1}{2}as$. Thus, for the entire octagon, $A = 8(\frac{1}{2}as)$.

Now suppose that we had a regular polygon of n sides rather than eight sides. The area of each triangle would still be $\frac{1}{2}as$, but instead of multiplying by 8 we would multiply by n. Thus the area of the regular n-sided polygon would be given by the formula $A = n(\frac{1}{2}as)$, or $A = \frac{1}{2}a(ns)$. But ns is the perimeter of the figure. Thus we have the following formula for the area of a regular polygon.

Formula for the Area of a Regular Polygon

$A = \frac{1}{2}aP$, where a is the apothem and P is the perimeter of the regular polygon.

Example 12.11 demonstrates how to apply the formula for the area of a regular polygon.

Example 12.11

Finding the Area of a Regular Polygon

Find the area of the regular polygon shown on the left. (Note: If the apothem is 12, the side length is close to, but not exactly, 10.)

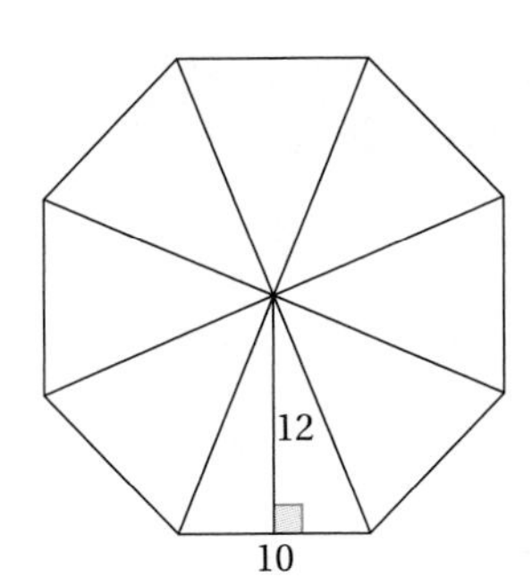

Solution

There are eight sides, so $P = 8(10)$, or 80 units. Because the apothem is 12, $A = \frac{1}{2}(12)(80)$, or 480 square units.

Your Turn

Practice: Find the area of a regular 10-sided figure if the apothem is 24 units and each side has a length of 15.6 units.

Reflect: What would happen to the area of a regular polygon in the example problem if the apothem remains the same but the number of sides keeps increasing? ■

Circumference and Area of Circles

In the preceding sections we discussed ways of finding the perimeters and the areas of various polygons. In this section we demonstrate how to find similar measures for a circle, namely, the distance around a circle and the area of a circle.

Circumference of a Circle. The distance around a circle is called the **circumference** of the circle. The concept of circumference is similar to that of perimeter except that circumference deals with curved lines and perimeter deals with straight lines. Figure 12.14 shows that, as the number of sides of an inscribed regular polygon increases, the perimeter of the regular polygon more closely approximates the circumference of the circle. The perimeter of the 12-sided regular polygon more closely approximates the circumference of the circle than does the perimeter of the regular octagon or the regular pentagon.

In Mini-Investigation 12.7 you are asked to examine the basis for the number π, which is used in finding the circumference of a circle.

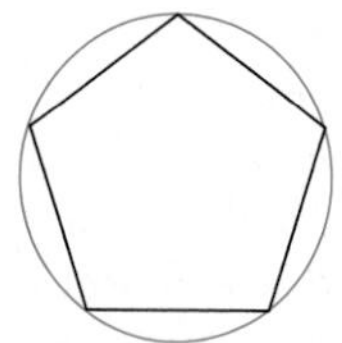
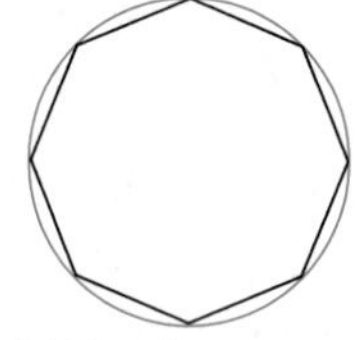

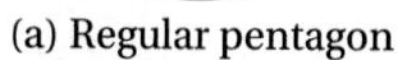

(a) Regular pentagon (b) Regular octagon (c) Regular dodecagon

FIGURE 12.14 | Circumference of a circle.

Write a general statement that you think would apply to the result of dividing the circumference of any circle by the diameter of that circle.

Mini-Investigation 12.7 Finding a Pattern

What can you conclude about the quotient of the measured circumferences of three different-sized cans divided by their measured diameters?

The result of Mini-Investigation 12.7 suggests that the ratio of the circumference of any circle to its diameter is always a number that is a little more than 3. This number is called **pi** and is written π. Pi is an irrational number, which means that it can't be written as an exact decimal or fraction. Note that π is the limiting value for the perimeter of regular polygons inscribed in circles with a diameter of 1 unit. Some commonly used approximations for π are 3.14, 3.1416, and $\frac{22}{7}$. Because $\frac{C}{d} = \pi$ for any circle, we have the following formula for the circumference of a circle.

Formula for the Circumference of a Circle

$C = \pi d$, where d is the length of the diameter of the circle, or $C = 2\pi r$, where r is the length of the radius of the circle.

Example 12.12 applies the procedure for finding the circumference of a circle to a real-world problem.

Example 12.12 **Problem Solving: The Highway Bypass**

A circular interstate highway bypass around a city has a radius of 10 mi from the city center. How many miles long is the bypass?

Solution

The radius of the circle is 10 mi. Thus the distance around or the circumference of the circular bypass is $C = 2\pi r = 2\pi(10) = 20\pi$, or about 62.8 mi.

Your Turn

Practice: Suppose that the bypass was 20 mi from the city center. How long would this bypass be?

Reflect: Suppose that only 300° of the 360° of the bypass with a radius of 10 miles has been completed. What would be the length of the completed part of the bypass? ■

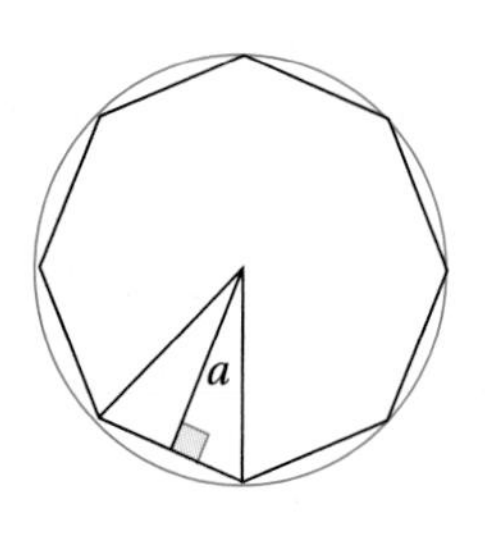

FIGURE 12.15 | Using polygon area to find circle area.

Area of a Circle. You can think of the area of a circle in much the same way you thought of the area of a regular polygon. Consider the regular polygon inscribed in a circle in Figure 12.15.

The area of the regular polygon is given by the formula $A = \frac{1}{2}aP$, where a is the apothem and P is the perimeter. If we think of the circle as the limiting figure of continually increasing the number of sides of an inscribed regular polygon, the area of the circle will be the limiting value of the area of the polygon. Also, the radius of the circle is the limiting value of the apothem and the circumference is the limiting value of the perimeter. These relationships suggest that we can make the following substitutions, beginning with the area of a regular polygon:

$A = \frac{1}{2}aP$	Regular polygon
$= \frac{1}{2}rC$	Replace apothem with radius and perimeter of polygon with circumference
$= \frac{1}{2}r(2\pi r)$	Replace circumference with $2\pi r$

Simplifying the last expression, we can produce the following formula for the area of a circle.

Formula for the Area of a Circle

$A = \pi r^2$, where r is the radius of the circle.

Another way to develop the formula for the area of a circle is to divide the circle, as shown in Figure 12.16, and rearrange the pieces to form what looks like a parallelogram. Here we divided the circle into eight parts, but if we divide the circle into many more parts, the figure on the right would look even more like a parallelogram.

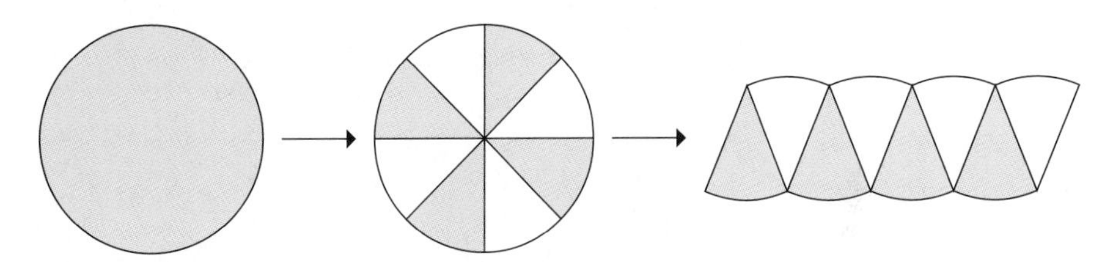

FIGURE 12.16 | Area of a circle: Using a model.

As we increase the number of partitions of the circle, the base of the parallelogram-like figure approximates one-half the circumference and the height approximates the radius of the circle. Beginning with the formula for the area of a parallelogram, $A = bh$, we can make the following substitutions:

$A = bh$	Area of parallelogram
$= (\frac{1}{2}C)r$	Replace b with $\frac{1}{2}C$ and h with r
$= (\frac{1}{2} \times 2\pi r)r$	Replace C with $2\pi r$
$= \pi r^2$	Simplify

Example 12.13 illustrates how to apply the formula for the area of a circle.

Example 12.13 **Problem Solving: Sprinkler Coverage**

A sprinkler has a range of 12 m and rotates with a circular motion. If the sprinkler covers the entire circular area, how many square meters does it cover?

Solution

The area covered is $A = \pi r^2 = \pi(12)^2 = 144\pi$, or about 452.4 m^2.

Your Turn

Practice: How much area would a sprinkler with radius 20 ft cover?

Reflect: If the sprinkler only rotated about 120° with a range of 12 m, how could you determine the area watered? ■

Students need to understand that formulas are shortcuts for finding areas and are tools for solving problems. Further, they should perceive formulas to be a logical result of other types of reasoning—similar to those presented in this section. It isn't productive for students to develop the notion that mathematical thinking is primarily a matter of memorizing formulas and using them in some abstract sense. Students should develop a sense of how formulas can be developed should they be forgotten.

Comparing Perimeters and Areas of Similar Polygons

In Mini-Investigation 12.8 you are asked to examine the relationship between the perimeters and areas of two squares of different sizes.

Write a general statement about what you discovered.

Mini-Investigation 12.8 Making a Connection

If the ratio of the side lengths of squares $AEFG$ and $ABCD$ is 2:1, what is the ratio of their

a. perimeters? **b.** areas?

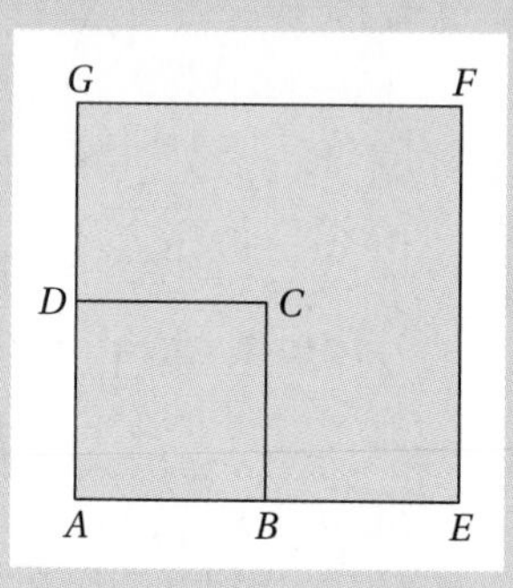

Recall that similar figures have the same shape but aren't necessarily the same size. The **scale factor** is the ratio of the lengths of the corresponding sides of two similar figures. For the rectangles shown in Figure 12.17(a), the scale factor is 2 (larger figure to smaller figure), or

$$\frac{EF}{AB} = \frac{GF}{BC} = \frac{GH}{CD} = \frac{HE}{AD} = \frac{2}{1}.$$

For the triangles shown in Figure 12.17(b), the scale factor is 3 (larger to smaller), or

$$\frac{DE}{AB} = \frac{FE}{BC} = \frac{FD}{CA} = \frac{3}{1}.$$

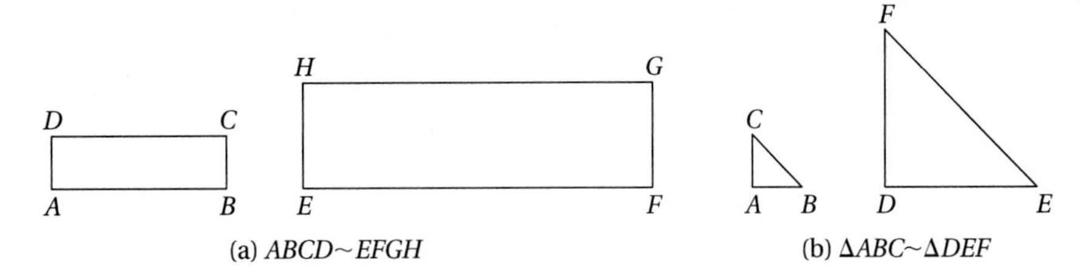

FIGURE 12.17 | Similar figures.

If the sides of rectangle *EFGH* are two times as long as those of rectangle *ABCD*, the perimeter of the larger rectangle must be twice that of the smaller rectangle. Similarly, the perimeter of triangle *DEF* must be three times the perimeter of triangle *ABC*. These conclusions suggest the following generalization.

Generalization about the Ratio of Similar Polygon Perimeters

If the scale factor of two similar polygons is k, then the ratio of their perimeters is k:1. That is, the ratio of the perimeters is the same as the ratio of any two corresponding sides.

To understand better the ratio of the areas of the rectangles or triangles in Figure 12.17, consider the number of smaller figures needed to cover the larger figures, as shown in Figure 12.18.

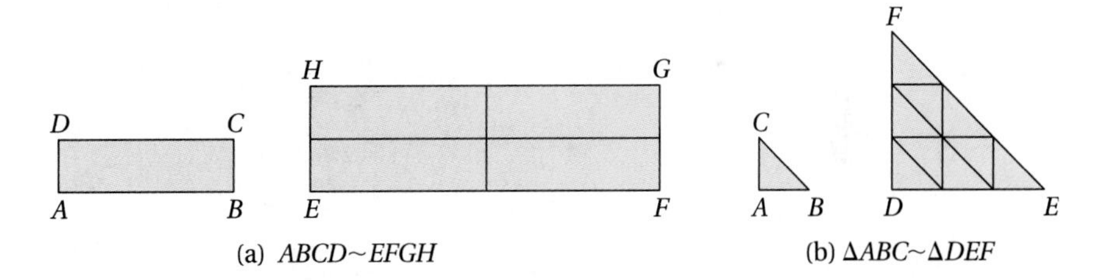

FIGURE 12.18 | Comparing areas of similar figures.

Four of the smaller rectangles are needed to make the larger rectangle, and nine of the smaller triangles are needed to make the larger triangle. Note that 4 and 9 are the squares of the scale factors, 2 and 3. This observation suggests the following generalization.

Generalization about the Ratio of Similar Polygon Areas

If the scale factor of two similar polygons is k, then the ratio of their areas is k^2:1.

Example 12.14 shows how to apply the generalizations about the ratios of the areas and perimeters of similar polygons.

Example 12.14 Problem Solving: The Air Duct Problem

A building contractor has been using a 12-in. square air duct in her buildings, but in the next building she wants to increase the air flow by 30%. What size should the new duct be?

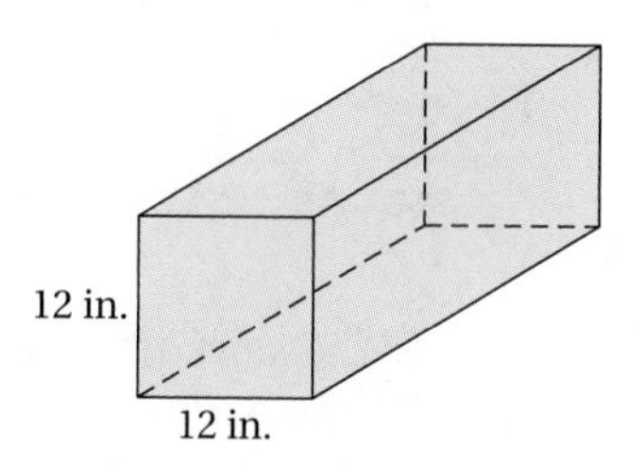

Working toward a Solution

Understand the problem	*What does the situation involve?*	Increasing the amount of air flow in an air duct by 30%
	What has to be determined?	The size of the new duct
	What are the key data and conditions?	The amount of air flow depends on the cross-sectional area of the duct. The cross section of the duct is a 12-in. × 12-in. square. We want the amount of air flow to increase by 30%.
	What are some assumptions?	The air flow through the air duct is constant. The new duct has a square cross section.
Develop a plan	*What strategies might be useful?*	Write an equation; use reasoning.
	Are there any subproblems?	Yes. Find the ratio of the new cross-sectional area to the old cross-sectional area.
	Should the answer be estimated or calculated?	Calculated
	What method of calculation should be used?	Mental math or pencil and paper calculation
Implement the plan	*How should the strategies be used?*	Set up a proportion comparing duct side lengths to duct areas and solve the proportion. If the increase is 30%, the ratio of the new area to the old area is 1.3:1. The proportion is $\frac{x^2}{(12)^2} = \frac{1.3}{1}$, where x is the side of the desired duct. Solving, we get $x^2 = (144)(1.3) = 187.2$, or $x \approx 13.7$ in.

	What is the answer?	The new duct should be about 13.7 in. on a side.
Look back	*Is the interpretation correct?*	The side length of the new duct has increased, suggesting that the interpretation is correct.
	Is the calculation correct?	Substituting 13.7 for x in the proportion $\frac{x^2}{(12)^2} = \frac{1.3}{1}$, and cross multiplying indicates that the calculation is correct.
	Is the answer reasonable?	30% of 144 is 43.2. Thus the new duct should have an area of 187.2 in.^{2}. A square with side length 13.7 in. produces about this much area. The answer seems reasonable.
	Is there another way to solve the problem?	Because the area of the new duct should be 1.3 times the area of the old duct, the side length of the new duct should be about $\sqrt{1.3}(12)$, or about 13.7 in. on a side.

Your Turn

Practice: What should be the size of the duct if the contractor wanted to *decrease* the airflow by 20%?

Reflect: If the sides of the ducts increase by 10%, 20%, and 30%, does the air flow increase by 10%, 20%, and 30%? Explain why or why not. ■

Problems and Exercises | *for Section 12.2*

Use $\pi = 3.14$ and a calculator, as necessary.

A. Reinforcing Concepts and Practicing Skills

1. Suppose that the dimensions of a rectangle are 12 cm × 40 mm. What do you need to do first in order to find the area of the rectangle?

Indicate whether the tasks in Exercises 2–6 involve perimeter or area.

2. Determining the amount of paneling to wall a den
3. Determining the amount of baseboard to finish a den
4. Determining the number of tiles needed to floor a bathroom
5. Determining the amount of paint to paint a bedroom
6. Determining the number of fence posts needed to make a dog pen

Find the perimeters and areas of the squares and rectangles in Exercises 7–10.

7.

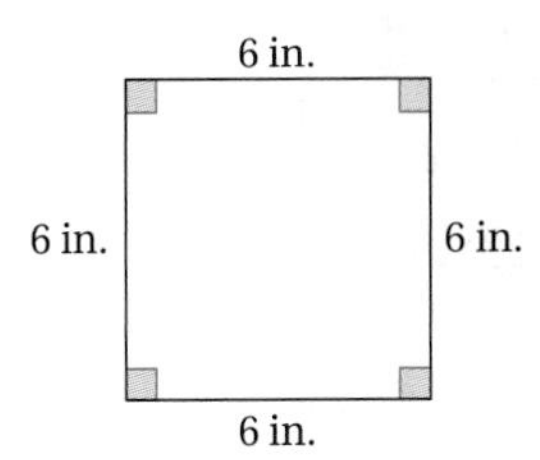

8.

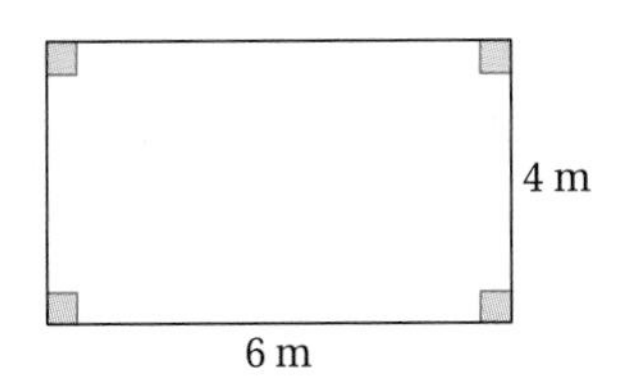

9.

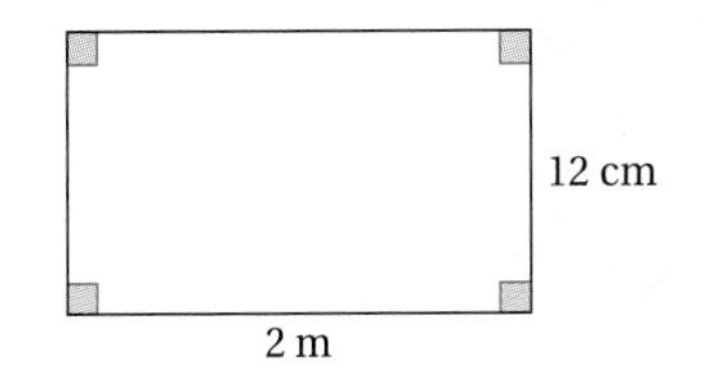

10.

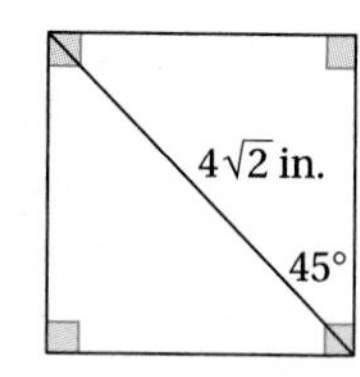

Find the perimeters and areas of the parallelograms in Exercises 11–14.

11.

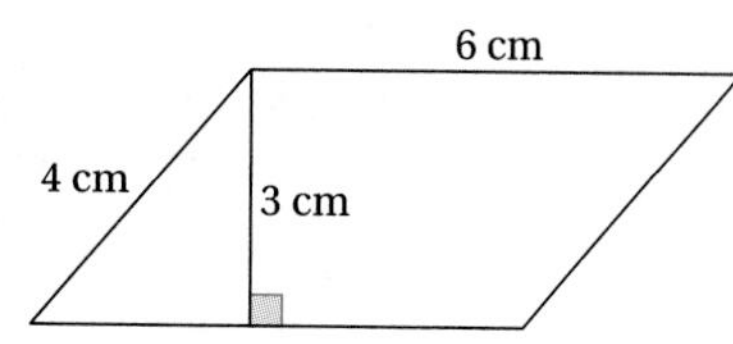

12.

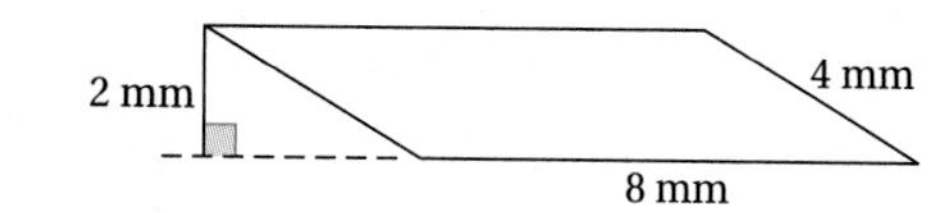

13.

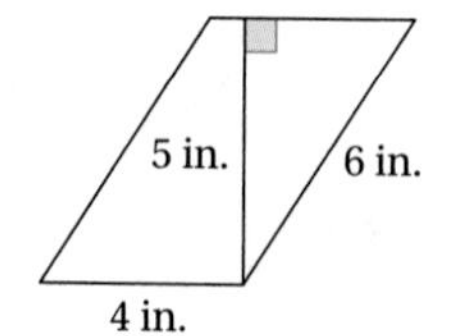

14.

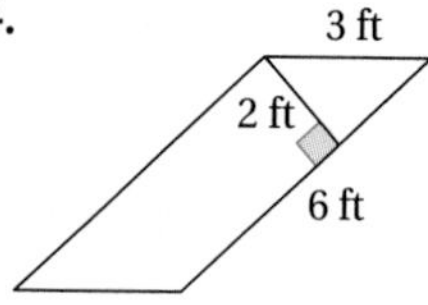

Find the areas of the trapezoids in Exercises 15–18.

15.

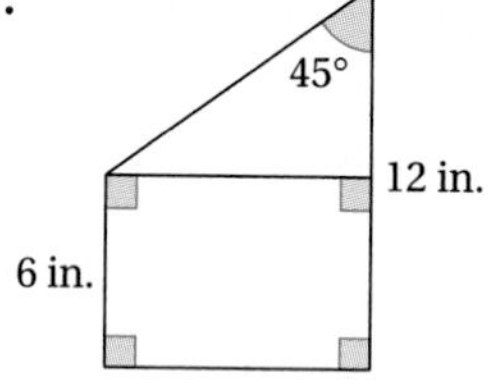

16.

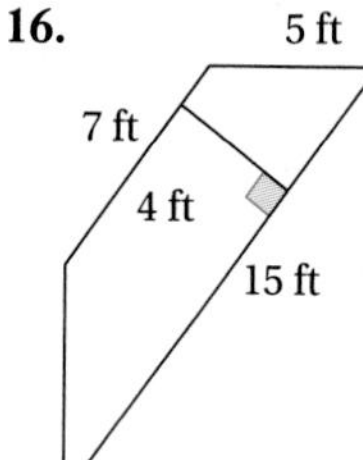

17.

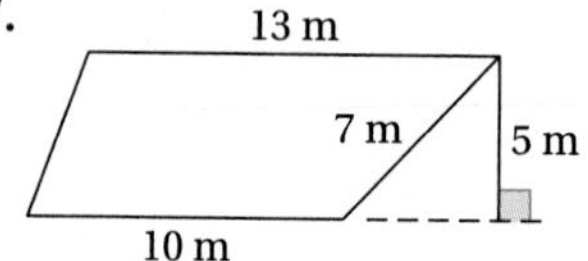

18.

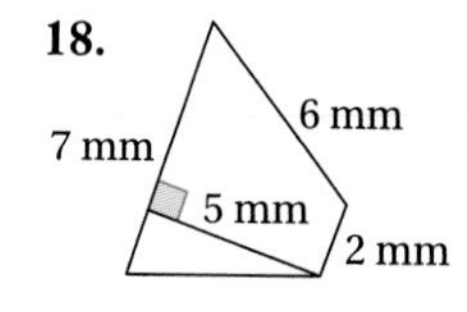

Find the perimeters and areas of the regular polygons in Exercises 19–20.

19.

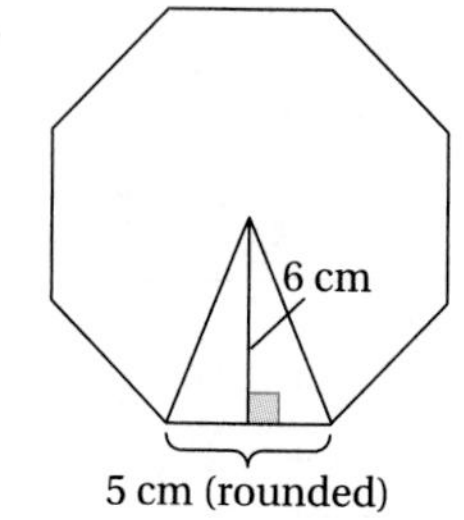

20.

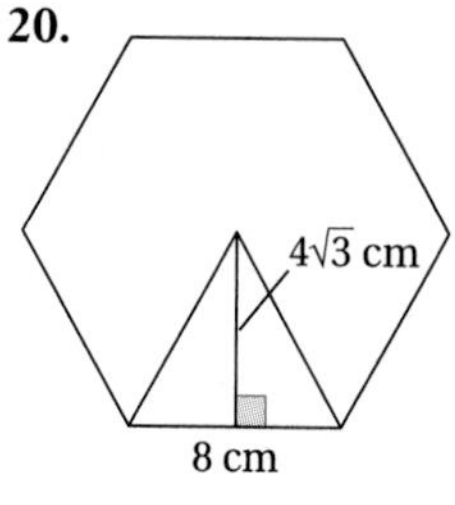

Find the circumferences and areas of the circles in Exercises 21–24.

21.

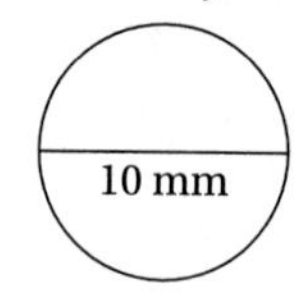

22.

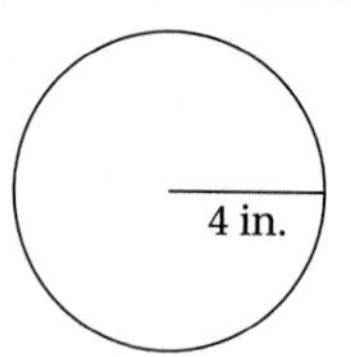

23.

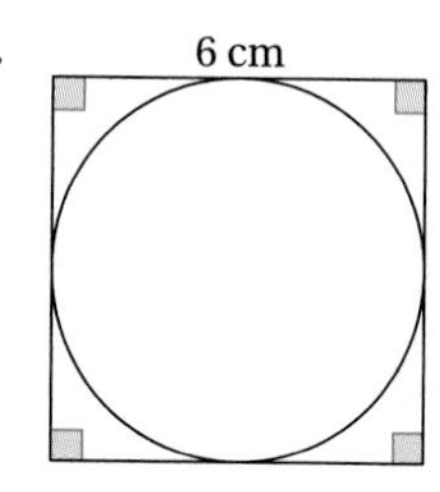

24.

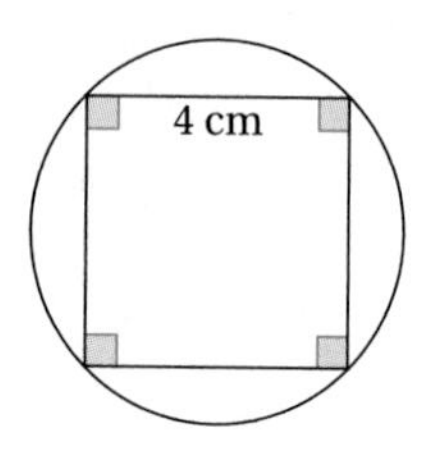

25. A small rug is 2 m long. A similarly shaped rug is 6 m long. If the cost of the first rug is $110 and the price is directly proportional to the area of the rug, what would you expect the cost of the larger rug to be?

26. Two corresponding sides of similar polygons are in the ratio of 5:4. If the perimeter of the smaller figure is 20, what is the perimeter of the larger figure?

27. The areas of two similar polygons are 25 mm^2 and 121 mm^2. What is the ratio of their perimeters?

28. Michelle wants to plant a new yard that is 90 ft $\times$ 60 ft. Each bag of seed covers 200 ft^2 of lawn. How many bags of seed does she need to buy?

29. Find the area of the following figure by dividing the figure into more familiar polygons.

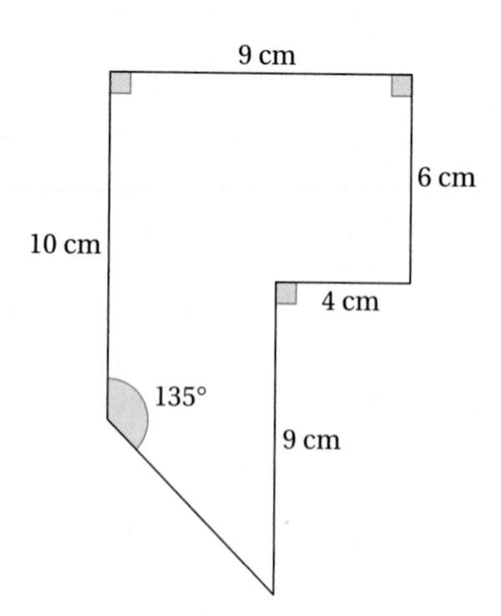

B. Deepening Understanding

30. Argue that the three triangles shown on the geoboard on the next page do [or do not] have the same area.

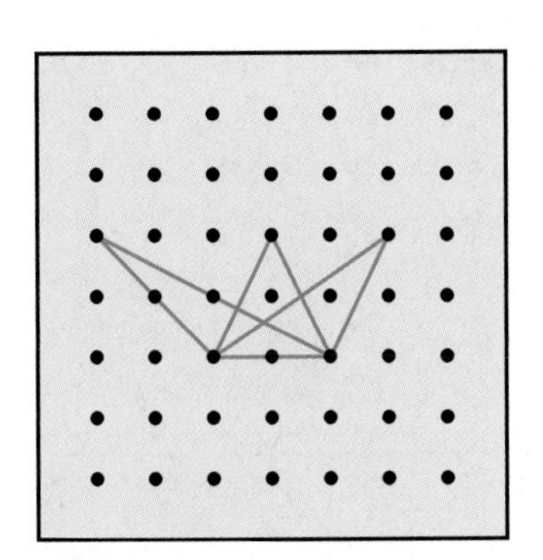

Find the areas of the triangles in Exercises 31–34. Congruent sides are marked.

31.

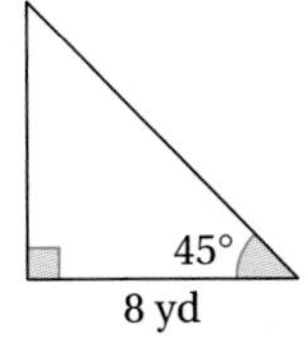

32.

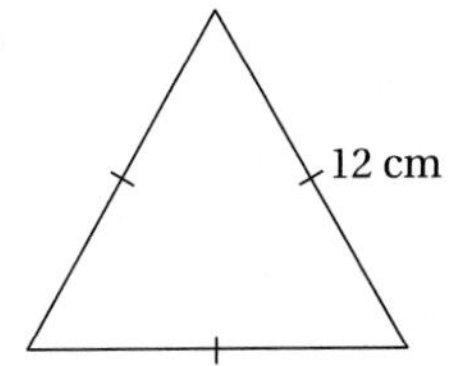

33.

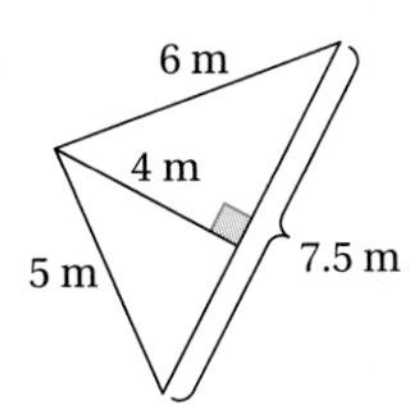

34.

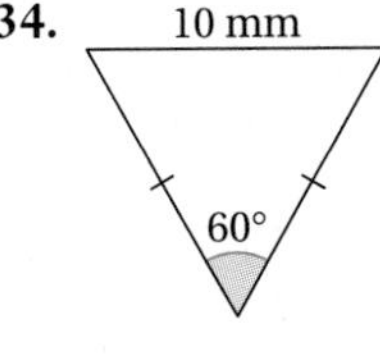

35. Find the arc lengths in the following figures. (Use the formula for the circumference of a circle and the fact that a circle has 360°.)

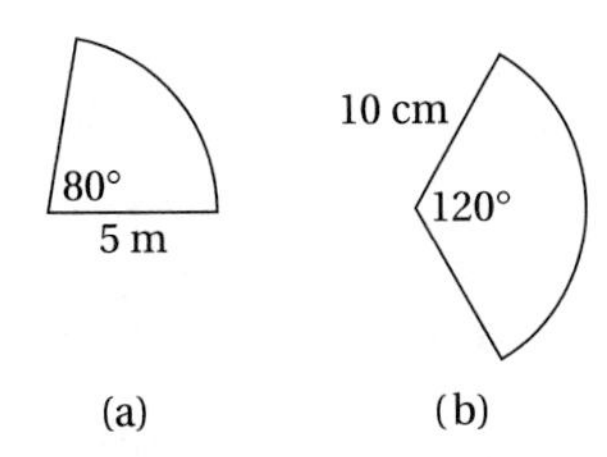

36. Find the areas of the following shaded figures. (Use the formula for the area of a circle and the fact that a circle has 360°.)

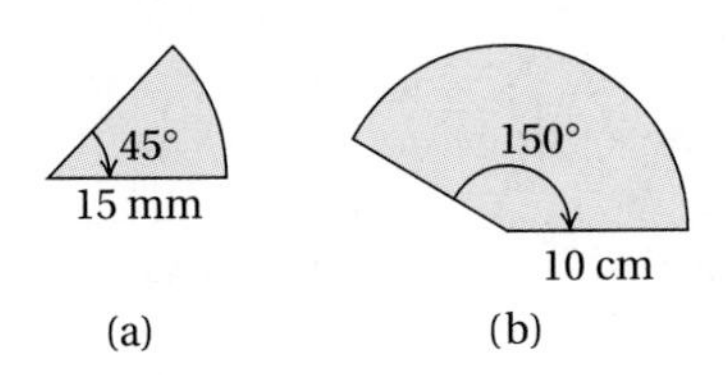

37. In $\triangle ABC$, $\overline{DE}$ bisects the altitude $\overline{AF}$. What is the ratio of the area of $\triangle ADE$ to the area of $\triangle ABC$? Assume that $\overline{DE} \parallel \overline{BC}$.

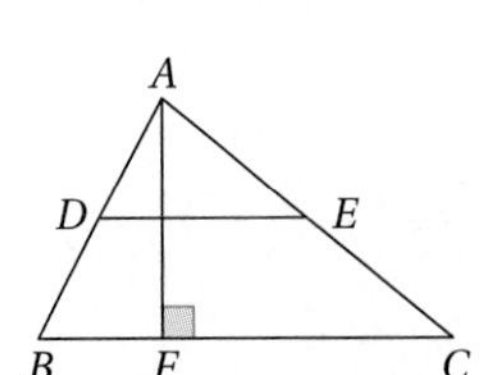

38. Making Connections. Consider the sequence of figures shown and describe what you discover as the relationship between the formulas for the areas of a rectangle, a trapezoid, and a triangle.

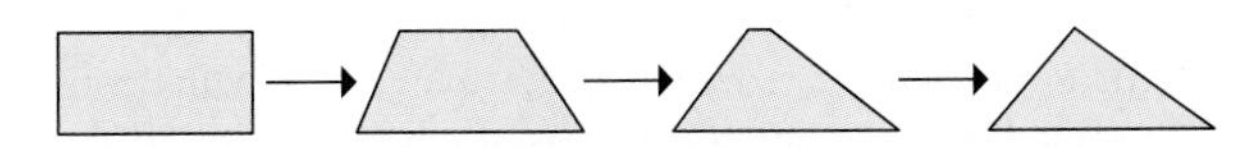

39. Suppose that a parallelogram is drawn similar to *ABCD* so that each side is four times that of the corresponding sides of *ABCD*. How many parallelograms *ABCD* would be needed to cover the larger parallelogram? Draw a diagram to show that you are correct.

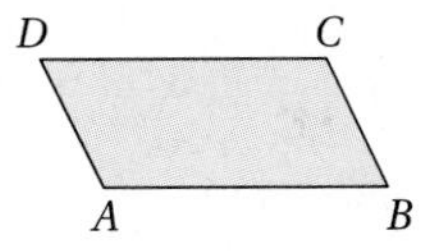

40. If the ratio of the areas of two similar polygons is 2:1, what is the ratio of their corresponding sides? What is the ratio of their perimeters?

41. Consider rectangles *ABCD* and *EFGH*. Argue that the ratio of their areas is [or is not] equal to the square of the ratio of their perimeters.

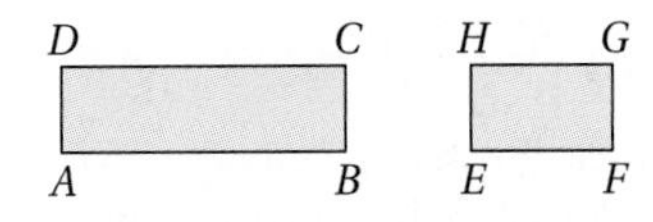

42. Find the length of a side of a square that has the same area as a 12-in. × 27-in. rectangle.

43. Identify the dimensions of a square and a rectangle that satisfy the following conditions.

a. They have equal perimeters, but the square has the greater area.

b. They have equal areas but the rectangle has the greater perimeter.

44. What is wrong with the following reasoning?

Rectangle I is similar to rectangle II. One side of rectangle I is 2 units and one side of rectangle II is 5 units. Therefore the ratio of their areas is 4:25.

C. Reasoning and Problem Solving

45. What would be the best formula to use to find the area of cross section ABC of the cube shown if the length of each edge of the cube is S? Justify your choice.

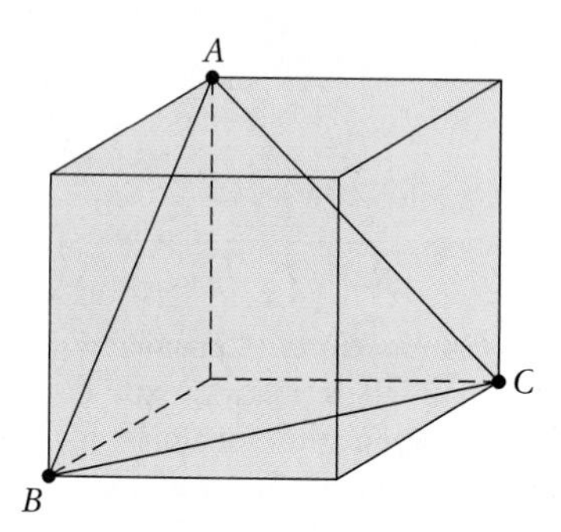

46. If a rectangle is redrawn so that its area remains constant but its perimeter increases, what happens to the shape of the rectangle?
47. If a rectangle is redrawn so that its perimeter remains constant but its area increases, what happens to the shape of the rectangle?
48. **The Dart Problem.** Suppose that you are playing a carnival game in which you throw a dart at a board 6 ft × 7 ft having 63 balloons attached that are 6 in. in diameter. What are your chances of winning?

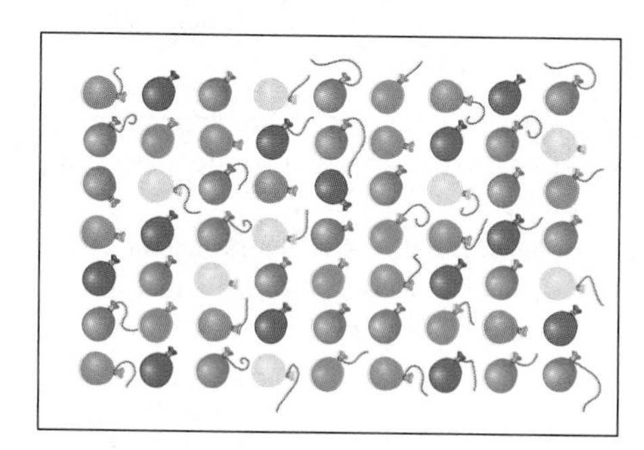

49. **The Track Problem.** Suppose that a running track consists of a 100-yd × 50-yd rectangle with semicircles at the ends, as shown in the next column. If Billie runs in a lane 3 yd farther out than Alan all the way around the track, how much farther will Billie have run after one lap?
50. **The Timer Problem.** Carlos notes that the sand in the hourglass takes 4 min to pass from one side to the other. If the diameter of the opening is 1 mm, what should be the diameter of the opening if the sand is to pass through in 2 min?
51. Figure $ABCD$ is a trapezoid with $DC = \frac{1}{4}AB$. What is the ratio of area $ABCD$ to area DCE? Provide an argument to support your answer.

Figure for Exercise 49

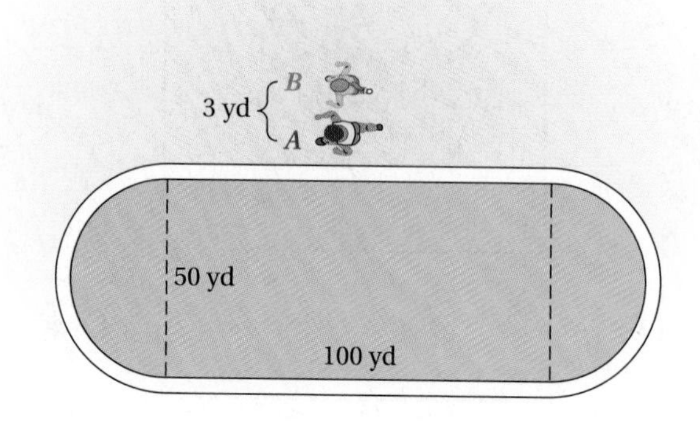

Figure for Exercise 50

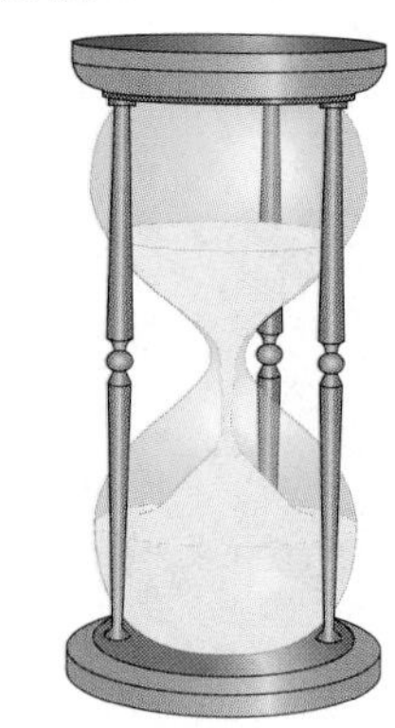

Figure for Exercise 51

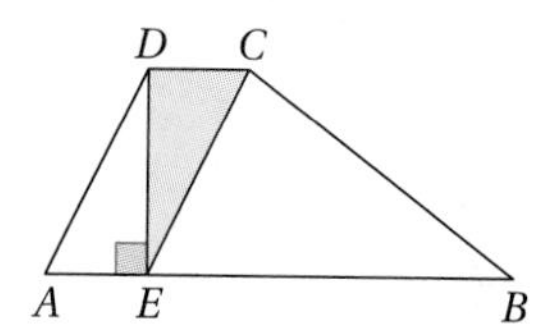

52. Figure $ABCD$ is a parallelogram and x is any point inside the parallelogram. What fraction of the area of the parallelogram is represented by the shaded areas? Provide an argument to support your answer.

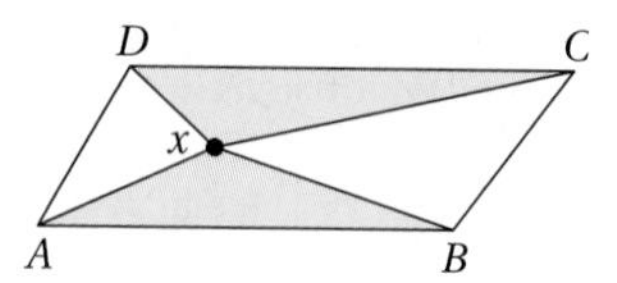

Find the areas of the shaded parts of the figures in Exercises 53–55.

53. Assume that both figures are squares.

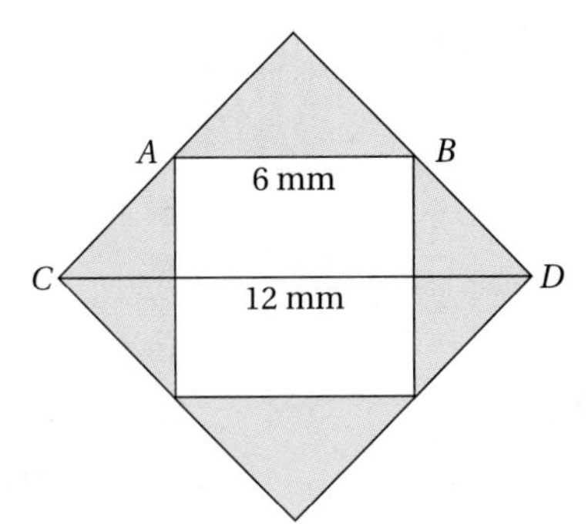

54.

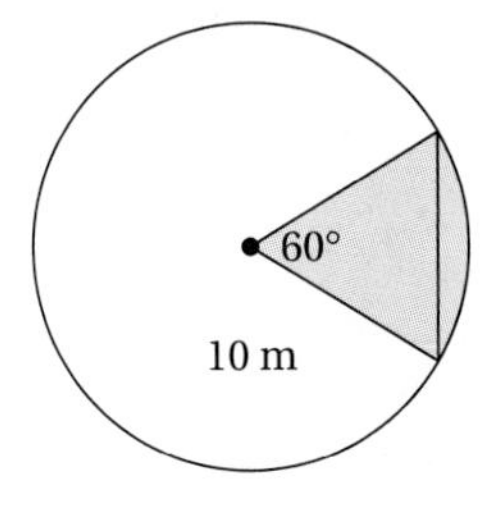

55. Each side of the square is 6 cm.

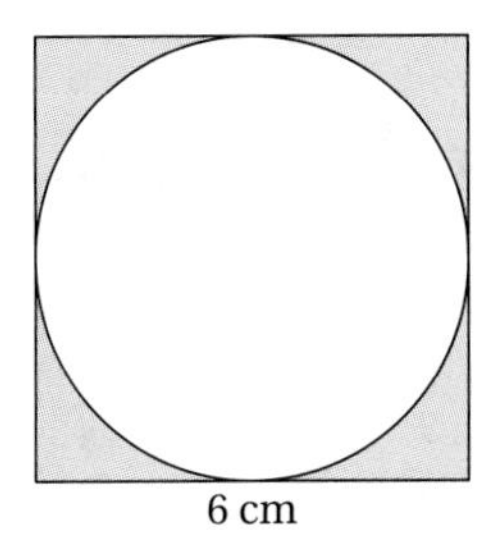

56. **The Rental Problem.** A businessman owns space at a local shopping mall. The sizes of the stores are shown in the following layout. If he charges $0.25/ft²/mo and if each store front is 50 ft wide, how much income would he have received after 1 mo if all four stores were rented?

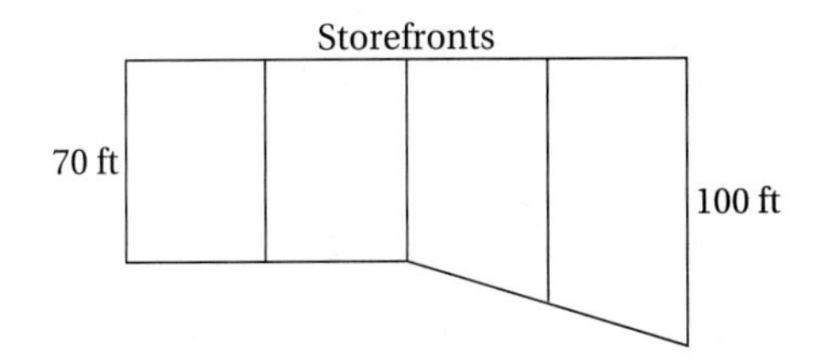

57. **The Lot Problem.** Form a group of students and find the area of the lot shown to the nearest square foot by dissecting the figure and estimating where necessary. Compare this area with those obtained by other groups.

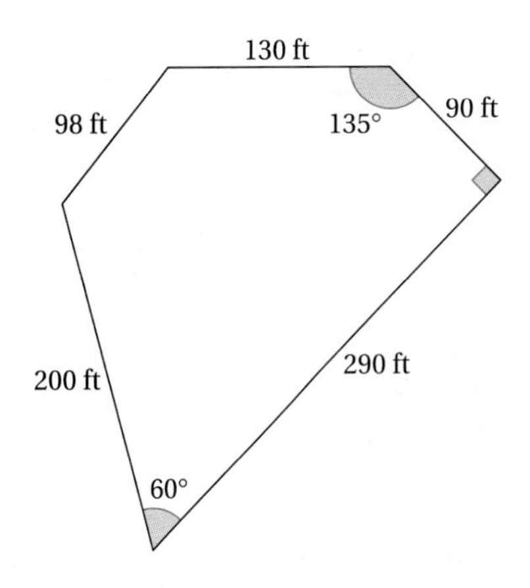

58. **The Land Purchase Problem.** A developer can buy a 90-yd × 50-yd plot of land for $15,000 or a 180-yd × 100-yd plot of land for $55,000. If all other things are equal, which is the better buy? Why?

D. Communicating and Connecting Ideas

59. Cris and Tony are arguing about which has more surface water, Shorty's pond or Elmo's pond.

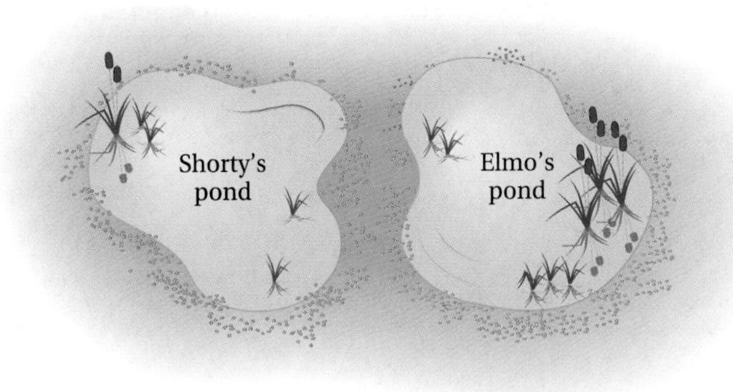

Cris says that if they walk around each pond, they can determine which is the longer distance and that will be the pond with more surface water. Tony claims that this method won't work. Write an explanation that supports either Cris or Tony.

60. **Historical Pathways.** The ancient Babylonians used the following formula for finding the area of quadrilaterals:

$$A = \frac{(a + c)(b + d)}{4},$$

where a and c are opposite sides and b and d are opposite sides. In what context does the formula work? Conjecture why you think that the Babylonians came up with this formula.

61. The Disturbed Motel Owner Problem. A motel owner is adding a swimming pool to the grounds. She told the contractor that she wants the pool area to be twice as large as her competitor's pool area, which is a 5-m × 7-m rectangle. When the contractor finished, the new pool covered 70 m^2, but the owner was upset. She claimed that she had asked that the competitor's pool be doubled but that the contractor hadn't done that. What do you think is the basis for the miscommunication between the owner and the contractor?

62. Making Connections. A homeowner's gutters aren't large enough to carry the water from a heavy rain away from his house. He is considering installing larger gutters. What is the percent increase in the amount of water that can be carried off with the larger gutters compared to the smaller gutters? (See figure in next column.)

Figure for Exercise 62

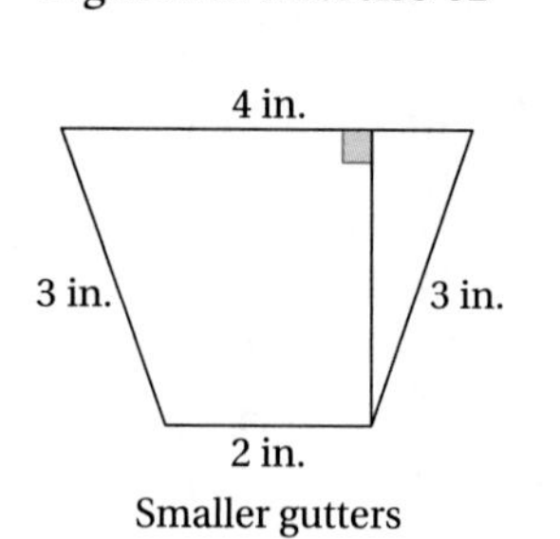

Smaller gutters

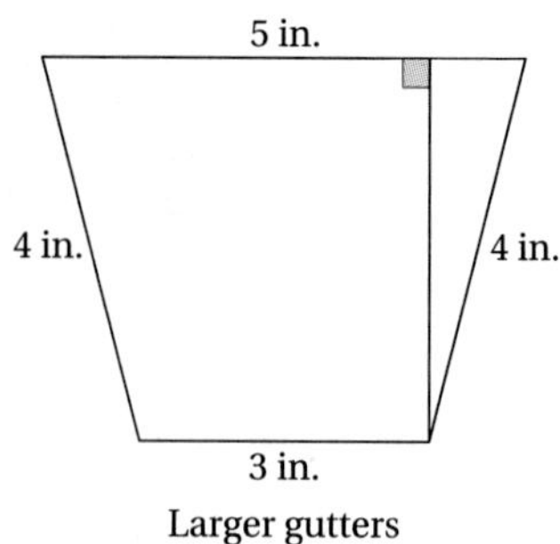

Larger gutters

63. Making Connections. An approximation for π can be obtained by dividing the perimeter of a regular polygon inscribed in a circle by the diameter of the circle. The perimeter of an inscribed regular hexagon is 6 when the radius is 1 (why?), which gives an approximate value for π of 3. The following formula can be used to find the perimeter of regular polygons of sides 6, 12, 24, 48, 96, 192, 384, and so on, assuming that the radius is 1 unit. For example,

$$s_{(12)} = \sqrt{2 - \sqrt{4 - s_{(6)}^2}},$$

where $s_{(12)}$ is the side length of a 12-sided regular polygon and $s_{(6)}$ is the side length of a 6-sided regular polygon. We can replace $s_{(12)}$ by $s_{(24)}$ and $s_{(6)}$ by $s_{(12)}$ to find $s_{(24)}$. Use a calculator or a spreadsheet to complete Table 12.2 to obtain an approximation for π.

TABLE 12.2

Number of Sides	Length of One Side	Perimeter	$P \div D$
6	1	6	3
12	$s_{(12)} = \sqrt{2 - \sqrt{4 - (1)^2}} = 0.51763809$	6.21165708	3.10582854
24	$s_{(24)} = \sqrt{2 - \sqrt{4 - (0.51763809)^2}} = 0.26105238$	6.26525722	3.13262861
48	$s_{(48)} = \sqrt{2 - \sqrt{4 - (0.26105238)^2}} = 0.13080626$	6.27870041	3.13935020
96	?	?	?
192	?	?	?
384	?	?	?

Section 12.3 MEASURING THE SURFACE AREA AND VOLUME OF SOLIDS

- Surface Areas of Solids
- Volumes of Solids
- Comparing Surface Areas and Volumes of Similar Solids

In this section we develop formulas for finding the surface areas and volumes of the following common solids: prisms, cylinders, spheres, pyramids, and cones. We also consider the relationship between the dimensions of similar solids and their surface areas and volumes. In Mini-Investigation 12.9 you are asked to compare the concepts of surface area and volume.

Talk about the reasons for your decisions and suggest other examples of finding surface area or volume.

Mini-Investigation 12.9 **Using Mathematical Reasoning**

Which of the following tasks involve finding area and which involve finding volume?

a. Finding the amount of leather needed to make a soccer ball
b. Finding the amount of frozen yogurt in a waffle cone
c. Finding the amount of concrete needed to make a sidewalk
d. Finding the amount of wrapping paper needed to wrap a gift box

Surface Areas of Solids

In Section 12.2, we discussed finding the area of different planar figures. In this subsection we extend this idea to include the **surface area,** *SA*, of a solid. Even though we are dealing with solids, their surface areas are measured in square units, as are the areas of planar figures. We consider the surface areas of selected polyhedra, cylinders, cones, and spheres.

Surface Areas of Rectangular Prisms and Cubes. We can develop a method for finding the surface areas of rectangular prisms by studying Figure 12.19. When the 2-unit (height) × 3-unit (width) × 5-unit (length) rectangular prism shown in Figure 12.19(a) is unfolded, it gives the figure shown in Figure 12.19(b).

Figure 12.19(b) contains two each of the following faces.

2×3	$(h \times w)$ rectangle
3×5	$(w \times l)$ rectangle
2×5	$(h \times l)$ rectangle

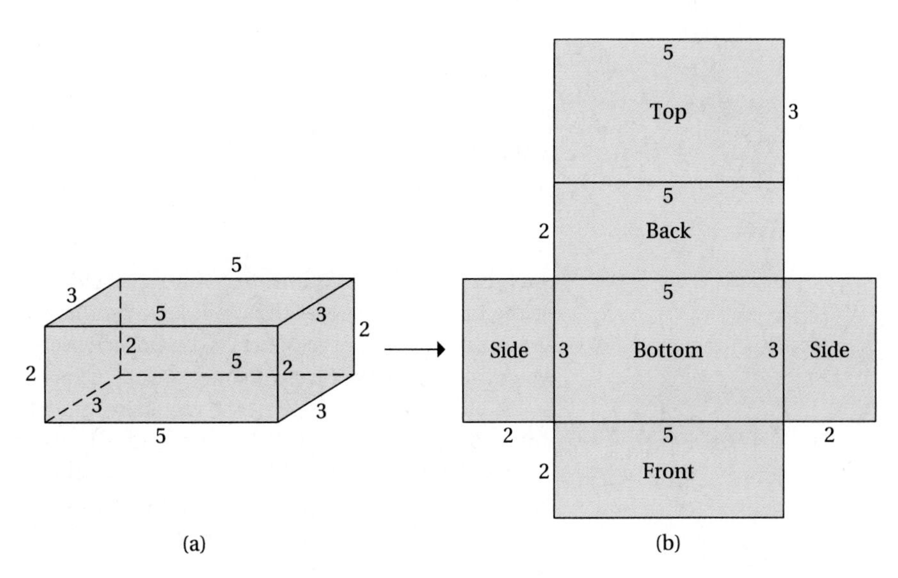

FIGURE 12.19 | Surface area of a rectangular prism.

The area of this specific rectangular prism then is $SA = 2(2 \times 3 + 3 \times 5 + 2 \times 5)$, or 62 sq units. Recall that a cube is a rectangular prism with six congruent square faces, which means that the edges of a cube (length, width, and height), denoted e, are all the same length. Thus the surface area of the cube is six times the area of one of its faces. These ideas suggest the following formulas.

Formulas for the Surface Areas of Rectangular Prisms

General Formula: $SA = 2(hw + wl + hl)$, where h, w, and l are the height, width, and length of the prism.

Cube: $SA = 6e^2$, where e is the length of each edge of the cube.

Example 12.15 applies finding the surface area of a rectangular prism.

Example 12.15

Problem Solving: Buying Wrapping Paper

Alexandria is buying wrapping paper that measures 18 in. wide and 48 in. long. She needs to wrap a box that is 8 in. high, 10 in. wide, and 16 in. long. Assuming that she has 20% overlap and/or waste when wrapping the box, how many sheets of wrapping paper does she need to buy?

Solution

The surface area of the box is $2(8 \times 10 + 8 \times 16 + 10 \times 16)$, or 736 in.^{2}. For the 20% waste or overlap she needs an additional $736 \times .2$, or 147.2 in.^{2} for a total of 883.2 in.^{2} of paper. Each sheet contains 18×48, or 864, in.^{2}. Therefore she needs to buy two sheets of wrapping paper.

Your Turn

Practice: If the box had been a cube 12 in. on each edge, how many square inches of wrapping paper would she need, assuming 20% waste and/or overlap?

Reflect: If each dimension of the 8-in. × 10-in. × 16-in. rectangular prism is reduced by one-half, is the surface area reduced by one-half? Would your answer apply to all possible sizes of boxes? Explain your reasoning. ■

Surface Area of a Cylinder. The surface area of a cylinder consists of the area of the curved surface of the cylinder plus the area of the two bases. We can think about finding the curved part of the surface area in the following way. Suppose that we cut the label off a can of soup and lay it out, as shown in Figure 12.20.

Note that the label forms a rectangle, the area of which is found by multiplying its length times its width. The length is also the circumference of the circular base of the can, or $2\pi r$. The height of the label is the height of the cylinder. Thus the area of the curved part is $2\pi rh$. The bases are circles, each of which has an area of πr^2. These observations suggest the following formula for the surface area of a cylinder.

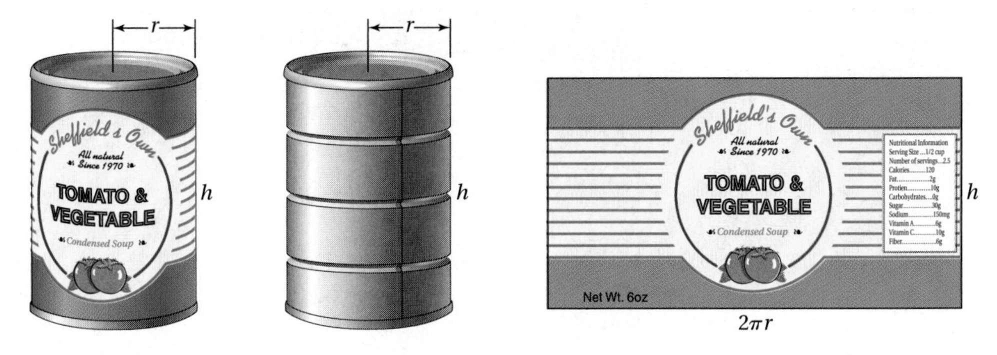

FIGURE 12.20 | Modeling the surface area of a cylinder.

Formula for the Surface Area of a Cylinder

$SA = 2\pi rh + 2\pi r^2$, where r is the radius of the circular base and h is the height of the cylinder.

This formula may also be written as $SA = 2\pi r(h + r)$, although the first formula is probably easier to understand. Example 12.16 demonstrates the use of the formula to find the surface area of a cylinder.

Example 12.16 Problem Solving: Manufacturing a Soup Can

How many square inches of tin are required to make a soup can that has a diameter of 4 in. and a height of 8 in. (Use $\pi = 3.14$.)

Solution

Substituting $h = 8$ and $r = 2$ in the formula for the surface area of a cylinder, we obtain $SA = 2(3.14)(2)(8) + 2(3.14)(2^2)$, or 125.6 in.^{2}. Thus 125.6 in.^{2} of tin are needed for one can.

Your Turn

Practice: Suppose that the can had an 8-in. diameter and a 4-in. height. How many square inches of tin would be required to make this can?

Reflect: Suppose that the size of a can with a 4-in. diameter and an 8-in. height were enlarged to a can having an 8-in. diameter and a 16-in. height. How many times greater would the surface area of the larger can be? Would this result hold in general if each dimension of the cylinder were doubled? ■

Surface Areas of a Right Regular Pyramid and a Right Regular Cone. Mini-Investigation 12.10 encourages you to explore finding the surface area of a right square pyramid.

Write a statement about how to find the surface area of a right square pyramid if each edge of the pyramid is s units long.

Mini-Investigation 12.10 **Solving a Problem**

If the following pattern can be used to make a right square pyramid with each edge 8 units long, what is the surface area of that pyramid?

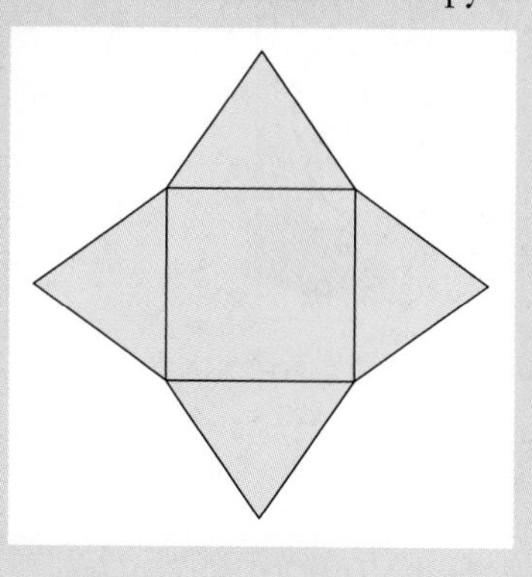

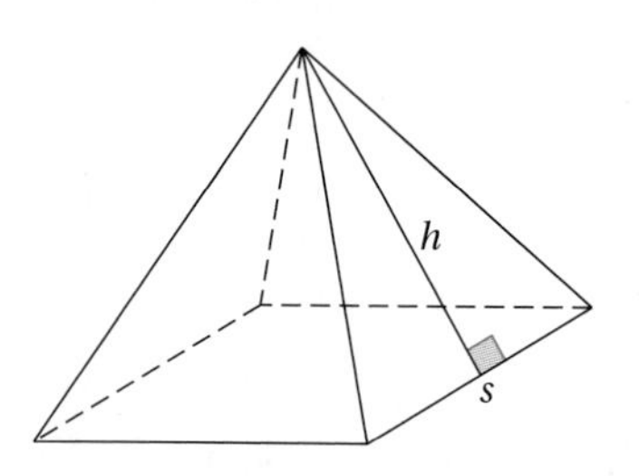

FIGURE 12.21

As you may have discovered in Mini-Investigation 12.10, the surface area of a pyramid is the sum of the areas of the triangular faces plus the area of the base. In this subsection we consider only right regular pyramids with regular polygons as bases and all the triangular faces congruent. For example, the surface area of the square pyramid shown in Figure 12.21 can be found by summing the area of the base (s^2) and four times the area of one of the triangular faces ($\frac{1}{2}sh$), where h is the height of the triangular face.

When the base of the pyramid is a regular polygon of n sides, the sum of the areas of the triangular faces is $n \times \frac{1}{2}sh$, where s is the length of the side and h is the height of the triangular face. But ns is the perimeter, P, of the pyramid's base. So the total surface area of the triangular faces is $\frac{1}{2}Ph$, and $SA(\text{pyramid}) = B + \frac{1}{2}Ph$, where B is the area of the pyramid's base.

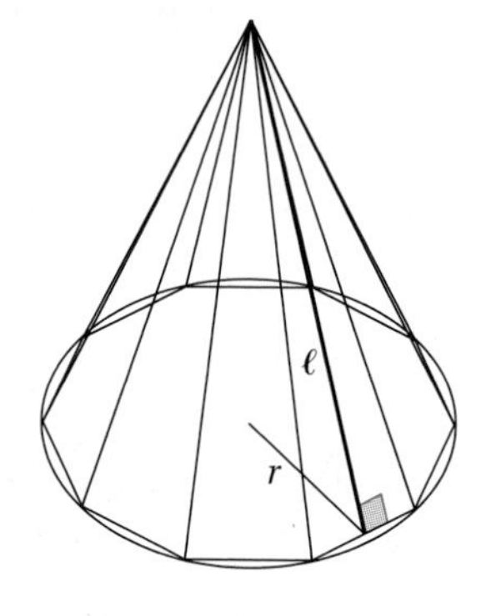

FIGURE 12.22

The surface area of a right circular cone can be thought of as the limiting value of the surface area of an inscribed right regular pyramid, as suggested in Figure 12.22. The area of the base of the cone is the limiting value of the area of the base of the pyramid, and the slant height of the cone, ℓ, is the limiting value of the height, h, of a triangular face of the pyramid. These relationships suggest the following substitutions.

$SA = B + \frac{1}{2}Ph$	Surface area of a regular pyramid
$SA = \pi r^2 + \frac{1}{2}Ph$	Replace area of pyramid base with area of cone base
$SA = \pi r^2 + \frac{1}{2}(2\pi r)\ell$	Replace P with the circumference, and h with ℓ

Simplifying the last expression, we produce the formula for the surface area of a cone. The two formulas are summarized as follows.

Formulas for the Surface Areas of a Right Regular Pyramid and a Right Circular Cone

Right Regular Pyramid: $SA = B + \frac{1}{2}Ph$, where B is the area of the base, P is the perimeter of the base, and h is the height of the triangular faces.

Right Circular Cone: $SA = \pi r^2 + \pi rl$, where r is the radius of the base and l is the slant height of the cone.

Example 12.17 illustrates the application of these formulas.

Example 12.17 Finding the Surface Area of Pyramids and Cones

A right circular cone and a square pyramid are inscribed in a cube, as shown.

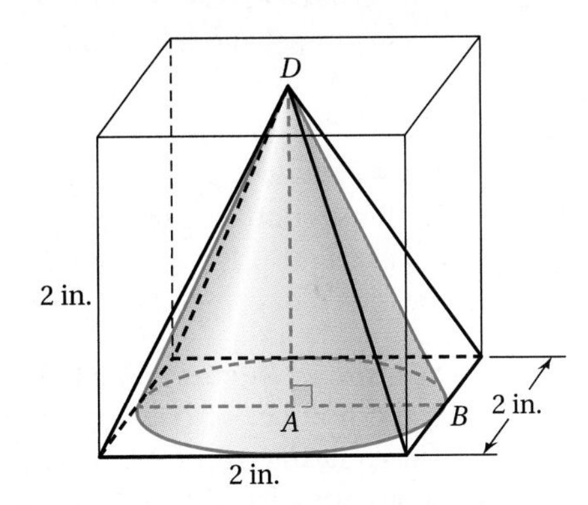

Find the surface area of the pyramid and cone. (Use $\pi = 3.14$.)

Solution

Cone: The radius of the cone is 1 in. and its height is 2 in. By the Pythagorean theorem, $BD^2 = 1^2 + 2^2$. Solving, we see that the slant height, BD, $= \sqrt{5}$. So $SA = \pi(1)^2 + \pi(1)\sqrt{5}$, or 10.2 in.^{2}.

Pyramid: The area of the square base of the pyramid is 4 in.^{2}. The height of each triangular face is $\sqrt{5}$, as determined in the first part of the solution. Thus $SA = 4 + (\frac{1}{2})(4)(2)(\sqrt{5})$, or about 12.9 in.^{2}.

Your Turn

Practice: Find the surface areas of the cone and the square pyramid if each edge of the cube is 4 in.

Reflect: Based on the example and practice problems, what do you think the surface area of the cone would be if each edge of the cube were doubled again to 8 in.? Doubled again to 16 in.? Use the formula to determine whether your guesses were correct. ■

Surface Area of a Sphere. The formula for the surface area of a sphere is difficult to verify but is interesting and simple to use. By using calculus, which is beyond the scope of this book, we can verify that the formula is as follows.

Formula for the Surface Area of a Sphere

$SA = 4\pi r^2$, where r is the radius of the sphere.

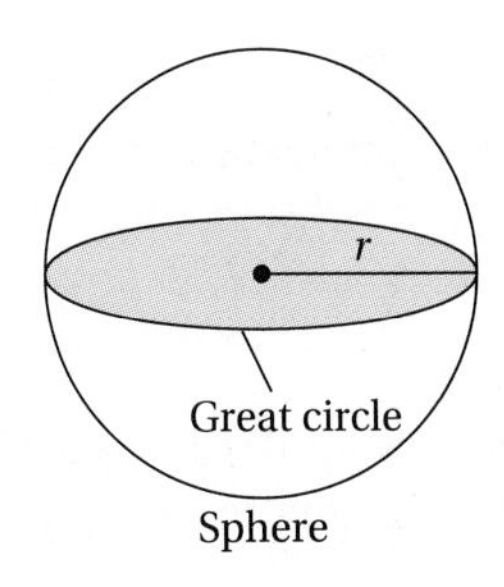

FIGURE 12.23

This formula is interesting in that it essentially states the unexpected result that the surface area of a sphere is four times the area of a great circle of the sphere. Note that a *great circle* of a sphere, as shown in Figure 12.23, is a circular cross section of the sphere with maximum area.

In Mini-Investigation 12.11 you are given the opportunity to apply ideas about the surface area of a sphere to a real-world situation.

Talk about the different procedures you and your classmates used.

Mini-Investigation 12.11 Solving a Problem

How would you find the surface area of a basketball, as accurately as possible?

Volumes of Solids

In this section we consider the volumes of the following solids: rectangular prisms, cubes, right prisms, cylinders, pyramids, cones, and spheres. Recall that volume is measured in cubic units, such as cubic inches, cubic feet, or cubic centimeters.

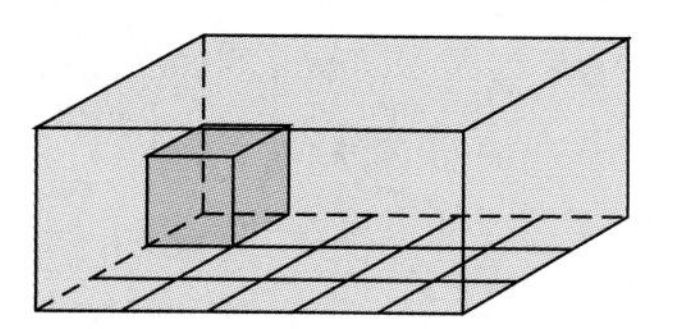

FIGURE 12.24

Volume of Rectangular Prisms and Cubes. The volume of a rectangular prism is the number of cubic units required to fill the prism. For the $2 \times 3 \times 5$ prism shown in Figure 12.24, we can determine its volume by finding out how many unit cubes are needed to fill the prism. On the "floor," or first layer of the prism, that would be 3×5, or 15, unit cubes (one is shown for illustrative purposes). But, as there are two layers, the total number of unit cubes would be 30. Thus the volume of the prism is $2 \times 3 \times 5$, or 30 cu units.

We can also think of finding the volume by multiplying the area of the base (15 sq units) by the height (2 units) to obtain the volume of 30 cubic units. That is, the product of length times width, $l \times w$, can be thought of as the area of the base, B. Multiplying B by the height, h, gives the volume. To find the volume of the cube we have to remember only that the length, width, and height of the cube are the same; that is, $l = w = h$. These ideas suggest the following formulas.

Estimation Application

The Giant Flush

Estimate how much water is used in 1 day in the United States for flushing toilets. Show the data you used to make the estimate, the source of the data, and how you made the estimate.

Formulas for the Volumes of a Rectangular Prism and a Cube

Rectangular Prism: $V = lwh$, where l is the length, w is the width, h is the height of the rectangular prism. $V = Bh$, where B is the area of the base and h is the height of the rectangular prism.

Cube: $V = e^3$, where e is the length of each edge of the cube.

Example 12.18 shows how to apply these formulas.

Example 12.18 Using the Formulas for a Rectangular Prism and Cube

A rectangular prism and a cube have the same volume. If the dimensions of the rectangular prism are 12 cm × 9 cm × 2 cm, what is the length of each edge of the cube?

Solution

The volume of the prism is $V = 12 \times 9 \times 2$, or 216 cm^3. Using the formula for the volume of a cube, $V = e^3$, we get $216 = e^3$, or $e^3 = 216$. Solving for e (by guess and revise, using a calculator), we obtain $e = 6$ cm.

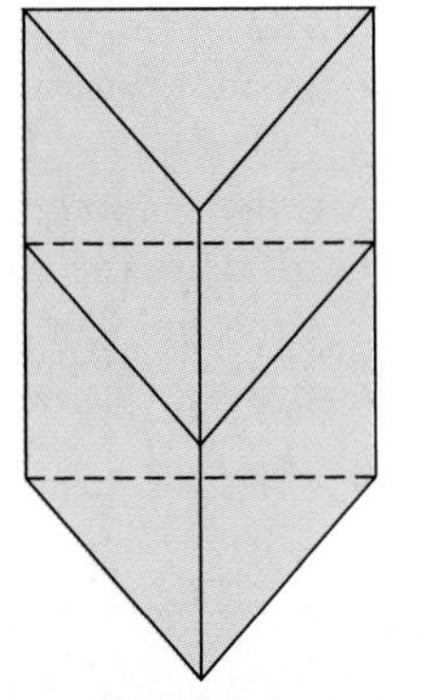

(a) Right triangular prism

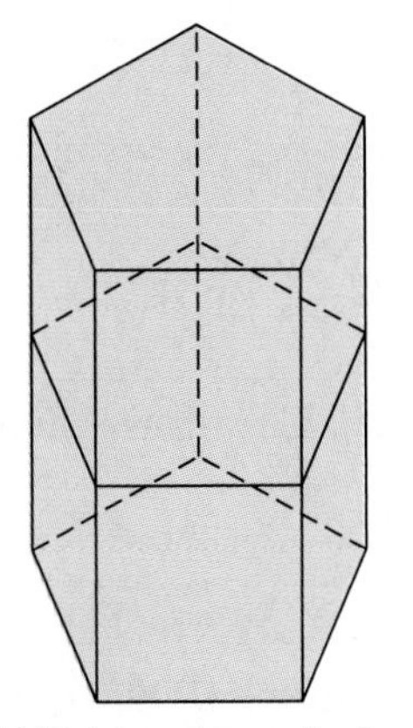

(b) Right pentagonal prism

FIGURE 12.25

Your Turn

Practice: Suppose that the volume of a cube is 64 cu units. Identify the dimensions of two different sizes of rectangular prisms that would give this same volume.

Reflect: Suppose that the volume of the cube was doubled to 128 cu units. What has to happen to the dimensions of the two rectangular prisms that you created if they are to have this same volume? ■

Volume of a Right Prism. Recall that a right prism is a prism in which the two bases are parallel and the other faces are perpendicular to the bases. A right prism is named by its base. For example, a right triangular prism and a right pentagonal prism are shown in Figure 12.25(a) and (b), respectively.

In both parts of Figure 12.25, a cross section is drawn parallel to the bases. If we think of this cross section as a very thin slice of the prism, we can then think of summing these very thin slices of the prism from one base to the other to find the volume of the prism. This summing process suggests the following formula.

Formula for the Volume of a Right Prism

$V = Bh$, where B is the area of the base and h is the height of the prism.

This formula is the same as that for finding the volume of a rectangular prism because a rectangular prism is a special kind of right prism. Example 12.19 demonstrates the application of the formula for the volume of a right prism.

Example 12.19

Problem Solving: The Waterfall

Experts estimate that 8000 cubic feet per second (cfs) of water were falling over a certain waterfall during a recent storm. Suppose that the water from the waterfall fills a prism that has a base of 100 yd × 50 yd, or about the size of a football field. How many yards high would the prism be filled in 1 minute from the falls?

Solution

In 1 minute 480,000 ft^3 of water will go over the falls. The base of the prism contains 5000 yd^2, or 45,000 ft^2. Substituting into $V = Bh$ yields $480{,}000 = 45{,}000h$. Hence $h \approx 10.67$ ft, or about 3.56 yd.

Your Turn

Practice: How deep will the water be in the prism if the water falls at the normal rate of 800 cfs?

Reflect: What effect does doubling the amount of water going over the falls have on the height of the water in the prism? ■

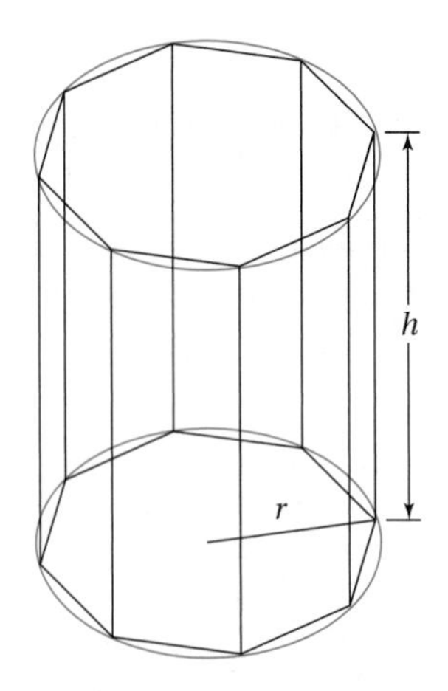

FIGURE 12.26

Volume of a Cylinder. We can think of the volume of a cylinder in the following way. Suppose that a regular right prism, such as an octagonal prism, is inscribed in a cylinder, as shown in Figure 12.26.

As the number of sides of the regular polygon increases, the volume of the prism becomes closer and closer to the volume of the cylinder. Because the volume of a right prism is given by the formula $V = Bh$, the volume of the cylinder also is determined by $V = Bh$, where B is the area of the circular base. Thus we have the following formula.

Formula for the Volume of a Cylinder

$V = \pi r^2 h$, where r is the radius of the cylinder and h is its height.

Example 12.20 illustrates the application of the formula for the volume of a cylinder.

Example 12.20 Problem Solving: Storage Tank Capacity

Suppose that an oil storage tank is 10 m wide and 10 m high. How many kiloliters of oil can it hold?

Solution

The storage tank is a cylinder with a height of 10 m and a radius of 5 m. Hence the volume is $V = \pi r^2 h$ or $V \approx 3.14(5^2)(10)$ or 785 m^3. A cubic meter contains 1000 l, or 1 kl, so the storage tank can hold 785 kl of oil.

Your Turn

Practice: What would be the volume of the storage tank if the height were doubled to 20 m and the radius were doubled to 10 m?

Reflect: Which do you think would have the greater impact on the volume of the cylinder in the example problem—doubling the radius or doubling the height? Why? ■

Volume of a Right Pyramid. In Mini-Investigation 12.12 you are asked to compare the relationship between the volumes of a pyramid and a rectangular prism when they have congruent bases and the same height. We revisit this relationship later in this section.

Write a generalization about what you think the relationship is between the volumes of a rectangular prism and a square pyramid that have congruent bases and the same heights.

Mini-Investigation 12.12 Finding a Pattern

How many square pyramids full of water would be needed to fill a rectangular prism that has the same area base and height as the pyramid? (Test your estimate by making the solids out of aluminum foil and pouring water.)

Compare your generalization in Mini-Investigation 12.12 with the following formula.

Formula for the Volume of a Pyramid

$V = \frac{1}{3}Bh$, where B is the area of the base and h is the height of the pyramid.

Example 12.21 shows how to apply the formula for the volume of a pyramid.

Example 12.21 Finding the Volume of a Pyramid

Find the volume of the square pyramid shown on the left.

P

$PC = 6$ cm
$AB = 2$ cm

6 cm

C

A 2 cm B

Solution

The area of the base is 4 cm^2, and the height is 6 cm. Thus $V = \frac{1}{3}(4 \times 6)$, or 8 cm^3.

Your Turn

Practice: Find the volume of the pyramid in the example problem if the height is doubled but the base remains the same. Find the volume if the length of each side of the base is doubled but the height remains the same.

Reflect: If you had a square pyramid and were to double only one dimension, which one would you double in order to maximize the volume? ■

Volume of a Right Circular Cone. In Mini-Investigation 12.13 you are asked to consider the volume of a cone.

Write a statement describing the relationship, identifying the conditions that must hold.

Mini-Investigation 12.13 Using Mathematical Reasoning

What do you think the relationship is between the volume of a cone and the volume of a cylinder, assuming they have the same radius and height?

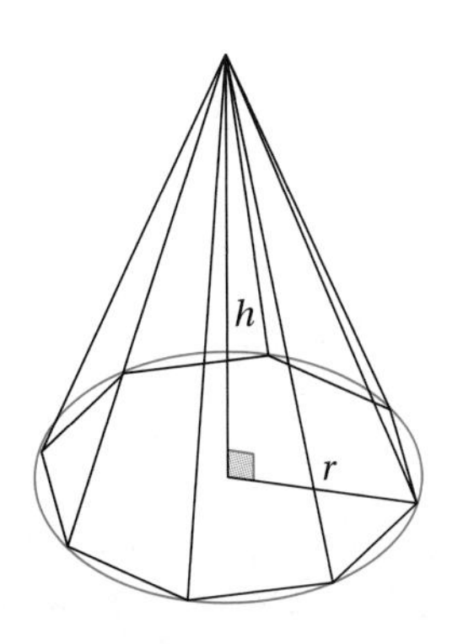

FIGURE 12.27

We can think of the volume of a cone in much the same way that we thought about the volume of a cylinder. That is, we can inscribe a pyramid with a regular n-gon as its base in the cone. As the number of sides of the n-gon base increases, the base more closely approximates the circular base of the cone and the pyramid more closely resembles the cone, as shown in Figure 12.27.

The volume of a pyramid is expressed by the formula $V = \frac{1}{3}Bh$, which suggests that the volume of a cone can also be expressed by the formula $V = \frac{1}{3}Bh$, where B is the area of the circular base and h is the height of the cone. The base of the cone is a circle, so we can express this relationship in the following way.

Formula for the Volume of a Cone

$V = \frac{1}{3}\pi r^2 h$, where r is the radius of the base and h is the height of the cone.

Volume of a Sphere. Suppose that we had a sphere and very small cone-like figures whose vertices are at the center of the sphere and whose bases are on the surface of the sphere, as shown in Figure 12.28.

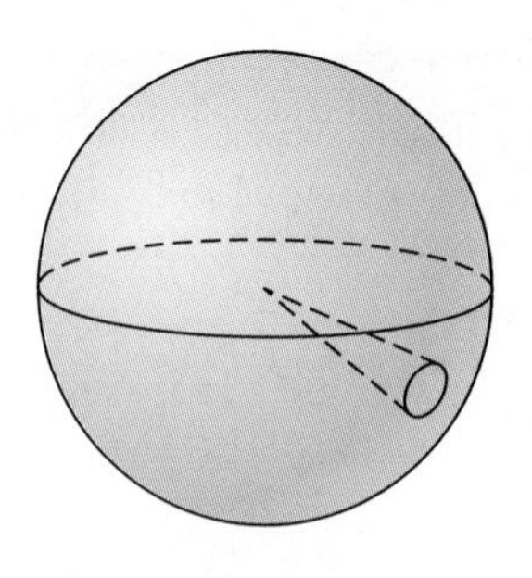

FIGURE 12.28

If the cone-like figure has a very small base that sweeps over the entire surface of the sphere, we can think of the volume of the sphere in the following way. The volume of a small cone-like figure is approximately $V = \frac{1}{3}Bh$, where B is the area of the base and h is the height of the cone. If the cone-like figure is small enough, h approximates the radius of the sphere and the base approximates a circle. By summing the volumes of the cone-like figures as they cover the entire surface of the sphere, the volume of the sphere can be expressed as one-third times the surface area of the sphere times the radius of the sphere. This result gives the following formula.

Formula for the Volume of a Sphere

$V = \frac{4}{3}\pi r^3$, where r is the radius of the sphere.

Example 12.22 illustrates how to apply the formulas for the volumes of a cylinder, cone, and sphere.

Example 12.22

Problem Solving: Volume of a Silo

What is the volume of a silo if the silo's diameter is 8 ft and height is 30 ft?

Solution

We can solve the problem by breaking it into two problems—finding the volumes of the cylinder and the hemisphere that forms the roof.

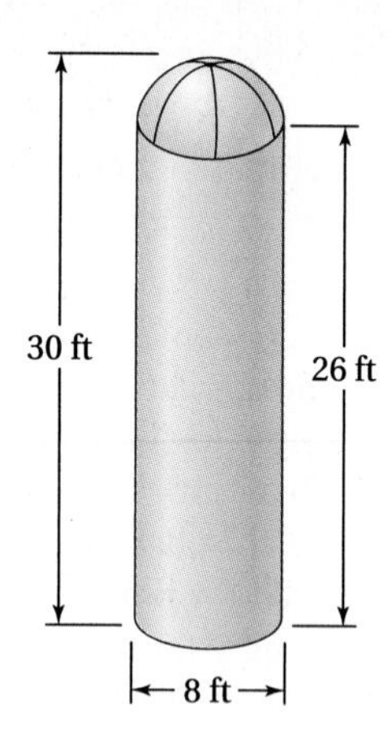

With a diameter of 8 ft, the radius of the silo's base is 4 ft and the sphere's radius also is 4 ft. If the silo's overall height is 30 ft, the cylinder's height is 26 ft. Thus $V_{\text{cylinder}} \approx 3.14(4^2)(26)$, or 1306 ft^3 (to the nearest cubic foot); $V_{\text{hemisphere}} \approx (\frac{1}{2})(\frac{4}{3})(3.14)(4^3)$, or 134 ft^3 (to the nearest cubic foot), so the volume of the silo is 1440 ft^3 (to the nearest cubic foot).

Your Turn

Practice: What would be the volume of the silo if the overall height were increased by 1 ft and the other dimensions remained the same?

Reflect: Which do you think would have the greater effect on the volume of the silo in the example problem, increasing the overall height by 1 ft or increasing the width by 1 ft? Justify your answer. ■

Comparing Surface Areas and Volumes of Similar Solids

In this section we consider the relationship between the sizes of similar solids. In Mini-Investigation 12.14 you are asked to identify the relationship between the surface areas and the volumes of two cubes of different sizes.

Write general statements about how doubling the edge length of a cube affects
a. the cube's surface area.
b. the cube's volume.

Mini-Investigation 12.14 Finding a Pattern

How does the surface area and volume of a cube made from a pattern of 1-in. squares compare with the surface area and volume of a cube made from a pattern of 2-in. squares? (Make the cubes, if needed, from a pattern like the one shown.)

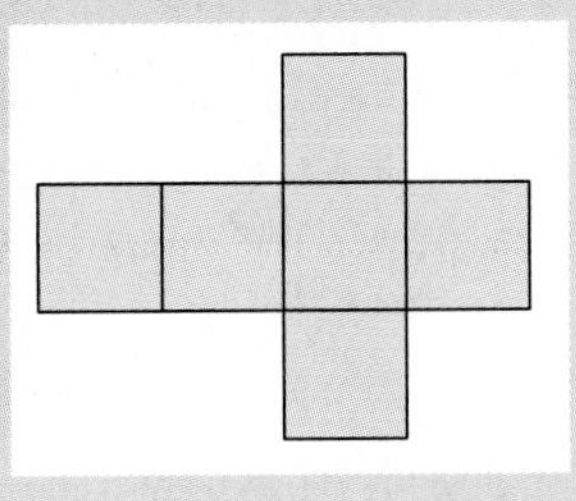

As suggested in Mini-Investigation 12.14, the surface areas and volumes of similar cubes are related in a particular way. Now consider the two rectangular prisms shown in Figure 12.29, which have the same shape but the edges of one are three times as long as the edges of the other.

The surface areas and volumes of the two rectangular prisms are as follows.

Figure 12.29(a)

$SA_{(a)} = 1 + 1 + 2 + 2 + 2 + 2$
$= 10$ sq units
$V_{(a)} = 1 \times 1 \times 2 = 2$ cu units

Figure 12.29(b)

$SA_{(b)} = 9 + 9 + 18 + 18 + 18 + 18$
$= 90$ sq units
$V_{(b)} = 3 \times 3 \times 6 = 54$ cu units

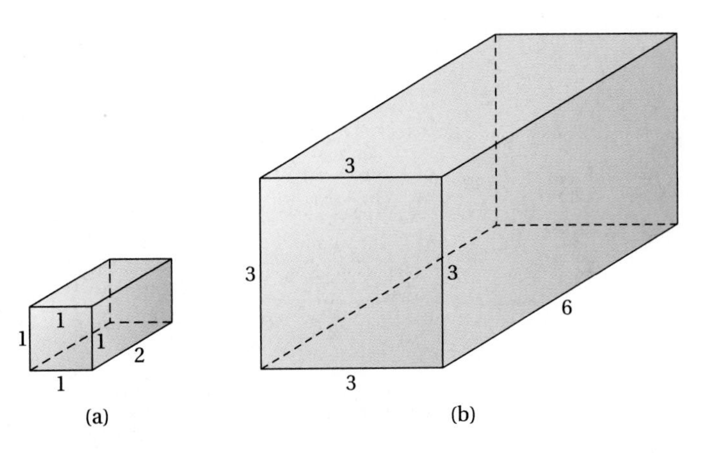

FIGURE 12.29 | Similar rectangular prisms.

Forming the ratios of the corresponding surface areas and volumes, we obtain the following.

Surface Area

$$\frac{SA_{(a)}}{SA_{(b)}} = \frac{10}{90} = 1:9, \text{ or } 1:3^2.$$

Volume

$$\frac{V_{(a)}}{V_{(b)}} = \frac{2}{54} = 1:27, \text{ or } 1:3^3.$$

This example and Mini-Investigation 12.14 suggest the following relationship between the lengths of the corresponding parts of similar solids and their surface areas and volumes.

Generalization About the Ratios of Surface Areas and Volumes in Similar Solids

If the lengths of corresponding parts of two similar solids are in the ratio of $a{:}b$, then

- the ratio of their surface areas is $a^2:b^2$ and
- the ratio of their volumes is $a^3:b^3$.

In Example 12.23 we consider the effect on surface area as the dimensions of a cylinder change.

Example 12.23 Problem Solving: The Cost of Producing Tuna Cans

Suppose that a manufacturer has determined that the cost of materials for making a cylindrical can to hold tuna is 10% of the $1 that the can of tuna sells for. The manufacturer is considering doubling each dimension of the can and selling the larger can for $6. As a percentage of the selling price, would the cost of materials for making the larger can go up or down? Why?

Working toward a Solution

Understand the problem	*What does the situation involve?*	Finding out how using a larger can would change the cost of materials as a percentage of the selling price
	What has to be determined?	The cost of the materials in the larger can compared to its price
	What are the key data and conditions?	The dimensions of the larger can are twice those of the smaller can. The cost of materials is 10% of the selling price for the smaller can. The smaller can sells for $1 and the larger can sells for $6.
	What are some assumptions?	The cost of materials is a factor in determining the cost of the can and is proportional to the amount of materials used.
Develop a plan	*What strategies might be useful?*	Use reasoning.

	Are there any subproblems?	Determine the ratio of the two surface areas. Find the cost of materials for the larger can.
	Should the answer be estimated or calculated?	Calculated
	What method of calculation should be used?	Use a calculator.
Implement the plan	*How should the strategies be used?*	We can reason in the following way. If the diameter and the height of the cylinder are doubled, the surface area will be increased fourfold. The cost of materials for the smaller can is 10¢, the cost of the materials for the larger can will be 40¢, or 6.7% of the cost of the larger can.
	What is the answer?	The manufacturer would decrease the relative cost of materials by using the larger can.
Look back	*Is the interpretation correct?*	We determined the cost of the materials in the larger can and compared it to the cost of $6, which is what the problem asks for.
	Is the calculation correct?	Four times 10¢ is 40¢; 40¢ divided by $6 is about 0.067.
	Is the answer reasonable?	Yes. The price increased faster than the amount of materials used. This outcome suggests that the percent cost of materials should decrease.
	Is there another way to solve the problem?	We could reason that the cost of materials increased fourfold but that the cost increased sixfold, making the materials relatively less costly.

Your Turn

Practice: In the example problem, if the manufacturer had tripled each dimension, would that have increased or decreased the cost of materials relative to the selling price as compared to doubling the dimensions?

Reflect: In the example problem the cost of materials for the can is 10% of the selling price. Would a different percent change the answer to the question? Why or why not? ■

Connection to the PreK–8 Classroom

Connecting Changes in Dimensions to Changes in Volume

When comparing the effect of increasing the dimensions of solids on their surface areas and volumes, students should use models to make actual comparisons. Having solids that can be filled with water is helpful in determining the effects on the solids if, for example, the dimensions are doubled or tripled. Use of such models, as illustrated in Figure 12.30, can enable students to visualize the effect of increasing the size of the solid rather than just thinking of the effect abstractly.

- **What** examples would you use to help students determine the effect of doubling the radius of a cylinder on the volume of the cylinder? Justify your choice.
- **Is** it possible for two cylinders to have the same lateral surface area but different volumes? Why or why not?
- **If** the dimensions of a solid each increases 50%, what percent does the volume of the solid increase? Explain.

12-3 Volume of Cylinders

What you'll learn
You'll learn to find the volume of cylinders.

When am I ever going to use this?
Many food containers are shaped like cylinders.

Word Wise
cylinder

The stack of quarters is a model of a **cylinder**. A cylinder is a solid figure that has two congruent, parallel circles as its bases.

To find the value of a stack of quarters, you multiply the value of one quarter, 25¢, by the number of quarters in the stack. You can use a similar method to find the volume of a cylinder.

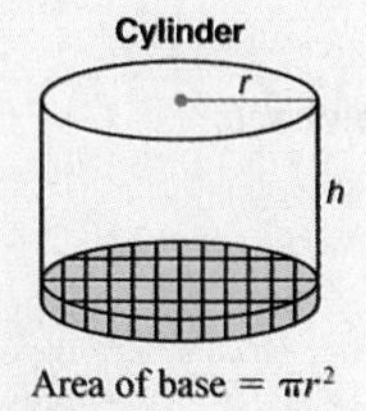

Area of base = πr^2

The area of the base tells how many unit cubes cover the base. The height tells how many layers there are.

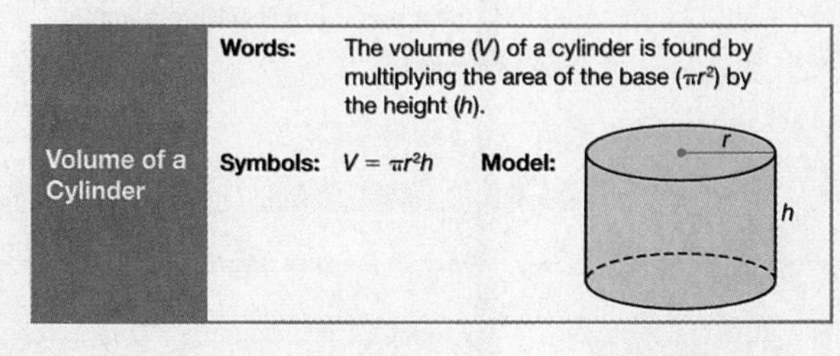

Volume of a Cylinder	
Words:	The volume (V) of a cylinder is found by multiplying the area of the base (πr^2) by the height (h).
Symbols:	$V = \pi r^2 h$ **Model:** (r, h)

Example 1 Find the volume of a cylinder with a diameter of 6 inches and a height of 4 inches.

The diameter of the cylinder is 6 inches. Therefore, the radius is 3 inches.

Estimate: $3^2 \times 3 \times 4 = 108$

$V = \pi r^2 h$ *Use 3.14 for* π.

$V \approx 3.14 \cdot 3^2 \cdot 4$ *Replace r with 3 and h with 4.*

$V \approx 113.04$

The cylinder has a volume of about 113 cubic inches.

Study Hint
Estimation You can estimate the volume of a cylinder by squaring the radius and multiplying the answer by 3 (the approximate value of π) and the height.

FIGURE 12.30 | Excerpt from an elementary school textbook.

(*Source:* From *Mathematics: Applications and Connections, Course 2*, by Collins, William, et al. Copyright © 2001, Glencoe/McGraw-Hill, p. 503. Reprinted by permission of Glencoe/McGraw-Hill.)

Problems and Exercises | *for Section 12.3*

Use $\pi = 3.14$ and a calculator, as necessary. In this exercise set, if lines or planes appear parallel or perpendicular, assume that they are.

A. Reinforcing Concepts and Practicing Skills

Find the surface area and volume of the solids in Exercises 1–13.

1.

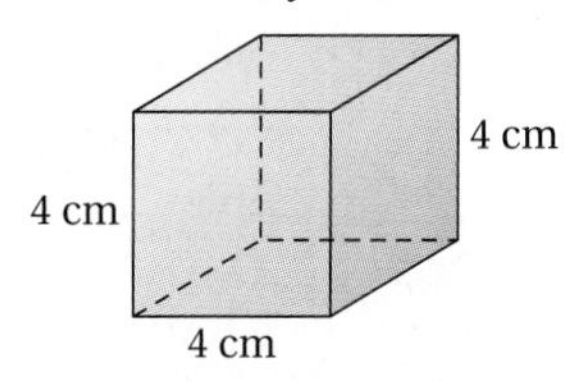

2.

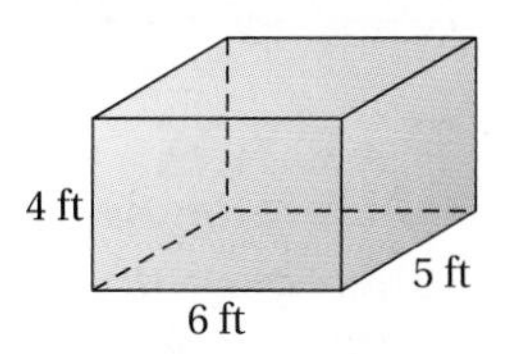

3.

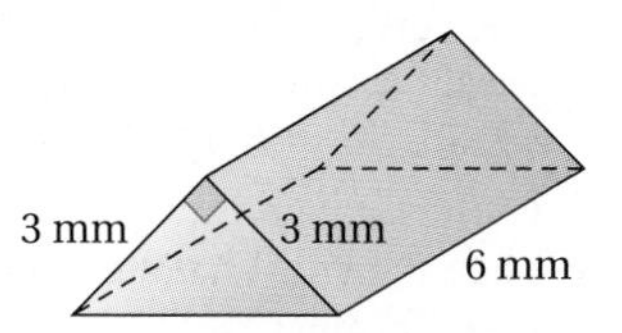

4.

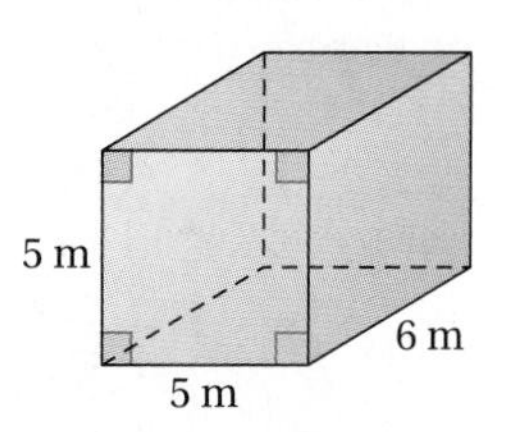

5.

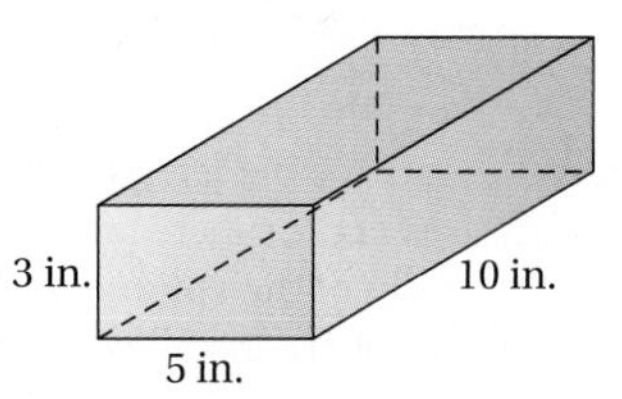

6.

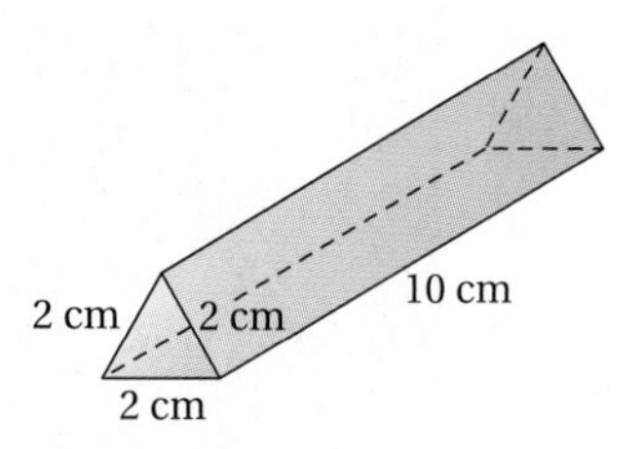

7.

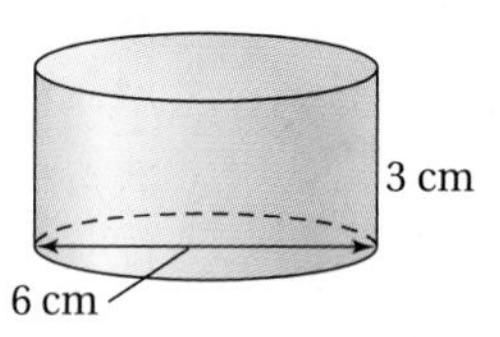

8.

9 cm
2 cm

9.

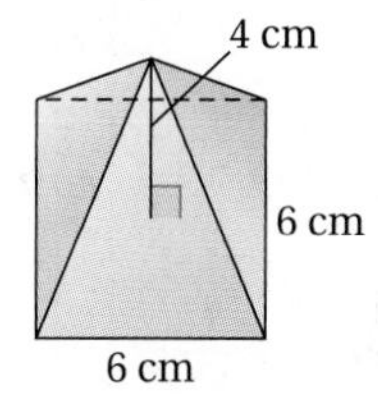

10.

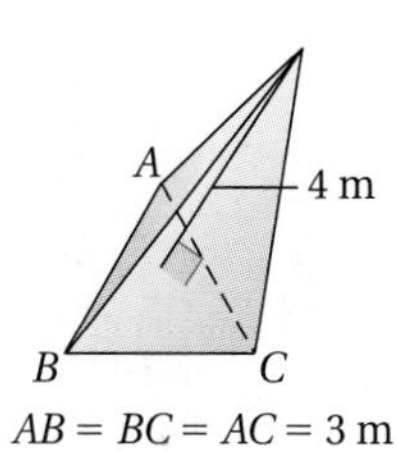

$AB = BC = AC = 3$ m

11.

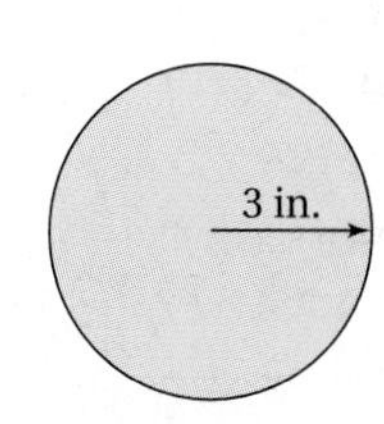

12.

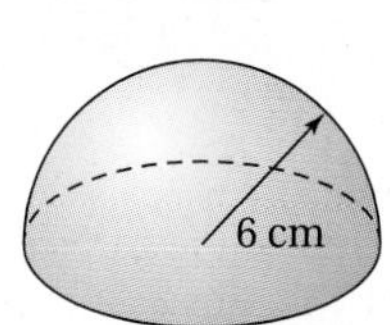

13.

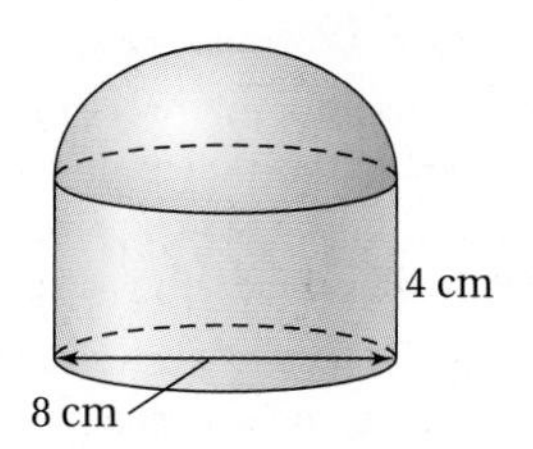

Find the volume of the solids in Exercises 14–18.

14.

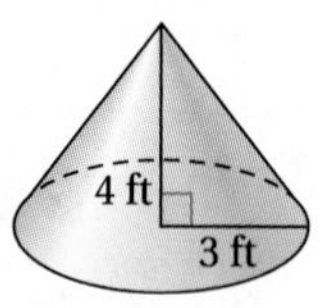

15.

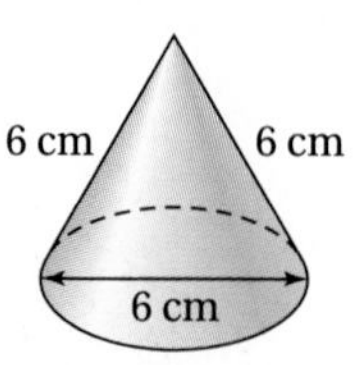

16.

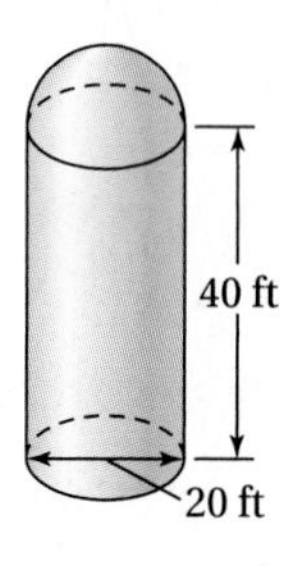

17.

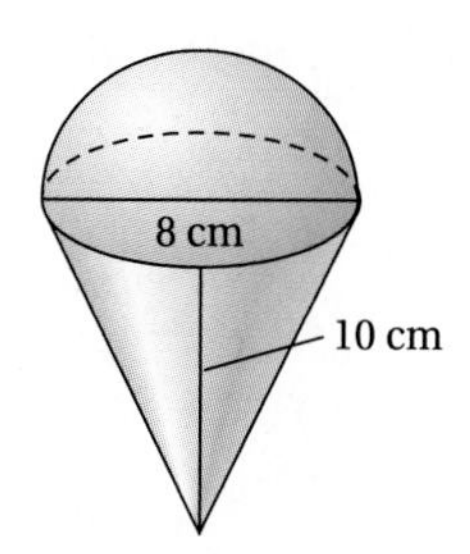

18.

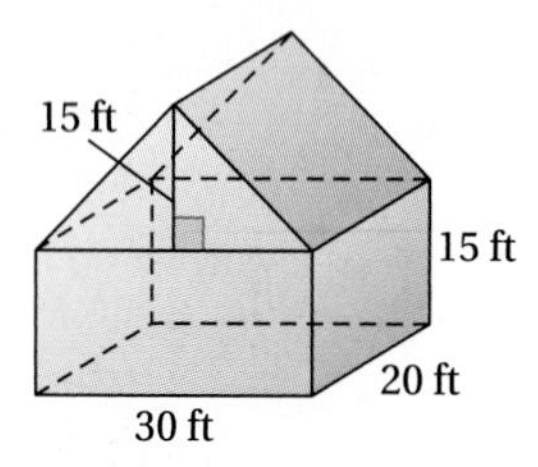

19. Determine the ratio of the surface areas of the pairs of solids in the next column.

20. Determine the ratio of the volumes of the pairs of solids shown in Exercise 19.

21. Experts agree that about three-fourths of the earth's surface is water. If the radius of the earth is approximately 3850 mi, how many square miles of water are there? (Assume that the earth is a sphere.)

22. How many cubic feet of topsoil should Mr. Jackson order if he wants to cover a rectangular area of his lawn that measures 60 ft $\times$ 40 ft to a depth of 3 in.?

Figures for Exercise 19

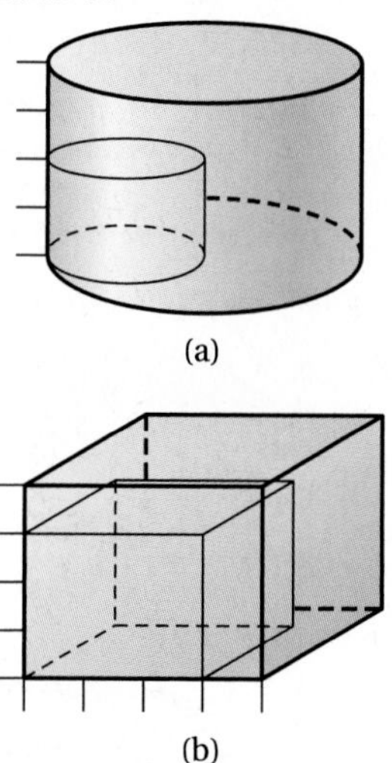

(a)

(b)

B. Deepening Understanding

23. Identify the corresponding parts for the following solids.
a. Two similar cylinders **b.** Two similar spheres
c. Two similar cones **d.** Two similar pyramids

24. Why does the formula for the volume of a pyramid, $V = \frac{1}{3}Bh$, give cubic units but the formula for the surface area of a pyramid, $SA = B + \frac{1}{2}Ph$, gives square units?

25. Fred wonders how much canvas it will take to make a tent like the one shown. What measures does he need to make and what formula would he find most helpful? What adjustment would he need to make in using the formula if the tent had a dirt floor?

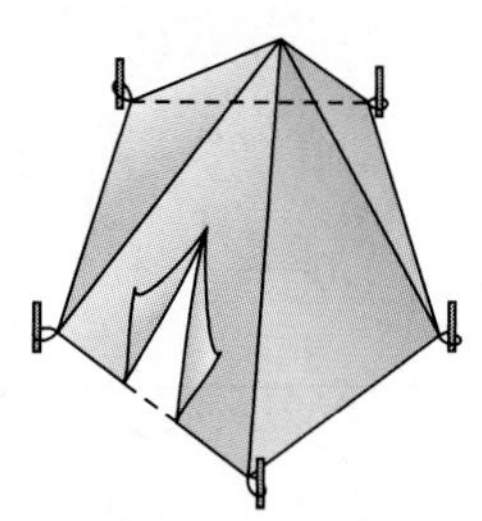

26. **a.** What is the ratio of the lateral area of the cylinder to the surface area of the sphere inscribed in the cylinder?
b. What is the ratio of the volume of the cylinder to the volume of the sphere inscribed in the cylinder?

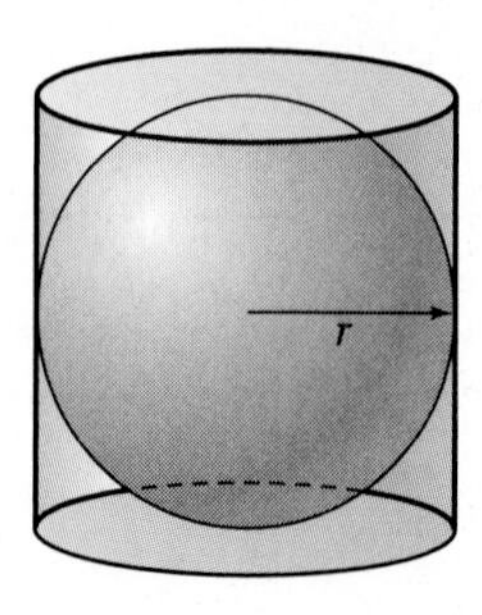

27. If the volume of a liquid in a sphere with radius 5 cm is transferred to a cylinder with the same radius in its base, to what height would the cylinder be filled?

28. A cylindrical can of cleaner holds 1 l of liquid. What is its height if the diameter of the base is 10 cm?

29. A container of popcorn holds 216 in.^{3} of popcorn. Give the dimensions of a cubical container that could hold the popcorn and three other dimensions of rectangular prisms that could hold the popcorn.

30. Ms. Jackson is considering building a 3000-ft^2 house. She is debating whether to have 8-ft ceilings or 9-ft ceilings. If she builds the house with the higher ceilings, how many more cubic feet of air do the furnace and air conditioner have to warm and cool?

31. Katrina has a fish tank that has a 30-cm × 60-cm base. The water is 20 cm deep. When she puts decorative rocks in the water, the water level rises 2 cm. What is the volume of the rocks?

32. What volume of wood remains if a 2-cm hole is bored out of a cube with edge lengths of 4 cm?

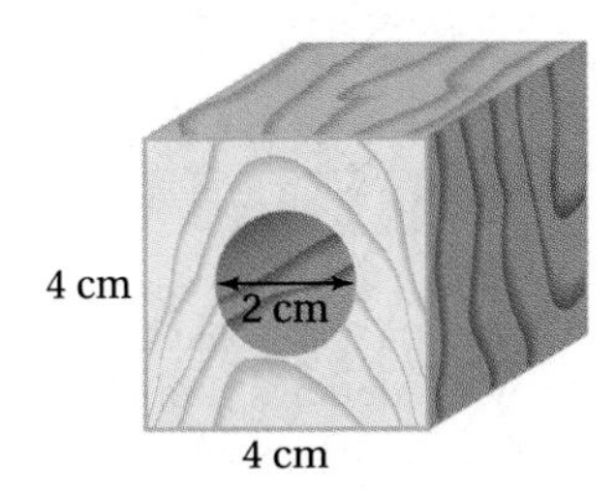

C. Reasoning and Problem Solving

33. Tracy says that if the length and the width of a 2-in. × 3-in. × 5-in. rectangular prism are doubled and the height is tripled, the volume will increase by a factor of 12. Jason says that you can't tell unless you specify which dimensions are the length, width, and height. Who is correct? Why?

34. Consider the square pyramid shown. If the size of the base remains the same and the height of the pyramid varies, what are the minimum surface area and minimum volume the pyramid can have?

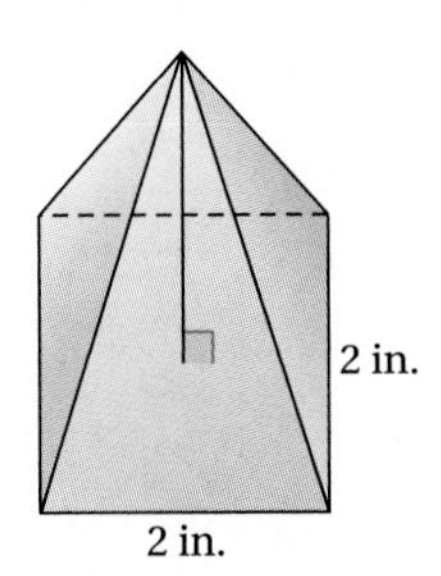

35. A copper pipe is 5 m long. It has an inside diameter of 1.5 cm and an outside diameter of 1.75 cm. How many cm^3 of copper are in the pipe?

36. The two cereal boxes shown have corresponding edges in a ratio of 2:3. If the smaller box sells for $2.50 and the larger box for $4.00, which is the better buy? Why? What assumption(s) do you have to make when solving the problem? Estimate, then check.

37. What percent of the volume of a can of three tennis balls consists of the balls? Assume that the balls are tangent to the sides, top, and bottom of the can.

38. Frankie and Johnnie are sharing a snow cone. If Frankie eats the hemisphere part of the cone (down to the level part), what percent of the cone has he eaten?

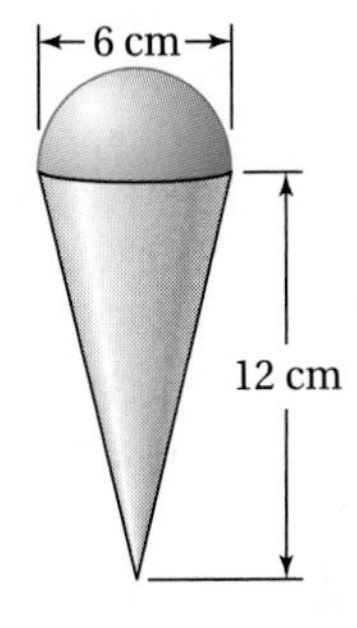

39. A concrete vat for storing chemicals has a length of 9 ft, a width of 5 ft, a depth of 4 ft, and a thickness of 2 in. Sketch the vat. How much concrete is needed to make the vat?

40. A cone is sliced into two pieces, as shown. Find the volumes of the pieces.

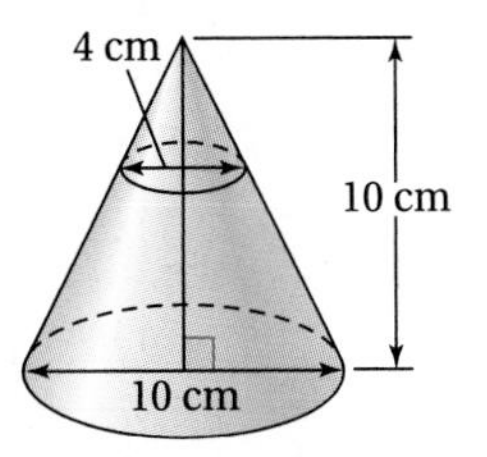

Figure for Exercise 40

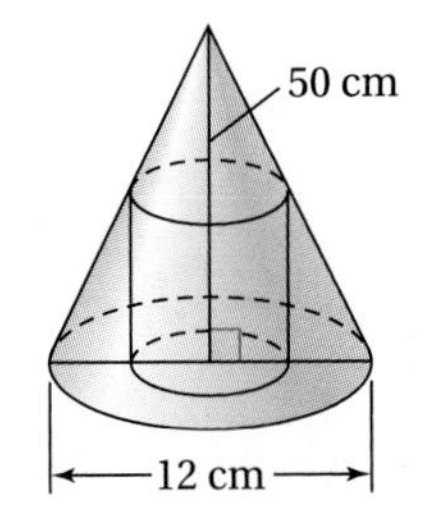

Figure for Exercise 41

41. A cylinder is inscribed in a cone, as shown. What percent of the cone is the cylinder if the diameter of the base of the cylinder is one-half the diameter of the cone?

42. Find the volume of a regular tetrahedron in which each edge is 4 in. long.

43. **The Packing Problem.** A box is packed with six soup cans, as shown in the following figure. What percent of the box is filled with the cans? If the same size of box contains eight cans rather than six, does the percent of the volume of the box occupied by the cans increase or decrease? Explain.

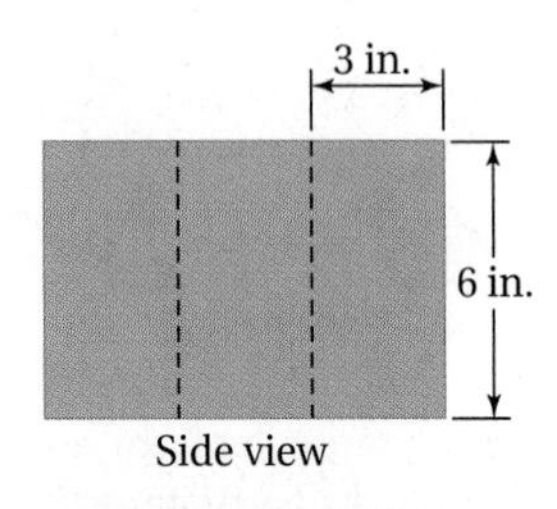

Side view

Top view

44. **The Canal Problem.** An engineer's plan shows a canal with a trapezoidal cross section that is 8 ft deep and 14 ft across at the bottom with walls sloping outward at an angle of 45°. The canal is 620 ft long. A contractor bidding on the job estimates the cost to excavate the canal at $\$175/\text{yd}^3$. If the contractor adds 10% profit, what should the bid be?

45. Find the surface area and volume of the solid casting shown.

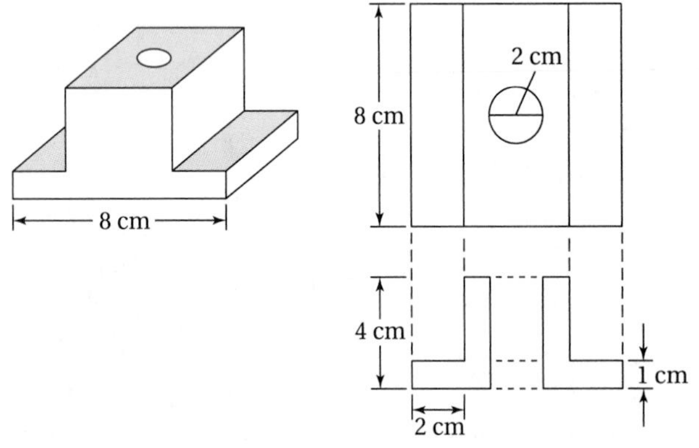

Source: From *Addison-Wesley Geometry* by S. Clemens, P. O'Daffer, T. Cooney, and J. Dossey. Copyright © 1994 by Addison-Wesley Publishing Company, p. 510.

46. Find the volume of the solid that was cut on a slant from the cylinder shown in the next column.

Figure for Exercise 46

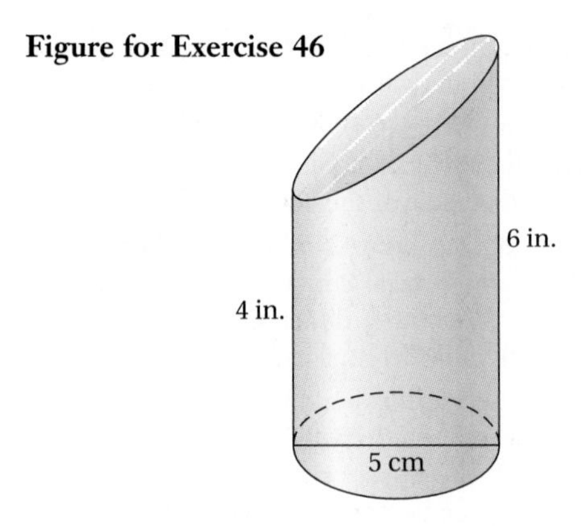

D. Communicating and Connecting Ideas

47. **Communication.** Consider the three cylinders shown.

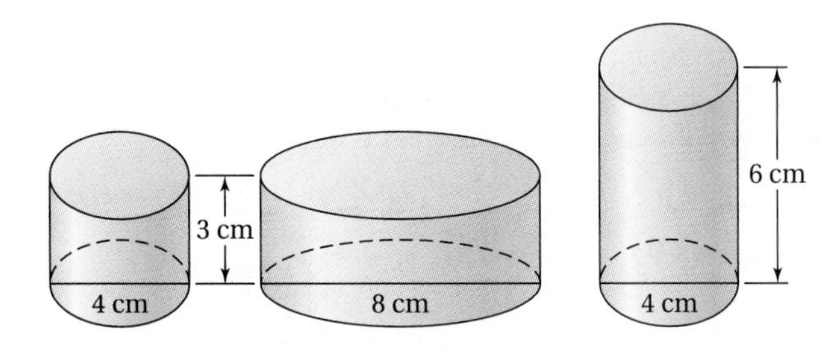

Find the surface area and volume of these three cylinders, analyze your findings, and write statements about the following relationships involving cylinders.

a. The effect on surface area if the radius is doubled but the height remains the same

b. The effect on volume if the radius is doubled but the height remains the same

c. The effect on surface area if the height is doubled but the radius remains the same

d. The effect on volume if the height is doubled but the radius remains the same.

48. **Making Connections.** Use a calculator to solve the following problem. Suppose that a 20-cm × 30-cm piece of tin has its corners cut out and the resulting piece is folded into an open box. The x quantities are whole numbers. When the cut is 1 cm ($x = 1$), the resulting box is as shown. What value of x will give the maximum volume of the box?

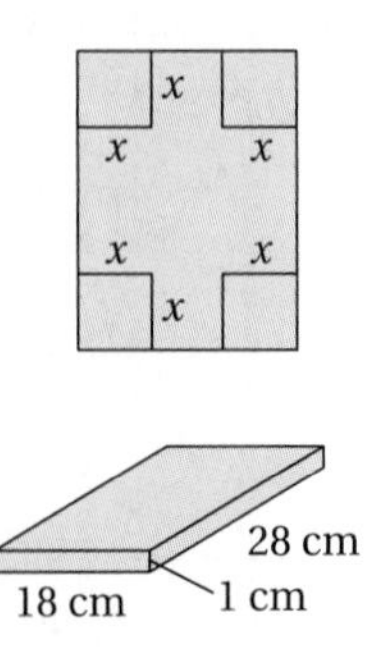

49. Making Connections. Use a calculator to solve the following problem. A 15-in. × 15-in. piece of construction paper has square corners cut out and then is folded to make an open box. For the integers 1–7, determine which size of cut maximizes the volume.

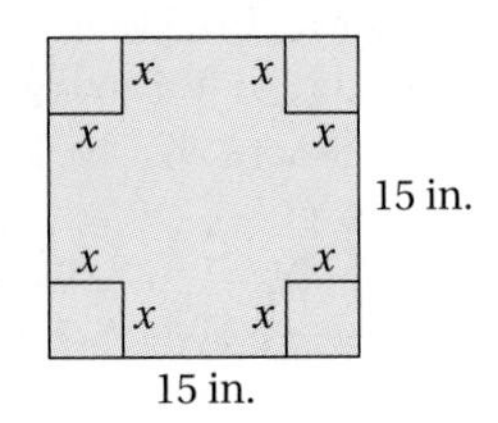

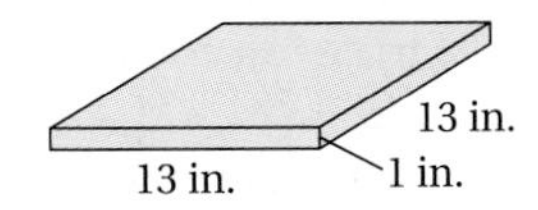

50. How do the volumes of the two solids shown seem to compare? What assumptions do you have to make to arrive at your conclusion?

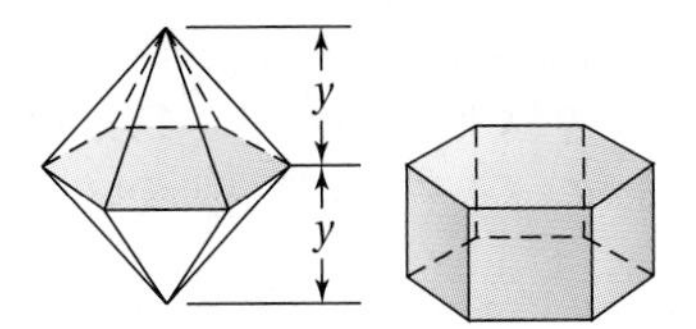

51. Historical Pathways. One of the most famous mathematics problems of antiquity is how to construct with compass and straightedge a cube that would have twice the volume of a given cube. Some mathematical historians trace the problem back to the Pythagoreans who considered the problem in about 540 B.C. A cube with an edge of 1 in. has a volume of 1 in.^{3}. Determine to the nearest 0.01 in. the length of the edge of a cube if the volume of the cube is twice that of the unit cube.

52. Historical Pathways. The following theorem is called Cavalieri's theorem, after Bonaventura Cavalieri (1598–1647), an Italian mathematician and disciple of Galileo.

Let S and T be two solids and let X be a plane. If every plane parallel to X that intersects S or T intersects both S and T in a cross section having the same area, then the volume of S equals the volume of T.

Any plane parallel to the bases of the three pyramids below cuts off sections of equal area on each pyramid. Using Cavalieri's theorem, what conclusions can you draw?

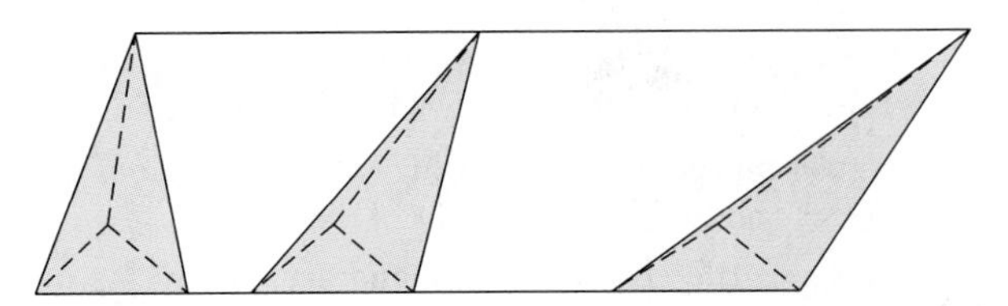

CHAPTER SUMMARY

Key Ideas: Questions and Answers

Section 12.1

- **What is the measurement process?** *(p. 648)* The measurement process has three components: the object to be measured, a unit of measure, and the determination of the number of units needed to cover or fill the object.
- **What are some types of measures and units used?** *(pp. 648–650)* Various types of measures include length, area, volume, mass, temperature, and time.
- **What are nonstandard units of measure?** *(pp. 650–651)* Nonstandard units of measure possess the same characteristics as any standard units of measure but vary in their sizes and are not used by everyone.
- **What are some standard units of measure?** *(pp. 651–654)* Standard units of measure in the English system include the inch, foot, yard, and mile for length; the square foot, square yard, acre, and square mile for area; the ounce, pint, quart, and gallon for capacity; and the pound for weight. Standard units of measure in the metric system include meter for length, square meter for area, liter for capacity, and gram for mass. The various divisions of the basic units for the metric system include kilo, hecto, deka, deci, centi, and milli.
- **How are time and temperature measured?** *(pp. 655–656)* The common units for measuring time include second, minute, hour, day, week, month, and year. The common units of measure for temperature are the Celsius and Fahrenheit units.

Section 12.2

- **How can the perimeters of common polygons be found?** *(pp. 659–660)* The perimeter of a polygon can be found by summing the lengths of each side of the polygon. Formulas for finding the perimeter of common polygons are as follows.

  Rectangle: $P = 2l + 2w$, where l and w are the length and width of the rectangle.

  Square: $P = 4s$, where s is the length of each side of the square.

  Parallelogram: $P = 2b + 2s$, where b and s are the lengths of the consecutive sides of the parallelogram.

  Triangle: $P = a + b + c$, where a, b, and c are lengths of the three sides of the triangle.

Equilateral triangle: $P = 3s$, where s is the length of each side of the triangle.

Regular polygon: $P = ns$, where n is the number of sides and s is the length of each side of the polygon.

- **How can the areas of common polygons be found?** *(pp. 661–667)* The area of a figure is found by determining the number of square units required to cover the region being measured. Formulas for finding the area of common polygons are as follows.

Rectangle: $A = lw$, where l is the length and w is the width of the rectangle.

Square: $A = s^2$, where s is the length of each side of the square.

Parallelogram: $A = bh$, where b is the base and h is the height to the base of the parallelogram.

Triangle: $A = \frac{1}{2}bh$, where b is the base of the triangle and h is the height from the apex to the base.

Equilateral triangle: $A = (\frac{\sqrt{3}}{4})s^2$, where s is the length of each side of the triangle.

Trapezoid: $A = \frac{1}{2}(a + b)h$, where a and b are the lengths of the two bases and h is the height of the trapezoid.

Regular polygon: $A = \frac{1}{2}aP$, where a is the length of the apothem and P is the perimeter of the polygon.

- **How can the circumference and area of circles be found?** *(pp. 667–670)* The circumference of a circle is the distance around the circle. The circumference and area of a circle can be found by using the following formulas.

$C = \pi d$ or $C = 2\pi r$, where d and r are the diameter and the radius of the circle.

$A = \pi r^2$, where r is the radius of the circle.

The number pi (π) is the ratio of the circumference of a circle to the diameter of the circle.

- **How do the perimeters and areas of similar polygons compare?** *(pp. 670–673)* If the scale factor of two similar polygons is k, the ratio of their perimeters is k:1 and the ratio of their areas is k^2:1.

Section 12.3

- **How can the surface area of solids be found?** *(pp. 679–684)* The surface area, SA, of a polyhedron is the sum of the areas of the faces of the polyhedron. The following formulas can be used to find the surface areas of common solids.

Rectangular prism: $SA = 2(hw + wl + hl)$, where h, w, and l are the height, width, and length of the prism.

Cube: $SA = 6e^2$, where e is the length of each edge of the cube.

Cylinder: $SA = 2\pi rh + 2\pi r^2$, where r is the radius of the circular base and h is the height of the cylinder.

Right regular pyramid: $SA = B + \frac{1}{2}Ph$, where B is the area of the base, P is the perimeter of the base, and h is the height of the triangular faces of the pyramid.

Right circular cone: $SA = \pi r^2 + \pi rl$, where r is the radius of the base and l is the slant height of the cone.

Sphere: $SA = 4\pi r^2$, where r is the radius of the sphere.

- **How can the volume of solids be found?** *(pp. 684–689)* The following formulas can be used to find the volume of common solids.

Rectangular prism: $V = lwh$, where l is the length, w is the width, and h is the height of the prism.

Cube: $V = e^3$, where e is the length of each edge of the cube.

Right prism: $V = Bh$, where B is the area of the base and h is the height of the prism.

Cylinder: $V = \pi r^2 h$, where r is the radius and h is the height of the cylinder.

Pyramid: $V = \frac{1}{3}Bh$, where B is the area of the base and h is the height of the pyramid.

Cone: $V = \frac{1}{3}\pi r^2 h$, where r is the radius of the base and h is the height of the cone.

Sphere: $V = \frac{4}{3}\pi r^3$, where r is the radius of the sphere.

- **How do surface area and volume of similar solids compare?** *(pp. 689–692)* If the lengths of corresponding parts of two similar solids are in the ratio of a:b, the ratio of their surface areas is a^2:b^2 and the ratio of their volumes is a^3:b^3.

Key Terms, Concepts, and Generalizations

Section 12.1

Measurement process (p. 648)
Length (p. 648)
Area (p. 648)
Volume (p. 648)
Mass (p. 648)
Temperature (p. 648)
Time (p. 648)
English system (p. 651)
Metric system (p. 652)
Celsius scale (p. 655)
Fahrenheit scale (p. 655)

Section 12.2

Perimeter (p. 659)
Perimeter of a square (p. 660)
Perimeter of a rectangle (p. 660)
Perimeter of a parallelogram (p. 660)
Perimeters of triangles (p. 660)
Perimeter of a regular polygon (p. 660)
Area of a rectangle (p. 661)
Area of a square (p. 661)
Area of a parallelogram (p. 662)
Areas of triangles (p. 664)
Area of a trapezoid (p. 665)
Area of a regular polygon (p. 667)
Pi (π) (p. 668)

Section 12.3

CHAPTER REVIEW

Concepts and Skills

1. How many square centimeters are in a square meter?
   a. 0.0001 **b.** 0.01
   c. 100 **d.** 10,000
2. How many liters are contained in a cube with each edge 1 m?
   a. 0.01 **b.** 10
   c. 100 **d.** 1000
3. A house is advertised as having 2700 ft^2. How many square yards does it contain?
   a. 100 **b.** 300
   c. 900 **d.** 8100
4. The perimeter of a square is 12 in. What is the area of the square?
   a. 3 in.^{2}. **b.** 9 in.^{2}.
   c. 16 in.^{2}. **d.** 144 in.^{2}.
5. Triangle *ABC* is a right triangle.

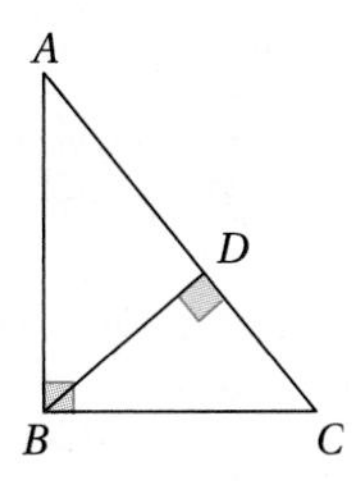

   If $AB = 16$ in., $BC = 12$ in., and $AC = 20$ in., what is the length of $\overline{BD}$?
   a. 4.8 in. **b.** 9.6 in.
   c. 12 in. **d.** None of these
6. If the edge of a cube is doubled, the volume is
   a. one-half as much. **b.** doubled.
   c. quadrupled. **d.** None of these
7. The trapezoid shown in the next column has bases $\overline{AB}$ and $\overline{CD}$, where $\overline{AB}$ is twice as long as $\overline{CD}$. What is the ratio of the area of the trapezoid to the area of triangle *CED*?

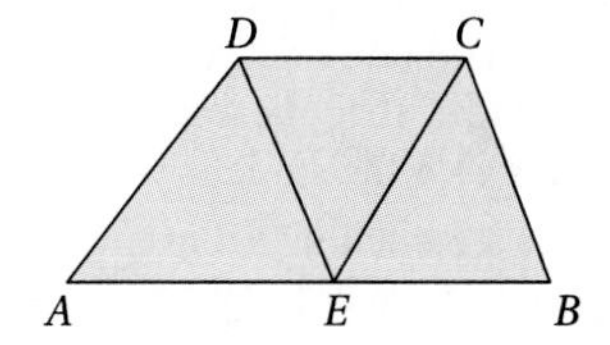

   a. 3:1
   b. 2:1
   c. 1:3
   d. Not enough information is given to decide.
8. What is the area of parallelogram *ABCD*?

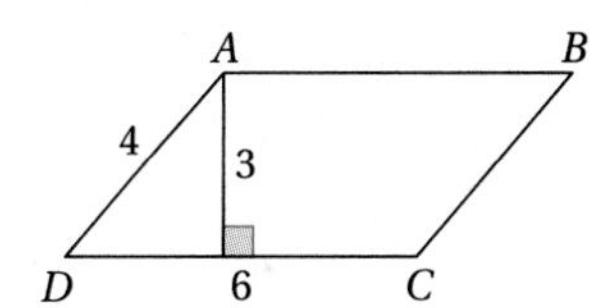

   a. 9 sq units **b.** 18 sq units
   c. 20 sq units **d.** 24 sq units
9. If the ratio of the perimeters of two similar figures is 4:1, what is the ratio of their areas?
   a. 2:1 **b.** 4:1
   c. 16:1 **d.** None of these
10. Find the volume of a sphere inscribed in a cube with each edge 2 cm.
11. If the cone and cylinder shown have the same base and the same volume, what is the height of the cone?

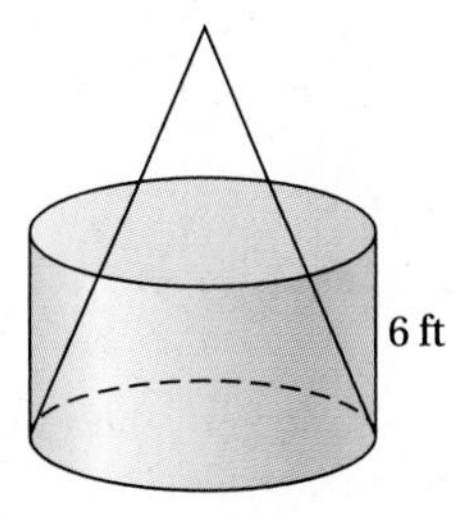

12. A cylinder has a diameter of 6 ft and a height of 3 ft.
    a. Find its surface area. **b.** Find its volume.
13. For the prism shown,
    a. find its surface area. **b.** find its volume.

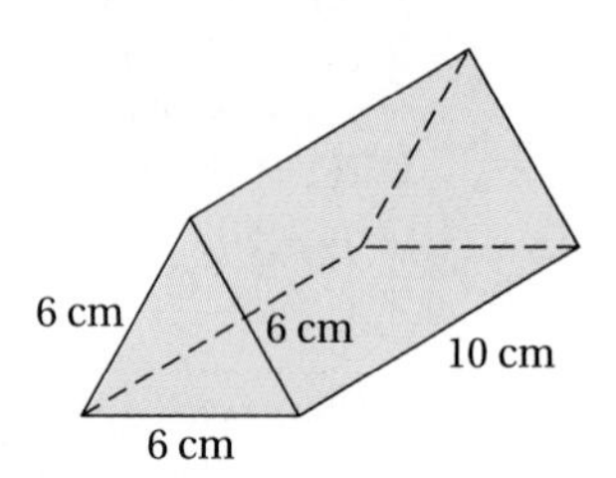

14. Triangles *ABC* and *DEF* are similar. One side of triangle *ABC* is 2 units and one side of triangle *DEF* is 4 units. A student concludes that the ratio of their areas is 4:1. What's wrong with the student's reasoning?
15. Ms. Jones has a garden plot that measures 40 ft × 20 ft. If she increases the garden by 2 ft on each side, by what percent has she increased the plot?

Reasoning and Problem Solving

16. Find the shaded area if the rectangle inscribed in the circle has dimensions 2 ft × 4 ft.

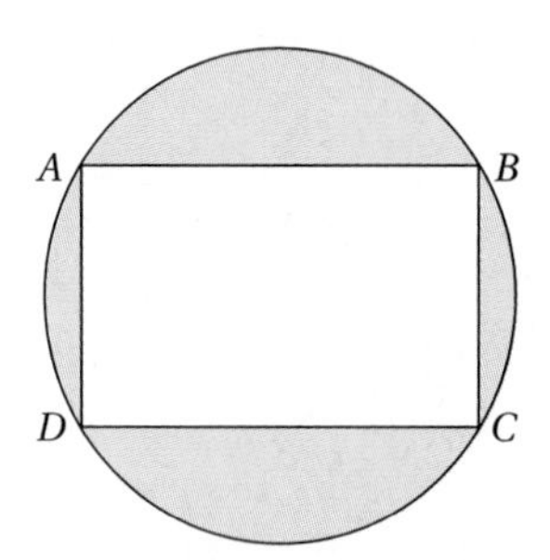

17. Figure *ABC* is an equilateral triangle measuring 4 units on a side. Altitude $\overline{CF}$ is bisected by $\overline{ED}$, which is parallel to $\overline{AB}$. What is the ratio of the area of trapezoid *ABDE* to the area of triangle *CED*?

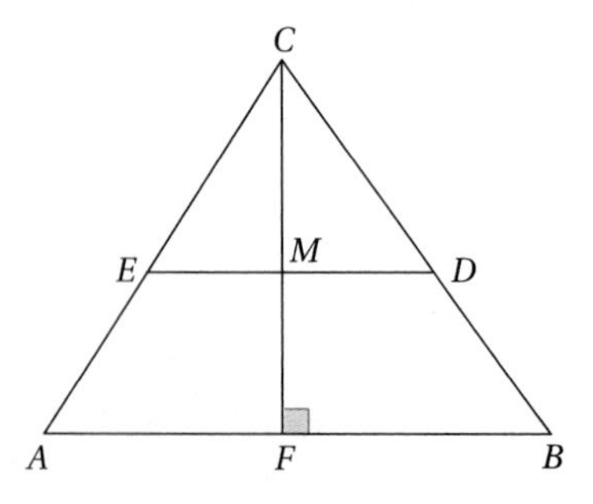

18. Consider various integer values for the radii of different sized spheres. Calculate the ratio of the numerical values of the surface areas to the volumes. What is the lowest integer value of the radius for which this ratio is less than 1? Less than one-half? Write a statement about the ratio of the numerical values of the surface area to the volume of spheres as the value of the radii increases.
19. **The Lot Problem.** A couple are considering buying the lot shown and building a house on it. The real estate agent says that the lot contains nearly 3 acres. (1 acre = 43,560 ft^2). Is the agent correct? Explain.

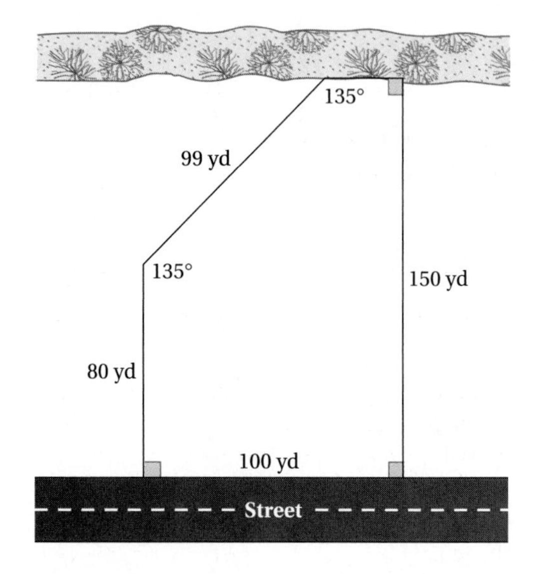

20. Cut out regions of circles so that the central angles are 90°, 180°, and 270°. Fold each region to form a cone. How do the volumes of the three cones compare? If the radius of the circles (slant height of the cones) is 1 unit, determine the diameter of the base of each new cone and its height. Calculate the volumes of the three cones. What will happen as the angle of the sector ranges from 0° to 360°?
21. A swimming pool 25 m long, 15 m wide, and 150 cm deep is being filled at the rate of 20 l/min. How many hours will be needed to fill the pool?
22. How many 4-ft × 8-ft wood panels are needed to panel a 16-ft × 12-ft den that has a 7-ft × 3-ft door in the narrow part of the room?
23. Sony TV sets have screens that approximate similar rectangles. If one is a 20-in. TV and the other is a 27-in. TV (measured on the diagonals), what is the ratio of their areas?
24. **The Leaky Roof Problem.** Maureen's office has a leaky roof that drips once every 2 s. She collects the water in a cylindrical bucket that is 30 cm across and 30 cm high. In about how many days will the bucket be filled, requiring Maureen to empty it? (Assume that one drop has a capacity of 10 mm^3. Use $\pi \approx 3$.)

25. **The Equator Problem.** Assume that the earth has a radius of 4000 mi. A rope 24,000 mi long is strung around the equator. If 1 yd is added to the rope, which of the following animals could *not* crawl under the rope? (Select all that apply. Use $\pi = 3$.)
   a. Ant **b.** Small dog
   c. Large dog **d.** Pig
   e. Cow

Alternative Assessment

26. Frank thinks that, as the perimeter of a rectangle increases, the area will also increase. He uses the examples shown in the figure to verify his conjecture.

If Frank is correct, argue why he is correct. If Frank is wrong, argue why he is wrong. Form a conjecture similar to Frank's regarding the surface area and volume of rectangular prisms. Argue whether your conjecture is correct or incorrect.

27. Create a house plan that has 1500 ft^2 with two bedrooms, one bath, a kitchen, and a living room.

Examples for Exercise 26

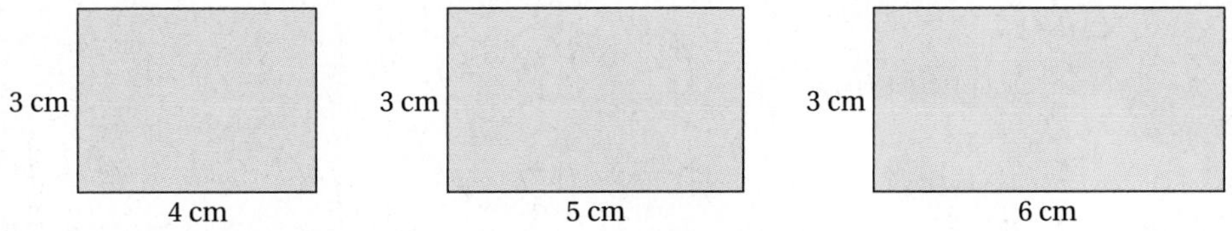

3 cm
4 cm

3 cm
5 cm

3 cm
6 cm

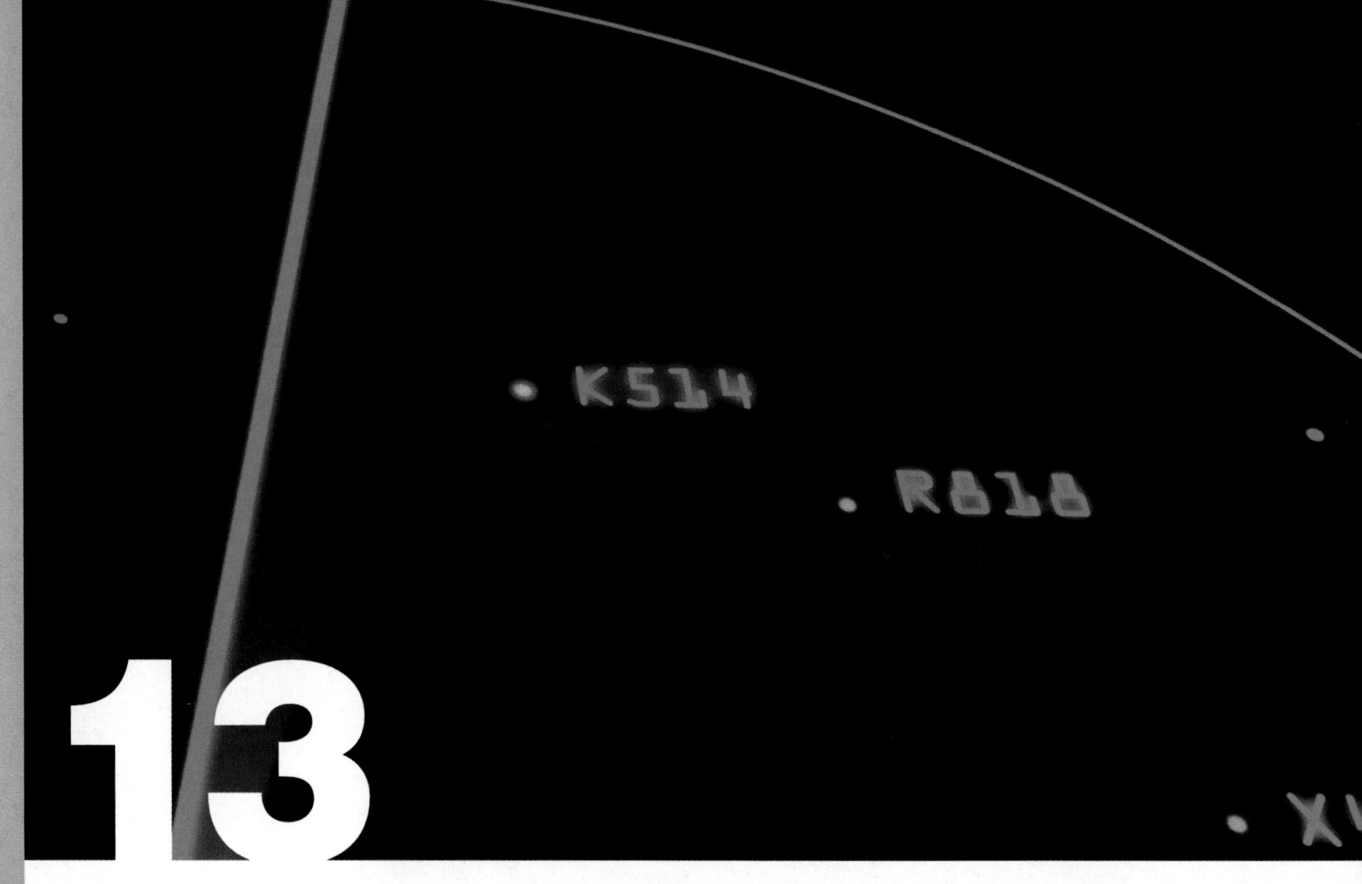

13

Exploring Ideas of Algebra and Coordinate Geometry

Chapter Perspective

Understanding, representing, and applying relationships between real-world quantities are involved in almost every occupation you can think of today and are the essence of algebra. The radar screen above demonstrates how the position of airplanes or other moving objects can be represented and then analyzed to give an air traffic controller crucial information. Graphs and equations are two of the most useful ways to show relationships between quantities and to help people use these relationships to make predictions and solve problems. Using data tables and models are other practical ways to do the same thing.

From a mathematical perspective, functional relationships, in which the value of one variable depends on the value of another or several other variables, are extremely important. In this chapter you will learn ways to represent relationships between quantities and ways to use these representations to solve problems. In particular, you will learn how to represent relationships using equations and graphs. You will learn techniques for solving linear and quadratic equations and you will learn why these techniques make sense. You will also learn how algebra and geometry can be connected through the coordinate plane.

Connection to the NCTM Principles and Standards

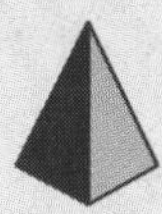

The NCTM *Principles and Standards for School Mathematics* (2000) recommend that the mathematics curriculum at grades PreK–8 include ideas related to algebra at all grades so students can

- understand patterns, relations, and functions;
- represent and analyze mathematical situations and structures using algebraic symbols;
- use mathematical models to represent and understand quantitative relationships; and
- analyze change in various contexts (p. 37).

Connection to the PreK–8 Classroom

In grades PreK–2, children work with various kinds of patterns, including simple shape patterns and numerical patterns. They explore both repeating patterns and growth patterns. Whole-number meanings and operations are the context used to develop an understanding of properties such as commutativity. The meanings of addition and subtraction developed at these grades provide the foundation for understanding relationships between real-world quantities.

In grades 3–5, work with patterns is extended using tables, words, and graphs. Children now analyze patterns and functions to make and test predictions. Mathematical relationships are represented using equations and properties of whole numbers are now used with multiplication. Children also explore how changes in one variable relate to changes in another variable.

In grades 6–8, children encounter a more formal development of algebra. In fact, in many states and cities, children are now required to take algebra in grade 8. A more formal analysis of functions occurs in these grades with extensive work on representing functional relationships and using these representations to solve problems. Both linear and nonlinear functions are explored in various representations. Graphs are used extensively at these grades to analyze changes in quantities in linear relationships.

Section 13.1 VARIABLES, EXPRESSIONS, AND EQUATIONS

- Understanding How Letters Are Used in Mathematics
- Understanding Numerical and Algebraic Expressions
- Understanding Equations
- Analyzing Different Types of Equations

In this section we examine different ways that letters are used in mathematics, ways to write algebraic expressions for both mathematical and real-world situations, and the meaning of an equation. Mini-Investigation 13.1 will help you think about some of the ways that you may have used letters in mathematics.

Talk about which of these equations involve the use of letters to make a statement about all numbers in a set, show a relationship between a pair of quantities that vary, represent a specific unknown number, or show a relationship among three quantities.

Mini-Investigation 13.1 **Communicating**

How does the use of letters differ in these four equations?

a. $d = rt$
b. $a + b = b + a$
c. $12 = 3x - 6$
d. $y = 3x - 6$

Understanding How Letters Are Used in Mathematics

Before discussing the different ways that letters are used in mathematics, let's first think about the idea of a **variable,** or a quantity that varies, and the idea of a **constant,** or a quantity that doesn't vary, as they relate to the real world.

Some examples of real-world constant quantities are

- the number of hours in a day,
- the number of cm in a m,
- the number of faces on a cube, and
- the distance around the earth at the equator.

Some examples of real-world variable quantities are

- the number of seconds a dropped ball takes to hit the floor,
- the areas of roofs,
- the costs of new cars, and
- the costs of paperback books.

In your earlier work with algebra, you have probably used the word *variable* to talk about *any* letter used in an equation. However, letters are used in different ways, not always as variables. Understanding algebra requires understanding the difference between quantities that vary and quantities that are constant. Let's examine the equations from Mini-Investigation 13.1 and the different ways that letters are used.

$d = rt$. You perhaps recognize the equation $d = rt$ as a formula relating distance, d, rate, r, and time, t. A formula is a way to show a relationship between quantities. Relationships expressed as equations that are used frequently are called formulas. Other formulas you undoubtedly have seen are $I = PRT$ and $A = \pi r^2$. English letters used in formulas represent quantities that vary. In all the formulas in this paragraph, the quantities represented by the English letters vary and thus can be called variables.

$a + b = b + a$. Statements such as the one represented by the equation $a + b = b + a$ are generalizations from arithmetic. This particular one, of course, is the commutative property of addition. It is usually stated as: $a + b = b + a$ for all real numbers a and b. In such generalizations, different values for a and b can be used in the equation, so these letters are used as variables. However, it is more appropriate to think of this generalization as: for *particular* real numbers a and b, $a + b = b + a$. So, although there is an element of variability in this use of letters, the quantities don't vary in the same sense as they do in formulas.

$12 = 3x - 6$. If you examine the equation $12 = 3x - 6$ in the absence of any context, the letter x isn't a variable because its value doesn't vary. Only one value for x will make it a true equation (the value on the left of the equals sign is the same as the value on the right of the equals sign). In this case x is an *unknown* value. However, x might be considered a variable here if you think of the unknown value for x that makes a true equation as one of many possible values that could replace x.

$y = 3x - 6$. The equation $y = 3x - 6$ is similar to the formula $d = rt$; it shows a relationship between two quantities whose values vary. Those quantities are represented by the letters x and y and thus are variables.

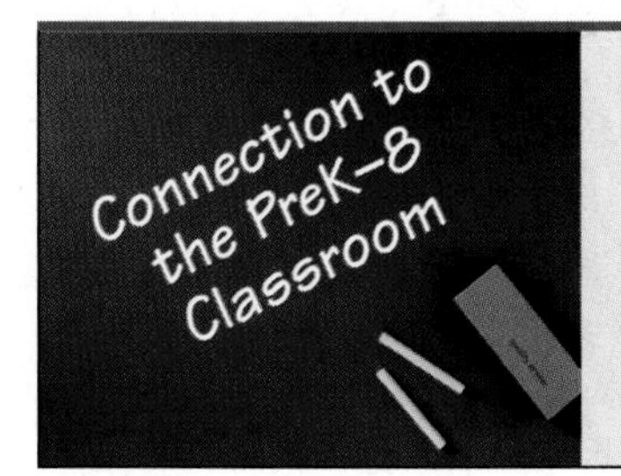

The idea of a symbol used to represent an unknown quantity is introduced early in some elementary school mathematics programs. Geometric figures are often used to write simple number sentences such as $\Delta + \bigcirc = 9$. Students find the values that can be written in the triangle and circle to make a true addition fact. Letters are used as unknowns in later grades when students are working with multiplication equations such as $a \times b = 24$, or as variables when introducing number properties such as $a + b = b + a$.

Understanding Numerical and Algebraic Expressions

Meaning of Numerical and Algebraic Expressions. A **numerical expression** involves one or more operations and represents a single number. For example, $3 + 2$ is a numerical expression, as is $(50 - 15) \div 7$. Each expression represents, or names, the number 5. Sometimes you can find a pattern in a sequence of numerical expressions and use variables to generalize this pattern, as in Mini-Investigation 13.2.

Talk about what helped you decide how to use variables to write the expression.

Mini-Investigation 13.2 Finding a Pattern

What pattern can you discover in the following table? Write an expression that represents the total cost of the movie tickets using a, the number of adults, and c, the number of children.

Number of Adults	Number of Children	Total Cost
2	1	\$6(2) + \$4(1)
1	4	\$6(1) + \$4(4)
4	6	\$6(4) + \$4(6)
5	0	\$6(5) + \$4(0)
12	18	\$6(12) + \$4(18)
a	c	?

In Mini-Investigation 13.2, the number of adults' and children's tickets varies, and the values are represented by the letters a and c. The expression you wrote is called an algebraic expression because it contains at least one variable.

Definition of an Algebraic Expression

An **algebraic expression** is a mathematical phrase involving variables, numbers, and operation symbols.

In using an algebraic expression to represent the values of a quantity, you must first define what the variable represents. For the situation in Mini-Investigation 13.2, the variables could be defined in this way:

a = Number of adults' tickets sold.

c = Number of children's tickets sold.

Example 13.1 further shows how to write algebraic expressions. It also involves some simplifying of algebraic expressions that you should recall from your previous work in algebra.

Example 13.1 Writing Algebraic Expressions

Define the variable and write an algebraic expression for each situation.

a. The perimeter of the rectangle shown.

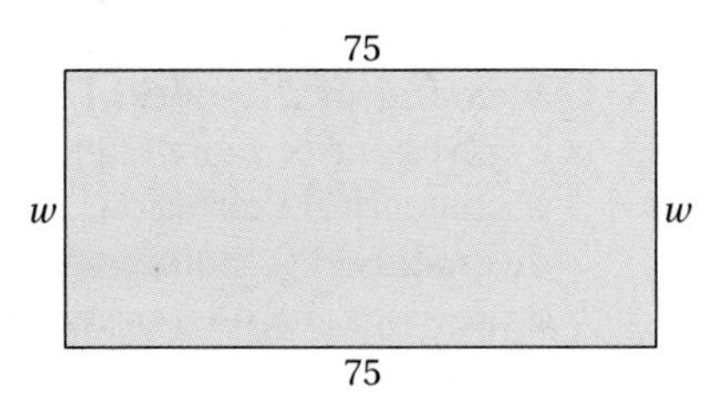

b. The volume of each cube shown.

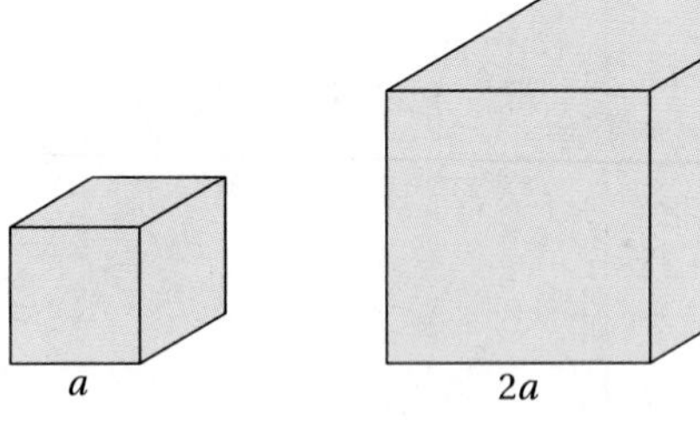

c. Marie has saved \$1200 toward the purchase of a new car. She estimates that she can save \$75 each week. Write an expression for the amount of money she will have saved after x weeks.

Estimation Application

The Great Perimeter
Estimate the perimeter of the mainland United States by estimating the perimeter of a rectangle that you think has about the same perimeter as the United States. Explain your procedure.

Solution

a. Let w = the width of the rectangle.

$$\text{Perimeter} = w + 75 + w + 75, \quad \text{or} \quad 2w + 150$$

b. Let a = the length of the side of the smaller cube and $2a$ = the length of the side of the larger cube.

$$\text{Volume of small cube} = (a)^3, \quad \text{or} \quad a^3$$
$$\text{Volume of large cube} = (2a)^3 = (2a)(2a)(2a), \quad \text{or} \quad 8a^3$$

c. Let x = the number of weeks. Then $75x$ is the amount of money Marie will save in x weeks, and $75x + 1200$ is the total amount of money she will have saved.

Your Turn

Practice: Define the variable(s) and write the algebraic expression(s) for the quantity or quantities involved in each situation.

a. The area of the right isosceles triangle shown.

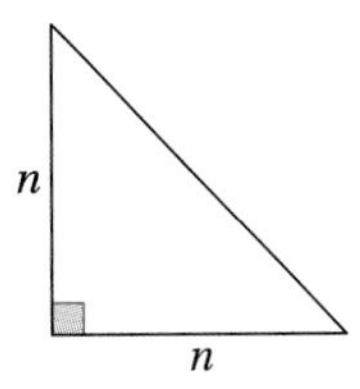

b. Micki earns $8 an hour for regular time and $12 an hour for overtime. The expression should show the total earned for working some regular time and some overtime.

Reflect: Which is the better way to define the variable for the number of hours of regular time that Micki worked in the practice problem?

h = Hours or h = Number of hours.

Explain. ■

Evaluating Algebraic Expressions. Have you ever used a formula such as $I = PRT$ by substituting numerical values for P, R, and T and calculating the value of I? If you have, you have evaluated an algebraic expression.

Procedure for Evaluating Algebraic Expressions

To **evaluate an algebraic expression,** replace the variable(s) with the given value(s) and do the calculation(s).

Example 13.2 demonstrates an application of the procedure for evaluating algebraic expressions.

Example 13.2

Evaluating Algebraic Expressions

Evaluate each algebraic expression.

a. $4x^2 - 5x$, for $x = -2$

b. $\sqrt{2x} + 8$, for $x = 50$

c. $P(1.0825)^{10}$, where $P = 2000$ = the amount of an initial deposit in a savings account paying annual compound interest of 8.25%. The final calculated amount is the value in the savings account after 10 years.

d. $P = 2l + 2w$, for $l = 8$ and $w = 5$, where l = length, w = width, and P = perimeter of a rectangle.

Solution

a.

$$\begin{aligned} 4x^2 - 5x &= 4(-2)^2 - 5(-2) && \text{Replace } x \text{ with } -2. \\ &= 16 + 10 && \text{Simplify.} \\ &= 26 && \text{Add.} \end{aligned}$$

b.

$$\begin{aligned} \sqrt{2x} + 8 &= \sqrt{2(50)} + 8 && \text{Replace } x \text{ with } 50. \\ &= \sqrt{100} + 8 && \text{Multiply.} \\ &= 10 + 8 && \text{Find the square root.} \\ &= 18 && \text{Add.} \end{aligned}$$

c. $P(1.0825)^{10} = 2000(1.0825)^{10}$.

i. Use a calculator to evaluate $(1.0825)^{10}$.

Key Sequence	Display
1.0825 [y^x]	y^x 1.0825
10 [=]	2.2094239

ii. Then complete the evaluation of the expression:

$$\begin{aligned} 2000(1.0825)^{10} &= 2000(2.2094) \\ &\approx 4419 \end{aligned}$$

The saver would have about $4419 after 10 years.

d. The following graphing calculator screen illustrates the use of stored values and expression evaluation.

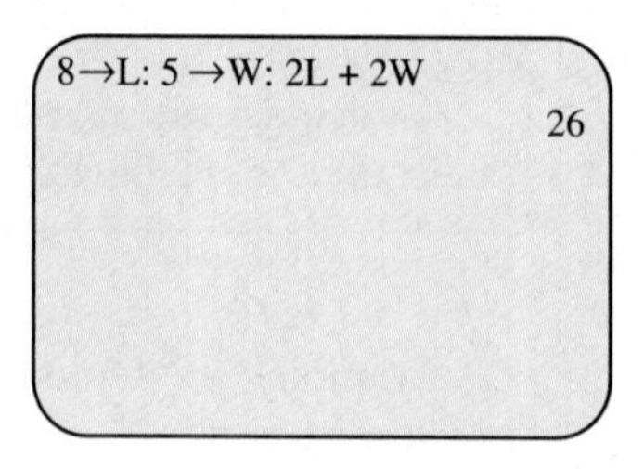

The perimeter is 26 units.

Your Turn

Practice: Evaluate each expression for the values given.

a. $6m^3 + 5$, for $m = 2$ and for $m = 1$
b. $\frac{2x + 6}{x - 9}$, for $x = 12$ and $x = 0$

Reflect: When evaluating an expression such as $4x^2 - 5x$, can you replace the x in $4x^2$ with a 3 and the x in $5x$ with a -2 to evaluate the expression? Explain. ■

As Example 13.2 clearly shows, calculators can help you evaluate algebraic expressions. In Mini-Investigation 13.3, you are asked to explore their use further.

Talk with your classmates about the variety of storage options available for different calculators.

Mini-Investigation 13.3 **The Technology Option**

How can you utilize your calculator to store values of one or more variables to be used in evaluating expressions?

Solving Problems Using Algebraic Expressions. Example 13.3 illustrates how expressions can be used in solving problems. The problem-solving strategies *make a table* and *look for a pattern* play a role in the solution to the problem presented.

Example 13.3 Problem Solving: Car Rental Charges

The charge for a rental car for one day is $25 plus $0.10 per mile (no tax). Make a chart showing the total driving charges for 50, 100, 150, and 200 mi. About how many miles can someone drive for $75?

Working toward a Solution

Understand the problem	*What does the situation involve?*	Analyzing car rental charges for distances traveled
	What has to be determined?	The total driving charge for different distances traveled; the number of miles that a person can drive a rental car for a $75 charge
	What are the key data and conditions?	The charge for a rental car is $25 + $0.10/mi. The maximum amount to be spent is $75.
	What are some assumptions?	The same schedule of charges applies to all distances.
Develop a plan	*What strategies might be useful?*	Total charges can be recorded in a table. Patterns in the table can be used to help answer the question.
	Are there any subproblems?	Yes, finding the cost for each of the total miles driven.
	Should the answer be estimated or calculated?	The exact amount for the miles given is needed. An estimate can be used to answer the question and then checked with a calculation.

	What method of calculation should be used?	A calculator since several calculations are needed.
Implement the plan	*How should the strategies be used?*	Write an algebraic expression that shows how to find the total cost. That expression can be used to find the total charges for the miles given. The data in the table can be used to estimate the answer to the question. Let m = the number of miles driven. Then total cost = $0.10m + 25$. **Number of Miles** — **Cost** 50 — \$30 100 — \$35 150 — \$40 200 — \$45
	What is the answer?	The cost increases by \$5 for each 50 miles driven. Six more increases of \$5 each will give \$75. So, $6 \times 50 = 300$ and $200 + 300 = 500$. A person can drive 500 miles for \$75.
Look back	*Is the interpretation correct?*	Yes.
	Is the calculation correct?	All calculations check. For m = 500 mi, $0.10(500) + 25 = \$75$.
	Is the answer reasonable?	Yes. In the table the total cost doesn't double when the number of miles driven doubles. So doubling 200, the last entry in the table, gives 400 miles, and the total cost of \$75 is less than $2 \times \$45 = \$90$. The answer seems reasonable.
	Is there another way to solve the problem?	Algebraic expressions can be used to find values that can be organized in tables. Patterns can then be used to help answer related question.

Your Turn

Practice: A basic cheese pizza at a certain restaurant costs \$9.89. Each additional topping costs \$0.25. Make a table showing the total cost for 1 to 7 toppings. Find the total cost for "The Works," a pizza with 12 toppings.

Reflect: Why might you say that a formula is based on an algebraic expression? Give an example. ■

Understanding Equations

The Meaning of an Equation. You used equations in your early days in elementary school when you wrote simple number sentences like $5 + 3 = 8$. You also have worked with equations in your study of algebra when you solved equations like $3x - 5 = 18$ or $x^2 + 2x - 8 = 0$ and used formulas such as $C = \frac{5}{9}(F - 32)$. In the following definition, note that an equation may or may not involve the use of variables.

Definition of an Equation

An **equation** is a mathematical sentence that uses an equals sign to state that two expressions represent the same number or value.

Equations Showing Functional Relationships. The equations of most interest in the study of algebra represent a **functional relationship** between two quantities. For example, the total monthly cost for cable television, C, is a function of the number of connections in the home, n. The total cost (the **dependent variable**) depends on the number of connections (the **independent variable**). The equation representing this relationship in one community is $C = 5.25n + 15$. In a functional relationship, the value of one quantity (the dependent variable) depends on the value of the other quantity (independent variable), and the value of the dependent variable is unique for each value of the independent variable.

The following are some other situations for which a functional relationship can be defined between two quantities and an equation written to represent this relationship. Note that only the first quadrants of the graphs apply to the real-world situations described.

- The distance a car travels, d, in 2 hr is a function of the speed of the car, r. Here, r is the independent variable and d is the dependent variable:

$$d = 2r.$$

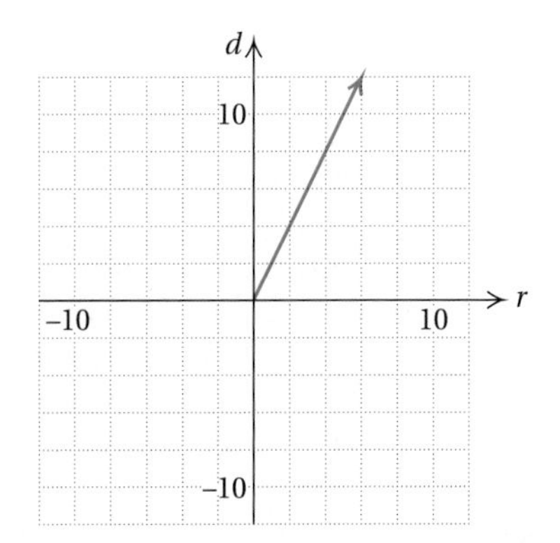

- The amount of light at a certain depth in water expressed as a percent of the surface light, y, is a function of the depth, x. Here, x is the independent variable and y is the dependent variable:

$$y = 100(0.975)^x.$$

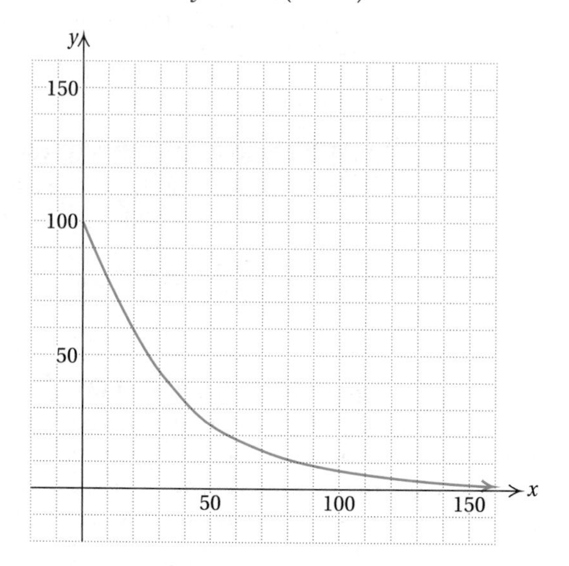

- The distance an object travels when dropped, d, is a function of the time it has been falling, t. Here, t is the independent variable and d is the dependent variable:

$$d = 5t^2$$

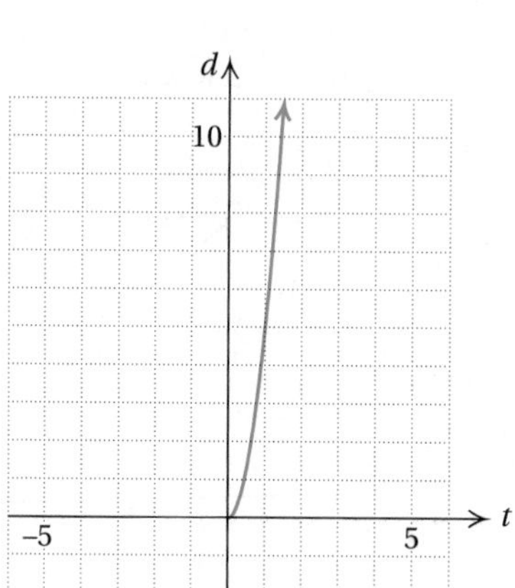

In some situations you will know the value of one variable and can also write an equation. If $d = 86$ mi in $d = 2r$, you can write the equation $86 = 2r$. If $y = 50$ in $y = 100(0.975)^x$, you can write the equation $50 = 100(0.975)^x$. If $d = 45$ m in $d = 5t^2$, you can write the equation $45 = 5t^2$. Example 13.4 illustrates how to write expressions and equations for relationships.

Example 13.4 **Writing Expressions and Equations**

a. The circumference of a circle, C, is a function of the radius of the circle, r. The circumference is 2π times the radius of the circle.

i. Write an expression in terms of r that shows the circumference.
ii. Write an equation showing the relationship between C and r.
iii. Write an equation in terms of r when the circumference is 20 cm.

b. The cost of a telephone call, c, at a certain time during the day is a function of the time spent talking. The cost is $0.25 for the first minute and $0.10 for each additional minute, m.

i. Write an expression in terms of m that gives the total cost of a call.
ii. Write an equation showing the relationship between c and m.
iii. Write an equation in terms of m when the total cost of the call is $5.

Solution

a. **i.** $2\pi r$ gives the circumference of the circle.
ii. C also represents the circumference, so $C = 2\pi r$.
iii. If $C = 20$ cm, we can write $20 = 2\pi r$.

b. **i.** $0.10m + 0.25$ gives the total cost of the call.
ii. c also represents the total cost of the call, so $c = 0.10m + 0.25$.
iii. If $c = \$5$, we can write $5 = 0.10m + 0.25$.

Your Turn

Practice: The sale price of leather jackets, p, is 10% off the regular price, r, plus an additional $5 off if you buy before the Christmas holiday.

a. Write an expression in terms of r that gives the sale price of the jacket. (*Hint:* 10% off is the same as 90% of the original price).
b. Write an equation showing the relationship between p and r.
c. Write an equation in terms of r when the sale price of the jacket is $85.

Reflect: Can the equation for the practice problem be written differently? Explain. ■

In Mini-Investigation 13.4 you are asked to think more about independent and dependent variables in real-world situations.

Talk with a classmate and explain why the values of one quantity in these situations depend on the values of the other quantity.

Mini-Investigation 13.4 **Communicating**

What three real-world situations can you identify in which the value of one quantity (the dependent variable) depends on the value of the other quantity (the independent variable)?

Analyzing Different Types of Equations

Throughout the remainder of this chapter you will encounter various types of equations. However, equations for linear and quadratic relationships are emphasized.

For example, the relationship between the number of gallons of gasoline purchased, n, and the total cost of the purchase, t, can be represented by a linear function such as $t=1.30n$. The relationship that gives the area, A, of a pizza in terms of its radius, r, can be represented by the quadratic function $A = \pi r^2$. The general forms for equations that represent linear functions and equations that represent quadratic functions are shown in Figure 13.1.

Definitions of Linear and Quadratic Functions

- An equation of the form $y = ax + b$, where a and b are real numbers, represents a **linear function.**
- An equation of the form $y = ax^2 + bx + c$, where a, b, and c are real numbers and $a \neq 0$, represents a **quadratic function.**

The graph of a linear function is a straight line, such as the one shown in Figure 13.1(a). The graph of a quadratic function is not a straight line; it is a U-shaped curve, such as the one shown in Figure 13.1(b). We further explore the graphs of functions in Section 13.3.

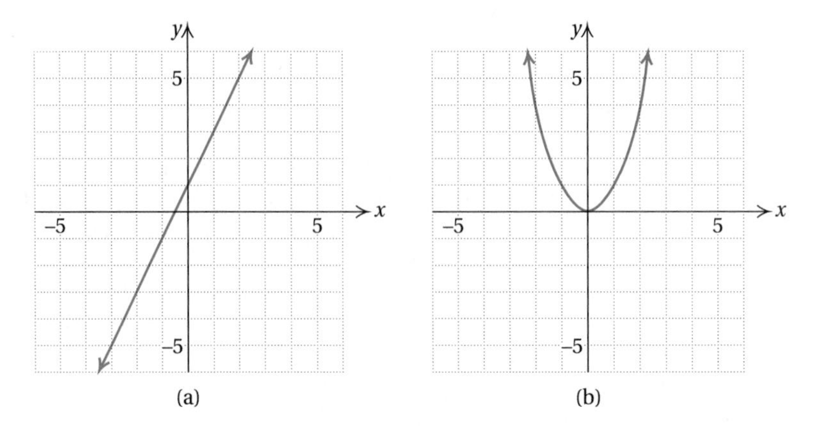

FIGURE 13.1 | (a) Graph of the linear function $y = 2x + 1$; (b) graph of the quadratic function $y = x^2$

In linear functions, a constant change in the independent variable, x, results in a constant change in the dependent variable, y. The amounts of change for the values of the variables aren't necessarily the same, but they are constant. In the graph of the linear function shown in Figure 13.1(a), a horizontal change of 1 always results in a vertical change of 2. For quadratic functions, a constant change in the independent variable doesn't result in a constant change in the dependent variable. The graph of the quadratic function shown in Figure 13.1(b) illustrates this condition. Example 13.5 applies the ideas of linear and quadratic functions.

Example 13.5

Identifying Linear and Quadratic Functions

Decide whether each equation represents a linear function, a quadratic function, or neither—and whether it represents a straight line. Explain.

a. $y = 3x^2 - 8$

b. $y = \frac{1}{x}$

c. $y = 12 - 2x$

Solution

a. The equation $y = 3x^2 - 8$ is quadratic. It can be written as $y = 3x^2 + 0x - 8$, which is the form for a quadratic function. The graph of this equation isn't a straight line.

b. The equation $y = \frac{1}{x}$ is neither linear nor quadratic. While it can be written as $xy = 1$, it cannot be put into either the $y = ax + b$ or $y = ax^2 + bx + c$ forms.

c. The equation $y = 12 - 2x$ is linear. It can be written as $y = -2x + 12$. The graph of the equation is a straight line.

Your Turn

Practice: Tell whether each equation is linear, quadratic, or neither and how you decided.

a. $3x + 7 - y = 0$

b. $5x^2 + 3x = y$

c. $y = 5$

Reflect: Explain how an equation such as $y = 5$ in the practice problem represents a linear function. ■

Mini-Investigation 13.5 gives a table in which there isn't a constant change in the *independent* variable. You are asked to decide whether the function is linear.

Give a convincing argument that the relationship shown in the table is or is not linear.

Technology Extension: For the equations given in the practice problem in Example 13.5, explore how you can use your graphing calculator to (a) graph each function represented by the equations and (b) create tables of values from the equations. Write a brief paragraph summarizing the graphing and table-building processes for your calculator and then compare it with those of your classmates.

Mini-Investigation 13.5 **Using Mathematical Reasoning**

Do you think that the function given by the following table of values is linear?

x	1	3	7	10	20
y	−7	−21	−49	−70	−140

Another equation encountered frequently in mathematics is an exponential function. Figure 13.2 shows the graphs and related tables of values of two exponential functions.

Definition of an Exponential Function

An equation of the form $y = a^x$, where a is some positive real number different from 1, is called the **exponential function, base *a*.**

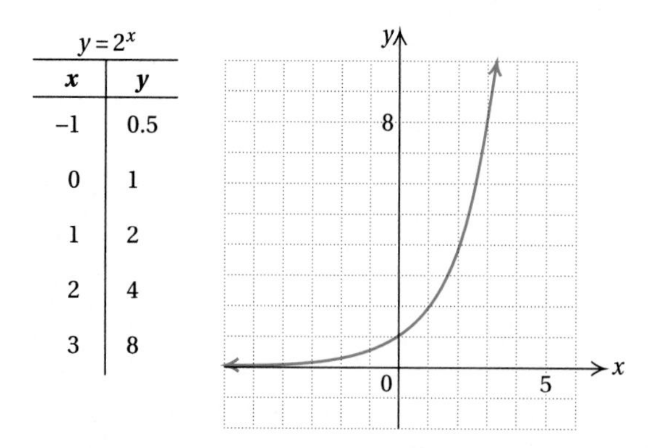

$y = 2^x$

x	y
–1	0.5
0	1
1	2
2	4
3	8

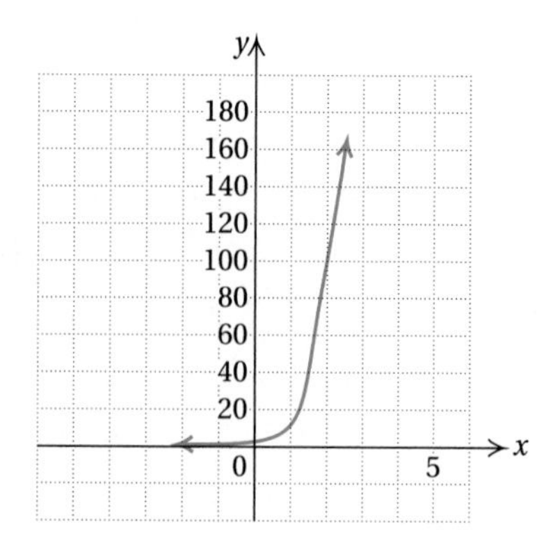

$y = 10^x$

x	y
–2	0.01
–1	0.1
0	1
1	10
2	100

FIGURE 13.2 | Examples of exponential functions.

Note that by definition, the independent variable, x, in an exponential function is always in the exponent. The following are *not* exponential functions because the independent variable is not the exponent.

$$y = x^3 \qquad y = x^{\frac{1}{2}} \qquad y = x^{0.5}$$

Many real-world growth phenomena can be represented using an exponential function. For this reason, exponential functions are often called **growth functions.**

The equation $y = A(1 + r)^x$ shows the relationship between the number of years, x, a certain amount of money is invested at a particular interest rate and the total amount of money, y, at the end of that time. The variable A represents the amount of money invested initially, sometimes called the principal, and r represents the annual interest rate when the interest is compounded annually. Suppose the amount of money invested is \$10,000 at an interest rate of 7% for 5 years. The exponential function then looks like $y = 10{,}000(1 + 0.07)^5$. Doing the calculations shows that $y = 14{,}025.51$ (rounded to the nearest tenth). So, the total amount of the investment at the end of five years would be about \$14,025.51.

Talk about how the graphs of $y = 2^x$ and $y = x^2$ are alike and how they are different.

Mini-Investigation 13.6 **Finding a Pattern**

Graph the exponential functions $y = 2^x$ and $y = 2^{-x}$ on the same coordinate grid. Describe in writing how the two graphs are alike and how they are different. You can use a graphing calculator if available.

Here are some other examples of real-world exponential (growth) functions.

- $A = 500 \cdot 2^{2t}$ shows the number of a certain bacteria, A, that will exist after t hours starting with 500 bacteria when this particular bacteria divides every half hour to produce two bacteria.
- $P = 14.7e^{-0.21h}$ shows an approximation for the atmospheric pressure, P (pounds per square inch), where h is the altitude above sea level in miles. The letter e represents an irrational number. The irrational number e to eight decimal places is $e \approx 2.71828183$.

Problems and Exercises | for Section 13.1

A. Reinforcing Concepts and Practicing Skills

Tell whether each quantity is constant or variable. Explain.

1. The distance from the earth to the sun
2. The number of the earth's moons
3. The number of days in a year
4. The total cost of tickets to a movie
5. The time required for a ball to roll down a ramp
6. The number of months in a year

The following table for Exercises 7–9 shows taxicab fares for various trip lengths

$\frac{1}{5}$ miles driven	Fare (f)
0	\$0.75
1	\$0.20(1) + \$0.75
2	\$0.20(2) + \$0.75
3	\$0.20(3) + \$0.75
4	\$0.20(4) + \$0.75
x	?

7. Define the variable quantity or quantities, using the letter given.
8. Write an algebraic expression for the last entry in the table, using the variables defined in Exercise 7.
9. Use the expression written in Exercise 8 to find the fare for a 5-mi ride. (*Hint:* How many $\frac{1}{5}$ mi are in 5 mi?)

The following table for Exercises 10–14 shows the rates of pay for various amounts of sales at an insurance company.

	A	B (t)
1	Sales (\$ thousands)	Total Earnings
2	0	\$5000
3	10	\$0.1(10) + \$5000
4	20	\$0.1(20) + \$5000
5	40	\$0.1(40) + \$5000
6	100	\$0.1(100) + \$5000
7	s	?

10. Define the variable quantity or quantities, using the letter given.
11. Write an algebraic expression for the last entry in the spreadsheet, using the variables defined in Exercise 10.
12. Use the expression written in Exercise 11 to find the total earnings for sales of \$1 million.
13. Use spreadsheet notation to write a formula for cell B2.
14. If the table were to continue, what formula would apply to cell B22? Explain.

Evaluate the algebraic expression in Exercises 15 and 16 for the value(s) given.

15. $18 - \frac{x}{2}$, for $x = 0, 2, -8, 36$, and 60
16. $2x^3$, for $x = -3, 0, 1$, and 4
17. Write an expression for the amount of walkway yet to be paved in the illustration below.

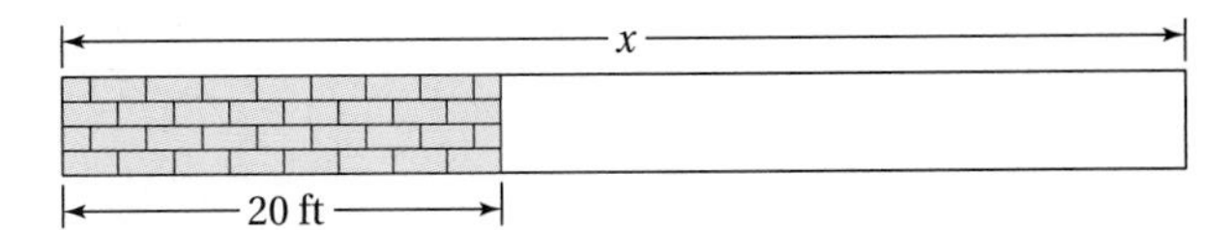

18. Write an expression for the area of the shaded region.

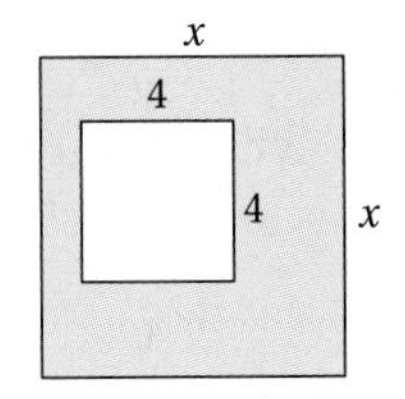

For Exercises 19–21, the total cost of purchasing from a mail order catalog, y, is a function of the cost of the items purchased, x. In addition, there is a fixed shipping charge of \$2.75 per order, and sales tax is 8%.

19. Write an expression in terms of x that gives the total cost of a purchase.
20. Write an equation that shows the relationship between the quantities.
21. Find the total cost for purchasing an item priced at \$125.

For Exercises 22–24, the amount of money remaining to be paid on a loan, y, is a function of the number of months, x. Suppose that monthly payments of \$225 are to be made on an original loan amount of \$3000.

22. Write an expression in terms of x that gives the amount of money remaining to be paid.
23. Write an equation that shows the relationship between the quantities.
24. Find the amount remaining after 8 mo.

State whether each equation in Exercises 25–30 represents a linear, quadratic, or exponential function. Tell how you decided.

25. $y = -34 + 7x$

26. $3x^2 = 5$

27. $y = -8x^2$

28. $y = 10^{3x}$

29. $y = 3x^2 + 12$

30. $y = 2x + 23 - 14x$

Copy each equation in Exercises 31–33. Replace the boldface italic letter with a number that shows the relationship indicated in the table. Then copy each table and complete it by using the corresponding equation.

31. $y = \boldsymbol{a}x + 4$

x	y
1	6
3	10
5	14
10	?
50	?
100	?

32. $y = -x + \boldsymbol{b}$

x	y
1	0
2	−1
8	−7
10	?
20	?
40	?

33. $y = \boldsymbol{a}x^2 + 5$

x	y
2	1
4	−11
6	−31
10	?
20	?
50	?

B. Deepening Understanding

34. Give five possible dimensions for the rectangle shown with the perimeter given.

Perimeter = $12x + 20$

35. Two students evaluated the expression $7 + 2 \times x - x - 4$ for $x = 12$ and found two different answers. Student A got 24, and Student B got 28. Which, if either, is correct? Explain how each value was found.

36. Describe a real-world functional relationship. Identify the independent variable and the dependent variable.

37. Write an expression that gives the volume of the box shown.

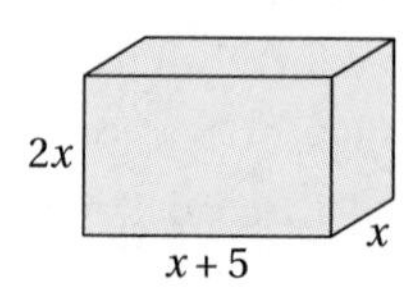

38. Write an expression for the volume of the box shown in Exercise 37 if the dimensions of each side were doubled.

For which table(s) of values in Exercises 39–42 is the relationship linear? Tell how you decided. Write the equation for each linear relationship.

39.

x	1	2	3	4	5	6
y	−8	−16	−24	−32	−40	−48

40.

x	1	2	5	8	10	11
y	1.5	2.5	5.5	8.5	10.5	11.5

41.

x	2	5	8	11	14	17
y	0	3	6	9	12	15

42.

x	5	10	15	20	25	30
y	20	30	40	50	50	50

43. Water is being poured into two containers, A and B, at the same rate and starting at the same time. It takes 3 times as long to fill container A than to fill container B. What can you conclude about these containers?

44. Is every equation a function? Explain and give a counterexample if possible.

The graph on the next page shows the relationship between the number of years (t)*, that one owns a complete set of Topps baseball cards, from 1971 to 1995, and the value,* V*, of the set. Note that* t = *0 for the year 1971. The equation* $V = 20(1.222)^t$ *also represents this relationship. Use the graph to estimate the answers to Exercises 45–48.*

45. The value of the set in 1976

46. The value of the set in 1991

47. The year when the set is worth \$500

48. The year when the set is worth \$1500

Figure for Exercises 45–48

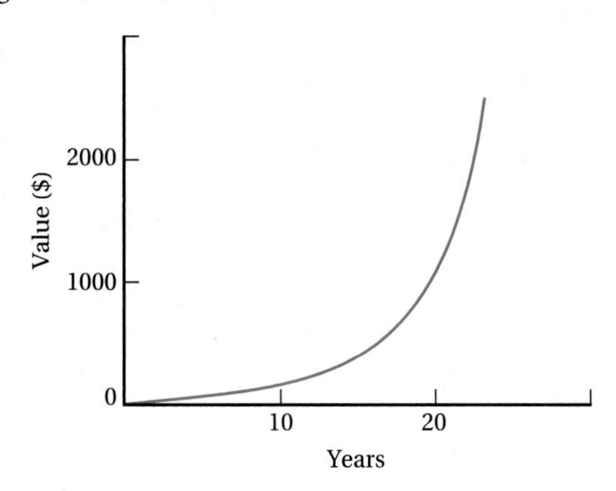

C. Reasoning and Problem Solving

49. Based on the patterns shown, how many blocks would be needed to build a patio with 10 blocks in the middle row? How many would be needed to build a patio with n blocks in the middle row? Explain your solution in writing.

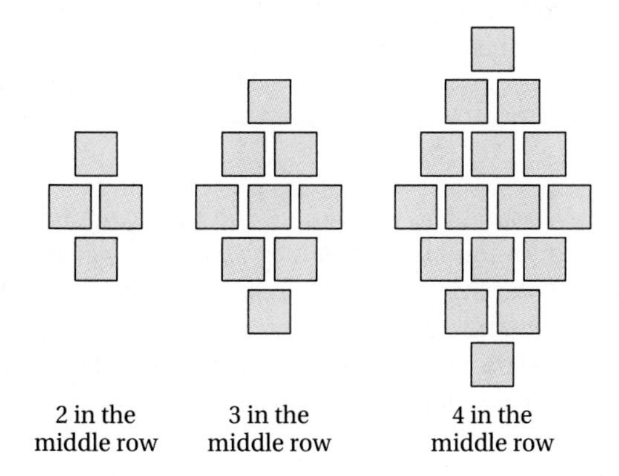

50. The Seating Problem. How many square tables seating 1 person at each side and arranged end to end are needed to seat 50 people? How many are needed to seat n people? Explain your solution in writing.

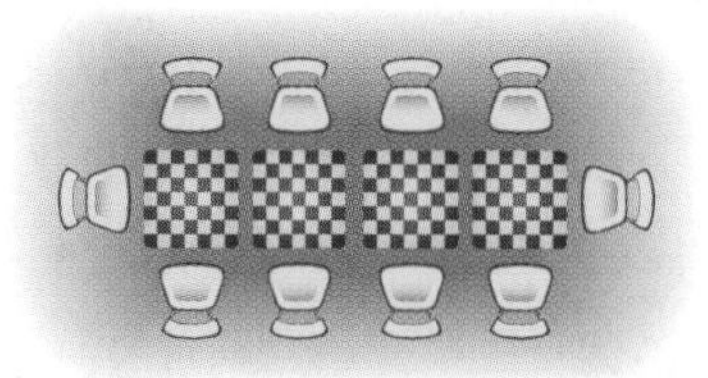

51. The Box Stack Problem. How many boxes would be needed altogether if there are 8 boxes on the bottom row? How many would be needed if there are n boxes on the bottom row? Explain your solution in writing.

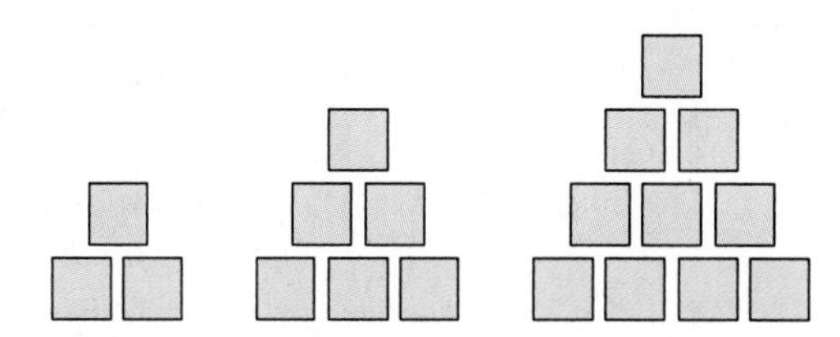

D. Communicating and Connecting Ideas

52. Making Connections. Name three ways of representing a function that were illustrated in this section. Show the same functional relationship between quantities represented in each of these ways.

53. Historical Pathways. The notation used today for expressions and equations was developed over hundreds of years. In the year 1631, the equation $x^3 - 3b^2x = 2c^3$ was written as $xxx - 3bbx$ ═══ $+2ccc$. Write the equation $2x^4 + 3bx^3 - x = -4c^2$, using the 1631 notation.

Section 13.2 SOLVING EQUATIONS

- Understanding the Meaning of a Solution to an Equation
- Developing Ways to Solve Linear and Quadratic Equations

In this section we explore the meaning of a solution to an equation and various ways of solving equations. We focus on linear equations such as $y = -12x + 18$ and quadratic equations such as $y = x^2 + 3x - 5$, although many of the techniques discussed can be applied to solving other types of equations. In this section we present techniques for solving equations and help you make sense of these different techniques. Mini-Investigation 13.7 will help you think about the meaning of a solution to an equation and recall ways you might go about solving equations.

Write a description of the techniques you used to solve each equation and compare your techniques with those that others in your class used.

Mini-Investigation 13.7 Communicating

How would you solve the following equations?

a. $12x - 5 = 67$

b. $x^2 = 125$

c. $2x^2 + 6x = 0$

d. $3x^2 + x - 5 = 0$

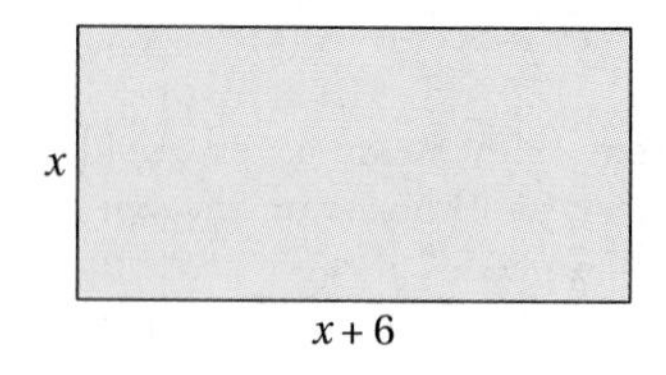

FIGURE 13.3 | A representation for $x(x + 6) = 112$.

Understanding the Meaning of a Solution to an Equation

Suppose that the area for the rectangle shown in Figure 13.3 is 112 sq units. You can write the equation $x(x + 6) = 112$ to show this relationship. To find the length and width of the rectangle, you could try to find a value or values that, when substituted for x, make a true equation. In doing so you would be finding a solution to the equation.

Description of a Solution to an Equation

A **solution** to an equation is a value for the variable or values for the variables that make the equation true. A **true equation** is one in which the value on the left of the equals sign is the same as the value on the right. To **solve** an equation means to find all of its solutions.

When an equation contains two variables, you should look for pairs of numbers that make the equation true. For example, suppose that you rent a car from an agency that charges a flat rate of $10 plus $25 each day you keep the car. The equation $y = 25x + 10$, where x represents the number of days the car is rented and y represents the total cost, shows the relationship between these quantities. An infinite number of ordered pairs (x, y) make this equation true. The following table of values contains five whole-number ordered pairs that are solutions to the equation $y = 25x + 10$.

x	0	1	2	3	4
y	10	35	60	85	110

Estimation Application

The Car Rental Challenge

The equation $y = 25x + 10$ shows the relationship between the total cost of renting a car, y, and the number of days the car is rented, x. Estimate the total cost of renting a car for the month of February.

Suppose, however, that your budget allows $340 for renting a car, y, and you want to know the number of days, x, that you can rent the car. The equation can be written as $340 = 25x + 10$. For this equation you are looking for values of x that make the equation true. In this case there is but one solution, $x = 13.2$. To check, calculate to determine that $340 = 25(13.2) + 10$ is a true equation. Although 13.2 is the solution to the equation, the answer that makes sense for the problem is 13 days.

Developing Ways to Solve Linear and Quadratic Equations

Most equations in mathematics may be solved in various ways. The technique you choose should depend on the complexity of the equation to be solved and the tools you have at your disposal, as well as your ability as an equation solver. In this section we explore some ways of solving equations that can be written in the form $y = ax + b$ (linear functions) and some ways to solve equations that can be written in the form $y = ax^2 + bx + c$, where $a \neq 0$ (quadratic functions).

Solving Linear Equations by Using Properties of Equality. Recall that equations such as

$$y = x + 34, \qquad y = 12 - 3x, \qquad \text{and} \qquad y = \frac{x}{3} + 23$$

represent linear functions.

When the value of y is known, as in

$$78 = x + 34, \qquad -26 = 12 - 3x, \qquad \text{and} \qquad 0 = \frac{x}{3} + 23,$$

the equations are called **linear equations** because they can be associated with linear functions.

The following properties of equality, together with the idea of inverse operations developed in Chapter 2, can be used to solve linear equations.

Properties of Equality

Addition and Subtraction: For all numbers a, b, and c, if $a = b$, then $a + c = b + c$ and $a - c = b - c$.

Multiplication and Division For all numbers a, b, and c, if $a = b$, then $ac = bc$; and when $c \neq 0$, if $a = b$, then $\frac{a}{c} = \frac{b}{c}$.

The addition and subtraction properties of equality say that you can add or subtract the same number on both sides of an equation and that the two sides will remain equal. The multiplication and division properties of equality say that you can multiply or divide both sides of an equation by the same non-zero number and that the two sides will remain equal. The other idea involved in solving equations by using the properties of equality is *inverse operations*. Recall that addition and subtraction are inverse operations because addition can undo subtraction and vice versa. Similarly, multiplication and division are inverse operations because multiplication can undo division and vice versa.

Example 13.6 illustrates how to use the properties of equality and inverse operations to solve linear equations. Always remember to check the values you find for the variable in the original equation to be sure that they really are solutions. Also, when a real-world situation is associated with the equation, be sure to check whether the solution makes sense in the real-world context.

Example 13.6 Solving Linear Equations

Solve and check the following equations. Give answers to the nearest hundredth.

a. $538 = 65x + 18$

b. $125.5 = \frac{x}{12.5} - 34.75$

c. $45 = \frac{2x + 18}{7x - 24}$

Solution

a.

$538 - 18 = 65x + 18 - 18$ Subtract 18 from both sides so that they remain equal.

$\frac{520}{65} = \frac{65x}{65}$ Divide both sides by 65 so that they remain equal.

$x = 8$

Check:

$538 = 65x + 18$

$= 65(8) + 18$ Replace x with 8.

$= 538$ The solution is 8.

b.

$125.5 = \frac{x}{12.5} - 34.75$

$125.5 + 34.75 = \frac{x}{12.5} - 34.75 + 34.75$ Add 34.75 to both sides so that they remain equal.

$(160.25)(12.5) = (\frac{x}{12.5})12.5$ Multiply both sides by 12.5 so that they remain equal.

$2003.13 = x$ Rounded to the nearest hundredth.

Check:

$125.5 = \frac{x}{12.5} - 34.75$

$125.5 = \frac{2003.13}{12.5} - 34.75$ Substitute 2003.13 for x.

$= 125.5$ It checks. The solution is 2003.13.

c.

$45 = \frac{2x + 18}{7x - 24}$

$45(7x - 24) = (\frac{2x + 18}{7x - 24})(7x - 24)$ Multiply both sides by $7x - 24$ so that they remain equal.

$315x - 1080 = 2x + 18$ Simplify.

$315x = 2x + 1098$ Add 1080 to both sides. Simplify.

$313x = 1098$ Subtract $2x$ from both sides.

$x = \frac{1098}{313}$ Divide both sides by 313.

$= 3.51$ Rounded to the nearest hundredth.

Check:

$45 = \frac{2x + 18}{7x - 24}$

$= \frac{2(3.51) + 18}{7(3.51) - 24}$

≈ 45 It checks, taking into account rounding. The solution is 3.51.

Your Turn

Practice: Solve and check.

a. $32 = \frac{2}{3}x - 24$

b. $64 = \frac{44 - x}{18 - 5x}$

c. $45 = \frac{x}{4} + \frac{5}{8}$

Reflect: How do you decide what the first step is in solving a linear equation? Use the example problem to explain your answer. ■

Solving Quadratic Equations by Using a Graph. Recall from Section 13.1 that equations such as the following represent quadratic functions.

$$y = x^2 + 2x + 1, \quad y = 4x^2 + 5, \quad \text{and} \quad y = -\frac{x^2}{3}$$

When the value of y in these equations is 0, as shown, the resulting equation is called a **quadratic equation** in standard form:

$$0 = x^2 + 4x + 1, \quad 0 = 4x^2 + 5, \quad \text{and} \quad 0 = -\frac{x^2}{3}$$

A quadratic equation such as $3x^2 + 5x - 4 = 7$ can be written in standard form by adding -7 to both sides to produce $3x^2 + 5x - 11 = 0$. A variety of techniques are available for solving quadratic equations. One very useful technique is to use the graph of the related quadratic function. To do so easily, you can use a graphing calculator. The following Spotlight on Technology feature focuses on the use of a graphing calculator.

Spotlight on Technology: **Graphing Calculators**

Graphing calculators provide fast and effective tools for graphing mathematical relationships, as well as carrying out numerical and symbolic calculations. Graphing calculators also provide visual representations of data sets and have extensive programming features.

One of the first graphing calculators was the fx7000G, created in 1985 by Casio. Several generations of graphing calculators have been produced since then, including some that graph in more than one color, some that split the calculator screen into two or more parts, and some that download applications from the Internet.

To use a graphing calculator to *graph a mathematical relationship*,

1. input a relationship between two variables,
2. establish a viewing rectangle, and
3. display the graph.

Let's consider each step in turn.

Input a relationship between two variables. We graph the relationship between the radius of a circle and its area, $A = \pi r^2$. With a graphing calculator, we use x to represent the radius length and y to represent the area, and input $y = \pi x^2$ into the calculator. Every calculator has a menu or screen for function input. On some, you can access the menu by pressing a key marked Y = ; on others you press a key or access a command marked GRAPH or PLOT. The screen shown in Figure 13.4 depicts the radius-area relationship as it was entered on a graphing calculator.

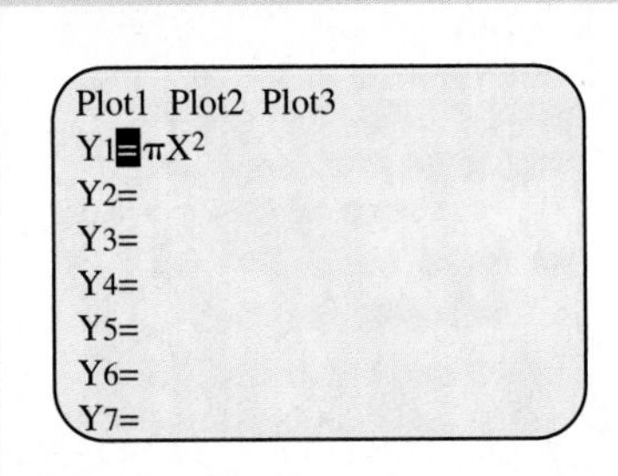

FIGURE 13.4 | Entering $y = \pi r^2$ in a graphing calculator.

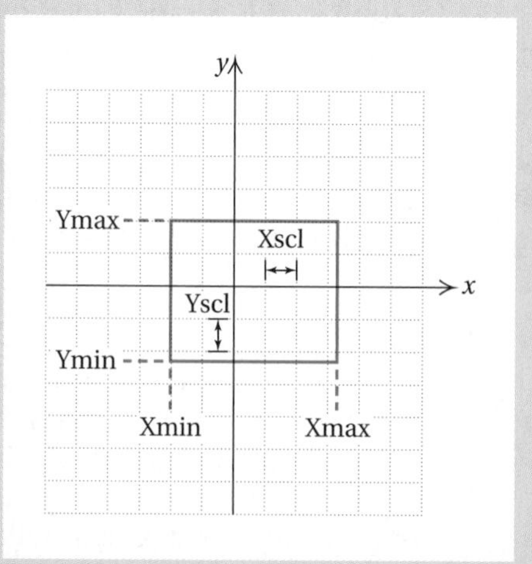

FIGURE 13.5 | The viewing rectangle on a graphing calculator.

Establish a viewing rectangle. The graphing plane continues infinitely in two dimensions. You need to record the portion of this infinite plane that you want to appear on your calculator's screen, called the *viewing rectangle* or the *window*. You must input the left and right boundaries (Xmin and Xmax), the bottom and top boundaries (Ymin and Ymax), and the spacing of the tick marks for each axis (Xscl and Yscl), as shown in Figure 13.5.

Every calculator has a menu or screen for window dimension input. On some, you access the menu by pressing a key marked WINDOW; on others you press a key or access a command named GRAPH, RANGE, or PLOT. The screen shown in Figure 13.6 indicates that the viewing rectangle stretches from 0 to 10 units on the x-axis, and from 0 to 25 units on the y-axis. The x-axis will be scaled with tick marks 1 unit apart, and the y-axis will have tick marks 5 units apart.

Display the graph. Having input a function and a viewing rectangle, you need to instruct the calculator to display the graph. Every calculator has a button or command that does so. On some, you simply press GRAPH; on others you press a PLOT key or access a command. The screen shown in Figure 13.7 contains the graph of the relationship $y = \pi x^2$ viewed through a window measuring from 0 to 10 units horizontally and from 0 to 25 units vertically.

Graphing calculators' capacity to display graphic representations of relationships allows you to concentrate on the global behavior of the relationships without getting bogged down in repetitive calculations and point plotting. Additional

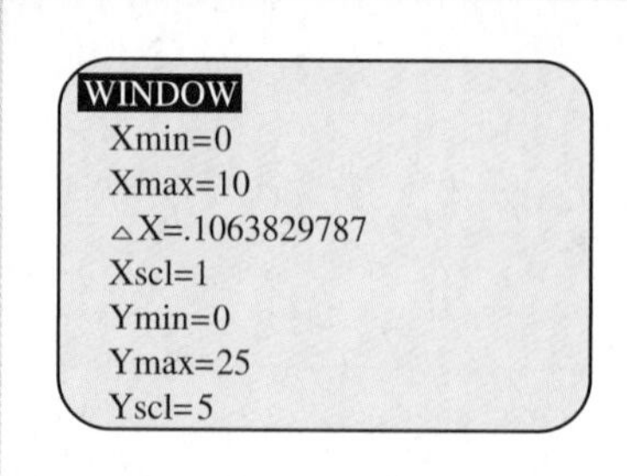

FIGURE 13.6 | Setting the window dimensions on a graphing calculator.

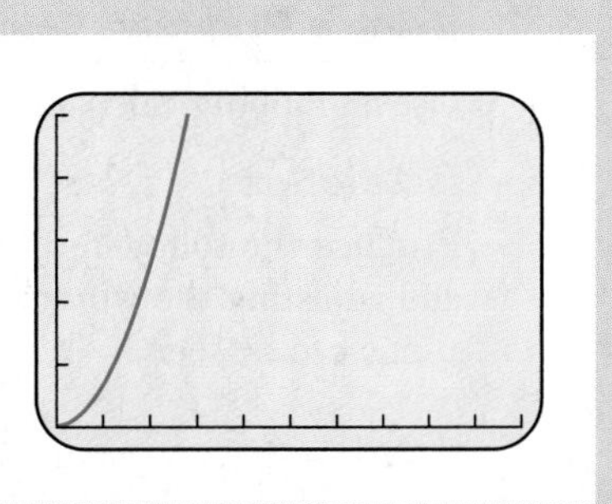

FIGURE 13.7 | Graph of the equation $y = \pi x^2$.

calculator commands allow you to manipulate the graph, zooming in to identify values that may be solutions to problems that you are exploring. These and many other features of today's graphing calculators provide powerful tools for mathematical investigations and analyses.

To illustrate the technique of solving quadratic equations by using graphs, let's start with the equation $x^2 - 3x - 70 = 0$. To solve this quadratic equation by using a graph, first think of it as a special case of the quadratic function $x^2 - 3x - 70 = y$. It's a special case because the value of y is 0. In other words, you are trying to find what, if any, values for x make the algebraic expression have a value of 0.

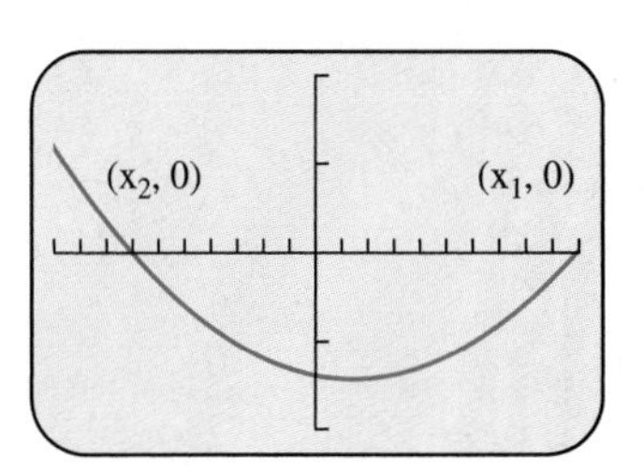

FIGURE 13.8 | Graph of the equation $y = x^2 - 3x - 70$.

The graph of $y = x^2 - 3x - 70$ is shown in Figure 13.8. The points $(x_1, 0)$ and $(x_2, 0)$ are the points at which the graph intersects the x-axis—the points at which the value of y is 0—and are called the ***x*-intercepts.** They also are the values of x that are solutions to the quadratic equation. Because the points of intersection are integers, the solutions to $x^2 - 3x - 70 = 0$ are $x = 10$ and $x = -7$.

The following summarizes the procedure used.

Procedure for Using a Graph to Solve a Quadratic Equation

When a Graph Might Be Used
Use if the quadratic equation isn't easily factored and a graphing calculator is available.

How to Use This Technique

1. Write the quadratic equation in standard form, $0 = ax^2 + bx + c$.
2. Graph the related function, $y = ax^2 + bx + c$.
3. Find the x-coordinate of the point(s) at which the graph intersects the x-axis.
4. Substitute the value(s) of x into the original quadratic equation to check whether each is a solution.

Example 13.7 illustrates how the zoom feature on most graphing calculators can be used in solving an equation.

Example 13.7 Using a Graphing Calculator to Solve a Quadratic Equation

Use a graphing calculator to solve $2x^2 + 9x - 56 = 0$.

Solution

Produce the following graph by inputting the relationship, displaying the graph, and adjusting the window so that the two points at which the graph intersects the x-axis can be seen.

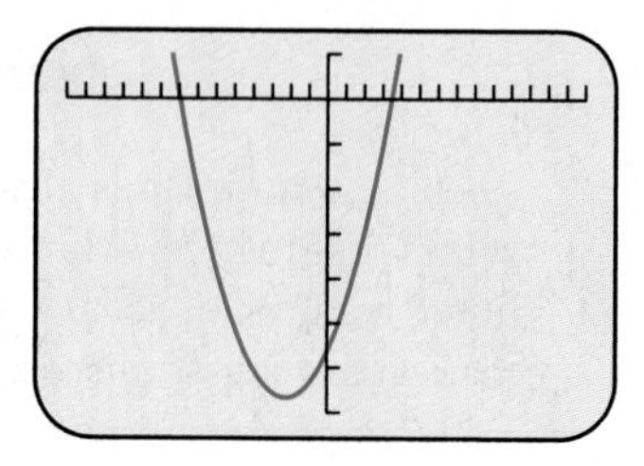

Use the ZOOM and TRACE on a graphing calculator to produce another view of the graph and find one value of x. Even though it may be difficult to move the cursor so y is exactly 0, this view is close enough to find the x value. Because the horizontal scale in this view is marked in increments of 0.1 unit, the value of x is 3.5.

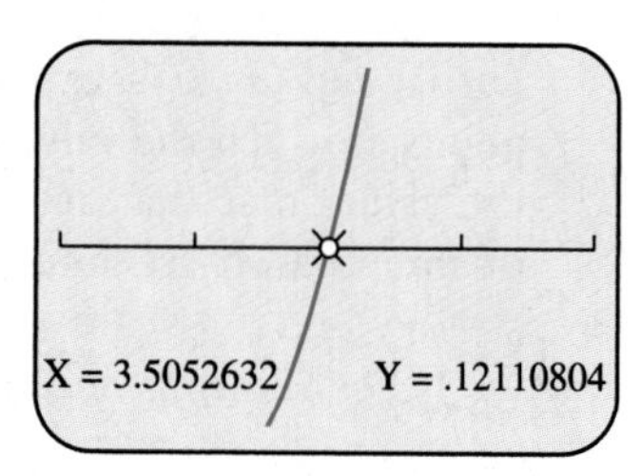

Use the same process to find the other value of x: -8.

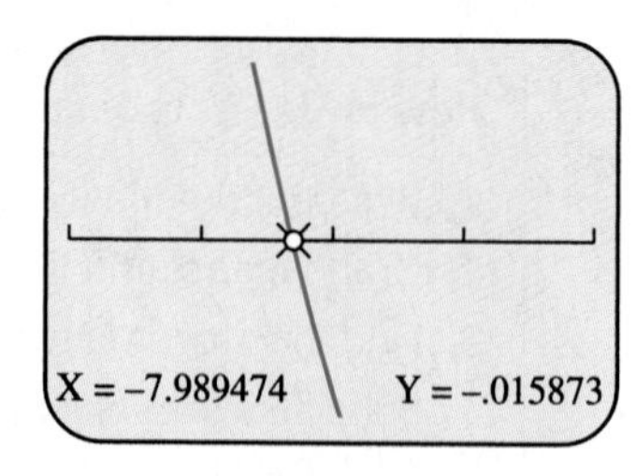

Substituting these values into the original equation gives the following results.

If $x = 3.5$, then $2x^2 + 9x - 56 = 0$.
If $x = -8$, then $2x^2 + 9x - 56 = 0$.

This outcome tells us that $x = -8$ and $x = 3.5$ are the exact values that make the expression equal 0 and that these are the solutions to the quadratic equation.

Your Turn

Practice: Solve by using a graphing calculator.

a. $x^2 + 11x + 18 = 0.$

b. $-3x^2 - 6.35x - 3.25 = 0.$

c. $9x^2 - 20.5x - 1 = 4.$ (*Hint:* Write in standard form.)

Reflect: Suppose that the value you get for x when you substitute into the expression from the quadratic equation makes the value of the expression close to, but not exactly, 0. Does this result mean that you made a mistake? Explain. ■

Solving Quadratic Equations by Using Numerical Zoom-In. A procedure sometimes called *numerical zoom-in* can also be used on a graphing calculator to solve equations. Consider, for example, the quadratic equation $3x^2 + 5x - 4 = 7$. One way to solve this equation is first to subtract 7 from each side, resulting in $3x^2 + 5x - 11 = 0$. Now enter $y = 3x^2 + 5x - 11$ under the [Y=] menu and press [GRAPH]. On a standard-viewing window, the graph appears as shown in Figure 13.9. Note that the x-axis intercepts appear near $x = -3$ and $x = 1$.

FIGURE 13.9 | Graph of the equation $y = 3x^2 + 5x - 11$

Now press [2nd] [TBLSET] and enter -5 and 1, as shown in Figure 13.10. Doing so builds a table from the currently selected function, evaluating it first at $x = -5$ and then incrementing x by 1 unit.

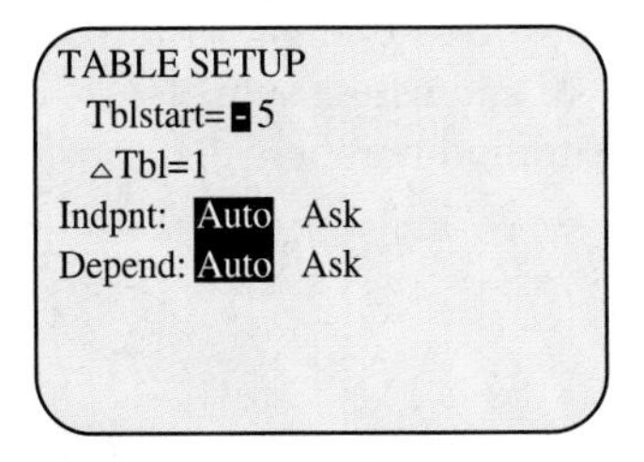

FIGURE 13.10 | Setting up a table on a graphing calculator.

Press [2nd] [TABLE] to show the ordered pairs. In Figure 13.11, Y1 is positive when $x = -3$ but negative when $x = -2$. Because Y1 is continuous, we know that $Y1 = 0$ between $x = -3$ and $x = -2$. We can return to the table setup screen and adjust the starting x-value and the increment.

Figure 13.12(a) shows a starting value of $x = -3$ and increments of 0.1. Here, Y1 is positive when $x = -3$ and negative when $x = -2.9$. Hence the value for x lies between -3 and -2.9. We can continue this numerical zoom-in procedure until we exhaust the accuracy capacity of the calculator or until we are satisfied with the numerical approximation generated by a particular ordered pair.

Figure 13.12(b) shows the zoom-in process after two more iterations. Here, you can see that Y1 is positive for $x = -2.922$ and negative for $x = -2.921$, so the value of x associated with $Y1 = 0$ lies between these two numbers. Using this graphing calculator technique to solve some of the quadratic equations given in this section is instructive.

X	Y1
-5	39
-4	17
-3	1
-2	-9
-1	-13
0	-11
1	-3

X=-5

FIGURE 13.11 | Some ordered pairs for the equation $y = 3x^2 + 5x - 11$

X	Y1
-3	1
-2.9	-.27
-2.8	-1.48
-2.7	-2.63
-2.6	-3.72
-2.5	-4.75
-2.4	-5.72

X=-3

(a)

X	Y1
-2.924	.02933
-2.923	.01679
-2.922	.00425
-2.921	-.0083
-2.92	-.0208
-2.919	-.0333
-2.918	-.0458

X=-2.918

(b)

FIGURE 13.12 | Some ordered pairs after zooming.

Solving Quadratic Equations by Factoring. The **factoring** technique for solving quadratic equations is based on the zero product principle. Consider the equation $ab = 24$. What values are possible for a and b? Obviously, many values are possible for both a and b. Now consider the equation $ab = 0$. For the product to be 0, a must equal 0, b must equal 0, or both a and b may equal 0. This relationship is the zero product principle.

Zero Product Principle

For any real numbers a and b, if $ab = 0$, then $a = 0$ or $b = 0$; and if $a = 0$ or $b = 0$, then $ab = 0$.

To solve a quadratic equation by factoring we need to determine whether it can be written as the product of two factors that equal 0. Consider the quadratic equation $x^2 + 3x - 40 = 0$. The goal is to rewrite the quadratic equation in the form $(x + a)(x + b) = 0$. Then either $x + a = 0$ or $x + b = 0$, and you can easily find the values of a and b. Remember that the values of a and b will be the x-intercepts when you graph the function $y = x^2 + 3x - 40$.

To help you understand factoring, work backward, starting with the end result you want to achieve, $(x + a)(x + b) = 0$, and write it in standard form for a quadratic equation. Using the distributive property results in an equation in standard form.

$$
\begin{aligned}
(x + a)(x + b) &= 0 \\
x(x + a) + b(x + a) &= 0 \quad \text{Distributive property} \\
x^2 + ax + bx + ba &= 0 \quad \text{Distributive property} \\
x^2 + (a + b)x + ab &= 0, \quad \text{Distributive property}
\end{aligned}
$$

When you start with an equation such as $x^2 + 3x - 40 = 0$, the goal for factoring is to find two numbers whose product, ab, is -40 and whose sum, $a + b$, is 3. You have to apply number sense and trial and error to find numbers that satisfy these conditions. In this situation, write

$$x^2 + 3x - 40 = 0.$$

Because

$$8(-5) = -40 \quad \text{and} \quad 8 + (-5) = 3,$$

write

$$(x + 8)(x - 5) = 0$$

The zero product principle tells you that

$$x + 8 = 0 \quad \text{or} \quad x - 5 = 0.$$

Solving for x gives

$$x = -8 \quad \text{or} \quad x = 5.$$

Substituting each of these values into the original equation and calculating shows that they are the correct solutions to the quadratic equation. We summarize this procedure as follows.

Procedure for Using Factoring to Solve Quadratic Equations

When This Technique Might Be Used

1. Use if the value of $c = 0$ in $ax^2 + bx + c = 0$.
2. Use if the coefficient of x^2 is 1 and the coefficient of x and the constant term are relatively small integers.

How to Use This Technique

1. If $c = 0$ in the quadratic equation, factor an x and, if possible, a constant from each term. Set each factor equal to 0 and solve for x.
2. If $c \neq 0$ and $b \neq 0$, look for two numbers whose product is the constant term and whose sum is the coefficient of x. Set each expression equal to 0 and solve for x.
3. Check the value(s) for x in the original equation.

Example 13.8 applies these ideas for solving quadratic equations.

Example 13.8

Using Factoring to Solve Quadratic Equations

Solve $x^2 - 3x - 70 = 0$ by factoring.

Solution

The pairs of integer factors of 70 are $1 \cdot 70, 2 \cdot 35, 5 \cdot 14$, and $7 \cdot 10$. As $-10 \cdot 7 = -70$ and $-10 + 7 = -3$, the numbers needed are -10 and 7. In other words, $(x - 10)(x + 7) = 0$, so, because of the zero product principle, we conclude that $x = 10$ or $x = -7$.

Check:

$$\begin{aligned} x^2 - 3x - 70 &= 0 \\ (10)^2 - 3(10) - 70 &= 0 \\ 100 - 30 - 70 &= 0 \\ 0 &= 0 \end{aligned} \qquad \begin{aligned} x^2 - 3x - 70 &= 0 \\ (-7)^2 - 3(-7) - 70 &= 0 \\ 49 + 21 - 70 &= 0 \\ 0 &= 0 \end{aligned}$$

The check verifies that 10 and -7 are solutions.

Your Turn

Practice: Solve the following equations by factoring and check.

a. $x^2 + 10x - 24 = 0$

b. $x^2 + 18x = 0$

Reflect: If $b = 0$ in $ax^2 + bx + c = 0$, can the equation be solved by factoring and using the zero product principle? Explain. ■

Mini-Investigation 13.8 involves a special case of quadratic equations and helps you make a generalization about solutions to these types of equations.

Write an explanation of your answer and compare it with that of a classmate.

Mini-Investigation 13.8 Using Mathematical Reasoning

In equations of the form $ax^2 + bx = 0$, what will always be one solution to the equation?

Solving Quadratic Equations by Using the Quadratic Formula. You probably have used the **quadratic formula,**

$$x = \frac{-b \pm \sqrt{b^2 - 4ac}}{2a}$$

to solve quadratic equations of the form $ax^2 + bx + c = 0$, $a \neq 0$. The procedure for using the quadratic formula is summarized as follows.

Procedure for Using the Quadratic Formula to Solve Equations

When This Technique Might Be Used

Consider using this technique if the coefficient of the x^2 term is not 1 and if most of the coefficients and constants are integers.

How to Use This Technique

1. Write the equation in the form $ax^2 + bx + c = 0$.
2. Identify the values of a, b, and c.
3. Substitute into the quadratic formula and calculate to find the value(s) for x.
4. Check by substituting the value(s) for x into the original equation and calculating.
5. If you are solving a real-world problem, evaluate the values for x to determine whether they make sense for the context of the problem.

Example 13.9 demonstrates the use of this procedure to solve a quadratic equation.

Example 13.9 **Using the Quadratic Formula to Solve Equations**

Use the quadratic formula to solve $x^2 + 4x + 1 = 0$.

Solution

For the equation $x^2 + 4x + 1 = 0$, $a = 1$, $b = 4$, and $c = 1$. Substituting

$$\begin{aligned}\frac{-b \pm \sqrt{b^2 - 4ac}}{2a} &= \frac{-4 \pm \sqrt{4^2 - 4(1)(1)}}{2(1)} \\ &= \frac{-4 \pm \sqrt{12}}{2} \\ &= \frac{-4 \pm 2\sqrt{3}}{2} \qquad \text{Note that } \sqrt{12} = \sqrt{4 \cdot 3} = 2\sqrt{3} \\ &= -2 \pm \sqrt{3}\end{aligned}$$

So

$$x = -2 + \sqrt{3} \quad \text{and} \quad x = -2 - \sqrt{3}$$

The decimal approximations for the x values are $x = -0.27$ and $x = -3.73$.

Your Turn

Practice: Solve $0 = -2x^2 + 10x - 7$ by using the quadratic formula and check.

Reflect: Must the values for a, b, and c in the quadratic formula always be positive? Explain. ■

Always check the x-values by substituting into the original equation and calculating. Also, if you are solving a real-world problem, one or both of the values obtained for x may not make sense in that context. For example, the quadratic function $s = -4.9t^2 + v_0t + h$ shows the relationship between the time an object is in the air, t, for an object thrown upward with an initial velocity, v_0, from a height, h, and the actual height, in meters, it obtains, s. Solving this quadratic equation gives both positive and negative values for t. The time that an object is in the air can't be negative, so the negative solution to the quadratic equation doesn't make sense in this real-world context.

Solving Problems That Involve a Quadratic Function. The complexity of the numbers involved, the nature of the situation (do you need an exact answer or is an estimate sufficient?), and the tools at your disposal (do you have a graphing calculator available?) influence the method you choose for solving a quadratic equation. Sometimes, as in Example 13.10, a problem involves creating, but not solving, a quadratic function. In this example we utilize the problem-solving strategies *simplify the problem*, *make a table*, and *look for a pattern*.

Example 13.10 Problem Solving: The Stair Steps

Suppose that squares of the same size were used to draw a stair-step design like the one shown. Give a function written as an equation that shows how the number of steps is related to the total number of blocks.

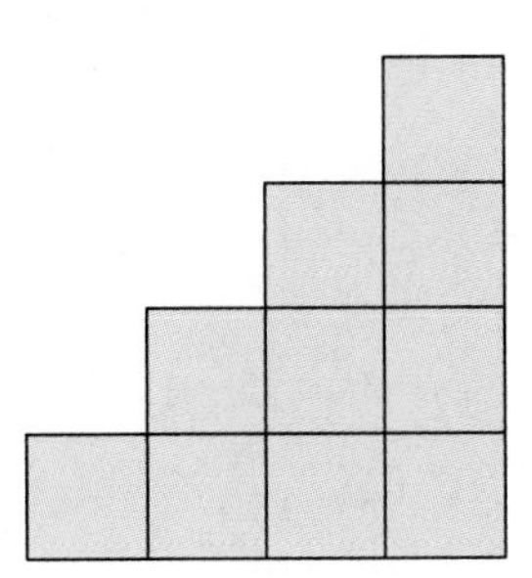

Working Toward a Solution

Understand the problem	*What does the situation involve?*	Showing the relationship between the number of steps and the number of blocks in a stair-step design

	What has to be determined?	We are trying to find a general rule that tells how the total number of squares in the design is related to the number of steps.
	What are the key data and conditions?	The data are shown in the figure.
	What are some assumptions?	We must assume that the design continues to grow as shown in the figure. The number of blocks needed to build each additional step is one more than the number needed to build the preceding one.
Develop a plan	*What strategies might be useful?*	Simplify the problem, make a table, and look for patterns.
	Are there any subproblems?	First, we look at the number of blocks for 1 step, then 2 steps, and so on.
	Should the answer be estimated or calculated?	The problem asks for a generalization, so exact calculations should be used for the subproblems in order to find a correct generalization.
	What method of calculation should be used?	Mental math is sufficient.
Implement the plan	*How should the strategies be used?*	Use a smaller number of steps and then build up. Record the data in a table and look for a pattern.

Number of Steps	Total Number of Blocks
1	1
2	3
3	6
4	10
5	15

	What is the answer?	At least two patterns appear in the table. One is that the total number of squares needed for n steps is the sum of the numbers $1 + 2 + 3 + \cdots + n$. The other pattern is that for n steps, multiplying n by 1 more than n and dividing the product by 2 gives the total number of squares. That is, $\frac{n(n+1)}{2} = T$, where n is the number of steps and T is the total number of squares.
Look back	*Is the interpretation correct?*	Yes. The interpretation fits the data given.

Is the calculation correct?	Several of the known values for n and T may be substituted into the equation to verify that it holds. It does.
Is the answer reasonable?	The equation $\frac{n(n+1)}{2} = T$ can be written as $\frac{1}{2}n^2 + \frac{1}{2}n = T$, which is a quadratic function. This result seems reasonable because for each increase of 1 step the change in the total number of blocks isn't constant.
Is there another way to solve the problem?	The expression may be written in different forms, but there is only one rule. Such growth problems can often be solved by making tables and looking for patterns to help generalize the approach.

Your Turn

Practice: Suppose that 78 squares of the same size are used to build a stair-step design like the one in the example problem. How many steps are in the design?

Reflect: How do you know that the equation $\frac{n(n+1)}{2} = T$ represents a quadratic function? ■

Problems and Exercises | *for Section 13.2*

A. Reinforcing Concepts and Practicing Skills

1. Which of the values, $x = 0, -3, 9, 3$, are solutions to the equation $8x^2 = 72$?

State the first step you would take to solve each equation in Exercises 2–6.

2. $2x - 18 = -34$

3. $\frac{x}{4} = -12$

4. $\frac{x + 13}{-6} = 24$

5. $5x + 18 = -7x - 6$

6. $4x^2 - 16 = 0$

Give five ordered pairs that make each equation true in Exercises 7 and 8.

7. $x^2 + 3x = y$

8. $y = 20 + \frac{x}{3}$

Copy and complete each table in Exercises 9 and 10, using the given linear equation.

9. $15 - 2x = y$

x	y
0	15
1	13
−1	17
−5	25
?	−3

10. $\frac{2x}{5} + 12 = y$

x	y
0	12
5	14
−5	10
10	16
?	4

Solve each equation in Exercises 11–16 by using the properties of equality. Check.

11. $6x - 25 = 35$

12. $\frac{3x}{4} + 6 = -12$

13. $2.5x + 32 + 1.5x = 20$

14. $3x + 12 = -7x - 28$

15. $\$395 = \$15x + \$12.50$

16. $78 = \frac{5x}{6} + 18$

Solve each quadratic equation in Exercises 17–22 using any technique you wish. Name the technique. Check.

17. $x^2 - 17x - 84 = 0$

18. $12x^2 + 10x = 0$

19. $-2x^2 - 13x = 7$

20. $x^2 - 4.5x + 1.75 = 0$

21. $12x^2 - 11x - 2 = 3$

22. $-4x^2 = 0$

State the solutions, or estimated solutions, to each quadratic equation in Exercises 23–26 by examining the graph of the related quadratic function.

23. Equation: $x^2 + x - 12 = 0$
Graph: $y = x^2 + x - 12$

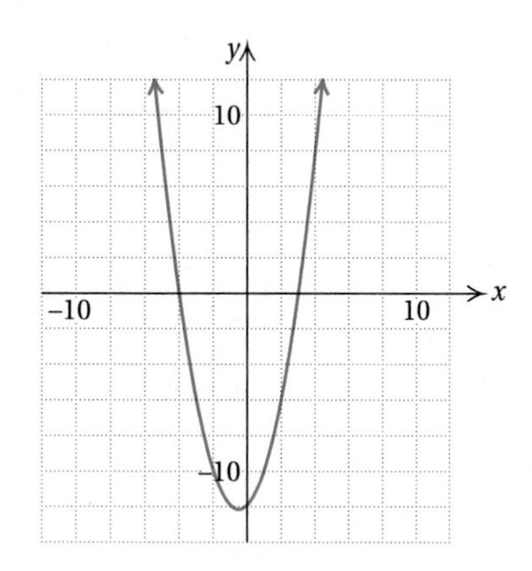

24. Equation: $x^2 - 9x + 20.25 = 0$
Graph: $y = x^2 - 9x + 20.25$

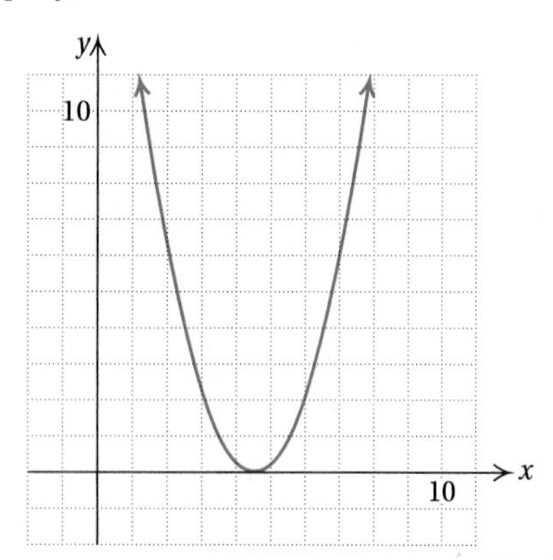

25. Equation: $4x^2 + 3x + 2 = 0$
Graph: $y = 4x^2 + 3x + 2$

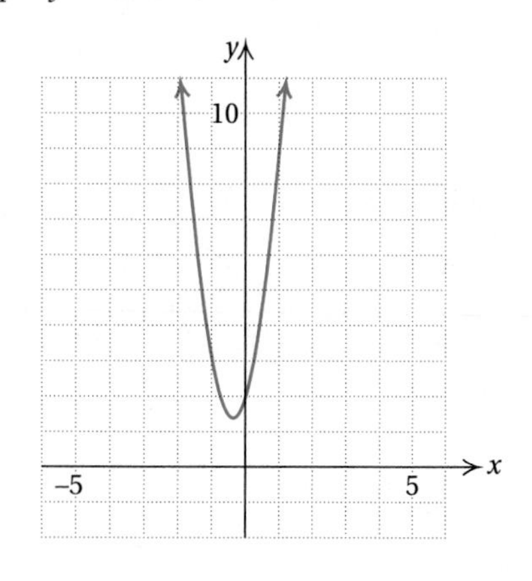

26. Equation: $2x^2 - 3x - 4 = 0$
Graph: $y = 2x^2 - 3x - 4$

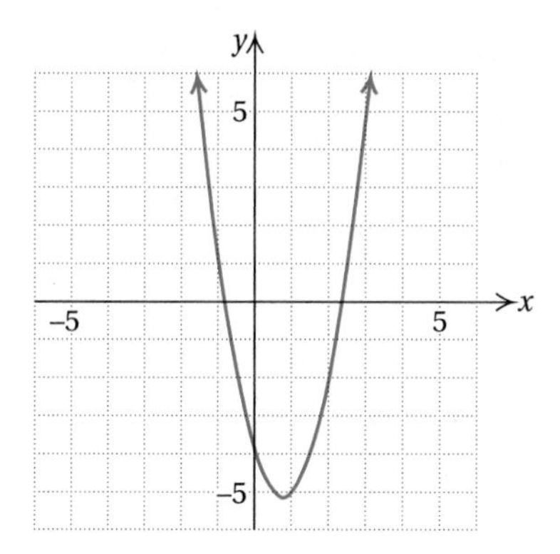

B. Deepening Understanding

27. Give two ways to solve the equation

$$\frac{1}{2}(x + 12) = -18$$

Tell the key sequence you would use on your calculator to solve each equation in Exercises 28–30.

28. $4x - 17 = 83$

29. $\frac{x - 13}{24} = 6$

30. $\frac{2x}{3} - 15 = 32$

31. Make up a linear equation. Then tell the calculator key sequence you would use to find the solution.

The graphs of quadratic functions in Exercises 32 and 33 are shown. How many solutions are there to the related quadratic equations?

32. $y = 2x^2 - 4x + 2$

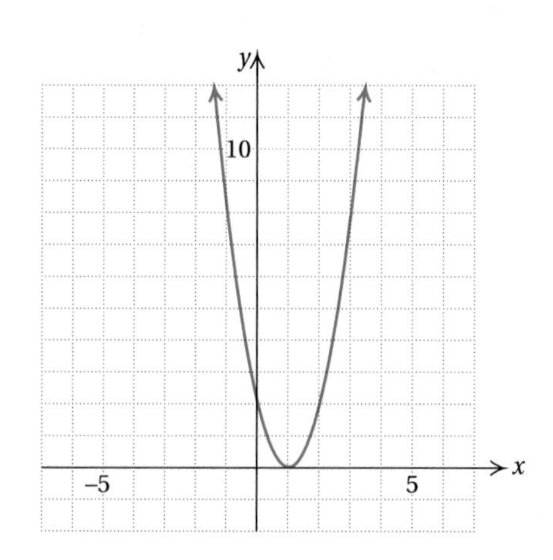

33. $y = x^2 - 6x - 7$

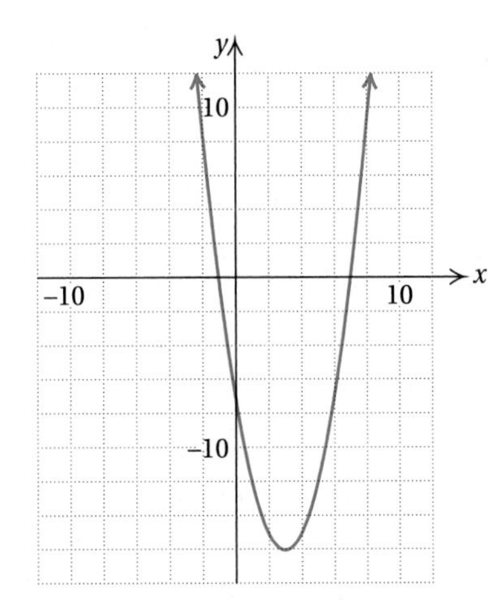

34. The quantity $b^2 - 4ac$ in the quadratic formula is called the *discriminant.* Describe how the value of the discriminant is related to the graph of the related function and to the number of solutions to the quadratic equation. Use quadratic equations from this section to support your argument.

35. Maurice solved the linear equation $8 = -4x + 28$ by graphing two lines. Use his graph to give the solution and explain how it works.

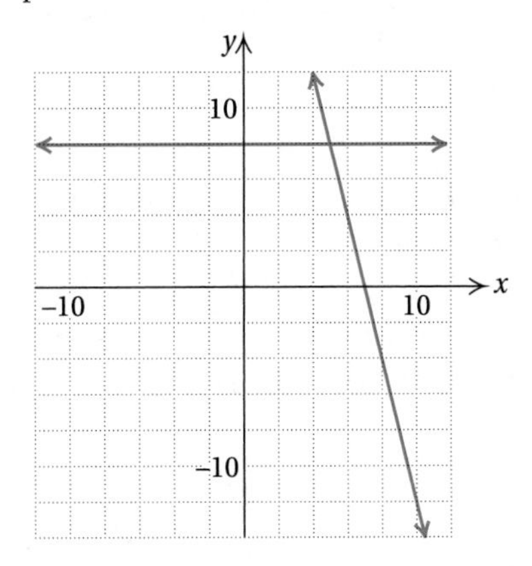

36. The graph of $y = x^3 - x^2 - 20x$ is shown in the next column. How many solutions are there to the equation $x^3 - x^2 - 20x = 0$? Explain, using the graph.

37. Write a quadratic equation that has two real-number solutions. Write a quadratic equation that has one real-number solution.

C. Reasoning and Problem Solving

38. The Windchill Problem. The function $y = -0.15x^2 + t$ is sometimes used to approximate the relationship between the wind velocity, x, and the windchill factor, y, at a given temperature, t, for low-velocity winds. If this formula were accurate, would a change of 5° in temperature or a change of 5 mph in wind speed have a greater effect on the windchill factor? Explain.

Figure for Exercise 36

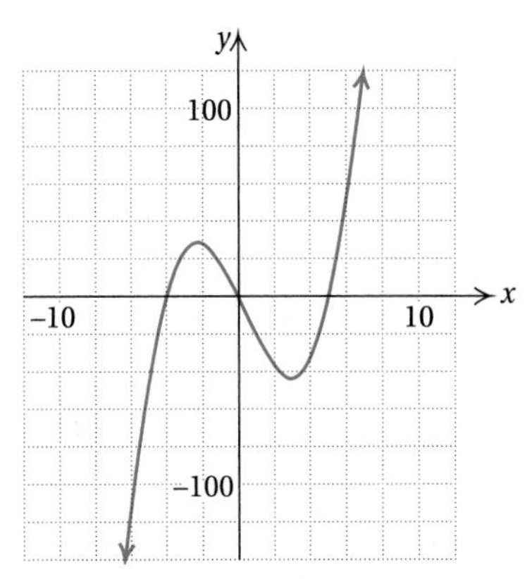

39. The Aesthetic Garden Problem. A landscape architect wants the length of a rectangular-shaped flower garden to be twice the width. The garden should have a total area of 288 ft^2. What are the dimensions of the garden?

40. The Maximum Garden Problem. A farmer has 230 ft of fence to enclose a rectangular garden. What is the largest garden area that can be enclosed with the 230 ft of fence? Explain your work.

41. The Skydiver Problem. A skydiver is free-falling at a rate of -9.8 m/s^2. The equation $d = -9.8t^2$ shows the relationship between the time the person has been falling, t, and the total distance traveled, d. If the ground is 1200 m below and the chute needs 5 s to open safely, what is the greatest number of seconds that the parachutist can wait to pull the rip cord?

D. Communicating and Connecting Ideas

42. Making Connections. Explain how a linear equation is related to a linear function and how a quadratic equation is related to a quadratic function.

43. Historical Pathways. Greek algebra was developed by Pythagoras and Euclid from 540 B.C. to 300 B.C. In those times, algebra was done through geometry. The following drawing illustrates how the Greeks showed that $(a + b)^2 = a^2 + 2ab + b^2$. Draw a picture like this one to find $(2x + 3)^2$. (*Hint:* Each side of the rectangle must be $2x + 3$ units long.)

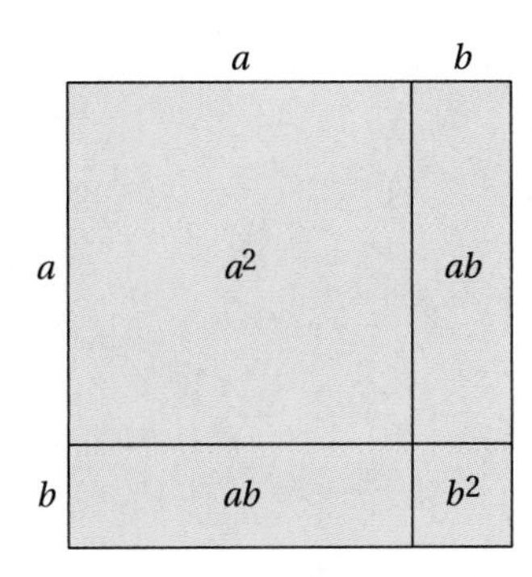

44. Work with a partner to discuss situations in which each of the techniques for solving quadratic equations illustrated in this section would be particularly useful and situations where they wouldn't be particularly useful. Explain your decisions.

45. The solutions to quadratic equations are sometimes called the "zeros" of the equation. Explain in writing why this phrase makes sense.

Section 13.3 EXPLORING GRAPHS OF LINEAR EQUATIONS

- Developing Techniques for Graphing Linear Equations
- Understanding How Graphs and Equations Are Related
- Understanding How Slope Is Related to Horizontal and Vertical Lines

In Section 13.2 we showed that a graph is one way to represent a functional relationship between two quantities. In this section we explore the graphs of linear equations further.

You have learned how to find solutions to equations such as $48 = \frac{x}{2} + 12$. A **solution to a linear equation** such as $y = \frac{x}{2} + 12$ is any ordered pair (x, y) that gives a true equation. Mini-Investigation 13.9 helps you think about finding solutions to linear equations.

Make a table to show your choices for the values for x and the related values of y.

Mini-Investigation 13.9 Using Mathematical Reasoning

What are five values of x that enable you to find the related values for y, using mental math, in the linear equation $y = \frac{x}{2} + 12$?

Developing Techniques for Graphing Linear Equations

One way to graph a linear equation is to create a table of ordered pairs and plot them on the Cartesian coordinate grid. Remember that the Cartesian coordinate grid is divided into four quadrants, as indicated on Figure 13.13. However, the real-world examples in this section have only positive values, so the graphs are in the first quadrant.

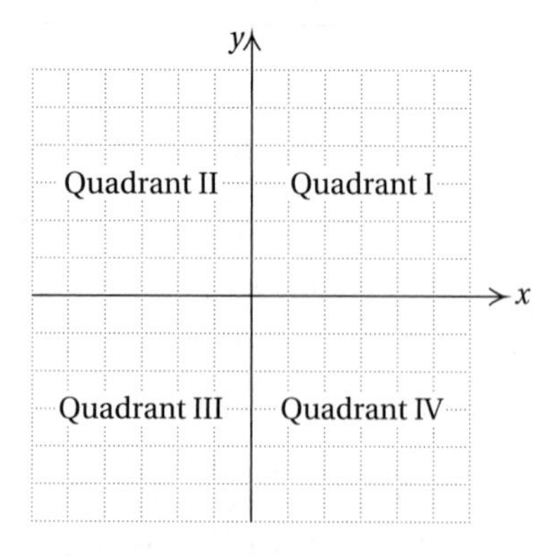

FIGURE 13.13 | Quadrants of the Cartesian coordinate grid.

Show the independent variable on the horizontal axis when you graph a function. Because x is often used to represent the independent variable, the horizontal axis is called the ***x*-axis.** Show the dependent variable on the vertical axis. Because y is usually used to represent the dependent variable, the vertical axis is referred to as the ***y*-axis.**

Consider the airline promotional plan whereby a special in-state ticket may be purchased for the regular price and a companion ticket for 50% off plus a $10 processing fee. The linear equation that represents this relationship is $y = 1.5x + 10$; x represents the cost of the regular ticket and y represents the total cost of the two tickets. The graph of this situation will represent all the values of x and y, (x, y), that make the equation true.

To graph the equation, we first make a table that includes at least three ordered pairs. Although any of the possible replacement values for x can be used in building the table, we decide to use values that will enable us to do the calculations mentally. Otherwise, we could choose any values and use a calculator. For building the following table, we used the values of 0, 10, 100, and 200 for x, which made the mental calculations straightforward. For the equation $y = 1.5x + 10$,

If $x = 0$, $\quad 1.5(0) + 10 = 10; \quad y = 10$

If $x = 10$, $\quad 1.5(10) + 10 = 25; \quad y = 25$

If $x = 100$, $\quad 1.5(100) + 10 = 160; \quad y = 160$

If $x = 200$, $\quad 1.5(200) + 10 = 310; \quad y = 310$

x	0	10	100	200
y	10	25	160	310

The next step is to decide on a scale for the graph. We don't really know what values of x make sense for this promotional deal, so we used a range of 0 to 200 for the x-axis. When $x = 200$, $y = 310$, so a range of 0 to 400 makes sense for the vertical axis. Constructing the scales and plotting the ordered pairs produces the graph shown in Figure 13.14. We drew a line through the points plotted to show that the solutions all lie on a straight line and that any point on the line is a solution to the equation $y = 1.5x + 10$. Of course, all possible real numbers can't be used for the price of a ticket, but there are so many possible values for x in the range that it's reasonable to use a solid line to connect the points.

Mini-Investigation 13.10 helps you think more about the points that you select to graph a linear equation.

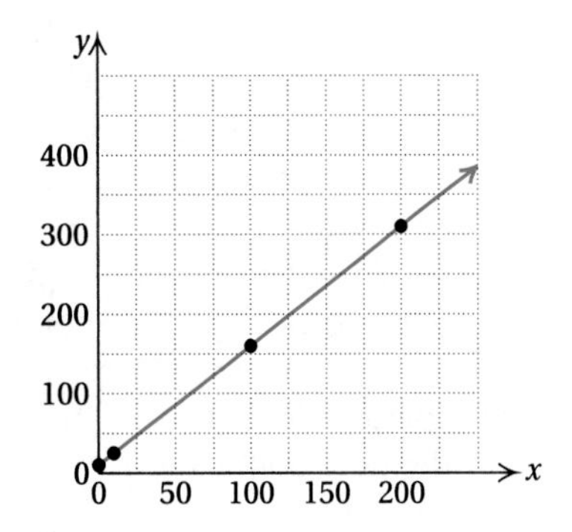

FIGURE 13.14 | First quadrant graph of the equation $y = 1.5x + 10$.

Talk about your answer with a classmate.

Mini-Investigation 13.10 Using Mathematical Reasoning

Why was the recommendation made that three or more points be used in building a table of ordered pairs to construct a graph?

Example 13.11 further demonstrates how to make and use a graph of a linear equation.

Example 13.11 Problem Solving: Summer Camp Enrollment

A summer camp prides itself in having a low children-to-counselor ratio: 8 to 1. The function $y = \frac{1}{8}x$ shows the relationship between the number of children who enroll, x, and the number of counselors needed, y. Make a graph that can be used for an advertising brochure that shows this relationship and use the graph to estimate the number of counselors needed if 125 children enroll in the camp.

Solution

Let x = the number of children who enroll and y = the number of counselors needed. Make a table of ordered pairs for $y = \frac{1}{8}x$

x	0	8	64	160	240
y	0	1	8	20	30

x-axis range: 0 to 300
y-axis range: 0 to 30
x-axis scale: 50
y-axis scale: 5

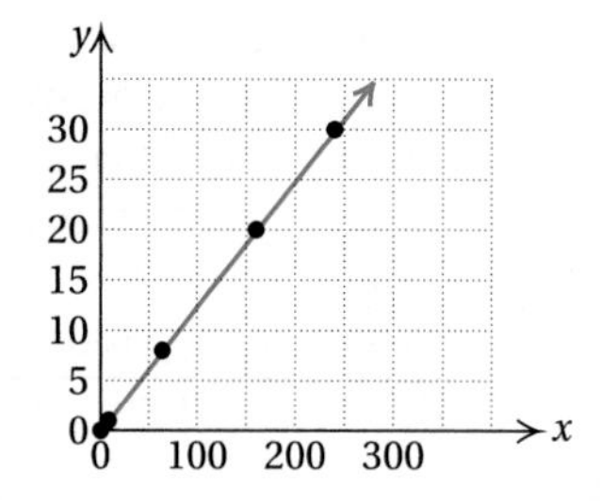

Connecting the points with a solid line is appropriate because all points on the line are solutions to the equation. However, we could connect the points with a dashed line to indicate that all the points don't make sense in the real-world situation (a fractional part of a child or counselor isn't possible).

Moving vertically from the point on the x-axis where $x = 125$ to a point on the line and then moving horizontally from this point on the line to the point on the y-axis gives the value for y when $x = 125$. The value for y is between 15 and 16 and appears closer to 16. The camp would have to have at least 16 counselors to meet the ratio goal.

Your Turn

Practice: Graph each equation. Use the graph to estimate the value for y when $x = 25$. Assume that the replacements for x and y are all real numbers.

a. $y = 12x - 2$
b. $y = -2x + 18$

Reflect: Why do the graphs of most real-world functional relationships fall only in the first quadrant of the coordinate grid? ■

Understanding How Graphs and Equations Are Related

A clear understanding of linear functions involves the ability to connect the graph representing the function to the linear equation representing the function. When you look at a linear equation, you should be able to create a visual image in your mind of what the graph looks like. Conversely, when you see the graph of a straight line, you should be able to create an image in your mind of what the linear equation might be. In order to create visual images of the equation or the graph, you must understand the relationship between the quantities.

Experience in making connections between a graph and an equation, such as those in Mini-Investigation 13.11, can help you create these visual images.

Write a description of the connections you found between characteristics of the linear equations and their graphs.

Technology Extension: Use a graphing utility to answer the question in the Mini-Investigation.

Mini-Investigation 13.11 Making a Connection

What can you discover about connections between linear equations and their graphs by graphing each of the following sets of equations on the same coordinate grid?

Set 1	Set 2
$y = 3x + 6$	$y = -3x + 2$
$y = -2x + 6$	$y = -3x - 3$
$y = \frac{x}{5} + 6$	$y = -3x$

Mini-Investigation 13.11 should have reminded you of some important connections between a graph and a linear equation. The following are some important connections to remember.

Description of How Lines Relate to Their Equations

1. Lines that slant up, moving from left to right, have a **positive slope.** The value of m when the equation is in the form $y = mx + b$ gives the **slope** of the line.
2. Lines that slant down, moving from left to right, have a **negative slope.**
3. The y-coordinate of the point at which a line intersects the y-axis is called the **y-intercept.** The constant term, b, when the equation is in the form $y = mx + b$, is the y-coordinate for the y-intercept $(0, b)$.
4. Equations written in the form $y = mx + b$ are said to be written in **slope-intercept** form.
5. **Parallel lines** have the same slope.
6. The product of the coefficients of x is -1 in equations whose graphs are **perpendicular lines.**

Slope and the Equation of a Line. A useful way to think about the slope of a line is as a rate. Recall from Chapter 7 that a rate is a ratio of two quantities where the second is a unit measure, such as 35 mi/(1) hr, or as a unit price, such as \$1.29/(1) lb.

We can use a simple situation to illustrate thinking about slope as a rate: $d = 55t$ represents the distance traveled, d, after traveling t hours at 55 mi/hr. The graph of this function is a straight line with a slope of 55.

The table in Figure 13.15(a) shows that, as the time increases by 1 hr, the distance increases by 55 mi, or at the rate of 55 mi/hr. The graph in Figure 13.15(b) is labeled to show that, as we move 1 unit from left to right, the vertical change is 55, which gives the rate and the slope of the line.

Now, look at the graph of $y = 75x + 1200$ shown in Figure 13.16. Note that, even though it doesn't pass through the origin, we can think of it as a "shifted" rate. For every 1 unit of horizontal change, as we move from left to right, the vertical change is 75, which is the rate or slope. In the real-world situation for this function, 75 was the number of dollars saved per week, starting with \$1200 in a savings account.

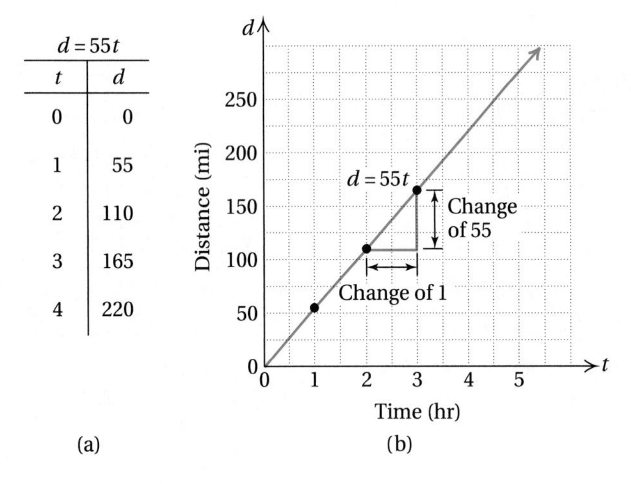

$d = 55t$

t	d
0	0
1	55
2	110
3	165
4	220

(a)

FIGURE 13.15 | (a) Table of ordered pairs for and (b) graph of $d = 55t$.

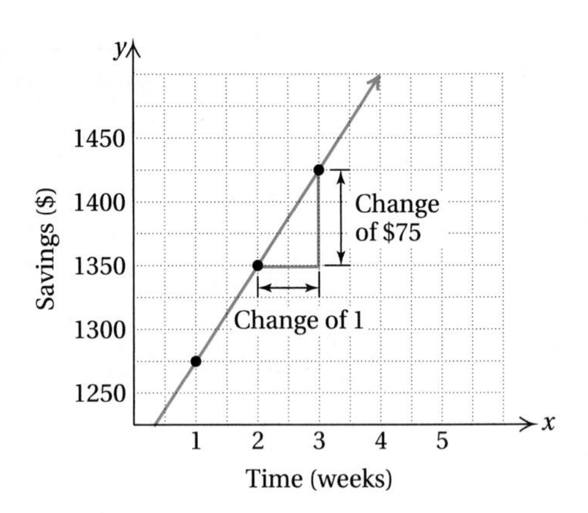

FIGURE 13.16 | First quadrant graph of the equation $y = 75x + 1200$.

Estimation Application

The Savings Race

Use the equation $y = 75x + 1200$ to estimate how much money you would have saved at the end of 1 yr. Here, x represents the number of weeks and y the total amount saved.

Now you can use the idea of a unit amount of change to develop a way to calculate the rate or slope. Again, a simple situation is helpful to show how slope can be thought about as rate.

Suppose that the total amount of a telephone charge for a certain type of call is a linear function of the number of minutes the call lasts. Also, suppose that you have a telephone bill showing a charge of \$1.00 for a 2-min call and a charge of \$3.50 for a 7-min call. Figure 13.17(a) shows these two points and the line passing through them. The amount of horizontal change for these two points isn't 1 unit, so finding the difference in the vertical change won't give the rate. Figure 13.17(b) shows that the amount of vertical change for 1 unit of horizontal change (1 min) is 0.50, which is the rate or slope of this line. Hence the cost is \$0.50/min.

Figure 13.17(c) shows that the small right triangle with a horizontal change of 1 min is similar to the larger right triangle drawn using the two original points. Recall from your work with geometry that the ratio of the legs of similar right triangles is constant. Therefore the ratio of the legs for the large triangle should also be 0.50.

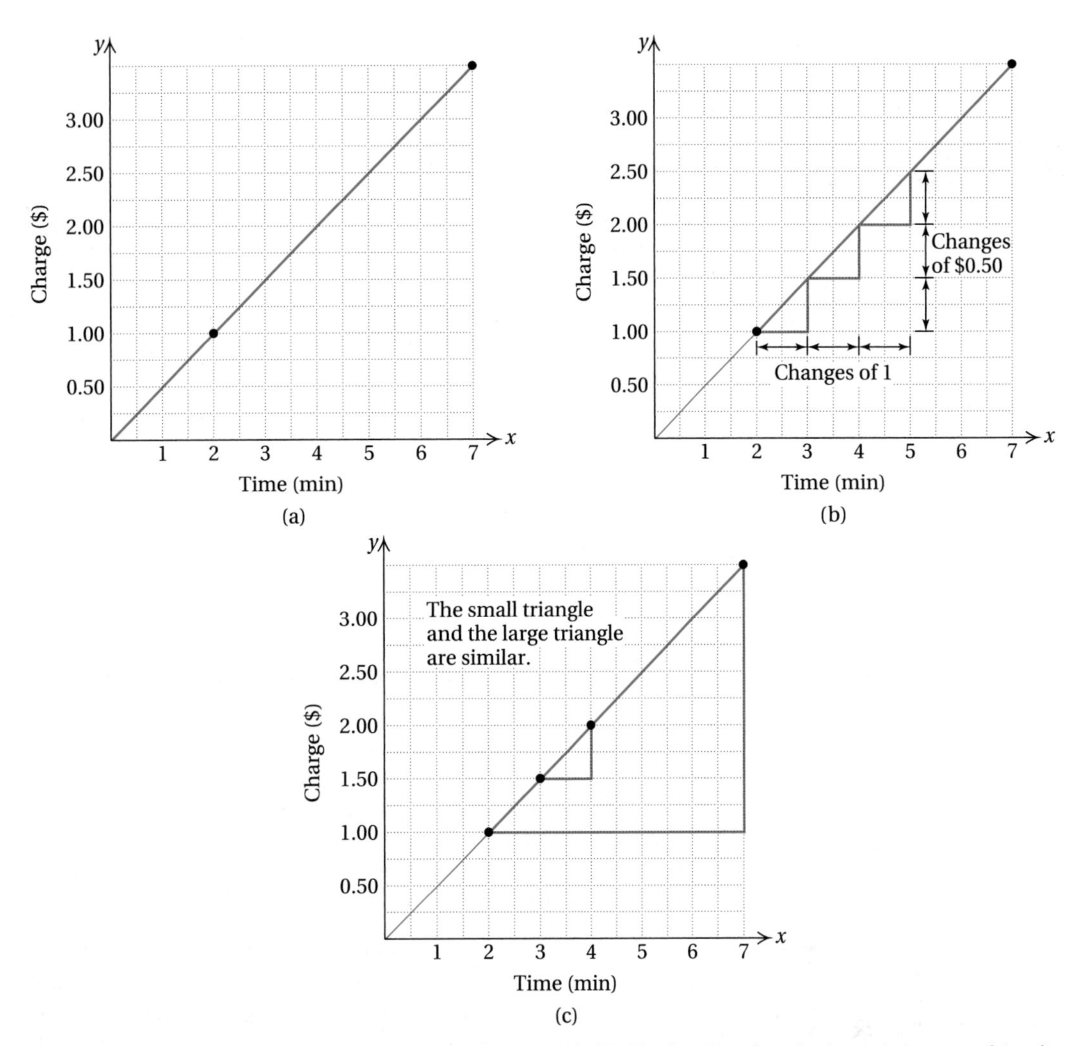

FIGURE 13.17 | (a) Using only two points to find a line, (b) finding its slope from horizontal changes of 1 unit, and (c) finding its slope from similar triangles.

In equation form,

$$\frac{\text{Vertical Change}}{\text{Horizontal Change}} = \frac{3.50 - 1.00}{7 - 2}$$
$$= \frac{2.5}{5}$$
$$= 0.50$$

We can summarize with variables this method of finding the slope or rate of change when any two points on a line are known.

Procedure for Finding Slope by Using Two Points

If (x_1, y_1) and (x_2, y_2) are any two points on a line, then the slope of the line, m, is

$$m = \frac{y_2 - y_1}{x_2 - x_1}$$

Example 13.12 shows how real-world paired data associated with a linear relationship can be used to find the slope of a linear equation. It also demonstrates why the slope is meaningful.

Example 13.12

Problem Solving: Community Planning

A new community of homes was opened in 1990. At the end of the third year of building homes, there were 360 homes in the community. At the end of 10 years, there were 1200 homes in the community. Assume that the relationship is linear and give an equation showing the relationship between the number of years that the community had been open for home construction, x, and the number of homes that had been built, y. Explain the meaning of the slope in this situation.

Solution

The two data points for this situation are (3, 360) and (10, 1200). The slope is

$$m = \frac{1200 - 360}{10 - 3} = \frac{840}{7}, \quad \text{or} \quad 120$$

Let x = the number of years the community had been open for home construction and y = the number of homes in the community. Then $y = 120x$. The slope, 120, means that 120 new homes were being built each year.

Your Turn

Practice: Suppose that a new coffee shop made \$2000 profit the first day it opened. After 6 mo it was making \$2900/day profit. After 15 mo the profit was \$4250/day. Assume that the relationship is linear and write the equation that shows the relationship between the number of months the coffee shop has been open and the daily profit. Explain the meaning of the slope in this situation.

Reflect: In the example problem, one of the data points was (10, 1200). The ratio of the y-coordinate to the x-coordinate for this one ordered pair gave the slope of the line. Does this process work in all situations? Explain. (*Hint:* Compare the results of the example problem to the practice problem.) ■

Understanding How Slope Is Related to Horizontal and Vertical Lines

We have discussed the meaning of slope and how to find the slope of a line with different techniques. Mini-Investigation 13.12 will help you further understand the connection between the equation of a line and its graph.

Mini-Investigation 13.12 **Looking for Patterns**

What patterns can you find in the slope of a line as it moves from (a) horizontal to vertical and (b) vertical to horizontal?

Write a description of what you discovered.

Technology Extension: Use a graphing calculator to explore the patterns you found.

You can use the formula for finding the slope of a line when two points are given to describe the slopes of horizontal and vertical lines. The points (2, 1) and (6, 1) lie on the same horizontal line. Using the formula to find the slope of the line containing the points, we write

$$m = \frac{1 - 1}{6 - 2} = \frac{0}{4}, \quad \text{or} \quad 0$$

This result shows that the slope of a horizontal line is 0. That makes sense because the vertical distance doesn't change for any change in horizontal distance.

The points (2, 4) and (2, 5) are on the same vertical line. Using the formula for slope we write

$$m = \frac{5 - 4}{2 - 2} = \frac{1}{0}, \quad \text{or} \quad \text{undefined}$$

We know that we can't divide by 0. Therefore we say that a vertical line has no slope, or that the slope is not defined.

Problems and Exercises for Section 13.3

Throughout the exercises, use a graphing calculator when appropriate.

A. Reinforcing Concepts and Practicing Skills

Graph each linear equation in Exercises 1–5. Make a data table with at least three ordered pairs.

1. $y = -5x + 7$

2. $y = -1$

3. $y = \frac{3x}{8}$

4. $y = 5$

5. $y = 2(x + 6) - 8$

The graph of $y = -4x + 5$ *is shown. Use it to find the answers to Exercises 6–8. Check your work.*

6. The value of y when $x = 1$

7. The value of y when $x = 0$

8. The value of x when $y = -3$

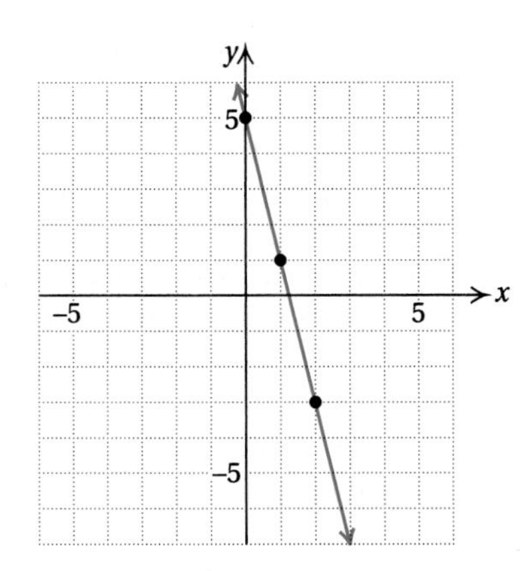

Without graphing, identify the slope and the y-intercept for the line associated with each equation or table of values in Exercises 9–15.

9. $y = 3 - x.$

10. $y = \frac{x}{2}$

11. $2x - 1 = -2y$

12. $5x + 20y = 60$

13. $16x - 120 = y$

14.

X	Y1
0	6
1	10
2	14
3	18
4	22
5	26
6	30

X=0

15.

	A	B	C	D	E	F
1	x	5	10	15	20	25
2	y	–80	–40	0	40	80

Which pair of equations in Exercises 16 and 17 represents lines that are parallel? Tell how you decided.

16. $y + x = -5$ and $y - x = -5$.

17. $y = -10x + 3$ and $y = -10x + \frac{1}{3}$

Give the equation of a line that is perpendicular to the graph of each equation in Exercises 18–21.

18. $y = 3x + 8$

19. $2x + y = -9$

20. $y = 12$

21. $y = -5(3x + 1)$

Find the slope of the line passing through each pair of points in Exercises 22–25.

22. $(3, 7)$ and $(6, -11)$

23. $(-2, -1)$ and $(4, 8)$

24. $(-12, 14)$ and $(12, 14)$

25. $(8, 5)$ and $(5, 8)$

B. Deepening Understanding

26. Use two points on any horizontal line to illustrate why a horizontal line has a slope of 0.

27. Use two points on any vertical line to illustrate why a vertical line has no slope.

28. The equation $y = 12x + 5$ shows the total cost for ordering tickets on the phone for a certain outdoor concert. Tickets are \$12 each, and there is a one-time service fee of \$5. Can the slope of this line be thought of as a rate? Explain.
29. Draw a line that passes through the second and third quadrants only. Give the equation of the line.
30. Find the equation of a line that passes through the point (1, 2) and has a slope of $\frac{1}{3}$. Show how you found the equation.

Write an equation that satisfies the condition stated in Exercises 31–35.

31. Slope $= -3$ and a y-intercept of 3
32. Parallel to the line $y = -4x + 1$
33. The same y-intercept as the line $y = 8 - x$
34. Intersects the line $y = 5x + 6$
35. Intersects the line $y = 12x - 2$ at infinitely many points
36. What happens to the value obtained for the slope if the coordinates of the two ordered pairs aren't subtracted in the same order? Give an example as part of your explanation.

C. Reasoning and Problem Solving

37. Make up a relationship between two real-world quantities that might be represented by the following graph. Describe in writing the relationship illustrated.

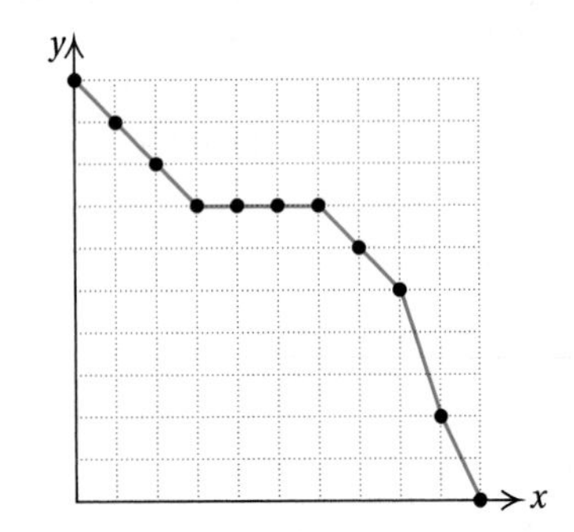

38. **The Juice Problem.** A store sells a 1-liter container of juice for 75¢ and a 2-liter container for \$1.40. Assume that the cost varies linearly with the number of liters, and find the slope of the line that shows the relationship. How many liters would be in a container that cost \$3.35?

D. Communicating and Connecting Ideas

Making Connections. *The following graph for Exercises 39 and 40 shows the relationship between the cost of a gallon of gasoline and the average number of gallons of gasoline used per year for several countries.*

39. What is the slope of this line?
40. Explain the meaning of the slope in this situation, using the quantities named and the idea of rate.

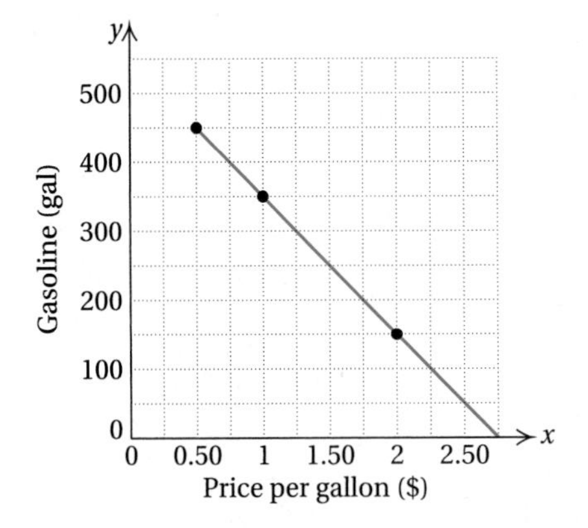

41. Work in a group to plan a display that summarizes the key ideas that relate to finding the slopes and equations of lines.
42. **Historical Pathways.** The following statement is in *Ray's New Elementary Algebra* book from 1866. Write an equation, using variables, that demonstrates this generalization. Show why it is true and then illustrate it with a numerical example.

    The square of any 2-digit number is composed of the square of the tens plus a quantity consisting of twice the tens plus the units, multiplied by the units.

Section 13.4 | CONNECTING ALGEBRA AND GEOMETRY

- Finding the Midpoint of a Line Segment
- Finding the Distance Between Two Points
- Finding the Equation of a Line
- Using Coordinate Geometry to Verify Geometric Conjectures
- Developing the Equation of a Circle
- Describing Transformations Using Coordinate Geometry

In this section we use coordinates and algebra to explore ideas about geometry. First, we extend techniques for analyzing and describing line segments and lines. Then we develop techniques for finding the equation of a line. We use these techniques to analyze intersecting lines and to revisit some geometric relationships explored earlier in this book but now from the perspective of coordinate geometry. We conclude this section by developing the equation of a circle.

Mini-Investigation 13.13 gives you a chance to review how points on the coordinate grid can be named and how geometric figures and positions can be described with coordinates.

Write your observations clearly and share them with a classmate for a reaction.

Mini-Investigation 13.13 Communicating

What observations can you make about the coordinates of the vertices of the quadrilateral formed by intersecting lines and the coordinates of the vertices of its image made by flipping the quadrilateral across the y-axis?

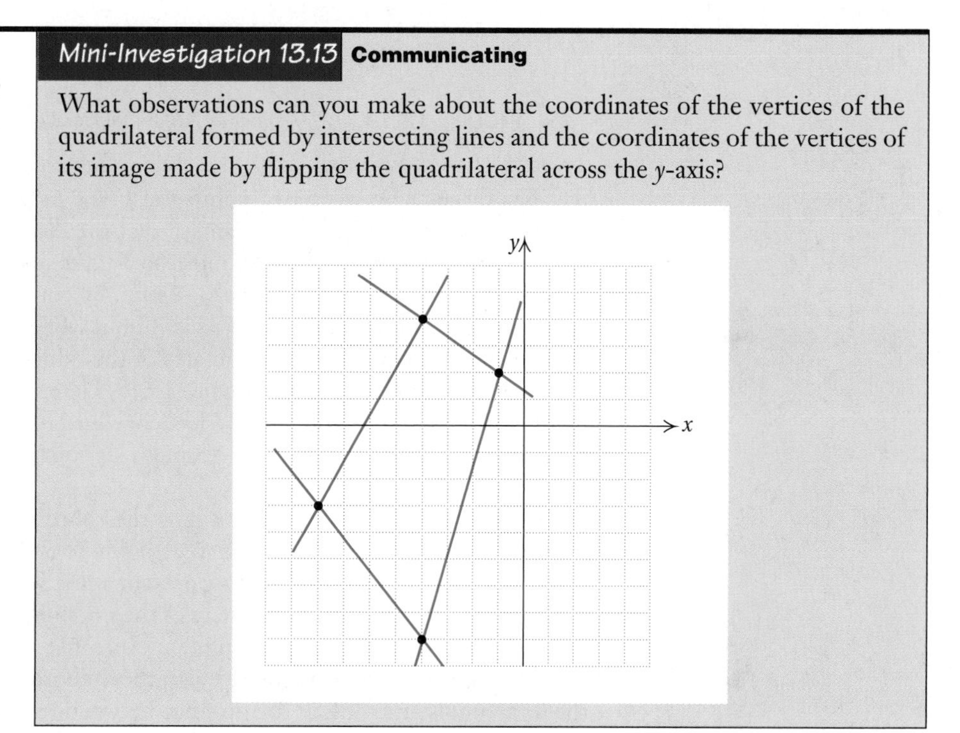

Finding the Midpoint of a Line Segment

Mini-Investigation 13.13 illustrates that geometric figures can be placed on coordinate grids and that ordered pairs can be used to name points on the figures. Remember that, to locate the point on a grid associated with an ordered pair of numbers, say, $(-4, 6)$, the first number tells you how far to move left or right and the second number tells you how far to move up or down on that grid. In this case the first number, -4, tells you to move 4 units to the left from the origin, $(0, 0)$. The second number, 6, then tells you to move 6 units up.

Several discoveries about geometry made in an earlier chapter involve the midpoint of a line segment. We can also use coordinates to develop a procedure for finding the midpoint of a line segment. For example, look at the two line segments shown on the grids in Figure 13.18. Finding the midpoint of $\overline{AB}$ and the midpoint of $\overline{CD}$ is straightforward. To find the total length of each segment, count units on the grid. Then find half the distance between the endpoints by counting to determine the midpoint.

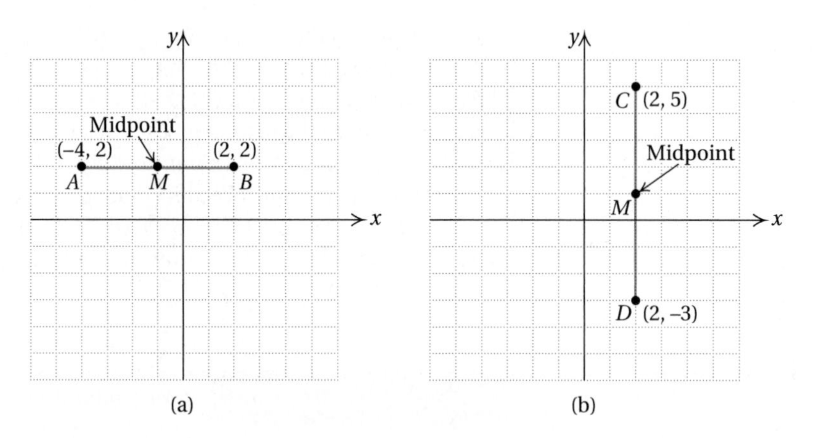

FIGURE 13.18 | (a) Midpoint of $\overline{AB}$; (b) midpoint of $\overline{CD}$.

The distance between two points must be 0 or positive; you can't have a negative distance. By counting, you can see that the distance between the endpoints of $\overline{AB}$ is 6 units. The midpoint, M, must be 3 units from the endpoints, so the coordinates of the midpoint are $(-1, 2)$. Again, by counting, you can see that the distance between the endpoints of $\overline{CD}$ is 8 units. The midpoint, M, must be 4 units from each endpoint, so the coordinates of the midpoint are $(2, 1)$.

Now look at $\overline{EF}$, shown in Figure 13.19. Determining that point M is the midpoint of this segment and that M has coordinates $(2, 1)$ also is straightforward. Note that $\overline{EF}$ forms the diagonal through six squares, so the midpoint is half that distance.

Drawing a line segment and finding the coordinates of the midpoint by counting squares on the grid is appropriate in simple cases, but in other applications it is inaccurate and impractical. We now present a method that works for all cases. Note in Figure 13.19 that the x-coordinate of the midpoint is half the horizontal distance between the endpoints of the segment. In other words, the horizontal distance extends from the x-coordinate of -1 to the x-coordinate of 5. The x-coordinate of the midpoint is at 2, or half the distance between -1 and 5. Another way to get this result is to treat the x-coordinate of the midpoint as the average of the x-coordinates of the endpoints of the line segment; 2 is the average of -1 and 5. Similarly, the

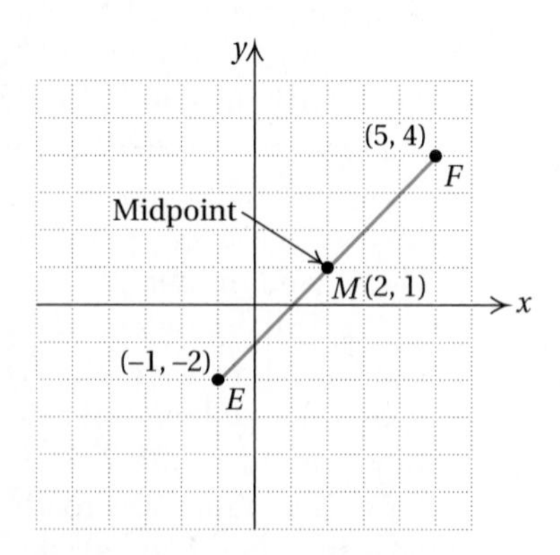

FIGURE 13.19 | Midpoint of $\overline{EF}$.

y-coordinate for the midpoint is the average of the y-coordinates of the endpoints of the line segment. The following theorem expresses the idea that the coordinates of the midpoints are the averages associated with the x- and y-coordinates for the endpoints of the segment.

Theorem for Coordinates of the Midpoint

If the coordinates of the endpoints of $\overline{PQ}$ are (x_1, y_1) for P and (x_2, y_2) for Q, then **the coordinates of the midpoint,** M, of $\overline{PQ}$ are

$$(x_M, y_M) = \left(\frac{x_1 + x_2}{2}, \frac{y_1 + y_2}{2}\right).$$

Example 13.13 shows how to use this theorem to find the coordinates of the midpoint of a segment.

Example 13.13

Finding the Coordinates of the Midpoint of a Segment

Find the coordinates of the midpoint for $\overline{AB}$ when the coordinates for A are $(6, -3)$ and the coordinates for B are $(-4, 8)$.

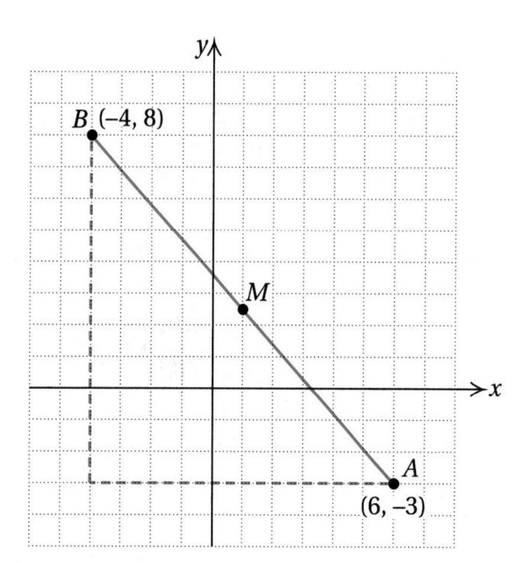

Solution

Let

$$(6, -3) = (x_1, y_1) \quad \text{and} \quad (-4, 8) = (x_2, y_2)$$

Then

$$\begin{aligned}(x_M, y_M) &= \left(\frac{x_1 + x_2}{2}, \frac{y_1 + y_2}{2}\right)\\ &= \left(\frac{6 + (-4)}{2}, \frac{-3 + 8}{2}\right)\\ &= \left(\frac{2}{2}, \frac{5}{2}\right)\\ &= (1, 2.5)\end{aligned}$$

Thus the coordinates of the midpoint of $\overline{AB}$ are $(1, 2.5)$.

Your Turn

Practice: Find the coordinates for the midpoint for each pair of points.

a. $(9, 12)$ and $(-5, 8)$
b. $(4.5, 7)$ and $(-8, -15)$

Reflect: How is finding the coordinates of the midpoint like finding the average? ■

Example 13.14 shows how the theorem for coordinates of a midpoint also can be used to find the coordinates of an endpoint of a line segment if the coordinates of the midpoint and the other endpoint are known.

Example 13.14 Finding the Coordinates of an Endpoint of a Segment

Find the coordinates of D for $\overline{CD}$, where the coordinates of C are $(-2, -4)$ and the coordinates of the midpoint are $(-3.5, 2)$.

Solution

Let C have coordinates $(x_1, y_1) = (-2, -4)$. The coordinates for D aren't known. Let (x_2, y_2) be the unknown coordinates for D. Then

$-3.5 = \frac{-2 + x_2}{2}$ The average of the x-coordinates of the endpoints is the x-coordinate of the midpoint.

$2 = \frac{-4 + y_2}{2}$ The average of the y-coordinates of the endpoints is the y-coordinate of the midpoint.

Solving each equation gives the coordinates for D:

$$-3.5 = \frac{-2 + x_2}{2} \quad \text{and} \quad 2 = \frac{-4 + y_2}{2}$$
$$-7 = -2 + x_2 \qquad 4 = -4 + y_2$$
$$x_2 = -5 \qquad y_2 = 8$$

Thus the coordinates of D, (x_2, y_2), are $(-5, 8)$.

Your Turn

Practice: Find the coordinates of the endpoint of a line segment whose one endpoint has coordinates $(-1, 0)$ and whose midpoint has coordinates $(1.5, 4.5)$.

Reflect: Does it matter which endpoint represents (x_1, y_1) and which endpoint represents (x_2, y_2) in the formula used in the practice problem? Explain. ■

Finding the Distance Between Two Points

If you are driving from one place to another in a city, you probably can't drive in a straight line from one point to the other. In other situations, the distance from one point to another can be the straight-line distance. Mini-Investigation 13.14 will help you think about ways of finding the distance between two points in the coordinate plane.

Talk about the procedures that can be used to find this distance.

Technology Extension: Explore your GES coordinate geometry options for measuring the length of a segment or the distance between two points, and use this technique to verify your answer to the question.

Mini-Investigation 13.14 Communicating

What is the distance between the fire tower and the fire? (Assume that the horizontal and vertical distances between each pair of grid lines is 1 km.)

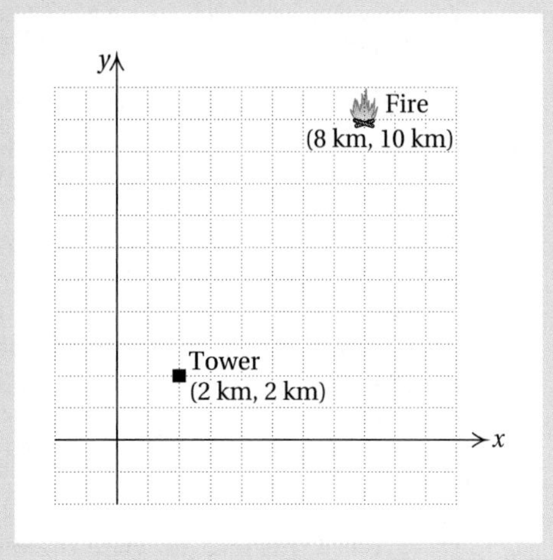

As you may have discovered in Mini-Investigation 13.14, the Pythagorean theorem can be used to find the distance between any two points in the plane. The distance between the tower and the fire is the hypotenuse, c, of a right triangle with one leg 6 km long and the other leg 8 km long, so $c^2 = 6^2 + 8^2$, or $c = \sqrt{100} = 10$ km. To find a general formula for the distance between points (x_1, y_1) and (x_2, y_2) in Figure 13.20(a), you can use the **Pythagorean theorem.** Figure 13.20(b) shows that a right triangle can be constructed and the coordinates of the vertex of the right angle named, (x_2, y_1).

If m and n are the lengths of the sides of the right triangle, we can use the Pythagorean theorem to write, $m^2 + n^2 = d^2$. Here, m is the distance between (x_2, y_1) and (x_1, y_1). Again, because distance must always be 0 or greater, we can take the absolute value of the difference of x_2 and x_1:

$$m = |x_2 - x_1|$$

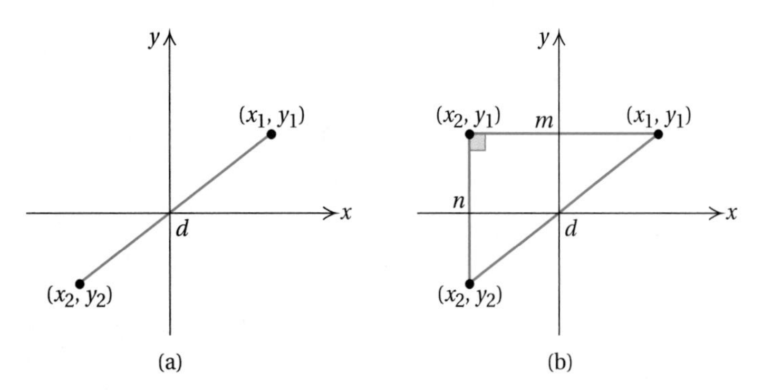

FIGURE 13.20 | Developing the formula for distance.

Similarly, we can express n in terms of the y-coordinates in (x_2, y_1) and (x_2, y_2):

$$n = |y_2 - y_1|$$

Substituting into the equation $m^2 + n^2 = d^2$ gives

$$(|x_2 - x_1|)^2 + (|y_2 - y_1|)^2 = d^2, \quad \text{or} \quad d^2 = (|x_2 - x_1|)^2 + (|y_2 - y_1|)^2$$

Taking the square root of each side yields

$$d = \sqrt{(|x_2 - x_1|)^2 + (|y_2 - y_1|)^2}$$

We can simplify this expression further. Consider the following absolute value expressions. The one on the left is the square of the absolute value of the difference. The one on the right is the square of the difference without first taking the absolute value. Thus, because the difference between the two values is squared, we don't have to take the absolute value of the difference.

$$(|2 - 5|)^2 = |-3|^2 = 9 \quad \text{and} \quad (2 - 5)^2 = (-3)^2 = 9$$

We use this idea to write the distance formula in a simpler form.

Distance Formula

If d is the distance between points (x_1, y_1) and (x_2, y_2), then

$$d = \sqrt{(x_2 - x_1)^2 + (y_2 - y_1)^2}.$$

Example 13.15 shows how to use the distance formula to solve a real-world problem.

Example 13.15 Problem Solving: Industrial Park Planning

A city development team might use a grid like the one shown to lay out a new industrial park. Each grid unit represents 50 m. Use the distance formula to find the distance across the park from point A to point B.

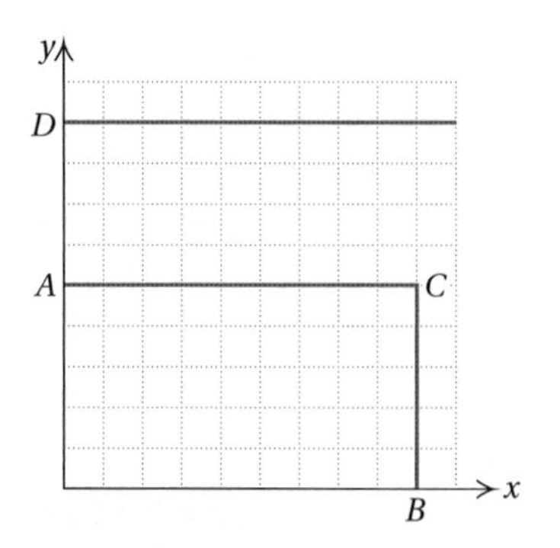

Solution

The coordinates of points A and B are $(0, 5)$ and $(9, 0)$. That is, $x_1 = 0, y_1 = 5$, $x_2 = 9$, and $y_2 = 0$. Substituting into the distance formula, we get

$$\begin{aligned} d &= \sqrt{(9 - 0)^2 + (0 - 5)^2} \\ &= \sqrt{9^2 + 5^2} \\ &= \sqrt{106}, \quad \text{or} \quad 10.30 \text{ (rounded to the nearest hundredth)} \end{aligned}$$

As each unit represents 50 m, the actual distance is approximately 10.30×50, or about 515 m.

Your Turn

Practice: Find the distance between the corner of the proposed industrial park at point C and the electric substation at point D. Each grid unit still represents 50 m.

Reflect: Does it matter which ordered pairs are used for (x_1, y_1) and (x_2, y_2) in the distance formula? Explain. ■

Finding the Equation of a Line

Just as geometric figures such as quadrilaterals and triangles can be placed on a coordinate grid and points on them identified by ordered pairs, so too can such polygonal figures be described by lines. The triangle shown in Figure 13.21(a) has sides that are line segments. But it also can be thought of as a figure formed by parts of three intersecting lines, as shown in Figure 13.21(b).

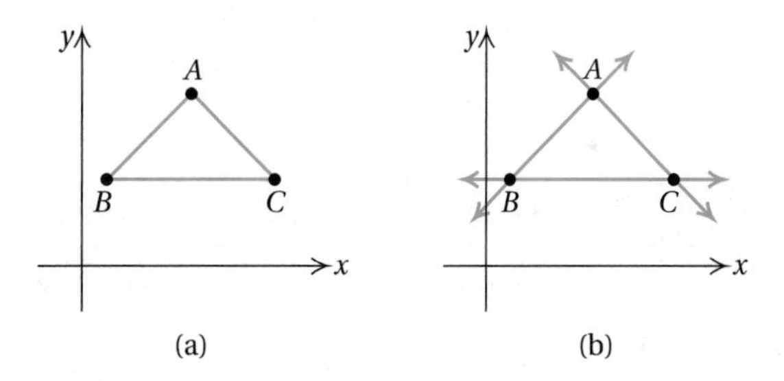

FIGURE 13.21 | Triangle formed by line segments or by intersecting lines.

To construct a triangle on a coordinate gird, we can give the three equations for lines through the sides of the triangle. There are two methods for finding the equation of a line: the **slope-intercept method** and the **two-point method.**

Slope-Intercept Method. Recall that the **slope** of a line can be thought of in terms of the steepness of the line. In the real world, you encounter the idea of slope in many situations.

- An expert skier likes hills with a steep slope.
- The slope of a roof indicates how steep the roof is.
- Wheelchair ramps must have a certain maximum slope.
- Truck drivers must be careful descending a road that has a steep slope.

Recall that the slope of a line can also be thought of in terms of rate of change involving the ratio of the change in vertical distance to the change in horizontal distance. We might say, for example, that a wheelchair ramp rises 1 ft for each 5 ft of horizontal distance. Sometimes we simply say that the rise-to-run ratio is 1 to 5.

Recall also that any line can be represented by an equation in the form $y = mx + b$ where m is the slope of the line and b is the y-intercept, or the y-coordinate of the point where the line crosses the y-axis. The slope of the line can be found using any two points on the line. Therefore we can calculate slope for the line passing through points (x_1, y_1) and (x_2, y_2) as

$$\text{Slope } (m) = \frac{\text{Change in } y\text{-coordinate}}{\text{Change in } x\text{-coordinate}} = \frac{(y_2 - y_1)}{(x_2 - x_1)}$$

For example, if points $(4, 1)$ and $(-2, 3)$ are points on a line, its slope is

$$\frac{3 - 1}{-2 - 4} = \frac{2}{-6}, \quad \text{or} \quad -\frac{1}{3}$$

Note that the change in the y- and x-coordinates must be calculated in the same order, or the sign for the slope will be incorrect. Note also that, for a line with slope $-\frac{1}{3}$, you can start at any point on the line and move 1 unit down and 3 units to the right and be at another point on the line. Or you can start at any point and move 1 unit up and 3 units to the left and be at another point on the line.

Example 13.16 shows how to use the slope-intercept method when the scales on the two axes are not the same.

Example 13.16 Finding the Equation of a Line by Using the Slope-Intercept Method

Find the equation of the line shown by using the slope-intercept method.

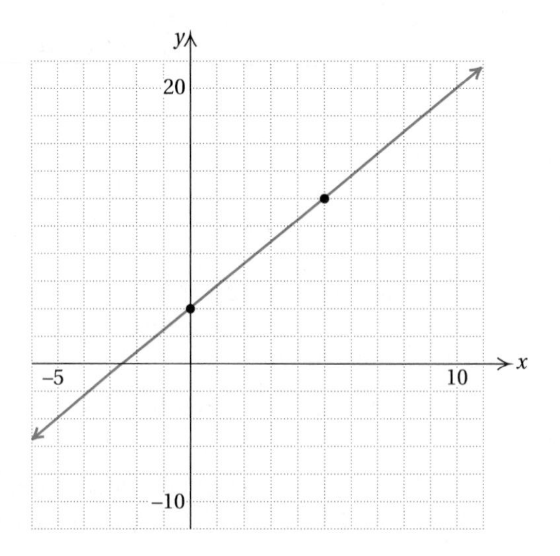

Solution

The coordinates of the two points are $(0, 4)$ and $(5, 12)$. Using these points, we calculate the slope as follows:

$$m = \frac{12 - 4}{5 - 0} \quad \text{or} \quad \frac{8}{5}$$

The y-intercept is 4, so the equation of the line is

$$y = \frac{8}{5}x + 4$$

Your Turn

Practice:

a. Use the slope-intercept method to find the equation of the line shown.

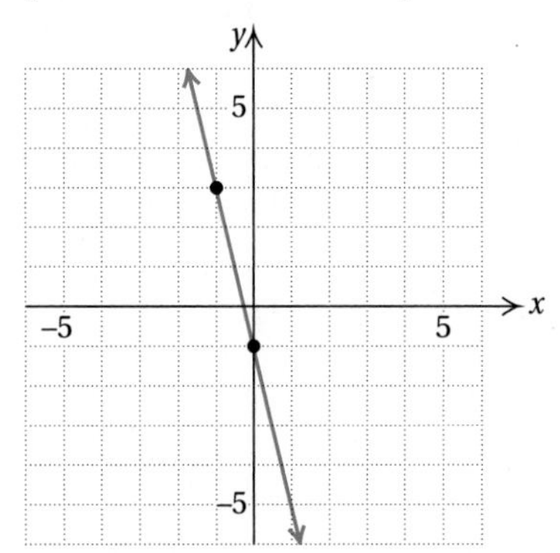

b. Suppose that the scale on the x-axis was 2 units rather than 1. Now find the equation of the line.

Reflect: Can you think of a situation in which the slope-intercept method might easily be used to find the equation of a line? Explain. ■

Estimation Application

Roof Construction

The design for a roof is placed on a grid as shown, where the scales of the axes are the same. Estimate whether the slope of segment AB is greater or less than 1. Tell how you know.

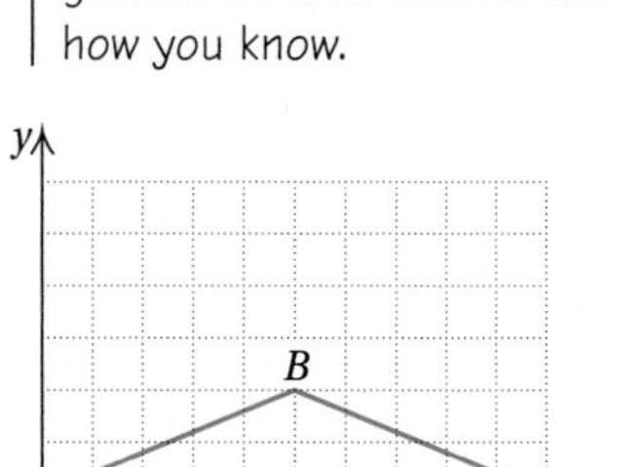

Two-Point Method. The second method for finding the equation of a line, the two-point method, is a slight variation of the first. We begin with two points on a line and first calculate the slope of the line by using the coordinates of those two points. That gives us the value of m in $y = mx + b$. To find the value of b, the y-intercept, we use either of the two points and substitute for the x- and y-coordinates. Example 13.17 shows how to use this method.

Example 13.17 Finding the Equation of a Line by Using the Two-Point Method

Find the equation of the line passing through the points $(-4, 7)$ and $(8, 3)$.

Solution

First find the slope of the line. Let $(x_2, y_2) = (8, 3)$ and $(x_1, y_1) = (-4, 7)$. Then

$$m = \frac{3 - 7}{8 - (-4)}$$
$$= \frac{-4}{12} \quad \text{or} \quad -\frac{1}{3}$$

So $y = \frac{1}{3}x + b$. We can use either $(8, 3)$ or $(-4, 7)$. Let's use $(x, y) = (8, 3)$:

$$y = -\frac{1}{3}x + b$$

$$3 = -\frac{1}{3}(8) + b$$

$$3 + \frac{8}{3} = b$$

$$b = 5\frac{2}{3}$$

So the equation of the line is

$$y = -\frac{1}{3}x + 5\frac{2}{3}$$

Your Turn

Practice:

a. Redo the example problem, using $(-4, 7)$ rather than $(8, 3)$ for (x, y).
b. Find the equation of the line passing through points $(5, 0)$ and $(-4, 9)$.

Reflect: Can you use the two-point method if you are given the graph of a line? Explain. ■

Example 13.18 demonstrates how to utilize the problem-solving strategies of *use a picture*, *find the equations of lines*, and *look for a pattern* to make a generalization about the slopes of perpendicular lines.

Example 13.18

Problem Solving: Water Lines

Two underground water lines are needed for a new housing development. The coordinate grid shown represents the layout of the new development. Proposed repair access tubes are shown on the grid as points A, B, C, and D. Lines ℓ and m represent the water lines, which meet to form a right angle (they are perpendicular lines). Suppose that you are developing a computer model for this situation and want to describe the locations of the new water lines with equations. Give the equation for each line. How are the slopes of the lines related?

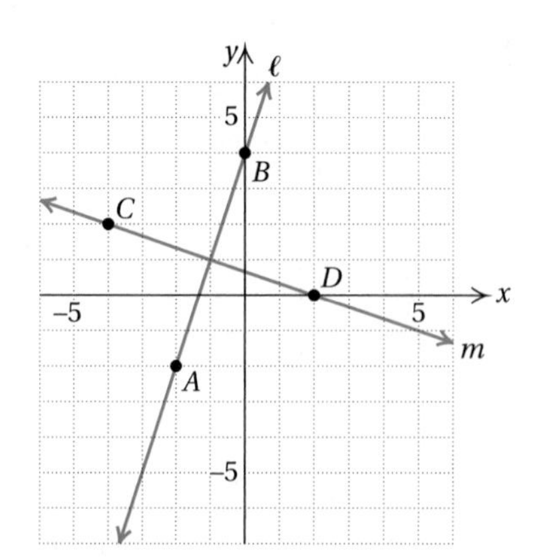

Working toward a Solution

Understand the problem	*What does the situation involve?*	Two water lines intersect at a 90° angle on a coordinate grid that represents a new development.
	What has to be determined?	The equation of each line and how the slopes of the lines are related.
	What are the key data and conditions?	The lines intersect at $(-1, 1)$ at a 90° angle. The coordinates of points A and B on line ℓ are $(-2, -2)$ and $(0, 4)$. The coordinates of points C and D on line m are $(-4, 2)$ and $(2, 0)$.
	What are some assumptions?	The water lines are essentially straight, as depicted on the graph.
Develop a plan	*What strategies might be useful?*	Draw a picture, write an equation, and look for a pattern.
	Are there any subproblems?	Find the equation of each line.
	Should the answer be estimated or calculated?	The exact equation of each line is needed in order for us to look for a pattern in the slopes.
	What method of calculation should be used?	The numbers involved are small. The use of paper and pencil makes sense.
Implement the plan	*How should the strategies be used?*	We can use the picture and either the slope-intercept or the two-point method to find the equation of each line. Writing the equations in the slope-intercept format will help identify patterns in the slopes.
	What is the answer?	Line ℓ: the y-intercept, 4, is easy to read, so we can use the slope-intercept method. The slope is $\frac{6}{2} = 3$. The equation for this line is $y = 3x + 4$. Line m: The y-intercept isn't easy to read, so we use the two-point method. The slope is $-\frac{1}{3}$, so $y = -\frac{1}{3}x + b$. We can use any point on the line, but $(2, 0)$ might make calculations easy. Substituting yields $0 = -\frac{1}{3}(2) + b$, or $b = \frac{2}{3}$. The equation for this line is $$y = -\frac{1}{3}x + \frac{2}{3}$$ The slopes are 3 and $-\frac{1}{3}$. Hence the slopes of the perpendicular lines have a product of -1.

Look back	*Is the interpretation correct?*	We can use the fact that the slopes of perpendicular lines have a product of -1 to write the equations of lines that we know are perpendicular.
	Is the calculation correct?	The correct numbers from the graph were used. One slope is negative and the other is positive, so a negative product makes sense. Also, the steeper one line is, the flatter the other must be. Therefore the relative sizes of the slopes make sense.
	Is the answer reasonable?	The signs of the slopes make sense, as do the y-intercepts. The equations seem reasonable.
	Is there another way to solve the problem?	The equations of the lines are unique, even though they can be written in different formats.

Your Turn

Practice: Use a grid and draw any two lines that you think are perpendicular. Find the equation of each line and the product of the slopes to test whether the lines in fact are perpendicular.

Reflect: Can the product of the slopes of two perpendicular lines ever be $+1$? Explain. ■

Example 13.18 shows that the product of the slopes of **perpendicular lines** is -1. In Mini-Investigation 13.15, you are asked to find the equations of **parallel lines** in order to verify the relationship of their slopes.

Talk with a classmate about how the equations of parallel lines written in slope-intercept form are alike and how they are different.

Mini-Investigation 13.15 **Looking for Patterns**

After writing equations for the parallel lines p and q, what pattern do you see in their slopes?

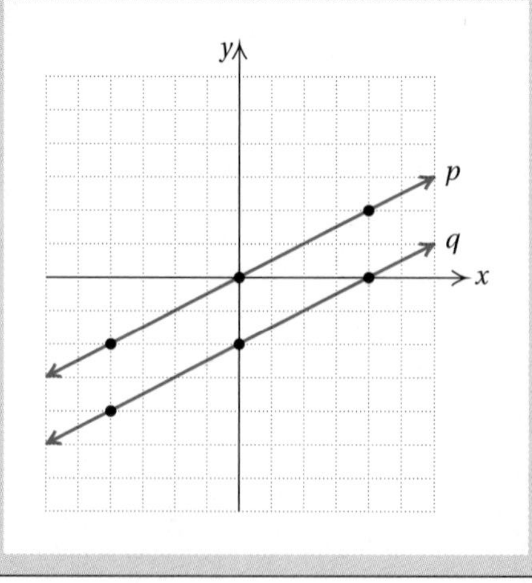

y
p
q
x

Finding the Point of Intersection of Two Lines. Example 13.19 shows how to use what we know about finding the equation of a line and solving equations to find the coordinates of the point at which two lines intersect. Such a point is a solution to the system of the linear equations of the lines. The coordinates of this point make both linear equations true.

Estimation Application

Shipwreck Search

Tracking methods located a sunken ship at the point shown on the grid. Estimate the coordinates of the point at which the shipwreck was located.

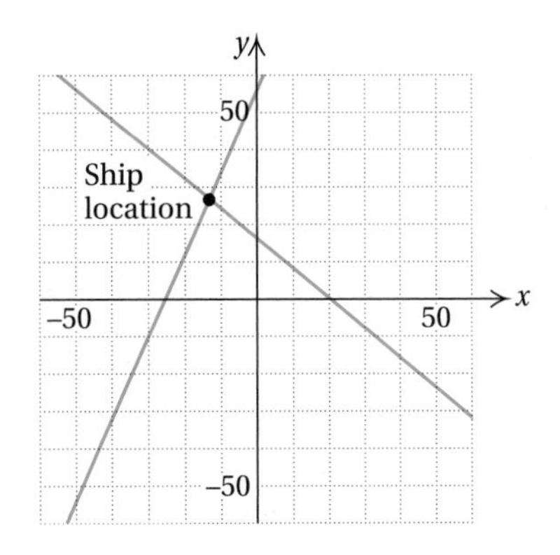

Example 13.19 Finding the Coordinates of the Point of Intersection of Two Lines

Find the coordinates of the point at which the lines shown intersect.

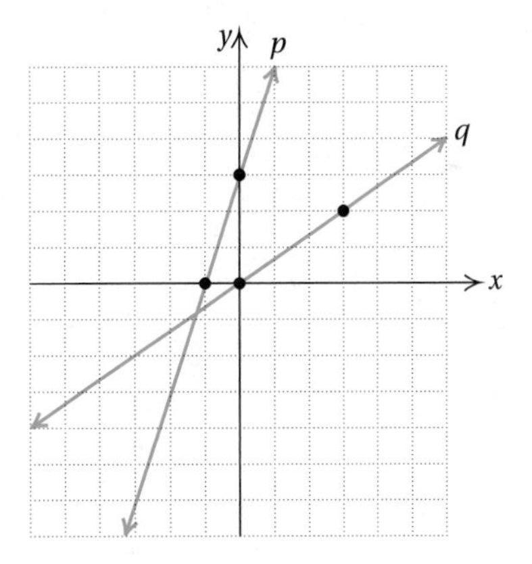

Solution

First find the equation for each line.

Line p	**Line q**
Use $(-1, 0)$ and $(0, 3)$.	Use $(0, 0)$ and $(3, 2)$.
$m = \frac{3 - 0}{0 - (-1)} = 3$	$m = \frac{2 - 0}{3 - 0} = \frac{2}{3}$
y-intercept $= 3$.	y-intercept $= 0$
Line p: $y = 3x + 3$	Line q: $y = \frac{2}{3}x$

Since $y = 3x + 3$ and $y = \frac{2}{3}x$, we can write and solve an equation to find the value of the x-coordinate for the point of intersection.

$$y = 3x + 3$$

$$\frac{2}{3}x = 3x + 3 \qquad \text{Substitute } \frac{2}{3}x \text{ for } y.$$

$$\frac{2}{3}x - 3x = 3$$

$$-\frac{7}{3}x = 3$$

$$\left(-\frac{3}{7}\right)\left(-\frac{7}{3}x\right) = -\frac{3}{7}(3)$$

$$x = -\frac{9}{7}$$

Using $y = 3x + 3$ and substituting $x = -\frac{9}{7}$ yields

$$y = 3x + 3$$

$$= 3\left(-\frac{9}{7}\right) + 3$$

$$= -\frac{27}{7} + \frac{21}{7} \qquad \text{Replace 3 with } \frac{21}{7}$$

$$= -\frac{6}{7}$$

The point of intersection appears to be $\left(-\frac{9}{7}, -\frac{6}{7}\right)$. (Note: Either of the original equations could have been used.) We can check by substituting into each of the linear equations.

Your Turn

Practice: Use paper and pencil or a graphing calculator to find the coordinates of the point of intersection for the two lines shown.

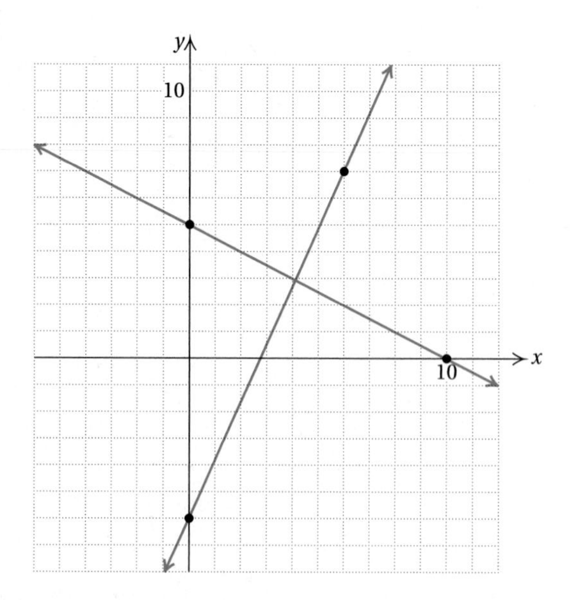

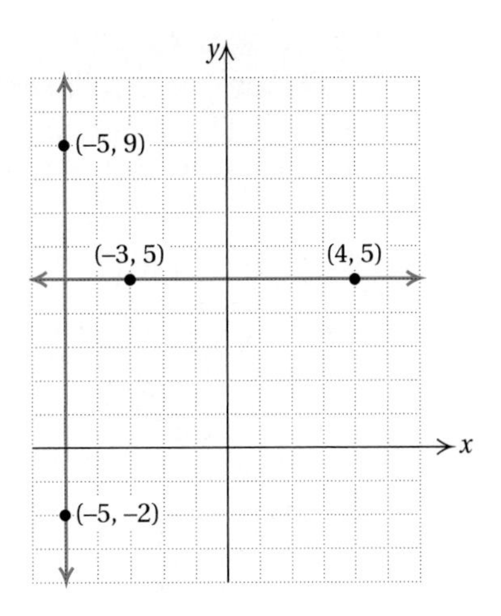

FIGURE 13.22 | Slopes for horizontal and vertical lines.

Reflect: Why does checking by substituting into each equation verify the point of intersection? ■

Horizontal and Vertical Lines. We can use the formula for finding slope to review the meaning of slope for horizontal and vertical lines. In Figure 13.22, two points are marked on both the horizontal and vertical lines. Let's look first at the slope for the **horizontal line.**

Using the formula for slope for the two points on the horizontal line, we get

$$m = \frac{5 - 5}{4 - (-3)} = \frac{0}{7} = 0$$

The y-coordinates will be the same for any two points on a horizontal line. In other words, the change in the y-coordinates—the numerator in the formula—will always be 0, so the quotient will always be 0. *The slope of any horizontal line is 0.*

Now let's look at the **vertical line.** Using the points shown, we get

$$m = \frac{9 - (-2)}{-5 - (-5)} = \frac{11}{0}$$

The x-coordinates will be the same for any two points on a vertical line. Therefore the denominator in the formula for slope will always be 0. Division by 0 isn't permitted, so the slope of any vertical line is said to be undefined. *A vertical line has no slope.*

Using Coordinate Geometry to Verify Geometric Conjectures

Mini-Investigation 13.16 provides an opportunity for you to revisit relationships involving polygons that you explored in Chapter 11, but now from the perspective of coordinate geometry.

Write a convincing argument that verifies your answer and compare it to an argument made by a classmate.

Mini-Investigation 13.16 Using Mathematical Reasoning

What is the relationship between the segment joining the midpoint of $\overline{AB}$ to the midpoint of $\overline{BC}$ and $\overline{AC}$?

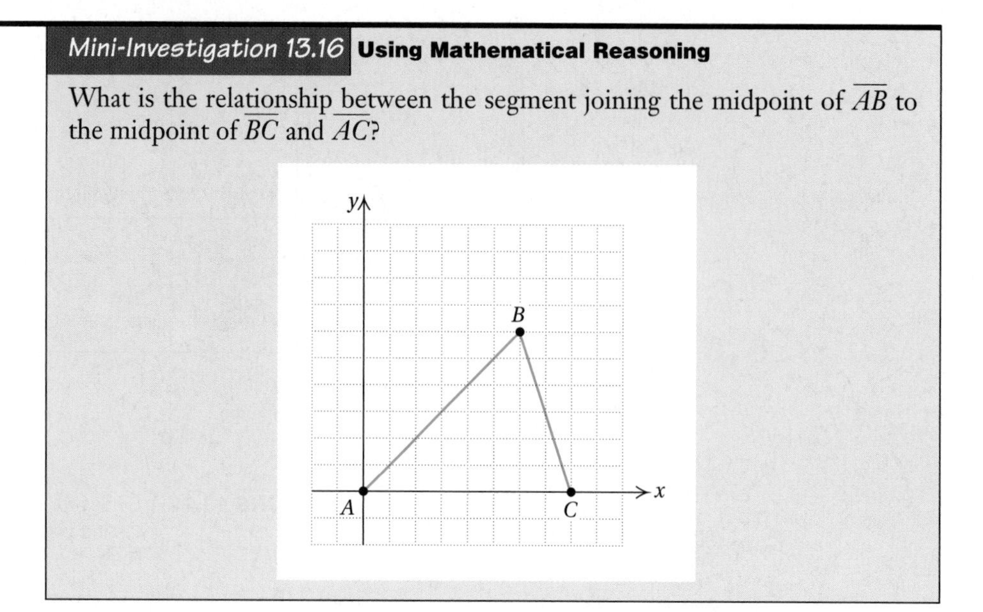

Geometric relationships such as the one in Mini-Investigation 13.16 can be proved with techniques involving coordinate geometry. Consider quadrilateral *ABCD* depicted in Figure 13.23(a). If the midpoints of its sides are found and connected in order as vertices of a new quadrilateral, what type of quadrilateral is formed? The midpoint formula yields the coordinates of the midpoints of the sides: $(2, 4)$, $(7, 7)$, $(9, 3)$, and $(4, 0)$. Connecting these points in order gives the quadrilateral shown in Figure 13.23(b).

The two-point method can be used to find the equations of the lines, parts of which form the sides of the new quadrilateral. The equations for these lines are as follows.

Through M and Q,	$y = -2x + 8$
Through N and P,	$y = -2x + 21$
Through M and N,	$y = \frac{3}{5}x + \frac{14}{5}$
Through Q and P,	$y = \frac{3}{5}x - \frac{12}{5}$

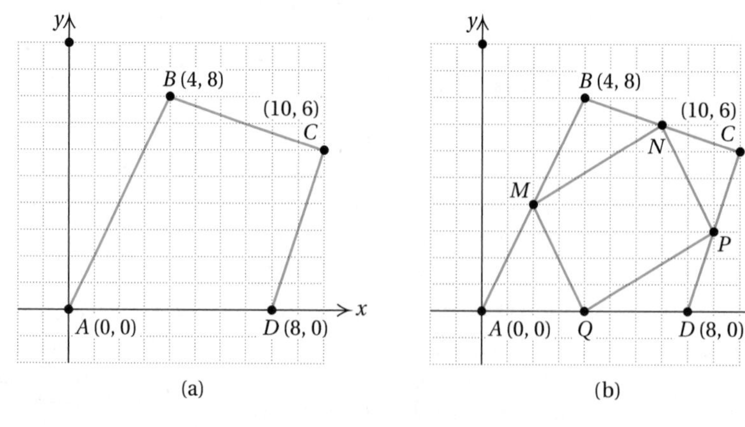

FIGURE 13.23 | Constructing a quadrilateral inside another quadrilateral.

Note that the slopes of the lines forming the opposite sides of the quadrilateral are the same. Thus the two sides are parallel and the quadrilateral *MNPQ* is a parallelogram.

You can use GES to explore in general the example just presented. With GES, create any quadrilateral on a coordinate grid. Locate the midpoints of the sides of the quadrilateral and connect them consecutively to form another quadrilateral. Then determine the slopes of the sides of the new quadrilateral. Finally, move one or more parts of the original quadrilateral (a vertex, a side) to produce a different quadrilateral. Observe what happens to the slope relationships in the quadrilateral you created from the midpoints, and how such changes provide inductive examples for this generalization.

Example 13.20 deals with another geometric relationship that can be verified by using coordinate geometry.

Example 13.20 Verifying Generalizations about the Diagonals of a Square

Show that the diagonals of a square intersect in a right angle and bisect each other. Use the figure shown to explain.

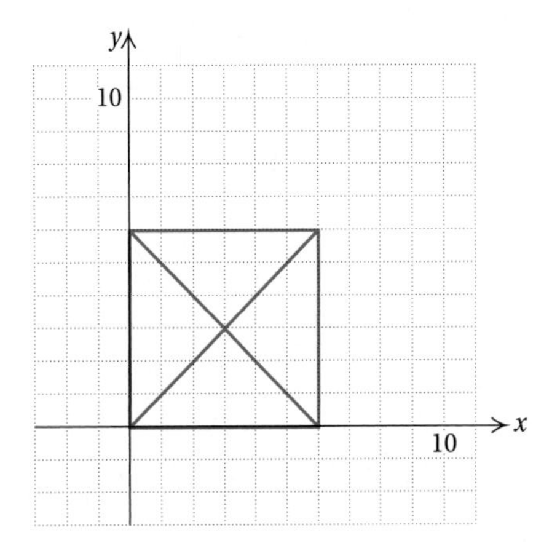

Solution

The equation of the line forming one diagonal of the square is $y = x$. The slope is 1, and that line passes through the origin. The equation of the line forming the other diagonal of the square is $y = -x + 6$. We can tell by inspection that the slope is -1 and that the y-intercept is 6. The product of the slopes of the two lines is -1, so the lines are perpendicular (they meet at a 90° angle).

Next, we solve the system of equations $y = x$ and $y = -x + 6$ to find the point at which the two lines intersect: $(3, 3)$. Now, we find the distance from the midpoint to each vertex. From $(3, 3)$ to $(0, 0)$,

$$\begin{aligned} d &= \sqrt{(3-0)^2 + (3-0)^2} \\ &= \sqrt{3^2 + 3^2} = \sqrt{18} \\ &\approx 4.24 \end{aligned}$$

Similarly, the distance from $(3, 3)$ to each of the other vertices is about 4.24. Because the distance from the point of intersection to each vertex is the same, the diagonals bisect each other.

Your Turn

Practice: Draw an isosceles trapezoid and find the midpoints of each side. Connect these midpoints in order. Determine the type of quadrilateral formed by connecting the midpoints. (*Hint:* Recall that an isosceles trapezoid has nonparallel sides of equal length.)

Reflect: Which method for finding the equation of a line, slope-intercept or two-point, was used most often in the example and practice problems? ■

Developing the Equation of a Circle

In this section so far, we have dealt with lines, line segments, and polygons. The coordinate grid can also be used to describe circles. Mini-Investigation 13.17 will help you connect distance on the coordinate grid and the meaning of a circle.

Make a graph that shows all the points you found and talk about the shape formed by connecting these points.

Technology Extension: Explore your GES coordinate geometry options for determining the equation of a circle that's been drawn on a coordinate grid.

Mini-Investigation 13.17 Looking for Patterns

How many points can you find on the grid that are exactly 3 units from the given point and what are their coordinates?

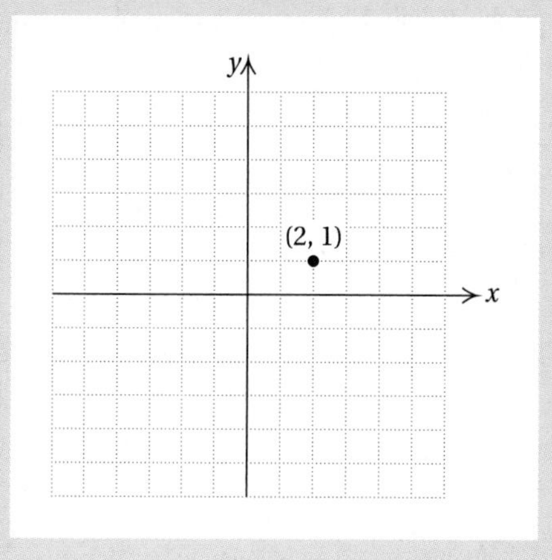

In Mini-Investigation 13.17 you may have observed that all points 3 units from point (2, 1) can be connected to form a circle. That is, a circle is defined as the set of all points in the plane the same distance from a given point. The point (2, 1) is the **center** of the circle and the **radius** of the circle is the distance each point on the circle is from the center.

We can use this information to create a procedure for finding the **equation of a circle.** But first, we need to think about certain relationships in a circle. Suppose that (x, y) is any point on a circle, as shown in Figure 13.24.

Any point on a circle is the same distance from the center, namely, the radius, r. Note that a right triangle is formed with sides of length x and y and the hypotenuse of length r. Using the distance formula gives $r = \sqrt{x^2 + y^2}$. We can rewrite this formula to get the general equation of a circle whose center is at the origin, (0, 0).

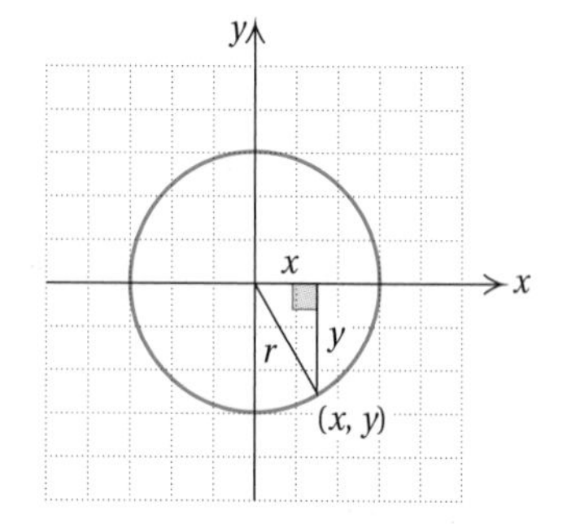

FIGURE 13.24 | Developing the equation of a circle with its center at the origin.

Circle Equation Theorem—Center at the Origin

The standard form for an equation of a circle with a center at the origin and radius r is: $x^2 + y^2 = r^2$.

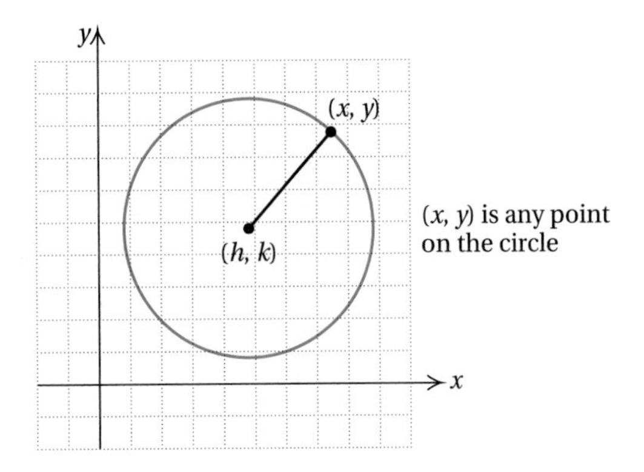

FIGURE 13.25 | Developing the equation of a circle with its center at (h, k).

What if the center of a circle weren't at the origin? How would the equation of a circle change? In Figure 13.25, the center of the circle is the point (h, k).

If r represents the distance from (h, k) to any point (x, y) on the circle, we can use the distance formula to write

$$r = \sqrt{(x - h)^2 + (y - k)^2} \qquad \text{or} \qquad (x - h)^2 + (y - k)^2 = r^2$$

Circle Equation Theorem—Center at (h, k)

The standard form for an equation of a circle with center (h, k) and radius r is $(x - h)^2 + (y - k)^2 = r^2$.

Example 13.21 shows how to use the circle equation theorem to find the equation of a circle.

Example 13.21 Finding the Equation of a Circle

Find the equation of a circle with its center at $(-4, -5)$ and a radius of 2.

Solution

Here, $(h, k) = (-4, -5)$ and $r = 2$. Hence

$$r = \sqrt{(x - h)^2 + (y - k)^2}$$
$$2 = \sqrt{(x + 4)^2 + (y + 5)^2}$$
$$4 = (x + 4)^2 + (y + 5)^2$$

Your Turn

Practice:

a. Find the equation of a circle with its center at $(1, 8)$ and a radius of 4.

b. Write the equation of a circle whose center is at $(0, -4)$ and whose radius is $\sqrt{5}$.

Reflect: When substituted into the formula for a circle, is there any point for the center of a circle that will give a value of r that's negative? Explain. ■

You can also use the circle equation theorem to describe a circle. Consider the circle whose equation is $(x + 1)^2 + (y - 3)^2 = 2$. Using $r = \sqrt{(x - h)^2 + (y - k)^2}$ gives $(h, k) = (-1, 3)$ and $r^2 = 2$. Hence the center is at $(-1, 3)$ and the radius is $\sqrt{2}$, or ≈ 1.414.

Describing Transformations Using Coordinate Geometry

Transformations can be described using coordinates and algebraic rules. The algebraic rules tell how the coordinates from the original figure move to corresponding points on the image. For example, the following notation is used to describe an example of a translation, a rotation, and a reflection.

a. Translation: $T_{AA'}(x, y) = (x + 3, y - 2)$
b. Rotation: $R_{(0,0),\,180°}(x, y) = (-x, -y)$
c. Reflection: $M_y(x, y) = (-x, y)$

In (a), if the coordinates of a point A are $(4, 5)$, the coordinates of its translation image, A', are $(4 + 3, 5 - 2)$, or $(7, 3)$. In (b) if the coordinates of a point A are $(4, 5)$, the coordinates of its rotation 180° about the origin image, A', are $(-4, -5)$. In (c) if the coordinates of a point A are $(4, 5)$, the coordinates of its reflection image in the y-axis, A', are $(-4, 5)$. The relationship between a point and its image for the transformations in (a), (b), and (c) can also be indicated using arrows as $(x, y) \rightarrow (x + 3, y - 2)$, $(x, y) \rightarrow (-x, -y)$, and $(x, y) \rightarrow (-x, y)$. Exercises 57–62 at the end of this section give additional opportunities to explore the use of coordinates to describe transformations.

Problems and Exercises | *for Section 13.4*

A. Reinforcing Concepts and Practicing Skills

1. Find the midpoint of each line segment in the graph below.

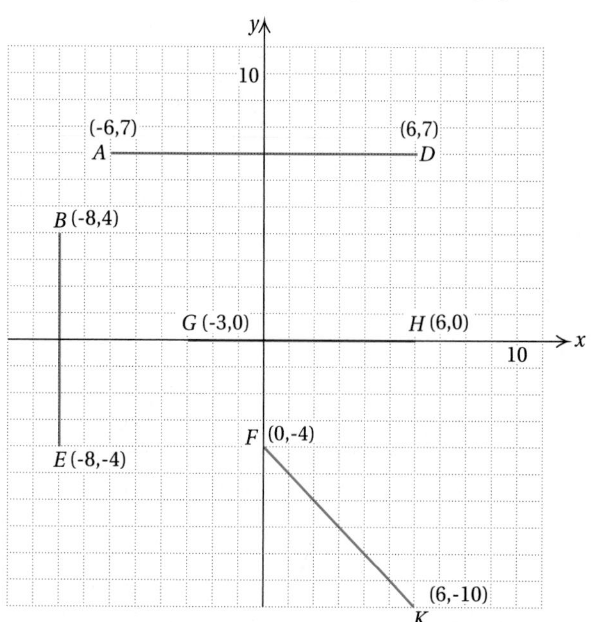

In Exercises 2–5, find the midpoint of the line segment having the given endpoints.

2. $(0, 5)$ and $(0, -8)$
3. $(-6, 0)$ and $(6, 0)$
4. $(8, 1)$ and $(2, 9)$
5. $(4, 3)$ and $(-5, -8)$

In Exercises 6–9, find the other endpoint of a line segment with the given midpoint and one endpoint.

6. Endpoint: $(6, 4)$; midpoint: $(3, 5)$
7. Endpoint: $(-5, 3)$; midpoint: $(-2.5, -1)$
8. Endpoint: $(0, 0)$; midpoint: $(4, -3)$
9. Endpoint: $(7, 7)$; midpoint: $(0, 0)$
10. Find the length of each line segment in the graph on the next page.

Find the distance between the given points in Exercises 11–14.

11. $(0, 6)$ and $(0, -5)$
12. $(3, 5)$ and $(8, 2)$
13. $(0, 0)$ and $(12, -4)$
14. $(-3, -7)$ and $(-4, 15)$

Find the slope of the line through the given points in Exercises 15–18.

15. $(5, 3)$ and $(12, -11)$
16. $(-2, 8)$ and $(2, 8)$

Figure for Exercise 10

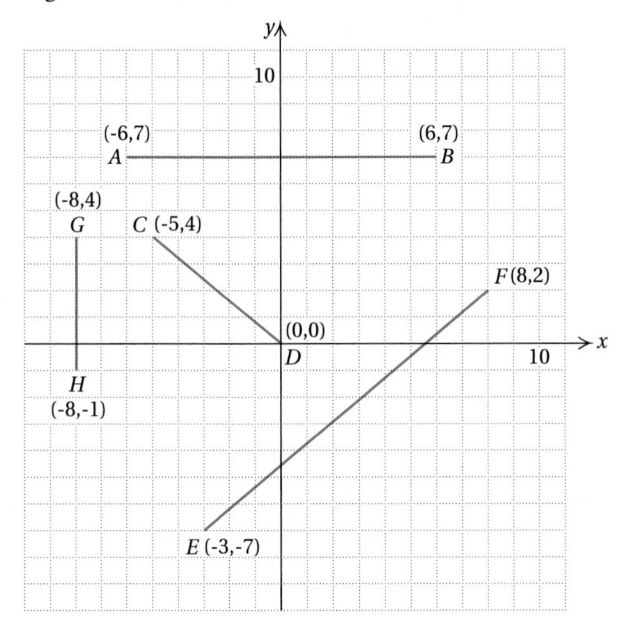

17. $(3, 7)$ and $(-5, -6)$
18. $(12, 18)$ and $(14, 2)$

Write the equation of a line having the given slope and y*-intercept in Exercises 19–22.*

19. Slope $= 4$; y-intercept $= -2$
20. Slope $= \frac{1}{2}$; y-intercept $= \frac{3}{4}$
21. Slope $= -1$; y-intercept $= 0$
22. Slope $= 0$; y-intercept $= 5$

Give the slope and y*-intercept for the line represented by each equation in Exercises 23–26.*

23. $y = -3x - 7$
24. $y = 2x + 12$
25. $y = 5$
26. $y = \frac{3x}{2}$

Find the equations of the lines shown in Exercises 27–29 by using the slope-intercept method. Be sure to check the scale of the graph.

27.

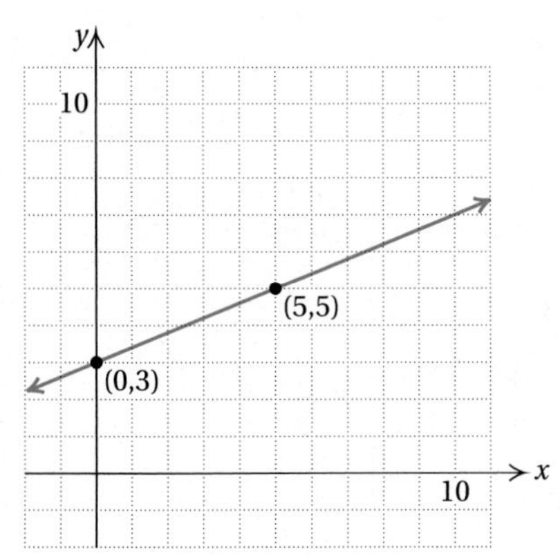

28.

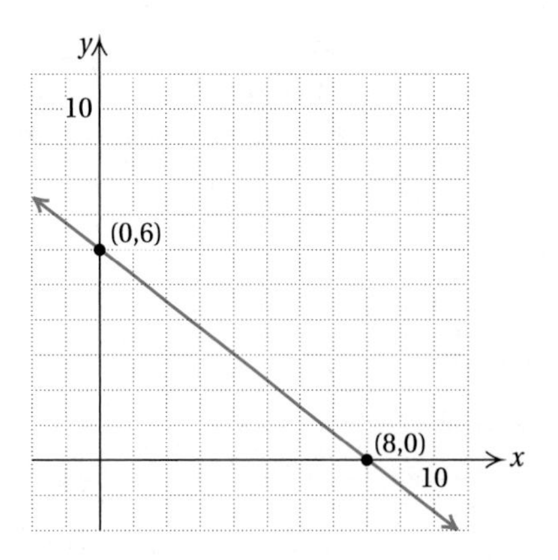

29.

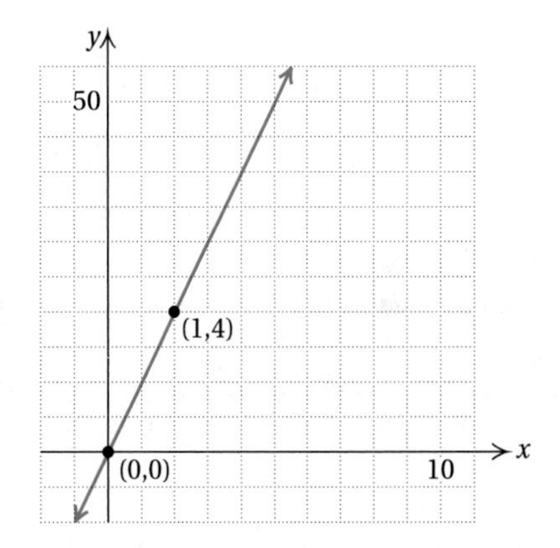

30. Solve the following system by first finding the equations of the two lines.

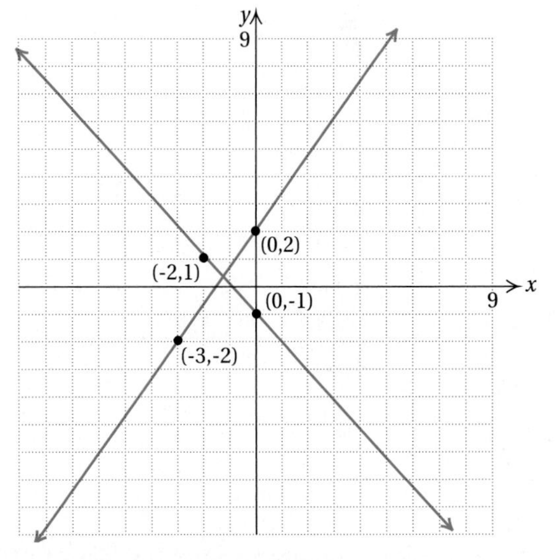

Give the equation of the circle with the given center and radius in Exercises 31–34.

31. Center at $(0, 0)$; radius $= 6$
32. Center at $(1, 2)$; radius $= \frac{1}{2}$

33. Center at $(-3, 4)$; radius $= 2$
34. Center at $(-2, -1)$; radius $= 5$

Give the center of the circle and the radius for each circle in Exercises 35–38.

35. $x^2 + y^2 = 36$
36. $(x - 3)^2 + (y + 5)^2 = 12$
37. $x^2 + (y - 6)^2 = 100$
38. $(x - 1)^2 + (y + 2)^2 = 1$

B. Deepening Understanding

39. Give an equation of a line that is parallel to the line $y = -2x - 3$.
40. Give an equation of a line that has the same y-intercept as the line $y = 24x + 15$.
41. Give the slope of any line that is perpendicular to $y = -\frac{x}{2} + 7$.
42. Sketch the line $x = 6$. How is it related to the line $y = 6$?
43. Tell why $x^2 + y^2 = -49$ doesn't make sense for the equation of a circle.
44. Give the equation of the line that passes through point $(5, 1)$ and is parallel to the line $y = 3x - 1$.
45. Give the equation of the line that is perpendicular to the line $y = -x + 2$ and passes through point $(2, 0)$.

C. Reasoning and Problem Solving

46. Determine the type of triangle formed by the base of an isosceles triangle and the altitudes drawn from the base angles to the congruent sides, as represented by the shaded area. Justify your conclusion.

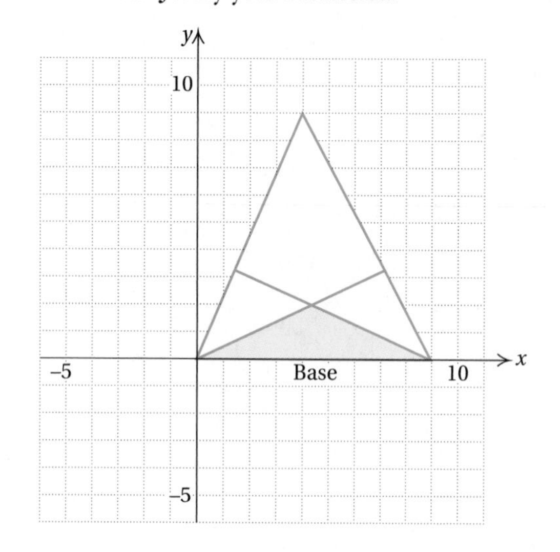

47. Suppose that a quadrilateral is formed by points $A\ (2, 4)$, $B\ (5, 8)$, $C\ (10, 8)$, and $D\ (7, 4)$. Use slope and the distance between these points to determine and verify the type of quadrilateral represented.

48. Give the equation of the line that is the perpendicular bisector of the line segment shown.

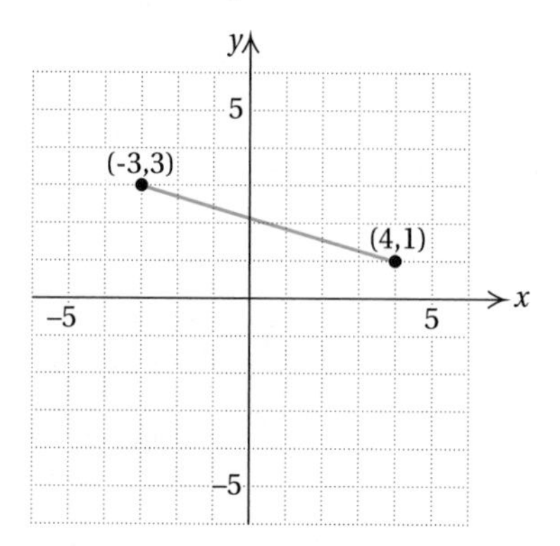

49. Give the center and radius of the circle represented by $2x^2 + 2y^2 = 36$.
50. **Garden Area Problem.** A designer created a garden from two concentric circles whose equations are

$$(x + 2)^2 + (y - 6)^2 = 16 \text{ and } (x + 2)^2 + (y - 6)^2 = 81$$

The area between the circles will be covered with grass. What is the area of that section?

Taxicab Distances Problem. *Imagine that the grid lines on a coordinate grid are streets and that the distance between two points must be measured by the number of "blocks" a taxicab would have to travel horizontally and vertically to get from one point to the other. The sum of the horizontal and vertical distances is called the taxicab distance. A cab driver estimates that traveling one block during rush hour takes about 5 min. Use this information to answer Exercises 51 and 52.*

51. How long would it take the driver to travel from point A to point B for the city represented by the grid?
52. What general rule can you use to find the taxicab distance between two points on a grid?

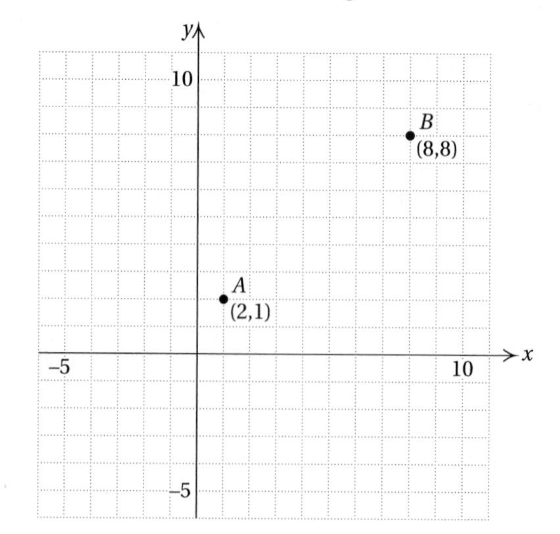

53. Give the coordinates of five points that are on the circle shown.

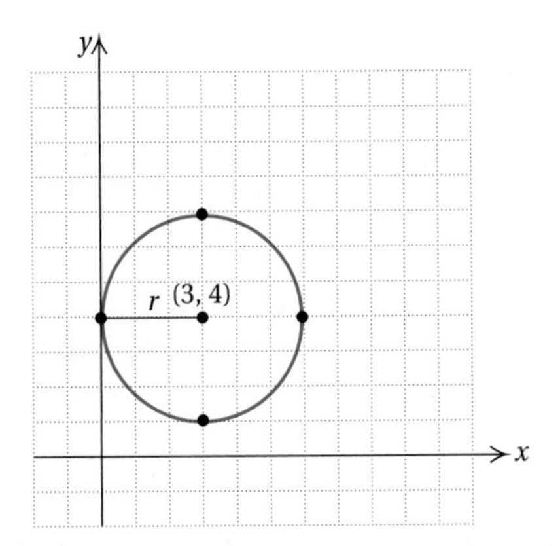

54. Suppose that someone said that the equation connecting points $(-1, 1)$ and $(2, 3)$ is $y = \frac{2}{3}x + \frac{5}{3}$. This person wants to know whether the line represented by this equation will contain point $(6, 9)$. Identify two ways by which this person can decide whether the graph of the equation contains the point.

55. Real-world images such as faces can be captured on the computer by representing the sides of the image with lines. Give the equations of the lines that are used to form the face shown and the coordinates of the points of intersection.

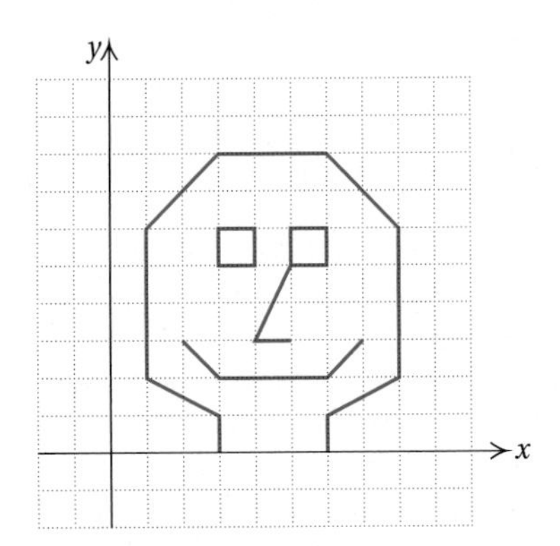

56. House Design Problem. Copy the following house elevation view started on a coordinate grid. Draw the other half of the house and use the tools of coordinate geometry to prove that the two halves of the house are identical.

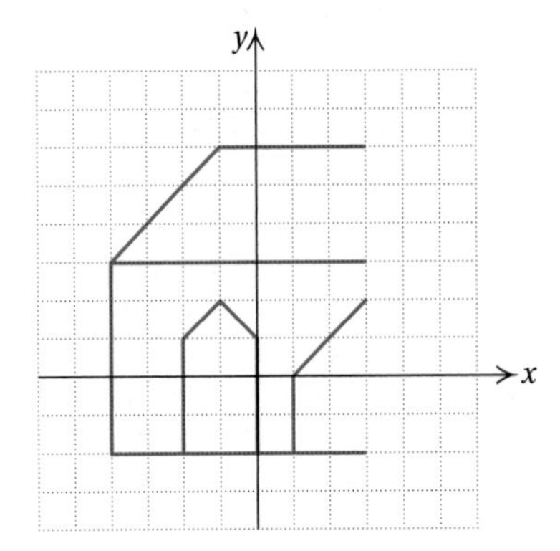

Use actual coordinates of the vertices of triangles to give examples of each of the following transformations.

57. $(x, y) \rightarrow (x + a, y + b)$ is a translation over a, up b.

58. $(x, y) \rightarrow (-x, -y)$ is a 180° rotation about the origin.

59. $(x, y) \rightarrow (-y, x)$ is a 90° rotation about the origin.

60. $(x, y) \rightarrow (y, x)$ is a reflection about the line $y = x$.

61. $(x, y) \rightarrow (x, -y)$ is a reflection across the x-axis.

62. $(x, y) \rightarrow (-x, y)$ is a reflection across the y-axis.

D. Communicating and Connecting Ideas

63. Work with a partner to describe in writing a procedure for using coordinates to analyze a polygon to determine whether it is a parallelogram.

64. Making Connections. Explain the different ways developed in this section for finding the equation of a line. Explain when each might be an appropriate method to use and why.

65. Historical Pathways. The Egyptian Rhind papyrus (c. 1650 B.C.) contains several problems related to the construction of a circle, including the statement that the area of a circle is the same as the area of a square with sides $\frac{1}{9}$ less than the diameter of the circle. Using grid paper, construct several circles and squares to test this statement. Do you agree with the Egyptian rule? Explain.

CHAPTER SUMMARY

Key Ideas: Questions and Answers

Section 13.1

- **How are letters used in mathematics?** *(pp. 704–705)* Letters can be used to represent an unknown number, a constant quantity, a generalized number from a particular set of numbers, or a quantity whose values vary (called variables).
- **What are numerical and algebraic expressions?** *(pp. 705–710)* A numerical expression is a name for a

number and involves one or more operations and numbers. An algebraic expression is a mathematical phrase that involves variables, numbers, and operation symbols.

- **What are the meaning and characteristics of an equation?** *(pp. 710–713)* An equation is a mathematical sentence that contains an equals sign to state that two expressions represent the same number or value.
- **What are some different types of equations?** *(pp. 713–716)* Quantities can be related in different ways. Two common relationships are a linear function and a quadratic function. An equation of the form $y = ax + b$, where a and b are real numbers, is one way to represent a linear function. A quadratic function can be represented by an equation of the form $y = ax^2 + bx + c$, where a, b, and c represent real numbers and $a \neq 0$. An equation of the form $y = a^x$ for $a > 0$ and $a \neq 1$ is an exponential function.

Section 13.2

- **What is a solution to an equation?** *(p. 720)* A solution to an equation is a value for the variable or values for the variables that make the equation true. A true equation is one in which the value on the left of the equals sign is the same as the value on the right. To solve an equation means to find all its solutions.
- **What are some ways to solve linear and quadratic equations?** *(pp. 720–733)* Using the properties of equality in conjunction with inverse operations is the most efficient method for solving complex linear equations. Some techniques for solving quadratic equations are graphing, factoring and the zero product property, and the quadratic formula. The nature of the equation being used can suggest which technique is most appropriate.

Section 13.3

- **How is a linear equation graphed?** *(pp. 736–739)* A linear equation is graphed by creating a table of ordered pairs that make the equation true and plotting them on the Cartesian coordinate grid. Although a line can be determined by two points, at least three ordered pairs should always be used.
- **How are graphs and equations related?** *(pp. 739–742)* The value of m in the linear function $y = mx + b$ is the slope of the line. The value of b is the y-coordinate of the point at which the line intersects the y-axis, called the y-intercept.
- **How is slope related to horizontal and vertical lines?** *(pp. 742–743)* The slope of a horizontal line is 0; the y-coordinates of all points on a horizontal line are the same, so the change in the y-values is 0. Vertical lines are said to have no slope; the x-coordinates of all points on a vertical line are the same, so the change in the x-coordinates is 0. Because slope is defined as the change in the y-coordinate divided by the change in the x-coordinate and division by 0 isn't permitted, the slope is not defined.

Section 13.4

- **How is the midpoint of a line segment determined?** *(pp. 745–748)* The x-coordinate of the midpoint of a line segment is the average of the x-coordinates for the endpoints of the segment. The y-coordinate of the midpoint is the average of the y-coordinates of the endpoints.
- **How is the distance between two points determined?** *(pp. 748–751)* The distance, d, between points (x_1, y_1) and (x_2, y_2) can be found using this formula:

$$d = \sqrt{(x_2 - x_1)^2 + (y_2 - y_1)^2}$$

- **How is the equation of a line determined?** *(pp. 751–759)* Use the slope-intercept method when you can easily find the y-intercept and you know two points on the line. Use the two points on the line to find the slope of the line, m. Substitute this value for m and the value for the y-intercept for b into $y = mx + b$. This gives the equation of the line. Use the two-point method when you know two points but cannot easily find the y-intercept. Start as above by finding the slope. Then using any point on the line, substitute values for x and y into $y = mx + b$ and solve for b, the y-intercept. You then have the values for m and b and can write the equation of the line. Horizontal lines have a slope of 0 and vertical lines are said to have no slope.
- **How can coordinate geometry be used to verify geometric conjectures?** *(pp. 759–762)* As an example, you can find the equations of lines and analyze these to determine if lines are parallel or perpendicular. You can use the distance formula to analyze line segment lengths in polygons. Describing how lines and line segments are related and finding the lengths of segments enables you to prove many geometric conjectures.
- **How can the equation of a circle be determined?** *(pp. 762–764)* To find the equation of a circle, substitute the values for the coordinates of the center of the circle, (h, k), and the value of the radius, r, into the equation below. Simplify as needed.

$$(x - h)^2 + (y - k)^2 = r^2$$

- **How can transformations be described with coordinate geometry?** *(p. 764)* The following notation, using coordinates and algebraic rules, can be used to describe a translation, a rotation, and a reflection. (a) Translation: $T_{AA'}(x, y) = (x + a, y + b)$, (b) Rotation: $R_{(0, 0), 180°}(x, y) = (-x, -y)$, (c) Reflection: $M_y(x, y) = (-x, y)$.

Key Terms, Concepts, and Generalizations

Section 13.1
Variable (p. 704)
Constant (p. 704)
Unknown (p. 705)
Numerical expression (p. 705)
Algebraic expression (p. 706)
Evaluate an algebraic expression (p. 707)
Equation (p. 711)
Functional relationship (p. 711)
Dependent variable (p. 711)
Independent variable (p. 711)
Linear function (p. 714)
Quadratic function (p. 714)
Exponential function, base a (p. 715)
Growth function (p. 716)

Section 13.2
Solution (p. 720)
True equation (p. 720)
Solve an equation (p. 720)
Linear equations (p. 721)
Addition and subtraction properties of equality (p. 721)
Multiplication and division properties of equality (p. 721)
Quadratic equation (p. 723)
x-intercept (p. 725)
Factoring (p. 728)
Zero product principle (p. 728)
Quadratic formula (p. 730)

Section 13.3
Solution to a linear equation (p. 736)
x-axis (p. 737)
y-axis (p. 737)
Positive slope (p. 739)
Slope (p. 739)
Negative slope (p. 739)
y-intercept (p. 739)
Slope-intercept (p. 739)
Parallel lines (p. 739)
Perpendicular lines (p. 739)

Section 13.4
Coordinates of the midpoint of a line segment (p. 747)
Pythagorean theorem (p. 749)
Distance formula (p. 750)
Slope-intercept method (p. 751)
Two-point method (p. 751)
Slope (p. 751)
Perpendicular lines (p. 756)
Parallel lines (p. 756)
Horizontal lines (p. 759)
Vertical lines (p. 759)
Center (p. 762)
Radius (p. 762)
Equation of a circle (p. 762)

CHAPTER REVIEW

Concepts and Skills

Tell whether each quantity in Exercises 1 and 2 is constant or variable. Explain.

1. The number of hours of sunlight in a day

2. The number of hours in a day

Evaluate the algebraic expression for the values given in Exercises 3 and 4.

3. $4(2x + 6)$, for $x = 0, 2, -5, 20$, and -40

4. $3x^3 - 5$, for $x = -5, 0, 2$, and 10

For Exercises 5–7, the total cost of hiring a certain plumber is a flat rate of \$35 plus \$9.50/hr.

5. Write an expression in terms of x that gives the total cost of hiring the plumber.

6. Write an equation that shows the relationship between the quantities.

7. Find the total cost for hiring the plumber for 8 hr.

Give five ordered pairs that make each equation true in Exercises 8 and 9.

8. $2x^2 - 5x = y$

9. $y = \frac{2x}{5} - 8$

Solve each equation in Exercises 10 and 11 by using the properties of equality. Check.

10. $4x - 18 = 65$

11. $3x + 15 = -4x - 20$

Solve each quadratic equation in Exercises 12 and 13 using any technique you choose. Name the technique you used. Check.

12. $x^2 - 5x - 84 = 0$

13. $8x^2 + 12x = 0$

14. Give the solutions to the quadratic equation by examining the graph of the related quadratic function.

Graph: $y = x^2 + 2x - 15$

Equation: $x^2 + 2x - 15 = 0$

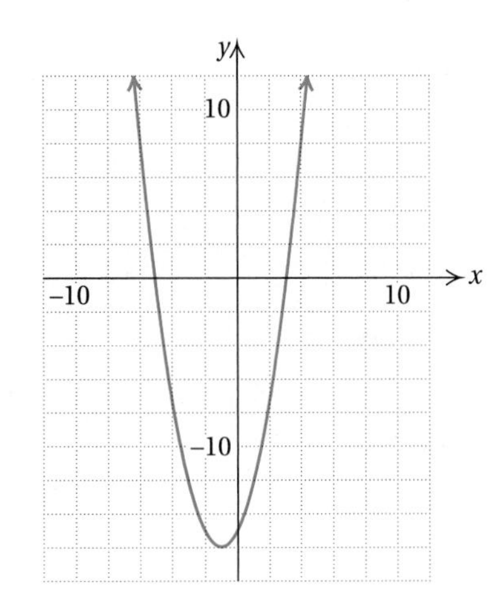

15. Find the slope and y-intercept of the line $y = -6x + 14$.

16. Graph the line given in Exercise 15.

17. Use the graph from Exercise 16 to find the value of y when $x = 3$.

18. Find the slope of the line passing through points $(3, 11)$ and $(18, 20)$.

Find the midpoint of the line segment with the given endpoints in Exercises 19 and 20.

19. $(0, 6)$ and $(6, -9)$

20. $(3, 4)$ and $(-6, 5)$

Find the distance between the given points in Exercises 21 and 22.

21. $(12, 6)$ and $(4, 5)$

22. $(-5, 5)$ and $(0, 12)$

Write the equation of a line with the given slope and y*-intercept in Exercises 23 and 24.*

23. Slope $= 3$; y-intercept $= -2$

24. Slope $= \frac{1}{2}$; y-intercept $= \frac{1}{4}$

25. Give the equation of the circle with its center at $(4, -3)$ and a radius of 3.

Reasoning and Problem Solving

26. Give the equation of a line that is parallel to $y = -4x + 7$.

27. Give the equation of a line that is *not* parallel to $y = 4x + 11$ but has the same y-intercept.

28. Give the equation of a line that is perpendicular to the line $y = \frac{x}{-3} + 7$.

29. Are all equations either linear or quadratic? If not, give an example and explain why it is neither linear nor quadratic.

30. The graph of an equation in the form $y = mx + b$ is a straight line. Can the equation of every straight line be written in the form $y = mx + b$? (*Hint:* What is the equation of a vertical line?)

31. The diagrams below show how a box was made by cutting a square of side length x from each corner of an 8-in. × 10-in. sheet of paper and then folding it to make the sides of the box. Write an equation that shows the volume of the box. Find the volume for side lengths of the cut-out square of 1 in., 1.5 in., 2 in., 2.5 in., and 3 in. Describe the patterns you see in the volume as the value of x changes.

32. Park Dimensions Problem. A large park is to be built in the shape of a rectangle. The length is to be 5 m shorter than 3 times the width. The perimeter of the park is to be 750 m. What will be the length and width of the park?

33. Give an equation of a line parallel to the line $y = -3x + 9$.

34. Give an equation of a line perpendicular to the line $y = 4x$ having the same y-intercept.

35. Give the coordinates of the original triangle when its image was created by reflecting across the x-axis and has coordinates $(-2, 1)$, $(-5, 1)$, and $(-2, 3)$.

Alternative Assessment

36. Give an example of a linear function. Represent the function in your own words as a table of ordered pairs, as a graph, and as an equation.

37. Make up a quadratic equation with the solutions 3 and 4. State how you know that these are the solutions.

38. Sketch the graph of a quadratic function for which the related quadratic equation has only one solution. Then sketch one that has two solutions. Explain how you know by looking at the graph the number of solutions there are.

39. Give examples and write an explanation for a classmate for how to find the equation of a line.

Diagrams for Exercise 31

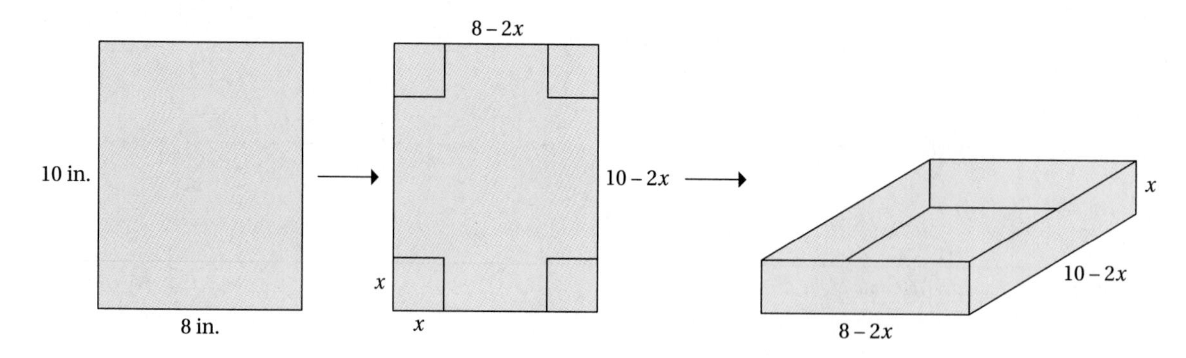

Appendix A

Graphing Calculator

INTRODUCTION

The graphing calculator has opened many opportunities for students and teachers to explore number concepts, algebraic relationships, and patterns in data. In the first two sections of this appendix we show how to use the graphing calculator to examine algebraic equations and their graphs. In the remaining sections we focus on use of the graphing calculator to examine data and statistics.

The illustration on page 772 shows the keyboard for the Texas Instruments TI-73, but most of the comments we make apply equally to TI-81, TI-82, and TI-83 calculators. Familiarize yourself with this keyboard. The top row contains keys related to the graphic displays. The key in the first column of the second row initiates the second function feature on most keys, indicated by the print on the keyboard above most keys. The four keys with arrowheads on them at the upper right of the keyboard move the cursor around on the screen.

Two other important keys are the [WINDOW] and [MODE] keys in the second column of the first and second rows, respectively. The first sets parameters for the display of information in the window of the calculator, and the second sets the actual processing mode for use in an application of the calculator.

To turn the calculator on, press the [ON] key at the lower left. To turn the calculator off, press the key sequence [2nd] [OFF], where the word OFF is the second function on the [ON] key at the lower left. To clear the screen of information no longer wanted, press [CLEAR] in the fifth column of the fourth row. To quit an application, press the sequence [2nd] [QUIT], where the word QUIT is found as the second function on the [MODE] key in the second column of the second row.

Section A.1 GRAPHING EQUATIONS

Entering the Equation in the Y= List

Enter the equation desired in the Y= list. For you to do so, the equation must be available in a form $y =$ some expression in x. You can let the x and y variables be the independent and dependent variables, respectively.

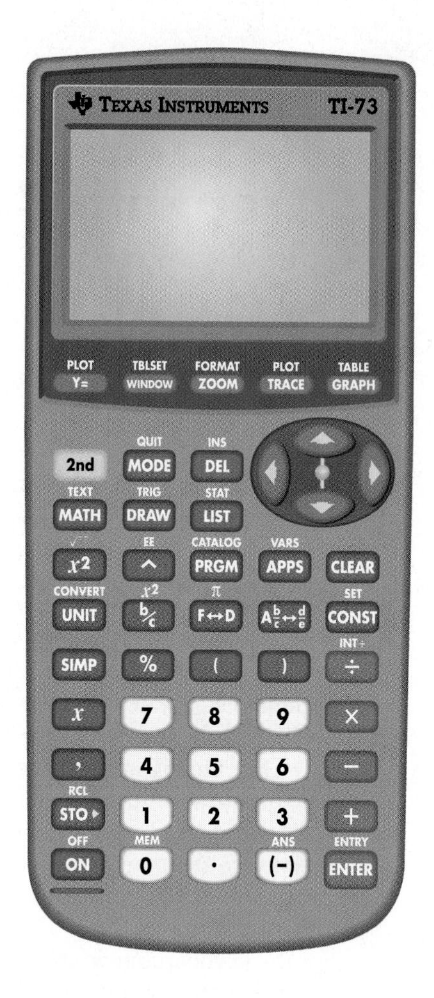

Press [Y=] to display the Y= edit screen. If any other equations are already on the screen, remove them by pressing [CLEAR]. Then enter, for example, the equation $y = 3x + 2$ with the key strokes [3] [x] [+] [2]. Note that you enter the variable x by pressing [x] in the first column of the seventh row. When you have completed the key strokes for the equation, press [ENTER] to complete entry of the equation. Your screen should appear as shown.

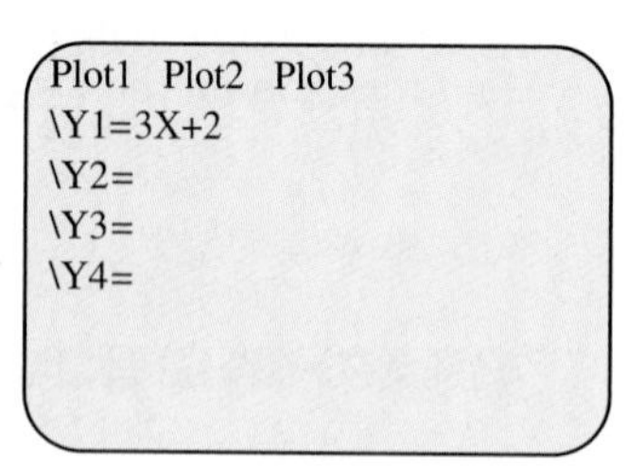

Defining the Viewing Window

Next press [WINDOW] in the first row. Doing so gives you the window on the following page. Use the appropriate cursor keys, including [(-)] in the fourth column of the bottom row for the negative sign, to set the values for the minimum and maximum x-values and minimum and maximum y-values, as shown. The scale settings indicate the positions of the tic marks along the axes. You can also set the screen display size by using the ZOOM key in the first row. Pressing ZOOM gives a number of options and the sixth option called Z Standard will also give this screen the dimensions $[-10, 10]$ by $[-10, 10]$, where the first range indicates the size of the x-interval graphed and the second range the size of the y-interval graphed.

```
WINDOW
  Xmin=-10
  Xmax=10
  ΔX=.2127659574...
  Xscl=1
  Ymin=-10
  Ymax=10
  Yscl=1
```

Displaying the Graph

Press GRAPH in the first row to display your graph. You should get a display similar to that shown for the equation $y = 3x + 2$. If you have trouble getting your graph to show, you may need to clear a statistics plot. To do so, press 2nd [PLOT]. Use the arrow keys to highlight the word OFF in the first line of each PLOT and press ENTER after each one. Be sure that all are switched to OFF and then try your graph again.

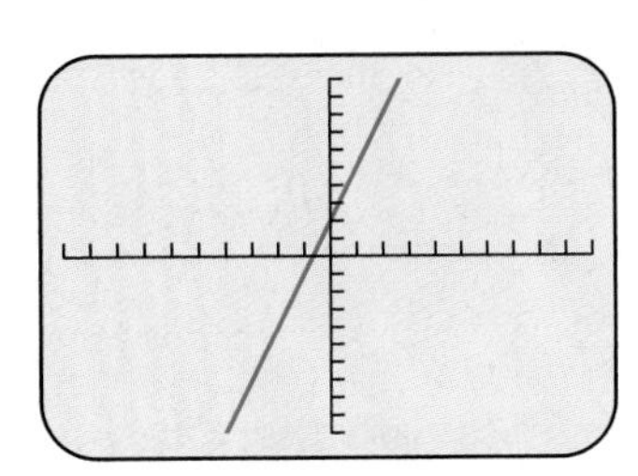

Clearing the Screen

To clear the screen, press CLEAR. To graph a second equation at the same time, you can press the Y= key, go to the equation list, and enter the equation on the second line. If you just want to see another equation by itself, use CLEAR to remove the first equation and type in your new equation.

Try These

1. Enter and examine the graphs of the following expressions. Make sketches of the graphs. How are they alike and how are they different?
   - **a.** $3x$
   - **b.** $3x + 1$
   - **c.** $3x + 2$
   - **d.** Predict the graph for $3x + 5$ and then graph the expression to check out your prediction.
2. Enter and examine the graphs of the following pairs of expressions. Make sketches of the graphs. How are they related?
   - **a.** x and $-x$
   - **b.** $2x$ and $-0.5x$
   - **c.** $3x$ and $-0.33x$
   - **d.** $10x$ and $-0.1x$
3. Enter and examine the graphs of the following expressions. Note that x^2 can be entered with the special sequence X x^2 or by X ^ 2.
   - **a.** 3
   - **b.** x
   - **c.** -2
   - **d.** x^2
   - **e.** $x^2 + 1$
   - **f.** $\sqrt{3x + 2}$
   - **g.** $\sqrt{9 - x^2}$

Section A.2 EXPLORING EQUATIONS

Entering the Equation(s) Desired

Key in the equations $y = 3x + 2$ and $y = x + 4$ for exploration as $Y_1 = 3x + 2$ and $Y_2 = x + 4$. Then graph them on the standard window. Your graph should appear as follows.

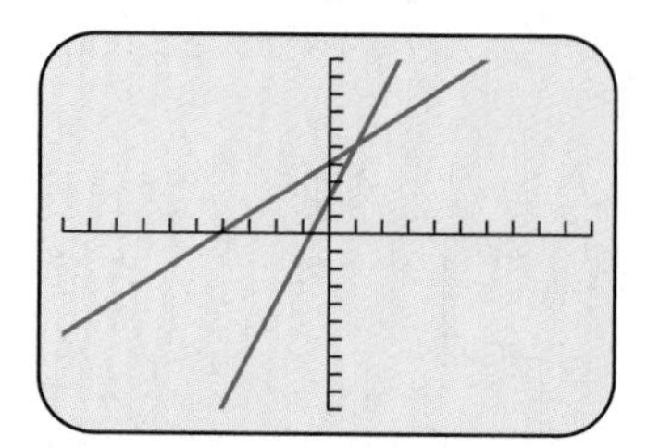

Activating the Free-Moving Cursor

Next, press TRACE in the first row. Doing so will activate a free-moving cursor on the screen that will travel, at your command, along the graphs selected. Note that the coordinates of the cursor are shown at the bottom of the screen.

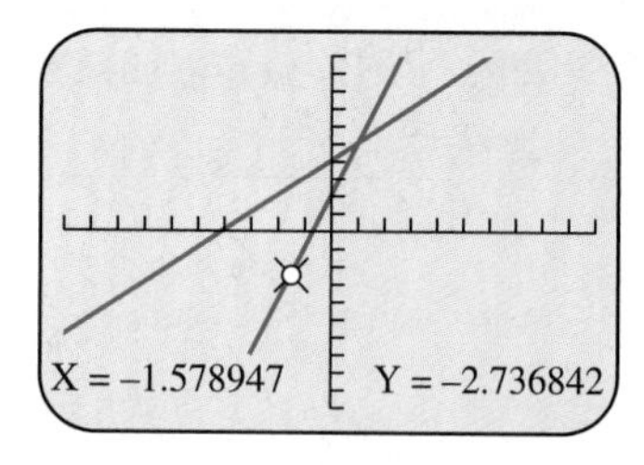

Moving the Cursor

You can move the cursor by using the cursor keys (keys with arrowheads) in the upper right hand corner of the keyboard. Pressing ◄ and ► will move the cursor left or right along the graph on which it is currently located. Pressing ▲ and ▼ will move the cursor to the second graph above or below the graph on which it is currently located. Use these keys to move the cursor about on the screen and approach the point of intersection of the two graphs.

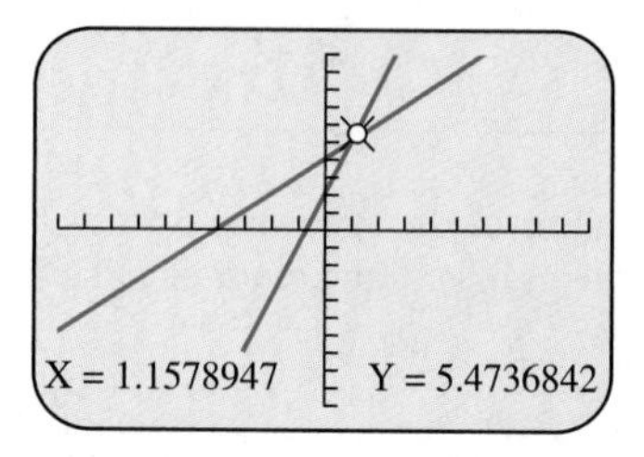

Using the TRACE Coordinates to Explore the Graphs

The coordinates shown on the preceding screen, $x \approx 1.1$ and $y \approx 5.4$, and the coordinates shown when the cursor is moved back to the left suggest that the intersection of the lines may be at $x = 1$ and $y = 5$, or the point $(1, 5)$. Testing these values in the equations shows that this point is the intersection of the two graphs, or the solution to the system of equations formed by the two equations.

Using TRACE to Find the Solution to an Equation

Suppose that you need to solve the equation $3x + 7 = 22$. You can consider the problem as asking the question, Where does the graph of $Y_1 = 3X + 7$ intersect the graph of $Y_2 = 22$? That is, when does the expression $3x + 7$ take on the value of 22? Display the graphs on a window of dimensions $[0, 10]$ by $[0, 25]$ and trace with the cursor to the intersection point. Your display should appear as shown.

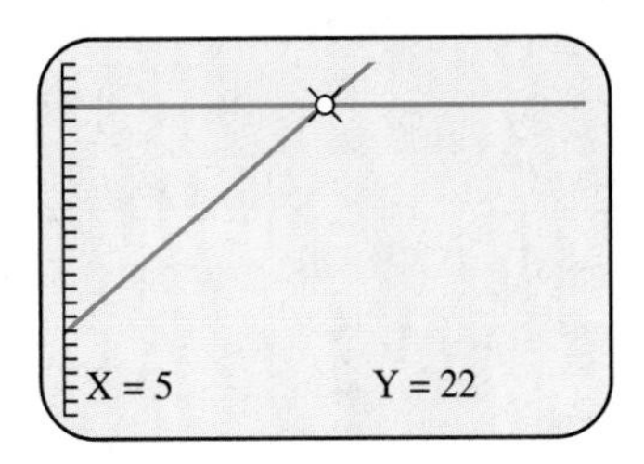

The solution appears to be $x = 5$. Testing it in the original equation shows that it is the solution. Obtaining a solution in this manner may not always be possible, but you can usually find a close decimal approximation to the solution.

Try These

1. Examine the coordinates of points along the graphs of the following expressions.
   a. $3x + 2$ **b.** $5x - 7$
   c. -2 **d.** x^2
   e. $\sqrt{x}$
2. Find a solution to each equation and check your results.
   a. $4x - 1 = 7$ **b.** $4x - 1 = 23$
   c. $5x + 3 = 23$ **d.** $x^2 = 4$
   e. $\sqrt{x} = 9$
3. Find solutions for the following systems of equations.
   a. $y = 2x + 2$ and $y = 3x - 1$
   b. $y = 4x + 5$ and $y = 10x - 1$

Section A.3 ENTERING AND EDITING DATA

One of the powerful applications of the calculator is that of analyzing data relating to statistics or numerical and algebraic patterns. In this section we illustrate the building and editing of data lists. Use the following data values: 158, 176, 150, 165, 158, 158, 166, 168, 168, 171, 152, 164, 165, 156, 151. This data set represents the height, in centimeters, of 15 female students who were enrolled in a summer mathematics program in California.

Selecting a List Location for Your Data

First, press [LIST] to open the list display. Use [◄] and [►] to move among the lists L1, L2, ..., L6. In this case select list L1 for the data.

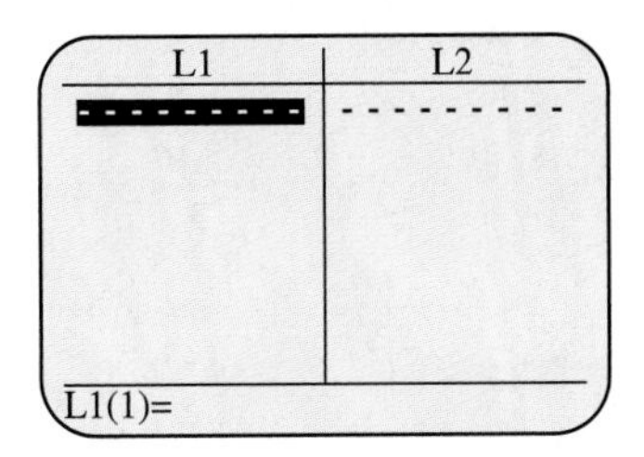

Entering Data into the Selected List

Using the data given, key in each value and press ENTER. The data will be entered on rows in list L1.

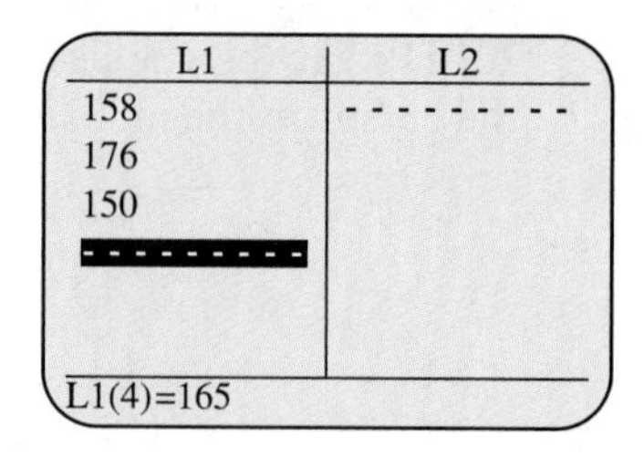

L1 | L2
152
164
165
156
151
L1(16)=

Here are some quick tips:

- As you enter a data value, it is echoed on the bottom line of the screen.
- Other information at the bottom of the screen shows you the current location and contents of the cursor. In the first data-entry screen above, L1(4) = 165 indicates that the cursor is in list L1 and is positioned on line 4. The value 165 will be stored in location L1(4) when you press ENTER.

Editing Values in a List

Press LIST and use the arrow keys to move the cursor to the value to be edited. In this case move the cursor to the value 165 in location L1(4).

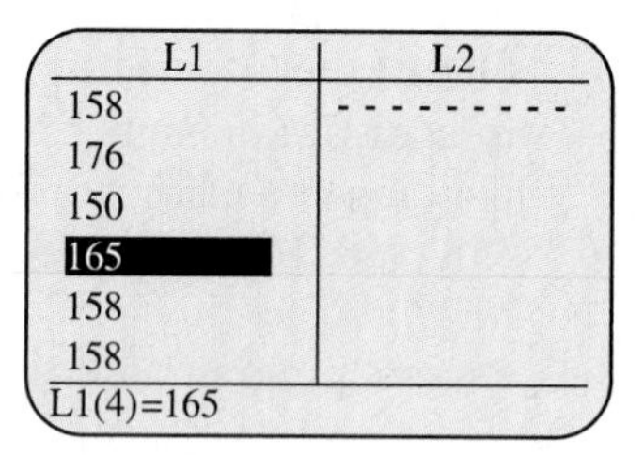

You now have two ways to edit the value.

- Key in an entire value to replace the original value and press ENTER. The cursor then highlights the next value in the list, as shown.

L1 | L2
158
176
150
166
158
158
L1(5)=158

- Press ENTER. A flashing cursor now appears on the editing line at the bottom of the screen, as shown. Use the arrow keys, as well as INS and DEL, for insert and delete, to modify the original data value. Press ENTER when you have finished editing.

L1	L2
158	---------
176	
150	
165	
158	
158	

L1(4)=165

Clearing a List

Press LIST and use the arrow keys to move the cursor to the top of the list to be cleared, in this case to the top of list L1. Note that the entire list is echoed on the bottom line of the screen.

L1	L2
158	---------
176	
150	
166	
158	
158	

L1=(158,176,150...

Press CLEAR. Note that the expression on the right-hand side of the formula at the bottom of the screen disappears.

L1	L2
158	---------
176	
150	
166	
158	
158	

L1=

Press ENTER. The selected list is now empty, with the cursor on the first line of the empty list.

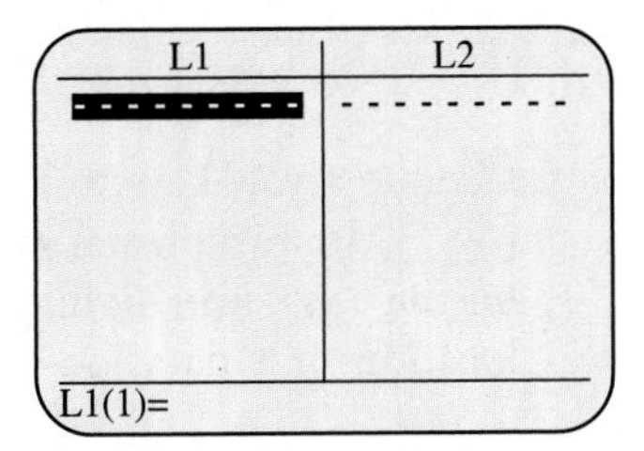

Try These

1. Create a list in L1 for the following measurements of bolt lengths (in centimeters): 23.5, 22.9, 22.1, 23.5, 24.5, 22.2, 23.9, 22.7, 21.8, 23.3, 22.0.
2. Edit your list in (1) by replacing the 23.9 measurement with 23.8.
3. Place the following measurements in L2: 4.5, 3.9, 4.2, 3.9, 4.5, 6.8, 4.2, 4.4, 5.3, 2.3, 4.4.
4. Clear your list in L1. Clear your list in L2.

Section A.4 CREATING A SCATTERPLOT

Deselecting Functions That You Don't Want Plotted

To begin, press [Y=]. If the equals sign of a function is highlighted, its graph will be displayed when you press [GRAPH]. For each function you don't want displayed, move the cursor onto its equals sign and press [ENTER] to deselect that function's graph.

Entering Data Pairs in Two Lists

First, review the discussion in Section A.3 regarding data entry. Then use list L1 for the first variable data (x) and list L2 for the second variable data (y). The ordered pairs (x, y) represent the heights and arm spans (in centimeters) of 15 female students enrolled in a summer mathematics program in California. Use the heights of 158, 176, 150, 165, 158, 158, 166, 168, 168, 171, 152, 164, 165, 156, 151. Use the arm spans of 155, 164, 142, 156, 157, 150, 162, 170, 168, 167, 154, 160, 166, 150, 148.

L1	L2
158	155
176	164
150	142
165	156
158	157
158	150

L1(1)=158

Turning on the Stat Plot, Indicating Plot Type, and Identifying Data Location

Press [2nd] [PLOT].

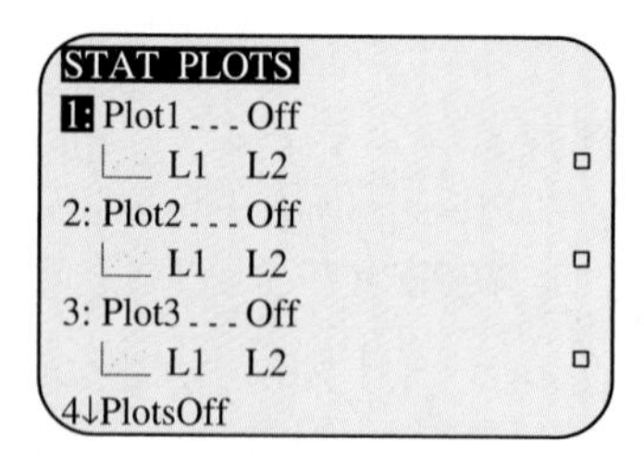

Use the arrow keys to highlight 1:PLOT1... and press [ENTER] or, alternatively, press [1]. To turn on stat plot 1, highlight ON and press [ENTER]. To indicate plot type, on the TYPE line highlight the first scatterplot icon and press [ENTER]. To indicate location of x data, on the Xlist line highlight L1 and press [ENTER]. To indicate location of y data, on the Ylist line highlight L2 and press ENTER. On the MARK line, highlight your choice of point marker to appear on the scatterplot and press ENTER.

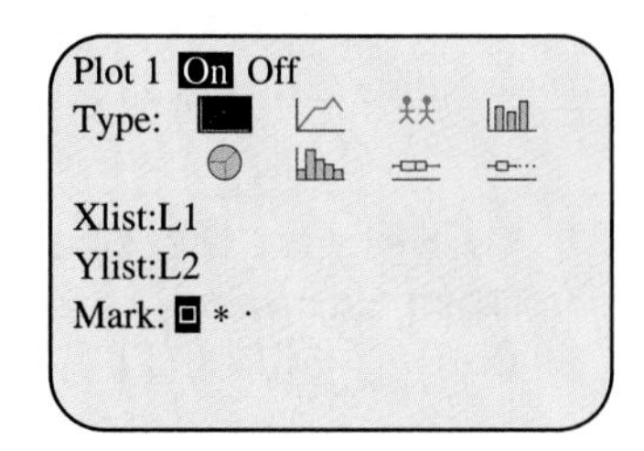

Adjusting Window Dimensions

Press WINDOW. Key in values for XMIN, XMAX, XSCL, YMIN, YMAX, and YSCL. Use the smallest and largest values in the two lists to help determine XMIN, XMAX, YMIN, and YMAX. Press ENTER after each entry to move to the next line.

Displaying the Scatterplot

Press GRAPH to display the scatterplot.

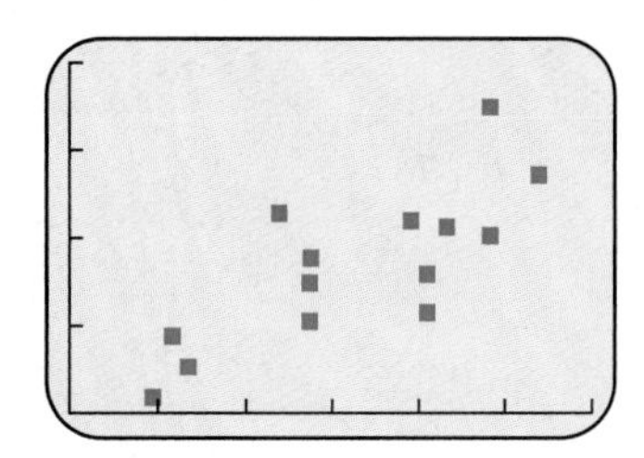

Try These

1. Make a scatterplot for the following data pairs where L1 is the number of grams of protein in five fast-food sandwiches and L2 is the number of grams of fat in the same sandwiches. L1: 42, 35, 27, 44, 39; L2: 52, 39, 22, 48, 50.
2. Make a scatterplot of the following matches of targets and measurements for five bolt lengths (in centimeters) on a vertical lathe assembly. Targets: 4.5, 4.3, 6.2, 8.1, 5.6; actual measurements: 4.4, 4.5, 6.1, 7.9, 5.5.

Section A.5 | CREATING A BOX-AND-WHISKERS PLOT

Deselecting Functions That You Don't Want Plotted

Press Y=. If the equals sign of a function is highlighted, its graph will be displayed when you press GRAPH. For each function you don't want displayed, move the cursor on to its equals sign and press ENTER to deselect that function's graph.

Entering Data Values into a List

First, review the data entry discussion in Section A.3. Then use list L1 for the data set used in Section A.4, representing heights of 15 female students enrolled in a summer mathematics program in California.

L1	L2
158	- - - - - - - - -
176	
150	
165	
158	
158	

L1(6)=158

Turning on the Stat Plot, Indicating Plot Type, and Identifying Data Location

Press [2nd][PLOT].

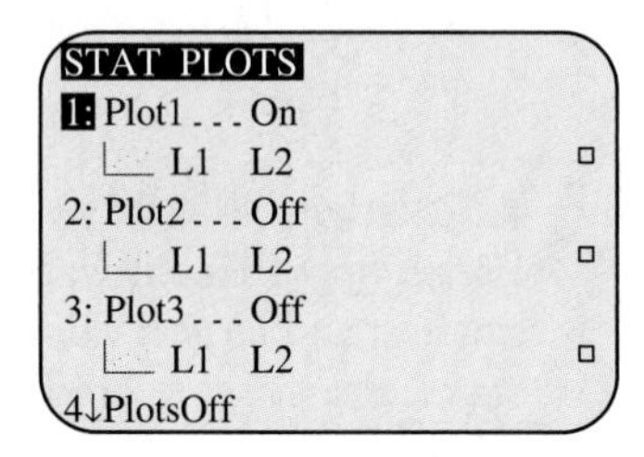

Use the arrow keys to highlight 1:PLOT1. . . and press [ENTER] or, alternatively, press [1].

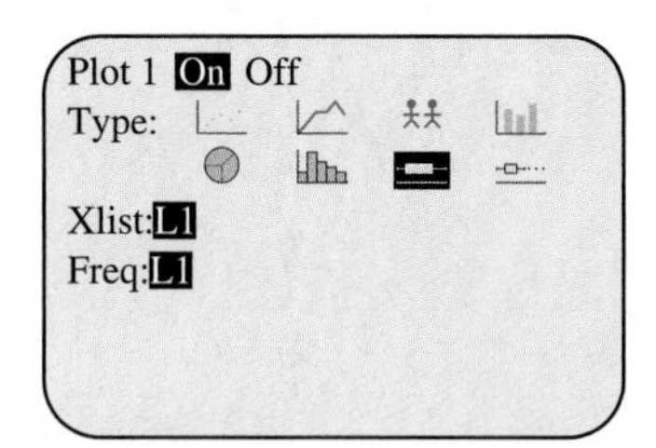

- To turn on stat plot 1, highlight ON and press [ENTER].
- To indicate plot type, on the TYPE line highlight the box-and-whiskers icon, and press [ENTER].
- To indicate location of the data, on the Xlist line highlight L1 and press [ENTER].
- To indicate the frequency of occurrence of each data value, on the Freq line highlight 1 and press [ENTER].

Adjusting Window Dimensions

Press [WINDOW] to display the adjustment controls. Key in values for XMIN, XMAX, and XSCL. For XMIN and XMAX, refer to the smallest and largest values in the list. In this case choose 150 and 180. The value for XSCL represents the distance between tic marks for the horizontal scale accompanying the box plot.

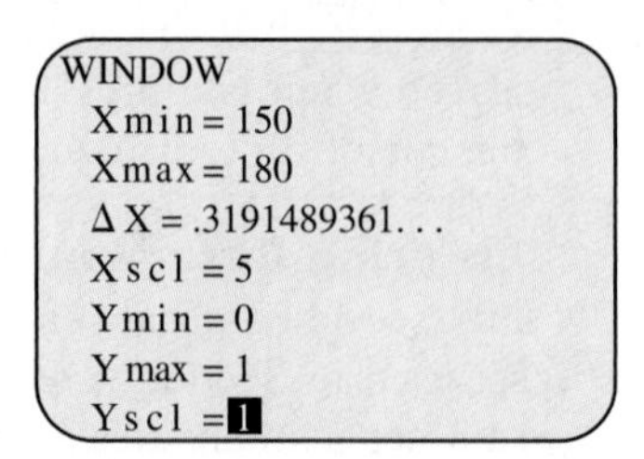

Key in values for YMIN, YMAX, and YSCL. These values will *not* change the location or size of the box plot, but they do determine whether a horizontal axis will be displayed. When you use 0 for YMIN, the horizontal axis is placed at the bottom of the screen.

Displaying the Box Plot

Press GRAPH.

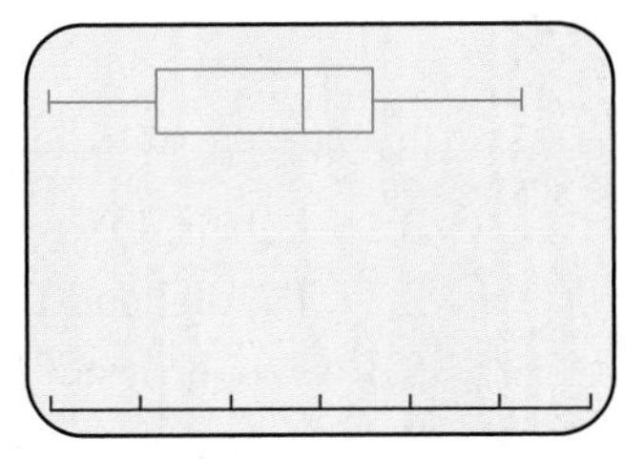

Try These

1. Make a box-and-whiskers plot for the following data representing the boiling points, in °C, for 12 different samples of a liquid: 194.5, 194.3, 197.9, 192.2, 195.4, 190.1, 200.8, 201.3, 199.4, 198.7, 197.6, 204.6.

2. The following measurements of the thickness of a pipe in centimeters were made with different instruments. Make a box-and-whiskers plot of the measurements showing the variations observed: 0.31, 0.45, 0.46, 0.43, 0.36, 0.29, 0.40, 0.23, 0.22, 0.21, 0.18, 0.23.

Appendix B

Geometry Exploration Software

INTRODUCTION

Geometry Exploration Software (GES) is a great tool for use in constructing geometric shapes and exploring their properties and relationships. The software extends traditional tools of Euclidean geometry—the compass and straightedge—by providing construction, measurement, and calculation tools that are accurate, efficient, and user friendly. GES also helps you investigate *consistency within change.* By moving one or more components of a construction—a vertex or a segment, for instance—you can begin to identify aspects of the construction that change and those that remain constant. GES allows you to examine numerous cases, which would hardly be possible with only pencil and paper.

We used *The Geometer's Sketchpad,* one of many GES packages now available, to produce the techniques and activities demonstrated in this appendix. Additional GES software, such as *CABRI: The Interactive Geometry Notebook,* the *Geometric Supposer* series (including the *Geometric SuperSupposer*), and others, provide many of the same capabilities we illustrate here.

In each of the seven sections in this appendix we present *key techniques* to help you become familiar with important features of *The Geometer's Sketchpad.* Throughout this appendix, step-by-step instructions accompany each key technique. You are encouraged to learn more about *Sketchpad* through its reference manuals, resource materials, and the sample files that accompany the software.

Section B.1 | LABELS, OBJECTS, AND MEASURES

In this section we introduce fundamental tools of *The Geometer's Sketchpad* to create, label, and measure geometric objects.

Pointer (Selection) Tool

In *The Geometer's Sketchpad,* use the pointer that appears on the screen to select tools and menu options. You can also use the pointer on any point of a figure, hold the mouse button down, and drag the point to change the size and shape of the figure. If the pointer isn't showing, click on the top icon, the pointer tool, in the **Toolbox** along the left margin of your screen.

Labels

Text Tool

Sketchpad labels every geometric figure you create—points, straight objects, circles, arcs, polygon interiors—but you must choose whether or not to show the labels.

To automatically show object labels

1. pull down the **Display** menu,
2. choose **Preferences,** and
3. under **Autoshow Labels,** check all three boxes.

To move and edit labels on the screen

1. select the text tool (pointing finger) from the **Toolbox,**
2. click and hold a label to move it around, and
3. double-click a label to edit it.

Objects

Begin constructions by using the **Toolbox** or the **Construct** menu.

The Toolbox. In the **Toolbox,** three tools create points, circles, and straight objects.

Point Tool

When you click on the point tool, the pointer changes to a crosshair: + .

To create a point

1. move the point tool anywhere on your screen and
2. click the mouse button to plot a point.

Circle Tool

When you click on the circle tool, the pointer changes to a crosshair surrounded by a circle: ⊕.

To create a circle

1. move the circle tool anywhere on your screen,
2. hold down the mouse button,
3. drag the mouse and watch the circle expand, and
4. release the mouse button to plot the circle.

Line Tool

When you click and hold on the line tool, three boxes extend from the line tool icon. The boxes represent the straight objects that you can create:

- a line segment (two endpoints),
- a ray (one endpoint), or
- a line (no endpoints).

Hold down the mouse button and move to any of the three boxes to select the type of straight object you want to construct. The line tool crosshair appears when you release the mouse button.

To create lines, rays, or segments

1. move the line tool anywhere on your screen,
2. hold down the mouse button to place one point,
3. drag the mouse and watch the straight object appear, and
4. release the mouse button to place a second point.

Try These

1. Draw a figure, using points, circles, segments, rays, and lines.
2. Construct a drawing similar to the one shown on the next page. You'll need to use all the construction tools in the **Toolbox.**

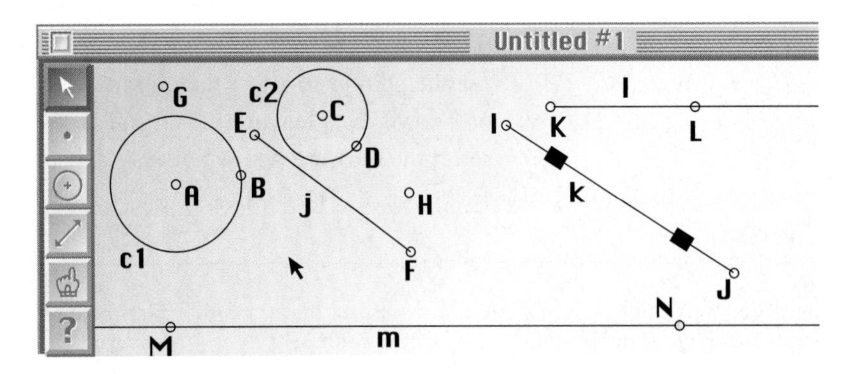

The Construct Menu. Each choice under the **Construct** menu requires that you select one or more objects that you have already constructed. Suppose that you want to construct the midpoint of a line segment. You first must select the segment. A selected object, such as segment **k** in the preceding exercise, shows its *selection handles.*

To select an object

1. be sure that the pointer tool is active and
2. click on the object with the pointer.

Sketchpad knows what you have selected and tells you the constructions that are possible. With a segment already selected when you pull down the **Construct** menu, only two constructions are possible: **Point on Object** and **Point at Midpoint.** These are the only choices highlighted on the **Construct** menu.

To complete a construction from the **Construct** Menu

1. use the pointer tool to select an object,
2. pull down the **Construct** menu, and
3. click on the desired construction, which must be highlighted.

Some constructions require that you select more than one object. To construct a perpendicular line, for instance, you must select a straight object and a point.

To select more than one object at a time

1. hold down **[SHIFT]** as you click the mouse button to select objects and
2. with **[SHIFT]** held down, every object you click on remains selected.

Try These

3. Construct a perpendicular line through a selected point and line.
4. Experiment with other constructions shown on the **Construct** menu. If you need help deciding what you must select, choose **Construction Help,** the last item on the **Construct** menu.

Measures

Use the **Measure** menu to measure the objects you construct and select. The measurement will appear on your screen.

To measure an object

1. use the pointer to select the object to be measured,
2. pull down the **Measure** menu, and
3. select the measurement you desire, which must be highlighted.

Try These

5. Construct a triangle and its interior. Measure the perimeter of the triangle and the area of the interior region.
6. Experiment with other measurements shown on the **Measure** menu.

Section B.2 PARALLEL AND PERPENDICULAR LINES

Many geometric shapes and properties involve parallel or perpendicular lines. The key techniques in this section help you create parallel and perpendicular lines. You also learn how to construct the intersection point of two lines and to bisect an angle and check its measure.

To construct two parallel lines

1. select a line and a point not on the line,
2. pull down the **Construct** menu and choose **Parallel Line.**

To construct two perpendicular lines

1. select a line and a point (the point can be either on or not on the line), and
2. pull down the **Construct** menu and choose **Perpendicular Line.**

When you drag any points on the screen, the lines should maintain their relative positions. The intersection point of two lines remains intact no matter how you drag the two lines.

To construct the intersection point of two lines

1. select two lines and
2. pull down the **Construct** menu and choose **Point at Intersection.**

To create an angle bisector and verify the angle measures

1. select an angle by selecting three points, *the second of which must be the vertex*, and
2. pull down the **Construct** menu and choose **Angle Bisector;**
3. then choose **Angle** under the **Measure** menu and
4. move to a point on the original angle (the two smaller angle measures will always sum to the measure of the original angle).

Try These

1. Create two perpendicular lines and verify that each of the four angles created measures 90°.
2. Create a parallelogram and measure opposite side lengths and opposite angle measures. How do those measures change as you drag points to change the shape of the parallelogram?
3. Create a right triangle and verify the Pythagorean theorem.

Section B.3 PROPERTIES OF GEOMETRIC OBJECTS

An important feature of the *Sketchpad* and other GES is the ability to create geometric figures with specific characteristics and explore what happens to those objects when components of the figure are changed. The key techniques presented here illustrate that feature. To construct the perpendicular bisectors of the sides of a triangle

1. create a triangle,
2. select each side of the triangle,
3. pull down the **Construct** menu and choose **Point at Midpoint,**
4. check to see that the **Toolbox** line tool is set to create lines,
5. select a side of the triangle and the midpoint on that side,
6. pull down the **Construct** menu and choose **Perpendicular Line,** and
7. repeat steps (5) and (6) for the other two sides of the triangle.

The following images illustrate the steps of the construction.

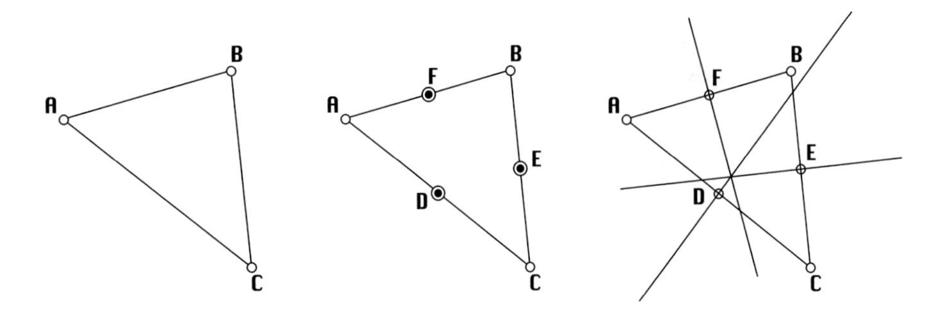

To explore the properties of these bisectors drag one or more of the vertices of the triangle to distort the original figure. What do you observe about the perpendicular bisectors? The following images illustrate the result.

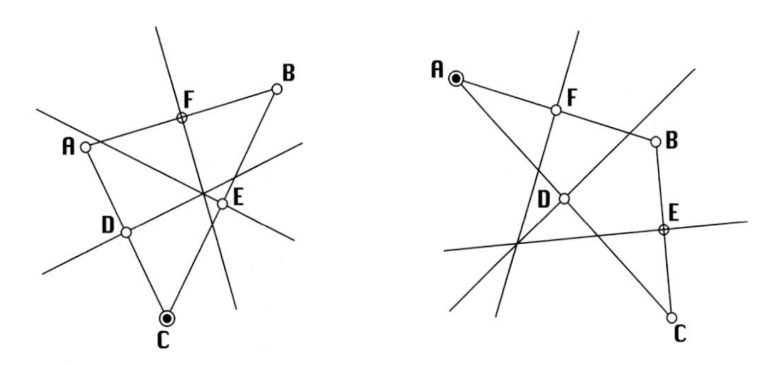

- What determines where the perpendicular bisectors intersect?
- Can you distort the figure so that each perpendicular bisector intersects a vertex of the triangle?
- What type of triangle is created?

Try These

1. Create the angle bisectors of the three interior angles of a triangle. Distort the triangle to explore what happens to the angle bisectors.
2. Create a square and measure its side lengths and angles. Test your construction by dragging the vertices and sides of the square. Does the figure remain a square?

Section B.4 TRANSFORMATIONS

The *Sketchpad* also helps you create transformations and investigate their properties. The key techniques in this section generate reflections, rotations, and translations.

To reflect an object

1. construct the object to be reflected;
2. construct a straight object for the mirror of reflection;
3. mark the reflection mirror, select it, and choose **Mark Mirror** in the **Transform** menu;
4. select the object to be reflected; and
5. choose **Reflect** under the **Transform** menu.

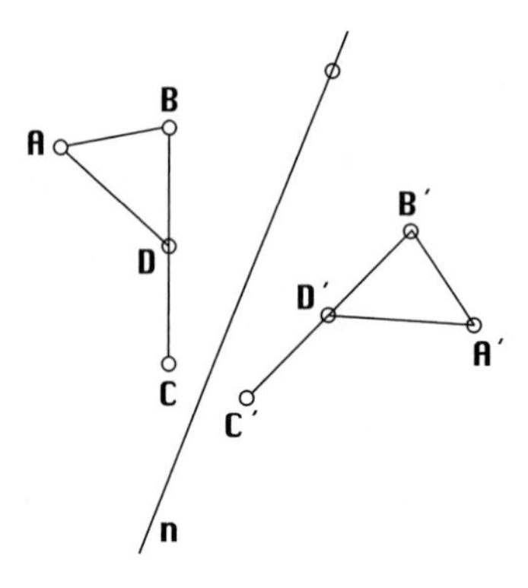

Figure **ABCD** was reflected across line **n**. *Sketchpad* automatically labels the reflected objects to correspond to the original image.

To rotate an object

1. construct the object to be rotated;
2. construct a point for the center of rotation;
3. mark the center of rotation, select it, and choose **Mark Center** in the **Transform** menu;
4. select the object to be rotated;
5. choose **Rotate…** under the **Transform** menu;
6. enter the degrees of rotation in the dialogue box that appears; and
7. click on **Okay.**

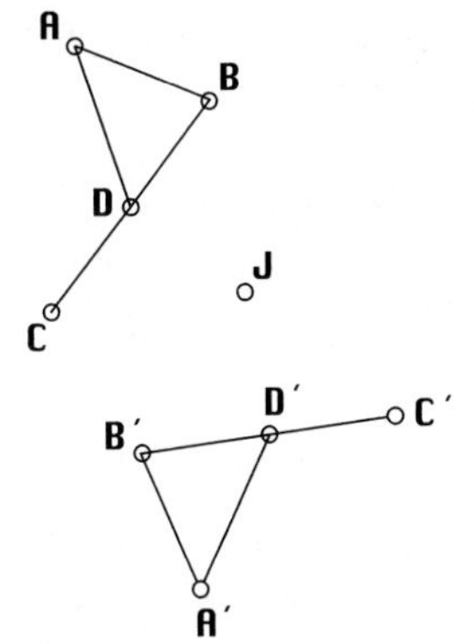

Figure **ABCD** was rotated counterclockwise 135° about point **J**.

To translate an object

1. construct and select the object to be translated,
2. choose **Translate...** under the **Transform** menu,
3. choose **By Rectangular Vector** at the bottom of the dialogue box that appears,
4. enter the **horizontal** and **vertical** translation distances in the appropriate boxes, and
5. click on **Okay.**

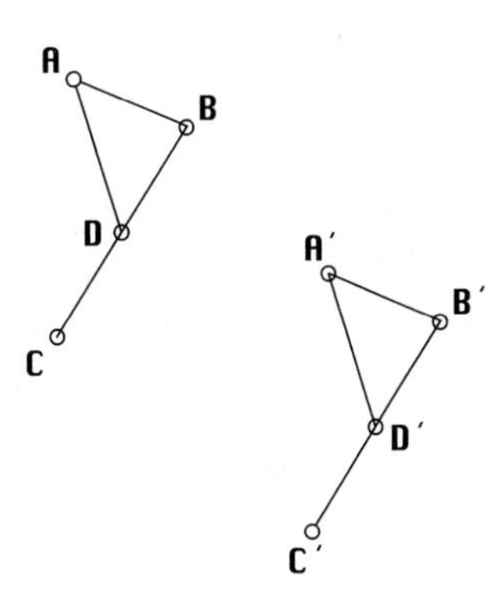

Figure **ABCD** was translated 2.2 centimeters horizontally and -1.6 centimeters vertically.

Try These

1. Practice using the **Reflect, Rotate,** and **Translate** constructions.
2. Create triangle ABC and reflect it across line m. Now create a segment with endpoints A and A' and label point P as the intersection of $\overline{AA'}$ with line m. What relationship seems to hold between $\overline{AA'}$ and line m? Distort triangle ABC to check your conjecture.

Section B.5 MORE EXPLORATIONS

The key techniques in this section reinforce and extend the use of GES as a tool to investigate relationships and propose conjectures. First, construct a quadrilateral and then create a new figure by using the midpoints of the first. After completing the construction, distort the quadrilateral to investigate the new figure. In the exercises, two more explorations are proposed.

To construct a new figure within a quadrilateral

1. create a quadrilateral,
2. select each side of the quadrilateral,
3. pull down the **Construct** menu and choose **Point at Midpoint** to construct the midpoint of each side of the quadrilateral,
4. check to see that the **Toolbox** line tool is set to create segments,
5. select all four midpoints you've constructed, and
6. pull down the **Construct** menu and choose **Segment.**

The following images illustrate the steps of the construction.

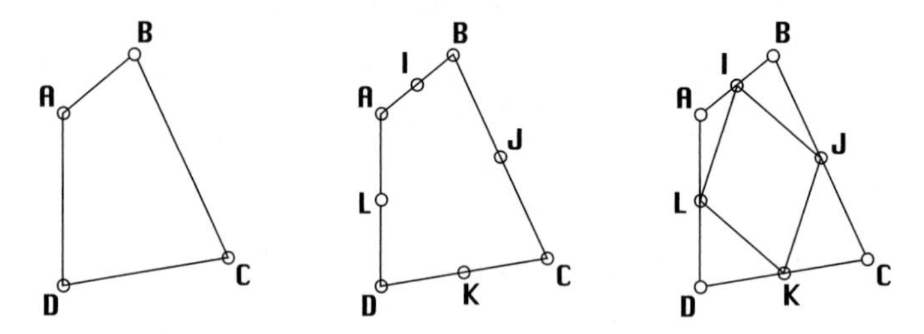

Now drag one or more of the vertices of the original quadrilateral to distort the figure. What do you observe about the new quadrilateral within the original? The following images illustrate the result.

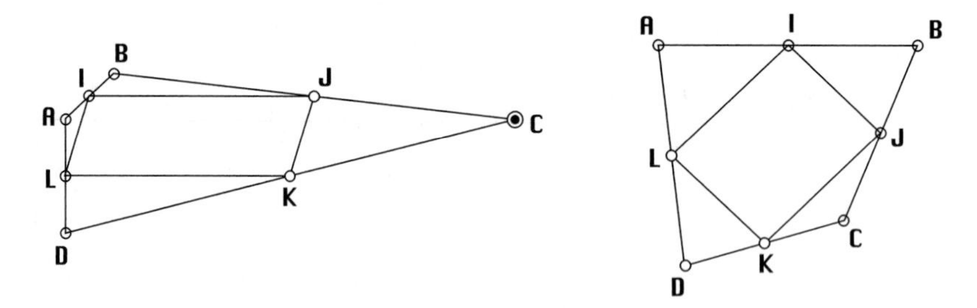

Which type of quadrilateral seems to be created within the original figure? Make some measurements to help you check your conjecture.

Try These

1. Create a circle and a diameter of the circle. Measure the *circumference* of the circle and the *length* of the diameter. Use **Calculate...** under the **Measure** menu to compare the circumference to the length of the diameter. Distort the circle and watch your calculation. What is the result?

2. Create a triangle and two intersecting lines away from the triangle. Reflect the triangle across one line and then reflect the new image across the second line. Distort the original triangle and the intersecting lines and observe the results. Do some measurements and make some conjectures about the relationships among the three triangles.

Section B.6 COORDINATE CONNECTIONS

The key techniques in this section introduce the coordinate geometry features of *The Geometer's Sketchpad.* First, plot two points by using ordered pairs and then draw a line through the points. Then determine the equation of the line. Finally, in the exercises, distort the coordinate geometry images and investigate the resulting change.

To plot points on a coordinate grid

1. choose **Plot Points…** under the **Graph** menu,
2. enter the (x, y) coordinates for one or more points, and
3. click on **Add,** then click on **Plot.**

The following image shows point **C** at $(2, 1)$ and point **D** at $(3, 5)$.

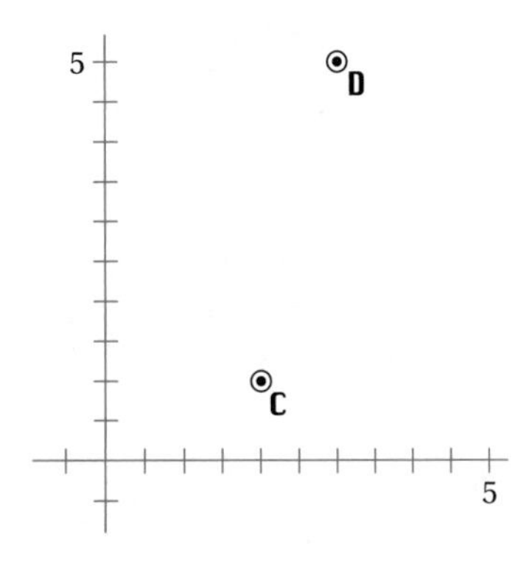

To determine the equation of a line on a coordinate grid

1. select a line and
2. pull down the **Measure** menu and choose **Equation.**

The equation will appear on the screen. You can specify whether the equation appears in slope-intercept form $(y = mx + b)$ or in standard form $(ax + by = c)$.

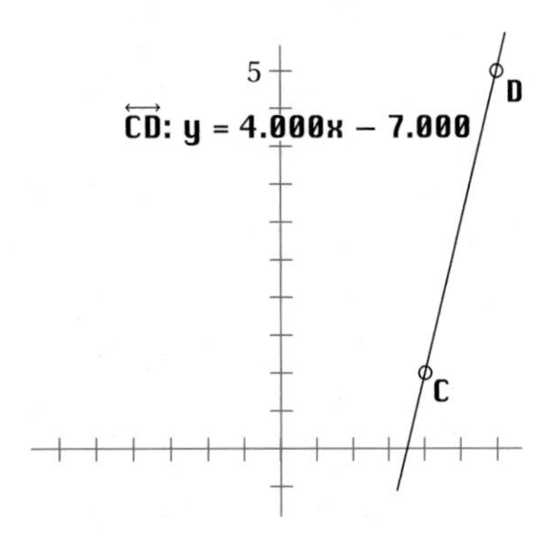

To change the form of a displayed equation

1. pull down the **Graph** menu and choose **Equation Form** and
2. under Lines choose **Slope/Intercept** or **Standard.**

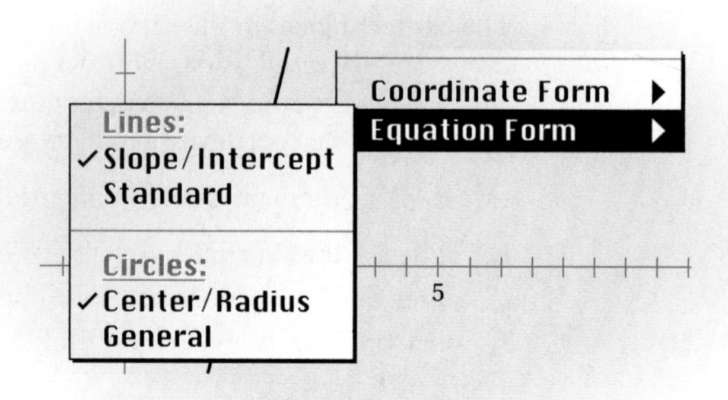

Try These

1. Plot one point on a coordinate grid and then create a circle with its center at that point. Select the circle and choose **Equation** under the **Measure** menu. Now change the size of the circle. What happened to the equation?
2. Plot two points on a coordinate grid, taking care to choose **Free Points** when you enter the points to be plotted. After constructing a line containing the two points, select it and determine its equation in slope-intercept form. Now select one of the two points plotted on the line and move it. What happens to the equation? What form does the equation take if the line is horizontal? Vertical?

Section B.7 TABULATE AND TRACE

The key techniques in this section highlight two advanced features of the *Sketchpad.* The **Tabulate** option records measurements and allows you to examine them as you distort a figure. The **Trace** feature plots a trail for one or more selected objects as you distort a figure.

To use the **Tabulate** option

1. make one or more measurements on an existing construction,
2. highlight one or more measurements to be recorded,
3. under the **Measure** menu, choose **Tabulate** (a table appears on your screen),
4. distort the construction,
5. select the table of values,
6. pull down the **Measure** menu, choose **Add Entry,** and
7. repeat steps (4) through (6) to continue adding new measures to your table.

The following example shows a table with a circle's circumference, its diameter length, and the ratio of those two values. Four entries in the table show that the ratio remains constant.

Circumference $\odot$AB = 3.826 cm
m $\overline{CB}$ = 1.218 cm

$$\frac{\text{Circumference } \odot AB}{m\ \overline{CB}} = 3.142$$

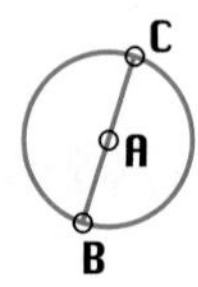

Circumference (Circle AB)	10.676	5.541	24.006	3.826
Length (Segment CB)	3.398	1.764	7.641	1.218
Circumference (Circle AB)/Leng...	3.142	3.142	3.142	3.142

To use the **Trace** feature

1. select one or more objects to be traced on an existing construction,
2. choose **Trace Objects** (or **Trace Point, Trace Segment,** etc.) under the **Display** menu,
3. distort the construction (the selected objects are traced as you distort the figure), and
4. click once on the screen to remove the trace.

In the figure shown point **G** is equidistant from points **C** and **D,** with $\overline{\textbf{DG}}$ perpendicular to $\overline{\textbf{AB}}$. Point **G** is traced as point **D** is moved along $\overline{\textbf{AB}}$.

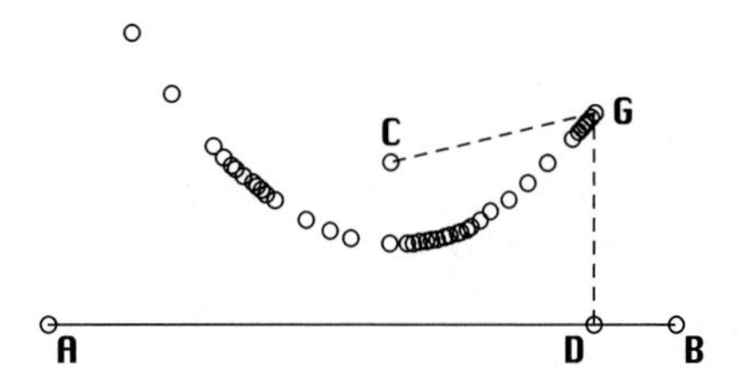

Try These

1. Create a triangle and a segment joining the midpoints of two sides of the triangle. Measure the length of this midsegment and the length of the side parallel to the midsegment. Use **Tabulate** to record these measures and their ratio. What do you observe as you distort the figure?
2. Construct a *Sketchpad* drawing that uses the **Trace** feature to create an ellipse.

Appendix C

Glossary of Geometric Terms[†]

acute angle An angle with measure less than 90°.

acute triangle A triangle with three acute angles.

alternate exterior angles Two exterior angles with different vertices on opposite sides of a transversal.

alternate interior angles Two interior angles with different vertices on opposite sides of a transversal.

altitude of a triangle A segment from a vertex to a point on the opposite side (perhaps extended) that is perpendicular to that opposite side.

angle The union of two noncollinear rays that have the same endpoint.

axiom A statement that is accepted as true without proof.

bisector of a segment Any point, segment, ray, line, or plane that contains the midpoint of the segment.

bisector of an angle The bisector of $\angle ABC$ is a $\overrightarrow{BD}$ in the interior of $\angle ABC$ such that $\angle ABD \cong \angle DBC$.

centroid of a triangle The point of intersection of the medians.

chord A segment that joins two points on a circle.

circle The set of all points in a plane that are a fixed distance from a given point in the plane.

circumcenter of a triangle The point equidistant from the vertices of a triangle.

circumference of a circle The distance around a circle, represented by the number approached by the perimeters of the inscribed regular polygons as the number of sides of the regular polygons increases.

circumscribed circle A circle that contains the three vertices of a triangle. The center of the circle is the point of intersection of the perpendicular bisectors of the sides of the triangle.

collinear points Points that lie on the same line.

complementary angles Two angles whose measures have a sum of 90°.

concurrent lines Three or more coplanar lines that have a point in common.

cone A solid figure with a vertex and a circular base.

congruent angles Angles that have the same measure.

congruent segments Segments that have the same length.

congruent triangles Triangles that have a correspondence between the vertices such that each pair of corresponding sides and angles is congruent.

conjecture A generalization that is hypothesized to be true.

convex polygon A polygon is convex if all of the diagonals of the polygon are in the interior of the polygon.

coplanar points Points that all lie in one plane.

corresponding angles Two angles on the same side of a transversal. One of the angles is an exterior angle; one is an interior angle.

counterexample A single example that shows a generalization to be false.

cross section of a solid A region common to the solid and a plane that intersects the solid.

cube A regular polyhedron with six square faces.

cylinder A solid figure having congruent circular bases in a pair of parallel planes.

deductive reasoning Starting with a hypothesis and using logic and definitions, postulates, or previously proven theorems to justify a series of statements or steps that lead to the desired conclusion.

degree measure The real number between 0 and 180 that is assigned to an angle.

diameter of a circle A chord that contains the center of a circle.

distance from a point to a line The length of the segment drawn from the point perpendicular to the line.

equiangular triangle A triangle with three congruent angles.

equilateral triangle A triangle with all sides congruent to one another.

Euler line of a triangle The line containing the circumcenter, centroid, and orthocenter of a triangle.

Euler's formula for a polyhedron The formula $V + F = E + 2$, where V is the number of vertices, F the number of faces, and E the number of edges.

[†]Adapted from O'Daffer, P.G., and Clemens, S.R. *Geometry: An Investigative Approach,* 2nd ed. Reading, Mass.: Addison-Wesley, 1992.

exterior angle of a triangle An angle that forms a linear pair with one of the angles of the triangle.

generalization A statement thought to be true, arrived at through inductive reasoning.

inductive reasoning Observing that an event gives the same result several times in succession, then concluding that the event will always have the same outcome.

intersecting lines Two lines with a point in common.

intersecting planes Planes that share at least a line in common.

isosceles triangle A triangle with at least two sides congruent to one another.

kite A quadrilateral with two distinct pairs of congruent adjacent sides.

line of symmetry A line in which a figure coincides with its reflection image.

line reflection A transformation that maps a figure into its reflection image about a line.

linear pair of angles A pair of angles with a common side such that the union of the other two sides is a line.

median of a triangle A segment joining a vertex to the midpoint of the opposite side.

midpoint of a segment The midpoint of $\overline{AB}$ is a point C between A and B such that $\overline{AC} \cong \overline{CB}$.

nonconvex quadrilateral A quadrilateral for which there exists a segment with endpoints on the quadrilateral, but with other points of the segment outside the quadrilateral.

obtuse angle An angle with measure greater than 90°.

obtuse triangle A triangle with an obtuse angle.

orthocenter of a triangle The point of intersection of the lines containing the altitudes.

parallel lines Lines in the same plane that do not intersect.

parallel planes Planes that do not intersect.

parallelogram A quadrilateral with both pairs of opposite sides parallel.

Pascal's triangle A triangular array of numbers in which 1's are written along the sides of an imagined isosceles triangle. Each other element filling up the triangle is the sum of the two elements directly above it to its left and its right.

perimeter of a polygon The sum of the lengths of the sides of the polygon.

perpendicular bisector of a segment A line that is perpendicular to the segment and contains its midpoint.

perpendicular lines Two lines that intersect to form congruent right angles.

perpendicular planes Planes that intersect in a line so that every line in one plane that is perpendicular to the line of intersection is also perpendicular to the other plane.

polygon A closed plane figure formed by the union of segments meeting only at endpoints such that (1) at most two segments meet at the point and (2) each segment meets exactly two other segments.

polyhedron A closed solid figure formed by a finite number of polygonal regions called faces. Each edge of a region is the edge of exactly one other region. If two regions intersect, then they intersect in an edge or vertex.

postulates A basic generalization accepted without proof.

prism A polyhedron such that (1) there is a pair of congruent faces that lie in parallel planes and (2) all other faces are parallelograms.

quadrilateral The union of four segments determined by four points, no three of which are collinear. The segments intersect only at their endpoints.

radius of a circle A segment whose endpoints are the center and a point on a circle.

ray A subset of a line. $\overrightarrow{AB}$ contains a given point A and all points on the same side of A as B.

rectangle A parallelogram with four right angles.

reflectional symmetry A figure $\mathcal{R}$ has reflectional symmetry if there is a line ℓ such that the reflection image over ℓ of each point P of $\mathcal{R}$ is also a point of $\mathcal{R}$. The line ℓ is called the line of symmetry.

regular polygon A polygon with all sides congruent to each other and all angles congruent to each other.

regular tessellation A tessellation made of regular, congruent, convex polygons such that each vertex figure is a regular polygon.

remote interior angles Two angles of a triangle with respect to an exterior angle that are not adjacent to the exterior angle.

rhombus A parallelogram with four congruent sides.

right angle An angle that measures 90 degrees.

right triangle A triangle with a right angle.

rotation A transformation involving a turn about a point O through an angle $x°$ that associates each point P of the plane with an image point P'. The measure of angle POP' is $x°$ and $OP = OP'$.

scalene triangle A triangle with no congruent sides.

segment $\overline{AB}$ is the set of points A and B and all the points between A and B.

semiregular tessellation A tessellation made of regular polygons of two or more types so that the arrangement of polygons at each vertex is the same.

skew lines Two nonintersecting lines that do not lie in the same plane.

square A rectangle with four congruent sides.

star polygon A polygon formed by joining every *d*th point of a set of *n* equally spaced points on a circle $(d < n)$.

supplementary angles Two angles whose measures have a sum of 180°.

tangram puzzle A famous puzzle consisting of seven pieces formed by dissecting a square in a special way.

tessellation or tiling A complete covering of the plane with polygons with no holes and no overlapping.

theorem A generalization that can be proved to be true using definitions, postulates, and the logic of deductive reasoning.

topological equivalence Two surfaces or curves are topologically equivalent if one can be transformed into the other by distorting, stretching, shrinking, or bending.

translation Given an arrow AA', the translation image of a point P for the arrow AA' is the point P', where $AA' = PP'$ and arrows AA' and PP' have the same direction.

transversal A line that intersects two coplanar lines in two different points.

trapezoid A quadrilateral with exactly one pair of parallel sides.

triangle The union of three segments determined by three noncollinear points.

undefined term A basic term at the simplest level that is not defined using other terms.

vertex of a polygon The endpoint of a side of a polygon.

vertex of an angle The common endpoint of the two noncollinear rays forming the angle.

vertical angles Two angles that are formed by two intersecting lines, but that are not a linear pair of angles.

volume A measure of the amount of space occupied by a solid. Each solid is assigned a unique positive number called its volume.

Appendix D

Constructions

INTRODUCTION

- D.1 Copying Figures
- D.2 Constructing Bisectors
- D.3 Constructing Perpendicular and Parallel Lines
- D.4 Constructing Special Polygons

In order to discover relationships in geometry, it is often helpful to draw accurate pictures. The Greek philosopher Plato used only the *compass* and *unmarked straight edge* as the tools for making these pictures (Figure D.1). To Plato, a *ruler-compass construction* was performed when "an accurate picture of a geometric idea was produced using only the unmarked ruler and compass according to agreements about the use of these tools." This restriction is historically interesting, and some very important ideas of mathematics have been discovered as people have attempted to make certain constructions using only these tools. We use these constructions occasionally in this text to discover or clarify geometric relationships.

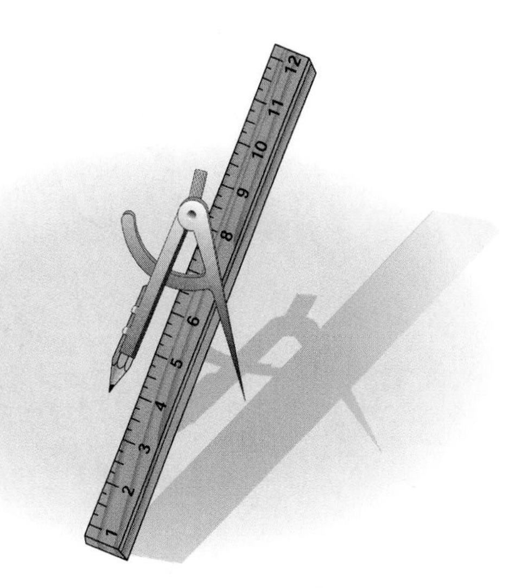

FIGURE D.1

Section D.1 | COPYING FIGURES

Copy a Circle

Using a compass (Figure D.2)

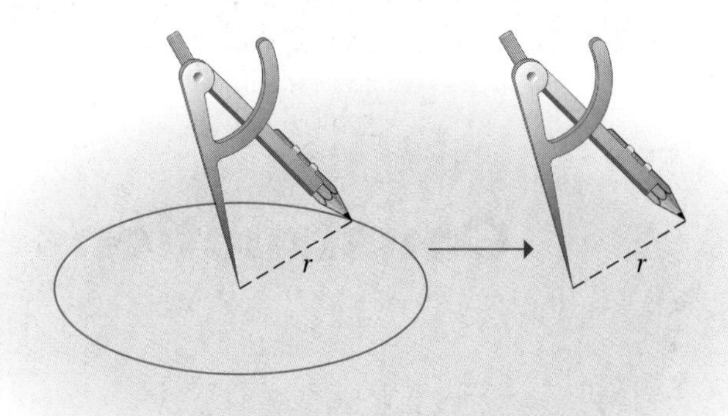

FIGURE D.2

1. measure the radius of the given circle with the compass, and
2. use this radius to draw a copy of the given circle.

Copy a Line Segment

Using a ruler and a compass (Figure D.3)

1. open your compass the length of the given segment (Figure D.3a),
2. draw a ray longer than the given segment (Figure D.3b), and
3. use the same compass to mark a copy of the segment on the ray (Figure D.3c).

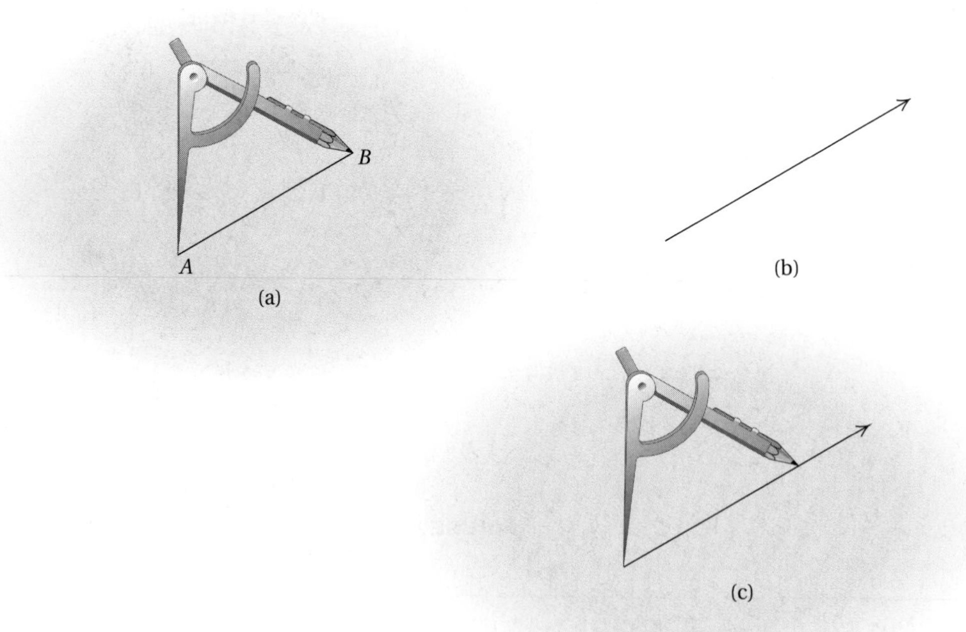

FIGURE D.3

Copy an Angle

Using a ruler and compass (Figure D.4)

1. draw an arc intersecting both rays of the given angle (Figure D.4a),
2. draw a ray to serve as one side of the copy (Figure D.4b),
3. with the same compass opening as in step 1, draw an arc crossing the ray (Figure D.4c),
4. open the compass to measure the opening of the given angle (Figure D.4d),
5. use the same opening as in step 4 and draw an arc (Figure D.4e), and
6. draw $\overrightarrow{AC}$ to complete the copy of the given angle (Figure D.4f).

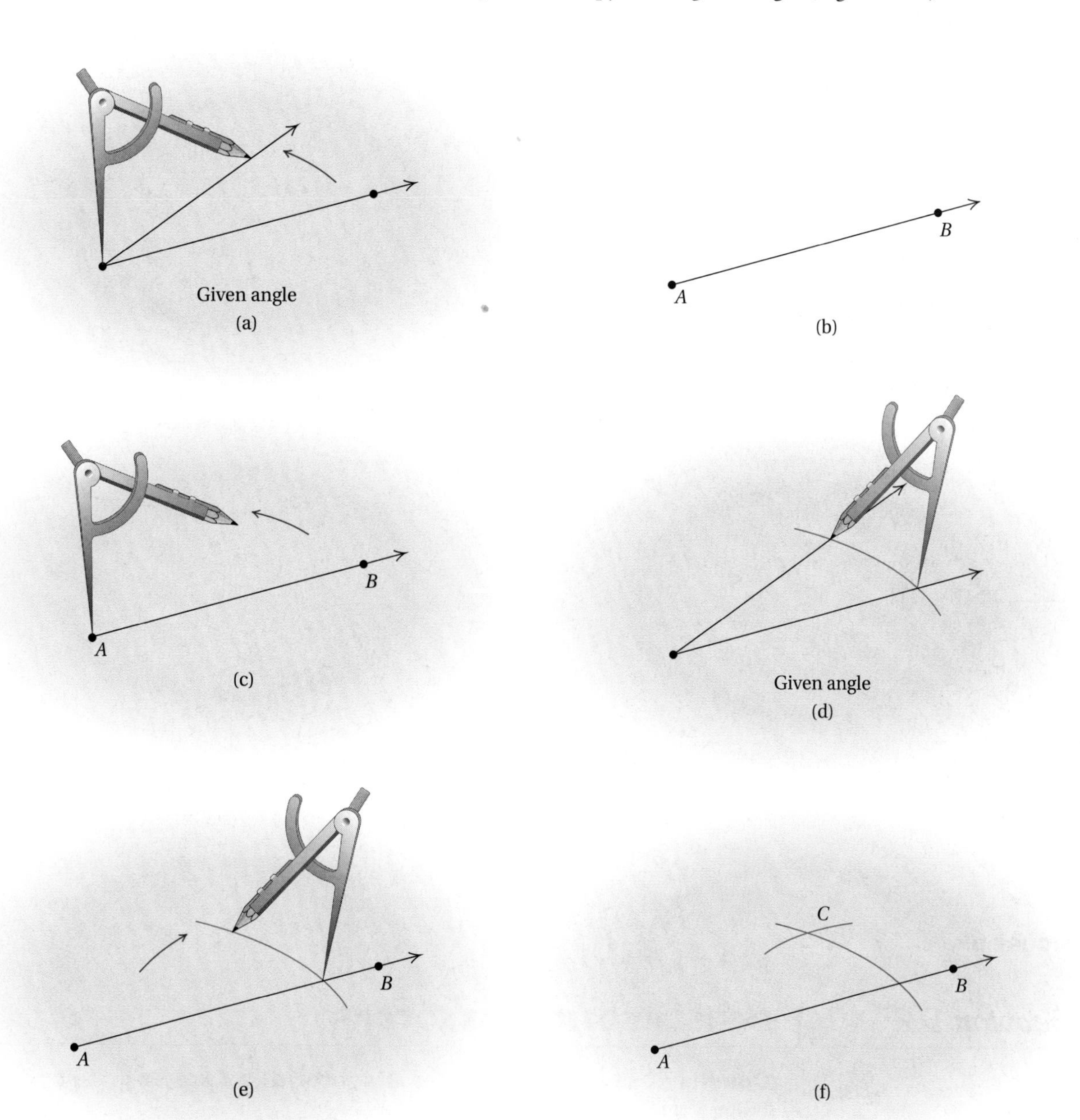

FIGURE D.4

Copy a Triangle

Using a ruler and compass (Figure D.5)

1. given a triangle *ABC* (Figure D.5a), draw a ray and copy $\overline{AB}$ of the triangle (Figure D.5b),
2. draw an arc with center A' and compass opening *AC* (Figure D.5c),
3. draw an arc with center B' and compass opening *BC* (Figure D.5d), and
4. use the ruler to complete the drawing of the triangle (Figure D.5e).

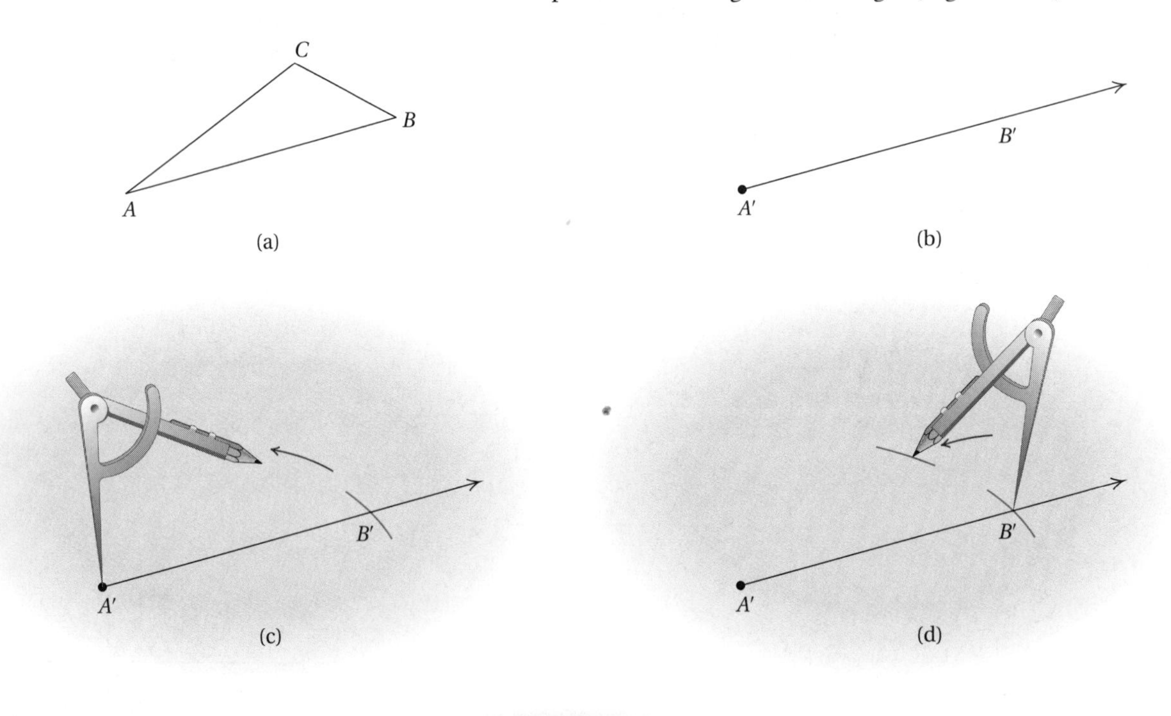

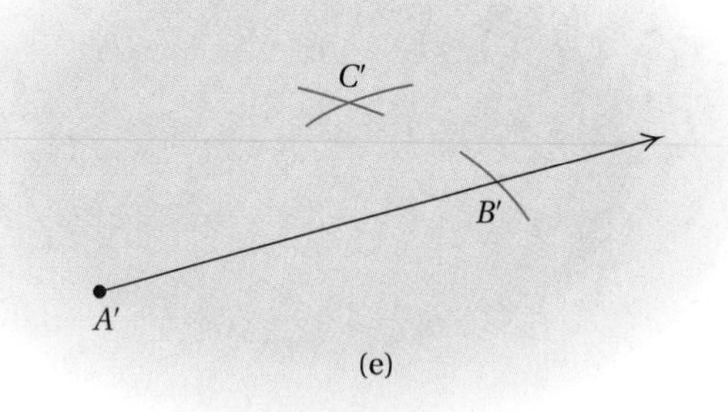

FIGURE D.5

Section D.2 CONSTRUCTING BISECTORS

Construct the Perpendicular Bisector of a Line Segment

Using a ruler and compass (Figure D.6)

1. given $\overline{AB}$ (Figure D.6a) and with *A* as center and a compass opening greater than half of *AB*, draw a semicircular arc (Figure D.6b),

2. with B as center and the same opening as in step 1, draw a semicircular arc that intersects the first arc (Figure D.6c), and
3. connect the two points of intersection to complete the construction of the perpendicular bisector of $\overline{AB}$ (Figure D.6d).

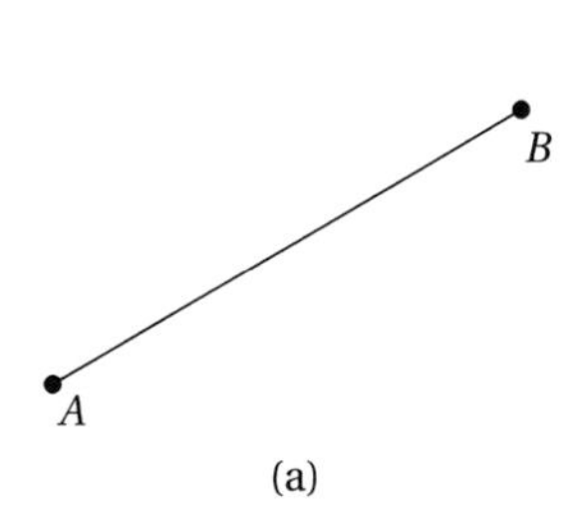

(a)

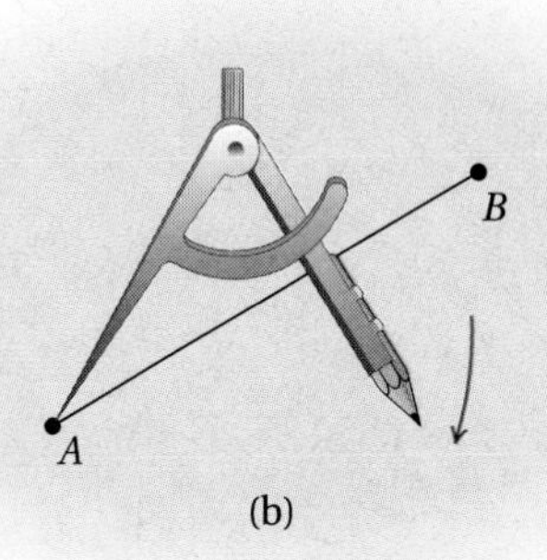

(b)

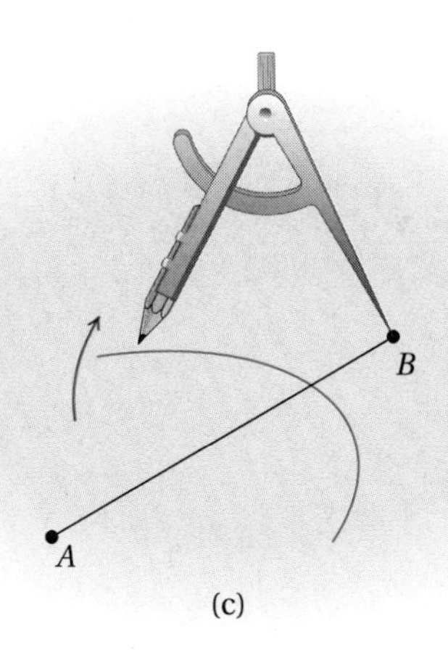

(c)

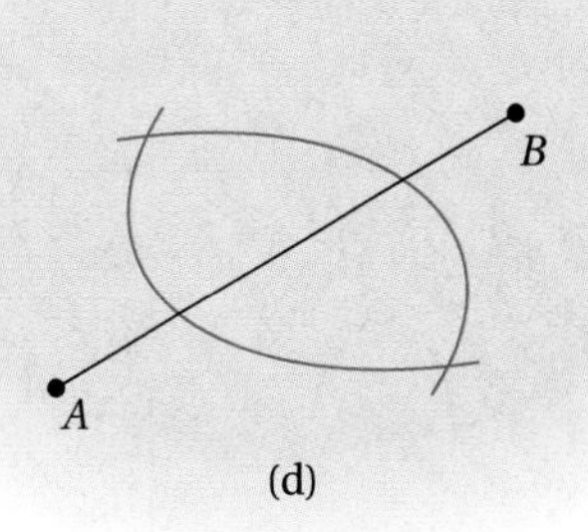

(d)

FIGURE D.6

Bisect an Angle

Using a ruler and compass (Figure D.7)

1. given $\angle ABC$ (Figure D.7a) and with B as center, draw an arc that intersects both sides of the angle, at F and G (Figure D.7b);
2. with F as center, draw an arc in the interior of the angle (Figure D.7c);
3. with G as center, and the same opening as in step 2, draw an arc that crosses the first arc (Figure D.7d); and
4. connect B and the point of arc intersection to produce the bisector of the angle (Figure D.7e).

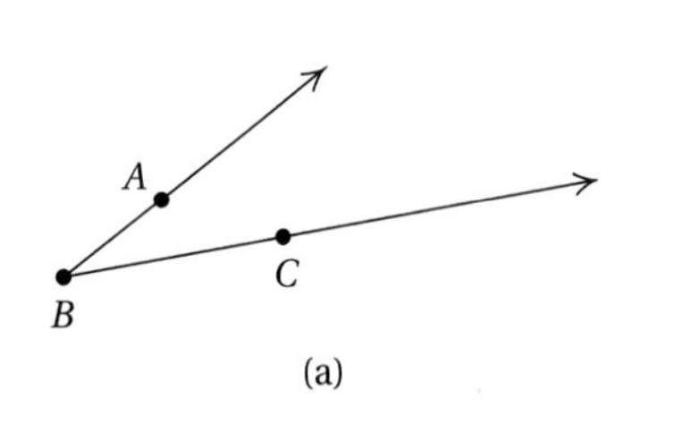

(a)

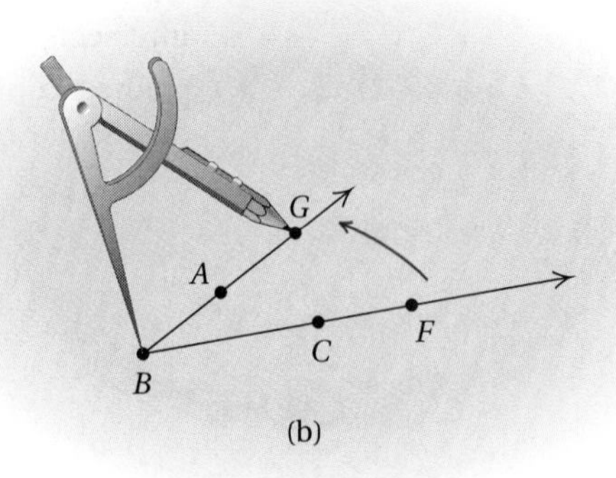

(b)

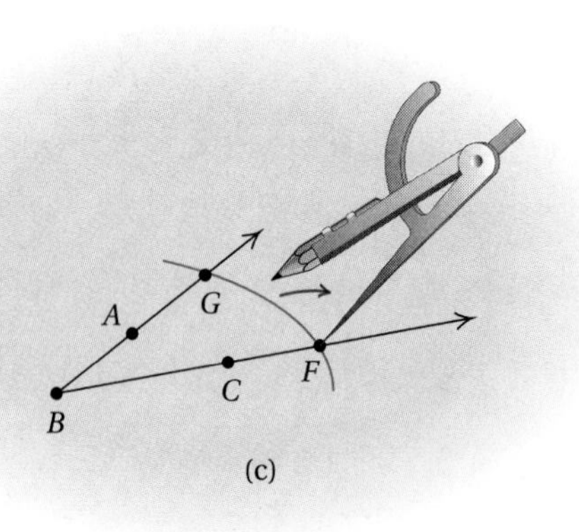

(c)

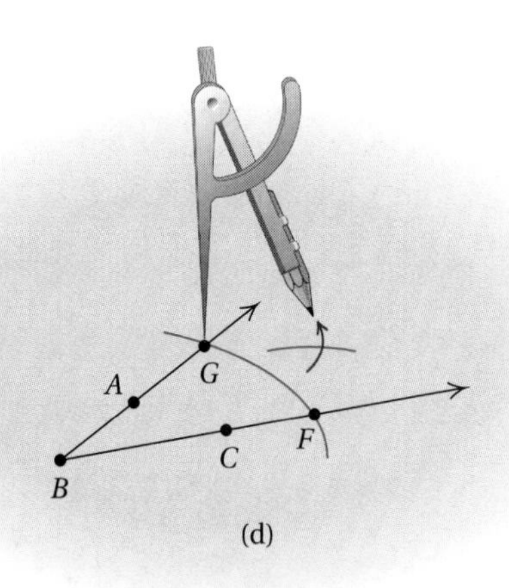

(d)

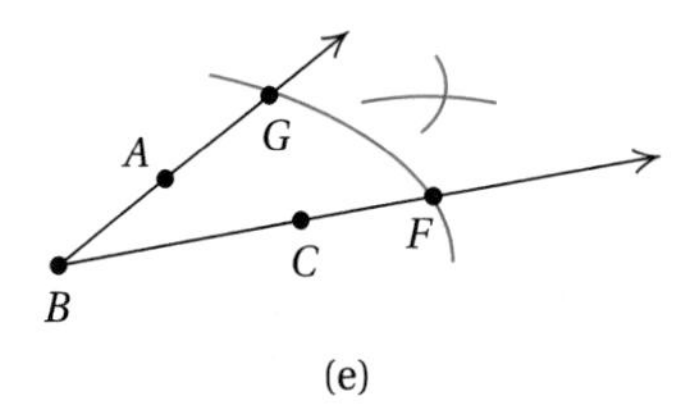

(e)

FIGURE D.7

Section D.3 CONSTRUCTING PERPENDICULAR AND PARALLEL LINES

Construct a Perpendicular to a Line Through a Given Point on the Line

Using a ruler and compass (Figure D.8)

1. given a line ℓ and a point P on ℓ (Figure D.8a), draw arcs on each side of P (Figure D.8b),
2. draw crossing arcs above line ℓ (Figure D.8c), and
3. draw $\overleftrightarrow{PQ}$, the perpendicular bisector of $\overline{AB}$ (Figure D.8d).

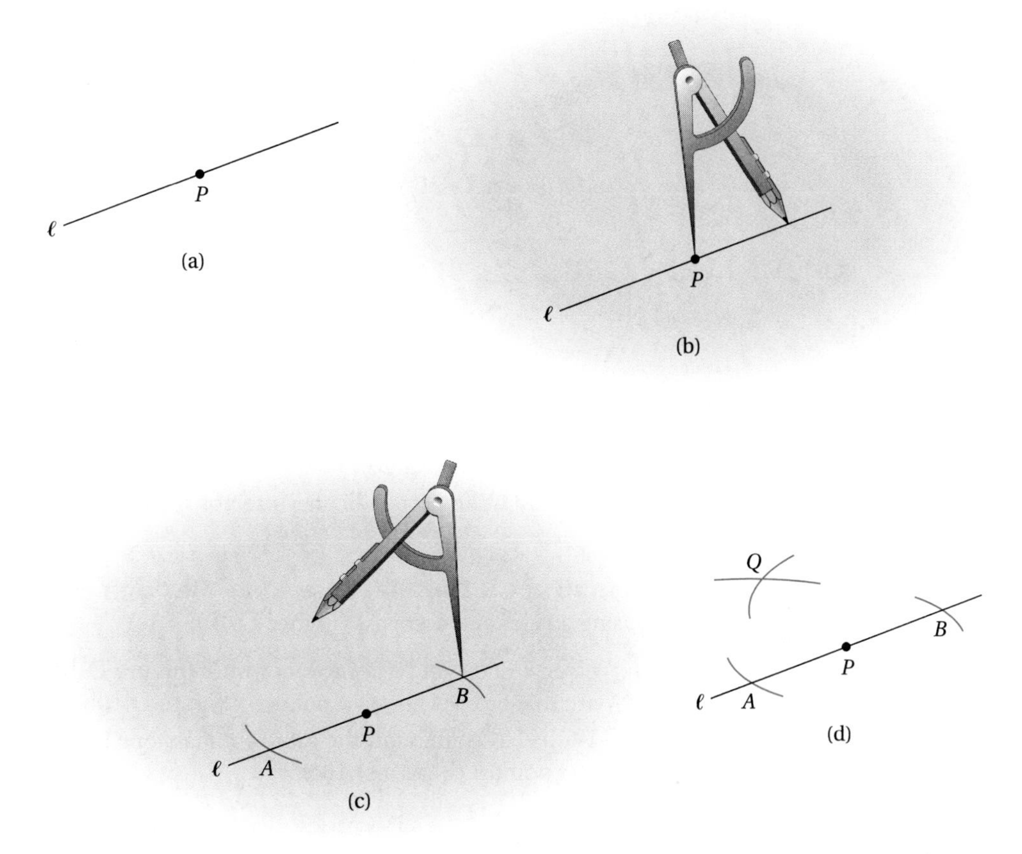

FIGURE D.8

Construct a Perpendicular to a Line Through a Given Point Not on the Line

Using a ruler and compass (Figure D.9)

1. given a line ℓ and a point P not on ℓ (Figure D.9a), draw two arcs cutting line ℓ (Figure D.9b),

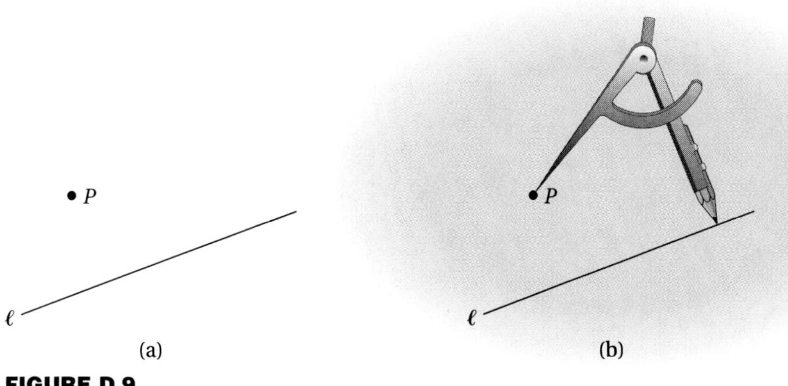

FIGURE D.9

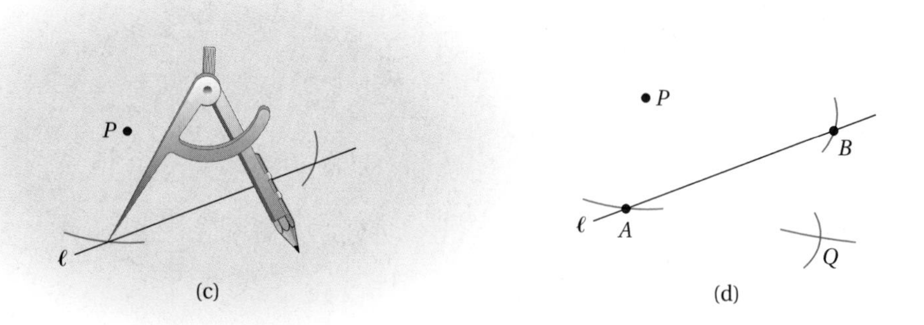

FIGURE D.9 *(continued)*

2. draw two crossing arcs below line ℓ (Figure D.9c), and
3. draw $\overleftrightarrow{PQ}$, the perpendicular bisector of $\overline{AB}$ (Figure D.9d).

Construct a Parallel to a Line Through a Point Not on the Line

Using a ruler and compass (Figure D.10)

1. given a line ℓ and a point P not on ℓ (Figure D.10a), with P as center, draw an arc that crosses line ℓ at point A (Figure D.10b);
2. With A as center and the same compass opening, draw an arc that crosses line ℓ at point B (Figure D.10c);

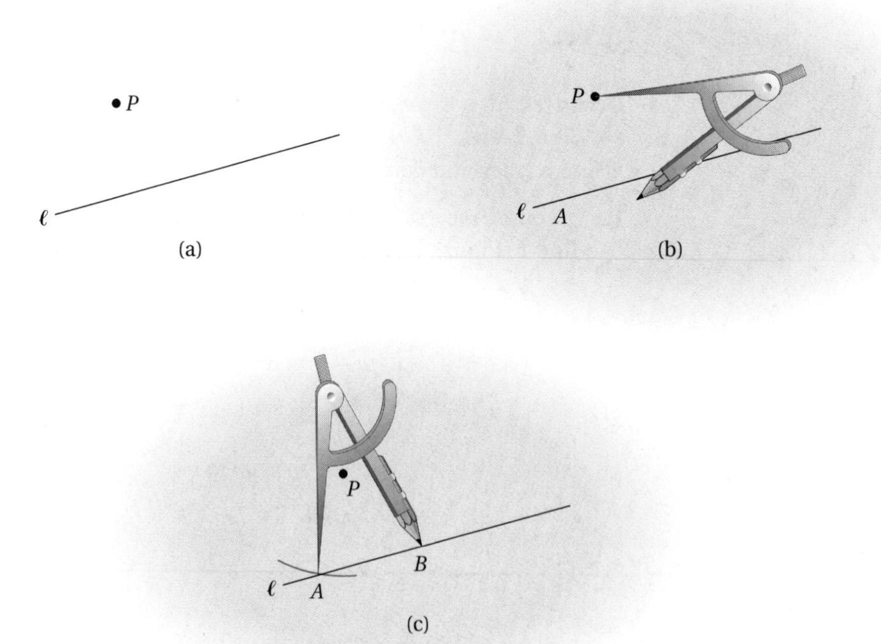

FIGURE D.10

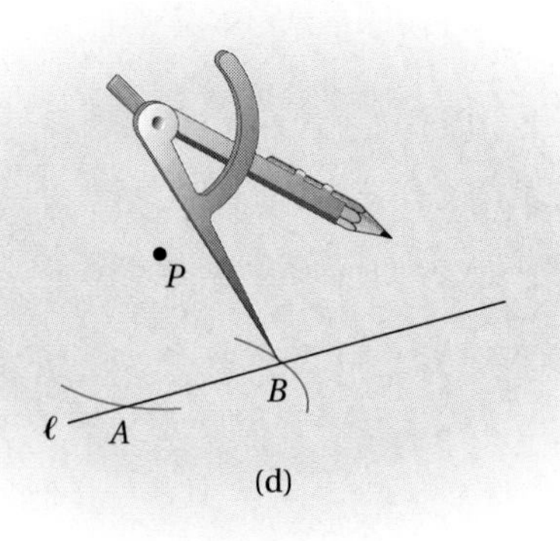

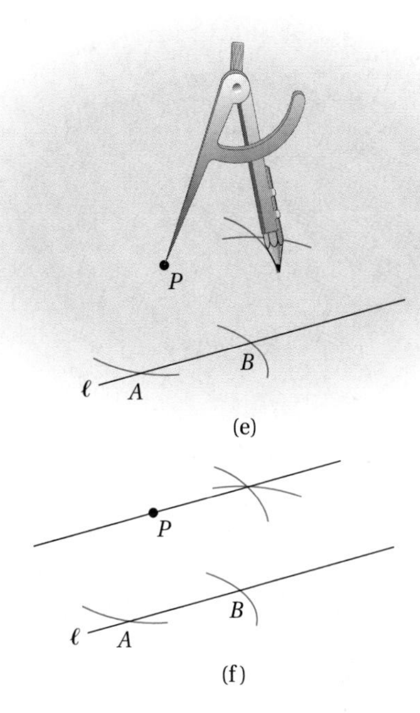

FIGURE D.10 *(continued)*

3. With B as center and the same compass opening, draw an arc above line ℓ (Figure D.10d);
4. with P as center and the same compass opening draw an arc crossing the arc you drew in step 3 (Figure D.10e); and
5. draw the line through P and the intersection of the arcs. This line is parallel to line ℓ (Figure D.10f).

SECTION D.4 | CONSTRUCTING SPECIAL POLYGONS

Construct an Equilateral Triangle, Given the Length of One Side

Using a ruler and compass (Figure D.11)

1. given side $\overline{AB}$ of an equilateral triangle (Figure D.11a), with A as center and compass opening AB, draw an arc centered above $\overline{AB}$ (Figure D.11b);

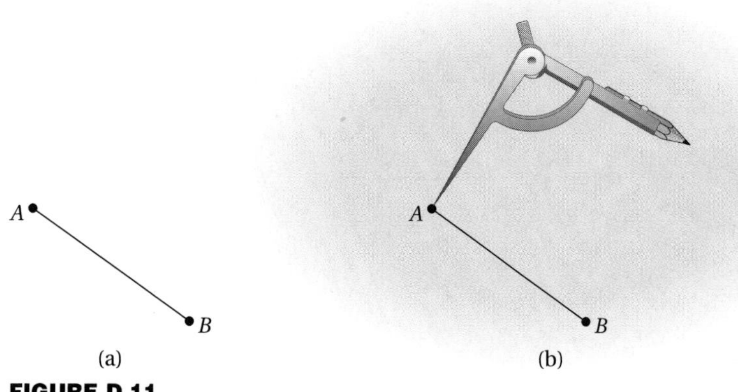

FIGURE D.11

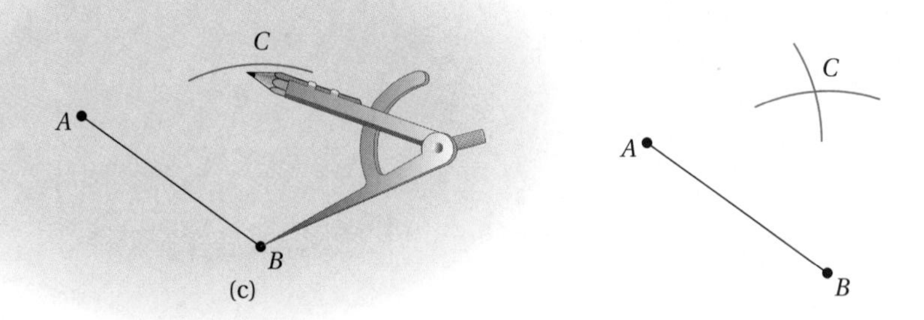

FIGURE D.11 *(continued)*

2. with B as center and compass opening BA, draw an arc that crosses the arc you drew in step 1 at point C (Figure D.11c); and
3. draw sides $\overline{AC}$ and $\overline{BC}$ to complete the construction of the equilateral triangle (Figure D.11d).

References

Amundson, H. E. Capsule 42: Percent. In *Historical Topics for the Mathematics Classroom* (thirty-first yearbook). Washington, D.C.: National Council of Teachers of Mathematics, 1969.

Boyer, C. B. *A History of Mathematics*, 2d ed. New York: John Wiley & Sons, 1991.

Escher, M. C. *The Graphic Work of M.C. Escher.* New York: Hawthorn, 1960.

Harvey, L. D. *Harvey's Practical Arithmetic: Book Two.* New York: American Book Company, 1909.

Hein, Piet. *Grooks II.* Garden City, N.Y.: Doubleday, 1969.

Henderson, K., and Pingry, R. Problem-solving in mathematics. In H. Fehr (ed.), *The Learning of Mathematics: Its Theory and Practice* (twenty-first yearbook). Washington, D.C.: National Council of Teachers of Mathematics, 1953.

Katz, V. J. *A History of Mathematics: An Introduction.* New York: HarperCollins, 1993.

Mikami, Y. *The Development of Mathematics in China and Japan*, 2d ed. New York: Chelsea, 1974.

O'Daffer, P., and Clemens, S. *Geometry—An Investigative Approach*, 2d ed. Reading, Mass.: Addison-Wesley, 1992.

Pólya, G. *How to Solve It.* Princeton, N.J.: Princeton University Press, 1945.

Professional Standards for School Mathematics. Reston, Va.: National Council of Teachers of Mathematics, 2000.

Ray, J. *New Elementary Algebra: Primary Elements of Algebra for Common Schools and Academies.* Cincinnati: Van Antwerp, Bragg, & Company, 1866.

Read, R. *Tangrams, 330 Puzzles.* New York: Dover, 1965.

Rees, N. (ed.). *Brewer's Quotations: A Phrase and Fable Dictionary.* London: Cassell, 1994.

Schmandt-Besserat, D. *Before Writing: From Counting to Cuneiform.* Austin: University of Texas Press, 1992.

Smith, D. E. *History of Mathematics*, vol. II. New York: Dover, 1953.

Smith, D. E., and Ginsburg, J. From numbers to numerals and from numerals to computation. In J. R. Newman (ed.), *The World of Mathematics*, vol. I. Redmond, Wash.: Tempus Books of Microsoft Press, 1988.

The Thirteen Books of Euclid's Elements (transl. from the text of Heiberg), with Introduction and Commentary by Sir Thomas L. Heath, vol. 2. New York: Dover, 1956.

Answers

This section contains answers to problems and exercises at the end of each section in the text and in the Chapter Review. It also contains answers to all of the exercises in the Appendices.

Chapter 1

Section 1.1

1. Answers will vary.

2. Answers will vary.

3. Answers will vary.

4. Answers will vary.

5. Answers will vary.

6. Answers will vary.

7. Answers will vary. It could be a three-cornered figure.

8. Answers will vary.

9. Answers will vary. Possible answers include π, $\div$, and $\sqrt{}$.

10. Point on the number line; shaded part of a rectangle; 6/10

11. Four groups of four; $20 - 4$; 2^4

12. Answers will vary.

13. Answers will vary.

14. one self-exam per month

15. fifty-three percent

16. Answers will vary.

17. **(a)** B2 = 0.10 * A2 **(b)** C2 = 0.90 * A2

18. fraction calculator

19. programmable scientific calculator

20. graphing calculator

21. Answers will vary.

22. Answers will vary.

23. Answers will vary.

24. Answers will vary.

25. The segment joining the midpoints of two sides of a triangle has a length equal to one-half the length of the remaining side; descriptions may vary.

26. **(a)** A3 = 1, B3 = 0.8 * B2 = 16,000; A4 = 2, B4 = 0.8 * B3 = 12,800; A5 = 3, B5 = 10,240; A6 = 4, B6 = 8,192; A7 = 5, B7 = 6,553.60.
(b) 7 rows

27. Answers will vary.

28. 11 or more hours

29. Estimates will vary.

30. Answers will vary.

31. Answers will vary.

32. 30 hens and 20 rabbits

33. Angles inscribed in semicircles are right angles.

Section 1.2

1. Any number plus zero is equal to that number.

2. Answers will vary.

3. Answers will vary.

4. Answers will vary.

5. Answers will vary.

6. **(a)** arithmetic sequence **(b)** geometric sequence

7. 16, 19, 22: arithmetic ($d = 3$)

8. $\frac{1}{2}, \frac{1}{4}, \frac{1}{8}$: geometric ($r = \frac{1}{2}$)

9. 36, 49, 64: neither

10. 26, 42, 68: neither

11. 63, 127, 255: neither

12. 81, 243, 729: geometric ($r = 3$)

13. 11, 16, 22: neither

14. 17, 21, 25: arithmetic ($d = 4$)

15. Answers will vary. One possibility: 121 is 11 squared and it ends in 1.

16. Answers will vary.

17. Hypothesis: an odd number is added to an even number. Conclusion: the sum is an odd number.

18. Hypothesis: a figure has four connected sides. Conclusion: it is a quadrilateral.

19. Hypothesis: it rains on Tuesday. Conclusion: it will be nice on Wednesday.

20. Hypothesis: the square of a number is even. Conclusion: the number is even.

21. inductive reasoning

22. Jack didn't tell the truth in statement (b).

23.

p	q	$p \rightarrow q$
T	T	T
T	F	F
F	T	T
F	F	T

The statement will be false when p is true and q is false.

24. The statement would be true under any of the following three conditions: inflation remains constant and salaries increase; inflation does not remain constant and salaries increase; inflation does not remain constant and salaries do not increase.

25. denying the conclusion

26. affirming the hypothesis
27. denying the conclusion
28. affirming the hypothesis
29. denying the conclusion
30. proportional
31. inductive
32. invalid deductive reasoning
33. inductive
34. **(a)** 25′ **(b)** 1.5″ **(c)** proportional
35. valid
36. invalid; assuming the converse
37. invalid; assuming the inverse
38. invalid; assuming the converse
39. only c
40. 1333332
41. Answers will vary.
42. She used inductive reasoning; her generalization is false because 14 is a counterexample.
43. Answers will vary.
44. Answers will vary.
45. **(a)** Conclusion: *ABCD* is a rectangle; affirmed the hypothesis. **(b)** Conclusion: *ABC* is a polygon; affirmed the hypothesis. **(c)** Conclusion: 15 is an odd number; affirmed the hypothesis.
46. 30 more blocks
47. 9 edges
48. Statement is reasonable in light of the given statements; true.
49. Statement is reasonable in light of the given statements; false ($5^2 - 1 = 24$ is a counterexample).
50. valid application of Rule B; denying the conclusion
51. invalid reasoning
52. valid application of Rule B; denying the conclusion
53. valid application of Rule A; affirming the hypothesis
54. invalid reasoning
55. Kevin is correct.
56. Answers will vary. If Mick is a dog, then Mick can fly.
57. Answers will vary. If you work for Sleezy, you will make a fortune.
58. **(a)** 36 **(b)** n^2
59. Frieda didn't convey the ad correctly; her first statement assumes the converse, and her second statement assumes the inverse.
60. Answers will vary.
61. 125 pounds
62. Answers will vary.
63. Answers will vary.
64. Answers will vary.
65. 3, 5, 11, 9, 6, 12, 15 (left to right; top to bottom)

Section 1.3

1. Answers will vary.
2. Answers will vary.
3. Six boxes on the bottom row.
4. Answers will vary.
5. One solution (two quarters, two dimes, one nickel)
6. understand the problem, develop a plan, implement the plan, look back over your work
7. Answers will vary. One possibility is to make a table.
8. Answers will vary. One possibility is to draw a diagram.
9. Answers will vary. One possibility is to make a table.
10. Answers will vary. One possibility is using a list-making approach.
11. Answers will vary. One possibility is to use reasoning to match and eliminate possible matches.
12. Answers will vary. One possibility is to represent the situation with an equation and solve the equation.
13. Answers will vary.
14. Answers will vary.
15. Answers will vary. Room is 8 ft by 12 ft.
16. The two numbers are 52 and 25.
17. 20 different crews
18. Answers will vary. **(a)** \$1.19 **(b)** 69 cents **(c)** 24 cents **(d)** 9 cents **(e)** 4 cents
19. Answers will vary. The minute hand passes the hour hand 9 times.
20. Answers will vary. One strategy is to *draw a diagram.*
21. Row 1 = 1 2 3 and row 2 = 5 6 4 is a solution to the problem.
22. Answers will vary. Probably *guess–check–revise* is the most popular strategy.
23. Answers will vary.
24. Answers will vary.
25. Problem 7: Phil is 19, Bill is 20, and Jill is 21.
Problem 8: Bella must pass 14 persons to be in third place.
Problem 9: 820 members
Problem 10: 24 possible arrangements
Problem 11: Beth coaches swimming; Deidre coaches soccer; Anton coaches volleyball; Cal coaches basketball.
Problem 12: \$27
26. Answers will vary.
27. Answers will vary.
28. 32 singles, 3 sets of twins, 5 sets of triplets. Strategies will vary.
29. Answers will vary.

30. Answers will vary.

31. Answers will vary.

32. Sixth day

33. **(a)** 10 **(b)** Answers will vary.

34. Answers will vary.

Section 1.4

1. They all represent the same number.

2. Answers will vary.

3. Answers will vary. Different uses of the number one-half suggest that different representations be used.

4. 24 is the answer to both problems.

5. Both areas can be found by multiplying the distance from the center of the figure to its edge, or side, by the distance around the figure by $\frac{1}{2}$.

6. Answers will vary. The relations here are coincidental.

7. Answers will vary. One answer is a circle inside a square so that the center of the circle is the center of the square.

8. Answers will vary. One answer is that the cities could be represented by the endpoints of a diameter of a circle. The two semicircles and the diameter represent the highways.

9. A square or rectangle could be used with the vertices representing the tennis players.

10. Answers will vary.

11. Answers will vary. A balanced view of mathematics involves doing computations and solving problems that involve concepts and relationships.

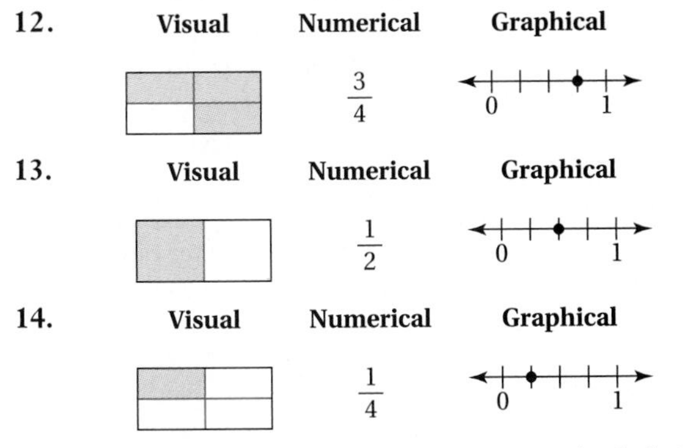

15. Answers will vary. One possibility is a circular hole in a cubical piece of wood.

16. Answers will vary. Place value is perhaps the most important because understanding of numeration is required for the other processes.

17. Answers will vary.

18. Answers will vary.

19. Answers will vary.

20. Answers will vary.

21. 2 inches

22. Answers will vary. Two possibilities are: *ABCDE* and *AEBCD*.

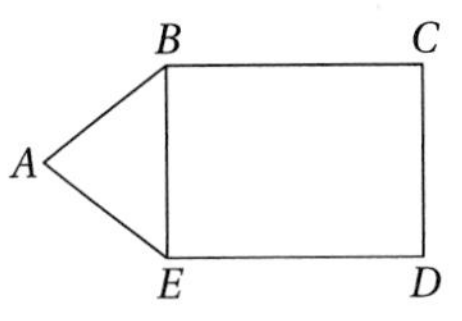

23. Answers will vary.

24. **i.** 0.285714 **ii.** 0.428571 **iii.** 0.571428 **iv.** 0.714286 **v.** 0.857143

25. 1376

Chapter 1 Review Exercises

1. Answers will vary.

2. C2 = A2 × B2 and C3 = A3 × B3.

3. **(a)** 0.2 (10 ÷ 0.2 = 50, which is larger than 10) **(b)** If the length and the width aren't the same length, the figure won't be a square. **(c)** 6 (or any even number)

4. **(a)** Hypothesis: Bulls score 100 points. Conclusion: They will win. **(b)** Hypothesis: He accepts the role in the proposed movie. Conclusion: Tom Hanks will win the Oscar.

5. Angle *C* is a right angle; affirming the hypothesis.

6. Alice didn't heat her coffee for 100 seconds; denying the conclusion.

7. Responses will vary. Some examples are: **(a)** My aunt lives in Austin, therefore she lives in Texas. **(b)** I don't know anyone who lives in Texas, thus I know no Austinites. **(c)** Since Joe lives in Texas, Joe lives in Austin. **(d)** Since I do not live in Austin, I don't live in Texas.

8. **(a)** The statement would be false if a person lived in Texas but did not live in Austin. **(b)** The statement would be false if I spent my money for concert tickets but I went to see the game.

9. proportional reasoning

10. **(a)** Yes, the pattern will fold to make an open-top box. **(b)** Spatial reasoning

11. **i.** Arithmetic ($d = 11$); 110, 121, 132 **ii.** Neither; 45, 55, 56 **iii.** Geometric ($r = 3$); 4374; 13,122; 39,366

12. **(a)** 7 ÷ 0.5 = 14 **(b)** 5 × 64 = 320

13. **(a)** first column is an arithmetic sequence with $d = 1$; second column is an arithmetic sequence with $d = 2$; the next three rows are: 7 13; 8 15; 9 17 **(b)** 25 49

14. Answers will vary.

15. Answers will vary. The ball cost $34 and the shoes cost $49.

16. Answers will vary.

17. These two problems are essentially the same because they can be represented by a common geometric model. The vertices in the model can represent either the people or the desks; the segments can represent the cables or a pairing of riders.

Figure for Exercise 17

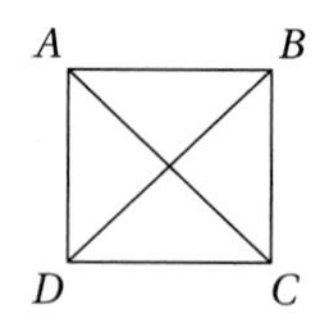

18. Answers will vary. Possibilities include $\frac{1}{2}$ or

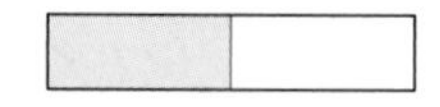

19. Answers will vary.
20. Answers will vary.
21. Answers will vary.
22. **(a) i.** 24, 22 **ii.** 22, 26 **iii.** 14, 30
(b) The areas decrease. **(c)** The perimeters increase.
(d) Answers will vary. One possibility: as the areas decrease, the perimeters increase.

23. **(a)** The common aspect of the figures is that the circle must be inside the square. **(b)** Answers will vary.

24. Answers will vary. The total is \$31.25; Andy will have enough money if sales tax is 2.4% or less.

25. The break-even point would be five shorts or shirts; more than five purchases would favor the discount plan.

26. Answers will vary.

27. Answers will vary.

28. Answers will vary. Inductive reasoning rests upon experimentation. Deductive reasoning is based on argument.

29. Caitlin is correct. She could pass the same person more than once and someone she passed could repass her to finish ahead of her.

30. Answers will vary.

Chapter 2

Section 2.1

1. A and D; B and C

2. **(a)** A, B; B, C; B, D **(b)** No solution; all equal sets are also equivalent. **(c)** A, C; A, D **(d)** C, D

3. **(a)** $n(S) = 5$ **(b)** $n(S) = 1$ **(c)** $n(S) = 0$
(d) $n(S) = 1$ **(e)** S is an infinite set

4. $A = \{a, b, c, d, e, f, g, h, i, j, k, l\}$ $n(A) = 12$

$B = \{A, B, C, D, E, F, G, H, I, J\}$ $n(B) = 10$

5. **(a)** A, C; B, A; B, C **(b)** No solution; every proper subset of S is also a subset of S. **(c)** A, A; B, B; C, C
(d) A, B; C, A; C, B

6. $2^3 = 8$: $\{c, d, e\}, \{c, d\}, \{d, e\}, \{c, e\}, \{c\}, \{d\}, \{e\}, \varnothing$

7. $S = \{a, b, c\}$, $T = \{a, b, c\}$, $R = \{a\}$

8. $P = \{a, b\}$ and $Q = \{a, b\}$; $P = Q$

9. **(a)** Every element of $\{z\}$ is also an element of $\{w, x, y, z\}$. **(b)** There is an element of $\{a\}$ that is not an element of $\{w, x, y, z\}$.

10. **(a)** $\{\}$ is a subset of $S = \{w, x, y, z\}$; since there are no elements in $\{\}$, there are no elements in $\{\}$ not in S.
(b) The argument above holds for any set S including $S = \{\}$.

11. **(a)** $S = \{w, x, y, z\}$ is a subset of itself because there is no element of S that is not an element of S. **(b)** The argument may be applied to any set S.

12. $\{\}, \{a\}, \{b\}, \{c\}, \{a, b\}, \{a, c\}, \{b, c\}$; the whole numbers associated with each of the proper subsets of S are less than the whole number associated with S: 0, 1, and 2 are less than 3.

13. **(a)** $\{0, 1, 2\}$; yes **(b)** $\{0, 2, 4, 6, \ldots\}$; no **(c)** No; the whole numbers can't get infinitely smaller, only infinitely larger.

14. Every subset of the whole numbers does have a least member.

15. $\{\}, \{a\}, \{b\}, \{c\}, \{d\}, \{a, b\}, \{a, c\}, \{a, d\}, \{b, c\}, \{b, d\}, \{c, d\}, \{a, b, c\}, \{a, b, d\}, \{a, c, d\}, \{b, c, d\}$ $\{a, b, c, d\}$

16. The counting numbers are a proper subset of the whole numbers; the whole numbers contain the element 0 and the counting numbers do not.

17. All sets that are in one-to-one correspondence are defined as equivalent sets and are associated with the same whole number.

18. If some set is equivalent to a proper subset of some other set, then the number associated with the first set is less than the number associated with the second set.

19. $\{1, 2, 3, 4\}$, $\{\&, *, \wedge, \%\}$, and so on **(a)** No **(b)** No **(c)** Equivalent sets must have the same number of elements.

20. $\{\}, \{a\}, \{b\}, \{c\}, \{d\}, \{a, b\}, \{a, c\}, \{a, d\}, \{b, c\}, \{b, d\}, \{c, d\}, \{a, b, c\}, \{a, b, d\}, \{a, c, d\}, \{b, c, d\}, \{a, b, c, d\}$

21. **(a)** $\{\}$ **(b)** There are none. **(c)** 0 is less than every whole number other than itself.

22. **(a)** Yes **(b)** No **(c)** Equal sets contain exactly the same elements and therefore the same number of elements. Equivalent sets contain the same number of elements, but not necessarily the same elements.

23. **(a)** False; suppose $A = \{a, b\}$ and $B = \{c, d\}$ **(b)** True; if two sets don't contain the same number of elements, they cannot contain exactly the same elements. **(c)** True; if two sets are equal, they have the same elements and therefore the same number of elements. **(d)** False; suppose $A = \{a, b\}$ and $B = \{c, d\}$

24. **(a)** False; suppose $A = \{a, b\}$ and $B = \{c, d, e\}$
(b) False; suppose $A = \{a, b\}$ and $B = \{c, d, e\}$ **(c)** True; since A is a proper subset of B, then all elements of A are in B and B contains elements not in A; therefore there are more elements in B than in A. **(d)** True; if A is a subset of B then either A and B contain exactly the same elements, in which case the cardinal numbers are equal, or all of the elements of A are also in B and B has some elements not in A. In this case the cardinal number of A is less than that of B.

25. The librarian means that they may check out only some, not all, of the newly acquired books.

26. (a) One subset: P itself **(b)** 2 subsets: $\{\}$ and $\{a\}$ **(c)** 4 subsets: $\{\}$, $\{a\}$, $\{b\}$, $\{a, b\}$ **(d)** A set with n elements has 2^n subsets.

27. There is a one-to-one correspondence between W and N because, for every whole number w in W, there is $n = w + 1$ in N to match it.

28. Given two sets, A and B, either there exists a one-to-one correspondence and $n(A) = n(B)$, or there are elements left over in set A and $n(A) > n(B)$, or there are elements left over in set B, and $n(A) < n(B)$.

29. $2^4 - 1 = 16 - 1 = 15$

30. Answers will vary. Discrete situations might include numbers of students in a class, cards in a hand, persons invited to a party, etc. Continuous situations might include the metric measures of length, mass, time, force, etc.

31. Answers will vary. Possibilities include: sets of students and sets of desks, persons invited to a party and party favors, plates and dinner guests. Some problems are trivial such as more Christmas cards than stamps; others are more serious such as more children than doses of vaccine.

Section 2.2

1. (a) $C \cup D = \{$people who are more than 20 years old or who are enrolled in college$\}$. **(b)** $A \cup B = \{10, 20, 30, 40\}$ **(c)** $E \cup F = \{0, 1, 2, 3, 4, \ldots, 100, 102, 104, \ldots\}$

2. (a) $C \cap D = \{$people who are more than 20 years old and who are enrolled in college$\}$. **(b)** $A \cap B = \{30\}$ **(c)** $E \cap F = \{\}$

3. (a) $\{e\}$ **(b)** $\{s, e, n, d, m, o, y\}$ **(c)** $\{m, o, r, e, n, y\}$ **(d)** $\{e\}$

4. (a) {@, @, @, @, @, @, @, @} ∪ {\$, \$, \$, \$, \$, \$, \$, \$} = {@, @, @, @, @, @, @, @, \$, \$, \$, \$, \$, \$, \$, \$} **(b)** {} ∪ {&, &, &, &, &, &, &, &, &, &} = {&, &, &, &, &, &, &, &, &, &}. **(c)** Show the union between the two disjoint sets: one with 123 elements; one with 324 elements.

5. (a)

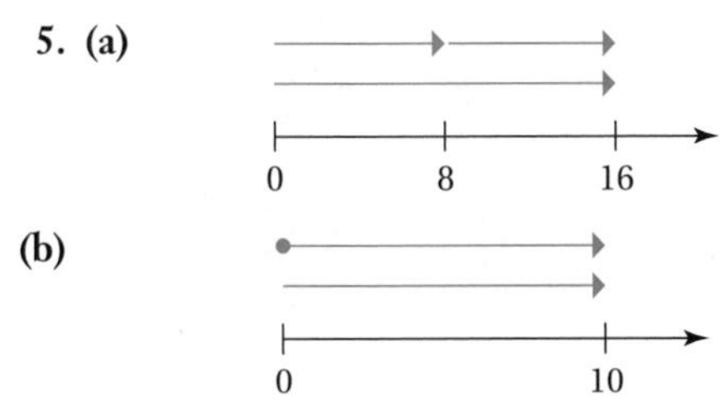

(b)

(c)

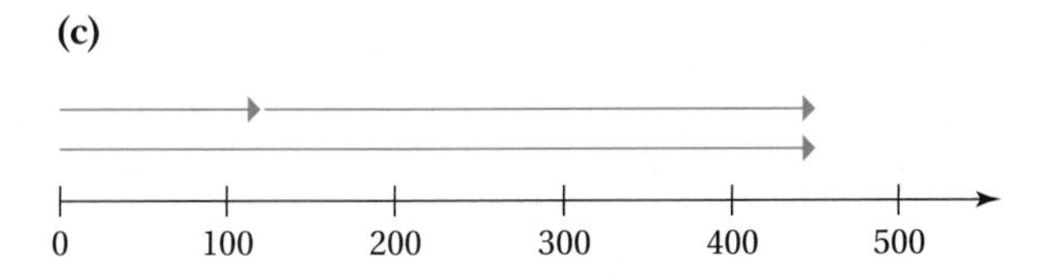

6. If 2000 tickets to a concert were presold and another 356 were sold at the door, how many tickets total were sold?

7. If 2000 miles were traveled on the interstate and 356 miles on secondary roads, what total distance was covered?

8. No. The set of those students bringing apples and the set of those students with peanut butter sandwiches may not be disjoint; some students may have brought both.

9. Examples will vary. **(a)** Commutative property of addition: $2 + 8 = 8 + 2$ **(b)** Identity property of addition: $a + 0 = 0 + a = a$ **(c)** Associative property of addition: $8 + 7 = (1 + 7) + 7 = 1 + (7 + 7) = 1 + 14 = 15$ **(d)** Closure property of addition

10. (a) There is a counting number, 2, such that $3 + 2 = 5$. **(b)** There is a counting number, 1362, such that $2500 + 1362 = 3862$. **(c)** $4 \times 103 = 412$, $3 \times 103 = 309$, and $412 > 309$ because there is a counting number, 103, such that $309 + 103 = 412$

11. (a) Draw a set of 16 objects with 8 of them crossed out. **(b)** Draw a set of 10 objects with all 10 crossed out. **(c)** Draw a set of 447 objects with 324 of them crossed out.

12. (a)

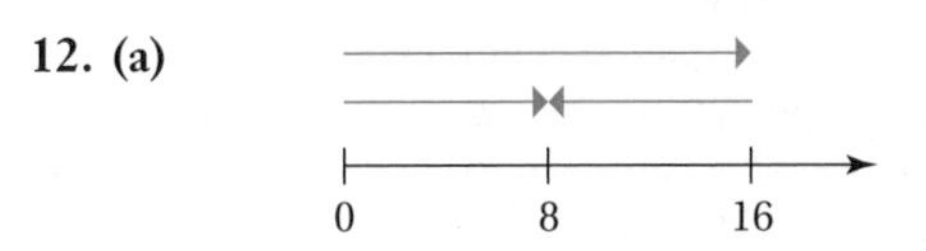

(b)

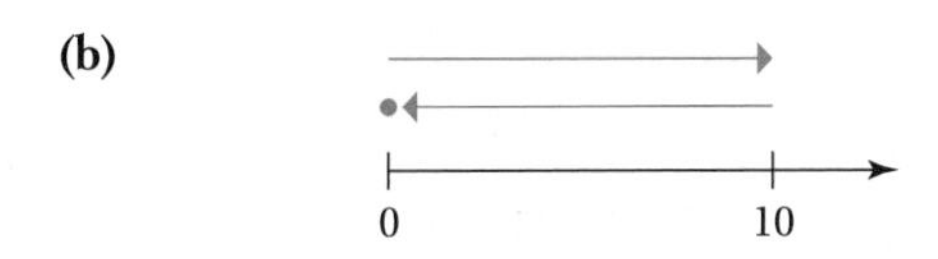

(c)

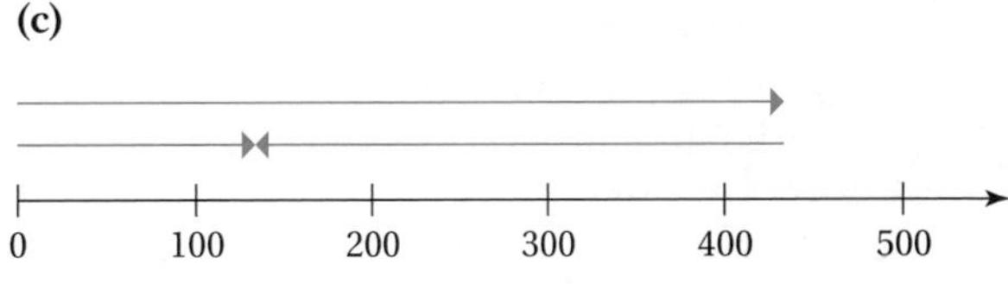

13. (a) $18 = n + 8$ **(b)** $25 = 15 + x$ **(c)** $y = 129 + 83$ **(d)** $a^2 = 30 + a$

14. (a) $8 = n - 7$ and $7 = n - 8$ **(b)** $14 = 25 - x$ and $x = 25 - 14$ **(c)** $r = t - s$ and $s = t - r$

15. A student had 12 lottery tickets. Five of these were known losers. How many tickets remained?

16. To complete a project, Sue cuts a 5-foot board from a 12-foot board. How long is the piece of board that she has left over?

17. A professor gives a quiz with 12 questions. If 5 of the questions are essay questions, how many of the questions are not essay questions?

18. Lashonda is 12 miles from school. Ross is 5 miles from school. How much farther does Lashonda have to travel to school than Ross?

19. The professor has written 5 essay questions for a quiz. How many more questions does the professor need to write so that the quiz will have 12 questions?

20. Answers will vary.

21. (a) $+1 = \quad = \quad + 10 = \quad = \quad = \quad = \quad + 100 = \quad = \quad =$ **(b)** 1, 2, 12, 22, 32, 42, 142, 242, 342 **(c)** Commutativity and associativity of addition

22. (a) Yes **(b)** Yes; zero **(c)** Yes **(d)** Yes

23. (a) Odd whole numbers **(b)** Natural numbers **(c)** Does not exist **(d)** Does not exist

24. Symbolically, the student is saying:
$(20 - 10) - 5 = (20 - 5) - 10$. This is true. However, commutativity would be expressed:
$20 - (10 - 5) = 20 - (5 - 10)$. This is not correct.
$20 - (10 - 5) = 15$ and $20 - (5 - 10) = 25$.

25. (a) Associativity of addition and base-ten place value **(b)** One possibility: A pattern in the 9s facts, i.e., $9 + 3 = 12, 9 + 4 = 13$, etc.

26. (a) If a set, S, is closed under an operation, the result of the operation on any element(s) of S is a unique element in S. Each cell in the table contains exactly one whole number. **(b)** The 0 row and 0 column are identical to the row and column representing the other addend. **(c)** Since the top right of the table is the mirror image of the bottom left, $a + b = b + a$. **(d)** Answers will vary.

27. $24_{five} < 30_{five}$ because $24_{five} + 1_{five} = 30_{five}$

28. Answers will vary. **(a)** Addition: How many students had lunch at school on Monday? Subtraction: How many more students had lunches from home on Monday than Tuesday? **(b)** $11 + 12 = ?$ $11 - 6 = ?$ Subtraction problems can also be written as $11 = 6 + ?$

29. Answers will vary.

30. The application of the mathematical relation $14 - 8 = 6$ to a variety of physical situations demonstrates the generality of mathematics.

31. (a–e) If a increases, e increases; if a decreases, e decreases; if c increases, e decreases; if c decreases, e increases; if a increases and c decreases, e increases.

32. (a) Not closed **(b)** Closed **(c)** Closed **(d)** Not closed **(e)** Closed

33. False

34. (a) Yes **(b)** No **(c)** No **(d)** No

35. Answers will vary. The following table shows one possibility.

#	a	b	c	d
a	a	b	c	d
b	b	c	d	a
c	c	d	a	b
d	d	a	b	c

36. Answers will vary.

37. (a) If the condition $k > 0$ were omitted from the definition of greater than, it would be possible for $k = 0$; if so, by the revised definition, $2 > 2$ because there would be a $k = 0$ such that $2 + 0 = 2$. **(b)** For a definition of less than, $a < b$ only if a $k > 0$ exists such that $a + k = b$ or such that $b - k = a$.

38. Answers will vary. You could remove objects from a set or move left on a standard number line.

39. Answers will vary. Scribes would have to look in the center of the table for the appropriate sum in the row or column of the addend that is being subtracted and then read the other addend from the row or column that forms the intersection at that sum.

40. Answers will vary. Some include deficits, losses, temperatures below zero, and elevations below sea level.

Section 2.3

1. (a) Show the union of 12 disjoint sets each with three elements. **(b)** Show the union of 14 empty sets. **(c)** Show the union of three disjoint sets each with 100 elements.

2. (a)

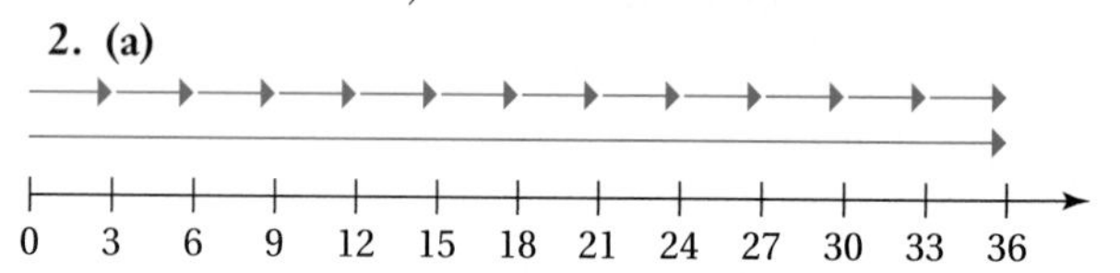

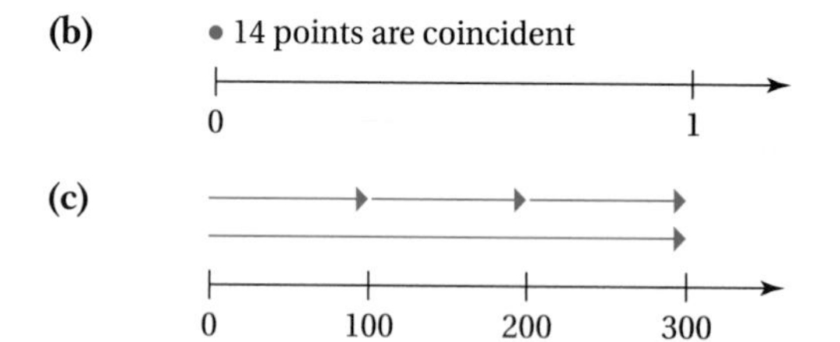

3. (a)

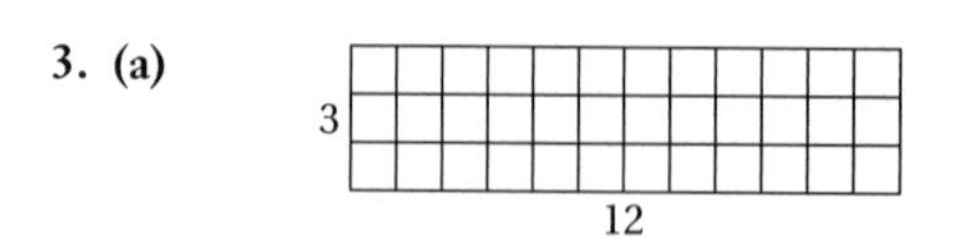

(b) No array, or rectangle, of 0 length **(c)** See answer below.

Answer for Exercise 3(c)

3

100

4. If each of 4 children received 30 pieces of candy on Halloween and put all the candy in a bag to share with their classmates, how many pieces of candy are in the bag?

5. If 4 students are running a relay in which each student runs 30 meters, how long is the race?

6. (a) $A = 18$ units $\times$ 15 units $= 270$ square units **(b)** 1 unit $\times$ 256 units $= 256$ square units **(c)** b units $\times$ b units $= (b \times b)$ square units

7. 5 soups $\times$ 8 salads $= 40$ soup-salad combinations

8. 8 blouses $\times$ 4 skirts $\times$ 4 vests $= 128$ blouse-skirt-vest combinations

9. w wrapping paper types $\times$ c bow colors $= 72$ paper-bow combinations

10. (a) See Exercises 7–9 for examples. **(b)** How many stamps are in a book with 25 pages of 4 stamps each? **(c)** What is the area of a garden 25 feet by 4 feet?

11. (a) Commutative property of multiplication **(b)** Zero property of multiplication **(c)** Distributive property of multiplication over addition **(d)** Closure property of multiplication **(e)** Associative property of multiplication

12. (a) 54 **(b)** 120 **(c)** $2a^2 + 11a + 12$

13. A teacher has 24 geoboards and wants to put them in boxes that will hold 6 each. How many boxes does she need?

14. A shop teacher cut a 24-inch piece of wire into 6-inch segments. How many pieces were made?

15. A teacher has 24 geoboards and wishes to share them evenly between 6 groups of students. How many geoboards will each group get?

16. A shop teacher had a piece of wire 24 inches long. He bent the wire into 6 pieces of equal length. How long was each straight piece?

17. A teacher made 6 tests, each with the same number of questions. The total number of questions was 24. How many questions were on each test?

18. (a) Subtract 15 nine times **(b)** Subtract 7 fourteen times **(c)** Subtract 26 fifty-four times

19. (a) $18 = 6 \times n; n = 3$ **(b)** $25 = 5 \times x; x = 5$ **(c)** $y = 42 \times 126 = 5292$ **(d)** $0 = b \times c; c = 0$

20. (a) $15 = n \div 3; 3 = n \div 15$ **(b)** $9 = 9 \div y; y = 9 \div 9$ **(c)** $r = t \div s; s = t \div r$ **(d)** $0 = 0 \div 8$

21. (a) 6 R 1; $6(3) + 1 = 19$ **(b)** 12 R 16; $20(12) + 16 = 256$ **(c)** 0 R 2; $0(8) + 2 = 2$

22. (a) i. The identity symbol for addition is the zero symbol for multiplication.
ii. Multiplication and addition are associative, are commutative, have closure, and have an identity property.
iii. Multiplication has a distributive and zero property; addition does not.

(b) Neither subtraction nor division is closed, associative, commutative, has an identity element, or has a zero element. Division does distribute over both addition and subtraction.

23. (a) $\{a, b, c\} \times \{r\} = \{(a, r), (b, r), (c, r)\}$; three ordered pairs. **(b)** $\{1, 2, 3, 4\} \times \{a, b\} = \{(1, a), (1, b), (2, a), (2, b), (3, a), (3, b), (4, a), (4, b)\}$; eight ordered pairs. **(c)** $\{\} \times \{3, 4, 5, 6, 7, 8\} = \{\}$; no ordered pairs. **(d)** $3 \times 1 = 3; 4 \times 2 = 8; 0 \times 6 = 0$.

24. No; if $S = \{a, b\}, T = \{a, b\}, S \times T = \{(a, a), (a, b), (b, a), (b, b)\}$.

25. It is used to generate the numerical value of each digit according to the place it is in; for example, $123 = (1 \times 100) + (2 \times 10) + (3 \times 1)$.

26. (a) Closure for multiplication; distributive property of multiplication over addition **(b)** $12 \times 64 = (10 + 2) \times (60 + 4) = (10 \times 60) + (10 \times 4) + (2 \times 60) + (2 \times 4) = 600 + 40 + 120 + 8 = 768$.

27. (a) $2(a + b)$ **(b)** $2r(7 + 9r)$ **(c)** $3(2c + 4d + 5e)$

28. (a) If a set, S, is closed under an operation, the result of the operation on any element(s) of S is a unique element in S. Each cell in the table contains exactly one whole number. **(b)** The 1 row and 1 column are identical to the row and column representing the other factor. **(c)** Since the top right of the table is the mirror image of the bottom left, $a \times b = b \times a$. **(d)** The 0 row and 0 column contain all 0s. **(e)** For example, if you add the 2 column to the 3 column, you get the 5 column. **(f)** Answers will vary.

29. Find the dividend (the product) in the interior of the table in the row or column of the divisor (a factor); then read the quotient (the other factor) from the row or column that forms the intersection at that product.

30. (a) The first type of problem asks "How many groups?" and the second asks "How many in one group?" The first type is acted out by removing a "bunch" at a time; the second is dealing out one at a time, etc. **(b)** They each can be thought of as subtracting d (the divisor) over and over again. Each represents separating into groups of equal size, etc.

31. (a) The (q, r) pairs are: $(0, 95), (1, 80), (2, 65), (3, 50), (4, 35), (5, 20), (6, 5)$ **(b)** $q = 6$ and $r = 5$ **(c)** There could be more than one possible quotient and remainder for a given problem.

32. (a) $(4 + 6) \div 2; (10 + 15) \div 5; (6 + 9) \div 3; (14 + 49) \div 7$ **(b)** a and b must be multiples of c because of the restriction of whole number quotients for each division expression. **(c)** The property is valid in the format in (b) because of the whole-number quotient restriction.

33. Answers will vary.

34. Answers will vary.

35. Yes, it could be true, because with 7 candy dishes, the size of the remainder can be greater.

36. Answers will vary.

37. All these sets are closed under multiplication.

38. False, consider $\{1, 2, 4, 7, 11, 16, \ldots\}$.

39. **(a)** $0 \div a = 0$ because, for all whole numbers a, $0 \times a = 0$. **(b)** $a \div 0$ is undefined because no whole number b exists such that $b \times 0 = a$, unless $a = 0$ (see part c). **(c)** $0 \div 0$ is undefined because no unique whole number b exists such that $b \times 0 = 0$; every whole number could satisfy this statement.

40. Answers will vary.

41. Answers will vary.

42. Answers will vary.

43. Fractions; additional responses will vary.

Section 2.4

1. **(a)** Numeral **(b)** Number **(c)** Numerals

2. **(a)** 17 **(b)** 10001_{two} **(c)** 32_{five}

3. **(a)** 408 = 4 hundreds squares, 0 tens sticks, 8 units cubes **(b)** $3699 = 3(1000) + 6(100) + 9(10) + 9(1)$ **(c)** $5{,}280{,}492 = 5(10^6) + 2(10^5) + 8(10^4) + 0(10^3)$

$$+ 4(10^2) + 9(10^1) + 2(10^0)$$

4. **(a)** 3 groups of 100 **(b)** 3 groups of 25 **(c)** 3 groups of b^2

5. **(a)** 99 **(b)** 20 **(c)** 83

6. **(a)** 455_{six} **(b)** $12\text{E}_{\text{twelve}}$ (where E represents eleven) **(c)** 10110011_{two}

7. **(a)** 34_{six} **(b)** 202_{three} **(c)** $1103_{\text{four}} = 123_{\text{eight}}$

8. 962, 980, 985, 1000, 2222, 2245

9. One-to-one correspondence; generation of equivalent sets

10. **(a)** 1405 **(b)** 260 **(c)** 119

11. **(a)** 70 **(b)** 59 **(c)** 4261

12. **(a)** 903 **(b)** 59 **(c)** 97

13. See answers below.

14. It contains place value if it uses a finite number of symbols, and the same symbol can represent different values, depending on its position. For example, in 12, 2 means two; in 21, 2 means twenty.

15. **(a)** Hindu–Arabic has only 10 symbols, but Egyptian has potentially infinitely many; Hindu–Arabic has a symbol for zero, but Egyptian doesn't; Hindu–Arabic uses place value, but Egyptian doesn't; Egyptian uses tallying, but Hindu–Arabic doesn't. **(b)** Babylonian involves tallying, but Hindu–Arabic doesn't. Hindu–Arabic has a symbol for zero, but Babylonian doesn't. **(c)** Hindu–Arabic has 10 symbols, but Roman has potentially infinitely many; Hindu–Arabic has a symbol for zero, but Roman doesn't; Roman uses some tallying, but Hindu–Arabic doesn't; Roman uses subtraction, but Hindu–Arabic doesn't.

16. For example, the tens digits increase by 1 and the ones digits remain the same as you move vertically.

17. **(a)** Go down one row in the same column. **(b)** Go up one row in the same column. **(c)** Move one space to the right. **(d)** Move one space to the left. **(e)** Move one row down and one space right. **(f)** Move one row up and one space left. **(g)** Move one row down and one space left. **(h)** Move one row up and one space right.

18. **(a)** 44_{five} **(b)** 444_{five}

19. $r(b^3) + s(b^2) + t(b^1) + u(b^0)$

20. **(a)** 58 miles per hour **(b)** \$85; \$0 **(c)** 2 days **(d)** 2 times as much **(e)** 85 to 52 **(f)** 8 checks **(g)** \$2500 **(h)** 5000 stamps

21. **(a)** Determine the value of each symbol, record the value for each symbol, and then add all the values. **(b)** Determine the place value of each symbol, multiply to determine the value of each symbol, and then add all the values. **(c)** Determine the value of each symbol, look at the order of symbols, and then add or subtract the values, as indicated by the order of the symbols.

22. **(a)** XXXVIII, XXXIX, XL, XLI **(b)** 98, 99, 100, 101 **(c)** ∩∩∩∩∩∩|||||||||, ∩∩∩∩∩∩∩, ∩∩∩∩∩∩∩|. **(d)** ▼ ‹‹‹ ▼▼▼▼▼▼▼▼; ▼ ‹‹‹ ▼▼▼▼▼▼▼▼▼; ▼▼; ▼▼▼ **(e)** XCVIII, XCIX, C, CI **(f)** 69, 70, 71, 72

23. It could be a square of 1000×1000 centimeters, or it could be a cube of $100 \times 100 \times 100$ centimeters.

Answer for Exercise 13

	100	61	608	94
(a) Egyptian	𓍢	∩∩∩∩∩∩|	𓍢𓍢𓍢𓍢𓍢𓍢||||||||	∩∩∩∩∩∩∩∩∩||||
(b) Babylonian	▼ ‹‹‹‹	▼ ▼	‹ ▼▼▼▼▼▼▼▼	▼ ‹‹‹ ▼▼▼▼
(c) Roman	C	LXI	DCVIII	XCIV

24. **Base-Five Numerals**

1	2	3	4	10	11	12	13	14	20
21	22	23	24	30	31	32	33	34	40
41	42	43	44	100	101	102	103	104	110
111	112	113	114	120	121	122	123	124	130
131	132	133	134	140	141	142	143	144	200
201	202	203	204	210	211	212	213	214	220
221	222	223	224	230	231	232	233	234	240
241	242	243	244	300	301	302	303	304	310
311	312	313	314	320	321	322	323	324	330
331	332	333	334	340	341	342	343	344	400

For example, if a numeral in base 10 ends in a 5 or 0, then the base five numeral for that same number ends in a 0.

25. The patterns identified in the two charts should be essentially the same.

26. The patterns predicted in the 8 by 8 base-eight chart should be essentially the same as those identified in the 5 by 5 base-five chart in problem 25.

Base-Eight Chart

1	2	3	4	5	6	7	10
11	12	13	14	15	16	17	20
21	22	23	24	25	26	27	30
31	32	33	34	35	36	37	40
41	42	43	44	45	46	47	50
51	52	53	54	55	56	57	60
61	62	63	64	65	66	67	70
71	72	73	74	75	76	77	100

27. Answers will vary.

28. Answers will vary.

29. **(e)** Higher than the distance to the moon (roughly 25 times that distance); estimate the height of a hamburger, how many can be stacked in a mile, and so on.

30. Answers will vary.

31. Answers will vary.

32. Answers will vary. One response should include the difficulty of using a place-value system lacking a zero symbol.

33. **(a)** 645_{eight} **(b)** $1A5_{\text{sixteen}}$ where A represents ten

Chapter 2 Review Exercises

1. **(a)** {!, @, #, $, %} **(b)** $A = \{1, 2, 3, 4, 5, 6, 7, 8, 9\}$, $B = \{1, 2, 3, 4, 5, 6, 7, 8, 9, 10, 11\}$; A is a proper subset of B **(c)** { } **(d)** { } is a proper subset of every nonempty set

2. See answers below.

3. **(a)** Egyptian system: tallying, additive, and new symbols based on groups of 10 **(b)** Babylonian system: tallying, additive, place value, and only two symbols **(c)** Roman system: some tallying, additive/subtractive (depending on position of symbols), and new symbols based on groups of 5 and 10 **(d)** Hindu–Arabic system: only 10 symbols, including one for zero, place value, and additive

4. **(a)** 1111_{five} **(b)** 10011100_{two}

5. **(a, b)** One possibility: To represent 1208, use 1 thousands cube, 2 hundreds flats, no tens sticks, and 8 units blocks. **(c)** $1(1000) + 2(100) + 0(10) + 8(1)$ **(d)** $1(10^3) + 2(10^2) + 0(10^1) + 8(10^0)$

6. **(a)** {10} **(b)** {3, 6, 8, 9, 10, 12} **(c)** { } **(d)** {9, 10, 11, 12}

7. Answers will vary.

8. Answers will vary.

9. Answers will vary.

10. Answers will vary.

11. A student may use from one to four types of flowers, which is the same as finding the number of nonempty subsets of a set with four elements: $2^4 - 1 = 15$; therefore there are enough different combinations of flowers for each student to have a different one.

12. $1 \times b^3 + 0 \times b^2 + 4 \times b^1 + 5 \times b^0$

13. $180 - 59 = 121$ because $121 + 59 = 180$

14. $123 > 85$ because $123 = 85 + 38$

15. **(a)** S is closed under addition because the sum of multiples of 10 is a multiple of 10. **(b)** The additive identity is 0 because any element of the set added to 0 results in the original element. **(c)** The set is closed under multiplication because the set is multiples of 10 and if a multiple of 10 is multiplied by any whole number not 0, then the product contains a factor of 10 and thus is a multiple of 10.

Answer for Exercise 2

Hindu–Arabic	200	55	398	120
Egyptian	𓍢𓍢	∩∩∩∩∩|||||	𓍢𓍢𓍢∩∩∩∩∩∩∩∩∩||||||||	𓍢∩∩
Babylonian	▼▼▼ <<	<<<<< ▼▼▼▼▼	▼▼▼▼▼▼ <<< ▼▼▼▼▼▼▼▼	▼ ▼
Roman	CC	LV	CCCXCVIII	CXX

(d) There is no multiplicative identity in S because there is no element t in S such that $t \times 10 = 10$. **(e)** The associative, commutative, and distributive properties of addition and multiplication do apply to S.

16. $120 \div 40 = 3$ because $3 \times 40 = 120$

17. $125 \div 40 = 3$ R 5 because $3 \times 40 + 5 = 125$

18. Answers will vary.

19. Answers will vary.

20. Answers will vary.

21. Answers will vary.

22. Answers will vary.

Chapter 3

Section 3.1

1. 140

2. 360

3. 2800

4. 290

5. 2207

6. 396

7. 960

8. 989

9. 548

10. 478

11. 680

12. 897

13. 36

14. 91

15. 152

16. 1450

17. 115

18. 280

19. 18

20. 286

21. 2822

22. 1325

23. 2400

24. 140

25. 79

26. 225

27. 139

28. 153

29. 109

30. 518

31. 429

32. 660

33. Responses will vary. A possible response is: $7 + 9 + 3 + 1 = 7 + 3 + 9 + 1 = 10 + 10 = 20$.

34. Responses will vary. A possibility is: $542 + 45 = 540 + 2 + 40 + 5 = 580 + 7 = 587$.

35. Responses will vary. A possible response is: 25 is $20 + 5$. So $11 \cdot 25$ is $11 \cdot 20$, which is 220, plus $11 \cdot 5$, which is 55. So the product is $220 + 55 = 275$.

36. Carly's thinking is not correct. When she subtracted 40, she subtracted 2 too many. Thus to compensate she must add 2, not subtract 2 more. So she should reason: to obtain the difference between 126 and 38, I'll first subtract 40, which is taking away 2 too many, getting 86. Now to compensate for taking away 2 too many, I'll add 2, getting 88.

37. $857

38. $672

39. Responses will vary. An example is: If you bought one of each item from Royalty Premier at this sale, what would be the total savings? It might be solved: $(799 - 699) + (599 - 499) + (279 - 229) + (199 - 179) = (800 - 700) + (600 - 500) + (280 - 230) + (200 - 180) =$ $270.

40. $44.99

41. $7

Answers will vary for Exercises 42–47. Possible responses are:

42. First use an underestimate: 28 times 20 is 560, more than 500. So the answer is "yes".

43. An exact answer is required. Although mental computations could be used, either pencil and paper or calculator computations are appropriate to find the difference: $365.25 - 48.89 = 316.16$.

44. An exact answer is called for. The data are such that mental computation by break apart into compatible numbers is an appropriate method. $34 + 46 + 52 + 38 = (30 + 40) + (50 + 30) + (4 + 6) + (2 + 8) = 170$.

45. An exact answer is called for. It may be determined as: $30 \times 150 = (3 \times 15) \times 10 \times 10 = 4500$.

46. Because the actual question asks ". . . about how much . . .", only an estimate is required. Mental computations are appropriate. Estimate the cost per tire as $50, and the cost for four as $4 \times 50 = 200$.

47. An exact answer is called for. Mental math computations or a fraction calculator are appropriate for finding the product $(10)(3\frac{1}{4} + 2\frac{1}{2} + 2\frac{3}{4})$.

48. Responses will vary. A representative response is: For a fund-raiser dinner for funds for cancer research, 48 tables of six have been pledged. How many dinners must be provided? The problem can be solved: $(48)(6) = (50)6 - 2(6) = 300 - 12 = 300 - 10 - 2 = 290 - 2 = 288$.

49. Responses will vary. A possible response is: Persons with poor understanding of place value may have difficulty in separating a number into place values such as when determining the product of of 8 and 28, one might improperly use the technique of adding 8 times 2 and 8 times 8 rather than adding 8 times 20 and 8 times 8.

50. Responses will vary.

Responses will vary for Exercises 51 and 52. Possible responses are:

51. One might have a symbolism that directly permits 42 to be separated into 40 + 1 + 1 just by spacing the symbols. One "1" could then be appended to the "39" resulting in the addition of 40 and 40. The remaining "1" could then be appended.

52. A similar symbolism might permit the appending of a "5" symbol to each numeral producing an equivalent but simpler computation: 2940 − 1500.

Section 3.2

1. 6780

2. 26.1

3. 210

4. 3000

5. 900

6. 750

Responses will vary for Exercises 7–10. Possible estimates are:

7. 7500

8. 7500

9. 18,000

10. 40

Responses will vary for Exercises 11–14. Possible estimates are:

11. 950

12. 1000

13. 200

14. 60

Responses will vary for Exercises 15–18. Possible estimates are:

15. 1500

16. 5000

17. 20

18. 13,000

Responses will vary for Exercises 19–22. Possible estimates are:

19. 1060

20. 38,000

21. 310

22. 4800

Responses will vary for Exercises 23–26. Possible estimates are:

23. 500

24. 280

25. 40

26. 1600

Responses will vary for Exercises 27–30. Possible estimates are:

27. 1500 to 2400

28. 1500 to 1700

29. 500 to 700

30. 180 to 240

Responses will vary for Exercises 31 and 32. Possible responses are:

31. About 12,000. The exact answer is greater than the estimate because both factors were rounded to smaller numbers.

32. About 8400. The exact answer is less than the estimate because both addends were rounded to greater numbers.

Responses will vary for Exercises 33–40. Possible estimates are:

33. 2400

34. 8000

35. 3200

36. 23,100

37. 5

38. 150

39. 62,000

40. 0 (exact answer)

Responses will vary for Exercises 41–46. Possible responses include:

41. About 91,000. The exact answer is greater.

42. About 60,000. The exact product is greater.

43. About 3200. The exact result is less.

44. About 1000. The exact product is greater.

45. About 10. The exact answer is greater.

46. About 260. The exact product is greater.

47. Responses will vary. A possible addend is any number between 400 and 449.

48. Responses will vary. A possible factor is 50.

49. Clustering and then rounding

50. Compatible numbers

51. 861 > 653

52. Responses will vary. The same if we round to tens.

53. Responses will vary. 238 + 273 is about 500. 21 × 26 is about 500.

54. The actual number falls between 90 bulbs and 120 bulbs.

55. Overestimate

56. Underestimate

57. Overestimate

58. Underestimate

59. Responses will vary. About 88 million.

60. Responses will vary. About 600,000.

61. Responses will vary. About 52%.

62. Responses will vary. Headline: 53 MILLION POOCHES IN U.S.

63. Responses will vary.

64. Responses will vary.

65. Responses will vary.

66. Responses will vary.

Section 3.3

1. 8 hundreds, 8 tens, 15 ones

2. 4 hundreds, 8 tens, 4 ones

3. Associative, commutative, associative, distributive

4. Correct

5. Incorrect: $(145 - 67) - 28 = 145 - (67 + 28)$.

6. Incorrect: $6[2(3.5 + 8.6)] = 12(3.5) + 12(8.6)$.

7. Correct

8. Correct

Responses will vary for Exercises 9–12 in the manner in which the error is described. Possible responses include:

9. 4 from 8 in the ones

10. Not changing 6 tens to 5 tens

11. There are several errors.

12. Place value was ignored.

13. 225

14. 146

15. 376

16. 266

17. Jorge neglected the parentheses.

18. $516

19. $38

20. Responses will vary.

21. $245

22. 20 tens = 19 tens + 10 ones

23. 625 and 348

24. 359

25. 824

26. 1465

27. 2123

28. 113

29. 259

30. 611

31. 4129

32. 72

33. 51

34. 16

35. 23

36. 272

37. 265

38. 1656

39. 2688

40. Separate the addends into compatible numbers. Add the leftovers.

41. $(75 + 52) + (100 + 300) = 127 + 400 = 527$

42. This algorithm shows another method of keeping track of regroupings.

43. The workers will pass the 875-meter depth sometime on the eighteenth day.

44. A possible response is: MORE THAN 1,000 ILL

45. There are 18 routes.

46. ABGFEJK has a distance of 1339 miles.

47. Another problem could be: What is the difference in distance between the shortest and longest routes?

48. A possible response is: A student received a rebate of $65 on purchases of $235 and $112. What was the net cost to the student?

49. $474

50. 10 months

51. Regrouping and trading reflect the concrete operations more than borrowing and carrying.

52. Responses will vary.

53. Responses will vary.

54. Responses will vary.

55. The missing addend is 13: $200 - 99 - 88 = 13$.

56. The largest possible addend is 99 and the sum is 286.

Section 3.4

1. 3 groups, 4 tens and 3 ones in each

2. 2 rows, 3 tens and 6 ones in each

3. **(a)** 368 and 1380 **(b)** You are multiplying by 30, not 3. **(c)** About 2,000

4.

$$\begin{array}{r} 34 \\ \times 28 \\ \hline 272 \\ 680 \\ \hline 952 \end{array}$$

10 10 10 4

10

10

8

= 680

= 272

5. $12 \times 22 = 264$

6. Find the sum of 6 42's or the sum of 42 6's.

7. 32, 240, 40, 300

8. In the first calculation, the multiplication was done starting with the tens digit. In the other, products were found by starting with the ones digit.

9. 11–20

10. 31–40

11. 21–30

12. 41–50

13. 61–70
14. 81–90
15. 51–60
16. 51–60
17. Possible estimate: 100
18. Possible estimate: 17
19. 8
20. 714
21. 147
22. Juanita left out the parentheses.
23. Highest = 200 points. Lowest = 50 points.
24. $75 \div 5$
25. 74×23 is about 1400, an estimate close to the (erroneously) computed value. 48×32 is about 1500, which is not close to 220.
26. Place 1 hundreds block in each of the six sets. Regroup the remaining hundreds block. Add these 10 tens to the 5 tens. We have 15 tens to distribute. Placing 2 tens in each of the six sets leaves 3 tens to regroup with the 8 units. The 38 units are distributed 6 in each of the six sets, leaving 2 units undistributed. So the quotient is 126 R 2.
27. distributive property
28. 3420
29. 7290
30. 6944
31. 5980
32. Responses will vary.
33. Since $32 \div 8 = 4$, $32 = 4 \times 8$.
34. About 40 years
35. $300,000
36. 11.4 years
37. $20,500
38. About $60 per month. This is below the actual payments.
39. $221 \times 226 = 49{,}946$
40. $358 \times 348 = 124{,}584$
41. $631 \times 542 = 342{,}002$
42. $145 \times 236 = 34{,}220$
43. $114,286 per year
44. $64,000 per year
45. $S = 75{,}000 + 200(2^0 + 2^1 + \cdots + 2^{29})$ or about 200 billion dollars
46. $12 \times W = 276$ or $276 \div 12 = W$
47. $450 \div 15 = 30$. The remaining 36 can be separated into 2 groups of 15, giving 32 bottles per carton with 6 left over.

Responses will vary for exercises 48 and 49. Possible responses are:

48. Use repeated addition.
49. Use repeated subtraction.
50. In repeated subtraction, how many 12's in 156? In sharing, if 156 things are equally shared among 12 individuals, how many does each individual receive?

Responses will vary for exercises 51–53. Possible answers are:

51. 328 pencils are to be given away, 6 at a time, to students. How many students will receive a group of 6 pencils?
52. $328, collected in a charity fund raiser, is to be given to 6 organizations. How much money does each organization receive?
53. Responses will vary.
54. The intermediate multiplications and subtractions are done mentally with only the differences recorded above and to the left of the next place to be considered in the division.
55. Responses will vary.

Chapter 3 Review Exercises

1. 84
2. 99
3. 448
4. 878
5. 2600
6. 127
7. 480
8. 100
9. 809
10. 701
11. 862
12. 324
13. 156
14. 369
15. 311
16. 203
17. 21
18. 130
19. 122
20. 104
21. 5700
22. 320
23. 7000
24. 600
25. About 900
26. About 200
27. About 800
28. About 400
29. About 3000
30. About 1300
31. About 300

32. About 900

33. About 1170

34. About 6700

35. About 250

36. About 89,000

37. About 4800

38. About 2750

39. About 7000. Exact, 6883.

40. About 1800. Exact, 1904.

41. About 240. Exact, 232.

42. About 200. Exact, 191.25.

43. 1500 and 2400

44. 800 and 1000

45. 150 and 160

46. 1200 and 1400

47.

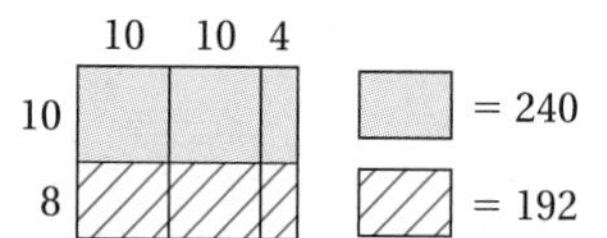

48. The 5 hundreds are divided into 3 groups. 1 hundred is placed in each group with 2 hundreds remaining. The 2 hundreds are traded for 20 tens. The 20 tens are combined with the 4 tens to give 24 tens. The 24 tens are divided into 3 groups. 8 tens are placed in each group with no tens left over. The 8 ones are divided into 3 groups. 2 ones are placed in each group so there are 2 ones left over.

49. Compatible numbers are numbers that are easy to compute with mentally. Different persons have different compatible numbers.

50. $34 \times 17 = 578$

51. The sums are (in thousands): 179, 904, 157.5, 775, 28.5, 753.5. The greatest difference is $875.5K.

52. Responses will vary.

56. Tennis vacation, golf clubs, camera: $407 left.

57. Responses will vary.

58. Responses will vary.

59. Responses will vary.

Chapter 4

Section 4.1

1. 7 is a factor of 87 if $7 \mid 87$. We can divide 87 by 7, and if there is a remainder 7 is not a factor of 87.

2. 1, 2, 3, 4, 6, 8, 12, 24.

3. 35. The Factor Test theorem

4. **(a)** 1, 2, 3, 6, 13, 26, 39, 78 **(b)** 1, 2, 3, 4, 6, 12, 13, 26, 39, 52, 78, 156 **(c)** 1, 2, 3, 4, 6, 7, 9, 12, 14, 18, 21, 36, 42, 63, 84, 28, 126, 252

5. Answers will vary. Factors of 253, in addition to 1, are 11, 23, 253. Multiples of 253, in addition to 0, are 253, 506, 759, etc.

6. You can multiply every whole number by a whole number (namely, 0) to get the product 0, so by the definition of a multiple, 0 is a multiple of every whole number.

7. $299 \div 13 = 23$ with no remainder, so 13 is a factor of 299. $13 \times 23 = 299$, so 13 is a factor of 299.

8. factor

9. multiple

10. multiple

11. factor

12. factor

13. multiple

14. multiple

15. factor

16. $0, 3n, 6n, 9n, 12n, \ldots$

17. **(a)** if it ends in an even number **(b)** if the sum of its digits is divisible by 3 **(c)** if the number represented by its last two digits is divisible by 4 **(d)** if it ends in 0 or 5 **(e)** if it is divisible by 2 and by 3

18. **(a)** divisible by 3 **(b)** divisible by 2, 3, 4, 5, and 6 **(c)** divisible by 5 **(d)** divisible by 2, 3, 4, and 6

19. The number is also divisible by any of the factors of 24, i.e., 1, 2, 3, 4, 6, 8, 12, 24.

20. Answers will vary. Possible answer: 4,632

21. Answers will vary. Possible answer: 2,736

22. Answers will vary. Possible answer: 72,294

23. Answers will vary. Possible answer: 5,142

24. Answers will vary. Possible answer: 24,168

25. Answers will vary. Possible answer: 82,194

26. Answers will vary. Possible answer: 8,532

27. Answers will vary. Possible answer: 73,656

28. Answers will vary. Possible answer: 62,595

29. **(a)** True, since $4 \times 5 = 20$. **(b)** False, since no whole number times 12 equals 6. **(c)** True, since $24 \times 1 = 24$. **(d)** False, since there is no whole number n such that $0 \times n = 8$. **(e)** True, since $4 \times 0 = 0$. **(f)** True, since $15 \times 4 = 60$. **(g)** False, since there is no whole number n such that $24 \times n = 8$. **(h)** False, since the definition of divisibility excludes 0 being divisible by 0.

30. 6 is a perfect number since the sum of its proper factors, $1 + 2 + 3$, equals 6.

31. 15 is a deficient number since the sum of its proper factors, $1 + 3 + 5$, is less than 15.

32. 18 is an abundant number since the sum of its proper factors, $1 + 2 + 3 + 6 + 9$, is greater than 18.

33. The smallest even, abundant number is 12, since $1 < 2$, $1 + 2 < 4$, $1 + 2 + 3 = 6$, $1 + 2 + 4 < 8$, $1 + 2 + 5 < 10$, but $1 + 2 + 3 + 4 + 6 > 12$.

34. **(a)** 9 is a square number. Its three factors are 1, 3, and 9. **(b)** 16 is an even square number. 25 is an odd square number. **(c)** 36 is even, square, and abundant. The sum of the proper factors of 36 is $1 + 2 + 3 + 4 + 6 + 9 + 12 + 18 = 55$, and $55 > 36$.

35. Divisibility by 3 and by 5. Yes.

36. If 4 divides 1516 then the persons can be seated as desired. The group can be seated.

37. No.

38. 1 by 84, 2 by 42, 3 by 28, 4 by 21, 6 by 14, and 7 by 12.

39. Responses will vary among students. These may include: 28 is a whole number. It is an even number. The factors of 28 are 1, 2, 4, 7, 14, and 28. Because the sum of the factors less than 28 sum to 28, it is called a perfect number.

40. $19{,}271{,}927 \div 73 = 263{,}999$, so $73 \mid 19{,}271{,}927$. This is not unusual, since any 8-digit number formed in the way described is divisible by 73.

41. **(a)** Not a leap year. 700 is divisible by 4, but as a multiple of 100 is not divisible by 400. **(b)** A leap year, since it is divisible by 4. **(c)** A leap year, since it is divisible by 4. **(d)** A leap year, since it is divisible by 4 and by 400. **(e)** Not a leap year, since it is not divisible by 4.

42. **(a)** Yes; since 5 is a factor of 10, a number divisible by 10 is also divisible by 5. **(b)** No; 15 is divisible by 5, but not by 10.

43. **(a)** True. **(b)** False. Consider 12. Both 2 and 4 are factors of 12 and thus divide 12 but 8 is not a factor of 12. **(c)** True. **(d)** False. Both 4 and 6 divide 12 but 24 is not a factor of 12.

44. All square numbers with prime square roots greater than 1 have 3 factors.

45. **(a)** always **(b)** sometimes **(c)** never **(d)** sometimes **(e)** always

46. **(a)** False. Let n be 6, a be 3, and b be 4. It is true that $6 \mid 3(4)$ but $6 \nmid 3$ and $6 \nmid 4$. **(b)** False. Consider that 6 divides $(7 + 5)$ but 6 does not divide 5 or 7.
(c) True. As an example, 3 divides 6 and 3 divides 9. 3 also divides $(6 + 9)$. **(d)** True. 4 divides 8 and 8 divides 16. 4 also divides 16. **(e)** False. Consider that 6 divides $(7 + 5)$ but 6 does not divide 5 or 7. **(f)** True. $6 \mid 48$, $6 \mid 18$, and $6 \mid 30$.

47. The proper factors of 220 are: 1, 2, 4, 5, 10, 11, 20, 22, 44, 55, and 110. The sum of these factors is 284. The proper factors of 284 are 1, 2, 4, 71, and 142. The sum of these factors is 220.

48. The proper factors of 496 are: 1, 2, 4, 8, 16, 31, 62, 124, 248. The sum of these factors is 496. So 496 is a perfect number.

49. Only 36 is abundant. The other numbers are deficient.

50.

Numbers	Operation	Result	Numbers	Operation	Result
even, even	+	even	even, even	×	even
even, odd	+	odd	even odd	×	even
odd, odd	+	even	odd, odd	×	odd

51. Answers will vary. Possible responses include 12 and 18. Since $12 = 2 + 4 + 6$, 12 is a semiperfect number. Since $18 = 3 + 6 + 9$, 18 is a semiperfect number.

52. \$4.56 is, in pennies, 456. If 456 is to be represented in multiples of 10 plus multiples of 25, then 456 must be separated into 2 parts, one a multiple of ten, the other a multiple of 25. The multiple of 10 will end in 0 leaving the 6 to be part of the number to be a multiple of 25. But 25 does not divide any number ending in 6. Thus \$4.56 cannot be represented in only quarters and dimes.

53. 301. Yes. 721 and 1141 satisfy the conditions.

54. 5717 is not divisible by 3. Thus 5717 tennis balls cannot be packed in multiples of 3. And since they can't be packed in multiples of 3 they can't be packed in multiples of 6.

55. **(a)** Begin by adding the factors from largest to smallest, checking the sum after each addition. The sum of 78 and 52 is 130. The sum of 130 and 39, 169, is greater than 156. So 156 is abundant. **(b)** Janie might apply some estimation techniques. Replace all the factors with the next higher number of 10's: 10, 10, 20, 30, 40, 100. The sum of these, which is an overestimate of the exact sum, is seen to be 210, less than 273. Thus 273 is deficient.

56. Responses will vary. Possible responses are:
(a) 452,452 divided by 13 is 34,804. **(b)** 4664 divided by $11 = 424$. **(c)** Consider 50, the sum of 20 and 30. The proper factors of 20 are 1, 2, 4, 5, 10, which sum to 22. The proper factors of 30 are 1, 2, 3, 5, 6, 10, 15, which add to 42. Both 20 and 30 are abundant. **(d)** Consider 38. Reversing and subtracting, we get $(83–38) = 45$. 45 divided by 9 is 5.

57. $4 \mid (12 + 20)$ and $4 \mid 20$, so $4 \mid 12$.

58. Suppose $2 \mid n$ and $5 \mid n$. We know that 2 and 5 have no common factors except 1, so, using the Divisibility by Products Theorem, we conclude that 2(5) or $10 \mid n$.

59. $4 \mid r$. The Divisibility by Sums Theorem.

60. When a number like 7648 is divided by 1000, the remainder is a 3-digit number. That is, $7648 = 1000 \times 7 + 648$. Consider any number n of the form $n = 1000q + r$. Since $8 \mid 1000$, we know by the Divisibility by Sums Theorem that if 8 also divides r, then $8 \mid n$.

61. $3 \mid 15$ and 2 is a natural number, so $3 \mid 15(2)$. To verify, since $a \mid b$, we know from the definition of divisibility that $ax = b$, where x is a whole number. Then $acx = bc$, $a(cx) = bc$, and since cx is a whole number, by the definition of divisibility we know that $a \mid bc$.

62. The error can be applied to 26/65, 19/95, and 49/98.

63. Daughter, 11 and Mother, 35.

64. 123654 and 321654 are the possible license plate numbers.

65. Responses will vary. One argument is since 43 is not divisible by 3, and since the scoring supposedly happened in 3's or in multiples of 3's, the score of 43 is not possible.

66. The address is 338.

67. The cost of one notebook is $2.39.

68. Responses depend on interactions among students.

69. The proper factors of 120 are: 1, 2, 3, 4, 5, 6, 8, 10, 12, 15, 20, 24, 30, 40, 60. The sum of these is 240, or 2 times 120.

The proper factors of 672 are: 1, 2, 3, 4, 6, 7, 8, 12, 14, 16, 21, 24, 28, 32, 42, 48, 56, 84, 96, 112, 168, 224, 336. The sum of these numbers is 1344, which is 2 times 672.

70. Responses will vary. Divisibility is used to reduce fractions to lowest terms. Multiples are used to find other fractions equivalent to given fractions.

Section 4.2

1. 2

2. 2, 3, 5, 7, 11, 13, 17, 19, 23, 29

3. A composite number has at least 3 distinct factors.

4. 32, 33, 34, 35, 36, 38, 39

5. **(a)** Not prime. 8 has more than 2 factors. **(b)** Prime. 11 has only 2 factors. **(c)** Neither prime nor composite. 1 cannot be expressed as the product of two or more unique factors. **(d)** Not prime. 51 is the product of 3 and 17. **(e)** Not prime. 221 is 13×17.

6. Responses will vary. 25, for example, is odd but not prime.

7. The order of the factors may vary.

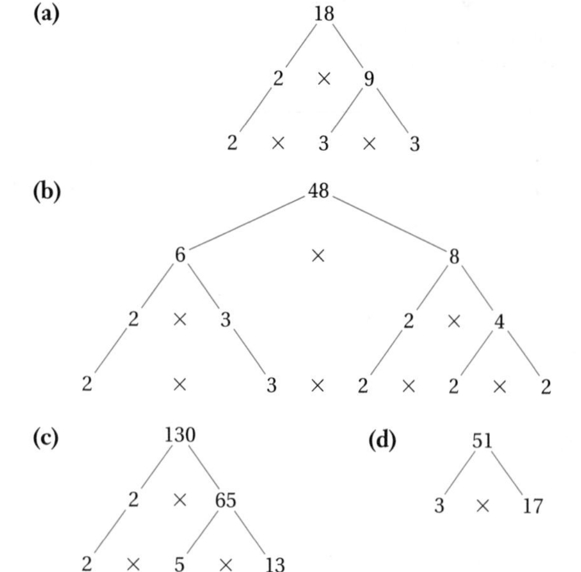

8. **(a)** $504 \div 2 = 252$, $252 \div 2 = 126$, $126 \div 2 = 63$, $63 \div 3 = 21$, $21 \div 3 = 7$. The prime factorization $= 2 \times 2 \times 2 \times 3 \times 3 \times 7$. **(b)** $1176 \div 2 = 588$, $588 \div 2 = 294$, $294 \div 2 = 147$, $147 \div 3 = 49$, $49 \div 7 = 7$. The prime factorization $= 2 \times 2 \times 2 \times 3 \times 7 \times 7$. **(c)** $2600 \div 2 = 1300$, $1300 \div 2 = 650$, $650 \div 2 = 325$, $325 \div 5 = 65$, $65 \div 5 = 13$. The prime factorization $= 2 \times 2 \times 2 \times 5 \times 5 \times 13$. **(d)** $3675 \div 5 = 735$, $735 \div 5 = 147$, $147 \div 3 = 49$, $49 \div 7 = 7$. The prime factorization $= 3 \times 5 \times 5 \times 7 \times 7$.

9. There are 8 pairs of twin primes, 3-5, 5-7, 11-13, 17-19, 29-31, 41-43, 59-61, 71-73.

10. The square root of 541 is between 23 and 24 so only natural numbers less than 24 need be tested for divisibility. Trial divisions on a calculator show no natural numbers less than 24 divide 541. Thus 541 has no factors other than 1 and 541, and is prime. Testing factors, we see that $437 = 19 \times 23$ and is not prime.

11. $16{,}731 \div 3 = 5577$, $5577 \div 3 = 1859$, $1859 \div 11 = 169$, $169 \div 13 = 13$. So $16{,}731 = 3 \times 3 \times 11 \times 13 \times 13$.

12. $42 = 2 \times 3 \times 7$
$28 = 2 \times 2 \times 7$
$\text{GCF} = 2 \times 7 = 14$

13. GCF (12, 20) = 4.

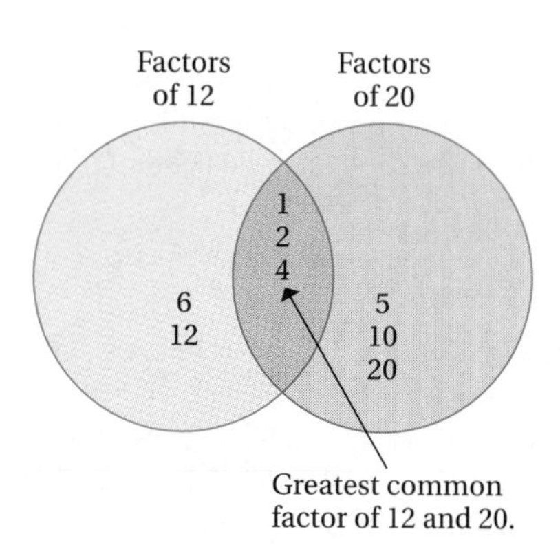

14. A = set of factors of 18 = {1, 2, 3, 6, 9, 18}. B = set of factors of 24 = {1, 2, 3, 4, 6, 8, 12, 24}. $A \cap B$ = set of common factors of 18 and 24 = {1, 2, 3, 6}. 6 is the greatest common factor of 18 and 24.

15. Answers will vary. A possible procedure is as follows. $80 = 2 \times 2 \times 2 \times 2 \times 5$. $124 = 2 \times 2 \times 31$. GCF of 80 and 24 is 2×2, or 4.

16. **(a)** $28 = 2 \times 2 \times 7$. $42 = 2 \times 3 \times 7$. Thus the GCF is $2 \times 7 = 14$. **(b)** $45 = 3 \times 3 \times 5$. $60 = 2 \times 2 \times 3 \times 5$. Thus the GCF is $3 \times 5 = 15$. **(c)** $36 = 2 \times 2 \times 3 \times 3$. $54 = 2 \times 3 \times 3 \times 3$. Thus the GCF is $2 \times 3 \times 3 = 18$.

17. **(a)** GCF (259, 888) = 37 **(b)** GCF (84, 308) = 28 **(c)** GCF (1232, 7560) = 56

18. Non-zero multiples of 9 = 9, 18, 27, 36, ... Non-zero multiples of 12 = 12, 24, 36, ... The least common multiple of 9 and 12 is 36.

19. The intersection of the set of non-zero multiples of 12 and the set of non-zero multiples of 18 is {36, 72, ...}. The least common multiple of 12 and 18 is the smallest number in this set, namely 36.

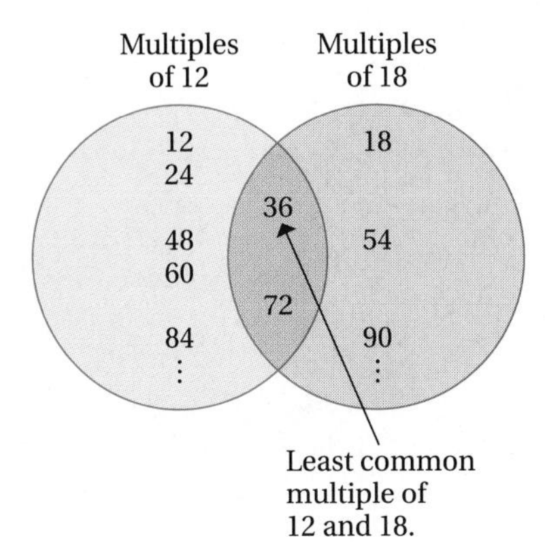

20. C = set of non-zero multiples of 6 = {6, 12, 18, 24, 30, 36, 42, ... 84, ...}. D = set of non-zero multiples of 21 = {21, 42, 63, 84, ...}. $C \cap D$ = set of common non-zero multiples of 6 and 21 = {42, 84, ...}. The least common multiple of 6 and 21 is 42.

21. Answers will vary. Non-zero multiples of 18 are 18, 36, 54, 72, ... Non-zero multiples of 24 are 24, 48, 72, ... The least common multiple of 18 and 24 is 72.

22. **(a)** LCM (27, 36) = 108 **(b)** LCM (42, 60) = 420 **(c)** LCM (28, 40) = 280

23. 3

24. Procedures will vary. Possible solutions are: **(a)** 4 **(b)** 6

25. Procedures will vary. Possible solutions are: **(a)** 180 **(b)** 280

26. 6

27. 90 ft

28. The siren and the alarm will sound at the same time again after 72 hours.

29. 8 stamps can be put on each page. 68 pages.

30. 60 min

31. 35 ft by 35 ft

32. (n, p) pairs are (1, 11), (2, 13), (3, 17), (4, 23), (5, 31), (6, 41), (7, 53), (8, 67), (9, 83) (10, 101), and (11, 121). All are prime but 121.

33. When $n = 41$, $P = 41^2$, and is not prime.

34. 2, 5, 17, 37

35. 3, 7, 31, and 127 are all primes. 2047, the next outcome, isn't prime.

36. Responses will vary.

37. The primes are all in the 1st and 5th columns, and are of the form $6n + 1$ or $6n - 1$.

38. **(a)** All ones digits are odd numbers. Only one prime has ones digit 5. **(b)** Groups of three consecutive primes are rare. **(c)** 13, 31; 17, 71; 37, 73; 107, 701; and 11, 11.

39. $5773 = 23 \times 251$, and is not prime.

40. Reponses will vary. Computers have a limit on their capacity.

41. GCF is always less than or equal to the LCM. GCF (17, 51) is no larger than 17, while LCM (17, 51) is no smaller than 51.

42. LCM is the product of the two numbers, using the GCF–LCM Product Theorem.

43. **(a)** It has no divisors other than 1 and 6. **(b)** 2, 6, 10, 14, 18 **(c)** Primes have the form $4n + 2$, where n is a whole number. **(d)** Composites have the form $4n$, where n is a natural number. **(e)** Yes **(f)** Yes. There are infinitely many numbers in E of the form $2 + 4n$. **(g)** $c = 2 \times e$, where e is an element of E greater than 1. **(h)** No, except for $e = 1$ **(i)** No. The minimum difference between two primes in E is 4. If a twin prime is defined as two primes separated by one composite, there are an infinite number of twin primes, in E. **(j)** Goldbach's conjecture is not true in E. $6 = 2 + 4$, but 4 is composite in E.

44. **(a)** $61 = 4(15) + 1$, and $61 = 25 + 36$. **(b)** 17 is a prime between 12 and 24. **(c)** $2^2 - 1 = 3$, a prime. $(2^{2-1})(2^2 - 1) = 6$, a perfect number. **(d)** For 16: $16 \div 2 = 8$, $8 \div 2 = 4$, $4 \div 2 = 2$, $2 \div 2 = 1$. For 7: $3(7) + 1 = 22$, $22 \div 2 = 11$, $3(11) + 1 = 34$, $34 \div 2 = 17$, $3(17) + 1 = 52$, $52 \div 2 = 26$, $26 \div 2 = 13$, $3(13) + 1 = 40$, $40 \div 2 = 20$, $20 \div 2 = 10$, $10 \div 2 = 5$, $3(5) + 1 = 16$, $16 \div 2 = 8$, $8 \div 2 = 4$, $4 \div 2 = 2$, $2 \div 2 = 1$.

45. All even numbers greater than 2 are divisible by 2. So they would have at least 3 unique factors and would not be prime.

46. Three consecutive whole numbers $2x$, $2x + 1$, $2x + 2$, or $2x - 1$, $2x$, $2x + 1$, would have to contain at least one even number, and there is no prime other than 2 that is even.

47. A square number has its square root as a factor, making it not prime because it has at least 3 factors.

48. GCF (9, 12) = 3
GCF [(9 + 12), LCM (9, 3)] = GCF (21, 9) = 3
GCF (12, 18) = 6
GCF [(12 + 18), LCM (12, 18)] = GCF (30, 36) = 6
Not a proof because it isn't generalized to all whole numbers.

49. The sum of the proper factors >1 of one of the numbers equals the other number.
Sum of proper factors of 140 = 2 + 4 + 5 + 7 + 10 + 14 + 20 + 28 + 35 + 70 = 195. Sum of proper factors of 195 = 3 + 5 + 13 + 15 + 39 + 65 = 140

50. Each time one or more of the three numbers are divided by a number, that number is established as a factor of the number. The LCM of the three numbers must contain all the factors of the numbers. So the product of all the divisors of the three numbers is their LCM.

2	24	28	45
2	12	14	45
2	6	7	45
3	3	7	45
3	1	7	15
5	1	7	5
7	1	7	1
	1	1	1

So LCM $= 2 \times 2 \times 2 \times 3 \times 3 \times 5 \times 7 = 2520$.

51. **(a)** $24 = 2^3 \times 3^1$ **(b)** $4 \times 2 = 8$, so there are 8 factors of 24.

52. 7

53. He ordered 19 watches at $23 each. $C \times N = 437$, where C is the cost and N is the number of watches. Other than 1 and 437, 19 and 23 are the only factors of 437.

54. The possibilities are: 21, 33, 39, 77, or 91.

55. Indiana, with zip prefix 317

56. 21,252 days, or a little over 58 years old.

57.

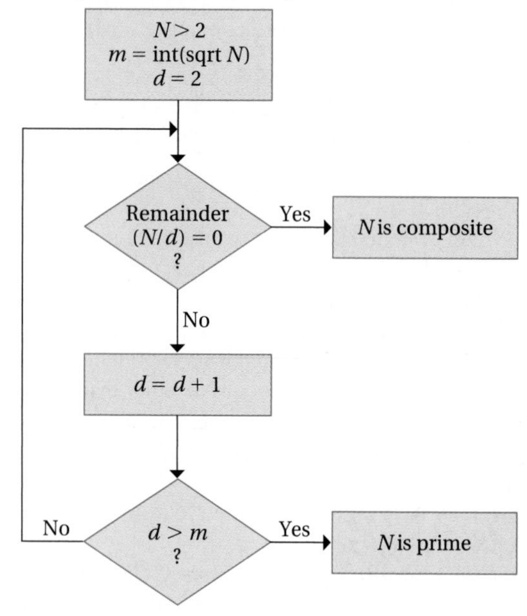

58. **(a)** 23, 9, 13 **(b)** 7 pairs of each **(c)** 7 pairs of each **(d)** yes

59. Responses will vary. One such program is: INPUT X, INPUT Y, IF X < Y, GOTO 1, X → A, Y → B, GOTO 2, LBL 1, X → B, Y → A, LBL 2, REMAINDER (A, B) → R, IF R = 0, GOTO 3, B → A, R → B, GOTO 2, LBL 3, DISP "GCF OF", DISP X, DISP Y, DISP "IS", DISP B

60. For $n = 0$, $P_n = 3$. For $n = 1$, $P_n = 5$. For $n = 2$, $P_n = 17$. For $n = 3$, $P_n = 257$. For $n = 4$, $P_n = 65{,}537$. All of the above are primes. For $n = 5$, $P_n = 4{,}294{,}967{,}297 = 641(6{,}700{,}417)$, a composite number.

61.

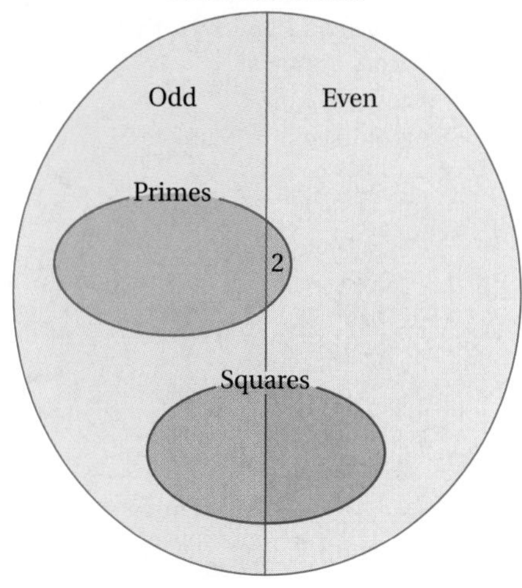

Chapter 4 Review Exercises

1. **(a)** 1, 2, 3, 6, 9, 18 **(b)** 1, 2, 4, 8, 16, 32 **(c)** 1, 2, 3, 4, 6, 8, 12, 16, 24, 48 **(d)** 1, 3, 5, 7, 15, 21, 35, 105

2. $\sqrt{625}$, or 25.

3. **(a)** 1, 7, 13, 91 **(b)** 1, 11, 13, 143 **(c)** 1, 3, 13, 17, 39, 51, 221, 663 **(d)** 1, 13, 23, 299

4. **(a)** A number is divisible by 2 if its ones digit is 0, 2, 4, 6, or 8. **(b)** A number is divisible by 3 if the sum of its digits is divisible by 3. **(c)** A number is divisible by 4 if the number formed by its last 2 digits is divisible by 4. **(d)** A number is divisible by 5 if it ends in 0 or 5. **(e)** A number is divisible by 6 if it is divisible by 2 and by 3.

5. **(a)** 1436, 4674, 5580 **(b)** 987, 4674, 5580 **(c)** 1436, 5580 **(d)** 5580 **(e)** 4674, 5580

6. **(a)** False. 3 divides 24 but 9 does not. **(b)** True. **(c)** False. 2 and 4 both divide 12 but 8 does not divide 12.

7. 7 is a prime because the only two numbers that divide 7 are 1 and 7. 6 is composite because in addition to being divisible by 1 and 6 it is divisible by 2 and 3.

8. **(a)** 89 is prime because it has only 2 divisors. **(b)** 39 is not prime because it is divisible by 1, 3, 13, and 39. **(c)** 137 is prime because it has only two divisors, 1 and 137. **(d)** 217 is composite because it is divisible by 1, 217, 31, and 7.

9. **(a)**

90	420
2×45	2×210
$2 \times 3 \times 15$	$2 \times 5 \times 42$
$2 \times 3 \times 3 \times 5$	$2 \times 5 \times 2 \times 21$
	$2 \times 5 \times 2 \times 3 \times 7$

(b) $90 \div 2 = 45$; $45 \div 3 = 15$; $15 \div 3 = 5$. $420 \div 2 = 210$; $210 \div 5 = 42$; $42 \div 2 = 21$; $21 \div 3 = 7$.

10. 36 has 9 factors.

11. (a) The set of factors of 48 is {1, 48, 2, 24, 3, 16, 4, 12, 6, 8}. The set of factors of 108 is {1, 108, 2, 54, 3, 36, 4, 27, 6, 18, 9, 12}. The intersection of these sets is {1, 2, 3, 4, 6, 12} and the largest element of the intersection is the GCF, 12. **(b)** The set of prime factors of 48 is {2, 2, 2, 2, 3} and the set of prime factors of 108 is {2, 2, 3, 3, 3}. The intersection of these sets is {2, 2, 3}. So the GCF is $2 \times 2 \times 3 = 12$. **(c)** $108 \div 48 = 2$ R 12. $48 \div 12 = 4$ R 0. Thus the GCF is 12.

12. (a) The set of multiples of 24 is {24, 48, 72, 96, 120, 144, 168, 192, …}. The set of multiples of 32 is {32, 64, 96, 128, 160, 192, …}. The smallest element of the intersection of these sets is the LCM, 96. **(b)** The two prime factorizations are $24 = 2 \times 2 \times 2 \times 3$ and $32 = 2 \times 2 \times 2 \times 2 \times 2$. The LCM is $2 \times 2 \times 2 \times 2 \times 2 \times 3 = 96$.

13. (a) Not relatively prime. 7 is a factor of both. **(b)** Relatively prime. No common prime factors.

14. By applying the GCF–LCM Product Theorem—the product of the GCF and LCM of two numbers equals the product of the numbers—we have $72 \times \text{GCF} = 432$. Thus the GCF = 6.

15. Any even number n other than 2 would have at least 1, 2, and n as factors, and thus couldn't be prime.

16. Because 4 and 2, unlike 3 and 2, are not relatively prime.

17. 420. This is the smallest such number because only the numbers necessary to ensure that the number is divisible by the given numbers were multiplied to produce 420.

18. 8, 10, 15, and 26 all have exactly 4 factors. The other numbers have more or less than this. Two additional such numbers are 21 and 55.

19. (a) 12 hours **(b)** 4 **(c)** Answers may vary. The siren could be set to sound every 80 minutes.

20. 70% of the first 10 numbers generated by $6n + 1$ are prime.

65% of the first 20 such numbers are prime.
63% of the first 30 such numbers are prime.
60% of the first 40 such numbers are prime.
54% of the first 50 such numbers are prime.

Since the percentage is decreasing and the continuation is infinite, the percent may well drop below 50%.

21. (a) 16 m × 16 m **(b)** 1155 **(c)** Smaller plots could have an edge length that is a factor of 16: 1, 2, 4, or 8.

22. Responses will vary. A chart of individual tests might be as shown in the next column.

Potential divisor, *d*	*d* divides dividend, *D*, if
2	*D* ends in 0, 2, 4, 6, or 8
3	The sum of the digits of *D* is divisible by 3
4	The number formed by the last 2 digits of *D* is divisible by 4
5	*D* ends in 0 or 5
6	*D* is divisible by both 2 and 3
7	The number formed by subtracting twice the last digit from the number formed by all but the last digit is divisible by 7
8	The number formed by the last 3 digits is divisible by 8
9	The sum of the digits is divisible by 9
10	The number ends in 0

Responses might include: tests for 2 and 5 are related to the test for 10; the test for 6 is related to the tests for 2 and 3; test for 3 and 9 both use the sum of the digits; tests for 4 and 8 both use some of the last few digits of the original number. The tests for 7 and 8 are not very efficient.

23. Responses will vary.

24. Descriptions will vary. The methods are alike in that one can find both of them by selecting factors from the prime factorizations of the original numbers, and multiplying. The methods are different in that for the GCF a factor is selected the smallest number of times it appears in *both* prime factorizations, and for the LCM a factor is selected the greatest number of times it is found in *either* of the prime factorizations.

Chapter 5

Section 5.1

1. (a) −5 **(b)** −6 **(c)** +8500

2. negative

3. positive

4. zero

5. positive

6. (a) Start with 8 blacks. Put in 3 reds. Cancel red–black pairs. There are now 5 blacks.

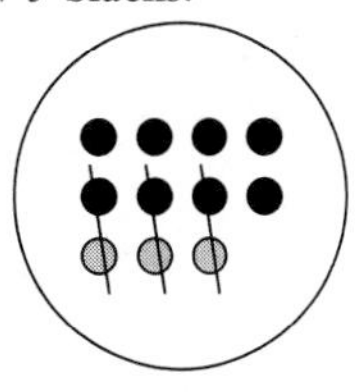

$8 + (-3) = 5$

(b) Start with 7 reds. Put in 4 blacks. Cancel red–black pairs. There are now 3 reds.

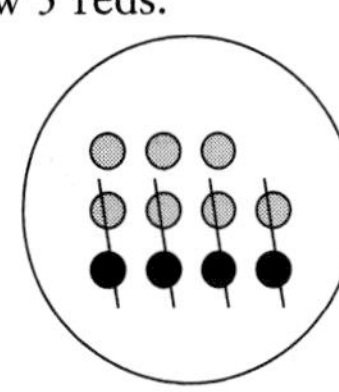

$-7 + 4 = -3$

(c) Start with 2 reds. Put in 6 reds. Cancel red–black pairs. There and now 8 reds.

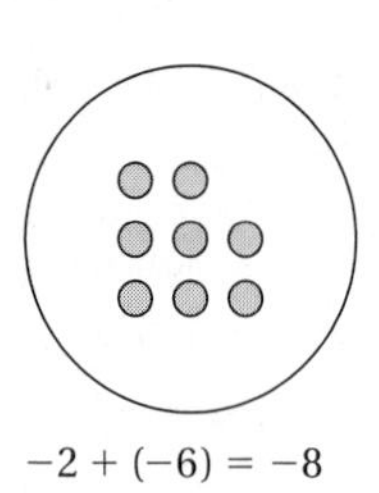

Put in 3 negative charges.

Put in 5 negative charges.

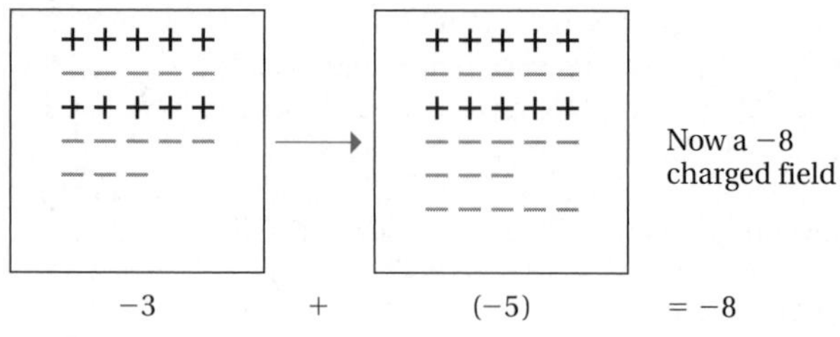

7. (a)

Start with a 0 charged field.

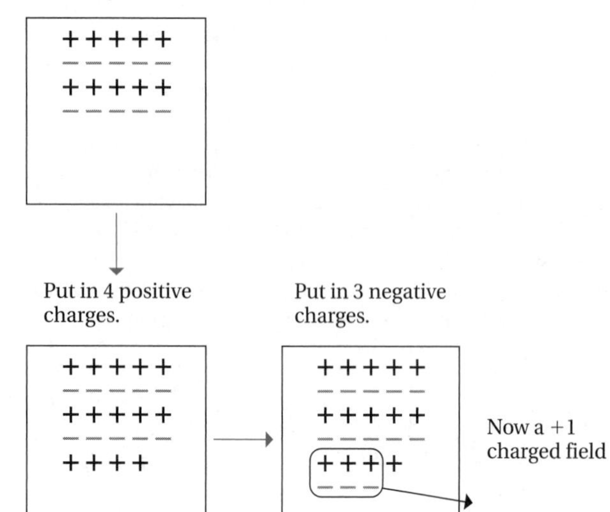

$4 + (-3) = 1$

(b) Start with a 0 charged field.

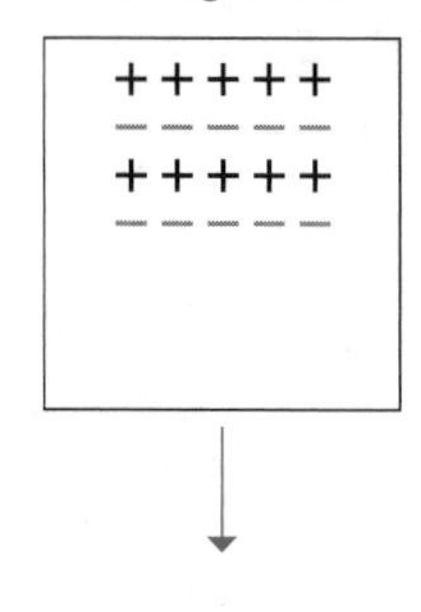

(c)

Start with a 0 charged field.

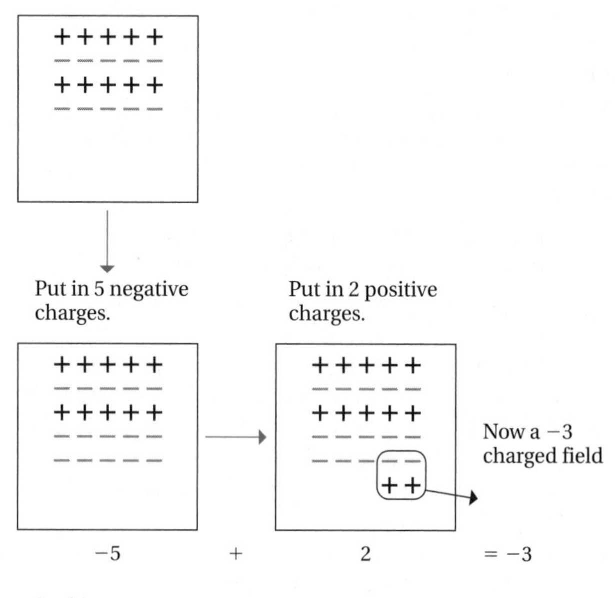

8. (a)

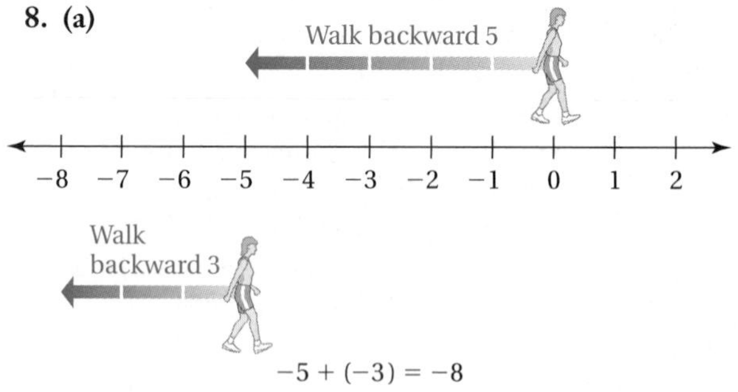

(b)

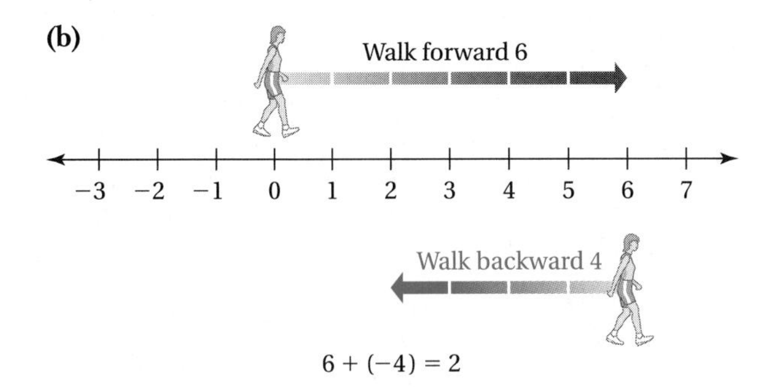

$6 + (-4) = 2$

(c)

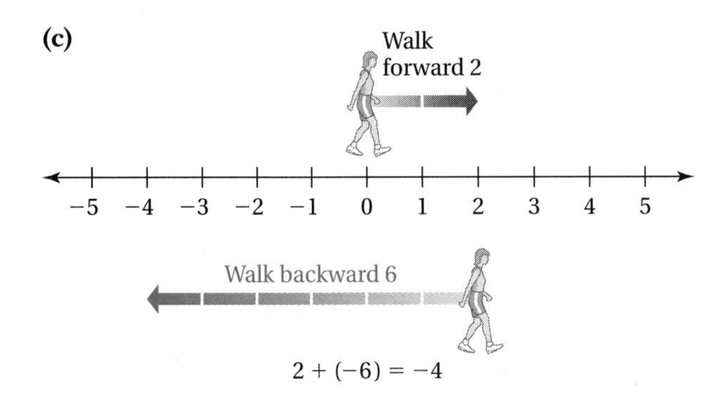

$2 + (-6) = -4$

9. Descriptions will vary. **(a)** 6 **(b)** 2 **(c)** 2

10. Responses will vary. A possibility is: For every number that is not 0 in the integers there is another number such that when the two numbers are added you get 0 for a sum. Either one of the numbers is called the additive inverse of the other.

11. **(a)** Associative property of addition: $(5 + 4) + (-4) = 5 + [4 + (-4)]$ **(b)** Additive inverse property: $[4 + (-4)] = 0$ **(c)** Additive identity property: $5 + 0 = 5$

12. **(a)** Take 5 black counters from a counter model for -2.

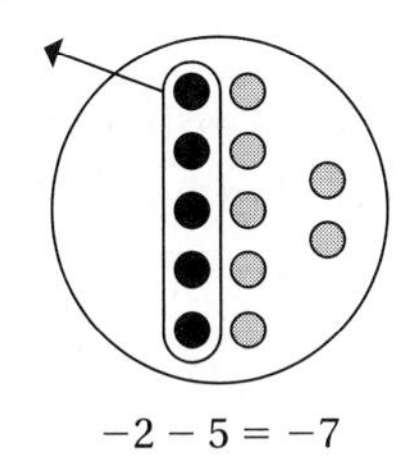

$-2 - 5 = -7$

(b) Take 5 red counters from a counter model for -8.

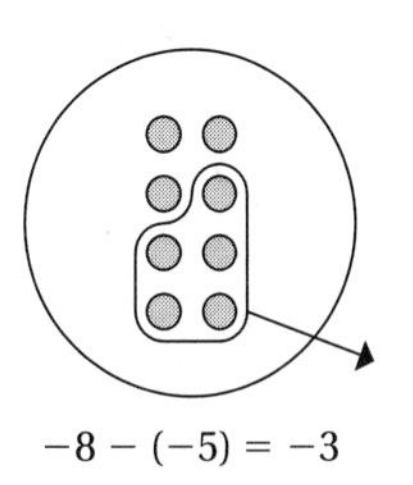

$-8 - (-5) = -3$

(c) Take 3 red counters from a counter model for 9.

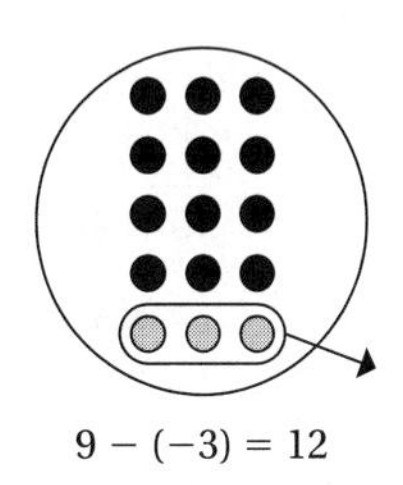

$9 - (-3) = 12$

13. **(a)**

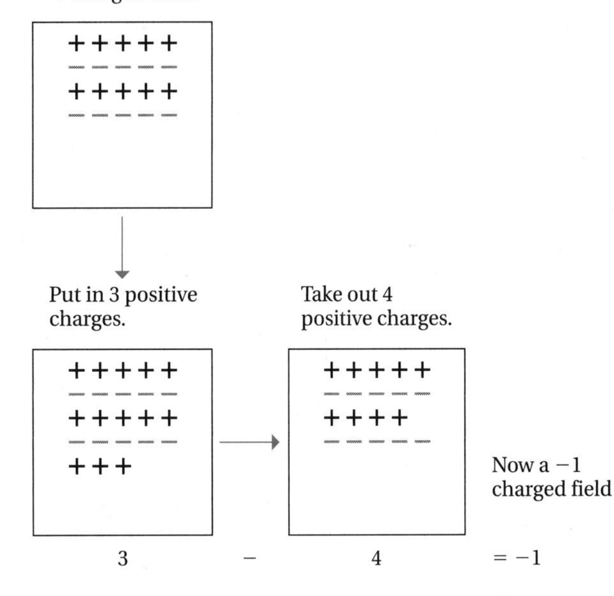

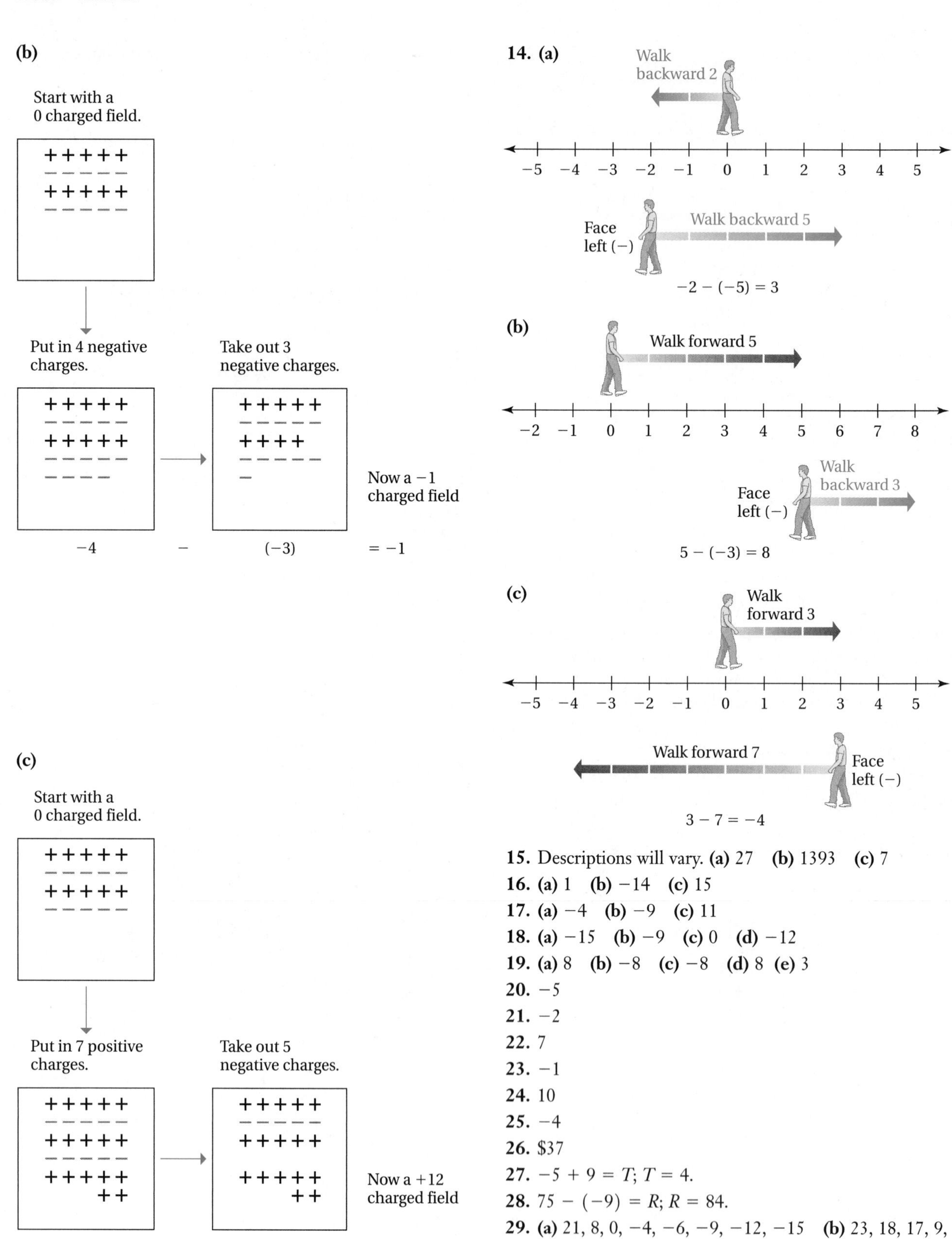

15. Descriptions will vary. **(a)** 27 **(b)** 1393 **(c)** 7

16. **(a)** 1 **(b)** −14 **(c)** 15

17. **(a)** −4 **(b)** −9 **(c)** 11

18. **(a)** −15 **(b)** −9 **(c)** 0 **(d)** −12

19. **(a)** 8 **(b)** −8 **(c)** −8 **(d)** 8 **(e)** 3

20. −5

21. −2

22. 7

23. −1

24. 10

25. −4

26. $37

27. $-5 + 9 = T$; $T = 4$.

28. $75 - (-9) = R$; $R = 84$.

29. **(a)** 21, 8, 0, −4, −6, −9, −12, −15 **(b)** 23, 18, 17, 9, −2, −12, −15, −18 **(c)** −13,570; −13,999; −14,000

30. By definition an integer b is greater than an integer a if there is a positive integer p such that $a + p = b$. Now, $-5 + 3 = -2$. So $-2 > -5$ or $-5 < -2$.

31. Since -13 is to the right of -19 on the number line, $-13 > -19$.

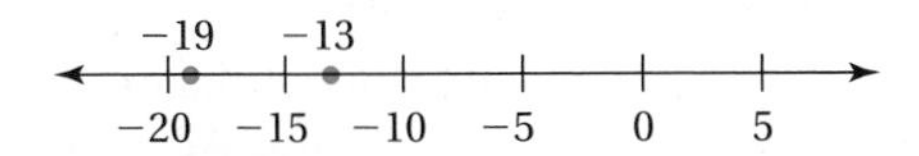

32. $t - 23 = -12$; $t = -12 + 23$; $t = 11$.

33. $a - (-459) = 9236$; $a = 9236 + (-459)$; $a = 8777$.

34. G (gain) $= -9 + 12$; $G = 3$.

35. **(a)** -1353 **(b)** 10,264 **(c)** 393 **(d)** 2012

36. Counterexamples will vary. Possibilities include: **(a)** False. $|0| = 0$, and 0 is neither positive nor negative. **(b)** Not true. Suppose $q = -2$. Then $-(-2)$ is positive 2. **(c)** True. By definition the sum of an integer and its opposite is 0. **(d)** Not true. The equation has a solution for integers $p = 2$ and $q = 6$.

37. Possible answers are: **(a)** 4, 8, 12 **(b)** $-6, -13, -20$ **(c)** $20, -30, 30$ **(d)** $-4, 4, -5$

38. 2301 years ago

39. **(a)** 7 **(b)** -1 **(c)** -7 **(d)** 2 **(e)** 30

40. Answers will vary. **(a)** $(-48 + 48) + 1 + (37 - 37) + 999 = 0 + 1 + 0 + 999 = 1000$. **(b)** $48 - 49 + 75 - 74 = 0$.

41. **(a)** k **(b)** $-(j + k)$ **(c)** $-j$ **(d)** $-(j - k)$, or $k - j$

42. No. Consider $|3 + (-7)|$, which is equal to 4. But $|3| + |-7| = 10$.

43. 5° above 0° to 0° is represented by $-5°$. 0° to 3° below 0° is represented by $-3°$. $-5° + (-3°) = -8°$.

44. Examples will vary. Some possibilities are: **(a)** No. $6 + (9 - 4) = 11$ but $(6 + 9) - (6 + 4) = 15 - 10 = 5$. **(b)** No. $9 - 6 = 3$, but $6 - 9 = -3$. **(c)** Yes. **(d)** No. $-(6 + 8)$ has a value of -14, but $(8 - 6)$ has a value of 2. **(e)** Yes. **(f)** Yes. **(g)** No. $(6 - 4) - 3 = -1$, but $6 - (4 - 3) = 5$. **(h)** No. $6 - (-2) = 8$, but $|6| - |-2| = 6 - 2 = 4$.

45. Methods of solution will vary. Possibilities include: **(a)** $x = 1$. **(b)** $n = -5$. **(c)** $y = 20$. **(d)** $z = 12$. **(e)** $t = -7$. **(f)** $r = 7$.

46. $8 + C = -4$ (beginning temperature plus change gives final temperature). $-4 - 8 = C$ (the difference between the final temperature and the initial temperature is the change). In either case $C = -12$.

47. The final position of the car is 3 miles west of the station.

48. **(a)** The range should be represented by a positive integer. So $|-415.469 - (-410.894)| = 4.5750$°F. **(b)** $16{,}188 - (-37.8921) = 6225.892$°F.

49. You might begin this exercise by ordering the numbers: $-61, -49, -29, -21, 39, 45, 56$. **(a)** $-49 + 45 = -4$. **(b)** $-61 + (-49) = -110$. **(c)** $39 + 56 = 95$. **(d)** $39 - (-61) = 100$. **(e)** $-21 - (-29) = 8$ **(f)** $39 - 45 = -6$.

50. -9

51. Estimates and methods will vary. Possibilities include: **(a)** Estimate by rounding: $700 - 500 = 200$. **(b)** All of the values cluster around -500. So $-500 - 500 - 500 - 500 - 500 = 5(-500) = -2500$. **(c)** Using compatible numbers and breaking apart the result is approximated by: $-300 - 50 + 50 + 1200 = 900$. **(d)** Front-end estimation gives the approximation $-800{,}000 + 300{,}000 = -500{,}000$. **(e)** Rounding to estimate gives $-400 - 800 + 200 + 1000 - 700 = -700$.

52. $4 \times A = 3$. There is no integer value for A to make this sentence true.

53.

−2	3	−7	4
0	−3	7	−6
5	−8	2	−1
−5	6	−4	1

54.

−2	3	2
5	1	−3
0	−1	4

55. Here are two solutions.

```
    1              3
  −2  2          −3  0
 3  0 −1        2  1 −1
```

56. In the model, in order to take away 2 red counters (subtracting -2), you must put in 2 black–red pairs.

After taking away the reds, the final effect is adding 2 blacks (adding 2). The statement on p. 246 is "to subtract an integer, add its opposite," which essentially describes the process in the model.

57. $-a + (-b)$ adds to $a + b$ to give the sum 0; i.e., $[-a + (-b)] + (a + b) = -a + a + (-b) + b = 0$. But $-(a + b)$ also adds to $a + b$ to give the sum of 0; i.e., $-(a + b) + (a + b) = 0$. Since there is a unique (only one) number that adds to $a + b$ to give the sum 0, $-a + (-b) = -(a + b)$.

58. **(a)** \$30 over budget **(b)** Dec. = −(Jan. + Feb. + Mar. + ⋯ + Oct. + Nov.).

59. Home Team: 80 net yards of offense. Visitors: 30 net yards including the kick-off return. Stories will vary. The home team scored on its first possession with 80 net yards of offense. After a successful conversion the home team led 7–0. The visitors received the ball on the kickoff and returned it 24 yards. On their first possession the visitors punted after a net gain of 6 yards.

60. **(a)** 75 revolutions clockwise, or -75 revolutions. **(b)** -45 rev/min. **(c)** $+160$ revolutions.

61. With 11 holes yet to be scored, Puttgood is 10 strokes ahead of Drivefar. But Drivefar can gain 1 stroke on each of the 11 holes. He has the potential to win by 1 stroke.

62. Responses to this question will vary. Some of the sentences that might be included are: Sat.–Thurs. $= -2 - 0 = -2°$, that is 2° lower on Saturday than on Thursday. Sat.–Fri. $= -2 - (-7) = 5°$ higher on Saturday than on Friday. To obtain the average temperature for the 3-day weekend, she will have to compute $(12 + -7 + -2) \div 3 = 1°$.

63. The arrow model is like the counters model (black and red counters cancel each other out) and the charged field model (negative and positive charges cancel each other out).

64. Responses will vary. A possibility is to represent the data with bar graphs with appropriate ranges of months (say 6 categories below the average and 6 above) and two colors, one to represent months above average, the other months below average.

65. Responses will vary. From the definition of subtraction, $a - b$ is the unique integer that adds to b to give a. From the adding-opposites theorem, $a + (-b)$ is also the unique integer that adds to b to give a. Hence $a - b = a + (-b)$.

66. Responses will vary. The elevator is similar to the number line. Designate some floor as 0, and agree that going upward is moving in the positive direction and downward is the negative direction. Use this model in the same way as the number line to model addition.

67. Yes, the method is correct. For $-7 + 4$, $|-7| - |4| = 3$; since 7 is larger than 4, the sum is -3. For $-3 + (-2)$, $|-3| + |-2| = 5$, so the sum is -5.

68. (a) Responses will vary. Arnauld's reasoning had some validity if one does not accept the properties of integers. He was possibly thinking of a false proportion such as $\frac{2}{3} = \frac{3}{2}$ when claiming that it is incorrect to think that a "smaller to greater" could equal a "greater to smaller." But this proportion has terms that are whole numbers. Because of the properties of integers, proportions with integers don't match with our intuition about whole number proportions. **(b)** There are two values for x for which the sentence $x^2 + 2x - 8 = 0$ is true. These are 2 and -4. Bhaskara was telling his students to ignore the negative solution for "it is inadequate," perhaps to represent a real-world application deemed important in that place at that time.

69. Responses will vary. A slash through a group of vertical tally marks anywhere within the numeral indicates that the number is negative. The numeration system is place value based on powers of 10. The initial tally may be either horizontal or vertical. Tallying continues through 5 within a place. After that, a single line—horizontal if tallying was done with vertical lines, vertical if tallying was done with horizontal lines—is used to represent 5. This line is not added until the face value becomes 6.

70. Responses will vary. They may include:
$I = P \cup N \cup R$; $I = W \cup N$; $E = P \cap N$;
$P \cap R = E$; $N \cap R = E$.

Section 5.2

1. **(a)** 0 **(b)** 5 **(c)** 10 **(d)** 15

2. Start with a bag containing several counters.
(a) Action A: Put 2 counters in the bag. Action B: Do action A a total of 5 times. There are 10 more counters in the bag than at the beginning, so $2(5) = 10$.
(b) Action A: Put 2 counters in the bag. Action B: Do the opposite of Action A a total of 5 times. There are 10 less counters in the bag than at the beginning, so $2(-5) = -10$. **(c)** Action A: Take 2 counters from the bag. Action B: Do the opposite of Action A a total of 5 times. There are 10 more counters in the bag than at the beginning, so $-2(-5) = 10$.

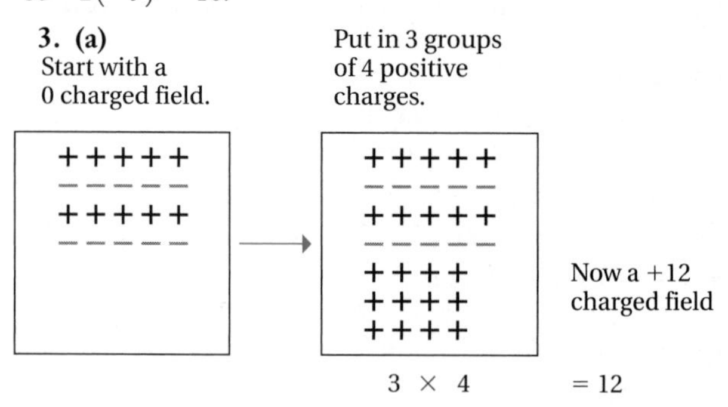

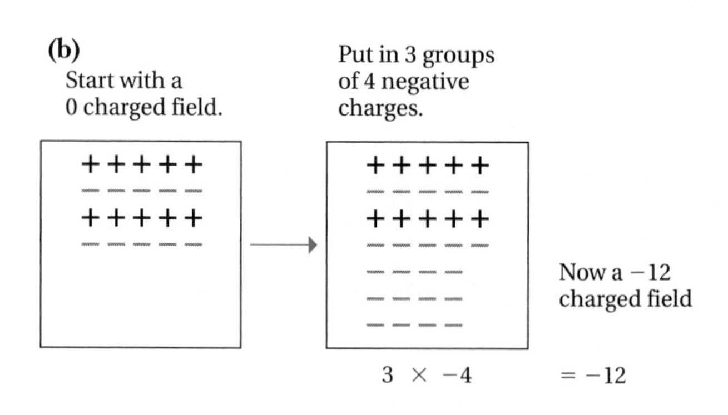

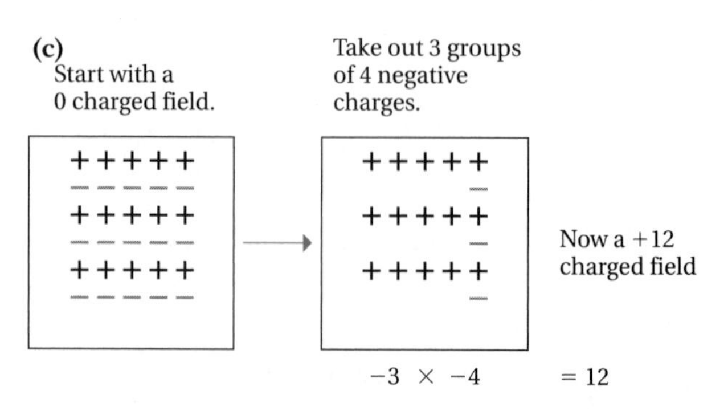

4. Start at zero on a number line. **(a)** Now at 0, walking east at 3 miles an hour. 2 hours from now you will be 6 miles to the right of zero, so $2(3) = 6$. **(b)** Now at 0, walking west at 3 miles an hour. 2 hours from now you will be 6 miles to the left of zero, so $2(-3) = -6$ **(c)** Now at 0, walking west at 3 miles an hour. 2 hours ago you were 6 miles to the right of zero, so $-2(-3) = 6$.

5. Methods will vary. **(a)** -63 **(b)** 506,898 **(c)** -80

6. -288

7. -104

8. 240

9. -512

10. 64

11. -280

12. -121

13. (a) -4370 **(b)** -8460 **(c)** 2278

14. -60

15. 105

16. -144

17. 64

18. negative

19. positive

20. (a) $-2[3 + (-4)] = -2(-1) = 2$; $-2(3) + [-2(-4)] = -6 + 8 = 2$. **(b)** $5[-3 + (-2)] = 5(-5) = -25$; $5(-3) + 5(-2) = -15 + (-10) = -25$. **(c)** $-6[-5 + (-4)] = -6(-9) = 54$; $-6(-5) + [-6(-4)] = 30 + 24 = 54$.

21. (a) $-2[3(-4)] = -2(-12) = 24$; $[-2(3)](-4) = -6(-4) = 24$. **(b)** $5[-3(-2)] = 5(6) = 30$; $[5(-3)](-2) = -15(-2) = 30$. **(c)** $-6[-5(-4)] = -6(20) = -120$; $[-6(-5)](-4) = 30(-4) = -120$.

22. (a) $-n - 4$ **(b)** $-6s$ **(c)** $192 - 8n$ **(d)** n

23. (a) -7 **(b)** 8 **(c)** -4

24. -8

25. -30

26. 3

27. -36

28. -289

29. -63

30. 81

31. 3

32. D (drop in temperature) $= 5(-7) = -35°$.

33. A (average change) $= -\$12M \div 4 = -\$3M$ per month.

34. T (total change) $= 5(-\$4) = -\$20$.

35. (a) 48 **(b)** $-80 \div 20 = -4$ **(c)** $128 \div 32 = 4$

36. \$229. $(-17 \times 9) + (-19 \times 4)$.

37. (a) Enter 81 in the calculator, push $\div$, enter -27, and push $=$. **(b)** Repeatedly subtract 27 from a beginning number of 81 until 0 is reached. The quotient is the number of subtractions, with the appropriate sign, or -3.

38.

-2	-9	12
-36	6	-1
3	-4	-18

39. $10(-12) = -120$.

40. (a) 104 **(b)** 18,096 **(c)** 160 **(d)** 16

41. Responses will vary. Possibilities include: **(a)** Property of opposites, $-a(-b) = -ab$ **(b)** Distributive property for opposites over addition, $-(a + b) = -a + (-b)$ **(c)** Property of opposites, $-1a = -a$

42. (a) Subtracting an integer by adding the opposite **(b)** Distributive property **(c)** A property of opposites **(d)** Subtracting an integer by adding the opposite

43. Positive integer exponents represent the number of times a number is to be used as a factor in multiplication **(a)** 125 **(b)** 4 **(c)** -125 **(d)** -3

44. n is a non-zero integer: **(a)** C **(b)** P **(c)** C **(d)** N **(e)** C **(f)** P **(g)** C **(h)** P **(i)** C **(j)** P

45.

Pos	Neg	0
Neg	Pos	0
0	0	0

46. (a) True for all integers, $c \neq 0$. **(b)** Not true. $6 \div (1 + 2)$ is not equal to $6 \div 1 + 6 \div 2$. **(c)** True. **(d)** Not true. $-8 < -4$ but $(-8)(-8) > (-4)(-4)$. **(e)** Not true. $-8 < -4$ but $10 - (-8) > 10 - (-4)$.

47. (a) 10, -20, 12 **(b)** 64, -128, 256 **(c)** -486, 1458, -4374 **(d)** 149, -299, 597

48. Responses will vary. Possibilities include: **(a)** Substituting compatible numbers, we obtain the estimate $(-100) \times (50) = -5000$. **(b)** Clustering the absolute values around 20, we obtain the estimate $-(20)(20)(-20)(-20) = -160{,}000$. **(c)** Substituting compatible numbers: $250 \div (-50) = -5$. **(d)** Rounding both factors we obtain: $-700(-200) = 140{,}000$. **(e)** Using front-end estimation we get: $-3600 \div 400 = -9$.

49. $-a + a = 0$, by the opposites property. $a(-1) + a = a(-1) + a(1) = a(-1 + 1) = 0$. Thus both $-a$ and $a(-1)$ are additive inverses of a. But the additive inverse of an integer is unique. So $-a = a(-1)$.

50. $-(ab) + ab = 0$ (opposites property). $a(-b) + ab = a(-b + b) = 0$ (distributive property, opposites property, zero property). $-ab + ab = (-a + a)b = 0$ (distributive property, opposites property, zero property). So $-(ab) = a(-b) = -ab$.

51. Because we want a unique answer to a division, and the argument that 5 is the quotient could be made for any other integer, such as 2, -4, 9, etc.

52. The arguments are intuitively valid. Because of the

student's understanding of multiplication, it seems reasonable that $3(-2)$ is $-2 + (-2) + (-2) = -6$. Since -3 is the opposite of 3, it also seems reasonable that $-3(-2)$ might be the opposite of $3(-2)$, and that $-3(-2) = 6$.

53. Integer multiplication is associative, while integer division is not: $(24 \times 6) \times 2 = 24 \times (6 \times 2) = 288$ but $(24 \div 6) \div 2 = 2$ and $24 \div (6 \div 2) = 8$. Integer multiplication is commutative, division is not. $12 \times 2 = 2 \times 12$ but $12 \div 2 \neq$ to $2 \div 12$.

54. Use maximum and minimum values and give the area of the field as between 13,685 and 15,717 square yards.

55. **(a)** Additive identity. **(b)** Distributive property. **(c)** Additive identity. **(d)** Since both 0 and $a \times 0$ add to $a \times 1$ to yield $a \times 1$, they must be different representations of the same number and thus be equal.

56. A spreadsheet could be used to produce the table below. The values predicted by the equation are in close agreement with the actual recorded values.

57. 1218 lbs.

58. Responses will vary. Assuming 2 packs/day, 365 days/year for 50 years at an average price of \$5/pack, the expense for cigarettes is \$182,500.

59. The argument refers to the absolute value of the numbers, not the numbers themselves.

60. The student did not prove the generalization. A model uses a specific example to *suggest* a correct procedure. A proof uses accepted assumptions and *logic* to *prove* that the generalization is true.

61. Responses will vary. A Korean mathematician may have laid out 4 black rods horizontally to represent 4 and, under each of the black rods, laid 3 red rods vertically to represent -3. By counting all the red rods, the product was -12.

62. Responses will vary

63. Responses will vary. The following ideas might be considered. Multiplication can be performed by repeated addition and division by repeated subtraction. Multiplication and addition are commutative and associative but subtraction and division are neither. The integers are closed under addition, multiplication, and subtraction but not under division.

Chapter 5 Review Exercises

1. **(a)** True. **(b)** False. The sum of a positive and a negative integer may be positive, negative, or 0. **(c)** False. The absolute value of 0 is 0 and 0 is not positive. **(d)** False. The product of 2 negative integers is always positive. **(e)** False. The integer 0 is neither positive nor negative.

2. **(a)** -2 **(b)** 7 **(c)** -9 **(d)** 3 **(e)** 14

3. **(a)** -7 **(b)** -4 **(c)** -8 **(d)** 9

4. **(a)** Use 3 red counters and 6 black, leaving 3 unpaired black counters. $6 + (-3) = 3$. **(b)** Start with a 0 charged field. Add 6 positive charges and 3 negative charges. End with a $+3$ charged field. $6 + (-3) = 3$. **(c)** Begin at 0, move 6 units right, then move 3 units left. End at 3. $6 + (-3) = 3$. **(d)** Subtract the absolute values, $6 - 3$, and use the sign of the number with the larger absolute value. $6 + (-3) = 3$.

5. **(a)** Represent 2 with 10 black and 8 red counters. Remove 5 red counters. $2 - (-5) = 7$. **(b)** Start with a 0 charged field. Put in 2 positive charges. Take out 5 negative charges. End with a $+7$ charged field. $2 - (-5) = 7$. **(c)** Begin at 0, move 2 units right, face left, move 5 units backward. End at 7. $2 - (-5) = 7$. **(d)** Subtract by adding the opposite. $2 - (-5) = 2 + 5 = 7$.

6. **(a)** $-12, -5, -3, 0, 1, 2, 8, 15$ **(b)** $-60, -56, -34, -17, 9, 27, 34, 46$

Answer for Exercise 56

Year	Actual min:sec	Actual sec	Predicted sec	High/Low (H/L)	Difference (sec)
1954	3:59.4	239.4	238.4	L	−1
1955	3:58.0	238	238	—	—
1957	3:57.2	237.2	237.2	—	—
1958	3:54.5	234.5	236.8	H	2.3
1962	3:54.4	234.4	235.2	H	0.8
1964	3:54.1	234.1	234.4	H	0.3
1965	3:53.6	233.6	234	H	0.4
1966	3:51.3	231.3	233.6	H	2.3
1967	3:51.1	231.1	233.2	H	1.1
1975	3:49.4	229.4	230	H	0.6
1980	3:48.8	228.8	228	L	−0.8
1985	3:46.3	226.3	226	L	−0.3
1993	3:44.4	224.4	222.8	L	−1.6

7. The larger of two integers is farther to the right on a number line.

8. (a) $-7 < -5$ because $-7 + 2 = -5$. **(b)** $1 > -5$ because $-5 + 6 = 1$. **(c)** $0 > -3$ because $-3 + 3 = 0$.

9. (a) Start with a bag and pile of counters. Do the opposite of taking 4 counters out of the bag 3 times. The bag now has 12 more counters than at the beginning. $-3(-4) = 12$. **(b)** Start with a 0 charged field. Take out 3 groups of 4 negative charges. The result is a +12 charged field. $-3(-4) = 12$. **(c)** You are at 0 walking west at 4 mph. 3 hours ago, you were at 12 miles east of 0. $-3(-4) = 12$.

(d) $3(-4) = -12$
$2(-4) = -8$
$1(-4) = -4$
$0(-4) = 0$
$-1(-4) = 4$
$-2(-4) = 8$
$-3(-4) = 12$

10. (a) $12 \div (-4)$ is the factor that multiplies by -4 to get 12. So $12 \div (-4) = -3$. **(b)** Dividing the absolute values we get $12 \div 4 = 3$. Only one of the dividend/divisor pair is negative, so the quotient is negative, and $12 \div (-4) = -3$.

11. (a) Distributive property of multiplication over addition **(b)** Associative property of multiplication **(c)** Additive inverse property **(d)** Distributive property **(e)** For all non-zero integers b, $b \div b = 1$. **(f)** The associative property of multiplication and the property of opposites.

12. Let $n = -12$. Substituting -12 for n, $-n = -(-12) = 12$.

13. (a) Jan. = −5, Mar. = −3, May = −1, Aug. = 2, Oct. = 4, Dec. = 6. **(b)** November.

14. $a - b$ and $b - a$ are opposites since their sum is 0. $(a - b) + (b - a) = (a - a) + (b - b) = 0$.

15. (a) Associative property of addition **(b)** Additive inverse property **(c)** Additive identity property

16. Examples will vary. Possibilities include: **(a)** $2 \div 4$ does not represent an integer. **(b)** $12 \div (2 + 4) = 2$, but $12 \div 2 + 12 \div 4 = 9$. **(c)** $6 \div 2 = 3$ but $2 \div 6$ does not represent an integer. **(d)** $(24 \div 6) \div 2 = 2$ but $24 \div (6 \div 2) = 8$. **(e)** $(12 - 6) - 4 = 2$, but $12 - (6 - 4) = 10$. **(f)** $6 - 2 = 4$, but $2 - 6 = -4$.

17. Low = −94°. High = 98°.

18. Next-to-last score: 5 under par. Last round score: 2 over par.

19. Responses will vary. The company is getting the income for 800 bags when selling 799 bags' worth of fertilizer.

20. Responses will vary. The counters model mirrors the more formal procedures. For example, in using the counters model for subtracting integers, the effect of subtracting 2 reds is the same as adding 2 blacks; i.e., to subtract −2, add its opposite.

21. The tables are:

+	P	N
P	P	S
N	S	N

−	P	N
P	S	P
N	N	S

×	P	N
P	P	N
N	N	P

÷	P	N
P	P	N
N	N	P

The discussions will vary. Responses may include the ideas that the tables are valuable as summary devices but are limited by the S entries. Also, the tables help to compare and contrast related operations, such as multiplication and division.

22. Diagrams and explanations will vary. They should show that integer subtraction is finding a missing addend in an addition, and that integer division is finding a missing factor in a multiplication.

Chapter 6 Answers

Section 6.1

1. Models will vary.
2. Models will vary.
3. Models will vary.
4. Models will vary.
5. Models will vary.
6. Models will vary.

7. The thousands cube represents 10; the tens stick represents 0.1; the units cube represents 0.01.

8. (a) $-\frac{2}{3}$ **(b)** $\frac{18}{4} = \frac{9}{2}$ **(c)** 0 **(d)** $\frac{386}{1000} = \frac{193}{500}$

9. (a) $\frac{1}{2}$ **(b)** $-\frac{1}{18}$ **(c)** $\frac{1}{2}$ **(d)** $\frac{386}{1000}$

10. Answers will vary. Possibilities include **(a)** $-\frac{12}{14} = -\frac{18}{21} = -\frac{24}{28}$ **(b)** $\frac{0}{10} = \frac{0}{1} = \frac{0}{1000}$ **(c)** $\frac{3}{5} = \frac{6}{10} = \frac{9}{15}$ **(d)** $\frac{2c}{2d} = \frac{3c}{3d} = \frac{4c}{4d}$

11. (a) 6.4 **(b)** $\frac{386}{1000}$ **(c)** $\frac{2}{3}$ **(d)** 0.3636...

12. (a) $3(100) + 8(10) + 5(1) + 1(\frac{1}{10}) + 9(\frac{1}{100}) + 2(\frac{1}{1000})$ **(b)** $0(1) + 0(\frac{1}{10}) + 0(\frac{1}{100}) + 6(\frac{1}{1000})$ **(c)** $6(1) + 0(\frac{1}{10}) + 8(\frac{1}{100}) + 5(\frac{1}{1000})$

13. (a) 3.48 09 or 3.48 E9 **(b)** 9.08 −05 or 9.08 E−5 **(c)** 6.085 00 or 6.085

14. Models will vary.

15. (a)

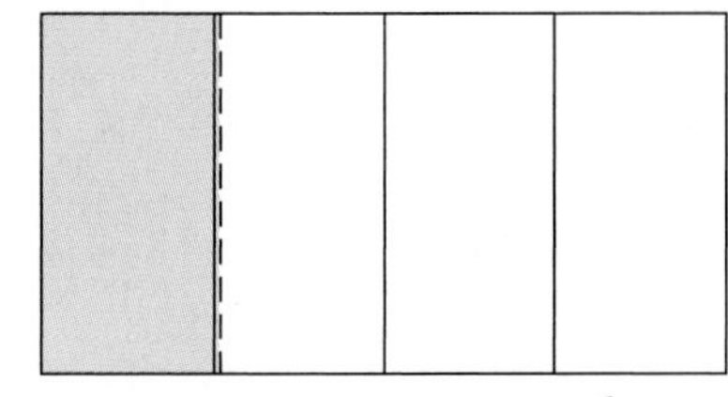

0.28 is a little more than $\frac{1}{4}$.

(b)

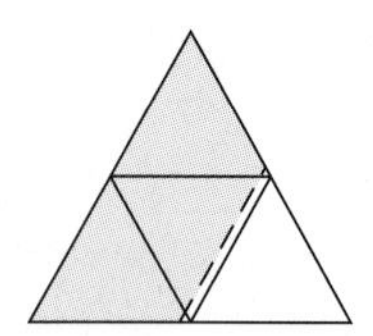

0.72 is a little less than $\frac{3}{4}$.

(c)

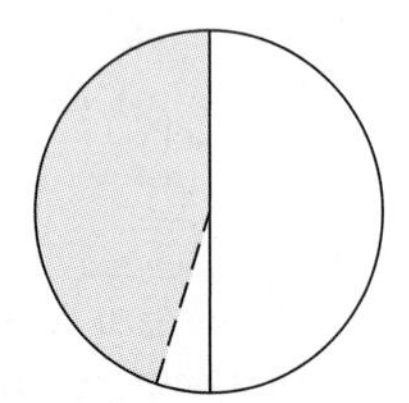

0.45 is a little less than $\frac{1}{2}$.

(d)

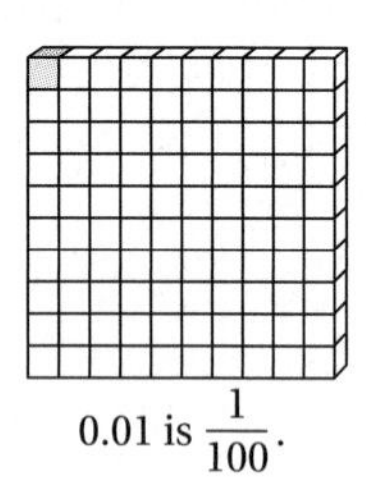

0.01 is $\frac{1}{100}$.

(e)

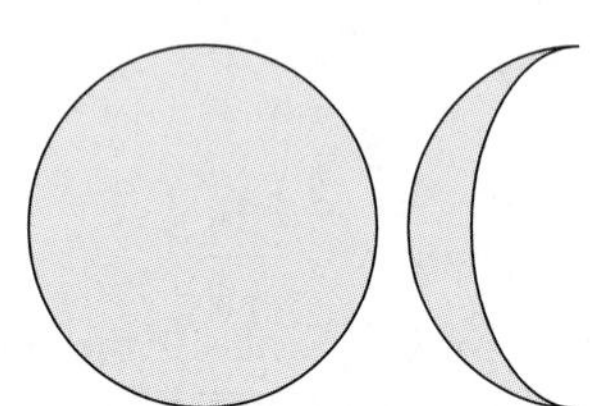

1.18 is a little more than 1.

(f)

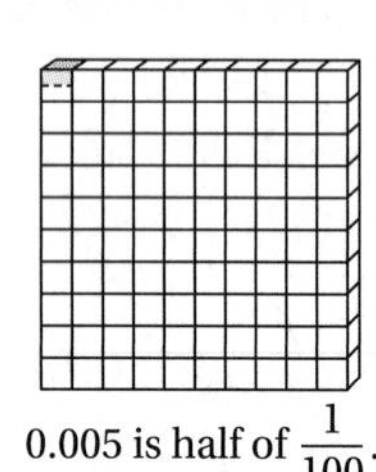

0.005 is half of $\frac{1}{100}$.

16. The part selected to represent 0.005 should be half the area selected to represent 0.01.

17. **(a)** about $\frac{1}{4}$ **(b)** about $3\frac{1}{10}$ **(c)** about $\frac{4}{100} = \frac{2}{50} = \frac{1}{25}$ **(d)** 1

18. **(a)** 0.666… **(b)** approximately $\frac{6}{100} = \frac{3}{50}$ **(c)** about 11 **(d)** 0.00333…

19. $\frac{3}{10} = \frac{30}{100}$

20. **(a)** $a = b = \frac{1}{4}, d = f = g = \frac{1}{8}, c = e = \frac{1}{16}$ **(b)** $a = b = 1, d = f = g = \frac{1}{2}, c = e = \frac{1}{4}$

21. If the project is anticipated to take h hours, each student should plan on working $\left(\frac{1}{n}\right)(h)$ hours.

22. Let a represent some nonzero integer. If a is divided by 0, $a \div 0 = b$ or $a = 0 \times b$. But $b \times 0 = 0$, thus $a = 0$. But a was declared to be nonzero.

23. Answers will vary.

24. **(a)**

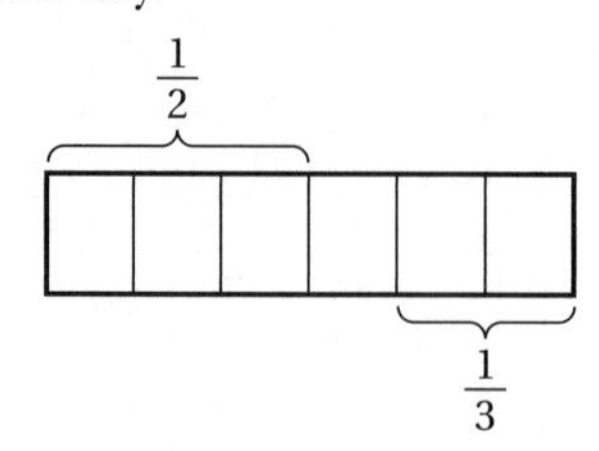

(b) Let X X X X X X X X X X X X represent 1. Then X X X X X X X X X represents $\frac{9}{12} = \frac{3}{4}$ and X X X X X X X X represents $\frac{8}{12} = \frac{2}{3}$. **(c)** Let 0 0 0 0 0 0 0 0 0 0 represent 1. Then 0 0 0 0 represents $\frac{4}{10} = \frac{2}{5}$ and 0 0 0 0 0 0 0 represents $\frac{7}{10}$. **(d)** The model of the whole needs to have a number of parts equal to the LCM of the denominators.

25. The facts are that $\frac{1}{13}$ of the women on one campus were polled and $\frac{48}{1000}$ responded that they would prefer to attend all female mathematics classes. On the other campus $\frac{1}{5}$ of the women students were polled and $\frac{48}{1000}$ of those polled said yes.

26. **(a)** $\frac{2}{5}, \frac{1}{25}, \frac{7}{8}$ can be written as terminating decimals; $\frac{2}{6}, \frac{6}{13}, \frac{9}{29}$ cannot. **(b)** A simplified fraction with a denominator that has prime factors of only 2 or 5 can be written as a terminating decimal. **(c)** $\frac{6}{12} = \frac{1}{2} = 0.5$, terminating; $\frac{6}{15} = \frac{2}{5} = 0.4$, terminating; $\frac{28}{140} = \frac{1}{5} = 0.2$, terminating; $\frac{0}{7} = \frac{0}{1}$, terminating. **(d)** For $\frac{a}{b} \times \frac{c}{c} = \frac{ac}{10^n}$, where a, b, and c are integers, b cannot have prime factors other than 2 or 5, because 10^n has prime factors of only 2 and 5.

27. Annexing a 0 to the right of a decimal number is like multiplying both the numerator and denominator of the fraction by 10: $0.3 = \frac{3}{10}$, $0.30 = \frac{30}{100}$.

28. Missing displays: 0.3333333; 0.6666667; 1; 1.3333333 **(a)** No; the calculator is rounding $0.\overline{3} + 0.\overline{3}$. **(b)** No; the calculator is adding $1 \div 3 = \frac{1}{3}$ three times, just displaying it in decimal form.

29. Missing displays: 0.142857142; 0.285714285; 0.428571428; 0.571428571; 0.714285714; 0.857142857; 1. The decimal representations of $\frac{2}{7}, \frac{3}{7}, \frac{4}{7}, \frac{5}{7}, \frac{6}{7}$, and $\frac{7}{7}$ are not the corresponding multiples of the decimal display for $\frac{1}{7}$. The quotient of 1 divided by 7 is an infinitely repeating decimal; when a number is entered from the keyboard it is necessarily finite.

30. The fractions $\frac{1}{5}$ and $\frac{1}{20}$ may refer to different wholes.

31. The circled numbers are negative exponents for powers of 10, as in our expanded notation.

32. Answers will vary.

33. **(a)** The women's wage, w, is $\frac{3}{4}$ of the men's wage, m: $w = \left(\frac{3}{4}\right)m$. Now, $w + \left(\frac{1}{4}\right)w = \left(\frac{5}{4}\right)w$, a 25% increase in pay. But $\left(\frac{5}{4}\right)w = \left(\frac{5}{4}\right)\left(\frac{3}{4}\right)m = \left(\frac{15}{16}\right)m$. This is still less than the men's wage. **(b)** The second strategy equalizes the hourly wages. But the gross earnings of women will still be

$\frac{3}{4}$ of those of men.

(c)

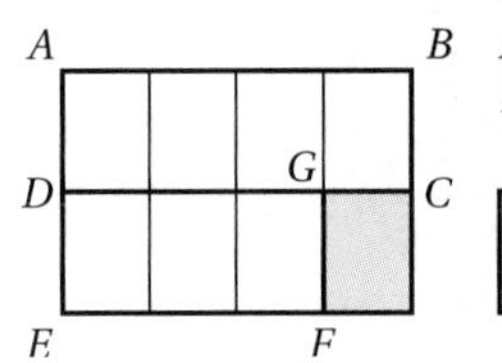

Section 6.2

1. (a)

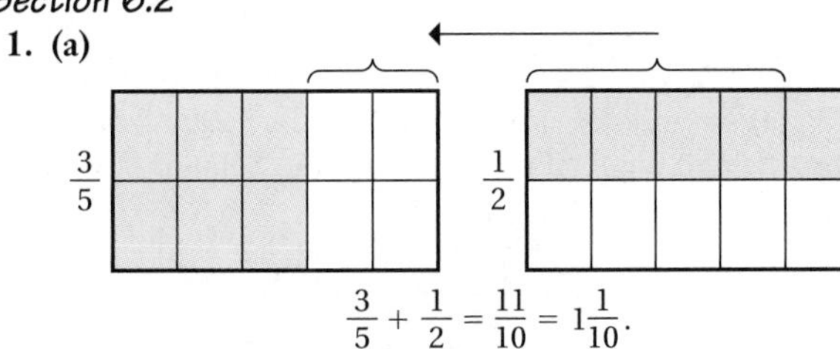

$\frac{3}{5} + \frac{1}{2} = \frac{11}{10} = 1\frac{1}{10}$.

(b)

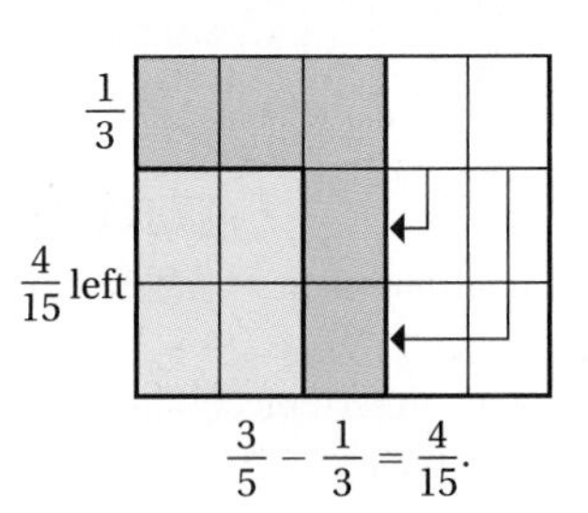

$\frac{3}{5} - \frac{1}{3} = \frac{4}{15}$.

(c)

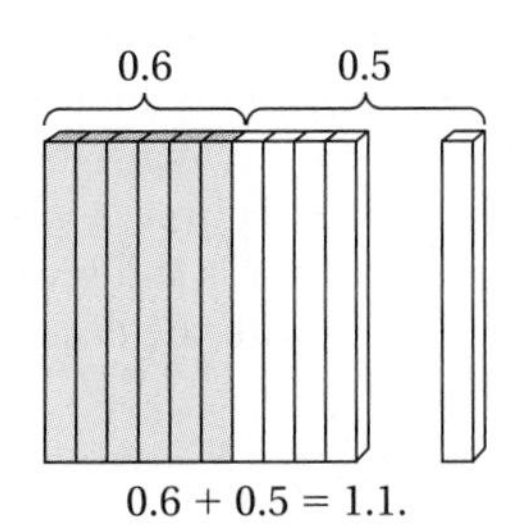

$0.6 + 0.5 = 1.1$.

(d)

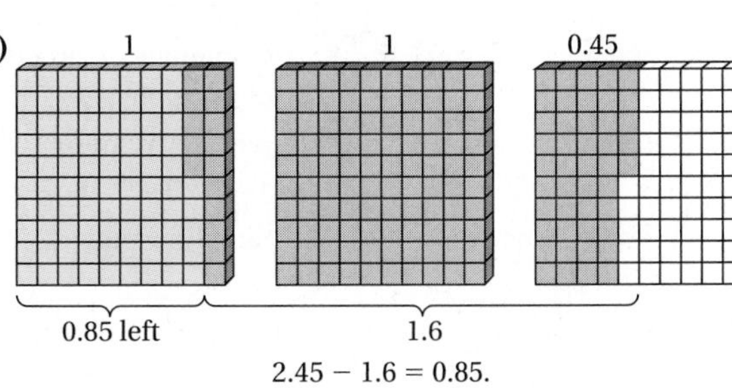

$2.45 - 1.6 = 0.85$.

2.

(a) → $\frac{1}{2}$ → $\frac{3}{5}$ → $1\frac{1}{10}$

(b) → $\frac{3}{5}$ ← $-\frac{1}{3}$ → $\frac{4}{15}$

(c) → 0.5 → 0.6 → 1.1

(d) → 2.45 ← − 1.6 → 0.85

3. **(a)** $\frac{20}{3}$ **(b)** $-\frac{19}{4}$ **(c)** $\frac{17}{2}$ **(d)** $\frac{109}{10}$

4. **(a)** $-3\frac{3}{5}$ **(b)** $3\frac{1}{8}$ **(c)** $4\frac{3}{10}$ **(d)** $7\frac{4}{7}$

5. **(a)** $(\frac{1}{3} + \frac{2}{3}) + \frac{3}{4} = 1\frac{3}{4}$ **(b)** $(\frac{5}{6} + \frac{4}{6}) + \frac{3}{8} = (\frac{9}{6}) + \frac{3}{8} = \frac{3}{2} + \frac{3}{8} = \frac{12}{8} + \frac{3}{8} = \frac{15}{8} = 1\frac{7}{8}$

6. **(a)** commutative and associative properties of addition **(b)** commutative and associative properties of addition; the fundamental law of fractions

7. **(a)** $2\frac{1}{3} - 1\frac{2}{3} + 3\frac{2}{3} = (2\frac{1}{3} + 3\frac{2}{3}) - 1\frac{2}{3} = 6 - 1\frac{2}{3} = 4\frac{1}{3}$
(b) $2\frac{1}{3} - 1\frac{2}{3} + 3\frac{2}{3} = 2\frac{1}{3} + (3\frac{2}{3} - 1\frac{2}{3}) = 2\frac{1}{3} + 2 = 4\frac{1}{3}$
(c) $2\frac{1}{3} - 1\frac{2}{3} + 3\frac{2}{3} = (2 - 1 + 3) + (\frac{1}{3} - \frac{2}{3} + \frac{2}{3}) = 4\frac{1}{3}$

8. **(a)** $1\frac{3}{8} + 3\frac{3}{4} - 1\frac{1}{4} = 1\frac{3}{8} + 2\frac{2}{4} = 1\frac{3}{8} + 2\frac{4}{8} = 3\frac{7}{8}$
(b) $1\frac{3}{8} + 3\frac{3}{4} - 1\frac{1}{4} = (1\frac{3}{8} + 3\frac{6}{8}) - 1\frac{2}{8} = 4\frac{9}{8} - 1\frac{2}{8} = 3\frac{7}{8}$
(c) $1\frac{3}{8} + 3\frac{3}{4} - 1\frac{1}{4} = (1 + 3 - 1) + (\frac{3}{8} + \frac{6}{8} - \frac{2}{8}) = 3\frac{7}{8}$

9. **(a)** $10\frac{7}{30}$ **(b)** $\frac{143}{60} = 2\frac{23}{60}$ **(c)** $2\frac{7}{24}$ **(d)** $\frac{1}{24}$ **(e)** 4.768 **(f)** 3.414

10. **(a)** $32.19 - 14.8 = 17.39$ because $17.39 + 14.8 = 32.19$ **(b)** $\frac{15}{16} - \frac{1}{4} = \frac{11}{16}$ because $\frac{11}{16} + \frac{1}{4} = \frac{11}{16} + \frac{4}{16} = \frac{15}{16}$

11. **(a) i.** $24.6 + 3.09 = \frac{246}{10} + \frac{309}{100} = \frac{2460}{100} + \frac{309}{100} = 27.69$

ii.
$$\begin{array}{r} 24.60 \\ +\ 3.09 \\ \hline 27.69 \end{array}$$

(b) The sums are the same.

12. **(a) i.** $24.6 - 3.09 = \frac{246}{10} - \frac{309}{100} = \frac{2460}{100} - \frac{309}{100} = 21.51$

ii.
$$\begin{array}{r} 24.60 \\ -\ 3.09 \\ \hline 21.51 \end{array}$$

(b) The differences are the same.

13. (a) greater than 1 **(b)** greater than 1 **(c)** less than $\frac{1}{2}$ **(d)** between $\frac{1}{2}$ and 1 **(e)** greater than 1 **(f)** between $\frac{1}{2}$ and 1

14. From least to greatest: $(\frac{9}{11} - \frac{3}{4})$, $(\frac{5}{6} - \frac{1}{5})$, $(\frac{3}{7} + \frac{4}{10})$, $(4\frac{4}{5} - 2\frac{1}{2})$, $(6\frac{7}{8} + 3\frac{2}{7})$

15. (a) $1 + \frac{1}{2} + \frac{1}{2} + \frac{1}{4} + \frac{1}{4} = 2\frac{1}{2}$ **(b)** $1 + \frac{1}{2} + \frac{1}{4} = 1\frac{3}{4}$ **(c)** The large square (all the pieces) = 4; as all the pieces but a and g have been used, the value of the shape is $4 - a - g = 4 - 1 - \frac{1}{2} = 2\frac{1}{2}$.

16. (a) $\frac{1}{12}$ **(b)** 8.79 **(c)** $-\frac{1}{6}$

17. (a) $\frac{3}{4} - \frac{1}{4} = \frac{2}{4}$ but $\frac{1}{4} - \frac{3}{4} = -\frac{2}{4}$ **(b)** $(\frac{7}{8} - \frac{4}{8}) - \frac{1}{8} = \frac{2}{8}$ but $\frac{7}{8} - (\frac{4}{8} - \frac{1}{8}) = \frac{4}{8}$

18. (a) -0.06; $0.06 + (-0.06) = (-0.06) + 0.06 = 0$ **(b)** $-\frac{3}{7}$; $\frac{3}{7} + (-\frac{3}{7}) = -(\frac{3}{7}) + \frac{3}{7} = 0$ **(c)** $2\frac{5}{8}$; $-2\frac{5}{8} + 2\frac{5}{8} = 2\frac{5}{8} + -2\frac{5}{8} = 0$ **(d)** 0; $0 + 0 = 0$

19. (a) 17.51 and 57.52; 9.23 and 65.8; 87.98 and −12.95 all add to 75.03. **(b)** The first number was randomly chosen; the other pair came from subtracting the first from 75.03.

20. (a) 67.4 and 23.8; 95.5 and 51.9; −79.8 and −123.4 **(b)** The first number chosen (listed second in the pairs) was randomly selected; the second number came from adding 43.6 to the first.

21. (a) $\frac{1}{6}, \frac{1}{30}, \frac{1}{60}, \frac{1}{60}, \frac{1}{30}, \frac{1}{6}$; $\frac{1}{7}, \frac{1}{42}, \frac{1}{105}, \frac{1}{140}, \frac{1}{105}, \frac{1}{42}, \frac{1}{7}$ **(b)** Any two adjacent fractions have a sum equal to the fraction between them in the row above. **(c)** The patterns are still there, but they aren't as visible because of the nonterminating, repeating decimals needed for $\frac{1}{3}, \frac{1}{6}, \frac{1}{7}$, and so on.

22. (a) $\frac{12}{5}$ **(b)** 21 times

23. (a) It is the sum of $\frac{1}{3}$ and $\frac{1}{4}$ in fraction form, and it is in simplest form. **(b)** $\frac{19}{12}$ **(c)** 10 times, because you press the equals sign once to get the first display of $\frac{7}{12}$ and then 9 more times because 34 − 7 is 27; 3 fourths are added each time the equals sign is pressed, and there are nine 3's in 27. **(d)** No; you can't add just $\frac{3}{4}$ to $\frac{7}{12}$ and ever get to $\frac{69}{12}$ because 69 − 7 is 62, and 62 isn't a multiple of 3.

24. (a) $-\frac{5}{6}$ **(b)** 14 times **(c)** N/D → n/d does not appear when the numerator and denominator are relatively prime.

25. (a) It is the difference of $\frac{1}{5} - \frac{2}{9}$ in fraction form. **(b)** $-\frac{31}{45}$, then $-\frac{41}{45}$ **(c)** 9 times because the number of equals signs pushed is always 1 less than the number of tens in the denominator of the fraction. **(d)** N/D → n/d will appear when the numerator and denominator have a common factor other than 1.

26. (a) One possibility is about $\frac{1}{2}$. The sum of $\frac{1}{10}$ and $\frac{1}{10}$ is a little less than $\frac{1}{4}$ and $\frac{1}{3}$ is a little bigger than $\frac{1}{4}$.

(b)

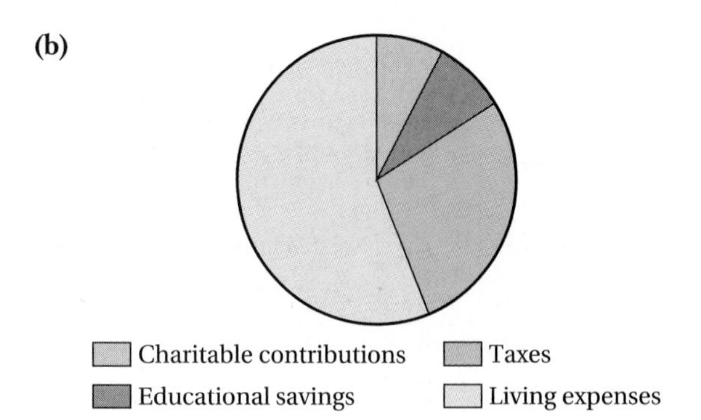

(c) Let E represent expenses as fraction of earnings and let 1 represent earnings. Then $1 - \frac{1}{10} - \frac{1}{10} - \frac{1}{3} = E$; $E = \frac{7}{15}$.

27. (a) Because $\frac{1}{3} + \frac{1}{4} + \frac{1}{6}$ is about $\frac{3}{4}$, the \$600 is about $\frac{1}{4}$ of the total. So the total was about 4 times \$600 or \$2400.

(b)

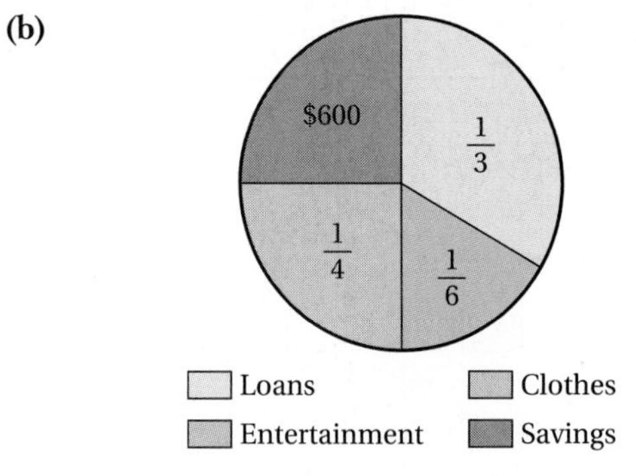

(c) Let I represent the inheritance. Then $[1 - (\frac{1}{4} + \frac{1}{6} + \frac{1}{3})]I = 600$; $I = 2400$.

28. $5\frac{1}{2}$ hours

29. Lining up the decimal points is like writing each of the decimals with common denominators.

30. For example, both algorithms are based on finding common denominators. For decimals, this is done by lining up the place-value positions. However, exact sums can be found for fractions such as thirds and sevenths with the fraction algorithm, but not with the decimal algorithm.

31.

$2\frac{1}{4} + 3\frac{1}{3} = 2 + \frac{1}{4} + 3 + \frac{1}{3}$	Symbolism for mixed numbers
$= 2 + (\frac{1}{4} + 3) + \frac{1}{3}$	Associative property of addition
$= 2 + (3 + \frac{1}{4}) + \frac{1}{3}$	Commutative property of addition
$= (2 + 3) + (\frac{1}{4} + \frac{1}{3})$	Associative property of addition
$= 5 + \frac{7}{12}$	Simplification
$= 5\frac{7}{12}$	

32. No; the fractional raises are with respect to different wholes.

33. **(a)** $\frac{4}{1} + \frac{3}{2} = 5\frac{1}{2}$ **(b)** $\frac{8}{5} + \frac{7}{6} = \frac{83}{30} = 2\frac{23}{30}$ **(c)** $\frac{9}{2} + \frac{8}{5} = \frac{61}{10} = 6\frac{1}{10}$ **(d)** If $0 < a < b < c < d$, then $\frac{d}{a} + \frac{c}{b}$ = greatest sum.

34. **(a)** $\frac{1}{3} + \frac{2}{4} = \frac{10}{12} = \frac{5}{6}$ **(b)** $\frac{5}{7} + \frac{6}{8} = \frac{82}{56}$ **(c)** $\frac{2}{8} + \frac{5}{9} = \frac{58}{72} = \frac{29}{36}$ **(d)** If $0 < a < b < c < d$, then $\frac{a}{c} + \frac{b}{d}$ = smallest sum.

35. **(a)** $\frac{4}{1} - \frac{2}{3} = 3\frac{1}{3}$ **(b)** $\frac{8}{5} - \frac{6}{7} = \frac{26}{35}$ **(c)** $\frac{9}{2} - \frac{5}{8} = \frac{31}{8}$ **(d)** If $0 < a < b < c < d$, then $\frac{d}{a} - \frac{b}{c}$ = greatest difference.

36. **(a)** If we restrict the problem to the smallest positive difference, we get $\frac{2}{4} - \frac{1}{3} = \frac{1}{6}$; if we drop the restriction, we get $\frac{2}{3} - \frac{4}{1} = -3\frac{1}{3}$. **(b)** $\frac{6}{8} - \frac{5}{7} = \frac{1}{28}$ **(c)** $\frac{5}{9} - \frac{2}{8} = \frac{11}{36}$ **(d)** If $0 < a < b < c < d$, then $\frac{b}{d} - \frac{a}{c}$ = smallest positive difference.

37. **(a)** Any integer equal to 0 couldn't be used as a denominator. **(b)** One negative integer in the numerator or denominator would cause the relative value of the fraction to be exactly the opposite of the relative value with the positive integer; for example, if $\frac{9}{2}$ were the largest possible fraction that could be made, $-\frac{9}{2}$ would be the smallest.

38. Answers will vary.

39. Answers will vary.

40. **(a)** One possibility: $\frac{7}{12} = \frac{1}{3} + \frac{1}{4}$ **(b)** $\frac{2}{7} = \frac{1}{4} + \frac{1}{28}$ **(c)** One possibility: $\frac{3}{5} = \frac{1}{2} + \frac{1}{10} = \frac{1}{3} + \frac{1}{5} + \frac{1}{15}$

41. **(a)** If $b = d$ then either is the least common denominator. **(b)** If either b or d is a factor of the other, then the larger of the two denominators is the least common denominator. **(c)** If b and d aren't factors of one another, but have common factors $f_1, f_2, f_3, \ldots$, the least common denominator is $bd/[(f_1)(f_2)(f_3)\ldots]$. **(d)** If b and d are relatively prime, the least common denominator is bd.

Section 6.3

1. Models will vary.

2. Models will vary.

3. Models will vary.

4. Models will vary.

5. $\frac{a}{b} \times \frac{c}{d} = \frac{ac}{bd}$

6. **(a)** The shaded area represents $\frac{1}{5}$ of $\frac{1}{4}$, a product of $\frac{1}{20}$. **(b)** The shaded area represents $\frac{2}{3}$ of $\frac{2}{3}$, a product of $\frac{4}{9}$.

7. **(a)** 1 **(b)** $\frac{17}{6} = 2\frac{5}{6}$ **(c)** 6.41 **(d)** $-\frac{21}{10} = -2\frac{1}{10}$ **(e)** $\frac{a^2bc^2}{bc}$ **(f)** $3\frac{3}{5}$

8. a and c may be represented by $\frac{1}{2} \times 10$.

9. Example: If $\frac{1}{3}$ of a class is over 21 and $\frac{1}{4}$ of those students age-eligible are registered to vote, what fraction of the class can vote in the next election?

10. Example: Find the area of a surface 2.5 by 3.5 meters.

11. Example: If $\frac{1}{3}$ of your stock portfolio has decreased by 150% and the other holdings remain constant, what is the change in the value of your holdings?

12. Example: Seven brothers had 7 dollars to give equally to seven brides. How much money did each bride receive?

13. Example: If the tax rate is \$0.125 per dollar, how much tax is there on \$8.00?

14. **(a)** **i.** $24.6 \times 3.09 = 24\frac{6}{10} \times 3\frac{9}{100} = \frac{246}{10} \times \frac{309}{100} = \frac{76{,}014}{1000} = 76.014$

ii.

$$\begin{array}{r} 24.6 \\ \times\ 3.09 \\ \hline 2214 \\ 0000 \\ 738 \\ \hline 76.014 \end{array}$$

(b) The products are the same because the same two numbers are being multiplied.

15. **(a)** $\frac{8}{5}$ because $\frac{5}{8} \times \frac{8}{5} = \frac{8}{5} \times \frac{5}{8} = 1$ **(b)** $-\frac{11}{36}$ because $-3\frac{3}{11} = -\frac{36}{11}$ and $-\frac{36}{11} \times -\frac{11}{36} = -\frac{11}{36} \times -\frac{36}{11} = 1$ **(c)** $\frac{100}{15}$ because $0.15 = \frac{15}{100}$ and $\frac{15}{100} \times \frac{100}{15} = \frac{100}{15} \times \frac{15}{100} = 1$ **(d)** $\frac{100}{918}$ because $9.18 = \frac{918}{100}$ and $\frac{918}{100} \times \frac{100}{918} = \frac{100}{918} \times \frac{918}{100} = 1$ **(e)** $\frac{36}{35}$ because $(\frac{1}{2} + \frac{1}{3})(\frac{1}{2} + \frac{2}{3}) = \frac{35}{36}$ and $\frac{36}{35} \times \frac{35}{36} = \frac{35}{36} \times \frac{36}{35} = 1$ **(f)** If $x \neq 0$, the multiplicative inverse of $-x$ is $-\frac{1}{x}$ because $(-x)(-\frac{1}{x}) = (-\frac{1}{x})(-x) = \frac{x}{x} = 1$; if $x = 0$, $-x = 0$, and 0 has no multiplicative inverse.

16. **(a)** between $\frac{1}{2}$ and 1 **(b)** less than $\frac{1}{2}$ **(c)** very close to 1 **(d)** greater than 1 **(e)** between $\frac{1}{2}$ and 1 **(f)** greater than 1

17. **(a)** $\frac{3}{4}$ times about 20 is about equal to 16 because $1 \times 16 = 16$ and $\frac{3}{4} < 1$ **(b)** $2 \times 0.25 = 0.5$ because $2 \times 25 = 50$ **(c)** $-\frac{4}{3}$ times $-\frac{1}{2}$ is equal to $\frac{2}{3}$ because $-4 \times -\frac{1}{2} = 2$ **(d)** $\frac{5}{6}$ times about 1.5 is about equal to 1.5 because $1.5 \div 1.5 = 1$ and $\frac{5}{6}$ is close to 1

18. The fundamental law of fractions states that for $c \neq 0$, $\frac{a}{b} = (\frac{a}{b})(\frac{c}{c}) = \frac{ac}{bc}$. The fundamental law of fractions is an application of the identity element of multiplication, since $\frac{c}{c} = 1$ and $\frac{a}{b} \times 1 = \frac{a}{b}$.

19. False; 0 is a rational number and has no multiplicative inverse because $\frac{1}{0}$ is not defined.

20. **(a)** $(\frac{4}{3})x = 16$. $(\frac{3}{4})(\frac{4}{3})x = x = (\frac{3}{4})16 = 12$. **(b)** $9y = 24$. $(9)(\frac{1}{9})y = (\frac{1}{9})24 = y = \frac{24}{9} = \frac{8}{3}$. **(c)** $-\frac{p}{6} = \frac{92}{5}$. $(-6)(-\frac{p}{6}) = p = (-6)(\frac{92}{5}) = -\frac{552}{5}$ **(d)** $\frac{4}{x} = \frac{12}{5}$. $(\frac{4}{x})(x) = 4 = (\frac{12}{5})x$. $(4)(\frac{5}{12}) = \frac{20}{12} = \frac{5}{3} = (\frac{12}{5})(\frac{5}{12})x = x$.

21. It is always valid because multiplication is commutative.

22. **(a)** $31.8 \times 42 = 1335.6$ because $30 \times 40 = 1200$. **(b)** Because 0.3 of 4 is just over 1, $0.318 \times 4.2 = 1.3356$ **(c)** $318 \times 0.42 = 133.56$ because $300 \times 0.5 = 150$. **(d)** Because $3 \times 4 = 12$, the product of 3.18 and 4.2 is 13.356.

23. Answers will vary.

24. **(a)** $\frac{256}{625}$ **(b)** 6 presses **(c)** It switches to the decimal display. **(d)** The display would be 0 because $0 \times \frac{4}{5} = 0$.

25. **(a)** The fraction in the display is not in its simplest form. **(b)** N/D → n/d $\frac{2}{48}$ and N/D → n/d $\frac{2}{96}$. **(c)** 8 presses **(d)** No; $\frac{2}{246}$ is not equivalent to the product $\frac{2}{3} \times (\frac{1}{2})^x$ for any integer value of x.

26. Answers will vary.

27. **(a)** $\frac{2}{55}$ each year **(b)** \$3090.91 to the nearest cent

28. **(a)** $S = 24.95 + 0.35m$, where S = standard cost and m = minutes. **(b)** Let E represent the economy cost; then, $E = 30.00 + (m - 30)(0.38)$ **(c)** The chart displays costs under the two plans as a function of time. **(d)** If the phone is to be used for calls other than emergencies, the economy plan is better.

29. **(a)** Experience with natural numbers; for example, the product of 3×4 is greater than either 3 or 4. **(b)** The conjecture is true if multiplication is restricted to the set of natural numbers > 1. **(c)** The conjecture is false when multiplying whole numbers $(0 \times 6 = 0)$, integers $(-2 \times 3 = -6)$, and rational numbers $(\frac{1}{2} \times \frac{1}{3} = \frac{1}{6})$.

30. The product in each case is $\frac{3}{7}$. The pattern is $(\frac{3}{a})(\frac{a}{7}) = \frac{3}{7}$, which can be generalized to suggest that multiplications of the form $(\frac{a}{b})(\frac{b}{c})$ always result in a product $\frac{a}{c}$.

31.

$2\frac{3}{5} \times 1\frac{1}{4} = (2 + \frac{3}{5})(1 + \frac{1}{4})$	Representation of mixed numbers
$= 2(1) + 2(\frac{1}{4}) + \frac{3}{5}(1) + \frac{3}{5}(\frac{1}{4})$	Distributive property
$= 2 + \frac{1}{2} + \frac{3}{5} + \frac{3}{20}$	Rational number multiplication
$= 2 + \frac{10}{20} + \frac{12}{20} + \frac{3}{20}$	Fundamental law of fractions
$= 2 + \frac{25}{20}$	Addition
$= 3\frac{1}{4}$	

32. **(a)** $(\frac{3}{1})(\frac{4}{2}) = 6$ **(b)** $(\frac{7}{5})(\frac{8}{6}) = \frac{28}{15}$ **(c)** $(\frac{8}{2})(\frac{9}{5}) = \frac{36}{5}$ **(d)** If $0 < a < b < c < d$, the greatest rational number product is $\frac{cd}{ab}$

33. **(a)** $(\frac{1}{3})(\frac{2}{4}) = \frac{1}{6}$ **(b)** $(\frac{6}{7})(\frac{5}{8}) = \frac{15}{28}$ **(c)** $(\frac{2}{8})(\frac{5}{9}) = \frac{5}{36}$ **(d)** If $0 < a < b < c < d$, the least product is given by $\frac{ab}{cd}$.

34. **(a)** The product will be 0 if the 0 is placed in a numerator position or undefined if placed in a denominator position. **(b)** If one of the integers is negative, but the absolute values are ordered $a < b < c < d$, then the greatest product is $\frac{ab}{cd}$. The product will be negative but closest to 0. The least product would be $\frac{cd}{ab}$. This product will also be negative but farthest from 0. **(c)** If three of the integers are negative the relation in (b) holds because the product will be negative in all cases. If two or all are negative, the argument is the same as if all are positive because the product of any placement will be positive. If the absolute values are ordered $a < b < c < d$, the greatest product is $\frac{cd}{ab}$ and the least product is $\frac{ab}{cd}$.

35. **(a)** $0.5 \times 0.25 = 0.125 = \frac{1}{8}$ **(b)** $\frac{2}{9}$; the multiplication algorithm cannot be applied to the decimal representations of $\frac{1}{3}$ and $\frac{2}{3}$.

36. Answers will vary.

37. **(a)** $\frac{1}{6} \times \frac{1}{6} = \frac{1}{36}$ **(b)** $\frac{1}{50} \times \frac{1}{49} = \frac{1}{2450}$ **(c)** $0.1 \times 0.1 \times 0.1 \times 0.1 = 0.0001$

Section 6.4

1.

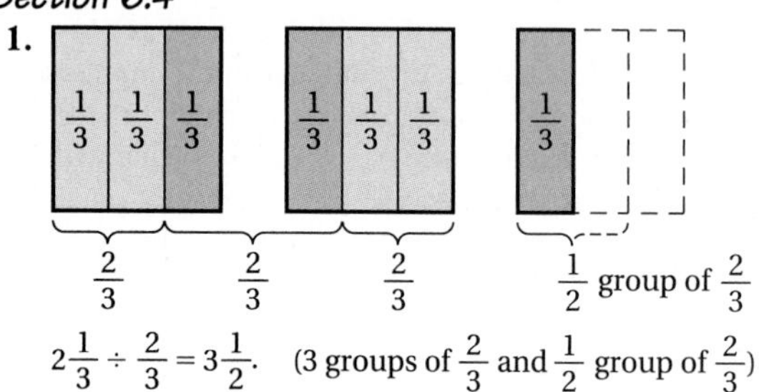

$2\frac{1}{3} \div \frac{2}{3} = 3\frac{1}{2}$. (3 groups of $\frac{2}{3}$ and $\frac{1}{2}$ group of $\frac{2}{3}$)

2.

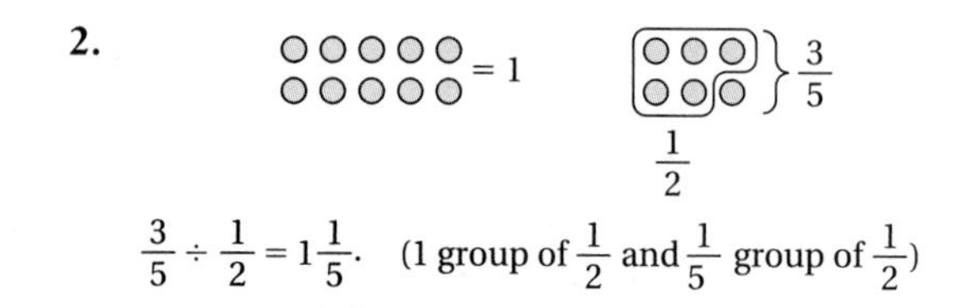

$\frac{3}{5} \div \frac{1}{2} = 1\frac{1}{5}$. (1 group of $\frac{1}{2}$ and $\frac{1}{5}$ group of $\frac{1}{2}$)

3. See answer below.

4.

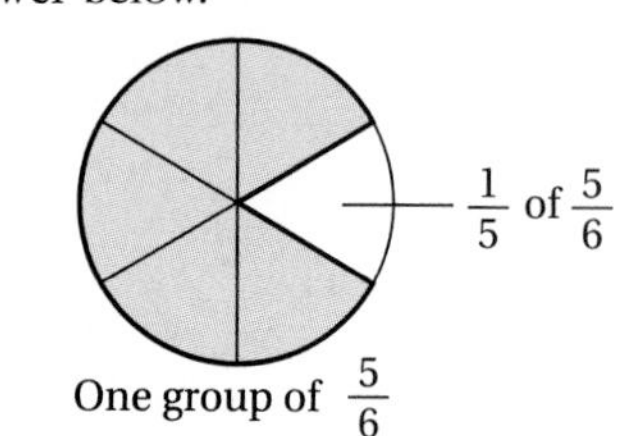

$1 \div \frac{5}{6} = 1\frac{1}{5}$ (or $\frac{6}{5}$).

5. $3\frac{1}{4} \div \frac{3}{4} = 4\frac{1}{3}$

6. $1\frac{2}{6} \div \frac{2}{6} = 4$

7. Only (b) and (c) are determined by $4 \div \frac{3}{4}$.

8. If one student takes, on average, $\frac{1}{2}$ hour to do a mathematics problem and another student takes $\frac{1}{3}$ hour, what is

Answer for Exercise 3

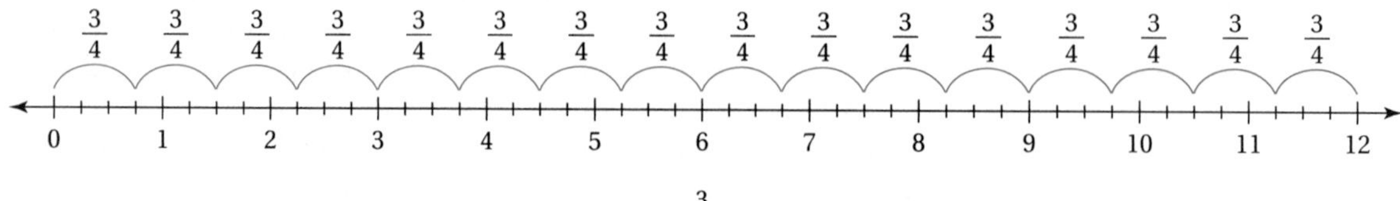

$12 \div \frac{3}{4} = 16$

the ratio of the times taken to do a mathematics assignment, first student to second?

9. Suppose the area of a rectangular surface is 1 m^2 and the length is $1\frac{1}{2}$ m. What is the width of the surface?

10. Suppose the NYSE dropped 16 points and the Nasdaq dropped $\frac{3}{4}$ of a point. What is the ratio of the stock market measure drops, NYSE to Nasdaq?

11. Suppose a runner covered 9.48 m in .96 sec. What is her speed in m/sec?

12. $\frac{3}{4} \div \frac{2}{5} = \frac{15}{20} \div \frac{8}{20} = 15 \div 8 = \frac{15}{8}$ or $1\frac{7}{8}$

13. $\frac{3}{4} \div \frac{2}{5} = (\frac{3}{4})/(\frac{2}{5}) = (\frac{3}{4} \times \frac{5}{2})/(\frac{2}{5} \times \frac{5}{2}) = (\frac{15}{8})/1 = \frac{15}{8}$ or $1\frac{7}{8}$

14. $\frac{3}{4} = \frac{2}{5} \times f$; so $f \times \frac{5}{2} \times \frac{2}{5} = \frac{3}{4} \times \frac{5}{2} = \frac{15}{8}$ or $1\frac{7}{8}$

15. $\frac{3}{4} \div \frac{2}{5} = \frac{3}{4} \times \frac{5}{2} = \frac{15}{8}$ or $1\frac{7}{8}$

16. **(a) i.** $0.9 \div 0.45 = \frac{9}{10} \div \frac{45}{100} = \frac{9}{10} \times \frac{100}{45} = \frac{10}{5} = 2$ **ii.** $0.9 \div 0.45 = 90 \div 45 = 2$ **(b)** The quotients are the same.

17. Methods will vary. **(a)** $\frac{27}{20}$ **(b)** $\frac{6}{7}$ **(c)** $-7.726315789\ldots$ **(d)** $\frac{5}{4} = 1\frac{1}{4}$

18. **(a)** $\frac{3}{8} < \frac{19}{24}$ **(b)** $-\frac{8}{3} < -\frac{8}{5}$ **(c)** $0.3232 < 0.3232\ldots$ **(d)** $35.01 < 35.1$

19. **(a)** $\frac{5}{6} < \frac{81}{96} < \frac{82}{96} < \frac{83}{96} < \frac{7}{8}$ **(b)** $\frac{13}{15} < \frac{53}{60} < \frac{54}{60} < \frac{55}{60} < \frac{14}{15}$ **(c)** $-0.2 < -0.199 < -0.198 < -0.197 < -0.1$ **(d)** $-\frac{1}{4} < -\frac{1}{8} < 0 < \frac{1}{8} < \frac{1}{4}$

20. **(a)** $\frac{3}{99} = \frac{1}{33}$ **(b)** $9\frac{41}{333}$ **(c)** $\frac{144}{990} = \frac{72}{495} = \frac{8}{55}$ **(d)** $\frac{1186}{225}$

21. Greater than 1: (b), (c), and (d); between $\frac{1}{2}$ and 1: (a); less than $\frac{1}{2}$: (e); equal to $\frac{1}{2}$: (f)

22. **(a)** $\frac{21}{10}$ **(b)** $-\frac{5}{2}$

23. **(a)** 123.400 **(b)** 1.23400 **(c)** 1.23400 **(d)** 12.3400

24. Answers will vary.

25. **(a)** $\frac{256}{81}$ **(b)** six times in all **(c)** 0, because the calculator begins with a display of 0, and $0 \div (\frac{3}{4}) = 0$.

26. **(a)** $\frac{3}{4}$ is the result of dividing $\frac{1}{2}$ by $\frac{2}{3}$. **(b)** $\frac{81}{32}$ and $\frac{243}{64}$ **(c)** Seven times in all **(d)** No; the numerator and the denominator are relatively prime.

27. **(a)** 0.1, 0.1313..., 0.2 **(b)** −4.74, −4.732121..., −4.73 **(c)** $\frac{3}{5}$, 0.7979..., $\frac{4}{5}$ **(d)** $-\frac{1}{4}$, 0.123123..., $\frac{1}{4}$

28. **(a)** $5 \div 2\frac{2}{3} < 1 \div \frac{1}{7}$ **(b)** $2\frac{2}{3} \times \frac{1}{4} < 6 \div \frac{3}{4}$ **(c)** $5.26 \times 0.23 < 0.63 \times 2$ **(d)** $-\frac{1}{2} \times \frac{4}{3} < -1.3 \times 0.5$

29. The rational numbers that are greater than 1.

30. One possibility: Use a calculator to subtract the numbers. If the difference is negative then the first number entered is less than the second. If the difference is 0 the numbers are equal. If the difference is positive then the first number is greater than the second.

31. **(a)** In the system of whole numbers, for $a \div b = c$, c is always less than or equal to a (but not necessarily less than b). **(b)** When the dividend > 0 and the divisor > 1, the conjecture in (a) is true. **(c)** The conjecture in (a) is false when the dividend is negative.

32. **(a)** $x \div x = 1$ because $x = 1(x) = x$ **(b)** $x \div 1 = x$ because $x = 1(x) = x$ **(c)** $(x \div a)(a) = x$ because $(\frac{x}{a})(\frac{a}{1}) = \frac{xa}{(a)1} = \frac{ax}{(a)1} = (\frac{a}{a})(\frac{x}{1}) = 1x = x$

33. **(a) i.** $\frac{4}{3}$ **ii.** $\frac{6}{5}$ **iii.** $\frac{7}{3}$ **iv.** $\frac{9}{4}$ **(b)** The quotient 1 divided by any nonzero rational number is the reciprocal of that rational number. **(c)** For rational number $\frac{a}{b}$, where $a \neq 0$, $1 \div \frac{a}{b} = 1 \times \frac{b}{a} = \frac{b}{a}$.

34. 21.7 miles/hr.

35. **(a)** Total square feet = Asking price ÷ Price per square foot. **(b)** 1532 square feet, 1198 square feet, 1689 square feet, 1280 square feet, 1598 square feet, 3190 square feet **(c)** E3 = B3 ÷ D3, E4 = B4 ÷ D4, ..., E8 = B8 ÷ D8 **(d) i.** \$52.80/square foot, \$53.84/square foot, \$42.92/square foot, \$57.03/square foot, \$53.75/ square foot, \$55.77/square foot **ii.** F3 = (B3 − 2000) ÷ E3, F4 = (B4 − 2000) ÷ E4,... **(e)** B9 = (B3 + B4 + B5 + B6 + B7 + B8) ÷ 6, B9 = average B3:B8, or whatever command the spreadsheet software has for averaging a column of numbers. In this case, the average is too high for a representative cost because the asking price for one house is much higher than that for all the others, which raises the average.

36. **(a)** $2325.68 = g + sD$

(b)

g	s	$D = (2325.68 - g) \div s$
117.50	200	11.04
60.21	500	4.53
19.52	750	3.08

(c) addition and multiplication **(d)** subtraction and division

37. **(a)** $\frac{4}{1} \div \frac{2}{3} = 6$ **(b)** $\frac{8}{5} \div \frac{6}{7} = \frac{28}{15}$ **(c)** $\frac{9}{2} \div \frac{5}{8} = 7\frac{1}{5}$ **(d)** For integers a, b, c, and d, where $0 < a < b < c < d$, $\frac{d}{a} \div \frac{b}{c}$ (or $\frac{c}{b} \div \frac{a}{d}$) = greatest quotient.

38. **(a)** $\frac{1}{6}$ **(b)** $\frac{15}{28}$ **(c)** $\frac{10}{72} = \frac{5}{36}$ **(d)** For integers a, b, c, and d, where $0 < a < b < c < d$, $\frac{b}{c} \div \frac{d}{a}$ (or $\frac{a}{d} \div \frac{c}{b}$) = least quotient.

39. **(a)** 0 couldn't be used as any denominator or as the numerator of the divisor; the numerator of the dividend would have to be 0, and 0 would be the greatest and least quotient. **(b)** The quotient becomes negative and the strategies for finding the least and greatest quotients reverse. **(c)** The strategies would depend on the absolute values of the negative numbers and whether two of them are multiplied to make a positive number.

40. $\frac{156}{723}$ is approximately $\frac{150}{750}$ or $\frac{1}{5}$; 285 to 1208 is about 300 to 1200 or $\frac{1}{4}$. So, the larger school had the better participation.

41. $\frac{2}{3}$ cup/6 servings $= \frac{2}{18} = \frac{1}{9}$ cup per serving; $\frac{3}{4}$ cup/8 servings $= \frac{3}{32}$ cup per serving; since $\frac{3}{32} < \frac{1}{10} < \frac{1}{9}$, the first recipe has more sugar per serving.

42. **(a)** $0.22\ldots = 2(\frac{1}{9}) = \frac{2}{9}$, $0.33\ldots = \frac{3}{9}$, $0.44\ldots = \frac{4}{9}$, $0.55\ldots = \frac{5}{9}$. It appears that $0.aa\ldots = \frac{a}{9}$. **(b)** $0.\overline{12} = \frac{12}{99}$, $0.\overline{23} = \frac{23}{99}$, and $0.\overline{34} = \frac{34}{99}$. It appears that $0.\overline{ab} = \frac{ab}{99}$.

43. **(a)** $3(\frac{1}{3}) = (\frac{3}{1})(\frac{1}{3}) = \frac{(3)(1)}{(1)(3)} = 1 = 3(0.3333\ldots) = 0.999\ldots$ **(b)** $9(\frac{1}{9}) = 1 = 9(.111\ldots) = 0.9999\ldots$
(c) $0.99\ldots = 9(\frac{1}{9}) = 1$

44. Answers will vary.

45. $$\frac{\frac{a}{b} + \frac{c}{d}}{2} = \frac{ad + bc}{2bd}$$
$$\frac{a}{b} = \frac{2ad}{2bd}$$
$$\frac{c}{d} = \frac{2bc}{2bd}.$$

Because $\frac{a}{b} < \frac{c}{d}$, $ad < bc$ and $\frac{2ad}{2bd} < \frac{(ad + bc)}{2bd} < \frac{2bc}{2bd}$; therefore $\frac{a}{b} < \frac{\frac{a}{b} + \frac{c}{d}}{2} < \frac{b}{d}$. Examples will vary.

46. increasing

47. **(a)** $\frac{h + 3}{b + 3}$ **(b)** $\frac{h + 3}{b + 3} \geq \frac{h}{b}$ because $b \geq h$; her "hit to at-bat" record will increase unless she has had a hit every time at bat ($h = b$); then it will remain the same.

48. For $\frac{a}{b} < 0$, the square of any number < 0 is greater than the number. If $0 < \frac{a}{b} < 1$ then $(\frac{a}{b})^2$ is less than $\frac{a}{b}$. If $\frac{a}{b} > 1$ then $\frac{a^2}{b^2}$ is greater than $\frac{a}{b}$.

49. **(a)** $\frac{2}{3} \div \frac{5}{6} = \frac{4}{5}$; because the decimal representations of both rational numbers are repeating, there is no algorithm for the computation. **(b)** $\frac{1}{2} \div \frac{3}{5} = \frac{5}{6}$; $0.5 \div 0.6 = 0.833\ldots$ (by calculator); both representations are equivalent because $\frac{5}{6}$ is the repeating decimal $0.833\ldots$.

50. **(a)** $\frac{2}{5} \div \frac{7}{9} = \frac{2}{5} \times \frac{9}{7} = \frac{18}{35}$ **(b)** $\frac{a}{b} \div \frac{c}{d} = \frac{ad}{b} \div c = \frac{ad}{bc}$ **(c)** $\frac{5}{8} \div \frac{2}{3} = \frac{15}{8} \div 2 = \frac{15}{16}$ **(d)** The method will not work when b, c, or d is zero.

51. **(a)** $1.4 \times 1.4 = 1.96$ **(b)** Since $1.96 < 2$, raise the estimate to, say, 1.41; $1.41 \times 1.41 = 1.9881$. **(c)** Revise upwards: $1.412 \times 1.412 = 1.993744$; try $1.414 \times 1.414 = 1.999396$; try $1.4145 \times 1.4145 = 2.00081025$. **(d)** Because the square root of 2 is not a rational number; i.e., there is no pair of integers a and b such that $a \div b$ is equal to the square root of 2. See Mini-Investigation 6.5 on p. 292.

52. Interest rates, mortgage rates, points on loans, Olympic times, etc.

Chapter 6 Review Exercises

1. **(a)**

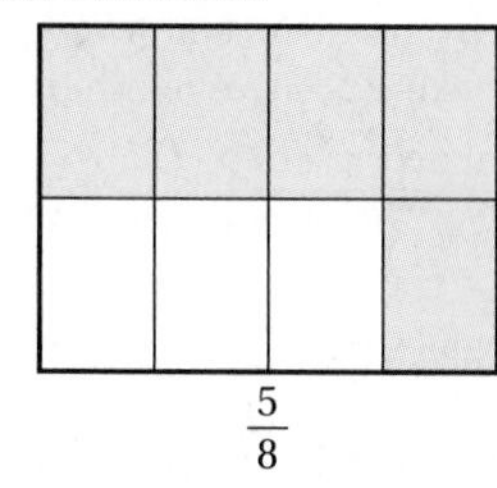

(b)

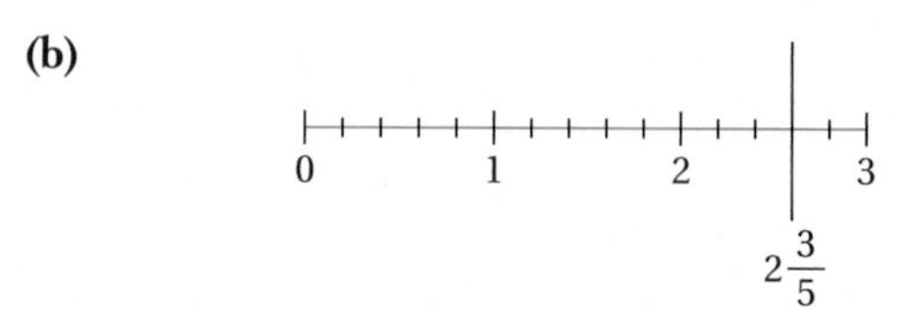

(c) If the thousands cube = 1, then 0.125 is

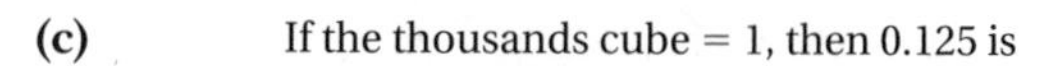

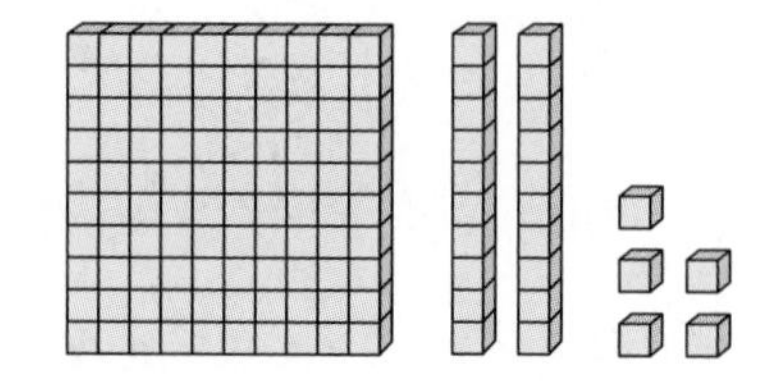

2. Examples will vary. One possibility: The fundamental law of fractions states that for every rational number $\frac{a}{b}$ and integer $c \neq 0$, $\frac{a}{b}$ is equivalent to $\frac{ac}{bc}$. Let $\frac{a}{b} = \frac{2}{3}$ and $c = 4$; by the fundamental law, $\frac{2}{3} = \frac{2(4)}{3(4)} = \frac{8}{12}$.

3.

Fraction	Decimal	Method of Solution
$\frac{5}{12}$	$0.41\overline{6}\ldots$	by calculator
$\frac{125}{999}$	$0.\overline{125}$	$n = 0.\overline{125}$, $1000n = 125.\overline{125}$, $999n = 125$, $n = \frac{125}{999}$
$\frac{56}{100}$	0.56	definition of decimal notation

4. **(a)** 1×10^{-10} **(b)** 1.1×10^{7} **(c)** 3.6754×10^{-2} **(d)** 2.4000059×10^{7}

5. **(a)** $-\frac{2}{3}$; $\frac{3}{2}$ **(b)** $\frac{4}{3}$; $-\frac{3}{4}$ **(c)** -0.25; $\frac{1}{.25}$ **(d)** 0; not defined

6. **(a)** 39.341 **(b)** $\frac{17}{15}$ **(c)** 27.117 **(d)** $-\frac{187}{24}$

7. **(a)** $-17\frac{1}{2}$ **(b)** $44\frac{17}{32}$ **(c)** 0.24004 **(d)** 6.63

8. **(a)** $-\frac{2}{27}$ **(b)** $\frac{32}{93}$ **(c)** 11.25 **(d)** 17.1

9. Responses will vary.

10. **(a)** $2.33\ldots$, $2\frac{1}{2}$, 2.51, $\frac{6000}{29}$ **(b)** $3.12\ldots$, 3.33, $3\frac{5}{8}$, $\frac{577}{154}$ **(c)** $2.71\ldots$, $\frac{32}{11}$, $\frac{477}{154}$, 22

11. Yes, because it is a repeating decimal.

12. Each rational number is defined by $\frac{a}{b} = a \div b$, where a and b are integers and b is not equal to 0.

13. **(a)** $(\frac{1}{2} + 6\frac{3}{4}) + 2\frac{3}{4} + 2\frac{1}{8} + 1$ is approximately $7 + 3 + 2 + 1 = 13$, so 2 sacks are needed. **(b)** The ham weighs about half of the approximately 13 pounds, so put the ham in one sack and all the other groceries in the second sack.

14. Exact processes will vary. **(a)** 5 **(b)** 96 **(c)** 8 **(d)** $-\frac{5}{9}$

15. **(a)** False; her salary was reduced by $\frac{1}{4}$. **(b)** Her salary would need to be increased by a third to again be equal to her original salary.

16. **(a)** \$70.45 **(b)** 7 hours

17. Both are correct; 32 R 2 with a divisor of 5 is the same as $32\frac{2}{5} = 32\frac{4}{10} = 32.4$.

18. Answers will vary.

19. Answers will vary.

20. Answers will vary.

21. Answers will vary.

22. Answers will vary.

Chapter 7

Section 7.1

1. 3 : 4; 7 : 4; 7 : 3

2. 5 : 1

3. 20 : 1

4. 1 : 3

5. 1 : 6

6. 2 : 3

7. 20 : 3

8. 2 : 5

9. 4 : 15

10. 12; 16; 60

11. Pete has the higher average.

12.

c	6	9	10	27	21	12
d	2	3	$3\frac{1}{3}$	9	7	4

13.

x	6	9	1.5	12	4.5	7.5
y	16	24	4	32	12	20

14.

x	1	3	3.2	4	4	10
y	2.5	7.5	8	10	10	25

15. 6 or 7

16. 0.857

17. Worse; $\frac{2}{6} > \frac{2.5}{9}$

18. 1 cup of nonfat dry milk and $3\frac{3}{4}$ tablespoons of baking powder

19. 2.5 feet

20. $\frac{3}{8}$: 1

21. 2 : 1

22. 4 : 1

23. $\frac{5}{24}$: 1

24. 6 beats per 5 seconds. 12 beats per 10 seconds.

25. $\frac{31}{57}$

26. $1.50 per liter

27. $.40 each

28. 43 cents each

29. $3.20 per kilogram

30.

Total lbs	8	16	24	40
Blue grass (lbs)	5	10	15	25
Clover seed (lbs)	3	6	9	15

31. 0 : 20

32. Alisha gets better mileage.

33. 5 hits

34. Marge is correct.

35. 10 : 8 or 5 : 4

36. $1.13 or less

37. **(a)** Juanita is correct. **(b)** The problem can be solved by comparing unit prices. **(c)** Answers will vary.

38. There is the same number of girls in both classes.

39. 1 : 3

40. 2 : 3

41. Each is approximately 25 cents per ounce. The actual price of the large tube is 25.6 cents per ounce; the small tube is 24.6 cents per ounce, so the small tube is the better buy.

42. 7.5 : 30; 10 : 40

43. Jackie's results are acceptable.

44. Answers will vary.

45. Answers will vary.

46. The headline should not be accepted.

47. Answers will vary.

48. 7 : 5

Section 7.2

1. Missing values: 4.5, 6, 7.5, 9, 10.5

2. Cross-product property

3. Reciprocal property

4. Reciprocal property

5. Cross-product property

6. $x = 9.8$

7. $x = 13.5$

8. $x = 17.6$

9. **(a)** $x = \frac{4}{3}$ **(b)** $\frac{28}{25}$ **(c)** $x = 8$

10. $x = 54$

11. $x = 63$

12. $x = \frac{8}{3}$ or $2\frac{2}{3}$

13. $x = \frac{16}{5}$ or $3\frac{1}{5}$

14. $x = \frac{15}{2}$ or $7\frac{1}{2}$

15. $x = \frac{28}{3}$ or $9\frac{1}{3}$

16. $x = 30$

17. $x = \frac{45}{4}$ or $11\frac{1}{4}$

18. $x = 10$

19. $x = 3$

20. $x = 1.5$

21. $x = \frac{13}{12}$

22. $x = 6$

23. $x = \frac{14}{9}$

24. $x = 36$

25. $x = 170$

26. $x = \frac{43}{12}$

27. $x = \frac{16}{5}$

28. $x = 5.5$

29. $x = -35$

30. The length to width ratio is the same for both; $\frac{44}{88} = \frac{35}{70}$.

31. The length to length ratio is not equal to the corresponding width to width ratio; $10:5 \neq 20:15$.

32.

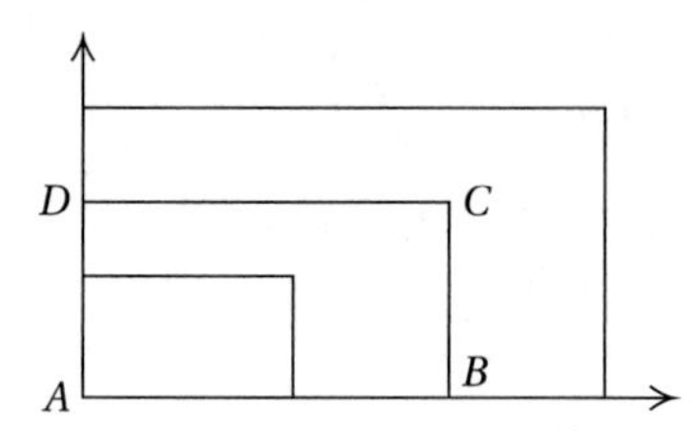

33. Answers will vary. For the following table, $k = -2$.

x	1	2	3	4	5	6
y	−2	−4	−6	−8	−10	−12

34. **(a)** No **(b)** Yes **(c)** No

35. $11\frac{2}{3}$ feet

36. 60 compact spaces

37. $\frac{14}{3}$ or $4\frac{2}{3}$ pounds of nickel

38. \$3.92

39. \$7.20

40. \$480

41. 90 minutes or $1\frac{1}{2}$ hours

42. 350

43. Answers will vary. The population is not proportional to the number of days.

44. Answers will vary. $N = 33$ tickets.

45. Reasons for disagreement will vary; one reason for agreement is that the ratio between all x, y pairs is constant.

46. Answers will vary. The ratios of the dollars and marks received are not proportional.

47. Barbara is correct.

48. Kay is correct.

49. 1.176470588

50. $O = 5.12$ ounces

51. No; the correct solution is 0.9524 (to four decimal places).

52. Correct

53. Dan: \$2000; Shelly: \$1250; Kirby: \$750

54. 9600 square miles

55. Yes; the cost is \$18.

56. 18 ounces of water with 12 ounces of glycerin.

57. Answers will vary.

58. Answers will vary.

59. Answers will vary.

60. Answers will vary; the number of corners is not proportional to the number of squares.

61. Answers will vary. The method will always work for proportions of the form $\frac{a}{x} = \frac{b}{c}$ in which a, b, and c are nonzero numbers.

62. Answers will vary.

63. Yes; the force and mass are proportional when v is held constant. No; the velocity is squared, so it won't vary proportionally with the force even if the mass is held constant.

64. Answers will vary: **(a)** Correct **(b)** Generally this is correct **(c)** Yes, but there are other factors as well **(d)** Correct **(e)** Correct

65. Answers will vary. Richie is correct; Martha's reasoning is also correct.

66. Answers will vary.

67. Answers will vary.

68. Answers will vary. The difference is 1500 marriages.

69. Answers will vary.

70. Answers will vary. **(a)** 16 men **(b)** 12 men

Section 7.3

1. 25 of 100 squares should be shaded.

2. One entire grid should be shaded and 7 squares on a second grid should be shaded.

3. 50 squares and half of one additional square should be shaded.

4. Two entire grids should be shaded and 10 squares on a third grid should be shaded.

5. Three-fifths of 1 square should be shaded.

6. Two squares and half of another square should be shaded.

7. Missing values: $\frac{7}{20}$; 0.35

8. Missing values: 42%; $\frac{21}{50}$

9. Missing values: 175%; 1.75

10. Missing values: 245%; $\frac{49}{20}$

11. Missing values: $\frac{1}{8}$; 0.125

12. Missing values: 0.5%; 0.005

13. Missing values: 0.25%; $\frac{1}{400}$

14. Missing values: $\frac{6}{5}$; 1.2

15. $200

16. $1440

17. $1300

18. $412.50

19. 100% increase

20. 50% decrease

21. 400% increase

22. 20% increase

23. 67% decrease (rounded)

24. 75% decrease

25. 4800 students

26. 60%

27. 4500

28. 7%

29. $21,875

30. Luis: 83%; Felipe: 81%; Maria: 80%

31. $125

32. 200%

33. About 114.29% increase

34. 5% price drop

35. Todd: 13.6%; Cassidy: 10%

36. Answers will vary.

37. **(a)** 19.2 **(b)** 161.28 **(c)** 0.625 **(d)** 577.5

38. About $300

39. About $6.00

40. About 530 points

41. About 11 people

42. Yes; the percent of increase can increase without limit.

43. Yes

44. $29,700

45. Less

46. About $900

47. 8% simple interest

48. Answers will vary. The costs are the same, $540.60.

49. **(a)**

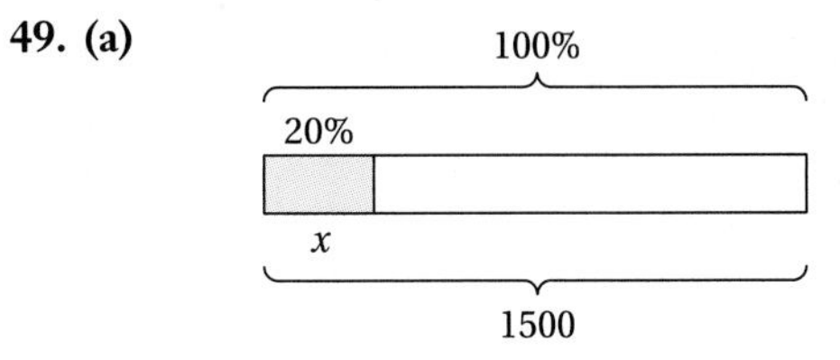

(b)

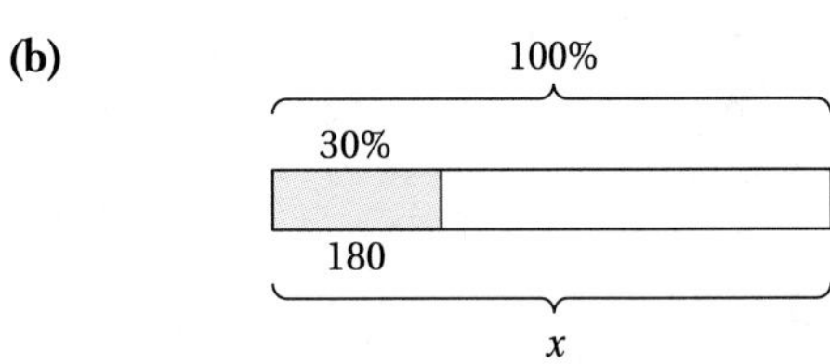

(c)

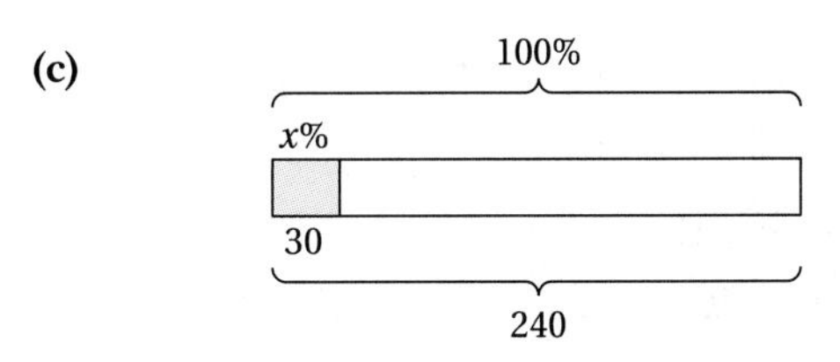

(d)

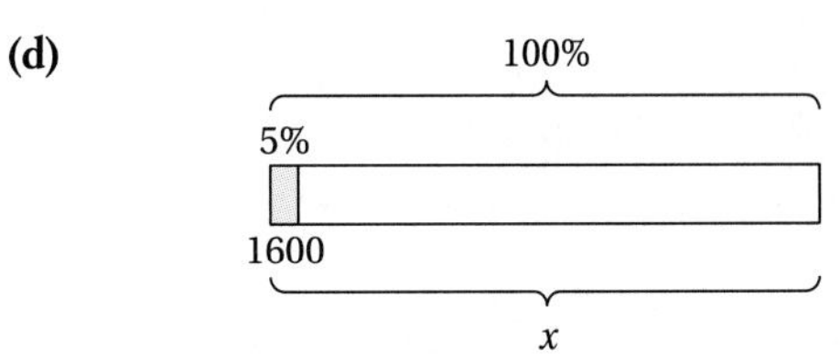

50. (c) is the better investment.

51. $1083.28; 8.33%

52. 1.5% per month is better.

53. Selected answers are: Year 6: 1419; 1587; 1974; Year 9: 1689; 1999; 2773; Year 18: 2854; 3996; 7690.

54. The rule seems accurate.

55. About 351 people

56. $11,810

57. $25,480.40

58. $226,842.23

59. 4.76%

60. 25%

61. 50%

62. Six years

63. Mark is correct when the interest is simple interest; Monica is correct when the interest rate is compounded.

64. 15.5%

65. The average rate $\frac{(a+b)}{2}\%$ is better.

66. Answers will vary.

67. Answers will vary.

68. Answers will vary.

69. About 770 couples.

70. Answers will vary.

Chapter 7 Review Exercises

1. **(a)** 25 : 8 or 25 : 17 **(b)** 8 : 17 or 17 : 8 **(c)** 8 : 25 or 17 : 25

2. Car B gets better mileage.

3. (c), (a), (b)

4. About $5.10

5. Missing values: 3.75, 5, 6.25, 7.5

6. *MNOP*: 3 by 8; *XYTZ*: 9 by 24

7. **(a)** True **(b)** False **(c)** False **(d)** True

8. Answers will vary.

9. **(a)** $x = 12$ **(b)** $x = 0.6$ **(c)** $x = 12$ **(d)** $x = 10\frac{2}{3}$

10. Estimates will vary.

11. $7.00

12. 3 and one-half squares should be shaded.

13. Missing values: 0.0025; $\frac{1}{400}$; 150%; $\frac{3}{2}$; 9.5%; 0.095

14. $127,500

15. 1479

16. 100

17. 25.4%

18. 11.2%

19. $242

20. Compounded value is $498.26 greater.

21. About $200

22. 550 francs

23. 150

24. Barney

25. 460 responses

26. 49

27. $\frac{a}{1000}$; $\frac{401b}{200}$

28. 123.1%

29. 5.13%

30. Rob: correct; Jan: correct; Michael: incorrect; Michelle: incorrect; Bill: correct

31. Discussions will vary. A 1 by 10 grid would serve as a model rather than our 10 by 10 grid. 0.5% would be $\frac{0.5}{10}$ and 5% would be $\frac{5}{10}$. A 5% increase would be half again as much. A 50% increase would mean 5 times as much as the original.

Chapter 8

Section 8.1

1.

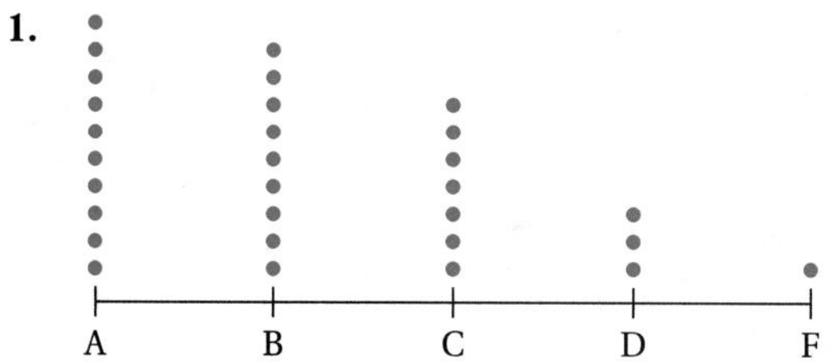

2. Models will vary.

3.

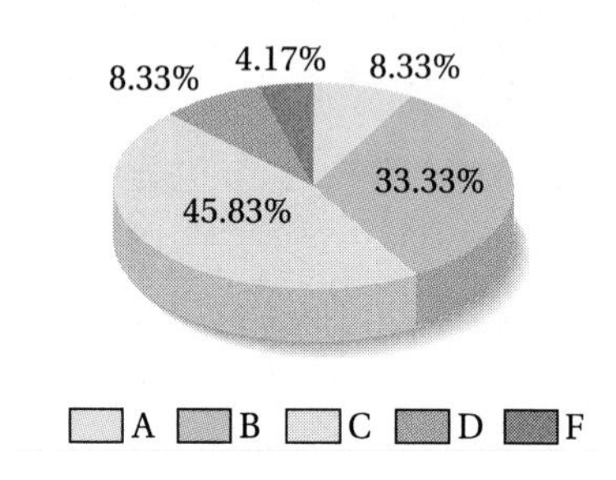

4. The graph would be unchanged.

5.

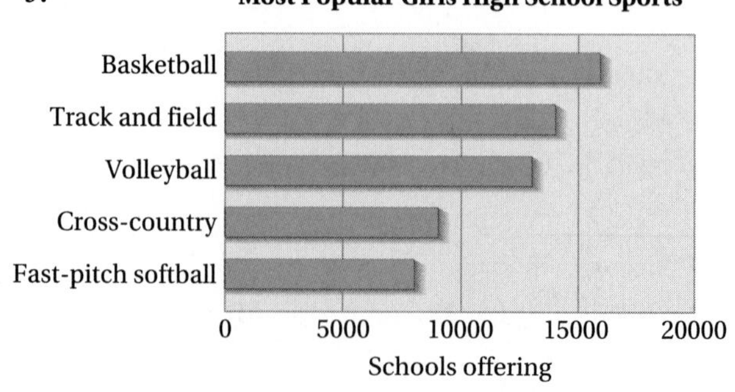

6.

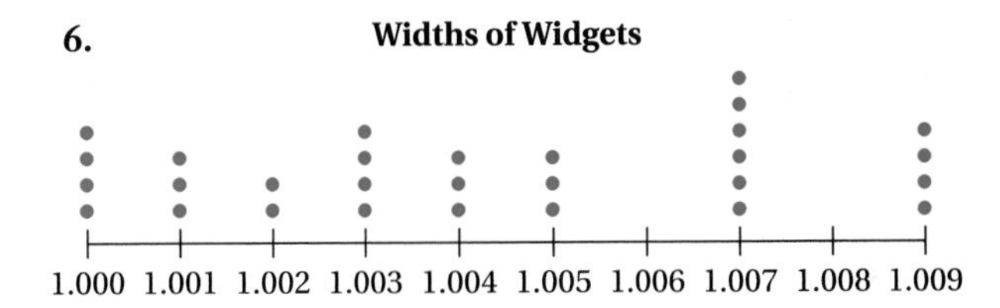

7. (a)

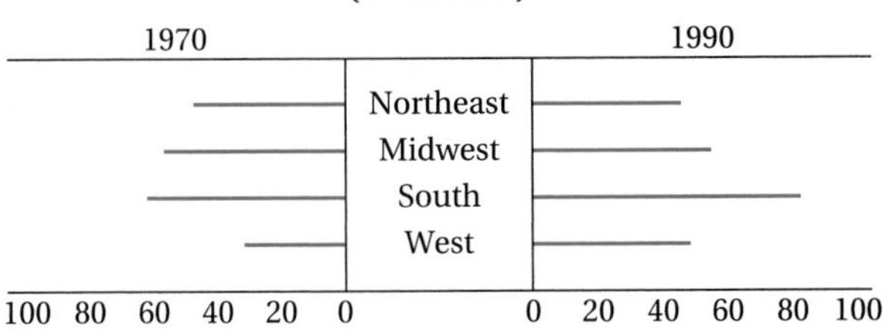

(b) The West grew most rapidly with a 51.5% growth.

8. Answers will vary.

9. (a)

199	001222
198	223456679
197	56899

(b) 17 years **(c)** 1992

10. Answers will vary.

11. The students in the class have scores from 60 to 100. The scores are spread out with 3 getting 100s, and several students from 99 to 91. Then there are students every 2 to 3 points on the scale until a cluster at 72 to 74.

12. The students in the first-hour class are bunched in the interval from 94 to 83 with another cluster from 71 to 75. The last-hour class has a cluster at 99 to 100, then the rest are spread along the entire scale.

13. Answers will vary. A bar graph would be the most appropriate.

14. Answers will vary. A bar graph would be the most appropriate.

15.

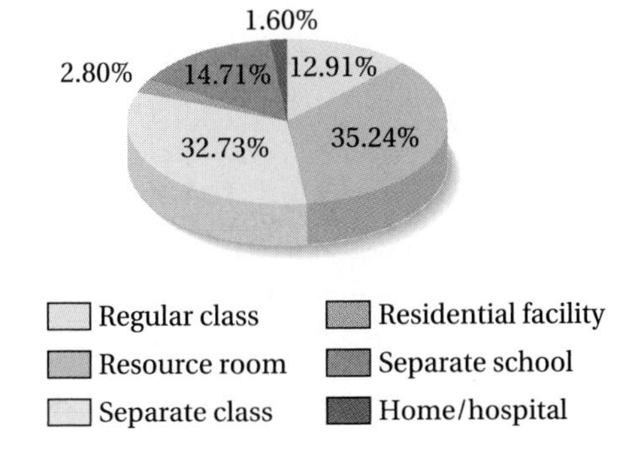

A circle graph was selected to show parts to the whole (100%).

16. Answers will vary. One possibility is a stem-and-leaf plot with the data rounded to hundreds of dollars.

17.

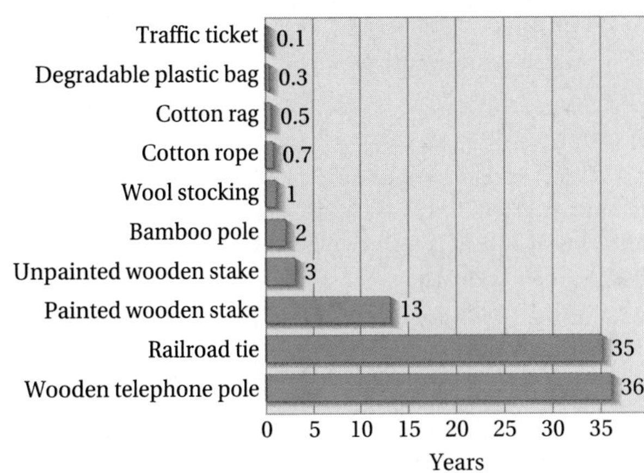

18. Answers will vary. A stem-and-leaf plot is appropriate.

19. (a) Highest: 30–39-year-old age group; lowest: under 12 years of age **(b)** Rate of exposure from sexual activity and time for disease to move from HIV to AIDS status. **(c)** The percentage of AIDS deaths increases to age group 30–39 and then declines. **(d)** Older people weren't exposed in their early years, and the rate of fatality is greatest at early ages.

20. Answers will vary. A scatterplot would be appropriate.

21. (a)

Number of Suicide Deaths in the United States, 1989

Females | Age | Males

Under 15
15–24
25–34
35–44
45–54
55–64
65–74
75–84
85 & Older

4000 3000 2000 1000 | | 1000 2000 3000 4000 5000 6000

(b) For each age, more men than women commit suicide. Both peak in the 25–34 age range. **(c)** As before, more men than women at each age, but the ratio of men to women is fairly constant.

22. Answers will vary.

23. (a)

Comparison of Test Scores for Class A and Class B

Class A		Class B
100	14	01112
552100	13	00122578
987650	12	1678
985	11	8

7|14 means 147   14|7 means 147

(b) Class B had slightly higher performance on the test.
(c) In class A, outliers appeared at 115 and 140–141, with clusters at 115–120, 125–132, and 140–141; in class B, outliers appeared at 118–121, with clusters at 126–132 and 137–142.

24. Answers will vary.

25. Answers will vary.

26. Answers will vary.

27. Answers will vary.

28. Answers will depend on the extent of student research.

Section 8.2

1. The projections of costs show increases each year from 1993 through 1998, with the exception of 1994. The costs decreased 1 percent in 1994 and have increased since, with a 10 percent increase predicted in 1998.

2.

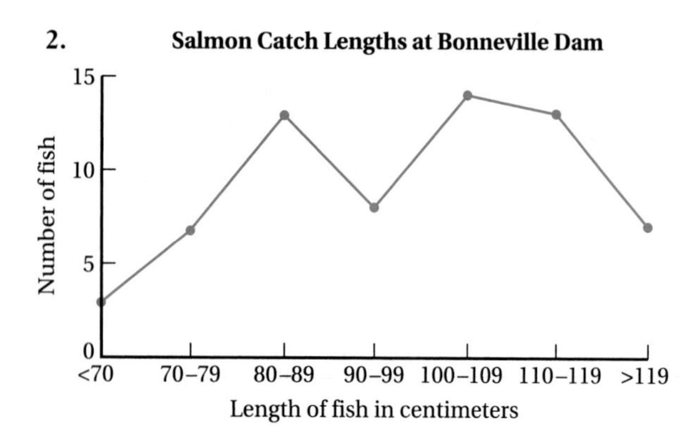

3.

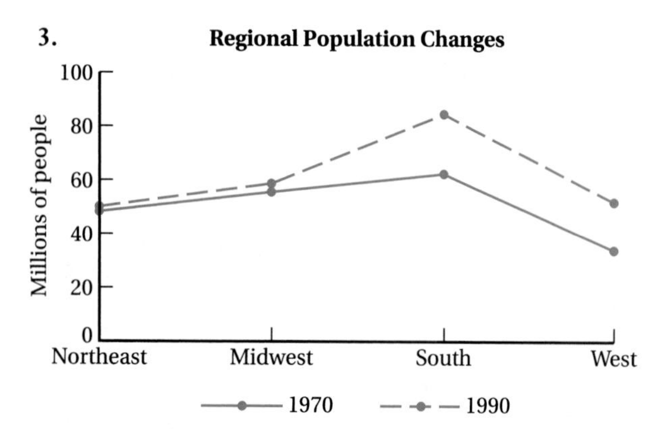

4. Answers will vary.

5. Answers will vary

6. 80; 15

7. **(a)** 0 **(b)** + **(c)** − **(d)** mild +

8. Answers will vary. Some may select (a) because of the axes' scales and the graph fills most of the space.

9. (b)

10. Lengthening it emphasizes changes in the vertical variable; shortening it smooths or de-emphasizes changes in the vertical variable.

11. No; 34 is beyond the range of scores on which the graph is based.

12. Answers will vary.

13. (a) There appears to be a high positive correlation between minutes played and points scored; a person has to play in order to have a chance to score points. **(b)** You could make the relationship appear less positive by elongating the horizontal axis.

14. Answers will vary.

15.

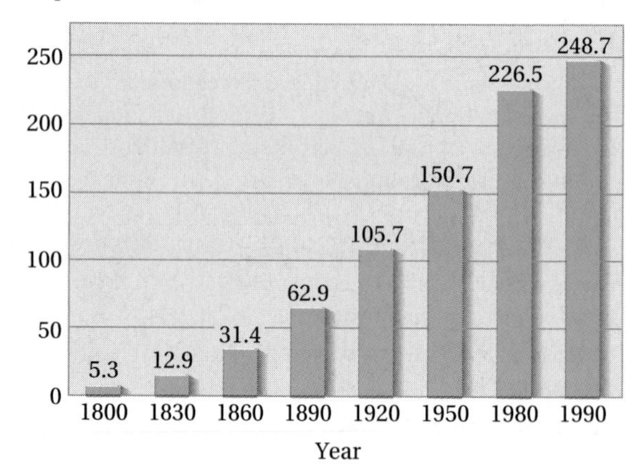

The column graph is used for the contrast it shows for adjacent columns.

16. Answers will vary.

17. Answers will vary.

Section 8.3

1. mean: 3; median: 2; mode: 0, 2; midrange: 5

2. mean: 13; median: 11; mode: 15; midrange: 17

3. mean: 84; median: 82; mode: 80; midrange: 84

4. mean: 77; median: 73; mode: 68; midrange: 77

5. mean: 83; median: 84; mode: none; midrange: 80.5

6. mean: 2; median: 0; mode: 0, 7; midrange: −.5

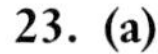

7. mean: $27,000; median and mode: $25,000; if the largest salary is dropped, the mean is $25,000 and the median and mode are unchanged.

8. **(a)** All three measures equal 80. **(b)** Any one of them would do. **(c)** 100; it appears unlikely, based on his other scores.

9. range: 5; variance: 2.166; standard deviation: 1.472; IQR: 2

10. range: 2; variance: 0.50; standard deviation: 0.7071; IQR: 1

11. range: 22; variance: 41.8; standard deviation: 6.47; IQR: 7

12. range: 24; variance: 53.71; standard deviation: 7.33; IQR: 10

13.

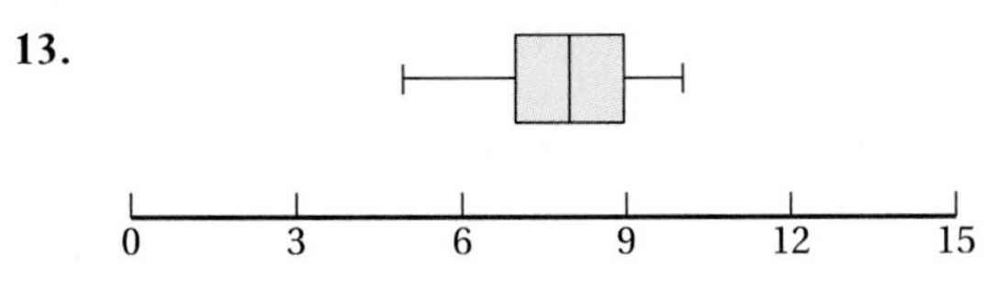

14.

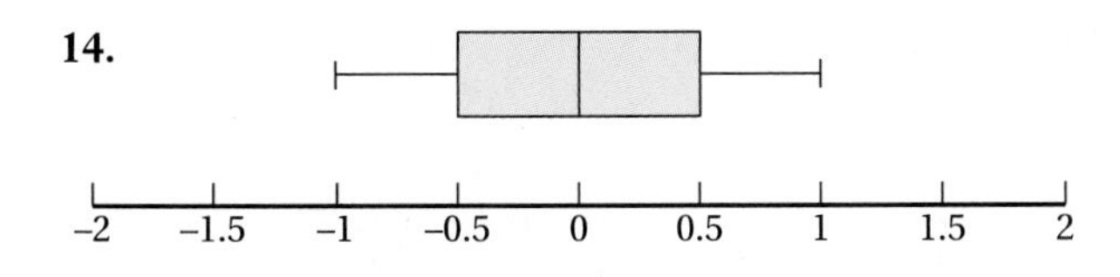

15.

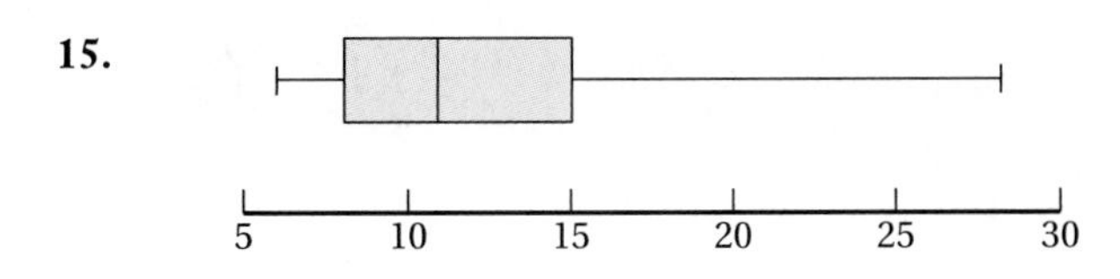

16.

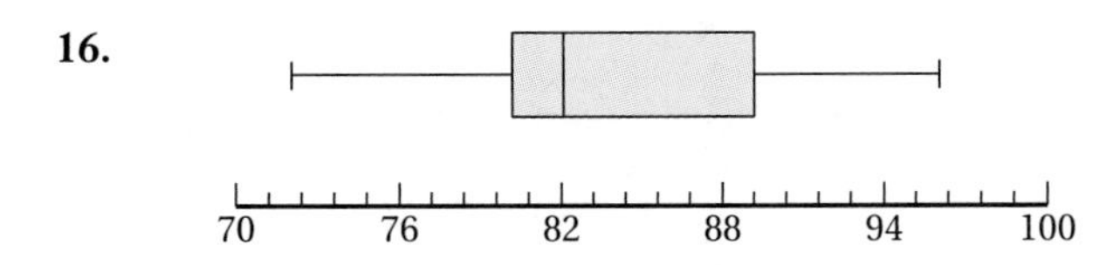

17. Answers will vary.

18. **(a)** Team C; team B **(b)** Team C; team B **(c)** Team B; symmetry **(d)** Team A; more high scores farther from the median

19. 288

20. 28 years of age

21. 76.481

22. Range = 43; One possible data set is {78, 79, 80, 81, 82, 83, 84}.

23. **(a)**

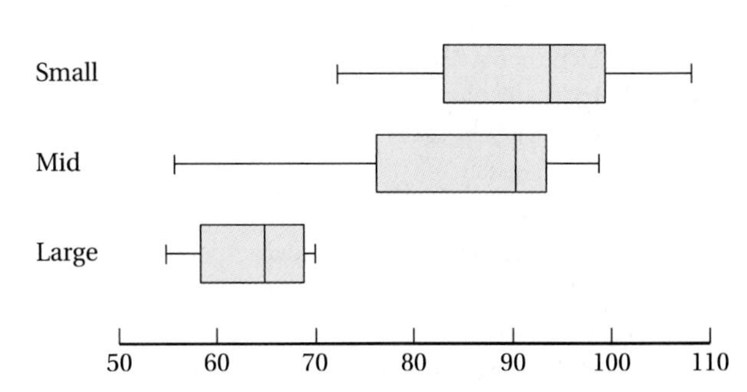

(b) Answers will vary.

24. 73.8¢

25. He cannot average 60 miles per hour no matter how fast he drives the second 2 miles.

26. **(a)** Answers will vary. **(b)** Flyers

27. Consider the data set with nine 0s and one 10 in it. The mean will be 1 and it will fall outside the box in the box-and-whisker plot.

28. No; consider what happens when the value of the variance is 0.49; its square root is 0.7, which is larger, and that only occurs when $0 < \text{variance} < 1$.

29. If the absolute value of the deviations were not used the average deviation would be 0. The average deviation and the standard deviation give similar information about the spread of data.

30. The mean, median, and mode are all 10 points larger; the range is the same, as is the standard deviation.

31. Not always; consider the data set 10, 10, 15, 16, 17, 20, 20. The IQR and the range both equal 10 in this case.

32. The mean increases by $10 \div 8$, or 1.25; the median, mode, and midrange are unchanged.

33. Example data sets will vary.

34. The mean, median, and midrange may or may not be, depending on their values. The mode always is a member of the data set, as it is the most frequently occurring member of that set.

35. Answers will vary.

36. Answers will vary.

37. Answers will vary.

38. Answers will vary.

39. Answers will vary.

40. Answers will vary.

Section 8.4

1. What type of survey? Who responded?

2. Correlation is not causation.

3. Bias resulted from collecting data at the ballpark, a nonrepresentative sample.

4. Maybe there is no other choice; where are connections made?

5. This may be the mean and may be distorted by high outliers. What is the data source?

6. What constitutes twice as much? Who were competitors?

7. Answers will vary.

8. Yes; it would be better to use a nonperspective pie chart for this presentation.

9. Answers will vary.

10. Yes; a poultry dog at the top of the range could have 150% of the calories of the dogs at the bottom of the other two dogs' ranges of scores.

11. The total minority enrollment has increased from 948,000 to over 2 million, or from 10% to almost 16% of total college enrollment.

12. The number of minority students has increased steadily, but the overall gap has increased over time, except in 1985 when it narrowed to 9645 units difference.

13. Answers will vary; there is a correlation, perhaps even a causation, between study time and grades, but it is not the same for all students.

14. It is a negative image in that it suggests that statisticians can manipulate data to arrive at the answer they want.

15. Answers will vary.

Chapter 8 Review Exercises

1. (a) Mean = 5.448; median = 5.46; modes = 5.29 and 5.34; range = 0.97.

(b)

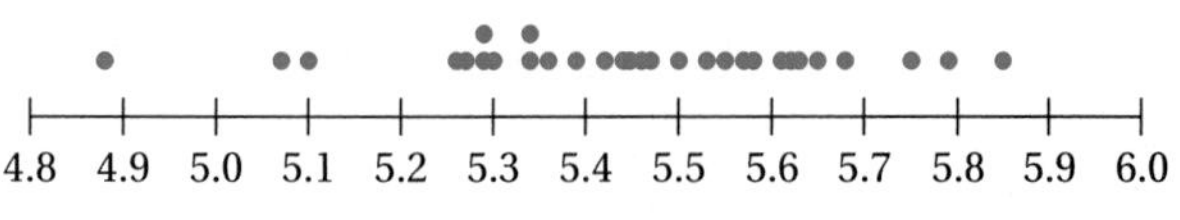

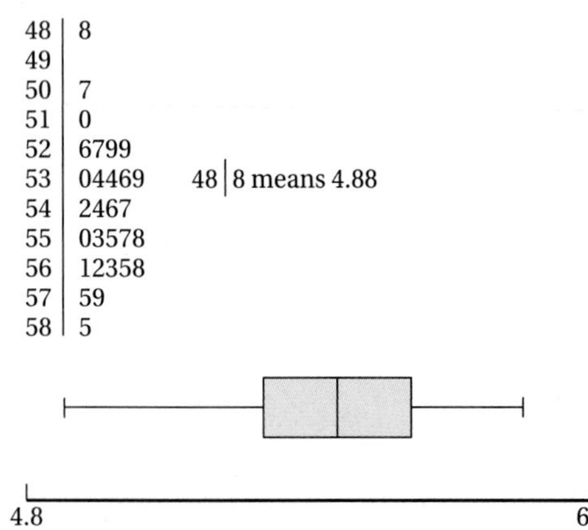

Stem	Leaf
48	8
49	
50	7
51	0
52	6799
53	04469
54	2467
55	03578
56	12358
57	59
58	5

48 | 8 means 4.88

(c) Variance = 0.049 and Standard Deviation = 0.221

(d) Use the median, or 5.46, as the data are slightly skewed to the left because of a low outlier.

2. Models will vary.

3. (a) **Comparison of Male and Female Scores**

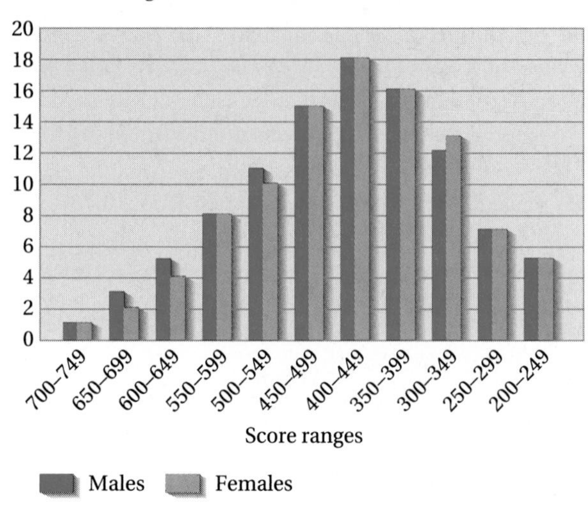

(b) **Male and Female Performance by Range**

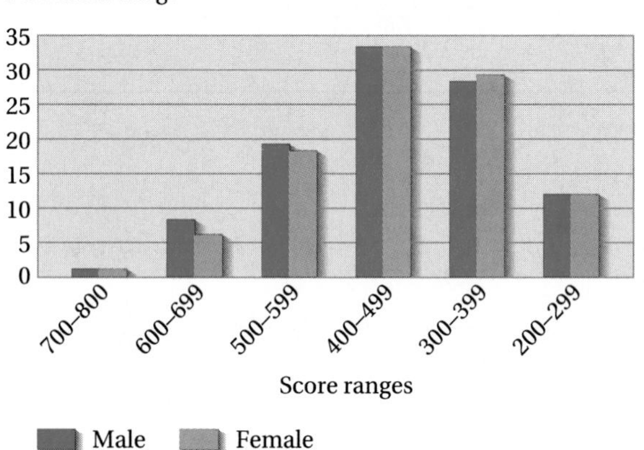

This graph shows less difference in the performance by gender per score range.

4. (a)

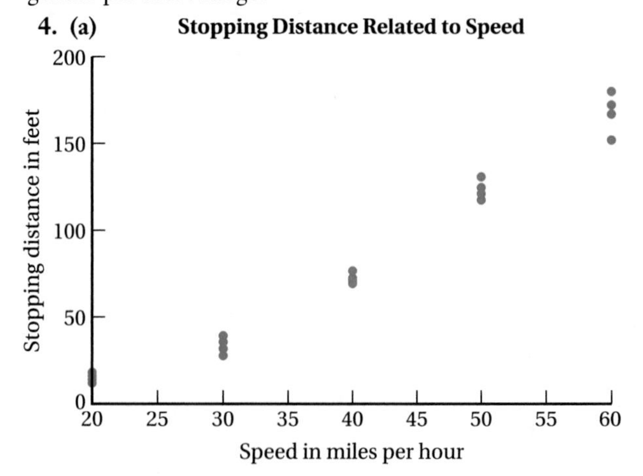

(b) The graph suggests a positive correlation between stopping distance and speed because as the speed increases, the stopping distance also increases.

5. (a) 9; 8 **(b)** positively **(c)** The number of data points is close to the tornado population for any day in May, so extending the data to larger numbers of tornadoes is misleading.

6. The sample is not randomly selected. It is possible that only strong advocates for either side are likely to call. Further, someone could make repeated calls biasing the results further.

7. No; those purchasing food at the snack bar probably don't dine at the area's best restaurants; also, sampling at one location only is dangerous.

8. If the number of households is small or the distribution of incomes contains an outlier, use the median. Otherwise either median or mean would probably work.

9.

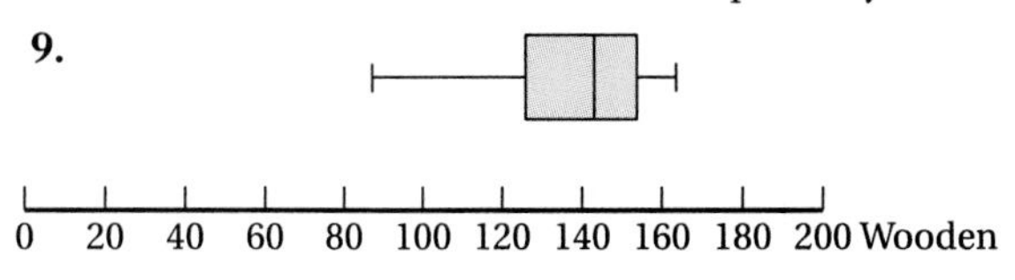

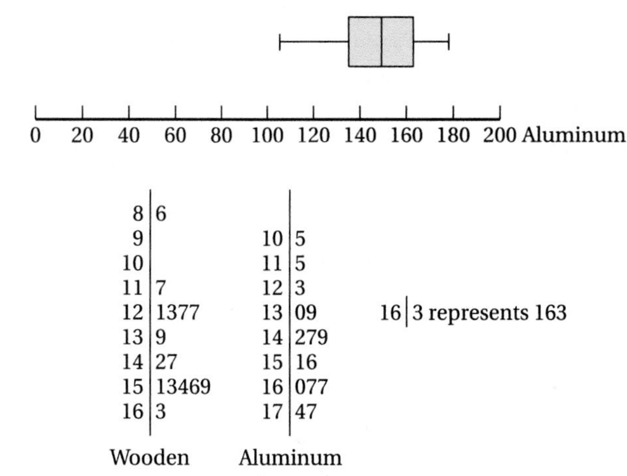

Wooden stem	Wooden leaves	Aluminum stem	Aluminum leaves
8	6		
9		10	5
10		11	5
11	7	12	3
12	1377	13	09
13	9	14	279
14	27	15	16
15	13469	16	077
16	3	17	47

Based on the two different analyses, the aluminum bat yields a slightly higher median distance and higher middle range of distances, as shown by the box.

10. Answers will vary.

11. Reading Scores

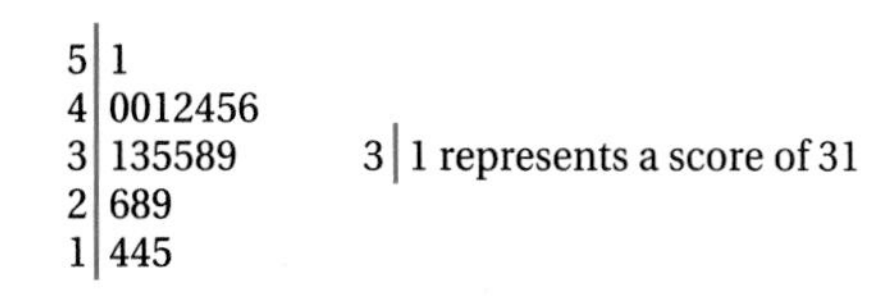

Stem	Leaves
5	1
4	0012456
3	135589
2	689
1	445

3 | 1 represents a score of 31

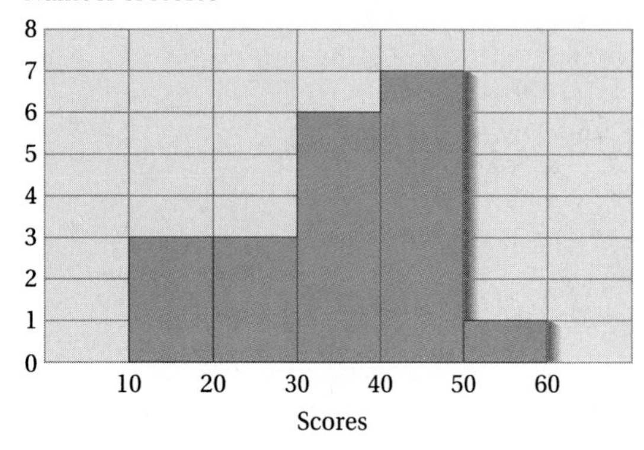

12. Ridership peaks over June, July, and August, and December and January. These are prime vacation times.

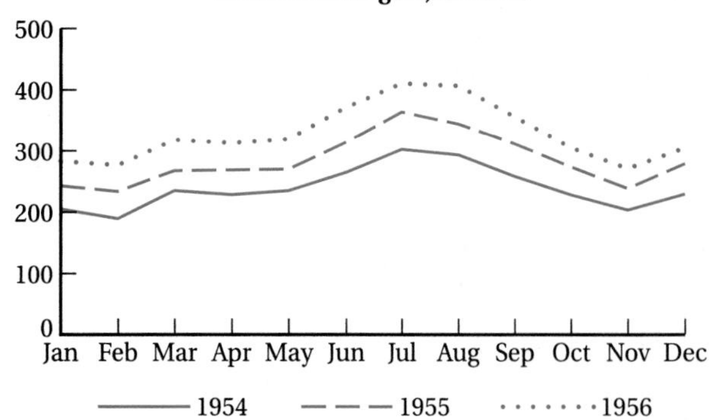

13. Select a random sample of individuals, balanced for right- and left-handers and by gender; use a set of hand-exercise grips equipped to measure the foot-pounds of force exerted by each hand; graph and chart the results individual by individual; analyze the results.

14. Answers will vary.

15. Graph (b) is most likely in that most students have pockets; in graph (a) that isn't the case; the 10 pockets for people and the rectangular pattern in graph (c) aren't very likely.

16. Answers will vary.

17. Answers will vary.

Chapter 9

Section 9.1

1. 1

2. $\frac{1}{2}$
3. $\frac{1}{3}$
4. $\frac{1}{2}$
5. $\frac{5}{6}$
6. $\frac{2}{3}$
7. $\frac{1}{2}$
8. $\frac{1}{6}$
9. $\frac{4}{13}$
10. $\frac{7}{13}$
11. $\frac{9}{13}$
12. $\frac{9}{13}$
13. $\frac{1}{13}$
14. $\frac{4}{13}$
15. $\frac{2}{13}$
16. $\frac{1}{13}$
17. $\frac{1}{3}$
18. **(a)** space has 25 elements **(b)** $\frac{9}{25}$ **(c)** $\frac{4}{25}$ **(d)** not complementary events
19. **(a)** If PH = penny heads, PT = penny tails, NH = nickel heads, NT = nickel tails: S = {PH NH, PH NT, PT NH, PT NT}. **(b)** $\frac{1}{2}$ **(c)** $\frac{1}{4}$ **(d)** $\frac{1}{2}$
20. **(a)** $\frac{21}{26}$ **(b)** $\frac{15}{26}$ **(c)** $\frac{7}{26}$
21. **(a)** 0.488 **(b)** 0.250 **(c)** 0.49 **(d)** 0.509
22. 0.3
23. $\frac{2}{5}$
24. $P(A) = P(B) = \frac{1}{4}; P(C) = \frac{1}{2}$
25. 0.81; based on her record, she should make approximately $\frac{34}{42}$ of her attempts.
26. **(a)** $\frac{23}{40}$ **(b)** $\frac{3}{40}$ **(c)** one with a prior in-school suspension
27. **(a)** $\frac{1}{4}$ **(b)** $\frac{1}{4}$ **(c)** $\frac{1}{4}$ **(d)** $\frac{3}{4}$
28. 0.245
29. $\frac{1}{4}$
30. $\frac{1}{6}$
31. $\frac{5}{6}$
32. **(a)** $P(A \text{ or } B \text{ or } C) \leq \frac{13}{13}$. **(b)** $P(A \text{ or } B \text{ or } C) = \frac{13}{13}$.
33. You would choose machine A in either case since it has the greatest output whether the player is alone or with other players.
34. **(a)** $\frac{1}{10}$ **(b)** $\frac{4}{25}$
35. Answers will vary.
36. $\approx$0.093
37. $\approx$0.01543
38. $\approx$0.73
39. The two experimental estimates are 0.507 and 0.501. Combining the data, we get 0.501. This is not an improvement.
40. a small hospital

Section 9.2

1. 6
2. 24
3. 180
4. 1620
5. 24
6. 24
7. 120
8. 36
9. $\frac{1}{17}$
10. $\frac{12}{51}$
11. $\approx$0.26
12. $\approx$0.06
13. $\frac{3}{4}$
14. $\frac{3}{4}$
15. **(a)** $\frac{1}{7}$ **(b)** $\frac{1}{3}$ **(c)** $\frac{2}{3}$
16. $\frac{2}{11}$
17. $\frac{13}{67}$
18. **(a)** $\frac{47}{69}$ **(b)** $\frac{20}{51}$ **(c)** $\frac{67}{120}$
19. **(a)** 0.3 **(b)** 0.25 **(c)** 0.19 **(d)** 0.81
20. **(a)** 0.0625 **(b)** 0.11 **(c)** 0.25 **(d)** 1
21. 0.75
22. **(a)** $\frac{1}{16}$ **(b)** $\frac{1}{1024}$
23. 0.614
24. **(a)** $\frac{1}{2}$ **(b)** $\frac{1}{4}$ **(c)** Each would decrease.
25. **(a)** $\frac{1}{3}$ **(b)** $\frac{1}{2}$
26. 19 games
27. **(a)** 60 **(b)** 15 **(c)** 5
28. $\frac{16}{625}$
29. **(a)**

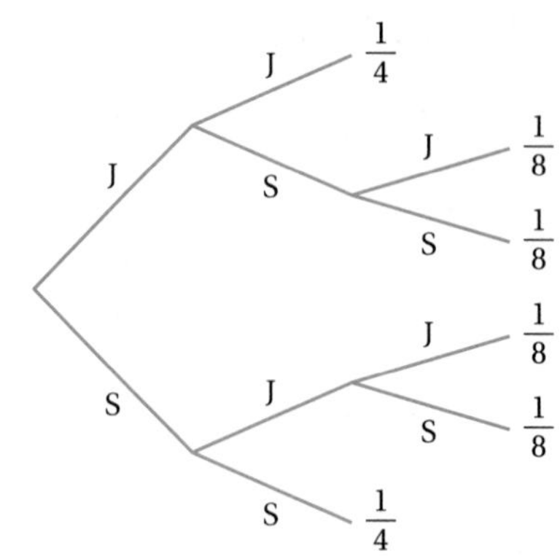

(b) $\frac{1}{2}$ **(c)** $\frac{1}{4}$ **(d)** $\frac{1}{4}$ **(e)** $\frac{1}{2}$

30. $\frac{2}{3}$

31. 0.76

32. 0.25, 0.50

33. Answers will vary.

34. $1.50 to A, $0.50 to B, a ratio of 3:1

Section 9.3

1. 0.25

2. Answers will vary.

3. 0.5625

4. Answers will vary.

5. 0.279

6. 0.45

7. Answers will vary. One is that it matches fairly closely.

8. 0.57

9. 1 through 6

10. 11 through 20

11. 1 through 1000

12. 1 through 20

13. About 0.7

14. 6

15. 5

16. About 0.004

17. About 0.65

18. 0.623

19. **(a)** 0.0256 **(b)** 0.1296 **(c)** 0.9744 **(d)** 0.096

20. estimated value of π: 3.35

21. Answers will vary, but about 1 match per trial.

Section 9.4

1. 156:2496 or 0.0625

2. 13:191 or 0.06806 ...

3. 16:1

4. 220:1

5. 16:1

6. 5:3

7. 7:1

8. 7:1

9. 5:4

10. 1:2

11. 6:23

12. 0:1

13. 23:14

14. 4:1

15. 670:1

16. 1:1

17. 5:31

18. 3:33

19. 36:0

20. 10:26 or about 0.3846:1

21. 5:1

22. **(a)** $20 **(b)** $8 **(c)** $0.002 or 0.2 cents

23. 51:1

24. 6:17

25. **(a)** 1:18 **(b)** 11:27 **(c)** 9:10 **(d)** 9:10

26. 2.5; 7.5

27. 3:2

28. $13,351

29. $20

30. EV = $1.60, so the price was not fair for Sam to pay.

31. 1:1325

32. 29:7

33. A is worst; C is best.

34. (c); (b) and (d); (a)

35. −5 cents

36. yes

37. $3.50

38. 0.78; 0.98; 0.99995

39. About 0.50

40. The game is not fair.

41. 8:3

42. **(a)** $14 **(b)** $4.29 **(c)** $6 **(d)** $40

Section 9.5

1. 625; 120

2. **(a)** 13,260,000 **(b)** 9,360,000

3. 240,240

4. 720

5. 210

6. 56

7. 19

8. 132

9. 1:6

10. 1:1,086,008

11. 31

12. 1:64

13. 167,960

14. 779,220

15. 2,598,960

16. 720

17. 300; 300

18. 11

19. 360

20. 0.0045

21. 2.039×10^{14}

22. 5040

23. 24

24. Answers will vary.

25. Yes, both are equal to $\frac{n!}{(n-r)!r!}$

26. $(1+1)^n$ is a geometric progression with a common ratio of 2.

27. 120; it doubles the number of ways.

28. Yes, there are 1152 different varieties.

29. About 60,000. One can multiply $10 \times 30 \times 20 \times 10$.

30. 24, assuming different positions at the table

31. 2 hours

32. 32,760; 1,365

33. Answers will vary.

34. 4.8×10^9 possible connections; bundling would help.

Chapter 9 Review Exercises

1. **(a)** {HHH, HHT, HTH, HTT, THH, THT, TTH, TTT} **(b)** {HTH, HTT, TTH, TTT} **(c)** {HHH, HTH, THH, TTH} **(d)** {HHT, THT}

2. **(a)** 1:16 **(b)** 1:16 **(c)** 1:8 **(d)** 9:16

3. **(a)** $\frac{1}{104}$ **(b)** $\frac{1}{3}$ **(c)** $\frac{2}{39}$

4. Answers will vary.

5. **(a)** 21:36 **(b)** 2:36 **(c)** 9:36 **(d)** 30:36

6. **(a)** 2:9 **(b)** 5:9 **(c)** 38:90

7. **(a)** 0 **(b)** 1:12 **(c)** 1:6 **(d)** 1:2

8. **(a)** 12:40 **(b)** 36:16 **(c)** 1:51 **(d)** 32:20

9. $9,000

10. **(a)** 225:360 **(b)** 75:360 **(c)** 270:360

11. **(a)** assumes uniform distribution **(b)** events aren't mutually exclusive **(c)** probabilities don't sum to 1

12. 96

13. 15!

14. 210

15. If mutually exclusive, they share no common outcomes; if independent, the outcome for one doesn't affect the outcome for the other.

16. 1.16×10^{12}

17. Three cheeses

18. 1:25

19. 0.0639

20. 1 person per flight

21. Answers will vary; there are 45 key designs.

22. between 2 and 3 cards

23. 1:15

24. 90

Chapter 10

Section 10.1

1. **(a)** point **(b)** line segment **(c)** ray **(d)** parallel lines **(e)** line **(f)** octagon **(g)** angle **(h)** circle

2. Responses will vary. **(a)** $\overleftrightarrow{PS}$ and $\overleftrightarrow{PQ}$ **(b)** $\overleftrightarrow{PS}$ and $\overleftrightarrow{QR}$ **(c)** $\overleftrightarrow{PS}$ and $\overleftrightarrow{UT}$ **(d)** plane *PSVU* and plane *QRTW* **(e)** plane *PSUV* and plane *WTUV* **(f)** $\overleftrightarrow{PV}$, $\overleftrightarrow{WV}$, and $\overleftrightarrow{UV}$ **(g)** points *P*, *R*, and *S* **(h)** points *Q*, *X*, and *R*

3.

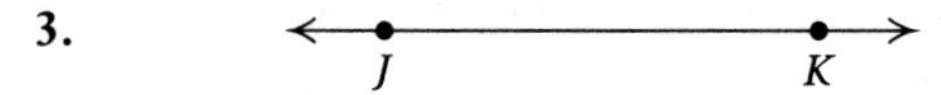

4.

R S

5.

Y X

6.

G H I

7.

N M O

8.

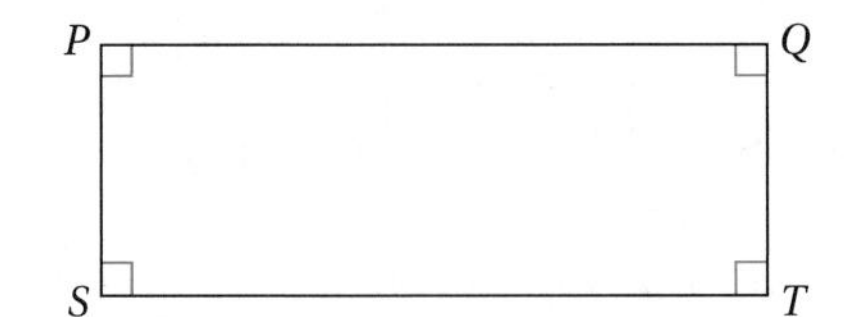

Responses for Exercises 9–15 will vary.

9. $\overline{KD}, \overline{FD}, \overline{MD}$

10. $\angle GHD, \angle FHM$

11. $\overleftrightarrow{DK}, \overleftrightarrow{KM}$

12. $\angle LJH$

13. *DKM* (right), *DHF* (scalene), *HIJ* (equilateral)

14. *DELK* (square), *DFMK* (rectangle), *GKMF* (trapezoid)

15. *GHJLK*

16.

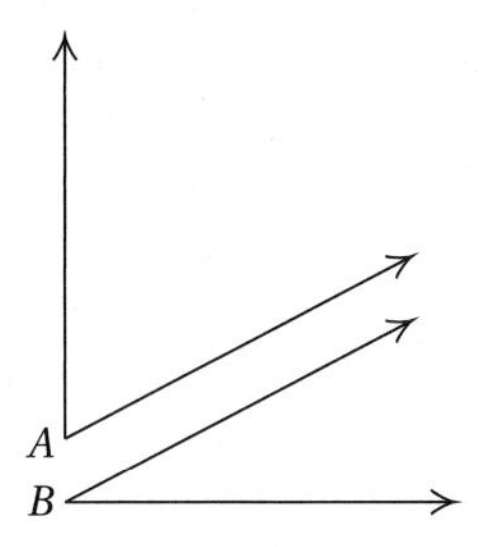

17.

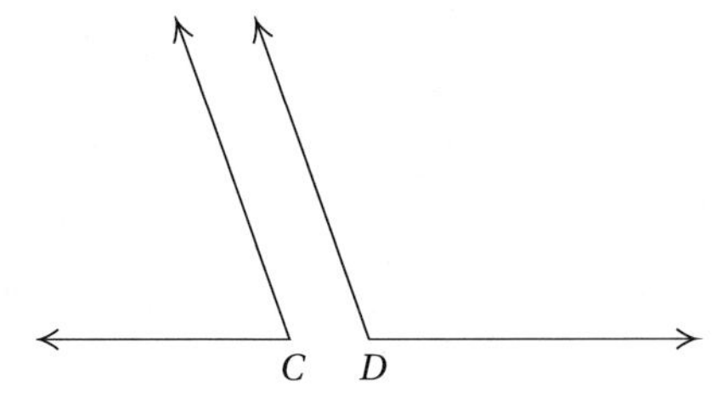

18.

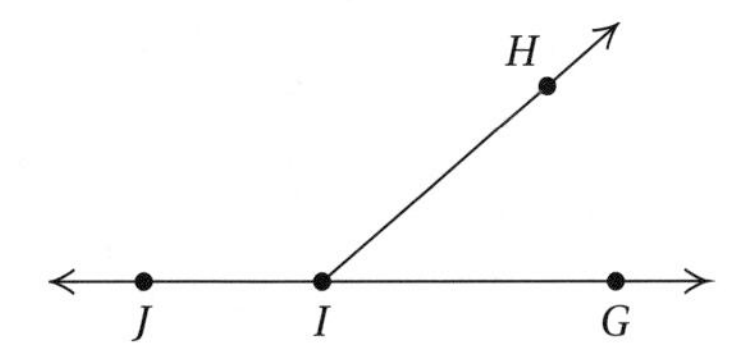

19.

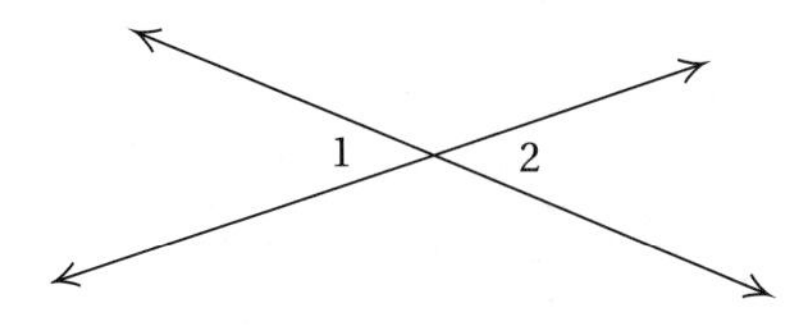

20. $m(\angle CBD) = 60°$. A bisector divides an angle into two congruent angles.

21. Since the lines are concurrent, they intersect in a point and hence cannot be parallel.

22. *D*, *E*, and *F* are on the same line, and there are infinitely many planes that contain a single line.

23. No. If they intersected they would be containable in a single plane, and skew lines are not in the same plane.

24. *A*, *B*, and *C* are collinear, and *A* is between *B* and *C*.

25. *A*, *B*, and *C* are collinear, and *B* is between *A* and *C* or *C* is between *A* and *B*.

26.

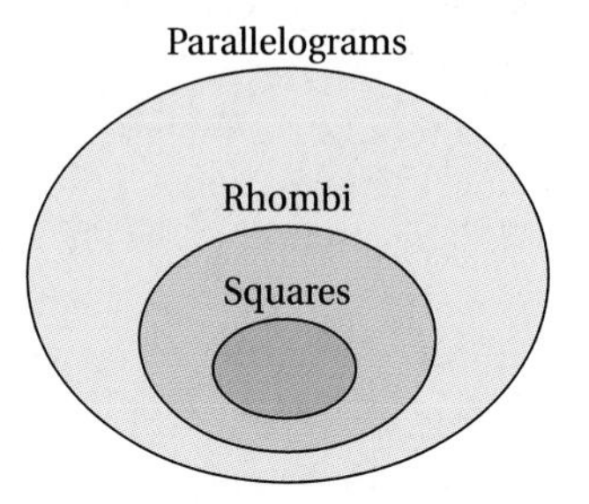

27. Every square is a rhombus, but some rhombi are not squares. Every rhombus is a kite, but not every kite is a rhombus.

28. An equilateral triangle has at least two congruent sides, which qualifies it as isosceles.

29.

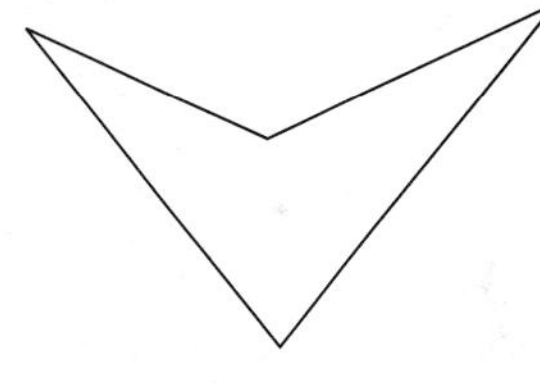

30.

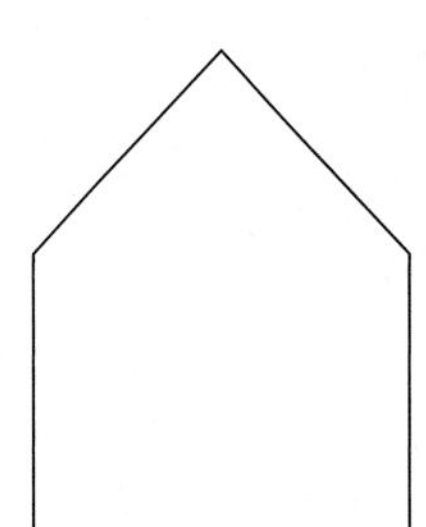

31. A linear pair with two obtuse angles is impossible, since each has measure greater than 90° so that their sum exceeds 180°. A linear pair with two acute angles is impossible, since each has a measure less than 90° so that their sum is less than 180°.

32. Responses will vary.

33. Responses will vary.

34. The lines are perpendicular.

35. A 5-by-8-inch card.

36. The $4\frac{1}{2}$-ft × $7\frac{1}{4}$-ft window.

37. 1, 1, 2, 3, 5, 8, 13, 21, 34, 55, 89, 144, 233, 377, 610. Add two preceding terms to get the next.

38. The ratios are: 1:1 = 1; 2:1 = 2; 3:2 = 1.5; 5:3 = 1.66; 8:5 = 1.6; 13:8 = 1.625; 21:13 = 1.615; 34:21 ≈ 1.619; 55:34 ≈ 1.6176; 89:55 ≈ 1.618; 144:89 ≈ 1.617978; 233:144 ≈ 1.61806; 377:233 ≈ 1.61802575; 610:377 ≈ 1.61803714. The golden ratio is 1.618033989. . . .

39. The square of a term of the Fibonacci sequence plus 1 (if it's an even number term) or minus 1 (if it's an odd number term) is equal to the product of the preceding and succeeding terms.

40. About $\frac{1}{3}$ of Fibonacci numbers are even. So about 8 of the first 25 Fibonacci numbers are even.

41. 1.618, .618 are approximations of the numbers. It appears that x is approaching the golden ratio, and y is its reciprocal.

42. The reciprocal of the golden ratio is 1/1.618033... ≈ 0.618033. 1.618033 − 1 = 0.618033. These are the numbers found in Exercise 41, numbers such that their difference and their product both equal 1.

43. Never. The sum of the angles about a point is 360°. Two intersecting lines form four angles. If they are congruent, then each must be 90°, a right angle.

44. Always. By definition, the angles formed upon bisections are congruent.

45. Always. The exterior sides of linear angles form a straight angle. If the sum of two angles is a straight angle, then the angles are supplementary.

46. Always. Vertical angles are formed by intersecting lines that form four linear pairs of angles. Angles in a linear pair are supplementary.

47. Sometimes. Linear angles are supplementary angles with a common vertex. But any pair of angles with a sum of 180° are supplementary, whether they are part of the same figure or completely distinct.

48. Sometimes. If the legs of a right triangle are congruent, then the right triangle is isosceles.

49. Sometimes. If the rhombus has a right angle, then it is a square.

50. Always. Since all three angles of an equilateral triangle measure 60°, the triangle is acute.

51. Never. If a kite had a pair of parallel sides as does a trapezoid, then the kite would have two pairs of parallel sides, making it a parallelogram. But no trapezoid is a parallelogram.

52. Never. A right triangle contains a right angle. The sum of the remaining two angles is 90°, thus neither can be greater than 90°. So the triangle has no angle greater than 90° and is not obtuse.

53. Sometimes. A parallelogram that is both a rectangle and a rhombus is a square. But not all parallelograms are squares.

54. Always. Since an equilateral triangle has all three sides congruent it necessarily has two sides congruent. An isosceles triangle is not restricted to two and only two congruent sides.

55. Never. Because a trapezoid has one and only one pair of parallel sides, no parallelogram is a trapezoid.

56. Yes. The planes W and X don't intersect, so the lines r and s in those planes can't intersect.

57. $m(\angle KPL) = 90° - m(\angle JPK) = 20°$. $m(\angle KPM) = 90°$, since the lines are perpendicular. $m(\angle LPM) = 90° - m(\angle KPL) = 70°$. $m(\angle JPM) = 90° + m(\angle LPM) = 160°$. $m(\angle NPM) = 180° - m(\angle JPK) = 20°$.

58.

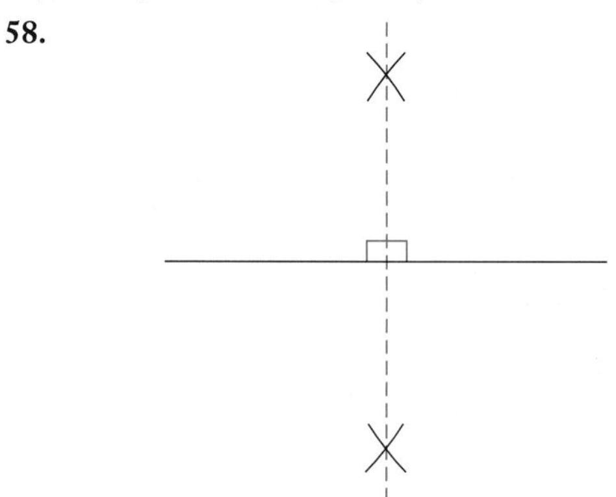

59.

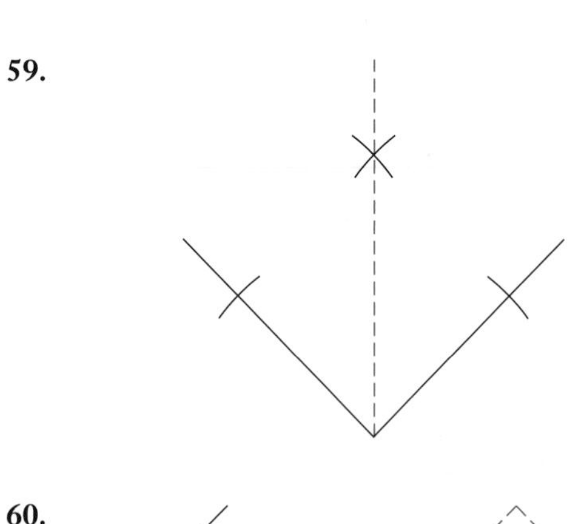

60.

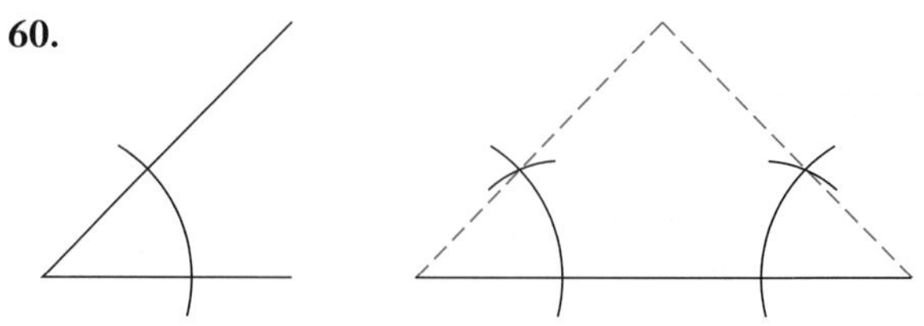

61. 

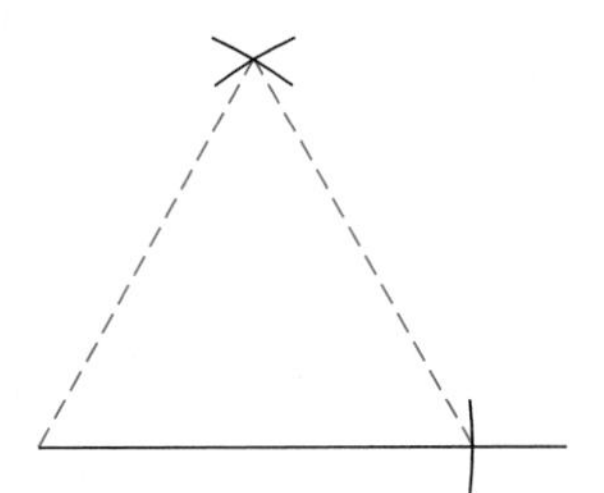

62.

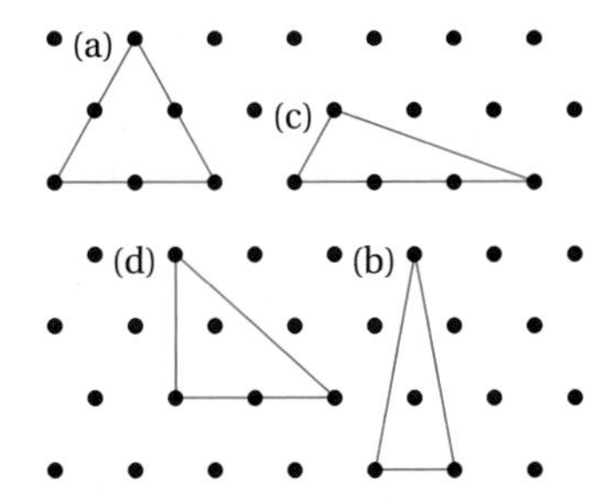

63.

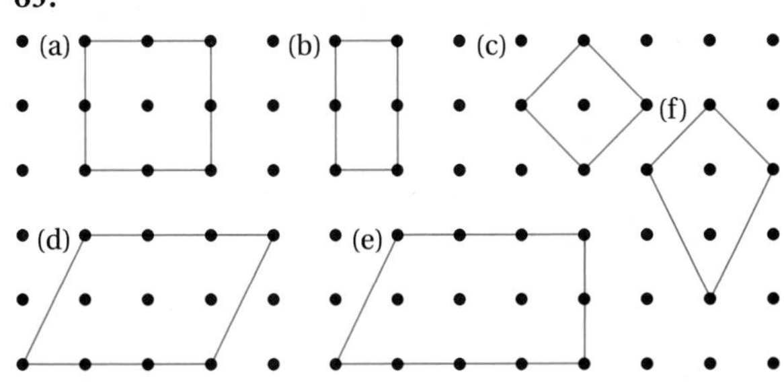

64. **(a)** The diagram is not correct. The difficulty in the diagram is with the relations among squares (all sides equal, all angles right), rectangles (all angles right), and rhombi (all sides equal). Squares are rhombi with right angles or rectangles with equal sides. The relations may be shown as in (b) below.

(b)

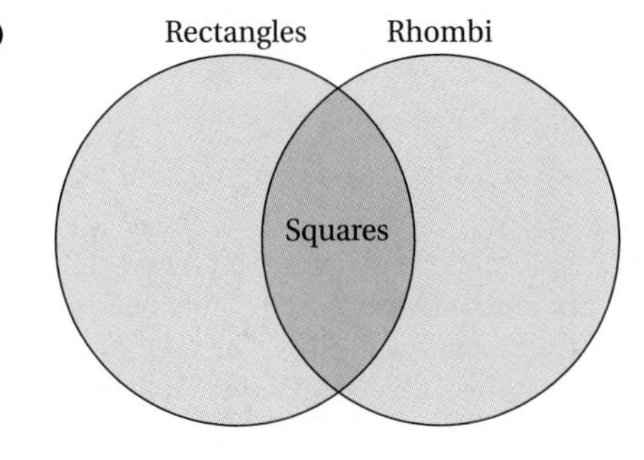

65. 108°

66.

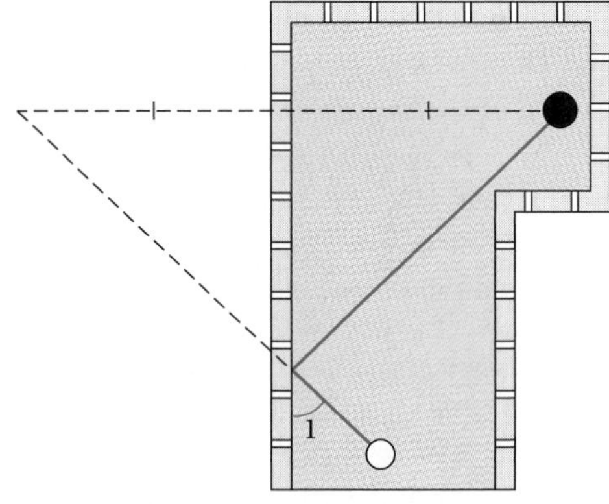

Angle 1 = approximately 52°

67.

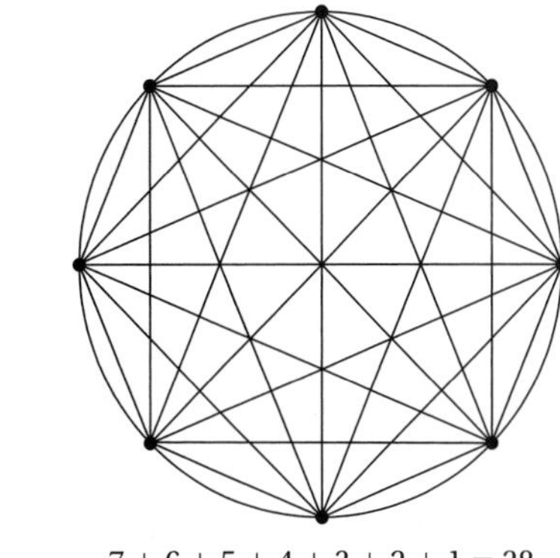

$7 + 6 + 5 + 4 + 3 + 2 + 1 = 28$

68. Responses will vary.

69. Students do Appendix B activities here.

70. **(a)** This is an activity. **(b)** Since it is a surface with exactly one edge and one side, when cut down the middle it remains in one piece, as a loop with two twists in it.

71. Designs will vary.

72. **(a)** A3 = A1 + A2 **(b)** After entering "= A1 + A2" in cell A3, highlight cells A3 through A15 and use "Fill Down" to automatically place the correct formula in each cell A4–A15.

Section 10.2

1. True. Three collinear points are on a line, and a line is in a given plane.

2. False. Concurrent lines intersect in a point, and are not parallel.

3. False. If they intersect at right angles, they are perpendicular.

4. False. If they were perpendicular, they would be in the same plane, and not skew.

5. True. A triangle contains only segments, not lines.

6. False. Only radii on the same line make a diameter.

7. True. An obtuse triangle can't have a right angle.

8. False. An equiangular triangle also has all sides congruent and is equilateral.

9. True. They all divide a figure into two congruent parts.

10. False. It takes all three angles of a triangle to sum to 180°.

11. The network has exactly two odd vertices, and is traversable type 2. The hiker would start at the rest area and end at the ranger station or vice versa.

12. Construct a trail from the rest area to the ranger station.

13. Start and return to point A.

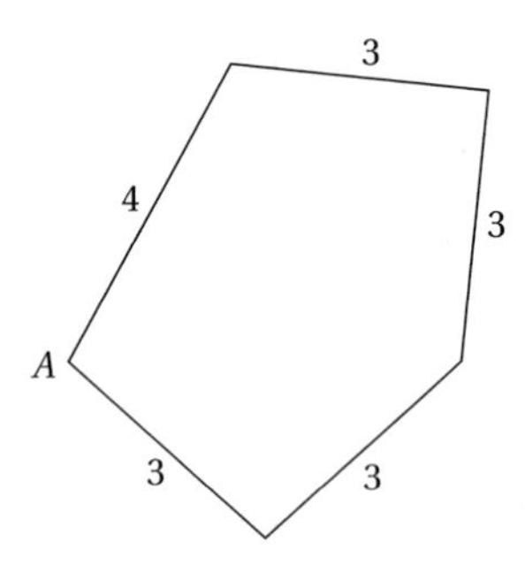

14. Yes. The network has all even vertices and is traversable type 1.

15. Yes. You can go from room A to room B, or vice versa. The network has exactly two odd vertices and is traversable type 2.

16. Examples will vary. The triangles should be different shapes.

17. Examples will vary.

18. Examples will vary.

19. They are collinear, on the Euler line.

20. They are the same point.

21. $\overline{GH}$ is twice as long as $\overline{OG}$. Use GES to construct the figure and measure the segments.

22. 850 feet long. The segment joining the midpoints of two sides of a triangle is half the other side.

23. 247 yards. If two angles of a triangle are congruent, the triangle is isosceles.

24.

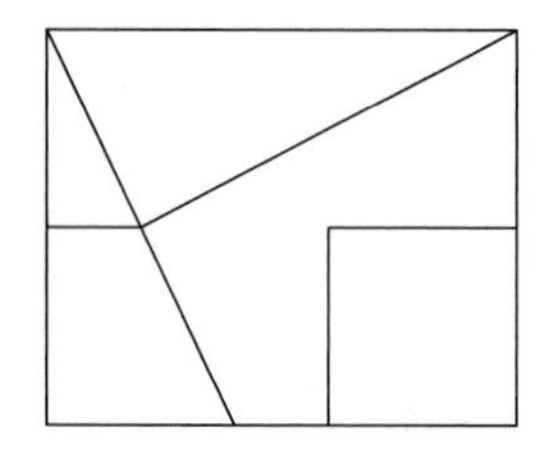

25.

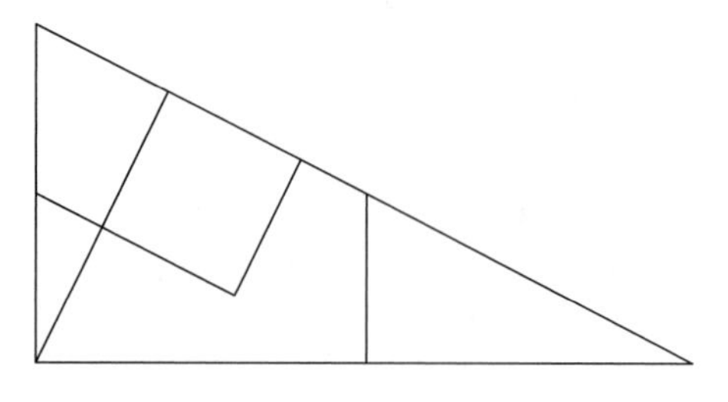

26.

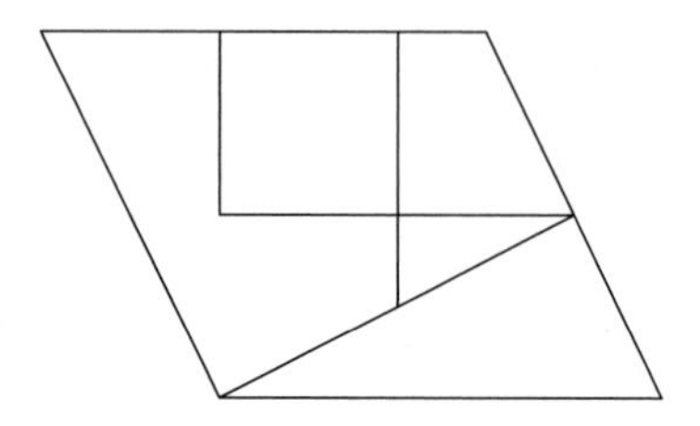

27.

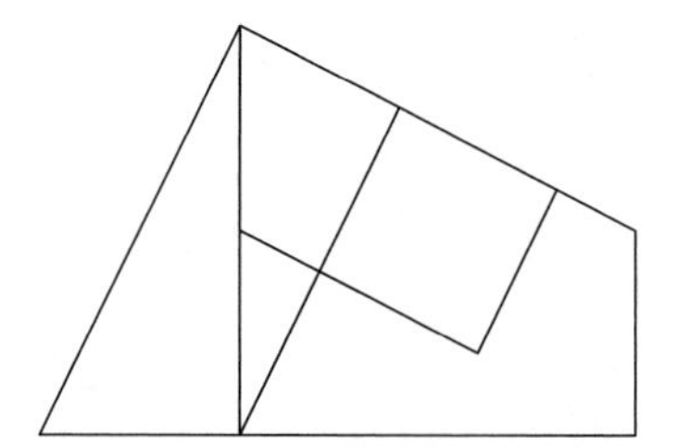

28.

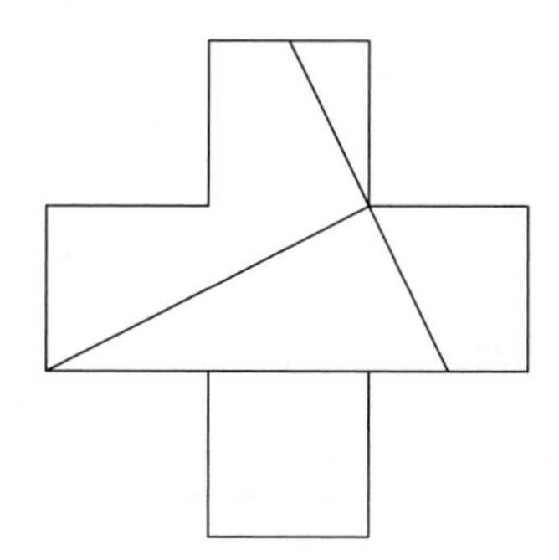

29. Designs will vary.

30. Designs will vary.

31. Constructions will vary. See descriptions in Appendix D.

32. The processes will depend upon the GES used. In general, **(a)** Congruent figures may be obtained by using a form of the COPY command. **(b)** The COPY command applies to all types of figures, angles as well as segments. **(c)** Most GES have a BISECT command that will determine midpoints and angle bisectors. **(d)** Same as (c). **(e)** Once a line and point have been drawn, GES has a construct PERPENDICULAR command. **(f)** Same as (e).

33. (a)

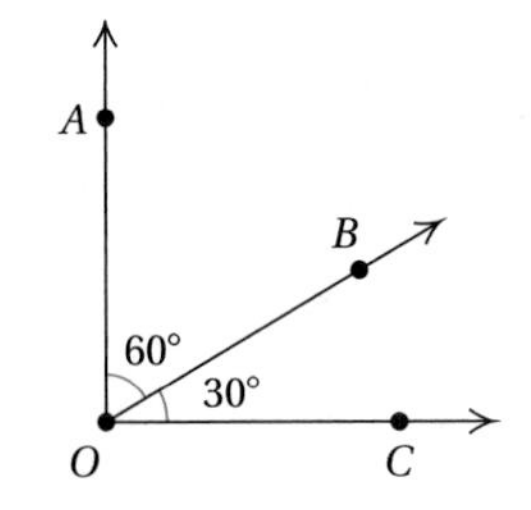

(b)

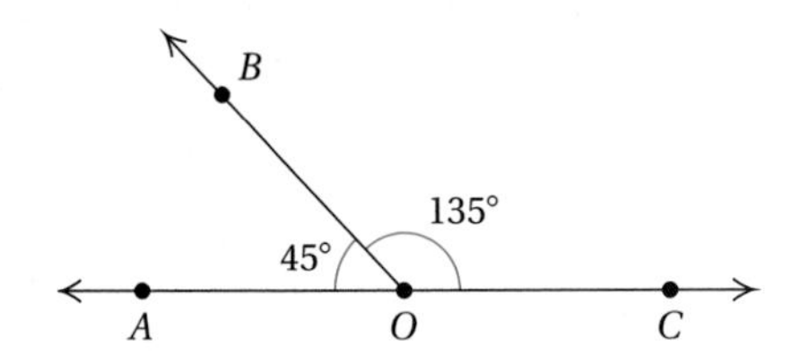

34. (a) It is not possible for four lines to intersect in exactly two points.

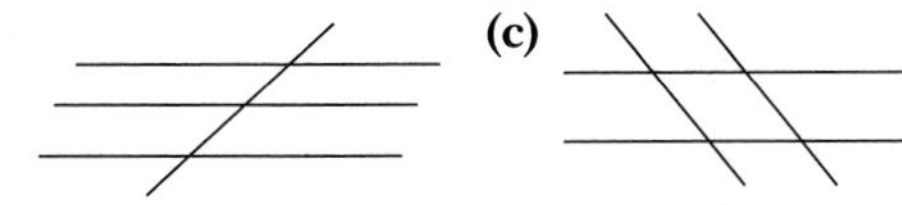

(d) **(e)**

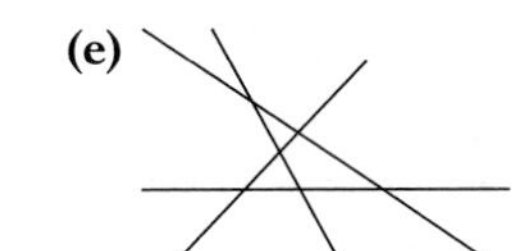

35. A triangle has six exterior angles; two per vertex. Usually the exterior angles are considered in groups of three as shown in the figures.

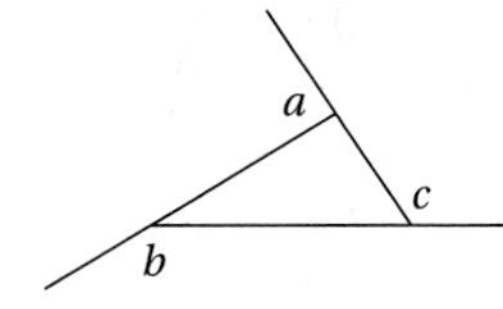

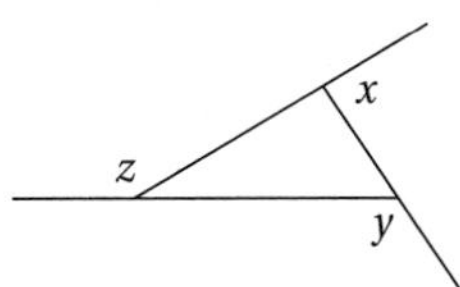

36. Six lines form a maximum of ten regions. You may note that n lines form a maximum of $(n^2 - 3n + 2)/2$ regions.

37. Responses will vary. Three such paths are: *ABCDEF BDFACEA*; *AFEDCBFDBAECA*; *AECAFDBFEDCBA*.

38. Responses will vary.

39. (a) 0.5 **(b)** 1 **(c)** 2 **(d)** 1

40. Responses will vary.

41. 20 lines

42. Responses will vary. All the triangles can be formed on a 3 by 3 Geoboard.

43. (a) 1 sq. unit **(b)** 16 sq. units **(c)** squares with area 2, 4, 5, 8, 9, and 10 sq. units **(d)** no

44. If D and F are trisection points of $\overline{AC}$ and E and G are trisection points of $\overline{AB}$, $\overline{DE} = \frac{1}{3}\,\overline{CB}$ and $\overline{FG} = \frac{2}{3}\overline{CB}$.

45. The area of the large square is 100. The area of the small square is 50, so the length of one side is $\sqrt{50}$ or a little over 7 feet.

46. If the first square has side 10, the first six squares have area 100, 50, 25, 12.5, 6.25, and 3.125.

47. Construct the perpendicular bisector of two chords of the arc. the intersection of these two bisectors is the center of the circle. You could make duplicate pieces of the arc and place them together to reconstruct the circle, then measure the diameter.

48. Since there are more than two odd vertices, there is no way to traverse the network where the vertices are the islands and the bridges the edges.

49. (a)

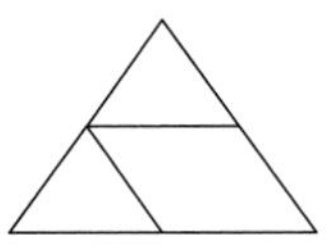

(b)

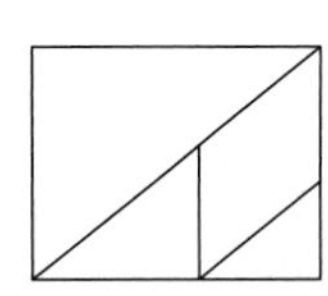

(c)

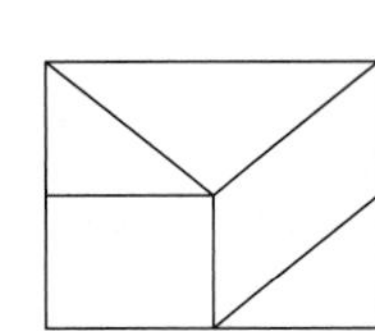

(d)

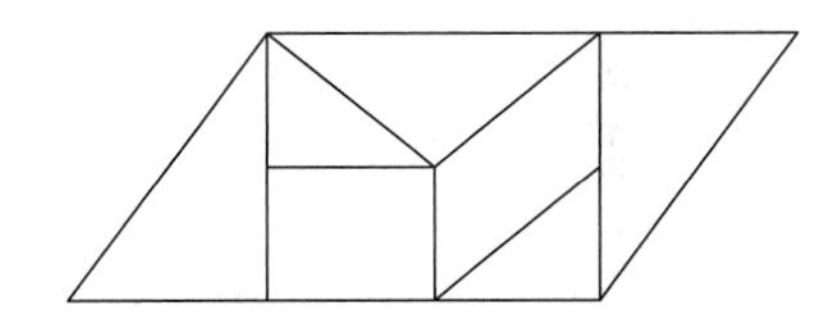

(e)

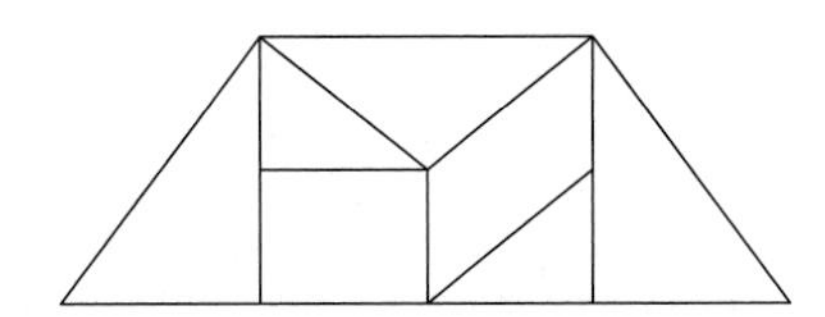

(f)

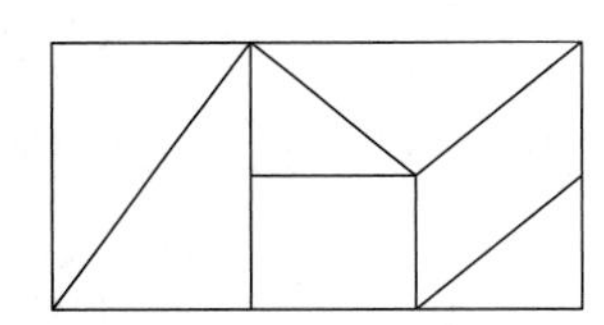

(g)

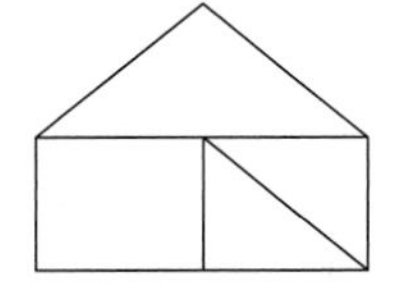

50. Lot *A*: \$142,857; Lot *B*: \$238,095; Lot *C*: \$238,095; Lot *D*: \$238,095; Lot *E*: \$47,619; Lot *F*: \$47,619; Lot *G*: \$47,619

51. (a) Responses will vary. One approach would be to take the cut-out triangles and use the smallest as the unit of area. The areas of the other triangles could be estimated in terms of this unit. **(b)** The triangles could be overlaid with the grid and the number of squares within each triangle estimated. Full squares inside the triangles are, of course, counted. Squares half or more in are counted; squares considered to be more than half out are not counted.

52. A *triquad* is a figure formed from a triangle and a quadrilateral such that the two figures share either a common vertex or a common side.

53.

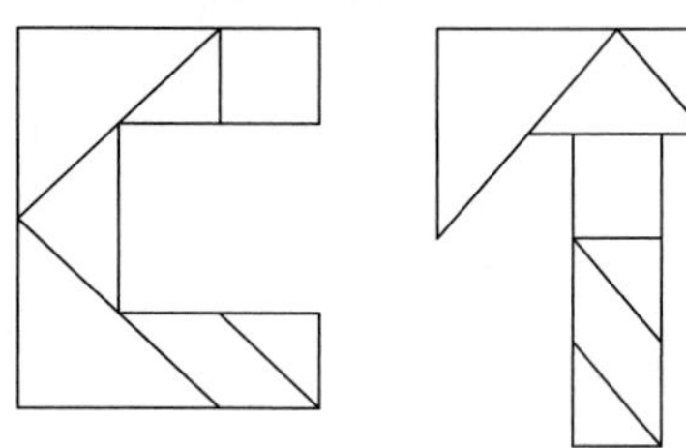

54. Responses will vary.

Section 10.3

1. (a) *d, f* and *c, e* **(b)** *a, e* and *d, g* and *b, f* and *c, h* **(c)** *a, h* and *b, g* **(d)** *d, e* and *c, f* **(e)** Answers will vary. *b, g* and *a, e* and *d, f* **(f)** Answers will vary. *d, e* and *a, g*

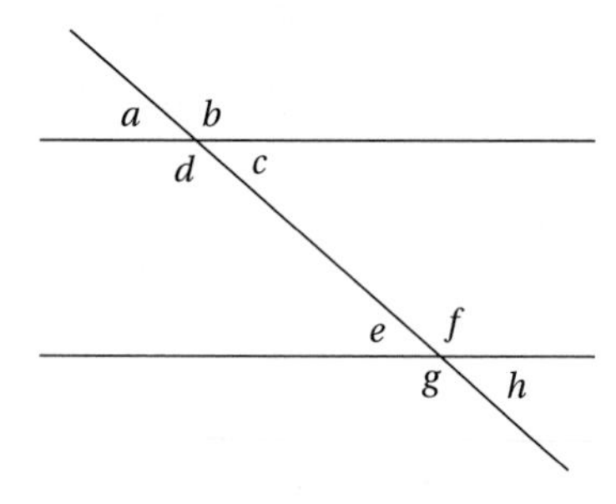

2. (a) $m\angle 1 = 52°$: vertical angles are congruent. **(b)** $m\angle 2 = 128°$: supplementary angles have a sum of 180°. **(c)** $m\angle 4 = 128°$: angles 2 and 4 are vertical and therefore congruent. **(d)** $m\angle 5 = 52°$: angles 52° and 5 are alternate interior angles, and so are congruent. **(e)** $m\angle 6 = 128°$: angles 6 and 2 are corresponding angles and are congruent. **(f)** $m\angle 7 = 52°$: corresponding angles 7 and 52° are congruent. (g) $m\angle 8 = 128°$: corresponding angles 8 and 4 are congruent.

3. $m\angle 5 = 80°, m\angle 4 = 100°$

4. $m\angle 2 = m\angle 8 = 120°$

5. When two parallel lines are cut by a transversal, alternate interior angles are congruent.

6. $m(\angle 3) = 180 - 52 = 128°$, $m(\angle 1 + \angle 2) = 180 - 52 = 128°$, so $m(\angle 3) = m(\angle 1 + \angle 2)$

7. $m(\angle 1) = m(\angle 2) = m(\angle 3) = m(\angle 4) = 90°$, since a square has four right angles. $\angle 5$ is a right angle, since $\angle 3$ and $\angle 5$ are a linear pair. $m(\angle 6) = 180 - 90 - 42 = 48°$, since the sum of the measures of the angles of a triangle is 180°.

8. Sum (interior angles of a polygon) $= (n - 2)180$, where n is the number of sides of the polygon.

9. $m(\text{interior angle of a polygon}) = [(n - 2)180]/n$

10. 108, 72, 72

11. 120, 60, 60

12. 144, 36, 36

13. 150, 30, 30

14. 140, 40, 40

15. The exterior angle and the interior angle are supplementary. The exterior angle and the central angle are congruent.

16. The measures of the interior angles get larger and larger, approaching 180°. The measures of the central angles get smaller and smaller, approaching 0°.

17. interior: 17,640° exterior: 360°

18. 84°

19. 99°

20. 135°

21. Since angles *S* and *Q* are inscribed angles, that is, they have vertices on the circle and their sides are chords, their measures are equal in degrees to half the measures of the arcs they intercept. Since they both intercept semicircles and the degree measure of a semicircle is 180°, the degree measure of each angle is 90°, the measure of a right angle.

22. 35°

23. 135°

24. 27°

25. 30°

26. 41°

27. They are congruent.

28. $m(\angle DAC) = \frac{1}{2}m(arc\ DC)$ or $\frac{1}{2}(360 \div 5) = 36°$

29. An angle inscribed in a semicircle is a right angle. When the angle between his sightings is a right angle he is at the danger semicircle.

30. The angle should be set at 135°, since a stop sign is an octagon and the measure of an interior angle of an octagon is 135°.

31. Responses will vary.

32. (a) The sum of the three angles of any triangle is exactly 180°. By definition the measure of an obtuse angle is greater than 90° and the sum of two obtuse angles is greater than 180°. Thus no two obtuse angles may be in the same triangle. **(b)** The sum of any obtuse angle and a right angle is a greater than 180°. But the sum of the angles of a triangle is exactly 180°. So no triangle may contain both a right and an obtuse angle.

33. They are complementary.

34. $m\angle A$ is 42°, since its supplement is 138°. $m\angle B$ is $180 - (42 + 83) = 55°$, since the sum of the measures of a triangle is 180°.

35. The sum of the measures of the angles of the red triangle + green triangle + blue rectangle is $180 + 180 + 360 = 720°$.

36. An interior angle of a regular pentagon is 108°. Three regular pentagons around a point would take 324° and leave a gap. Four regular pentagons around a point would take 432° and create an overlap.

37. Yes. The measures of their interior angles are 108°, 108°, and 144°, respectively. These measures sum to 360°.

38. The measures of their interior angles are 60°, $128\frac{4}{7}°$, and $171\frac{3}{7}°$, respectively. These measures sum to 360°.

39. 18 sides

40. 9 sides

41. 16 sides.

42. **(a)** The bisectors are concurrent. **(b)** It is always true. Theorem: In any triangle, the bisectors of an interior angle at one vertex and the exterior angles at the other two vertices are concurrent. **(c)** The excenter of a triangle is the intersection of the bisectors of two exterior angles of the triangle. It is the center of the escribed circle of the triangle, that is, the circle tangent to one side of a triangle and tangent to the extensions of the other two sides.

43. a square

44. **(c)** sometimes perpendicular

45. The measure of the exterior angle is equal to the sum of the measures of the two nonadjacent interior angles.

46. Angles H and I are complementary.

47. a parallelogram

48. The sum of the measures of the opposite angles of a quadrilateral is 180°.

49. **(a)** 5 points: 16 regions **(b)** Generalization, 6 points: 32 regions **(c)** The generalization is incorrect. In actuality, 6 points: 31 regions.

50. 6 sides

51. The ten arrangements, using the number of sides, are:
6, 6, 6;
5, 5, 10;
4, 8, 8;
4, 5, 20;
4, 6, 12;
3, 7, 42;
3, 8, 24;
3, 9, 18;
3, 10, 15;
3, 12, 12;

52. Place the square as shown and mark points A and B and points C and D. Draw $\overline{AB}$ and $\overline{CD}$. Because X and Y are right angles, $\overline{AB}$ and $\overline{CD}$ are diameters. Diameters intersect at the center of a circle.

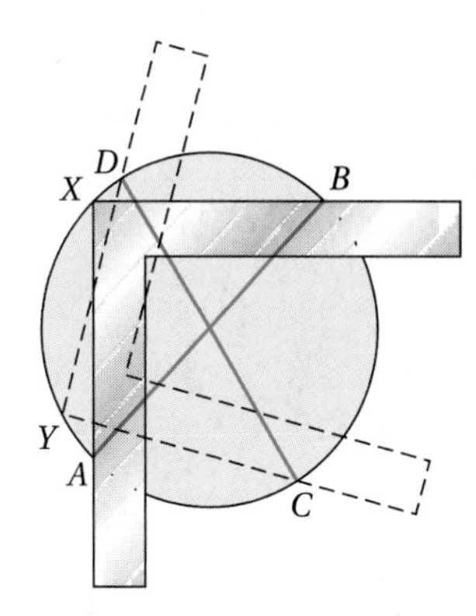

53. The measure of the central angle is $\frac{360}{n}$. The measure of an interior angle is $[(n - 2)180]/n$. This is equivalent to $180 - \frac{360}{n}$, so the book's statement was correct.

54. The exterior angles make up a rotation of 360°. As their number increases, they get smaller to compensate, and their sum is constant. However, as the number of interior angles increases, the interior angles get larger, and their sum increases.

55. Responses will vary.

56. 7.5°: 500 = 360°:C. So $C = 24{,}000$. The circumference of the Earth is given as 24, 902 miles. Both Erathosthenes' linear and angular measure lacked the precision available today.

57. For each of the n sides there is a triangle with an angle sum of 180°. So the sum of all the angles is $180n$. Subtracting the sum of all the angles at the point inside the polygon, 360°, we are left with $180n - 360 = 180n - 2(180) = (n - 2)180$. This sum is the sum of all the remaining angles of the n triangles that make up the interior angles of the polygon.

Section 10.4

1. $\triangle ABC \cong \triangle DEF$, SAS

2. $\triangle PGO \cong \triangle JAH$, SSS

3. $\triangle SRO \cong \triangle BAT$, AAS

4. $\triangle JRT \cong AGO$, HL

5. not necessarily congruent

6. $\triangle LAT \cong \triangle OPM$, ASA

7. A counterexample would be to show two triangles with all three corresponding angles congruent, but with the sides of one twice the length of the corresponding sides of the other.

8. A counterexample would be to show two triangles with a pair of two adjacent corresponding sides congruent, but with the angle between these sides twice the size of the corresponding angle in the other triangle.

9. $\triangle BDM \cong \triangle FDM$, SAS

10. $\triangle MKI \cong \triangle MLI$, SSS

11. $\triangle BAJ \cong \triangle FGH$, AAS

12. $\triangle CDB \cong \triangle EDF$, ASA

13. $\triangle KBM \cong \triangle LFM$, HL, or SSS, or SAS

14. $\triangle KJI \cong \triangle LHI$, AAS

15. No. The corresponding sides of two equilateral triangles are not necessarily congruent.

16. $5\sqrt{2}$

17. $5\sqrt{3}$

18. $3\sqrt{3}$

19. $3\sqrt{2}$

20. 26

21. $a^2 + b^2 = c^2$, where a and b are lengths of legs and c is the length of the hypotenuse of a right triangle. Dealing with lengths of sides of a right triangle, the hypotenuse squared equals the sum of the legs squared.

22. If $a^2 + b^2 = c^2$, where a and b and c are the lengths of the sides of the triangle, then the triangle is a right triangle. Since $5^2 + 12^2 = 13^2$, the triangle is a right triangle.

23. $d^2 = 120^2 + 120^2 = 120\sqrt{2} \approx 169.7$ feet

24. 3.464; 4

25. 25 inches

26. 17 inches

27. 5.66 ft.

28. **(a)** $13^2 = 12^2 + 5^2$, so the triangle contains a right angle. **(b)** $6^2 < 5^2 + 4^2$, so the triangle is not right. **(c)** $17^2 = 15^2 + 8^2$, so the triangle is a right triangle. **(d)** $13^2 > 11^2 + 7^2$ so the triangle does not contain a right angle. **(e)** $25^2 = 24^2 + 7^2$, so the triangle is a right triangle.

29. The diagonal of the doorway is $\sqrt{(6.5^2 + 3^2)} \approx$ 7.16 feet. So the 7-foot square piece of plywood can be carried through the doorway.

30. 127 feet, 3 inches

31. 413.26 feet

32. 39 feet

33. **(a)** $\sqrt{2}$ **(b)** 2 **(c)** $\sqrt{3}$ **(d)** $\frac{\sqrt{3}}{3}$

34. $h = 16$ $s = 8\sqrt{3}$

35. $d = 10$

36. $x = 10$ $y = 5\sqrt{3}$

37. $\frac{\sqrt{2}}{2}$

38. $2\frac{\sqrt{3}}{3}$

39. $\frac{1}{2}$

40. SAS

41. ASA

42. AAS

43. 10.77 inches

44. **(a)** Never: No obtuse triangle contains a right angle. **(b)** Sometimes. Some isosceles triangles are right. **(c)** Sometimes. Some scalene triangles are right triangles. **(d)** Always. The Pythagorean theorem specifically addresses a relationship among the sides of right triangles.

45. For three numbers a, b, and c, c the largest, if $a^2 + b^2 = c^2$, then the numbers can represent the sides of a right triangle. So: **(a)** $(\sqrt{3})^2 + (\sqrt{4})^2 = (\sqrt{7})^2$: yes. **(b)** $0.3^2 + 0.4^2 = 0.5^2$: yes. **(c)** $24^2 + 10^2 = 26^2$: yes. **(d)** $3^2 + 2^2 < 4^2$: no.

46. ≈ 4.47

47. $d^2 = s^2 + s^2$, so $d = \sqrt{2s^2} = s\sqrt{2}$

48. $r \approx 10.19$; distance $\approx 10.19\sqrt{2} \approx 10.19$

49. an equilateral triangle

50. an equilateral triangle

51. The nine points lie on a circle. The midpoint of the segment on the Euler line joining the orthocenter and the circumcenter is the center of the circle.

52. **(a)** $\frac{1}{2}$ as long **(b)** The four interior triangles are congruent to each other.

53. Estimates will vary.

54. Responses will vary. Statements should be similar to the following. **(a)** $\Delta\ ABC$ and $\Delta\ DEF$ are right triangles, $\overline{EF} \cong \overline{BC}$, $\angle C \cong \angle F$ (given data) **(b)** $\angle A \cong \angle D$ (all right angles are congruent) **(c)** $\Delta ABC \cong \Delta DEF$ (AAS)

55. **(a)** $a = m^2 - n^2$; $b = 2mn$; $c = m^2 + n^2$ **(b)** Yes, increase the upper limit of the loop of statement 10. **(c)** (3, 4, 5); (5, 12, 13); (7, 24, 25) **(d)** The program could be

```
10 FOR N = 1 to 5
20 LET A = 2N + 1
30 LET B = 2N^2 + 2N
40 LET C = 2N^2 + 2N + 1
50 PRINT A;",";B;",";C
60 NEXT N
70 END
```

The first program produces all integer triples within the limits of the loops. The program above produces only primitive triples.

56. In the regular septagon, the angle size of the sharp points of the cutter is found by subtracting the two angles of the white right triangle from the measure of the interior angle of the septagon ($128\frac{4}{7}°$). B ut the two angles of the right triangle are congruent at each point because the right triangles are all congruent (because of the HL Congruence theorem). So the same amount is subtracted each time.

57. The area of the large square is c^2. Each of the triangles has an area $(\frac{1}{2})ab$. So the area of the four triangles taken together have an area of $2ab$. Now, the small square has an area $(b - a)^2$ or $b^2 - 2ab + a^2$. So summing all the pieces we have $c^2 = 2ab + b^2 - 2ab + a^2 = b^2 + a^2$. Since a and b are the legs of a right triangle with hypotenuse c, we have the Pythagorean relation.

58. The two large squares both have sides of length $a + b$ and so have the same area. Two of the triangles cover one of the rectangles so if all four triangles are removed from one large square and both rectangles are removed from the other large square the remaining areas, a square of side c, and two areas, squares of sides a and b, are equal.

59. The student claims that $a^2 = (c + b)(c - b)$. Now, $(c + b)(c - b) = c^2 - cb + cb - b^2 = c^2 - b^2$.
So $a^2 = c^2 - b^2$ and $c^2 = a^2 + b^2$, the Pythagorean relation. So the student is correct.

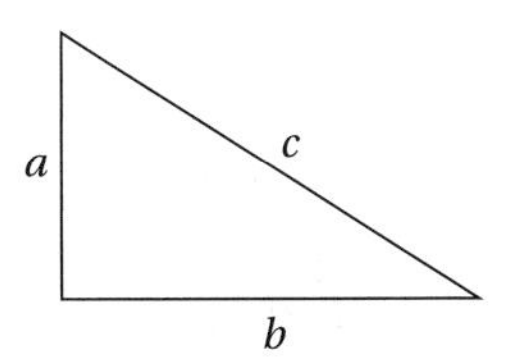

60. **(a)** $a^2 + b^2 > c^2$ **(b)** $a^2 + b^2 < c^2$

61. Yes. Substitute $(b + 1)$ for c in the Pythagorean theorem, simplify, and solve for the other leg, a.
$a = \sqrt{[b + (b + 1)]}$. **(a)** 4 **(b)** 4.36 **(c)** 9 **(d)** 11

62. **(a)** For m an even number greater than 2, the longest side is $(\frac{m^2}{4}) + 1$ and because $m^2 + [(\frac{m^2}{4}) - 1]^2 = m^2 + \frac{m^4}{16} - (\frac{2m^2}{4}) + 1 = \frac{m^4}{16} + (\frac{2m^2}{4}) + 1 = [(\frac{m^2}{4}) + 1]^2$ the statement is true. **(b)** $(2n + 1)^2 + (2n^2 + 2n)^2 = 4n^4 + 8n^3 + 8n^2 + 4n + 1 = (2n^2 + 2n + 1)^2$, so the statement is true. **(c)** $x^2 + (x + 1)^2 = (x + 2)^2$ is equivalent to $x^2 + (x + 1)^2 - (x + 2)^2 = 0$ which is also equivalent to $x^2 - 2x - 3 = 0$. The solutions to this last equation are 3 and -1. So the only Pythagorean triple with consecutive integers is 3, 4, 5 because -1 cannot be the length of the side of a triangle.

63. 3.46 feet

64. The statements will still be true. The conclusion can be supported by expressing the areas on each leg and proving algebraically that the sum of the areas on the legs equals the area on the hypotenuse. Or GES could be used to support the conjecture.

65. The area of the trapezoid, A, is the sum of the areas of the three triangles. So $A = (\frac{1}{2})ab + (\frac{1}{2})ab + (\frac{1}{2})c^2$. The area of a trapezoid is also $(\frac{1}{2})$ altitude(base 1 + base 2). So $A = \frac{1}{2}(a + b)(a + b) = \frac{1}{2}(a^2 + 2ab + b^2)$. Thus $(\frac{1}{2})ab + (\frac{1}{2})ab + (\frac{1}{2})c^2 = \frac{1}{2}(a^2 + 2ab + b^2)$ and $a^2 + b^2 = c^2$.

66. Because the two squares have equal area, we use areas to conclude that $a^2 + b^2 + 4(\frac{1}{2}ac) = c^2 + 4(\frac{1}{2}ac)$. Subtracting $4(\frac{1}{2}ac)$ from both sides of the equation, $a^2 + b^2 = c^2$. In the geometric argument, we removed four triangles from each square figure, leaving a square with side a and a square with side b equal to a square with side c. In the algebraic argument, we symbolized these actions with an equation and simplified it.

Section 10.5

1. $\overline{AB} \parallel \overline{CD}, \overline{AC} \parallel \overline{BD}, \overline{AB} \cong \overline{CD}, \overline{AC} \cong \overline{BD}$, $\angle CAB \cong \angle BDC$, $\angle ACD \cong \angle ABD$. $\angle CAB$ and $\angle ABD$ are supplementary, $\angle ABD$ and $\angle BDC$ are supplementary, $\angle BDC$ and $\angle ACD$ are supplementary, and $\angle ACD$ and $\angle CAB$ are supplementary.

2. $\overline{AB} \parallel \overline{CD}$, $\overline{AC} \parallel \overline{BD}$, $\overline{AB} \cong \overline{CD}$, $\overline{AC} \cong \overline{BD}$, $\angle A \cong \angle D$, $\angle C \cong \angle B$. $\angle A$ and $\angle B$ are supplementary, $\angle B$ and $\angle D$ are supplementary, $\angle D$ and $\angle C$ are supplementary, and $\angle C$ and $\angle A$ are supplementary. $\angle A$, $\angle B$, $\angle C$, and $\angle D$ are right angles.

3. $\overline{AB} \parallel \overline{CD}$, $\overline{AC} \parallel \overline{BD}$, $\overline{AB} \cong \overline{BD}$, $\overline{CD} \cong \overline{AC}$, $\angle A \cong \angle D$, $\angle C \cong \angle B$. $\angle A$ and $\angle B$ are supplementary, $\angle B$ and $\angle D$ are supplementary, $\angle D$ and $\angle C$ are supplementary, and $\angle C$ and $\angle A$ are supplementary. $\overline{AD}$ is the perpendicular bisector of $\overline{BC}$. $\overline{AD}$ is the bisector of $\angle A$ and $\angle D$. $\overline{BC}$ is the bisector $\angle C$ and $\angle B$.

4. rectangle

5. rhombus

6. rhombus

7. square

8. None. It could be a trapezoid.

9. None. It could be a general quadrilateral, with two sides congruent.

10. rhombus

11.

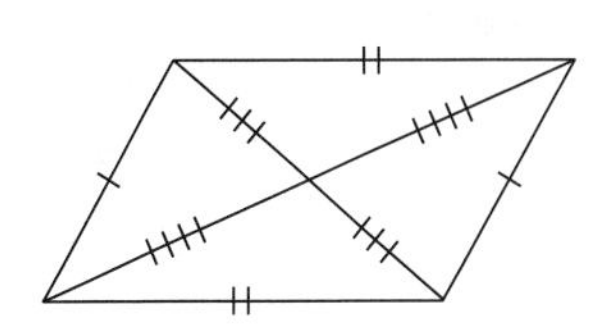

12.

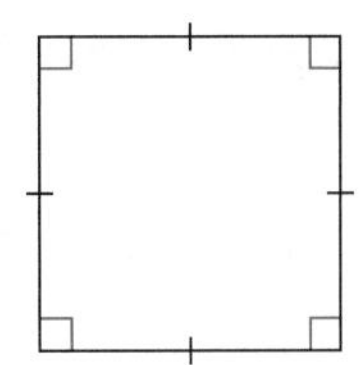

13.

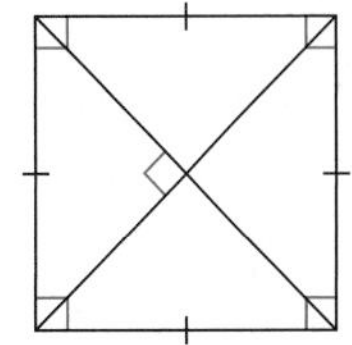

14. not possible

15. not possible

16.

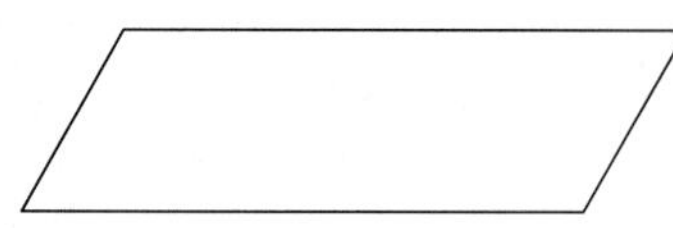

17. not possible

18. **(a)** yes **(b)** no **(c)** no **(d)** yes **(e)** yes **(f)** yes

19. **(a)** no **(b)** no **(c)** no **(d)** yes **(e)** no **(f)** no

20. **(a)** yes **(b)** no **(c)** no **(d)** no **(e)** no **(f)** yes

21. **(a)** no **(b)** no **(c)** no **(d)** no **(e)** no **(f)** no

22. **(a)** yes **(b)** no **(c)** no **(d)** no **(e)** no **(f)** yes

23. Responses will vary. Opposite roads are parallel. Opposite roads are congruent. Consecutive angles between the roads are supplementary.

24. Responses will vary. Its sides meet at right angles. Opposite sides are congruent and parallel.

25. The diagonals of a rectangle are congruent.

26. If the diagonals of a quadrilateral bisect each other, then the quadrilateral is a parallelogram.

27. If the opposite sides of a quadrilateral are congruent, then the quadrilateral is a parallelogram.

28. The sides of the nonsquare rectangle are not all congruent. The diagonals of the nonsquare rectangle are not perpendicular.

29. The rhombus has all four sides congruent. The diagonals of the rhombus are perpendicular bisectors of each other.

30. It is a rhombus because the compass openings used ensure that all four sides are congruent.

31. a parallelogram

32. a parallelogram

33. Not necessarily a parallelogram. The pair of parallel sides may not be the same pair as the pair of congruent sides.

34. a parallelogram

35. a parallelogram

36. Not necessarily a parallelogram. Could be a nonparallogram kite.

37. a parallelogram

38. Not necessarily a parallelogram. It is possible to have a quadrilateral with a pair of sides congruent, but with no pairs of sides parallel.

39. Not necessarily a parallelogram. It could be a trapezoid.

40. Not necessarily a parallelogram. It could be a trapezoid.

41. The opposite angles of the parallelogram are congruent and the consecutive angles are supplementary, so you can use these properties to show that all angles of the parallelogram are right angles, and it is a rectangle.

42. If the adjacent sides are congruent, you can use this idea to show that all sides are congruent, and the parallelogram is a rhombus.

43. A parallelogram with at least one right angle is a rectangle. If adjacent sides of the rectangle are congruent, then you can show that all sides are congruent and the rectangle is a square.

44. Given that line r is parallel to line s, choose a point A on line r and a point B on line s such that $\overline{AB}$ is perpendicular to r and s. Then mark a point E on r and a point F on s such that $\overline{AE} \cong \overline{BF}$. $AEFB$ is a parallelogram, since it has a pair of sides $\overline{AE}$ and $\overline{BF}$ that are congruent and parallel. It follows that $\overline{AB} \cong \overline{EF}$, since these are opposite sides of a parallelogram.

45. No. Consider a generalization such as "A rectangle can have a pair of adjacent sides that are not congruent." This generalization would not be true if the word "rectangle" were replaced with the word "square."

46. A rectangle must have at least one right angle.

47. A rhombus must have all sides congruent.

48. **(a)** rhombus, square, kite **(b)** parallelogram, rectangle, rhombus, square **(c)** parallelogram, rectangle, rhombus, square **(d)** rhombus, square, kite

49. The area of each figure in (a)–(g) is always $\frac{1}{2}$ the area of the original figure. **(a)** square **(b)** rhombus **(c)** parallelogram **(d)** rectangle **(e)** rectangle **(f)** parallelogram **(g)** parallelogram

50. Estimates will vary. GES gives the ratio between the shaded area and the total area as $\frac{1}{6}$.

51. The segment joining the midpoints of the sides of a trapezoid is equal to $\frac{1}{2}$ the sum of the lengths of the bases.

52. The bisectors are perpendicular.

53. Use the fact that the diagonals of a rhombus are perpendicular.

54. a square

55. If x is the measure of each angle in one pair of opposite congruent angles, y is the measure of each angle in the other pair, $2x + 2y = 360$, so $x + y = 180$, and consecutive angles of the quadrilateral are supplementary.

56. **(a)** The factory can be located at any point inside the rectangle. **(b)** No matter where F is placed inside the rectangle, $a^2 + b^2 = c^2 + d^2$.

57. **(a)** The ratio of the original quadrilateral to the shaded quadrilateral is 5 to 1. **(b)** Questions will vary. Questions might be "What if the vertices are connected in the counterclockwise direction, will the ratio be the same?" or "What if the vertices are connected to trisection points instead of midpoints?" Responses will also vary.

58. Twelve parallelograms can be formed.

59. The perimeter of the Varignon quadrilateral is equal to the sum of the lengths of its diagonals.

60. Responses will vary.

Chapter 10 Review Exercises

1. 1, 1, 2, 3, 5. The ratio of a term and its predecessor approaches the golden ratio as the number of terms increases. The golden ratio appears as the ratio of the length to the width in aesthetically pleasing rectangles.

2. Responses will vary. An appropriate size might be 3 feet by 1 foot 10 inches $\left(\frac{36}{22} = \frac{1.64}{1}\right)$.

3. **(a)**

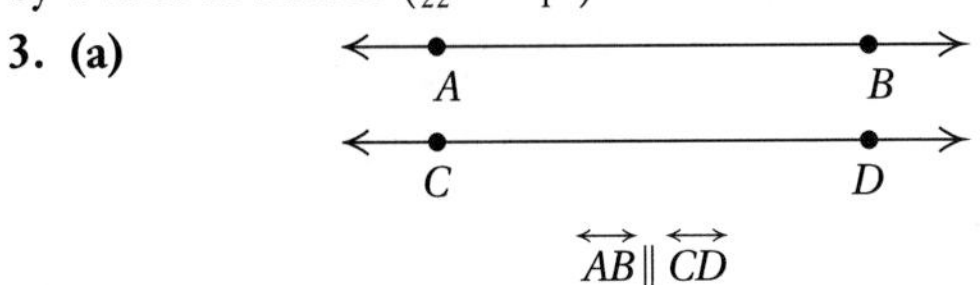

(b)

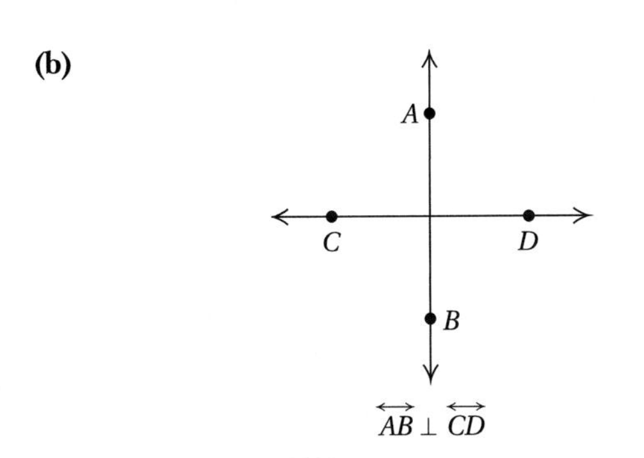

(c)

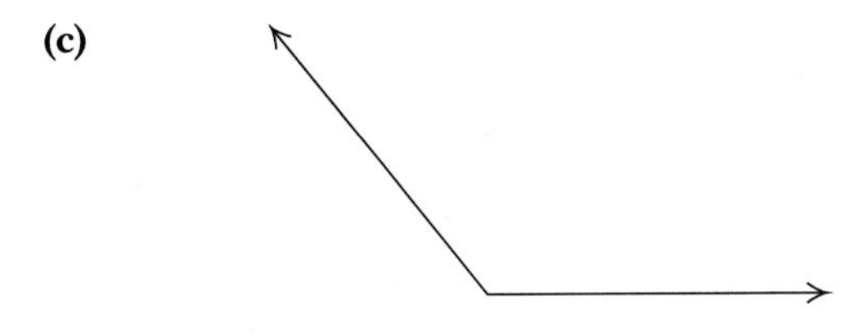

(d)

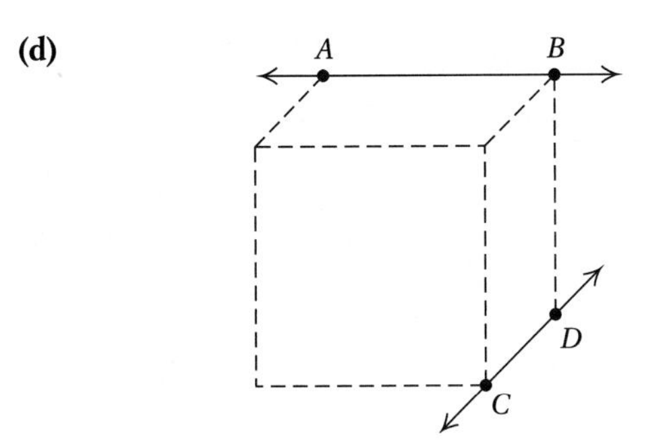

(e)

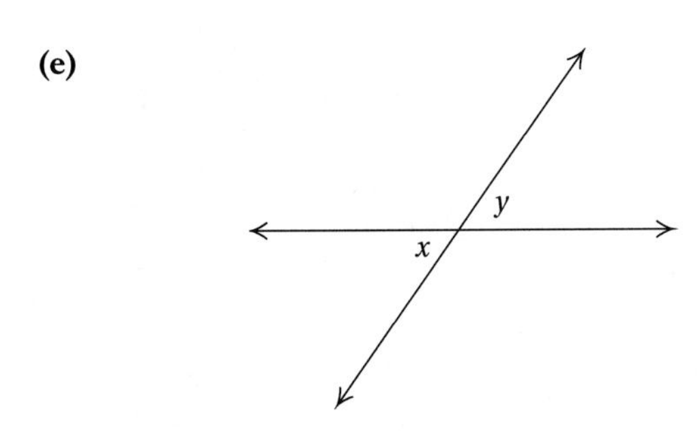

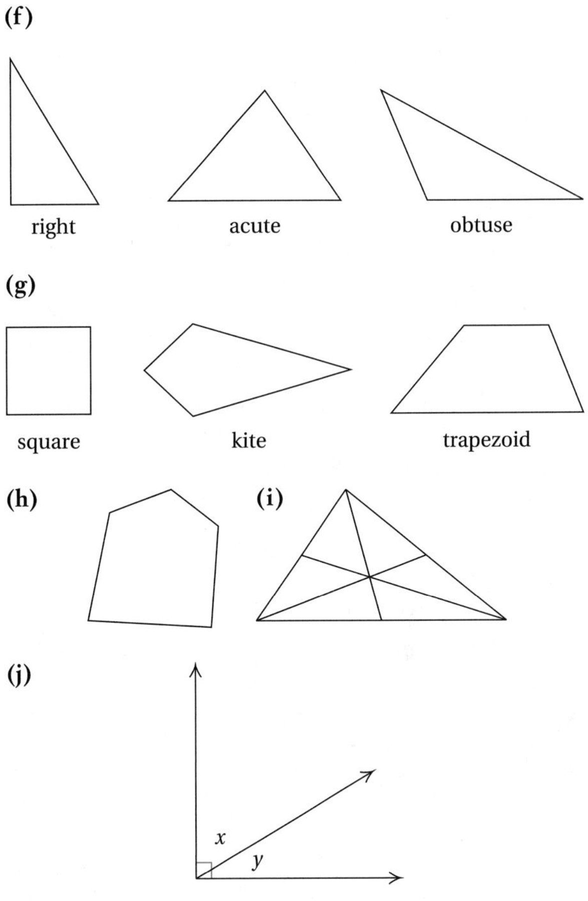

4. **(a)** Complementary angle measures add to 90°. Supplementary angle measures add to 180°. **(b)** A perpendicular bisector is a line that bisects a segment and is perpendicular to it. An angle bisector divides an angle into two congruent parts. **(c)** An isosceles triangle has at least two sides congruent. An equilateral triangle has all three sides congruent. **(d)** A rhombus has all sides congruent and opposite sides parallel. A kite doesn't necessarily have all sides congruent nor opposite sides parallel. **(e)** A rhombus has all sides congruent and diagonals perpendicular. A parallelogram does not have to have all sides congruent or its diagonals perpendicular. **(f)** A square is a special case of a rectangle that has all sides congruent. **(g)** A trapezoid has only one pair of sides congruent. A parallelogram has two pairs. **(h)** A right triangle has a right angle. An obtuse triangle has an obtuse angle. **(i)** A simple closed curve does not intersect itself. A nonsimple closed curve does. **(j)** A segment connecting points on two sides of a convex polygon lies entirely in or on the polygon. In a nonconvex polygon, such a segment can be found that lies outside the polygon.

5. Responses will vary.

6. Yes. The network has exactly two odd vertices, *P* and *Q*, and so is traversable type 2.

7. interior angle: 108°; exterior angle: 72°; central angle: 72°

8. Since the sum of the angles of any triangle is 180°, $m\angle C = 180 - 54 - 78 = 48°$.

9. 540°

10. 135°

11. 35°

12. **(a)** The incenter is the intersection of the angle bisectors of the triangle. **(b)** The circumcenter is the intersection of the perpendicular bisectors of the sides of the triangle. **(c)** The orthocenter is the intersection of the altitudes of the triangle. **(d)** The centroid is the intersection of the medians of the triangle.

13. The Euler line is the line that contains the centroid, orthocenter, and circumcenter of a triangle.

14. The Pythagorean theorem. If a and b are legs of a right triangle and c is the hypotenuse, $a^2 + b^2 = c^2$.

15. 12 feet

16. Responses will vary.

17. **(a)** $16\sqrt{2}$ **(b)** ≈ 10.39, or $6\sqrt{3}$

18. No. If it did, the sum of the angles of the quadrilateral would be greater than 360°, which is impossible.

19. As the number of sides of a regular polygon increases, the sum of the exterior angles remains 360°, so since there are more exterior angles, the measure of each exterior angle must decrease.

20. Since the sum of the exterior angles is 360°, the measure of one exterior angle, E, is $\frac{360°}{n}$. And since the interior and exterior angles are supplementary, the measure I of the interior angle would be $I = 180° - E = 180 - \frac{360°}{n}$.

21. Yes. It has four sides, and could be identified as a nonsimple, nonconvex quadrilateral.

22. 25 feet

23. **(a)** Responses will vary. You could cut the angles from a cardboard triangle and arrange them so that one is adjacent to the other two and show that the exterior sides form a line. You could also rotate an arrow through each of the angles, and show that you had rotated 180°. **(b)** Use the same procedure as in (a), only with a quadrilateral. You could also rotate an arrow through each of the angles, and show that you had rotated 360°.

24. Responses will vary. Answers should address the methods of dissection, motions, measurement and computation, and algebra.

Chapter 11

Section 11.1

1. direction and distance of the slide

2. The pair of letters in the subscript identifies the segment (distance) describing the slide, and the order of the letters identifies the direction.

3. the point to turn around, the direction of the turn, and the turn angle in degrees

4. The letter in the subscript identifies the point of rotation, and the angle in the subscript identifies the size and direction of the turn.

5. the line to flip the plane around

6. The subscript identifies the line of reflection.

7. reflection

8. glide–reflection

9. rotation

10. 180° rotation or translation

11.

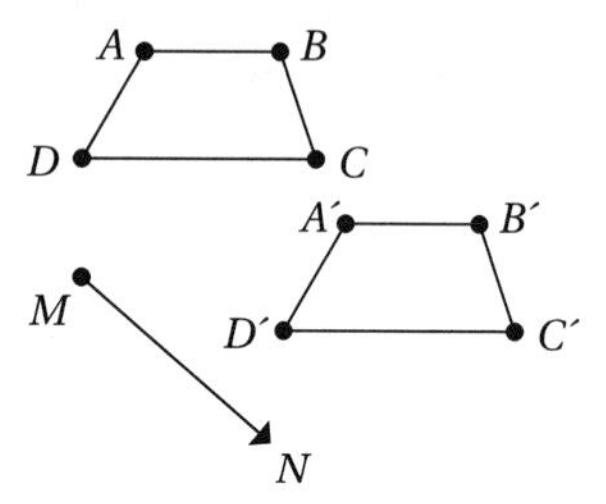

12.

13.

14.

15.

16.

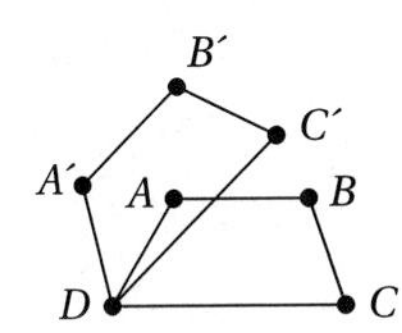

17.

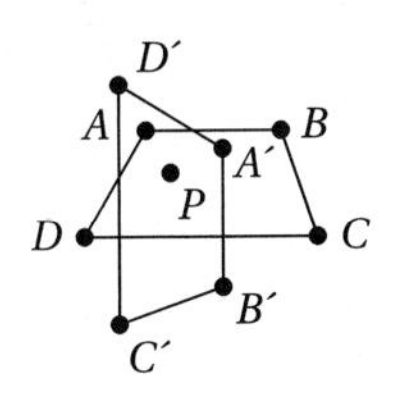

18.

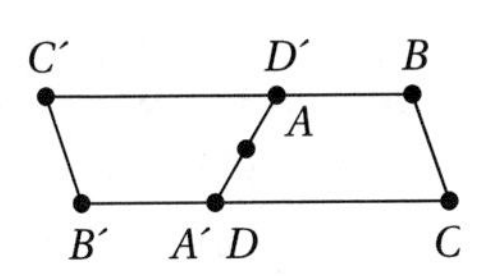

19.

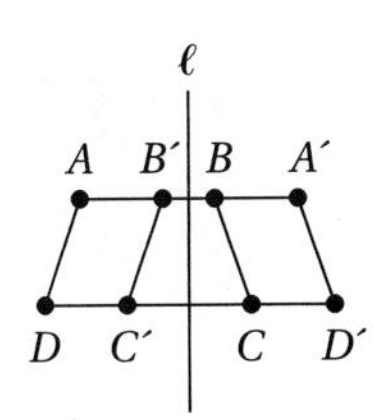

20.

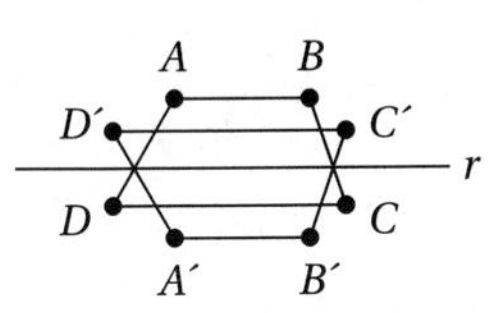

21.

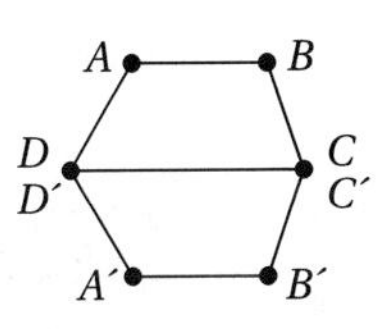

22.

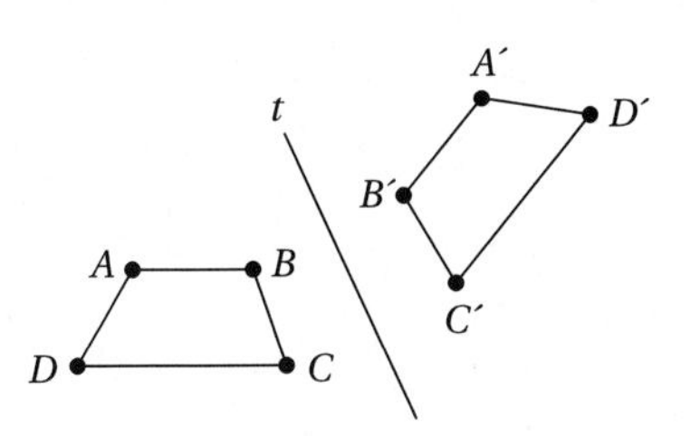

23. **(a)** 120° and 240° rotational symmetry **(b)** two lines of reflectional symmetry, 180° rotational symmetry **(c)** 180° rotational symmetry

24. **(a)** Yes, each can be straightened into a segment without cutting. **(b)** No, the first one has two "ends" and the second has five.

25. **(a)** a closed box, a sealed bottle **(b)** a coffee cup, a section of pipe **(c)** a two-handled tray, a bushel basket (with two handles)

26. {A, R}, {C, I, L, M, N, S, U, V, W, Z,}, {E, F, G, Y, T}, {H, I}, {K, X}, {O, D}, {P, Q}

27. **(a)**

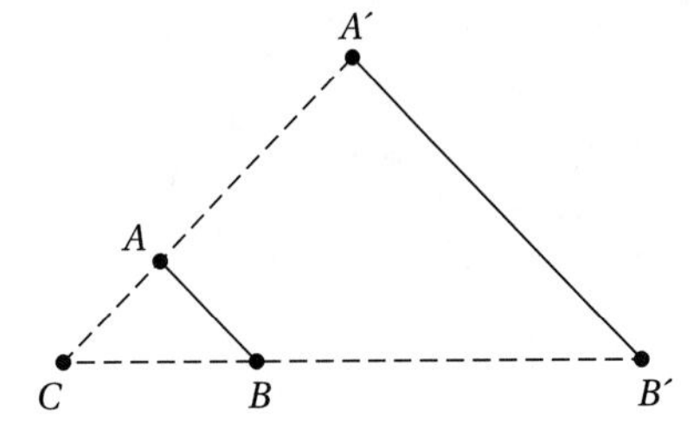

(b) $\frac{3}{1}$ **(c)** $\frac{3}{1}$ **(d)** $\frac{3}{1}$

(e)

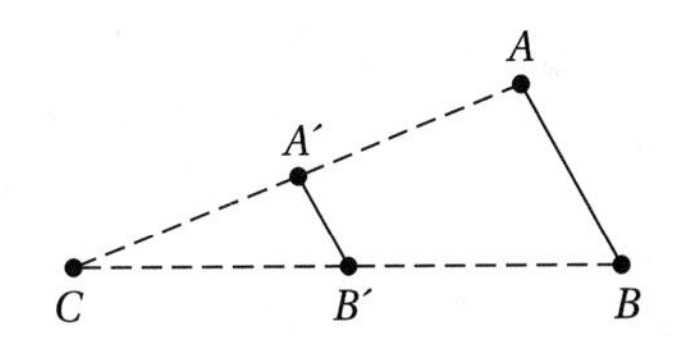

(f) $\frac{1}{2}$ **(g)** $\frac{1}{2}$ **(h)** $\frac{1}{2}$

28. Responses will vary.

29. **(a)** lines of reflection symmetry on the three point/dent lines; rotational symmetry in multiples of 120° about its center **(b)** rotational symmetry about its center in multiples of 90° **(c)** a line of reflection symmetry on each diameter; rotational symmetry around the center of any number of degrees from 0° to 360° **(d)** a line of reflection symmetry along the long diagonal

30.

	Translation	Reflection	Rotation
(a)	No change in orientation	There is a left-right exchange or a top-bottom, but not both.	Change in orientation
(b)	No point maps onto itself.	Points on the line of reflection map onto themselves.	The center of rotation maps onto itself.
(c)	Direction and distance	Line of reflection	Center and angle
(d)	Measures unchanged	Measures unchanged	Measures unchanged

31. Responses will vary.

32. It doesn't matter which transformation you do first.

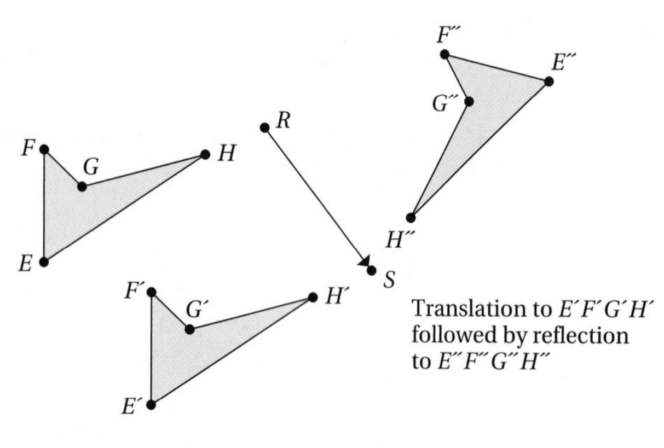

Translation to $E'F'G'H'$ followed by reflection to $E''F''G''H''$

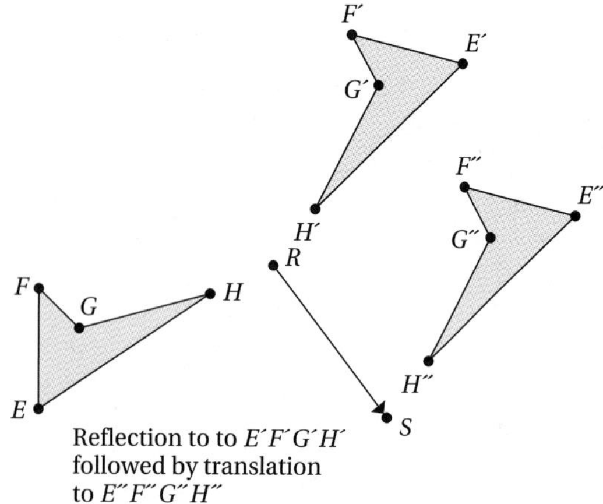

Reflection to to $E'F'G'H'$ followed by translation to $E''F''G''H''$

33. (a)

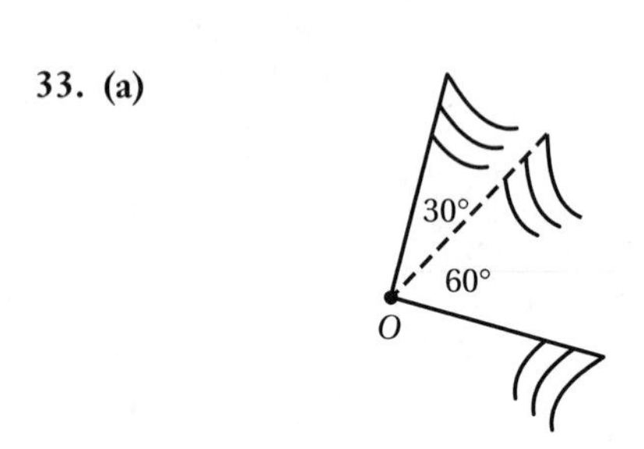

(b)

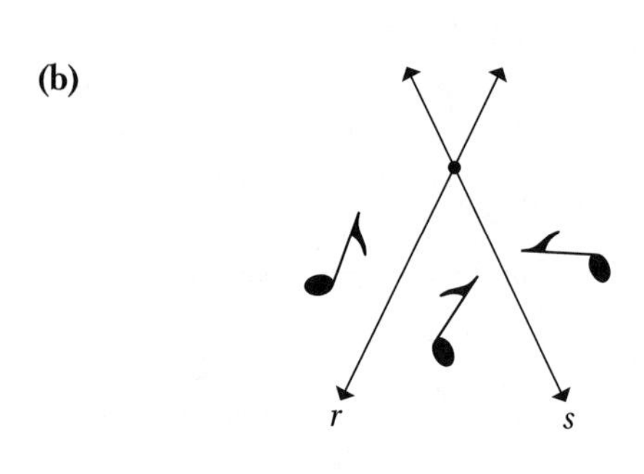

(c)

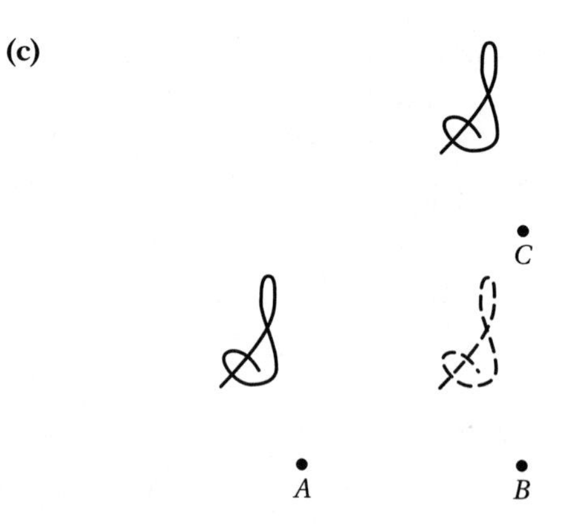

34. (a) 180° rotational symmetry **(b)** 120° and 240° rotational symmetry; three lines of reflectional symmetry **(c)** one line of reflectional symmetry **(d)** no symmetry

35. Responses will vary.

36. (a) a kite that is not a rhombus **(b)** a parallelogram that is not a rectangle **(c)** a rectangle

37. (a) **(b)** **(c)**

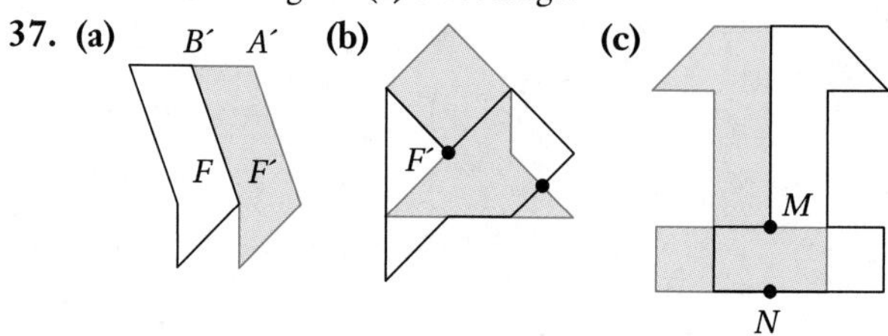

38. Gina used a reflection in line m followed by a reflection in line n; translation through twice the distance between lines m and n gives the same image.

39. Responses will vary.

40. (a)

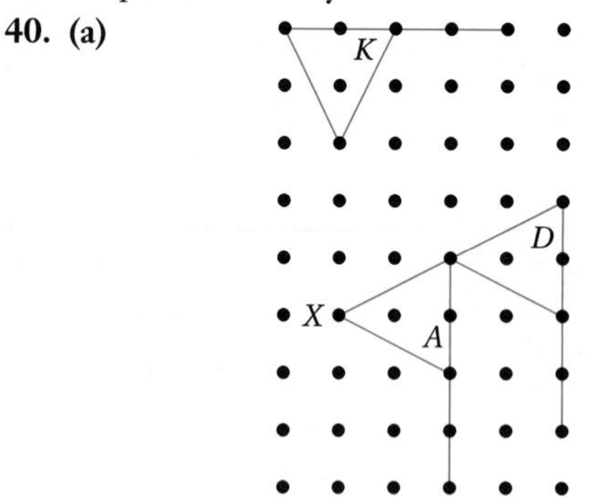

(b) Yes, because the transformations are isometries.

41. No, they are not congruent, because the scale factor of the size transformation is not equal to 1. (See diagram.)

42. No, because the ratios of the sides in the two rectangles would not be equal.

43. Half the height of the person hung at eye-level. (See diagram.)

Answer for Exercise 41

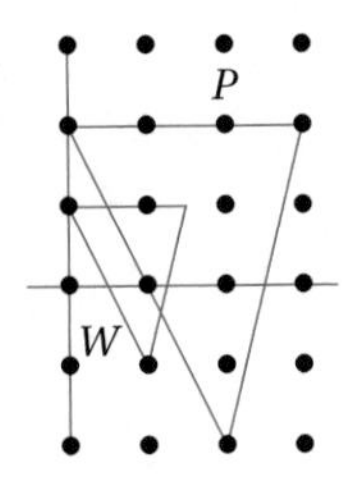

Answer for Exercise 43

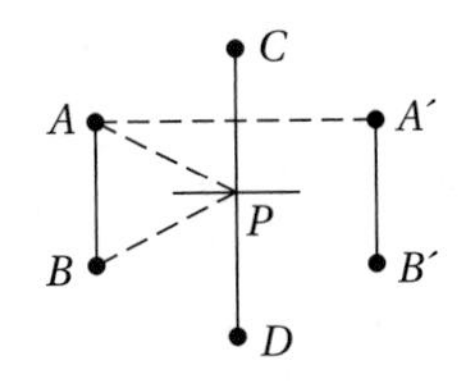

44. Not necessarily; the old lot might have been a rectangle, and the new lot could be a parallelogram that is not a rectangle.

45. No. Yes.

46. Responses will vary.

47. One possibility is:

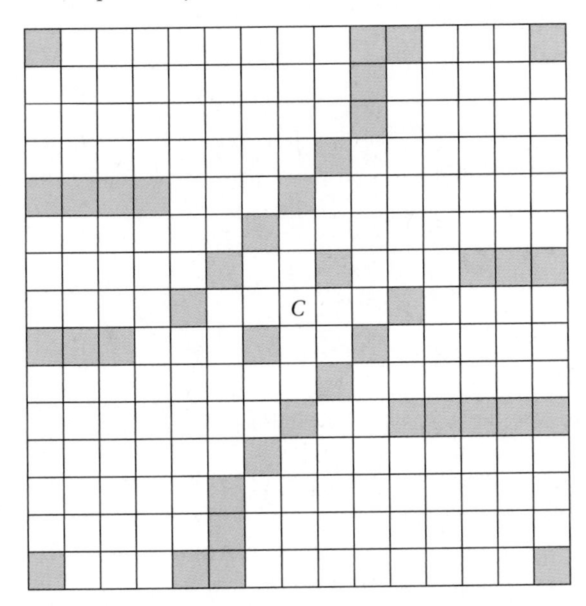

48. Responses will vary.

49.

Polygon	Lines of Reflectional Symmetry	Angles of Rotational Symmetry
Trapezoid	none	none
Parallelogram	none	180°
Isosceles trapezoid	one	none
Kite	one	none
Rhombus	two	90°, 180°, 270°
Rectangle	two	180°
Square	four	90°, 180°, 270°
Scalene triangle	none	none
Isosceles triangle	one	none
Equilateral triangle	three	120°, 240°

50.

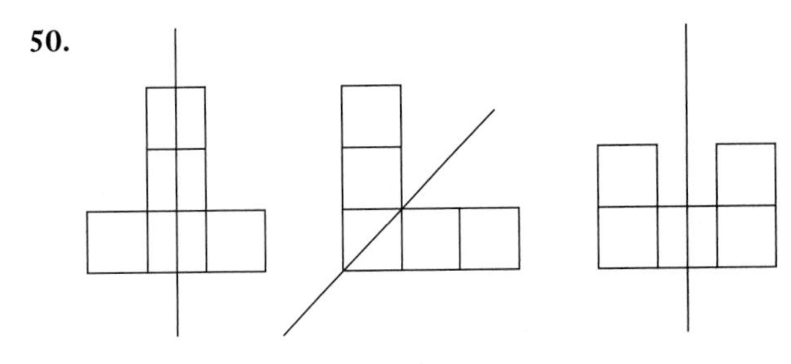

51. **(a)** yes **(b)** no **(c)** no **(d)** yes

52. **(a)** A figure with line symmetry can be created by joining a figure and its reflection image. **(b)** A figure with rotational symmetry can be created by joining a figure and its rotation image(s).

Section 11.2

1. Responses will vary.

2. Responses will vary.

3. 6, because $6 \times 60° = 360°$.

4. 4, because the sum of the four angles of a quadrilateral $= 360°$.

5. Two of the triangles form a parallelogram, which then can be used to tessellate the plane.

6. three, one with equilateral triangles, one with squares, and one with regular hexagons

7.

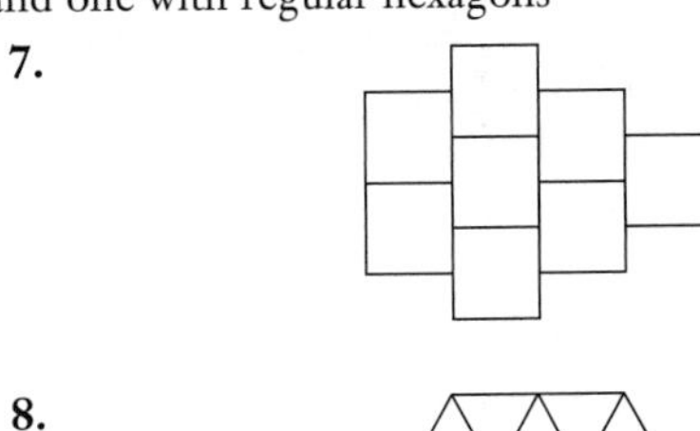

8.

9. Because the 108° angle of a regular pentagon is not a divisor of 360°.

10. No, because it is not made of regular polygons.

11. Yes, because the 4 angles of the kite add up to 360°; all convex quadrilaterals tessellate.

12. no

13. The angle measures of the polygon are greater than $\frac{1}{3}$ of 360°.

14. The equilateral triangles would tessellate the plane, but the colors of the parts of the triangles would make it look like a tessellation of regular hexagons.

15. a regular dodecagon

16. 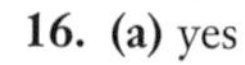 **(a)** yes 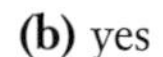 **(b)** yes

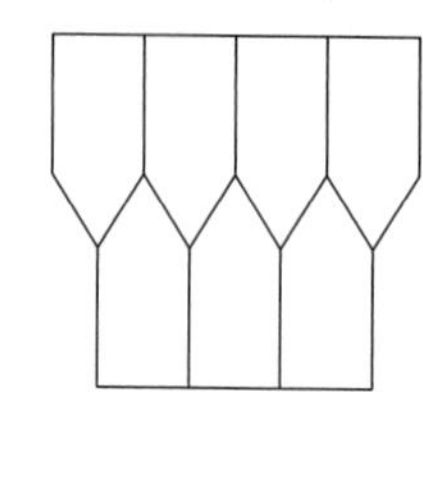

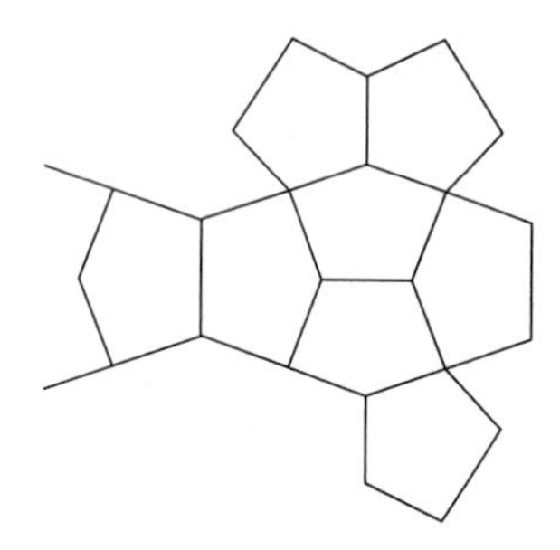

17. The angles are too large for three to fit around a point. Regular octagons with squares will tessellate.

18. **(b)** 3, 3, 3, 4, 4, or $3^3 4^2$ **(d)** 3, 4, 4, 6, or $34^2 6$ **(g)** 4, 6, 12

19. eight

20. Because the hexagons are not regular hexagons.

21. Because not every vertex has the same arrangement of polygons.

22. It will not tessellate because the four polygons form a concave polygon.

23. 180° rotational symmetry

24. no symmetry

25. **(a)** Translations from the vertex of any square to the corresponding vertex of a square of the same size; or rotated 90°, 180°, or 270° about the center of any square. **(b)** Translations from the center of a hexagon to the center of another hexagon; rotations of 180° about the center of any square; 120° or 240° about the center of any triangle; or 60°, 120°, 180°, 240°, or 300° about the center of any hexagon; lines of symmetry of the hexagons can be used as lines of reflection.

26.  **(a)**

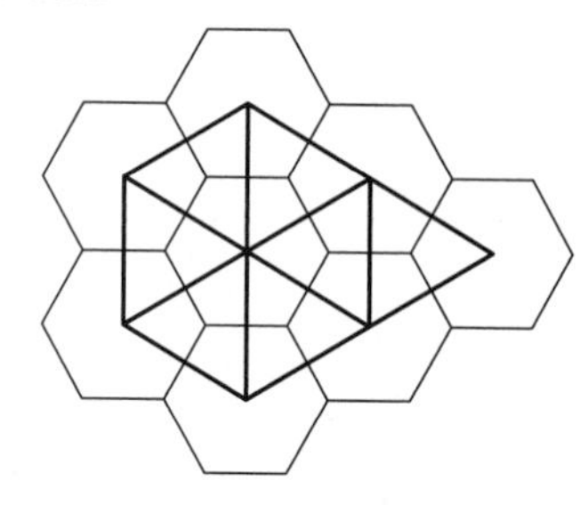

(b)

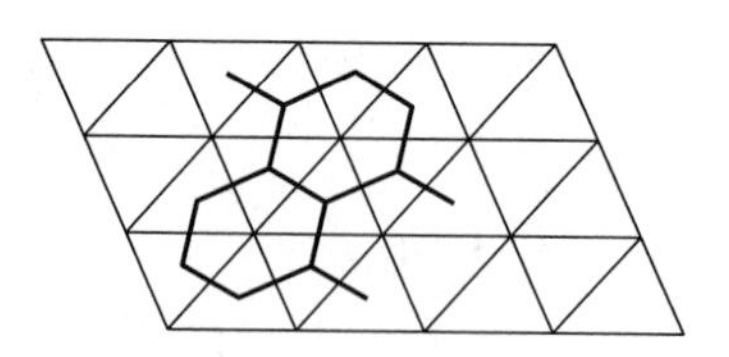

(c) The tessellations are duals of each other.

27.

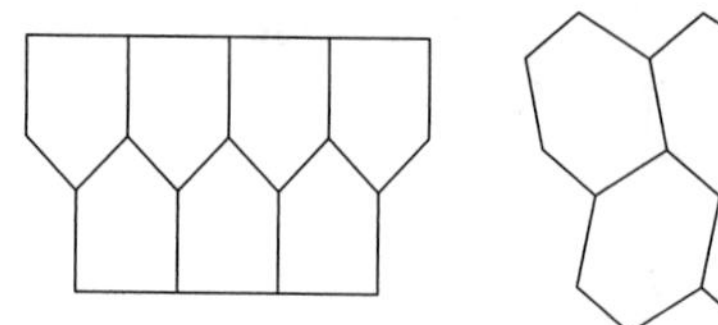

28. **(a)** **(b)** **(c)**

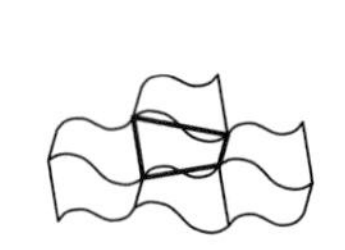

29. A semiregular tessellation, 3, 6, 3, 6, that combines hexagons and equilateral triangles.

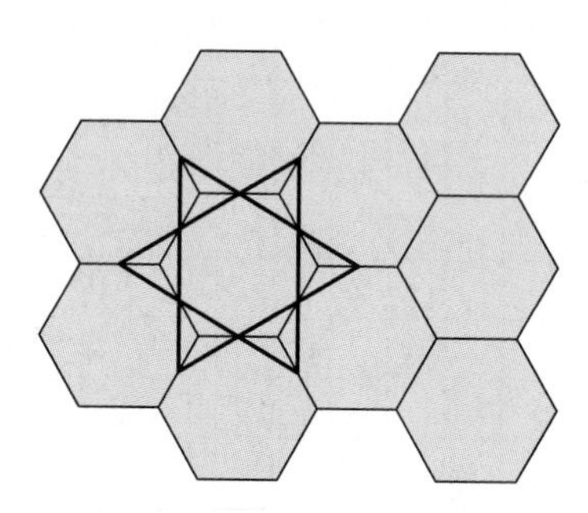

30. Responses will vary. Possibilities for each are:

(a)
```
TO SQUARE :SIDE
    REPEAT 4[FD :SIDE RIGHT 90]
    TO TESSQ
    PENUP BACK 80 PENDOWN LEFT 90
    REPEAT 4[SQUARE 30 FD 30]
    LEFT 180 REPEAT 4[SQUARE 30 FD 30]
```

(b)
```
TO EQTRI :SIDE
    REPEAT 3[FD :SIDE RIGHT 120]
    TO TESSTRI
    PENUP BACK 80 PENDOWN RT 30
    REPEAT 4[EQTRI 30 FD 30]
    RIGHT 120 FD 30 RT 60
    REPEAT 4[EQTRI 30 FD 30]
```

31. Responses will vary. One possibility is

```
TO TRISQ
RT 30 REPEAT 2[FD 30 RT 120] FD 30
REPEAT 3 [LEFT 90 FD 30]
END
```

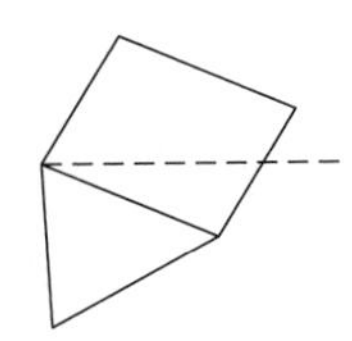

32. All pentominoes will tessellate.

33.

34.

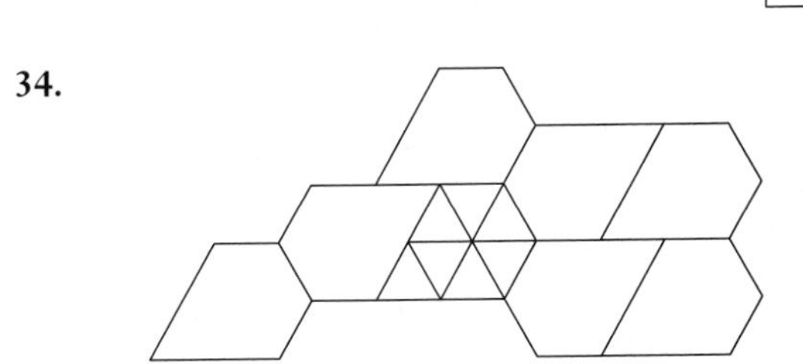

35. The design will make the tessellation of squares, hexagons, and dodecagons shown.

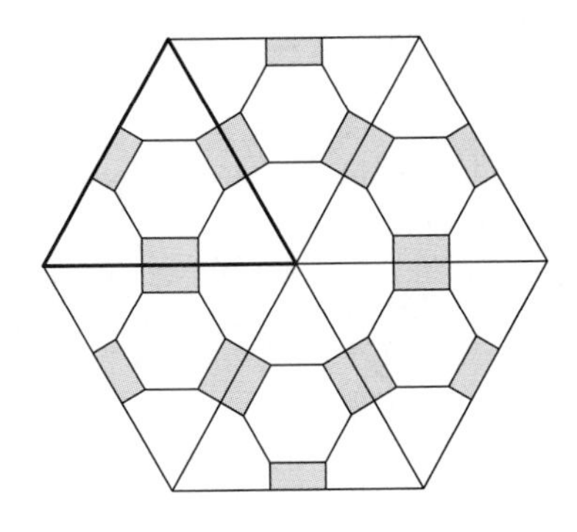

36. Responses will vary.

37. Responses will vary.

38. The geometer is correct.

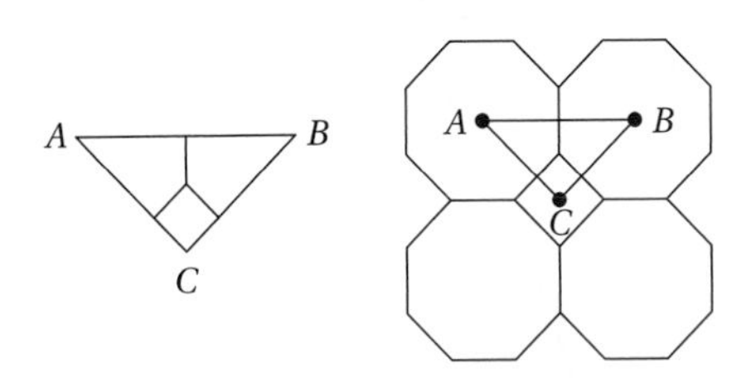

39. Regular pentagons, regular decagons, 5-pointed stars, and figures made up of two 8-sided portions of decagons, fitted together to make a 16-sided concave polygon.

40. The supposed tessellation has vertices of 6, 6, 7, which would be a sum of 120°, 120°, and $128\frac{4}{7}$°, which is not equal to 360°.

41. Responses will vary.

42. **(a)** yes **(b)** One vertical line of symmetry passes through two points of the stars and bisects the rectangles. Another line of symmetry passes through two dent angles of the stars and divides the rectangles in half lengthwise.

43. The area of a grid line square is c^2. The area of the larger square in the tessellation is b^2 and the area of the smaller is a^2. Now $1 + 2 + 3 + 4$ make up a square of area b^2. Thus $a^2 + b^2 = c^2$.

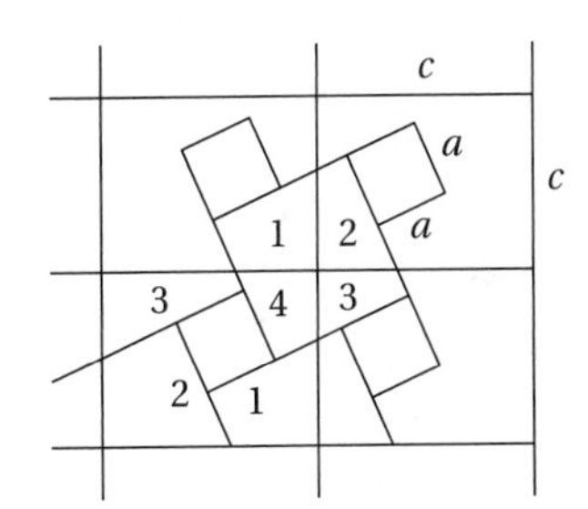

Section 11.3

1. $5 \div 3$ is 1.666 . . ., and $8 \div 5 = 1.6$, which are both close to the golden ratio.

2. Responses will vary. The ratio should be close to 1.6.

3. Responses will vary.

4. The golden triangle is an isosceles triangle with angles of 72°, 72°, and 36°; the ratio of the length of a longer side to the shorter side is the golden ratio.

5. Possible answers are *QT*:*TS* and *QD*:*DE*.

6. 72°, 72°, and 36°

7. five sides

8. It is a triangle; not all of the six points are vertices of the polygon.

9. **(a)** a segment **(b)** a quadrilateral **(c)** a triangle **(d)** a hexagon

10. The figures are identical; $\{\frac{5}{3}\}$ is a $\{\frac{5}{2}\}$ drawn counterclockwise.

11. **(a)** 36° **(b)** 45° **(c)** 100°

12. a segment

13. a quadrilateral

14. an octagon

15. a hexagon

16. Responses will vary.

17. Responses will vary.

18. 36°

19. 45°

20. 108°

21. Responses will vary.

22. The star polygon $\{\frac{9}{4}\}$ has 9 vertices and 9 sides. The star shaped polygon $9_{20°}$ has 18 sides.

23. 50°

24. 68°

25. 39°

26. 10°

27. 92°

28. 70°

29. 80°

30. 106°

31. 60°

32. $\{\frac{5}{2}\},\{\frac{7}{2}\},\{\frac{7}{3}\},\{\frac{8}{3}\},\{\frac{9}{2}\},\{\frac{9}{4}\}$

33. No

34. $\{\frac{12}{5}\}$

35. $\{\frac{10}{3}\}$

36. The sum of the measures of all point angles in any nonsymmetric five-point star is 180°.

37. The measure of the dent angles is equal to the sum of the measure of the point angles and the quotient of 360 divided by the number of point angles.

38. $CD/GH = 2.8$ cm/1.1 cm. The square root of this ratio is approximately 1.60.

39. Responses will vary. The dent angle is 90°.

40. $\{\frac{36}{5}\}, \{\frac{36}{7}\}, \{\frac{36}{11}\}, \{\frac{36}{13}\}, \{\frac{36}{17}\}$

41. Point angle 20°, dent angle 80°; point angle 30°, dent angle 90°; point angle 40°, dent angle 100°. There are infinitely many different six-pointed star-shaped polygons.

42. The sum of the measures of the exterior angles *a*, *e*, *d*, *c*, and *b* is 360°. The sum of the measures of the exterior angles *g*, *h*, *i*, *j*, and *f* is 360°. The sum of the measures of these two sets of angles plus the measures of the point angles *p*, *q*, *r*, *s*, and *t* is $5(180°) = 900°$ (the sum of the angles of 5 triangles). Subtracting out the two 360° sums leaves the sum of the point angles equal to 180°.

43. **(a)** (i) squares, equilateral triangles, regular hexagons (ii) dent angles of 60°, point angles of 30° (iii) 3, 6, 4, 3; 3, star **(b)** (i) equilateral triangles (ii) dent angles of 120°, point angles of 30° (iii) 3, star, 3, star, 3, star; 3, 3, star

44. For the star, dent angles of 140°, point angles of 80°

45. The triangles are similar isosceles triangles. If $\frac{a}{b}$ is the golden ratio, then the ratio of the length to the width of the rectangle (corresponding sides of the two triangles) is also the golden ratio.

46. Consider the rectangle with width, w, equal to 1 unit and length, l, such that $l/w = (l + w)/l$. So $l/1 = (l + 1)/l$ and $l^2 = l + 1$. Applying the quadratic formula to $l^2 - l - 1 = 0$ we have $l = [-(-1) \pm \sqrt{[(-1)^2 - 4(1)(-1)]}]/2(1) = [1 \pm \sqrt{5}]/2$. Since l has positive length, $l = [1 + \sqrt{5}]/2$.

47. Since the sum of the angles about a point on one side of a line is 180°, we have $\frac{360}{n} + x + \alpha + x = 180$ and $\frac{360}{n} + \alpha + 2x = 180$. Since the sum of the angles of a triangle is 180°, we also have $\beta + 2x = 180$. So, $\frac{360}{n} + \alpha + 2x = \beta + 2x$. Hence $\frac{360}{n} + \alpha = \beta$.

48. Responses will vary.

49. Responses will vary.

50. Responses will vary. The pentagon star has point angles of 36° and dent angles of 108°. The 10-point star has point angles of 36° and dent angles of 72°. The hexagons have one angle of 72°, 3 of 144°, and 2 of 108°.

51. Responses will vary.

52. The length of the golden rectangle, DC, is $\ell + \frac{1}{2}$, and the width is 1. By the Pythagorean theorem,

$$\ell = \sqrt{1 + \frac{1}{4}} = \sqrt{\frac{5}{4}} = \frac{\sqrt{5}}{2}$$

Hence the ratio of the length to the width could be expressed in simplified form as $\frac{(\sqrt{5} + 1)}{2}$.

Section 11.4

1. **(a)** The faces of a regular octahedron must be congruent regular polygons, and all the vertices of the polyhedron must have the same configuration. These conditions are not necessary for a nonregular octahedron. **(b)** Eight congruent, equilateral triangles, with four edges at each vertex.

2. **(a)** regular tetrahedron **(b)** regular octahedron **(c)** cube, or regular hexahedron **(d)** regular dodecahedron **(e)** regular icosahedron

3. **(a)** a six-sided die **(b)** a cereal box **(c)** a Toblerone box, or a pencil gripper **(d)** a gemstone **(e)** an olive jar **(f)** the tip of a pencil **(g)** a basketball

4. any rectangular prism

5. **(a)** three **(b)** any quadrilateral prism, right or oblique (leaning)

6. **(a)** a cylinder about 3 times as tall as it is wide **(b)** a cylinder about 3 times as wide as it is tall **(c)** a cone about

3 times as tall as it is wide **(d)** a cone about 3 times as wide as it is tall

7. **(a)** $F = 5, V = 6, F + V = 11, E = 9; 11 = 9 + 2$ **(b)** $F = 6, V = 6, F + V = 12, E = 10; 12 = 10 + 2$ **(c)** $F = 5, V = 5, F + V = 10, E = 8; 10 = 8 + 2$

8. For the truncated octahedron: $F = 14, V = 24, F + V = 38, E = 36; 38 = 36 + 2$. For the cube octahedron: $F = 14, V = 12, F + V = 26, E = 24; 26 = 24 + 2$.

9. Five planes of symmetry and three axes of rotational symetry.

10. **(a)** Planes of symmetry pass through pairs of opposite vertices. These planes contain the vertices that create a square cross-section. Axes of rotational symmetry contain opposite vertices, midpoints of opposite edges, or centers of opposite faces. **(b)** Five planes of symmetry and 13 axes of rotational symmetry **(c)** A cube has 7 planes of symmetry and 13 axes of rotational symmetry.

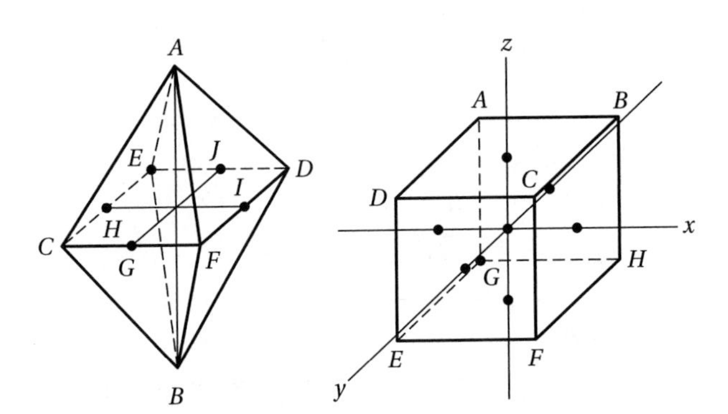

11. A regular tetrahedron

12. It must have a base with rotational symmetry.

13. Patterns **(a)** and **(b)**

14. eight

15. **(a)** **(b)**

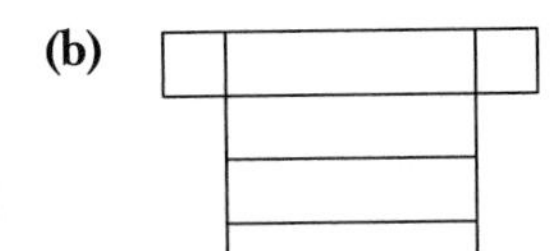

16. **(a)**

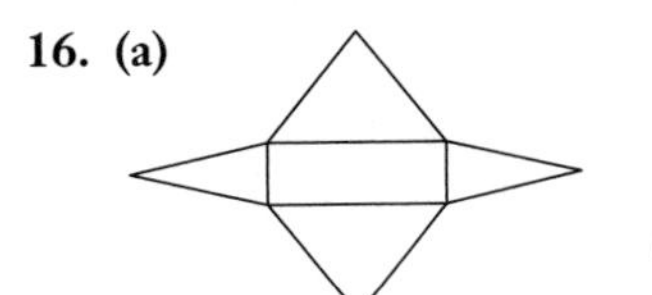

(b)

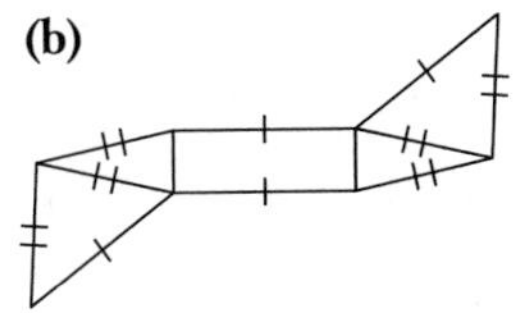

17. **(a)** an ellipse **(b)** a rectangle

18. From the top view, the bottom left square represents a stack of 3 cubes, and the bottom right square represents a stack of 1 cube. Therefore, the top square represents a stack of 1 cube.

19. a stack of 1 cube and a stack of 4 cubes or a stack of 2 cubes and a stack of 3 cubes

20.

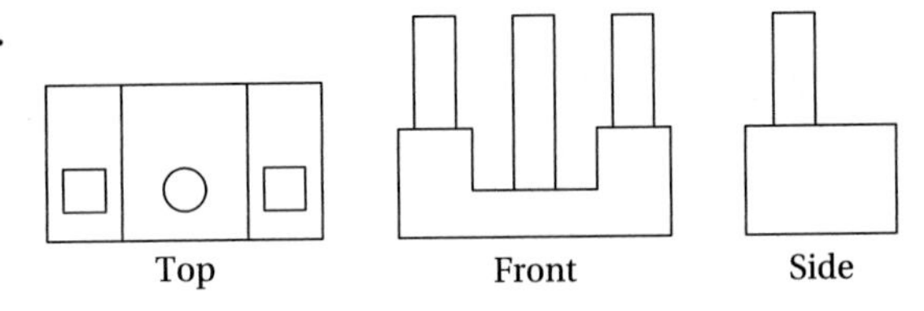

21. a square, nonsquare rectangles, a regular hexagon, nonregular hexagons

22.

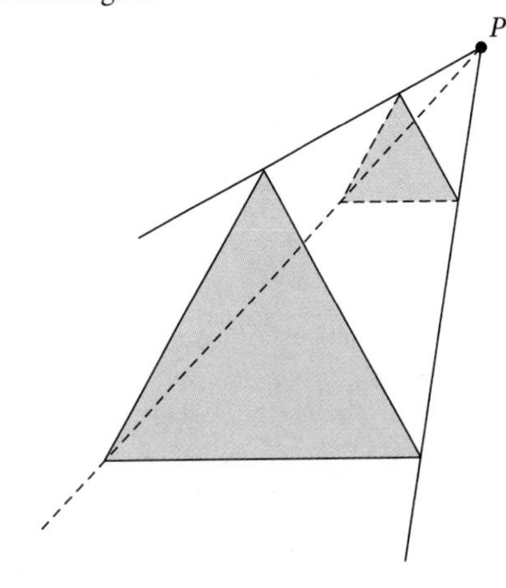

23.

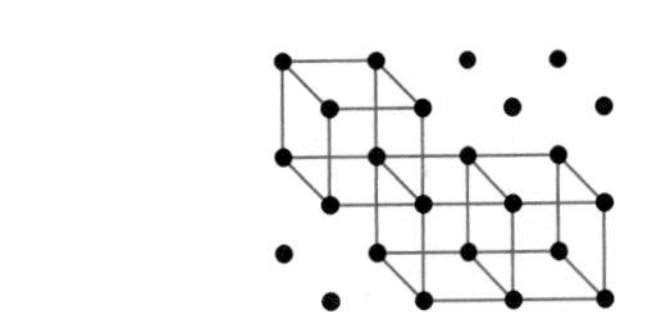

24. **(a)** a cube **(b)** another tetrahedron

25. Since the faces in a regular polyhedron are all congruent, and the angles at each vertex are all congruent, then each face of the regular polyhedron has an equally likely chance of falling up (or down, in the case of the regular tetrahedron).

26. **(a)** Both are made of polygons. **(b)** The prism pattern has two congruent bases and all the other faces are rectangles. The pyramid has only one base, and all the other faces are triangles. **(c)** A prism is a polyhedron with two congruent bases and lateral faces that are parallelograms (rectangles for a right prism). A pyramid is a polyhedron with one base and an apex and its lateral faces are triangular.

27. **(a)** For the square antiprism, $F = 10, V = 8, F + V = 18$, and $E = 16; 18 = 16 + 2$. **(b)** For the triangular antiprism, $F = 8, V = 6, F + V = 14, E = 12; 14 = 12 + 2$.

28.

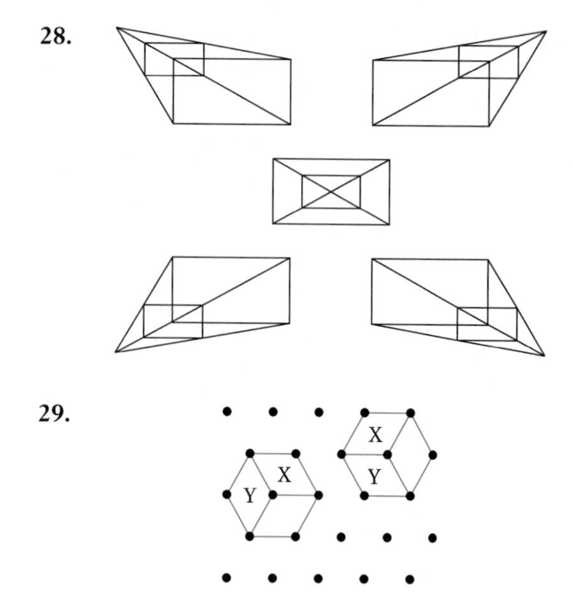

29.

30. **(a)** 6, 8, 12 **(b)** gain 4 and lose 1, gain 1, gain 4 **(c)** +3, +1, +4 **(d)** +4 **(e)** Since the change in $V + F$ is equal to the change in E, then Euler's formula still holds.

31. When you slice off a corner of a prism, there is a gain of $n - 1$ vertices, a gain of 1 face, and a gain of n edges. Therefore, there is a change of $(n - 1) + 1 = n$ in $V + F$, which is equal to the change of n in E, and Euler's formula still holds.

32.

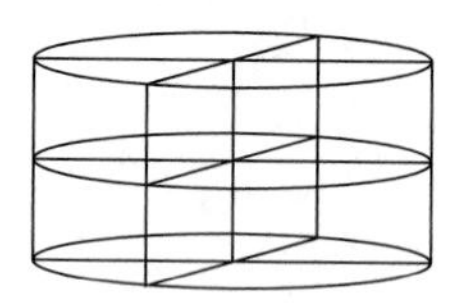

33. a regular hexagon

34. **(a)** nine cuts **(b)** 6,400 unpainted blocks and 19,200 blocks painted on 2 sides

35. **(a)** cubes **(b)** Yes, because the sum of their dihedral angles is 360°.

36. Student constructions will vary.

37. **(a)** regular tetrahedra, regular octahedra, and regular icosahedra **(b)** a double triangular pyramid

38. no

39. No, responses will vary.

40. Responses will vary.

41. When you slice off a corner of a cube, there is a gain of 3 − 1 vertices, a gain of 1 face, and a gain of 3 edges. Therefore, there is a change of $(3 - 1) + 1 = 3$ in $V + F$, which is equal to the change of 3 in E, and Euler's formula still holds.

42. Responses will vary.

43. **(b)** an octahedron **(c)** a tetrahedron

44. Responses will vary.

Chapter 11 Review Exercises

1. A regular tetrahedron has 4 congruent faces that are equilateral triangles; a cube (regular hexahedron) has 6 congruent faces that are squares; a regular octahedron has 8 congruent faces that are equilateral triangles; a regular dodecahedron has 12 congruent faces that are regular hexagons; and a regular icosahedron has 20 congruent faces that are equilateral triangles.

2. One possibility is 3 feet by 1 foot 10 inches, because $\frac{36}{22} \approx 1.64$.

3. $V = 5$, $F = 5$, $V + F = 10$, and $E = 8$; $10 = 8 + 2$.

4. One possibility is:

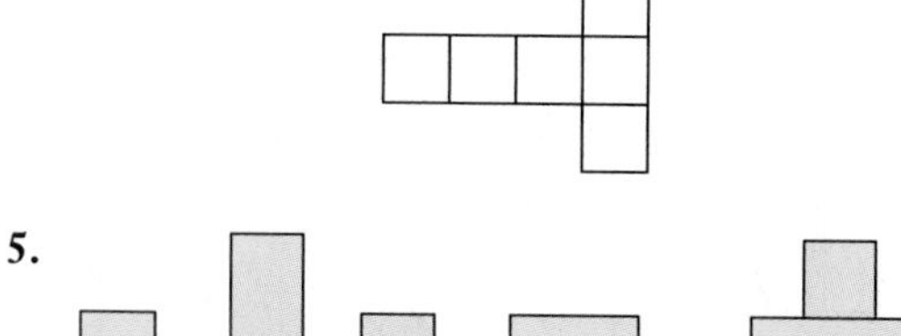

5.

Front view

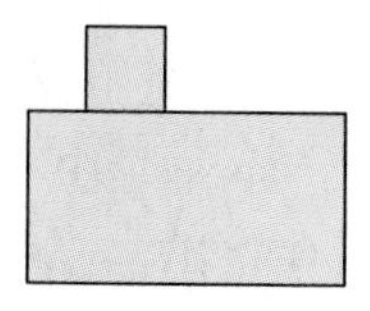

Side view

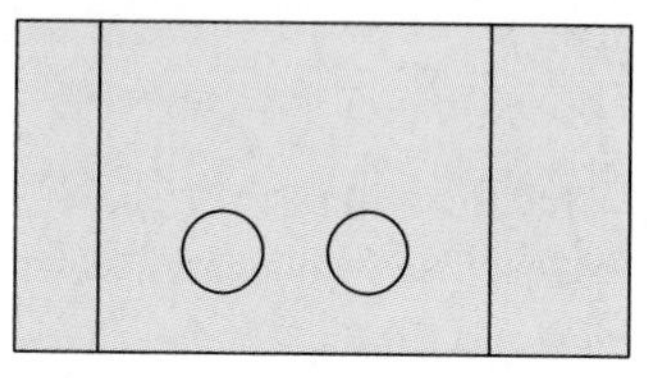

Top view

6.

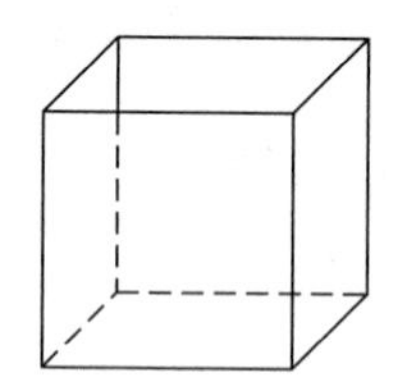

7. $\{\frac{7}{4}\}$; it is a regular star polygon if the points are evenly spaced on the circle, but it is not a regular polygon.

8. 105°

9. **(a)** The figure has reflection symmetry with respect to a vertical line that bisects the square and passes through the point of the section in the lower middle of the design. **(b)** The figure has rotational symmetry with respect to its center and for angles that are multiples of $\frac{360}{16} = 22.5°$.

10. The cylinder has an axis of symmetry through the centers of its bases, and infinitely many axes that bisect the segment joining the centers of the two bases. The cylinder has a plane parallel to the bases and bisecting the segment joining their centers, plus it has infinitely many planes of symmetry perpendicular to the bases and containing their centers.

11. an equilateral triangle, a square, and a regular hexagon

12. **(a)** The tessellation is made up of two types of regular polygons, and the arrangement at each vertex is the same. **(b)** The arrangement of the polygons is not the same at each vertex.

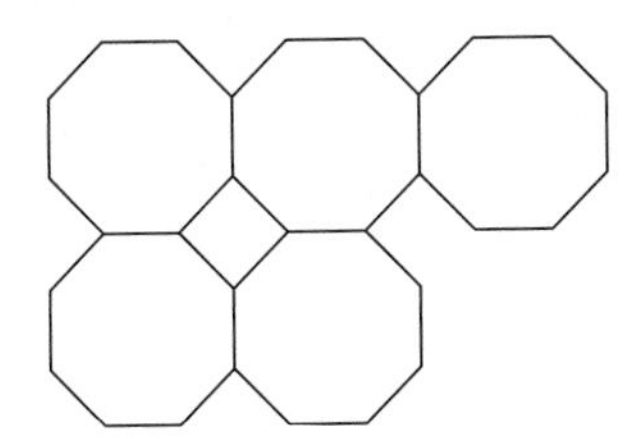

13.

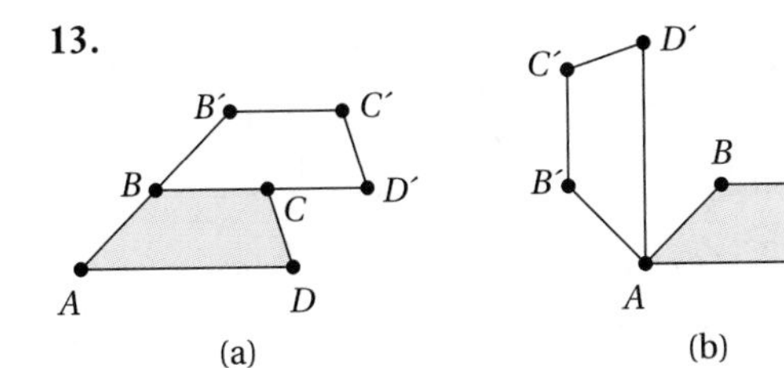

(a) (b)

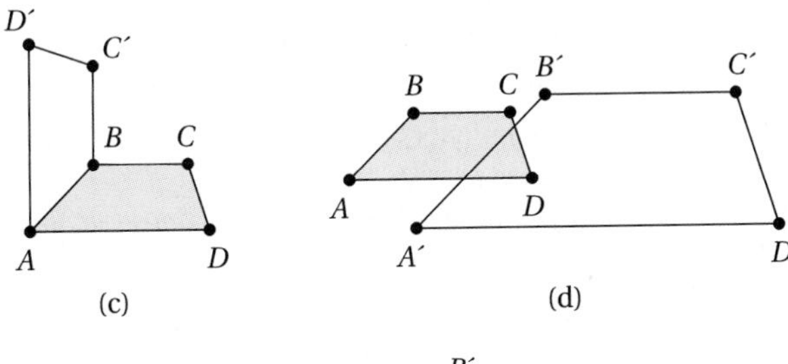

(c) (d)

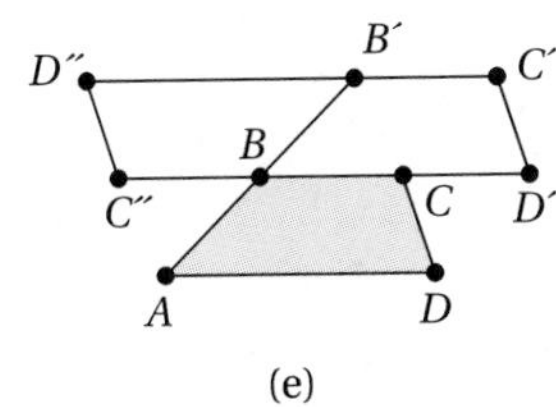

(e)

14. Answers will vary. The following are possibilities.

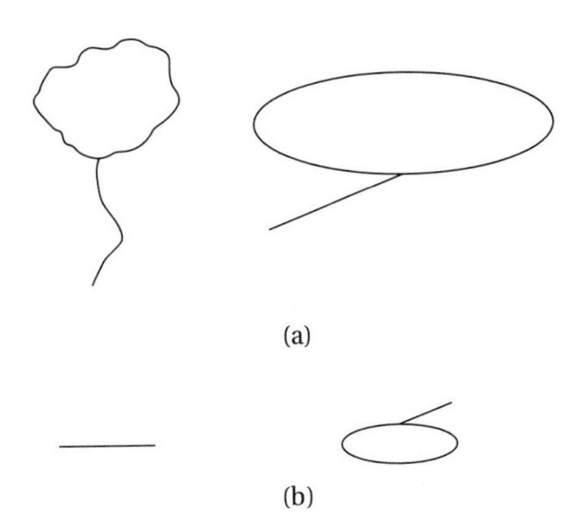

(a)

(b)

15. 8 faces that are equilateral triangles

16. Yes, it is true for a regular star polygon.

17. For the figure given, the axes of rotational symmetry are exactly the same.

18. The dent angles of the 12-pointed star polygons are 60° and the point angles are 30°.

19. **(a)** Both preserve shape, betweenness, straightness. **(b)** A size transformation does not preserve absolute size, only relative size.

20. Responses will vary.

21. See diagram for one possibility: $\frac{AB}{EF} = \frac{AC}{EC}$.

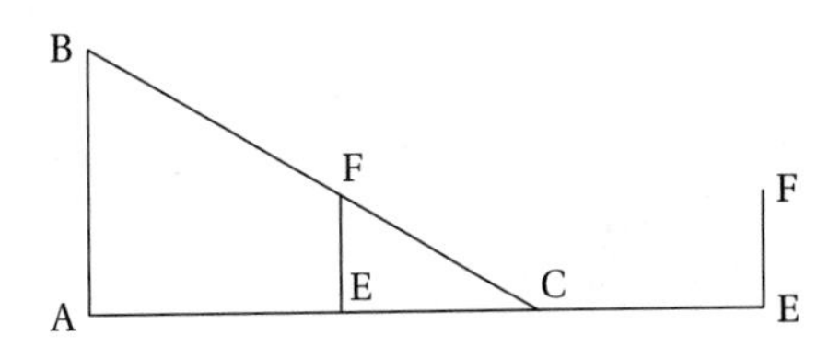

22. Only one:

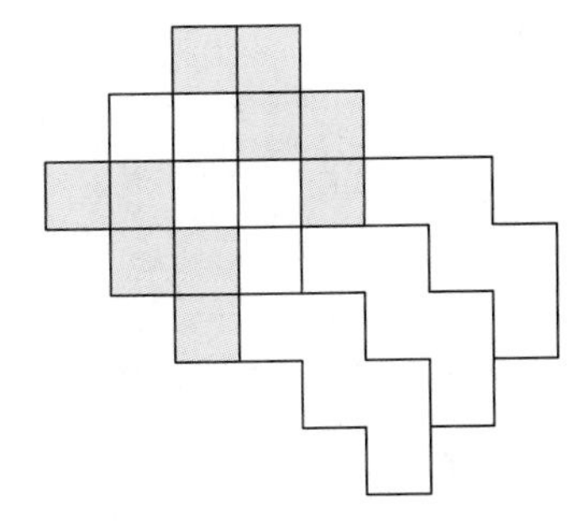

23. Responses will vary.

24. Responses will vary. The ratio of length to width of the rectangles should be close to the golden ratio.

Chapter 12

Section 12.1

1. Responses will vary; about $3\frac{1}{4}$ paperclips or $1\frac{1}{5}$ index fingers.

2. Responses will vary; about $8\frac{3}{4}$ paperclips or $3\frac{1}{4}$ index fingers.

3. gram

4. kilogram

5. either grams or (tenths of) kilogram

6. kilogram

7. kilogram

8. grams

9. kilograms

10. kilograms

11. milligrams

12. kilograms

13. grams

14. 12 pt

15. $1\frac{1}{4}$ gal

16. 576 sq in.

17. $1\frac{1}{3}$ sq yd

18. 15 cm

19. 25,000 ml

20. 5200 g

21. 0.015 km

22. 230 cm

23. 12 ml

24. 210 m

25. 0.002 kg

26. 0.6 liters

27. 1000 sq mm

28. Responses will vary.

29. Responses will vary.

30. Responses will vary.

31. Responses will vary.

32. Responses will vary.

33. Responses will vary.

34. Responses will vary.

35. Responses will vary.

36. 86° F

37. −4° F

38. 203° F

39. 32° F

40. 248° F

41. 68° F

42. −18° C

43. −31° C

44. 21° C

45. 4° C

46. 49° C

47. 30° C

48. Responses will vary. 16 cm

49. Responses will vary. 5 m

50. Responses will vary. 250 ml

51. Responses will vary. 4 kg

52. Responses will vary. 100 g

53. not reasonable

54. reasonable

55. not reasonable

56. reasonable

57. reasonable

58. 37° C

59. **(a)** 80 units of area **(b)** 40 units of area **(c)** 160 units of area

60. 16 units of area

61. 8 units of area

62. 15.5 units of area

63. not reasonable

64. reasonable

65. reasonable

66. not reasonable

67. reasonable

68. Estimates will vary; about 700 sq mm.

69. −10° F is colder

70. 4 times as many units of area with 1-in. × 1-in. squares

71. As unit of length doubles, measures of length are one-half as much.

72. 18°

73. $2.80/gal

74. 1612.9 g

75. $3/400 g is the better price

76. 5 kg/ream

77. About −40° C or −40° F

78. Multiply kilometers by 0.6 to get miles.

79. Responses will vary; object doesn't change, only measures of its length change.

80. Responses will vary.

81. Responses will vary.

Section 12.2

1. First change measures to the same units.

2. area

3. perimeter

4. area
5. area
6. perimeter
7. $P = 24$ in., $A = 36$ sq in.
8. $P = 20$ m, $A = 24$ sq m
9. $P = 4.24$ m, $A = 0.24$ sq m
10. $P = 16$ in., $A = 16$ sq in.
11. $P = 20$ cm, $A = 18$ sq cm
12. $P = 24$ mm, $A = 16$ sq mm
13. $P = 20$ in, $A = 20$ sq in
14. $P = 18$ ft, $A = 12$ sq ft
15. 54 sq in.
16. 44 sq ft
17. 57.5 sq m
18. 22.5 sq mm
19. $P = 40$ cm, $A = 120$ sq cm
20. $P = 48$ cm, $A = 96\sqrt{3}$ sq cm
21. $C = 31.4$ mm, $A = 78.5$ sq mm
22. $C = 25.12$ in., $A = 50.24$ sq in.
23. $C = 18.84$ cm, $A = 28.26$ sq cm
24. $C = 17.76$ cm, $A = 25.12$ sq cm
25. $990
26. 25
27. $\frac{5}{11}$
28. 27 bags
29. 86.5 sq cm
30. The triangles have the same area. They each have the same base and height. Supporting arguments will vary.
31. 32 sq yd
32. 62.35 sq cm
33. 15 sq m
34. 43.3 sq mm
35. **(a)** 6.98 m **(b)** 20.9 cm
36. **(a)** 88.3 sq mm **(b)** 130.8 sq cm
37. 1:4
38. Responses will vary.
39. 16
40. $\sqrt{2}$:1 or about 1.41:1 for both corresponding sides and perimeters
41. Responses will vary.
42. 18 in.
43. Responses will vary.
44. Responses will vary.
45. $A(ABC) = (s^2/2)\sqrt{3}$ where s is the length of an edge of the cube.
46. The rectangle becomes more elongated.
47. The rectangle becomes more "square-like" in its shape.
48. Probability of success is 0.29.
49. 18.84 yd
50. about 1.4 mm
51. Ratio of areas is 5:1.
52. $\frac{1}{2}$. Arguments will vary..
53. 36 sq mm
54. 52.3 sq m
55. 7.74 sq cm
56. $3875 per month
57. 38,470 sq ft
58. The 180-yd × 100-yd plot is the better buy.
59. Tony is correct. Arguments will vary.
60. The formula works if $a = c$ and $b = d$.
61. Responses will vary.
62. Larger gutters carry about 82% more water.
63. Selected answers:

No. of Sides	Length of One Side	Perimeter	$P \div D, D = 2$
96	0.06543817	6.28206639	3.14103195
384	0.01636228	6.28311522	3.14155761

Section 12.3

1. $SA = 96$ sq cm, $V = 64$ cu cm
2. $SA = 148$ sq ft, $V = 120$ cu ft
3. $SA = 70.5$ sq mm, $V = 27$ cu mm
4. $SA = 170$ sq m, $V = 150$ cu m
5. $SA = 190$ sq in., $V = 150$ cu in.
6. $SA = 63.46$ sq cm, $V = 17.3$ cu cm
7. $SA = 113$ sq cm, $V = 84.8$ cu cm
8. $SA = 62.8$ sq cm, $V = 28.26$ cu cm
9. $SA = 96$ sq cm, $V = 48$ cu cm
10. $SA = 22.3$ sq m, $V = 5.2$ cu m
11. $SA = 113$ sq in, $V = 113$ cu in.
12. $SA = 339$ sq cm, $V = 452$ cu cm
13. $SA = 251.2$ sq cm, $V = 334.9$ cu cm
14. 37.7 cu ft
15. 48.9 cu cm
16. 14,653 cu ft
17. 301.44 cu cm
18. 13,500 cu ft
19. **(a)** 4:1 **(b)** 16:9
20. **(a)** 8:1 **(b)** 64:27
21. 139,627,950 sq mi
22. 600 cu ft

23. **(a)** similar cylinders: radii, diameters, heights, and circumferences **(b)** spheres: radii, diameters, and great circles **(c)** similar cones: base radii, base diameters, heights, slant heights, base circumferences **(d)** similar pyramids: corresponding segments of the bases, heights, and corresponding parts of the faces

24. The volume involves the product of three dimensions; the surface area involves the product of only two dimensions.

25. formula for pyramid is most helpful

26. **(a)** 1:1 **(b)** 3:2

27. 6.67 cm

28. 12.7 cm

29. Responses will vary.

30. 3000 cu ft

31. 3600 cu cm

32. 51.4 cu cm

33. Tracy is correct.

34. Minimum surface area approaches 8 sq in. Minimum volume approaches 0.

35. 318.9 cu cm

36. The large box is the better buy.

37. $66\frac{2}{3}\%$

38. $33\frac{1}{3}\%$

39. 25.16 cu ft

40. 16.7 cu cm and 245 cu cm

41. 37.5%

42. 7.56 cu in.

43. 78.5% of the box filled with either 6 cans or 8 cans.

44. $777,985

45. $SA = 251$ sq cm, $V = 147$ cu cm

46. 98.13 cu cm

47. **(a)** Doubling the radius with the same height more than doubles the area. **(b)** Doubling the radius with the same height quadruples the volume. **(c)** Doubling the height with the same radius less than doubles the surface area. **(d)** Doubling the height with the same radius doubles the volume.

48. The maximum volume (1056 cu cm) occurs when $x = 4$.

49. The maximum volume (243 cu in.) occurs when $x = 3$.

50. 3:2

51. 1.26 in. (1.26 approximates the cube root of 2)

52. The pyramids have the same volume.

Chapter 12 Review Exercises

1. (d)

2. (d)

3. (b)

4. (b)

5. (b)

6. (d)

7. (a)

8. (b)

9. (c)

10. 4.19 cu cm

11. 18 ft

12. **(a)** 113 sq ft **(b)** 84.8 cu ft

13. **(a)** 211 sq cm **(b)** 156 cu cm

14. Responses will vary. The given lengths may not represent corresponding sides.

15. 15.5%

16. 7.7 sq ft

17. 3:1

18. 4; 7; as the radius of a sphere increases, the ratio of the surface area to volume decreases.

19. The total area is about 2.6 acres.

20. 90 degrees: 0.0633 cu unit, 180 degrees: 0.227 cu unit; 270 degrees: 0.389 cu unit

21. 469 hr or 19.5 days

22. 14 panels

23. 1.82:1

24. almost 47 days

25. The rope would be about 1 foot high. An ant and a small dog could crawl under it.

26. Frank is not correct.

27. Responses will vary.

Chapter 13

Section 13.1

1. variable

2. constant

3. variable

4. variable

5. variable

6. constant

7. $x =$ the number of $\frac{1}{5}$ miles; $f =$ the total fare

8. $f = 0.75 + 0.20x$

9. $5.75

10. $s =$ sales in thousands of dollars; $t =$ total earnings

11. $t = 5000 + 0.1s$

12. $5100

13. B2: 0.1*A2 + 5000

14. B22:0.1*A22 + 5000

15. 18, 17, 22, 0, −12

16. $-54, 0, 2, 128$
17. $x - 20$
18. $S = x^2 - 16$
19. $1.08x + 2.75$
20. $y = 1.08x + 2.75$
21. \$137.75
22. $3000 - 225x$
23. $y = 3000 - 225x$
24. \$1200
25. linear
26. quadratic
27. quadratic
28. exponential
29. quadratic
30. linear
31. $y = 2x + 4$
32. $y = -x + 1$
33. $y = -x^2 + 5$
34. Responses will vary.
35. $x = 15$; order of operations
36. Responses will vary.
37. $2x^3 + 10x^2$
38. $16x^3 + 80x^2$
39. linear, $y = -8x$
40. linear, $y = x + 0.5$
41. linear, $y = x - 2$
42. nonlinear
43. The volume of container A is 3 times the volume of B.
44. no, $y^2 = x$
45. about \$50
46. about \$1100
47. about 1987
48. about 1993
49. 100 blocks, n^2
50. 24, $(\frac{n}{2}) - 1$
51. 36, $\frac{n^2}{2} + \frac{n}{2} = \frac{n(n + 1)}{2}$
52. graphs, equations, tables
53. $2xxxx + 3bxxx - x ==== -4cc$

Section 13.2

1. -3 and 3

Responses will vary for Exercises 2–6. Possible responses are:

2. add 18 to both sides
3. multiply both sides by 4
4. multiply both sides by -6
5. add $7x$ to both sides
6. add 16 to both sides
7. (0, 0) (1, 4) (2, 10) (3, 18) (4, 28)
8. (0, 20) $(1, 20\frac{1}{3})$ $(2, 20\frac{2}{3})$ (3, 21) $(4, 21\frac{1}{3})$
9. 9
10. -20
11. 10
12. -24
13. -3
14. -4
15. 25.5
16. 72
17. 21, -4
18. 0, $-\frac{5}{6}$
19. $-5.9075, -0.5925$
20. 4.07, 0.429973
21. $\frac{5}{4}, -\frac{1}{3}$
22. 0
23. 3, -4
24. 4.5
25. no solutions
26. 2.4, -0.9
27. guess, check, and revise or use the properties of equality
28. 25
29. 157
30. 70.5
31. example: $2x - 6 = 12$; $12 + 6$, $\div 2$
32. one
33. two
34. +: 2 real solutions; 0: 1 real solution; −: no real solutions
35. When $y = 8$, $x = 5$.
36. 3: $-4, 0, 5$
37. Possible answers: $x^2 + 3x + \frac{9}{4}$ has one real-number solution; $x^2 + 3x + 2$ has two real-number solutions.
38. At low speeds, temperature changes have a greater effect; at greater wind speeds, wind speed has a greater effect.
39. $w = 12$ ft; $\ell = 24$ ft
40. A square with sides 57.5 ft gives area of 3306.25 square ft.
41. about 6 seconds
42. Responses will vary.
43. $4x^2 + 12x + 9$
44. Responses will vary.
45. The solutions are the values of x for which the y value is 0.

Section 13.3

1.

x	y
0	7
1	2
2	–3

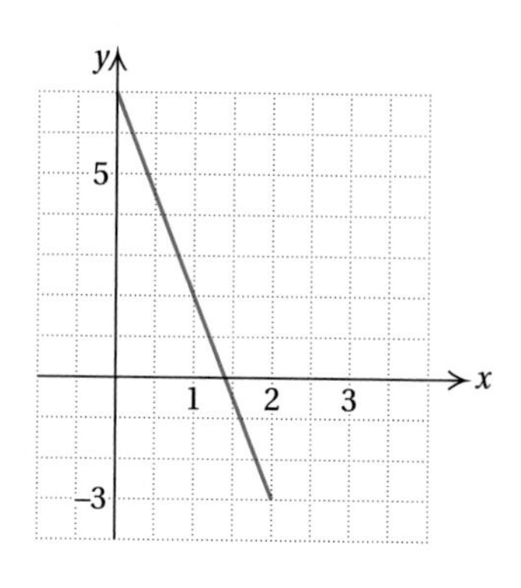

2.

x	y
0	–1
1	–1
2	–1

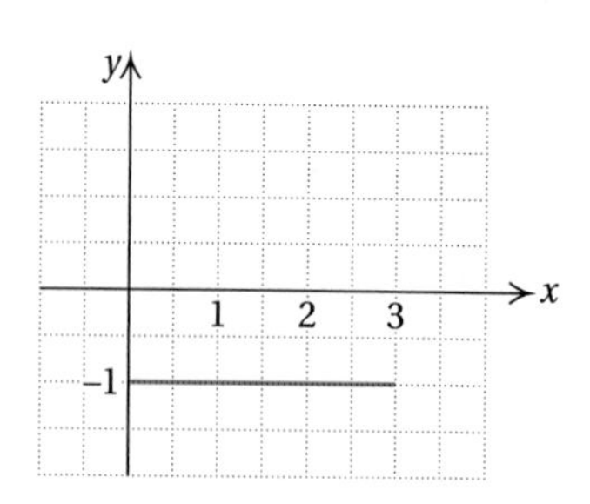

3.

x	y
0	0
1	.375
2	.75

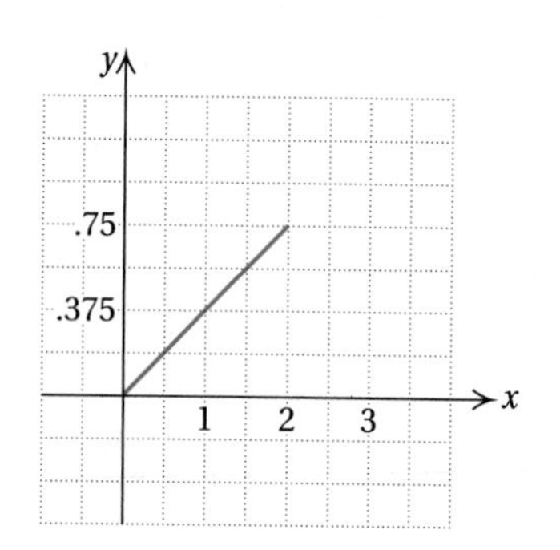

4.

x	y
0	5
1	5
2	5

5.

x	y
0	4
1	6
2	8

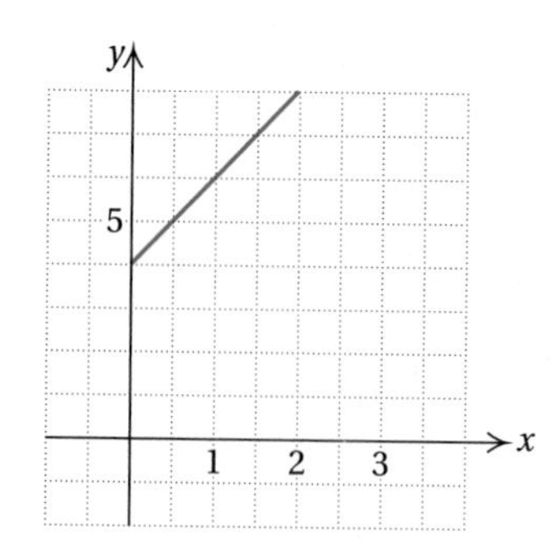

6. $y = 1$

7. $y = 5$

8. $x = 2$

9. $m = -1; 3$

10. $m = \frac{1}{2}; 0$

11. $m = -1; \frac{1}{2}$

12. $m = -\frac{1}{4}; 3$

13. $m = 16; -120$

14. $m = 4; 6$

15. $m = 8; -120$

16. not parallel, different slopes

17. parallel, slope $= -10$

Responses will vary for Exercises 18–21. Two lines are perpendicular if the product of their slopes is -1 or if one is vertical and the other horizontal.

18. $y = (-\frac{1}{3})x + 1$

19. $y = (\frac{1}{2})x - 3$

20. $x = 6$

21. $y = (\frac{1}{15})x$

22. slope $= -6$

23. slope $= \frac{3}{2}$

24. slope $= 0$

25. slope $= -1$

26. The y values do not change, or have a change of 0, as the x values change. Thus the slope is 0.

27. As y changes value, the corresponding change in x is 0. To calculate the slope we would be dividing by 0, an undefined operation.

28. Yes

29. $x = -1$

30. $y = \frac{1}{3}x + \frac{5}{3}$

31. $y = -3x + 3$

32. Possible answer: $y = -4x + 4$

33. Possible answer: $y = 8 - 3x$

34. Possible answer: $y = -2x + 3$

35. $y = 12x - 2$

36. The result will be the opposite of the correct value.

37. Responses will vary.

38. 5 liters

39. -200

40. The graph suggests that as gasoline usage decreases by 200 gallons, the price of 1 gallon will increase by one dollar.

41. Responses will vary.

42. Let tu represent a 2-digit number. So, $(tu)^2 = (10t)^2 + u(2 \cdot 10t + u)$.

Section 13.4

1. $\overline{AD}$: $M(0, 7)$; $\overline{BE}$: $M(-8, 0)$; $\overline{GH}$: $M(1.5, 0)$; $\overline{FK}$: $M(3, -7)$

2. $M(0, -1\frac{1}{2})$

3. $M(0, 0)$

4. $M(5, 5)$

5. $M(-\frac{1}{2}, -2\frac{1}{2})$

6. $(0, -6)$

7. $(0, -5)$

8. $(8, -6)$

9. $(-7, -7)$

10. AB: 12; CD: $\sqrt{41}$; GH: 5; EF: ≈ 14.2

11. 11

12. ≈ 5.83

13. ≈ 12.64

14. ≈ 22.02

15. -2

16. 0

17. 1.625

18. -8

19. $y = 4x - 2$

20. $y = \frac{1}{2}x + \frac{3}{4}$

21. $y = -x$

22. $y = 5$

23. $m = -3$, y-intercept $= -7$

24. $m = 4$, y-intercept $= 12$

25. $m = 0$, y-intercept $= 5$

26. $m = \frac{3}{2}$, y-intercept $= 0$

27. $y = (\frac{2}{5})x + 3$

28. $y = -(\frac{3}{4})x + 6$

29. $y = 10x$

30. $(-1.29, 0.29)$

31. $y^2 + x^2 = 36$

32. $(x - 1)^2 + (y - 2)^2 = \frac{1}{4}$

33. $(x + 3)^2 + (y - 4)^2 = 4$

34. $(x + 2)^2 + (y + 1)^2 = 25$

35. $(0, 0)$; 6

36. $(3, -5)$; ≈ 3.46

37. $(0, 6)$; 10

38. $(1, -2)$; 1

39. Example: $y = -2x + 2$

40. Example: $y = -2x + 15$

41. Example: $y = 2x + 5$

42. $x = 6$ is vertical through $(6, 0)$; perpendicular to $y = 6$

43. -49 is not possible for r^2

44. $y = 3x - 14$

45. $y = x - 2$

46. isosceles

47. nonrectangular parallelogram

48. $y = (\frac{7}{2})x + \frac{1}{4}$

49. $(0, 0)$; $3\sqrt{2}$

50. 204.1 sq units

51. 13 blocks, 65 min.

52. From (a, b) to (c, d): $|(d - b)| + |(c - a)|$

53. $(0, 4)$, $(4, 7)$, $(4, 1)$, $(7, 3)$

54. $(6, 9)$ is not on the line

55.

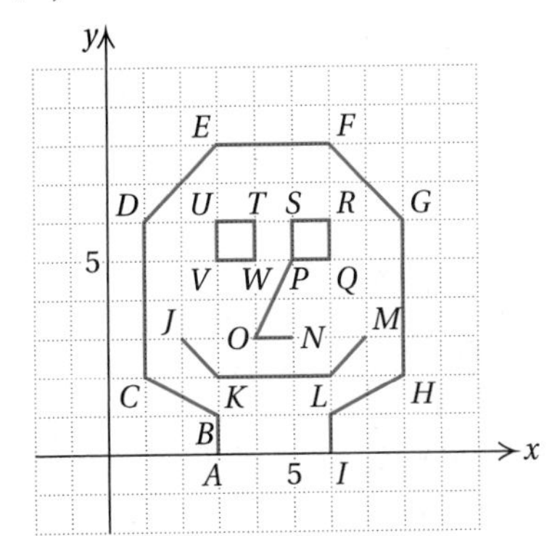

The equations are:

$\overleftrightarrow{AB}$: $x = 3$	$\overleftrightarrow{LM}$: $y = x - 4$
$\overleftrightarrow{BC}$: $y = -\frac{1}{2}x + 2\frac{1}{2}$	$\overleftrightarrow{NO}$: $y = 3$
$\overleftrightarrow{CD}$: $x = 1$	$\overleftrightarrow{OP}$: $y = 2x - 5$
$\overleftrightarrow{DE}$: $y = x + 5$	$\overleftrightarrow{PQ}$: $y = 5$
$\overleftrightarrow{EF}$: $y = 8$	$\overleftrightarrow{QR}$: $x = 6$
$\overleftrightarrow{FG}$: $y = -x + 14$	$\overleftrightarrow{RS}$: $y = 6$
$\overleftrightarrow{GH}$: $x = 8$	$\overleftrightarrow{SP}$: $x = 5$
$\overleftrightarrow{HI}$: $y = \frac{1}{2}x - 2$	$\overleftrightarrow{UV}$: $x = 3$
$\overleftrightarrow{LI}$: $x = 6$	$\overleftrightarrow{VW}$: $y = 5$
$\overleftrightarrow{JK}$: $y = -x + 5$	$\overleftrightarrow{WT}$: $x = 4$
$\overleftrightarrow{KL}$: $y = 2$	$\overleftrightarrow{TU}$: $y = 6$

56. Responses will vary.

Responses will vary for Exercises 57–62. Possible responses are:

57. $(3, 4) \to (3 + 2, 4 + 5) \to (5, 9)$ by translating 2 over and 5 up.

58. $(3, 4) \to (-3, -4)$ by rotating 180 degrees about the origin.

59. $(3, 4) \to (-4, 3)$ by rotating 90 degrees about the origin.

60. $(3, 4) \to (4, 3)$ by reflecting over the line $y = x$.

61. $(3, 4) \to (3, -4)$ by reflecting over the x-axis.

62. $(3, 4) \to (-3, 4)$ by reflecting over the y-axis.

63. Responses will vary.

64. Responses will vary.

65. yes, approximately

Chapter Review

1. variable

2. constant

3. 24, 40, −16, 184, −296

4. −380, −5, 19, 2995

5. $\$35 + \$9.50x$

6. $C = 35 + 9.50x$

7. $111

8. possible answer: (0, 0), (1, −3), (2, −2), (3, 3), (4, 12)

9. possible answer: (0, −8), (5, −6), (10, −4), (15, −2), (20, 0)

10. 20.75

11. −5

12. 12 and −7

13. 0 and $-\frac{12}{8}$ or $-\frac{3}{2}$

14. 3 and −5

15. slope = −6; y-intercept = 14

16. Straight line passing through (0, 14) and (2, 2)

17. −4

18. $\frac{3}{5}$

19. $(3, -\frac{3}{2})$

20. $(-\frac{3}{2}, \frac{9}{2})$

21. about 8.06

22. about 8.60

23. $y = 3x - 2$

24. $y = \frac{x}{2} + \frac{1}{4}$

25. $(x - 4)^2 + (y + 3)^2 = 9$

26. any equation of form $y = -4x + a$

27. any equation of form $y = mx + 11$. $(m \neq 4)$

28. any equation of form $y = 3x + a$

29. $y = 2x^3 + 2x - 3$ not linear or quadratic

30. $x = 5$ is a vertical line and not in the $y = mx + b$ form.

31. $V = X(8 - 2X)(10 - 2X)$

X	1	1.5	2	2.5	3
V	48	52.5	48	37.5	24

The volume is a maximum in the vicinity of $x = 1.5$.

32. $w = 95$ m, $\ell = 280$ m

33. any equation of form $y = -3x + a$

34. $y = (-\frac{1}{4})x$

35. (−2, −1) (−5, −1) (−2, −3)

36. any equation of form $y = mx + b$

37. possible answer: $x^2 - 7x + 12 = 0$

38. any quadratic function whose graph intersects the x-axis in exactly one point

39. Possible response: If you know the slope, m, and y-intercept, b, the equation is $y = mx + b$.

Appendix A Answers

Section A.1

1. (a)

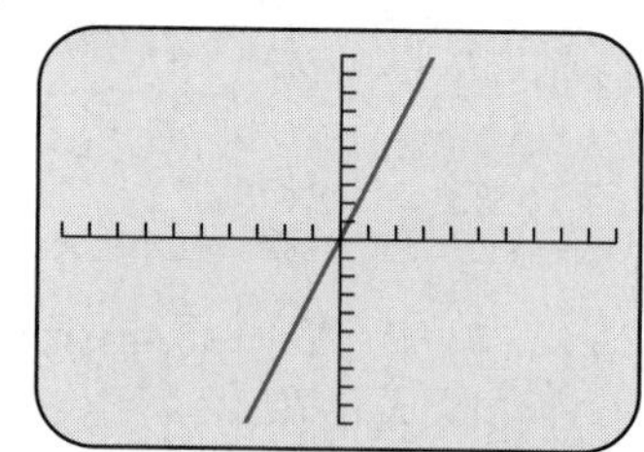

(b)

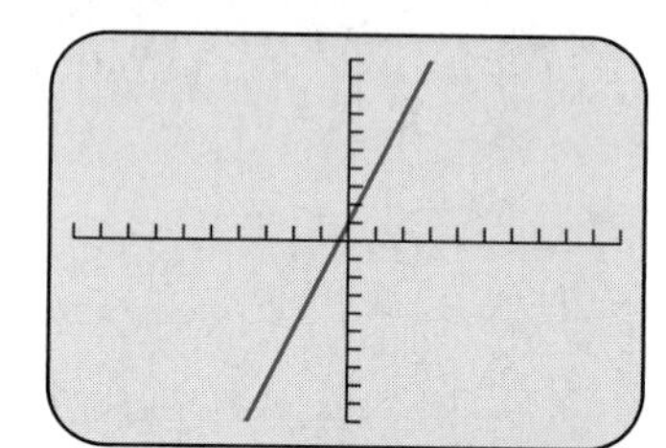

(c)

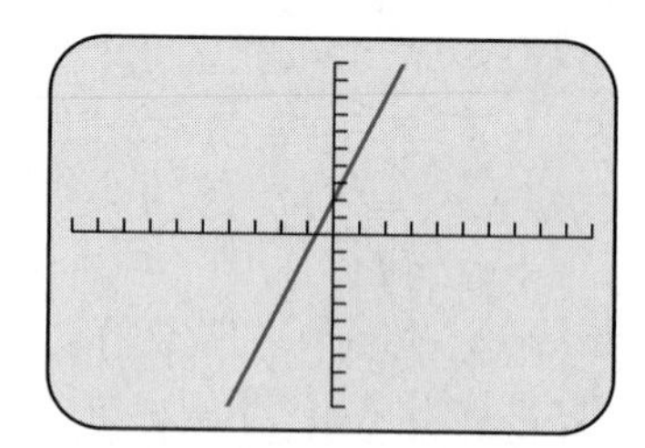

(d) It should be a line with a slope of 3 and a y-intercept at (0, 5).

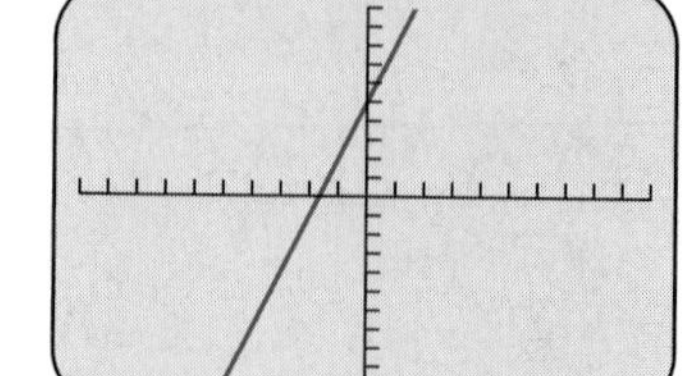

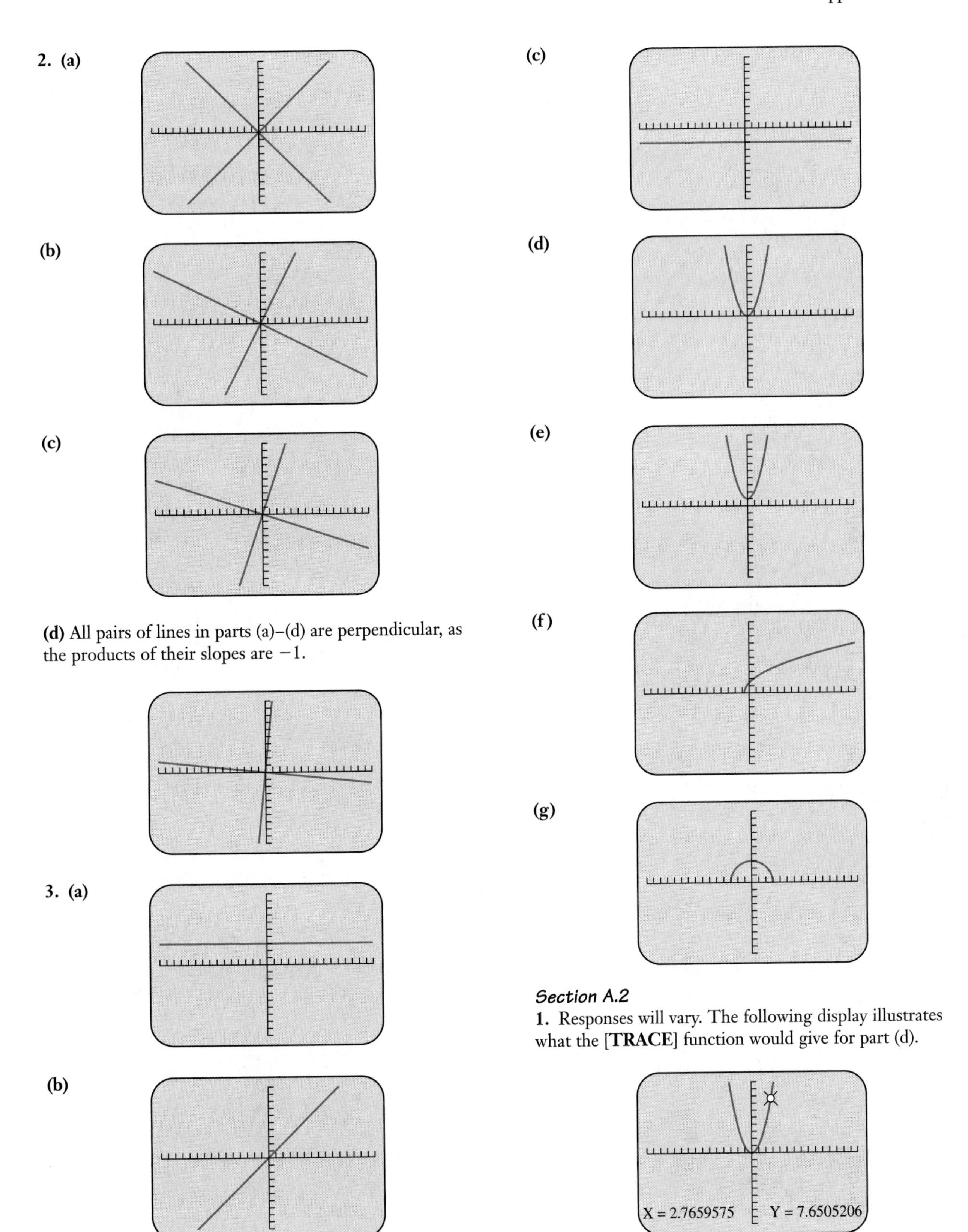

(d) All pairs of lines in parts (a)–(d) are perpendicular, as the products of their slopes are -1.

Section A.2

1. Responses will vary. The following display illustrates what the [**TRACE**] function would give for part (d).

2. **(a)** The graphs of the pair are shown; the intersection point is $(2, 7)$, so the solution is $x = 2$.

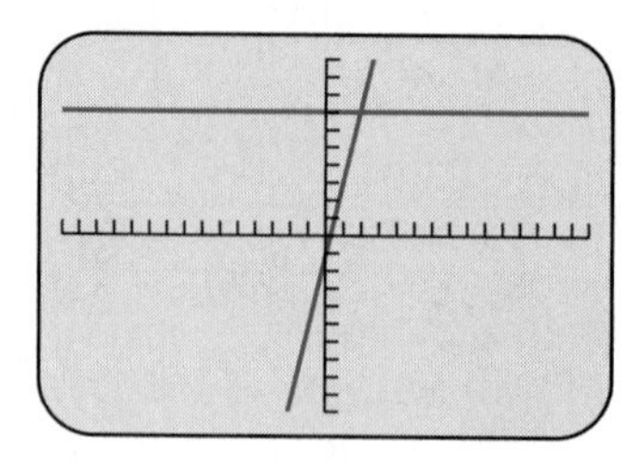

(b) $x = 6$. **(c)** $x = 4$. **(d)** $x = 2$ or $x = -2$.
(e) $x = 81$.

3. **(a)** The graphs of the equations are shown; the intersection point is $(3, 8)$, so the solution to the system of equations is $x = 3$ and $y = 8$.

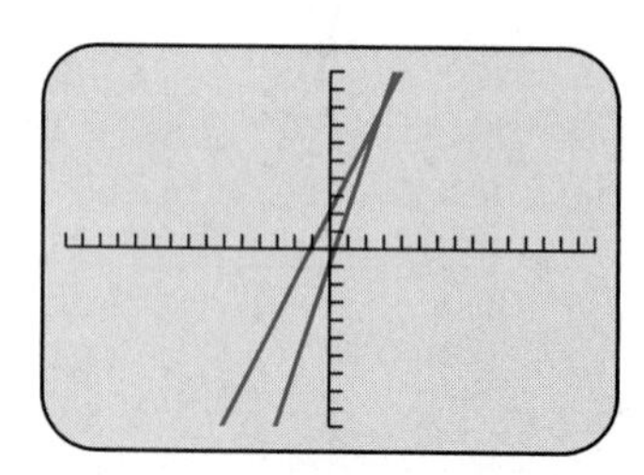

(b) $x = 1$ and $y = 9$; the intersection point is $(1, 9)$.

Section A.3

1. The numbers should appear in the table under the L1 heading.

2. The end of your list should appear as shown in the answer to Exercise 4.

3. The end of your list should appear as shown in the answer to Exercise 4.

4. The display should have a list end that looks like that shown.

L1	L2	L3
22.2	6.8	
23.8	4.2	
22.7	4.4	
21.8	5.3	
23.3	2.3	
22	4.4	
------	------	

L2 (12)=

Section A.4

1. The table is shown in (a); the scatterplot shown in (b) is on a $[0, 60] \times [0, 60]$ plot.

L1	L2	L3
42	52	------
35	39	
27	22	
44	48	
39	50	
------	------	

L2 (6)=

(a)

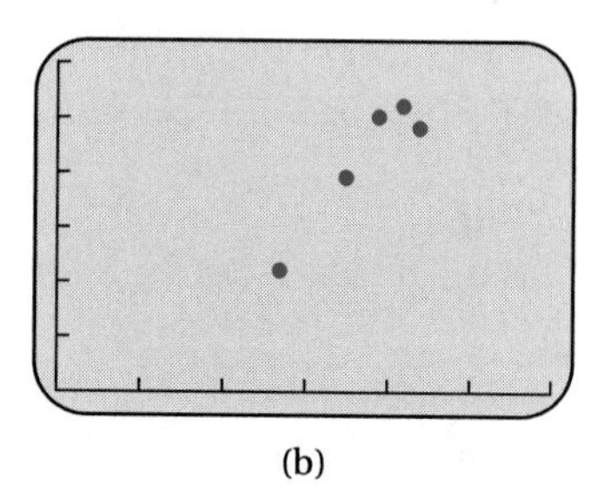

(b)

2. The scatterplot shown is on a $[0, 10] \times [0, 10]$ plot.

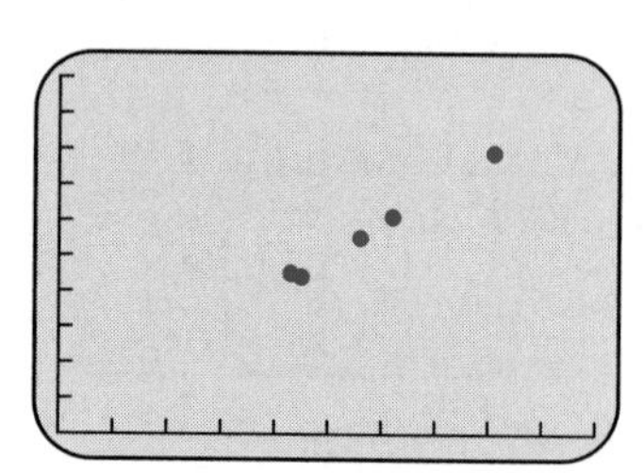

Section A.5

1. Using a window with a horizontal scale from 190 to 210 with tic marks at units produces the following display.

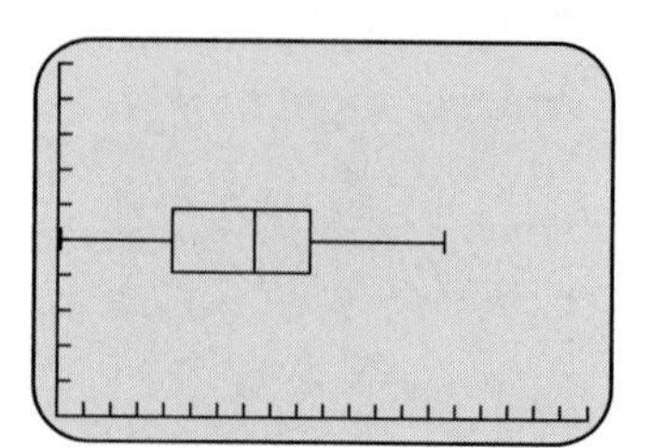

2. Using a window with a horizontal scale from 0 to 0.50 with tic marks at tenths produces the following display.

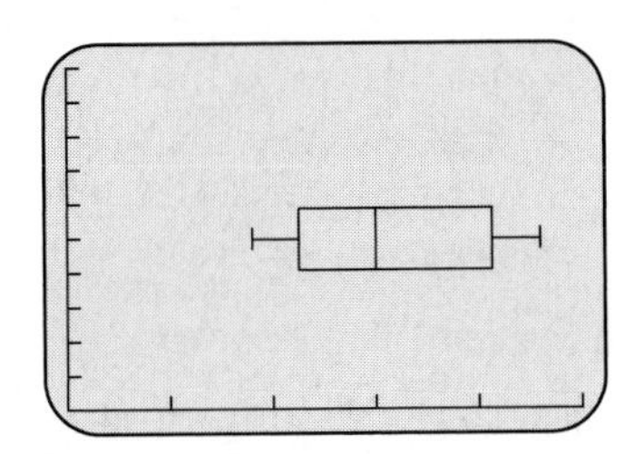

Appendix B Answers

Section B.1

1. Responses will vary. Examples include the figures shown in Exercise 2.

2. The drawing should look similar to the ones you're copying, as shown at the bottom of the page.

3.

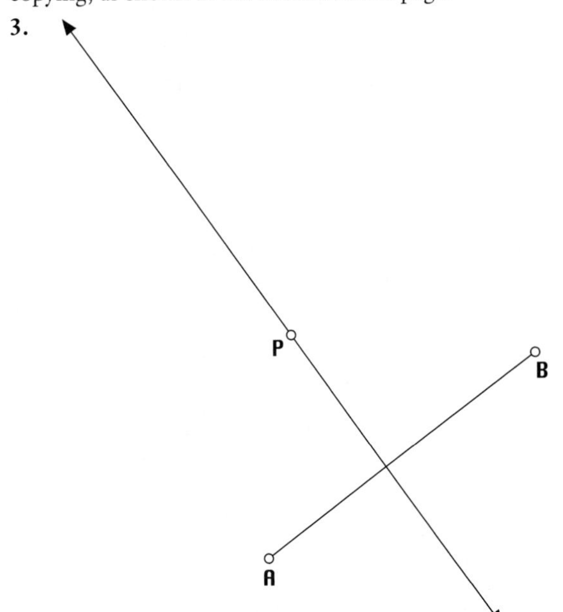

4. Constructions will vary. One possibility is constructing the bisector of an angle, as shown.

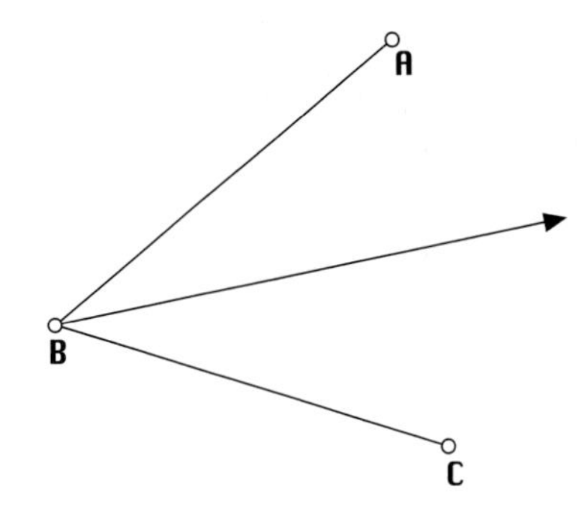

5. Area(Polygon 1) = 1.27 square inches
Perimeter(Polygon 1) = 5.40 square inches

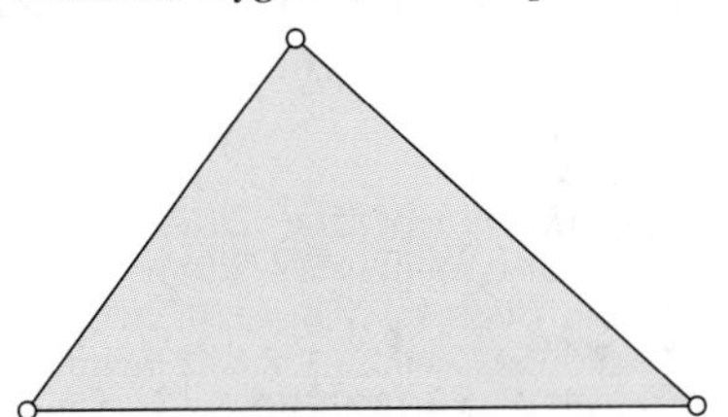

6. Responses will vary. One example is measuring the slope of a segment, as shown.

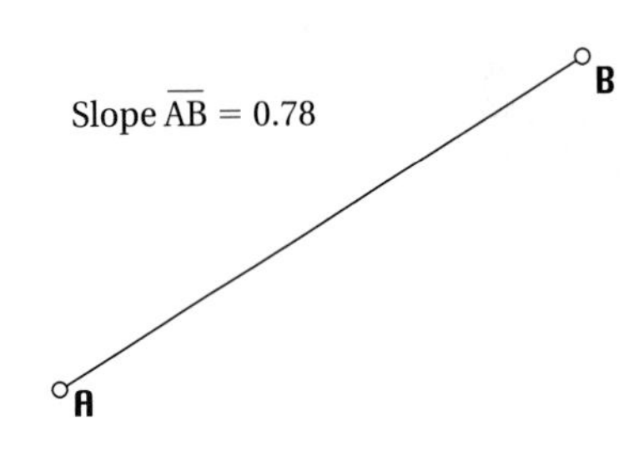

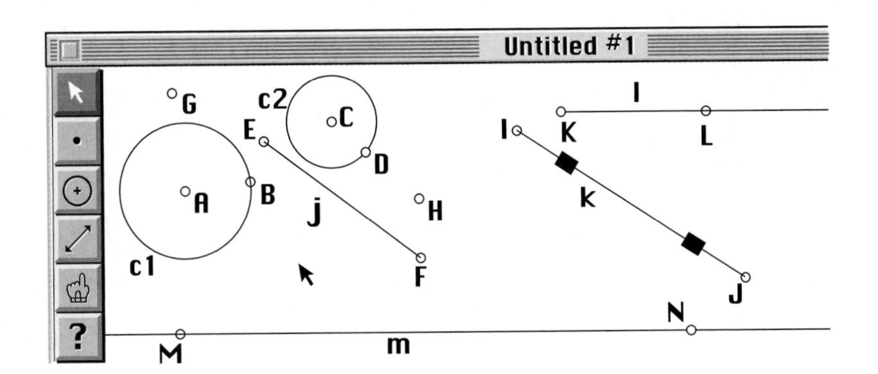

Section B.2

1. The lines are perpendicular, and all four angles measure 90°, as shown.

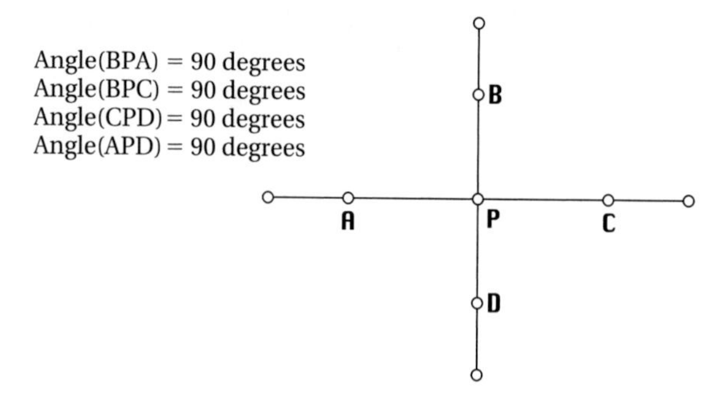

2. The display shown in (a) depicts the data of opposite sides and angles for a parallelogram; the display shown in (b) depicts the same data for a parallelogram of a different shape; repeating the process several times clearly shows that the lengths of the opposite sides are equal and that the measures of opposite angles are equal.

Length(Segment AB) = 2.52 inches
Length(Segment DC) = 2.52 inches
Length(Segment AD) = 1.42 inches
Length(Segment BC) = 1.42 inches
Angle(DAB) = 109 degrees
Angle(DCB) = 109 degrees
Angle(ADC) = 71 degrees
Angle(ABC) = 71 degrees

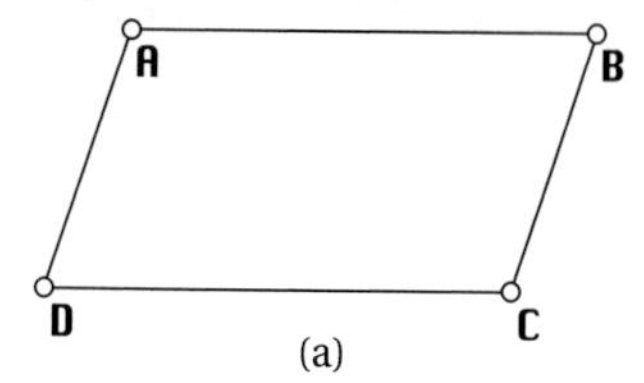

(a)

Length(Segment AB) = 2.97 inches
Length(Segment DC) = 2.97 inches
Length(Segment AD) = 1.62 inches
Length(Segment BC) = 1.62 inches
Angle(DAB) = 125 degrees
Angle(DCB) = 125 degrees
Angle(ADC) = 55 degrees
Angle(ABC) = 55 degrees

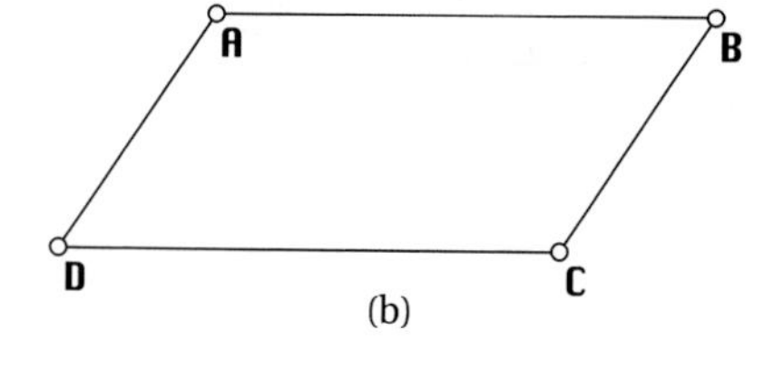

(b)

3. When the size and shape of a triangle are varied, as in displays (a) and (b), $a^2 + b^2$ always gives the same result as c^2 [note in the fifth line that the GES abbreviated the sentence and showed three dots, an ellipsis, instead of the last factor, Length(Segment b)].

Length(Segment a) = 1.06 inches
Length(Segment b) = 1.96 inches
Length(Segment c) = 2.23 inches

Length(Segment a)*Length(Segment a)+
Length(Segment b)*...= 4.98
Length(Segment c)*Length(Segment c) = 4.98

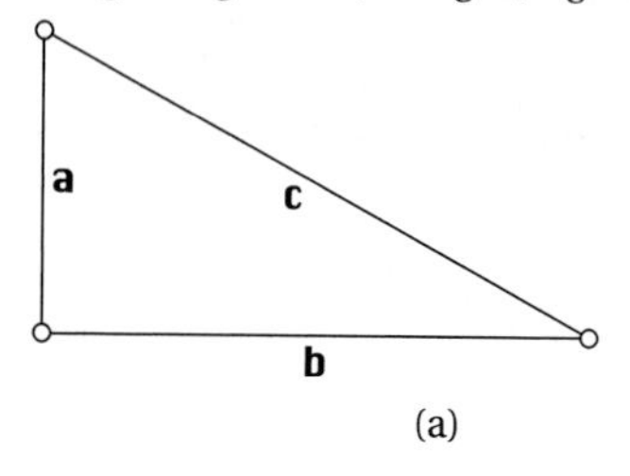

(a)

Length(Segment a) = 0.53 inches
Length(Segment b) = 2.91 inches
Length(Segment c) = 2.96 inches

Length(Segment a)*Length(Segment a)+
Length(Segment b)*...= 8.75
Length(Segment c)*Length(Segment c) = 8.75

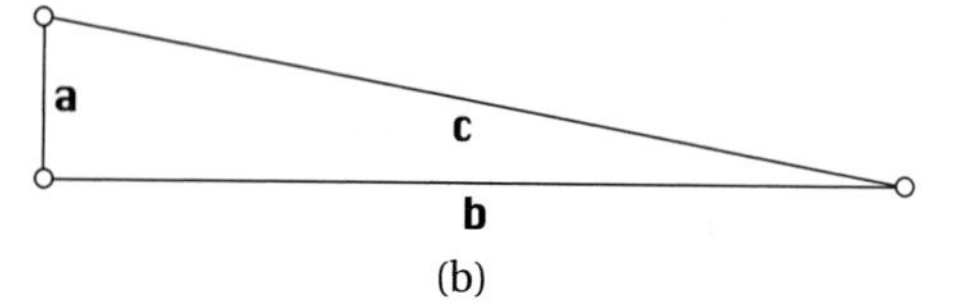

(b)

Section B.3

1. In the displays shown in (a) and (b) the angle bisectors of the interior angles of a triangle appear always to be concurrent.

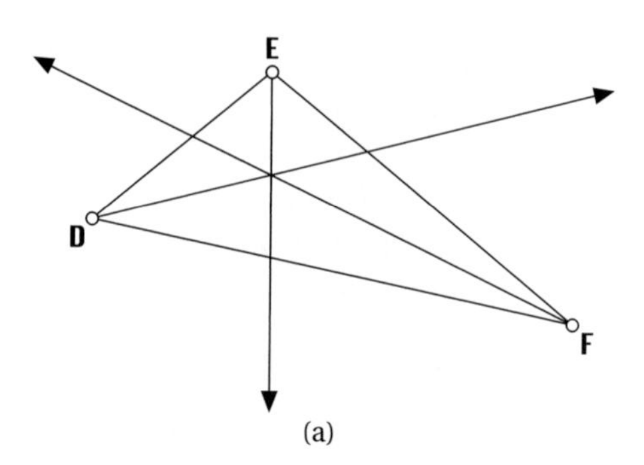

(a)

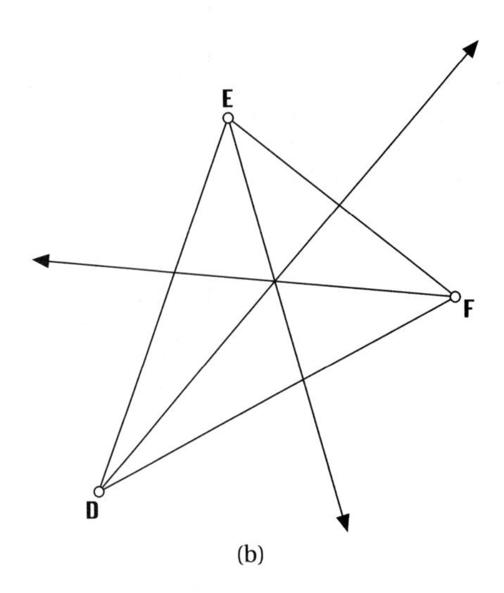

(b)

2. The display shown in (a) depicts a square (equal sides and angle measures). In the display shown in (b) vertex B was moved, and the angle measures indicate that the figure is no longer a square; a square of a different size could be produced by moving the sides and vertices in a different way so that the lengths and angle measures again would be equal.

Length(Segment AB) = 1.53 inches
Length(Segment BC) = 1.53 inches
Length(Segment CD) = 1.53 inches
Length(Segment DA) = 1.53 inches

Angle(DAB) = 90 degrees
Angle(ABC) = 90 degrees
Angle(BCD) = 90 degrees
Angle(CDA) = 90 degrees

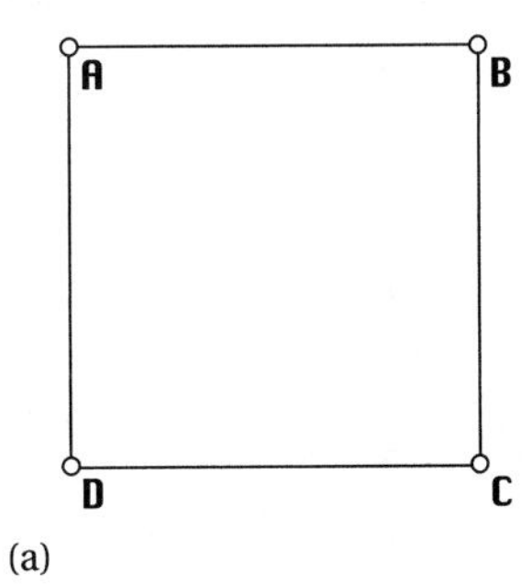

(a)

Length(Segment AB) = 1.82 inches
Length(Segment BC) = 1.33 inches
Length(Segment CD) = 1.53 inches
Length(Segment DA) = 1.53 inches

Angle(DAB) = 83 degrees
Angle(ABC) = 86 degrees
Angle(BCD) = 102 degrees
Angle(CDA) = 90 degrees

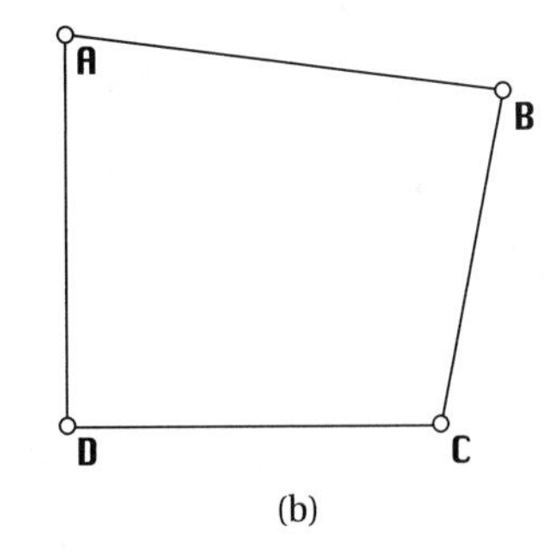

(b)

Section B.4

1. Responses will vary. Possibilities include drawing a polygon (of any type), then reflecting it in a chosen line, rotating it about a chosen point, or translating it according to a chosen vector.

2. In the display shown in (a) $\overline{\mathbf{AA}'}$ is perpendicular to line **m** and **P** is the midpoint of $\overline{\mathbf{AA}'}$; in the display shown in (b) triangle **ABC** has been distorted, but the original relationships hold.

Angle(APS) = 90 degrees
Angle(A′PS) = 90 degrees

Distance(P to A) = 1.22 inches
Distance(A′ to P) = 1.22 inches

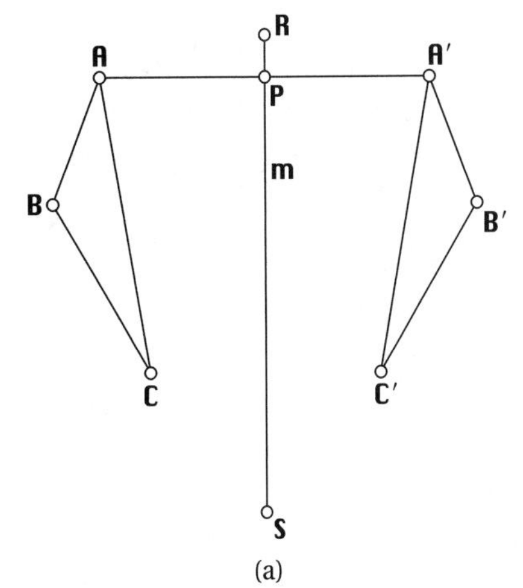

(a)

Angle(APS) = 90 degrees
Angle(A′PS) = 90 degrees

Distance(P to A) = 0.90 inches
Distance(A′ to P) = 0.90 inches

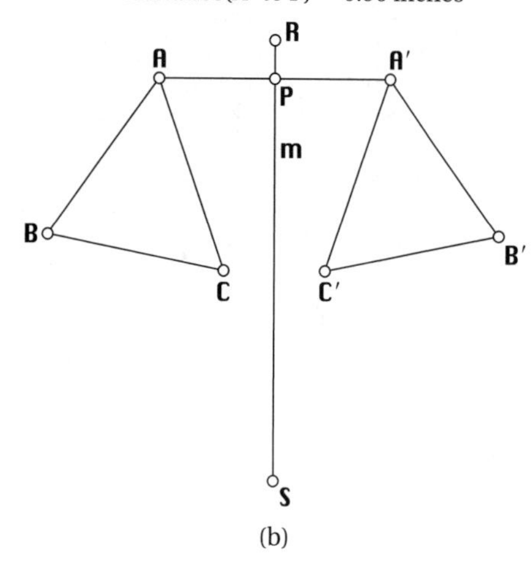

(b)

Section B.5

1. In the displays shown the ratio of the circumference to the diameter of the circles of different size is always 3.14; this relationship holds for all circles, regardless of size.

Length(Segment AB) = 1.66 inches
Circumference(Circle 3) = 5.21 inches
Circumference(Circle 3)/Length(Segment AB) = 3.14

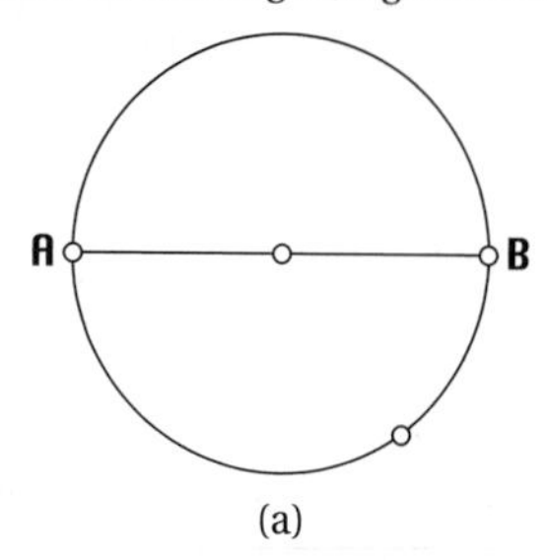

(a)

Length(Segment AB) = 2.55 inches
Circumference(Circle 3) = 8.01 inches
Circumference(Circle 3)/Length(Segment AB) = 3.14

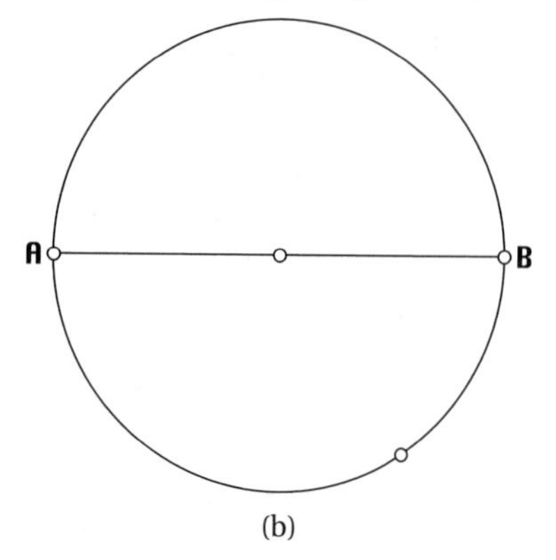

(b)

2. The arrangement is shown; the image resulting from the two successive reflections could have been produced by a single rotation about point **E**, through twice the measure of ∠**DEF**; note that the first figure goes from **A** to **B** to **C** clockwise, the reflected image goes counterclockwise, and the final image again goes clockwise.

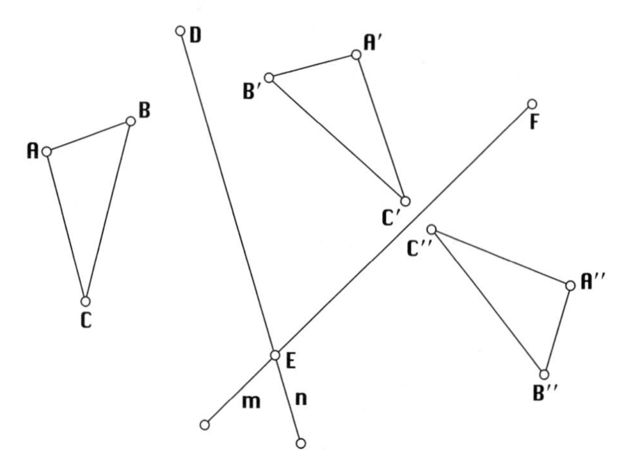

Section B.6

1. The image shown in (a) is the result of the first construction, and that shown in (b) depicts a change in the size of the circle; the only term in the equation that changes is the value of the squared constant, which is the length of the radius; the other values, 2.000 and 3.000, represent the coordinates of the center of the circle (unchanged).

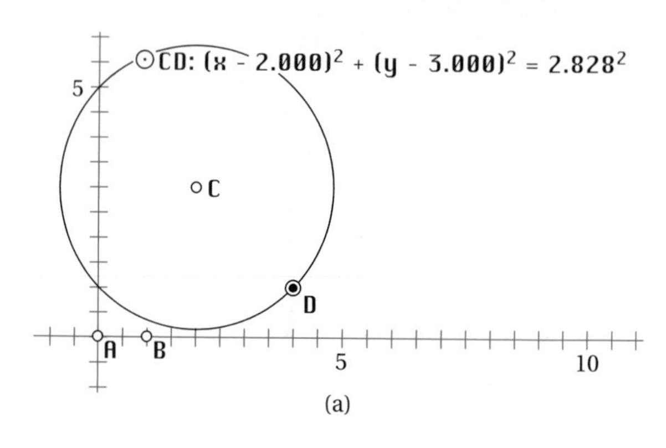

(a)

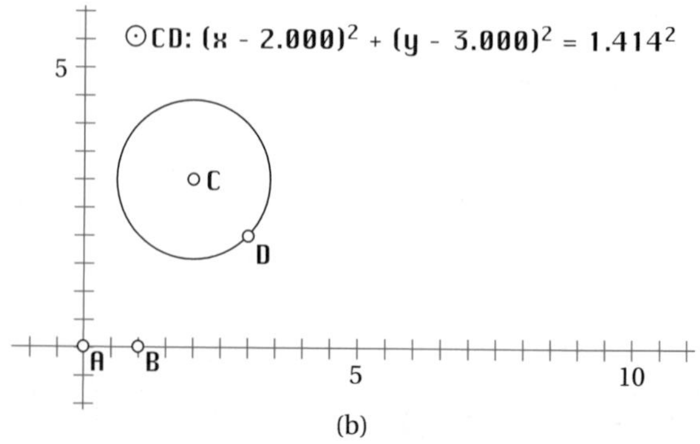

(b)

2. The image shown in (a) is the result of the first construction, and that shown in (b) is the result of moving point **D**; two values in the equation have changed: the slope, from 2.500 to −4.000, and the y-intercept, from −0.500 to 6.000.

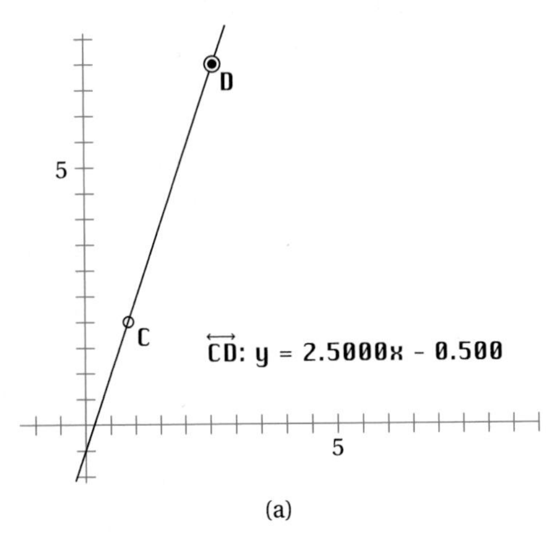

(a)

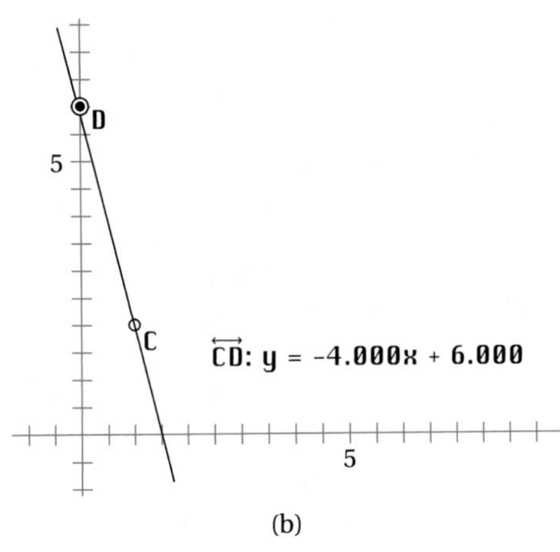

(b)

The following images show graphs of equations for (c) horizontal and (d) vertical lines. Note that the equation for a horizontal line shows only the y-intercept (the slope of a horizontal line is 0; the equation could also be written as $y = 0.000x + 2.000$ to show that the slope is 0). The equation of a vertical line changes significantly and is expressed here as $x = 1.000$; this equation shows the value of the x-coordinate for every point on the line; for any vertical line any change in the vertical direction results in *no change* in the horizontal direction; the slope, m, would be the ratio

$$\text{Slope, } m = \frac{\text{Change in } y}{\text{Change in } x} = \frac{\text{Some positive or negative number}}{0}$$

Because no real number is associated with this ratio, the equation can't be written in the form $y = mx + b$; thus a vertical line has no slope.

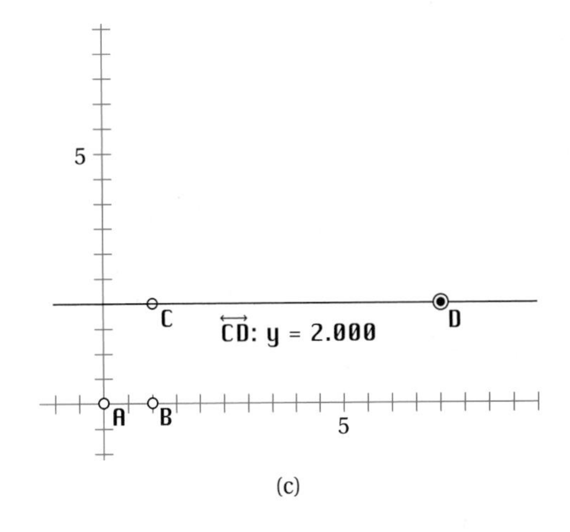

(c)

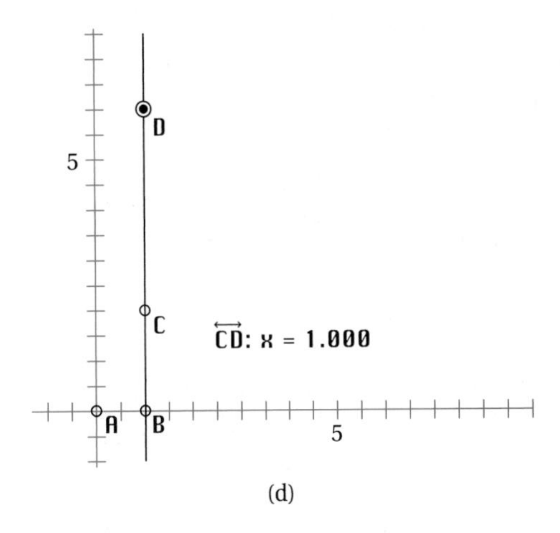

(d)

Section B.7

1. The image shown in (a) is the result of an original construction, and that shown in (b) is the result of moving point **A** to a new location; the table shows that lengths **CB** and **EF** haven't changed and therefore that their ratio hasn't changed, which holds for any movement of point **A**.

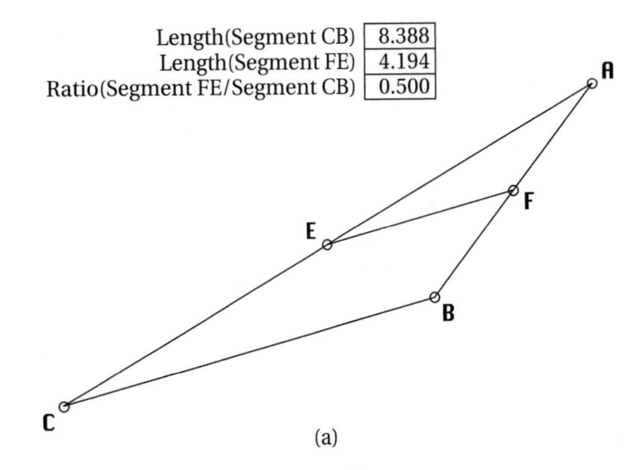

(a)

Length(Segment CB)	8.388	8.388
Length(Segment FE)	4.194	4.194
Ratio(Segment FE/Segment CB)	0.500	0.500

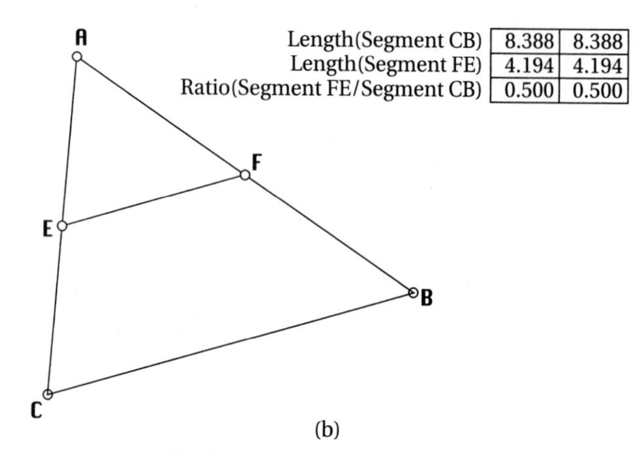

(b)

In the images shown in (c) and (d) point **C** has been moved; lengths **CB** and **EF** change, but their ratio remains 0.500; for any triangle created, this ratio of the lengths of a midsegment and its corresponding side of the triangle is constant.

Length(Segment CB)	8.388	8.388	2.892
Length(Segment FE)	4.194	4.194	1.446
Ratio(Segment FE/Segment CB)	0.500	0.500	0.500

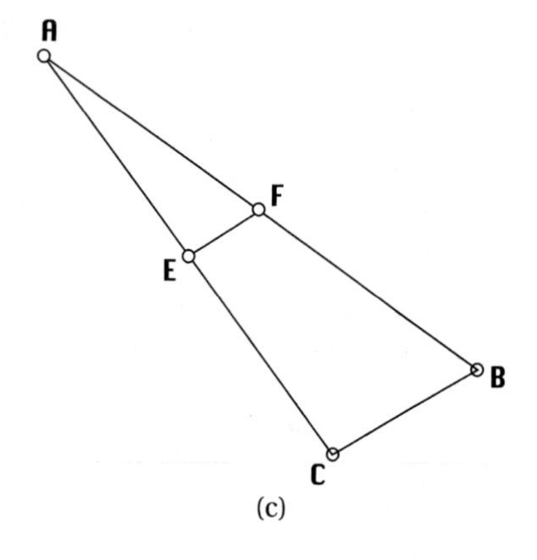

(c)

Length(Segment CB)	8.388	8.388	2.892	11.249
Length(Segment FE)	4.194	4.194	1.446	5.624
Ratio(Segment FE/Segment CB)	0.500	0.500	0.500	0.500

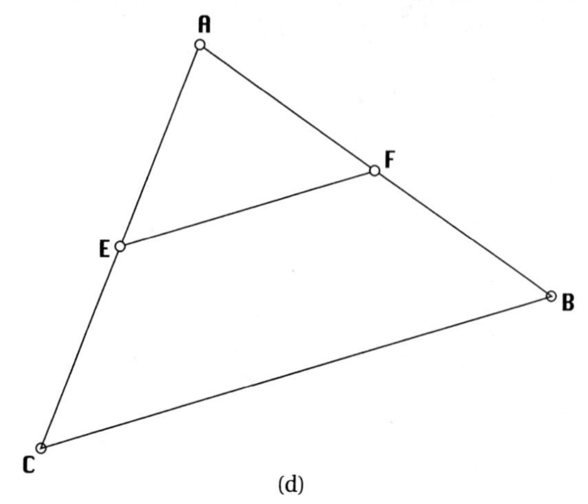

(d)

2. In the following display the circle with its center at **F1** and having a radius of **AP** intersects the circle with its center at **F2** and having a radius of **BP** in two points; select these two points and choose **Trace Locus;** then move point **P** along $\overline{\mathbf{AB}}$, and as the radii of the two circles change, the two points will trace an ellipse because the sum of their distances from the two fixed points remains constant.

Length(Segment AP) = 0.66 inches
Length(Segment BP) = 1.29 inches
Length(Segment AP) + Length(Segment BP) = 1.95

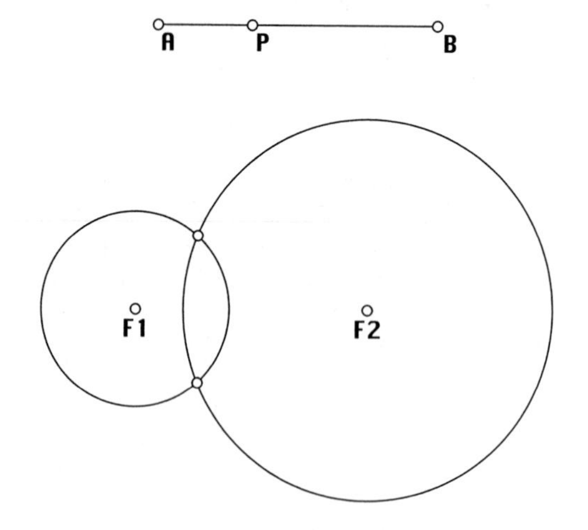

Index